3rd International Symposium on LAPAN-IPB Satellite for Food Security and Environmental Monitoring 2016

IOP Conference Series: Earth and Environmental Science
Volume 54

Bogor, Indonesia
25 – 26 October 2016

Editor:

Yudi Setiawan

ISBN: 978-1-5108-3664-8
ISSN: 1755-1307

Printed from e-media with permission by:

Curran Associates, Inc.
57 Morehouse Lane
Red Hook, NY 12571

Some format issues inherent in the e-media version may also appear in this print version.

Printed by Curran Associates, Inc. (2017)

For permission requests, please contact the Institute of Physics
at the address below.

Institute of Physics
Dirac House, Temple Back
Bristol BS1 6BE UK

Phone: 44 1 17 929 7481
Fax: 44 1 17 920 0979

techtracking@iop.org

Additional copies of this publication are available from:

Curran Associates, Inc.
57 Morehouse Lane
Red Hook, NY 12571 USA
Phone: 845-758-0400
Fax: 845-758-2633
Email: curran@proceedings.com
Web: www.proceedings.com

Table of contents

Volume 54

The 3rd International Symposium on LAPAN-IPB Satellite For Food Security and Environmental Monitoring 2016

25–26 October 2016, Bogor, Indonesia

Accepted papers received: 19 January 2017
Published online: 11 February 2017

Preface

RS Application in Agriculture

RS Application in Land Use Land Cover

Climate Change and Variability

Dynamic Modelling in Climate System

Deforestation and Forest Degradation

Forest Fire Assessment

Biodiversity Conservation

RS Application in Coastal Zone and Fisheries

RS Application in Marine

Sensor and Satellite Platform

UAV and Microsatellite Technology

SAR Detection and Satellite Application

IOP Conf. Series: Earth and Environmental Science **54** (2017) 011001 doi:10.1088/1755-1315/54/1/011001

Foreword

Bogor Agricultural University (IPB) and National Institute of Aeronautics and Space of Indonesia (LAPAN) have committed to work together in developing microsatellite namely LAPAN-A3/ LAPAN-IPB Satellite (LISAT) in order to support the food security and environmental monitoring program especially in Indonesia. This satellite has been launched on June 22, 2016 aboard the India-owned Polar Satellite Launch Vehicle (PSLV-C34) in Sriharikota, India. Moreover, the series of international symposium is a part of the agenda that will support the LISAT satellite programs and it is also intended to support the research related to agriculture, forestry, marine and fisheries, and climate change.

The 3rd International Symposium on LAPAN-IPB Satellite for Food Security and Environmental Monitoring 2016 (LISAT-FSEM 2016), to which this issue of IOP Conference Series: Earth and Environmental Science (EES) is dedicated, was held on October 25-26, 2016 at the IPB International Convention Center, Bogor, Indonesia. This symposium is jointly organized by the Bogor Agricultural University (IPB), Center for Environmental Research, Research and Community Services Institution, IPB (PPLH-LPPM, IPB), Directorate of Research and Innovation, IPB (DRI-IPB) and the National Institute of Aeronautics and Space of Indonesia (LAPAN), Indonesia. This symposium was focused on: (1) enhancing beneficial use of satellite technology in the fields of agriculture, forestry, climate change, and marine resources, in terms of food security and environmental monitoring and (2) to prepare and complete the processing chains of the LISAT satellite data utilization from upstream to downstream as well as the existing environmental satellite data of the selected locations in accordance with the focus study.

This symposium was being participated by around 300 participants that come from various government institutions such as: Ministry of Agriculture, Ministry of Environment and Forestry, Indonesian Agency for Meteorology, Climatology and Geophysics (BMKG), Indonesian Institute of Science (LIPI), Geospatial Information Agency (BIG) and some local government's agencies; from diverse universities such as: Bogor Agricultural University (IPB), University of Indonesia (UI), Gadjah Mada University (UGM), Bandung Institute of Technology (ITB), Diponegoro University, Udayana University, University of Tsukuba, Yamaguchi University, Hokkaido University of Education, Japan, National Autonomous University of Mexico (UNAM) and many more. Several international research institute and organization also gave their enthusiasms such as: International Rice Research Institute (IRRI), Center for International Forestry Research (CIFOR), Center for Remote Sensing and Ocean Sciences (CReSOS), Institute of Northern Environment, Institute for Marine Research and Observation, Indonesian Society for Remote Sensing (MAPIN), and many more.

The committee has received 207 research titles for both category of oral presentation (120 titles) and poster presentation (87 titles), which come from various scientists and students from different institutions. These research papers have been reviewed by the scientific committee of LISAT-FSEM 2016. Regarding the peer review result, 106 papers have been selected to be published in this issue, which mostly focused on enhancing the beneficial use of satellite technology in terms of food security in the fields of agriculture, forestry, climate change, and fisheries/marine resources. Indeed, the result from this symposium will provide useful information from satellite technology in order to support the government's programs for food security and environmental monitoring. Moreover, this symposium will strengthen the collaboration between multi-institutions and develop a better cooperation among the different research groups, either for basic research or technology developments.

We wish to thank the authors who have contributed to yield a high scientific standard to this issue. We are also grateful to IOP Publishing, especially Anete Ashton (IOP Conference Commissioning Editor), for allowing us to share some of the symposium results with a large audience of colleagues through the IOP Conference Series: Earth and Environmental Science. We would like to thank Ministry of Research, Technology and Higher Education (Ristekdikti) for giving us opportunity to get the additional funding for our international symposium. This symposium was supported also by the IPB World Class University Program for fiscal year 2016.

Yudi Setiawan
Editor-in-Chief
Proceeding LISAT-FSEM 2016

Lilik Budi Prasetyo
Conference Chair
The 3rd International Symposium on LISAT-FSEM 2016

Editorial Committee

Scientific Committee

Assistant Editors

LAPAN
THE 3RD
INTERNATIONAL SYMPOSIUM
ON LAPAN-IPB SATELLITE (LISAT)
2016 OCTOBER 25-26TH 2016
venue: IPB International Convention Center, Bogor
INDONESIA
Strengthening the Role of Environmental Satellites
in Support of Food Security and Environmental Monitoring
publication partner IOP Conference Series
a Scopus and Thomson Reuters indexed proceedings

IOP Conf. Series: Earth and Environmental Science **54** (2017) 011001 doi:10.1088/1755-1315/54/1/011001

IOP Conf. Series: Earth and Environmental Science **54** (2017) 011001 doi:10.1088/1755-1315/54/1/011001

Peer review statement

All papers published in this volume of *IOP Conference Series: Earth and Environmental Science* have been peer reviewed through processes administered by the proceedings Editors. Reviews were conducted by expert referees to the professional and scientific standards expected of a proceedings journal published by IOP Publishing.

Identification of rice field using Multi-Temporal NDVI and PCA method on Landsat 8 (Case Study: Demak, Central Java)

Abdi Sukmono*, **Ardiansyah**
Department of Geodesy Engineering, Faculty of Engineering, Diponegoro University Semarang, Indonesia

E-mail : sukmono35@gmail.com

Abstract. Paddy is one of the most important agricultural crop in Indonesia. Indonesia's consumption of rice per capita in 2013 amounted to 78,82 kg/capita/year. In 2017, the Indonesian government has the mission of realizing Indonesia became self-sufficient in food. Therefore, the Indonesian government should be able to seek the stability of the fulfillment of basic needs for food, such as rice field mapping. The accurate mapping for rice field can use a quick and easy method such as Remote Sensing. In this study, multi-temporal Landsat 8 are used for identification of rice field based on Rice Planting Time. It was combined with other method for extract information from the imagery. The methods which was used Normalized Difference Vegetation Index (NDVI), Principal Component Analysis (PCA) and band combination. Image classification is processed by using nine classes, those are water, settlements, mangrove, gardens, fields, rice fields 1st, rice fields 2nd, rice fields 3rd and rice fields 4th. The results showed the rice fields area obtained from the PCA method was 50 ,009 ha, combination bands was 51,016 ha and NDVI method was 45,893 ha. The accuracy level was obtained PCA method (84.848%), band combination (81.818%), and NDVI method (75.758%).

1. Introduction

Rice is one of the main crops in Indonesia because most Indonesian people consume rice as a staple food. Based on data from the BPS national socio-economic survey, Indonesia's population of rice consumption reached 78.82 kg per capita per year [1]. It is assessed by the Ministry of Agriculture is fairly high consumption. In fact, more than world rice consumption is only 60 kg per capita per year. In Asia, Indonesia's rice consumption is the highest. The high number of national rice consumption because the grains are separated part of the food culture of Indonesia [2].
Indonesian government wants to realize Indonesia became self-sufficient in 2017. The Indonesian government therefore should be able to seek the stability of the fulfillment of basic needs for food. The first step that must be done is to map the rice fields. Map of rice lands can provide comprehensive information on the development of paddy fields. The development of paddy land area can be used as an estimate of rice production.

Many studies have proved that images such as Landsat and NOAA Advanced Very High Resolution Radiometer (AVHRR) has potential to identify paddy rice fields [3]. After the launch of Landsat 8 satellite image, many researchers are trying to explore its use for identification of land cover. Landsat 8 had the privilege of both the specification and efficiency. It is launched in 2013 and

has a spatial resolution of 30 meters, without the cost and easy to obtain [4]. The temporal resolution of the Landsat 8 is every 16 days, so it can be possible to perform analysis in time series objects such as identification of rice field which has 4 major phases in its development phase, i.e. water, vegetative, generative and fallow [4]. The identification of rice lands using satellite images often have difficulty. This is because the rice land have the same colour and texture by fields, moor or weeds. In this study used multi-temporal satellite images Landsat 8. Multi-temporal methods have been tried are used for identification of rice fields cover effectively on a satellite image of Modis by zhang et all [5]. This method is used by assuming that the multi time to change the hue of rice land more quickly and significantly than other land. Multi temporal or time series data h identification of multi-temporal rice can be combined with several classification methods such as Normalized Difference Vegetation Index (NDVI), Principal Component Analysis (PCA) and band combination. The objective of this research are testing the best multi-temporal combination with that some method on Landsat 8.
Based on the above explanation, the study was done by identifying the rice fields as an object of research, several methods are compared, i.e. the method of Normalized Difference Vegetation Index (NDVI), Principal Component Analysis (PCA) and band combination. Using multi-temporal Landsat 8 based on Rice Planting Time 1st in Demak that range between the end of October 2013 to early March 2014.

2. Methodology

2.1. Description of Study Area
The study was conducted in Demak of the Province of Central Java, Indonesia. Centred at latitude 6° 43' 26"–7° 09' 43" and longitude 110° 27' 58 "-110° 48' 47" and is located approximately 25 km east of Semarang City (figure 1). The longest distance from West to East is along 49 km and from North to South along the 41 km [2].

Figure 1. Location of study area

2.2. Method
Method this study was followed the flowchart in figure 2. In this study is using 4 images multi-temporal with reference to Rice Planting Time 1st in Demak 2013/2014, i.e. the image on October 30, December 1, 2013, January 2 and March 7, 2014.

2.2.1 Geometric correction
Geometric correction geo-referenced image to real world coordinates. The geometric correction method of the image used image to map rectification with reference Indonesian Topographical map (RBI) of Demak, scale 1:25,000. GCP on this research taken as many as 16 points of each image. Root mean square error (RMSE) in this process were required maximum 1 pixel.

2.2.2 Radiometric correction

This stage corrected image using the conversion of digital number to Top of Atmosphere (TOA) Reflectance. OLI band data of Landsat 8 can be converted to TOA planetary reflectance using reflectance rescaling coefficients provided in the product metadata file (MTL file). MTL is the format of metadata file for Landsat 8 imagery. The following equation is used to convert digital number (DN) or pixel values to TOA reflectance for operational land imager (OLI) data as follows [4]:

$$\rho\lambda' = M\rho Qcal + A\rho \quad (1)$$

TOA reflectance with a correction for the sun angle is then [6]:

$$\rho\lambda = \frac{\rho\lambda'}{\cos(\theta SZ)} = \frac{\rho\lambda'}{\sin(\theta SE)} \quad (2)$$

where:

$\rho\lambda'$ = TOA planetary reflectance, without correction for solar angle. Note that $\rho\lambda'$ does not contain a correction for the sun angle

M_ρ = Band-specific multiplicative rescaling factor from the metadata (REFLECTANCE_MULT_BAND_x, where x is the band number)

A_ρ = Band-specific additive rescaling factor from the metadata (REFLECTANCE_ADD_BAND_x, where x is the band number)

Q_{cal} = Quantized and calibrated standard product pixel values (DN)

$\rho\lambda$ = TOA planetary reflectance

θSE = Local sun elevation angle. The scene center sun elevation angle in degrees is provided in the metadata (SUN_ELEVATION)

θ_{SZ} = Local solar zenith angle; $\theta_{SZ} = 90° - \theta_{SE}$.

2.2.3 Highlight cloud

Clouds and shadows in the image is a problem that can't be avoided. This is an obstacle in the use of medium resolution imagery such as Landsat 8, especially in equatorial regions. This causes the loss of some of the information contained in the images. Highlight Cloud is used to discriminate or distinguish clouds with land cover [6]. Because of the clouds in the application or interpretation of land use is considered as interference to other land cover reflectance values, so that should be eliminated in the analysis. In this study, the clouds separated by digitization after sharpening cloud value.

2.2.4 Image Cropping

Cropping images do reduce the file size of the image so that the processing of data to be lighter and faster. It are made in accordance with the required area or region of interest (ROI). Moreover, in this study uses data multi-temporal for classification process, so cropping of the image is needed in order to accelerate the image processing.

2.2.5 Transformation method

In this research, transformation method by using Normalized Difference Vegetation Index (NDVI), Principal Component Analysis (PCA) and band combination (7-6-4) were used in each image. The results of image transformation at several different times are combined for a single image, it is intended to create new combinations of image so that the pattern of changes in the value of this index is used to identify rice fields and other land. This index pattern is used as a reference in the classification process. There are four image results from the transformation of the NDVI, namely on NDVI October 30, 2013; NDVI December 1, 2013; NDVI January 2 and NDVI March 7, 2014 are combined in a single layer (the layer stack). At PCA method, there were 7 PCs results from the transformation of the PCA combined 28 band (where prior to transformation, 7 band each image are combined to produce a 28-band). While in band combination (7-6-4) there were 12 image from the combined results of band combination 7-6-4 each image.

2.2.6 Supervised Classification
Classification is performed using the supervised classification method, with type maximum likelihood. Class classification is namely water, settlements, mangrove, gardens, fields, rice fields 1st, rice fields 2nd, rice fields 3rd and rice fields 4th. The taking of ground data is done by field survey. It was used to knowing the field rice in actual area. That can be used to identification the spectral pattern on image combined.

2.2.7 Analysis of the cloud-covered area of rice fields
Analysis of the cloud-covered area of rice fields to find out the information that is lost due to clouds and shadows. It was assisted by using land use map of the Land Office of Demak 2014 that updated using Google image. The results processing digitation using Google Image 2013 as well as field survey to check the suitability of the map. Based on the result of processing, the rice fields covered area acquired cloud 6.307 ha.

2.2.8 Test the Accuracy
The classification accuracy were tested by conducting field survey (33 points), and compare the results with rice fields area is available from the Department of Agriculture in 2014[7].

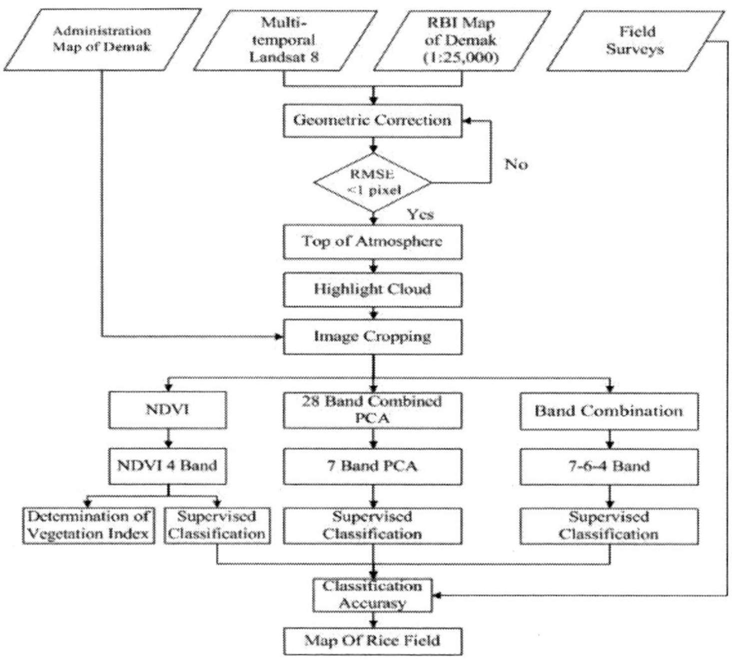

Figure 2. Flowhchart of the research

3. Result and Discussion
Identification of rice field in the image get four type of rice field. This type is based on the growth stage of paddy in the field that is different phase. Determination of the pattern of vegetation index for rice fields utilize the result of a combination of NDVI in different time. Fifty of ground data with field surveys were taken to get the pattern of rice field in four types. Fifty field data was distributed by 18 in the first type of rice field, 15 in the second type of rice field, 12 in the third type of rice field and 5 in

fourth type of rice field. The value of NDVI at the four types of wetland in detail shown in table 1, table 2, table 3 and table 4.

Table 1. Vegetation index for rice fields 1st

Month	Min.	Max.	Average
October 30, 2013	-0.041361	0.208411	0.110164
December 1, 2013	0.409407	0.744219	0.595514
January 2, 2014	0.587388	0.806071	0.735100
March 7, 2014	-0.229917	0.280647	0.075846

Table 2. Vegetation index for rice fields 2nd

Month	Min.	Max.	Average
October 30, 2013	0.129808	0.227061	0.173888
December 1, 2013	0.209410	0.429601	0.295438
January 2, 2014	0.501472	0.807062	0.684695
March 7, 2014	0.038147	0.501958	0.231346

Table 3. Vegetation index for rice fields 3rd

Month	Min.	Max.	Average
October 30, 2013	0.218776	0.645476	0.393743
December 1, 2013	0.095961	0.304005	0.185014
January 2, 2014	0.415126	0.805949	0.671330
March 7, 2014	0.235618	0.495557	0.358231

Table 4. Vegetation index for rice fields 4th

Month	Min.	Max.	Average
October 30, 2013	0.463593	0.723882	0.565158
December 1, 2013	0.271310	0.442523	0.340143
January 2, 2014	-0.124463	0.292501	0.157182
March 7, 2014	0.286817	0.639328	0.514898

The value of NDVI multi temporal indicates the condition of a growth phase at the time of image acquisition. The four table based on four types NDVI values seen that the four types of wetland have different patterns of NDVI, for example in the first field rice's NDVI approach zero at October 2013. It shown that growth stage of paddy were first phase because dominated with water. The value of NDVI increase at December 2013 and at January 2014, it showed that growth stage of rice were also increase. At March 2014, the value of NDVI were null, it show that growth stage of rice in bera phase. Bera is the phase where the fields are not planted after the harvest.

Generate 4 patterns of vegetation index for rice fields in the four type of rice field shown in the figure 3, figure 4, figure 5 and figure 6.

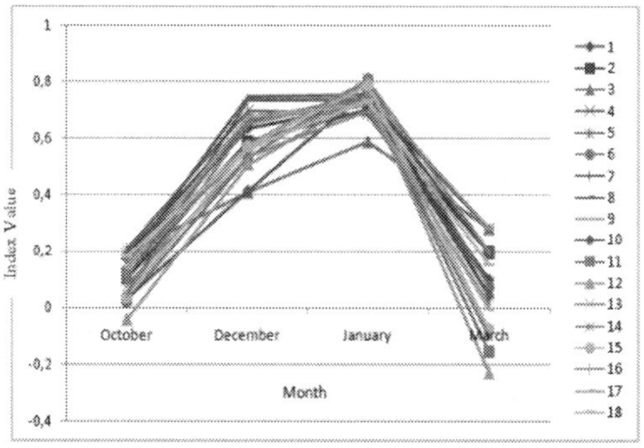

Figure 3. Vegetation Index for rice fields 1st

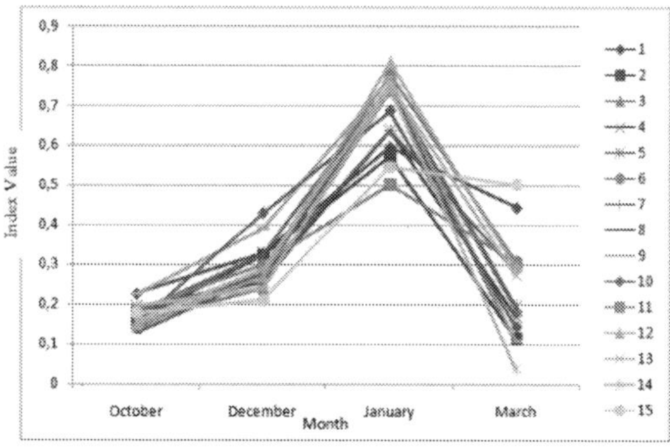

Figure 4. Vegetation Index for rice fields 2nd

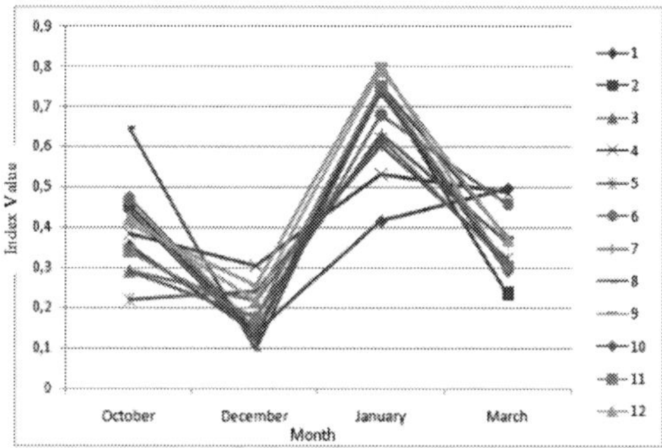

Figure 5. Vegetation index for rice fields 3rd

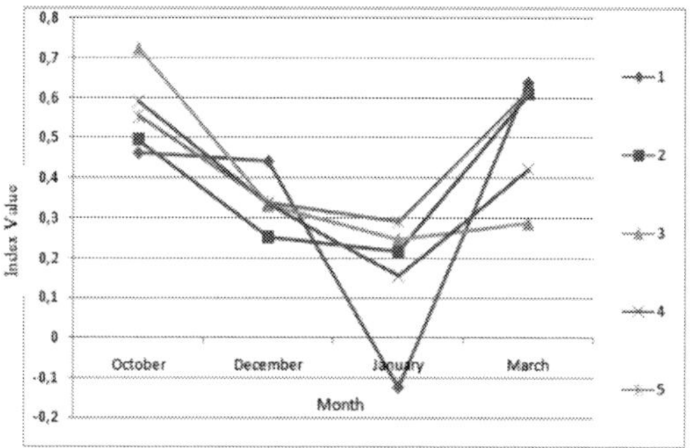

Figure 6. Vegetation index for ricefields 4rd

The four pattern were used to identification of rice field using supervised classification. PCA multi-temporal clasification and band combination clasification also used the same proces in NDVI multi-temporal. The results of supervised classification can be seen in figure 7, figure 8 and figure 9.

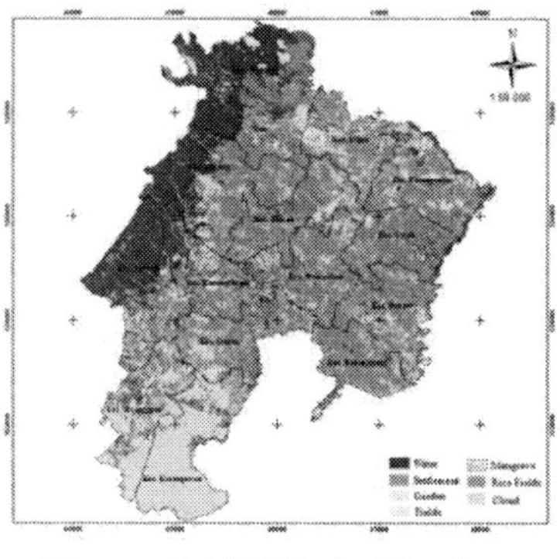

Figure 7. NDVI clasification

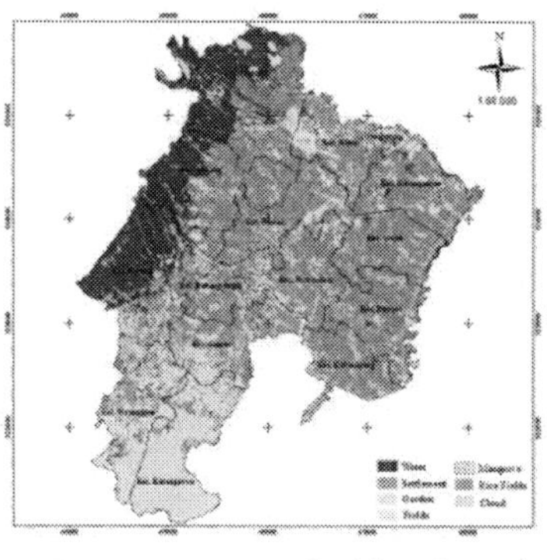

Figure 8. PCA clasification

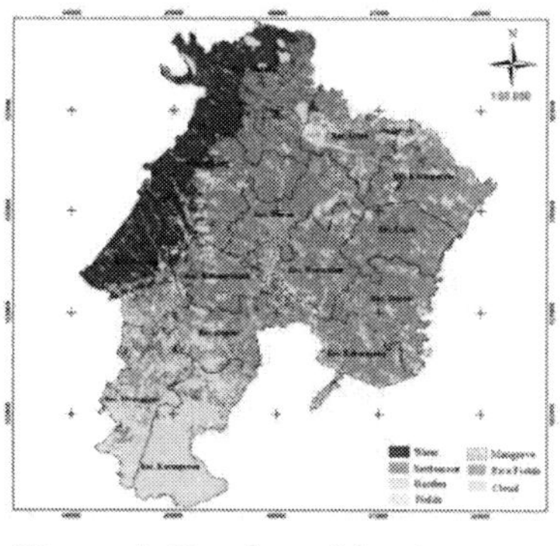

Figure 9. Band combination

PCA multi-temporal clasification has producer accuracy of 86.67%, user accuracy of 86.67% and Overall acuracy of 84,85 %. Then NDVI multi-temporal clasification has producer accuracy of 80 %, user accuracy of 92.31 % and Overall acuracy of 75,76 %. The band combination 7-6-4 clasification has producer accuracy of 86.67%, user accuracy of 81.25% and Overall acuracy of 81,82 %. For more details can be seen in table 5, table 6 and table 7. While the area of rice fields clasification were compared to the methods of approaching data Department of Agriculture of 49.657 Ha [4], PCA clasification has difference of 328 ha (0.71%) , band combination 7-6-4 of has difference of 1,359 ha (2.73%) and lastly NDVI of 45,893 ha with a difference of -3,764 ha (7.58 %).

Table 5. Matrix acuration of multi-temporal PCA 7 band

Land use	Water	Settle-ment	Plantation	Moor	Mang-rove	Rice Field	Total	UA (%)	CE (%)
Water	4	0	0	0	0	0	4	100,00	0,00
Settlement	0	4	0	0	0	0	4	100,00	0,00
Plantation	0	0	3	0	1	0	4	75,00	25,00
Moor	0	0	0	3	0	2	5	60,00	40,00
Mangrove	0	0	0	0	1	0	1	100,00	0,00
Rice Field	0	0	1	1	0	13	15	86,67	13,33
Total	4	4	4	4	2	15	33		
PA (%)	100,00	100,00	75,00	75,00	50,00	86,67			
OE (%)	0,00	0,00	25,00	25,00	50,00	13,33			
Overall Accuracy(%)	84,848								
Kappa Coefficient	0,792								

Table 6. Matrix acuration of multi-temporal NDVI

Land use	Water	Settle-ment	Plantation	Moor	Mang-rove	Rice Field	Total	UA (%)	CE (%)
Water	4	0	0	0	1	0	5	80,00	20,00
Settlement	0	3	0	0	0	0	3	100,00	0,00
Plantation	0	0	2	0	0	0	2	100,00	0,00
Moor	0	1	1	3	0	3	8	37,50	62,50
Mangrove	0	0	1	0	1	0	2	50,00	50,00
Rice Field	0	0	0	1	0	12	13	92,31	7,69
Total	4	4	4	4	2	15	33		
PA (%)	100,00	75,00	50,00	75,00	50,00	80,00			
OE (%)	0,00	25,00	50,00	25,00	50,00	20,00			
Overall Accuracy(%)	75,758								
Kappa Coefficient	0,677								

Table 7. Matrix acuration of multi-temporal band combination 7-6-4

Land use	Water	Settle-ment	Plantation	Moor	Mang-rove	Rice Field	Total	UA (%)	CE (%)
Water	4	0	0	0	0	1	5	80,00	20,00
Settlement	0	4	0	0	0	0	4	100,00	0,00
Plantation	0	0	4	0	1	0	5	80,00	20,00
Moor	0	0	0	1	0	1	2	50,00	50,00
Mangrove	0	0	0	0	1	0	1	100,00	0,00
Rice Field	0	0	0	3	0	13	16	81,25	18,75
Total	4	4	4	4	2	15	33		
PA (%)	100,00	100,00	100,00	25,00	50,00	86,67			
OE (%)	0,00	0,00	0,00	75,00	50,00	13,33			
Overall Accuracy(%)	81,818								
Kappa Coefficient	0,747								

IOP Conf. Series: Earth and Environmental Science **54** (2017) 012001 doi:10.1088/1755-1315/54/1/012001

4. Conclusion

In this study can be concluded that PCA 7 band multi-temporal classification has better accuracy for all land use type. Over all accuray of PCA 7 band up to 84, 85 %. This qualifies a clasification accuracy of 80 %. However when viewed from the user accuracy of rice field, NDVI multi-temporal has better accuracy. The user accuracy of NDVI multi-temporal for rice field up to 92,31 %.

5. References

[1] Center Statistic of Indonesia (BPS) 2014 *Demak in Figure 2014* (Demak: BPS)
[2] Bratani, Gigih 2014 *Growth Stage of Rice* (Bandung: Aneka)
[3] Genc L, Inalpulat M, Kizil U, and Aksu S 2014 Determination of Paddy Rice Using Landsat 8 *Proceeding International Conference on Biological, Civil and Enviromental Engineering* (BCEE-2014) (Dubai UEA, Marc 17-18)
[4] USGS 2013 Using the USGS Landsat 8 Product http://Landsat.usgs.gov. (accessed on10 April 2015)
[5] Zhang G, Xiao X, Dong J, Kou W, Jin C, Qin Y, Zhou Y, Wang J, Menarguez MA, and Biradar C 2015 Mapping paddy rice planting areas through time series analysis of MODIS land surface temperature and vegetation index data *ISPRS Journal of Photogrammetry and Remote Sensing* **106** 157-171
[6] Putra E H 2011 *Remote Sensing Using ERMapper* (Yogyakarta: Graha Ilmu)
[7] Department of Agriculture 2015 *Conditions of Land Agriculture in Demak Th 2014-2015* (Demak: Department of Agriculture)

The method for detecting biological parameter of rice growth and early planting of paddy crop by using multi temporal remote sensing data

D D Domiri
Remote Sensing Application Center, LAPAN

Email: dede_dirgahayu03@yahoo.com

Abstract. Rice crop is the most important food crop for the Asian population, especially in Indonesia. During the growth of rice plants have four main phases, namely the early planting or inundation phase, the vegetative phase, the generative phase, and bare land phase. Monitoring the condition of the rice plant needs to be conducted in order to know whether the rice plants have problems or not in its growth. Application of remote sensing technology, which uses satellite data such as Landsat 8 and others which has a spatial and temporal resolution is high enough for monitoring the condition of crops such as paddy crop in a large area. In this study has been made an algorithm for monitoring rapidly of rice growth condition using Maximum of Vegetation Index (EVI Max). The results showed that the time of early planting can be estimated if known when EVI Max occurred. The value of EVI Max and when it occured can be known by trough spatial analysis of multitemporal EVI Landsat 8 or other medium spatial resolution satellites.

1. Introduction

Application of Remote Sensing (RS) activities for monitoring the growth conditions of rice paddy fields in Java – Bali have undertaken since 2005 by RS Application Center, LAPAN, because Java and Bali island still contribute ± 55 % of Indonesian rice production. Monitoring about growth condition of main crop need to be done by using satellite technology. Like as reported by some research, ie Takeuchi [1], Sakamoto [2], Xiao [3] and Uchida [4,5] about ability of low or medium spatial satellite data, such as MODIS and Landsat for monitoring rice growth. While Dirgahayu [6] has been provided about useful of soil moisture in paddy field which can be derived from MODIS data for predicting drought condition. Now, the activity is part of research focus in the field of food security in Indonesia has be done operationally for whole Indonesian area by cooperating between LAPAN with ministry of Agriculture. Factors that affect food security in a region associated with spatial aspects such as: food availability and vulnerability. Spatial information that shows the actual condition of the indicators of food availability and vulnerability can be derived from RS data which has high spectral and temporal resolution, like as Landsat 8, MODIS data. Spatial information include, ie the greenness rice crop, rice growth phase, early growing season, rice crop growth stage, harvest failures caused by drought, flood, pets and diseases in the paddy field, harvest area, and productivity can be estimated by RS data. Those information are needed to predict the rice production. Based on the background above, the purpose of research are to estimate when EVI maximum as one of biological parameter of rice growth occurred and the early planting time of rice crop in paddy field.

2. Materials and Method

The study site is Paddy field in Subang District, West Java. Geographically, the study area is located between 787438 m to 794519 m of easting and between -694463 m to -704894 m of Northing with NUTM Zone 48 projection as shown in Figure 1. Primary data used in this research is surface reflectance of Landsat 8 data, P/R 122/064. In order To obtain a rice plant growth profile completely, so needed multitemporal data as many as 16 scenes in the period from May to August, 2015. Stages of research done can be seen in the flowchart in Figure 2. Two parameter will be used in this research based on result by Dirgahayu [7], ie NDWI and EVI. NDWI (Normalize Difference Water Index) to detect wetness condition can be created by using the following formula :

$$NDWI = (R3 - R7)/(R3 + R7) \quad (1)$$

This formula is different from the method used by Uchida [8] and also Li, et all [9], Zhou, et all [10], because based on investigation result by Dirgahayu [7] show that peak high value of water body generally occur at Green Reflectance (R3) of Landsat 8 data, while SWIR2 (R7) reflectance always the lowest value for water body. Vegetation Index to detect land greeness using Enhanced Vegetation Index (EVI) equation :

$$EVI = 2.5 * (NIR - Red)/(1 + NIR + 6 * Red - 7.5 * Blue) \quad (2)$$

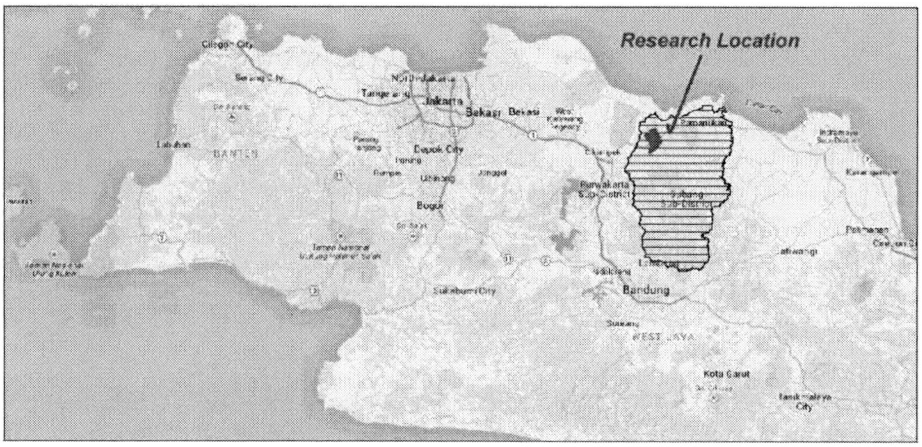

Figure 1. Research location at Paddy area, Subang District, West Java

3. Results and Discussions

3.1. Visual Interpretation based on Colour Composite Image of Landsat 8

Most of the paddy area shows still look of rice crops based on RGB color composite image 653 dated August 31, 2015 (Figure 3) and some areas occurred harvesting. Object water showed is also seen in some areas. While vegetation object in paddy field looks dark green to light green and yellowish green colour.

IOP Conf. Series: Earth and Environmental Science **54** (2017) 012002 doi:10.1088/1755-1315/54/1/012002

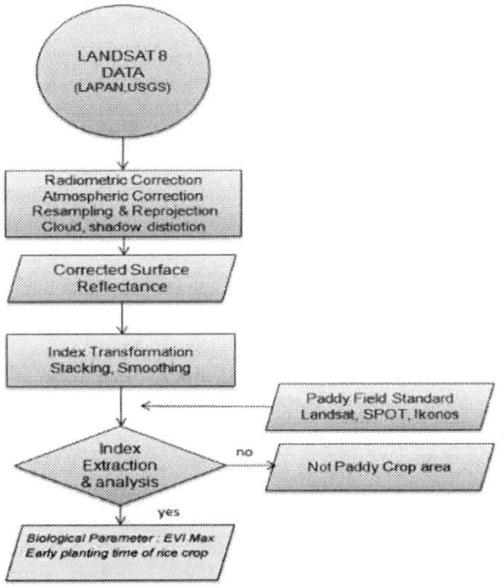

Figure 2. The Flowchart of Research Stage

Figure 3. Natural Color Composite band 653 (SWIR1,NIR, Green) of Landsat 8 on August 31, 2015

The condition paddy field has been harvested is shown by the appearance of the object of open area (bare land) in paddy fields with gradation of dark brown and magenta to light gradation. In order to know the rice plant in vegetative or generative growth phase, then it must be known change of the vegetation index (EVI) compared with the previous 16 days of data. Vegetative phase is indicated by the increase in the vegetation index or a trend of positive changes. While the generative phase occurs otherwise is indicated by negative changes or a trend of decline in the vegetation index.

3.2. *Extraction of Maximum EVI as Biological Growth Parameter of Rice Crop*

Statistical analysis of EVI multitemporal using Landsat 8 from January to September 2015 is shown in Image Composite RGB (Mean EVI, EVI Max and Min EVI) in Figure 4. The paddy field area in Subang appear light green with a smooth texture that is different from the object of other land cover. While the spatial distribution of paddy field classification based on Maximum of EVI in the region can be seen in Figure 5. Based on the spatial distribution of the maximum EVI in Figure 5 shows that the dominant class is the class that has the EVI Maximum range: 0.55-0.60. The next occurred in class 5 and class 3.

The percentage of rice area in Subang based on EVI maximum grade is shown in Table 1 and the graph in Figure 6.

Table 1. Distribution of Paddy Field Area in Subang District

Class	EVI_Max	Area (Ha)	Area (%)
1	0.40-0.45	1,832	2.5%
2	0.46-0.50	5,741	7.7%
3	0.51-0.55	15,551	20.9%
4	0.56-0.60	29,943	40.2%
5	0.61-0.65	17,356	23.3%
6	0.66-0.70	3,789	5.1%
7	> 0.70	237	0.3%

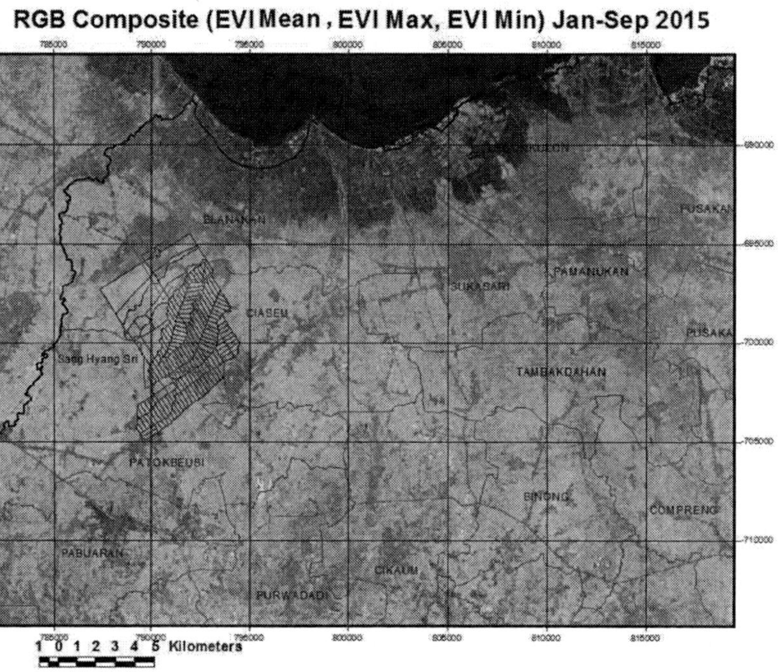

Figure 4. Color Composite RGB (EVI Mean, Max, EVI Min) of EVI Multitemporal

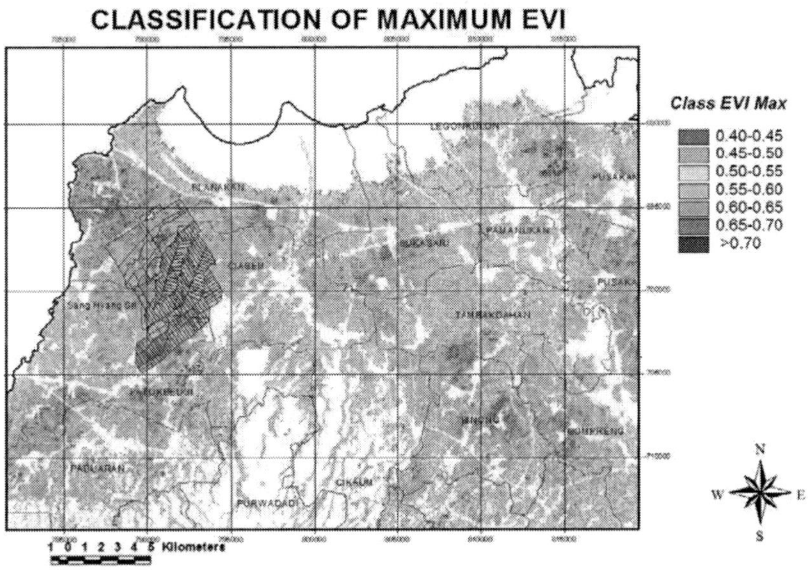

Figure 5. Spatial distribution of Paddy Field based on Maximum EVI in Subang District

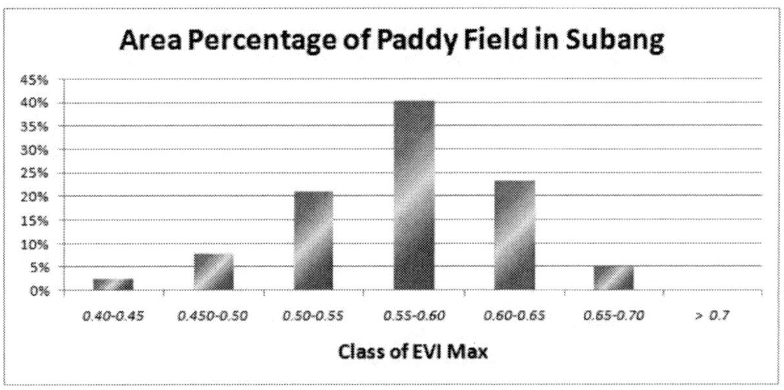

Figure 6. Histogram of Percentage Paddy Field Area Based on EVI Max in Subang District

3.3. Determining the Early Planting of Rice Crop in Paddy Field

Rice growth profile can be known by using the 16 days data from multi temporal vegetation index (EVI). For an example of the extraction result of the EVI Multi temporal Landsat 8 from early planting to harvest time can be viewed in Figure 7. Based on Figure 7 shows the peak value of vegetation index occurred at about the middle of growth, that is when the rice plant ages ranged between 60-64 days after planting (DAT). Early planting is characterized by EVI value (≤ 0.188) lower than at the harvest time (100-120 days) and bare land. Based on the time when EVI Maximum has occur (t_EVI Max, in Julian date), so it can be created the spatial distribution of the planting time of rice and also can be predicted when and how amount area of harvesting. Early planting with still inundation of water can be decided by the value of EVI ≤ 0.188 and NDWI ≥ 0.226. When the early planting occurred (TEP) can be estimated by using the equation :

$$TEP = t_{-EVI}Max - \frac{1}{2} * (\frac{LP}{Period}) \quad (3)$$

Where LP is length period (120 days). Period is temporal resolution of Landsat 8 (16 days). The implementation result can be seen in Figure 8.

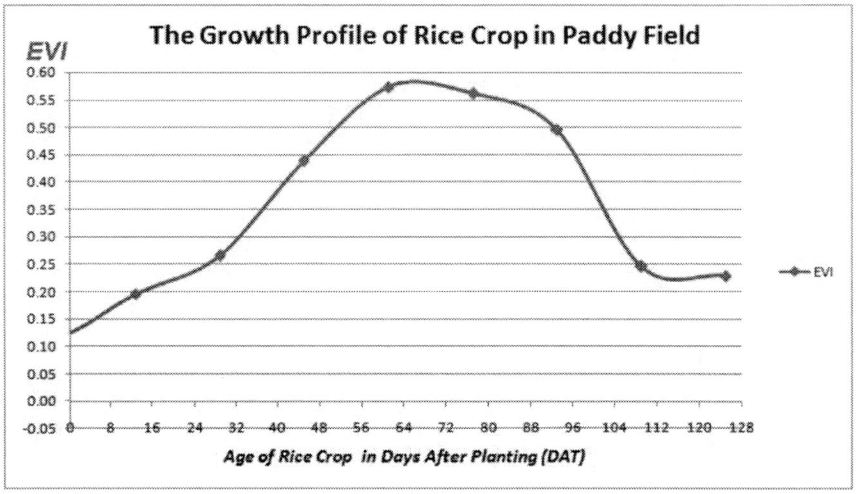

Figure 7. The Growth Profile of Rice Crop in Paddy Field (EVI Maximum 0.55-0.60)

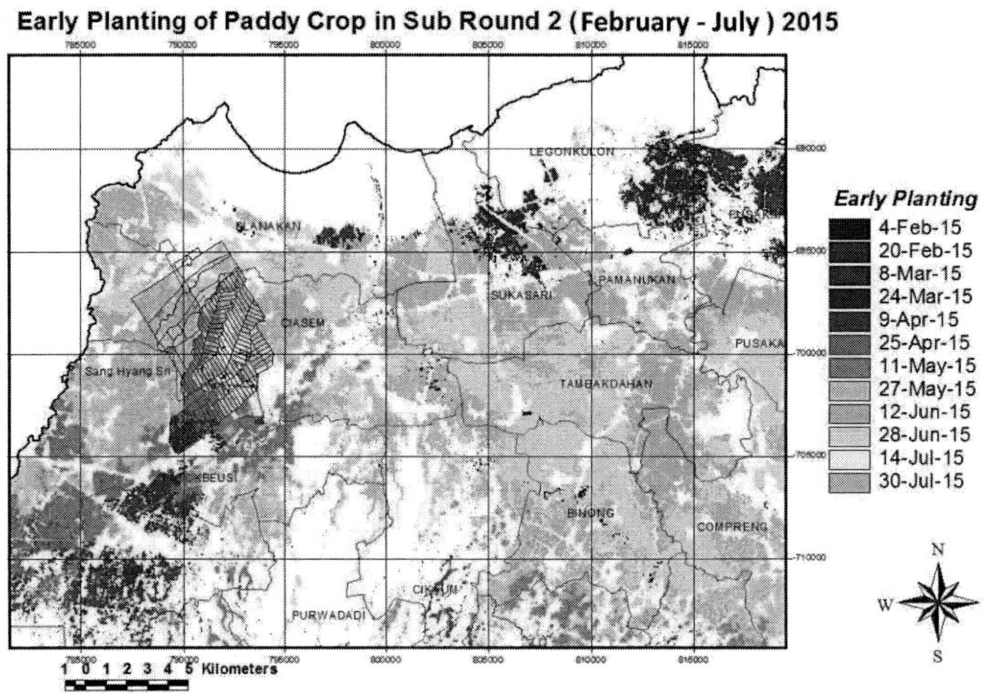

Figure 8. Spatial Distribution of The Early Planting Time of Rice Crop in Paddy Field at Subang District, from February to July 2015

4. Conclusions
The research has successfully demonstrated the ability of Landsat 8 to estimate when the early planting time of rice crop have been occurred. The early planting can be determined based on the maximum of EVI as biological growth parameter and using combination EVI and NDWI parameters.

Acknowledgements
The author would like to thank to PT Sang Hyang Sri, Subang, West Java, for their cooperation and giving of information about the schedule of planting and harvesting of rice plants at the sites.

References
[1] Takeuchi W and Yasuoka Y 2004. Estimating spatio-temporal patterns of paddy fields over Southeast Asia using MODIS time series, *25th Asian Conference on Remote Sensing (Chiang Mai, C-8.7)*

[2] Sakamoto T, Nguyen N V, Ohno H, Ishitsuka N and Yokozawa M 2006 Spatio-temporal distribution of paddy phenology and cropping systems in the Mekong Delta with special reference to the seasonal water flow of the Mekong and Bassac rivers *Remote Sensing of Environment* **100** p 1-16.

[3] Xiao X, Boles S, Frolking S, Li C, Babu J Y, Salas W and Moore B 2006. Mapping paddy agriculture in South and Southeast Asia using multi-temporal MODIS images *Remote Sensing of Environment* **100** p 95-113.

[4] Uchida S, 2010 a Monitoring of planting paddy with complex cropping pattern in the tropical humid climate region using Landat and MODIS data -A case of West Java Indonesia *International Archives of the Photogrammetry, Remote Sensing and Spatial Information Science* **38** p 477-481.

[5] Uchida S, 2010 b. Variation of paddy planted time in recent years for West Java in Indonesia observed by using MODIS data *31st Asian Conference on Remote Sensing (Hanoi, TS05-4)*

[6] Dirgahayu D, 2013. Development of land moisture estimation model using modis infrared, thermal, and EVI to detect drought at paddy field. *International Journal of Remote Sensing* **10** 1 June 2013

[7] Dirgahayu D, 2016 The new method for detecting early planting and bare land condition paddy field by using vegetation-bare-water index. *2nd International Conference of Indonesian Society for Remote Sensing (Yogyakarta, 17-20 October 2016)*

[8] Uchida S, 2007 Monitoring of paddy rice planting with complex cropping pattern using satellite remote sensing data -A case of West Java Indonesia *28th Asian Conference on Remote Sensing, Kuala Lumpur*

[9] Li P et all 2014. Cross-Comparison of Vegetation Indices Derived from Landsat-7 Enhanced Thematic Mapper Plus (ETM+) and Landsat-8 Operational Land Imager (OLI) Sensors. *Journal of Remote Sens.* **6** p 310-329

[10] Zhou Y et all 2016 Mapping paddy rice planting area in rice-wetland coexistent areas through analysis of Landsat 8 OLI and MODIS images *International Journal of Applied Earth Observation and Geoinformation* **46** (Elsevier) p 1-12

Vegetation type classification and vegetation cover percentage estimation in urban green zone using pleiades imagery

Bambang Trisakti*
Remote Sensing Application Center, LAPAN-Pekayon

E-mail: btris01@yahoo.com, bambang.trisakti@lapan.go.id

Abstract. Open green space in the urban area has aims to maintain the availability of land as a water catchment area, creating aspects of urban planning through a balance between the natural environment and the built environment that are useful for the public needs. Local governments have to make the green zone plan map and monitor the green space changes in their territory. Medium and high resolution satellite imageries have been widely utilized to map and monitor the changes of vegetation cover as an indicator of green space area. This paper describes the use of pleaides imagery to classify vegetation types and estimate vegetation cover percentage in the green zone. Vegetation cover was mapped using a combination of NDVI and blue band. Furthermore, vegetation types in the green space were classified using unsupervised and supervised (ISODATA and MLEN) methods. Vegetation types in the study area were divided into sparse vegetation, low-medium vegetation and medium-high vegetation. The classification accuracies were 97.9% and 98.9% for unsupervised and supervised method respectively. The vegetation cover percentage was determined by calculating the ratio between the vegetation type area and the green zone area. These information are useful to support green zone management activities.

1. Introduction

In general, the development of urban areas resulted in an increasing number of settlement areas and the decreasing of open green space areas. In the regulation of minister of Public Work No.05 of 2008, it was explained that the open green space is elongated or clustered area, that its utilization is more open, where the plants grow naturally or planted by human [1]. Open green space has some functions including the function of ecological, social and cultural, aesthetic and economic. Law No. 26 of 2007 on spatial planning explains that 30% of the urban area has to be open green space, where 20% of open green space for public area and 10% of it for private area [2].

Currently, the Jakarta government are pursuing the target to provide open green space area up to 30%, where the city government responsible to 16% of the open green space, and privat parties responsible the remaining 14% of it. To achieve the target, the Jakarta government will annually spend 5 billion to increase 6% of open green space in Jakarta until 2030, and it was already stated in Regulation No.1 of 2012. In connection with the implementation of this program, City Planning Department of East Jakarta has coordinated with Remote Sensing Application Center (Pusfatja) LAPAN to be able to take advantages of satellite remote sensing data utilization to monitor the open green space in East Jakarta region. The monitoring activities using satellite imagery will be very helpful to know the last condition of vegetation cover, and to find out the land cover change occured in the green zone area.

Utilizations of remote sensing data for monitoring changes in the open green space and to map land cover in urban areas have been conducted using medium and high spatial resolution remote sensing imagery. Some research studies were carried out using medium resolution satellite data (Landsat TM, Landsat ETM + and Landsat 8 OLI) for monitoring the changes of green space area [3], vegetation greeness level [4] as well as land use mapping [5] in urban areas. On the other hand detail scale mapping needs high spatial resolution satellite imagery, such as IKONOS, Quicbird or Pleiades. Some research studies were also used those high resolution satellite imageries to classify land cover and then used the land cover information to analysis urban green space [6], to map greenways in the city [7], and to digitally extract urban green space [8] [9]. Furthermore various types of data such as digital aerial photograph, satellite imagery and time series data were used to get more detail and accurate information related to green space and urban environment [10], and also the combination of LIDAR data and multi spectral imagery [11] were used to produce information of green space and urban building distribution. But the common method to map green space based on remote sensing data were to identify the vegetation cover, where the vegetation cover can be extracted using various classification methods or extracted using various indexes such as Normalised Different Vegetation Index (NDVI), Modified Soil Adjusted Vegetation Index (MSAVI) and chlorophyll index [4, 9, 12]. In contrast to the object appearance on the medium spatial resolution imagery, the object appearance on the high spatial resolution imagery up to submeter unit (less than 1 meter) becoming very detail, so the separation of vegetation from other objects around the vegetation (such as building or open land) becomes more difficult.

In this research activity, the high spatial resolution Pleiades imagery was used to classify vegetation types and estimate vegetation cover percentage in the green zone. Pleiades imagery has 0,5 meter spatial resolution, so this satellite imagery can be used is to identify tree canopy and single house building accurately, and it will be a suitable data for monitoring the vegetation cover in urban areas.

2. Research Method

The study area of this research was covered area surrounding Halim Perdana Kusuma Airport in East Jakarta, as shown in figure 1. This study area was selected due to various vegetation types and vegetation density. The use of Pleaides imagery with high spatial resolution was expected to be able to identify the various vegetation types in the study area accurately. The Pleiades imagery used was Pleiades-1A obtained from Technology and Data Center, LAPAN. The imagery was recorded on May 12th 2015, and it was already corrected for ortho systematic and processed for pansharpening (data fusion between 2 meter spatial resolution of multi spectral band and 0,5 meter spatial resolution of panchromatic band). So the Pleiades imagery had already 0,5 meter spatial resolution in multi spectral band (Blue, Green, Red and Near Infrared). Detailed information of Pleiades-1A is shown in table 1.

Tabel 1. Detail information of Pleaides-1A imagery used in this Research

Data	Processing level	Waktu perekaman	Spectral resolution	Spatial resolution
Pleiades-1A	Ortho Sistematic Without ground control points: 3m (CE90)	12 May 2015	Pan: 480-830 nm	0.5 m
			Blue: 430-550 nm Green: 490-610 nm Red: 600-720 nm Near Infrared: 750-950 nm	2 m

Source: http://www.satimagingcorp.com/satellite-sensors/pleiades-1/

Research flowchart is shown in figure 2. The whole data processing can be divided into three main processes, as follow: 1) vegetation and non vegetation separation process (vegetation cover mapping),

2) vegetation types classification process, and 3) estimating vegetation cover percentage in the green zone. Based on the results from the previous research conducted at the same location [9], vegetation and non vegetation pixels in the Pleiades imagery can be separated through two separation stages. In the first stage, the combination algorithm of Normalized Difference Vegetation Index (NDVI) and blue band was performed to the Pleiades imagery, then the histogram was adjusted to make the pixel value of non vegetation becoming zero, and then the result was saved as a new data. In the second stages, threshold values were determined to separate between pixels of vegetation and pixels-like vegetation, that they were still mixing from the first stage's result. This method made the process becoming easier to determine threshold values for accurately separating between vegetation pixels and non vegetation pixels.

Figure 1. Pleiades imagery of Halim Perdana Kusuma Airport

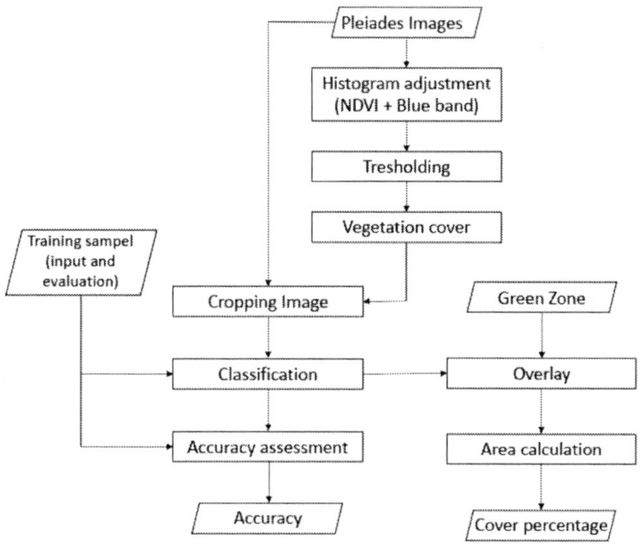

Figure 2. Reserach Flowchart

Vegetation pixels or vegetation cover area obtained from the first process was used to crop Pleaides imagery for obtaining the imagery inside the vegetation cover area. Vegetation types inside the vegetation cover area were identified using composite RGB 321 image (true color composite) and composite RGB 423 (false color composite). There area several types of vegetation were recognized inside the area. Spectral patterns for all vegetation type were plotted and analyzed to determine whether there is any differences of pixel value at each vegetation type. The difference of the pixel values can be used to separate the vegetation type in the study area. The vegetation types were classified using the unsupervised classification method (classification without using training samples) and supervised classification method (classification using training samples). The number of classifed classes were determined by identifying the vegetation types found in the study area. The classes of vegetation types in this research were divided into 3 classes: sparse vegetation, low-medium vegetation and medium-high vegetation.

Unsupervised classification was done using ISODATA unsupervised method [13], and it used the following setting parameters: maximum iteration 100, maximum classes 50 and pixel unchanged threshold 98%. The classification result produced 50 classes, which were then reclassed and labelled into three classes of vegetation types that previously determined. Supervised classification was done by collecting training samples for three classes of vegetation types. Each class used around 30 training samples, then classification process was performed using Maximum Likelihood Enhanced Neighbor (MLEN) method. According to the previous result, MLEN method has a higher accuracy than other maximum likelihood methods [14]. The classification results of vegetation type were tested using confusion matrix. Training samples were needed in the confusion matrix, the training samples were different with the training samples for classification input. The result from the confusion matrix was consist of user accuracy, producer accuracy and overalls acuracy of the classification result.

Furthermore, the classification results was overlaid with polygon of green zone. In this research, polygons of green zone was simulation polygon that created as examples to calculate the percentage of vegetation cover in the green zone. The percentage of vegetation cover was calculated by the following equation:

$$P_{Veg(i)} = \frac{A_{Veg(i)}}{A_{GZ}} \times 100\% \quad (i = 1,2,3) \tag{1}$$

$$P_{Total} = \frac{A_{Veg1}+A_{Veg2}+A_{Veg3}}{A_{GZ}} \times 100\% \tag{2}$$

Where, P, A, Veg 1, Veg 2, Veg 3 and A_{GZ} are vegetation cover percentage (%), area, sparse vegetation, low-medium vegetation, medium-high vegetation, and green zone area respectively.

3. Result and Discussiom

Figure 3a shows the separation result of vegetation and non vegetation in the study area, where green indicates vegetation area while black indicates non vegetation area. The vegetation area is also called vegetation cover area. The result of vegetation cover in figure 3a was verified by doing visual comparison between appearances of vegetation cover in figure 3a with appearances of vegetation in figure 1 (Composite RGB 321 of Pleiades imagery). By this comparison, it can be known that vegetation can be separated well with non vegetation around it (such as runway, settlement, water and open land). Figure 3b shows an example result of the vegetation cover in urban areas. Left side is Pleiades imagery in urban area, and right side is the overlay picture of vegetation cover area and Pleiades imagery. Separation of vegetation from non vegetation was done accurately along the main roads and among the dense building. The information of vegetation cover area was then used to crop the Pleaides imagery for obtaining the Pleiades imagery inside the vegetation cover area.

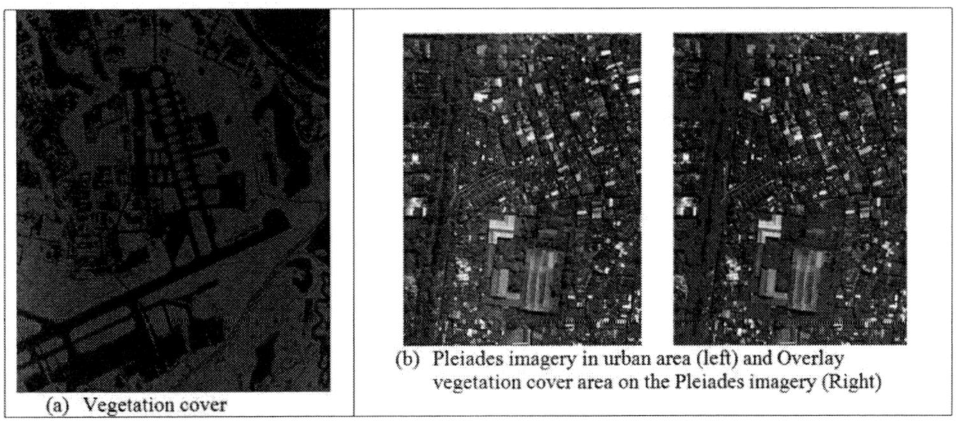

(a) Vegetation cover

(b) Pleiades imagery in urban area (left) and Overlay vegetation cover area on the Pleiades imagery (Right)

Figure 3. Vegetation cover result and Its visual verification

Based on the results from visual observation by using a composite RGB 321 (true color composite) and RGB 423 (false color composite) of Pleiades imagery in the vegetation cover area, vegetation types could be divided into four classes, those were: sparse vegetation (vegetation mixed with soil), low vegetation (such as grass field and other low plants), medium vegetation (like bushes and other plants with a height approximately equal to house building), and high vegetation (dense trees with heights exceeding house building). But in practice it is quite difficult to determine the boundary of medium vegetation, so it causes trouble to separate between low vegetation and medium vegetation, or between medium vegetation and high vegetation. Therefore, the classes of vegetation types were divided only into three vegetation types, those are: sparse vegetation, low-medium vegetation and medium-high vegetation.

Classification of vegetation types was done using two methods, ISODATA unsupervised and MLEN supervised methods. ISODATA unsupervised method was conducted using fixed setting parameters: maximum iteration 100, Maximum classes 50 and pixel unchanged threshold 98%. This classification resulted in 50 unlabelled classes, then the 50 unlabelled classes were reclassified and labelled into three classes of vegetation types (sparse vegetation, low-medium vegetation and medium-high vegetation). While MLEN method required training samples from the three classes of vegetation types as input data. Each class used 30 training samples. The MLEN classification directly produced three classes of vegetation types (sparse vegetation, low-medium vegetation and medium-high vegetation). The classification results using unsupervised and supervised classification method are shown in figures 4a and 4b. Furthermore, the accuracy testing was carried out using the confusion matrix, and the testing results are shown in table 2. The level of accuracy (producer accuracy, user accuracy and overall accuracy) for both the classification results did not differ significantly, and the accuracy of all classification were high accuracy more than 90%. Overall accuracy for unsupervised classification was around 97.9%, while the supervised classification was around 98.9%. Therefore, both classification methods performed good results, and can be used to classify vegetation types accurately from Pleiades imagery.

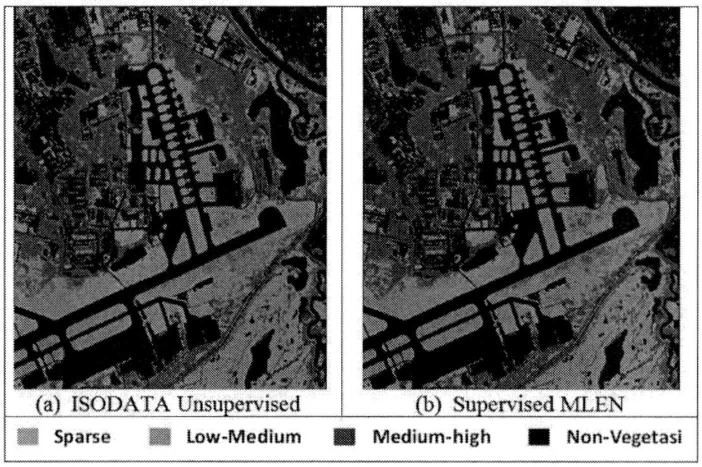

Figure 4. Vegetation type classification using unsupervised and supervised methods

Tabel 2. Accuracy result of vegetation type classification

Method	User Accuracy (%)	Producer Accuracy (%)	Overal Accuracy (%)
ISODATA Unsupervised	98,5	97,7	97,9
MLEN Supervised	99,1	98,7	98,9

The last proces is to estimate vegetation cover percentage inside the green zone. In this resarch, two polygons of green zone (GZ1 and GZ 2) were created in the study area as a simulation to calculate the vegetation cover percentage inside the green zone. The calculation of vegetation crop percentage can be done for all vegetation types. Figure 5 shows two simulation polygons of the green zone (GZ1 and GZ2) which are dioverlaid on Pleiades imagery and the classification result of vegetation types. By doing the visual observation, we can know that GZ1 and GZ2 are consisting of vegetation cover and non vegetation cover (building on GZ1 and open areas on GZ2).

Figure 5. GZ and GZ2 are overlaid on imagery(left) and classification result (Right)

The calculation results of vegetation type area, percentage of each vegetation types cover, and percentage of total vegetation cover and non vegetation cover inside GZ1 and GZ2 are shown in table 3. GZ1 had vegetation cover percentage around 84.8%, which is consisting of sparse vegetation (2.9%), lower-medium vegetation (32.4%) and medium-high vegetation (49.5%). While GZ2 had vegetation cover percentage around 93.5%, which is consisting of sparse vegetation (29.3%), low-medium vegetation (53.1%) and medium-high vegetation (11.1%). This kinds of information can be used by local governments to monitor the last condition vegetation cover in the green zone, to evaluate the existing plan map of the green zone, and to make a better policy for green zone management activities.

Tabel 3. Accuracy result of vegetation type classification

Green Zone	Sparse	Low-Medium	Medium-High	Non Vegetation	Vegetation
GZ1					
Area (ha)	0,145	1,607	2,454	0,756	4,206
Percentage (%)	2,9	32,4	49,5	15,2	84,8
GZ2					
Area (ha)	0,292	0,528	0,110	0,065	0,930
Percentage (%)	29,3	53,1	11,1	6,5	93,5

4. Conclusion

Very high sapatial resolution Pleiades imagery was used to map vegetation cover, classify vegetation types and estimate vegetation cover percentage in the green zone, and the results were verified qualitatively and quantitatively. Vegetation cover can be obtained accurately using combination of NDVI and blue band. Vegetation types in the study area were classified into sparse vegetation, low-medium vegetation and medium-high vegetation using ISODATA unsupervised and MLEN supervised methods. The accuracy of both methods did not differ significantly, where the overall accuracies were 97.9% and 98.9% for unsupervised and supervised method respectively. The vegetation cover percentage was determined by calculating the ratio between the vegetation type area and green zone area. This information will be useful to support green zone management activities.

Acknowledgement
The Author thank to officials of Remote Sensing Application Center, LAPAN, who have facilitated this research activity, and the author also thank to official of Technology and Data Center who has fulfilled the request of the author to provide Pleiades data at the study area.

5. References

[1] Kemen PU 2008 *Pedoman penyediaan dan pemanfaatan ruang terbuka hijau di kawasan perkotaan* (Dirjen Penataan Ruang, Departemen Pekerjaan Umum)
[2] Rahmy W A, Faisal B and Soeriaatmadja A R 2012 Kebutuhan Ruang Terbuka Hijau Kota pada Kawasan Padat, Studi Kasus di Wilayah Tegallega, Bandung *Jurnal Lingkungan Binaan Indonesia* **1**
[3] Mahmoodzadeh H 2007 Digital Change Detection Using Remotely Sensed Data for Monitoring Green Space Destruction in Tabriz *Int. J. Environ. Res.* **1** 35-41
[4] Rao Y, Zhu X, Chen J and dan Wang J 2015 An Improved Method for Producing High Spatial-Resolution NDVI Time Series Datasets with Multi-Temporal MODIS NDVI Data and Landsat TM/ETM+ Images *Remote Sens.* **7** 7865-7891
[5] Hu T, Yang J, Li X, and Gong P 2016 Mapping Urban Land Use by Using Landsat Images and Open Social Data *Remote Sens.* **8** 151
[6] Simarmata M 2012 Kajian Ketercukupan Ruang Terbuka Hijau Berdasarkan Jumlah Penduduk Kota Pematangsiantar *Jurnal Akar* **1** 91-103
[7] Andriyani E P 2012 Pemetaan Greenways Menggunakan Citra Quickbird Di Kota Surakarta *Jurnal Bumi Indonesia* **1**
[8] Trisakti B 2016 Utilization Very High Spatial Resolution Pleiades Data to Identify Vegetation and Its Density *Seminar Nasional Penginderaan Jauh* Depok Indonesia 27 July 2016
[9] Zylshal, Sulma S, Yulianto F, Nugroho J T and Sofan P 2016 A support vector machine object based image analysis approach on urban green space extraction using Pleiades-1A imagery, *Model. Earth Syst. Environ* **2** 54
[10] Caccetta P, Collings S, Hingee K, McFarlane D and Wu X 2011 Fine-scale monitoring of complex environments using remotely sensed aerial, satellite, and other spatial data, *2011 International Symposium on Image and data fusion (ISIDF 2011)* Tengchong Yunnan China 9-11 August 2011
[11] Li X, Meng Q, Li W, Zhang C, Jancso T and Mavromatis S 2014 An explorative study on the proximity of buildings to green spaces in urban areas using remotely sensed imagery *Annals of GIS* **20** 193–203
[12] Nichol J dan Lee C M 2005 Urban vegetation monitoring in Hong Kong using high resolution multispectral images *International Journal of Remote Sensing* **26** 903
[13] Adam S, Vitse L, Johannsen C and Monbliu J 2006 Sediment Type Unsupervised Classification Of The Molenplaat, Westerschelde Estuary, The Netherlands *EARSeL eProceedings* **5** 2
[14] Trisakti B 2012 Peningkatan Akurasi Hasil Klasifikasi Penutup Lahan menggunakan Metode Maximum Likelihood *Prosiding PIT MAPIN XIX* Makasar Indonesia 7-8 Juni 2012

IOP Conf. Series: Earth and Environmental Science 54 (2017) 012004 doi:10.1088/1755-1315/54/1/012004

Transfer function control strategy of Subak rice field land and agricultural development in Denpasar city

Indayati Lanya[*], N. Netera Subadiyasa, Ketut Sardiana, Gst. Putu Ratna Adi
Study Program of Agroecotechnology, Faculty of Agicultural, Udayana University

E-mail: indahnet@yahoo.co.id

Abstract. The success of tourism development in Bali gave a negative impact on Subak rice fields, especially on land convertion over 2579 ha year^{-1} (2002-2013) to the area awakened. Denpasar city has lost rice fields 185 ha year^{-1} and six Subak, as well as potentially losing 10 Subak, as a result of the allocation of space in the region in the Spatial Planing. UNESCO, in 2012 the establishment of Subak as a cultural heritage. Most Subak rice fields designated as an Urban Green Open Space (UGOS). Satellite image Iconos 2002, World 2015 View Coverage of Denpasar, and ArcGIS 10.3 software used for mapping the balance of rice field and violation of land use in the area of UGOS. The control strategy over the convertion of spatial land-based environment is done through zoning map. Land conversion of rice fields for 13 years (2002-2015) in Denpasar (572.76 ha), comes standard acreage of rice fields in 2015. Denpasar city has experienced of food deficits, even in the UGOS has awakened 96.04 ha (24.04 ha year^{-1}). A period of 50 years into the future, rice fields which needs to be protected 872.83 ha, buffer area 984.77 ha, and can be converted 499.81 ha.

1. Introduction

The National Central Bureau of Statistics, released the mass media: conversion of agricultural land to non-agricultural area of 110,000 ha year^{-1}. Impact destabilize national food security. The results of population census of Indonesia as many as 238 million people. Indonesia has experienced a food deficit. Results of research in Bali, obtaining data in the last 10 years there is a reduction of wetland of 800 ha year^{-1} [1]. Subsequent research going over the land 4734 ha (68.61 ha year^{-1}) in Denpasar. Data showed land use as a negative impact of successful development and infrastructure supporting tourism.

Subak was agreed as the indigenous people of Bali, nationally recognized as a world cultural heritage [2]. Along with advances in the tourism sector and the development of population in urban areas, conflicts of interest utilization of water resources. Originally springs as a source of irrigation water, has changed functions for household needs, and tourism have an impact on the lack of irrigation water and wetland function transfer.

Conversion of agricultural land to non-agricultural needs to be controlled by law to provide protection of rice fields Subak as urban green open space, normalization of the urban ecosystem, and the preservation natural and agrarian culture. Regional Regulation concerning Regional Spatial Planning (RSP) and the Detailed Spatial Planning (DSP), not enough to conserving rice fields water control system (Subak), due to punishment and law enforcement is weak. In Act 41 of 2009 [3], the provincial government and the city district is required to have a Regional Regulation (Perda), and the regulation of Regional Head (Perwali) to the protection of sustainable agricultural land (LP2B).

The products will be very applicable law, when equipped with spatial data in the form of a map, such as the location of the fields which need to be protected / preserved, as a buffer, and the fields may be converted to development needs. It is necessary for the analysis of spatial data and the physical condition of the environment, so that the geographical position of the fields that need to be protected and can be converted quickly be known location through information technology in order to facilitate the public service in land use.

The use of satellite imagery of natural resources is a means of high-resolution remote sensing can be used to obtain data quickly with high accuracy [4]. Local Regulation or Regulation mayor, has been mandated by the Act 41 of 2009, Government Regulation (GR) and the Regulation of the Minister of Agriculture (Permentan), related protected LP2B. In other words, the study of control over the land to support agricultural development in urban areas based remote sensing and Geographic Information Systems (GIS) is very urgent, information in order to increase national food security and sustainable agricultural development.

Results of previous studies Land Transfer Function Control Strategy [4] analysis of satellite imagery Quich Bird, 2002, Aster satellite imagery in 2013, and Landsat satellite imagery 8 in 2012 obtained the data transfer funggsi rice field in the province of Bali. In 2002 108,336.9942 ha rice fields area, changed into dry land area of 3530,3569 ha and into other uses 24,839.4552 ha, so that the rice area in 2013 only 79,967.1821 ha. The level of accuracy satellite image analysis 98% of 81,165 ha The results of academic studies related to LP2B been delivered to Bappeda and Department of Agriculture Bali Provincial. At this year's Legislative Assembly of the Province of Bali to discuss the Draft Regulation on LP2B.

Along with the policy direction of agricultural development in the improvement of food supplies. Rice field conversion in urban areas, particularly in Denpasar continue to occur on the existing land supply only form rice field. Design of agricultural land for urban development, especially in the tourism sector and its supporting facilities; Plus a with the lack of oversight of the regional spatial planning, such as violations of UGOS, requires a legal framework to transfer function control of Subak rice fields. Control study of land conversion, in the form of academic studies to determine Subak for conservation of nature and culture, need to be strengthened by the rule of law about control of land conversion and rice fields protection as sustainable agricultural Land.

This research aims to: (1) analysis of rice field area data and mapping of land use in Denpasar in 2002 and 2015 based on remote sensing, (2) to get projection data and the potential food needs until 2015, and (3) create a zonation map rice fields (zoning map: protection, buffer, and converted) through classification of numerical-based remote sensing and GIS.

2. Material and Methods

The research was conducted in Denpasar with an area of 127.78 km^2. located between 08^0 35' 31"- 08^0 44' 49" south latitude and between 115^0 10' 23 "- 115^0 16' 27" east longitude. The north and west by the Badung Regency, east by Gianyar and Badung Strait, south by the Gulf of Benoa (figure 1). Composed of four of the 37 districts and 36 villages and 6 village with 404 banjar/village. The area of West Denpasar District 2,407 ha, East Denpasar District 2,230 ha, South Denpasar District 4993 ha, and 3,139 ha of North Denpasar District.

The research location in Denpasar with a flat-sloping shape of the region, rainfall between 1000-1750 mm year^{-1}, the population of 788 589 people and a land area of 2506 ha rice fields in 2013. Materials used include: topographic maps, Topographic maps for the analysis: watershed, high places, relief, distance from the town center, and irrigation canals. Thematic maps (land suitability, rainfall map, map of rice production, the distance from the city center) and maps of Regional Spatial Plan (RSP) Denpasar 2011-2031 [5]. Satellite images: Iconos 2002, World View Denpasar Coverage 2015. Remote sensing technology is used to create spatial maps of land use, statistical data [6] and [7]. This research tool in the form of specific set of computers with the ArcGIS 10.3 program, is used for the digitization and analysis of thematic maps.

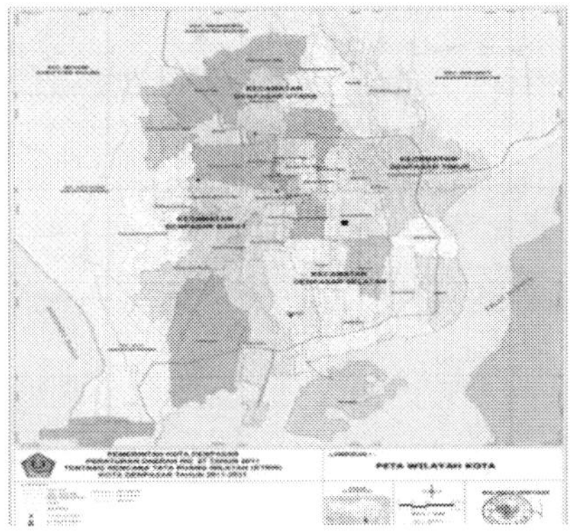

Figure 1. The map of Denpasar City(PerdaRSP, 2011-2031)

Methods used include: (1) wetland function transfer analysis of topographic maps and statistical data for various periods of development, (2) the interpretation of satellite imagery for mapping land use, (3) field survey, (4) Subak rice fields overlay maps with maps of UGOS, (4) digital mapping thematic maps based on GIS, (5) the classification of numerical rice fields, conducted through weighting and scores 10 parameter to get a zoning map using ArcGIS 10.3 program using the overlay method, intersect and query analysis, and (6) the preparation of rice fields classification criteria (sustainable, buffer, and can be converted).

3. Result and Discussion

3.1. Subak Rice Field

Rice area in Denpasar in 1943 is analysis results of topographic maps scale of 1:50000. Rice field in that year was Subak standard acreage of rice fields, because it has not happened transfer function. Rice fields converted into the center of government office center area of Bali provincial level in the early 1970s, happen in Yangbatu Subak. Conversion to residential areas and urban infrastructure happen in Sanglah, Tegalinjung, and Tunggulaji Subak, contained Dauh Puri Kaja in the Dauh Puri, and Pemecutan village.

Denpasar has lost five of Subak in a radius of two km from the city center, and seven area subak <5 ha located in a radius <5 km from the city center potentially converted. The causes of inventory of land for urban development is only a rice field. Recapitulation rice area in 1943, area rice fields of Bali Provincial Cultural Office [8], and statistical data of 1996-1912 (Denpasar CBS) Rice file data after the year 2012 do not suitable to existing conditions in the field. Chart decrease in rice filed area in each district is illustrated in figure 2, the regression equation in figure 3.

Figure 2 shows the land use was highest in 1997-2000. The time is before the Bali bombing and after the riots in Jakarta. In 2002-2007 conversion rice fields flat due to the Bali bombings in 2002 and 2005. The impact of the Bali bomb influence on land use change, indicates the future of tourism has not recovered. Year 2007-2012 rice fields area declined more steeply compared to previous years, during this period tourism has recovered. Denpasar City does not have Spatial Plan of 2005-early 2011. Denpasar has lost five Subak (Sanglah, Tegal Injung, Buluh, Tunggulaji, and Tegal Injung) and the potential loss of seven Subak (Yangbatu, Serogsogan, Peraupan West, Ubung, Petangan, and Kedaton Subak) established as a non-agricultural cultivation area and the extent of <5 ha.

Figure 2 and 3 show that the reduction in rice area is highly significant from year to year, with more than R^2 = 0,75-0.85. The coefficient of the highest decrease in South Denpasar District, followed by East Denpasar. Both areas directly adjacent to the tourist center of Sanur and Kuta. Figure 4 provides information on average over the land 68.61 ha $years^{-1}$ (1943-2012).

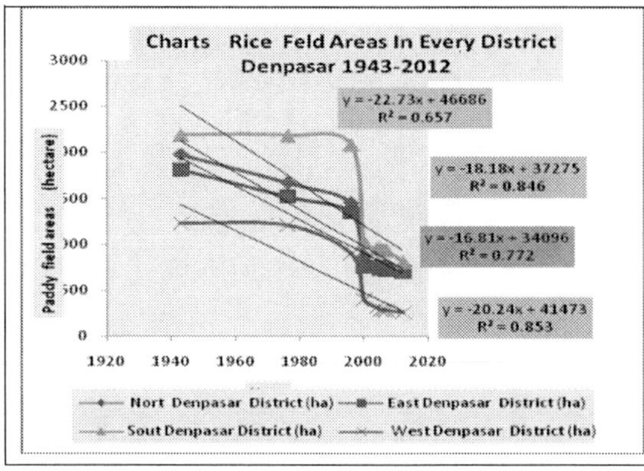

Figure 2. Graph decrease in rice field area and regression in each sub-district in Denpasar (analysis of statistic data, 2016)

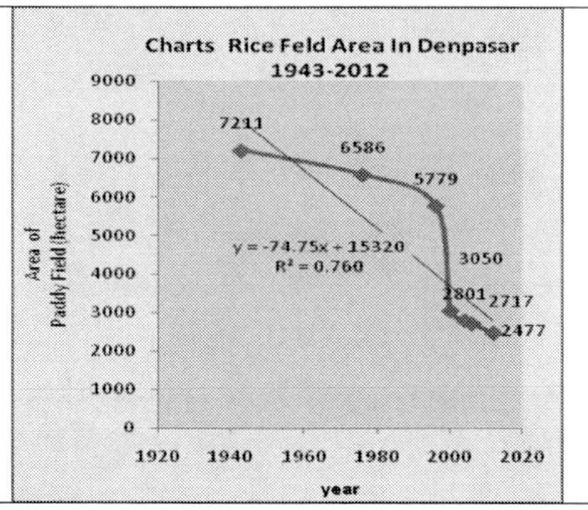

Figure 3. Graph decrease in rice field area and the regression in Denpasar (analysis of statistic data, 2016)

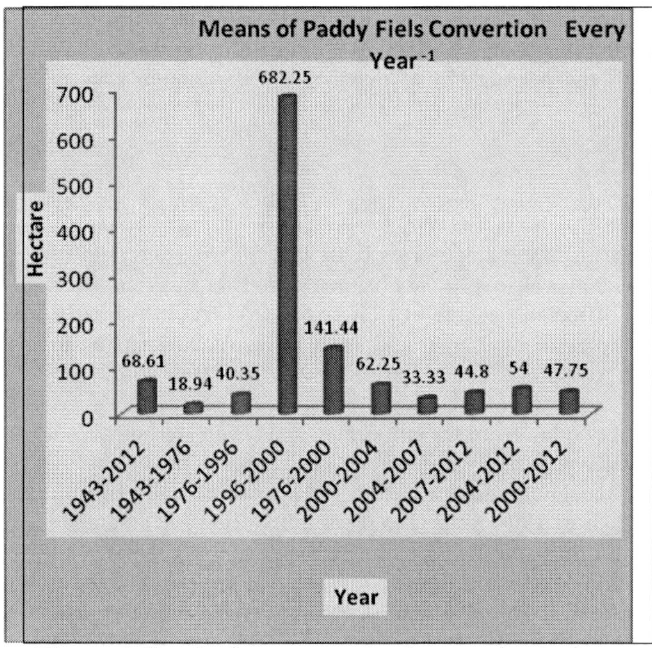

Figure 4. Graph of average wetland conversion in the Old Order-Order Reform (analysis of statistic data, 2016)

Description of Figure 5.

years	Period	Paddy fiels conversion (hectare/year)
1943-2012	Data analysis	68.61
1943-1976	Development of the old order to New Order	18.94
1976-2000	Development of the New Order	141.44
1976-1996	Early until towards the end of New Order (tourism development)	40.35
1996-2000	The transitional period from New Order to Reform (unrest in Java)	682.25
2000-2004	Early Reform Order, there RTRW I	62.25
2004 -2007	TRTW II (Tourism Bombing Bali I & II)	33.33
2007-2012	The status quo SP II	44.80
2004-2012	RTRW beginning to RTRW III (tourism recovered)	54.00
2000-2012	Reform Order RTRW II to RTRW III (average)	47.75

The highest rice field conversion (682.25 ha) in the 1996-2000 period. During that period, a transition period of the New Order to Reform. At that time there were riots in Jakarta and Solo, impact on the exodus to Bali and tourism is growing rapidly. Bali Bombing I and II have an impact on the data transfer decrease in rice field function (only 33.33 ha). In addition the RSP Denpasar City in 2004-2007 can control land conversion. Rice field conversion increases after recovering tourism sector (2007-2012). The data showed: tourism, urban infrastructure and spatial planning influence on rice field conversion in Denpasar. Earlier research 4], the conclusion of population growth and development of urban infrastructures means a negative impact on rice field conversion.

IOP Conf. Series: Earth and Environmental Science 54 (2017) 012004 doi:10.1088/1755-1315/54/1/012004

3.2. *Transfer Function*

Analysis of satellite imagery Iconos 2002, and WorldView Coverage in 2015 acquired data conversion rice fields in Denpasar in table 1, for each village and decrease in of each sub-district in in tables 2. Rice fields in 2002 covering an area of 2,784.24 ha, in 2015 acquired 2,211.48 ha of rice area experienced conversion be undeveloped land 572.76 ha in the period of 13 years (44.06 ha year^{-1}). The highest in South Denpasar District (219.4 ha), followed by West Denpasar (156.62 ha), East Denpasar (105.62 ha) and lowest in North Denpasar (91.12 ha).

Table 1. Area land use change in 2002-2015 in Denpasar City

Satellite Imagery Iconos 2002	Satellite Imagery World View Coverage 2015			
	Land Use (ha)	Rice Fields (ha)	Other Use (ha)	Total (ha)
	Rice	2.211,48	572,76	2.784,24
	Other Use		9.822,95	9.822,95
	Total	2.211,48	10.395,70	12.607,19

Table 2. Area and rice field conversion in each village For 13 years on the analysis of satellite imagery in 2002 and 2015, as well as regression to the district level in Denpasar (Analysis, 2015)

Rural / urban villages in Nort Denpasar District	Area of Rice Fiels 2002	Area of Rice Fiels 2015	Converted rice field
1. Peguyangan Kaja	168.28	149.51	18.78
2. Peguyangan Kangin	159.74	137.11	22.63
3. Peguyangan	153.96	139.87	14.09
4. Ubung Kaja	114.91	86.67	28.24
5. Tonja	10.95	5.66	5.29
6. Ubung	2.76	0.83	1.93
7. Pemecutan Kaja	0.16	-	0.16
Total area	**610.75**	**519.63**	**91.12**

Rural / urban villages in East Denpasar District	Area of Rice Fiels 2002	Area of Rice Fiels 2015	Converted rice field
1. Penatih	210.55	189.06	21.49
2. Penatih Dangin Puri	188.76	176.68	12.08
3. Kesiman Kertalanggu	157.47	134.76	22.70
4. Kesiman	113.01	89.63	23.38
5. Kesiman Petilan	75.18	65.45	9.73
6. Sumerta Kelod	31.03	21.91	9.12
7. Dangin Puri Kelod	13.97	7.83	6.14
8. Sumerta	5.96	4.98	0.97
9. Sumerta Kauh	0.01	0.01	(0.00)
Total area	**795.93**	**690.31**	**105.62**

Rural / urban villages in South Denpasar District	Area of Rice Fiels 2002	Area of Rice Fiels 2015	Converted rice field
1. Pemogan	255.97	202.06	53.92
2. Pedungan	176.96	142.95	34.01
3. Renon	135.01	100.39	34.62
4. Sanur Kauh	111.26	93.42	17.84
5. Sanur Kaja	73.39	52.11	21.28
6. Sidakarya	69.47	51.10	18.37
7. Panjer	52.75	23.58	29.17
8. Sesetan	39.08	28.89	10.19
Total area	**913.89**	**694.49**	**219.40**

Rural/urban villages in West Denpasar District	Area of Rice Fiels 2002	Area of Rice Fiels 2015	Converted rice field
1. Pemecutan Kelod	207.12	155.23	51.89
2. Padang Sambian Kelod	105.15	63.19	41.96
3. Padang Sambian	96.83	50.87	45.96
4. Padang Sambian Kaja	45.32	34.43	10.89
5. Dauh Puri Kauh	5.16	2.91	2.25
6. Pamecutan	3.41	0.42	2.99
Total area	**463.67**	**307.05**	**156.62**

Rice field conversion each years in each district indicated by the regression equation (figure 5). The average conversion of rice field highest in South Denpasar District (16.88 ha year^{-1}), West Denpasar (12.05 ha year^{-1}), East Denpasar (8,12 ha year^{-1}) and the smallest in North Denpasar (7.01 ha year^{-1}). Denpasar West and East Denpasar bordering tourism centers, such as Kuta, Sanur and Nusa Dua Access. Denpasar west directly adjacent to the center of the city tourism. South Denpasar there are a tourism area of Sanur and Nusa Dua are on track access to Kuta, Sanur and area of town.

This research is in accordance with previous studies, that the negative impact of tourism, the Subak rice field [10]. While North Denpasar is an upstream area, rice fields designated as an area UGOS. The data show a negative impact on tourism progress toward rice fields conversion in Denpasar. The existence of Denpasar RSP 2011-2031 are expected to control land use change. Results of the analysis of satellite imagery in 2015 rice fields allocated to UGOS has occurred land use change, especially in the area of South Denpasar.

To determine the cause of rice field conversion to non-agricultural use suitability analysis between rice fields map 2015 (2211.48 ha) and map rice fields in the UGOS (1563.52 ha) set forth in spatial planning regulation of Denpasar 2011-2031. That is until the end of the regional spatial plan (2031), rice fields which can be converted 647.96 ha (40.59 ha year^{-1}). Results of statistical analysis of data rice field area in 2002 and 2015 (the results of satellite image analysis) showed a greater drop is 44.06 ha year^{-1}. This shows that the rice field conversion faster than planned in spatial planning. This phenomenon can be seen from the building violations in the UGOS (figure 6) covering an area of 96.04 ha for 4 years (2011-2015).

Figure 5 gives information on the spread of yellow rice fields located outside the area of UGOS; While green is the spread of wetland defined as an area of green space terbukan. The pattern of the spread of violations of building in the area of UGOS (figure 6) shows that the greatest offense (40.26 ha), in East Denpasar, followed by violations of UGOS in South Denpasar area of 37.1 ha. Similarly rice field area that is outside the UGOS, the largest in East Denpasar, followed by South Denpasar and West Denpasar. This shows that the conversion rice field into non-agricultural and location of such violations were near Kuta-Sanur tourism track. The pace of reduction rice field between 42.34 -74.75 ha year^{-1}, obtained from statistical data analysis rice filed area in 1943-2012 with the regression equation $y = -74.752x + 153\ 205$ $R^2 = 0.7608$. Rice field area on the analysis of satellite imagery in 2002 (2784.20 ha), in 2013 (2355.60 ha), and 2015 (2211.48 ha), with the regression equation $y = -42.341x + 87\ 555$ $R^2 = 0.9898$. When we review the aspects of the legal aspects, the existence of Subak in the RSP, 2011-2031, Subak rice field in Denpasar will be reduced in 913.48 ha.

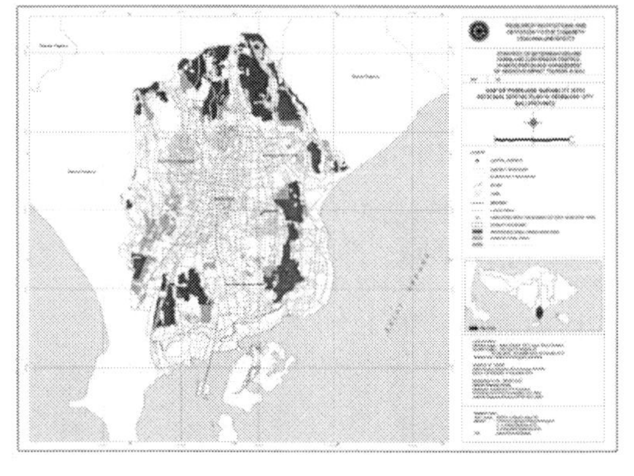

Figure 5. Map suitability of rice field with UGOS in Denpasar City RSP 2011-2031

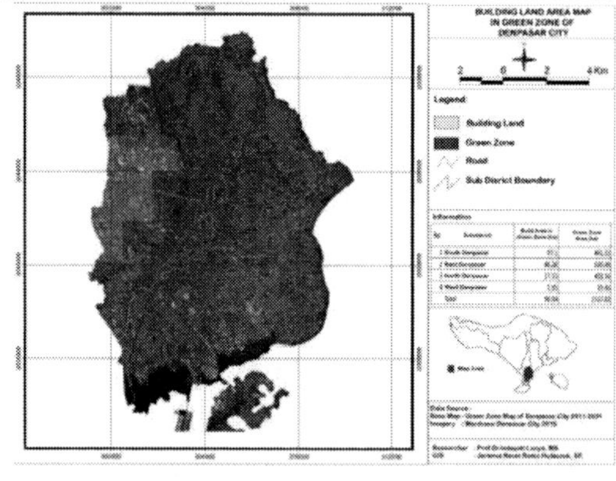

Figure 6. Map of Violations of buildings in the area of UGOS

3.3. Food Balance Sheets

The parameters of food balance sheet includes data rice field area, planting area, harvested area, rice production, population growth, and food needs. Previous research [9] gain Data that food deficits in Denpasar occurred since 2000. Projection rice fields area, food supply, population and food needs in 2015-2050 in figure 10 shows the rice fields in 2050 only 1503 ha (96.13%) of 1563.52 ha allocated to UGOS rice fields in RSP Denpasar City 2011-2031. The regression equation decline in rice field, food supplies, the increase in population and food needs are listed in figure 7. Figure 8 shows between supply and food needs is inversely. Inventories decreased food highly significant, while the increasing food needs are very highly significant. Assumptions 3.4% increase in population every year and the conversion of rice field land 26.78 ha year^{-1}. The regression equation in figure 11 shows the increase in food needs an average of 3630 tons every year. Conversely of food supplies was reduced 87 053 tons every year.

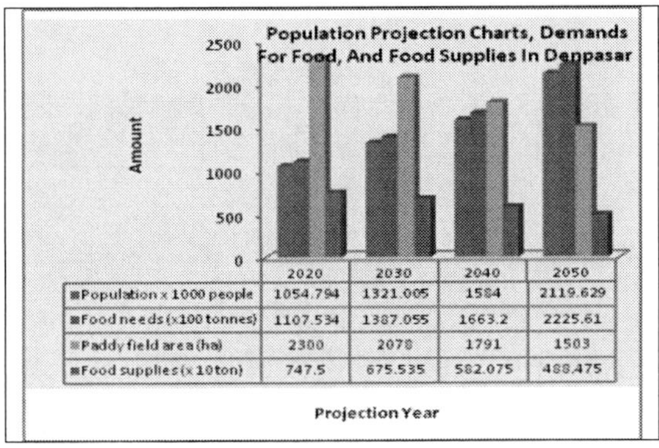

Figure 7. Projected land area rice fields, food supply, population and food supplies years 2020, 2030, 2040, 2050 in Denpasar City.

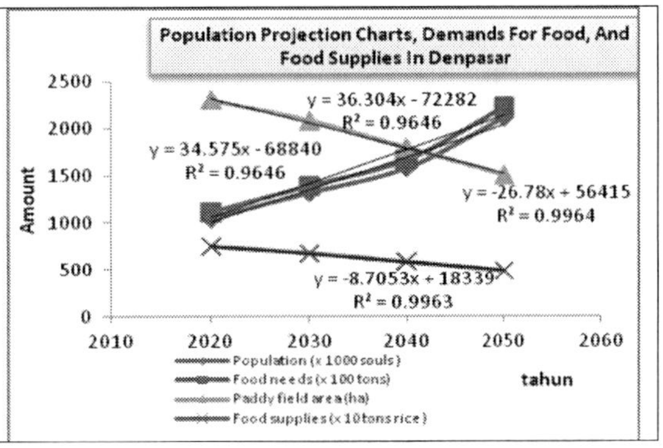

Figure 8. Graph and the regression equation of population, need food, rice area and food supplies years 2020, 2030, 2040, in Denpasar City.

3.4. Zoning of Subak Rice Field

Problems decline in food supplies as a result of land conversion and increased food needs of the population increase has been anticipated by the government with various strategies for agricultural development in Long Term Development Plan (LTDP) and medium-term Development Plan (MTDP). The existence of various regulations, Act No. 41 of 2009, about Sustainable Food Agricultural Land Protection (PLP2B), equipped with a variety of Government Regulation (GR) and the Regulation of the Minister of Agriculture (Permentan), which is associated with PLP2B , and food security. In these regulations, local governments (LGs) are required to have the relevant legislation PLP2B. In these regulations, local governments (LGs) are required to have the rules relevant legislation.

Land for development in the city of Denpasar is very high (42-76 ha year^{-1}). RSP allocating non-agricultural development needs, in rice fields outside the region RSP. That is legally for a period of 20 years, allowed to converted rice fields covering an area of 913.48 ha for non-agricultural development. RSP local regulation negatively impact the the existence rice field (figure 8). While there RSP, urging rapid urban development and oversight of weak local regulation, impact on the violation of UGOS area 1563.51 ha, which is done by the community (figure 7). Although there are RSP, the insistence rapid development of of urban and weak regulatory oversight regions, an impact on violations of UGOS 1563.51 ha which is done by the community (figure 7). Additionally the status of the land of individual ownership and indigenous land that have been certificated, triggered the land conversion.

Therefore, the strategy of the government of Denpasar to conserve the Subak rice fields are set out in the RPJP and RPJM, in the form of tax subsidies in the area of UGOS, revitalizing agriculture, urban agriculture, and household-based farming, advisory services to farmers and most of the fields, in the area of green open space as Agro-ecotourism. Based on the above research results, Subak rice field zoning as a sustainable food agricultural land, is the answer to controlling land use change, and conserve of subak rice fields in Denpasar as a cultural heritage. This requires the Regional Regulation (RR) Protection of Subak rice fields as PLP2B Region, and set forth in the Detailed Spatial Plan (DSP).

The results of numerical classification using the ArcGIS 10.3 overlay 10 thematic maps that have been weighted and scored, as parameters in accordance with its role in the preservation of agricultural land resources based environments. Menu intersect and query analysis is used in the process of GIS to create maps of Subak rice field zoning (zoning map), sustainable (total value> 125), buffer (limited conversion) (between 100-125), and can be converted (total value <100) presented in figure 9.

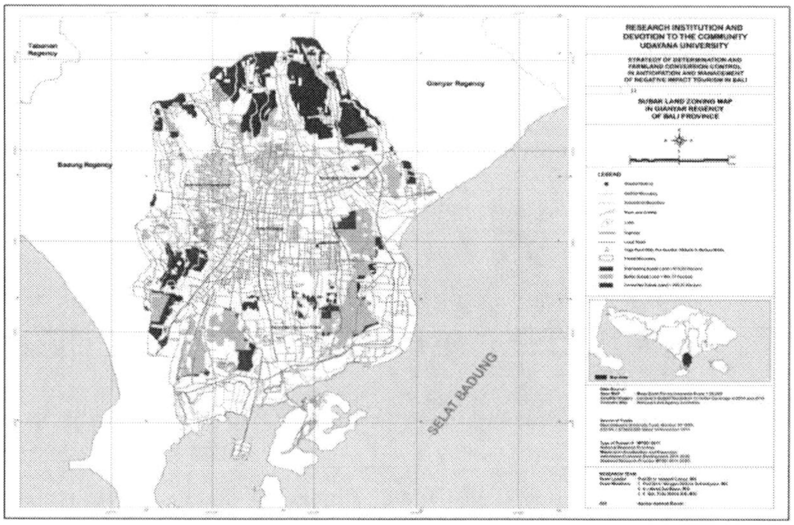

Figure 9. Map Subak sustainable zoning, buffer and can be converted in Denpasar

The numerical classification is the result of the simplification of statistical analysis of modeling, that Subak Lestari = > 125 (> average + 1 standard deviation), Subak Buffer = 107-125 (average-1x standard deviation) to (mean + 1standar deviation), and Subak Converted = <107 (average - 1x standard deviation).

Results of qualifying to get the data rice fields that need to be protected in Denpasar up to 2050 covering an area of 872.83 ha. Rice field as a buffer (limited protection 984.77 ha), and rice fields that can be converted 499.81 ha. Rice field need to be protected and limited protection (buffer) up to 2050 covering an area of 185.6 ha. The modeling results obtained different data, comprehensive sustainable Subak 694.99 ha, spacious Subak buffer = 926.516 ha, and extensive water control system that can be converted 383.269 ha. Classification of different modeling methods will get different data. However, the determination of this classification data required land for urban development needs that require urban green open space minimum 30%.

In RSP set UGOS region covering 1563.52 ha rice fields that need to be protected until 2031. In other words, the result of numerical classification in urban areas can help to define rice fields UGOS needs to be protected for the next 20 years (one LTDP), is a rice field that needs to be protected, and limited protection. For the time of the next 40 years (twice LTDP in 2055), protected rice field (872.83 ha) as UGOS, rice fields buffer/conversion limited 984.77 ha. Rice fields can be converted to non-agricultural construction supplies in urban areas 499.81 ha.

Protected area of rice fields, the total value ≥ 125 of 10 parameters, covering an area of 872.83 ha. There is upstream the city of Denpasar, as water catchment areas, rain catchment, and maintaining the balance of urban ecosystems. Denpasar spread in East, South and West Denpasar a small part. Rice fields protected as UGOS, technical irrigation water, is located> 5 km from the city center. Spread in

North Denpasar five Subak (Kedua, Lungatad, Pakel I, Pakel II, and Subak Sembung), East Denpasar eight Subak (Poh Manis, Anggabaya, Umalayu, Paang ,Taman, Saba), South Denpasar 10 Subak (Sanur, Intaran West, East Intaran, Renon, Panjer, Sesetan, Sidakarya, Kerdung, Cuculan, Kepaon). West Denpasar five Subak (Margaya, Tegal Buah, Tegal Lantang, and Lange Subak). Existence, the necessary legislation or put in the Perda RSP and revised Detailed Spatial Plan. Denpasar already experiencing food deficit since 2013 total – 70,279.38 tons of rice increased food deficit in 2020 total 137,124 tons of rice, and 2050 deficit-275,555.20 tons of rice. The real condition of many rice field that only one of planting due to shortage of water in the dry season and limited irrigation water.

Rice fields of limited conversion/buffer. The total value of between 100-125, the extent of 984.77 ha. Location are 2,5- 5 km from the city center and government services. Its presence in accordance with theRSP as cultivation area of agriculture (crops, horticulture). Integrated farming systems, and /or farming on-farm and off-farm, bailouts and tax subsidies production facilities, and in violation of liable to sanctions. Required Local regulations (Perda) as mandated by the Act. Land conversion only in the public interest so that the necessary permits restrictions on the transfer function according to the GR No.1/2011 [10]. Land conversion must be in accordance with DSP, zoning, and the applicable laws and regulations.

Rice field conversion. Total value of <100, covering an area of 499.81 haa, contained in the RSP and DSP non-agricultural cultivation areas, is only allowed in a radius of <2.5 km from the city center, a radius of 500 m from the center of the capital district, and a radius of 100 m from the left as the right of the provincial road. Land conversion is carried out through licensing of land use changes, based on technical considerations of land use planning and regulations.

4. Conclusion and Recomendation

4.1. Conclusion

Conversion of rice fields is caused by: the allocation of space in the Regional Spatial Regulation, the period of development, need for land for urban development and infrastructure, development of tourism centers, the housing needs of residents, and law enforcement Provincial Regulation on the regional spatial plan.

Analysis of Ikonos satellite image 2002 and image Worldview 2015 coverage, data obtained over rice field function for 13 years an area of 572.76 ha (44.06 ha year^{-1}). The high land conversion in South Denpasar, West and East as the negative impact of tourism development in the area of Kuta and Sanur. Local Regulations Denpasar RSP 2011-2031, establishes rice fields permitted for conversion area of 913.48 ha (45.67 ha year^{-1}). Violations of the building, in the area of UGOS, for 4 years covering an area of 96.0 ha, due to pressure by the community development and weak enforcing the law RSP Local Regulation.

Projected increases in population (+34575 person/year), is inversely proportional to decreasing of rice fields (-26.78 ha year^{-1}). Food supplies in Denpasar, projected to 2050 decreased -87 053 tonnes year^{-1}, is inversely proportional to the needs of food is projected to increase +3630.4 tonnes year^{-1}.

Denpasar has lost five Subak and the potential loss of seven Subak. Food deficit in 2013 as -70,279.38 tons of rice. In 2020, food deficit -137124 tons of rice, and 2050, the deficit -275,555.20 tons of rice. As cultural heritage, of Subak rice fields needs to be protected and conserved. The main objective to produce food, and the normalization of urban ecosystems.

Numerical classification and mapping of rice field based zoning sustainable land resources (zoning map) using remote sensing and GIS technology of the 10 parameters in Denpasar until 2050 with a total value of> 125 as class rice fields to be protected / preserved area of 872.83 ha; The total value of between 100-125 as class rice fields buffer / conversion limited area of 984.77 ha, and the total value of <about 100 as a class can be converted rice fields area of 499.81 ha for the supply of non-agricultural development in urban areas.

Rice fields protected contained in the upstream region (Denpasar North and East), technical irrigated, is in accordance with the UGOS and regional agricultural cultivation, relief slope, land suitability agroekosistem appropriate, harvest index more than 2 times, distance from the town center more than 5 km. Class rice field as a buffer and can be converted, each located at a distance of 2.5 - 5 km and <2.5 km from the city center.

4.2 Recommendation
Control of land conversion in urban areas, can be done with the establishment of of rice fields as an area of UGOS in RSP and DSP, followed by law enforcement. Determination of of rice fields spatially: protected area of rice fields, as a buffer and can be converted to urban development, can use the appropriate technology, namely: PLP2B mapping technology, based on the physical condition of agricultural land and the environment, through the use of remote sensing and GIS technologies.

Regional regulation on PLP2B, equipped with maps of of rice fields area (conservation, buffer, and can be converted), will easier in control land conversion and the calculation of a regional food supply.

Acknowledgments
Thanks and appreciation goes to: Rector and Chairman of the Institute for Research and Community Services University of Udayana, with the approval and guidance in this study; Dean of the Faculty of Agriculture, University of Udayana, with the approval and assistance in implementation research; Head of the Department of Agriculture for the help of data and information provided.

5. References
[1] Lanya Indayati, N. Netera Subadiyasa, Ketut Sardiana and Gst Putu Adi Ratna 2014 Numerical clasification, Subak zoning and land transfer function rice field in Province of Bali on *Remote Sensing and GIS. Procedia Enverimental Sciences*. Vol. 24 pp 47-55
[2] UNESCO 2012 Subak System as Manifestation of the Tri Hita Karana Philosophy (Rusia: St Petersburg)
[3] Republik Indonesia. 2009. Act No. 41 Year 2009 about Sustainable Food Agricultural Land Protection State Gazette of the Republic of Indonesia Year 2009 Number 149. Gazette of the Republic of Indonesia Number 5068. Secretariat of State of the Republic of Indonesia. Jakarta. [cited 2014 March. 7]. Available from: URL: www.dpr.go.id/dokjdih/document/uu/UU_2009_41.pdf.
[4] Lanya I 2010 Research Technology Applications in Remote Sensing and GIS Inventory and mapping of Natural Resources and Environmental Potential of Bali Province *The final report of the research results* (Bali: Udayana University, Bukit Jimbaran)
[5] Denpasar Government 2011 *Regional Regulation No. 27 In 2011 the Spatial Plan (RTRW) Denpasar City from 2011 to 2031* (Denpasar: Denpasar government)
[6] Central Bureau of Statistics (BPS) Denpasar 1996-2013 *Denpasar in Figures* (Denpasar: BPS)
[7] Department of Agriculture and Horticulture Denpasar 2014 *Books Agricultural Statistics Denpasar* (Denpasar: Department of Agriculture and Horticulture)
[8] Bali Provincial Cultural Office 1976 The seminar Subak Provincial Development Planning Agency Bali, Denpasar
[9] Lanya I, N. Netera Subadiyasa, Ketut Sardiana and Gst Putu Adi Ratna 2015 Transfer fungsion control strategy Subak rice field through zoning map and zoning teks. *Proceedings Senastek (National Seminar on Science and Technology)* (Badung Bali: Institute for Research and Community Services, Bukit Jimbaran)
[10] President of the Republic of Indonesia. 2011. Government Regulation No. 1 Year 2011 About Determination and Sustainable Food Agricultural Land Transfer Function. State Gazette of the Republic of Indonesia Year 2011 Number 2. Gazette of the Republic of Indonesia Number 5185. Secretariat of State of the Republic of Indonesia. Jakarta. [cited 2014 March. 7]. Available from: URL: www.hukumonline.com/ pusatdata/ downloadfile/lt4d467fa097aad/ /lt4d46795fea065

Deriving vulnerability indicators for crop production regions in Indonesia

Perdinan[1,2,4]* , Tri Atmaja[2,4], Ujang Sehabuddin[3], Yon Sugiarto[1,2], Lina Febrianti[1] and Ryco Farysca Adi[2,4]
[1] Department of Geophysics and Meteorology, Bogor Agricultural University
[2] Generasi Hijau Indonesia
[3] Department of Natural Resources and Enviromental Economics, Bogor Agricultural University
[4] PI AREA

E-mail: perdinan@gmail.com

Abstract. Food supply is considered as one of the most vulnerable to the effects of climate change. Higher temperature and changes in rainfall patterns and intensity may adversely impact crop production, which will eventually affect the food supply. Consequently, adaptation strategies should be devised to minimize the potential adverse impacts and maximize its potential benefits. The adaptation strategies should be devised by considering factors contributed to causing vulnerability following the concept of food supply chain, starting from production to consumption. This study focuses on identifying the contributed factors to vulnerability of crop production regions in Indonesia. The contributed factors were identified by defining indicators for each component of the food supply chain using an example of crop production centers in Indonesia, the West Java Province. The identification considers existing issues of the food supply chain, covering aspects of production, post-harvest and storage, distribution, and consumption, based on the field surveys conducted in Indramayu district of the West Java, the main grower of paddy production, and Garut district of the West Java, the main grower of corn production. The selection of the vulnerability indicators was also considered the data availability for the study area. The analysis proposed a list of indicators classified into production, post-harvest and storage, distribution and consumption that are proposed to assess the regional vulnerability of crop production regions in Indonesia. This result is expected to contribute in understanding the process of devising climate change adaptation intended for enhancing food supply resilience to climate change.

1. Introduction

Food supply chain is a complex process, composing of several steps starting from production, post-harvest dan storage, distribution, and consumption. Many factors control each step of the supply chain, which can be generally categorized into biophysical factors (e.g., climate, land) and socio-economic condition (e.g., access to seeds, fertilizers, machinery, farming technology, farmers' network, storage system, transportation, and etc). Climate fluctuation has been widely acknowlegded affecting crop productivity over a crop growing region, assuming the land fertility is maintained. This climate sensitivity has drawn attention worldwide in anticipating the potential impacts of climate change on crop production [1-3], which can eventually affect the food supply. The socio-economic condition contributes to create enabling condition, namely: the farming activities (i.e., the production process), the access to crop production, and the consumption to the food supply.

As an agricultural country, Indonesian government is already paid a serious attention to stabilize food supply, particullary staple food (i.e., rice), in meeting the domestic demand for food. Two major challenges associated with the crop production process are climate extremes and agricultural land conversion. Climate extremes, which may be exarcebated by the global climate change, can hamper crop growth and development [4, 5], resulting in declining farm production [6, 7]. Climate extremes, leading to climate hazards (e.g., droughts and floods), have been reported posing serious challenges in crop production over central growing areas in Indonesia, such as West Java Province, one of the national production centers of food crops (i.e., paddy and corn). The province contributes about 15% and 4.9% for the national rice and corn production, respectively [8]. Two main crop growing areas in the West Java are the Indramayu district for paddy (about 14.3% of the provincial paddy production) and the Garut district for corn (about 59.2% of the provincial corn production) [8]. Climate extremes associated with drought caused paddy production failure of about 16 thousand hectares and paddy production damage called 'puso' of about a thousand hectare in Indramayu [8, 9]. In addition, drought occurred in the district of Garut reported on 25/9/2012 damaged thousand hectares of corn production ('puso') such as those occurred in the sub-district of Banyuresmi [10].

Conerning the impacts, it is important to learn factors caused the impacts. Theoritically, the potential impacts may be different spatially with regard to the level of regional vulnerability, which is defined as "*the propensity or predisposition to be adversely affected, encompasses a variety of concepts and elements including sensitivity or susceptibility to harm and lack of capacity to cope and adapt*" [11]. At the national level, Kementerian Lingkungan dan Kehutanan (Ministry of Environment and Forestry) already launched a tool called Sistem Informasi Data Indeks Kerentanan (SIDIK) (System Information on Vulnerability) in 2015 [12]. This tool is currently purposed to measure the level of regional vulnerability based on the components of risk indicators [11, 10], calculated using available socio-economic data. Thus, further development can be proposed to expand the capability of the SIDIK. This research is proposed to develop vulnerability indicators specifically for food security derived based on the concept of supply chain.

The district of Indramayu and Garut are chosen as the case study. This case was employed particularly to explore contributed factors (i..e, indicators) to measure the vulnerability level that should be considered at each stage of the supply chain (i.e., production, post harvest and storage, distribution, and consumption). The exploration was performed through literature review and stakeholders' engagement, i.e., discussions and meetings with key players of the supply chain in the two crop production centers. The derived indicators can provide inputs to improve the SIDIK to measure the vulnerability of agriculture sector. The indicators can also provide valuable information on what strategies should be devised, concerning the contributed factors to the regional vulnerability. In the context of climate change, the anticipated strategies are called as adaptation strategies directed to minimize the potential negative impacts of climate change and to maximize its positive impacts [11].

2. Related works

The development of vulnerability indicators for agriculture sector in Indonesia is not really new; although, the indicators derived based on the concept of food supply chain is still rare. As an example, there is on-going works on the development of vulnerability indicators for complex areas called as Agropolitan in Malang. This work aims at developing vulnerability indicators for the components of Agropolitan, namely: agriculture, economy, industry, and tourism [13]. Another study [14] analyzed the factors affected corn production in Garut, identifying landarea and seeds significantly controlled corn production. Current work conducted by research and development division of Ministry of Agriculture developed vulnerability indicators in accordance with the definition of food security of the FAO (1996) and Law No. 7 in 1996. The food security defined criteria that should be fulfilled, namely 1) availability; 2) supply stability; 3) accessibility/ affordability, and 4) quality/safety [15].

3. Methodology
Concerning the focus of this research on crop production center, spatial variability of crop production areas in the study region (i.e., West Java) were evaluated. Information on paddy and corn production obtained from the National Statistical Agency were mapped using spatial tools to provide information on crop growing center in the study region. For the selected production centers (i.e., Indramayu for paddy and Garut for corn), spatial variability of productivity and harvested area for paddy and corn was also evaluated to identify major crop growing areas in the centers. This information in accordance with climate classification over the region provides information on which climate type is suitable for crop growing areas. The climate classification was completed using climate classification techniques (i.e., Koppen, Oldeman, and Schmidt – Ferguson) applied to climatological gridded rainfall data of WorldClim 1971-2000 [13, 16]. The development of the vulnerability indicators were completed based on the concept of supply chain and the vulnerability components (i.e., "Sensitivity" and "Capacity") and the other risk component namely "Exposure" proposed in the IPCC report [11]. The concept of supply chain has been applied to identify factors and its associated issues within each stage of the supply chain, i.e., production, post-harvest and storage, distribution, and consumption, under the exposure of climate change [17]. Field survey was conducted at the two crop growing areas (Indramayu and Garut) to interview key players of each stage of the supply chain. The interview was conducted through meetings or group discussions to explore factors and issues within each stage of the supply chain. This field survey provided information on what indicators should be derived to measure the vulnerability of crop production centers. We grouped the indicators contributed to each stage of the supply chain into the vulnerability components ("Sensitivity" and "Adaptive Capacity") and "Exposure", following the definition of each vulnerability component [11]. The major source of data employed for identifiying the vulnerability indicators in this study were obtained from the National Statistical Agency named *Badan Pusat Statistik* (BPS), i.e., *BPS Dalam Angka* (BPS in Numbers) 2015, *Sensus Penduduk* (Population Census) 2010, *Survei Potensi Desa* (Village Potency Survey) 2014 and *Sensus Pertanian* (Agricultural Census) 2013.

4. Results and discussion
West Java province is one of the national production centers of paddy and corn. Production areas of paddy are spatially distributed across the northern and southern areas of the province, with the largest production centers of over 800 thousand tons/year located along the northern coast (i.e., Karawang, Subang and Indramayu). For corn, the largest corn production centers located in Garut with a total production reaching to 600 thousand tons/year (figure 1).

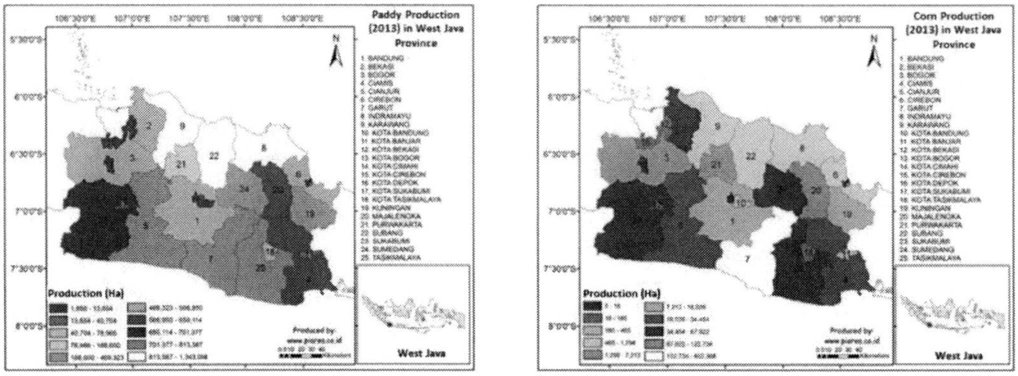

Figure 1. Regional variability of paddy (left) and corn (right) production areas in West Java 2013 [23]

Further exploration in the case study area is directed to understand the diversity of regional climate types. This information can provide information on the climate type for each crop production center. The application of Oldeman climate classification technique suggested that climate types for West Java can be categorized into 12 climate zones. This classification is a little bit different with the classification of Badan Meteorologi, Klimatologi, and Geofisika (BMKG) – Meteorology, Climatology, and Geophysics Agency, whose categorized the West Java into 46 seasonal climate zonation or called as ZOM [18]. The paddy production center of Indramayu comprises three types of Oldeman classification, namely: D3, C2 and C3; whereas, with refer to the BMKG classification, the region can be classified into ZOM 77, 78, 79 and 80. For Garut, the Oldeman climate types are A, B1, B2, C1 and C2, and the BMKG categorizes the region into ZOM 88 and 89 for the year of 2015/2016. Generally, Indramayu and Garut have monsoonal climate pattern, which is characterized by the unimodial monthly precipitation over a year. This pattern suggests that the months of June, July, and August are the dry months, while December, January, and February are the wet months.

The different climate types for the district of Indramayu and Garut; although, the general pattern (i.e., the monsoonal pattern) are the same, reveals that spatial variability of crop productivity can also be explained by the climate types. The Oldeman climate type is applied considering the technique is purposed to classify climate for crop production region [17, 19]. For the purpose of comparison, the Koppen and Schmidt-Ferguson classification types were also applied to the same gridded climate data for the West Java. The classification shows a little different because of the criteria applied for each climate classification; however, the general pattern shows latitudinal direction of climate groups where the southern and norther part of the region exhibit distinct climate types. The climate classification also tells that the district of Indramayu located in the northern area and the district of Garut located in the southern area has different climate types (figure 2).

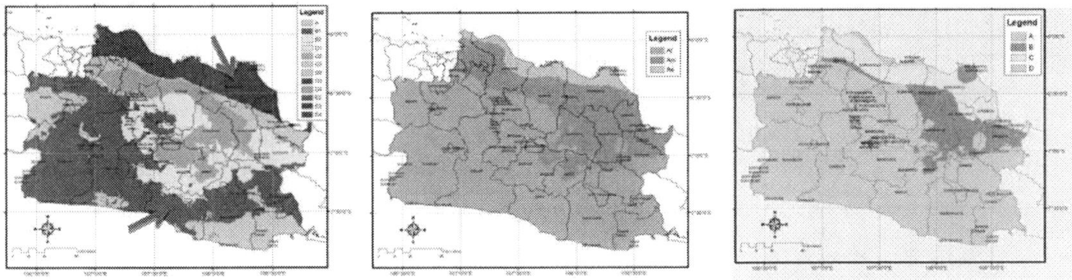

Oldeman Koppen Schmidt - Ferguson

Figure 2. Climate zonation of the West Java determined by the classification methods of Oldeman (left), Koppen (middle), and Schmidt – Ferguson (right) applied to gridded climate data of WorldClim 1971-2000. The legend in the right corner informs about the climate types showed in different colors.

The diversity of climate classification can be the reason for Indramayu and Garut as the major growing areas for paddy and corn, respectively. For Indramayu, the major growing areas of paddy, the areas with the largest productivity are in the sub-district of Widasari (a total production of 78 to 80 kw/ha), and the lowest productivity are in the sub-district of Kandanghaur, Indramayu and Sindang (a total prodiction of ≤ 62 kw/ha). The harvested area in Indramayu are considered lower on the east and the north coast (figure 3); which is reasonable as the two areas are the main areas for industry and goverment offices. However, the highest productivity areas such as the sub-district of Widasari does not always automatically reflect the highest cropping intensity (CI), calculated based on the proportion of harvested area to the area of paddy fields. The Widasari has the value of CI ~2.03, which means two planting times a year. This CI value is lower than that for the District Gantar whose CI is ~3.18.

As for corn, the productivity of corn production at each sub-district of Garut district ranges from 60 to 80 kw/ha, with the highest productivity in the northern areas. The harvested area of corn in the district are generally considered low to moderate (0-3000 ha), only several areas recorded relatively high harvested area such as the sub-district of Blubur Limbungan (7000 ha) (figure 4). The CI of corn in the district of Garut on average is about 2.82, which means about three planting times a year.

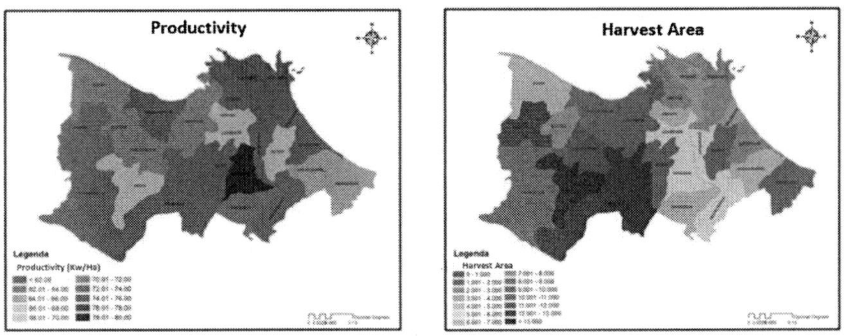

Figure 3. Productivity and harvested area of paddy in the district of Indramayu in 2014. Data source: BPS Indramayu dalam Angka 2015 [22].

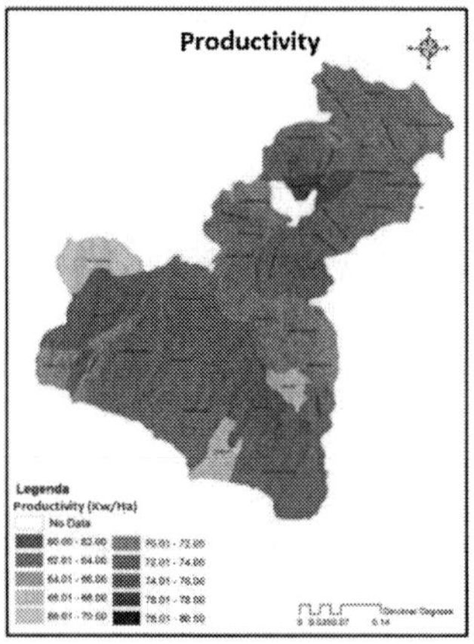

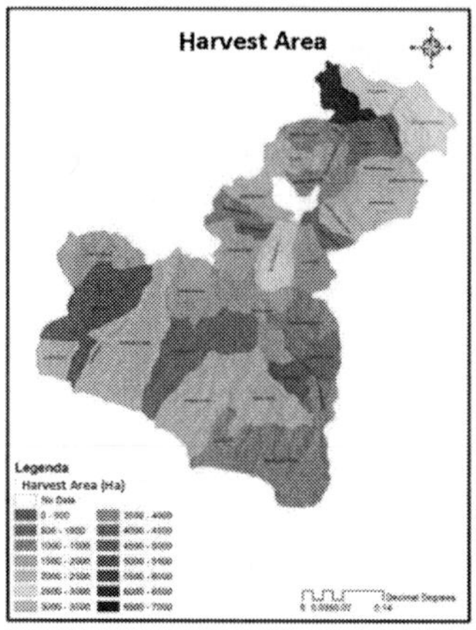

Figure 4. Productivity and harvested area of corn in the district of Garut. Data source: BPS Garut dalam Angka 2015 [8]. Legend color explains the level of productivity / harvested area from high (blue) to low (orange)

Unfortunately, these production centers face serious challenges due to land conversion and erratic rainfall causing difficulty in determining the begining of growing season. It was reported that the largest conversion rate occured in 2000 to 2001; about 189,813 hectares of paddy fields were converted into non-agricultural uses [8]. This decline in planted areas is subseted with better farming management and technology purposed to increased crop productivity. However, unpredictable growing season can cause difficulty to farmers and may result in delaying the sowing date. Climate extremes, leading to climate hazards (e..g, drought and floods), also exarcebate the challenges. This challenge, hereafter frequently known as climate risk, is a function of "Hazards", "Vulnerability" and "Exposure". Hazards in this context is a function of climate variability and change; whereas, Vulnerability and Exposure are a function of enabling condition surrounding human activities such as socio-economic condition [11].

Understanding the concept of risks, the development of indicators for measuring the component of risks, namely: Vulnerability and Exposure [11], was completed by considering contributed factors to the food supply chain identified in the two crop production centers, i.e., Indramayu and Garut. The identification considers factors contributed to each stage of the supply chain (i.e., production, post-harvest and storage, distribution, and consumption) and the related studies on vulnerability assessment [13, 15, 20, 21]. For example, the statement of Agriculture Minister, Bapak Andi Amran Sulaiman, who said that population densisty including agricultural workers, water resources, and access to agricultural inputs are challenges that should addressed [9] for sustaining crop production. The previous study [13, 20] suggested that crop production is determined by factors such as the sources of family income and agricultural workers to measure vulnerability. The ratio of paddy/corn production per planted area and the ratio of crop production area to agricultural land should be considered as the vulnerability indicators [15]. This identification produced a list of indicators categorized into each stage of the supply chain as presented in figure 5.

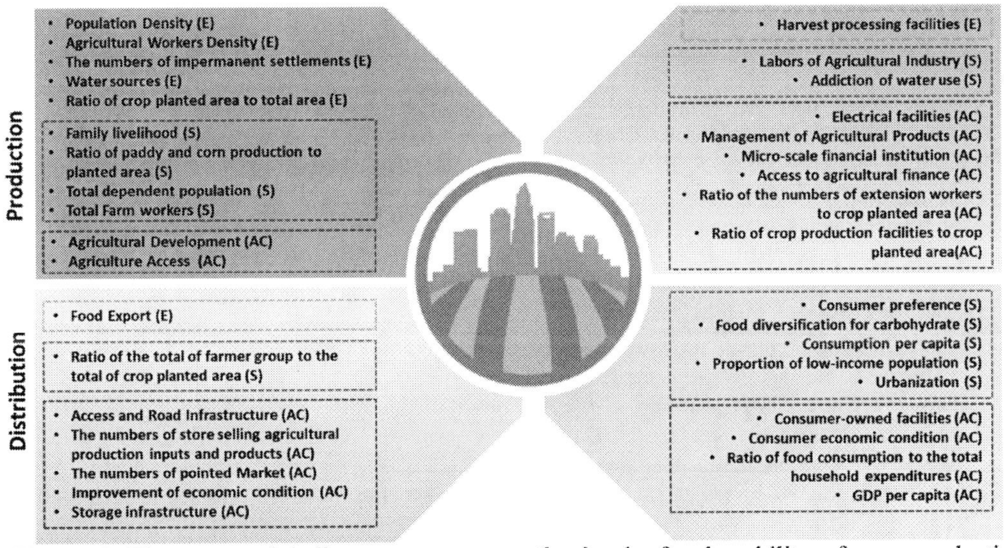

Figure 5. The proposed indicators to measure the level of vulnerability of crop production regions. The abbreviation in the brackets indicated the risk components of E (Exposure), S (Sensitivity) and AC (Adaptive Capacity) as defined by IPCC [11]

5. Conclusion

Agriculture is a sector considered mostly influenced by climate fluctuation. Classification of climate types for the case study (i.e., the West Java) identified that differences in climate type might dictate the commodity grown in the crop production centers. The district of Indramayu, located in the nothern area of West Java, and the district of Garut, located in the southern area of West Java, have differences in climate type and major crop commodity grown, namely: paddy for Indramayu and corn for Garut. Unfortunately, the production centers are currently facing challenges due to land conversion and unpredictable growing season. Future climate change may exarcebate this challenge. As an approach to devise proper anticipation strategies, this research derived indicators for Vulnerability and Exposure based on contributed factors to the supply chain of the commodities in Indramayu and Garut. The indicators were developed based on literature review, meetings and discussions with key players of each stage of the supply chain, i.e production, post-harvest and storage, distribution, and consumption. Each stage of the supply chain can be associated with specific vulnerability indicators as proposed in this study presented in figure 5. This information can help further studies focussed on the regional vulnerability assessments, which can be used as a reference study to identify proper interventions required to support the targets of sustaining or enhancing food security.

References

[1] Cline, William R 2007 *Global Warming and Agriculture: Impact Estimates by Country* (Washington, DC: Center for Global Development and Peterson Institute for International Economics)
[2] Fischer, Shah G M, Tubiello F N and Velhuizen H v 2005 Socio-economic and climate change impacts on agriculture: an integrated assessment, 1990-2080. *Philosophical Transactions of the Royal Society B-Biological Sciences* **360** 2067-2083
[3] Motha R P and Baier W 2005 Impacts of present and future climate change and climate variability on agriculture in the temperate regions: North America *Climatic Change.* **70** 137-164
[4] Hoogenboom G 2000 Contribution of agrometeorology to the simulation of crop production and its applications *Agricultural and Forest Meteorology.* **103** 137-157

[5] Parry M C, Rosenzweig and Livermore M 2005 Climate change, global food supply and risk of hunger. Philosophical Transactions of the Royal Society B-Biological Sciences **360** 2125-2138

[6] Santoso, Heru and Nur Layli 2011 Dampak Perubahan Iklim Terhadap Produksi Dan Pendapatan Usahatani Jagung (*Zea mays* L) *J Agrise.* **11** 151

[7] Tubiello F N C, Rosenzweig R A, Goldberg S, Jagtap and Jones J W 2002 Effects of climate change on US crop production: simulation results using two different GCM scenarios. Part I: Wheat, potato, maize, and citrus *Clim. Res.* **20** 259-270

[8] Statistics Indonesia 2015 Kabupaten Garut Dalam Angka 2015. (Garut: Badan Pusat Statistik Kabupaten Garut)

[9] Tempo 2016 [cited 2016 31 August]; Available from: https://m.tempo.co/read/news/2015/08/07/058689825/kekeringan-di-jawa-barat-ini-dampaknya-terhadap-produksi-padi

[10] [PEMKAB Garut] 2015 Ribuan Hektare Tanaman Jagung Gagal Panen [cited 2016 16 October]; Available from: http://www.garutkab.go.id/pub/news/detail/8435-Ribuan-Hektare-Tanaman-Jagung-Gagal-Panen.html

[11] IPCC 2014 Climate Change 2013: The Physical Science Basis. Contribution of Working Group I to the Fifth Assessment Report of the Intergovernmental Panel on Climate Change. Cambridge University Press, Cambridge, United Kingdom and New York, NY, USA, 1535 pp, doi:10.1017/CBO9781107415324.

[12] Ministry of Environment and Forestry 2015 Sistem Inventori Data dan Indeks Kerentanan Online. Retrieved from http://apps.cs.ipb.ac.id:8008/administrator/dashboard on 17 October 2015.

[13] Perdinan, Sugiarto Y, Santikayasa I P, Dasanto B D, Adi R F, Atmaja T and Arini E Y 2016 Analisis Adaptasi Perubahan Iklim Kawasan Agropolitan Kabupaten Malang. UNDP – IPB. Bogor

[14] Perdinan, Kartikasari K and Malahayati M 2015 Promotion of Climate Resilience in Rice and Maize Indonesia National Study. Deutsche Gesellschaft für Internationale Zusammenarbeit (GIZ) GmbH, Bonn and Eschborn - Germany.

[15] Estiningtyas W 2015 Kerentanan Pangan dan Risiko Iklim Berbasis Sumberdaya Lahan, Iklim dan Air. Balai Penelitian Agroklimat dan Hidrologi, Badan Litbang Pertanian – Bogor [personal communication]

[16] Hijmans R J, Cameron S E, Parra J L, Jones P G and A Jarvis 2005 Very high resolution interpolated climate surfaces for global land areas *International Journal of Climatology* **25** 1965-1978

[17] Agustian A 2015 Analisis Faktor-Faktor yang Mempengaruhi Produksi pada Usaha Tani Jagung di Kabupaten Garut, Provinsi Jawa Barat. Indonesia. Pusat Sosial Ekonomi dan Kebijakan Pertanian, Badan Penelitian dan Pengembangan Pertanian, Ministry of Agriculture.

[18] BMKG 2016 Prakiraan Awal Musim Hujan 2015/2016 Zona Musim di Banten, DKI Jakarta dan Jawa Barat. Badan Meteorologi Klimatologi dan Geofisika

[19] Handoko 1995 Klimatologi Dasar (Jakarta: Dunia Pustaka Jaya)

[20] Perdinan, Rizaldi B, Adi R., Anter R S, Mirna Z and Betty N 2015 Pilihan Adaptasi Perubahan iklim Nusa Tenggara Timur. UNDP-SPARC Project. Ministry of Environment and Forestry. Jakarta

[21] Perdinan, Sugiarto Y, Impron, Sehabudin U, Adi R F, Atmaja T, Edyanto, Dewi W S and Arini E Y 2015 *Pengembangan Metode Kajian Risiko Iklim Fokus Anak* (Bogor: UNICEF – IPB)

[22] Statistics Indonesia 2015 *Kabupaten Indramayu Dalam Angka 2015* (Indramayu: Badan Pusat Statistik Kabupaten Indramayu)

[23] Statistics Indonesia 2016 *Agricultural Cencus, National Statistical Agency 2013* (Jakarta: Badan Pusat Statistik Indonesia)

Assessment of the potential enhancement of rural food security in Mexico using decision tree land use classification on medium resolution satellite imagery

A Bermeo and S Couturier*
Laboratory for Geospatial Analysis (LAGE), Geography Institute,
National Autonomous University of Mexico (UNAM)
Circuito Exterior s/n, Del. Coyoacán, CP:04510, Mexico City, Mexico

E-mail: stephcamelo@mailcity.com

Abstract. Because of its renewed importance in international agendas, food security in sub-tropical countries has been the object of studies at different scales, although the spatial components of food security are still largely undocumented. Among other aspects, food security can be assessed using a food selfsufficiency index. We propose a spatial representation of this assessment in the densely populated rural area of the Huasteca Poblana, Mexico, where there is a known tendency towards the loss of selfsufficiency of basic grains. The main agricultural systems in this area are the traditional *milpa* (a multicrop practice with maize as the main basic crop) system, coffee plantations and grazing land for bovine livestock. We estimate a potential additional milpa - based maize production by smallholders identifying the presence of extensive coffee and pasture systems in the production data of the agricultural census. The surface of extensive coffee plantations and pasture land were estimated using the detailed coffee agricultural census data, and a decision tree combining unsupervised and supervised spectral classification techniques of medium scale (Landsat) satellite imagery. We find that 30% of the territory would benefit more than 50% increment in food security and 13% could theoretically become maize self-sufficient from the conversion of extensive systems to the traditional multicrop *milpa* system.

1. Introduction

Because of its importance in international agendas (e.g. the Zero Hunger Challenge by the United Nations), assessing food security has been the object of increased focus in many countries. A vibrant debate exists as to the contribution to food security of agroindustrial and capital intensive agricultural systems with respect to smallholder and traditional cultivation systems. To the advantage of the latter, it is claimed that environmentally harmful production practices are counterproductive since production in the long run decreases in marginalized areas [1, 2]. In this sense, production of food for future demand is a great international challenge since it should be constrained to existing farming systems without expanding extensive agriculture and grazing lands [3].

Following the hypothesis that smallholder agricultural systems are more sustainable than extensive agricultural systems and monoculture lands, research has recently emerged on the spatial characterization and analysis of smallholder rural territories. For example in Mexico, spatially explicit studies have focused on the extent of indigenous territories as fundamental biocultural entities [4], on the compatibility or not of smallholder agriculture with surrounding forest conservation [5], and on the availability, or lack of availability of land for traditional cultivation systems [6].

Food security is defined by FAO [7] as "the access to sufficient and nutritious food by all people at all times to lead a healthy and active life". We suggest that an innovating insight on the problem of food security in marginalized rural areas could be to estimate its potential improvement by intensifying traditional cultivation systems. Specifically we propose to assess food security benefits as a consequence of (re)establishing the milpa system on the land currently under extensive agriculture and grazing systems in Mexico. The balance between availability of food and population intake needs is an approach [7] involving food self-sufficiency, employed for example to build the Global Food Security Index of The Economist Intelligence Unit [8].

In this research, food security is approximated as self-sufficiency of basic foods (balance between basic food production and its consumption) in a typical, densely populated, rural area in Mexico. The area pertaining to capital intensive agricultural systems is first derived from landscape scale classification of satellite imagery and fine scale analysis of the agrarian structure. Then, the Potential Increment of Food Security (PIFS) is presented as a spatially explicit index, based on the hypothesis that this area is converted to the traditional milpa cultivation.

2. Study area and available spatial data

The study area covers approximately 1337 km^2 of the Sierra Madre Oriental eastern mountain range. It corresponds to the Xicotepec Rural Subdistrict (see figure 1) which is the tropical portion of the Rural Development District 01 of the Puebla State, in the United States of Mexico.

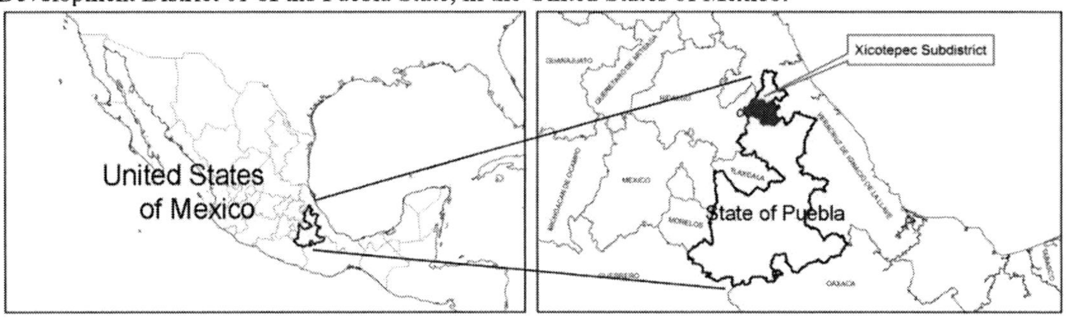

Figure 1. Location of the study area (area in red color)

The Sierra Madre Oriental is one of the main densely populated indigenous regions of Mexico. Maize cultivation was traditionally the main agricultural system in the region, however it was reported with a steadily decreasing production in the last century, while coffee and pasture based systems arose and developed till occupying the major portion of the landscape nowadays, namely the largest coffee region in the State of Puebla. Most livestock estates as well as large coffee plantations are clearly related to capital intensive agricultural activities. Original vegetation ranges from oak forests upland to medium tropical forest and include a rich set of cloud forest ecotones.

196 rural settlements are distributed throughout the study area, each of which received a coffee census detailed at the level of production units in 2004 [9]. Production data from most crops is also reported in the agricultural and pasture activity census ("Censo Agropecuario", [10]. The pasture land component of the landscape can be estimated from the land use national cartography (e.g. [11]) or in medium scale satellite imagery (Landsat, ASTER) available online from the GloVis USGS global viewer, for example.

3. Method

The method consisted in three steps, detailed in the following subsections: 1) assessment of the areas of capital intensive production systems, 2) estimation of the potential maize production using traditional cultivation systems in this area, and 3) calculation of the potential of food self sufficiency increment (the PIFS spatial index).

3.1. *Assessing the areas of capital intensive production systems*

Capital intensive production systems in the Sierra are grazing lands and coffee large plantations. The 2004 coffee census contains the surface distribution of agricultural production units, allowing to distinguish between smallholder cultivated surface (< 5ha) and large plantation surface. More recently, a polygon-based (parcel clusters) spatial representation with agricultural production data have been released (rural Geostatistical Units: [10]), however, this data does not discriminate small from large estates, which is a high limitation to smallholder agriculture studies such as ours; we could not make use of this recent database for this reason and instead, the large coffee plantation surface was derived on a locality point basis using the 2004 detailed coffee census, as plotted in figure 2.

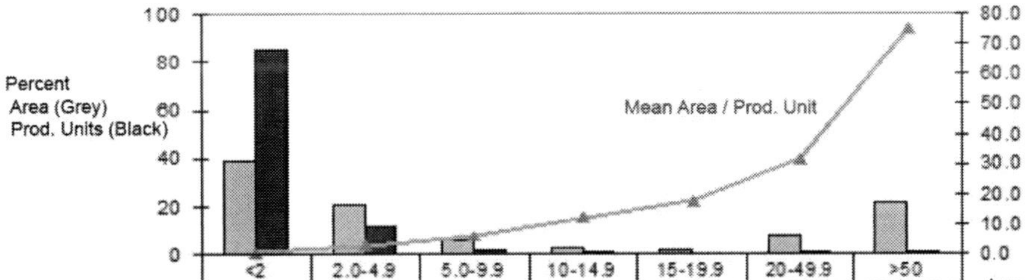

	<2	2.0-4.9	5.0-9.9	10-14.9	15-19.9	20-49.9	>50
% Total AREA	39	21	6	2	2	8	22
% Number PU	85	12	1.71	0.34	0.18	0.41	0.48
Mean AREA	0.5	2.9	6.2	12.0	17.6	31.5	74.8

Figure 2. Land distribution among coffee Production Units (PU) in the Xicotepec Subdistrict (derived from [9])

Official land use cartography may be useful for vast pasture surface estimation, as has been done in a previous study by Bermeo *et al.* [6]. With the food security focus of this study, an attempt was made to estimate not only large surfaces of grazing systems (>100ha), but instead a more exhaustive account of surfaces pertaining to medium (>10ha) and large land owners. Landsat imagery (year 2003, at the time of the coffee census) was acquired for the study area and geo-referenced to the INEGI 1:50,000 scale official Mexican topographic maps via the national database of ground control points by Mas *et al.* [12]. A decision tree, based on the combination of unsupervised spectral cluster detection and supervised Maximum Likelihood class assignation was applied to the Landsat corrected imagery, building on the hybrid classification scheme of Couturier *et al.* [13]; in a first step, ISODATA clustering algorithm was applied to the 7 bands of the Landsat imagery (with 40 to 60 clusters). Next, empirical "pure" spectral classes (representing a gradient of vegetation densities, grazing land, mixed land uses, rainfed agriculture, and forested land, with differing sun expositions and shadowed areas) were selected from the imagery. Then, a supervised ML classifier was applied and each pixel was assigned to the "grazing land" and other classes. Finally, a majority filter was applied to reduce the dispersion of the results, and patches of less than 10 ha were left out of the "grazing land" class, under the assumption that these patches pertained to smallholders.

The classification suffered from some omission of pasture parcels in the shadowed areas due to the reduced contrast. A simplified set of classes was attempted on topographically corrected Landsat imagery, however, because of the extreme topographic gradient of the study area, no significant improvement was obtained.

3.2. *Potential milpa production in capital intensive areas*

The calculations derived from the earlier subsection (point-based coffee area A_{coffee} and polygon-based grazing land area $A_{pasture}$) were assimilated and synthetized in hexagonal spatial units (HSU), because village spatial limits are largely undocumented. Hectagonal cells were preferred over square cells because the connectivity between cells is greater (one to six neighbors rather than one to four neighbors), and better suited to intercommunity-scale analysis (see [6]). The extent of each HSU was selected with an average of 4 localities per spatial unit according to point distribution analysis by Mitchell *et al.* [14] adapted by Bermeo *et al.* [6]. A preliminary analysis of A_{coffee} and $A_{pasture}$ was necessary to handle outlier values (unrealistically high areas of coffee due to the spatial uncertainty of the location of coffee plantations, for example), after mapping these variables over the entire study area (figure 4); a threshold of 70% of the HSU area was applied to constrain these outlier values.

Next, the potential production of maize ($Prod_{maize\_add}$) in these capital intensive areas was calculated according to the following equation:

$$Prod_{maize\_add} = (A_{coffee} + A_{pasture}) * Productivity_{milpa} \quad (1)$$

Where $Productivity_{milpa}$ (ton/ha), conceived as the productivity of the traditional practice of the milpa system, was estimated on the grounds of field interviews with elder indigenous inhabitants that practice traditional milpa in three localities of the study area. Accordingly, a 1 ha milpa plot typically yields 1.8 tons of maize (among yields of other crops in the plot), but implies 4 ha associated plots under fallow to maintain land fertility. Therefore, $Productivity_{milpa} = 0.36$ tons/ha.

3.3. Calcultating and mapping the Potential Increment in Food Security (PIFS)

Current Food Selfsufficiency (CFS) can be evaluated on the basis of the balance between local food production and food consumption [15]:

$$\text{CFS (\%)} = {Prod_{maize}}/{Cons_{maize}} \quad (2)$$

$Prod_{maize}$ was estimated at locality level according to the Agricultural Census [10] and $Cons_{maize}$ was estimated to 0.274 tons/year, according to a set of interviews in the field.

Finally, the Potential Increment of Food Security (PIFS) was estimated using the following equation:

$$\text{PIFS (\%)} = {Prod_{maize\_add}}/{Cons_{maize}} \quad (3)$$

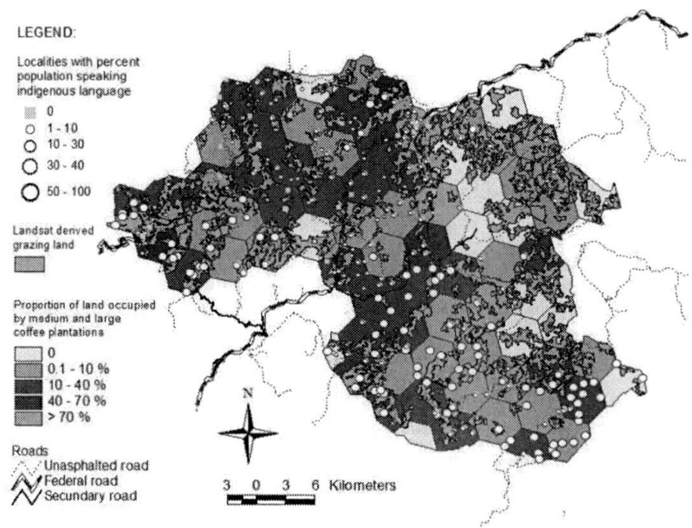

Figure 3. Mapped proportion of land occupied for capital intensive agricultural activities.

4. Results

A_{coffee} and $A_{pasture}$ were illustrated for preliminary analysis (figure 3). HSUs with very high surface proportion of coffee plantation (outlier values >70%) were probably affected by a saturation effect caused by the spatial uncertainty of the location of coffee plantations registered in the census.

Table 1. Value distribution of the Potential Increment of Food Security (PIFS)

Class	Description	N° cells
No population	Land occupied by large estates	11
Small increment	PIFS < 25%	27
Medium increment	25% < PIFS < 50%	16
Large increment	50% < PIFS < 100%	13
Very large incr. (over self sufficiency)	PIFS > 100%	11

Indeed, most of these HSUs contained small urban centers (head towns of municipios) where many plantations are registered, although their location is in nearby HSUs. To tackle these unrealistic data, it was assumed than no more than 70% of the HSU area could be assigned to coffee plantations.

Next, CFS and PIFS distributions were mapped using the HSU representation (figure 4). Table 1 reports an increment of food self-sufficiency of more than 50% for 24 HSUs out of a total of 77 (30% of the study area), and an increment of more than 100% for 11 HSUs (13% of the study area).

To the East of the study area (towards the coast), several HSUs registered null population because of the absence of localities (low altitude, easily accessible land occupied by very large coffee plantations). A set of HSUs with small PIFS values was detected in the Southern part of the study area, with a high density of indigenous localities and sharper relief. In contrast, a clear pattern of high PIFS values exists near to a circuit of roads in the vicinity of Xicotepec (center North of the study area), the main town of the region, where cattle breeders or coffee plantation farmers could establish easily in spite of the relatively sharp relief. Finally, to the North, a "mestizo" densely populated area, the pattern is more heterogeneous.

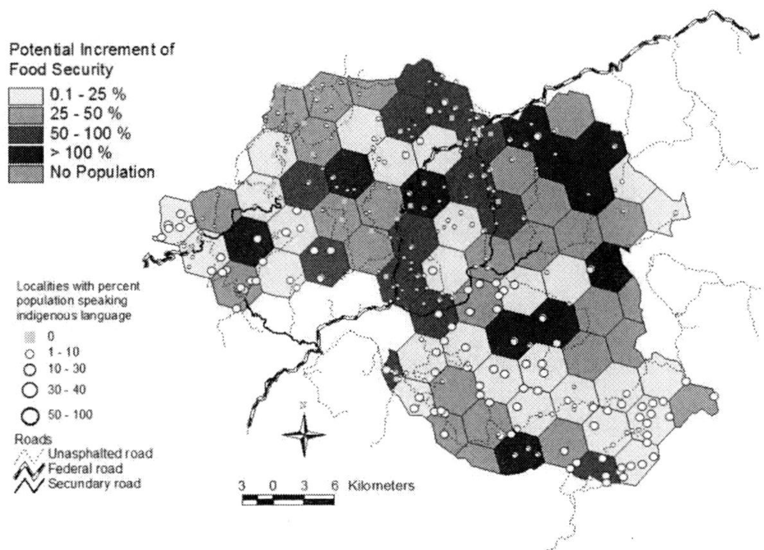

Figure 5. Mapped Potential Increment of Food Security (PIFS) through additional milpa activity in the study area.

5. Conclusions

In the context of increasing concerns for massive migration and environmental impacts of the agro-industry in rural territories, the resilience of traditional cultivation systems retains the attention of urban citizens and the scientific community; this situation in turn appeals for a renewal of spatial analysis tools to better grasp the role of these territories in the national and global food security issues.

The proposed approach provides a means of handling smallholder activities and livelihood strategies in spite of highly uncertain statistical data or poor documentation on agrarian structure in subtropical rural areas. Our approximation to food security to the Xicotepec Subdistrict suggests great potential towards more sufficient and more nutritious maize supply in these rural areas (increment of food self-sufficiency of more than 50% for 30% of the territory) if *milpa* cultivation were to be massively implemented back in the region. Because of the key role of the coffee detailed census of 2004 in our analysis, we recommend the Ministry for Agriculture (SAGARPA) and INEGI to adopt the design of the 2004 census for future census in rural areas.

The mentioned improvement of food self-sufficiency ignores many factors including cash income needs of smallholders, migration induced neglect of cultivation practices, effects from losses of interregional exchange of products from large farms, as well as the possible loss of agricultural production because of non-traditional practices, all of which should be studied further with a regional perspective.

Acknowledgments

This research was developed in the context of the academic PAPIIT project (DGAPA – UNAM financed) number IA300515, entitled "A multiscale territorial characterization of the contribution of the Mexican society to global environmental change".

References

[1] Verburg P, Mertz O, Erb K, Haberl H and Wu Wenbin 2013 Land system change and food security: towards multi-scale land system solutions *Current Opinion in Environmental Sustainability.* **5** 494-502

[2] Knoke T, Calvas B, Ochoa S, Onyekwelu J and Griess V 2013 Food production and climate protection- What abandoned lands can do to preserve natural forest *Global Environmental Change.* **23** 1064-1072

[3] Perfecto, Ivette, John V and Angus Wright 2009 *Nature's Matrix: Linking Agriculture, Conservation and Food Sovereignty* (London: Earthscan) pp. 40-43.

[4] Boege E 2008 *El patrimonio biocultural de los pueblos indígenas de México: hacia la conservación in situ de la biodiversidad y agrodiversidad en los territorios indígenas* (Mexico City: INAH/CNDPI) 342p.

[5] Roy Chowdhury R 2006 Landscape change in the Calakmul Biosphere Reserve, Mexico: Modeling the driving forces of smallholder deforestation in land parcels *Applied Geography.* **26** 129-152

[6] Bermeo A, Couturier S and M. Galeana-Pizaña 2014 Conservation of smallholder traditional cultivation systems in indigenous territories: mapping land availability for milpa cultivation in the Huasteca Poblana, Mexico *Applied Geography.* **53** 299-310.

[7] FAO (Food and Agriculture Organization) 2006 Food Security *Policy Brief Issue No. 2*. FAO Food Security Program, Rome.

[8] The Economist Intelligence Unit 2015 *Global Food Security Index*. Retrieved from: http://foodsecurityindex. eiu.com/Index (last access February 2015).

[9] SAGARPA 2005 *Padrón cafetalero del Estado de Puebla*. Consejo Poblano del Café. SAGARPA, Xicotepec de Juárez, Puebla, Mexico. Digital documents on CD ROM in Excel Format (56395 items). Individual items available at: http://www.spcafe.org.mx/wb3/wb/spc/spc_reportes_pnc

[10] INEGI 2014 *Censo Agropecuario 2007. Sistema de Consulta de Información Geoestadística*

Agropecuaria (SCIGA). Accessed at http://gaia.inegi.org.mx/sciga/viewer.html
[11] INEGI 2012 *Cartografía de Usos de Suelo y Vegetación, Serie V,* in: Series históricas del INEGI. INEGI, Aguascalientes, Aguascalientes, Mexico.
[12] Mas J.-F, Velázquez A, Palacio-Prieto JL, Bocco G, Peralta A and Prado J 2002 Assessing forest resources in Mexico: Wall-to-wall land use/ cover mapping *Photogrammetric Engineering and Remote Sensing*. **68** 966-969.
[13] Couturier S, Ricárdez M, Osorno J, y López-Martínez R 2011 Morpho-spatial extraction of urban nuclei in diffusely urbanized metropolitan areas *Landscape and Urban Planning*. **101** 338-348
[14] Mitchell A 2005 *The ESRI guide to GIS analysis, Volume 2: Spatial Measurements and Statistics.* Redlands, (CA: ESRI Press) 238 p.
[15] FAO, IFAD 2014 *The State of Food Insecurity in the World: Strengthening the Enabling Environment for Food Security and Nutrition*. (Rome: Food and Agriculture Organization of the United Nations)

IOP Conf. Series: Earth and Environmental Science 54 (2017) 012007 doi:10.1088/1755-1315/54/1/012007

Sensible heat flux of oil palm plantation: Comparing Aerodynamic and Penman-Monteith Methods

Nurul Amri Komarudin[1], Tania June[1]* and Ana Meijide[2]

[1] Department of Geophysics and Meteorology, Faculty of Mathematics and Natural Sciences, Bogor Agricultural University (IPB), Kampus IPB Dramaga, FMIPA Building, Wing 19, Level 4, Jalan Raya Dramaga, Bogor, Jawa Barat, Indonesia 16680

[2] CRC990, University of Goettingen

E-mail: tania.june@yahoo.com

Abstract. Oil Palm (*Elaeis guinensis* Jacq) has a unique morphological characteristics, in particular it has a uniform canopy. As the plant become older, its canopy coverage will completely cover the surface and influence characteristics of its microclimate. Sensible heat flux estimation of oil palm plantation could be used to identify the contribution of oil palm in reducing or increasing heat to its surrounding environment. Determination of heat flux from oil palm plantation was conducted using two methods, Aerodynamic and Penman-Monteith. The result shows that the two methods have similar diurnal pattern. The sensible heat flux peaks in the afternoon, both for two and twelve years oil palm plantations. Sensible heat flux of young plantation is affected by atmospheric stability (stable, unstable and neutral), and is higher than that of older plantation, with mean values of 0.52 W/m^2 (stable), 43.53 W/m^2 (unstable), 0.63 W/m^2 (neutral), with standard deviation of 0.50, 28.75 and 0.46 respectively. Sensible heat flux estimated by Penman-Monteith method in both young and older plantation was higher than the value determined by Aerodynamic method with respective value of 0.77 W/m^2 (stable), 45.13 W/m^2 (unstable) and 0.63 W/m^2 (neutral) and 0.34 W/m^2 (stable), 35.82 W/m^2 (unstable) and 0.71 W/m^2 (neutral).

1. Introduction

Oil Palm (*Elaeis guinensis* Jacq) is a plantation crop from West Africa, which can grow well in the humid tropics like Indonesia. It has a unique morphological characteristics, in particular it has a uniform canopy [1]. As the plant become older, its canopy coverage will completely cover the surface and influence characteristics of its microclimate, including the absorption of solar radiation, wind speed, air temperature, and relative humidity as well as other elements of microclimate.

Sensible heat flux is the energy required to heat the air, in addition to the heat flux was also associated with an increase and decrease in temperature and serves to keep the temperature of the environment in order to maintain at the optimum interval [2]. Several methods have been developed to Determine the value of heat flux feels one of them with the estimate or calculate the gradients of temperature and humidity at or near PBL *(Planetary Boundary Layer)*, this research compared two methods, Aerodynamic and Penman-Monteith, identify the contribution of oil palm in reducing or increasing heat to its surrounding environment. Since there is little difference in the needs of weather data and climate data for both methods, so it is important to do a comparison between the two methods. Research aims are to determine the sensible heat flux of different age-oil palm plantation and to compare Aerodynamic and Penman-Monteith Methods.

2. Methods

Measurement of solar radiation, wind velocity, air temperature, soil temperature and relative humidity in 2 year old oil palm plantation (Pompa Air Village) running for 7 months July 2013 – February 2014, sensor of solar radiation installed above oil palm plantation, sensor wind velocity (anemometer), air temperature (thermometer), soil temperature and relative humidity installed at three heights are 2.4 meters, 3.15 meters and 5.8 meters and 12 year old oil palm plantation (PTPN VI) march – October 2014, sensor of solar radiation installed above oil palm plantation , sensor wind velocity (anemometer), air temperature (thermometer), soil temperature and relative humidity installed at three heights are 13 meters, 15.4 meters and 21.7 meters. Measurements in 2 year old oil palm plantation (Pompa Air Village) and 12 year old oil palm plantation (PTPN VI) are research collaboration with *CRC 990 subproject A03, EFForts projects (University of Goettingen).* Because measurements must be made on the planetary boundary layer, so that the measurements were taken at three heights for both oil palm age.

2.1. *Stability atmosphere*

$$Ri = \frac{g(\frac{\delta\theta}{\delta z})}{T\prime(\frac{\delta u}{\delta z})^2} \quad (1)$$

where g is acceleration due to gravity (9.8 m/s^2), θ_a is the average potential temperature (K) at a height of z1 and z2. These measurements were performed at a position close to the surface (2 years old oil palm plantation, with the highest measurement position 5.8 meters and 12 years old oil palm plantation, with the highest measurement position 21.7 meters) [3].

2.2. *Aerodynamic methods*

$$H = -\rho_a C_p k^2 \frac{(u_2-u_1)(\theta_2-\theta_1)}{\left[ln\left(\frac{z_2-d}{z_1-d}\right)\right]^2 \varphi_m \varphi_s} \quad (2)$$

On the atmospheric conditions of stable and unstable or in addition to neutral, formula sensible heat flux (H) in equation (6) must be corrected, because the equation can only be used on the atmospheric conditions of neutral, correction equations can be done by using a correction factor Monin-Obukhov Stability parameter, Monin-Obukhov stability parameters at every condition of atmospheric stability is shown in the following equation [4]:

$\zeta = Ri$ for $Ri < 0$ (3)
$\zeta = Ri/(1\text{-}5Ri)$ for $0 \leq R \leq 0{,}1$ (4)
$\zeta = 0{,}2$ for $Ri > 0{,}1$ (5)

Monin-Obukhov value of parameter of stability is obtained, then the substitution to the following equation dimensionless wind shear [4]:

$\Phi_s = \Phi m^2 = (1-5\zeta)^{-1/2}$ for $\zeta < 0$ (6)
$\Phi_s = \Phi m = 1-5\zeta$ for $\zeta \geq 0$ (7)

where ρ is the density of air (Kg / m3), Cp is specific heat at constant pressure (1004.2 J / kg / K). and K is the Von Karman constant with a value (0.4). For level measurement is low (2 years old oil palm plantation) use absolute temperature, (T) in units of kelvin and for level measurement of high (12 years old oil palm plantation) use potential temperature (θ) in units of kelvin where θ = T - Γdz, with Γd is *dry adiabatic lapse* rate of -0.00976 K/m.

2.3. Penman-Monteith methods

$$L_E = \frac{\Delta(Rn-G)+ \rho_a C_p(e_s-e_a)/r_a}{\Delta+ \gamma[1+\frac{r_s}{r_a}]} \qquad (8)$$

where L_E is latent heat flux (MJ m^{-2} day^{-1}), Δ is the slope of the vapor pressure curve (kPa°C^{-1}), Rn is net radiation (MJ m^{-2} day^{-1}), G is the soil heat flux (W m^{-2}), (e_s - e_a) shows the deficit pressure of vapor of the air, ρa is the average air density at constant pressure, γ is the psychometric constant (kPa°C^{-1}), Cp is the specific heat capacity of air at constant pressure (1.013x 10^{-3} MJ kg^{-1}°C^{-1}), ra is the aerodynamic resistance (s m^{-1}) and r_s is canopy resistance (s m^{-1}) [5].

Residual of evapotranspiration using the Penman-Monteith method of equation above, can be used to estimate the sensible heat flux was using the energy balance equation, with the assumptions net radiation (Rn) is distributed to $L_E, H, G\ dan$ S, can use the following equation:

$$R_n - LE - G - H - S = 0 \qquad (9)$$

or

$$H = R_n - (L_E + G + S) \qquad (10)$$

Where:
H : Sensible Heat Flux (W/m²)
Rn : Net Radiation (W/m²)
L_E : Laten Heat Flux (W/m²)
G : Ground Heat Flux (W/m²)
S : *Storage* (W/m²)

3. Result

3.1 Study area

Pompa Air Village located in the Province of Jambi, Bajubang Sub District, Batang District, located at a latitude of 01° 52'06" south latitude and 103° 15'17" east longitude. With a height of between 11-500 meters above sea level, air temperature ranged between 20°C - 30°C and annual rainfall between 2265 to 2976 mm [6]. PTPN VI located in the province of Jambi, Kotabaru Sub District. Jambi District located between 01° 38'19" south latitude and 103° 34'42" east longitude. With a height of 0-60 meters above sea level. The average air temperature ranged between 22°C - 32°C with an annual rainfall average of 2296 mm [7].

3.2 Characteristics micrometeorology

3.2.1 Global radiation

Solar radiation is an energy been derived from thermonuclear process that occurs in the sun. The variation of the average diurnal solar radiation above 2 years old oil palm plantation (Pompa Air Village) and 12 years old oil palm plantation (PTPN VI). Shown in figure 2. From the morning towards midday, The incandescence of the air above 12 years old oil palm plantation is happening faster then the 2 years old oil palm plantation. Although the global radiation on average in the 12 years ol oil palm plantation is lower then the 2 years old oil palm plantation.

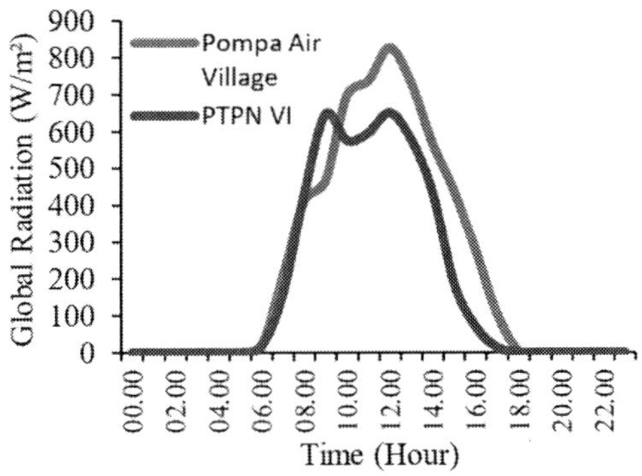

Figure 1. Average diurnal variation of global radiation in 2 year old plantation (Pompa Air Village) and 12 years old plantation (PTPN VI), Jambi

3.2.2 Air temperature and relative humidity

Air temperature is the manifestation of sensible heat flux. Based on figure 2 two research areas show that increasing the height of areas decreases its temperature tend to be declining. Based on theory, the air density near the surface during the day is higher than the layers above it and the acceptance of reflected radiation from the Earth's surface is more, so temperature near the surface is higher than the above layer, because the existence of air turbulence or mass movement such as the high wind speed, causes the mixture of air masses and air temperature tend to be homogeneous, so the difference air temperature between the height is relatively small .

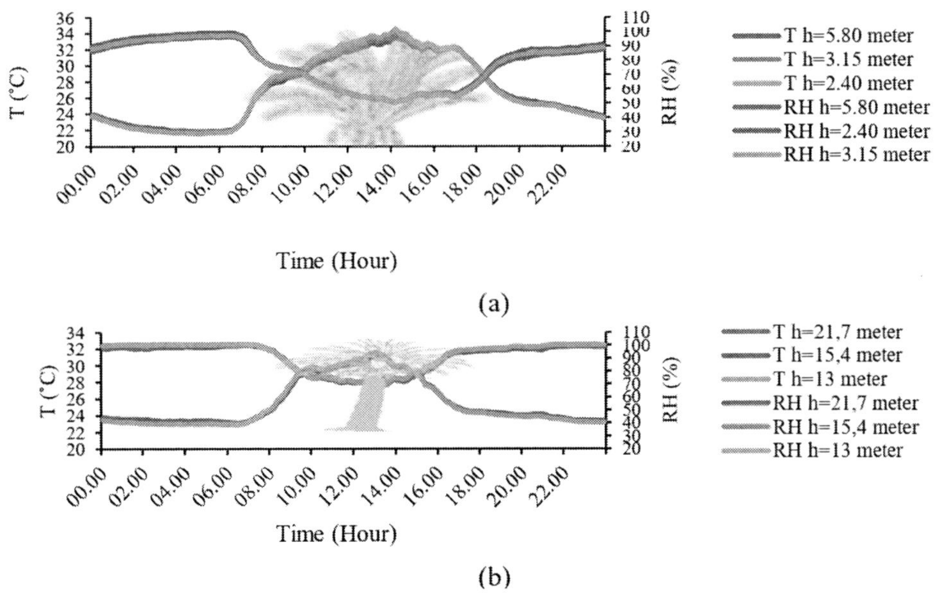

Figure 2. Average diurnal variation of temperature and relative humidity in 2 year old plantation (Pompa Air Village) and 12 years old plantation (PTPN VI), Jambi

3.2.3 Wind speed

Wind speed profile in the area of oil palm plantation, Pompa Air Village measured at height of 2.40 m, 3.15 m and 5.80 m have almost same diurnal variation. The same is also shown by the wind speed profile in the region of PTPN VI measured at a height of 13 meters, 15.4 m and 21.7 m. Based on figure 5, maximum wind speeds in the oil palm plantation of age 2 years (Pompa Air Village) happens at 12.00 WIB with a value 2.56 m/s and the maximum wind speed in the oil palm plantation of age 2 years (PTPN VI) happens at 13.00 WIB with a value 2.3 m/s. The higher altitude will increase the wind speed, because at the height near land surface the movement of air will be hampered by the existence of friction between the surface of the style that is caused by the interaction between the surface by the air above it, so experiencing barriers in the form of friction, that cause wind speed declining.

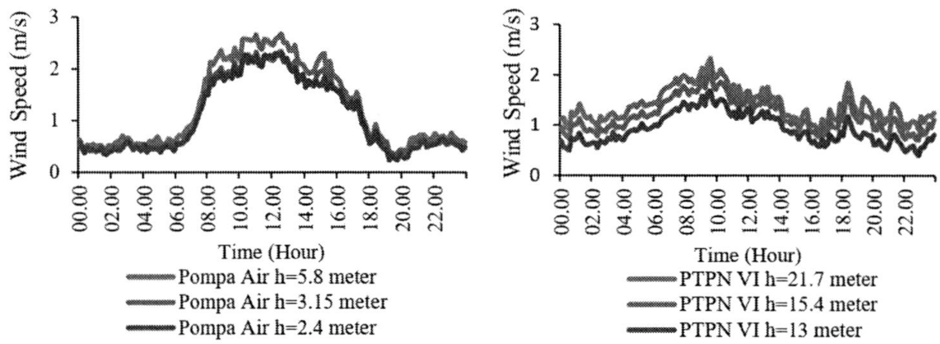

Figure 3. Average diurnal variation of wind speed on oil palm plantations (a) the age of 2 years (Pompa Air Village) and (b) the age of 12 years (PTPN VI)

3.3 Roughness characteristic parameters

Roughness characteristics parameters can be determined by three parameters, there are the zero-plane displacement (d), friction velocity (u *) and roughness length (z_0) [8] is obtained from the profile wind speed, because surface roughness can be known through data analysis wind speed at the top of the canopy on the state of the atmosphere is neutral [9], because the atmospheric conditions of neutral assumed to be able to represent the roughness condition in all conditions of atmospheric stability and the neutral condition, shear stress that occurs purely because of friction with the roughness parameters and no flux of vertical mass or buoyancy such as unstable atmospheric conditions [10].

Zero-plane displacement (d) is a function of the height and density and the mechanical state of the plant [11, 12]. Based on calculations, the value of the zero-plane displacement (d) of oil palm (Pompa Air Village) is lower than the oil palm plantations (PTPN VI) with value of 2.08 meter and 12.44 meter respectively. Zero-plane displacement (d) will be higher if the the roughness length increases and when it near the height of the canopy [4].

Table 1. The average zero-plane displacement (d), friction velocity (u *) and roughness length (z_0) in oil palm plantations age 2 years (Pompa Air Village) and age 12 (PTPN VI) Jambi

Location	Heigh (m)	d (m)	u* (m/s)	z_0 (m)
Pompa Air Village	2.5	2.08	0.03	0.0006
PTPN VI	13	12.44	0.17	0.006

The value of friction velocity (u*) and roughness length (z0) on 2 year old oil palm plantation (Pompa Air Village) and 12 year old oil palm plantation (PTPN VI) are different (Table 1). The value of friction velocity (u*) tends to increase with increasing wind speed [13].*Stability of atmosphere*

Based on the Richardson number (RI), there are three categories of atmospheric stability: neutral (Ri $\pm$ 0.01), stable (Ri < 0:01) and unstable (Ri > 0:01). The results show that the atmospheric stability in the plantation Pompa Air Village (3a) and PTPN VI Jambi (3b) are neutral, stable and unstable. However, for the stability of the atmosphere in the neutral condition it is very rare. Although the neutral atmospheric conditions at both areas are relatively small, the data in these conditions can be used to determine the roughness characteristic parameters (figure 4). That's because the atmospheric stability at neutral condition is considered to be representative of condition of roughness in all conditions of atmospheric stability, logarithmic wind profile and also lack of influence bouyancy, so purely the influence from surface characteristics itself [10].

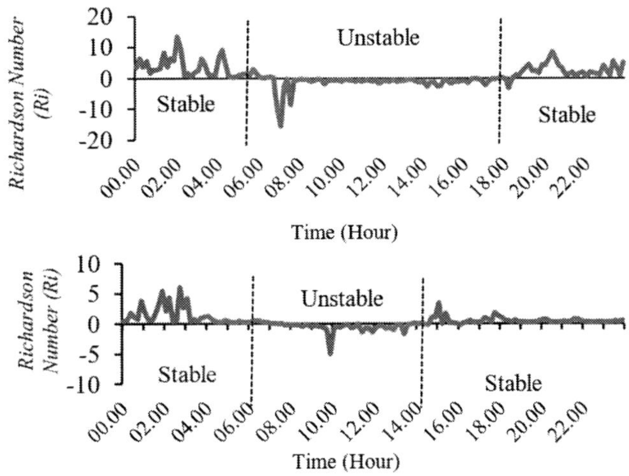

Figure 4. Variation diurnal of Richardson Number in oil palm plantations (a) the age of 2 years (Pompa Air Village) and (b) the age of 12 years (PTPN VI)

3.4 *Sensible heat flux profile*

Sensible heat flux is the energy required to heat the air. Expression of sensible heat flux is the temperature. Heat flux was a heat transfer that could be felt through the air from one place to another. The flux density is expressed as energy per unit time per unit surface area (W/m^2). Estimation of oil palm in reducing heat to its surrounding environment. Figure 1 shows similar diurnal variations. The peak of sensible heat flux happen in the afternoon, both for two and twelve years oil palm plantations. Sensible heat flux of young palm plantation is higher than the older plantations, with mean value 0.52 W/m^2 (stable), 43.53 W/m^2 (unstable), 0.63 W/m^2 (neutral), with standard deviation of 0.50, 28.75 and 0.46 respectively and the old plantation is 0.29 W/m^2 (stable), 35.38 W/m^2 (unstable) dan 0.47 W/m^2 (neutral), with standard deviation of 0.54, 23.39 and 0.46 respectively (table 1). Sensible heat flux estimated by Penman-Monteith method in both study areas was higher than the value estimated by Aerodynamic method with respective value of 0.77 W/m^2 (stable), 45.13 W/m^2 (unstable) dan 0.63 W/m^2 (neutral), with standard deviation of 0.96, 29.61 and 0.91 respectively dan 0.34 W/m^2 (stable), 35.82 W/m^2 (unstable) dan 0.71 W/m^2 (neutral), with standard deviation of 0.20, 24.31 and 0.90 respectively (table 2).

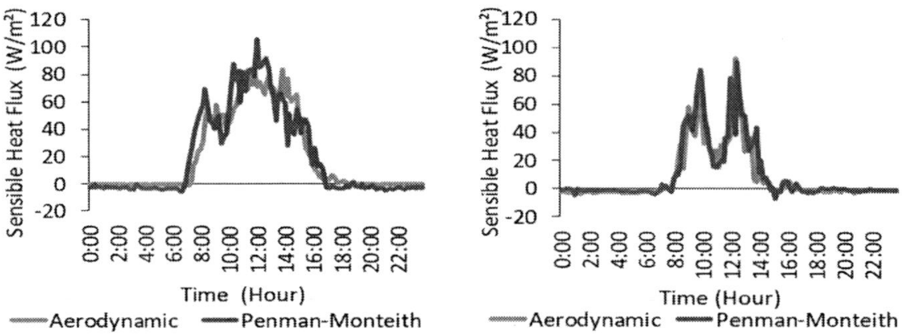

Figure 5. Sensible heat flux variation of oil palm plantation (a) age 2 years (Pompa Air Village) and (b) age 12 years (PTPN VI) by aerodynamic and Penman-Monteith methods.

Table 2. Average diurnal sensible heat flux of oil palm plantation in each situatuion of atmospheric stability in Pompa Air Village and PTPN VI, Jambi

Method	Pompa Air Village (2 years old) W/m²	Standard Deviation	PTPN VI (12 years old) W/m²	Standard Deviation
Aerodynamic				
Stable	0.27	0.24	0.24	0.88
Unstable	41.94	27.9	34.95	22.47
Neutral	0.27	0.01	0.23	0.02
Penman-Monteith				
Stable	0.77	0.76	0.34	0.20
Unstable	45.13	29.61	35.82	24.31
Neutral	0.73	0.91	0.71	0.90
Average				
Stable	0.52	0.50	0.29	0.54
Unstable	43.53	28.75	35.38	23.39
Neutral	0.63	0.46	0.47	0.46

RMSE (Root Mean Square Error) showed 8.9 between the method of Penman-Monteith and aerodynamic for plantation Pompa Air Village Jambi and 3.4 for observation at the plantation PTPN VI, can be said to be a match between the method Penman-Monteith with aerodynamics in both areas, which can also be seen from a high correlation coefficient between the two methods ($r = 1$ to the Village Pump and 0.99 for PTPN VI) and Agreement Index score between the two methods at two research sites close to the value 1. So the methods of aerodynamic and Penman-Monteith are a perfect fit to estimate the heat flux was (table 3)

Table 3. Statistical calculation of sensible heat flux by Aerodynamic and Penman-Monteith Methods

	Pompa Air Village	PTPN VI
Average of Sensible Heat Flux:		
Penman-Monteith	20.03	9.93
Aerodinamik	20.02	8.43
RMSE	8.9	3.4
I.A (*Index of Agreement*)	0.98	0.99
r^2	0.92	0.98
R	1.00	0.99

4. Conclusion

Sensible heat flux estimation using aerodynamic and penman-monteith methods show a similar diurnal variation in 2 years and 12 years aged oil plantation. Penman-monteith method show a higher estimation than aerodynamic method and 2 years aged oil plantation show a higher estimation than 12 years aged oil plantation. Sensible heat flux of young plantation is affected by atmospheric stability (stable, unstable and neutral), and is higher than that of older plantation. Sensible heat flux estimated by Penman-Monteith method in both young and older plantation was higher than the value determined by Aerodynamic method.

References

[1] Direktorat Jenderal Perkebunan 2014 Buku Statistik Perkebunan Tahun 2008 – 2012. Retrieved from http://www.pertanian.go.id/infoeksekutif/bun/isi_dt5thn_bun.php on January 24, 2016.
[2] Oke T R 1978 *Boundary Layer Climates* (London (UK): Methuen & Co Ltd)
[3] Kaimal J C and Finnigan J J 1994 *Atmospheric Boundary Layer Flows: Their Structure and Measurement* (New York: Oxford Univeristy Pr.)
[4] Arya S P 2001 *Introduction to Micrometeorology Ed ke-2.* (San Diego: Academic Pr.)
[5] Allen R G, Pereira L S, Raes D and Smith M 1998 Crop Evapotranspiration: Guidelines Computing Crop Water Requirements *FAO Irrigation and Drainage Paper.* **56**
[6] Pemerintah Provinsi Jambi 2015 *Sekilas tentang Provinsi Jambi.* Retrieved from *http://jambiprov.go.id/index.php?letluaswil* on May 15, 2016
[7] Badan Perencanaan Pembangunan Daerah Provinsi Jambi and [BPS Jambi] Badan Pusat Statistik Provinsi Jambi 2012 *Jambi dalam angka* 2011 (Jambi: Tim Penyusun)
[8] Montheith J L and Unsworth MH 2013 *Principle Environmental Physics Fourth Edition.* Polandia. **16** 290-218. ISBN: 978-0-12-386910-4.
[9] Tsai J L, Tsuang B J 2005 Aerodynamic roughness over an urban area and over two farmland in a populated are as determined by wind profiles and surface energy flux measurement, *J Agric Forest Meteorol.* **132** 154-170
[10] Kusuma A P 2014 Dinamika Karakteristik Lapisan Perbatas Permukaan *(Surface Boundery Layer)* Di Atas Perkebunan Kelapa Sawit. *Undergraduate Thesis.* (Bogor (ID) : Institut Pertanian Bogor)
[11] Chang J 1986 *Climate and Agriculture* (Chicago: University of Winconsin)
[12] June T 2012 *Modul Praktikum Mikrometeorologi: Pengukuran Profil Iklim Mikro, Fluks Momentum, Fluks Bahang dan Fluks Uap Air dari Permukaan Kanopi Tanaman.* [Unpublished]. Bogor : Jurusan Geofisika dan Meteorologi. FMIPA-IPB.
[13] Suciatiningsih F 2013 *Karakteristik Kekasapan Permukaan dan Pengaruhnya Terhadap Transfer Turbulen, Momentum dan Bahang. Undergraduate Thesis.* (Bogor (ID): Institut Pertanian Bogor)

Analysis of relationship between meteorological and agricultural drought using standardized precipitation index and vegetation health index

U Ma'rufah[1], R Hidayat[1*], I Prasasti[2]

[1]Department of Geophysics and Meteorology,Faculty of Mathematics and Natural Science, Bogor Agricultural University, Bogor, Indonesia
[2]Remote Sensing Application Center, Indonesian National Institute of Aeronautics and Space (LAPAN), Pasar Rebo, Jakarta, Indonesia

E-mail: rahmath@apps.ipb.ac.id

Abstract. Agricultural drought is closely related to meteorological drought in which the agricultural drought is an impact of meteorological drought. This study aim to understand the duration, spatial extent, severity and lag time of meteorological and agricultural drought during El Niño years. The data used in this study are monthly data of CHIPRS and MODIS. Meteorological drought and agricultural drought are intensified in the El Niño years. The duration of meteorological drought is different in each region but generally occurs during June to November. Agricultural drought mostly occurs from August to November. Spatially, meteorological drought and agricultural drought in 2015 has wider extent and higher severity (SPI <-2 and VHI <10) than in 2002. Agricultural drought generally intensified in areas that have monsoonal rainfall type such as Java, Bali, Nusa Tenggara, Lampung, southern part of Kalimantan, and southern part of Sulawesi. We found that VHI is significantly correlated with SPI-3 reach 58% of the total area of Indonesia. It means rainfall deficit during three months has a significant impact on agricultural drought in Indonesia. In general, SPI-3 and VHI clearly explain the relationship between meteorological drought and agricultural drought in Indonesia.

1. Introduction

El Niño is widely known as global-scale climate anomalies which clearly modulates rainfall anomalies in many parts of the world including Indonesia [1]. Decreasing rainfall during El Niño frequently causes dryness over major part of Indonesia. In Indonesia some extreme drought occured simultaneously with El Niño event such as in 1997/1998, 2002/2006, 2006/2007, and 2009/2010 [2]. Indonesian National Board for Disaster Management (BNPB) also mentioned that El Niño in 2015 resulted in Puso in 111,000 ha farming area [3].

Drought has a significant impact on the economic, agriculture, environment, and social. Drought event during El Niño 2002 caused a Puso in 42,000 ha of paddy fields in Java [4]. Monitoring drought needs to be done to minimize such drought impact. Various drought indices have been developed by a number of previous studies to determine the duration and severity of drought [5]. These indices have been developed based on different parameters so that each drought indices reflects different drought conditions.

Drought indices is widely applied for monitoring drought are Standardized Precipitation Index (SPI) and the Vegetation Health Index (VHI) [6-8]. SPI is a meteorological drought index that shows level of drought due to rainfall deficit while the VHI is an agriculture drought index based on remote sensing that shows stress level of crops due to drought. VHI is constructed by combining the Vegetation Condition Index (VCI) and the Temperature Condition Index (TCI) [6-10].

Agricultural drought is closely related to meteorological drought in which the agricultural drought is basically the impact of meteorological drought. In general, plants do not respond to the meteorological drought directly. It seems there is a lag time between meteorological drought and agricultural drought. Therefore, the objectives of this study are (1) to identify the duration, spatial extent, and severity of drought and (2) to determine lag time of meteorological drought and agricultural drought in Indonesia during El Niño events.

2. Methodology

2.1. Standardized Precipitation Index

SPI is calculated in different time scale 1,3,6,9, and 12 months (SPI-, SPI-3, SPI-6, SPI-9 and SPI-12) based on CHIRPS data. CHIPRS is rainfall data set incorporating infrared Cold Cloud Duration (CCD) observation and in-situ stasion data for trend analysis and seasonal drought monitoring [11]. SPI calculation requires historical precipitation data at least 30 years. So that, CHRIPS data from 1981 to 2015 are used to do the calculation. The SPI calculation is conductedby following McKee *et.al* (1993) [12]. Meteorological drought category based on SPI shown in Table 1.

Table 1. Classification of drought severity by WMO [13]

SPI	Category
$-1,0 < SPI \leq 1,0$	Normal
$-1,5 < SPI \leq -1,0$	Moderate
$-2,0 < SPI \leq -1,5$	Severe
$SPI < -2,0$	Extreme

2.2. *Vegetation Health Index*

MODIS data that consist of Enhanced Vegetation Index (EVI) and Land Surface Temperature (LST) are used to calculate VCI and TCI, respectively [8].We usetime series data from 2001 to 2015 to calculate their minimum and maximum values. The calculation of VCI and TCI is shown byfollowing equations:

$$VCI = 100 \text{ x } \frac{E - E_{min}}{E_{max} - E_{min}} \quad (1)$$

where E is EVI value of a given month. E_{min} and E_{max} denote the maximum and minimum EVI values, respectively, for the month from multiyear time series.

$$TCI = \frac{L_{max} - L}{L_{max} - L_{min}} \text{ x } 100 \quad (2)$$

where L is LST value of a given month. L_{min} and L_{max} denote the maximum and minimum LST values, respectively, for the month from multiyear time series. VHI are calculated based on VCI and TCI value using equation 3 [10]

$$VHI = (0.5 \text{ x } VCI) + (0.5 \text{ x } TCI) \quad (3)$$

Agricultural drought classified into five category shown in Table 2 [10].

Table 2. Agricultural drought category besed on VHI

VHI	Category
< 10	Extreme
< 20	Severe
< 30	Moderate
< 40	Mild
>40	No Drought

2.3. Correlation analysis between SPI and VHI

Correlation analysis is conducted spatially between SPI and VHI over Indonesia region.This analysis is used to determine the time lag between meteorological drought and agricultural drought over the region.

3. Result and Discussion

3.1. Meteorological Drought

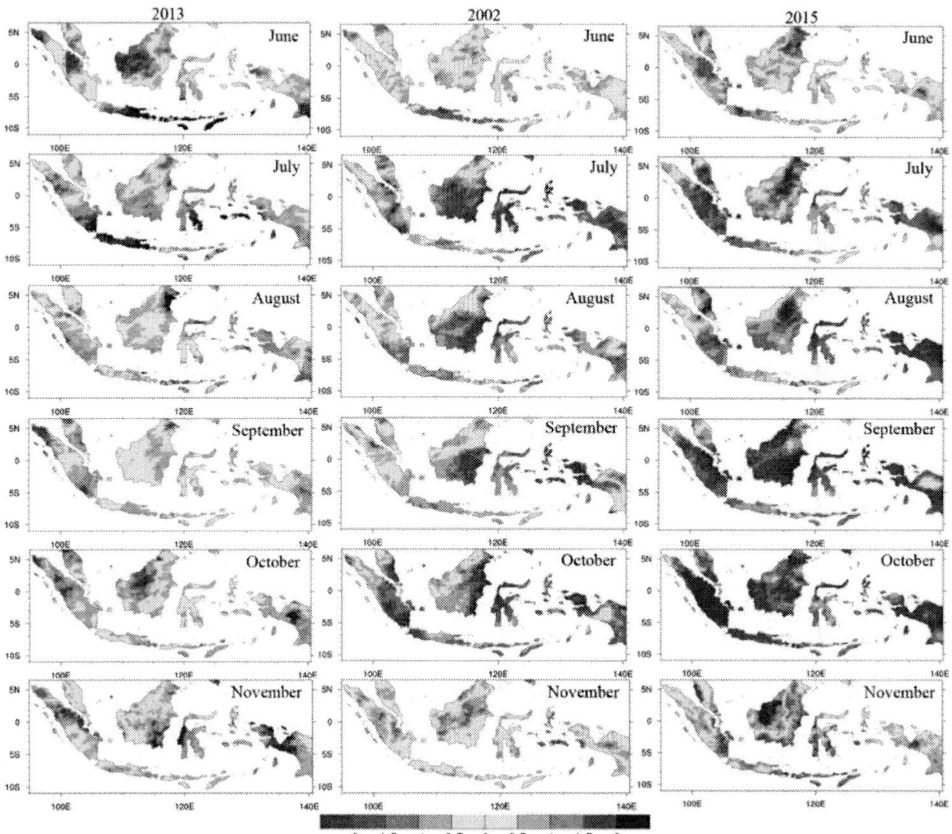

Figure 1. Spatial distribution of SPI in normal year(2013), weak El Niño (2002), and Strong El Niño (2015)

Meteorological drought is analyzed based on 1-month SPI (SPI-1). SPI-1 has significant correlation with rainfall so it is more appropriate to use it to analysis the meteorological drought [14]. The calculation of SPI-1 was done from 1981 to 2015. However, this study focuses on the drought conditions that occurred in 2002, 2013 and 2015. SPI-1 in 2002, 2013 and 2015 was chosen to explain the spatial distribution of drought in Indonesia during weak El Niño (2002), normal (2013), and a strong El Niño (2015).

El Niño has significant impact on the occurrence of meteorological drought in Indonesia [2]. In a normal year (2013), most of regions in Indonesiaare not affected by drought (see Figure 1). Drought occurs only in some parts of southern Indonesia (e.g. Java, Bali, Nusa Tenggara) and some part of northern Indonesia (e.g. north Sumatera and west Kalimantan). Southern region of Indonesia experienced drought from August to November, while the northern part of Indonesia such as Sumatera and Kalimantan experience drought from June to November. Moderate to extreme drought occured in this year.

Drought during weak El Niño (2002) was not spread uniformly across Indonesia (see Figure 1). Southern part of Indonesia experienced drought earlier than other region that is started from June. In addtion, Kalimantan, Sulawesi, and Papua experienced drought from July and some part of Sumatera experienced drought from August and reachits peakin October. In general, the drought is in the category of moderate to extreme drought with dominated by strong drought.

Drought in a strong El Niño (2015) occured almost over the regions during July to October as seen in Figure 1. In June, drought has occured only in Java and Sumatra, while in November drought occured only in the southern part of Indonesia. The peak of the drought observed in October. In general, the drought is in the category of moderate to extreme drought.

3.2. Agricultural Drought

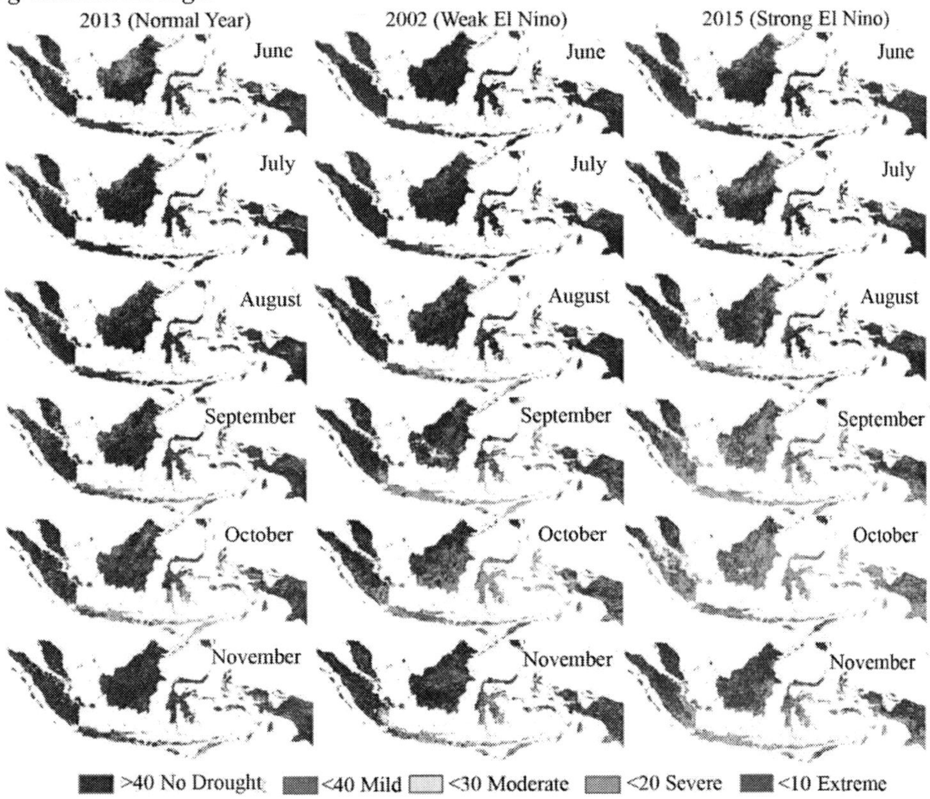

Figure 2. Spatial distribution of VHI in normal year(2013), weak El Niño (2002), and Strong El Niño (2015)

Decreasing levels of soil water or water stress can causes transpiration, photosynthesis, and retrieval of ions decrease that ultimately lead to impair growth and development of plants [15]. Reducing soil water content in constantly lead to agricultural drought [7]. This study used VHI to understand the duration, spatial distribution, and severity or category of agricultural drought. As well as SPI-1, VHI in 2002, 2013 and 2015 were chosen to determine the spatial distribution of agricultural drought in Indonesia during weak El Niño (2002), normal (2013), and strong El Niño (2015). Figures 2 show spatial distribution of VHI during normal year, weak El Niño, and Strong El Niño.

Agricultural drought in Indonesia is generally in the category of moderate to severe in the El Niño year. In 2013, drought observed from September to October in category mild to moderateespecially in the southern part of Indonesia. While in 2002, the drought started from August to November where drought dominantly occured in the southern part of Indonesia such as Java, Bali, Nusa Tenggara, southern part of Kalimantan, and southern part of Sulawesi. In this year, drought occurred in the category of mild to extreme in which extreme droughts observed in October. Moreover, drought occurred from August to November in mild to extreme category during 2015 which extreme drought is dominant in October. In all cases (2002, 2013, and 2015) drought were observed in the southern part of Indonesia. Due tomost of the region are dominanted by crops. Wang and Liu (2014) pointed out that crops area more quickly respond to drought than the forest [9].

3.3. Relationship between Meteorological and Agricultural Drought

Meteorological drought and agricultural drought generally has lagged in time [9]. It is exist due to soil water content is not reduced directly when rainfallgradually decreased. Spatially SPI-1, SPI-3, SPI-6, SPI-9, and SPI-12 are positively correlated with VHI in most Indonesian regionas seen in Figure 3. The correlation coefficient has increased from SPI-1 to SPI-3, then tend to be declined following the increase in time scale. Percentage of coveragewith significant correlation coefficient also increased from SPI-1 to SPI-3, then gradually decline following the increase in the time scale (Table 3).

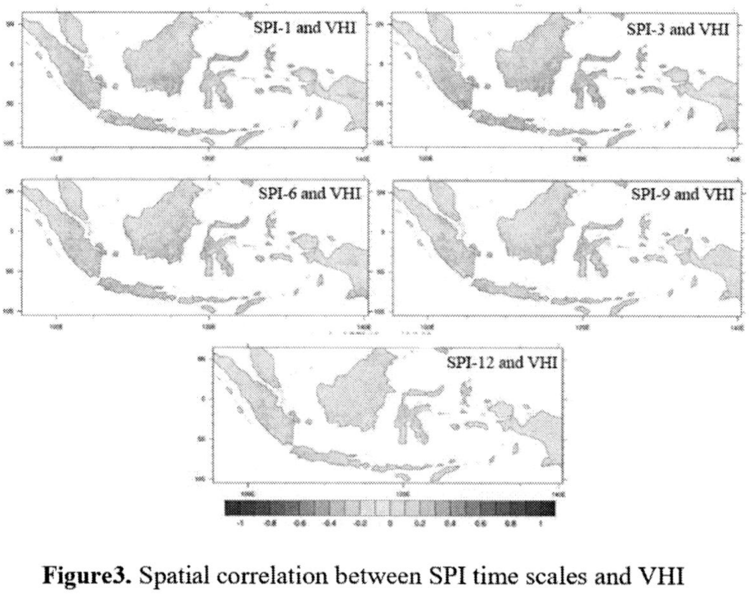

Figure3. Spatial correlation between SPI time scales and VHI

The coefficient correlation between the SPI-3 and VHI is 0.63 with significant at 99% level. The high coeficcient correlation covers about 58% of Indonesian region. It implies that the total deficit of rainfall for three months giving a significant impact on agricultural drought or it can be speculated that the lag time between meteorological and agriculture drought is about three months. Significant coefficient correlation values on SPI-3 and VHI are common in the southern part of Indonesia. This might be explained that the southern part of Indonesia are mostly center for crop production. This finding is agreed with previous studies (e.g. Wang et al, 2014; Huang et al, 2015). They pointed out crops respond to the drought after a rainfall deficit for three months [9,16].

Table 3. Data of correlation between SPI and VHI

Correlation	Maximum Correlation Value	Minimum Correlation Value	Percentage of coverage with significant correlation: Level 95%	Level 99%	Total
SPI-1 and VHI	0,55	-0,33	19%	38%	57%
SPI-3 and VHI	0,63	-0,32	15%	43%	58%
SPI-6 and VHI	0,58	-0,35	16%	31%	47%
SPI-9 and VHI	0,58	-0,47	15%	24%	39%
SPI-12 and VHI	0,61	-0,5	13%	15%	28%

Table 4. Coefficient correlation betweenSPI and VHI in some territory and containing note[a]

Territory	SPI-1 and VHI	SPI-3 and VHI	SPI-6 and VHI	SPI-9 and VHI	SPI-12 and VHI
Bali	0,34**	0,37**	0,32**	0,25**	0,21**
Demak	0,14	0,19*	0,16*	0,13	0,13
Kediri	0,25**	0,30**	0,21**	0,12	0,08
Cianjur	0,19*	0,23**	0,19*	0,19*	0,17*
Progo	0,29**	0,38**	0,32**	0,18*	0,17*
Lamongan	0,09	0,11	0,06	0,01	-0,01
Karawang	-0,01	0,05	0,07	0,05	0,09

[a]Note are referenced using alpha superscripts
*(**)correlation with significant level 95% (99%)

Mostly paddy fields in Indonesia is located in southern part and 40% centered on Java Island [17]. Territories of Java and Bali are choosen to get more detail about the correlation between SPI and VHI in paddy field. The correlation between SPI and VHI in paddy fields in most areas of Java and Bali (see Table 4) also show that the correlation increases in 3-months scale (SPI-3). Moreover, the coefficient correlation gradually decreased in SPI-6, SPI-9, and SPI-12. We noted that Karawang has a different pattern due towell-maintenance dirrigation system. Again, Table 4 shows the higher coefficient correlation between SPI dan VHI is in 3-months scale (SPI-3). It shows that the agricultural drought, especially in crop land, occurs after rainfall deficit for three months. We speculated that in generally agricultural drought in the majority of land crop has 3 months behind the occurrence of meteorological drought.

4. Conclusion

Meteorological drought and agricultural drought in Indonesia is more intensive during the El Niño years. The duration of meteorological drought is different in each region but generally observe during June to November. However, agricultural drought mostly occurs from August to November. In strong El Niño (2015), we found wider extent and higher severity of meteorological drought than in weak El Niño (2002). Areal extent and severity of meteorogical drought generally is increased in each month and reached its peak in October. Agricultural drought in strong El Niño (2015) also has wider areal extent and higher severity compared to weak El Niño (2002). The intensive agricultural drought mostly occurs in the areas with monsoonal precipitation such as Java, Bali, Nusa Tenggara, Lampung, southern part of Kalimantan, and southern part of Sulawesi. Moreover, VHI is significantly correlated with SPI-3 which accounted for 58% of the total area of Indonesia. It implies that the deficit of rainfall during three months has a significant impact on agricultural drought in Indonesia. In other words the agricultural drought is seen about 3-month behind the occurrence of meteorological drought. Therefore, SPI-3 and VHI clearly explain the relationship between meteorological drought and agricultural drought in Indonesia.

References

[1] Aldrian E and Susanto RD 2003 Identification of three dominant rainfall region within Indonesia and their relationship to sea surface temperature *International Journal of Climatology* **23** 1435-1452

[2] Adianingsih ES 2014 Tinjauan Metode Deteksi Parameter Kekeringan Berbasis Data Penginderaan Jauh *Seminar Nasional Pengindraan Jauh* 210-220

[3] BNPB 2015. Dampak EL-Niño Tahun 2015 terhadap Kekeringan di Indonesia. Retrieved from http://www.bnpb. go.id/berita/2554/dampak-el-nino-tahun-2015-terhadap-kekeringan-di-indonesia on May 27th, 2016

[4] Utami AW Jamhari and Hardyastuti S 2011 El Niño, La Niña, dan penawaran pangan di Jawa, Indonesia *Jurnal Ekonomi Pembangunan* **12** 257-271

[5] Zargar A, Sadiq R, Naser B, and Khan FI 2011 A review of drought indices *Environ Rev* **19** 333-349

[6] Bhuiyan C, Singh RP, and Kogan FN 2006 Monitoring drought dynamics in the Aravalli region (India) using different indices based on ground and remote sensing data *Elsevier* **8** 289-302

[7] Dalezios NR, Blanta A, Spyropolous NV, and Tarquis AM 2014 Risk identification of agriculture drought for sustainable agroecosystems *Natural Hazards and Earth System Science* **14** 2435-2448

[8] Roswiantiarti O, Sofan P, and Anggraini N 2011 Monitoring of drought-vulnerable area in Java Island, Indonesia using satellite remote-sensing data. *Jurnal Penginderaan Jauh* **8** 21-34.

[9] Wang H, Lin H, and Liu D 2014 Remotely sensed drought index and its responses to meteorogical drought in Southwest China *Remote Sensing Letters* **5** 413-422

[10] Kogan FN 1995 Application of vegetation index and brightness temperature for drought detection *Adv Space Res* **15** 91-100

[11] Funk C, Peterson P, Landsfeld M, Pedreros D, Verdin J, Shukla S, Husak G, Rowland J, Harrison L, Hoell A, and Michaelsen J 2015 The climate hazards infrared precipitation with stations-a new environmental record for monitoring extremes *Scientific Data* DOI: 10.1038/sdata.2015.66

[12] McKee TB, NJ Doesken, and J Kleist 1995 Drought monitoring with multiple time scales *American Meteorological Society* 233-236

[13] WMO 2012 Standardized Precipitation Index User Guide *Geneva (SY): Publication Board World Meteorogical Organization*

[14] Adeogun BK, Nwude MO, Mohammad YS, and Adie DB 2014 Evaluation of suitable Standardized Precipiation Index time scales for meteorological, agricultural, and hydrological drought analyses *FUTA Journal of Research in Science* **2** 140-149

[15] Arve LE, Torre S, Olsen JE, and Tanino KK 2011 Stomatal responses to drought stress and air humidity *Venkateswarlu*

[16] Huang S, Huang Q, Chang J, Leng G, and Xing L 2015The response of agricultural drought to meteorological drought and the influencing factors: A case study in the Wei River Basin, China *Agricultural Water Management* **159** 45-54

[17] Ministry of Agriculture 2014 Statistics of Agricultural Land 2009-2013 *Jakarta (ID): Center for Agricultural Data and Information System-Ministry of Agriculture*

Comparison between remote-sensing-based drought indices in East Java

Luisa Febrina Amalo[1*], Rahmat Hidayat[2] and Haris[3]
[1]Center for Environmental Research, Bogor Agricultural University (PPLH-LPPM), Bogor, 16680, Indonesia
[2]Department of Geophysics and Meteorology, Bogor Agricultural University, Bogor, 16680, Indonesia
[3]Gowa College of Agriculture (STTP), Gowa, Indonesia

E-mail: luisafebrina@gmail.com

Abstract. Drought is natural hazard which has causing several impacts, such as decreasing of air and water quality, land degradation, forest fire, decreasing of agricultural crops production. Drought assessment using drought indices have widely conducted for drought monitoring. Remote-sensing-based indices defined as an index which using remote sensing data for mapping the drought condition in particular area or region. This research aims to compare remote-sensing-based drought indices, namely TCI, VCI and VHI to obtain a better understanding about the differentiation between each index, and their application for monitoring drought in East Java on El Nino year 2015. LST and EVI data were used to construct the indices. The result showed, each index proved to be useful, quick, sufficient and inexpensive tool for drought monitoring. However, each index has its differences. TCI proved to be detected drought sensitively in dry season or months when high temperature occurred. While VCI detected drought more sensitive in wet season as well (December-January-February to May) than TCI and VHI. Meanwhile, VHI which the enhancement of TCI and VHI has combined two indicators to provide better comprehension about drought occurrence.

1. Introduction

Drought is natural hazard which frequently occurs in some regions caused by the deficiency of precipitation [1]. Drought has causing several impacts, such as decreasing of air and water quality, land degradation, forest fire, decreasing of agricultural crops production. It may led to water conflict and bring people to immigrate to better place [2][3]. Therefore, drought monitoring is an essential to prevent and minimalize the impact of drought occurrence, Drought monitoring can be assessed by using drought indices. Drought indices defined as a combination of drought indicators (temperature, precipitation, vegetation condition, evapotranspiration, etc.) and provide more comprehensive information for drought analysis than using raw data of each indicators [4].

Niemeyer [5] categorized drought indices into three types, i.e. comprehensive, combined and remote-sensing-based drought indices. Comprehensive drought indices using meteorology, hydrology, and vegetation indicators for describing the drought more widely. PDSI (Palmer Drought Severity Index) is one of comprehensive drought index calculated by using evapotranspiration, runoff, soil recharge, and precipitation indicators. Secondly, combination drought indices is an index combination of two or several drought indices used for drought mapping and monitoring. The example of this indices are USDM (US Drought Monitoring) which is a combination of several drought index, namely SPI (Standardized Precipitation Index), PDSI, vegetation, and hydrology variable. Moreover, remote sensing based indices defined as an index which using remote sensing data for mapping the drought condition in particular area or region [4].

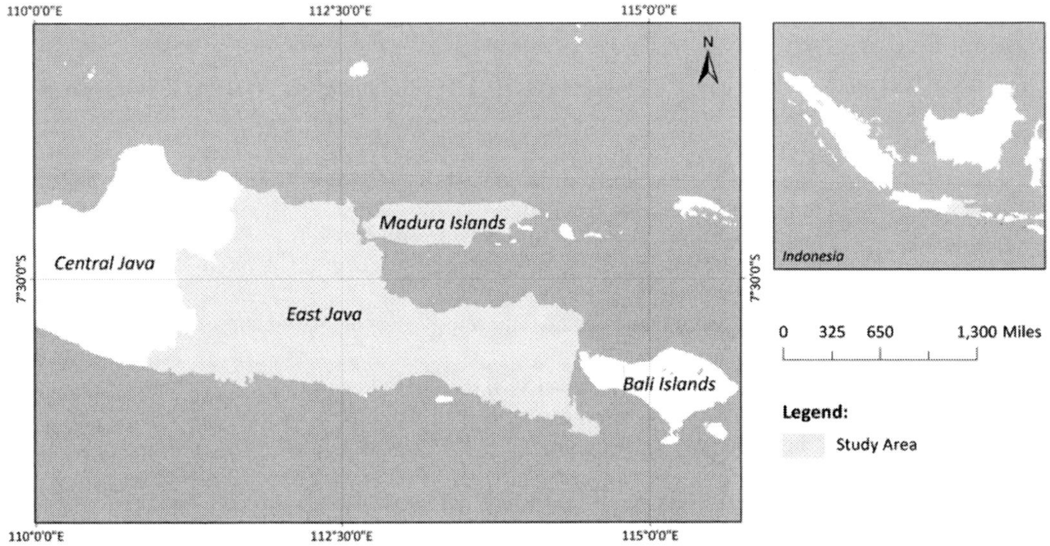

Figure 1. Geographical map of East Java Region

Remote-sensing-based drought indices are widely used because satellite data is more quick, inexpensive, and efficient tools for drought monitoring, than using observational data [1]. There are many remote-sensing-based drought indices which used for drought assessment, e.g. NDVI (Normalized Difference Vegetation Index), TCI (Temperature Condition Index), VCI (Vegetation Condition Index), VHI (Vegetation Condition Index). Beside as drought indices, TCI, VCI, and VHI are classified into vegetation index which describe the condition of vegetation in particular area and categorized them into drought classes, thus they are often to use as indices for drought monitoring [6]. TCI identifies vegetation stress caused by high temperature, as well as excessive wetness [7][8][9]. VCI is commonly used to identify the changes of vegetation from bad to optimum condition [7][8]. VHI describes vegetation health from the combination of TCI (temperature) and VCI (vegetation condition) [8][1].

TCI, VCI, and VHI are remote-sensing-based indices used in this research for drought monitoring in East Java as study area, which one of agricultural production centre in Indonesia [10] is shown in figure 1. Focus period of this research was year of 2015 as strong El Nino event [11]. Thus it may leading to drought occurrence in study area. This research aims to compare these three indices to understand the differentiation between each index, and its application for monitoring drought in East Java on El Nino 2015.

2. Data and Methods

The Moderate Resolution Imaging Spectroradiometer (MODIS) datasets were used to calculate remote-sensing-based drought indices in this research. MODIS datasets were obtained from U.S. Geological Survey (USGS). TCI was derived by using LST (Land Surface Temperature) from MOD11C3 monthly LST with spatial resolution of 0.05°, and calculated by the following equation (equation 1):

$$TCI_i = \frac{LST_{max} - LST_i}{LST_{max} - LST_{min}} \times 100 \quad (1)$$

where LST_i, LST_{min} and LST_{max} defined as LST of current month, and maximum and minimum LST value in multi-year. TCI value ranges 0-100, while low TCI indicates an unfavourable condition, and high TCI indicates an optimum condition. In several study, TCI is also produced from brightness temperature [1][7][8]. In addition, TCI derived from LST was conducted in several studies, e.g. [12][13].

VCI was constructed by using EVI (Enhanced Vegetation Index) which similar with NDVI but passed through the enhancement. EVI have corrected the aerosol in atmosphere and shadow captured by satellite to reduce the bias that can influenced EVI values [14]. In this study we used MODIS product MOD11A3 monthly EVI with spatial resolution of 1 km. TCI then calculated by the following equation :

$$VCI_i = \frac{EVI_i - EVI_{min}}{EVI_{max} - EVI_{min}} \times 100 \quad (2)$$

where VCI_i, VCI_{min} and VCI_{max} defined as VCI of current month, and maximum and minimum VCI value in multi-year. VCI also ranges from 0-100, while low VCI indicates an unfavourable condition, and high VCI indicates an optimum condition [1][9].

Lastly, VHI is composed from VCI and TCI provides a better comprehension about drought occurrence, than using only single drought index indicator. The detail of VHI calculation from TCI and VCI illustrated in Figure x (??). VHI calculation also given in following equation:

$$VHI = 0.5VCI + 0.5TCI \quad (3)$$

TCI, VCI, and VHI then classified into five severity classes (see Table 1).

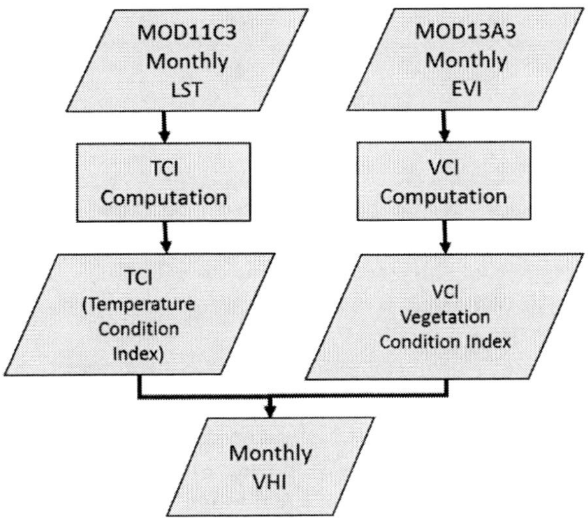

Figure 2. Flow diagram of VHI calculation

Table 1. Drought severity classes for TCI, VCI, and VHI [8][9]

Severity class	Values
Extreme Drought	< 10
Severe Drought	< 20
Moderate Drought	< 30
Mild Drought	< 40
No Drought	> 40

3. Result and discussion

The result showed TCI, VCI, and VHI generally has similar pattern from January and December in 2015, which has low areal extent in wet season (December-January-February) and March-April-May shift season from wet season to dry season (figure 3). East Java is influenced by Australian and Asian Monsoon system [15] that has been caused drought occurs in June-July-August and maximum in October in each year frequently [16]. However, result of strong El Nino event in 2015 caused a longer drought duration, which usually end in November, but in 2015 drought was ended in December with areal extent higher than normal year [16].

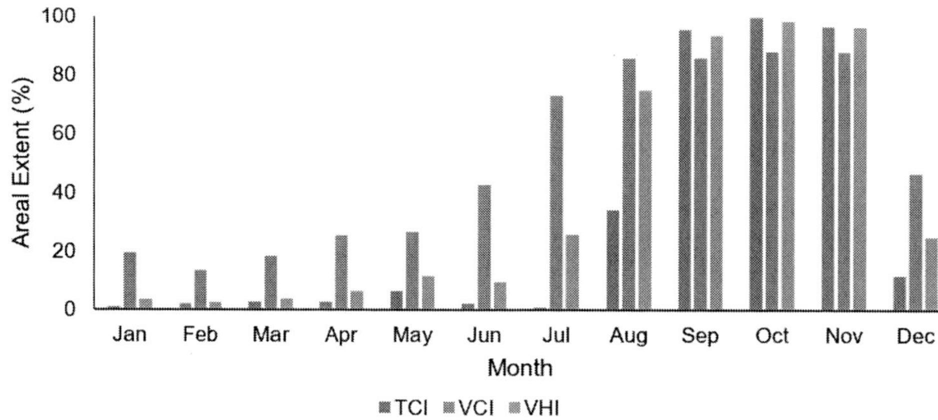

Figure 3. Drought areal extent (index value 0-40) of TCI, VCI, and VHI year 2015

The difference between TCI, VCI, and VHI showed higher areal extent of VCI than TCI and VHI in January to May and December. It clearly seen because areal extent of TCI and VHI were relatively low in those months. It can be due to the indicator used in VCI calculation is vegetation index (EVI) which only consider vegetation condition for monitoring drought. Then, if the harvest season appeared and land became empty, EVI will detect it as drought. Moreover, VCI has high spatial resolution thus it more detail than TCI and VHI.

In contrary, drought areal extent of TCI was tend to high in August to November (dry season). It may be due to indicators used for calculating the index was land surface temperature, which can led to high areal extent of drought if dry season or high temperature occurs. Drought areal extent of VHI also has most similar pattern with VHI (figure 3), however the drought area extent was range between drought areal extent TCI and VCI, caused by the equal weight of TCI and VCI in VHI calculation.

The drought map of TCI, VCI, and VHI were analysed to see any difference or similarity between each index spatially. Selected months in 2015, i.e. September, October, and November, were selected to show the drought condition before, during, and after maximum drought in October (Figure 3) El Nino 2015. Figure 4 showed that TCI drought map has similar with VHI drought map in term of spatial, and drought severity classes also similar. The two indices were dominant with moderate and severe drought. In contrast with drought map of VCI which has areal extent of extreme drought in some areas.

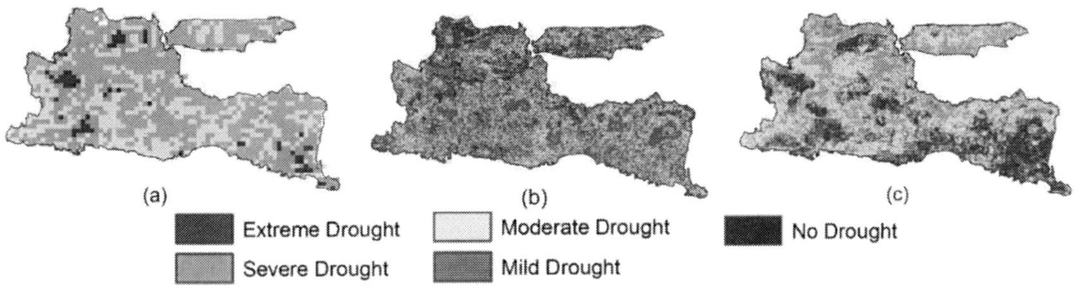

Figure 4. Drought map of September 2015 using (a) TCI, (b) VCI and (c) VHI

While figure 5 showed that TCI, VCI, and VHI drought map were seen similar. Maximum of drought occurrence in East Java is in October [16] which can be seen from Figure 3. The drought maps were dominated by extreme drought in each map. However, TCI has highest areal extent of extreme drought among VCI and VHI. While TCI has the lowest areal extent of extreme drought, and still has no drought areal extent remained in October.

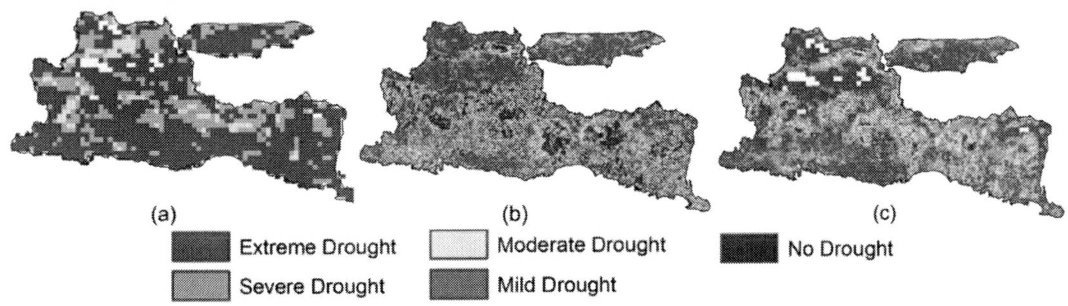

Figure 5. Drought map of October 2015 using (a) TCI, (b) VCI and (c) VHI

While figure 6 showed that TCI, VCI, and VHI drought map were seen also similar. The drought areal extent for three indices has decreased in term of severity. Extreme drought areal extent no longer dominated East Java as in October. In specific area, such as, East area of Madura islands was categorized as extreme drought on each index.

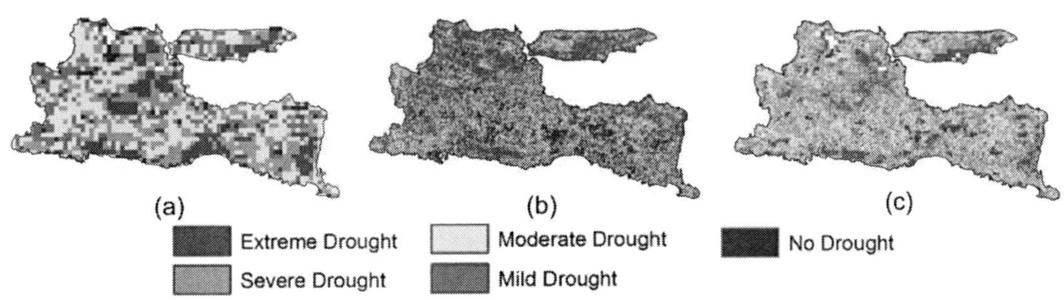

Figure 6. Drought map of October 2015 using (a) TCI, (b) VCI and (c) VHI

Each index proved to be useful, quick, sufficient and inexpensive tool for drought monitoring. However, each index has its differences. For example, in this study, TCI proved to be detected drought sensitively in dry season or months when high temperature occurred. Kogan et al. [8] showed that TCI provides information about vegetation stress caused by high temperature or extensive wetness. While VCI detected drought more sensitive in wet season as well (December-January-February to May) than

TCI and VHI. It may caused by the high resolution of data used for VCI calculation and bias from harvest season thus the index misinterpret unused land after harvest as drought area. But VCI is useful for determine the onset and duration of drought [8] and weather impact on vegetation health [6]. VCI also used for estimating crop productivity [17] because it uses vegetation condition indicator. Meanwhile, VHI which the enhancement of TCI and VHI has combined two indicators, namely temperature and vegetation to provide better comprehension about drought occurrence [8]. The results also showed VHI values were in the middle between TCI and VCI value because of its weight-equal calculation of TCI and VCI to produce VHI.

4. Conclusion

The difference between TCI, VCI, and VHI showed higher areal extent of VCI than TCI and VHI in January to May and December. In contrary, drought areal extent of TCI was tend to high in August to November (dry season). While, drought area extent of VHI was range between drought areal extent TCI and VCI, caused by the equal weight of TCI and VCI in VHI calculation. Each index proved to be useful, quick, sufficient and inexpensive tool for drought monitoring. However, each index has its differences. TCI proved to be detected drought sensitively in dry season or months when high temperature occurred. While VCI detected drought more sensitive in wet season as well (December-January-February to May) than TCI and VHI. Meanwhile, VHI which the enhancement of TCI and VHI has combined two indicators to provide better comprehension about drought occurrence.

Acknowledgment

The authors gratefully acknowledge the U.S. Geological Survey (USGS) for providing MODIS archive used in this research.

References

[1] Dalezios N R, Blanta A, Spyropoulos N V, Tarquis A M 2014 Risk identification of agricultural drought for sustainable Agroecosystems *Nat. Hazards Earth Syst. Sci.* **14** 2435-2448

[2] Amarsaikhan D, Battsengel V, Nergui B, Ganzorig M, Bolor G 2014 A study on air pollution in Ulaanbaatar city, Mongolia. *Journal of Geoscience and Environment Protection* **2** 123-128

[3] Neri C and Briones F 2010 *Assessing drought risk and identifying policy alternatives for drought risk management. Risks, impacts and social meaning of drought: Characterization of the vulnerability in Sonora, Mexico* Background Paper prepared for the 2010 Global Assessment Report on Disaster Risk Reduction (Geneva: UNISDR)

[4] Zargar A, Sadiq R, Naser B, Han F I 2011 A review of drought indices *Environ. Rev.* **19** 333–349

[5] Niemeyer S 2008 New drought indices *Options Méditerranéennes Série A: Séminaires Méditerranéens* **80** 267–274

[6] Jiao W, Zhang L, Chang Q, Fu D, Cen Y, Tong Q 2016 Evaluating and enhanced vegetation condition index (VCI) based on VIUPD for drought monitoring in the continental United States *Remote Sens.* **8** 1-21

[7] Singh R P, Roy S, Kogan F 2013 Vegetation and temperature condition indices from NOAA AVHRR data for drought monitoring over India *Int. J. Remote Sensing* **24** 4393-4402

[8] Kogan F N 1995 Application of vegetation index and brightness temperature for drought detection *Adv. Space Res.* **15** 91-100

[9] Ghaleb F, Mario M, Sandra A N 2015 Regional landsat-based drought monitoring from 1982-2004 *Climate* **3** 563–577

[10] Statistics Indonesia 2015 *East Java in Figures 2015* (Surabaya: CV Media Kontruksi)

[11] National Oceanic and Atmospheric Admnistration 2016 *ENSO years* Retrieved from http://www.cpc.ncep.noaa.gov/products/analysis_monitoring/ensostuff/ensoyears.shtml
[12] Wang H, Lin H, Liu D 2014 Remotely sense drought index and its responses to meteorological drought in Southwest China *Remote Sensing Letters* **5** 413-422
[13] Roswintiarti O, Sofan P, Anggraini N 2011 Monitoring of drought-vulnerability area in Java island, Indonesia using satellite remote-sensing data Jurnal Penginderaan Jauh **8** 21–34
[14] Xiao X, Braswell B, Xhang Q, Boles S, Frolking S, Moore B 2002 Sensitivity of vegetation to atmospheric aerosols: Continental-scale observations in Northern Asia. Remote Sensing of Environment **84** 385-392
[15] Bey A, Amien L I, June T, Pramudia A 2002 The analysis of climate and its impact on water availability in food production centers, Java Island [Unpublished Statements competitive research grant] (Bogor Agricultural University: Bogor)
[16] Amalo L F, Hidayat R, Sulma S 2016 *Analyses of Agricultural Drought in East Java Using Vegetation Health Index* Undergraduate thesis (Bogor: Bogor Agricultural University)
[17] Rahman A, Roytman L, Krakauer N Y, Nizamuddin M, Goldberg M 2009 Use of vegetation health data for estimation of Aus Rice yield in Bangladesh *Sensors* **9** 2968-2975

Conservation planning on eroded land based of local wisdom in Kintamani sub-district, province of Bali

Made Sri Sumarniasih[1*] **and Made Antara**[2]
[1] Department of Agroecotechnology, Faculty of Agriculture, Udayana University PB. Sudirman Street, Denpasar. Bali-Indonesia
[2] Department of Agribusiness, Faculty of Agriculture, Udayana University PB. Sudirman , Denpasar. Bali-Indonesia

E-mail: madesris@gmail.com

Abstract. Location determination is based on the compilation of soil type's map, land use map and slope map. Uniformity of soil type, slope and land use is classified into one unit of land, so that there are 48 units of land to be use as sample points. The purpose of this research are to identify patterns of land use, determine the amount of erosion, the amount of erosion that is tolerable and erosion control through a conservation plan based on local wisdom. The erosion prediction used USLE method, erosion of tolerated *(Edp)* using the formula Hammer.Results of laboratory and field observations having analyzed using USLE showed some level of erosion on land use in the Kintamani classified from very mild to very severe: 4.79 to 370.60 t ha-1yr-1, while Edp ranges from 30.00 to 48.00 t ha-1 yr-1. erosion Severe to very severe found on the use of mixed garden/citrus garden, dry land and shrubs/ reeds.The planned of use of land is intercropping annuals with horticultural crops (cabbage), citrus trees intercropped with flower of gumitir and bush land planting with elephant grass on a slope of less than 25% and planted of trees on land with a slope above 25%.

1. Introduction

In an effort to maintain the agriculture as a source of food and the preservation of an agrarian culture, the presence of agriculture need to be maintained towards the conservation and sustainable agriculture approaches. In relation to land use, the pattern of land management by local farmers, traditionally which are known as local wisdom that has been practiced by the farmers for a long time which need to be raised again to avoid production failures.

Local wisdom is an idea, action and truth that has been a tradition, traditional rules (*awig awig*) in an area, formed as a local culture of excellence, the value it contains is considered very universal. Local wisdom in an area different from other regions[1][2][3]. One of the local wisdom of Bali which has been recognized globbally is rice paddy farming system[4][5].

In nature actually impossible obviate erosion, especially lands that are cultivated on sloping ground, so need conservation action and conservation methods applied to be able to suppress the erosion rate becomes smaller till to the tolerated limit.Lowland rice farming system in Bali is one of the local wisdom, not just a source of food and income for thr farmers, but also the prevention of soil erosion, especially on sloping lands.

The use of land in the district of Kintamani is still less attention to the rules of conservation, in areas with a slope slopes above 25%, is still used for farming seasonal crops. As is known Kintamani is upstream of several rivers vital for the region Bali, one of which is the Watershed of Ayung. Overcome this, the necessary efforts to improve the condition of the watershed with the soil and water conservation, so that controlled the interrelationships of natural resources and environment of the watershed. In an effort to tackle the downturn is more severe on land resources, the necessary land use follow the rules of soil and water conservation to minimize erosion, and can increase farm production.

Based on the issues described earlier, then do research on the amount of erosion and erosion tolerated, at some certain pattern of land use and land conservation action based on local wisdom, if erosion exceeds the tolerable erosion. The purpose of this research is (1) to identify patterns of land use, (2) determine the amount of erosion, (3) the amount of erosion that is tolerable, (4) erosion control through a conservation plan based on local wisdom.

2. Methodology

2.1. Research location

The study was conducted in several land uses in the district of Kintamani, Bali. Kintamani District is geographically located on 115°18'27 " *BT* until 115° 23'00"*BT* and 08°10'00 " *LS* until 08°20'00"*LS*. Total rainfall is very high in this research area which can causes high potential for erosion. Erosivity value average is 1355.59 for Kintamani area near an observation station and to the observation post Catur 2764.80. Soil type is Regosol, these lands are vulnerable to erosion. Land use is dominated dry land. Most people live from farming.

2.2. Materials of the research

Materials used in this research are: map of the earth (1: 25,000)[6], a map of soil types (1: 25.000), land use maps (1: 25,000). [6] [7].

2.3. Research procedure

Research procedures are as follows:

1) Digitizing a map of the study area Sub District Kintamani. using the program GIS. [8].
2) Making a map of the land unit based overlaying map soil type, slope maps and land use maps, in order to obtain a map of the land unit using the program GIS (Figure 1.)
3) Based on the map of the land unit of each soil sample was taken one to represent, in order to get 48 samples, for analysis of physical (bulk dencity, permeability, soil structure), and chemical (organic matter content) properties of the soil. Laboratory analysis coducted in the departement soil and environment, faculty of agriculture, University of Udayana, Denpasar-Bali.
4) Identify land-use patterns is dry land/moor, mixed garden, bush, forest and settlement or open land.

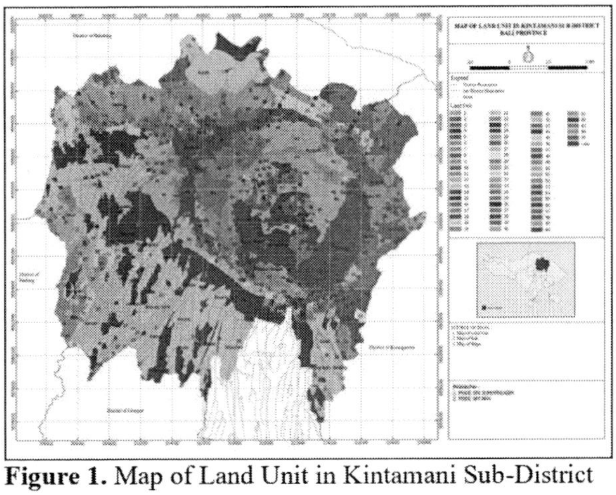

Figure 1. Map of Land Unit in Kintamani Sub-District

2.4. Method of data analysis

Erosion is calculated by using the formula of the *Universal Soil Loss Equation* (USLE) [9]. The amount of erosion is A = R.K.L.S.C.P, where A is the number of eroded soil (t ha^{-1} yr^{-1}), R is a factor rainfall or erosivitas (mm/ha/day/ yr). R = 6.119 (monthly rainfall in cm)$^{1.21}$ (number of rainy days per month)$^{-0.47}$ (maximum daily rainfall 24 hours)$^{0.53}$. K = soil erodibility (tonnes ha/day rain/ha), 100K=1.292 {2.1M^{1,14} (10-4) (12-a) +3.25 (ba) +2.5 (c-3)}. LS is the length and slope factor, LS = √L (0.0136+0.00965S+0.00138S2). C is a factor of soil and vegetation cover crop management, P is a factor of soil and water conservation measures. CP value is determined by the approach of the state in the field with the results of the Land Research Center, Bogor. CP value is the value of crop management factor (vegetation) and the value of land management or conservation measures

Erosion is tolerated *(Edp)* on some land use was analyzed according to the formula Hammer [10], calculated of *Edp*namely:

$$Edp = \frac{\text{Dept of the soil x factor of dept soil}}{\text{The useful life of land}} \quad (1)$$

If the existing conditions causing erosion exceeds the tolerable erosion it needs improvement in land and crop management(CP) with a few scenarios through simulation such as by changing the pattern of land use, so that the erosion rate is smaller than the tolerable erosion. Erosion is tolerated (*Edp*), if a maximum speed of soil loss per year is still tolerated in order to maintain the optimum use of soil productivity for a long period. In determining the value of *Edp* such as the thickness of the layer of soil, soil physical properties and organic matter decline should be considered.

3. Results and discussion

This research was conducted by looking atland use pattern in the district of Kintamani, which is the upstream part of the ayung watershed, one of the longest watersheds in the province of Bali. The followings are the factors that affect erosion:

1) Climate is a factor that affects the hydrological processes and is an important element in the process of hydrology. One of the climate elements that affect most of the hydrological processes is rainfall. Rainfall data in Kintamani Sub-district is obtained from *Tuban* Meteorology and Geophysics Agency (*BMKG*) based on rainfall observation stations located in the village of Kintamani and Catur.
2) The type of soil in the Kintamani Sub-district [7] is Regosol Humus, Brown Regosol and Gray Regosol. Each covers an area of 8729.74 ha; 12014.61 ha and 412.90 ha. This soil type is sensitive to erosion. It is means the soil is easily to be eroded particularly in a land uses that do not follow the rules of soil and water conservation. If the rainfall with hight intencity If it rains with high intensity, then a layer of surface soil (top soil) is easy to carry rainwater runoff into the form of erosion or landslides. The calculation of land value from 0.02 to 0.34 erodibilitas classified from very low to very high.
3) Topography factor of research area show the shape of the area, including differences in slope steepness. The length and slope are two elements that most influence on erosion. Topography affects the hydrological processes, especially on the transformation of rainfall into runoff. Increasingly steep slope, so the greater of rainfall becomes runoff or erosion caused by rainwater into the fall was not given the opportunity infiltrated. Length slopes in the study area is from 10 meters to 100 meters with a slope of 2% up to 41%.
4) The land use is dominated dry land area of 1960,55 ha; mixed garden is 994.73 ha; forest is 318.51 ha; bush 175.64 ha; open land 163.46 ha and undeveloped land is78.34 ha. The use of dry land, mixed garden/citrus area with a less dense planting, patio there is not maintained and without strengthening plants such as elephant grass or patio plants that can maintain the stability of aggregates. The use of dry land and mixed garden in the upper without the principles of

conservation and with high rainfall intensity will cause severe to very severe erosion. This means it is very necessary repairs and maintenance of existing porch with terrace strengthening plants that have dual functions as reinforcing the terrace so that erosion can be minimized and economic value, because it can be used as a means of *upakara* especially for Hindus (based on local wisdom). Value of crops and land management (CP) obtained at the existing condition is 0.005 to 0.28, the lowest CP value obtained on land use for forest land-use and highest in shrubs or overgrown with weeds

Results of laboratory analysis and field observations was analyzed using *USLE* showed tha some level of erosion on land use in the Kintamani district classified from very mild to very severe, namely: 4.79 to 370.60 t $ha^{-1}yr^{-1}$, while erosion is allowed (*Edp*) ranges from 30.00 to 48.00 t ha^{-1} yr^{-1}. Severe erosion to very severe erosion found on the use of mixed garden/citrus garden (land units 1, 6, 25, 29, 33, 35, 46, 52, 55, 61 and 64), dry land (land units 2, 11, 19, 23, 31, and 58), and shrubs/ reeds (land units 15, 41, 47, 53, 59 and 65) (Figure 2). Erosion analysis presented in Table 1. While the erosion of the result of improved land management and crop management (simulation result) is presented in Figure 3.

Based on the analysis at Table 1, erosivity and erodibility factors (R, K) is a natural factor that can not be repaired, while the factors of slope and slope length (L,S) can be repaired, but the costs necessary to make the building mechanically (terrace) is large and expensive, the factor which most can be corrected is the crop and land management (CP). In the planning of erosion control, so the most effective conservation action and acceptable by local communities is based on local wisdom, which is planted intercropped with plants that can live in the study site, but it has economic value that can increase farmers' income.

Table 1. Erosion, Tolerated Erosion (*Edp*), CP Value, and Erosion of the Improvement Result of CP Value (Simulation)

Unit of Land	Name of Village	Land Use	Erosion (t ha^{-1} th^{-1})	Edp (t ha^{-1} th^{-1})	CP Value After Improve ment	Erosion of Improvement Result of CP Value (t ha^{-1} th^{-1})
1	Belandingan	Mixed garden	19.85	48	0.003	0.74
2	Belandingan	Dry land	70.06	36	0.021	5.25
3	Belandingan	Mixed garden	21.60	48	0.003	0.81
4	Belandingan	Bush	17.19	30	0.005	4.09
5		Settlement				
6	Belandingan	Mixed garden	91.41	48	0.003	3.43
7	Subaya	Forest	14.48	48	0.004	11.58
8	Belandingan	Mixed garden	36.71	48	0.003	1.38
9		Settlement				
10	Belandingan	Bush	24.86	30	0.005	5.92
11	Subaya	Dry land	118.22	36	0.021	8.87
12	Belandingan	Forest	8.59	48	0.004	6.87
13	Belandingan	Mixed garden	38.60	48	0.003	1.45
14		Settlement				
15	Belandingan	Bush	37.04	30	0.005	8.82
16	Belandingan	Mixed garden	38.60	48	0.003	1.45
17	Catur	Mixed garden	15.98	48	0.003	0.60
18		Settlement				
19	GunungBau	Dry land	111.95	36	0.021	8.40
20	Buahan	Forest	4.97	48	0.004	3.97
21	A. Batu Dinding	Mixed garden	36.80	48	0.003	1.38
22		Settlement				
23	Ulian	Dry land	167.93	36	0.021	12.59
24	Buahan	Forest	10.97	48	0.004	8.78

25	Catur	Mixed garden	59.22	36	0.003	2.22
26		Settlement				
27	Buahan	Mixed garden	37.71	48	0.003	1.41
28	Sukawana	Forest	11.13	48	0.004	8.91
29	Mengani	Mixed garden	132.22	48	0.003	4.96
30		Settlement				
31	Subaya	Dry land	320.72	36	0.021	24.05
32	Sukawana	Forest	17.17	48	0.004	13.74
33	Bantang	Mixed garden	49.27	48	0.005	11.73
34		Settlement				
35	Pinggan	Mixed garden	126.01	48	0.040	63.01
36	Batur Selatan	Forest	4.79	48	0.004	3.83
37	Kintamani	Mixed garden	25.06	48	0.003	0.94
38		Open land				
39		Settlement				
40	Songan A	Dry land	73.95	36	0.021	5.55
41	Kintamani	Bush	62.41	30	0.005	14.86
42	Songan A	Forest	6.07	48	0.004	4.86
43	Suter	Mixed garden	61.22	48	0.003	2.30
44		Open land				
45		Settlement				
46	BayungGede	Mixed garden	59.69	48	0.003	2.24
47	Kintamani	Bush	48.19	30	0.005	11.47
48	Songan A	Forest	10.17	48	0.004	8.13
49	Suter	Mixed garden	48.97	48	0.003	1.84
50		Open land				
51		Settlement				
52	Manikliu	Mixed garden	73.19	48	0.003	2.74
53	Kintamani	Bush	51.35	30	0.005	12.23
54	Batur Selatan	Forest	20.67	48	0.004	16.53
55	Kintamani	Mixed garden	114.50	48	0.040	4.29
56		Open land				
57		Settlement				
58	Sukawana	Dry land	370.60	36	0.021	27.79
59	Kintamani	Bush	248.91	30	0.005	59.26
60	Buahan	Forest	32.85	48	0.004	26.28
61	Sukawana	Mixed garden	250.28	48	0.003	9.39
62		Open land				
63		Settlement				
64	Batur Tengah	Mixed garden	167.31	48	0.003	6.27
65	Batur	Bush	170.81	30	0.005	20.67

Source: result analysis
Unit of land : 5,9,18,22,26,30,34,38,39,44,45,50,51,56,57,62,63 is the land use settlement or open land.

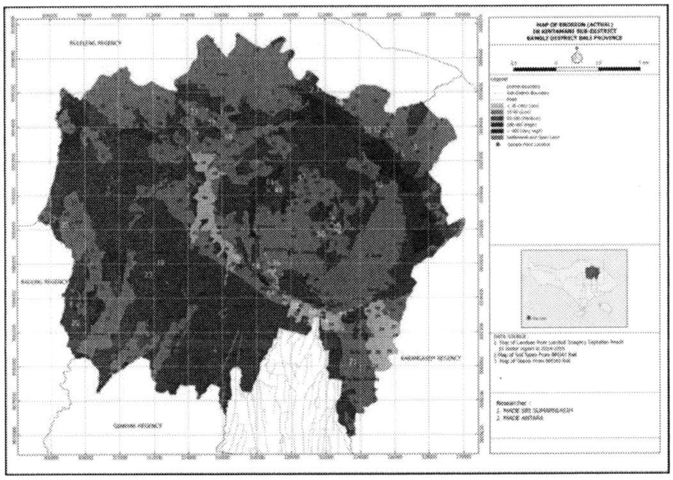

Figure 2. Map of The Erosion Actual in Sub-District Kintamani

The use of land for mixed garden/orange seemed to be filled plants sidelines due to lack of appropriate terms of land cover (crop factor or C factor), because the plant canopy at the time the plant was three years less dense, high rates of erosion caused among citrus crop there are still parts of land that is not covered by the canopy of citrus. This is when it rains with high intensity will cause erosion, because there is no canopy of leaves that withstand the kinetic energy of the rainfall. As for the use of the existing terrace moor land should be maintained by planting amplifier terrace, which has a dual function. On land overgrown with reeds or left bush/shrub should be planted with elephant grass, in addition for animal feed, can also prevents erosion, especially on steep slopes.

The use of dry land, mixed garden/citrus, and shrubs/reeds patio that there are not maintained properly, causing erosion exceed erosion tolerated, it needs management. While on land that is overgrown with shrubs/reeds should be managed well with land management and intercropping crops, maintaining terraces of *gulud* existing with planted by h elephant grass as a reinforcement of the terrace and can be used to feed livestock (has a dual function), so that erosion on land that is left alone with weeds can be minimized.

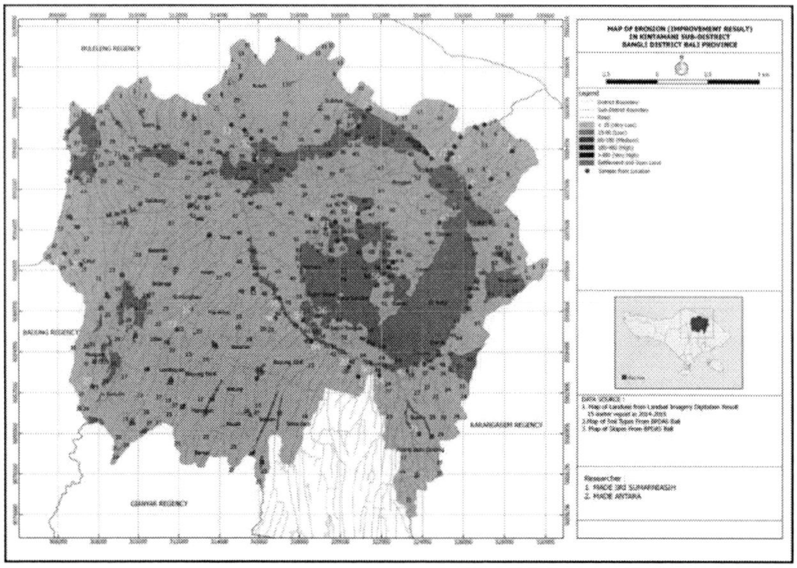

Figure 3. Map of The Erosion Result of Simulation in Sub-District Kintamani

Gumitir flower crops being planted in the hallway of citrus increase land cover so as to reduce erosion will happen, because the rainfall on the soil surface can be mitigated by a canopy of gumitir flowers. Keep in mind that the hall plants should be lower of the canopy than citrus trees so as not to disrupt the growth of the main crops (citrus trees). Patio that there should be maintained by planting plants such terrace amplifier: elephant grass plant, lemongrass, and *pandan*, it is suggested because the plant has a dual function. Elephant grass has a dual function as an amplifier terraces as well as fodder, lemongrass as the reinforcement terrace and can be used for herbs and pandan leaves as reinforcement terrace and means of *upakara* (Hindus). The plant has a socio-cultural values as a means of *upakara* and economic value because it can increase the income of farmers and protecting the environment towards sustainable agriculture.

On land shrubs/reeds, steep topography with slope above 25%, or the land is left without treatment, especially in the border river/gorge suggested planted woody trees such as pecan, *cempaka*, *sandat, beringin, bunut*, breadfruit tree, jackfruit, and others. Trees such as *upakara* plants produce leaves, flowers or fruit that can be harvested for sale, without cutting down the tree, so that the economy can increase revenue. In agro ecology can preserve the environment, the main function is to protect the steep areas to avoid landslides and springs that exist in these locations are well preserved because of the fountain in the upstream is very influential to the downstream part of the watershed.

The farming community actually know besides slope, land use is a factor affecting the surface runoff and erosion, but farmers still not yet apply the conservation techniques that right because of production costs to be incurred if the action is done. This can actually be solved by involving the community at every activity of government Provincial / District in accordance with Act No. 37 Year of 2014 Clause al 46 paragraph (4). The government also should regularly provide counseling, mentoring, greening along with the communities in which activities are conducted, and the provision of seeds required for free so that people care about and feel they have, want to keep the environment well in order to reach sustainable agriculture. In addition to growing crops interrupted the crop rotation patterns most likely be implemented is to plant annual crops with horticultural crops for rainfed paddy fields or upland. Bush land on the river banks and ravines improvements made is by planting trees or woody plants such as Albizia forest industry, Albizia, jackfruit and breadfruit. The tree is useful to keep the border river and the gorge in order not to landslides because it has deep and strong roots. Trees can also be cut down if the age is enough to be sold so as to increase farmers' income, because they are in areas with slope <25%, but as an important note is to replace the harvested crop of young plants. Preferably prior to clearing groomed young plants or seedlings that will be planted as a replacement.

4. Conclusions and suggestion

4.1. Conclusion

Based on the results and discussion, it can be formulated some conclusions, as follow:

1. Erosion happens in Kintamani District on some land uses and belong to very mild to very severe category (4.79 to 370.60 t $ha^{-1}yr^{-1}$), whereas the tolerable erosion is 30-48 t $ha^{-1}yr^{-1}$.
2. Counter measures on the eroded land through conservation planning based on local wisdom in the use of dry land, mixed garden/citrus garden, shrubs/reeds.
3. The planned land use is intercropping annuals with horticultural crops such as, citrus trees intercropped with *gumitir* flowers and planting the elephantgrass on bush land with a slope of less than 25% and growing trees on land with a slope above 25%.

4.2. Suggestion

1. Counseling could be done periodically to prevent the conversion of agricultural land into non-agricultural land
2. In order to increase farmers' income above a decent life apart from planting economically valuable commodity, it also can be done by raising livestock and looking for side jobs such as trader, labourer or tourism activities.

References

[1] Hertanto 2008 Local wisdom to improve of food security agrimandiri year to 4
[2] Ridwan N A 2007 Scientific grounding of local wisdom *IBDA Journal* Vol **5**, pp 27-38
[3] Salain P R 2007 *Role of local wisdom in spatial planning in Bali* In the local wisdom in environmental management (Denpasar: Udayana University Press)
[4] Sutrisno A 2011 *Existence of Subak in the regions of ex transmigration in North Bengkulu* Downloaded September 12, 2012
[5] Windia W 2006 *Subak irrigation system transformation based the concept of Tri Hita Karana* Bali Post Publisher Bali
[6] Bakosurtanal 2000 *Earth Visual Map of Indonesia* (Jakarta)
[7] BPDAS Unda Anyar 2009 *Integrated watershed management plan watershed swp pangi ayung* Report of Watershed Management Center of Unda Anyar (Denpasar)
[8] Barus B and Wiradisastra U S 2000 *Geographic Information System, Remote Sensing and Cartography Laboratories*, Departement of Soil (Bogor: Bogor Agricultural University)
[9] Wischmeier W M and Smith D D 1978 *Predicting rainfall erosion losses-a guide to conservation planning* Washington DC, US Gov.US Dep of Agriculture, Agric Print Off Handbook
[10] Hammer W I 1981 *Soil Conservation* Consultant Report AGOF INS/78/006 Technical Note No 7CSR Bogor

Comparison between wavelet transform and moving average as filter method of MODIS imagery to recognize paddy cropping pattern in West Java

Kreshna Dwi Nugroho[1*], Singgih Pebrianto[1], Muhammad Arif Fatoni[1], Alvin Fatikhunnada[1], Liyantono[1], Yudi Setiawan[2]
[1]Department of Mechanical and Biosystem Engineering, Faculty of Agriculture Technology and Engineering, Bogor Agricultural University, Lingkar Akademik Street, Campus of IPB Dramaga, Bogor 16680, West Java, Indonesia
[2]Center for Environmental Research (PPLH), Bogor Agricultural University, PPLH Building, 2th-4th floor, Lingkar Akademik Street, Campus of IPB Dramaga, Bogor 16680, West Java, Indonesia

E-mail: kreshna.dwinugroho40@gmail.com

Abstract: Information on the area and spatial distribution of paddy field are needed to support sustainable agricultural and food security program. Mapping or distribution of cropping pattern paddy field is important to obtain sustainability paddy field area. It can be done by direct observation and remote sensing method. This paper discusses remote sensing for paddy field monitoring based on MODIS time series data. In time series MODIS data, difficult to direct classified of data, because of temporal noise. Therefore wavelet transform and moving average are needed as filter methods. The Objective of this study is to recognize paddy cropping pattern with wavelet transform and moving average in West Java using MODIS imagery (MOD13Q1) from 2001 to 2015 then compared between both of methods. The result showed the spatial distribution almost have the same cropping pattern. The accuracy of wavelet transform (75.5%) is higher than moving average (70.5%). Both methods showed that the majority of the cropping pattern in West Java have pattern paddy-fallow-paddy-fallow with various time planting. The difference of the planting schedule was occurs caused by the availability of irrigation water.

1. Introduction

Indonesia is a country in fourth-highest rice consumption in the world after Myanmar, Vietnam, and Bangladesh, at 1.62 kg per person per weeks in 2014 until now [1]. Therefore needs government effort to keep rice supply to Indonesian people. West Java with a number of people in 2013 as many as 45.340.799 people [2] and increase every year, needs rice supply to fulfill food requirement. Information about agriculture land utilization (especially paddy field) is important things to support of sustainable agriculture and Indonesian food security. Understanding the potential of paddy field production in Indonesia is one way to increase effectiveness and efficiency of agriculture land development program.

Indonesia has big paddy field area. Characteristic change of paddy field vegetation at the time can give information about Indonesian paddy crop intensity. The way that used to knows about vegetation dynamic of paddy in a very wide area is using remote sensing. Utilization of remote sensing data with the geographic information system (GIS) to various environmental applications already many used, especially to detect land use change [3]. This study discusses remote sensing for paddy field monitoring based on MODIS time series data. In time series MODIS data, difficult to directly classify data due to temporal noise. Therefore wavelet transform and moving average are needed as filter methods. The objective of this study is to recognize paddy cropping pattern with wavelet transform and moving average in West Java using MODIS imagery (MOD13Q1) from 2001 to 2015 then compared between both of methods.

2. Method

2.1. Study Site

West Java Province is one of six Province in Java Island, Indonesia. Geographically, West Java Province located between 5°50'-7°50' South Latitude and 104°48'-108°48' East Longitude (figure 1). In 2014, West Java has paddy field area 924,307 ha which divided by 742,658 ha irrigated lands and 181,649 ha unirrigated lands. This paddy field area produced 7,306,009 tons [2].

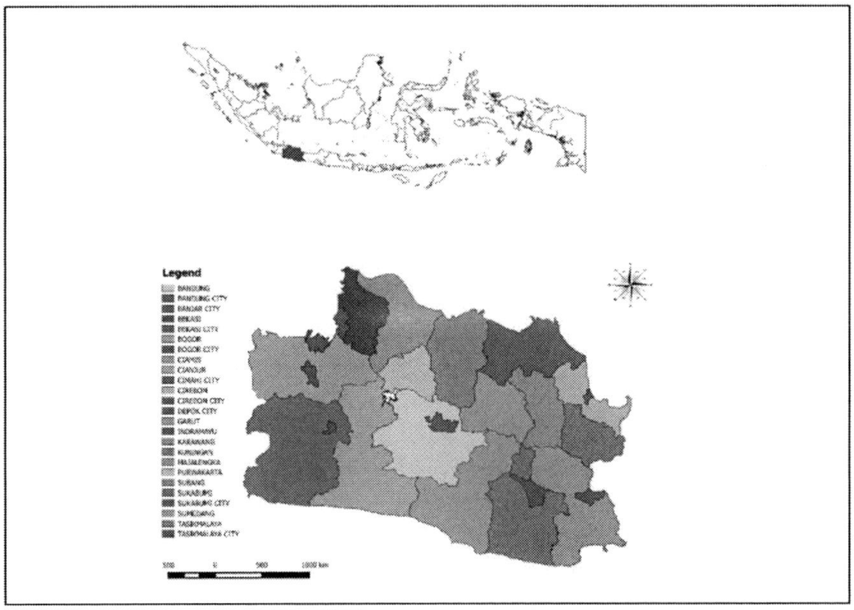

Figure 1. The study site

2.2. Method of analysis

2.2.1. Collecting data and pre-processing MODIS Image. Data that used was map paddy field land use, map of administration and MODIS data in West Java. Land use and administration map of West Java obtained from Geospatial Information Agency (BIG). MODIS data obtained, at no cost, from the NASA Land Processes Distributed Active Archive Center (LPDAAC) [4]. MODIS imagery has 36 bands with spatial resolution 250 m (band 1-2), 500 m (band 3-7), and 1000m (band 8-36) for each sensor. The MODIS data used in this study is the vegetation indices (VI) composite 16-day Global 250 m SIN Grid V005 or MOD13Q1 product [5], which is generated from MODIS data by using the maximum value composite (MVC) method and combined with the MODIS BRDF (Bidirectional Reflectance Distribution Function) in order to get a greater percentage of clear-sky data. The acquired datasets represented the period from January 2001 to December 2015. In this study, we used the MODIS EVI datasets, which were filtered by wavelet transforms and moving average, in order to reduce inevitable disturbances in the time-series datasets. Figure 2 illustrates the flow of data handling and image analysis in this study.

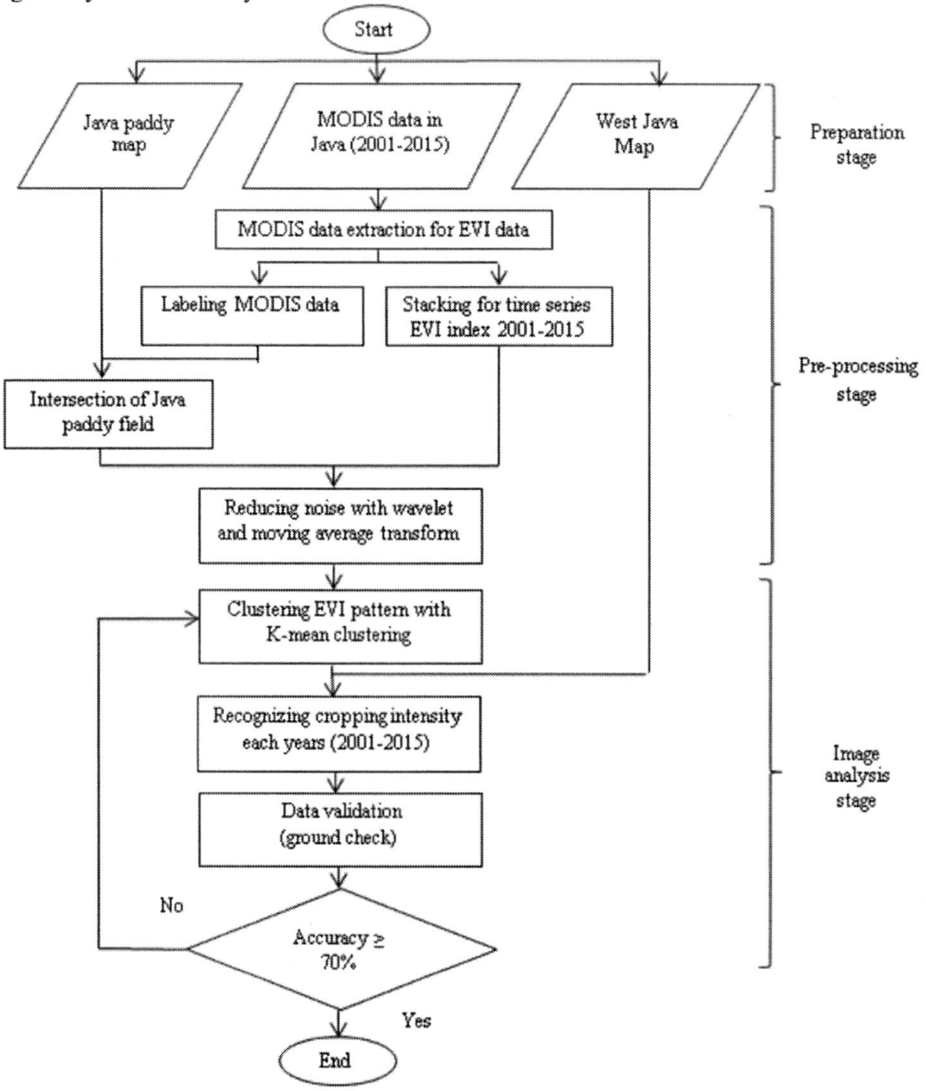

Figure 2. Study flowchart

The MODIS developed the Enhanced Vegetation Index (EVI) based on this equation below [6].

$$EVI = G\frac{\rho_{NIR}-\rho_{red}}{\rho_{NIR}+C_1\times\rho_{red}-C_2\times\rho_{blue}+L} \quad (1)$$

Where, ρ_{NIR}, ρ_{red} and ρ_{blue} are the remote sensing reflectance in the NIR, red, blue, respectively. L is a soil adjustment factor, and C1 and C2 describe the use of the blue band in a correction of the red band for atmospheric aerosol scattering. The coefficients C1, C2, and L are empirically determined as 6.0, 7.5, and 1.0. respectively. G is a gain factor set to 2.5.

2.2.2. *Noise reduction.* In time series MODIS data, it is difficult to direct classified of data, because of temporal noise. Temporal noise can be caused by something like clouds covers and atmosphere effect. Cloud covers become a problem where a research using MODIS data conducted in a tropical area like Indonesia. Because heavy cloud covers may affect the quality of the image and information of earth's surface can be lost. Besides cloud covers, the atmosphere also becomes a problem. Because the atmosphere can absorb reflectance energy from the earth surface, which causes reduced total energy. In this study to reduce such temporal noise, the wavelet transform and moving average transform were used. Wavelet transform and moving average are needed to reduction noise or as filter methods.

2.2.3. *Clustering and recognize the cropping pattern of paddy field.* The clustering method subdivides data sets into k clusters through an iterative process in order to optimize a criterion function [7]. Data clustering aim to easier of analysis. The cluster based on similarity of spectral profile each MODIS pixel. Data that already classified using k-mean clustering afterward determined the cropping pattern. Then classified data again based on similarity of the pattern each class based on Sakamoto *et al.* [8]. They were using wavelet and fourier transform, paddy growth period divided 3 phase, that is planting date, heading date, and harvesting date. In general, paddy fields are flooded before rice planting, then decreases during the period and increases again after rice planting. Furthermore, vegetation index will increase in growing period, and reach a peak in heading date (maximum vegetation index). After the heading date, vegetation index in paddy field begins to decrease as leaves wither and die. It then decreases abruptly because of harvesting. Figure 3 illustrates recognizing of the cropping pattern based on EVI [8].

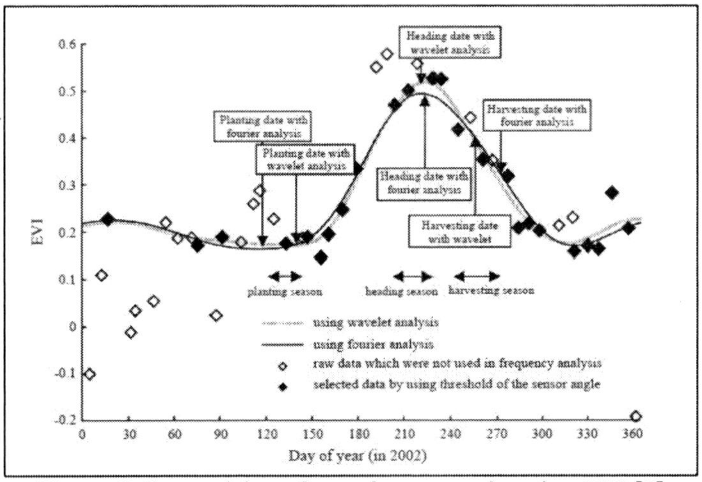

Figure 3. Recognizing of cropping pattern based on EVI [8]

2.2.4. *Validation of data.* Stage of data validation (ground check) held in some area in West Java to certain compatibility MODIS data and field condition. A number of observation points in this study are 139 points. Using 70 % threshold based on a study [9], He was analyzing of paddy field change dynamics in Java and generated 71.1% accuracy.

3. Result and Discussion

3.1. Pre-processing MODIS image

The first stage of this study was pre-processing data, which are divided into some stage such as EVI value extraction and stacking of MODIS imagery, cut image of administration and paddy field map of West Java, and noise reduction. At this stage, noise reduction using wavelet transform and moving average methods. Wavelet transforms using level 1, the approximate 1, and coiflet level 1. Coiflet is kind of wavelet transform. The higher level of the selected value cause the greater value reduced. Meanwhile, moving average methods was using window level 3. In selected level of the window, not too big and small, because if too big (window > 5), data will be too smooth and fluctuated of pattern not to recognize. But, if using window level is too small (window ≤ 2), the results still have an irregular pattern (noise).

3.2. Analysis of West Java MODIS image from 2000-2015

To analyze a number of big data, used k-mean clustering method. This method divided by 3 stage, first operator input the maximum class. Second based on input, randomly the computer will position average value as a centroid. Third based on a distribution of centroid, all pixel calculated and classified based minimum distance. A number of maximum class that chosen in this study was 25 class. When each centroid does not yet represent a number of class around then the computer will determine new centroid. The process going on until formed a class based on minimum distance from the centroid.

K-mean clustering also caused cropping pattern each class more smoothed. More and more class input then cropping pattern that formed more and can represent dynamic that happen in the field. But using more class caused long processing time, therefore must adjust with computer ability. Data that classified then analyzed by cropping pattern with directly seen from cropping pattern graph result clustering. Starting at 25 class, data classified again based on similarity of pattern and generated 10 main class.

3.3. Cropping pattern in West Java paddy field

The process of the pattern recognizing paddy field using EVI value, as described in methodology (point 2.2.3). Distribution of cropping pattern with wavelet transform and moving average are given in figure 4.

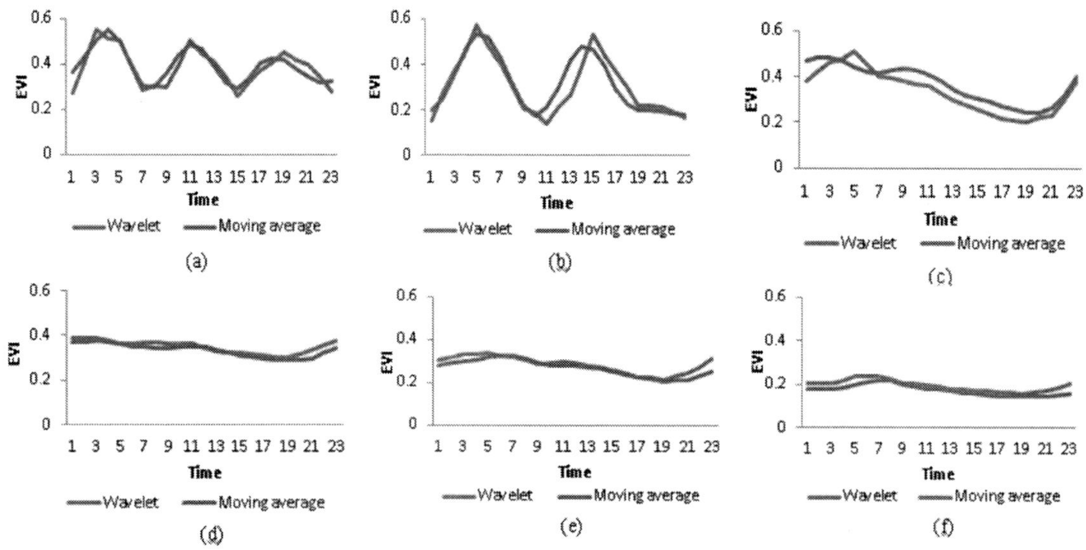

Figure 4. The sixth type of cropping pattern in West Java from recognizing the temporal vegetation pattern of EVI. (a) paddy triple cropping; (b) paddy double cropping; (c) paddy single cropping; (d) mix and other crops dominantly; (e) mix and building dominantly; (f) fishpond.

Spatial mapping of paddy cropping pattern with wavelet transform and moving average shown in figure 5.

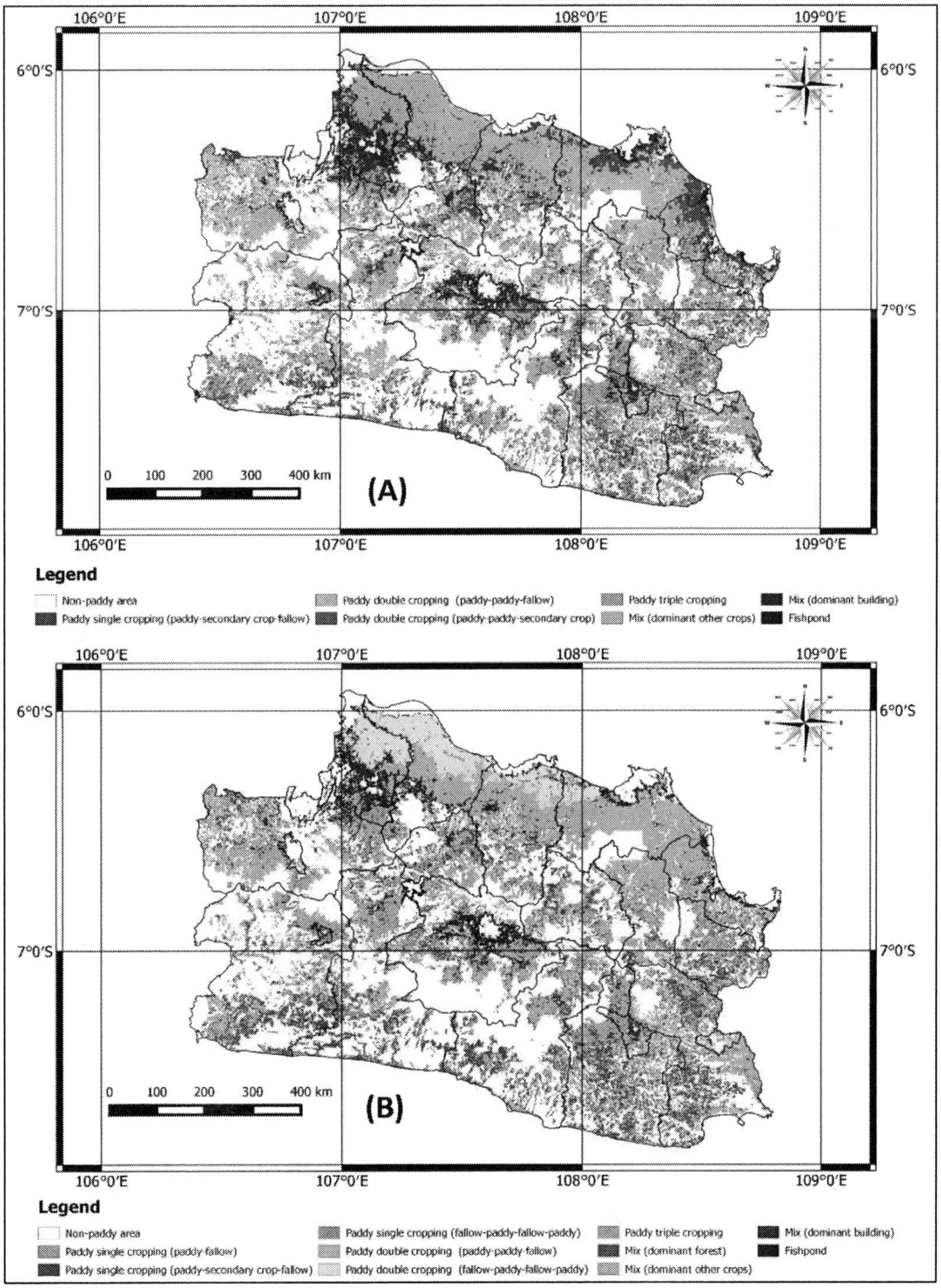

Figure 5. Comparison of paddy cropping pattern on West Java in 2015 with wavelet transform (a) and moving average (b).

3.4. *Accuracy of cropping pattern*

The cropping pattern was calculated based on ground check result using confusion matrix method. A number of reference sample as many as 139 points and generated overall accuracy to be 75.5% in wavelet transform and 70.5% in moving average method. The result of ground check refers to Table 1.

Table 1. Confusion matrix for paddy cropping pattern on West Java with wavelet transform (a) and moving average (b)

Cropping Pattern	Reference Data Paddy Field Type																				User's Accuracy (%)	
	PBB		PLB		PPB		BPP		PPL		P3		CH		CP		CT		T			
	(a)	(b)	(a)	(b)	(a)	(b)	(a)	(b)	(a)	(b)	(a)	(b)	(a)	(b)	(a)	(b)	(a)	(b)	(a)	(b)	(a)	(b)
PBB	0	4	0	0	0	0	0	0	0	0	0	0	0	0	0	0	0	1	0	0	-	80.0
PLB	0	0	1	4	2	4	0	0	0	0	0	0	0	1	0	0	0	1	0	0	33.3	40.0
PPB	0	3	1	1	61	49	0	0	0	0	2	1	0	0	0	0	1	1	0	0	93.8	89.1
BPP	0	0	0	0	0	3	0	10	0	0	0	1	0	0	0	0	0	1	0	0	-	66.7
PPL	0	0	0	0	1	0	0	0	1	0	0	0	0	0	0	0	0	0	0	0	50.0	-
P3	0	0	0	0	0	0	0	0	0	0	12	14	0	0	0	0	0	1	0	0	100.0	93.3
CH	0	0	0	0	0	3	0	0	0	0	0	0	0	1	0	0	0	2	0	0	-	16.7
CP	0	0	0	0	5	7	0	0	0	0	2	2	0	0	3	3	2	2	0	0	25.0	21.4
CT	0	0	2	2	12	5	0	0	0	0	4	1	0	0	0	0	24	11	0	0	80.0	64.7
T	0	0	0	0	0	0	0	0	0	0	0	0	0	0	0	0	0	0	3	2	100.0	100.0
Procedure's accuracy (%)	-	57.1	25.0	80.0	88.4	69.0	-	100.0	100.0	-	60.0	73.7	-	50.0	100.0	100.0	88.9	55.0	100.0	100.0		
Overall Accuracy : (a) = 75.5 % ; (b) = 70.5 %																						

Note:

PBB = Paddy single cropping (paddy-fallow)
PLB = Paddy single cropping (paddy-secondary crop-fallow)
PPB = Paddy double cropping (paddy-fallow-paddy-fallow /.paddy-paddy-fallow)
BPP = Paddy double cropping (fallow-paddy-fallow-paddy)
PPL = Paddy double cropping (paddy-paddy-secondary crop)
P3 = Paddy triple cropping (paddy-paddy-paddy)
CH = Mix (dominant forest)
CP = Mix (dominant building)
CT = Mix (dominant other crops)
T = Fishpond

Based on the result of the calculation, overall accuracy obtained with wavelet transform higher than moving average method with accuracy succession 75.5% and 70.5%. It means wavelet transform better to reduce noise than moving average. The higher user's accuracy both of method are PPB, P3 and T. It caused the pattern of PPB and P3 easier to difference, that showed 2 peaks and 2 valleys for PPB and 3 peaks and 3 valleys for P3. In fishpond pattern, the high accuracy caused EVI average of fishpond very low, and different from paddy pattern generally that has high EVI average. Low of user's accuracy in both of method showed at PLB and CP. In PLB, occur mistakes to recognize cropping pattern PPB (paddy-paddy-fallow) that identified as PLB (paddy-secondary crop-fallow), therefore accuracy is low. It caused many of second season paddy that crop with a different period, therefore pattern that identified is PLB. Low accuracy in CP pattern caused the inability of MODIS image to recognize the cropping pattern with pixel lower than MODIS resolution (250 m).

4. Conclusion

Based on the study, image satellite is used to recognize the cropping pattern in West Java and generated overall accuracy 75.5% and 70.5% in wavelet and moving average transform. Methods for recognizing the cropping pattern in West Java successfully, and generated ten cropping patterns with three main cropping patterns: paddy single cropping (paddy-secondary crop-fallow) with accuracy 33.3% and 40.0%, paddy double cropping (paddy-paddy-fallow) with accuracy 93.8% and 89.1%, and paddy triple cropping with accuracy 100.0% and 93.3%.

Acknowledgements
The authors would like to thank the NASA LP PAAC for making MODIS data available, also the Center for Environmental Research (PPLH IPB) for providing the facilities in this research.

5. Reference

[1] Badan Pusat Statistik 2014 Average consumption per person a week of important food, 2007-2014 *Pusat Data dan Sistem Informasi Pertanian* (Jakarta: Badan Pusat Statistik)
[2] Badan Pusat Statistik 2015 Statistics of agriculture land 2015 *Pusat Data dan Sistem Informasi Pertanian* (Jakarta: Badan Pusat Statistik)
[3] Liu Y, Wang X, Guo M, Tani H, Matsuoka N and Matsumura S 2011 Spatial and temporal relationships among NDVI, climate factors, and land cover changes in Northest Asia from 1982 to 2009 *GIS science and Remote Sensing* **48** 371-393
[4] US Geologic Survey Land Processes Distributed Active Archive Center 2010 MODIS Data Product Table Available online: http://lpdaac.usgs.gov/products/modis_products_ table (accessed on 3 December 2010)
[5] US Geologic Survey Land Processes Distributed Active Archive Center Data Pool Available online: https://lpdaac.usgs.gov/lpdaac/get_data/ data_pool (accessed on 10 August 2009)
[6] Huete AR, Didan K, Miura T, Rodrlquez EP, Gao X and Ferreira LG 2002 Overview of the radiometric and biophysical performance of the MODIS vegetation index *Remote Sensing of Environmental* 2003 **83** 195-213
[7] Jain AK, Murty MN and Flynn PJ 1999 Data Clustering: A Review, ACM Computing Surveys **31** 264–323
[8] Sakamoto T, Yokozawa M, Toritani H, Shibayama M, Ishitsuka N and Ohno H 2005 A crop phenology detection method using time-series MODIS data *Journal Remote Sensing of Environment* **96** 366–374
[9] Setiawan Y, Rustiadi E, Yushino K, Liyantono and Effendi H 2014 Assessing the seasonal dynamics of the Java's paddy field using MODIS satellite images *ISPRS International Journal of Geo-Information* **3** 110-129

Spatial change analysis of paddy cropping pattern using MODIS time series imagery in Central Java

Muhammad Arif Fatoni[1*], Kreshna Dwi Nugroho[1], Alvin Fatikhunnada[1], Liyantono[1], Yudi Setiawan[2].
[1]Faculty of Agricultural Engineering and Technology, Bogor Agricultural University, Fateta Building, 4th, Jl. Kamper, Kampus IPB Darmaga, Bogor 16680, West Java, Indonesia
[2]Center for Environmental Research (PPLH), Bogor Agricultural University, PPLH Building, 2 nd - 4 th, Jl. Lingkar Akademik, Kampus IPB Darmaga, Bogor 16680, West Java, Indonesia

E-mail: muhammadariffatoni@yahoo.co.id

Abstract: Central Java had the diverse paddy field cropping patterns and it was influenced by several factors such as water availability, land condition, paddy fields ownership, and local culture. This research was aimed to analyze dynamic changes of paddy cropping pattern using MODIS imagery (MOD13Q1 16-day composite from 2001 to 2015). This research used k-means clustering algorithm for classified cropping pattern in Central Java based on similarity pattern of annual data from vegetation index. The result of this research classified cropping pattern become a main class and produced 15 maps of distribution cropping patterns (from 2001 to 2015). The result also divided Central Java's paddy fields become 2 section (constant and change) based on cropping pattern that majority was caused by water availability. This research got the better accuracy (77.67%) of cropping pattern than long time series analysis from previous research. Although some classes successfully obtained upon annual time series analysis, MODIS still difficult to detect mixed crop pattern.

1. Introduction

Satellite technology for detecting land cover change has been used in several researches for the advancement of precision agriculture. Characteristics of spectral reflectance from vegetation were affected by the leaf pigment content, organic matter, water and leaf structural characteristics such as the shape of leaves and leaf area [1]. The spectral characteristics of vegetation could be divided into the visible spectrum and the infrared spectrum which could be combined to form vegetation index. According to Danoedoro [2], a vegetation index is a form of mathematical transformations involving multiple channels at the same time giving a new image that is more representative of presenting the changes in vegetation. Rice plants which were the majority staple food in Indonesia had the fields area that centralized in Java Island and could be detected using MODIS (moderate resolution imaging spectroradiometer) EVI (enhanced vegetation index) imagery. Setiawan [3] assessed Java's Paddy Fields became 8 specific types and produced distribution map by annual cropping intensity. Moreover, Muhammad [4] used a more specific object that was East Java for analyzing paddy fields dynamic land use. They used long time series MODIS EVI and showed paddy fields land use dynamic in Java region which happened in the last 15 years. The changes that could be occurred in every time and everywhere in Java region should be detected and analyzed because the changes of paddy cropping pattern could affect the paddy productivity. Many methods were used and developed to get the better result for paddy fields problem solving. Some of the research focused on a spatial problem (pixel of satellite image), and the others modified the method to process time series data. This research focused to analyzed the annual change, and it was intended to support the development of near real-time monitoring in the future. Central Java as the central Province in Java Island had the diverse paddy field cropping patterns and it was influenced by several factors such as water availability, land condition, paddy fields ownership, and local culture. This research analyzed the spatial change of cropping patterns in Central Java and produced the changes distribution map using annual time series MODIS EVI.

2. Methodology

2.1 Study Area

Central Java was located at 8° 30' – 50° 40' S and 108° 30' – 111° 30' E (figure 1) and had the area about 3254412 ha with the paddy fields area about 952525 ha at 2013[5]. Central Java which had 29 District and 6 City is the central Province in Java island. Setiawan[3] explained that Java Island comprises the most active volcanic island in the world, therefore it provides vast areas of fertile soils, which support intensive agricultural lands. The combination of high rainfall rate and fertile soils make many areas of the islands ideally suited for agriculture.

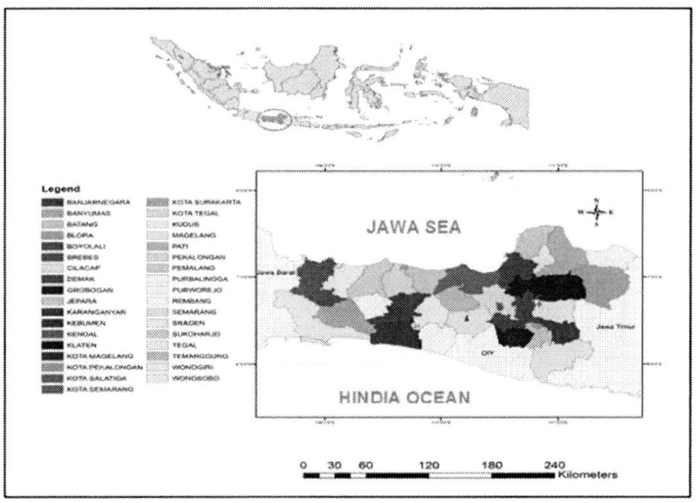

Figure 1. Central Java Province

2.2 Data

MODIS EVI (MOD13Q1) was provided for biomass change analysis and it was corrected from radiometric disruption in atmosphere, gas and aerosol scattering, and more sensitive to high biomass change. MODIS provided 16 composite images with the 250m spatial resolution. It was needed 23 images for representing one-year paddy crop phenology. MODIS EVI was expressed in an equation[6]:

$$EVI = G\frac{\rho_{NIR}-\rho_R}{\rho_{NIR}+C_1\times\rho_R-C_2\times\rho_{blue}+L} \quad (1)$$

Where, ρNIR, ρR, and ρblue are reflectance from near-infrared, red, and blue wave from electromagnetic energy, L (1.0) is a soil adjustment factor, C_1 (6.0) and C_2 (7.5) are aerosol scattering coefficient for blue band correction. This algorithm was optimization from NDVI (normalized difference vegetation index) which can't distinguish the change of vegetation index in the high biomass area. We used 23 MODIS EVI data for interpreting annual patterns, so we needed 345 MODIS images to produced annual patterns from 2001 to 2015. The method is describe in figure 2.

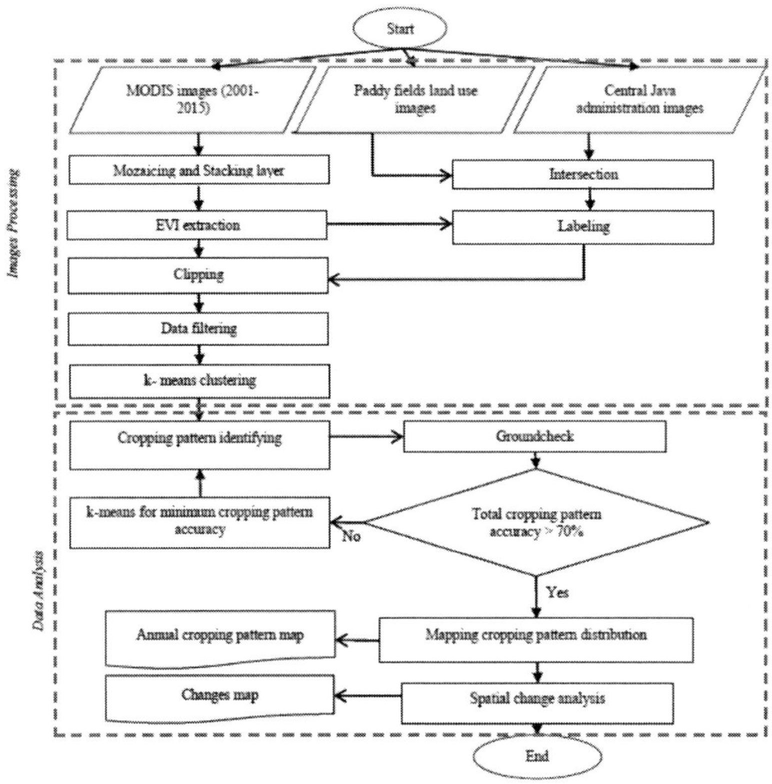

Figure 2. Research method

2.3 Data denoising

Temporal pattern of EVI still had the noise that couldn't be erased by EVI algorithm, and it will be filtered by wavelet transform. This small wave was generally used for MODIS noise reduction process in several previous researches like Setiawan [3] and Muhammad [4] that utilized MODIS EVI for analysis paddy fields in Java region. Sakamoto [8] conclude that the wavelet transform gave the best time profile of EVI for establishing crop phenology. The function of wavelet transform was expressed as follow:

$$Wf(a,b) = \int_{-\infty}^{+\infty} f(x)\frac{1}{\sqrt{a}}\,\Psi\left(\frac{x-b}{a}\right)dx \qquad (2)$$

where a is a scaling parameter, b is a shifting parameter, and Ψ implies a mother wavelet [8]. Because MODIS EVI was discrete data, this study used discrete wavelet transform (DWT) for filtering temporal EVI data that the scaling and shifting parameters transformed as follows:

$$(a,b) = (2^j, 2^j\,k) \qquad (3)$$

Scaling and shifting were two key parameters of the wavelet transform. DWT decomposed a signal into the different scale and through the signal into low pass filter (LPF) and high pass filter (HPF). Setiawan [7] explained these filters retain the large and small scale components of the signals also known as the approximation (A) and detail (D) series, respectively. This study used Coiflet mother wavelet for filtering MODIS EVI signal because previous researches (Muhammad [4] and Setiawan [7]) gave the good accurate for cropping pattern in Java region with the Coiflet mother wavelet. This wavelet shape also gives remarkably good results in determining phenological stages as pointed out by [8]. This study used DWT with Coiflet order 1, level 1, and approximation 1. The level number was representing the decomposition level that has two component, there are the approximation (A1) and detail (D1). Setiawan [7] explained that the function of the level 1 decomposition was F(t)= A1+D1, and then A1 was decomposed into A2 and D2 using LPF and HPF filtered in decomposition level 2. So, there are 3 component results from level 2 decomposition: A2, D1, and D2. This research used decomposition level 1 and approximation 1 (app1) that mean A1.

3. Result and Discussion

3.1 Cropping Pattern Classification

The annual filtered patterns were classified into 25 classes using k-means clustering. The number of the class was determined based on previous research in the Java region and 25 class is the near maximum number of categories that could be discriminated effectively; identification of other categories would likely be difficult to achieve [7]. K-means algorithm classified similar patterns become one class with the number of class was determined. Then, 25 classes were identified and classified into main classes in every year. Each year had the difference total of main classes, that mean it could be increased or decreased every year. For example, there are 12 main classes at 2015: Paddy-Paddy-Bare (PPB), Paddy-Bare-Paddy (PBP), Paddy-Secondary Crop-Bare (PPLB), Paddy-Paddy-Paddy (PPP), Paddy-Paddy-Secondary Crop (PPPL), Paddy single crop in mixed time cropping (P1), Paddy double crop in mixed time cropping (P2), Mixed crop (C), Forrest dominant (H), Pond dominant (T), Housing dominant (Pr), and Bare-Paddy (BP). It is not same with the total of main classes at 2014 that only have 11 main classes because there is no BP class on this year. It means that BP just the particular class which happened in 2015 as a result of changes cropping pattern (it can be abrupt, seasonal, stable, or gradual). The main classes were divided by two: clear class and mixed class. A clear class is the class that has same plant and planting schedule, so the pattern was clear between peak and valley (PPB, PBP, PPLB, PPP, PPPL, BP).

The accuracy of this cropping pattern was approach by confusion matrix (Table 1), which compared with ground check data. Amount 103 reference point in Central Java considered representing the main class for ground check. The total accuracy of this research (77.67%) was better than Setiawan [3] and Muhammad [4] which has the accuracy 74.43% and 71.11% with the long time series analysis. It means that annual time series analysis gave the better result for Pattern recognizing using MODIS EVI data. The difference accuracy also determined several factors like amount of reference point, how they classified the main class, and the variance data of the class. Although there was difference accuracy (77.67%, 74.43%, and 71.11%), the accuracy above 70% is the good accuracy, considering that MODIS had the moderate resolution (250m).

Table 1. Confusion matrix of cropping pattern in Central Java

Cropping pattern	PPB	PBP	PPLB	PPP	PPPL	P1	P2	C	H	T	Pr	*User accuracy (%)*
PPB	12	0	0	0	0	3	0	0	0	0	0	80
PBP	0	3	0	0	0	0	0	0	0	0	0	100
PPLB	0	0	10	0	0	0	0	0	0	0	0	100
PPP	0	0	0	9	2	0	1	0	0	0	0	75
PPPL	0	0	0	7	14	0	0	0	0	0	0	66.67
P1	0	0	0	0	0	8	1	0	0	0	0	88.89
P2	0	0	0	0	0	7	12	0	0	0	0	63.16
C	0	0	0	0	0	1	1	8	0	0	0	80
H	0	0	0	0	0	0	0	0	1	0	0	100
T	0	0	0	0	0	0	0	0	0	2	0	100
Pr	0	0	0	0	0	0	0	0	0	0	1	100
Procedural accuracy (%)	100	100	100	56.25	87.50	42.11	80	100	100	100	100	100
Total accuracy (%)							77.67					
Total data							103					

3.2 Annual change

This research produced annual distribution cropping pattern maps for analyzing the changes of paddy fields land use in Central Java. The cropping pattern distribution maps can be seen in **Error! Reference source not found.**. Spatial analysis was done to know where and when the changes happened based on the color change from cropping pattern distribution maps. From **Error! Reference source not found.**, we know that the cropping pattern changed annually. The map also presented the majority of cropping pattern in an area or District. The majority of cropping pattern was affected by water availability, land area, paddy fields ownership, and local culture. Paddy fields ownership and local culture could affect the cropping pattern in Central Java because the smaller paddy fields area which owned by farmers increasingly diverse the plant and planting schedule in one pixel. When each farmer planted difference crop, the temporal EVI pattern will be mixed. The others two factors (water availability and land area) are the factors which came from nature. The paddy fields that had the location near water sources like reservoir or lake would get more water support for irrigation, so they could grow rice in the dry season. The place that had the PPP (Paddy-Paddy-Paddy) cropping pattern marked in green as Karanganyar, Klaten, Sragen, and Wonogiri region which were close to Gajahmungkur reservoir or Kedungombo reservoir. From this fact, we know that water availability and the land area could affect the cropping pattern.

The area of cropping patterns every year didn't always remain, they can be increased or decreased. It will automatically affect harvest area in Central Java. Figure 3 give an estimation of harvest area in Central Java using MODIS images which were compared with harvest area from Statistics of Indonesia [11]. The estimation of harvested area was obtained by regional measurements using QGIS. From Figure 3 we know that there were several years (2010, 2011, 2014, 2015) which had quite a big difference between MODIS data and BPS data. Based on data reported by [12], from July 2010 until the end of 2011 there was La Nina phenomenon. In the western equatorial region like Indonesia, There would be the formation of more clouds than usual, so it automatically caused a lot of disturbance for MODIS images. It would scatter EVI value that should be patterned rice planting, so the pixel didn't count as rice harvest. Otherwise, from the end of 2014 to round 2015, there was El Nino phenomenon. Each farmer would plant rice with not uniform time because they tend to grew paddy when it rains (El Nino make the rainy season less than usual). Consequently, EVI temporal patterns became docked and difficult to distinguish between one or twice planting.

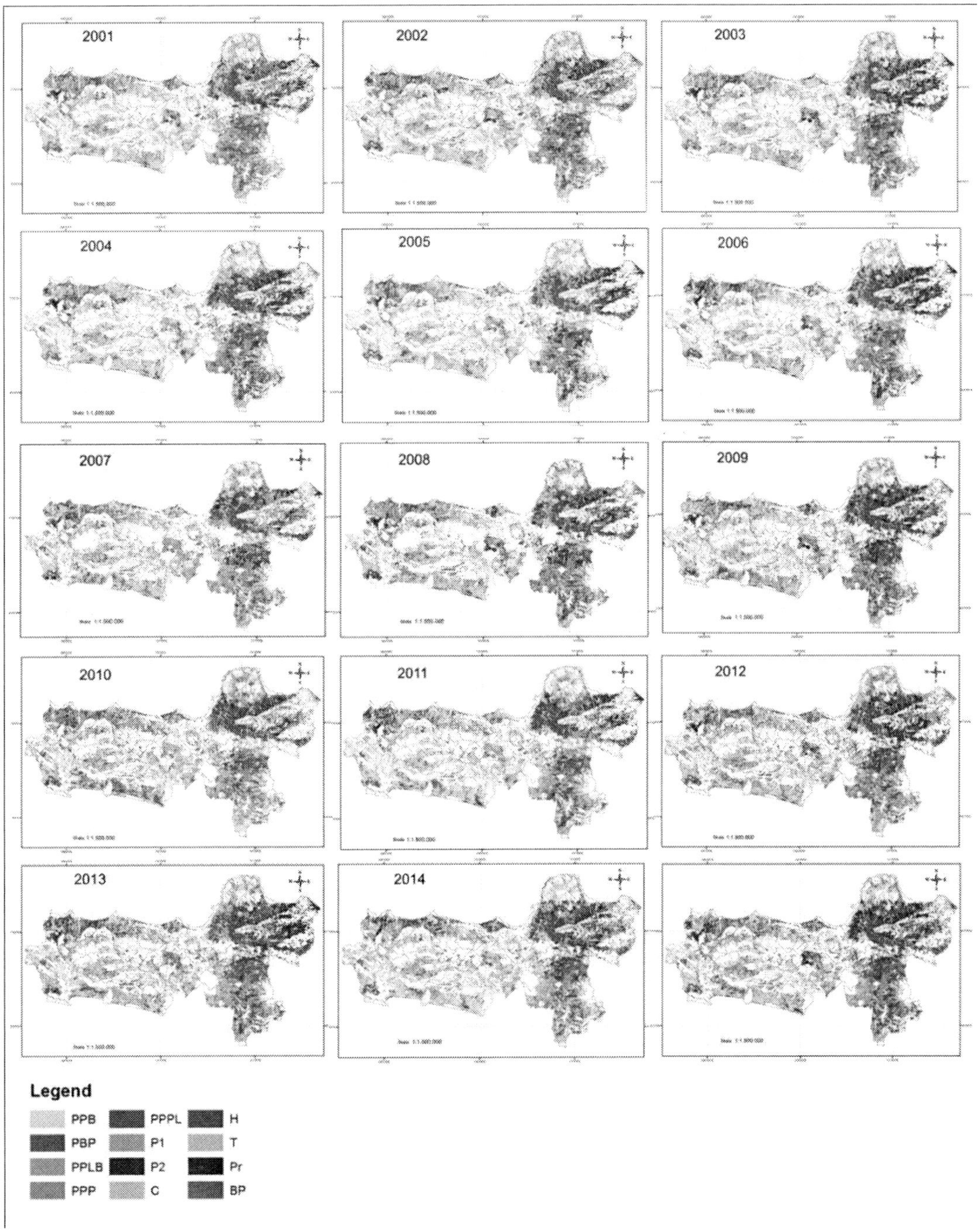

Figure 3. Annual distribution cropping pattern maps in Central Java

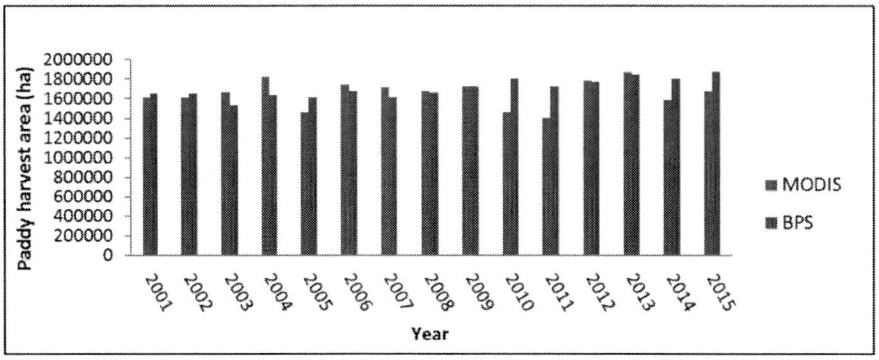

Figure 3. Paddy harvest area from MODIS and BPS

3.2 Change analysis

The change of paddy cropping patterns could be detected by clipping the annual cropping pattern map. The changes could be abrupt, seasonal, stable, and gradual. The spatial analysis had the objective for discussing where, when, and what was the changes happened. **Error! Reference source not found.** give the example gradual change that happened from 2001 to 2015 in Central Java that was decreasing of the area with the PPB pattern and gradually became PPPL every year. This changes happened in Undaan, Kudus District. Abrupt changes also happened in the beginning of 2008 that BP pattern suddenly widespread in Undaan, Kudus. The Bare-Paddy pattern indicated that there was a bare area at the beginning of the year. It could be caused due to flooding which happened in 2008. The stable changes also happened from PPB cropping pattern became PPPL cropping pattern start from 2009. The differenced types of change (abrupt and stable) could happen from difference factors. Abrupt changes are normally caused by disturbances such as fire, flood, urbanization, insect attack, or drought [9]. While stable changes usually occurred by water availability or the changes of cropping pattern culture.

The land cover change dynamic analysis produced the information about the stable cropping pattern map and shows in f. There were about 1262976.26 ha (5.2%) paddy fields that haven't change in last 15 years. This fact indicated that majority of paddy fields in Central Java ever changing cropping pattern. From Figure 4, we know that majority stable cropping pattern was PPPL (red color) pattern which gathered in Kudus, Demak, and Grobogan District. Paddy fields at those regions were not irrigation type, and according to [2] was paddy double cropping area. Based on ground check, the farmer prefers to plant secondary crops in dry season. This is also corroborated by the rainfall data in Kudus from 2011 to 2015[10] which had the small rainfall at June until October. The rainfall not enough to irrigated rice crops in that period.

These maps give the main information that Central Java always changing their cropping pattern and just little area that stable with their cropping pattern. This picture seemed to give the message that the near real-time monitoring system was very important for assessed this phenomenon. The majority of paddy fields area always changed each period, or the map explained that majority of paddy fields in Central Java was not stable. Variation of the changes time indicated the change was affected by time period. It could be from rainfall influence, topography (land area), or farmers habit (culture and ownership). It difficult for concluding the cause of changes, because the variation culture and large area for analysis made it became the complex problem.

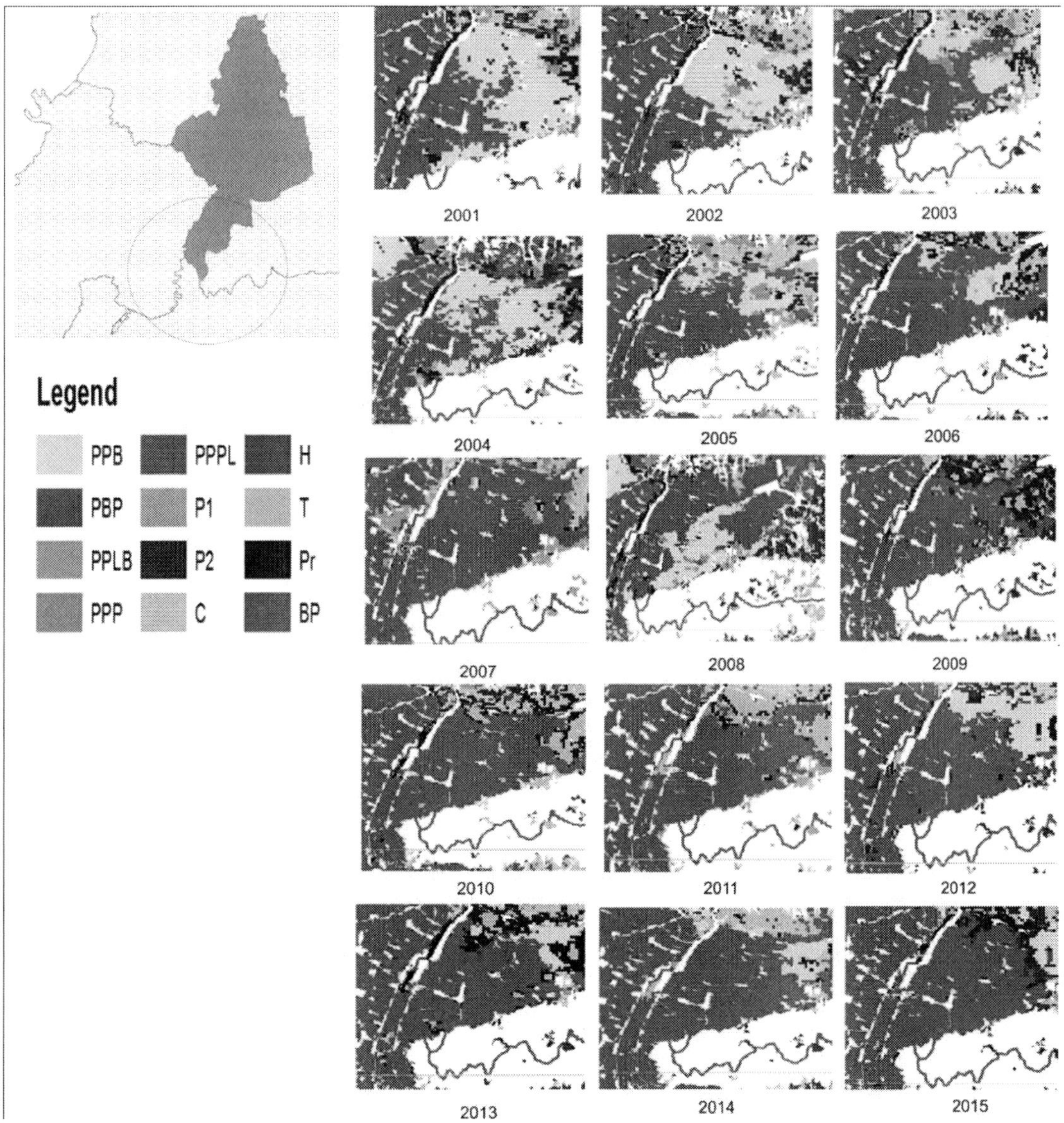

Figure 5. Gradual change in Undaan, Kudus District

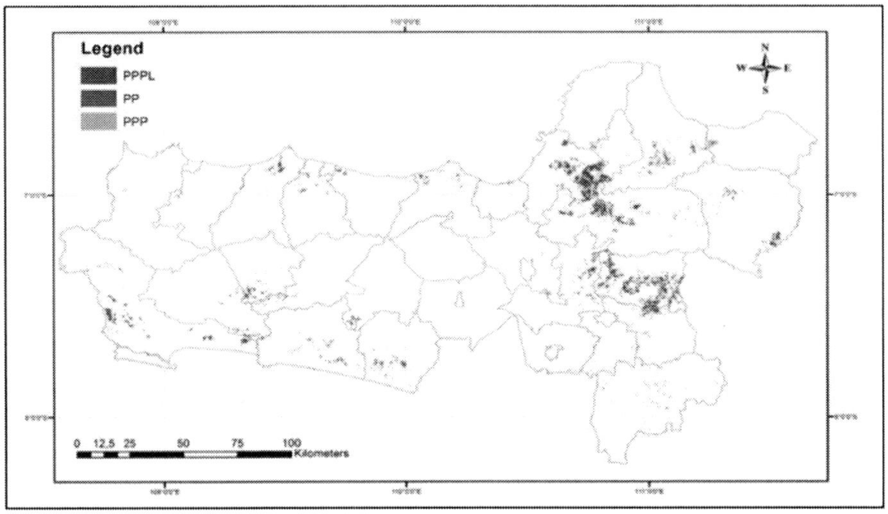

Figure 4. Stable cropping pattern map in last 15 years

4. Conclusion

This research got the better accuracy (77.67%) of cropping pattern than long time series analysis from Setiawan [3] and Muhammad [4]. Although some classes successfully obtained upon annual time series analysis, MODIS still difficult to detect mixed crop pattern. Spatial change in Central Java could be detected by annual time series analysis of MODIS EVI. The changes divided into abrupt, seasonal, stable, and gradual. Using annual data analysis, the spatial changes was more represented. The majority of paddy fields area (94.8%) in Central Java ever changing their patterns in last 15 years. Variation of the changes time indicated the change was affected by time period. Paddy double cropping pattern (PPPL and PP) was more stable than paddy triple cropping pattern (PPP). it means that Paddy in the dry season was only slightly which consistent for last 15 years. The changes could be affected by drought, flood, urbanization, or insect attack. The result of this research seemed to give the conclusion that paddy fields near real-time monitoring system was needed for distribution of more precise and accurate information.

Acknowledgments

The authors would like to thank the Center for Environmental research (PPLH-IPB) for facilitating this research. Gratitude also gives to NASA LP DAAC, Indonesian Geospatial Information Agency (BIG), and Indonesian central statistics agency for providing data of this research.

5. References

[1] Huete A and Glenn E 2011 *Remote Sensing of Ecosystem Structure and Function, Advance in Environtment Remote Sensing* (Boca Raton (USA: CRC Press)
[2] Danoedoro P 2012 *Pengantar Penginderaan Jauh* (Yogyakarta, Indonesia : ANDI)
[3] Setiawan Y, Rustiadi E, Yoshino K, Liyantono and Effendi H 2014 Assessing the seasonal dynamics of the Java's paddy field using MODIS satellite images *ISPRS Int. J. Geo-Information* pp 110-129
[4] Muhammad M, Liyantono, Setiawan Y and Fatikhunnada A 2015 Prediction of paddy field cropping pattern using time series MODIS image in East Java *Proc. Int. Symp. on Geoinformatics* pp 36-40
[5] Badan Pusat Statistik 2013 *Paddy fields area of Indonesia 2003-2013* (Jakarta, Indonesia: Pusat Data dan Sistem Informasi Pertanian)

[6] Huete A, Didan K, Miura T, Rodriguez E P, Gao X and Ferreira L G 2003 Overview of the radiometric and biophysical performance of the MODIS vegetation indices *Remote Sensing of Environment* **83** pp 195– 213
[7] Setiawan Y, Yoshino K and Philpot W D 2013 Characterizing temporal vegetation dynamics of land use in regional scale of Java Island, Indonesia *Journal of Land Use Science* **8** pp 1–30
[8] Sakamoto T, Yokozawa M, Toritan H, Shibayama M, Ishitsuka N and Ohno H 2005 A crop phenology detection method using time-series MODIS data *Remote Sens. Environ* **96** pp 366–374
[9] Kawabata A, Ichii K and Yamaguchi Y 2001 Global monitoring of interannual changes in vegetation activities using NDVI and its relationships to temperature and precipitation *Int. journal of remote sensing* **22** pp 1377-1382
[10] Badan Pusat Statistik 2015 *Monthly of rainfall in Kudus District 2011-2015* (Jakarta, Indonesia: Pusat Data dan Sistem Informasi Pertanian)
[11] Badan Pusat Statistik 2015 *Paddy harvest area 1993-2015* (Jakarta, Indonesia: Pusat Data dan Sistem Informasi Pertanian)
[12] National Oceanic and Atmospheric 2015 *Data of oceanic nino index* (Marryland, United States: Center of Weather and Climate Prediction)

Identification of potentially high yielding irradiated cassava 'Gajah' genotype with different geographic coordinates

I Subekti, N Khumaida* **and SW Ardie**
Department of Agronomy and Horticulture, Faculty of Agriculture, Bogor Agricultural University (IPB), Jl. Meranti Kampus IPB Darmaga, Bogor

E-mail: nkhumaida@yahoo.com

Abstract. Cassava is one of the main and important carbohydrate producing crops in Indonesia. Thus cassava production and its tuber quality need to be improved.'Gajah' genotype is a local genotypes cassava from East Kalimantan, has high potential yield (> 60 ton Ha^{-1}). However, the harvest time of this genotype is quite long (>= 12 months). The objective of this research was to identify the high yielding cassava mutants from the gamma rays irradiated 'Gajah' genotype at M1V3 population and potential yield at different location. Several putative cassava mutants (12 mutants) were planted in Cikabayan Experimental Field, IPB from March 2015 to March 2016 and the yields compared with the same genotype grown at different location by seeing its coordinates to observe the potential yield. Our result showed that the fresh tuber weight per plant of some putative mutants could reach more than 8 kg (yield potential of 64 ton Ha^{-1}). The harvested tubers also had sweet flavor, although the tubers of some putative mutants were bitter. Based on previous research study, the different geographic coordinate has resulted variability on fresh tuber yield. It seems that it needs to observe the stability of 'Gajah'- irradiated mutants in several location in Java Island.

1. Introduction

Cassava (*Manihot esculenta* Crantz) is one of the main carbohydrate sources after rice and maize. In Indonesia, mainly cassava is used as food (58%), as raw material for industry (28%), exported in form of cassava chips (8%), as feed (2%), and the rest (4%) is agricultural waste that has not been utilized [1]. As the main utilization of cassava is as food, it is important to match cassava quality with the demand. Important character of cassava as food are high productivity (cassava productivity of several national varieties is around 20-40 ton ha^{-1}), high starch content (starch content of several released varieties is 25-31%), and low cyanide acid (HCN) content [2]. HCN mostly found in cassava leaves and tuber, but the level of HCN content is depend on genotype and environmental factors (Salvador *et al.* 2014), with the maximum HCN content to be safe as food is 10 mg/kg [1]. Ferrero and Villegas [3] classified tuber based on HCN content, which were HCN $\leq$ 50 ppm as *innocuous*, 50-100 ppm as *moderately toxic* and $\geq$ 100 ppm as *dangerously toxic*, so it is better if consumed cassava has HCN content around 10 mg/kg. In irradiation research on cassava by Khumaida *et al.* [4], cassava mutants with bitter tuber taste and had 10-20 kg tuber mass per plant was resulted. Therefore, new genotype that adaptive and has sweet tuber taste should be developed so cassava tuber is suitable as raw material for food industry.

Cassava germplasm in Indonesia that has been started to be developed is 'Gajah' genotype. This genotype has been listed as variety with serial number of 183/PVHP/2013/18 November 2013, has high yield potential in its origin area, East Kalimantan (Samarinda) and even has higher yield potential than other local varieties. Beside of high yielding property, this genotype has quite sweet tuber taste, making 'Gajah' suitable for further research and development. However, 'Gajah' genotype gave different yield potential in different area, which suspected because the absent of adaptive property. One of means to improve adaptive property of a genotype is by improving diversity of potential genetic material and continued with selection. Schwartz [5] explained, improving genetic diversity and variability on performance of species through changing the DNA chain can be done by induction mutation approach using gamma ray irradiation. Furthermore, the yield potential information of the genotype and also other varieties in several areas is needed to understand the interaction between gen and environment in order to determine the best environment for the genotype. Fransisco *et al.* [6] tested *Brassica rapa* in several different areas to understand environment or genotype effect on secondary metabolite and production. The objective of this research is to analyze yield potential of Gajah genotype in different areas and identify several cassava mutants resulted from gamma ray irradiation on M1V3 generation.

2. Materials and methods

The planting material used in this study was 12 genotypes of mutant resulted from gamma ray irradiation of 'Gajah' genotype on M1V3 generation.

The research was conducted by planting 12 mutant genotypes and one background genotype (Gajah) using Completely Randomized Block Design with three replications in experimental field Cikabayan IPB. Maintenance was performed on the plant including fertilizing, pruning and weeds control. Observations were made on morphological characters refer to Fukuda et al. [7], as well as physicochemical characters analysis performed on harvested tubers of certain number genotypes which were genotypes with high yield potential. Analysis of variance was done on quantitative data and Kruskal-Wallis analysis based on qualitative data. To determine the yield potential of 'Gajah' cassava in different distribution areas, data collection of 'Gajah' genotype cassava was carried out, in several regions of Indonesia, followed by Pearson correlation analysis between the plantation locations coordinates and yield potential. Data analysis was perform using Statistical Tool for Agricultural Research (STAR) software.

3. Result and discussion

Distribution of 'Gajah' genotype cassava in Indonesia has not been spread evenly, because it was less than ten years since its release as new variety. However based on search result, this genotype has been spread in several area in Indonesia, namely in Kalimantan, Sumatera, and Java with yield potential per plant vary. When viewed in the map, the spread of 'Gajah' genotype cassava was in different coordinate points (Table 1) and had different yield potential. Yield potential in origin area (East Kalimantan – Samarinda) was the same with West Kalimantan – Pontianak, which was 35 kg/plant, followed by Central Java – Wonogiri, and North Sumatera.

Based on Pearson correlation analysis between coordinate points with yield potential, coordinate point was not correlated with yield potential (Figure 1), which meant that the change in coordinate point will not affect resulted tuber yield potential. However, based on analysis on Table 1, there was difference on harvested yield potential between planting area. This difference on yield based on environment showed that there was interaction between 'Gajah' genotype genetic and environment (GxE). Harvested yield is a quantitative character and this character is highly affected by environment. If the environment or location effect is significant, the genotype can be developed as location specific variety and if insignificant, the genotype can be developed as widely adaptive variety. Syukur *et al.* [8] stated that genetic potential and interaction between genotype x environment (GxE) are needed in developing plant variety in order to determine gen stability on tested genotype. Based on this study, 'Gajah' genotype was suspected as specific location variety, thus wide adaptation of 'Gajah' genotype needs to be improved. Khumaida et al. [9] conduct a research on cassava irradiation, and the study resulted that yield of cassava background and mutant in two different areas has not significant. It means, by irradiation a genotype can be change the level of adaptation. So that, the irradiation on 'Gajah' genotype expected to improve their adaptation and other important characters.

Table 1. Different coordinate points and yield potential of Gajah genotype cassava

Location	Latitude	Longitude	Harvest potential (kg/plant)*
Samarinda	0°28’ LS	117°11’ BT	35
Pontianak	0°05’ LS	109°22’ BT	35
Palembang	2°59’ LS	104°47’ BT	15
North Sumatera	3°38’ LU	98°38’ BT	20
Wonogiri	7°5’ LS	110°55’ BT	25
Surabaya	7°15’ LS	112°45’ BT	15
Bogor	6°37’ LS	106°48’ BT	8**

* Data was compilled from various sources; yield each plant based on information and news

**Data obtained from experiment using Completely Randomized Block Design

Latitude			
Longitude	0.6202		
Yield potential	-0.179	0.2855	
	Latitude	Longitude	Yield potential

Figure 1. Pearson correlation analysis result of coordinate point- yield potential

Research on ‘Gajah’ genotype from gamma ray irradiation has been conducted until M1V3 generation on both qualitative and quantitative character. This article contains results from observation on qualitative and quantitative character of harvested tuber. Quantitative character observed was tuber mass, number of tuber per plant and number of commercial tuber (tuber length > 20 cm). Results showed that there was no significant difference on quantitative character of mutant genotypes (Table 2).

Table 2. Average value of three quantitative characters on 13 cassava genotypes

Genotype	Tuber mass (kg)	Number of tuber	Number of commercial tuber
	P value		
	1.09ns	0.79ns	1.40ns
U1-0-2	2.7	7.22	2.69
U1-15-2	4.11	7.11	3.66
U1-15-4	3.23	8.08	4.5
U1-15-5	3.88	5.06	3.17
U2-15-1	3.67	7.33	4.33
U2-15-3	2.81	7.42	3.08
U2-15-4	3.89	7.38	4.64
U2-15-5	3.1	7.72	4.08
U3-15-1	2.94	5.83	3.72
U3-15-2	3.73	6.96	3.48
U3-15-3	2.43	6.22	3.33
U3-15-4	3.17	7.25	3.42
U3-15-5	3.31	6.86	3.31

*ns= no significant

On tuber mass per plant character, the highest tuber mass was 4.11 kg and the lowest was 2.43 kg, but on individual observation there were mutants with tuber mass of 8 kg per plant. That tuber mass was obtained on plant resulted from 15 Gy irradiation (mutant plant), which were U1-15-2, U2-15-1 and U3-15-4 genotypes. On number of tuber per plant, the highest number of tuber was 8 and the lowest was 5. But same as tuber mass, individual observation result showed that there were mutants with number of tuber > 10, even reach 14 tubers, namely on mutants U1-15-2, U1-15-4, U2-15-1, U2-15-3, U2-15-5, U3-15-2, and U3-15-4. Analysis result for number of commercial tuber showed that observed mutants had 2 – 4 commercial tubers per plant. On individual observation, there was mutant with 10 commercial tubers, which was U2-15-1 genotype (Figure 2). Khumaida *et al.* [4] reported that highest number of tuber from mutant plant was 35 tubers, which was higher than its background genotype of 18.78 tubers.

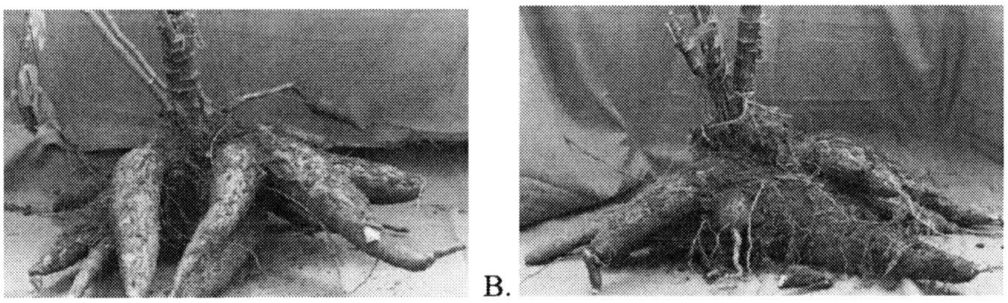
A. B.

Figure 2. Cassava genotype with superior quantitative characters. A. U2-15-1 genotype with 14 tubers, 10 commercial tubers and tuber mass of 5.8 kg; B. U3-15-2 genotype with 14 tubers, 8 commercial tubers and tuber mass of 7.6 kg.

Observation on qualitative character was conducted on characters: tuber level from stem, tuber constriction, tuber shape, and outer tuber color, tuber color, and cortex color, easiness to peel, tuber skin texture, tuber taste, and cortex thickness. Result of Kruskall-Wallis analysis on quantitative characters showed that there was no significant difference between mutant genotypes (Table 3). On tuber level from stem character, the most common was *pedunculate*, which meant tubers were not attached directly to stem, but had stalk. On tuber constriction, all tubers had no constriction or smooth. On tuber shape character, analysis result showed that overall tuber shape was *conical-cylindrical*, but individual observation still showed diversity among mutants. There were found mutants with tuber shape of *conical, cylindrical,* and *conical-cylindrical* (Figure 3).

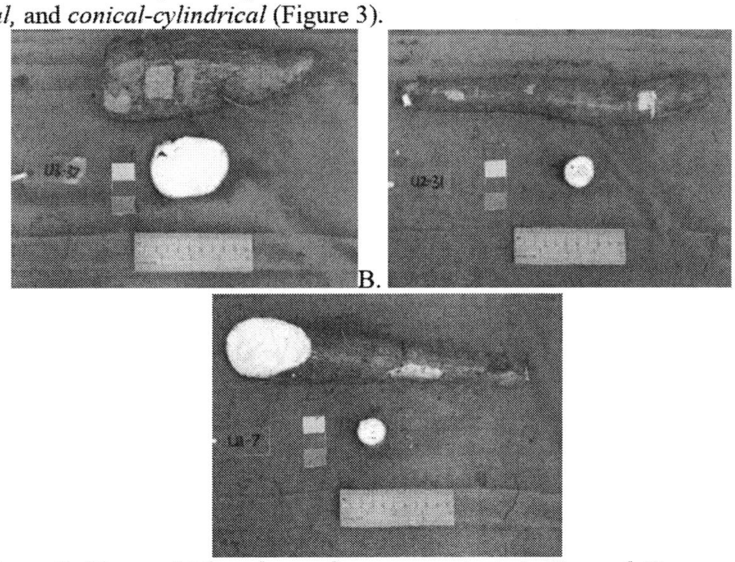
A. B. C.

Figure 3. Observed tuber shape of cassava mutant, A. *Conycal*; B. *Cylindrical*; C. *Conical cylindrical*

Next characters observed on all mutants were outer tuber color (dark brown), tuber color (white), cortex color (purple), easiness to peel (easy), and cortex tuber (intermediate). For tuber taste character, the most common was sweet and intermediate, but there were some numbers with bitter tuber taste. The change in tuber taste could wither be sweet to bitter like in this research or bitter to sweet, as shown in study from Khumaida *et al.* [4]. Roslim *et al.* [10] reported that partial gene mutation of *Meisal* (gen that responsible of coding *isoamylaseI* enzyme that involved in starch metabolism in cassava) happened in intron cannot affect tuber taste. Based on this study, it was suspected that occurred mutation was not partial and happened on exon, thus can affect tuber taste.

Tabel 3. Kruskall-Wallis analysis results of qualitative characters from 13 cassava mutant genotypes

Genotipe	Extent of root peduncle	Root constrictions	Root shape	External color of storage root	Color of root pulp (parenchyma)	Color of root cortex	Cortex: ease of peeling	Texture of root epidermis	Root taste	Cortex thickness
					P Value					
	0.81ns	0.63ns	0.51ns	0.20ns	0.44ns	0.16ns	0.44ns	0.25ns	0.13ns	NaN
U1-0-2	Sessile to pedunculate	Few to none	Conical-cylindrical	Dark brown	White	Purple	Easy	Intermediate to rough	Intermediate	Intermediate
U1-15-2	Pedunculate	Few to none	Conical-cylindrical	Dark brown	White	Purple	Easy	Rough	Sweet	Intermediate
U1-15-4	Pedunculate	Few to none	Conical-cylindrical	Dark brown	White	Purple	Easy	Rough	Intermediate	Intermediate
U1-15-5	Sessile to pedunculate	Few to none	Conical-cylindrical	Dark brown	White	Purple	Easy	Intermediate to rough	Intermediate	Intermediate
U2-15-1	Sessile to pedunculate	Few to none	Conical-cylindrical	Dark brown	White	Purple	Easy	Intermediate to rough	Sweet	Intermediate
U2-15-3	pedunculate to mixed	Few to none	Conical-cylindrical	Dark brown	White	Purple	Easy	Intermediate to rough	Intermediate	Intermediate
U2-15-4	Sessile to pedunculate	Few to none	Conical-cylindrical	Dark brown	White	Purple	Easy	Intermediate to rough	Intermediate	Intermediate
U2-15-5	Pedunculate	Few to none	Conical-cylindrical	Dark brown	White	Purple	Easy	Intermediate to rough	Intermediate	Intermediate
U3-15-1	Pedunculate	Few to none	Conical-cylindrical	Dark brown	White	Purple	Easy	Rough	Intermediate	Intermediate
U3-15-2	Sessile to pedunculate	Few to none	Conical-cylindrical	Dark brown	White	Purple	Easy	Intermediate to rough	Sweet	Intermediate
U3-15-3	Pedunculate	Few to none	Conical-cylindrical	Dark brown	White	Pink	Easy	Intermediate to rough	Intermediate	Intermediate
U3-15-4	Pedunculate	Few to none	Conical-cylindrical	Dark brown	White	Purple	Easy	Intermediate	Intermediate	Intermediate
U3-15-5	pedunculate to mixed	Few to none	Conical-cylindrical	Dark brown	White	Pink	Easy	Intermediate to rough	Intermediate	Intermediate
H	10.15	-	-	-	-	4.71	-	7.91	6.43	-

*ns= no significant

Observation of the physicochemical character of tubers was conducted to see the difference in chemical content of tubers in the mutant plant tubers and its background genotypes. Observations were made on some characters, such as starch content, amylose, amylopectin, moisture content, whiteness and HCN (Table 4). Mutant tuber starch content ranging from 77.17-83.01%, higher compared to the background genotype, which was 75.615%. Amylose content also found higher in the mutant plants ranging from 26.2-26.9%, while the background genotype was 23.5%. On amylopectin content, background genotype had higher levels, namely 53%, whereas in the mutant plants 51 and 55%. According Rawel and Kroll [11], the levels of amylopectin and amylose ranges from 83% to 17% of total starch tubers, while Keeling and Myers [12] states amylopectin ranges from 70-80% and amylose ranges from 20-30% of total starch content. The results of this study indicate amylose content was 30% and amylopectin was 70% of the total starch content. The highest water content observed was 8.06% and the lowest was 6.8%. Highest whiteness character of the observed in the mutant plants was 100%, but the background genotype plant has low degree of whiteness, which was 97.6%. Whiteness indicate the level of β-carotene, the yellow color of the tuber indicate higher β-carotene content. However, the content of β-carotene is negatively correlated with the production of tubers, although not very significant, as reported by Akinwale *et al.* [13], the greater the weight of the tuber, the lower levels of β-carotene. In Indonesia, the degree of tuber whiteness affects the level of consumer preferences, because consumers prefer white tubers, making mutants with white tuber potential to be developed in

Indonesia. The last character was the HCN content, which the lowest observed was 8.5% in mutant plants. This value was lower than the background genotype HCN content, which was 12.9%. Wilson and Dufour [14] states HCN content negatively correlated with starch content of the tuber, the lower the HCN content, the higher starch content. Thus, the low HCN levels causing sweet tuber taste.

Table 4. Physicochemical analysis results of severeal selected genotypes tubers

Genotype	Starch (%)	Amylose (%)	Amylopectin (%)	Water content (%)	HCN (ppm)*	Whiteness (degree)
U2-2	75.61	23.57	53.91	7.58	12.98	97.6
U3-19	77.17	26.56	51.72	8.06	14.02	96.2
U2-32	83.01	26.98	55.32	6.8	10.57	99.8
U3-49	78.11	26.27	51.85	7.28	8.57	100.05

*ppm= part per million, physicochemical analysis was done base on dry base

4. Conclusion

Planting location affect different yield potential, as in cassava 'Gajah' genotype showed different yields potential between regions, it suspected that ‘Gajah’ genotype was specific location variety. Observations on harvested yield of M1V3 population resulted from gamma ray irradiation, showed that some mutant genotypes had higher tuber weight with lower HCN content, and had sweet tuber taste. Thus, some mutant from 'Gajah' genotypes with high yield potential, low HCN and sweet tubers taste need to be further developed to support the food industry.

Acknowledgments

The authors wish to acknowledge The Ministry of Research, Technology and Higher Education, Republic of Indonesia, for the funding of this research at 2016.

References

[1] [Dirbudkabi] Direktorat Budidaya Kacang-kacangan dan Umbi-umbian 2007 *Budidaya Ubi Kayu* (Jakarta: Direktorat Jenderal Tanaman Pangan)
[2] Sudarmonowati E, Hartati N S and Amzal A 2012 *Perbaikan Sifat Ubi Kayu dan Pengembangannya untuk Ketahanan Pangan dan Nutrisi*. http://www.wnpg.org/frm_index.hp?pg=informasi/info_makalah.php&act=edit&id=73
[3] Ferrero M T and Villegas L 1992 *Proc. CBN (Colombia)* (Cali: CIAT) p 433-7
[4] Khumaida N, Ardie S W and Syukur M 2015 *Procedia Food Sci* **3** 27–34
[5] Schwartz S 2014 *'Fast-forward Genetics' Induces Mutations to Produce Higher-Yielding Crops* http://www.pbs.org/wgbh/nova/next/nature/fast-forward-genetics-induces-mutations-produce-higher-yielding-crops/
[6] Fransisco M, Cartea M E, Butron A M, Sotelo T and Velasco P 2012 *J. Agric. Food Chem.* **60** 5507-14
[7] Fukuda W M G, Guevara C L, Kawuki R and Ferguson M E 2010 *Selected morphological and agronomic descriptors for the characterization of cassava* (Nigeria: International Institute of Tropical Agriculture)
[8] Syukur M, Sujiprihati S and Yunianti R 2012 *Teknik pemuliaan tanaman* (Jakarta: Penebar Swadaya)
[9] Khumaida N, Evii, Ardie S W and Sopandie D 2016 *Procedia Env Sci.* **33** 70-7
[10] Roslim D I, Nisa F and Herman 2016 *J. Of Biology & Biology Edu., Biosaintifika* **8** 103-10
[11] Rawel H M and Kroll J 2003 *DLR* **99** 102-10
[12] Keeling P L and Myers A M 2010 *Annu. Rev. Food Sci. Technol.* **1** 271–303
[13] Akinwale M G, Aladesanwa R D, Akinyele B O, Dixon A G O and Odiyl A C 2010 *Intl. J. of Genetics and Mol. Biol* **2** 198-201
[14] Wilson W M and Dufour D L 2002 *Economic Botany* **56** 49–57

[15] FAO/WHO 1991 *XII Suppl. 4ed.* FAO 1-42
[16] Salvador E M, Steenkamp V and McCrindle C M E 2014 *J. Of Agric. Biotech. And Sustainable Dev.* **6** 29-38

Identification of paddy field using Landsat image in Karawang Regency, West Java

Bambang Riadi* and Ahmad Budiman Suriadi
Geospatial Information Agency, Jalan Raya Jakarta Bogor Km.46 Cibinong Indonesia

E-mail : bambang.riadi@big.go.id, briadi_jasinfo@yahoo.com

Abstract. The Government of the Republic of Indonesia has a mission to achieve food self-sufficiency in 2017, therefore it is necessary efforts to the stability of food needs. Karawang Regency as a granary states have a vital role in maintaining the national rice self-sufficiency, so indispensable information paddy field area. Paddy field accurate mapping can be done with a fast and efficient method of using remote sensing technology. This study aims to identify the paddy field using remote sensing technology. The data used is Landsat TM 2002 and Landsat-8 2015. The classification method using an approach Normalized Difference Vegetation Index (NDVI) and the Tasseled Cap Transformation (TCT). This method can be implemented to identify the fields that are still green or in the growing season. TCT produces three images of the six combinations, namely Brightness (BRT), greenness (GRN), and wetness (WET). BRT provides graytone gradation of non vegetation to water. GRN is indicated the gradation of vegetation cover, begin from densely vegetated until the most rare or non-vegetated areas. While wetness (WET) indicates the area associated with the presence of water. The main results of the classification is not a wetland and paddy. Based on existing data paddy land area of research area is ± 57% of the area of Karawang Regency.

1. Introduction

Food is a basic human need for survival and life, food is a central issue for people around the world, including Indonesia. So the human needs to food is a top priority the fulfillment of which cannot be postponed. The government is targeting self-sufficiency in the next two years, there is the possibility of constraints in achieving food sovereignty in the form of climate change, namely the existence of extreme weather such as droughts, floods, landslides, pest / disease with serious consequences to the decline in agricultural production, especially food crops. The conversion of agricultural land into non-agricultural land is currently increasing, and many factors influence the occurrence of land use both internally and externally.

Population growth and increasing infrastructure needs cause the needs of the land will increase also, over the function of agricultural land is the impact of the transformation of economic structures (agriculture to industrial) and demographics (rural to urban areas) which encourage land use change from agriculture to non-farm. In the last ten years, the conversion of the rice fields in central home rice-producing Indonesia i.e. Java Island, an average of more than 22,000 hectares/year [1]. Karawang as one of the highest rice contributor region especially in West Java until the currently still experiencing over the function of agricultural land especially the rice fields. The negative impact of the conversion of the rice fields is declining support national food resilience. Agricultural earnings decline and increasing poverty [2]. To know the over the function of agricultural land is to identify the widespread use of land of each type of land use on a particular year and then compared the next year.

The mapping of agricultural land in this research using the image Landsat TM 2002 and image Landsat-8 2015 using technology Remote Sensing.

Remote sensing is the science of obtaining information about an object, area, or symptoms by analysis of data obtained by using the tool without direct contact to the object, area, or symptoms [3]. Remote sensing data can be obtained through the recording sensors mounted either on aircraft, satellites, space shuttle, or other vehicle. The sensor will produce data that vary according to the location of the height sensor and the characteristics of the object being studied [4]. Remote sensing analysis based on the method of distinguishing an object to other objects using the properties of electromagnetic waves emitted. Sensing for agriculture generally uses light waves visible, near infra-red (NIR), and thermal infrared spectral approach to the character of the different vegetation cover which can indicate the condition of land and crops.

Information needs in food security programs associated with the conversion of land need to support rapid technological, informative and inexpensive. Remote sensing technology is a technology that can be utilized in monitoring changes in the use of agricultural land to non-agricultural with basic data with satellite imagery. Landsat image is one type of remote sensing imagery generated from passive systems remote sensing. Landsat TM imageries are widely used for mapping land cover, land use mapping, soil mapping, geological mapping, and mapping of sea surface temperatures. The accuracy of the land cover information will provide convenience in analysing the planning and development of a region. Landsat 8, 2015 in this study is used to identify the current condition of land cover that has a resolution of 30 m and 12 bit pixel, image data can be accessed free of charge, and is an opportunity for utilization as possible.

2. Method

Stage of growth and the size of agricultural land becomes very important for horticultural crops, because the vast canopy of leaves is the variable that is relatively easy to measure that became an indicator of light absorption in plants. In the studies linking the use of the leaf canopy to the need for watering the plants in agricultural areas [5]. Spectral vegetation index calculated from the data visible and near infrared reflectance (NIR), correlate linearly with the number of active radiation absorbed by the canopy of plants during photosynthesis [6].

2.1. Normalized Difference Vegetation Index (NDVI)

Vegetation index is a measure of vegetation greenness value obtained from the digital signal processing brightness value data (brightness) some channels satellite sensors. The absorption of red light by chlorophyll and reflectance of infrared light close by mesophyll tissue in the leaves will make brightness values received satellite sensors through the canals will be much different. On land without vegetation, including territorial waters and residential, vacant land is open or damaged vegetation conditions, cannot provide high-value ratio on a comparison of the value of these canals. And contrary to the mainland by dense vegetation and healthy conditions, comparisons of both channels provide a very high value (maximum). In mainland non-vegetation, including territorial waters, residential, vacant land is open, and areas with vegetation damaged condition, it will not show the value of a high ratio (minimum).

Vegetation index most commonly used is the Normalized Difference Vegetation Index (NDVI). The vegetation index value based on the difference between the maximum absorption of radiation in the red channel (red) as a result of the pigment chlorophyll and maximum reflectance in the near infrared spectral channels (near infra-red / NIR) as a result of the cellular structure of the leaves [7].

The NDVI formulation is as follows:

$$NDVI = (\rho_{NIR} - \rho_{Red}) : (\rho_{NIR} + \rho_{Red}) \quad (1)$$

IOP Conf. Series: Earth and Environmental Science **54** (2017) 012014 doi:10.1088/1755-1315/54/1/012014

Where:
ρ_{NIR} = reflectance value canal near infrared
ρ_{Red} = reflectance value of the red channel

2.2. Tasseled Cap Transformation (TCT)
Tasseled Cap Transformation (TCT) is a mathematical formula to calculate the brightness level (brightness), greenness (greenness), and moisture (wetness) of digital numbers in each band (band 1 to band 5 and band 7) on Landsat imagery. TCT was first introduced of the Landsat MSS, TCT subsequently refined using data from Landsat TM. That the values in the TCT ie Brightness, greenness, and wetness can be used to analyze the drought, and the bands that influence the growth phases of rice, among others, are blue, green, red, and Near-Infra Red (NON). The calculation of the level of greenness with TCT method can only be applied by the Landsat satellite data only. Vegetation index is a quantitative calculation that is used to calculate the biomass or vegetation conditions. Generally made by using a combination of several spectral bands.

Mathematical formulation third components TCT are expressed in the following three equations for Landsat 8:

$$Brightness = B2 * 0.3029 + B3 * 0.2786 + B4 * 0.4733 + B5 * 0.5599 + B6 * 0.508 + B7 * 0.1872 \quad (2)$$

$$Greenness = B2 * 0.2941 + B3 * -0.243 + B4 * -0.5424 + B5 * 0.7276 + B6 * 0.0713 + B7 * -0.1608 \quad (3)$$

$$Wetness = B2 * 0.1511 + B3 * 0.1973 + B4 * 0.3283 + B5 * 0.3407 + B6 * -0.7117 + B7 * -0.4559 \quad (4)$$

where B1 = Band1, B2= Band2, B3= Band3, B4= Band4, B5= Band5, B6= Band6, B7= Band7 [8].

Overhead perspective, using electromagnetic radiation in a few parts of the electromagnetic spectrum that is reflected or emitted from the earth's surface, the technology is able to generate information on the growth stage of rice plants by reflectance (reflection) electromagnetic waves generated. The growth phase of rice based International Rice Research Institute (IRRI) is divided into nine phases, whose determination requires a model of appropriate classifier to produce a high degree of accuracy [9].

Satellite Landsat-8 has a sensor Onboard Operational Land Imager (OLI) and the Thermal Infrared Sensor (TIRS) with the number of channels as many as 11 pieces. Among these channels, 9 channels (bands 1-9) are in the OLI and the other 2 (bands 10 and 11) on TIRS. Most channels have specifications similar to Landsat-7 [10]. Date acquired October 18, 2015. Target_WRS_PATH 122 and target_WRS_ROW 64. Date acquired Landsat-7 December 22, 2001. Here's the specs channels owned by the Landsat-8 are as follows table.1.

Table 1. Specifications spectral sensor canals imagers LDCM (Landsat-8)

Sensor	Band	Wavelength (μm)	Resolution (meter)	Sensor
1. Operational Land Imager (OLI)	*Band 1*	*0.43 - 0.45*	*30*	*Visible (Coastal aerosol)*
	Band 2	*0.45 - 0.51*	*30*	*Visible (Blue)*
	Band 3	*0.53 - 0.59*	*30*	*Visible (Green)*
	Band 4	*0.64 - 0.67*	*30*	*Visible Red (Red)*
	Band 5	*0.85 - 0.88*	*30*	*Near Infrared*
	Band 6	*1.57 - 1.65*	*30*	*SWIR 1*
	Band 7	*2.11 - 2.29*	*30*	*SWIR 2*

	Band 8	*0.50 - 0.68*	*15*	*Pancromatic*
2. Thermal	*Band 9*	*1.36 - 1.38*	*30*	*Cirrus*
Infra-red Sensor	*Band 10*	*10.60 - 11.19*	*100*	*TIRS 1 (Thermal Infrared)*
(TIRS)	*Band 11*	*11.50 - 12.51*	*100*	*TIRS 2 (Thermal Infrared)*

Source : Butler,2013 (*Obtained from USGS website)*

3. Result and discussion

Based on the results of this analysis also shows that the minimum and maximum values of NDVI in rice fields is lower than all the study sites. This is because the water will cover lowest NDVI values, whereas the vegetative phase in rice fields have NDVI values lower than the forest cover. Chlorophyll will absorb more light in the visible spectrum, especially the red spectrum (band 3) i.e. in a wavelength range of 0.63 to 0.690 microns. Therefore land with healthy plant will have a value of spectral band 3 which is lower than the land areas experiencing drought. In normal land spectral values that are in band 4 which has a wavelength range from 0.76 to 0.90 (near infrared) has a high sensitivity to the plant canopy, but not too sensitive to water. It is therefore more appropriate band 4 is used to identify vegetated land and open land and vegetation less able to reflect drought conditions. For dry land characterized by chlorophyll leaves dry and damaged / dead, were analyzed in band 5 (mid infra-red) wavelength range from 1.55 to 1.75 micron. Band 5 sensitive to the water content in the plant that will reflect the higher spectral dry plant (drought). Band 4 (near infra-red) has a high sensitivity to the plant canopy, but not too sensitive to water.

Colour composite composed of Landsat NDVI statistical value gives the appearance of a green color that contrasts indicating as paddy fields, paddy fields look very different than non paddy area. NDVI value in rice fields will follow a phase change (fallow, water and vegetation), wherein the water phase NDVI value would be very low, in the vegetative phase ranges from low to high. In the area of non-paddy changes in the value of NDVI is generally not too big, although NDVI vegetative be greater than NDVI vegetative rice. Cover water while the negative NDVI vegetation cover will reach its lowest NDVI if in an open condition.

Table 2. NDVI Landsat-8

NDVI value	Cover area
23 -- 57	*Wetland*
99 -- 124	*Settlemen*
185 -- 255	*Paddy filed*
230 -- 237	*Forest*

The research area is one of the districts in West Java who experienced conversion of agricultural land to non-agricultural (roads, settlements, industry, etc.). Since 1989 there were approximately 2,502 ha of land switch functions are 54% for industry and 36% for housing (Department of Agriculture). The results of the analysis of the largest agricultural land-use transitions occurred in the Cikampek District that many have compared to the industrial areas of agricultural land. Changes in land use may occur because of changes in spatial plans and change of policy direction and development as well as market mechanisms. The last two things happen more often in the past due to lack of understanding of the community and government officials regarding spatial. The transformation from agriculture to non-agriculture was widespread in line with the development policy that emphasizes the aspects of amenity growth through investment, both to local and overseas investors in the provision of land.

3.1. Land use in 2002

Land use types used consists of 5 that is wetlands, settlements, dry land, water bodies (swamp, pond), and paddy field. Total is the largest area are paddy fields. Dry land has an area of 28, 704 hectares or 14.9% of the total land area. The land area that is both comprehensive settlement which has 7,504 hectares or 3.8% of the land area. Then the paddy fields with an area of 116,054 hectares, or 60.3%, while for the water bodies (pond) have extensive percentage 23,496 with 3.8%, other area 17,226 hectares of the total land area in the Landsat images have been classified.

3.2. Integration NDVI and TCT for identification of drought

Use of TCT and NDVI has been done in the later study applied to map drought in the Citarum River Basin and the northern coast of West Java Province [11]. In assessing the level of dryness, TCT analysis is divided into five classes humidity of land as presented in table 3 and analysis of NDVI which produces plant greenness level 5 class in table 4.

Table 3. Classification of land surface moisture levels

Class	TCT value	water content %	moisture
1	*-295 – -30*	*<5%*	*very dry*
2	*-30 – -13*	*5 - 20*	*dry*
3	*-13 – 10*	*20 - 70*	*moist*
4	*10 – 35*	*70 - 100*	*very moist*
5	*35 –*	*>100*	*stagnant*

Source: Wahyunto *et al.* (2003).

Table 4. Greenness level classification of plants

Class	NDVI value	Level greenish / land cover conditions
1	*<-0.03*	*Land is not vegetated*
2	*-0.03 – 0.15*	*The greenery is very low*
3	*0.15 – 0.25*	*low greenish*
4	*0.26 – 0.35*	*The greenery was*
5	*0.36 – 0.61*	*high greenery*

Source: Wahyunto *et al.* (2003).

NDVI value analysis to identify plants experience drought conditions, in addition to visual analysis can use any combination of bands sensitive to vegetation and water. NDVI has a value of -1 to 1 whereas a negative value describes the conditions of open ground and greenery is low, the value of 1 is the high land with greenery. Use of the method of integration is the use of a single analysis using only the NDVI or TCT cannot be used to determine crop drought conditions. The results of the analysis of NDVI or TCT has the same appearance at a different plant drought conditions that need to be done integration by combining the value of greenness (NDVI) and surface moisture (TCT). Integration of greenness and moisture level of the land surface can be made that the land does not have a greenish vegetation and very low humidity that has a very low surface area categorized as experiencing severe drought. In contrast, no vegetated land but has a high moisture classified as land is not dry or watery (figure 1).

3.3. Spectral analysis of plant condition

Analysis of spectral patterns of soil conditions can be briefly described in figure 1. The analysis of Landsat satellite data Colour composite image with RGB composition TCB (Tasseled Cap Brightness) -NDVI (vegetation index) -TCW (Tasseled Cap Wetness), giving the appearance of the object is not vegetated red, green in areas with vegetation, and blue or dark the appearance of water or wet objects [11]. Gradation of redness depending on the level of vegetation and water shortages, more and more red tone sparse vegetation, as well as in the more dense vegetation NDVI increasingly green, and for TCW getting increasingly wet blue or the deeper the colour.

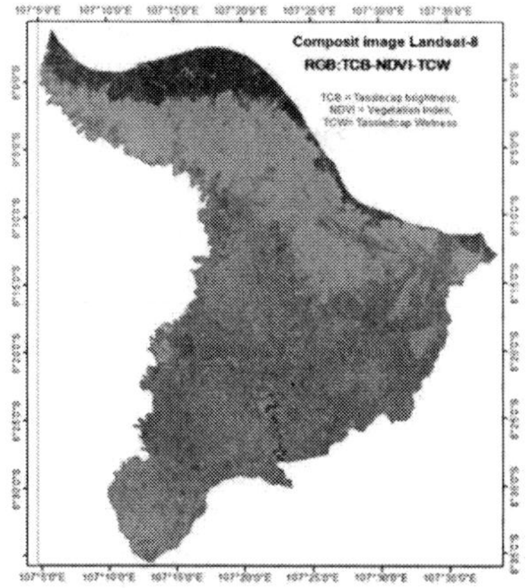

Figure 1. Composite image Landsat-8

3.4. Land use in 2015

Land use types used consists of 5 land use that is that is wetlands, settlements, dry land, water bodies (swamp, pond), paddy field. Total is the largest area paddy fields. Dry land has an area of 30,934 hectares or 16.1% of the total land area. The land area that is both comprehensive settlement which has 4,241 hectares or 2.2% of the land area. Then the paddy fields with an area of 112,878 hectares, or 58.7%, while for the water bodies have extensive percented 25,180 with 13.1% and other area 19,051 hectares with 9.0% of the total land area in the Landsat 8 images have been classified figure.2.

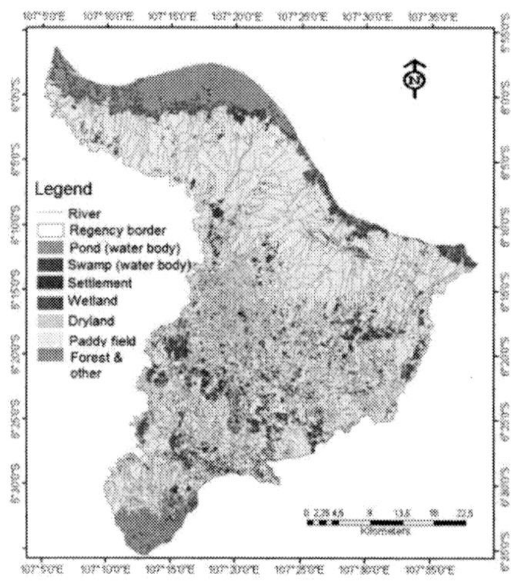

Figure 2. Landuse image Landsat-8

4. Conclusions

Remote sensing data on wave spectrum of visible light, near infrared, and thermal can be used to identify, assess, and monitor wetland. Analysis of soil surface moisture (wetness) using TCT (Tasseled Cap Transformation) and analysis of the level of greenness of plants using NDVI (Normalized Difference Vegetation Index) can be used to map the condition of agricultural land. TCW (Tasseled Cap Wetness) effectively used for monitoring the condition of the paddy fields. To enhance the interpretation of the results, the identification and classification of wetland, carried out the integration of two or more methods of analysis, because it can complement the advantages and disadvantages of each.

Identification of wetland by using remote sensing technology can help provide a comprehensive picture of the condition and the spatial distribution of wetland in the region. Wetland of Landsat TM rice cultivation area covers 60.3% of Karawang Regency, Landsat 8 2015 showed a decrease of acreage of paddy be 58.7%. The information obtained can be used to determine management measures in keeping with the food security in the Karawang Regency.

Acknowledgments

Thank you to the USGS Website for the ease of data access Landsat and Geospatial Information Agency for assistance data, so this article can be resolved.

References

[1] Sumaryanto and Tahlim S 2005 Pemahaman Dampak Negatif Konversi Lahan Sawah Sebagai Landasan Perumusan Strategi Pengendaliannya. *Prosiding seminar penanganan konversi lahan dan pencapaian pertanian abadi* (Bogor: Pusat studi Pembangunan Pertanian dan Pedesaan LPPM Institut Pertanian Bogor)

[2] Irawan, Supena, Friyanto, Supriatna, Anugrah, Kirom, Rahmanto and Wiryono 2001 Pencadangan lahan pertanian di Jawa *Buletin Agro Ekonomi*. **1** 1 – 6.

[3] Lillesand TM and Keifer RW 1979 *Remote Sensing and Image Interpretation* (New York: JohnWilley and Sons)

[4] Sutanto 1986 *Penginderaan Jauh* (Yogyakarta: Gadjah Mada University Press)

[5] Troutl T J, Johnson L F and Gartung J 2008 Remote sensing of canopy cover in horticultural crops *HortScience*. **43** 333-337

[6] Raun W R. 2001 In-season Prediction of Potential Grain Yield in Winter Wheat Using Canopy Reflectance *Agron. J.* **93** 131–138.

[7] Tucker C J 1979 Red and Photographic Infrared Linear Combinations for Monitoring Vegetation *Remote Sensing of Environment* **8** 127–150.

[8] Baigab M H A, Zhanga L, Shuaiab T and Tonga Q 2014 *Derivation of a tasselled cap transformation based on Landsat 8 atsatellite reflectance.* State Key Laboratory of Remote Sensing Sciences, Institute of Remote Sensing and Digital Earth (RADI), Chinese Academy of Sciences, Beijing, China University of Chinese Academy of Sciences, Beijing, China http://grindgis.com/blog/vegetation-indices-arcgis.

[9] Widjaja, Moeljono, Arief Darmawany and Sidik Mulyono 2012 *Fuzzy Classifier of Paddy Growth Stages Based on Synthetic MODIS Data*, ICACSIS.

[10] Butler K 2013 Band Combinations for Landsat 8. Retrieved from http://blogs.esri.com/esri/arcgis/2013 /07/24/band-combinations-for-landsat-8

[11] Wahyunto S, Ritung and Widagdo 2003 Laporan Akhir Teknologi Penginderaan Jauh untuk Efisiensi Inventarisasi dan Monitoring Sumberdaya Lahan. Balai Penelitian Tanah. Puslitbang Tanah dan Agroklimat. Badan Litbang Pertanian.

[12] Babar M A 2006 *Spectral Reflectance to Estimate Genetic Variation for In Season Chlorophyll, and Canopy Temperature in Wheat*, Crop Science, America.

Synthetic Aperture Radar (SAR)-based paddy rice monitoring system: Development and application in key rice producing areas in Tropical Asia

T D Setiyono[1*], F Holecz[2], N I Khan[1], M Barbieri[2], E Quicho[1], F Collivignarelli[2], A Maunahan[1], L Gatti[2] and G C Romuga[1]

[1]International Rice Research Institute (IRRI), Los Baños, Philippines
[2]Sarmap, Purasca, Switzerland

E-mail: t.setiyono@irri.org

Abstract. Reliable and regular rice information is essential part of many countries' national accounting process but the existing system may not be sufficient to meet the information demand in the context of food security and policy. Synthetic Aperture Radar (SAR) imagery is highly suitable for detecting lowland paddy rice, especially in tropical region where pervasive cloud cover in the rainy seasons limits the use of optical imagery. This study uses multi-temporal X-band and C-band SAR imagery, automated image processing, rule-based classification and field observations to classify rice in multiple locations across Tropical Asia and assimilate the information into ORYZA Crop Growth Simulation model (CGSM) to generate high resolution yield maps. The resulting cultivated rice area maps had classification accuracies above 85% and yield estimates were within 81-93% agreement against district level reported yields. The study sites capture much of the diversity in water management, crop establishment and rice maturity durations and the study demonstrates the feasibility of rice detection, yield monitoring, and damage assessment in case of climate disaster at national and supra-national scales using multi-temporal SAR imagery combined with CGSM and automated methods.

1. Introduction

Rice is a pivotal political commodity in many Asian countries with its price often serves as a key indicator for government performance [1]. It is therefore crucial for policymakers to control rice trade flow for domestic rice market to be stable. Reliable and regularly available sub-national information on rice area, seasonality, and yield is an essential part of many countries' national accounting process but existing system may not be sufficient to meet the information demand in the context of food security and policy [4,5]. Nevertheless, this same information is the basis of policy decisions related to imports, exports and prices, which directly impact food security, especially amongst the poor [6–8].

Remote sensing technology offers scalable and unbiased estimates of rice area to support, augment, and complement the existing system based on survey and statistical methods [5]. However, the following are technical challenges for the development of national-scale, operational, remote sensing-based rice crop monitoring systems in tropical monsoon Asia, region where 70% of rice is being produced: (i) Cloud-cover is extensive and pervasive during key rice growing season in this region [10,11], (ii) Wide range of conditions and environment where rice is grown in this region [1,10], (iii) rice cultivation in this region is mainly associated with small holders with relatively small cultivated paddy rice area (< 2 ha); (iv) Diverse and complicated cropping calendar practices exist and often a very short distance from field to field in this geography due to all-year suitable growing temperature [1,13].

The issue of cloud obstruction (i) can be addressed by using Synthetic Aperture Radar (SAR) imagery and the suitability of SAR for rice area mapping have been well documented [14–24]. Optical images, even with multi-temporal compositing, are still unreliable given many consecutive days or weeks data can be unusable due to cloud cover, which is particularly problematic for algorithms that rely on the detection of agronomic flooding at the start of rice growing season [4,9]. The wide geographic distribution of rice across Asia (i) necessitates wall-to-wall coverage to adequately capture the 144 million hectares [2] of rice area and suggests that automated or low-level supervised processing is required to do this. Automated processing would also be suited to low-cost cloud computing platforms and low-level operator supervision, which could offset the common bottlenecks of infrastructure, bandwidth and human capacity in emerging economies. The wide range of practices and environments (ii) means that rice detection algorithms should be generalizable and robust [9]. Such methods should show high skill for irrigated and rainfed rice, for short (90 day), medium (120 day) and long (150 day+) rice growth duration [1] and for different establishment practices such as direct seeding or transplanting. The spatial complexity (iii) of rice environments requires high-resolution imagery and the temporal complexity (iv) requires high-frequency acquisitions across many months of the year.

Remote-sensing requirements are non-trivial and they go some way towards explaining the dearth of operational rice crop monitoring systems. However, recent launches of Sentinel-1 mission by the European Space Agency (ESA) with SAR carrying sensors together with the state-of-the-art automated processing can provide sustainable solutions to this challenge to map and monitor one of the world's most important crops.

It is clear from the literature [14–24] that multi-temporal SAR data can be used to detect and map rice fields. However, most of these studies have focused on a limited number of observation locations per study and few studies can test the robustness of rice detection algorithms under different conditions. Varietal choice, crop establishment methods and crop management practices can have a significant effect on the structure of both plant and canopy, the duration and growth rate of the plant and water content. This contextual information on the characteristics of different rice cropping systems must be considered in the development of a robust rice detection algorithm for application over diverse rice environments.

This paper highlights recent initiative from the RIICE project – Remote Sensing-based Information and Insurance for Crops in Emerging economies to involve larger number of study sites (at least 13 sites x 2 seasons within phase 1 and 4 sites x 3 seasons within phase 2) and covers a larger total area (4.78 m ha in phase 1 and 17 m ha in phase 2) than any previous assessment of SAR-based remote sensing for rice mapping. In addition in this project the rice area, seasonality, and SAR back-scatter is coupled to crop model in order to also generate high resolution yield maps. Thus, it is a timely contribution to the literature on the use of SAR for national and sub-national mapping of rice in Asia for food security applications.

2. Methodology

The RIICE project – Remote Sensing-based Information and Insurance for Crops in Emerging economies – tested SAR-based rice monitoring across six countries (India, Thailand, Cambodia, Vietnam, Indonesia and the Philippines) between late 2012 and early 2014 (figure 1). These countries account for 51.5% of the world's rice area and 45.9% of production, and include both rice net exporters (Thailand, Vietnam, India and Cambodia) and rice net importers (Indonesia and the Philippines) [2]. Beginning from 2015, the project embarked in effort to upscale the SAR-based rice monitoring toward national/sub-national scale in Vietnam, Cambodia, Thailand, and Tamil Nadu, India (figure 1). This international collaborative effort used multi-temporal X-band (phase 1) and C-Band (phase 2) SAR data, semi-automated processing chains, in-season field monitoring, end-of-season validation points to map rice crops across the target areas.

IOP Conf. Series: Earth and Environmental Science **54** (2017) 012015 doi:10.1088/1755-1315/54/1/012015

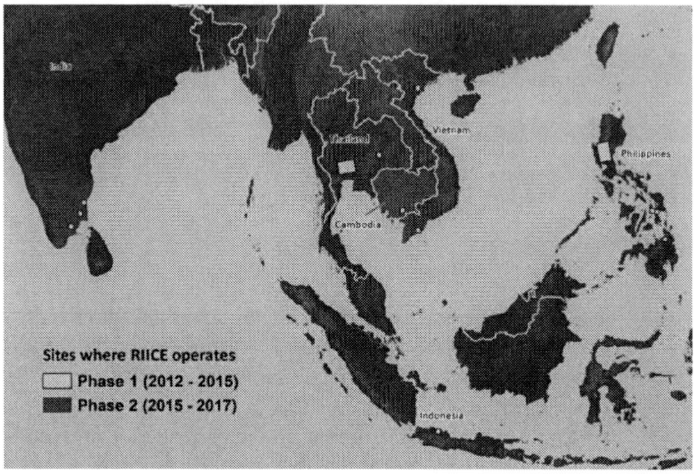

Figure 1. Study sites of the RIICE project during phase 1 and 2 across South and South East Asian Countries.

SAR-based paddy rice monitoring system deploys two core software: MAPscape-RICE and ORYZA Crop Growth Simulation Model (CGSM) [25] and involves SAR image acquisition, integration of SAR data into ORYZA, and generation of yield map (figure 2). MAPscape-Rice processes the SAR data into terrain-geocoded images (backscatter σ^{o} values) to produce the rice area estimates, start of season (SoS), phenological field status, and LAI. LAI values at approximately 33% maturity of the rice variety together with the SoS product are inferred from radar backscatter using cloud vegetation model with parameters calibrated with in situ LAI measurements. ORYZA uses daily weather, management practices, and crop and soil characteristics to estimate rice yield at 14% moisture content. Rice Yield Estimation System (Rice-YES) interfaces the assimilation process of SAR products (SoS and LAI) into the ORYZA crop growth model and facilitate provision of all the necessary inputs for the crop model.

Description of the processing chains to generate rice area using MAPScape-RICE has been previously documented in detailed [26]. Methodology description in this paper is therefore are focused more toward complementary modules such as the yield modeling component. Likewise, readers should review previous publication [26] for detailed descriptions the study sites and the SAR data used in the RIICE project phase 1.

In order to consider soil nitrogen dynamic processes, the CGSM uses soil data (https://sites.google.com/a/irri.org/oryza2000/) extracted from the World Inventory of Soil Emission potential (WISE) dataset (http://www.icasa.net/toolkit/wise.htm) and Harmonized World Soil Database (HWSD) (http://webarchive.iiasa.ac.at). Some assumptions on puddling effect on physical soil properties have been made. Daily solar radiation data are obtained from NASA Power dataset (http://power.larc.nasa.gov). Daily min and max temperature are obtained by interpolation of weather station data (http://www.ncdc.noaa.gov) using Shuttle Radar Topography Mission (SRTM) Digital Elevation Model (DEM) (http://www2.jpl.nasa.gov/srtm/) whereas daily rainfall were obtained from Tropical Rainfall Measurement Mission (http://trmm.gsfc.nasa.gov) until 2014 and from Climate Hazards Group Infra-Red Precipitation with Station data (CHIRPS) starting from 2015. The weather data were set at 15-arc minutes resolution and local data obtained from the National collaborators were also used whenever available.

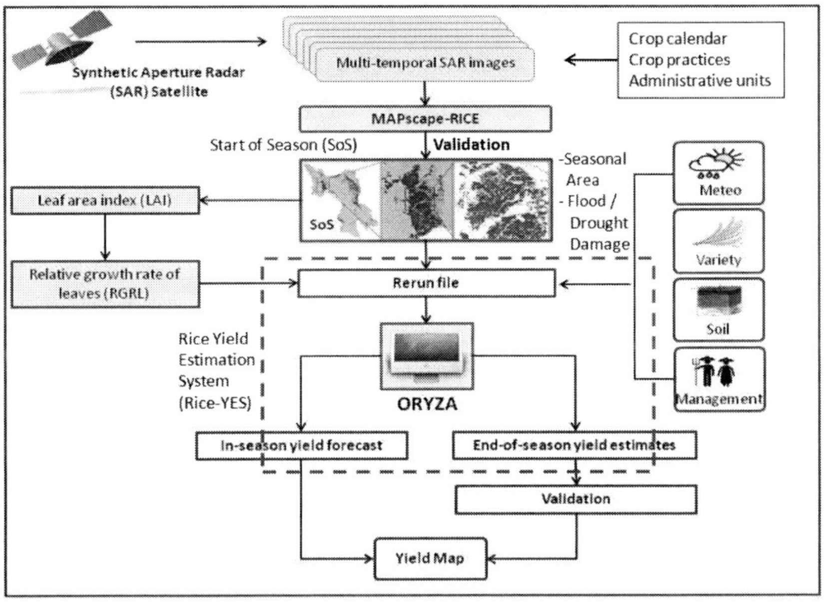

Figure 2. Operational diagram of the SAR-based paddy rice monitoring system implemented in the RIICE project.

The yield simulations account for water and nitrogen dynamics based on climatic, soil conditions and management rice practices. Irrigation and nitrogen fertilizer inputs are assumed as recommended for achieving attainable yield. Leaf Area Index (LAI) values at early leaf expansion stage (at 33% of rice maturity progress, roughly 40 days after start of season for 110 days rice) are inferred from radar backscatter using water cloud vegetation model [27] with parameters calibrated with in situ LAI measurements. Inferred LAI are finally used to calibrate the relative leaf growth rates parameters in ORYZA (figure 3).

LAI values at each simulation pixel were grouped in serveal claseees prior to running the CGSM in order to attain processing efficiency while maintaining high resolution of the output (same resolution of the SAR data being used) (figure 4). Post simulation, the yield results were remapped using the look up table containing geo-reference information of the original pixel classified according to the LAI values and in combination with SoS and weather data grid information.

3. Results

Figure 5 shows rice area maps derived from multi-temporal X-band SAR imagery during RIICE phase 1 in the Philippines, Indonesia, India, Cambodia, and Thailand. Summary of the related validation results is shown in table 1. Cosmo-SkyMed (CSK) Stripmap 3m resolution were the source SARdata for all of these maps except for in central Luzon, Philippines (TerraSAR-X, TSX Scansar 10m resolution) and Sivaganga, India (TSX Stripmap 3m resolution), and Central Plain Thailand (CSK Scansar 15m resolution). All of these sites were observed during the wet/monsoon season.

Rice map validation was conducted for each Area of Interest (AOI), led by the national partners given their local knowledge and easy access to the monitored areas. A rapid land-cover appraisal method has been adopted whereby partners travel through the footprint and collect GPS coordinates, photos and land-cover descriptions (and, when necessary, farmer interviews), with a good spatial coverage for approximate 50%-50% distribution across rice and non-rice areas (figure 6).

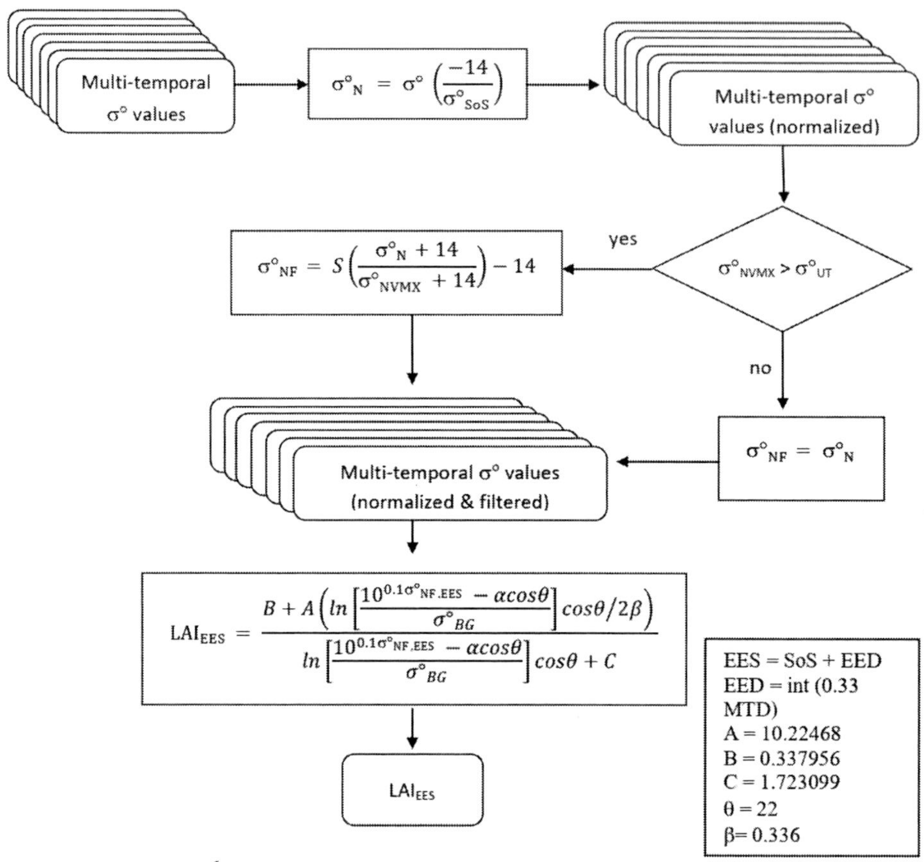

$$\sigma^{o}_{NF,EES} \begin{cases} = \sigma^{o}_{NF,i}\text{ , if i is within } \pm 2 \text{ days of EES date} \\ \text{else} \\ = \text{linear intrapolated } \sigma^{o}_{NF} \text{ between i} < EES \; and \; i > EES \end{cases}$$

σ^{o}_{UT} = -8.5 (for vv polarization), -9.2 (for vh polarization); S = 5.5 (for vv polarization), 4.8 (for vh polarization) ;σ^{o} = SAR backscatter; σ^{o}_{N} = Normalized SAR backscatter; σ^{o}_{SoS} = Normalized SAR backscatter at the start of season; σ^{o}_{NVMX} = Maximum value of normalized SAR backscatter during early vegetative phase, approximately SoS + 4 SAR S1A acquisition dates (12 days interval); σ^{o}_{UT} = Upper threshold value for normalized SAR backscatter during early vegetative phase; σ^{o}_{NF} = Normalized and filtered SAR backscatter for LAI calculation using vegetation cloud model; $\sigma^{o}_{NF,EES}$ = Normalized and filtered SAR backscatter for LAI calculation using vegetation cloud model at the early expansion stage; S = parameter for filtering SAR backscatter values for LAI calculation using vegetation cloud model; SoS = Start of season (day of year); EES = Early leaf expansion stage (day of year); EED = Duration to early expansion stage (days after start of season); MTD = Rice variety maturity duration (days); A,B,C = vegetation cloud model parameters: plant height & moisture parameters (Shen et al., 2009); α = vegetation cloud model parameter: backscattering coefficient at full canopy closure (m^2 m^{-2}); β = vegetation cloud model parameter: coefficient of attenuation per unit canopy water (m^2 kg^{-1}); θ = incident angle of radar beam (°); σ^{o}_{bg} = vegetation cloud model parameter: backscattering from canopy background (m^2 m^{-2})

Figure 3. Process diagream for conversion of multi-temporal backscatter into LAI at early expansion state as implemented in the SAR-based paddy rice monitoring sytstem.

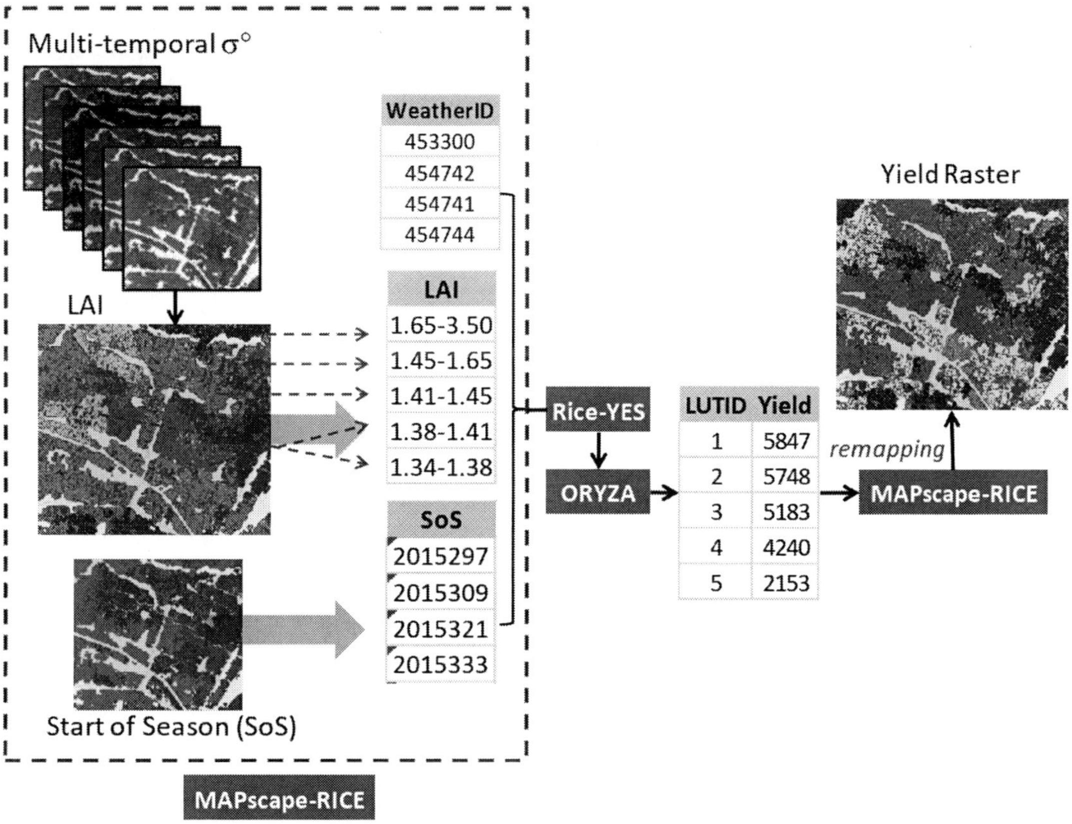

Figure 4. Key processing steps to allow efficient & effective yield map generation in the SAR-based rice monitoring system. MAPscape-RICE communicates back and forth with ORYZA and Rice-YES to ultimately generate yield raster. Programmatically LAI & SoS outputs are deconstructed from the raster forms into combination values of LAI x SoS x Weather pixel ID (thick gray arrows) with LAI grouped into several classes. This allows a more efficient processing of yield by running ORYZA only for unique combination of LAI x SoS x WeatherID whereas running yield simulation is avoided for pixels with near identical LAI values and that belongs to the same SoS and Weather ID. The simulation results can then remapped using the Look Up table approach since every pixel were marked (LUTID) prior to deconstruction process. The resulting outputs are simulated yield results (raster) at the same resolution of the backscatter & SoS inputs with relatively efficient processing requirement.

Table 3 shows a summary of the validation data, rice area and rice classification accuracy. Accuracy assessments in the field were generally conducted in-season, in the reproductive or ripening stage before harvesting, but in some cases the field assessment was conducted post-season and rice stubble and farmer surveys were used to confirm that the observed post-harvest situation reflected the presence of a rice crop during the monitored season.

The total classified rice area across these sites is more than 1.6 million hectares but the proportion of the footprint area that was classified as rice varied from 8% to 94% across footprints. The overall classification accuracy was consistently high (86% to 95%). There is no relationship between the classification accuracy and either the rice area or the proportion of the footprint classified as rice. Large, homogeneous and landscape-dominating rice areas and small, fragmented, heterogeneous rice areas are all classified equally well.

Although the classifier can properly detect most rice areas, some land-cover types can cause misclassifications. Wetland or seasonal water bodies that are subjected to drying followed by sudden vegetation growth can contribute significantly to an increase in commission errors. This behaviour is observed for some rivers that drain very quickly and also for water tanks used to store irrigation water (Sivaganga, for example). The temporal signature of this land cover is similar to the typical rice signature, and, if the timing of the event corresponds to the known rice crop calendar, then the discrimination of rice is extremely challenging. One solution is to use multi-temporal SAR or optical images acquired in suitable periods – often outside the rice-growing season – to develop such a mask. In some cases, but this depends on the rice environmental conditions, the use of additional polarizations or the combination of different frequencies (for instance, C- and L-band as shown in [28]) may help in the exclusion of some non-rice areas.

Many of the omission errors were associated with a lack of correspondence between the observed rice crop calendar in isolated areas within the footprint and the acquisition period. This is particularly problematic when rice is sown early with respect to the average crop calendar since the signature does not include the critical land preparation/agronomic flooding, which is the foundation of rice detection. A further problem relates to extremely short-duration varieties, around 90 days, which are transplanted as 15- or 20-day-old seedlings. This means that the remaining vegetative stage in which biomass increases substantially is of very short duration and can be hard to detect, especially if there are cancellations during that time of the season.

Rice classification accuracy was maintaind above 85% for the mapping activities using Sentinel S1A data (C-band, 20m resolution) starting from phase 2 of the RIICE project. An example of this is shown by the nation-wide rice area map for the 2015/16 dry season (recession rice) in Cambodia (figure 7) with rice classification accuracy at 87%. Correspondingly, the SAR-based yield estimates were at 93% agreement against reported yield data at district level as evaluated in four southern Cambodian provinces of Kandal, Svay Rieng, Prey Veng, and Takeo.

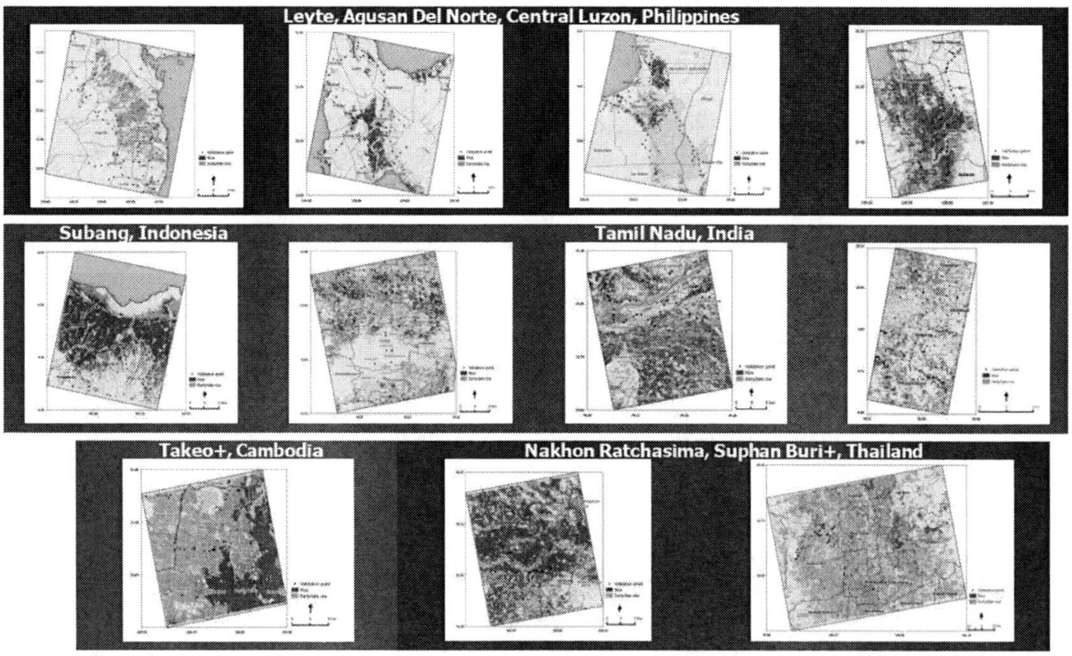

Figure 5. Rice area maps generated during the RIICE project phase 1. Rice is green, early/late rice is orange. See previous publication for more detailed version of these maps.

Table 3. Summary of accuracy assessments of the rice map outputs of RIICE project phase 1.

Site	Season	Period	Establishment	Maturity (days)	Water source	Rice area (ha)	Accuracy
Cambodia, Takeo	Dry	Oct to Apr	Direct seeding (DS)	95	Irrigated (IR)	150,026	85%
Philippines, Leyte East	Wet	May to Sep	Transplanting (TP)	114	IR	17,817	87%
Philippines, Leyte West	Wet	May to Sep	TP	110-112	IR	15,229	89%
Philippines, A. del Norte	Dry	May to Oct	TP & DS	107-123	IR & some rainfed (RF)	13,163	89%
Philippines, Nueva Ecija	Wet	Jul to Nov	TP	114	IR	424,801	86%
Vietnam, Soc Trang	Summer-autumn	Jun to Sep	TP & DS	95-120	IR	55,216	87%
Vietnam, Nam Dinh	Summer	Jul to Nov	TP	125-134	IR	108,733	89%
Indonesia, Subang	Wet	Nov to Apr	TP	115-135	IR	64,533	95%
India, Cuddalore	Samba	Jul to Jan	TP	130-160	IR	26,015	92%
India, Thanjavur	Samba	Aug to Dec	TP & DS	135-160	IR	83,871	91%
India, Sivaganga	Samba	Sep to Jan	TP & DS	100-110	Semi-dry rice	41,825	87%
Thailand, Muang Yang	Wet	May to Nov	DS	150-178	RF	91,908	86%
Thailand, Suphan Buri	Wet	Jun to Oct	DS	92-120	IR	555,317	87%
	Dry – 2 Wet – 11		DS – 7 TP - 10	92-178	IR – 12 RF – 3	1.65M ha	85-95 %

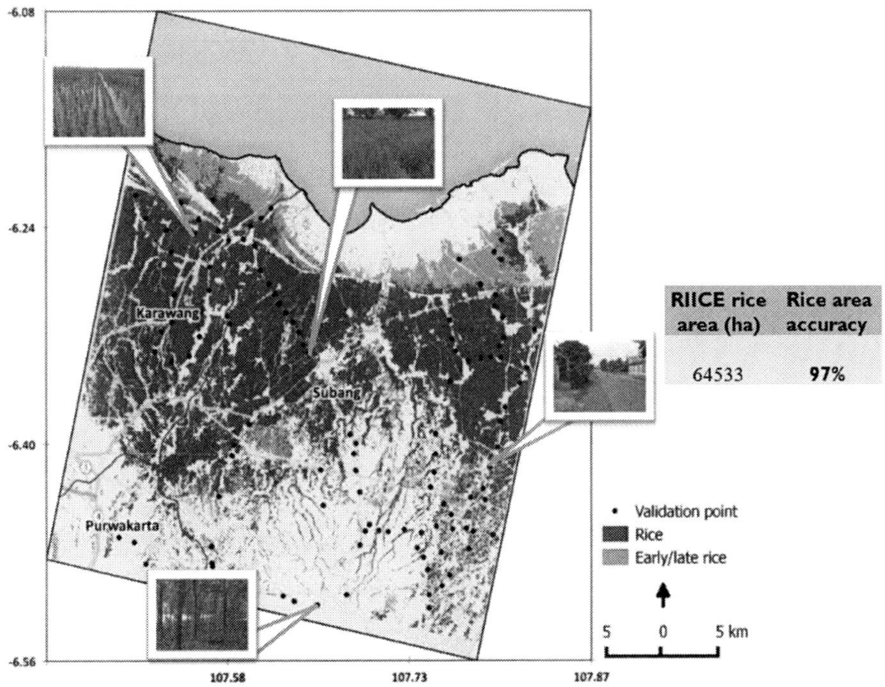

Figure 6. Rice area map for 2013/14 Wet Season in Subang Indonesia with validation points for assesing rice area map accuracy. The validation exercise involves ground verification of rice and non-rice of at least 50 points each.

While maintaing rice area classificaiton accuracy above 85%, the SAR-based monitoring system can effectively capture complexity in paddy rice cropping calendar such as the case in Mekong River Delta, Vietnam (figure 8). In this example, paddy rice start of season in An Giang and Dong Thap provinces were characterized with relatively long span of start of season from April to September 2016 with difference between field to fields even within a short distance. The map showed the general trend of relatively earlier rice season start in Don Thap as compared to An Giang an expression of difference in irringation water release in the area.

Responding to climate-induced disaster such as typhoon landing in the study area, the RIICE project monitored the extent of resulting floods using the SAR data. An example of this activity is shown in Figure 9. Tropical Storm Mirinae passed through the Red River Delta, Vietnam from 27 July 2016, 18UTC to July 28, 2016, 09UTC. Flood detection involved comparision of SAR images before and after the typhoon event (figure 9 bottom panel). The estimated total flooded areas are 8, 256 ha where 1,007 ha represented as flooded rice area. The area affected by flooding from this storm was relatively minor as detected on 1 August 2016 (4 days after the flood event). It is suggested therefore that rice cultivation in Red River Delta was spared from major damage from this tropical storm. Ground monitoring further confirmed that the extent of the flooding was relatively minor and the impact on the rice (at vegetative stage by the time the storm passed the area) was rather insignificant as flooding duration for the affected area was less than 2 days.

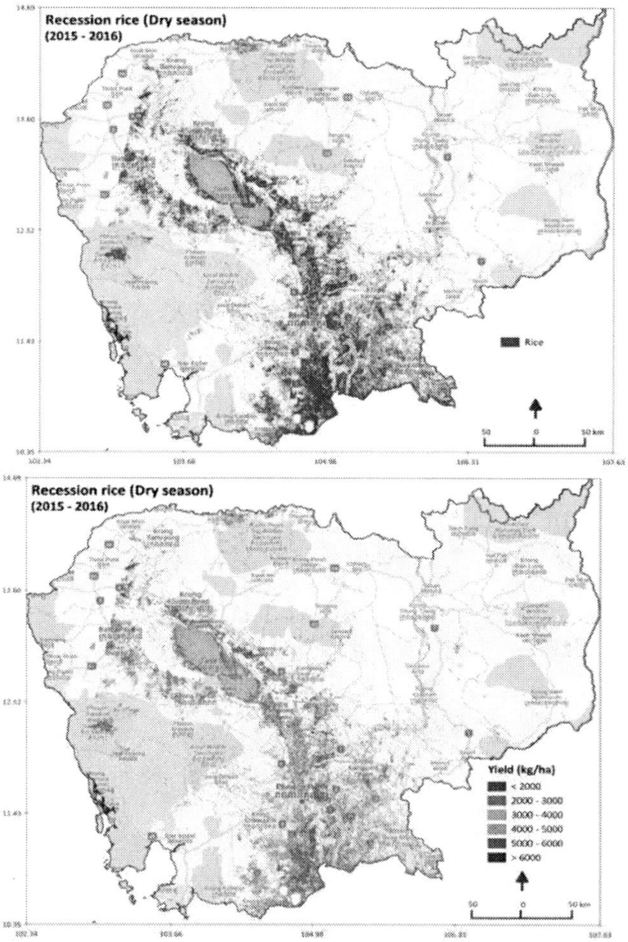

Figure 7. Rice area and yield maps for 2015/16 dry season (recession rice) in Cambodia generated during the phase 2 using SAR Sentinel-1 data. Accuracy of rice area classification is 87%. Yield estimates were at 93% agreement against official yield data at district level.

Yield estimate maps derived from C-band SAR data from Sentinel 1A for Mekong River Delta (2014/15 Winter-Spring season) and Red River Delta (2015 Summer season), Vietnam, Suphan Buri, Thailand (2015 Wet/Major season), Cauvery Delta, Tamil Nadu India (2014/15 Samba season), and Southern provinces of Cambodia (2015 Early Wet season) are showin in Figure 10. The map provided assessment of spatial distribution of yield and can be potentially used to target where interventions may be provided. Across these study sites, in these particilar seasons paddy rice yield ranged from less than 2.0 t/ha to greater than 6.0 t/ha.

Quality assesment of SAR-based yield estimate results is summarized in table 4 with simulated yields were compared against reported yield at district level from the goverment source. Aggregated SAR-based yield estimates at district level were within 81 to 93% agreement against official yield statistics. With such reasonable agreement level, this remote sensing-based rice yield monitoring system offers comparative advantage of providing: (i) timely yield information soon after the end of season and even before the season completed as yield forecast, (ii) high level of details (high resolution yield maps), and (ii) process-based yield information to allow further investigation for necessary interventions.

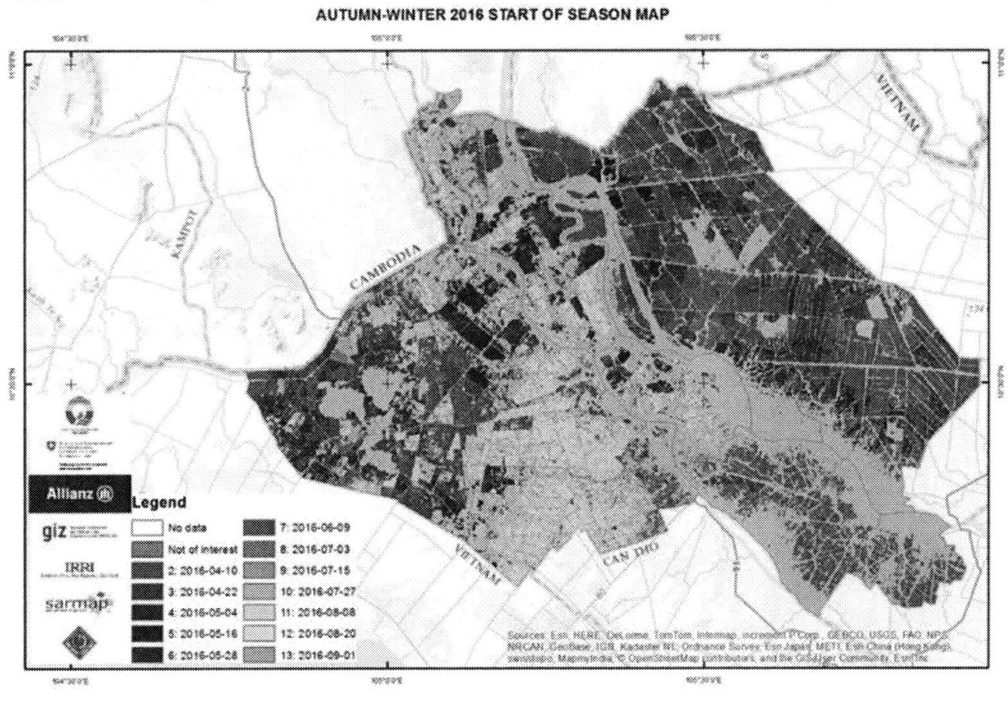

Figure 8. Start of Seeason (SoS) map for Autum-Winter season in An Giang and Dong Thap, Mekong River Delta, Vietnam.

IOP Conf. Series: Earth and Environmental Science 54 (2017) 012015 doi:10.1088/1755-1315/54/1/012015

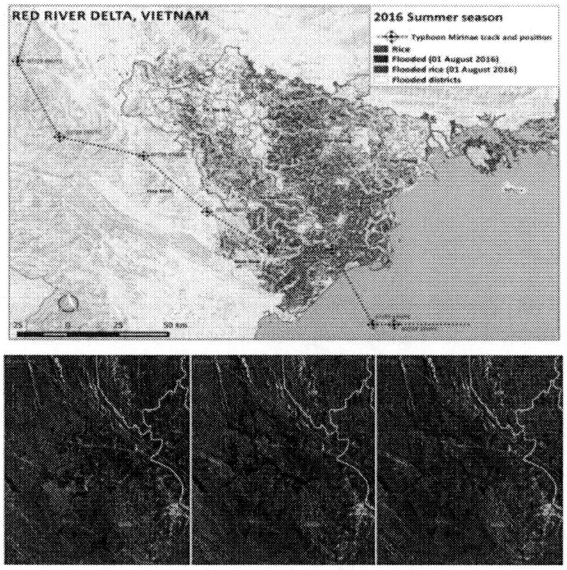

Figure 9. Assesmemt of flood damage in Red Rive Delta, Vietnam, responding to the typhoon Miriniae entering the study area on Jul 27-28, 2016.

Table 4. SAR-based yield estimate results compared against reported yields for RIICE phase 1 & 2.

Country or State*	Province or District*	Season	Period covered	Ecosystem	Source of SAR data	Modeled Yield	Reported Yield	Yield Agreement (%)
						----- kg/ha	-----	
Phippines	Leyte (East)	Wet	2013 May to Oct	Irrigated	CSK	5028	5620	84
Phippines	Leyte (West)	Dry	2013 May to Oct	Irrigated	CSK	3390	3483	91
Phippines	Leyte (West)	Wet	2013 Dec to 2014 May	Irrigated	CSK	3870	4133	89
Phippines	Nueva Ecija	Wet	2014 May to Oct	Irrigated	TSX	5478	5313	90
Vietnam	Soc Trang (Mekong River Delta)	Summer-Autumn	2013 Jun to Oct	Irrigated	CSK	5900	5539	90
Vietnam	Soc Trang (Mekong River Delta)	Summer-Autumn	2014 Jun to Sep	Irrigated	CSK	5478	5698	90
Vietnam	An Giang and Dong Thap (Mekong River Delta)	Winter-Spring	2015 Oct to 2016 Apr	Irrigated	S1A	7172	7385	86
Vietnam	Nam Dinh (Red River Delta)	Summer	2013 Jul to Nov	Irrigated	CSK	4818	4376	84
Vietnam	Nam Dinh (Red River Delta)	Summer	2014 Jul to Nov	Irrigated	CSK	4986	5170	88
Vietnam	Hanoi, Hai Duong, Hai Phuong, Hung Yen, Nam Dinh, Thai Binh, Back Ninh, Ha Nam (Red River Delta)	Summer	2015 Jun to Nov	Irrigated	S1A	5273	5516	88
Tamil Nadu, India	Cuddalore	Samba	2013 Jul to 2014 Jan	Irrigated	CSK	3748	4276	83
Tamil Nadu, India	Thanjavur	Samba	2013 Aug to Dec	Irrigated	CSK	4896	4523	85
Tamil Nadu, India	Sivaganga	Samba	2013 Sep to 2014 Feb	Semi-Dry	TSX	1827	1753	88
Thailand	Nakhon Ratchasima	Wet	2013 May to Dec	Rainfed	CSK	2748	2595	92
Thailand	Nakhon Ratchasima	Wet	2014 May to Dec	Rainfed	CSK	2628	2823	81
Thailand	Suphan Buri	Wet	2013 May to Oct	Irrigated	CSK	4964	4440	89
Thailand	Suphan Buri	Wet	2014 May to Oct	Irrigated	CSK	5086	4941	93
Thailand	Suphan Buri	Wet	2015 May to Oct	Irrigated	CSK	5394	5125	84
Cambodia	Takeo	Early Wet	2014 Apr to Aug	Rainfed	CSK	3760	3850	85
Cambodia	Takeo	Main Wet	2014 Jun to Oct	Rainfed	CSK	3123	3018	92
Cambodia	Kandal, Prey Veng, Svay Rieng, Takeo	Early Wet	2015 Apr to Aug	Rainfed	S1A	3395	3555	85
Cambodia	Kandal, Prey Veng, Svay Rieng, Takeo	Dry (Recession Rice)	2015 Oc to 2016 Apr	Irrigated	S1A	4490	4430	92

* Upper & lower administrative boundaries of interest for India

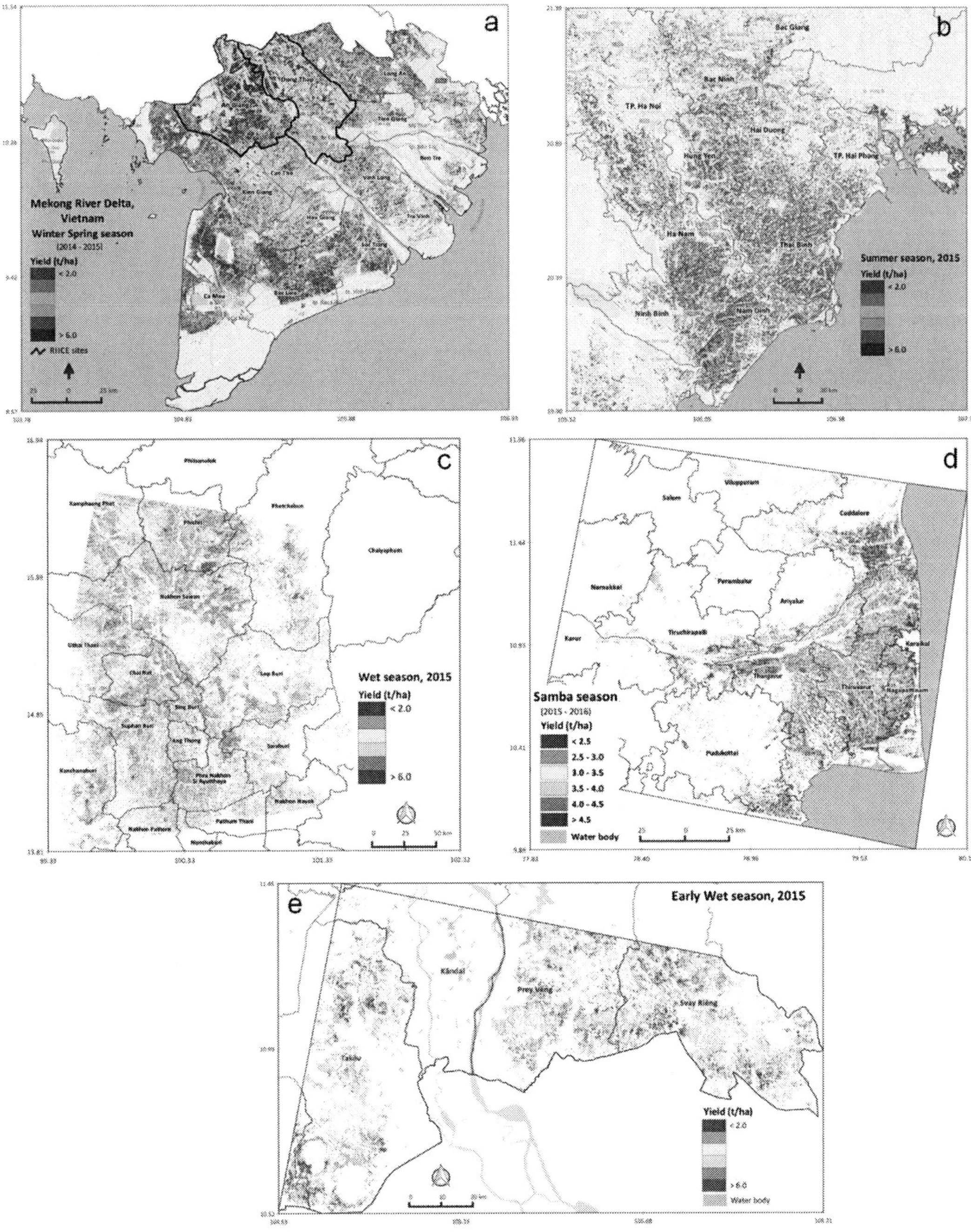

Figure 10. Yield estimates maps generated with SAR-based paddy rice monitoring system for (a) Mekong River Delta, Vietnam (2014/15 Winter-Spring), (b) Red River Delta ,Vietnam (2016 Summer) (c) Suphan Buri, Thailand (2015 Wet/Major) (e)Tamil Nadu, India (Samba), and (e) Southern Cambodia (Early Wet).

4. Conclusions

The consistently high accuracy of the rice area classification across different rice environments and crop management practices demonstrates that the methodology is appropriate for rice detection across the most common rice agro-ecologies in Asia. The classification is based on a temporal analysis of the σ° signature, including a detection of agronomic flooding at the land preparation and/or seedling stage followed by a rapid increase in biomass relative to the duration of the vegetative stage of the varieties in the SAR footprint. The ground monitoring data are critical for the correct interpretation of the σ° signature, especially for the distinct differences between transplanting and direct-seeding crop establishment methods. It is also demonstrated that this SAR-based monitoring technology can provide reliable, timely, and detailed yield estimates and thus allowing potential necessary intervention and an effective tool to respond to climate related disaster such as typhoon-induce flooding that has potential negative impact on rice production.

Acknowledgement

This work has been undertaken within the framework of the Remote Sensing-based Information and Insurance for Crops in emerging Economies (RIICE, www.riice.org) program, financially supported by Swiss Development Cooperation (SDC). It was also funded by the CGIAR Global Rice Science Partnership (GRiSP, www.grisp.net) program.

SAR data were provided by ASI/e-GEOS and GISTDA from COSMO-SkyMed and by InfoTerra GmbH from TerraSAR-X, and Sentinel-A from European Space Agency (ESA). The boundaries, colors, denominations and other information shown on any map in this work do not imply any judgment on the part of the authors or their institutes concerning the legal status of any territory or the endorsement or acceptance of such boundaries.

The authors would like to thank the following individuals for their contribution to the paper: Andy Nelson, Arnel Rala, Cornelia Garcia, Jenny Ravis, Wahyunto, Dwi Kuntjoro, Sri Murdiyati, Phuong Do, Vo Quang Minh, Vo Quoc Tuan, Duong Hoang Trinh, Quyen Huu Nguyen, Amornrat Intrman, Preesan Rakwatinh, Thana Setapayak, Sommai Lertna, Sirilux Noikeaing, Anyada Penporn, Jittima Chaimuanwong, Men Soothy, Touch Veasna, Lim Vanndy, Sellaperumal Pazhanivelan, Pandian Kanan, P. Christy Nirmala Mary, E. Subramanian, Eduardo Jimmy P. Quilang, Mary Rose O. Mabalay, Ninh H. Nguyen, Pongmanee Thongbai, Chharom Chin, and Manoj Yadav.

References

[1] GRiSP (Global Rice Science Partnership) 2013 *Rice Almanac 4th ed.* (Philippine: *International Rice Research Institute*) pp 283

[2] FAO FAOSTAT http://faostat3.fao.org/faostat-gateway/go/to/home/E (accessed Jun 1, 2014).

[3] Timmer C P 2010 Food Security in Asia and the Changing Role of Rice. *The Asia Foundation. Occasional Paper* **4**

[4] Xiao X, Boles S, Frolking S, Li C, Babu J Y, Salas W, Moore B 2006 Mapping paddy rice agriculture in South and Southeast Asia using multi-temporal MODIS images *Remote Sens. Environ.* **100** 95–113

[5] Gumma M K, Thenkabail P S, Maunahan A, Islam S, Nelson A 2014 Mapping seasonal rice cropland extent and area in the high cropping intensity environment of Bangladesh using MODIS 500m data for the year 2010. *ISPRS J. Photogramm. Remote Sens.* **91** 98–113.

[6] Balagtas, J V, Bhandari H, Cabrera E R, Mohanty S, Hossain M 2014 Did the commodity price spike increase rural poverty? Evidence from a long-run panel in Bangladesh *Agric. Econ.* **45** 303–312

[7] Mittal A 2009 The 2008 Food Price Crisis: Rethinking Food Security Policies (New York) pp 1–40.

[8] Dawe D, Timmer C P 2012 Why stable food prices are a good thing: Lessons from stabilizing rice prices in Asia. *Glob. Food Sec.* **1** 127–133

[9] Boschetti M, Nutini F, Manfron G, Brivio P A, Nelson A 2014 Comparative analysis of

normalised difference spectral indices derived from MODIS for detecting surface water in flooded rice cropping systems *PLoS One* **9** e88741

[10] Huke R E, Huke E H 1997 Rice area by type of culture: South, Southeast, and East Asia, A revised and updated data base p 32

[11] NASA Cloud Fraction (1 month Terra/MODIS) 1 FAO FAOSTAT http://faostat3.fao.org/faostat-gateway/go/to/home/E (accessed Jun 1, 2014)

[12] Pandey S, Byerlee D, Dawe D, Dobermann A, Mohanty S, Rozelle S, Hardy B editors. 2010 Rice in the Global Economy: Strategic Research and Policy Issues for Food Security p 477

[13] Nguyen T T H, De Bie C A J M, Ali A, Smaling E M A, Chu T H 2012 Mapping the irrigated rice cropping patterns of the Mekong delta, Vietnam, through hyper-temporal SPOT NDVI image analysis *Int. J. Remote Sens.* **33** 415–434

[14] Ulaby F T, Allen C T, Eger G, Kanemasu E 1984 Relating the microwave backscattering coefficient to leaf area index *Remote Sens. Environ.* **14** 113–133

[15] Le Toan T, Laur H 1988 *Multitemporal and Dual Polarisation Observations of Agricultural Crops by X-band SAR Images* Proc. of the 1988 Int. Geoscience and Remote Sensing Symp. IGARSS '88. Remote Sensing: Moving Toward the 21st Century International 3 1291–1294

[16] Inoue Y, Kurosu T, Maeno H, Uratsuka S, Kozu T, Dabrowska-Zielinska K, Qi J 2002 Season-long daily measurements of multifrequency (Ka, Ku, X, C, and L) and full-polarization backscatter signatures over paddy rice field and their relationship with biological variables *Remote Sens. Environ.* **81** 194–204

[17] Suga Y, Konishi T 2 October 2008 *Rice crop monitoring using X, C and L band SAR data Proc.* SPIE 7104 Remote Sensing for Agriculture, Ecosystems, and Hydrology X,710410 doi:10.1117/12.800051

[18] Bouvet A, Le Toan T, Lam-Dao N 2009 *Monitoring of the rice cropping system in the mekong delta using ENVISAT/ASAR dual polarization data* Geosci. Remote Sensing, IEEE *Trans.* 47 517–526

[19] Oh Y, Hong S Y, Kim Y, Hong J Y, Kim Y H 2009 *Polarimetric Backscattering Coefficients of Flooded Rice Fields at L- and C-Bands: Measurements, Modeling, and Data Analysis* Geosci. Remote Sensing, IEEE Trans 47 2714–2721

[20] Kim Y H, Hong S Y, Lee Y H 2009 Estimation of paddy rice growth parameters using L, C, X-bands polarimetric scatterometer *Korean J. Remote Sens.* **25** 31–44

[21] Lopez-Sanchez J M, Ballester-Berman J D, Hajnsek I 2011 *First results of rice monitoring practices in Spain by means of time series of TerraSAR-X Dual-Pol Images* Sel. Top. Appl. Earth Obs. Remote Sensing, IEEE J. 4 412–422

[22] Lopez-Sanchez J M. Cloude S R, Ballester-Berman J D 2012 *Rice Phenology Monitoring by Means of SAR Polarimetry at X-Band.* Geosci. Remote Sensing, IEEE Trans 50 2695–2709

[23] Inoue Y, Sakaiya E. 2013 *Relationship between X-band backscattering coefficients from high-resolution satellite SAR and biophysical variables in paddy rice* Remote Sens. Lett. 4 288–295

[24] Inoue Y, Sakaiya E, Wang C 2014 *Capability of C-band backscattering coefficients from high-resolution satellite SAR sensors to assess biophysical variables in paddy rice* Remote Sens. Environ. 140 257–266

[25] Bouman B A M, Kropff M J, Tuong T P, Wopereis M C S, Ten Berge H F M, van Laar H H 2011 ORYZA2000: modeling lowland rice International Rice Research Institute, Los Baños, Laguna

[26] Nelson A, Setiyono T, Rala A B, Quicho E D, Ravis J V, Abonete P J, Maunahan A A, Garcia C A, Bhatti H Z M, Villano L S, Thongbai P, Holecz F, Barbieri M, Collivignarelli F, Gatti L, Quilang E J P, Mabalay M R O, Mabalot P E, Barroga M I, Bacong A P, Detoito N T, Berja G B, Varquez F, Wahyunto, Kuntjoro D, Murdiyati S R, Pazhanivelan S, Kannan P, Mary P C N, Subramanian E, Rakwatin P, Intrman A, Setapayak T, Lertna S, Minh V Q, Tuan V Q, Trinh D H, Nguyen Q H, Kham D V, Hin S, Veasna T, Yadav M, Chin C, Nguyen N H 2014

Toward an operational SAR-based rice monitoring system in Asia: Examples from 13 demonstration sites across ASia in the RIICE project. *Remote Sensing.* **6** 10773-10812

[27] Attema E P W, Ulaby F T 1978 Vegetation modeled as a water cloud *Radio Science* **13** 357-364.

[28] Holecz F, Barbieri M, Collivignarelli F, Gatti L, Nelson A, Setiyono T D, Boschetti M, Manfron G, Brivio P, Quilang E, Obico M, Minh V Q, Kieu D P, Huu Q N, Veasna T, Intrman A, Wahyunto, Pazhanivelan S 2013 *An Operational Remote Sensing Based Services for Rice Production Estimation at National Scale* ESA Living Planet Symposium, Edinburgh

IOP Conf. Series: Earth and Environmental Science **54** (2017) 012016 doi:10.1088/1755-1315/54/1/012016

Monitoring of landscape change in paddy fields: Case study of Karawang District – West Java Province

E E Franjaya[1]*, Syartinilia[1] and Y Setiawan[2]

[1] Department of Landscape Architecture, Faculty of Agriculture, Bogor Agricultural University, Kampus IPB Darmaga, Bogor 16680, Indonesia
[2] Center for Environmental Research, Bogor Agricultural University, Kampus IPB Darmaga, Bogor 16680, Indonesia

E-mail: eduwinekofranjaya@yahoo.com

Abstract. Paddy field is an important agricultural land in Indonesia, as one of the largest rice producing-country in the world. At least 26 from 33 provinces in Indonesia are characterized by the existence of paddy field landscape. However, due to the increasing of population and development of infrastructure building, a conversion of paddy field rapidly occurs in many sites. This study aimed to examine the dynamics change in paddy field in Karawang District-West Java during the period of 1994-2015. The method used in this study mainly by the remote sensing technique using satellite images data. The result indicated that conversion of paddy fields to built area/infrastructure in Karawang is approximately 10.326,6 ha. It took up 56% from the paddy that were changed. Based on the result, the changes are likely to occur in the middle of karawang district, near the central city. This result showed the change of paddy field in 1994 converted into some built-up areas such as settlement or roads in 2015. However, about 85.597,56 ha paddy field is not changed during these period. The study showed that paddy fields landscape is facing a changes over the last two decades.

1. Introduction

Rice paddy is one of the important agricultural commodity in Indonesia. Even in the world, Indonesia is known as one of the top three rice paddy producers. This is because paddy is staple food for Indonesian people. Based on data from BPS [1], it is known that 26 of 33 provinces in Indonesia have paddy field landscape. But unfortunately, this paddy field landscape also facing the changes. In fact, paddy field landscape especially wet land paddy field can support in cultivating fishes (agro-fisheries), controlling the flood, filling the ground water reserve, and maintaining ecosystems and biodiversities[2]. In the other side, paddy field can also be cultural heritage, particularly at terraced paddy field and rice fish culture as founded in China, Japan, Republic of Korea, and Philippines [3]. They became part of site countries that are designated as GIAHS (Globally Important Agricultural Heritage Systems) in Asia and the Pacific region. In Indonesia, we call cultural landscape such paddy field and its water management as Saujana Heritage. We found it in Subak System, Bali region.

In this research, Java Island has chosen as the research area due to the facts as follow: First, based on BPS data [1], three provinces with the highest amount of crop yield were West Java, East Java, and Central Java, respectively. Those provinces are in Java Island. Second, based on Verburg, Veldkamp, and Bouma [4], the level of urbanization in Java Island is higher than the other with the density more than 800 people/km^2. The existence of Jakarta as the largest city in Indonesia also carry a huge influence. The increasing of socio-economic activities in this city have led to massive land use changes [5]. The changes are happened in its surrounding areas, not only Bodetabek area as usual but also Karawang District. Types of the changes are in the form of green areas such as forests, paddy fields, into settlements and the other built up areas. Certainly, Karawang as one of paddy's production center that was very close to Jakarta will get influence of the existence of that largest city. Third, based on Handayani [6], the growing and developing economy and social change were consequenced of urbanization. It indicates that higher the level of urbanization, the higher its impact on social and economy. Those facts encourage researcher to know whether there is a change or not in paddy field landscape in Karawang.

We need some satellite images data to know the change and how fast it changed. The result will be extracted to be land cover map and it applies the remote sensing technique. This research aimed to get land cover change map of paddy field in Karawang District-West Java. Using this land cover change map, we can give recommendations so that more changing would not be happen in the future.

2. Materials and methods

2.1. Study area

This study was conducted in Karawang District-West Java, which located latitude S5°56′ - S6°34′ and longitude E107°02′ - E107°40′. This area has also chosen as the research area based on some facts as follow: *First*, Based on BPS data [1], West Java is the province with the highest crop yield in Indonesia, and Karawang is one of paddy's production center. *Second*, Karawang is in the distance 112 km from Jakarta [1], and it is known that the area around Jakarta has faced the high level of urbanization [7].

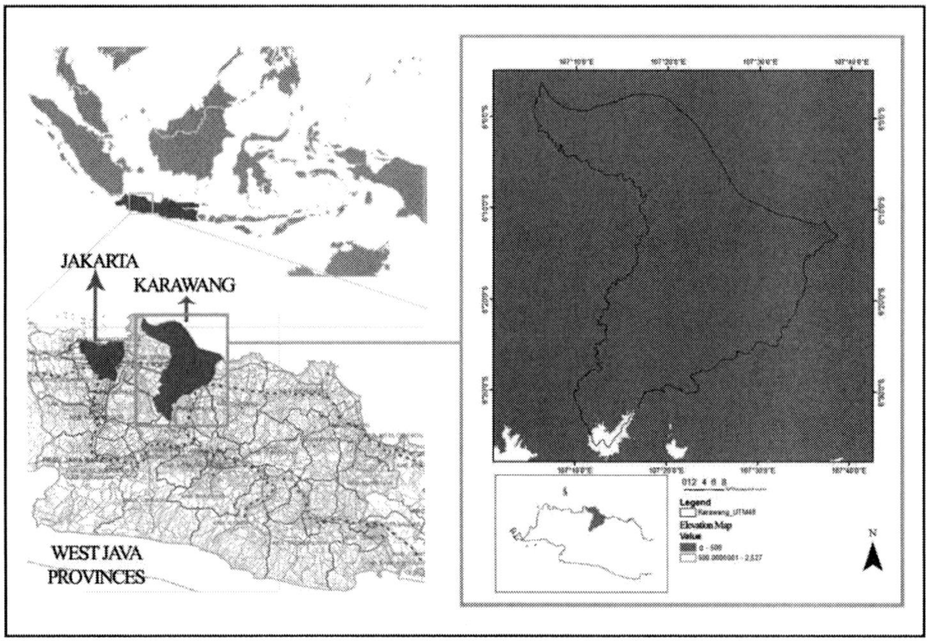

Figure 1. Study area

The figure 1 above is the location map of Karawang. Originally, it is an elevation map of this area. The elevation map was extracted from ASTER Global Digital elevation Model (ASTER GDEM) and was classified due to the height of planting paddy from *Balai Besar Padi* [8]. According to this classification, approximately 95% of Karawang District is in low land area (0-500meter above sea level) in which classified as S1 category (very suitable) for planting the paddy. Based on Integrated Crop Calendar, the average rainfall in Karawang is 272 mm/month [9].

2.2. *Data collection*

The main data used in this research is satellite images data. The images data are Landsat 5 (1994) and Landsat 8 (2015). Besides the satellite images, this research also need Indonesia Digital Map (RBI) at Karawang scene. This RBI data will be used for geometric correction and as a reference on supervised classification process. The tools that used in this research were varied. The main tools are ArcGIS software and Erdas Imagine 9.1. These tools serve to collect, to save, and to analyze data.

Table 1. Research's material, source, and its function.

Material	Unit	Source	Tool's function
Digital Map (RBI)	Vector	Geospatial Information Agency	Earth Surface Map
Satellite Images Path 122 row 064	Landsat 5 and Landsat 8 (Raster) 30x30 m pixel resolution	National Aeronautics and Space Administration	Land Cover Change Analysis
Aster GDEM	Raster (30x30m)	National Aeronautics and Space Administration	Elevation and Slope

2.3. *Land cover analysis*

This research follows the research process as illustrated at study flowchart (figure 2). Images data were selected using purposive sampling with considering the quality of images (cloud free) and time interval. The time interval is related to probability of change's time that is closely related to the change of technology, economy, and government policies.

Land cover analysis in this research use the Maximum Likelihood Classification (MLC method). This is a standard method that often used in supervised classification algorithm. It quantitatively evaluates the varian and covarians from spectral response of an unknown pixel [10]. The algorithm of MLC method is able to know the spectral characteristic of each class on an unknown data set using statistic data. This statistic data obtained from training site that had done before [10]. The algorithm is assumed normal distribution of multivarians from each spectral classes. The average vector and matrix covarians from a distribution can be used to explain it completely. With these parameters, it is possible to approximate the statistic opportunity from pixel value that given to be part of a spectral class [10]. Simply, this method facilitates us to categorize pixels into specific class. This MLC method delivers land cover map.

2.4. *Land cover change analysis*

On this stage, the result of MLC method will be processed further to detect the changes. But previously, we need to do accuracy assessment that can corrected the land cover map of each year. The input of accuracy assessment comes from the points that we take using GPS on each land cover through all area of Karawang District. The PCC method will be executed if the result of accuracy assessment reach 80% or more.

Change detection analysis that using this method is done at model maker in Erdas Imagine. Basically, this method is comparative analysis of classification result at different year using simply mathematical combination pixel per pixel.

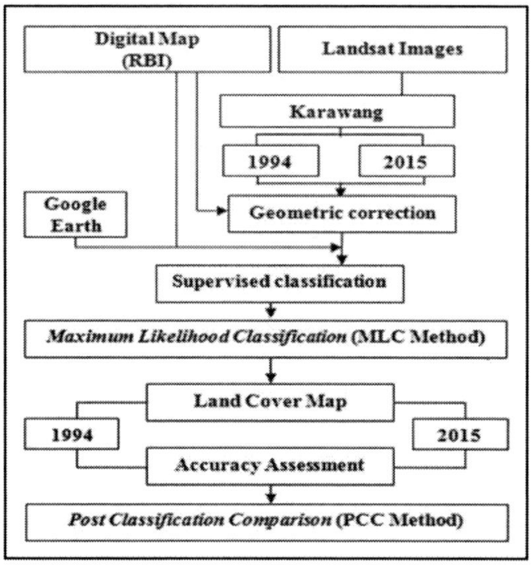

Figure 2. Study flowchart

3. Results and discussions

Based on the result at supervised classification process, the images were classified into 7 classes: bare land, dry land/mix garden, paddy field, built area, forest, fishpond, and water. This classification produces land cover map year of 1994 and 2015 (figure 3 & 4). The result of overall classification accuracy reaches 87.79 % for 1994's image and 89.43 % for 2015's image.

The paddy field landscape (figure 5 and 6) is still dominated the area in 1994 and 2015. Visually, the built area (in the middle of Karawang District) is increased in 2015 while the paddy field area is decreased. From those figures, we completely see that paddy field is decreased with amount of 7178.5 ha and built area is increased with amount of 6598 ha. In figure 7, we will see the paddy field landscapes that were changed and were not changed during 1994-2015 period.

At figure 7, there was 18460.8 ha paddy field landscapes that were changed from paddy to other classes. At this, we can focus on Paddy to Built Area (see the legend at figure 7). This area is paddy field landscapes in 1994 that were changed to built area in 2015, with a wide area reached 10326.6 ha. Please see figure 8 and 9 for detailed information.

From the figures 8 and 9, the change in the form of paddy to built area is the highest in wide area. It took up more than half. In summary, we can conclude that 56% of paddy field landscapes in 1994 have changed to settlements/infrastructures in 2015.

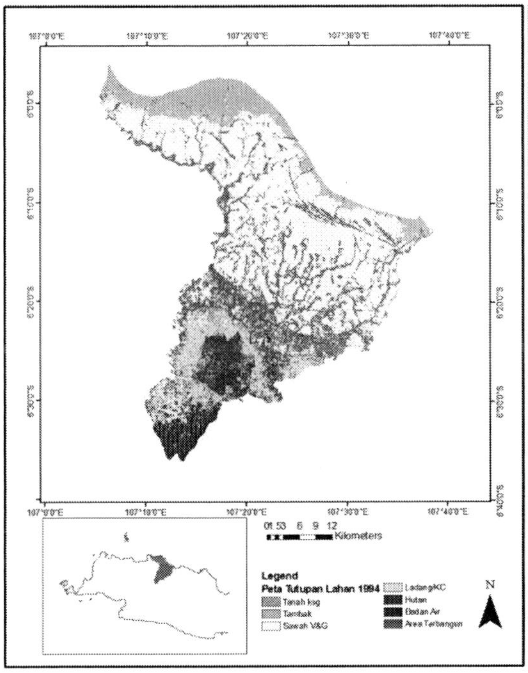

Figure 3. Land cover map 1994

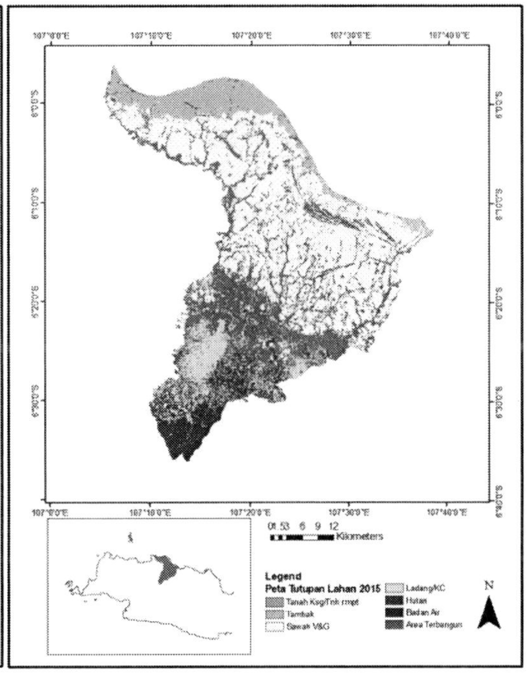

Figure 4. Land cover map 2015

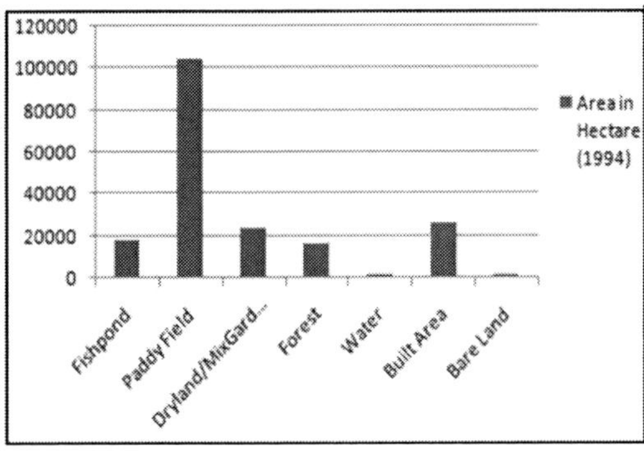

Figure 5. Wide area in 1994

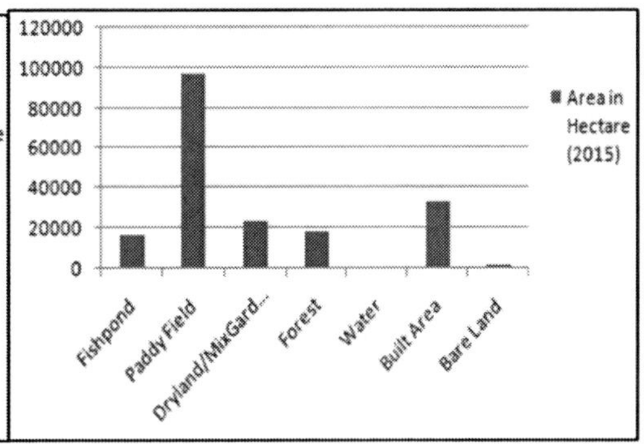

Figure 6. Wide area in 2015

Furthermore, based on land cover map 2015, Karawang District is faced to a very significant development. More clearly, at the land cover change map (1994-2015), the changes are dominantly take place in the center of the city. Through the ground check and RBI data, area in the center of the city is dominated by housing and commercial area. At this area, there are toll road and arterial road too. From those facts, it can be assumed that beside the population growth which characterized by the increasing of housing, the existence of road access and commercial buildings were estimated as driving factors of change. But, it needs a further research to prove it.

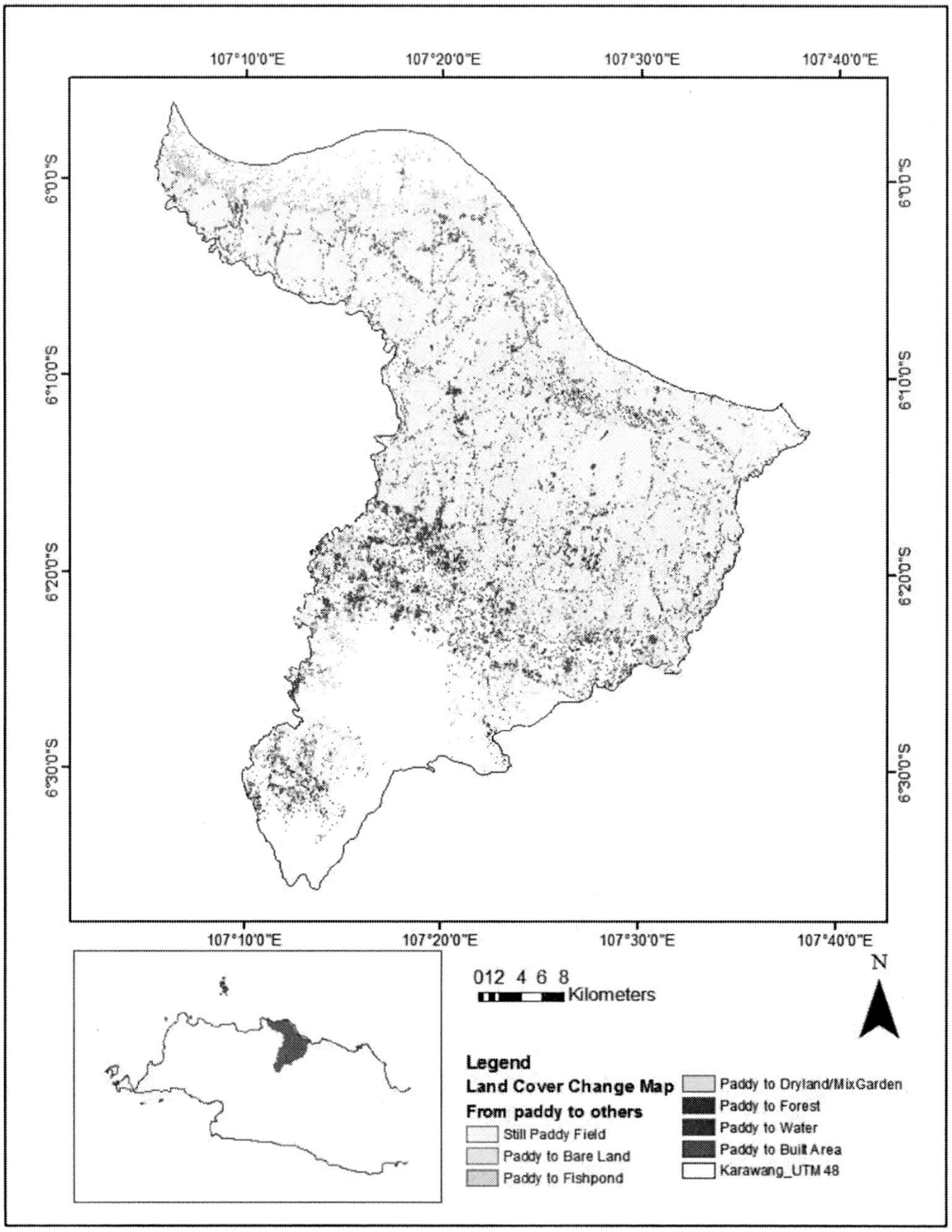

Figure 7. Land cover change map from paddy to other classes (1994-2015)

4. Recommendation

We need to maintain the paddy field area, especially near the area that was estimated as driving factors of change. Housing area, commercial building area, and economy area need to be maintained so that the paddy keep produce well. As mentioned before, we also need to do a further research especially in driving factors analysis. In the other side, a good supervision and spatial arrangement will be helpful in preventing the changes. The legislation that protects the area of food production is also needed.

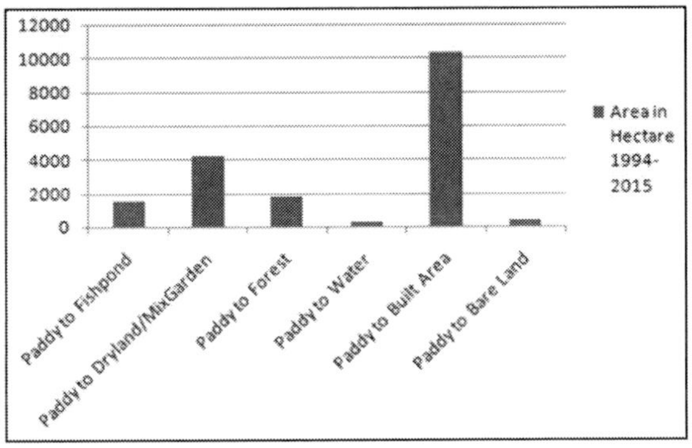

Figure 8. Wide area of paddy fields that were changed to other land cover classes

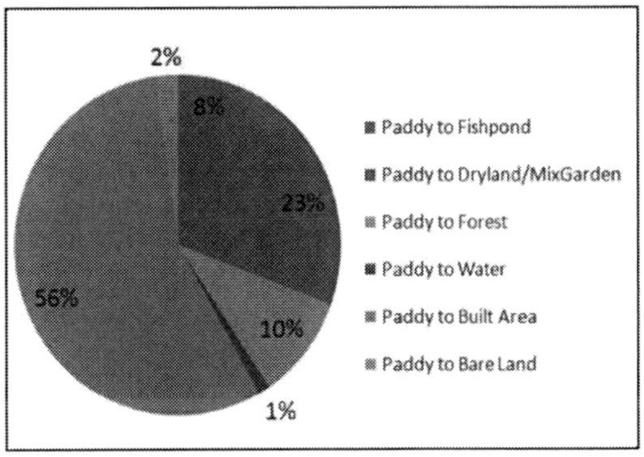

Figure 9. Pie chart of wide area of paddy fields that were changed

Acknowledgments
We gratefully acknowledged for the help that given by *Badan Informasi Geospasial* (en: Geospatial Information Agency, abbreviation: BIG), Staff of *Pusat Penelitian Lingkungan Hidup* (en: Center for Environmental Research, abbreviation: PPLH), and the government of Karawang so that this research is well done. We should also thanks to the other for the helpful suggestions.

References
[1] BPS 2013 *The table of crop area, productivity, production of paddy throughout Indonesia*. Retrieved from *http://www.bps.go.id.* on February 28th, 2014; Dated 2013 (in Indonesian)
[2] INWEPF Japanese Committee 2007 *Message from the INWEPF to the first asia-pacific water summit INWEPF workshop "Ecosystem services of paddy irrigated basin"* Beppu Japan. International Network for Water and Ecosystem in Paddy Fields
[3] FAO 2016 *Globally Important Agricultural Heritage System: Designated Sites Asia and The Pacific*. Retrieved from *http://www.fao.org/giahs/giahsaroundtheworld/designated-sites/asia-and-the-pacific/en/* on December 2nd, 2016 (in Indonesian)
[4] Verburg PH, Veldkamp TA and Bouma J 1999 Land use change under conditions of high population pressure: the case of Java *J. Global Environmental Change* **9** 303-312
[5] Indraprahasta GS 2013 The potential of urban agriculture development in Jakarta *Procedia Environmental Sciences* **17** 11-19
[6] Handayani W 2013 Rural-urban transition in central java: population and economic structural changes based on cluster analysis *Land* **2** 419-436
[7] Rustiadi E, Panuju DR and Trisasongko BH 2009 *Environmental impacts of urbanization in Jabodetabek area* S Uchida *et al* Bogor *Proc. of The ICALRD-JIRCAS Workshop on Enhancement of Remote Sensing and GIS Technologies for Sustainable Utilization of Agricultural Resources in Indonesia* Indonesian Center for Agricultural Land Resources Research and Development Ministry of Agriculture p 44-55
[8] BB Padi 2015 *The influence of elevation on physical quality of several aromatic rice*. Retrieved from *http://bbpadi.litbang.pertanian.go.id.* on January 17th, 2016 (in Indonesian)
[9] Balitbang Kementan 2014 *Integrated crop calendar of Karawang District, West Java Provinces*. Retrieved from *http://katam.litbang.pertanian.go.id.* on September 11th, 2014 (in Indonesian)
[10] Sallaba F 2009 *Potential of a Post-Classification Change Detection Analysis to Identify Land Use and Land Cover Changes* (Lund: Lund University)

Land conversion and economic development in Jawa Barat Province: *Trade off or Synergy?*

N Nuryartono[1,2*], A Tongato[1], S Yusdiyanto[1], S H Pasaribu[1,2], T Anggraenie[1]
[1] International Center for Applied Finance and Economics (InterCAFE), Bogor, Indonesia
[2] Departments of Economics, Faculty of Economics and Management, Bogor Agricultural University, Bogor, Indonesia.

E-mail: nuryartono@yahoo.com

Abstract. Jawa Barat province contributed 22 percent to the national rice production. Since land availability is one of important factors for food production, this must be addressed seriously. During the period of 2008-2013, there had been a shrinking of wetland area of 20,502 hectares (nearly 2.2 percent out of total wetland area in Jawa Barat). This condition may have implications on local and national food production and food security. This study analyzes influencing factors for land use changes of wetland in West Java using a panel of 5200 villages' level data in 2011-2014 by panel method. Important influencing factors that contribute to the land conversions mostly related to the development such as increasing number of new settlements and regional economic development. Spatially, using GIS map shows some areas have high rate of land conversion compared to other areas. The highest land conversion was in Kabupaten Bogor followed by Kabupaten Bekasi, while lowest was Bogor city. As there are tradeoff between economic development and land use, government policy should be addressed for economic development while it should be simultaneously preventing wetland conversion.

1. Introduction

The increasing in food security has always been the top priority of national food development scheme. It is logically understood that when there is a lack in food it could result in some great consequences economically, politically and socially. Land is one of the factors that plays an important role in agriculture production. Noting the importance of land in contributing the agriculture production and also the potential revenue of villages (PAD), increasing the potential revenue through agriculture production is becoming an important effort in achieving national food needs. Based on Agricultural Census (SUSENAS) in 2013, it is shown that Indonesian farming households in food crops dominated by the households that involved directly in revenue management of the villages. The number of Indonesian farming households in food crops dominated by plants that contributed to villages' income is 14 million households, with a percentage of 79.80 percent, or decreased by 58 thousand households (-0.41 percent) compared to achieved revenue in 2003. Villages' food crops had planting area of 94,478,528,364 m^2 and the average acreage of 6678 m^2 [1]. Jawa Barat is the third-largest province contributor to meet the food security needs of Indonesia. As for the year 2013, the agricultural area of paddy fields in Jawa Barat amounted to 925 thousand hectares.

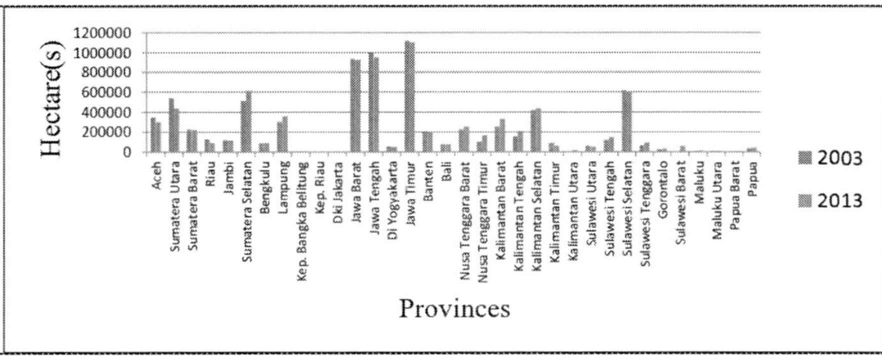

Source: [2]
Figure 1. Agricultural area of paddy fields in several provinces in Indonesia (Hectares) in 2003-2013

Based on the results of Agricultural Census in 2013, it is found that the number of farming household in food crops in Jawa Barat province amounted to 2 million households. Compared to 2003, the number decreased by 121 thousand households. Meanwhile, the households number of business plantation contributing to villages' revenue in Jawa Barat province in 2013 was 2 million households, or increased by 6 percent compared to 2003, based on BPS data [3].
However, the efforts of Jawa Barat province in contributing to the national granary to meet national food needs recently have some difficulties. This is reflected in the growth rate of rice production in Jawa Barat province which had been steadily declining from the year 2013 as many as 12 million tonnes to 11 million tonnes of villages' revenue in 2015.

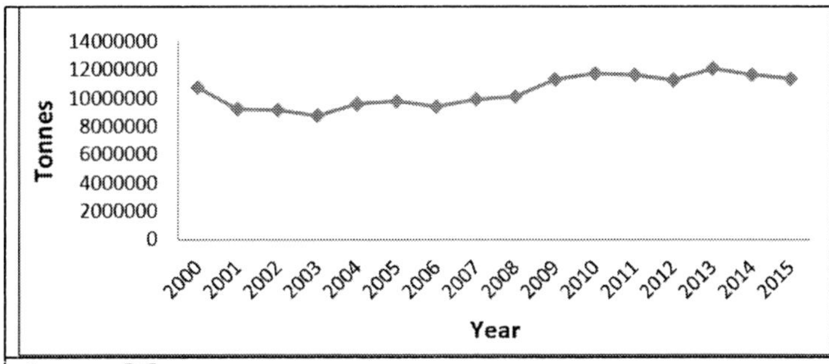

Source: [1]
Figure 2. Rice Production in Jawa Barat Province (Tonnes)

According to [1] showed that in 2013 the conversion of agricultural land into nonagricultural paddy fields in Jawa Barat province amounted to 12 331 hectares made it as the highest number compared to Jawa Tengah province that was equal to 8243 hectares. In contrast with Jawa Barat province which Jawa Timur province experienced an increase in the number of rice area as much as 2346 hectares. The impact of agricultural land conversion into non-agricultural land involves a very large dimensions like decreasing in food production as it is related to aspects that changes the orientation in economy, social-cultural and political community. The change direction, be it directly or indirectly, will affect the shift in economic conditions, spatial agriculture and agricultural development both at regional and national levels. Besides the economic aspect which significantly arising from conversion of agricultural land, the loss of long-term social aspect will also take place such as: (1) decreasing in food production; (2) degradation in agro-ecosystems; (3) degradation in farming tradition and culture; (4) limited arable farming; and (5) declining in farmers welfare, resulted by [4].

The increasing land conversion into non-agricultural land such as the expansion of housing, shops, industrial estates, and roads have led in reducing paddy field area and will ultimately reduce the production of rice from Jawa Barat province. According to [2], the level of agricultural land conversion had resulted in reducing paddy fields area in Jawa Barat province to 20 thousand hectares from 2008 which originally accounted to 945 thousand hectares and significantly decreased to 925 thousand hectares in 2013.

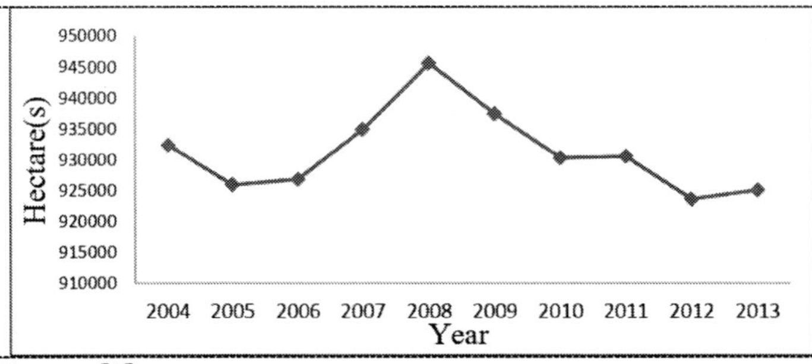

Source: [2]
Figure 3. Paddy Field Area in Jawa Barat province (ha) in 2004–2013

The increasing agricultural land conversion (paddy field area) into non-agricultural land has caused the reducing in rice production ultimately. This means a threat to national food security. Therefore, control of wetland conversion is an important effort that must be taken in order to support national food security. Based on the facts mentioned above, this study aims to answer the critical questions related to:

1. What is the pattern and development of agricultural land conversion into non-agricultural land in Jawa Barat province?
2. What are factors that affect the conversion of agricultural land into nonagricultural land in Jawa Barat province?

2. Methods

The observation of this study includes all districts in Jawa Barat province which consists of 20 districts namely: Bandung district, Bandung Barat district, Banjar district, Bekasi district, Bogor district, Ciamis district, Cianjur district, Cimahi district, Cirebon district, Depok district, Garut district, Indramayu district, Karawang district, Kuningan district, Majalengka district, Purwakarta district, Subang district, Sukabumi district, Sumedang district, and Tasikmalaya district. The data used in this study were obtained from the Central Bureau of Statistics (BPS). All type of the data used which is the combination of individuals (cross section) data and time series data used survey data from Potential Village Survey conducted by BPS in 2011 and 2014 in Jawa Barat province. The variable type used to answer the factors that affect the conversion of land at the village level is described in Table 1 below.

Table 1. Variables Type

Data	Unit
Micro, Small, and Medium Enterprises (MSME)	Unit
Convenient Stores	Unit
Hotels and Lodgings	Unit
DumTransportation	Transportation development 1=Exist; 0=Non
DumResidence	Residence development 1=Exist; 0=None
DumEconomy	Economy development 1=Exist; 0=None
Villages' Revenue (PAD)	Millions of Rupiah
Agricultural land conversion_non-agricultural lan	Percentage

The analysis methods of this study used descriptive statistics method and quantitative method. The using of descriptive statistics method is to draw the characteristics, pattern and the development of agricultural land conversion into non-agricultural land in Jawa Barat province. Meanwhile, the quantitative analysis method is panel data method. The process in selecting data observations into a dataset is used by eliminating a few data observations that not exist in village potentials data in 2011 but then exist in 2014, which accounted to 149 villages. Therefore, total observations in this study amount to 11.626 observations.

The purpose of using panel data method in this study is to analyze factors that affect agricultural land conversion into non-agricultural land. Panel data method is analysis method that uses time series data in 2011 and 2014. In stylized form, the estimated model of determinant factors of agricultural land conversion is written as follow:

$$Lc_{it}=\beta_0+\beta_1(MSME_{it})+\beta_2(\text{Convenient Stores}_{it})+\beta_3(\text{Hotels Lodgings}_{it})+\beta_4(\text{DumTransportation}_{it}) + \beta_5(\text{DumResidence}) + \beta_6(\text{DumEconomy}_{it}) + \beta_7(\text{Villages'sRevenue}_{it}) + \mu_{it} \quad (1)$$

Where:

β_0	= *intercept*,
MSME	= Total Micro, Small and Medium Enterprises (MSME) (unit),
Convenient Store	= Total Convenient Store (unit),
Hotels_Lodgings	= Total Hotels and Lodgings (unit),
DumTransportation	= Infrastructure Development in Transportation (dummy variable)
DumResidence	= Infrastructure Development in Residence, (dummy variable)
DumEconomy	= Development of Economy, (dummy variable)
Villages' Revenue (PAD)	= Total Villages' Revenue or PAD (Millions of Rupiah),
Land Conversion (Lc)	= Agricultural land conversion into non-agricultural land (percent),
β_n	= *slope* (n= 1, 2, ...),
μ	= *error*

The independent variables apply in the model has been modified to answer research questions follows some authors [5-8].

The operational definition of each variable in this study is described below:

a. Micro, Small and Medium Enterprise (MSME)
Total number of micro, small, and medium enterprises (workforce less than 20 workers) in the village according to main raw material which consist of leather industry (bags, shoes, slippers, etc.), wood industry (furniture and etc.), metal and metal materials industry (furniture and jewelry from metal, etc.), fabric and weaving industry (woven handicrafts, convection, etc.), food and beverage industry (processing and preservation of meat, fish, fruits, vegetables, oils and fats, milk and dairy foods, other food and beverage industry, and etc.), and other related industries.

b. Convenient Stores
The total number of convenient store (projected by self-service system, ability in selling various kinds of retail goods with given price tags, and floor area < 400 m^2).

c. Hotels and Lodgings
Hotels are interpreted as the total number of hotels which providing accommodation including a restaurant, and lodging certified license as a hotel while lodgings are interpreted as the total number of lodging area including hostel / motel / inn / guesthouse which providing accommodation and lodging with a business license.

d. DumTransportation
DumTransportation means all the programs and community development activities in the villages during the last 3 years including construction and the improvement of facilities and infrastructure in transportation (roads, bridges, etc.) in each village in Jawa Barat province.

e. DumResidence
DumResidence means all the programs and community development activities in the villages during the last 3 years including construction and the improvement of facilities and infrastructure in residence area, namely housing and health (sanitation, clean water, lighting, and posyandu) in each village in Jawa Barat province.
f. DumEconomy
DumEconomy means all the programs and community development activities in the villages during the last 3 years including construction and the improvement of facilities and infrastructure in terms of economy aspect (irrigation, markets, fish auction place (TPI), fish hauling harbor (PPI), trade facilities, and other facilities to support economy) in each village in Jawa Barat province.
g. Villages' Revenue (PAD)
Villages' revenue (PAD) means all the sources, forms and its value during 2010 and 2013: the revenue is measured in millions of Rupiah in each village in Jawa Barat province. The villages' revenue comes from village enterprises, wealth of the village (coming from village treasury, village markets and village buildings), the result of self-help and peoples' participation and also the result of mutual cooperation.
h. Conversion of agricultural land into nonagricultural land (Land Conversion)
The change in agricultural land (conversion) during the last year into non-agricultural land is in the form of percentage.

3. Results

3.1. Characteristics of Economic Development in Jawa Barat Province

Economic Characteristics of Jawa Barat province describes the development and economic growth in Jawa Barat province. The more rapid economic development of a region the more it indicates that the economic growth moves forward.

Table 2. Descriptive Statistics Variables Affecting the Conversion of Agricultural Land into Nonagricultural land in Jawa Barat province

Variable	Indicated Conversior	Standard Error
MSME (Unit)	37.331	92.406
Convenient Stores (Unit)	1.542	17.761
Hotels_Lodgings (Unit)	0.481	4.124
DumTransportation (1)	0.931	0.252
DumResidence (1)	0.686	0.463
DumEconomy (1)	0.430	0.495
Villages' Revenue (Millions Rupiah)	114.863	298.177

The highest number of MSME in 2011 was achieved by Gunungsari village in Cikatomas, Tasikmalaya which amounted up to 1,292 units. Fabric and weaving manufacturing (woven handicrafts, convection, etc.) was the most dominating industry in the area which amounted to 998 units. Meanwhile, the highest number of MSME in 2014 was achieved by Tanjungsari village in Rajadesa, Ciamis which amounted to 1,041 units. However, th woven industry (cane/bamboo equipment, grasses, rushes, papyrus, mats, bags, wall hangings, etc.) is the most dominating industry in the area which amounted to 998 units.

The highest number of convenient stores in 2011 was achieved by Cinunuk village in Wanaraja, Garut district; Singasari village in Singaparna, Tasikmalaya district; Sinagar village in Sukaratu, Tasikmalaya district; Sirnajaya village in Gununghalu, Bandung Barat district; and Pondok Petir village in Bojongsari, Depok district which contributed to 98 units. Meanwhile, in 2014, in Samudrajaya village, Tarumajaya, Bekasi district, total number of convenient stores amounted to 1,427 units. The high number of convenient stores in Bekasi happens not only because of the existing industrial area in this

location but also this location is a part of the mega Jabodetabek area. The urban development in Bekasi also gives an impact on the development of other facilities especially convenient stores in Bekasi.
The highest number of hotels and lodgings in 2011 and 2014 was achieved by Cisarua village, Bogor district with the total number of 109 units and 113 units respectively. Cisarua village which located in the peak tourist area has its own charming characteristics. Tourism development was also playing an important role in encouraging the growth of hotels and lodgings while at the same time can also have implications in the conversion of agricultural land into non-agricultural land.

Programs and community development activities basically can be grouped into: the development of transportation infrastructure, housing and health infrastructure, and economic infrastructure. Over the last 3 years, there were 10 831 villages that owned empowerment program for transportation infrastructure, 7981 villages owning housing and health infrastructure program and 5000 villages that received economic infrastructure program. Thus, within a period of 3 years, the majority of villages in Jawa Barat province had enjoyed government programs related to road and transportion infrastructure. With the opening of road accessibility will also affect the conversion of agricultural land as physical development of road infrastructure requires more land area.

The highest amount of villages' revenue (PAD) in 2011 was achieved by Cikalong village, Cimaung, district of Bandung. Meanwhile, in 2014, the Jelegong village in Cidolog, Ciamis district contributed to 9812 (in millions of Rupiah). The revenue comes from village enterprises, wealth of the village (coming from village treasury, village markets and village buildings), the result of self-help and peoples' participation and also the result of mutual cooperation. The average amount of PAD accounted to 115 thousand millions rupiah.

3.2. Pattern and Development of Agricultural Land Conversion into Non-Agricultural Land in Jawa Barat Province

Increasing land conversion is a serious threat to national food security. The increasing in agricultural land conversion has led in the decreasing of paddy field area every year. In the village potentials data, there were some indicated conversion of agricultural land into non-agricultural land in Jawa Barat province in 2011 and 2014.

Table 3 .The Conversion of Agricultural Land (Field Paddy Area) into Non-Agricultural Land in Jawa Barat Province in 2011 and 2014

Conversion to Non-Agricultural Land	2011 (villages)	2014 (villages)
Conversion Indicated	1.886 (32.44%)	1.769 (30.43%)
Conversion Unindicated	3.927 (67.56%)	4.044 (69.57%)
Total	5.813 (100%)	5.813 (100%)

Based on Table 3, the phenomenon of villages that suffered the conversion of agricultural land depicts a slight decreasing. As shown in 2011, 32 percent of villages in Jawa Barat province had been converted to non-agricultural land, while in 2014 it is shown that the percentage decreased to 30 percent. However, it is very mandatory be wary of this situation.

The general picture of land conversion in Jawa Barat province can be seen in figure 4. Some districts that had quite high increasing in land conversion from 2011 to 2014 were Kabupaten Bogor, Kabupaten Bekasi, Kabupaten Bandung, Kabupaten Cianjur, Kabupaten Garut and Kabupaten Cirebon. Particularly, the one that gaining serious concern is the district of Bekasi. In 2011, there were land conversion around 7 percent and rising to 11,2 percent in 2014. Within 3 years, the number increased and doubled. It certainly could not be separated from the fact that Bekasi district located in hinterland of the capital area of Jakarta as well as developing and expansion area of many industries. Not to mention, this situation also had further implications especially the decreasing of existing paddy field area.

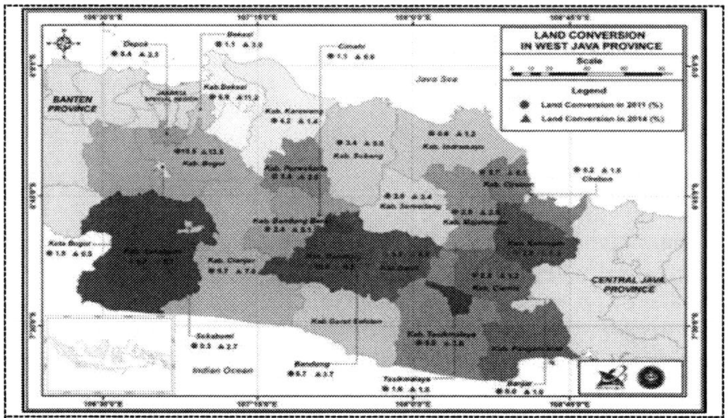

Figure 4. Total amount of Agricultural land (Paddy Field Area Converted into Non-Agricultural Land in Jawa Barat Province 2001 and 2014

3.3. Dynamic Changes in Agricultural Land Conversion into Non-Agricultural Land in Jawa Barat Province

The conversion of agricultural land into non-agricultural land in Jawa Barat province had changed in terms of conversion patterns in several villages in 2011 and 2014. A total of 1886 villages recorded to have converted their agricultural land into non-agricultural land in 2011. From those 1886 villages, it is then found that in 2014, around 893 villages (47%) happened to have converted agricultural land into non-agricultural land while the rest 993 villages had done no land conversion.

In 2011, 3927 villages did not experience any conversion of agricultural land into non-agricultural land. From those 3927 villages, a total number of 876 villages (22%) experienced the agricultural land conversion into non-agricultural land, while the other 3051 villages did not experience any of land conversion in 2014.

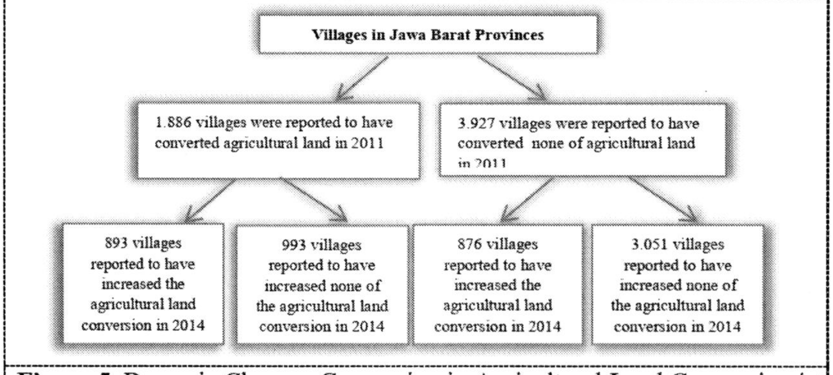

Figure 5. Dynamic Changes Conversion in Agricultural Land Conversion into Non-Agricultural Land in Jawa Barat Province

3.4. Analysis of Factors Affecting Land Conversion of Agricultural Land into Non-Agricultural Land

According to BPS (2015), the level of agricultural land conversion has caused reducing in paddy fields area in Jawa Barat province to 20 thousand hectares. The total area in 2008 was 945 thousand hectares, and later in 2013 it decreased to 925 thousand hectares. This also means a threat to national food security. Therefore, the analysis of the factors that affect the conversion of agricultural land into non-agricultural land is required in order to control the conversion of agricultural land.

The existence and development of convenient stores positively effect the conversion rate of agricultural land into non-agricultural land in Jawa Barat province and statistically significant at the 5 percent significance level. This indicates that an increasing of one unit convenient stores establishment resulting in an increasing of the level of agricultural land conversion into non-agricultural land in Jawa Barat province by 0.0002 percent. This result is also in line with the study by Jinyan *et al* (2010), which explains that the dominant factor affecting the conversion of agricultural land into non-agricultural land is the increasing of economic activities on non-farming industries. The increasing activity on retail trade is also portrayed by the more development and establishment of convenient stores in rural areas which implies the needs for land for convenient store construction. Later, this would put more pressure on the need for land.

Table 4. Analysis of factors affecting the conversion of agricultural land into non-agricultural land

Variable	Koef	Standard Error	t value	Prob
MSME	-7.16E-06	7.47E-06	0.957796	0.3382
Convenient Stores	0.000257***	9.10E-05	2.823262	0.0048
Hotels_Lodgings	0.000143*	8.51E-05	1.679298	0.0931
DumTransportation	-0.005604***	0.001355	-4.135869	0.0000
DumResidence	0.005929***	0.000661	8.975779	0.0000
DumEconomy	0.002653***	0.000295	8.979503	0.0000
Villages' Revenue (PAD)	-0.00000285***	6.05E-07	4.717230	0.0000
C	1.742871	0.000752	2318.479	0.0000

R-Square : 0,9869
Adj R-Square : 0.9826
Note: ***) Significant level at 1%; **) Significant level at 5%; *) Significant level at 10%

Hotels and lodgings positively effect the conversion rate of agricultural land into non-agricultural land in the in Jawa Barat province. persen. With the significant level of 10 percent, it indicates that an increasing in the construction for 1 unit of hotels and lodging resulting in an increasing of the level of agricultural land conversion into non-agricultural land in Jawa Barat province by 0.0001 percent.

From the estimation of factors that affect the conversion of agricultural land into non-agricultural land in in Jawa Barat province, the coefficient of Dummy Transportation is -0.0056. Dummy Transportion is worth 1 if there is a program or community development activity in the village during the last 3 years which includes construction or improvement of facilities and infrastructure such as roads, bridges, etc. in each village of the Jawa Barat province and worth 0, otherwise. This result implies that the difference in conversion of agricultural land into non-agricultural land between developed transportation and non-developed transportation is 0.0056 percent. When there is a construction for infrastructure, the conversion of agricultural land into non-agricultural land in in Jawa Barat province will decrease. This is in line with [8] which explains that access to market does negatively affect the agricultural land conversion. Better infrastructure access will increase the manpower and capital investment as it can increase the amount of output since input costs are relatively cheaper, results by [9]. Access to better infrastructure can also increase the high-value of agricultural production and the using of inputs to the fullest. Conversely, if there is an increase in production costs, it will affect the amount of production output, due to factors of production that will do better in better infrastructure, results by [10]. The better infrastructure transportation will reduce the conversion rate of agricultural land. This is as explained before that in terms of infrastructure development, the local government considers sustainable aspects of Nationals' Land Management Program (SLMP) which includes environmental aspects of sustainable development improvement on agricultural productivity through soil and water conservation in developing community infrastructure.

In addition, based on the estimation results of factors that affect the conversion of agricultural land into non-agricultural land in Jawa Barat province, the coefficient of Dummy Residence obtained is 0.0059. The Dummy Residence is worth 1 if there is a program or community development activity in the village during the last 3 years, which includes construction or improvement of facilities, and infrastructure in housing and health (such as sanitation, clean water, lighting, and posyandu) in each village of the Jawa Barat province and worth 0, otherwise. This result implies that the difference in conversion of agricultural land into non-agricultural land between developed residence infrastructure and non-developed residence infrastructure is 0.0059 percent. When there is a construction for residence area, the conversion of agricultural land into non-agricultural land in Jawa Barat province will increase. This is in line with [11], which explains that the dominant factor affecting land conversion when viewed from the landscape aspect is on the accessibility to the location area.

The estimation coefficient result of Dummy Economy is 0.0026. Dummy Economy is worth 1 if there is a program or community development activity in the village during the last 3 years which includes construction or improvement of facilities and infrastructure that generally supports the economy (such irrigation, markets, TPI / PPI, trade facilities, and other supporting economic tools) in each village in Jawa Barat province and is worth 0 otherwise. This result implies that the difference in conversion of agricultural land into non-agricultural land between the region that has active economic development and the one that does not have it is 0.0026 percent. When there is economic development, the conversion of agricultural land into non-agricultural land in Jawa Barat province will increase.

Villages' Revenue or PAD negatively affects the conversion rate of agricultural land into non-agricultural land in Jawa Barat province. The revenue itself comes from village enterprises, wealth of the village (coming from village treasury, village markets and village buildings), the result of self-help and peoples' participation and also the result of mutual cooperation. The result shows that any increase in PAD amount to 1 million rupiah, the rate of agriculture land conversion to non-agricultural land in Jawa Barat province decreases by 0.00000285 percent. This result is in line study by [7], which explains that the external factors that affect the conversion of agricultural land is that GDP negatively affects agricultural land conversion. In other words, if there is an increase in GDP in the agricultural sector, it will lead to a decrease in conversion of agricultural land. The high villages' revenue the more it will reduce the rate of agricultural land conversion.

4. Conclusion

The conclusions of this study are explained as written below:

1. The analysis results above of the increasing in agricultural land conversion into non-agricultural land in Jawa Barat province from 2011 to 2014 occurred in some districts such as: 2.62 percent in Bandung Barat district, 0.99 percent in Banjar district , 6.34 percent in Bekasi district, 1.25 percent in Cianjur district, 3.88 percent in Cirebon district, 0.93 percent in Garut district, 0.58 percent in Indramayu district, 0.39 percent in Majalengka, 1.16 percent in Purwakarta district and 0.57 percent in Sumedang district. The highest rate of agricultural land conversion into non-agricultural land in Jawa Barat province is reported in Bekasi district from 7.92 percent in the year 2011 and doubled to Rp 14.26 percent in 2014.
2. The analysis results using the panel data showing the variables that positively affect the rate of agricultural land conversion into non-agricultural land in Jawa Barat province are convenient stores, hotels and lodgings, Dummy Residence, and Dummy Economy. Meanwhile, variable Dummy Transportation and Villages' Revenue (PAD) showing negative effects. DummyRecidence is the variable that mostly affects the rate of agricultural land conversion into non-agricultural land in Jawa Barat province.

5. Suggestion

1. Economic factors that becoming proxy in agricultural land conversion explain the fact that economic development has some certain implications related to the conversion problem. In other hand, the government still should spur economic growth and develop the economic activities. The other related things which also should be concern is that economic development, the need for housing, industrial areas and trade will reduce the agricultural area. It is therefore necessary to establish a consistent policy areas in accordance with existing spatial allocation. The determination of perennial land in effort to ensure the agricultural production process should also

get more attention in its implementation. Industrial zones, residence area and trade are set and should be seen in the context of inter-regional connections.

2. Residence area is the variable that mostly affects the rate of agricultural land conversion into non-agricultural land in Jawa Barat province. Thus, when it comes to residence construction, it is suggested to use the concept of *vertical building* than *landed building*. It is suggested because the concept of *vertical building* will reduce the land use area.
3. The development of the region will affect the development of other areas. The phenomenon that occurred in Bekasi district is an important lesson noting that Bekasi district is also the hinterland of Jakarta. The land conversion Bekasi district is relatively high compared to other areas. Therefore, the government needs to pay attention to hinterland areas of Jakarta in order to prevent massive conversion of agricultural land to non-agricultural. However, coordination between regions becomes absolutely necessary for the survival of the agricultural land area.

Acknowledgement
The authors are sincerely grateful to the Ministry for Research Technology and Higher Education, Government of Indonesia for financial support this research under research grand for International Collaboration and Publication.

6. References

[1] BPS 2016 *Agricultural Census 2013* (Jakarta: Central Bureau of Statistics)
[2] BPS 2015 *The total area of Paddy Field Area by Province (Hectares) tahun 2003 dan 2013* (Jakarta: Central Bureau of Statistics)
[3] BPS 2016 *Production in Villages' Revenue (PAD) in Jawa Barat Province (Tonnes)* (Jakarta: Central Bureau of Statistics)
[4] Harini R, Nurdin M H and Aprilia L 2014 The Overview of Spatial Economic Analysis for Villages' Revenue (PAD) on Agriculture: The Case of Converted Agricultural Land in District of Sleman *Indonesian Journal of Geography* **44** 120 - 133
[5] Jinyan Z, Nana S, Shujin H E and Yingzhi L 2010 Factors and mechanism driving the land-use conversion in Jiangxi province *Journal of Geographical Sciences* **20** 525-539
[6] Quasem M D A 2011 *Conversion of agricultural land to nonagricultural uses in Bangladesh:extent and determinants* (Bangladesh Development Studies) **34**
[7] Harini R, Yunus H S, Kasto and Hartono S 2012 Agricultural land conversion: determinants and impact for food sufficiency in Sleman Regency *Indonesian Journal of Geography* **44** 120 – 133
[8] Girma H M and Hassan R M 2015 Drivers of land-use change in the Southern Nations, Nationalities and People's Region of Ethiopia *African Journal of Agriculture and Resource Economics* **9** 148-164
[9] Binswanger H and J Mcintire 1997 Behavioral and Material Determinants of Production relation in Land Abundant Tropical Agriculture *Economic Development and Cultural Change* **36** 73-99
[10] Pender J N, Ephraim J, Pamela,S, Dick S and Henry 2004 Strategies to increase agricultural productivity and reduce land degradation: evidence from Uganda *Agricultural Economics, Blackwell* **31** 181-195
[11] Benu N M, Maryunani, Sugiyanto and Kindangen P 2013 Analysis of land conversion and its impacts and strategies in managing them in City of Tomohon, Indonesia *Asian Transactions on Basic and Applied Sciences* **03** 65-72

Spatial modelling of land use/cover change (LUCC) in South Tangerang City, Banten

K Saifullah[1*], B Barus[2], and E Rustiadi [3]
[1] Graduate School of Regional Planning Studies, Bogor Agricultural University (IPB), Jl. Raya Darmaga, IPB Graduate School Building, IPB Darmaga Campus, Bogor, Indonesia
[2] Remote Sensing and Spatial Information Division, Department of Soil Sciences and Land Resources, Faculty of Agriculture, Bogor Agricultural University (IPB), Jl. Meranti, IPB Darmaga Campus, Bogor, Indonesia
[3] Regional Planning and Development Division, Department of Soil Sciences and Land Resources, Faculty of Agriculture, Bogor Agricultural University (IPB), Jl. Meranti, IPB Darmaga Campus, Bogor, Indonesia

E-mail : khalids240181@gmail.com

Abstract. Urban expansion in peri-urban region remains a major issue for regional planning and development, as it converts most of the land cover into built-up settlements rapidly and extensively beyond its spatial allocation plan (RTRW). South Tangerang City is one of municipalities in Jabodetabek with the most rapid annual population growth rate reached 6.87%. LUCC analysis from three LANDSAT land cover maps (1990, 2002, and 2014) shows that the built-up area has increased 8 times (8650 ha) in 1990-2014 with the average annual growth rate of 10.83% per sub-district (kelurahan). The most extensive land conversion type in 1990-2002 was vegetated land to open land (3605ha), while in 2002-2014 land conversion type was dominated by open land to built-up (3446ha). In general, the built-up area expansion in 1990-2002 shows irregular ribbon development pattern in district Serpong Utara, Pondok Aren, Ciputat Timur, Ciputat, and Pamulang, while district Serpong tends to represent leap frog development pattern. The expansion in 2002-2014 shows infill and ribbon development pattern continuation across the city. MLP was used to model LUCC sensitivity prediction. The model performance shows 73.16% accuracy with ROC validation of prediction output reaches 0.804 for 2014, which is qualified to predict LUCC sensitivity in 2032.

1. Introduction

South Tangerang City as a proliferated municipality from Tangerang Regency in 2008 can serve as an example of urban expansion phenomenon, which has been particularly triggered by peri-urbanization and regional agglomeration between Jabodetabek Metropolitan Region and Bandung Metropolitan Region to form a larger mega urban region called Jakarta-Bandung Region (JBR). According to Firman and Dharmapatni [1] and Firman [2], the development of JBR has been significantly affected by large scale land development projects of private new towns and industrial centers that are connected by inter-region highway (toll roads) infrastructure network. These major transportation corridors have stretched even further from the western part of Banten Province (Merak) to the eastern part of Bandung Regency (Cileunyi) and northern coast of Java regencies (i.e., Palimanan-Kanci, Cirebon Regency) as part of Trans Java toll roads development mega project. KEMENPUPR [3] statistics showed that there are already 544.58 km total lengths of active connected toll roads network that link Jabodetabek region to its surrounding outer suburbs. The JBR agglomeration has also been intensified with the construction of a 142 km long high speed railway line connecting the two regions [4]. The project is considered to cause a significant amount of land conversion not only for its infrastructure, but also for the urban settlements, new towns and other property sectors that will be attracted in along with the new railway as similar to induced ribbon development along the toll roads in Bodetabek regions. One of the studies that found this effect of main roads on urban expansion in Jabodetabek was conducted by Nagasawa and Fukushima [5], who statistically analyzed land cover category along main roads using multiple ring-buffer distance parameters.

The rapid and extensive growth of urban expansion in JBR have converted most of primary agricultural land, vegetated land (forest, plantation), dry land, and green open space into built-up areas as an environment and development trade-off. Studies on environmental and socioeconomic impacts of land conversion or land use/cover change (LUCC), its driving force factors and policy related subjects in JBR have generally evolved since 1990's. Firman and Dharmapatni [1] has reported the negative impact of land conversion on water resources (surface and ground water), air quality, and to some extent on soil erosion. Pravitasari *et al.* [6] and Rustiadi *et al.* [7] have analyzed the increasing of environmental degradation and anthropogenic disasters such as flood and landslides in the Jabodetabek region in terms of quantity, intensity and distribution. In addition, Firman [2, 8-11] have continually studied and updated the process and implications of land conversion for urban and regional development; the domestic and foreign direct investment driving factor; the speculative land-business behavior of large private developers; the excessive government land development permits; the social segregation and inequality; and the regional disparity and privatization phenomena as indirect impacts of rural to urban land conversion in the fringe of large cities, especially in JBR. These broad perspectives of studies have brought us to believe that land conversion / LUCC related topics continue to be prominent issues in response to solve the future challenges on environmental sustainability, food security, and socio-economic stability in urban planning and development.

In general, the Jabodetabek built-up area has increased 31 times (± 195,000 ha) in 1972-2012. Forest area was falling significantly with approximately 71 % or 152,000 ha; and the dry land has also reduced by 31 % or 71,000 ha during the same period [7]. In another LUCC study with different temporal high resolution satellite images by Nagasawa and Fukushima [5], the expansion of urban built-up area in Jabodetabek has increased by 11% (35,000 ha) in the 20 years between 1989 and 2009. In contrast, the vegetated area has decreased by approximately 12% through the period. Nagasawa and Fukushima [5] study also shows that the most significant land conversion pattern of urban built-up area in Bodetabek region occurred in Tangerang Regency, which is the parent region of South Tangerang before proliferation in 2008. The expansion of the urban built-up area in this city in 1989, 2001, and 2009 have respectively reached ≈ 7,000 ha, 11,000 ha, and 17,000 ha exceeding other municipalities and regencies in Bodetabek region.

South Tangerang has become one of municipalities in Jabodetabek with the most rapid annual population growth rate reached 6.87%. Based on land use map of 2011 in South Tangerang spatial planning documents (RTRW 2011-2031), this city has been dominated by settlement land use type as large as 9.095 ha (61,79% of the total area). This number exceeds the allocation area of housing (horizontal and vertical housing) in RTRW, that is only as large as 7.610,67 ha.

Spatial studies on urban expansion that utilize GIS (Geographic Information Systems) and remote sensing technology have been summarized by Bhatta [12], including the use of LUCC modelling with transition matrices, spatial metrics, and spatial statistics. These approaches now are not only capable to measure, analyze, and map the spatial pattern of LUCC, but also to model the relationship between LUCC and its driving factors. The model can then be used to simulate / predict the future LUCC by the inductive pattern-based approach that utilizes the transitional spatial pattern of the observed LUCC; by the deductive theory / expert knowledge-based approach that relies on the explicit descriptions of the process; by the emulation of individual decision makers / agent-based behavior; or by the hybrid combinations approach [13][14]. The spatial modelling of LUCC in this study belongs to inductive pattern-based approach that employs one of the machine learning algorithms (i.e., Artificial

Neural Network) called Multi Layer Perceptron (MLP) in LCM (Land Change Modeler) IDRISI platform. According to Eastman [15], MLP as a non-parametric statistic has the ability to model various forms of relationship between dependent and independent variables that are linear or complex non-linear without considering any form of data distribution (distribution free).

In general, the LUCC spatial modelling objectives are: 1). to measure the LUCC and explain its dynamics; 2). to identify the spatial pattern of LUCC and urban expansion rate; 3). to model the spatial relationship between LUCC and its driving factors; and 4). to predict the LUCC sensitivity in the future.

2. The LUCC Dynamics in South Tangerang

The first stage of LUCC modelling in LCM IDRISI is change analysis or change detection stage using post-classification comparison method [12]. In post-classification comparison, each date of rectified imagery is independently classified to fit a common land type schema. LCM requires equal number and type of land-cover classes in an exact order for comparing the land cover maps. This study uses 3 satellite images including the following: Landsat 5 TM (1990), Landsat 7 ETM (2002), and Landsat 8 (2014) derived from USGS (*United States Geological Survey*) at coordinate path 122 and row 64. All of the images were classified using combination techniques between NDVI (*Normalized Difference Vegetation Index*) and manual editing with the help of Google Earth high resolution images and ground truth survey to correct the NDVI classification errors in ArcMap. These techniques have divided land cover into 4 classes: 1). Built-up area; 2). Vegetated area; 3). Open / bare area; and 4). Water body (rivers and lakes). The resulting land cover maps are then overlaid and compared on a pixel-by-pixel basis in Change Analysis Tab in LCM.

The LUCC analysis in this study has been divided into two periods: 1990-2002 (LUCC I), and 2002-2014 (LUCC II). This division is designed not only to match LCM validation scheme recommended by Eastman [15], but also to understand the dynamics and the spatial process of LUCC in South Tangerang across all of the periods. The figure 1 and 2 below are the result of the first stage containing LUCC quantitative graphics (i.e., gains and losses, net change, and contributors to net change).

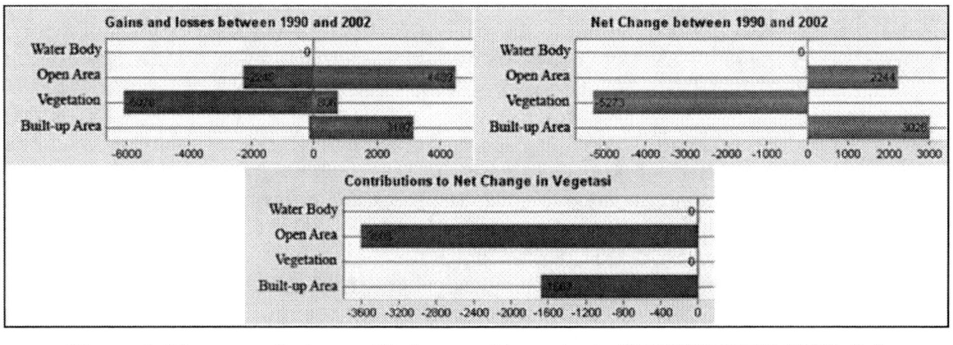

Figure 1. Change analysis quantitative graphics output of LUCC I (1990-2002) (ha)

The result of the overall LUCC analyses in 1990-2014 show that vegetated area is the most extensively converted land cover type with 56.01% of the total area reduction. In contrast, the built-up area is the most developed land cover type with 52.28% of the total area increase. In general, the expansion of the built-up area in South Tangerang between 1990 and 2014 has reached 8 times (8650 ha). According to Figure 1, the most extensive land conversion type in 1990-2002 was vegetated area to open area (3605 ha), followed by vegetated area to built-up area (1667 ha); while the land conversion type in 2002-2014 (Figure 2) was dominated by open area to built-up area (3446 ha), followed by vegetated area to built-up area (2176 ha). The water body in this study was classified to be constant since it is a protected area with insignificant area change that cannot be detected by Landsat spatial resolution.

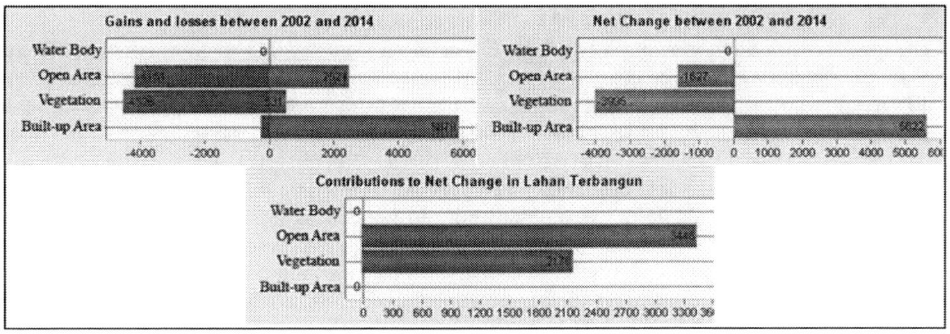

Figure 2. Change analysis quantitative graphics output of LUCC II (2002-2014) (ha)

The alteration of the dominant land conversion type between 1990-2002 and 2002-2014 has explained the spatial process dynamics of South Tangerang LUCC that are linearly related to the large scale land acquisition and subdivision projects to develop new towns and their supporting infrastructure in Jabodetabek region at the property boom bust period during mid-1980s until the economic crisis in 1997. There are many open / bare areas (i.e., ready to build areas) acquired by private developers that were left opened and undeveloped since the economic crisis collapsed the property sector. This condition was supported by Winarso and Firman [16] and Firman [8] studies that have noted the increase of urban land demand; the excessive land development and property sector projects that had been part of triggering economic crisis; and the expansive land conversion during the boom bust periods in Jabodetabek regencies and municipalities. It is particularly interesting to note that the amount of infertile (idle) land in the area increased by 1600 %, which partly indicated the increasing abandonment of land to speculation, acquisition and subdivision projects by the private developers. The disproportion between the proposed area of land development permits (i.e., location permits) and the actual realization of land development and housing production has promoted the increase of the unutilized open areas that were inefficient. At the mid-1980s to 1995, BPN had issued 345 land development permits as large as 81,200 ha, with the annual average reached 8000 ha/year, but then only 2/5 of the total area in the permits that were actually developed in Bodetabek region, where the remaining portion had become idle / unutilized land [8].

In the case study of South Tangerang, Winarso [17] has compiled 60 land development projects for new town in Tangerang Regency and municipality since 1980's, where 24 of them were developed in South Tangerang. There are 3 largest new town projects in South Tangerang, including the following: BSD City (6000 ha) in Serpong and Setu District; Bintaro Jaya (2300 ha) in Pondok Aren District; and Alam Sutera (700 ha) in Serpong Utara District. These new self sustained towns have attracted migrants from inside and outside Jabodetabek region [18], and become pull factors for other commercial, industrial, property, and service sectors to thrive around the regions. The districts within the new towns together with densely populated district in Ciputat Timur, Ciputat, and Pamulang have shaped *Cipasera* region as the origins of South Tangerang municipality.

The dominant land conversion type in 2002-2014 is largely attributed to the spatial process resumption of land development projects for property sector that had been halted for several years due to prolonged monetary crisis in 1997. The pattern of urban built-up area development in 2002-2014 tends to utilize the available open areas / idle areas that had been acquired by developer groups over the previous years. According to BSD [19] annual report from Sinar Mas Land developer group, the undeveloped land in BSD City new town at the end of 2015 have reached 2307 ha from the total areas (6000 ha) that have been acquired by the company since 1984. This kind of extensive oligopolistic land tenure is certainly contradictory to KEMENATR [20] regulation number 5/2015 that regulates the maximum area limit of the land tenure location permits hold by a residential developer company (i.e., 400 ha for 1 province and 4000 ha for the whole of Indonesia).

3. The Spatial Pattern of LUCC and Urban Expansion Rate

The spatial distribution mapping of each LUCC resulted from the first section will be the second stage of LUCC modelling. The change maps panel in Change Analysis Tab in LCM IDRISI is able to create different kind of maps, including the following: changes map; persistence map; gains and losses map; and transitions or exchanges map. However, this section will focus on the spatial trend of change panel (i.e., the last panel in Change Analysis Tab) to directly identify the actual spatial pattern of each dominant land conversion that has taken place in South Tangerang in 1990-2002 and 2002-2014. This panel works by the principle of polynomial order to generalize the spatial pattern / trend of LUCC between the two land cover maps. According to Eastman [15], the spatial trend of change panel in LCM is developed based on Trend Surface Analysis (TSA) as in the TREND module in IDRISI. This panel calculates 1 to 9 order trend surface polynomial equations for spatial data sets, and interpolates surfaces based on those equations. The generic equation for the polynomials fitted by TREND is:

$$Z = \sum_{i=0}^{k} \quad \sum_{j=0}^{i} b_{ij} x^{i-j} y^{j} \qquad (1)$$

In this equation, k is the maximum order to be fitted; b is a coefficient of the polynomial equation; both i and j are iteration variables associated to k, in which $i = 0,...k$ and $j = 0,...i$. For example, to fit a cubic trend $k = 3$, the equation is:

$$Z = b_{00} + b_{10}x + b_{11}y + b_{20}x^2 + b_{21}xy + b_{22}y^2 + b_{30}x^3 + b_{31}x^2y + b_{32}xy^2 + b_{33}y^3 \qquad (2)$$

The outputs of spatial trend of change panel in Figure 3, 4, and 5 are raster surface maps with numeric values that do not have any special significance nor measurement unit, other than to provide an indication of where the change was more intense (higher numbers) or less intense (lower numbers).

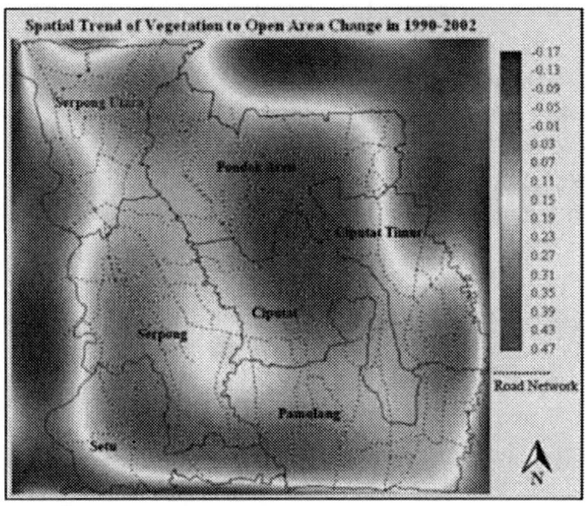

Figure 3. Spatial trend of vegetation to open area change in 1990-2002 (9th order trend).

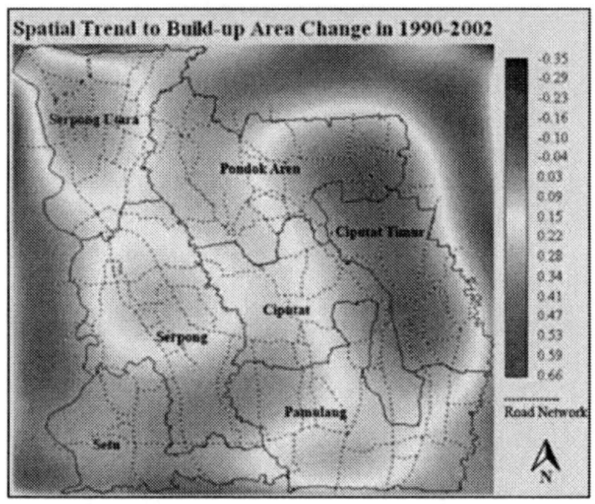

Figure 4. Spatial trend to built-up area change in 1990-2002 (9th order trend)

The mapping result in Figure 3 identifies the spatial trend of vegetation to open area change as the most extensive land conversion type in 1990-2002. The spatial trend pattern in Figure 3 also implies the locations of land acquisition, land clearing, and land subdivision with moderate to high intensity prior to the built-up area construction in all districts. The highest intensity of this transition occurred in Pondok Aren District (Jurang Mangu Barat, Pondok Jaya, and Pondok Pucung Sub-district); Ciputat District (Sawah Lama and Sawah Baru Sub-district), and Ciputat Timur District (Pondok Ranji Sub-district). The transition intensity in these regions showed radial / concentric spatial pattern that were centered in Pondok Jaya Sub-district, Pondok Aren District. This spatial concentration is matched to the actual condition of the land development performed by developer company (PT. Jaya Real Property) in preparing Bintaro Jaya New Town (sector 1 to 9) as large as ± 2300 ha since 1979. The developer has also designed the development of Bintaro X Change Mall CBD (Central Bussiness District) Superblock in part of Pondok Jaya Sub-district that was opened in 2013. The development of Bintaro CBD as one of economic strategic area in South Tangerang has been supported by the construction of JORR (Jakarta Outer Ring Road) network progressively since 1990s together with the commuter line railway network and stations that connect BSD City – Bintaro Jaya – Jakarta City.

The transition to open area in Setu District (Muncul, Kademangan, Setu, Babakan, and Bhakti Jaya Sub-district), the western part of Pamulang District (Pondok Benda and Pamulang Barat Sub-district) and the southern part of Serpong District (Serpong and Buaran Sub-district) have been influenced by the development of Taman Tekno Industrial Park in BSD City and PUSPIPTEK (Center for Research, Science and Technology) that were also induce the housing development for the workers in the surrounding area. PUSPIPTEK have been developed since 1976 with total investment of more than 500 million USD and covering 460 ha total area that contains 47 research, development and testing centers under non-ministerial government institutions, including LIPI, BPPT and BATAN [21]. The spatial trend concentration of vegetated area to open area transition also occurred from Serpong Utara District (Pakujaya and Pondok Jagung Timur Sub-district) to the western part of Pondok Aren District (Pondok Kacang Sub-district) which is the area of Kebayoran Regency (Graha Raya) housing development as large as 400 ha. This housing area is also directly adjacent to Alam Sutera new town development in Serpong Utara District that has been developed and expanded now to more than 800 ha since 1990s. In the southeastern part of South Tangerang, the transition to open area intensity in Pondok Cabe Ilir Sub-district (Pamulang District) has been triggered by the opening of Pondok Cabe airport by Pertamina in 1972 which attracted the urban settlement development in its surroundings, including Pisangan and Cirendeu Sub-district (Ciputat Timur District). These districts are also crossed by the main street network connecting South Tangerang with adjacent municipalities (i.e., South Jakarta and Depok) which in turn become densely populated districts.

According to Figure 4, the spatial pattern of urban built-up area expansion in 1990-2002 has distributed irregularly with the spatial concentration trend in some districts including the following: Serpong Utara, Serpong, Pondok Aren, Ciputat Timur, Ciputat, and Pamulang. The urban built-up area expansion in those districts except for Serpong, tend to represent ribbon development pattern which is linear to the main street network and the distance to the nearest regional service centers, while the urban built-up area expansion in Serpong District tend to show leap frog development pattern that is relatively detached (isolated) from other adjacent regions. This leap frog development was largely influenced by the development of BSD City since 1980s that was designed to become new self sustained city and economic center with powerful infrastructure and numerous employment and urban service centers.

To explain the leap frog development pattern in BSD City, Winarso [17] has studied the residential land developer behavior in selecting land for the location of their residential projects including in South Tangerang. The result of his study found that the low-priced land was a stronger determinant of developers' locational decisions than access to main roads. There is possibility of a higher profit margin if the development is located farther and isolated from Jakarta, where land prices are cheaper. Low priced land enables developers to achieve economies of scale. By acquiring low-cost land, developers were able to construct on-site major infrastructure and service facilities, including streets and roads, and even a toll road connecting their projects with Jakarta. With the construction of streets and roads developers boosted the land price and enjoyed not only the value added beyond costs to the land's selling price due to the possibilities for urban use after the installation of basic infrastructure, but also the value gained beyond building and site service costs in the selling price of a house caused by the demand for urban housing with the amenities provided by these developers [17].

The urban expansion in Serpong Utara in 1990-2002 (Figure 4) has been primarily triggered by Jakarta-Tangerang-Merak toll roads. It has also been affected by the urban development of adjacent Tangerang municipality, Alam Sutera, BSD City and Gading Serpong New Town in Tangerang Regency since 1990s. While, the urban expansion in the eastern part of South Tangerang, Pondok Aren, Ciputat Timur, Ciputat and Pamulang District were largely influenced by South Jakarta City; the development of Bintaro Jaya New Town in Pondok Aren District; the national housing project (Perumnas KPR BTN); and the government employees housing in the remaining districts.

The spatial pattern of built-up area expansion in 2002-2014 in Figure 5 has generally showed infill development pattern continuation across the city that utilizes the undeveloped open areas as well as intensifies the densification of the available built-up areas (i.e., vertical development). The ribbon development pattern of built-up area expansion have also still occurred in the intersection area, along the arterial and collector roads as well as JORR toll network that connect between the urban service centers. The development of JORR that built across Pondok Aren, Ciputat Timur, Ciputat and Serpong have played important roles in improving urban accessibility and accelerating the distribution of the built-up area expansion in regions around South Tangerang. Figure 5 shows the dense spatial concentrations in several districts comprising Serpong Utara (Pakualam, Pakujaya, Pakulonan, and Pondok Jagung Timur); Pondok Aren (Pondok Jaya, Pondok Aren, Jurang Mangu Barat, and Pondok Pucung); Ciputat (Serua Indah, Serua, Sawah Baru, Sawah Lama, and Jombang); Pamulang (Pondok Benda and Pamulang Barat); and Setu (Setu). The urban expansion in 2002-2014 has developed a government center strategic area located in Ciputat District and 3 economic strategic areas that serve as CBDs, including BSD City (Serpong), Living World Alam Sutera (Serpong Utara), and Bintaro Jaya Xchange (Pondok Aren).

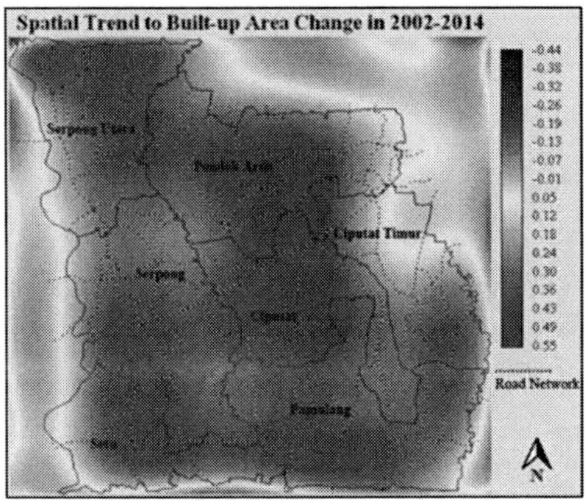

Figure 5. Spatial trend to built-up area change in 2002-2014 (9th order trend)

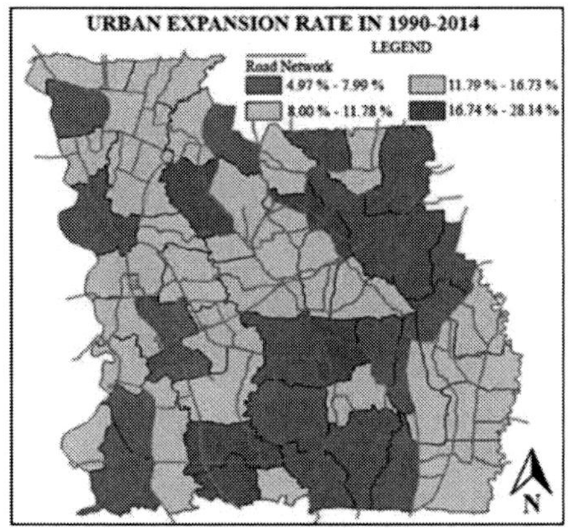

Figure 6. Urban expansion rate of each sub-district in 1990-2014

This study also uses choropleth mapping technique in ArcMap to map the urban expansion rate aggregated over each sub-district by adding a new built-up area growth rate field; inputting its data into the attribute table of sub-district polygon; and classifying the quantities of the growth rate field based on natural breaks. Prior the choropleth mapping, the analysis of built-up area growth rate were calculated based on exponential growth model with the assumption of large undeveloped potential areas acquired by developers and the fact that the growth rate of built-up area is still constantly increasing each year in South Tangerang. According to the urban expansion rate map of South Tangerang in 1990-2014 (Figure 6), Pondok Jaya (Pondok Aren), Babakan (Setu), Serua Indah (Ciputat), Parigi Baru (Pondok Aren), and Buaran (Serpong) are the top 5 sub-districts with the highest growth rate. While, Muncul (Setu), Pamulang Barat (Pamulang), Ciputat (Ciputat), Serua (Ciputat), and Rempoa (Ciputat Timur) are the top 5 sub-districts with the lowest growth rate. The annual growth rate average of each sub-district is 10.83%. To identify the spatial pattern of urban expansion rate objectively, this study employs Global Moran's I Spatial Autocorrelation analysis in ArcMap's spatial statistics toolbox. The Global Moran's I is capable to measure whether the spatial pattern of the urban expansion rate is distributed at random, dispersed, or clustered pattern based on statistical significance test using *z-score* and *p-value*. The result concludes that the urban expansion rate in 1990-2014 is distributed in random spatial pattern since the statistical test showed insignificant result (i.e., *Moran's Index* = 0.088; *z-score* = 1.32; *p-value* = 0.186).

4. The *Multi Layer Perceptron* (MLP) Modelling and The Prediction of LUCC Sensitivity

In order to predict sensitivity of LUCC in the future, the transition potential map of LUCC have to be created in LCM Transition Potential Tab. This study only used transitions to built-up area (i.e., vegetation to built-up area, and open area to built-up area transition) since the main concern of this study is the urban built-up area expansion. Hence, the actual final output of this study is the sensitivity prediction of built-up area transition from the urban expansion. There are 3 model options in Transition Potential Tab, including MLP, Sim Weight, and Logistic Regression. Eastman [15] recommends MLP because of its capability to run multiple transitions (up to 9) of each sub-model at once.

MLP as a non-parametric regression analysis also has both a dependent and independent (predictor) variables. The dependent variable used in MLP is all of the built-up area transitions in 1990-2002 and 2002-2014 that were reclassified into binomial classes (i.e., No change: Y=0; Change: Y=1). While, the independent variables are the driving factors of the urban expansion in South Tangerang including the following: 1). Population density (head/km^2) per sub-district in 2003 and 2014 (X_1); 2). Agriculture household per district in 2003 and 2013 (X_2); 3). Number of IMB (built-up area construction permits) per sub-district in 2014-2015 (X_3); 4). Total value of domestic and foreign investment (PMDN-PMA) per sub-district in 2014 (X_4); 5). Number of companies based on SIUP and TDP per sub-district in 2014-2015 (X_5); 6). Number of vehicles per sub-district in 2015 (X_6); 7). Regional development (hierarchy) index value per sub-district in 2003 and 2014 (X_7); 8). Land value zone in 2014 (X_8); 9). Distance to built-up area in 1990 and 2002 (X_9); 10). Vegetated area in 1990 and 2002 (X_{10}); 11). Open area in 1990 and 2002 (X_{11}); 12). Water body area (X_{12}); and 13). Distance to road network (X_{13}).

This set of independent variables was chosen based on preliminary investigations over the case study as well as academic literature study. All of the variables in the model had to be converted to raster format at 30 m resolution and at equal spatial extend (columns and rows) in the same coordinate reference system. In accordance with Arsanjani [13], all of the independent variables used in MLP were already calibrated and refined individually with IDRISI logistic regression test to pick the optimal variable set that had to reach the highest ROC (Relative operating characteristic) statistics value (0,5-1). ROC = 1 indicates a perfect fit, while ROC = 0.5 indicates a random fit and values in between have a degree of membership. The logistic regression with the sampling rate applications (i.e., random or systematic sampling) of each independent variable was also functioned to reduce the impact of spatial dependency (spatial autocorrelation) between observations [15]. This study also employed correlation test between the independent variables in order to measure the linear relationship between each variable and to check for severe multi-collinearity prior to running the MLP transition model. Table 2 below is the result of MLP transition model with 13 independent variables combination.

Table 2. The Result of MLP with backwards Stepwise constant forcing

Model	Variables included	Accuracy (%)	Skill measure
With all variables	All variables	73.16	0.6421
Step 1: var.[4] constant	[1,2,3,5,6,7,8,9,10,11,12,13]	73.23	0.6431
Step 2: var.[4,12] constant	[1,2,3,5,6,7,8,9,10,11,13]	73.23	0.6431
Step 3: var.[4,12,3] constant	[1,2,5,6,7,8,9,10,11,13]	72.97	0.6396
Step 4: var.[4,12,3,6] constant	[1,2,5,7,8,9,10,11,13]	72.61	0.6348
Step 5: var.[4,12,3,6,5] constant	[1,2,7,8,9,10,11,13]	72.20	0.6294
Step 6: var.[4,12,3,6,5,2] constant	[1,7,8,9,10,11,13]	71.76	0.6235
Step 7: var.[4,12,3,6,5,2,13] constant	[1,7,8,9,10,11]	71.44	0.6192
Step 8: var.[4,12,3,6,5,2,13,7] constant	[1,8,9,10,11]	70.07	0.6009
Step 9: var.[4,12,3,6,5,2,13,7,10] constant	[1,8,9,11]	68.89	0.5852
Step 10: var.[4,12,3,6,5,2,13,7,10,1] constant	[8,9,11]	66.49	0.5532
Step 11: var.[4,12,3,6,5,2,13,7,10,1,8] constant	[9,11]	62.78	0.5037
Step 12: var.[4,12,3,6,5,2,13,7,10,1,8,9] constant	[11]	49.94	0.3325

The backward stepwise MLP result in Table 1 was used to assess the best model combination of independent variables based on accuracy rate (%) and skill measure by consecutively eliminating the weakest independent variable one by one. The result in Table 1 and the rerunning MLP test prove that the 13 independent variable combination to be the best model combination with 73.16% of accuracy rate and 0.6421 of skill measure. MLP rerunning test with the elimination of variable 4 and 12 to confirm the higher accuracy rate of the two variables have proved to be imprecise and outperformed by the 13 independent variable result. This combination was then used to project sensitivity of urban built-up area expansion in 2014 and 2032 as shown in the Figure 7 and 8, by activating the create soft prediction option and running the model in the Change Allocation panel in LCM Change Prediction Tab.

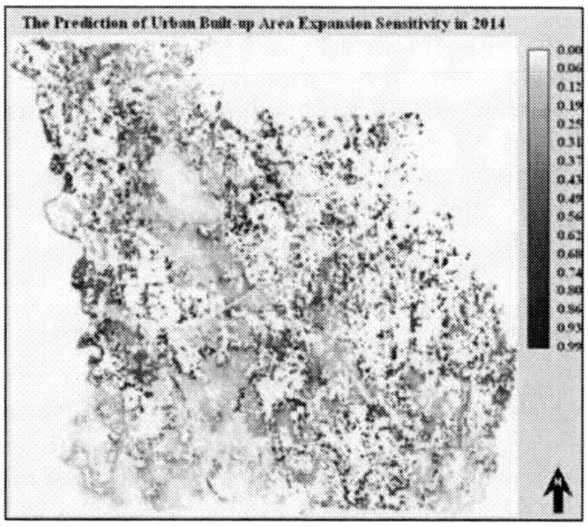

Figure 7. The Prediction of Urban Built-up Area Expansion Sensitivity in 2014

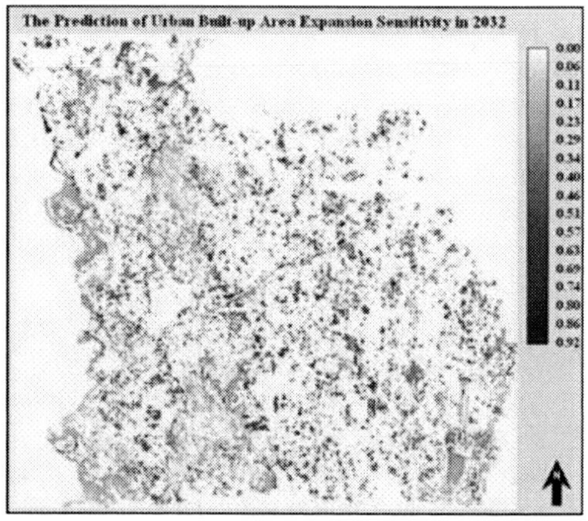

Figure 8. The Prediction of Urban Built-up Area Expansion Sensitivity in 2032

The sensitivity prediction map in 2014 (Figure 7) were validated with the actual transition to built-up areas map in 2002-2014 as the reference image in ROC module to assess its prediction accuracy. The validation result showed a ROC value of 0.804 that was considered as a good accuracy (i.e., ROC = 1 as a perfect fit). Hence, the MLP model was qualified to predict the sensitivity of urban built-up area transition in 2032 as shown in Figure 8. A raster cell with the probability value close to 1 (dark red) is more vulnerable to built-up areas transition than one that close to 0 (white cream). Zero probability of a raster cell inside the boundary means that the raster cell were already developed an intense built-up area.

5. Conclusions

In general, the paper is part of thesis research project on the topic of LUCC, urban expansion, and urban sprawl phenomenon in South Tangerang as part of Jabodetabek Mega Urban Region. This paper serves as introductory research on providing general measurement and analysis by spatial simulation models to better understand the LUCC as a broad phenomenon and the urban expansion as a more specific issue.

The result findings in LUCC dynamics provide the basic information on how large the urban expansion during the Jabodetabek's land development boom bust period to the post economic crisis affects the land cover and penetrates into South Tangerang interior region. The alteration of the dominant vegetated area to open area transition in 1990-2002 (LUCC I) and the open area to built-up area transition in 2002-2014 (LUCC II) have confirmed the dynamics of the excessive land development in the peri-urban regions of Jakarta especially in South Tangerang that were intensively studied by Winarso and Firman [16], Winarso [17], and Winarso *et al.* [18]. To account for the scale of urban expansion in South Tangerang, the change analysis in this study have found 8 times increase (8650 ha) of the built-up area in 1990-2014. The spatial pattern mapping of each LUCC have also confirmed the locations and the influences of large developer company's new town and housing projects together with their infrastructures that progressively accelerated and intensified the whole urban development in South Tangerang. The developer preferences on constructing the new large residential projects and infrastructure in the farther and isolated peri-urban region to gain super-normal profit [17] were captured in the spatial trend of change maps.

The spatial trend of vegetation to open area change map (Figure 3) has showed the land acquisition, land clearing, and land subdivision of the large land development projects in 1990-2002 with moderate to high change intensity especially in Pondok Aren, Ciputat, and Ciputat Timur District. It showed radial /concentric spatial pattern that were centered in Pondok Jaya Sub-district as a CBD of Bintaro Jaya new town. The spatial trend to built-up area change map in Figure 4 has also showed the leap frog urban development pattern in 1990-2002 that occurred relatively in Serpong (BSD City) and the inner part of Pamulang District; and the irregular ribbon development patterns that were also found in Serpong Utara, Pondok Aren, Ciputat Timur, Ciputat, and Pamulang district. The spatial trend to built-up area change in 2002-2014 showed infill and ribbon development pattern continuation across the city. In this study, the aggregated urban expansion rate of each sub-district in 1990-2014 was also mapped and spatially analyzed by the Global Moran's I Spatial Autocorrelation analysis to conclude the general spatial pattern of urban expansion rate in South Tangerang. The annual urban expansion rate average of each sub-district is 10.83% and the result of the Global Moran's I showed a random / irregular spatial pattern.

The last result finding to conclude is the MLP LUCC model with 13 independent variables that showed the best model performance with 73.16% of accuracy rate and 0.6421 of skill measure. This MLP model combination then was used to produce the sensitivity prediction map in 2014. The output map then was validated with the actual built-up transition map to assess its accuracy by ROC statistics. The result showed a ROC value of 0.804 that was considered as a good accuracy and qualified to predict the future urban built-up area transition sensitivity map in 2032.

As mentioned above, all of the modelling stages in this study are meant to provide basic information to be elaborated and explored in the future researches. The sensitivity prediction of urban built-up areas map in 2032 can be aggregated into administrative units to draw attention and prioritization for supporting the regional plan, policy and decision making. The sensitivity map can also be overlaid with the existing spatial plan to evaluate its consistency. The single use of MLP model should be validated with other LUCC models to understand the model behavior and compare their advantages and disadvantages in understanding the LUCC and urban expansion phenomena. Our next research will cover the logistic regression as another model option in LCM. We will also compare these raster based models in LCM with exploratory OLS and GWR regression as vector based models to assess the model performance, the independent variables selection process, and the final prediction results.

References

[1] Firman T and Dharmapatni I A I 1994 The challenges to sustainable development in Jakarta metropolitan region *Habitat International* (Elsevier Science Ltd) **18** 3 p 79-94

[2] Firman T 2009 The continuity and change in mega-urbanization in Indonesia: A survey of Jakarta - Bandung Region (JBR) development *Habitat International* (Elsevier Ltd) **33** p 327-339

[3] KEMENPUPR 2015 *The Book of Public Works and Housing Infrastructure Statistics Information*. (Jakarta: PUSDATIN-KEMENPUPR (Ministry of Public Works and Housing of the Republic of Indonesia)) chapter 4 p 11-17

[4] Salim W and Negara S D 2016 Why is the High-Speed Rail Project so Important to Indonesia. *Perspective* (Singapore, ISEAS-Yusof Ishak Institute) **16** p 1-10

[5] Nagasawa R and Fukushima A 2013 Urbanization and Its Influences on the Suburban Land Use Changes in JABODETABEK Region, Indonesia. *In*: Rustiadi E, Barus B, and Sunito S, editor. *International Seminar of Geospatial and Human-Dimensions on Natural Resource Management; (IPB Convention Center Bogor, 12-13 September 2011)* Crestpent Press p 17-24

[6] Pravitasari A E, Saizen I, Tsutsumida N and Rustiadi E. 2014 Detection of spatial clusters of flood and landslide prone areas using local moran index in Jabodetabek metropolitan area Indonesia *International Journal of Ecology and Environmental Sciences* **40** 4 p 233-243

[7] Rustiadi E, Pribadi D O, Pravitasari A E, Indraprahasta G S, and Iman L S 2015 Jabodetabek Megacity: From city development toward urban complex management system. *In*: Singh RB, editor. *Urban Development Challenges, Risks and Resilience in Asian Mega Cities.* Advances in Geographical and Environmental Sciences (Japan: Springer) p 421-445.
[8] Firman T 1997 Land conversion and urban development in the northern region of West Java, Indonesia *Urban Studies* **34** 7 p 1027-1046 (SAGE Publications Inc.)
[9] Firman T 2000 Rural to urban land conversion in Indonesia during boom and bust periods *Land Use Policy* **17** p 13-20 (Elsevier Science Ltd.)
[10] Firman T 2004 New town development in Jakarta Metropolitan Region: a perspective of spatial segregation. *Habitat International* **28** p 349–368 (Elsevier Ltd.)
[11] Firman T 2014 Privatization of the outskirts: The trends of "Post-Suburbanization" in Jabodetabek (Jakarta extended metropolitan region) Indonesia (*Population Association of America (PAA) 2014 Annual Meeting Program, May 1-3*)
[12] Bhatta B 2010 *Analysis of Urban Growth and Sprawl from Remote Sensing Data*. Advances in Geographic Information Science Book Series (Verlag Berlin Heidelberg: Springer) p 172
[13] Arsanjani J J 2012 *Dynamic Land-Use/Cover Change Simulation: Geosimulation and Multi Agent-Based Modelling* Springer Theses Book Series (Verlag Berlin Heidelberg: Springer) p 139
[14] Mas J F, Kolb M, Paegelow M, Olmedo M T C and Houet T 2014 Inductive pattern-based land use / cover change models: a comparison of four software packages. *Environmental Modelling and Software* **51** p 94-111 (Elsevier Ltd.)
[15] Eastman J R 2012 *IDRISI Selva Help System. IDRISI Selva Version: 17.* (US: Worcester, MA Clark Labs, Clark University)
[16] Winarso H and Firman T 2002 Residential land development in Jabotabek Indonesia: Triggering economic crisis?. *Habitat International* **26** 487–506 (Elsevier Ltd.)
[17] Winarso H 2002 Access to main roads or low cost land? Residential land developers' behaviour in Indonesia *Bijdragen tot de Taal, Land- en Volkenkunde*, **158** 4 p 653-676
[18] Winarso H, Hudalah D and Firman T 2015. Peri-urban transformation in the Jakarta metropolitan area *Habitat International* **49** p 221–229 (Elsevier Ltd.)
[19] B S D 2015 *Annual Report 2015*: *Footprints in Indonesia and Beyond* PT. Bumi Serpong Damai (BSD) Tbk. Sinar Mas Land Group p 227
[20] KEMENATR 2015 *Regulation Number 5 / 2015 on Location Permit.* Ministry of Agrarian and Spatial Planning / National Land Agency (KEMENATR / BPN) Jakarta
[21] KEMENRISTEK 2013. *Masterplan of PUSPIPTEK Revitalization and I-STP Development.* Ministry of Research and Technology (KEMENRISTEK) Jakarta p 111

IOP Conf. Series: Earth and Environmental Science 54 (2017) 012019 doi:10.1088/1755-1315/54/1/012019

Impact of oil and gas field in sugar cane condition using landsat 8 in Indramayu area and its surrounding, West Java province, Republic of Indonesia

Tri Muji Susantoro[1,2*], **Ketut Wikantika**[1], **Alia Saskia Puspitasari**[1], **Asep Saepuloh**[1]

[1] Center for Remote Sensing, Bandung Institute of Technology, Indonesia

[2] Research and Development Center for Oil and Gas Technology "LEMIGAS", Ministry of Energy and Mineral Resources, Indonesia

E-mail: trimuji_s@yahoo.com

Abstract. This study tried to monitor sugar cane condition surrounding of oil and gas field area. The spectral approaches were conducted for mapping sugar cane stress. As an initial stage Landsat-8 was corrected radiometrically and geometrically. Radiometric correction is an important stages for spectral approaching. Then all pixel values were transformed to the surface reflectance. Several vegetation indices were calculated to monitor vegetation stress surrounding of oil and gas field. NDVI, EVI, DVI, GVI, GRVI, GDVI and GNDVI were applied for generating tentative sugar cane stress images. The results indicated that sugar cane surrounding of oil and gas field has been influenced by oil and gas field.

1. Introduction

The presence of oil and gas beneath the earth can be identified through their seepage. There are three types of seepage; gas seepage, oil seepage and mudvolcano [1]. All types of seepage sometimes are associated each other. In mudvolcano sometimes contained gas and or oil seepage and also in oil seepage sometimes contained gas seepage [2]. Based on the appearance, seepage divided into macroseep and microseep. Macroseep is oil and gas seepage that can be seen by eye directly on the surface [3]. Macroseep in the surface can be seen as asphalt, oil pond or gas bubbles when associated with water or mudvolcano. Microseep is seepage that occurs vertically or nearly vertically from the reservoir to the surface, but can not be seen directly by eye [3].

Macroseep and microseep are common phenomenon occurs in oil and gas fields. More than 75% of oil and gas basin occur macroseep [4]. Around of 85% of oil and gas occur in the shape of microseep to the earth surface. Microseep could cause "hydrocarbon altered column" and hydrocarbon alteration halo" on the earth surface in over hydrocarbon reservoirs [5]. Microseep is basically pervasive to the earth surface. Hydrocarbon gas blows spread of the earth resulting from the migration of natural gas from reservoirs [6]. The presence of both, macroseep and microseep may affect to the vegetation and soil surface conditions. Stress on vegetation can occur which is characterizatized by slower growth and reduced chlorophyll [7] [8]. It is also reduce the vegetation density as the oxygen in the soil is reduced [9].

Remote sensing has the ability to detect the vegetation condition, whether healthy or stressed. This is because vegetation spectral reflectance is dependent on the chlorophyll and water absorption in the leaves, which get altered by oil and gas impact [10]. Healthy vegetation possess high spectral reflectance pattern at wavelengths near infrared (800-1100 nm). Healthy vegetation is characterized by high absorption of blue (450 nm) and red (670) wavelengths. A decline of the health vegetation due to interference from accumulation of seepage or other obstruction will change the spectral pattern. The near infrared spectral shows a drop value due to decreasing chlorophyll in the leaves [11] [7]. In addition, this will increase spectral reflection at red wavelength [12].

The purpose of this study is to identify the impact of the oil and gas fields on the sugar cane condition, Indramayu Regency and surroundings, West Java, Republic of Indonesia. Remote sensing data used were Landsat 8. It is a new generation of Landsat series that was launched. Initially known as the Landsat Data Continuity Mission (LDCM), it took the name Landsat 8 when handed over to the U.S. Geological Survey (USGS) for operational work on May 30, 2013 [13]. Spectral bands of Landsat 8 has high similarities with Landsat 7ETM+, except the three parameters as follows: the Landsat 8 had higher values of the near infrared band for vegetative landcover types, lower values for the shortwave infrared (2.11–2.29 μm) band for all land cover types and higher values for the shortwave infrared (1.57–1.65 μm) band for non-water land cover types [14].

Several vegetation indices namely Normalized Difference Vegetation Index (NDVI), Enhanced Vegetation Index (EVI), Difference Vegetation Index (DVI), Green Vegetation Index (GVI), Green Ratio Vegetation Index (GRVI), Green Difference Vegetation Index (GDVI), Green Normalized Difference Vegetation Index (GNDVI) are used to monitor the condition of sugar cane plantations (Table 1).

Tabel 1. Several Vegetation Indices used in the study

No	Veg. Index	Algoritm	References
1	Normalized Difference Vegetation Index (NDVI)	(NIR-Red)/(NIR+ Red)	[21]
2	Enhanced Vegetation Index (EVI)	(NIR-Red)/(NIR+6*Red-7.5*Blue+1)	[22]
3	Difference Vegetation Index (DVI)	(NIR-Red)	[23]
4	Green Vegetation Index (GVI)	(-0.2848*TM1)+(-0.2435*TM2)+(-0.5436*TM3)+(0.7243*TM4)+(0.0840*TM5)+(-0.1800*TM7)	[24]
5	Green Ratio Vegetation Index (GRVI)	(NIR/Green)	[25]
6	Green Difference Vegetation Index (GDVI)	(NIR-Green)	[25]
7	Green Normalized Difference Vegetation Index (GNDVI)	(NIR-Green)/(NIR+ Green)	[26]

2. Study Area

The study area is located in Indramayu and Majalengka Regency, West Java Province, Republic of Indonesia (Figure 1). Sugar cane plantation is part of Jatitujuh Company. The area is around of 11,921.56 hectares in Indramayu Regency and 5,673.04 hectares in Majalengka Regency [15].

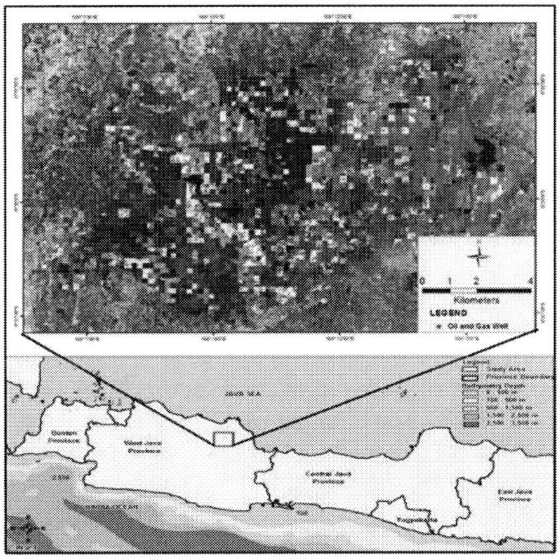

Figure 1. Study area presented by red rectangles with subset of 653 RGB of Landsat 8 overlaid with oil and gas well-head in red dots

3. Material and Method

Landsat 8 data used were collected from Indonesian National Institute of Aeronautics and Space (LAPAN). Landsat 8 data was acquired on September 25, 2015. The ancillary data used include: oil and gas well obtained from Research and Development Center for Oil and Gas Technology "LEMIGAS", Topographic Map from Geospatial Information Agency (BIG) and IKONOS Imagery from Ministry of Agriculture. The methods implemented in this study for observing the condition of sugar cane plantation is shown in Figure 2. The following tasks were performed: image pre-processing, NDVI, EVI, DVI, GVI, GRVI, GDVI and GNDVI calculation and analysis.

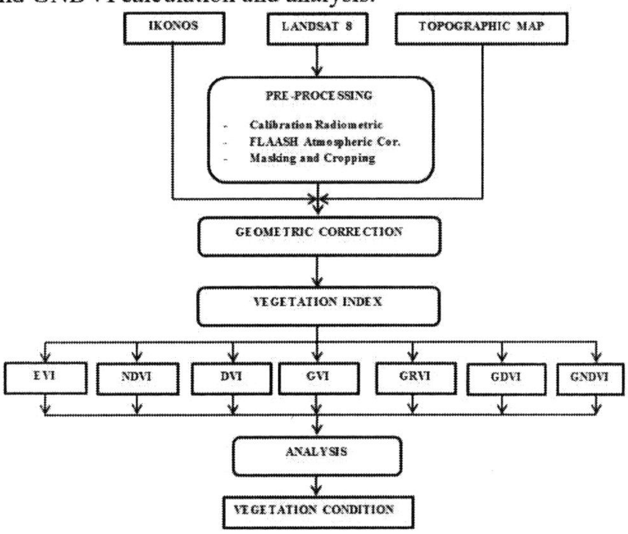

Figure 2. Work flows of analyzing vegetation condition at oil and gas field at Indramayu and Majalengka Regencies

3.1 Digital Image Processing

Radiometric calibration is done by calculation the digital number to Top of Atmosphere (ToA) radiance values. Calculation of at-sensor spectral radiance is the fundamental step in converting image data from multiple sensors and platforms into a physically meaningful common radiometric scale. Radiometric calibration of Landsat 8 sensors involves rescaling the raw digital numbers transmitted from the satellite to calibrated digital numbers which have the same radiometric scaling for all scenes processed on the ground for a specific period [16]. Converting digital number to ToA is very important because the light energy that comes from the sun, the radiance is often normalized into a reflectance values. Then FLAASH method was used to change the radiance values into surface reflectance and undertake atmospheric correction [10].

Geometric correction is done by image to image using IKONOS imagery and supporting by Topographic map. This correction is conducted to minimize bias of locations (Spatial displacement) [17]. Interpolation method of spectral values is using nearest neighbour [18]. Transformation of pixel positions is using polynomial orde 2. The minimum GCP of this method are 6 GCP. Consideration of using this method is based on the area condition that has undulating topography [19]. Residual Mean Square Error (RMSE) maximum of this correction is 0.5 X pixel resolution [20].

3.2 Analyisis of Vegetation Index

Several vegetation indices is used in this study to extract information related to sugar canes vegetation stress (Table 1). NDVI is a measure of healthy and green vegetation. The combination of its normalized difference formulation and use of the highest absorption and reflectance regions of chlorophyll make it robust over a wide range of conditions [21]. EVI was developed as a standard of MODIS product to improve the NDVI by optimizing the vegetation signal in LAI regions. It uses the blue reflectance region to correct for soil background signals and to reduce atmospheric influences, including aerosol scattering [22]. DVI used to distinguish between soil and vegetation [23].

GVI minimizes the effect of the background soil while emphasizing green vegetation. It uses global coefficients that weigh the pixel values to generate new transformed bands. It is also known as Tasseled Cap Green Vegetation Index [24]. GRVI is sensitive to photosynthetic rates in canopies [25]. GDVI was originally designed to predict nitrogen requirements for corn in color-infrared photography [25]. GNDVI is more sensitive to chloropyll concentration than NDVI [26].

Analysis of vegetation stress is based on the vegetation indices results. The vegetation index values will reflect the health of the vegetation condition. Low vegetation indices values can describe the condition of vegetation stress or dominated by soil and or water underneath. Buffer of oil and gas wells conducted to assess the change of vegetation index results as an indication of stress. Buffer is done with a distance of 100 meters, 250 meters and 500 meters.

4. Result and Discussion

The results of several vegetation indices in sugar cane plantations in Indramayu and its surroundings can be seen in Figure 3. NDVI maximum value is 0.59. this NDVI values is considered as moderate category [22]. At the location where there is no vegetation, NDVI value will be close to 0 to 0.1. It values typically occurs in baren land [27]. Based on growing phase of sugar cane, this condition occurs in post-harvest and or early planting (phase-1).

NDVI values describe the sugar cane condition in oil and gas field has no high density canopy. This should occur in three phases, namely stem elonging phase. This phase occurs in 4 to 9 months ages after planting. In this phase, especially after 9 months, sugar cane condition has no less than 12 of tillers, 3.8 meters of height, 3.2 cm of diameters and 20 segments [28]. NDVI with a high bulk density has values close to +1 or 0.6 to 0.9. But the dominant NDVI value in sugar cane ranges from 0.11 to 0.35 (95%) of the total area. The value is in low to moderate categories.

DVI value in sugar cane ranges from 0 to 0.4. The dominant values is ranging between 0 to 0.184. This index distinguishes between soil and vegetation. But it does not account for the difference between reflectance and radiance caused by atmospheric effects or shadows. Lower DVI value which means

IOP Conf. Series: Earth and Environmental Science **54** (2017) 012019 doi:10.1088/1755-1315/54/1/012019

sugar cane condition is not dominant vegetation cover the area. It is assumed as impact of oil and gas field and or sugar cane plantations dominated by phase-1 and or post-harvest phase. GVI value ranges from 0 to 0.0674. This indicates that the green level of sugar cane is very low. GVI maximum values is 0.02 (2%). GNVDI values are almost equal to the NDVI values. The maximum value is 0.5382 and dominant value ranges from 0.1339 to 0.3361 (83%). GNDVI function almost the same as the NDVI, but more sensitive to chlorophyll concentration.

EVI value ranges from 0 to 0.2901. The dominant value is between 0.058 to 0.174. Based on EVI values, sugar cane vegetation signal is very low. Actually this index is used to estimate foliage cover and to forecast crop growth and yield. GRVI used to assess the sensitivity of photosynthetic process. On sugar cane plantations GRVI values range from 0 to 3.5. However GRVI values is dominant at 1.25 to 2.5 which reached 98%. Based overlaying with NDVI values, GRVI in dominant values has low NDVI values. It means the photosynthetic rates is low. GDVI value ranges from 0 to 0.3. The dominant GDVI value ranges from 0.0376 to 0.1681. It means, the level nitrogen in sugar cane is low. Based on the analysis of EVI, DVI, GVI, GRVI, GDVI and GNDVI values tends to be low and correlated with NDVI values (Table 2).

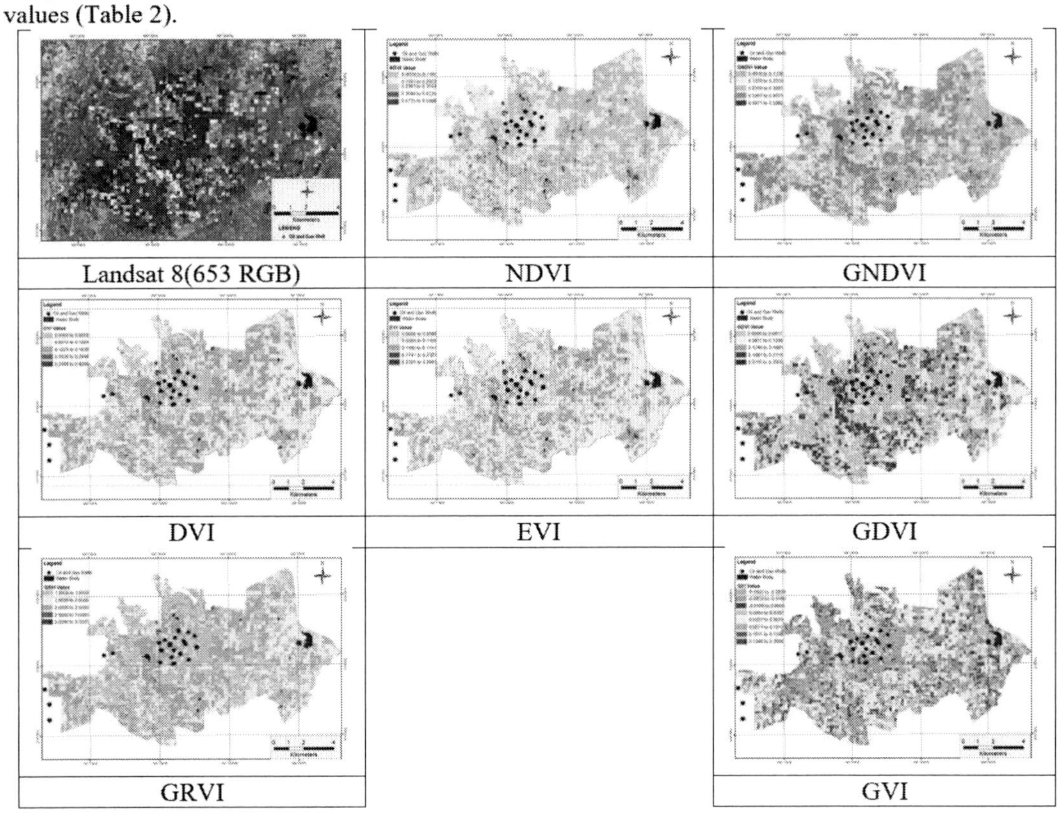

Figure 3. Vegetation indices were extracted from Landsat 8, Namely: NDVI, GNDVI, DVI, EVI, GDVI, GRVI and GVI

Table 2. Vegetation index results and their areas

NO	DVI VALUES	AREA (M2)	NDVI VALUES	AREA (M2)	GNDVI VALUES	AREA (M2)	EVI VALUES	AREA (M2)
1	-0.0461 to -0.0000	874.800	-0.2181 to -0.0000	879.300	-3.0000 to -0.0000	942.300	-0.0526 to -0.0000	879.300
2	0.0000 to 0.0612	25.483.500	0.0000 to 0.1181	943.200	0.0000 to 0.1339	2.646.000	0.0000 to 0.0580	17.091.900
3	0.0612 to 0.1224	58.519.800	0.1181 to 0.2363	52.874.100	0.1339 to 0.2350	32.053.500	0.0580 to 0.1160	76.552.200
4	0.1224 to 0.1836	38.133.000	0.2363 to 0.3544	65.103.300	0.2350 to 0.3361	51.752.700	0.1160 to 0.1741	28.467.900
5	0.1836 to 0.2448	769.500	0.3544 to 0.4725	3.831.300	0.3361 to 0.4371	36.071.100	0.1741 to 0.2321	793.800
6	0.2448 to 0.4000	156.600	0.4725 to 0.5906	291.600	0.4371 to 0.5382	459.000	0.2321 to 0.2901	143.100

NO	GVI VALUES	AREA (M2)	GRVI VALUES	AREA (M2)	GDVI VALUES	AREA (M2)
1	-0.1222 to -0.0500	902.700	0.4164 to 1.0000	942.300	-0.0600 to -0.0000	942.300
2	-0.0500 to -0.0250	6.280.200	1.0000 to 1.2500	1.330.200	0.0000 to 0.0376	3.703.500
3	-0.0250 to -0.0100	24.070.500	1.2500 to 1.6656	42.523.200	0.0376 to 0.0811	34.550.100
4	-0.0100 to 0.0000	15.266.700	1.6656 to 2.0000	41.407.200	0.0811 to 0.1246	40.324.500
5	0.0000 to 0.0337	38.355.300	2.0000 to 2.5000	37.098.000	0.1246 to 0.1681	29.156.400
6	0.0337 to 0.0674	36.060.300	2.5000 to 3.0000	557.100	0.1681 to 0.2116	14.170.500
7	0.0674 to 0.1011	2.634.300	3.0000 to 3.5000	80.100	0.2116 to 0.3000	1.070.100
8	0.1011 to 0.1348	252.000				
9	0.1348 to 0.2000	117.900				

Analysis of all vegetation indices showed low to moderate. This is presumably due to the oil and gas beneath that affect to vegetation in the surfaces [29]. Oil and gas beneath the earth precipitated to bring carbonate, pyrite, sulfur, pyrhotite, uranium, radon and others to the surface. In addition there is also oxidation process that forms microseeps in the surface [30] [31] [12] [32]. These conditions resulted vegetation anomaly that causes slow growth and the vegetation becomes sparse [11]. The presence of excess carbon dioxide gas (increased carbon content) will reduced chlorophyll content of leaves and the leaves become yellowish [8].

In healthy vegetation, near infrared reflectance will occur significantly spectral ranges from 0.7 to 1.2 μm. Only few are absorbed by the leaves. Healthy vegetation have 40 to 60% of reflectance, 40 to 60% of transmission and only 5 to 10% of absorption. The biggest adsorption is in blue and red channels. The presence of vegetation damaged/unhealthy will indicate the minimum of vegetation index values, but will be very high in healthy vegetation [33]. Vegetation may be affected by natural gas in many ways. The gas may be taken in the vegetation by the root system where it metabolized or passed in the transpiration stream [34]. Vegetation anomaly (stress) may exist in area where the soil contain unusually high concentration of the certain elements or compounds [35].

Based on buffer analysis of wells at a distance 100 m, 250 m and 500 m seen that sugar cane plantations around the wells has a NDVI, EVI, DVI, GVI, GRVI, GDVI and GNDVI values are likely to be low. Those values begin to rise at 50 m, but not significant (Figure 4). This indicates that the presence of oil and gas is expected to affect the sugar cane plantation. This is due to the long-term presence of oil and gas seepage continuosly cause anomalies mineral and chemical changes in the surface [36]. These condition altered the biophysical and chemical characteristics of vegetation. The biophysical and biochemical characteristics changes in vegetation induced by oil pollution would affect the spectral signature of vegetation in visible and near infrared wavelengths [10]. Hence the vegetation cover is affected by gas in the soil and it can be a key to predict hydrocarbon seeps [37]. The vegetation growing in areas of gas production have a different spectral signature from vegetation growing off-gas field. Vegetation growing off-gas field have higher reflectivity in the 750-1250 nm spectral region and an absorption feature at 1200 nm is much stronger in off–gas field [38]. The model of vegetation change condition of oil and gas field described where the vegetation at the center of field will grow up stunted and abnormal, as for getting away from center will get better and grow normally [11] (Figure 5).

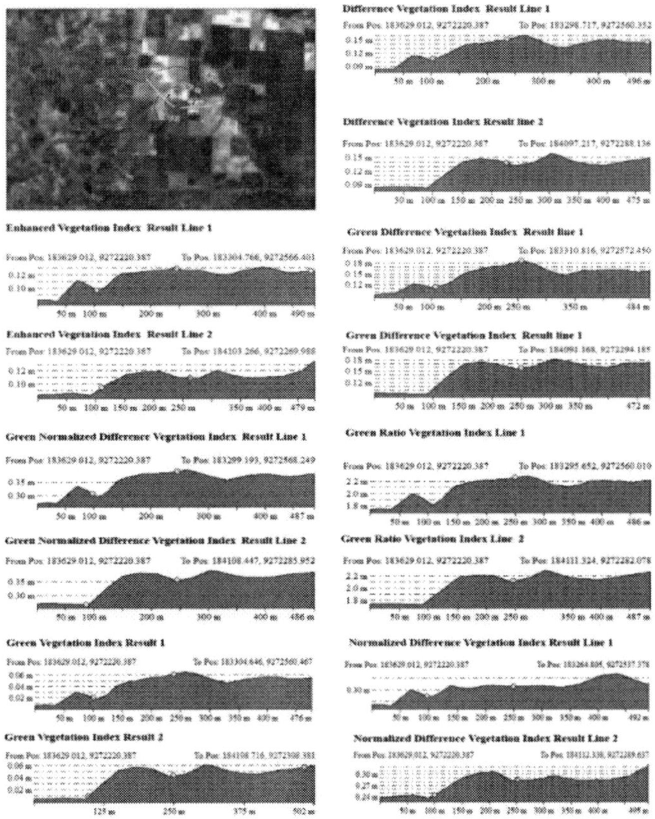

Figure 4. NDVI, EVI, DVI, GVI, GRVI, GDVI and GNDVI values around of oil and gas wells.

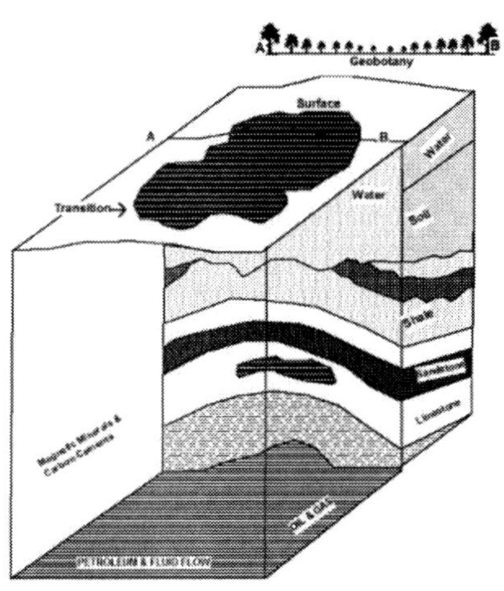

Figure 5. Conceptual model of vegetation change condition in oil and gas field that vegetation willgrow up stunted and abnormal in the center of field and away from center will get better and grow normally [11].

5. Conclusion

This study was conducted to analyze the impact of oil and gas beneath the earth to sugar cane plantations. The analysis was conducted by processing and reviewing NDVI, EVI, DVI, GVI, GRVI, GDVI and GNDVI values. Based on its vegetation indices values, oil and gas beneath the earth are affecting the condition of sugar cane plantations. The effect is analyzed based on NDVI, EVI, DVI, GVI, GRVI, GDVI and GNDVI values that has dominant low values in all of sugar cane plantations. Results of buffer analysis around of oil and gas wells of several vegetation indices tends to be low values and then increase away from the well. It can be concluded that, the seven vegetation indices from Landsat 8 corresponded to vegetation condition at oil and gas field and agreed with previous published references. Ground checking is crucial to validate the obtained vegetation indices for next step.

Aknowledgement

We would like to express our thank to head of Remote Sensing Technology and Data Center-LAPAN for supporting Landsat 8 Data, Geospatial Information Agency (BIG) for supporting Topographic Map and Ministry of Agriculture for supporting High Resolution Data for making this paper.

References:

[1] Etiope G 2009 A global dataset of onshore and oil seeps: a new tool for hydrocarbon exploration *Oil and Gas Business* Downloaded June 17th, 2016

[2] Susantoro T M, Puspitasari A S and Wikantika K 2016 Spectral pattern of several types mudvolcano using analytical spectral devices *National Seminar of Remote Sensing* LAPAN (Depok West Java)

[3] Van der M, Van Dijk P, Werff H V D and Hong Y 2002 Remote sensing and petroleum seepage: a review and case study *Terra Nova* **14** Blackwell science Ltd

[4] Clarke R H and Cleverly R W 1991 Petroleum seepage and post-accumulation migration. in: england, w.a., fleet, a.j. (eds.), petroleum migration *Geological Society Special Publication N. 59* (London : Geological Society of London) pp 265–271

[5] Tian Q 2012 Study on oil and gas reservoir detecting methods using hyperspectral remote sensing international archives of the photogrammetry *Remote Sensing and Spatial Information Sciences Volume XXXIX-B7. XXII ISPRS Congress*, August 25 – Sept 1, 2012 Melbourne Australia

[6] Etiope G and Klusman R W 2010 Microseepage in Drylands: flux and implications in the global atmospheric source/sink bugdet of methane *Global and Planetary Change* **72** pp 265-274

[7] Noomen M F 2007 Hyperspectral reflectance of vegetation affected by underground hydrocarbon *Dissertation* International Institute for Geo-information Science & Earth Observation. Enschede, The Netherlands (ITC)

[8] Lakkaraju V R, Zhou X., Apple M E, Chunningham A and Dobeck L M 2010 Studying the vegetation response to simulated leakage of sequestered co_2 using spectral vegetation indices *Economic Informatics* Elsevier

[9] Van der W H M A, Noomen M F, Van der M and Van D M F D 2006 Remote sensing of onshore hydrocarbon *Geological Society, London, Special Publication* **283** pp 125-133

[10] Adamu B, Tansey K and Ogutu B 2015 Using vegetation indices to detect oil pollution in the Niger Delta *Remote Sensing Letters* **6** pp 145-154

[11] Hong Y 1999 Imaging spectrometry for hydrocarbon *Dissertation* TU Delft Master of Science in Geology ITC Publication Number 76

[12] Tejada P J Z and Canto G S 2007 Remote sensing of vegetation biophysical parameters for detecting stress condition and land cover changes *Estudios de la Zona No Saturada del Suelo* Vol. **VIII**. J.V. Giráldez Cerveray F.J. Jiménez Hornero

[13] Flood N 2014 Continuity of reflectance data between Landsat-7 ETM+ and Landsat-8 OLI, for both top of atmosfphere and surface reflectance: a study in the Australian landscape. *Remote Sensing Journal* **6** pp 7952-7970

[14] Li P, Jiang L and Feng Z 2014 Cross-comparison of vegetation indices derived from Landsat-7 Enhanced Thematic Mapper Plus (ETM+) and Landsat-8 Operational Land Imager (OLI) sensors *Remote Sensing Journal* **6**: pp 310-329
[15] Fadilah R and Sumardjo 2011 patnership analysis between jatitujuh sugar factory with sugar cane factory in Majalengka, West Java *Transdisiplin Sociology, Communication and Human Ecology Journal* pp 159-172
[16] Chander G, Markham B L, and Helder D L 2009 Summary current radiometric calibration coefficients for Landsat MSS, TM, ETM+ and EO-1 ALI. *Sensors.* Vol. **113**
[17] Murti S H and Wicaksono P 2014 The analysis of the most effective spectral bands for coral reefs health identification: case study on Menjangan Besar and Menjangan Kecil Island, Karimunjawa Islands *Globe Journal* **16** pp 117-124
[18] Danoedoro P 1996 *Digital image processing, theory and application in remote sensing* (Yogyakarta : Faculty of geography, Gadjah Mada University)
[19] Setianto A 2003 Geological of front block mountain area, Central Sumatera Basin, Riau based on Landsat Thematic Mapper Imagery *Thesis* Postgraduate Program (Yogyakarta: Gadjah Mada University)
[20] Susantoro T M 2009 Optimalization of Landsat 7ETM+ and SRTM data for revising geological map of bojonegoro sheet *Thesis* Remote Sensing Postgraduate Program (Yogyakarta : Facultu of geography Gadjah Mada University)
[21] Rouse J, Haas R, Schell J and Deering D 1973 Monitoring vegetation systems in the Great Plains with ERTS *Third ERTS Symposium* NASA pp 309-317
[22] Huete A, Didan K, Miura T and Ferreira L G 2002 Overview of the radiometric and biophysical performance of the MODIS vegetation indices *Remote Sensing of Environment* **83** pp 195–213
[23] Tucker C 1979 Red and photographic infrared linear combinations for monitoring vegetation *Remote Sensing of Environment* **8** pp 127–150
[24] Kauth R and Thomas G 1979 The tasselled cap-A graphic description of the spectral-temporal development of agricultural crops as seen by Landsat *Proceedings of the LARS 1976 Symposium of Machine Processing of Remotely-Sensed Data*, West Lafayette, IN: Purdue University, pp. 4B41-4B51
[25] Sripada R P, Heinigerb R W, Whitec J and Meijer A D 2006 Aerial color infrared photography for determining early in-season nitrogen requirements in corn." *Agronomy Journal* **98** pp 968-977
[26] Gitelson A and Merzlyak M 1998 Remote sensing of chlorophyll concentration in higher plant leaves *Advances in Space Research* **22** pp 689-692
[27] Neigh C S R, Tucker C J and Townshend J R G 2008 American vegetation dynamic observed with multi-resolution satellite data *Remote Sensing Environment* **112** pp 1749-1772
[28] Pramuhadi G 2010 Climate factor at sugar cane culture at Dryland. *Food Journal* Vol. **19** pp 331-344
[29] Qianqian L, Chen X, Liu X, Mao B and Ni G 2012 Study on oil and gas exploration in sparse vegetation areas by hyperspectral remote sensing data *Chinese Optic Letter* Vol **10** (Suppl)
[30] Joshua J 2015 Hyperspectral remote sensing for oil exploration *Published in Science* Downloaded April 2, 2016
[31] NASA 2011 *Finding Oil and Gas from Space* Downloaded April 2, 2016
[32] Sounder D F, Burson K R and Thompson C K 1999 Model for hydrocarbon microseepages and related near-surface alteration *Bull. Am. Ass. Petrol. Geology* **83**. pp 170-185
[33] Sudiana D and Diasmara E 2008 Analysis of vegetation index using satellite data NOAA/AVHRR and TERRA/AQUA-MODIS *Seminar on Intelligent Technology and Its Applications*
[34] Smith K L, Steven M D and Colls J J 2004 Spectral respons of pot grown plants to displacement of soil oxigen *International Journal Remote Sensing* Vol. **25** pp 1-17
[35] Lillesand T M and Kiefer R W 1994 *Remote sensing and image interpretation*, 3rd ed. John Wiley and Sons, Inc.

[36] Schumacher D 1996 Hydrocarbon induced alteration of soil and sediments, hydrocarbon migration and its near-surface expression (D. Schumacher and M.A. Abrams, Eds.). *Mem. Am. Ass. Petrol. Geology* **66**, pp 71-89

[37] Noomen M F, Skidmore A K and Van der Meer F D 2003 Detecting the influence of gas seepage on vegetation, using hyperspectral remote sensing *Presented at the 3rd EARSel Workshop on Imaging Spectroscopy* Herrsching, May 13-16, 2003

[38] Hong Y, Van der Meer F D, Zhang J and Kroonenberg S B 2000 Direct detection of onshore hydrocarbon microseepages by remote sensing techniques *Remote Sensing Reviews* March 2000

IOP Conf. Series: Earth and Environmental Science 54 (2017) 012020 doi:10.1088/1755-1315/54/1/012020

The critical land in Komering watershed as a result of land use changes from 2000-2016 period

Kusratmoko E*, Dayanti S T, and Supriatna
Magister Program Departement of Geography, Faculty of Mathematics and Natural Science, Universitas Indonesia, Depok 16424

E-mail: eko.kusratmoko@ui.ac.id

Abstract. Land use changes in a watershed could affect the ecological system, hydrological system and water quality, meanwhile land use changes study is needed to conduct especially in Komering watershed. Land use is the main factor that determines level of critical land, so the land use changes should correspond with the capability through which degraded land can be controlled. This study aimed to calculate land use/ land cover changes from 2000 to 2016 and generate critical land criterions. The objective of the study was to determine the physical and social factors that influence the degree of criticality land in watershed. Physical and social characteristics were measured including land use changes, solum, slopes, outcrop, and land productivity. This study applied supervised classification-maximum likelihood algorithm in ENVI 5.1 imagine to detect land use changes observed in Komering watershed, South Sumatera province, using multispectral satellite data obtained from Landsat 7 and Landsat 8 for the years 2000 to 2016 respectively. Determination of the level of critical land and land use conditions are executed through a spatial approach by utilizing Geographic Information Systems (GIS) and statistical analysis.

1. Introduction

The phenomenon of land-use change is one of the major environmental changes occurring around the world today, including in Indonesia. The determining factors of land use change are the increase in the population which resulted in the need for land is also increasing [1]. According to Verburg [2] land use changes is a complex interaction between structural factors and human behavior associated with technological capabilities, needs, and social interaction. The complex interactions will affect the carrying capacity of the environment. The impact of land use changes will affect the ecology directly for both biodiversity and aquatic ecosystems. The land use such as uncontrolled deforestation, the expansion of agricultural land, and urban development will worsen the condition of the watershed [3]. Watershed conditions are poorly marked by a decline in the quality and availability of water [2], the intensity of flooding is high, reduced groundwater discharge, the extinction of some endangered species, reduced productivity of agricultural land, and an increase critical land [4].

The critical land is soil that function loss for growing plants [5] because the soil loses its fertility [6]. The critical land occurs due to the incompatibility of the soil's ability to use the land, resulting in soil physical damage, chemical, or biology [7]. Understanding some of the definitions of degraded soils have the same meaning, namely soil loss as a function of production factors and protective factors. The production function characterized the performance of the plants that can be the object of production (agriculture and forestry), or the production of services (comfort and beauty environmental), while the protection function regarding the protection of soil and hydrology. So, the critical land can be critical in terms of physical land, a critical socio-economic, and critical in terms of hydrology [8]. Land degradation has deteriorated the quality of land and it can be caused by the water erosion, wind erosion, ravines, salt-affected areas ,water logging, shifting cultivation, forests degraded and the other special issues [9].

This research is aimed to identify the impact of land use change over 15 years time periode at watershed with perspective spatial and temporal. However the specific objectives are (i) to identify and delineate different land use/ land cover categories and the pattern of land use change in watershed from 2000 to 2016 (ii) to form criteria of critical land in watershed with combining physical and social factors by using remote sensing data and spatial analysis with GIS, which is reinforced by quantitative analysis. Given the watershed as an ecosystem is an indicator of environmental damage, this study is considered important to do as a preventive measure the incidence of environmental problems that will be caused by human activity in utilizing land.

2. Methods

2.1 Studi Area

This study was conducted in Komering watershed from July to September 2016. Komering watershed is one of sub watersheds in Musi basin located on Sumatera island at South Sumatera province, Indonesia. The Komering watershed belongs to three administrative regency of South Sumatera, the first is Ogan Komering Ilir, second is Ogan Komering Ulu Timur, and third is Ogan Komering Ulu Timur. The absolute location of this study is between 103° 34' 12" – 105° 0' 36" E and 02° 58' 12" – 04° 59' 24" S with area coverage of 806,062 Ha. In general, thetopography of the watershed is relatively flat in the middle and lower with reaches of slope of between 0-2% and 2-15%. Whereas upper Komering watershed began undulating topography dominated by a slope of between 15-25% with the highest point is 2126 mean sea level. The type of land use in the watershed is dominated by plantation with an area 313,798 ha, or approximately 8.93% of the total area of research (Fig 1).

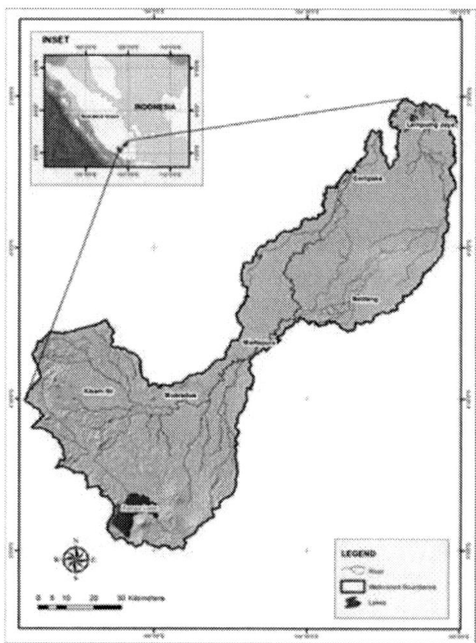

Figure 1. Map of Komering Watershed

2.2 Materials

- Satellite data
  Landsat images of path/row 124/62 and 124/63; Landsat 7 ETM for the year 2000, and Landsat 8 OLI for the year 2016. The satellite acquisition date for the images was July to September.
- Raster Data
  Digital Elevation Models Shuttle Radar Topography Mission (DEM SRTM) path-row 57-13 and 58-13
- Vector data
  1. Watershed boundary (Obtained from Musi Watershed Management Agency, "BPDAS Musi" –South sumatera province).
  2. Administrative district boundary of Indonesia (Obtained from Geospasial Information Agency, "BIG")

3. Landuse of South sumatera province (2000 and 2015), obtained from Ministry of Agricultural dan Spatial. [Scale:1:50.000].
4. Landsystem of South sumatera province obtained from Geospasial Information Agency, "BIG"[Scale 1 : 250.000]

- Tabular Data
  Production of crop (in units of Ton) at South sumatera province in 2015 obtained from Central Bureau of Statistic,
- Software
  ENVI 5.1, Arc GIS 10.1, and SPSS 17.

2.3 Data Analysis

- Landuse classification
  Classification of land use was conducted using supervised classification in ENVI 5.1 software. There are 180 traning samples were used to detect the type of land use/ land cover in Komering Watershead. The classification of land use/land cover in Komering watershed was produced in six classes. These included; forest, water, settlement, agriculture, bare-land and plantation. Assessment of classification accuracy of 2000 and 2016 images was carried out to determine the quality of information derived from the data. For the accuracy assessment of land use/ land cover maps extracted from satellite images, stratified random method was used to represent different land cover classes of the area. The accuracy assessment was carried out using 100 points, based on visual interpretation from high resolution images. In addition, a nonparametric Kappa test was also performed to measure the extent of classification accuracy as it not only accounts for diagonal elements but for all the elements in the confusion matrix [10]. Post-classification change detection technique, performed in ArcGIS 10 was employed by the study.
- Critical Land Classification
  In this study critical land classification in watershed developed based on physical dan social factors such as: percentage of land use changes, slope, solum, outcrop, and productivity. In order to generate sampels, the study area was divided into 2,5 km x 2,5 km grid and as many as 1.432 boxes. The number of sample points (n) is determined based on population-sample table that is 10% of total grid (N) used [5]. Samples will be processed using the SPSS software.

2.4 Research Prosedure

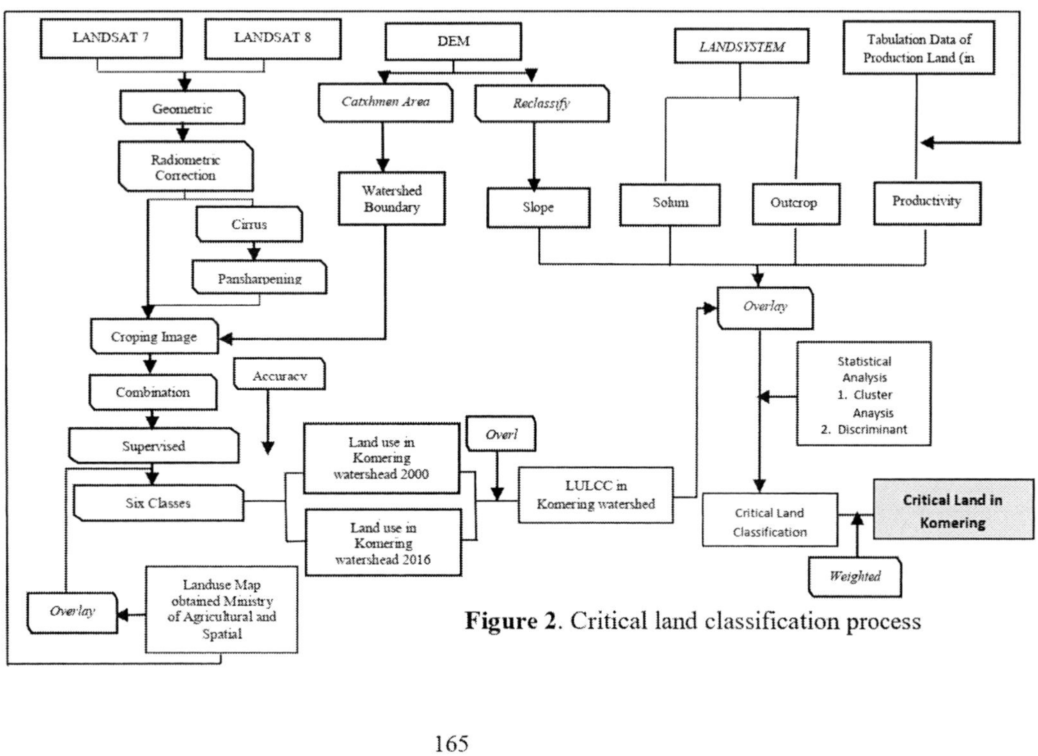

Figure 2. Critical land classification process

3. Result And Discussions

3.1 Landuse Change

The classified land-use map of Komering watershed of years 2000 and 2016 is given in Fig 3 and Fig 4. The achieved overall classification accuracies were 82.96% and overall kappa statistics were 0.7523 respectively for the classification images. The accuracy assessment reporting requires the overall classification accuracy more than 70% which were successfully achieved in the present research [11]. The cross-tabulation matrices (Table 1) show the nature of change of different land cover classes or in other words the shift in the land cover classes. According to the information revealed by classification results, the plantation showed 4.46% increase from 2000 to 2016. Out of the 300,256 ha that was plantation area in 2000, 299,999 ha was still plantation area in 2016 but 256 ha was converted to settlement, and then to agriculture and forest. At the same time the increased of plantation, from 2000 to 2016, was mainly from agriculture (13,270 ha). Forest also increase from 201,662 ha in 2000 to 201,812 ha in 2016, indicating a 0,07% of forest regrowth. Another type of LULC agricultural land, have been increased from 300,256 ha in 2000 to 313,798 ha in 2016, representing an 4.32 % increase. Bare land out of 3,054 ha in 2000 lost area mainly to plantation (25.26%), agriculture, forest and retained 2,088 ha of total in 2016. Bare land decreased rapidly (by 46.23%) between 2000 and 2016. Water decreased from 21,816 ha in 2000 to 21,814 in 2016, indicating an 0,01 % decrease.

Table 1 Cross tabulation of land use classes in Komering watershed in 2000 to 2016 (ha)

LU/LC type 2000	2016	Water	Forest	Agriculture	Plantation	Settlement	Bare land	**Total**
	Water	21,813				3		**21,816**
	Forest		201,661					**201,662**
	Agriculture			236,347	13,270	232		**249,850**
	Plantation				299,999	256		**300,256**
	Settlement					29,421		**29,421**
	Bare land		150	208	527	79	2,088	**3,054**
	Total	**21,814**	**201,812**	**236,556**	**313,798**	**29,992**	**2,088**	**806,062**

(Source : Data processing, 2016)

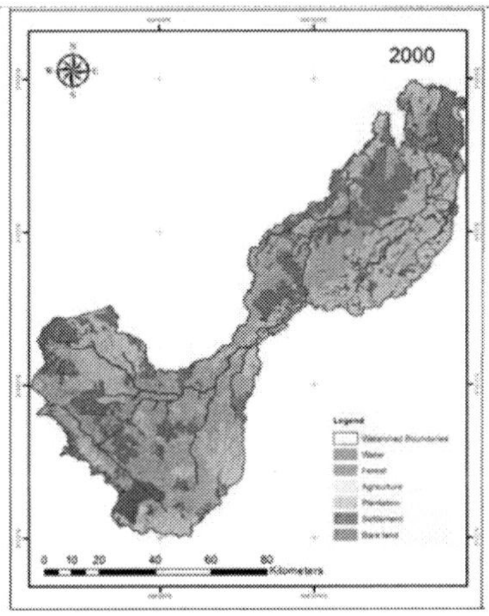

Figure 3. Landuse Classification of Komering Watershed in 2000

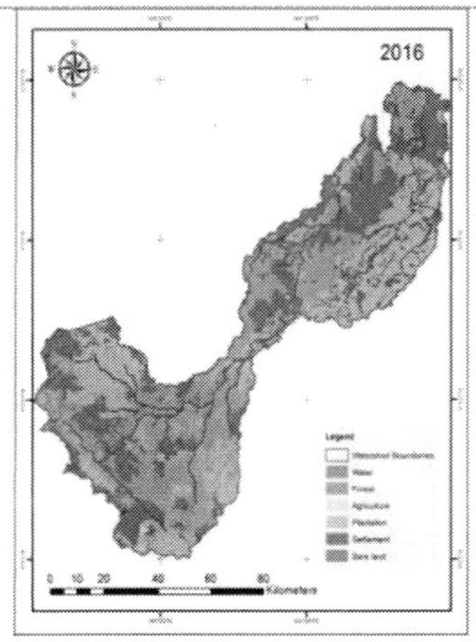

Figure 4. Landuse Classification of Komering Watershed in 2016

3.2 Critical Land Classification

The classes of critical land in watershed perform using cluster analysis by method Average Linkage (Between Groups) with Euclidean distance interval [12]. The results of an analysis from five variables used, shown that there are three classes of critical land like; critical, potential critical, and not critical. Preparation of critical land criteria based on the assumption that each population expected to have a homogenous covariance [1]. Box's M test results against factors that meet the homogeneity of covariance (P>0.05) obtained five factors that is: forest, water, settlement, agriculture, bare-land and plantation. These factors are analyzed stepwise discriminant method. Method stepwise secrete factors each detected collinear (multicollinearity), with sub methods mahalanobis distance to identify outliers (Sitorus *et al*, 2011). The coefficient obtained in a linear function is used to determine the factors that have biggest contribution to the difference between classes. so it can be used to perform weighted factors selected [13] [1].

Discriminant analysis results shown there are three variables that contribute to make three classes of critical land, namely: outcrop (X1); slope (X2), and productivity (X3). The contribution of each variable is obtained from the coefficients contained in the function generated by discriminant analysis (D = -4.70 + 0,189X1 - 0,138 X2 - X3 0.04). Based on the test results significant discriminant function has amounted to 57.30% ability to explain variation among classes of critical land. Besides this discriminant function has eigenvalue 1,340 or >1. This indicates that the discriminant function was valid [13] [14].

Based on the coefficient of the factors that contributed most to the difference between the classes is weighted each factor namely: outcrop (48%), slope (27%), and the productivity of the land (25%). Based on the weighting, critical land in the Komering watershed was influenced by the percentage of outcrop with values of 48%, while the rest influenced by slope of 27% and land productivity by the value of the effect of 25%. According to Ministry of Agrarian and Spatial Planning (2016) outcrop is one of the factors that caused a critical land. The main problems in regions that have an outcrop are serious soil erosion, mining activities in stony/rocky areas, landslides and it will cause critical land [9]. Furthermore, based on the analysis and previous description, criteria acquired critical land and indicators as shown in Table 2. By adding up the value (score) of the indicator, the cumulative amount of each class as shown in Table 3 was derived. The result of statistical analysis by weighting of each variable is presented in spatial data with overlay method.

Tabel 2. Criteria of Variables to make Classification of Critical Land

Variables	Classification of Critical Land		
	Not Critical	Critical Potential	Critical
Outcrop (weighted 48%)	<10%	10-20%	>20%
Slope (weighted 27%)	0-20%	20-40%	>40%
Land Productivity (weighted 25%)	> 4 Ton/ Ha	2-4 Ton/Ha	<2 Ton/Ha

Tabel 3. Scoring of indicator critical land

Variables (weigted %)		Indicator	Scoring
Outcrop		< 10%	1
	48%	10-20%	2
		>20%	3
Slope		0-20%	1
	27%	20-40%	2
		>40%	3
Land Productivity		>4 Ton/ Ha	3
	25%	2-4 Ton/ Ha	2
		<2 Ton/ Ha	1

The result of weighting main factors that influence critical land was shown in a critical land map, as illustrated in Fig. 4. Most of the study area was dominated by non critical area by 95,66% (771,079 ha), while area of critical detected by 0,6 % (4,836 ha). Based on the land use/ land cover in the Komering Watershed, forest have a most tendency as a potential critical land by 60,74% (18,990 ha) of the total area critical potential, and then plantation by 32,28 % (10,090 ha) of the total area critical potential Fig. 5

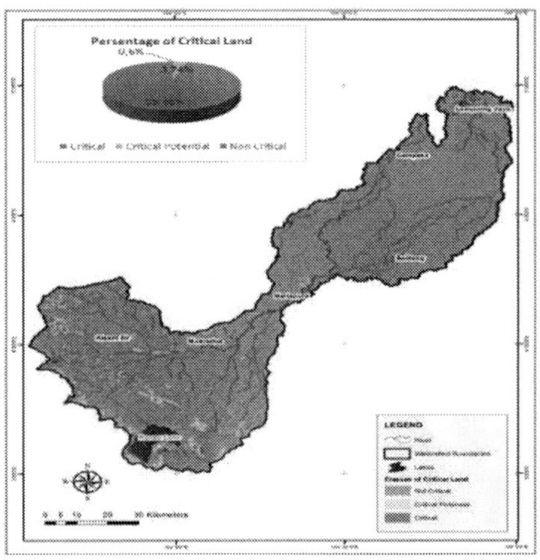

Figure 4. Map of critical land in Komering Watershed

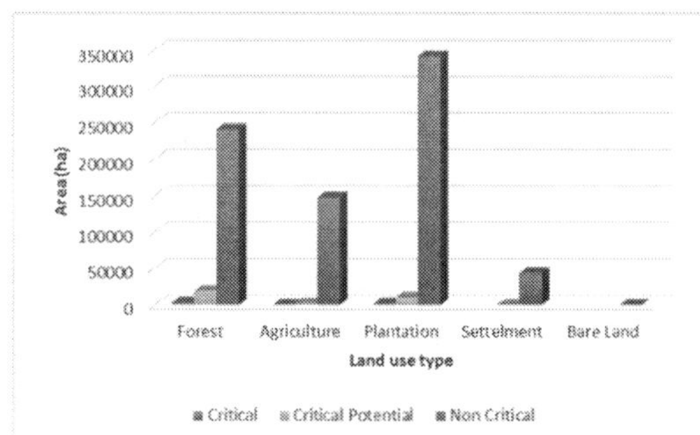

Figure 5. Comparison of Critical Land

4. Conclusions

Based on the results obtained by employment of GIS, Remote Sensing, and statistical analysis to achieve the specific research objectives, it is concluded that the increase in plantation areas from 37,25% to 38,93% is closed related with decrease in agriculture areas from 31,00% to 29,35%. There are three classes of critical land and the factors that contributed most to the difference between the classes namely; outcrop (48%), slope (27%), and the productivity (25%). The result of overlay analysis from the variables shown that study area was dominated by non critical area by 95,66% (771,079 ha), while area of critical detected by 0,6 % (4,836 ha). Based on the type of land use/ land cover, forest have a most tendency as a potential critical land by 60,74% of the total area critical potential.

References

[1] Sitorus S R P, Mashudi and Haridjaja O 2011 Pengembangan kriteria dan klasifikasi lahan kritis serta keterkaitannya dengan produktivitas lahan di kabupaten Bogor *Prosiding Seminar Nasional Sumberdaya Lahan Pertanian* Balai Besar Penelitian dan Pengembangan Sumberdaya Lahan Pertanian Kementerian Pertanian Bogor, November 30 – December 1, 2010
[2] Butt A, Shabbir R, Ahmad S S and Aziz N 2015 Land use change mapping and analysis using remote sensing and GIS: A case study of simly watershed Islamabad Pakistan. *The Egyptian Journal of Remote Sensing and Space Science* **18** p 251-259
[3] Agaton M, Setiawan Y and Effendi H 2015 Land use/land cover change detection in an urban watershed: a case study of upper citarum watershed West Java Province Indonesia *J.Procedia Environtmental Sciences* **33** p 654-660
[4] Haryanto E T, Herwanto T and Kendarto D R 2007 Perubahan bentuk penggunaan tanah dan implikasinya terhadap koefisien air larian DAS Citarum Hulu Jawa-Barat *Jurnal Bionatura* **9** p 1-15
[5] Irawan P 2006 *Penelitian kualitatif dan kuantitatif untuk ilmu-ilmu sosial* (Depok : Departemen Ilmu Administrasi UI)
[6] Sandi I M 1977 Penggunaan tanah (land use) di Indonesia Publikasi No 75 (Jakarta : Direktorat tata Guna Tanah Dirjen Agraria Departemen dalam Negeri)
[7] Arsyad S 1989 *Konservasi Tanah dan Air* (Bogor: Penerbit IPB Press)
[8] Sandi I M 1975 Tanah kritis sehubungan dengan usaha pertanian *Simposium Pencegahan-Pemulihan Tanah Kritis* Jakarta
[9] Bhan S 2013 Land degradation and integrated watershed management in India. *J. International Soil And Water Conservation Research* **1** p 49-57
[10] Rosenfield G H and Fitzpatirck L K 1986 *A coefficient of agreement as a measure of thematic classification accuracy.* Photogramm Eng Remote Sens **52** p 223–227
[11] Purwadi 2001 *Interpretasi Citra Digital* (Jakarta : Penerbit PT. Grasindo)
[12] Aswandi S R, Supriyo H and Hartono 2015 Faktor-faktor penentu kekritisan dan pengembangan kriteria indikator kekritisan ekosistem gambut tropika di Trumon dan Singkil Provinsi Aceh *Jurnal Manusia dan Lingkungan* **2** p 319-325
[13] Priyanto 2007 Penerapan analisa diskriminan dalam pembedaan kelas umur tegakan pinus *J. Manajemen Hutan Tropika* **13** p 155-165
[14] Owojori A and Xie H 2005 *Landsat image-based LULC changes of San Antonio, Texas using advanced atmospheric correction and object-oriented image analysis approaches* Paper Presented at the 5th International Symposium on Remote Sensing of Urban Areas Tempe, AZ
[15] Kadir S 2015 Penutupan tanah untuk pengendalian tingkat kekritisan DAS Satui Provinsi Kalimantan Selatan *Prosiding Seminar Nasional Masyarakat Biodiversitas Indonesia* **1** p 1190-1196
[16] Kementerian Agraria dan Tata Ruang 2016 *Petunjuk Teknis Inventarisasi Tanah Negara Bekas Hak, Bekas Kawasan, dan Tanah Kritis* (Jakarta: Direktorat Penatagunaan Tanah)
[17] Zulfikar 1999 *Aplikasi Penginderaan Jauh Dan Sistem Informasi Geografi (SIG) Untuk Mengidentifikasikan Dan Memetakan Lahan Kritis (Studi Kasus Pada Lahan Kritis Di Sub Das Bancak Propinsi Jawa Tengah)* (Bogor: Pascasarjana IPB)

Modelling of land use change in Indramayu District, West Java Province

L D W Handayani[1]*, M A Tejaningrum[2] and F Damrah[3]

[1] Center for Environmental Research Bogor Agricultural University, Jl. Lingkar Akademik, Bogor, 16680, Indonesia
[2] Soil Science and Land Resource Departement Bogor Agricultural University, Jl. Meranti, Bogor, 16680, Indonesia
[3] Forestry Departement West Sulawesi University, Jl. Prof. Dr. Baharuddin lopa, 91412, Indonesia

E-mail: handayani.luluk@gmail.com

Abstract. Indramayu District into a strategic area for a stopover and overseas from East Java area because Indramayu District passed the north coast main lane, which is the first as the economic lifeblood of the Java Island. Indramayu District is part of mainstream economic Java pathways so that physical development of the area and population density as well as community activities grew by leaps and bounds. Growth acceleration raised the level of land use change. Land use change and population activities in coastal area would reduce the carrying capacity and impact on environmental quality. This research aim to analyse landuse change of years 2000 and 2011 in Indramayu District. Using this land use change map, we can predict the condition of landuse change of year 2022 in Indramayu District. Cellular Automata Markov (Markov CA) Method is used to create a spatial model of land use changes. The results of this study are predictive of land use in 2022 and the suitability with Spatial Plan (RTRW). A settlement increase predicted to continue in the future the designation of the land according to the spatial plan should be maintained.

1. Introduction

Along with the increasing human activities in various sectors, especially the economic sector, hence the demand for land resources will also increase. The availability of land relatively fixed will lead to high competition of land use. Land use changes would be a problem if it occurs in areas that have a conservation function or agriculturally productive because it can cause food production decreased and environmental losses [1]. Land-use system is a natural, social and economic complex spatial dynamics with open, non-linear, comprehensive, random, hierarchical, dynamic, and uncertainty of complex characteristics. Dynamic evolution of urban land use is highly complex affected by the natural, social, economic, cultural, political and legal and other factors. At present, urban land use changed dramatically under the common influence of physical geography and human environmental system, affecting all aspects of sustainable urban development, the problems brought to mankind [2].

The CA-Markov land use change model can be applicable for spatial land use simulation and landcover reconstructions. Using satellite-derived land use or cover maps, Courage *et al.*[3] combined the Markov-cellular automata model to simulate land use changes in the Musana and

Masembura case in Zimbabwe. Although this study represented an important contribution to land use modeling by the integration of biophysical and socioeconomic data into a spatially explicit Markov cellular automata land use simulation model in a landscape, these simulation result lack of model validation and accuracy assessment. Some researchers in China used this robust approach to simulate the land use change and analysis the structure the land use in the future [4, 5].

In this paper, the study area was Indramayu District, West Java Province. Indramayu District into a strategic area for a stopover and overseas from East Java area because Indramayu District passed the north coast main line, which is the first as the economic lifeblood of the Java Island. Indramayu District is one district that included the Regional Development Ciayumajakuning (Indramayu-Cirebon-Majalengka-Kuningan) located in the northern part of West Java Province. This distric located on the northern coast of West Java Province and rear area (hinterland) from Cirebon City. Cirebon City be directed to the National Even Centre while Indramayu District directed into a Regional Activity Centre [6]. Indramayu District is part of mainstream economic Java pathways so that physical development of the area and population density as well as community activities grew by leaps and bounds. Growth acceleration raised the level of land use

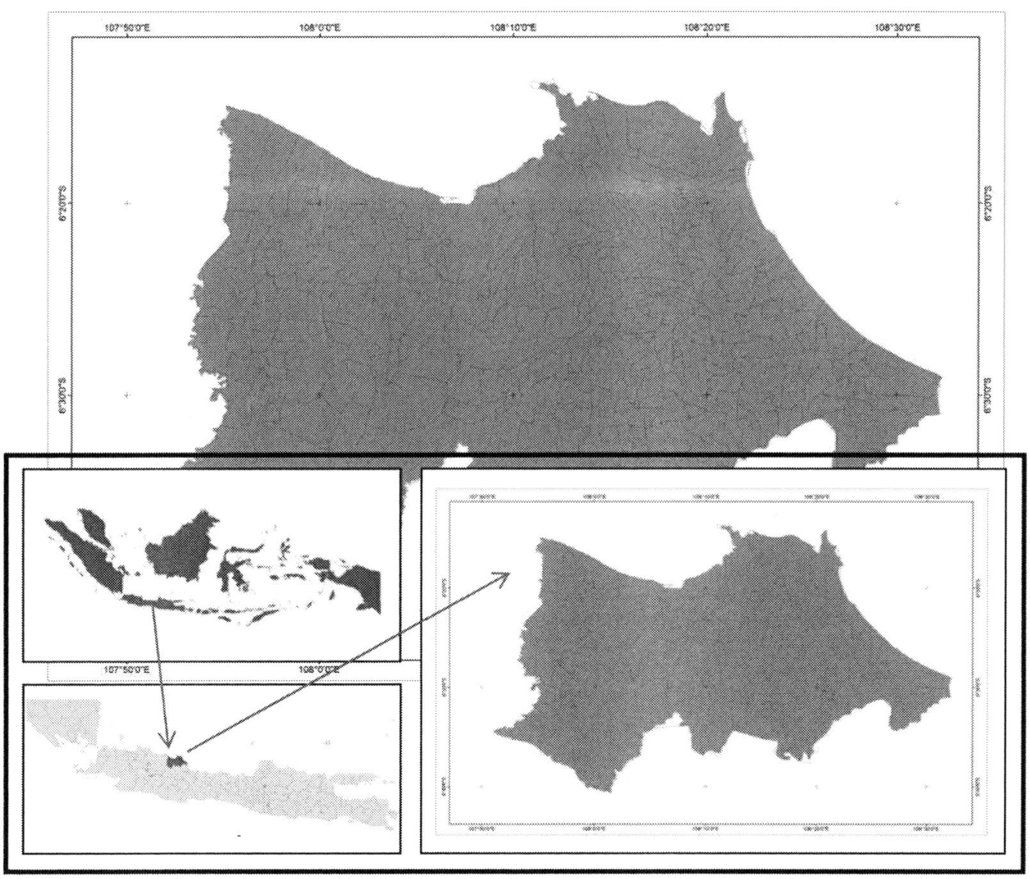

Figure 1. Study area

IOP Conf. Series: Earth and Environmental Science **54** (2017) 012021 doi:10.1088/1755-1315/54/1/012021

2.2. Data collection

The main data used in this research is satellite images data. The images data are Landsat 7 ETM (2000 and 2011). Besides the satellite images, this research also need Indonesia Digital Map (RBI) at Indramayu scene. This RBI data will be used for geometric correction. The tools that used in this research were varied. The main tools are ArcGIS 10.1 software and Idrisi Selva. These tools serve to collect, to save, and to analyze data.

Table 1. Research's material, source, and its function.

Material	Unit	Source	Tool's function
Digital Map (RBI)	Vector	Geospatial Information Agency	Earth Surface Map
Satellite Images Path 121 row 064 and 065	Landsat 7 ETM (Raster) 30x30 m pixel resolution	earthexplorer.usgs.gov U.S Geological survey	Land Use Change Analysis

2.3. Land use analysis

Satellite images obtained prior correction of geometry into place in accordance with the position of objects on the Earth's surface. The image that has been corrected and then a visual interpretation refers to the interpretation that the key hue, texture, pattern, size, shape, shade and sites [7]. Interpretation of results were then verified to determine the accuracy of interpretation by interpreters.

2.4. Land use suitability map

Land use suitability maps for each land use is one of the inputs in the process of CA-Markov. The series serves as a reference map of allocating a land use [8]. Land suitability map serves as a reference for land use allocation. Land suitability evaluation is an attempt to evaluate the use of land, where the land use should be tailored to the land characteristics (physical and chemical) [8]. In this study, land suitability analysis was limited to the physical characteristics of the land only, include: soil physical properties (rock material type, soil type and soil texture), slope, elevation, rainfall, lithology, and analysis of competitive commodities. Land suitability maps were made for eight land use. Land suitability classes consist of two classes of grade appropriate (S) and is not suitable (N). Conformance criteria based on the criteria [9].

2.5. Land use change analysis with Markov-Cellular Automata

Markov Chain is approach in order to get transition probability matrix from past periods of land use to predict land use in future periods. The main factor in Markov Chain is transition probability, a conditional probability for system to undergo transition to new state and gives current state of the system. Transition matrix is representing the changes among two land use occurred during that period. The changing happened in two different land use will form transition matrices. [10] explained that Markov Chain can be used to predict future matrix of vector (tn) based on past matrix of vector (t1) by transition probability matrix, while transition probability matrix is derived from probability changing between past matrix vector and previous (t1 and t0).

$$Q_{t+1} = Q_t^T P^n$$

Where:

$$\begin{matrix} q_1 \\ q_2 \\ \cdots \\ q_m \end{matrix} = \begin{matrix} q_1 \\ q_2 \\ \cdots \\ q_m \end{matrix}_t^{\;T} \begin{matrix} q_1 & q_1 & \cdots & q_1 \\ q_1 & q_1 & \cdots & q_1 \\ \cdots & \cdots & \cdots & \cdots \\ q_1 & q_1 & \cdots & q_1 \end{matrix}^{\;n} \qquad (1)$$

n = number of time steps
m = number of states
Q_t = vector initial states at an initial time t
Q_{t+1}= vector of states at the next time $t+1$
P = transition probabilities matrix

Cellular Automata (CA) is one of application in spatial dynamics modeling that mainly in land use change simulation [11]. Cellular Automata are neighborhood based cell while transition between it is determined by states in its neighborhood and probability transition. Standard cellular automaton can be defined as function that defines change of state from time past time to future and influenced by neighborhood of all cells [12]. Cellular Automata as modeling approaches has basic elements comprises of series of cells or grid, set of local states, a neighborhood and transition rules. There are several factors involved in Cellular Automata model such as Transition Probability which is from Markov Chain, Suitability factors and Neighborhood influences. The explanation of those factors explained as follows: suitability indexes and neighborhood indexes.

Inputs for Markov Chain were:
1. Land use image of t1=2000; t2=2011
2. Defining number of time periods between t1 and t2
3. Defining number of time periods to project forward from t2

Outputs from Markov Chain analysis were:
1. Transition probability matrix (TPM) which express probability of each land use category that will change to every other category.
2. Transition areas matrix which express the total area (number of pixels) which expected to change in next period.
3. A set of conditional probability images which express the probability that each pixel will belong to the designated class in next period.

Inputs for Cellular Automata model were
1. Basis land use image of t2
2. Transition Probability Matrix (TPM 2000-2011)
3. Land suitability
4. Defining the moving filter size as 5x5 where central pixel where this filter will move and weighted the pixel one by one.
5. Defining iteration of model: 2, 4, 8, 11, 22, 33

Outputs from Cellular Automata analysis were:
1. Predicted of land use of t=2022
   Firstly, CA-Markov modeling was conducted in order to validate the model at various iteration Validation was conducted by assessing predicted land use t=2022 with actual land use t=2011 using Kappa Accuracy assessment.
2. Predicted of land use change in future years t=2022
   Secondly, CA-Markov modeling was conducted to predict land use t=2022 by using iteration that had optimum accuracy result based on model validation. The result of predicted land use then was used to analyze future condition of land use.

The first model is done by executing input models with base year is 2000, and the matrix area trasition 2000-2011 resulting in a prediction map land use in 2011. Validation is done with the existing land use map in 2011 which is based on the value of kappa [13],

$$Kappa\ Accuracy = \frac{N\sum_{i=1}^{r} x_{ii} - \sum_{i=1}^{r}(x_{i+} \times x_{+i})}{N^2 - \sum_{i=1}^{r}(x_{i+} \times x_{+i})} \quad (2)$$

x_{+I} :wide of land use type-i observation result

x_{i+} : wide of land use type-i simulation result
x_{ii} : wide of land use type-i simulation result correspond with observation result
i : Path and row
r : number of land use type
N : total area all of land use type

Kappa value in validating the model is expected to get a percentage of more than 85%.

2.6. Control of land use change scenario

Preparation of land use change control scenarios in Indramayu District, carried out using land suitability based on the policies to be taken. The scenario in Indramayu geared to fit the pattern space Spatial Plan (RTRW) 2011-2031 Indramayu district. The scenario that has the smallest mismatch with space on spatial pattern is the most ideal scenario for controlling land use changes.

3. Results and discussions

3.1 Land use change 2000-2011

Based on the result at interpretation process, the images were classified into 8 classes: plantation, settlement, dryland farming, paddy field, scrub, fishpond, fishpond, open field, and water. Value kappa accuracy of Landsat imagery interpretation results is 85.01%. This classification produces land use map year of 2000 and 2011 (figure 2 and 3).

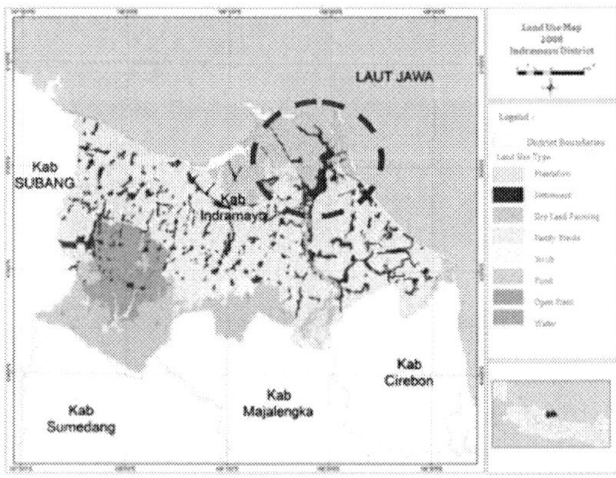

Figure 2. Land use map 2000

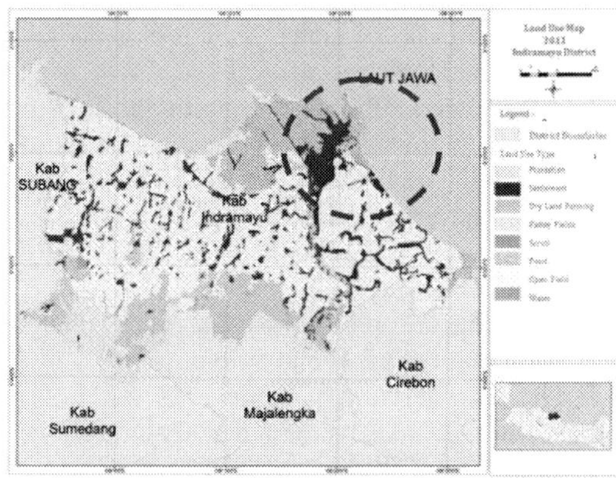

Figure 3. Land use map 2011

The paddy field landscape (figure 2 and 3) is still dominated the area in 2000 and 2011because Indramayu District dominated by alluvial deposits. Alluvial deposits came from sediment of river cause many sub watershed contained in Indramayu (32 sub watersheds), it suitable for paddy field that supported with rainfall ranging between 1900-4100 mm/year. Indramayu District dominantly used by slope class 1 (0-3%), that used for paddy field, settlement, and pond. The interpretation results showed the settlement increased from 2000 to 2011 (blue circle). Settlement area located in the northern part of the district, it is due to the northern coastal road into artery lane of transportation in Java Island. In the southern of the district bordering with Majalengka District, in 2000 was dominated by scrubs and dryland farming, this is due to have a slope grade 3 (8-15%) and 4 (15-45%), while in the year 2011 was dominated by dry land farming, and the scrub converted into a plantation. Indramayu District land use is dominated by paddy fields with an area of 146.115 ha is equivalent to 69.59% of total area. While the settlements area of 18.520 ha approximately 8.82% of the Indramayu District total area.

The prediction results of land use in 2022 compared to 2011 are presented in figure 4-9.

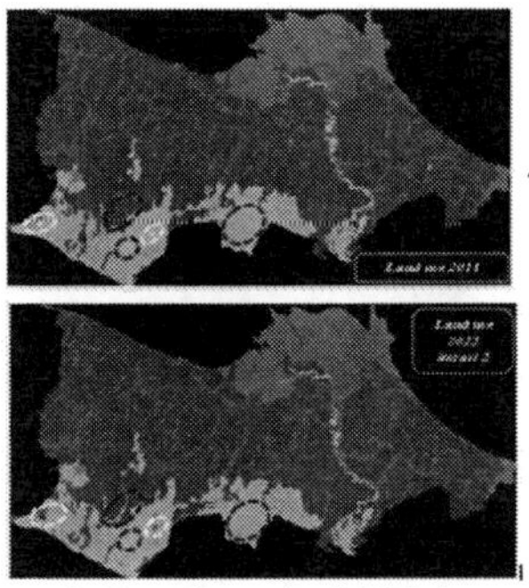

Figure 4. Land use in 2022 iteration 2

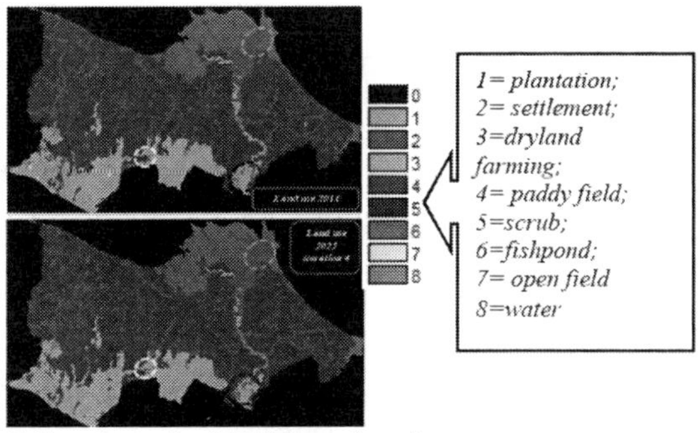

Figure 5. Land use in 2022 iteration 4

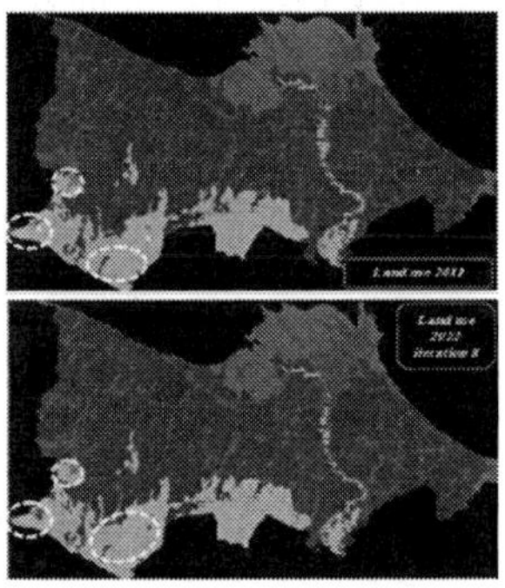

Figure 6. Land use in 2022 iteration 8

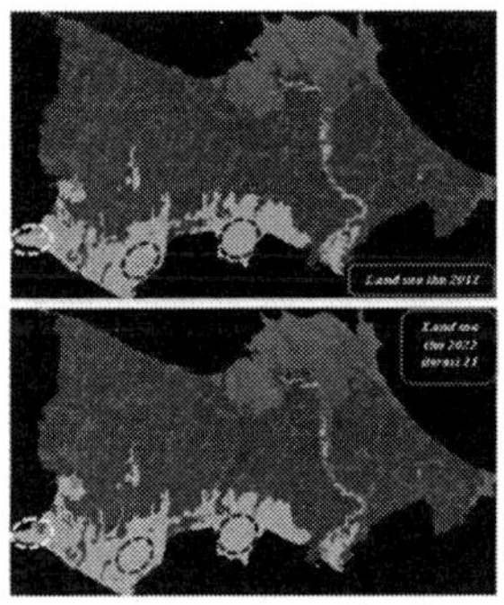

Figure 7. Land use in 2022 iteration 11

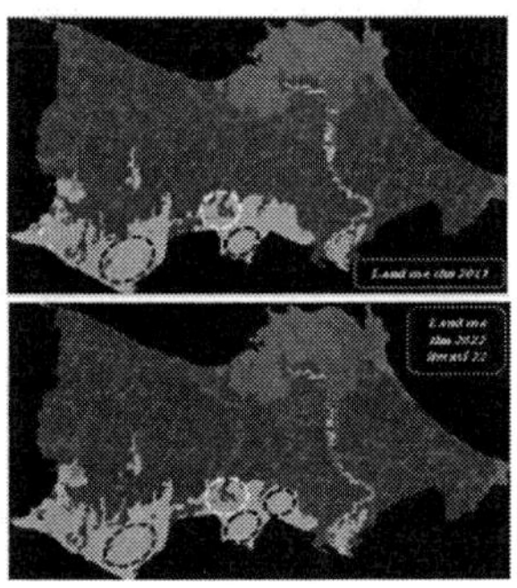

Figure 8. Land use in 2022 iteration 22

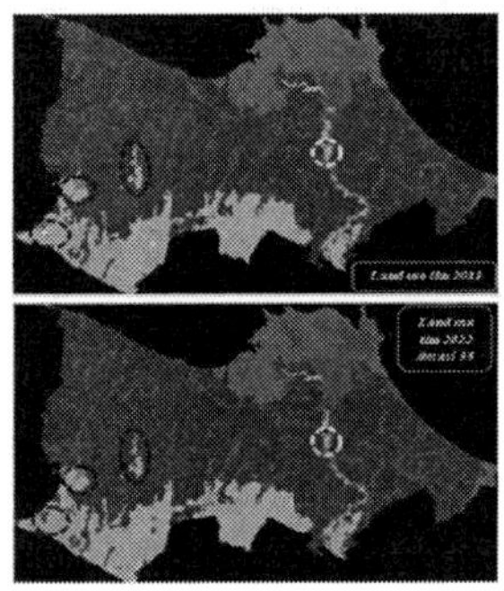

Figure 9. Land use in 2022 iteration 33

Based on the validation results is kappa values for all kinds of iterations shows that all products are estimated based on ca-Markov produced can be quite good for kappa values of all the results of the validation approach 1, where in kappa values of each iteration it is; iteration 2 is 0.9116 (91.16%), iteration 4 is 0.9079 (90.79%), iteration 8 is 0.9079 (90.79%), iteration 11 is 0.9086 (90.86%), iteration 22 is 0.9085 (90.85%) and iteration 33 is 0907 (90.7%). The highest kappa value validation results prediction of land use in 2011 obtained the iteration 2 is equal to 91.16%. Transitional Probababilty Matrix, land suitability, filter 5 × 5 and iteration 2 then used to make predictions of land use for a period of 11 years into the future is the use of land in 2022 (figure 8).

The matrix table (figure 9) shows that the numbers in red circles are a component of land use changes that seemed less appropriate, that occurs in class2-class4 (transition 236 cell of the settlement into paddy fields) and class2-class6 (transition 6 cell of the settlement into the pond) , Components class1-class1, class2-class2, class3-class3, class4-class4, class5-class5, class6-class6, class7-class7 and class8-class8 indicates that changes the class follow the types of land use before (predicted not easily turned into other land use class, for example will remain a paddy field).

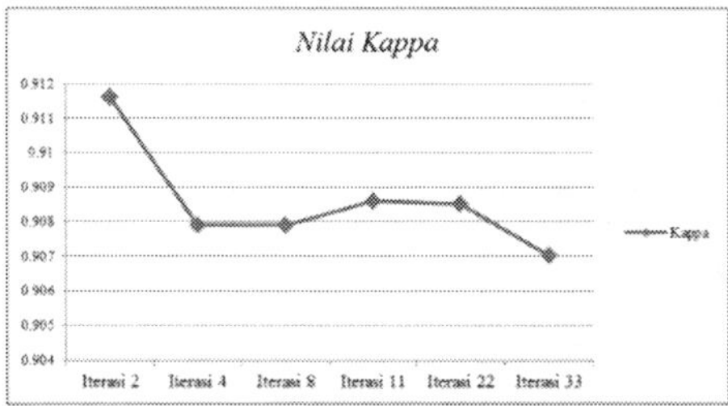

Figure 8. Comparison of Kappa value in transition the various iterations of use for estimations of land use for the year 2022

	Cl. 1	Cl. 2	Cl. 3	Cl. 4	Cl. 5	Cl. 6	Cl. 7	Cl. 8
Class 1 :	87207	0	0	0	0	0	0	0
Class 2 :	0	246206	17	236	0	6	0	0
Class 3 :	66629	8577	197882	14059	0	0	1	0
Class 4 :	0	2426	19390	1385778	0	1279	0	0
Class 5 :	328	0	1801	298	156	0	0	0
Class 6 :	0	24617	0	60	0	195536	0	0
Class 7 :	0	0	0	1888	0	0	20	0
Class 8 :	0	0	0	0	0	0	0	12046

Figure 9. Prediction number of cells undergoing

On the map, red circle shows unsuitability land use designation in the draft Spatial Plan (RTRW) Indramayu District, which in the north (coastal) is ponds land use while the allocation is protected areas (red circle).
in accordance with the analysis, it is known that unsuitability RTRW 2011-2031 with land use in 2022 in the amount of 67.576 ha and area suitibility with RTRW is 136.221 Ha.

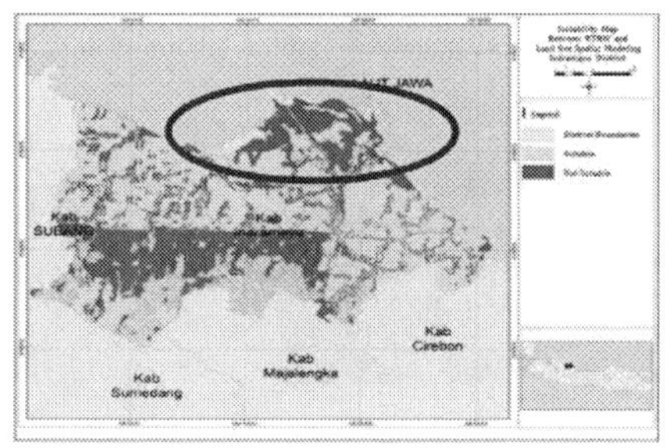

Figure 10. Suitability map

4. Conclusion

Land use changes in 2000-2011 increased in the total area of settlement (18 520 ha approximately 8.82% of the total area of Indramayu district). Based on modeling results, it is known that unsuitability RTRW 2011-2031 with land use in 2022 attained 67.576 ha. Hence necessary for government intervation related the control of Land functional shift. Thus, the designation of the land according to the spatial plan should be maintained. The intervention of the government can not directly change the previous land use, example pond as a protected area (mangrove). This is because the associated livelihoods of coastal residents. It is hoped that the restoration of the protected areas do not affect the population income and socio-economic conditions in the region.

References

[1] Pribadi D O, Shiddiq D and Ermyanyla M 2006 Model Perubahan Tutupan Lahan dan Faktor-Faktor yang Mempengaruhinya *Jurnal Teknologi Lingkungan* **7** 35-51

[2] Wang S Q, Zheng X Q and Zang X B 2012 Accuracy Assessment of Land Use Change Simulation Based on Markov-Cellular Automata Model *Procedia Environmental Sciences* **13** 1238-1245

[3] Kamusoko C, Aniya M, Adi B, Manjoro M and Eason G 2009 Rural sustainability under threat in Zimbabwe-Simulation of future land use/cover changes in the Bindura district based on the Markov-cellular automata model *Applied Geography* **29** 435-47

[4] Hou X Y, Chang B and Yu X F 2004 Land use change in Hexi corridor based on CA-Markov methods *Transactions of the CSAE* **20** 286-91.

[5] Yang G Q, Liu Y L and Wu Z F 2007 Analysis and simulation of land-use Temporal and spatial pattern based on CA-Markov model *Geomatics and Information Science of Wuhan University* **32** 414-7

[6] BAPPEDA 2012 Local Regulation No 1 *Spatial Plans Indramayu District 2011-2031.* Retrieved from *http://www.indramayukab.go.id.* on Desember 2015; Dated 2012 (in Indonesian)
[7] Lilesand M T and Kiefer R W 1993 *Penginderaan Jauh dan Interpretasi Citra [terjemahan]* (Yogyakarta (ID): Universitas Gadjah Mada)
[8] Munibah K 2008 Model Spasial Perubahan Penggunaan Lahan dan Arahan Penggunaan Lahan Berwawasan Lingkungan Studi Kasus DAS Cidanau, Provinsi Banten *Disertation* (Bogor: Sekolah Pascasarjana Institut Pertanian Bogor)
[9] Hardjowigeno and Widiatmaka 2007 *Evaluasi Kesesuaian Lahan dan Perencanaan Tataguna Lahan* (Yogyakarta: Gadjah Mada University Press)
[10] Ozah A P, Adesina F A and Dami A 2012 *A* Deterministic Cellular Automata Model for Simulating Rural Land Use Dynamics: A Case Study of Lake Chad Basin *IJEE* **2** 22-34
[11] White R and Engelen G 2000 High-Resolution Integrated Modelling of The Spatial Dynamics of Urban and Regional Systems *Comput Environ Urban* **24** 383- 400
[12] Li X and On Yeh AG 2001 Zoning Land for Agriculture Protection by The Integration of Remote Sensing, GIS and Cellular Automata *Photogramm Eng Rem S* **67** 471-477
[13] Jensen J R 1996 *Introductory digital image processing: A remote sensing perspective* 2nd Edition (New Jersey (US) Prenctice Hall)

Surface runoff estimation from various land use in Mapili Watershed using SCS Curve Number and Geographic Information System

Sri Malahayati Yusuf[1*], David Guluda[2], Trisumitra Jayanegara[3]
[1] Center for Environmental Research, Bogor Agricultural University, Indonesia
[2] Postgraduated Program, University of Tadulako, Central Sulawesi, Indonesia
[3] Agency for Watershed Management and Protected Forest of Lariang Mamasa, West Sulawesi, Indonesia

E-mail: malahayati10@gmail.com

Abstract. Mapili watershed which is located in West Sulawesi is dominated by secondary forest and agriculture land. The condition of the watershed is affected by land use and soil type. Land use is a form of interaction between man and land in order to meet their needs. Soil is function from organism, climate, topographic, material, and time. This information then will affect the hydrological condition in the watershed, especially surface runoff. The objective of study is to analyze the surface runoff in Mapili watershed. SCS curve number method that integrates with geographic information system was used in the analysis. The highest surface runoff value is about 887.6 mm and the highest of surface runoff coefficient is about 0.31.

1. Introduction

Land use is characterised by arrangements, activities, and inputs people undertake in a certain land cover type to produce, change or maintain it [1, 2]. Land as a natural resource is a utilized resource by humans to meet their needs. Land use that does not comply with the carrying capacity and soil and water conservation service will trigger land degradation [3-5]. Land degradation is one of the parameters that affect the hydrological aspects of the watershed, especially runoff [6-8].

Surface runoff is a part of rainfall that flowing on gound surface after the interception, depression storage, and infiltration takes place [9, 10]. The amount of measured runoff is a function of rainfall, soil and land use type [4, 9, 10]. The high of surface runoff is one cause of problems in a watershed [11, 12].

Mapili watershed which is located in Province of West Sulawesi, Indonesia is dominated by steep and very steep slope around 64%. Dominant land uses are upland agriculture mixed with shrub land (42.28%) and secondary forest (43.57%). The combination of these circumstances will contribute to the occurrence of surface runoff [13-15]. Therefore, it is necessary to estimate the surface runoff value in Mapili watershed so that the watershed condition can be determined.

Surface runoff value for Mapili watershed was estimated using SCS Curve Number method. The method is able to describe the effect of various type of land use on runoff in a watershed [16-19]. The integration between SCS Curve Number with Geographic Information System is able to provide the

IOP Conf. Series: Earth and Environmental Science **54** (2017) 012022 doi:10.1088/1755-1315/54/1/012022

information of runoff value spatially in the entire watershed, so the most problematic part of a watershed will be known [20-22].

2. Methodology

2.1 Watershed description

The research was conducted in Mapili watershed which is located in Mamuju and Polewali Mandar Region, Province of West Sulawesi (figure 1). The research area is dominated by secondary forest type (43.57%) and very steep slope (33%).

2.2 Research Procedure

The research procedure is described in Figure 2. From the figure, runoff value was estimated for each sub watershed in Mapili watershed base on soil, land use, topographic, and also rainfall data. Hydrology of a watershed is affected by soil data through hydrology soil group (HSG) information. HSG is defined by soil texture and minimum infiltration rate.

Effect of combination from various land use and soil type on watershed hydrology is described by curve number (CN). CN is determined based on HSG, land use type, and antecedent moisture condition (AMC).

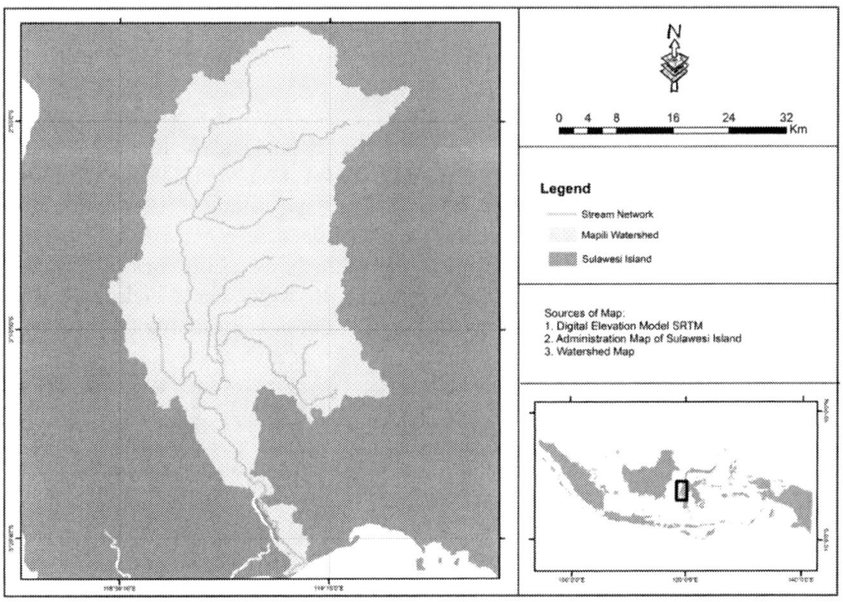

Figure 1. Study area

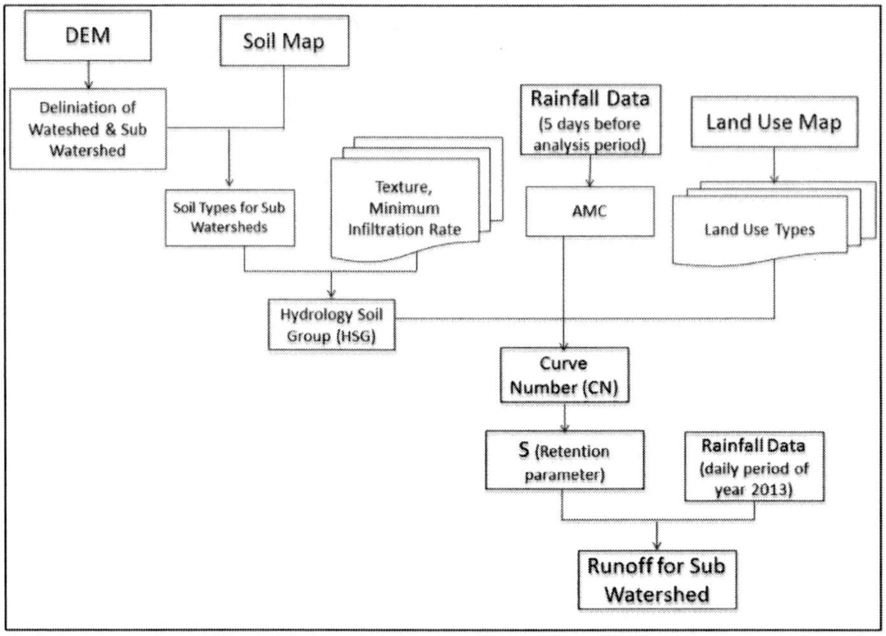

Figure 2. Procedure to estimate runoff value

3. Results and Discussion

Analysis of runoff for Mapili watershed was done in detail for each sub watershed. Sub watershed was divided using hydrological model with input data digital elevation model. Base on the process, there was 25 sub watersheds in Mapili watershed (figure 3). Then, hydrology soil group (HSG) for each sub watershed was defined base on soil texture data. From the data, it is known that Mapili watershed is dominated by HSG A (figure 4), where the soil in the watershed has lowest runoff potential [10, 16-18]. But, the total surface runoff also effected by slope in the watershed which is dominated by steep and very steep slope. Before water infiltrate to the soil profile, it will flowing on the ground surface as runoff.

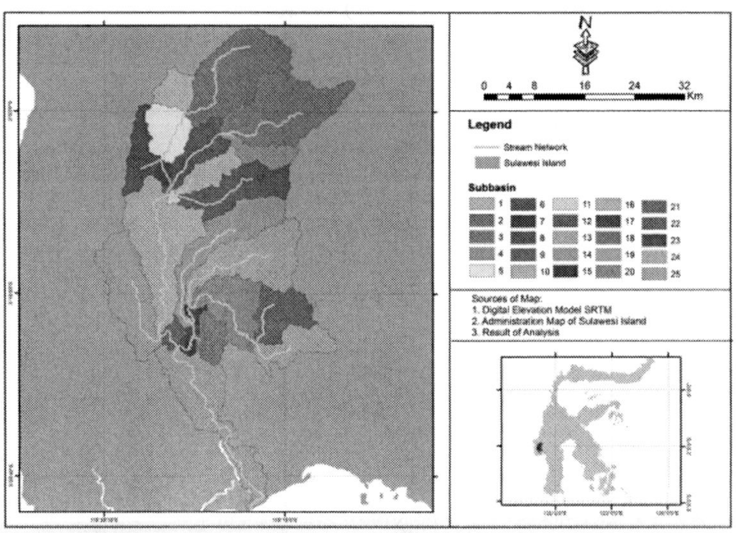

Figure 3. Map of sub watershed in Mapili watershed

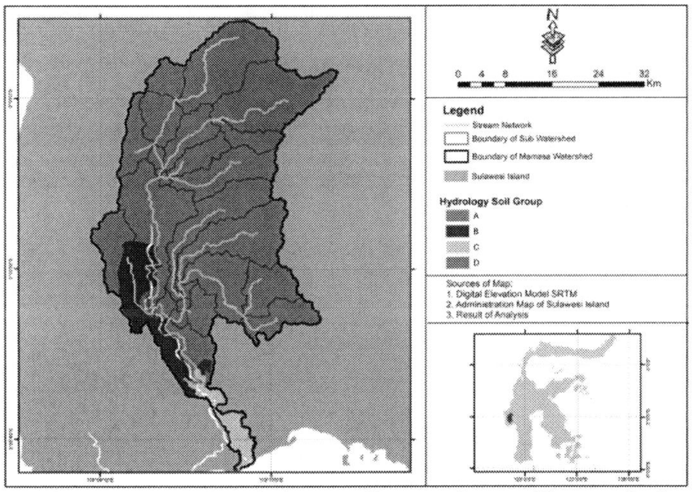

Figure 4. Map of Hydrology Soil Group in Mapili watershed

Data of HSG and land use map were used to define the curve number for each land use type in Mapili watershed. The curve numbers describes the condition of land use, soil and also the moisture condition [4, 18, 20]. The value of curve numbers in Mapili watershed are varies between 25 for forest area to 92 for water body (figure 5). It means that around 25 until 92% from rainfall will transform into surface runoff. The lowest surface runoff is 51.9 mm and the highest is 887.6 mm with the average value is around 398.90 mm (figure 6).

The status for each sub watershed can be determined using parameter of surface runoff coefficient. The coefficient of surface runoff is resulted from surface runoff value and rainfall data. From the data, Mapili watershed is dominated by coefficient surface runoff under 0.2 (figure 7). The highest coefficient surface runoff was derived from sub watershed number 4 as 0.31. This sub watershed is dominated by agriculture land and steep slope. Base on the Rule of Minister of Forestry of Republic Indonesia No. P. 61/MENHUT-II/2014, the coefficient surface runoff < 0.2 is categorized as low and very low class, while the highest value in Mapili watershed is categorized into medium class. It is mean that all sub watershed is still in good condition. But, all the stakeholders in watershed should keep this condition so the sustainable watershed can be reached.

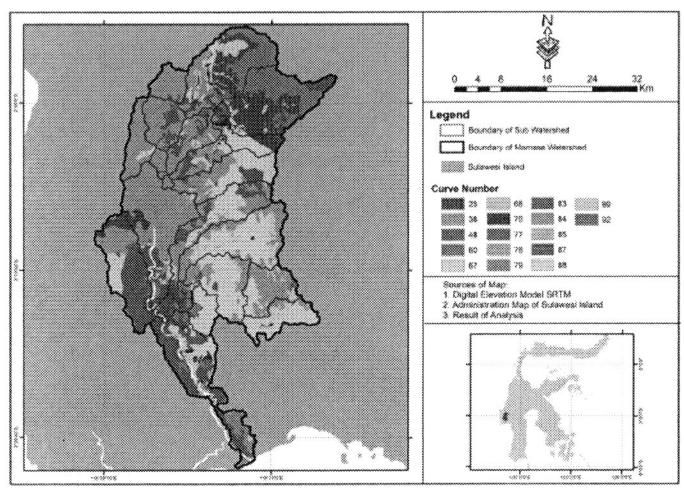

Figure 5. Map of curve number in Mapili watershed

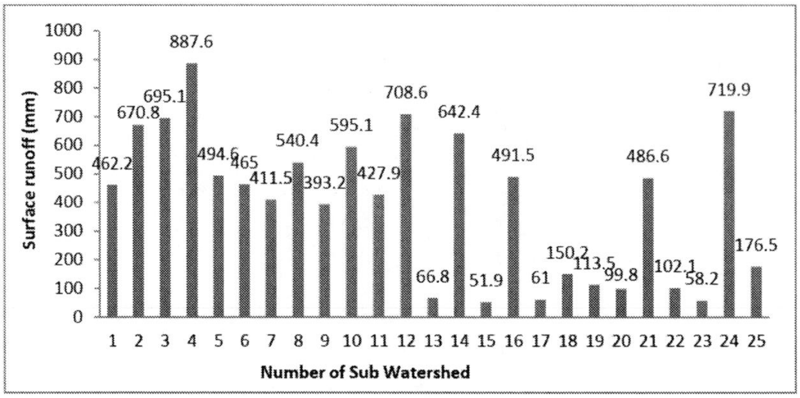

Figure 6. Surface runoff value for sub watersheds

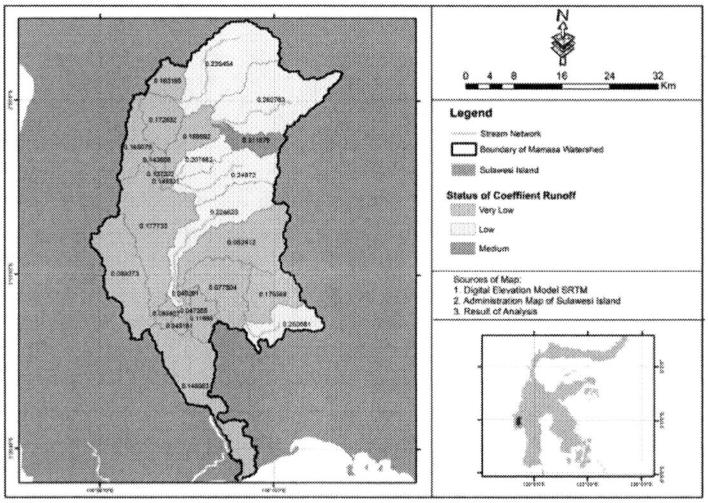

Figure 7. Map of surface runoff in Mapili watershed

4. Conclusion

The surface runoff value in Mapili watershed is range between 51.9 - 887.6 mm. The highest surface runoff derives the highest of surface runoff coefficient around 0.31. This value is classified into medium class of surface runoff condition.

5. References

[1] FAO 1999 *Land Resources* http:www.fao.org (accessed on July 10, 2016)

[2] Vink, A P A 1975 *Land Use in Advancing Agriculture* (Berlin Heidelberg, New York: Springer-Verlag)

[3] Agency for Watershed Management and Protected Forest of Citarum Ciliwung 2009 *Management planning of Citarum Ciliwung watershed* (Bogor: Agency for Watershed Management and Protected Forest of Citarum Ciliwung) (in Bahasa Indonesia)

[4] Arsyad S 2006 Konservasi tanah dan air (Bogor: IPB Press)

[5] Berry E and Levy J 1983 *Soil degradation problems and soil conservation techniques and management: use and implementation in Africa*, Resource Handbook **1** (New York: United State Agency for International Development)

[6] Ghimire C P 2014 *Hydrological impacts of reforesting degraded pasture land in the middle mountain zone of Central Nepal* (doctor dissertation) (Netherlands: ITC Printing Department)

[7] van Luijk G, Cowling R M, Riksen M J P M and Glenday J 2013 Hydrological implications of desertification: Degradation of South African semi-arid subtropical thicket *J. Arid Environ.* **91** 14-21

[8] Johansson J and Svensson J 2002 *Land degradation in the semi-arid catchment of Lake Baringo, Kenya: a minor field study of physical causes with a socioeconomic aspect* (Sweden: Earth Science Centre, Goteborg University)

[9] Linsley R K, Kohler M A and Paulhus J L H 1982 *Hydrology for engineers* 3rd Ed. (McGraw-Hill, Inc)

[10] Schwab G O, Frevert R K, Edminster T W and Barnes K K 1981 *Soil and water conservation engineering* 3rd Ed. (Canada: John Wiley & Sons, Inc)

[11] Valentin C, Agus F, Alamban F, Boosaner A, Bricquet JP, Chaplot V, de Guzman T, de Rouw A, Janeau JL, Orange D, Phachomphonh K, Phai DD, Podwojewski P, Ribolz O, Silvera N, Subagyono K, ′Baux JPT, Tran DT and Vadari T 2008 Runoff and sediment losses from 27 upland catchments in Southeast Asia: impact of rapid land use changes and conservation practices *J. Agriculture, Ecosystems and Environment* **128** 225–238

[12] Wang S, Kang S, Zhang L and Li F 2007 Modelling hydrological response to different land use and climate change scenarios in the Zamu River Basin of Northwest China *Journal of Hydrological Process* (Wiley Interscience)

[13] Pawitan H 2006 Land use change and its impact on watershed hydrology (Bogor: Hydrometeorology Laboratory, IPB) (in Bahasa Indonesia)

[14] Agency for Climate and Meteorology Research 2007 Analysis of land cover change and its effect on water balance and sedimentation in Lake Tempe http://balitklimat.litbang.deptan.go.id (accessed on 14 Mei 2012)

[15] Yusuf S M, Murtilaksono K, Harjianto M and Herlina E 2016 The utilization of satellite imagery data to predict hydrology characteristics in Dodokan Watershed *Procedia Environmental Sciences* **33** 36-43

[16] McCuen R H 1982 *A guide to hydrologic analysis using SCS method* (New Jersey: Prentice Hall, Inc)

[17] United States Department Agriculture, Natural Resources Conservation Service and Conservation Engineering Divison 1986 *Urban hydrology for small watersheds TR-55* (Washington: USDA)

[18] Natural Resources Conservation Service 1997 *Engineering hydrology training series: Module 205-SCS runoff equation* (Washington: NRCS)

[19] Tedela N H, McCutcheon S C, Rasmussen T C, Hawkins R H, Swank W T, Campbell J L, Adams M B, Jackson C R and Tollner E W 2012 Runoff curve numbers for 10 small forested watersheds in the mountains of the Eastern United States *J Hydrol. Eng.* **17** 1188-1198

[20] Nayak T, Verma M K and Hema B S 2012 SCS curve number method in Narmada basin International *Journal of Geomatics and Geosciences* **3** 219-228

[21] Gajbhiye S and Mishra S K 2012 Aplication of NRCS-SCS Curve Number model in runoff estimation using RS & GIS *IEEE-International Conference on Advances in Engineering, Science, and Management (ICAESM-2012)* (March 30-31, 2012)

[22] Ahmad I, VermaV and Verma M K 2015 Application of curve number method for estimation of runoff potential in GIS environment *2nd International Conference on Geological and Civil Engineering IPCBEE* **80** p 16-20

Assessing the ecological status of the Cisadane River's headwaters using benthic macroinvertebrates

M Krisanti[1], Y Wardiatno[1], Y M Anzani[2]*

[1] Department of Aquatic Resources Management, Faculty of Fisheries and Marine Science, Bogor Agricultural University, Kampus IPB Dramaga, Bogor 16680, Indonesia.

[2] Master Program in Aquatic Resources Management, Graduate School of Bogor Agricultural University, Kampus IPB Dramaga, Bogor 16680, Indonesia

E-mail: yunitamagrimaanzani@gmail.com

Abstract. Benthic macroinvertebrates are commonly used in river health biomonitoring. In monitoring program biotic indices are now widely established in water quality monitoring around the world, including in the tropical countries. The aim of this study was to reveal the ecological status of Cisadane River's headwaters in inside and outside of Mount Halimun-Salak National Park by using benthic macroinvertebrates. The research was conducted in the headwaters of Cisadane River located in Mount Halimun-Salak National Park. Macroinvertebrates were collected from four sites, i.e. inside the park (station 1, 2, 3, and 4) and from two sites outside the park (station 5 and 6). Collections were made twice a month, starting from April to June 2015 by means of Surber sampler (frame area 30x30 cm). A total of 65 genera from 38 families and 11 orders were found in the river. The results showed that based on diversity index, Lincoln Quality Index (LQI), Family Biotic Index (FBI), and Stream Invertebrate Grade Number Average Level 2 (SIGNAL 2), stations located within national park were ecologically better than those outside national park. Rivers with well-preserved riverside vegetation, as in the national park area have greater ecological status.

1. Introduction

The headwaters of Cisadane River are originally from Salak and Gede-Pangrango Mountains and flow into Teluk Naga coastal waters. The river has multiple uses to fulfill agricultural, domestic, and industrial needs [1] and provides economic values to people living in the cities where the river goes through [2].

As an aquatic ecosystem, headwaters of stream is favorable habitat for several benthic macroinvertebrate animals and is mostly dominated by aquatic insects [3, 4] and plays important food source for fish, birds, and other vertebrates [5]. These organisms has been widely used as biological indicators (bio-indicators) to assess the health status of the river [6]. For example, 49 of the 50 states in the United States use benthic macro invertebrate in water quality monitoring [7].

Monitoring of water quality can be done in various ways, such as with the physical and chemical analyses of water and biological analysis. The use of biological materials as bio-indicators in the management of waters is needed, because directly related to the ecological conditions or the health of aquatic ecosystems. Biological measurements provide a comprehensive and integrated assessment of the waters health from time to time [8]. Biota characteristics such as the presence or abundance can be an overview of the status change or an environmental condition. Because of these reasons these biotas serve as a tool to continuous monitor for the waters. One of the Cisadane River's headwater is located in Mount Halimun-Salak National Park (MH-SNP).

The aim of this study was to reveal the ecological status of Cisadane River's headwaters in inside and outside of Mount Halimun-Salak National Park by using benthic macroinvertebrates.

2. Material and methods

The research was conducted in the headwaters of Cisadane River. One of the Cisadane River's headwater is located in Mount Halimun-Salak National Park. Macroinvertebrates were collected from four sites inside the park (station 1, 2, 3, and 4) and from two sites outside the park (station 5 and 6) (figure 1). Ten replicates of samples were collected at each station. Collections were made twice a month, starting from April to June 2015 using Surber sampler (frame area 30x30 cm) and preserved using 70% alcohol. Macroinvertebrates were identified up to genus level following identification guides [9]. Some environmental characteristics were recorded, i.e. hydrological characteristics, velocity, temperature, dissolved oxygen (DO), and pH. Environmental characteristics were measured in situ at the time of biota sampling. Macroinvertebrate biotic indices were used is diversity index, Lincoln Quality Index (LQI), Family Biotic Index (FBI), and Stream Invertebrate Grade Number Average Level 2 (SIGNAL 2). Dendrogram of grouping stations in Cisadane's River headwater was used by using Bray-Curtis Index [10]. One-way ANOVA was performed to detect a significant difference in benthic macroinvertebrate diversity between stations inside and stations outside the national park at p= 0.05.

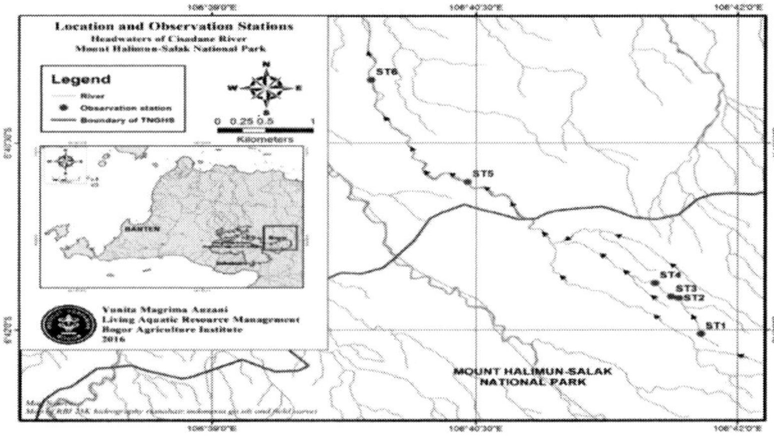

Figure. 1. Map of location and observation stations in the headwaters of Cisadane River, West Java - Indonesia.

2.1. Biotic indices

Biotic indices as numerical estimators of stream health, based on the tolerance or sensitivity of macroinvertebrates to an environmental gradient, are now widely established in monitoring regimes around the world. A few biotic indices have been developed for Southeast Asia, and more generally the tropics [11]. Diversity of species shows the number of organisms present in an area. To determine the species present within a community and the level of diversity can be seen with the diversity index (H') [12]. Evenness index (E) is the composition of individuals of each species contained in a community [12]. This is obtained by comparing the diversity Index to the maximum value of diversity index. If E value approaches 0, means there is a dominance of species. While if E value approaches 1, means number of individuals of each species was same or no dominance.

LQI method was used to determine the environmental criteria. Organisms that have been identified to the family level, given a score [13], then the scores were summed whole and of the values obtained BMWP. BMWP value divided by the number of taxa to get the value of Average Score Per taxon

(ASPT). The calculation of the value BMWP and ASPT used to determine the value of Overall Quality Rating (OQR). OQR value is used to provide Lincoln Quality Index (LQI) [13].
FBI method was used to determine which organisms are more sensitive to dissolved oxygen content in an aquatic ecosystem impacted by organic matter pollution [14]. Calculations were performed using the multiplication of the value indicator based on relative abundance of organisms with a score of each family. The total amount is divided by the total number of organisms, then the results adjusted to the water quality criteria [15]. SIGNAL 2 was developed for assessing water quality based on the presence of particular benthic macroinvertebrate to indicate the type of physics and chemistry pollution. The calculation is performed by using a weighting factor multiplied by the score of each family, and the multiplication result is summed. The sum of the multiplication is divided by the total number of weighting factors, and the value obtained SIGNAL 2. The value obtained plotted on the graph associated with a number of families [16].

3. Result and discussion

3.1. Diversity, evenness, and Simpson's dominance indices

A total of 65 genera from 38 families and 11 orders benthic macro invertebrate were found in the river. The value of diversity (H'), evenness (E), and Simpson's dominance index (C) in each station in Cisadane's River headwater can be seen in table 1. The value of diversity index was of relatively high at stations located inside the national park (station 1, 2, 3, and 4) than at stations outside the national park (station 5 and 6). Benthic macroinvertebrate diversity was significantly higher inside the national park than outside the park (ANOVA, $p<0.05$).

Table 1. Diversity, evenness, and Simpson's dominance indices of Cisadane's River headwaters

	St 1	St 2	St 3	St 4	St 5	St 6
Diversity (H′)	1.00	1.05	1.04	1.02	0.53	0.61
Eveness (E)	0.79	0.81	0.81	0.87	0.50	0.67
Dominance (C)	0.16	0.13	0.14	0.11	0.43	0.34

Riverside conditions at the stations within national parks are mostly forests and located in protected area. Well-preserved riverside vegetation, as in the National Park Area, can prevent soil erosion and nutrient release into adjoining rivers because it stabilizes the river banks. Consequently, rivers with preserved watersheds have greater biodiversity [17]. The diversity index was lower at stations that are outside the national park. The decrease of diversity index was partly due to changes in the composition and balance level of macroinvertebrate taxa abundance [18]. The values of dominance index were relatively higher at stations outside the national park compared with stations located inside the park. This indicates that there are one or a few organisms that dominate at stations outside the national park. Dominance may happen, because only a few types of organisms are able or tolerant to their lives environment [19].

3.2 LQI, FBI, and SIGNAL 2

The value of LQI in the headwaters of Cisadane's River can be seen in table 2. Based on table 2, the value of LQI in rivers located within national park (station 1, 2, 3, and 4), was A + up to A ++. While the stations outside the park (station 5 and 6) the value was A. All of those LQI values were still classified the rivers into the excellent water quality criteria.

Table 2. LQI values of Cisadane's River headwaters. Determination was in accordance to LQI water quality criteria [13]

	OQR	LQI	Water Quality Criteria
St 1	6.0	A++	*excellent*
St 2	6.3	A++	*excellent*
St 3	6.3	A++	*excellent*
St 4	5.9	A+	*excellent*
St 5	5.3	A	*excellent*
St 6	5.1	A	*excellent*

FBI values of Cisadane's River headwaters are shown in table 3. According to table 3, the water quality criteria of rivers located within national park (station 1, 2, 3, and 4) were classified as good to excellent. While the river that located outside the park (station 5 and 6) was moderate to quite bad.

Table 3. FBI values of Cisadane's River headwaters. Determination was in accordance to FBI water quality critera [15].

	FBI	Water quality Criteria
St 1	4.45	good
St 2	4.23	excellent
St 3	4.08	excellent
St 4	4.44	good
St 5	5.67	moderate
St 6	5.93	quite bad

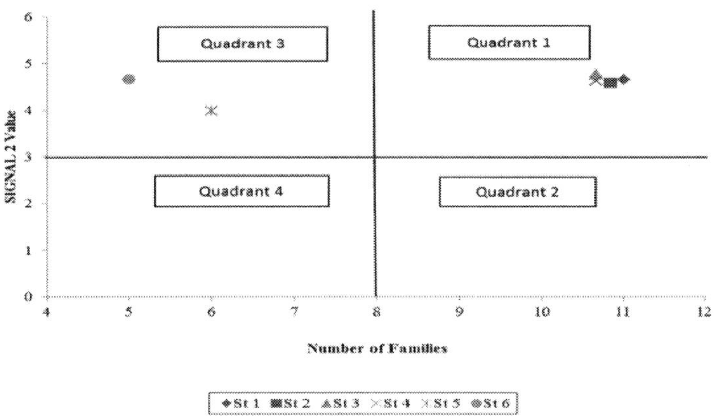

Figure 2. Graph of SIGNAL 2 scores based on the number of families that are found in the headwaters of Cisadane's River.

Position of SIGNAL 2's score of each station based on the number of families is seen in figure 2. As seen in figure 2, the point scattered in quadrants 1 and 3. Stations located within national park (station 1, 2, 3, and 4) are posted in quadrant 1. Quadrant 1 means that the diversity of benthic macroinvertebrate was high and there was no ecological stress factor in the stations. Quadrant 1 illustrates good water conditions (lack of interference) [16]. While the stations that located outside national park (station 5 and 6) are in quadrant 3. Quadrant 3 indicated those rivers have been contaminated. This quadrant showed that water quality criteria of river was moderate polluted waters [16].

Based on these three indices (LQI, FBI, and SIGNAL 2), the river was located within national park tend to have better water quality than the river was located outside national park. FBI is more advisable to use, as the index is known to be more specific in the assessment of the sensitivity of aquatic organisms to environmental conditions [20].

3.3 Grouping stations based on biological index
The level of similarity from the four biological indexes (Diversity index, LQI, FBI, and SIGNAL 2) was calculated and made groupings. Dendrogram of grouping station based on biological indices can be seen in figure 3.

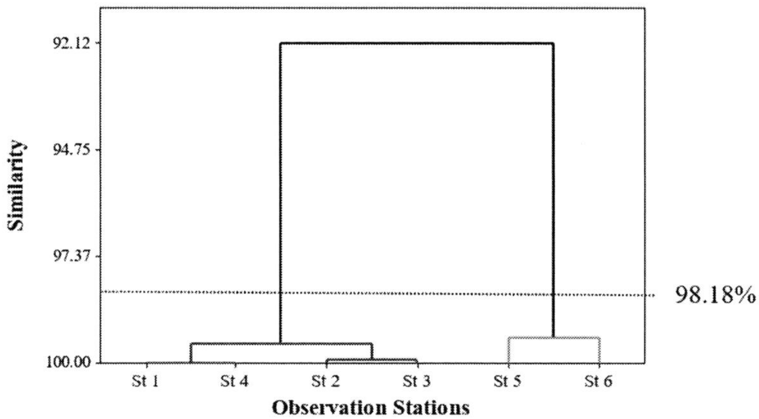

Figure 3. Dendrogram of grouping stations based on biology indices in the headwaters of Cisadane's River

At the similarity level of 98.18%, grouping stations are divided into two groups. Group 1 consisted of Stations 1, 2, 3, and 4 (located within national park). While the second group consisted of station 5 and 6 (located outside of national park). Some factors that affect the groupings are the type of families, number of families, and the abundance of benthic macroinvertebrate. In addition, at the similarity level of 92.12%, the sixth station is still in the same group. This is because the entire station was still on the order of 1-3 river, which means the sixth station is still part of headwaters [21].

3.4 Environmental characteristics
Benthic macroinvertebrates are affected by changes in the environmental characteristics of rivers such as width, depth, type of substratum, water velocity, and physicochemical variables [22]. The results of environmental characteristics measured in Cisadane's River headwaters given in table 4.

The velocity at four stations which located within national park was quite high, in contrast to that of the two stations outside the park. It is influenced by differences in gradient/slope of the river [23]. The water temperature at stations within national park area tended to be low, while outside were higher. This was presumably because stations within national park area were of higher elevation, canopy coverage was high, and with 100% green area coverage. In contrast, stations outside national park area were of lower elevation, canopy coverage was low, and with 16-55% green area coverage.

Table 4. Environmental characteristics of Cisadane's River headwaters

Parameter	Unit	St 1	St 2	St 3	St 4	St 5	St 6
River width	m	1.7	3.1	2.9	2.5	11.4	7.0
River body width	m	2.7	4.3	3.7	3.0	14.0	13.2
Depth	cm	20.7	19.4	20.3	19.2	18.1	18.5
Altitude	mdpl	1168	1090	1075	1020	717	584
Canopy coverage	%	90-100	60-80	40-60	0-20	0-10	0-10
Green area coverage	% (right)	100	100	100	100	55	16
	% (left)	100	100	100	100	20	27
Velocity	m/s	0.45	0.43	0.37	0.38	0.22	0.28
Temperature	^{0}C	20.08	20.33	20.48	22.57	24.15	27.78
DO	mg/L	7.08	6.53	6.70	5.67	5.18	4.80

Dissolved oxygen of the stations inside the national park area tend to be high, while those outside the park were lower. This is influenced by the low water temperature, causing the oxygen solubility is high. Besides that, the high velocity also increases oxygen in the waters through diffusion [24]. Benthic macroinvertebrate mostly dominated by aquatic insect prefers a cold and clean river with high dissolved oxygen concentration [25, 26]. Environmental characteristics stations located within the national park more support benthic macroinvertebrates that live in these waters. This is suitable with the results of the four indices and grouping stations based on these indices.

4. Conclusion

The value of diversity index in rivers located within national park has a range of 1-1.05, while in outside the park has a range 0.53-0.61. LQI value in rivers located within national park was A + up to A ++, while in outside the park the value was A. the water quality criteria of rivers located within national park were classified as good to excellent. While the river that located outside the park was moderate to quite bad. Based on graph of SIGNAL 2, river located within national park are posted in quadrant 1 (good water condition). While the river that located outside the park are posted in quadrant 3 (moderate polluted waters). The results showed that based on diversity index, LQI (Lincoln Quality Index), FBI (Family Biotic Index), and SIGNAL 2 (Stream Invertebrate Grade Number Average Level 2), stations located within national park were ecologically better than those outside national park. Rivers with well-preserved riverside vegetation, as in the national park area have greater ecological status.

References

[1] Siahaan R, Indrawan A, Soedharma D and Prasetyo L B 2012 Keanekaragaman makrozoobenthos sebagai indikator kualitas air Sungai Cisadane, Jawa Barat-Banten *Jurnal Bioslogos.* **2** 1-9

[2] Rahayu D M, Yoga G P, Effendi H, and Wardiatno Y 2015 Penggunaan makrozoobenthos sebagai indikator status perairan hulu Sungai Cisadane, Bogor *JIPI.* **20** 1-8

[3] Anzani YM, Effendi H, and Wardiatno Y 2013 Makrozoobenthos sebagai bioindikator kualitas perairan Sungai Ciambulawung *Prosiding Pertemuan Tahunan Masyarakat Limnologi Indonesia* 101-115.

[4] Yoga G P, Lumbanbatu D T, Riani E, and Wardiatno Y 2014 Secondary production of the net-spinning cadisfly, *Cheumatopsyche* spp. (Trichoptera: Hydropsychidae) inmercuric contaminated river. *J Trop Biol Conserv.* **11** 1– 12

[5] Myers L W, Kondraftieff B C, Mihuc T B and Rutter D E 2011 The mayflies (Ephemeroptera), stoneflies (plecoptera), and caddisflies (Tricoptera) of the Adirondack Park (New York State). *Trans Am Entomol Soc.* **137** 63-140

[6] Sudarso Y, Yoga G P, Suryono T, Syawal M S, and Yustiawati 2009 Pengaruh aktifitas antropogenik di Sungai Cikaniki (Jawa Barat) terhadap komunitas fauna makrobenthik *Limnotek.* **16** 153-166
[7] [US-EPA] U.S. Environmental Protection Agency 2002 *National Recommended Water Quality Criteria* (Washington DC (US): USEPA)
[8] Karr J R 1999 Defining and measuring river health *Fresh Biol.* **41** 221-234
[9] McCafferty W P 1983 *Aquatic entomology.* (London: Jones and Bartlett Publisher) p 488
[10] Bray J R and Curtis J T 1957 An ordination of the upland forest communities of Southern Wisconsin. *Ecol Mon.* **27** 325-349
[11] Blakely T J, Eikaas HS, and Harding JS 2014 The Singscore: a macroinvertebrate biotic index for assessing the health of Singapore's stream and canals *Raffles Bulletin of Zoology.* **62** 540-548.
[12] Krebs C J 1989 *Ecological methodology* (New York: Harper Collins Publisher Inc). p 654
[13] Mason C F 1991 *Biology of freshwater pollution Ed ke-2* (New York (US): Longman, Inc) p 351
[14] Cairns Jr J, Dickson K L 1971 A simple method for the biological assessment of the effects of waste discharge on aquatic bottom dwelling organism *J Water Pollut Control Fed.* **43** 755-765.
[15] Hilsenhoff W L 1988 Rapid field assessment of organic pollutionwith a family-level biotic index *J N Am Benthol Soc.* **7** 65-68
[16] Hauer F R and Lamberti G A 2007 *Methods in stream ecology Ed ke-2* (California: Academic Press) p 877
[17] Osborne L L and Kovacic DA 1993 Riparian vegetated buffer strips in water quality restoration and stream management *Fresh Biol.* **29** 243-258
[18] Sudarso Y, Wardiatno Y, Sualia I 2008 Pengaruh kontaminasi logam berat di sedimen terhadap komunitas bentik makroavertebrata: Studi kasus di Waduk Saguling-Jawa Barat *JIPPI.* **15** 49-59.
[19] Chessman B C 2003 New sensitivity grades for Australian river macroinvertebrates *Mar Fresh Res.* **54** 95-103
[20] Gitarama A M, Krisanti M and Agungpriyono D R 2016 Komunitas makrozoobenthos dan akumulasi kromium di Sungai Cimanuk Lama, Jawa Barat *JIPI.* **21** 48-55.
[21] Vannote R L, Minshall G W, Cummins K W, Sudell JR and Cushing CE 1980 The river continuum concept *Can J Fish Aquat Sci.* **37** 130-137
[22] Arimoro F O 2009 Impact of rubber effluent discharges on the water quality and macroinvertebrate community assemblages in a forest stream in the Niger Delta, Nigeria *Chemosphere.* **77** 440–449
[23] Effendi H, Romanto and Wardiatno Y 2015 Water quality status of Ciambulawung River, Banten Province, based on pollution index and NSF-WQI *Procedia Environmental Sciences.* **24** 228-237
[24] Hershey A E, Lamberti G A, Chaloner D T and Northington R M 2010 Aquatic insect ecology.Ecology and Classification of North American Freshwater Invertebrates *Elsevier.* 659-694.
[25] Bottova K, Derka T, Beracko P and de Figueroa J 2013 Life cycle, Feeding and secondary production of Plecoptera community in a constant temperature stream in Central Europe. *Limnologica.* **43** 27-33
[26] Anzani Y M, Krisanti M, Wardiatno Y 2016 The diversity and density of Plecoptera in the headwaters of Cisadane River, West Java, Indonesia *Journal of Biodiversity and Environmental Sciences.* **9** 126-132.

IOP Conf. Series: Earth and Environmental Science **54** (2017) 012024 doi:10.1088/1755-1315/54/1/012024

Carbon stock and plants biodiversity of *pekarangan* in Cisadane watershed West Java

Tatag Aisyah Filqisthi* **and Regan Leonardus Kaswanto**
Department of Landscape Architecture, Faculty of Agriculture, Bogor Agricultural University

E-mail: aisyahfilqist@gmail.com

Abstract. The presence of vegetation in *Pekarangan* can be proposed to mitigate global climate change impacts by CO_2 sequestration and at the same time to promote the availability of food for the community. The aims of this research is to calculate carbon stock and biodiversity in *pekarangan*, and to compare carbon stock and biodiversity on three levels of Cisadane Watershed. Four groups of *Pekarangan* defined on a purposive random sampling. Allometric models were developed to estimate aboveground biomass of vegetation, and an inventory was conducted in 48 *pekarangan*. Shannon Weiner Index (H') and Margalef Index (D_m) are used to evaluate biodiversity, averaged 2,84 and 5,10 (G1); 2,55 and 4,27 (G2); 2,56 and 4,52 (G3); 2,68 and 4,84 (G4), while carbon stock averaged 33,20 Mg Carbon/ha (G1); 29,97 Mg/ha (G2); 59,18 Mg/ha (G3); and 40,98 Mg/ha (G4). There is no relationship between biodiversity with carbon stock on pekarangan (R^2 = 0,02), or tree's biodiversity with carbon stock (R^2 = 0,23). High resolution satellite imagery can be used to extrapolate carbon stock and plants biodiversity of *Pekarangan* at watershed level.

1. Introduction

Pekarangan is one of traditional agroforestry systems that usually has biological, ecological, and economic interactions among the component. Its presence also important in households diet and social activities. Multistory level of vegetation structures and species richness of *pekarangan* can be promoted as supporting agricultural land for food security at the household level [1]. *Pekarangan* production contributed 11.5% of household's income and 12.9% of household's diet in term of food expense [2]. Biodiversity in pekarangan also contribute to diversification of products and diets. One of ecological functions of *pekarangan* is carbon sequestration. The existence of woody plants contribute in absorbing CO_2 in the atmosphere which increasing. Agroforestry pattern pure stands can absorb carbon dioxide 15.4-80.2 Mg / ha and mixture agroforestry 10.4-73.8 Mg / ha [3]. Based on these facts, existence of pekarangan play an important role to reduse GHGs emission, especially CO_2. Unlike others agroforestry systems, pekarangan not only sequester C in biomass and soil, but also reduce fossil-fuel burning by promoting woodfuel production, and conserve agrobiodiversity [4]. Other functions is conservation of genetic resources. The complex pattern with traditional land use systems such as tropical homegardens, have a potential to maintain high species richness and diversity.

Sequestration carbon from pekarangan potensial as a CO_2 mitigation and also build environment-friendly society. Landscape services provided by well-managed pekarangan will promote to establish low carbon society (LCS), means that environment friendly society and aware about concept of sustainability so that promote acceleration of agricultural productivity [5]. Therefore it is necessary to analyze plants biodiversity and carbon stored in pekarangan and also the correlation between them.

This study aimed to analyze and compare plants biodiversity and carbon stored in pekarangan in three levels Cisadane watershed. This watershed divided into three levels, upper stream, middle stream and downstream. Measurements carried out on those levels and then compared. Analysis of carbon stored conducted by estimation species biomass using alometric equations with different measured variables such as DBH, height, and others, while plants biodiversity analyzed with Shannon-Wiener index and Margalef index.

2. Study Site

This research site is Cisadane watershed, West Java. The objects are vegetation structures in pekarangan in upper stream, middle stream and downstream Cisadane watershed. One village were randomly selected as sample in each those levels are Tegal Waru on Ciampea district, Putat Nutug on Ciseeng district and Karawaci on Karawaci district (Figure 1).

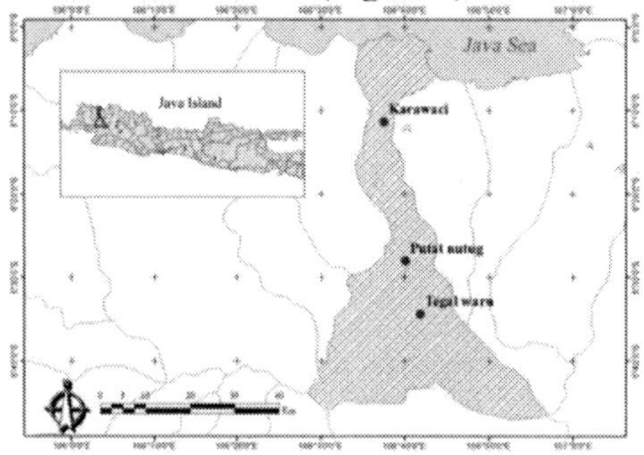

Figure 1 Study areas on Cisadane watershed

3. Method

3.1. Sampling design for selecting pekarangan

The pekarangan samples are divided into four groups, G1, G2, G3, and G4. G1 is pekarangan size less than 120 m^2 with no other agricultural land (OAL), G2 is pekarangan < 120 m^2 with OAL, G3 is pekarangan 120 – 400 m^2 with no OAL and G4 is pekarangan 120 – 400 m^2 with OAL (**Figure 2**) [2]. In each village four group were selected besed on the sample frame. In total, 48 pekarangan were investigated. The consideration is the size of pekarangan and the ownership of OAL. Representative samples were collected by (1) village selection, (2) pekarangan utilization, and (3) sample frame requirement.

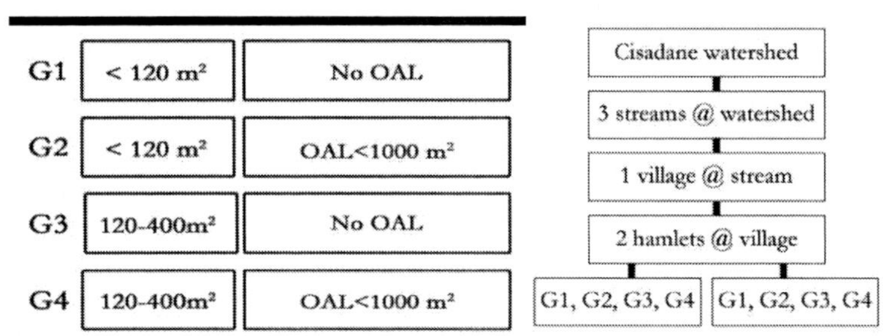

Figure 2 Schematic diagram of group criteria and flowchart of the research

3.2. Data analysis

Data were gathered are biodiversity and carbon stock. Biodiversity were calculated by 2 indices: Shannon-Weinner index, and the Margalef index. Carbon stored conducted by estimation species biomass using alometric equations. These two aspects correlated by simple linear regression.

3.2.1. Species diversity (Shannon-Weinner index). Shannon-Wiener index is formula for calculating diversity, which combination number of species and number of individuals of each type within a community [6].

The index is thus:

$$H' = \sum_{i=1}^{s}(pi\ lnpi) \quad (1)$$

where *S* is the total number of species, and *pi* is the frequency of the *i*th species (the probability that any given individual belongs to the species, hence *p*).

3.2.2. Species richness (Margalef index). The Margalef index is a species diversity index divided into two types of species richness (how many types are there) and assessment of species evenness or dominance (how individual species are distributed among the community). The formula of this index is [7]:

$$D = (S - 1)/\ln N \quad (2)$$

where S is the number of species in a sample and N is the total individuals in pekarangan.

3.2.3. Estimating biomass and carbon stocks. All plants on pekarangan considered for determining above ground biomass. Trees biomass that have more than 5 cm DBH were estimated by allometric model of Chave et al. [8], while shrub generalized by allometric model of Ali et al. [9] (Table 1). The above ground carbon stock was considered as 50% of above ground biomass. Results were then scaled from Mg/plot to Mg/ha.

Table 1 Allometric equations

Vegetation	Allometric model	Source
Teak (*Tectona grandis*)	$Y = 0,153\ D^{2,39}$	[10]
Banana (*Musa* sp.)	$Y = 0,0303\ D^{2,1345}$	[11]
Multispecies for shrub and trees D < 5 cm	Ln(AGB) = -3,50 + 1,65 Ln(D) + 0,842 Ln(H)	[9]
Others tree	$Y = 42,69 - 12,8\ D + 1,242D^2$	[8]

4. Result and discussion

4.1. Plants biodiversity in pekarangan

There are 244 species in 76 plant families that have collected in pekarangan in Cisadane watershed with 8 functions ie ornamental plants, industrial plants, starch crops, medicinal plants, herbs, vegetables, fruit, and more. This is larger than previous studies in four watersheds West Java, namely 214 species [2], in the courtyard of Santa Rosa in the Peruvian Amazon with 168 species [12] and in homegardens of northern Thailand with 230 species [13]. However, this result is smaller than other watershed in West Java, Cianjur watershed is about 440 species [1], the western part of Kenya (253: [14]), Zaire (272: [15]) and Nicaragua (324: [16]). Overall in West Java there are 602 species [17] and nearly 40% of plants were recorded from this study.

Average Shannon Wiener index (H ') on G1 and G2 in middle stream no higher than upper stream and downstream, but overall Shannon Weinner index in middle stream Cisadane watershed has a higher value compared to other stream levels (**Figure 3**). The average H' on middle stream is 2,85; upper stream and downstream are 2,49 and 2,63. So that the number of species and their relative abundance (number of individuals of each species) are higher on middle stream. The average of Margalef index (species richness) in middle stream is higher than upper stream and middle stream 5.45, the upper stream equal to 4,15, and downstream is 4,44 (**Figure 3**). Overall pekarangan in middle stream of watershed Cisadane is more heterogenity than others stream. Middle stream is a ecotone or intermediate region of mountain and lowland [1, 2], thats why diversity in middle stream higher than upper or downstream.

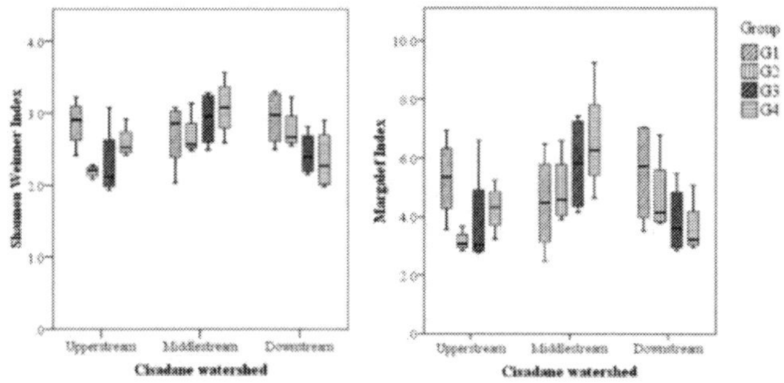

Figure 3 Shannon Weinner Index and Margalef Index in pekarangan in Cisadane watershed

Vegetation in pekarangan can be assigned in eight use categories or usually known as horizontal diversity, consisting of ornamental plant, medicinal plant, fruit plant, vegetable crops, spice crops, industrial plants, starchy crops, and other plants like fuel wood species, wood for handycraft or wood for building materials [1]. Ornamental plants are the most common plant (50,47%), the second is medicinal plant (23,05%), followed by fruits (9,66%), vegetables (5,92%), spices (4,36%), other plants (4,05%), industrial plants (1,25%), and strachy plants (1,25%) (**Figure 4**). Ornamental plants identical to the unique plants morphology not only function to beautify home but also give pleasure to its owner. Some plants such as rose, jasmine and cananga used indigenous people for their activities too such as traditional weddings and traditional ceremony. Pekarangan divided into three spaces, front, sides, and back [18]. Ornamental plant commonly planted in front yard while medicinal plants, spice and vegetable crops in back or side yard. The presence of ornamental plants related to the economic status of the owner of the yard. Families with higher economic status tend to plant ornamental plants and productive plants like clove [19].

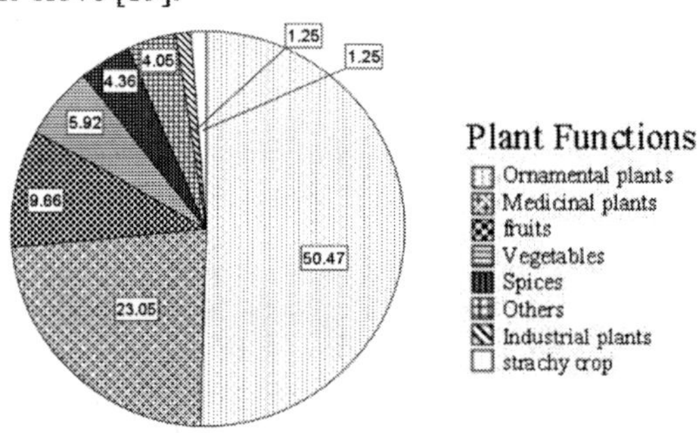

Figure 4 Percentage of plant function in pekarangan

Agavaceae and Araceae family are highest number of species with a percentage of 7.82% and 7.00%. Most of plants from these families fungsional as ornamental plant. Medicinal plants are usually found in Euphorbiaceae (6,17%) such as patah tulang (*Euphorbia tirucalli*), sambang darah (*Excoecaria cochinchinensis*), betadine (*Jatropha multifida*) and katuk (*Sauropus androgynus*), as well as Zingiberaceae (4,12%) such as temu kunci (*Boesenbergia pandurata*), temu item (*Curcuma aeruginosa*), turmeric (*Curcuma domestica*), and curcuma (*Curcuma zanthorriza*).

Based on plant height vertical diversity devided into five stratum, namely stratum I that plant less than 1 meter, stratum II 1 to 5 meters, stratum III 5 to 10 meters, stratum IV 10 to 15 meters and stratum V with more than 15 meters high. Plants in the first stratum are commonly found with 53,3%, then stratum II with 23,8%. The tree is more often found in stratum III with 12,3%. Trees in stratum IV and V only by 7,0% and 3,7% (**Figure 5**). In general, plant in pekarangan mostly are grasses, herbs or shrubs.

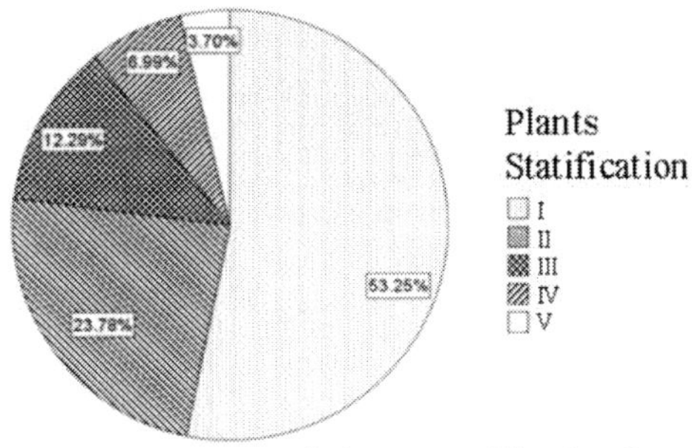

Figure 5 Percentage of plants stratification in pekarangan

Pekarangan <120 m^2, G1 and G2 were combined to represent a small pekarangan, while pekarangan 120-400 m^2 (G3 and G4) formed a large pekarangan. In addition, G1 and G3 were combined to non-OAL group, while G2 and G4 to OAL group. The average of Shannon Wiener index on small pekarangan (H '= 2,70) tend to higher than large pekarangan (H' = 2,62) (**Figure 6**). However numbers of Margalef index are similar on both small and large pekarangan (4,68). Previous research [2] showed G4 is more diverse than others group.

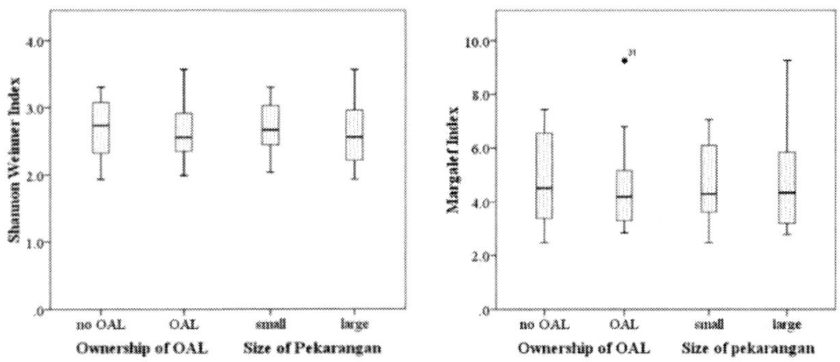

Figure 6 Shannon Weinner and Margalef Indexs based on both ownership of OAL and size of pekarangan

The average of Shannon Wiener index and Margalef index (**Figure 6**) in non-OAL pekarangan higher than OAL pekarangan. The absence of OAL, makes homeowners take advantage of pekarangan intensively, and thats make non-OAL pekarangan more diverse. Most of the non-OAL pekarangan cultivate food, spices, and medicinal plants that can be harvested as home use or market sales. Pekarangan without other agricultural land can be used to produce carbohydrates by planting strachy crops such as cassava and sweet potatoes [20]. Furthermore, it is mentioned that the product of pekarangan also contribute to produce calories, protein, vitamin A and C [2, 21, 22]. Homeowners may sell products of pekarangan including fruits, vegetables, and timber to increase their income. The income derived from pekarangan production contributed up to 11,5 % to total income derived from the main occupations of all family members [2].

4.2. Carbon stock in pekarangan

The value of carbon stock from 48 samples pekarangan is around 0,08 Mg / ha to 131,61 Mg / ha. The average carbon stock in pekarangan in middle stream Cisadane watershed tend to higher (Figure 6), as well as number of diversity index. Overall the average is 42,11 Mg / ha, while in upper stream is equal to 33,54 Mg / ha and in downstream is 38,92 Mg / ha (Figure 7). These results are higher than the previous research in the pekarangan of Java and Sumatra (35.3-58.6 Mg / ha) [23, 24], agroforestry in Sumatra was 9 years old (14 Mg / ha) [24], and homegarden in Central Kerala, India (16-36 Mg / ha)

[15]. However these are lower than agroforestry in Sumatra over the age of 35 years (101 Mg / ha) [24] and in the traditional agroforestry Panama (145 Mg / ha) [25]. Cisadane watershed is one of northern watershed in the west part of Java. Based on previous research,the carbon stock in the northern region is lower than in the southern region (ex. Cimandiri and Cibuni watershed). It is probably because of higher intensity of pekarangan utilization from rural community in the southern region [26].

Carbon stock on small pekarangan group (28,15 Mg / ha) lower than large pekarangan group (48.23 Mg / ha) (Figure 7). pekarangan with no OAL are higher than pekarangan with OAL. The average of carbon stock on non-OAL pekarangan are 43.54 Mg / ha, while pekarangan with OAL equal to 32.84 Mg / ha (Figure 6). The number of plants per area and tree growth rate affects the amount of carbon that can be stored [27]. The structure and composition of vegetation such as type, size, height, density and age of trees may also affects the above ground carbon stock [28, 29]. Wood of slower-growing species is usually have higher spesific gravity than faster-growing species, therefore the slow-growing species may accumulate more carbon in the long term [30]. Some researchers argue the existence of exotic plants related to the carbon stored in agroforestry. Local plants are more adaptable than exotic plants, because of their supposedly better adaptability to local conditions [28] besides exotic plants will be more dangerous especially if located near the conservation area [2, 29].

One of pekarangan on G3 in middle stream saving of 108.52 Mg / ha. Big trees of mango, rambutan and guava planted in this area that accounted for nearly 95.32% 103.44 Mg / ha. Tree component represented the most important C pool of the aboveground biomass of perennial plants that growing in pekarangan [27, 31]. Results from previous studies show the tree biomass potentially save 93.73% [24] and 55-81% [27] of the total carbon in above-ground.

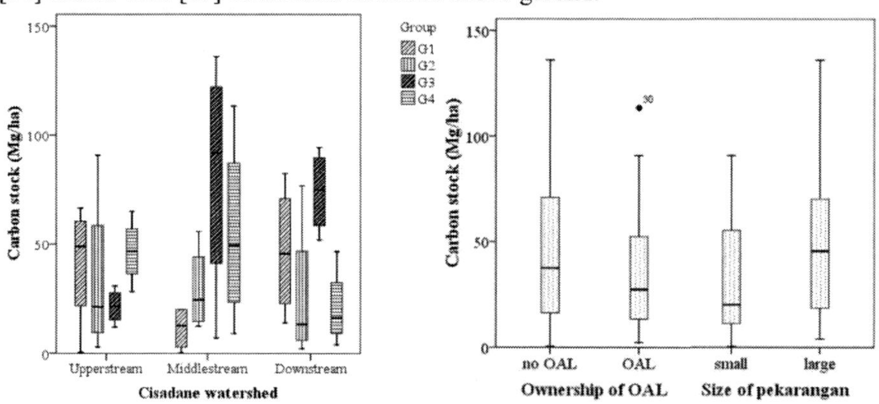

Figure 7 Carbon stock in pekarangan in Cisadane watershed and carbon stock based on ownership of OAL and size of pekarangan

4.3. Correlation between biodiversity and carbon stock

In tropical forests, carbon storage depends largely on species composition [32] and thus there may be exist relationship between them in agroforestry such as pekarangan. Results of linear regression between biodiversity and carbon stock is positively associated though not significant (R^2 = 0,024). Similarly, relationship between biodiversity and carbon stock in trees in pekarangan with R^2 = 0,234 (Figure 8). Variations of carbon stock does not affected by plants biodiversity. These results are consistent with several previous studies [25, 27, 33, 34]. Biodiversity can be seen as an independent pekarangan function even though C sequestration project that contribute to enhance biodiversity should be considered as more ethical in the long term than conventional afforestation project [27].

The composition of tree species and agroforestry is affects the quality and quantity of biomass on the ground and also carbon stored below ground [28]. Water availability, quality and quantity of litter, root composition, and distribution of carbon in the soil profile will influence both the quality and quantity of the biomass returned to the soil. Recent research found a positive relationship between species richness and underground carbon biomass in the West African region [35].

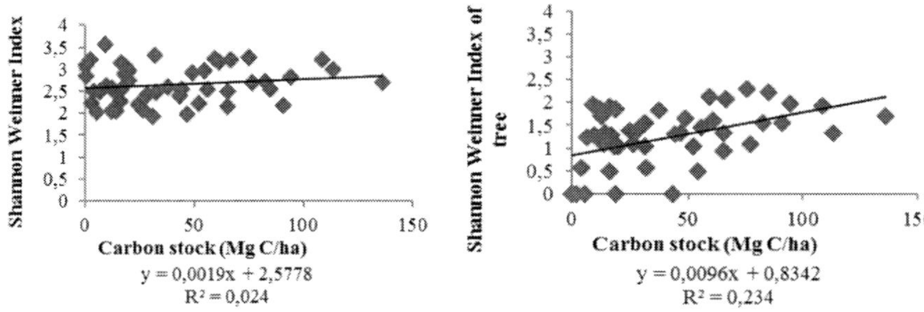

Figure 8 Relation between Shannon Weinner Index and Carbon stock in pekarangan

5. Conclusion and Recommendation

Plant diversity and carbon stock are examples of landscape services, in this case is pekarangan. Landscape reviews Reviews These services can be seem optimally by the owner if pekarangan is well managed. This study bounded by watershed to minimize others environmental variables that can affect plant diversity and carbon stock data. The diversity of plant species was recorded growing in pekarangan in Cisadane watershed. Diversity indexs are higher in middle stream, as well as carbon stock. Small pekarangan (<120 m^2) is more diverse, H '= 2,70; D = 4,68. However the higher carbon stock is in large pekarangan (120-400 m^2), 50.08 Mg / ha. A lot of small pekarangan are usually dominated by shrubs and grasses, while large pekarangan dominated by small and big tree which require more space. That's causes small pekarangan more diverse than large pekarangan. However large pekarangan can store more carbon because of existing trees. Non-OAL pekarangan is more diverse than OAL pekarangan (H '= 2.70; D = 4.81) as well as carbon stock (46.19 Mg / ha). Tree component are the most contributed to above ground carbon storage. Biodiversity and aboveground carbon have a weak positive trend relationship, and thus biodiversity significantly contribute to food security while C sequestration on pekarangan contribute to mitigate CO^2 concentrations in atmosphere. From this study we recommend to maximize pekarangan function, plant diversity and carbon stock by planting varied plants either vertically or horizontally. These information collected here may contribute to refining broader scale methods based on remote sensing. High resolution satellite imagery can be used to extrapolate carbon stock and plants biodiversity of *Pekarangan* at watershed level. The further researchs may aiming at analyzing on the long term resilience of plants diversity and C sequestration, and also the potensial of below ground carbon sequestration.

References

[1] Arifin H S, Kaswanto R L, Nakagoshi N 2014 Low carbon society through pekarangan traditional agroforestry practices in Java, Indonesia *Designing Low Carbon Societies in Landscapes* **8**: 129-143

[2] Kaswanto R L and Nakagoshi N 2014 Landscape ecology-based approach for assessing pekarangan condition to preserve protected area in West Java *Designing Low Carbon Societies in Landscapes* **8**: 289-311

[3] Rusolono T 2006 Model pendugaan persediaan karbon tegakan agroforestry untuk pengelolaan hutan milik melalui skema perdagangan karbon [Disertation] Bogor (ID): Sekolah Pascasarjana, Institut Pertanian Bogor

[4] Kumar B M and Nair P K R 2004 The enigma of tropical homegardens *Agroforestry System* **61-62**: 135-152

[5] Kaswanto R L, Baihaqi M, and Hadi A A 2013 Potensi pekarangan dalam peningkatan jasa lanskap di kasawan perdesaan In: *Prosiding lokakarya nasional dan seminar. forum komunikasi perguruan tinggi pertanian indonesia*

[6] Ponce-Hernandez R 2004 Assessing carbon stocks and modelling win-win scenarios of carbon sequestration through land-use changes *Food and Agriculture Organization of the United Nations* (Rome)
[7] Death R 2008 Margalef's index. Dalam: Sven Erik J, Brian F (eds) Encyclopedia of ecology Academic Oxford pp 2209-2210
[8] Chave J, Andalo C, Brown S, Cairns M A, Chambers J Q, Eamus D, Fölster H, Fromard F, Higuchi N, Kira T, Lescure J-P, Nelson BW, Ogawa , Puig , Riéra B, and Yamakura T 2005 Tree allometry and improved estimation of carbon stocks and balance in tropical forests *Oecologial 45: 87–99*
[9] Ali A, Xu M S, Zhao Y T, Zhang Q Q, Zhou L L, Yang X D and Yan E R 2015 Allometric biomass equations for shrub and small tree species in subtropical China *Silva Fennica* **15**: 1-10
[10] Hairiah K, Rahayu S 2007 Pengukuran karbon tersimpan di berbagai macam penggunaan lahan (Indonesia : ICRAF)
[11] Hairiah K, Sitompul S M, Van M N and Palm C 2001 Methods for sampling carbon above and below ground *ASB Lecture Note 4B* (Indonesia: ICRAF)
[12] Padoch C and de Jong W 1991 The house gardens of Santa Rosa : Diversity and variability in an Amazonian agricultural systems *Economic Botany* **45**:166-175
[13] Black G M, Somnasang P, Thamathawan S and Newman J M 1996 Cultivating continuity and creating change: women's homegarden practices in North-eastern Thailand – Multicultiral consideration from cropping to consumption *Agriculture Human* Vol **13**: 3-11
[14] Backes M M 2001 The role of indigenous trees for the conservation of biocultural diversity in traditional agroforestry land use systems: The Bungoma case study. *Agroforest. Syst.* **52** 119–132
[15] Kumar B M 2011 Species richness and aboveground carbon stock in the homegardens of central Kerala, India. *Agriculture, Ecosystem, and Environment* **140** 430-440
[16] Mendez V E, Lok R and Somarriba E 2001 Interdisciplinary analysis of homegardens in Nicaragua: Micro-zonation, Plant Use and Sosioeconomic Importance *Agroforestry System* **51** 85-96
[17] Karyono 1990 Homegardens in Jawa: their structure and function eds. Landauer K, Brazil M. *Tropical Homegardens* (Tokyo: United Nations. University Press)
[18] Arifin H S 1998 Study on the vegetation structure of Pekarangan and its changes in West Java, Indonesia [Disertasi] (Okayama ,Japan: Okayama University)
[19] Achmad H, Martadihardja A, Suharto, Wawan, Gunawan and Hadyana 1978 Sociocultural aspect of the homegarden .Seminar on the Ecology of Homegarden II, October 25-26, Institute of Ecology, (Bandung: Padjadjaran University)
[20] Karyono 1981 Structure of homegarden in the rural area of Citarum Watershed, West Java [Thesis] (Bandung: Padjajaran University)
[21] Ochse J J and G J A 1937 The economic aspect of the "Koetawinangun report." *Landbouw* 13:54
[22] Christanty L, Oekan S, Abdoellah, Marten G G, Iskandar J 1986 Traditional agroforestry in West Java *Traditional agriculture in Southeast Asia: a human ecology perspective*, Westview Press: 132-158
[23] Jensen M 1993 Soil conditions, vegetation structure and biomass of a javanese homegarden. *Agroforestry System* **24** 171-186
[24] Roshetko M, Delaney M, Hairiah K, Purnomoshidi P 2002 Carbon stock in Indonesian homegarden systems: can smallholder systems be targeted for increased carbon storage? *American Journal of Altern. Agric* **17** 125-137
[25] Kirby K R and Potvin C 2007 Variation in carbon storage among tree species: implications for the management of a small-scale carbon sink project *Forest Ecology and Management* **246** 208-221
[26] Kaswanto and Utami F N H. 2016 The disparity of watershed development between northern and southern region of Java Island *Procedia Environmental Sciences* **33**: 21-26

[27] Henry M, Tittonell P, Manlay R J, Bernoux M and Alberecht A 2009. Biodiversity, carbon stock and sequestration potential in aboveground biomass in smallholder farming systems of Western Kenya *Agriculture, Ecosystems and Environment* **129**: 238-252
[28] Nair P K R, Nair V D, Kumar B M and Haile S G 2009 Soil carbon sequestration in tropical agroforestry systems: a feasibility appraisal *Environmental Science and Policy* **12** 1099-1111
[29] Bajigo A, Mikrewongel T, Yitebitu M and Agena A 2015 Estimation of carbon stored in agroforestry practices in Gununi Watershed, Wolayitta Zone, Ethiophia *Ecosystem and Ecography* **5** 157-162
[30] Baker T L, Phillips O L, Malhi Y, Almeida S, Arroyo L, Fiore A D, Erwin T, Killeen T J, Laurance S G, Laurance W F, Lloyd J, Monteagudo A, Neill D A, Patino S, Pitman N C A, Silva J N M and Vasquez M R 2004 Variation in wood density determines spatial patterns in Amazonian forest biomass *Global Change Biol.* **10**: 1-18
[31] Beenhouwer M D, Geeraert L, Mertens J, Geel M V, Aerts R, Vanderhaegen K and Honnay O 2016 Biodiversity and carbon storage co-benefits of coffee agroforestry across a gradient of increasing management intensity in the SW Ethiopian highlands *Agriculture, Ecosystems and Environment* **222**: 193-199
[32] Bunker D E, DeClerck F, Bradford J C, Colwell R K, Perfecto I, Phillips O L, Sankaran M, and Naeem S 2005 Species loss and aboveground carbon storage in a tropical forest *Science* **310** 1029–1031
[33] Karna Y K 2012 *Mapping above ground carbon using world view satellite image and lidar data in relationship with tree diversity and forests* [Thesis] (The Netherlands : University of Twente)
[34] Mandal R A, Dutta I C, Jha P K and Karmacharya S 2013 Relationship between Carbon Stock and Plant Biodiversity Collaborative Forests in Terai, Nepal *Hindawi Publishing Corporation* Vol **2013** 1-7
[35] Dayamba S D, Djoudi H, Zida M, Sawadogo L and Verchot L 2016 Biodiversity and carbon stocks in different land use types in the Sudanian Zone of Burkina Faso, West Africa *Agriculture, Ecosystems and Environment* **216** 61-72

Land use cover changes and water quality of Cipunten Agung Watershed Banten

Afifa Karima* and Regan Leonardus Kaswanto
Department of Landscape Architecture, Faculty of Agriculture, Bogor Agricultural University, Meranti St. Dramaga, Bogor 16680, Indonesia

E-mail: karimaafifa@gmail.com

Abstract In general, the land and natural resources utilization of Cipunteng Agung Watershed could be classified into protected and cultivated region. Based on satellite imagery classification, protected region covers 885.76 ha (22.71%), while cultivated region occupied 3,041.69 ha (77.29%) in 2011. It means that the land use and covers change (LUCC) to be cultivated rapidly increase. Those conditions had impacted positively to the local economy development, however it had negatively degrades water quality in Ciputen Agung river. Therefore, the purpose of this research is to analyze LUCC impacts to the water quality (WQ) in Cipunten Agung watershed. Supervised classification method and Water Pollution Index (WPI) approach were correlated to define the optimal solution to reduce the rate of LUCC. The result shows that area of cultivation tree and paddy field are higher than others in midstream, and settlement is higher in downstream, particularly at riparian landscapes. The concentration of total phospate, nitrite, and nitrate which indicated from agriculture land were complies with WPI class 2 until 4 standart. The rate of LUCC of Cipunten Agung watershed and the density of settlement are potential to be the driving factor of water quality degradation. Regional planning with ecology approach was recommended for sustainable development of Cipunten Agung Wateshed.

1. Introduction

Land conversion into agriculture land or settlement was occurs gradually in many watershed in Indonesia. The one of negative impact from land conversion is water quality degradation in main watershed river. The impact of Land Use Cover Changes (LUCC) to water quality has been become the main issue in several landscape architecture dan hydroecology research within a few decades, such as Kaswanto (2012), Xia (2012), and Firdaus (2013). It also occurs in Cipunten Agung Watershed, Pandeglang Regency, Banten. Downstream community continues to build dense settlements and facilities for the fishing industry without proper planning. Improper planning can bring about watershed degradation such as barren land, critical land, and slope erotion. Eventually, the degradation process can occur such as flooding in rainy season, lower water discharge and less soil humidity, forest fires in dry season, sedimentation, and water quality degradation (Directorate of Forestry and Water Resources Conservation of Indonesia).

The evaluation of LUCC and water quality has been known as the critical point in deteremine pollution source and watershed management. Land use cover type and water quality degradation shows significant relation [1]. Point source and non-point source pollution which become the main factor of water pollution will affect water quality [2]. The greater the percentage of man made landscape such as construction and agricultural land, the water quality will become worse [3]. Cipunten Agung watershed environment problem become obstacles and challenge for stakeholders. Therefore, LUCC is necessary to analyze for determining the impact of its into water quality. So it can be considered on watershed management for the sustainability of landscape development.

2. Methodology

2.1 Location, tools, and materials

This study was located at Cipunten Agung watershed, Pandeglang Regency, Banten Province with coordinate of 6°17'30" to 6°22'30" Latitude and 105°50'0" to 105°55'0" Longitude, seluas 3979,9 Ha (figure 1). The tools used in this study consist of hardware, such as digital camera, GPS, and several Geography Information System (GIS) softwares. In water sampling process, this study use tools such as temprature meter, pH meter, DO meter, and water tube Materials of this study are Landsat imagery of year 195, 2005, 2015 (table 1). Sub-district administration map, watershed boundary map, precipitation map, digital elevation model (DEM), and soil type map. Water sampling was tested by Environmental Productivity Laboratorium of Bogor Agricultural University.

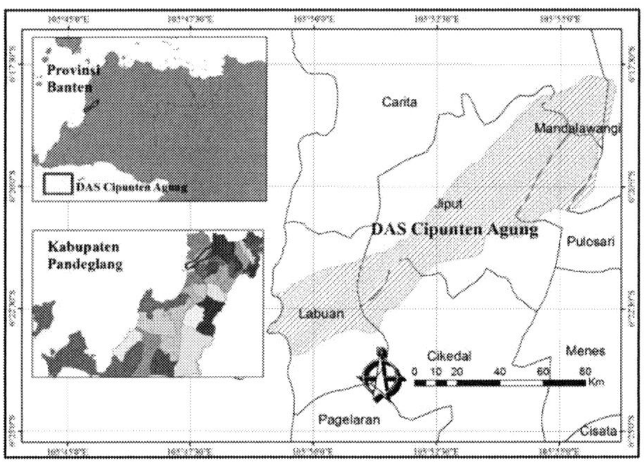

Figure 1. The study area of Cipunten Agung

Table 1. Landsat data acquisiton

Landsat data	Acquisition date
Landsat 5	Sept 16th, 1995 Path 123 Row 64
Landsat ETM + 7	May 30th 2005 Path 123 Row 64
Landsat 8	Sept 23th 2015 Path 123 Row 64
Landsat ETM + 7	July 31th 2004 Path 123 Row 64

2.2 Method

The main method used in this study was supervised classification for LUCC analysis and Water Pollution Index for water quality analysis. Land use and cover was classified into five classes, namely water bodies, forest, cultivated tree, settlement and paddy field. Water quality parameter that used in this study are Total Suspended Solid, Total Dissolved Solid, Temprature, Dissolve Oxygen (DO), Chemical Oxygen Demand (COD), Biologycal Oxygen Demand (BOD), Nitrite (NO_2), Nitrate (NO_3), Ammonium (NH_4), Phospate (PO_4), pH, and, Total Coliform. Water quality standart threshold which used in this study following Indonesian Goverment Decree No. 82 of 2001 for classifying water quality into four classes.

Water Pollution Index (WPI) analysis which used in this study, following Evironmental Ministry of Indonesia decree No. 115 of 2003, with some criteria of WPI such as :

$0 \leq WPI \leq 1.0 = not\ polluted$,
$1.0 \leq WPI \leq 5.0 = lightly\ polluted$,
$5.0 \leq WPI \leq 10. = moderate\ polluted$,
and $WPI > 10.0 = heavy\ polluted$.
The relationship between LUCC and WPI was identified by spatial correlation between LUCC map and water sample point at radius 1 km (figure 2).

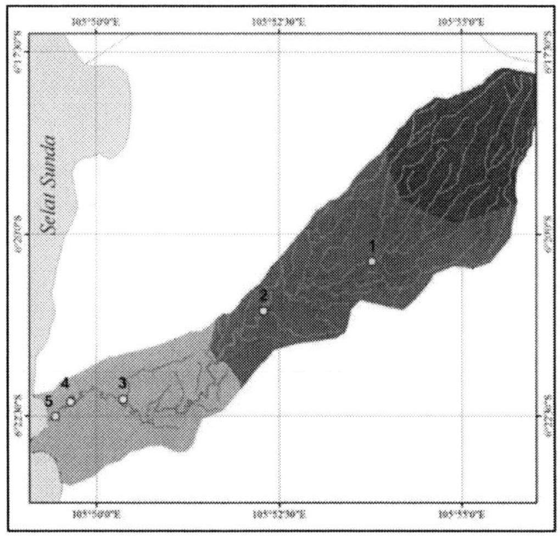

Figure 2. Water sampling station

3. Result and discussion

3.1. Land use cover changes

Land Use Cover classification use clases namely, water bodies, forest, cultivated tree, settlement, and paddy field. There were 75 training area which used for accuration test of classification. Accuracy standart that used in this classification is 75% following minimal standart of National Institute of Aeronautics and Space of Indonesia. Land Use Cover classification from Landsat Image of 1995 result (**Figure 3**) shows that, the widest class of land use cover is cultivated tree 2570 ha and paddy field 793,63 ha, while the smallest class is water bodies 38,98 ha or 0,98% of total area. Land utilization as cultivated tree and paddy field was intensive for daily needs supply. In 1995 at downstream, setllement area was developed 20 m near the river, which was one of inundation and flooding area.

Land Use Cover classification from Landsat Image of 2005 result (**Figure 34**) shows that the widest class of land use cover is cultivated tree 2286,01 ha. Cultivated tree area was decreasing while paddy field area was increasing to 969,9 ha or 24,37% of total watershed area. Settlement area not too expanded, the total area is 114,44 ha or 2,88% from total watreshed area. There was a stripping landsat image problem and it solved by gap filling. The material used for gap filling was landsat imagery of 2004,with assume there was no significant LUCC in one year. Gap filling process result was added two classes of Land Use Cover such as cloud and shade. The classification result shows people tend to built residential area at downstream since 1995.

Landsat imagery of 2015 Land Use Cover classification result (**Figure 5**), the widest area is cultivated tree 2420,5 ha or 61,41% of watershed total area, while the smallest area is water bodies 46,69 ha or 1,19% from watershed total area. Classification in 2015 was tested with accuracy assessment of classification in 2015 result is 82,67%, more than the minimum threshold 75%. The increases of paddy field and decreases of cultivated tree occur especially at Jiput Districs (Mid stream). LUCC into paddy field many occur at 1995 to 2005 period with 591,66 ha area, while LUCC into settlement many accour at 2005 to 2015 period with 195,21 ha area.

IOP Publishing
doi:10.1088/1755-1315/54/1/012025

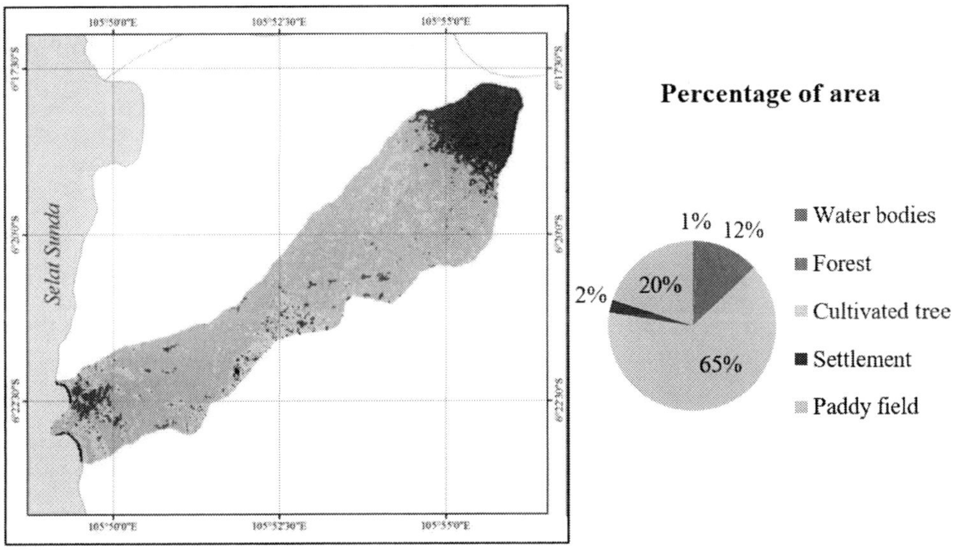

Figure 3. Land use cover of Cipunten Agung Watershed in 1995

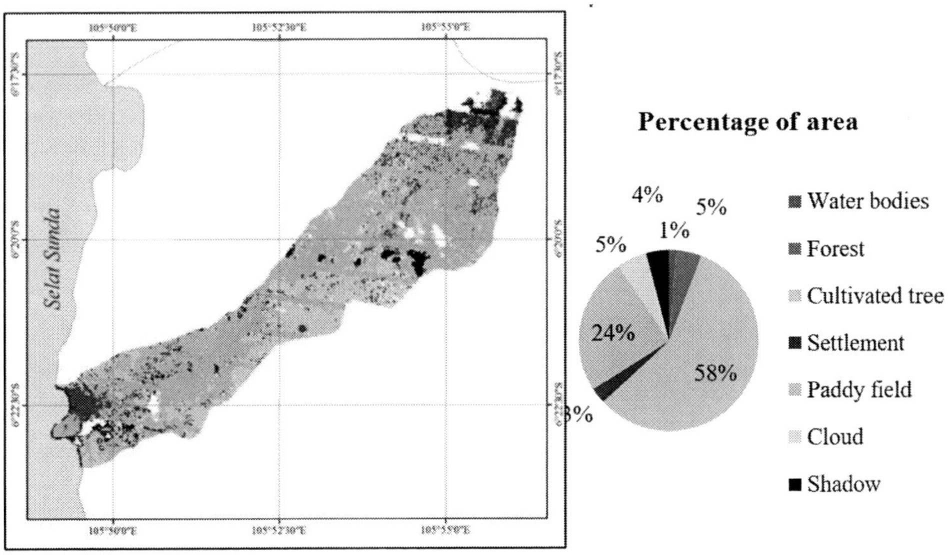

Figure 4. Land use cover of Cipunten Agung Watershed in 2005

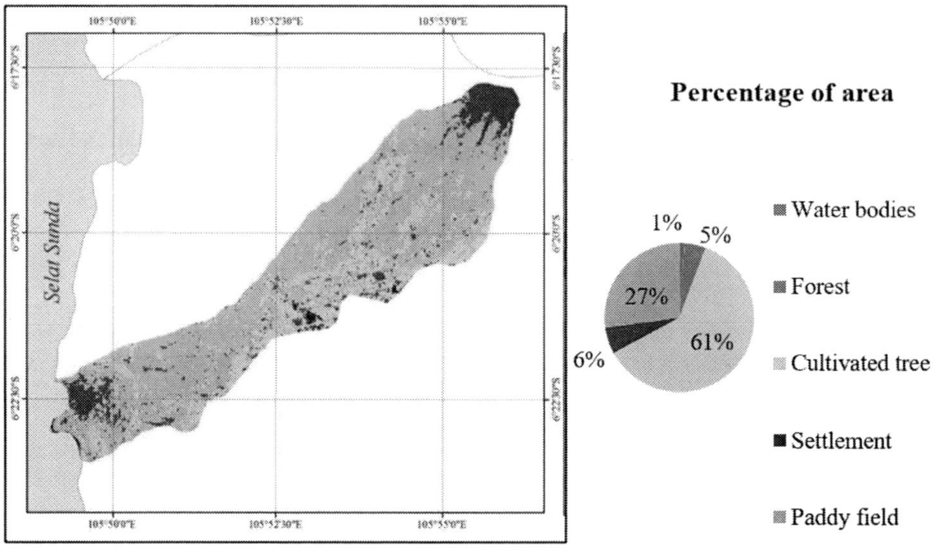

Figure 5. Land use cover of Cipunten Agung Watershed in 2015

3.2. *Cipunten Agung water quality*

Water sampling location was determine based on accessibility, watershed character, and land use cover of Cipunten Agung Watershed. Station (ST) 1 and 2 were located at the middle stream which surounded by paddy filed, precisely at Sikulan and Bangkuyung village. ST 3, 4, and 5 were located at down stream which surounded by settlement area, precisely at Sukawati, BTN, and Teluk village. The observation shows that, Cipunten Agung water recently was used for washing, irrigation, and livestock. Water quality sampling test result presented descriptively at **Table 2** it shows the minimum, maximum, and the standart deviation of eleven parameter.

Table 2. Descriptive statistics of Cipunten Agung water quality

Parameter	Unit	*Min*	*Max*	*Mean*	*St.dev*
Physics					
Temprature	°C	27,00	31,90	28,56	2,11
Total Suspended Solid (TSS)	mg/L	9,00	18,00	13,00	4,24
Total Dissolve Solid (TDS)	mg/L	20,00	4232,00	1012,80	1826,79
Chemical					
pH	-	6,35	7,18	6,91	0,35
Dissolve Oxygen	mg/L	4,60	6,90	5,66	0,99
Biologycal Oxygen Demand	mg/L	3,00	5,80	3,68	1,20
Chemical Oxygen Demand	mg/L	25,68	35,56	27,91	4,29
Total Phosphate	mg/L	0,03	0,35	0,12	0,14
Amonium (NH_4)	mg/L	0,01	0,45	0,11	0,19
Nitrat (NO_3-N)	mg/L	0,18	0,43	0,33	0,10
Nitrit (NO_2-N)	mg/L	0,01	0,02	0,01	0,00
Microbiolgy					
Total Coliform	MPN/100mL	17000	43000	29000	10319

The average value of physics parameters such as temprature and TSS are 28,56 mg/L and 13 mg/L were below the threshold of class 1 until 4. Meanwhile, the average value of TDS is 1012,8 mg/L overpass the threshold of class 3. It indicates, the Total Dissolved Solid concentration in water is still good, but its not drinkable. The average value of chemical parameter such as pH, DO, and BOD were below the threshold of class 1 until 4. pH of Cipunten Agung river is neutral from 6,35 to 7,18. DO, BOD, and COD never reach out the threshold standart. The DO concentration average is 5,66 mg/L, it indicates the water was good enough for human and aquatic organism. The BOD concentration averge is 3,68 mg/L. It indicates, the biologycal activity in water ecosystem are still in good condition.

Another chemical parameter such as Total Phosphate, Ammonium, Nitrate, Nitrite was measured in small quanitites with an average value are, 0,12; 0,11; 0,33, and 0,01 mg/L below the standart threshold from class 1 until 4. Regarding to the result, the concentration of those parameter are still in permitted level. The chemical parameter was important to analyze due to its possibility to predict the water pollution point source from agriculture land such as, chemical fertilizer, excrement, and erosion [4].

One of microbiolgy parameter is Total Coliform which has value from 17000 to 43000 and high average value 2900 MPN/100mL. That value had overpass the Total Coliform thershold standart at class 4. It indicates a high potential of pathogenic bacteria in Cipunten Agung River water. Total Coliform is a group of bacteria commonly found in the environment, e.g in soil or vegetation, mammals, including human. Coliform bacteria had small potential to become a source of disease, but the presence of these bacteria indicates that the water supply is vulnerable to contamination by the microorganism which more dangerous (www2.gnb.ca). So, its not recommended to use Cipunten Agung water for consumption.

3.3. LUCC and water quality correlation

The impact of LUCC to the water quality was identified by the land cover variety analysis which surrounding the sample stations in radius 1 km. Water Pollution Index analysis refers to Evironmental Ministry of Indonesia decree No. 115 of 2003. In WPI calculation process, each parameter in 5 station was compared with four water quality standart. ST 1 (**Figure 6**) and ST 2 (**Figure 7**) located at middle stream, 13.77 and 7.72 km from the estuary area. In radius 1 km, ST 1 was dominated with cultivated tree and paddy field with the size of area are 150,8 ha and 141,4 ha. The result shows, ST 1 has the highest concentration of COD 35,56 mg/L, it indicates the potential of the organic material which not degraded properly, and it potentialy occur muddy areas.

ST 2 was dominated by cultivated tree and paddy field with the size of area 146,6 ha and 104.9 ha. Nitrate concentration at ST 2 was increasing, it potentialy occur by the COD and Nitrate concentration which already high at ST 1 and the extenssion of paddy field area at the middle stream. Nitrate concentration at agriculture land significantly higher than at the riparian area [5]. Nitrate are actually not poisonous, but if it is reduced to Nitrite then it will be toxic digestion. Nitrite is one of carciogenic tumor disease source [6]. The content of N should be considered because it can damage health. The density of cultivated tree in ST 2 has potential to provide landscape services as biofilter to reduce the concentration of COD and Nitrates in ST 1 and 2 rbelow the quality standart limits. The biofilter also potentialy reduce the level of water pollution index which from agricultural land use.

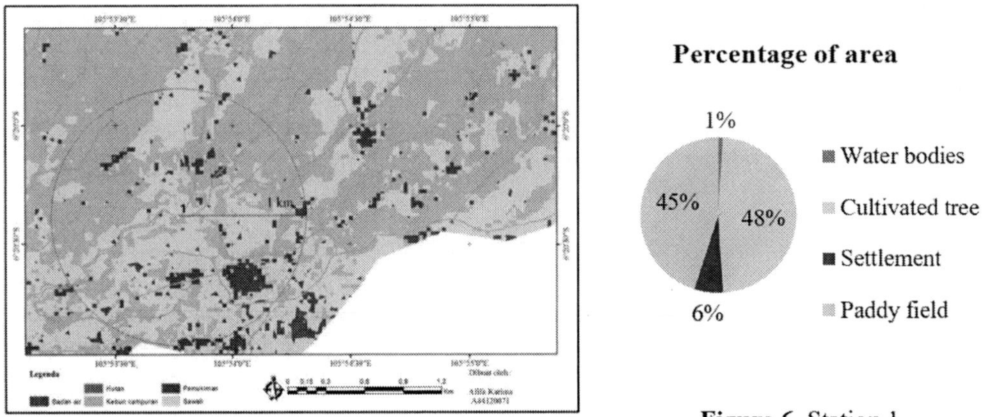

Figure 6. Station 1

ST 3, 4, and 5 was located at downstream area, with distances 3 km; 0,68 km; and 0,14 km from the estuary area. In radius 1 km, ST 3 (**Figure 8**) was dominated by cultivated tree and paddy field with the size of area 220 ha and 65,9 ha, smaller than at the middle stream. The concreation of Total Phosphate and Ammonium increase to 94 % and 55 % from ST 2. Those parameter is toxic metabolites due to the presence of nitrogen which can stimulate the growth of algae (algae bloom). Algae can form a layer on the water surface and prevent the penetration of oxygen and sunlight which needs by the aquatic ecosystem. When waters contain enough phosphate, algae accumulate it inside the cell and exceed its needs, its known as over consumption [7].

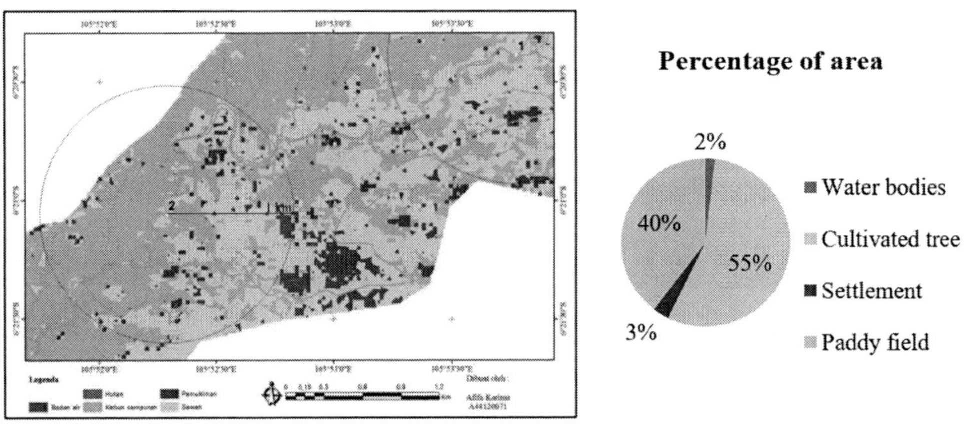

Figure 7. Station 2

Settlement and industry area was intensely developed at down stream area. In radius 1 km, ST 4 (**Figure 9**) was dominated by cultivated tree and settlement with the size of area 79,9 and 91,6 ha. BOD concentration in ST 4 is 5,8 mg/L, its higher than others. This indicates, the oxygen decreases rapidly due to the organic material degradation. LUCC with the increases of domestic activity, agriculture, and industry will influence and impact on the water quality condition, mainly the domestic activity that will increases BOD concentration [8].

In radius 1 km, ST 5 (**Figure 10**) was dominated by cultivated tree and settlement area with the size of the area 71,9 ha and 88,2 ha. Nitrate and Total Coliform was decreasing 45% and 63%, while the TDS concentration was higher than other ST. The TDS concentration was 4232 mg/L, its overpass the standart threshold at 2000 mg/L. This indicates, the solid material was dissolve rapidly and impacted the water taste. Its also indicate, the presence of toxic minerlas. Enviromental Proteciton Agency (EPA) USA suggested the maximum threshold water pollutant is 500 mg/L (500 ppm).

All stations of water samples has a high concentration of Total Coliform 17000-43000 MPN/ml overpass the standart threshold at 1000 MPN/ml at class 1 until 4. This indicates the water quality and sanitation were not in good condition. Eventhough, the potential of coliform bacteria to causing disease was small, the presence of its shows the vulneralibility of water for contaminated by more dangerous microorganism (www2.gnb.ca). Water quality degradation problems was impacted by LUCC, deforestation, monoculture planting system, urbanization, industrialization, and another development activity [9].

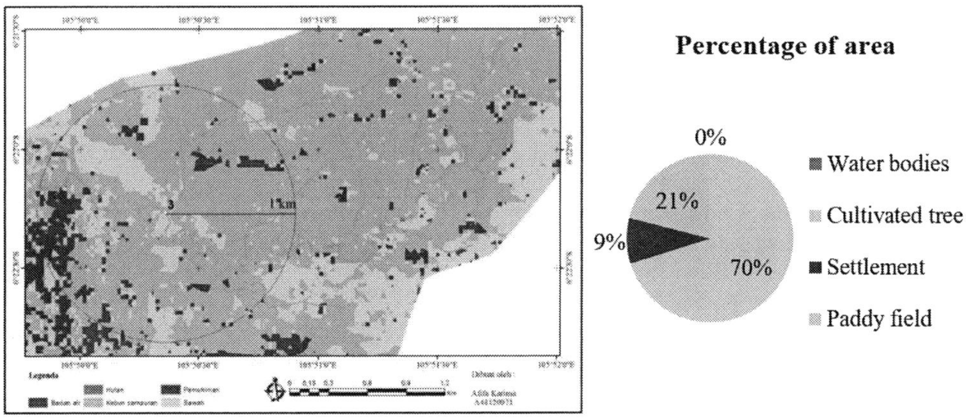

Figure 8. Station 3

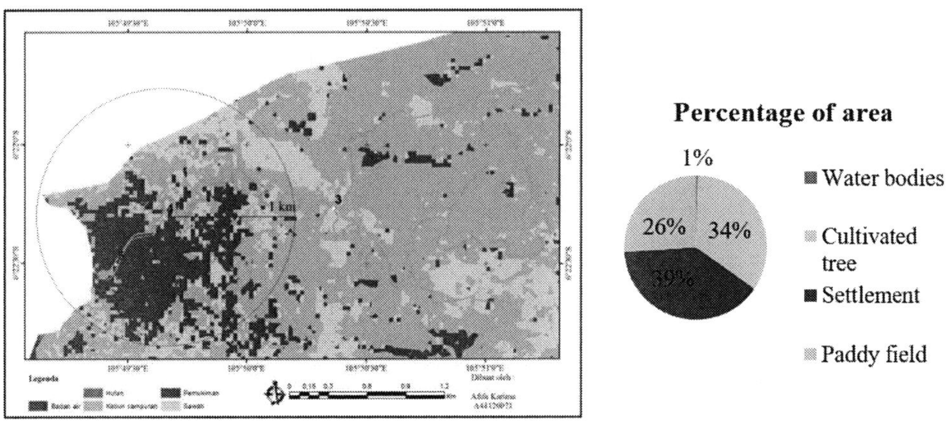

Figure 9. Station 4

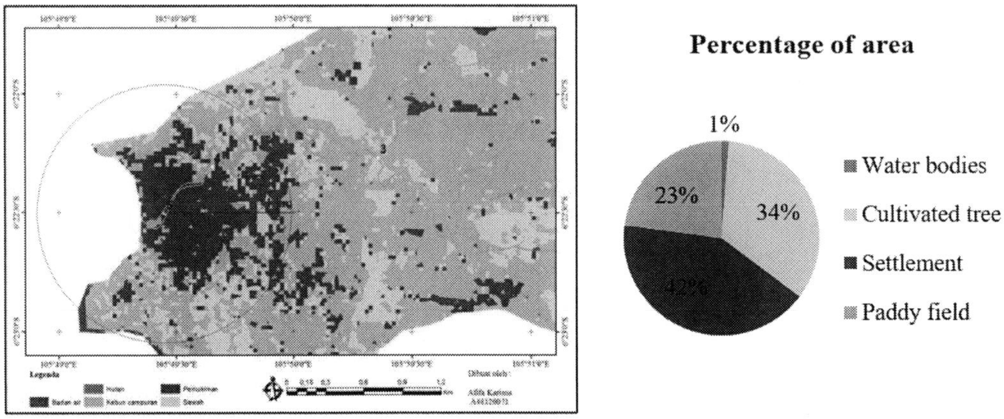

Figure 10. Station 5

The diversity of land use cover and water utilization from upstream to down stream will influence the water quality from the physics, chemical, and microbiology parameter (**Figure 11**). Water Pollution Index can determine the polluted degrees compared with standart threshold of class 2, based on the Cipunten Agung water utilization. Indonesia Government Regulation No. 82 year 2001 stated, water with the status of class 2, can be used as water recreation, fish cultivation, livestock, and irrigation. The pollution degrees was classified into $0 \leq WPI \leq 1.0 = not\ polluted$; $1.0 \leq WPI \leq 5.0 = lightly\ polluted$; $1.0 \leq WPI \leq 10.0 = moderate\ polluted$; $WPI > 10.0 = high\ polluted$. Water Pollution Index shown as Table 3.

Table 3. Water Pollution Index in Cipunten Agung Watershed

DAS Character	Station	WPI	Class	Status
Mid stream	1	5,05	2	Lightly polluted
Mid stream	2	4,03	2	Lightly polluted
Down stream	3	2,48	2	Lightly polluted
Down stream	4	6,18	2	Lightly polluted
Down Stream	5	3,31	2	Lightly polluted

Land use cover and water quality analysis in 1 km radius potentialy support for pollution degree prediction of Cipunten Agung river. Land use cover composition at ST 1,2, and 3 which dominatied by cultivated tree and paddy field determine the lower pollution degree at 5,05; 4,04; and 2,48 with WPI status is lightly polluted (**Figure 11**). The increases of nutrient at water surface was correlated with eutrophication, which occurs in the ecosystem due to the addition of artificial or natural substance from manure or waste into water system [4]. Meanwhile, land use cover composition at ST 4 and ST 5 which dominated by cultivated tree and settlement , determine the pollution degree at 6,18 and 3,31 with WPI status is lightly polluted. ST 4 was highlighted because it surrounded by the settlement with high density and poor sanitation. ST 5 which surrounded by fish industry and settlement has lower WPI score than ST 1, because of the influence of estaury salinity and affected by sea water. Water Pollution Index result of all station was at range of $1.0 \leq WPI \leq 5.0$ determine that 80 % Cipunten Agung main river area was ligthly polluted.

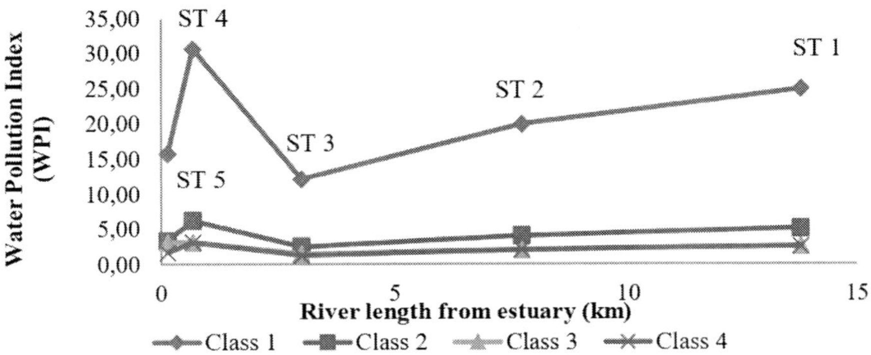

Figure 11. WPI based on river length

4. Recommendation

Material from many source will accumulated at river, it makes the role of the river is very crusial. Water pollutant concentration potentialy increased due to the land conversion, such as agricultural land and industry area. Based on WPI value, its important to arrange Cipunten Agung river water restoration for water quality improvement. Ecological approaches in various scales can be applied in regional water resource managemnent with cost and energy efficiencies for enviromental quality improvement. In meso scale such as Cipunten Agung watershed, agroforestry scenario can be applied for agriculture production and environmental sustainability in the same time. Other research shows agroforestry landscape provide landscape services, spesifically pollutant absorption from agriculture activities such as nitrogen and phosphorus [10].

In micro scale, natural landscape engineering such as biofilter potentialy improve water quality which used wet land plant such as *Typha latifolia*, *Scirpus grossus, Pistia straiotes*, and *Echinodorus palaefolius*. Biofilter can be applied at some point of Cipunten Agung main river near by the agriculture, settlement, and industry outlet area. *Typha latifolia* density variation will affected the COD reduction significantly 97,18%, it will be more efficient for BOD, COD, and TSS reduction with plant density value 1 g/cm^2 [11]. The combination of *Scirpus grossus* and *Pistia straiotes* efectively 70-100% for reduction the concentration of TSS, TDS, Nitrit, Ammonia, and COD [12]. Integrated emergent plant such as *Scirpus grossus* and submergent plant system were potentialy remove the biological oxygen demand (BOD5), 65.7%; chemical oxygen demand (COD), 40.8%; ammonium (NH4+-N), 74.8%; nitrate (NO3--N), 38.8%; phosphate (PO43-), 61.2%; total suspended solids (TSS), 65.8%; and fecal coliforms, 94.8% [13].

The abilities to maintain the local wisdom knowledge to maintain the environmental quality are more likely the best options for achieving sustainable management in rural landscape [4]. Low Carbon Society (LCS) potentialy applied to involve the communities in watershed resource management. The paradigm of water back landscape at many level of society and goverment should be restored to the water front landscape. The incentives should be given to LCS communities and disentive for the program offenders [14].

5. Conclusion

LUCC can support water quality and ecosystem sustainability prediction in Cipunten Agung Watershed. Paddy field was increases at 2005, its potentialy increase Nitrate concentration. Meanwhile, the cultivated tree landscape at 1995 potentialy influence the environmental rejuvenation, especially water quality. Cipunten Agung Watershed management with ecology approaches is important to ensure the sustainability of environment and social ecosystem.

Acknowledgements
This study was conducted under the project entitled Management of Agroforestry Landscape as Landscape Services Provider towards Low Carbon Societies of PUPT PB 2016. This study was supported by KEMENRISTEK DIKTI of Indonesia.

References
[1] Firdaus R 2013 Assessment of the realtionship between land use land cover and water quality status of the tropical watershed : a case of batang merao watershed, indonesia. *J of Biodiversity and Environmental Sci.* **3** 21-30
[2] Basnyat P, Teeter L D, Flynn K M and Lockaby B G 1999 Relationships between landscape characteristics and nonpoint source pollution inputs to coastal estuaries *J Environ. Manage.*, 539– 549
[3] Xia L L, Liu R Z and Zao Y W 2012 Correlation analysis of landscape pattern and water quality in baiyangdian watershed *Procedia Environmental Sciences* **13** 2188-2196
[4] Kaswanto, Arifin H S and Nakagoshi N 2012 Water quality index as a simple indicator for sustainability management of rural landscape in west java, indonesia *International J of Enviromental Protection* **2** 19.
[5] Osborne L L and Kovacic D A 1993 Riparian vegetated buffer strip in water-quality restoration and stream management *Freshwater Biology Journal.* **29** 243-258
[6] McDonald A T and Kay D 1998 *Water Resources: Issues and Strategies* (New York (US): Longman Scientific and Technical)
[7] Effendi H 2003 *Telaah Kualitas Air : Bagi Pengelolaan Sumberdaya dan Lingkungan Perairan*. (Yogyakarta: Kanisius)
[8] Agustiningsih D, Sasongko S B and Sudarno 2012 Analisis kualitas air dan strategi pengendalian pencemaran air sungai blukar kabupaten kendal *J Presipitasi.* **9** 65-71
[9] Arifin H S and Nakagoshi N 2011 Landscape ecology and urban biodiversity in tropical Indonesian cities *Landscape and Ecological Engineering.* **7** 33-43
[10] Harashina K, Takeuchi K, Tsunekawa K and Arifin H S 2003 Nitrogen flows due to human activities in the Cianjur-Cisokan watershed area in the middle Citarum drainage basin, West Java, Indonesia : a case study at hamlet scale *Elsevier Agriculture, Ecosystems and Environment* **100** 75-90
[11] Elystia S, Sasmita A and Purwanti 2014 Pengolahan kandungan COD limbah cair pabrik kelapa sawit oleh *Typha latifolia* dengan metode fitoremediasi *Jurnal Teknik Lingkungan UNAND* **11** 88-95
[12] Syafrani 2007 Kajian pemanfaatan media penyaringan dan tumbuhan air setempat untuk pengendalian limbah cair pada sub-DAS tapung kiri, provinsi riau. *Graduate Thesis* (Bogor: Sekolah Pascasarjana Insititut Pertanian Bogor)
[13] Tanaka N, Jinadasa K B S N, Werellagama I B, Mowjood M I M and Wun JN 2006 Constructed tropical wetlands with integrated submergent-emergent plants for sustainable water quality management. *Journal of Environmental Science and Health Part A* **41** 2221-36
[14] Arifin H S 2014 Revitalisasi ruang terbuka biru sebagai upaya manajemen lanskap pada skala bio-regional *Risalah Kebijakan Pertanian dan Lingkungan* **1** 172-180

The effect of land use change on water quality: A case study in Ciliwung Watershed

Prita Ayu Permatasari[1]*, Yudi Setiawan[1,2], Rahmi Nur Khairiah[1] and Hefni Effendi[1]

[1] Center for Environmental Research, Bogor Agricultural University, PPLH Building 2nd-4th Floor, Jl Lingkar Akademik, Kampus IPB Dramaga, Bogor, Indonesia

[2] Department of Forest Resource Conservation and Ecotourism, Faculty of Forestry, Bogor Agricultural University, Kampus IPB Dramaga, Bogor, Indonesia

E-mail: pritapermatasari@gmail.com

Abstract. Ciliwung is the biggest river in Jakarta. It is 119 km long with a catchment area of 476 km^2. It flows from Bogor Regency and crosses Bogor City, Depok City, and Jakarta before finally flowing into Java Sea through Jakarta Bay. The water quality in Ciliwung River has degraded. Many factors affect water quality. Understanding the relationship between land use and surface water quality is necessary for effective water management. It has been widely accepted that there is a close relationship between the land use type and water quality. This study aims to analyze the influence of various land use types on the water quality within the Ciliwung Watershed based on the water quality monitoring data and remote sensing data in 2010 and 2014. Water quality parameters exhibited significant variations between the urban-dominated and forest-dominated sites. The proportion of urban land was strongly positively associated with total nitrogen and ammonia nitrogen concentrations. The result can provide scientific reference for the local land use optimization and water pollution control and guidance for the formulation of policies to coordinate the exploitation and protection of the water resource.

1. Introduction

Jabodetabek (Jakarta, Bogor, Depok, Tangerang, and Bekasi) megacity covering an area of approximately 7,000 km^2 and had a total population of more that 26.7 million in 2010. After 1981, the rate of population growth in Jakarta and surrounding cities has changed significantly. Jabodetabek area can not be separated from its upstream area in Puncak and Cianjur since the source of water for Jabodetabek comes from those two regions. Therefore, the concept of Jabodetabekpunjur (Jabodetabek, Puncak, and Cianjur) was designed to control development in the metropolitan region and upstream of the metropolitan area [1].

Ecologically, Jabodetabekpunjur coverage is the area that includes three primary watersheds (DAS), namely Ciliwung, Cisadane, and Bekasi Watershed. Ciliwung watershed is one that has a very important ecological role because it crosses several administrative areas such as Bogor Regency, Bogor City, Depok City, and Jakarta. Ciliwung River upstream located in Puncak and flows along 119 km with an average monthly discharge 882 m^3 per second (in Manggarai) toward Jakarta Bay.

Total of forest area, settlements, agricultural area, and green space in Jabodetabek from 1970-2005 has changed drastically. In the last 35 years, Jabodetabekpunjur has lost 27% of green open space.

Built area such as settlement grew more than 12 times, causing the carrying capacity of the environment to be very limited. The uncontrolled growth of settlements and urban area along the watershed, the malfunction of the canals, and the absence of an adequate drainage system caused further delays in the flow of water into the sea, resulting in Jakarta and the area along the river banks were vulnerable to flood. The other problems is the decline in the quality and quantity of river water [2].

Land cover and land use are very important elements in relation to water quality. Different types of land use and land cover affect the quality of water. Agricultural and household fertilizers have different chemicals within them, such as nitrogen and phosphorus. These chemicals can potentially run off into nearby water sources such as groundwater, streams and larger bodies of water. In turn, this could damage the nutrient content within that water supply, affecting the overall water quality itself [3].

The high population growth in Jakarta, Bogor, Depok, and surrounding areas resulted in changes of land use and land cover. Changes in land use and land cover indirectly affect water quality in Ciliwung River. This study aim is to look at the effects of changes in land use and land cover to the Ciliwung River water quality in 2010 and 2014.

2. Material and methods

2.1. Study area

Ciliwung River (figure 1) is one of 13 river that traversed in Jakarta. Drainage area of Ciliwung River approximately 42,818.74 hectares area and the main river channel is ± 119 kilometers long with an average gradient of about 10-15%. The Ciliwung River has its source at Mandalawangi in Bogor Regency with the highest peak at 3,002 m. The river flows in a northern direction passing several active volcanoes, crosses three main cities Bogor, Depok, and Jakarta before finally flowing into the Java Sea through Jakarta Bay. Ciliwung River is divided into six segments. It crosses some administrative regions such as Bogor Regency, Bogor City, Depok City, East Jakarta, South Jakarta, Central Jakarta, and North Jakarta. Mean rainfall reaches 3,125 mm, with mean annual discharge of 16 m^3/s as measured at Ciliwung Ratujaya observation station (231 km^2). With such topographical, geological and hydrological features the Ciliwung River is often overflowing and inundating parts of Jakarta [4].

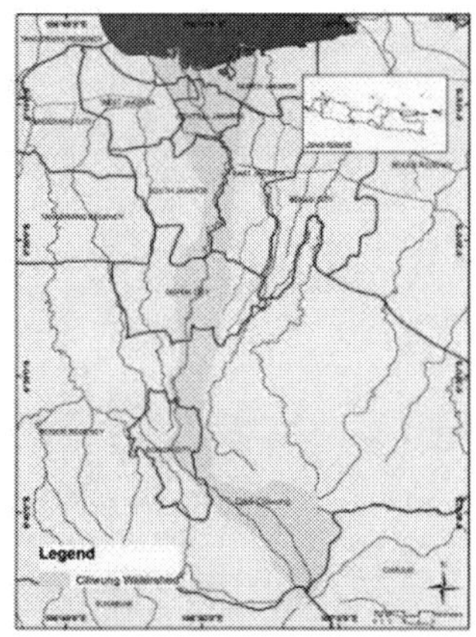

Figure 1. Study area

2.2. Sample collection and parameter measurement

A total of 12 points were sampled along the main stem of Ciliwung River. Water sampling was carried out in rainy season and dry season in 2010 and 2014. Sampling data in those two years will be used to compare the water quality from time to time and to see land cover change effect in 4 years to water quality. A combination of spatial and temporal analyses was applied to provide more robust results for the relationship between land use and water quality. Duplicate samples were taken out at each sampling site. Data of water quality were obtained from Ministry of Environment and Forestry Republic Indonesia. Sampling points along Ciliwung River can be seen in table 1.

Table 1. Ciliwung sampling points

No	Sampling Points	Location
1	Atta'awun	S 06°41'56.2 E 106°59'14.5"
2	Bendung Gadog	S 06°39'12.1" E 106°52'10.3"
3	Katulampa	S 06°37'59.7" E 106°50'12.6"
4	Sempur	S 06°35'25.9" E 106°47'56.6"
5	Pondok Rajeg	S 06°26'43.7" E 106°48'46.2"
6	Jembatan Panus	S 06°24'02.4" E 106°49'54.3"
7	Kepala Dua/Srengseng Sawah	S 06°20'56.2" E 106°50'18.2"
8	Intake PAM Condet/Kampung Gedong	S 06°17'34.2" E 106°51'12.5"
9	Sebelum Pintu Air Manggarai	S 06°12'45.7" E 106°51'28.0"
10	Jl. Kwitang	S 06°10'53.78" E 106°50'11.40"
11	Jl. Gajah Mada	S 06°09'34.64" E 106°49'08.48"
12	Jemb. Pantai Indah Kapuk/Muara Angke	S 06°07'00.04" E 106°46'23.03"

During and after sampling, six water parameters were chosen to be measured, which are important indicators of water pollution influenced by anthropogenic activities. Parameters that sampled consist of total suspended solids (TSS, mg/L), Total phosphorus (TP, mg/L), chemical oxygen demand (COD, mg/L), biological oxygen demand (BOD, mg/L), and dissolved oxygen (DO mg/L).

2.3. Spatial analysis
Landsat Thematic Mapper (TM) satellite imagery with 30 m resolution for 2010 and 2014 was obtained from Indonesian Planology Agency. The land use classes were generated using a supervised classification technique with a maximum likelihood algorithm in the Erdas Imagine 9.2 (ERDAS Corporation, Norcross, GA, USA). The Landsat images were also rectified using aerial photographs and field survey data. The overall accuracy for the classification was 88% and the kappa coefficient was 0.85 (i.e., 85% more accurate than a random classification) [5]. The land use classes were categorized into five groups: urban land, agricultural land, forest, bareland, and water body. Urban land use consists of built areas such as settlement and industry. Agricultural land including rice field, plantation, shrubs, and secondary crops field. While water bodies consist of fish pond and wetland.

3. Result

3.1. Land use site grouping and their distribution
From 2010 until 2014 urban land made up the vast majority of total area. The percentage of urban land has increased in 4 years from 66.10% to 66.85%. While agricultural land in the whole watershed showed degradation from 21.65% to 20.90% (table 2). There is no changes in total area of other land uses (bareland, forest, and water body) within those 4 years. Land use map of Ciliwung Watershed in 2010 and 2014 can be seen in figure 2.

Table 2. Land use characteristics in Ciliwung Watershed

Land Use	2010		2014	
	Area (Ha)	Percentage	Area (Ha)	Percentage
Agricultural land	9,269.15	21.65	8,947.64	20.90
Bareland	12.06	0.03	12.06	0.03
Forest	5,048.19	11.79	5,048.19	11.79
Urban	28,301.37	66.10	28,622.88	66.85
Water body	187.98	0.44	187.98	0.44
Total	42,818.74	100.00	42,818.74	100.00

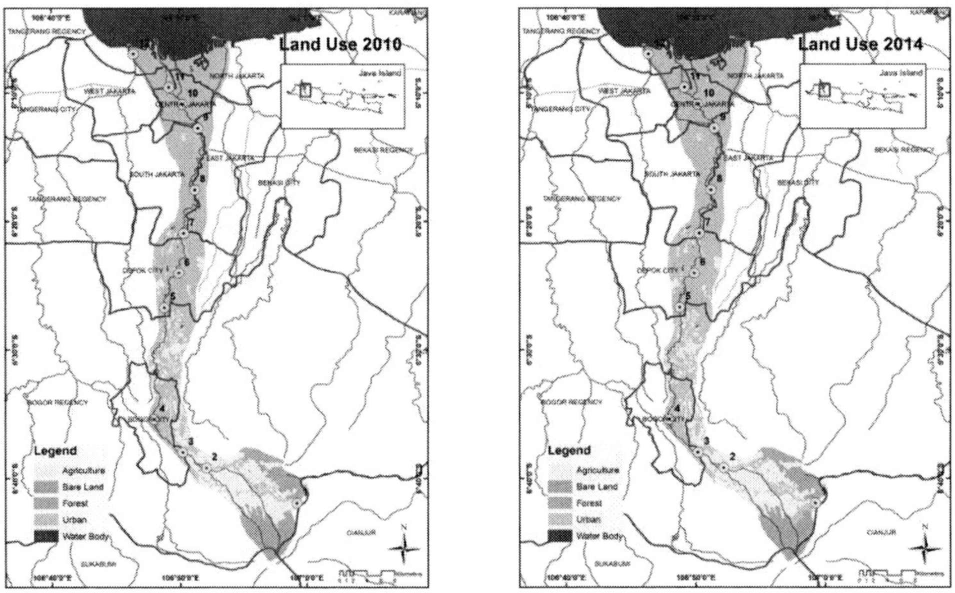

Figure 2. Land use in Ciliwung Watershed in 2010 and 2014

From the figure 2, it can be seen that land use changes from agricultural land to urban land occurred in the upstream of Ciliwung Watershed near sampling point 1. This indicates that land use changes in the Ciliwung Watershed not only occurred in an urban area or downstream of the river but also the conservation area located on the upper river. The rapid economic activity and development in Puncak is one of the factor. The land use within the watershed has great impacts on the water quality of rivers. The water quality of rivers may degrade due to the changes in the land cover patterns within the watershed as human activities increase of [6, 7]. The high attraction of Puncak as tourism object for the urban population causes local people to prefer to change agricultural land into buildings such as villas and hotels that have higher economic value than agricultural commodities. If this happens constantly then the Ciliwung upperstream would lose its function as a conservation area and ecological conditions along the Ciliwung will be degraded.

3.2. Temporal variations of water quality

The comparison of water quality in every Ciliwung sampling point in 2010 and 2014 can be seen in figure 3. According to Government Regulation No. 82/2001, Ciliwung River was categorized into water quality class 2. Indonesian Government Regulation No. 82 year 2001 stated that water quality class 2 can be used as water recreation, fish cultivation, livestock, and irrigation. The standard of water quality class 2 of each parameter can be seen from the horizontal blue line in the graphic. Based on class 2 water quality standards, the parameters of TSS, TP, BOD, and COD should be at or below the blue line while the DO parameter should be at or above the blue line.

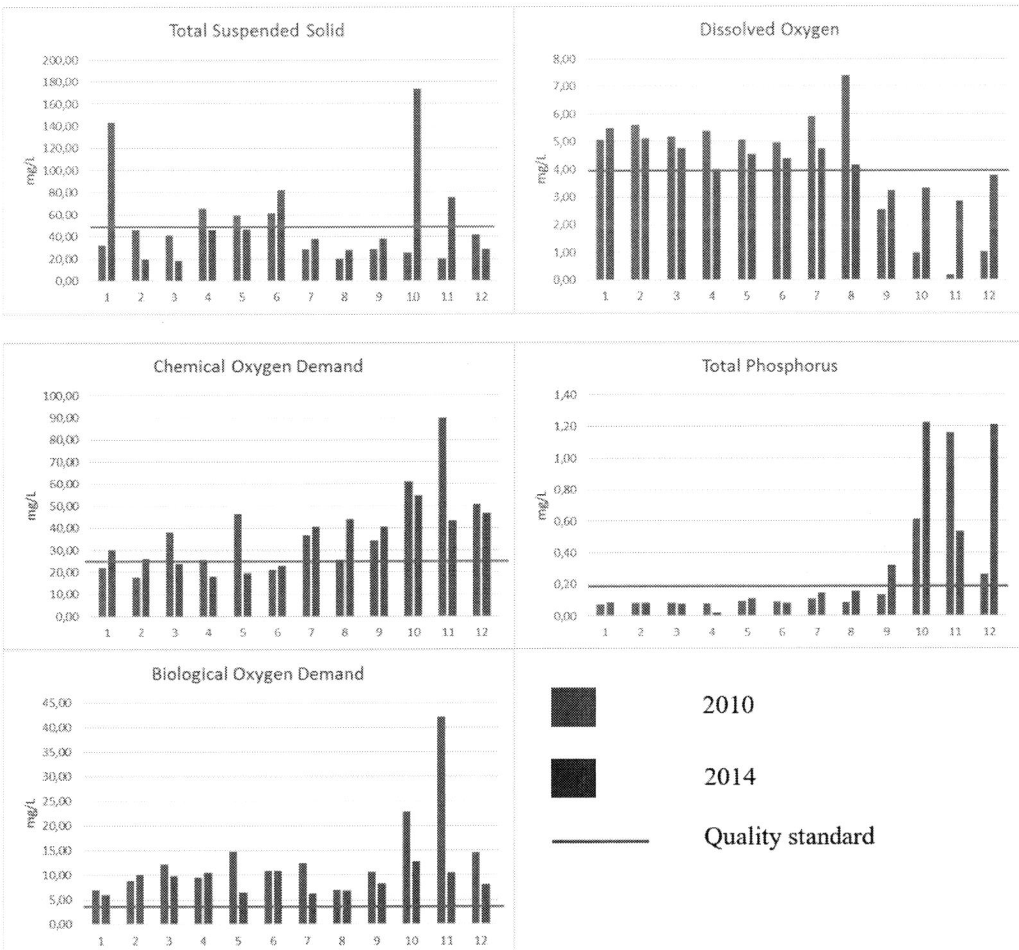

Figure 3. Graphic of Ciliwung River water quality in 12 sampling points

TSS value at point 2, 3, 4, and 5 showed a decline from 2010 to 2014, while the rest showed an increase. At some point, the value of TSS is already at quality standards. In general, the parameter DO showed a decline from 2010 to 2014. At the point 1-8, the value of DO meet the quality standards. However, in the downstream area of the river (point 9-12) DO value is below the quality standard. The decrease of DO value from upstream to downstream indicates water quality degradation.

Similarly with DO, TP values showed a decrease in water quality from upstream to downstream. TP value showed an increase on point 9 to 12. According to NGRDC (2015) [8], the value of TP and DO has negative correlation. The more TP value, the less DO value. Nutrients such as phosphorous and nitrogen are essential for the growth of algae and other plants. Excessive concentrations of nutrients, however, can overstimulate aquatic plant and algae growth. Bacterial decomposition of organic respiration and can use up dissolved oxygen, depriving fish and invertebrates of available oxygen in the water (eutrophication).

COD value shows an increase from upstream to downstream. Some sampling point showed a decrease COD value within 4 years. COD value at the sampling point in the downstream area is also above the quality standard of 25 mg/L. In contrast to COD, BOD value in every sampling points have already meet water quality standard for class 2 (3 mg/L). However, in general, the value of BOD in 2014 showed a decline from the 2010. This represents an improvement of water quality.

Biochemical oxygen demand (BOD) is a measure of the amount of oxygen that bacteria will consume while decomposing organic matter under aerobic conditions. Chemical oxygen demand (COD) does not differentiate between biologically available and inert organic matter, and it is a measure of the total quantity of oxygen required to oxidize all organic material into carbon dioxide and water. COD values are always greater than BOD values, but COD measurements can be made in a few hours while BOD measurements take five days. Generally, water quality parameters in Ciliwung River from 2010 to 2014 has degraded.

4. Discussion

4.1. Impact of different land uses on water quality

Generally land use types related to economic activity and development negative impacts on water quality, so positive correlations exist between percentages of these land use types and concentrations of water pollutants, while undeveloped areas such as natural forest area are related to good water quality [9]. Different land use types are associated with different water pollution problems. For example, Tong and Chen (2002) [10] examined the relationships of land use and water quality on regional scale in the watershed of the Ohio State, USA. They found that TP (Total Phosphorus) was significantly positively related to commercial, residential, and agricultural lands, but non-significant relationships with forest, and BOD had a significant positive correlation with residential and commercial lands, a significant negative correlation with forest, but non-significant correlation with agricultural land [11].

4.2. Impact of urbanization on water quality

Ciliwung river water quality degradation within 4 years shows that the change in land use, in this case the increase in urban land, generally have shown that urbanization provides a real influence on the degradation of water quality in all parameters. In addition, the land utilization in Jabodetabekpunjur area has increased from upstream to downstream. Urban land has a great correlation with water quality. Urban land use was identified as the strongest contributor of pollutants in the Ciliwung Watershed. This may have been highly influenced by point source as well as non-point source pollution.

Urban land consists of various land uses such as residential, industrial, office, commercial, and other built area. Compared with other land use, urban land produces more wastewater. In addition, urbanization has expanded impervious areas, which has led to faster storm flows and greater runoff volumes [12]. Impervious storm runoff washes all types of pollutants (e.g., non-point source and point source pollutants) into rivers, which increases concentrations of nutrients and other pollutants in surface waters [13]. Besides, a shortage of wastewater treatment infrastructure in suburban areas also contributes to increased nitrogen levels [14].

5. Conclusion

Based on the Ciliwung watershed land use map in 2010 and 2014, it is known that there is a land use change of agricultural areas into urban land over 4 years period. Land use change has great impact on water quality. Generally, Ciliwung river water quality has decreased in the past 4 years. The quality of river water also decreased from upstream to downstream. This is related to the increasing urban land along the Ciliwung river from upstream to downstream. The high level of economic growth have resulted in the rise urban land. While, the rising urban land has a positive correlation with increasing concentration of pollutants in the river.

References

[1] Rustiadi E, Pribadi D O, Pravitasari A E, Indraprahasta G S and Iman L S 2014 Jabodetabek Megaciy: From City Development toward Urban Complex Management System (Berlin: Springer)
[2] Djakapermana R D 2015 Jabodetabekpunjur Spatial Plan: Efforts on Balancing Economic Growth to Environmental Sustainability Rencana Tata Ruang. Retrieved from www.scribd.com on January 2, 2017 (Indonesian)
[3] Bowden C, Mike K, Josh A, Keelie C and Shane T 2015 Water Quality Assessment: The Effects of Land Use and Land Cover in Urban and Agricultural Land (Kansas: Natural Resources and Environmental Sciences (NRES) Kansas State University)
[4] DPRI Kyoto 2014 Ciliwung Disaster Prevention Technology and Policy Research (Kyoto: Kyoto University)
[5] Congalton R G 1991 A review of assessing the accuracy of classifications of remotely sensed data *Remote Sens. Environ.* **37** 35–46.
[6] Ngoye E and Machiwa J F 2004 The influence of land-use patterns in the Ruvu river watershed on water quality in the river system *Physics and Chemistry of the Earth* **29** 1161–1166
[7] Sliva L and Williams D D 2001 Buffer zone versus whole catchment approaches to studying land use impact on river water quality *Water Research*. **35** 3462–3472
[8] Northeast Georgia Regional Development Center 2015 Watershed Protection Plan Development Guidebook (Athens: the University of Georgia)
[9] Bu H, Meng W, Zhang Y and Wan J 2014 Relationships between land use patterns and water quality in the Taizi River basin, China. *Ecol. Indic.* **41** 187–197
[10] Tong S T Y and Chen W 2002 Modeling the relationship between land use and surface water quality. *J. Environ. Manag.* **66** 377–393
[11] Tu Q 1990 The Researches on the Eutrophication in Chaohu Lake. Heifei (China: University of Science and Technology of China Press) (Chinese).
[12] Waters E R, Morse J L, Bettez N D and Groffman P M 2014 Differential carbon and nitrogen controls of denitrification in riparian zones and streams along an urban to exurban gradient. *J. Environ. Qual.* **43** 955–963
[13] White M D and Greer K A 2006 The effects of watershed urbanization on the stream hydrology and riparian vegetation of Los Peñasquitos Creek, California. *Landsc. Urb. Plan.* **74** 125–138
[14] Sun R, Chen L, Chen W and Ji Y 2013 Effect of land-use patterns on total nitrogen concentration in the Upstream Regions of the Haihe River Basin, China. *Environ. Manag.* **51** 45–58

The Contribution of the Mesoscale Convective Complexes (MCCs) to total rainfall over Indonesian Maritime Continent

Trismidianto[1,2*], E Yulihastin[2], H Satyawardhana[2], J T Nugroho[3] and S Ishida[1]
[1]Meteorological Laboratory, Graduate School of Science and Technology, Hirosaki University, Hirosaki, Japan
[2]Center for Atmospheric Science and Technology, National Institute of Aeronautics and Space, Jakarta, Indonesia
[3]Remote Sensing Application Center, National Institute of Aeronautics and Space, Jakarta, Indonesia

E-mail: trismidianto@lapan.go.id

Abstract. The MCCs contribution is expressed as the ratio of MCCs precipitation to the total rainfall at each grid point where MCCs identified by using infrared satellite imagery. The rainfall obtained from the Real-Time Tropical Rainfall Measuring Mission's (TRMM) data. This study found that MCCs contribute to total rainfall during 15-years over Indonesian Maritime Continent up to 20% where the greatest contribution concentrated over Central Kalimantan, South China Sea, Indian Ocean and Papua Island. The contribution of the MCCs is slightly increased in seasonally and monthly up to 24% and 30%, respectively. The greatest contribution is more existence over continent than the ocean in each season and month, except in July. The greatest contribution of the oceanic MCC concentrated in the Indian Ocean almost in each season and month. The contribution of the MCCs over Java Island is almost small in each season and month, but the contribution a slight increase during JJA. The MCC not only contribute to rainfall in the MCC area but the MCC give a contribution to its surrounding area. The frequency distribution of MCCs contribution to the total rainfall is very similar and consistent with the geographic distribution of the MCCs over the IMC during 15-years.

1. Introduction

Mesoscale convective systems (hereafter MCSs) are the largest of the convective storms that form when clouds occurring in response to convective instability amalgamate and organize upscale into a single cloud system with a very large upper cirriform cloud structure and rainfall covering large contiguous rain areas. They account for a large proportion of precipitation in both the tropics and warmer midlatitudes. A special case of the MCS is the mesoscale convective complex (hereafter MCCs), defined by Maddox [1], as an MCS that maintained a large, contiguous cold cloud shield for at least 6 h. The MCC criteria were chosen to isolate those systems that relatively rounded, with minor axis/major axis 0.7 at the time of maximum extent. These large, long-lasting organized systems are comprised of an ensemble of thunderstorms that together, often yield intense rain rates within contiguously sizeable precipitation areas and can greatly influence the hydroclimate of a region [2]. Some of the papers examined showed that the passage of an MCC changes the surface conditions over a large region and produce the heavy rainfall must be regarded as a cause of several of the hazards (flooding, localized flash floods, mudslides, and property and crop damage) produced by MCCs [e.g., 1, 3]. MCCs play also a major role in the rainfall regimes of the tropics [4][5]. The convection related to sea breeze convergence is able to aggregate into MCCs later in the days, which move off land to give the greatest precipitation during the local morning time over the oceans [6].

Ashley et al. [7] and Fritsch et al. [8] showed that MCCs have contributed up to 60% o the total precipitation in a portion of the central USA. Laing [9] demonstrated that MCCs contributed an average of 22% of total rainfall during July - September 1987 across portions of the Sahel in Africa. Durkee et al. [10] determined that MCCs account for a considerable portion of total warm season precipitation in subtropical South America, with 11%-20% of the total rainfall in most warm seasons provided by MCCs in much of the study domain. Blamey and Reason [11] found that MCCs contributed up to 20% of the total summer rainfall in parts of the eastern region of southern Africa. Nesbitt et al. [12] have reported that MCSs (5-years data) produce as much as 90% of rainfall in certain land areas. For the global tropics, they produce more than 50% of the rainfall in heavily raining areas. Nesbitt et al. [12] were also documented that MCSs produce 50%–90%, are typical for heavy rain regions of the global tropics. MCSs contribute greater than 70% of the rainfall in the Tropics but make up a much smaller fraction of the cloud cover [2]. Yuan and Houze [13] showed that the active MCSs to be associated with 56% of tropical precipitation. The rainfall is strongly correlated with MCSs or MCCs, is highly spatially and temporally variable, and is heavily influenced by land-surface properties and large-scale circulations [14]. However, the contribution of MCC over Indonesian Maritime Continent (hereafter IMC) to total rainfalls is not yet well documented. Therefore, it is interesting to determine the contribution of MCCs to total rainfalls over IMC. In addition, previous studies raise important questions: what is the monthly, seasonal, and annual analyze of MCC contribution over IMC? So that, the overarching goals of this study is to determine the contribution of MCCs to rainfall over IMC for a period of 2001-2015.

2. Data and Method

To identify the MCC in this study, we use black body temperature (hereafter T_{BB}) from the Himawari geostationary satellites, operated by the Japan Meteorological Agency (JMA) that consists of Himawari-5/GMS-5 for data from January 2001- April 2003, Pacific GOES/GOES-9 for data from May 2003- June 2005, Himawari-6/MTSAT-1R for data from July 2005 - June 2010, Himawari-7/MTSAT-2 for data from July 2010 - June 2015, and Himawari-8 for data from July 2015 - December 2015. That satellite data has spatial and time resolutions 0.05°×0.05° and one hour, respectively. This paper adapts the method from Ismanto [15] and Trismidianto [16] that based on the maximum spatial correlation tracking technique (MASCOTTE) method by Carvalho and Jones [17] and almost similar way with Laing and Fritsch [18] to identify MCC with input the temperature, latitude, and longitude values for each cloud shield pixel that obtained from IR to a modified version of a computerized MCC program using MATLAB for measuring the satellite-observed characteristics of the MCCs [1] which stated that MCC have cloud shield with continuously $T_{BB} \leq -32$°C must have an area ≥ 100,000 km^2 and the interior cold cloud $T_{BB} \leq -52$°C must have an area ≥ 50,000 km^2. To distinguish MCCs from the other MCSs type, thus, the MCC will have a type of elliptical with eccentricity ≥ 0.7 and life cycle of MCC about ≥ 6 hours. This program computes the area, eccentricity, and the centroid positions of the -32°C and -52°C cold cloud shield associated with deep convection. The areal extent and eccentricity values were used to determine the duration and life cycle of the MCCs, while the centroid position indicated the path of propagation of the systems.

To determine the contribution of MCC to rainfall, we following the methods like Laing et al. [9], Durkee et al. [10], and Blamey and Reason [11] but a little different because in this paper we do not determine the storm area for MCC rainfall use the convex hull method but the MCC rainfall just determined from accumulated rainfall during the MCC event from initial stage until three or six hours after post-MCC stage at each grid point that occurs from 2001 - 2015 over the IMC and the total rainfall is accumulated rainfall over the IMC during 2001 - 2015. We make accumulated rainfall until around six hours after post-MCC stage due to MCCs are not only contribute to rainfall in the MCC area but the MCC give a contribution to its surrounding area after several hours dissipated in accordance with the results already reported by Trismidianto et al. [19]. The contribution of the MCC is expressed as the ratio of the rainfall during MCCs to the total rainfall at each grid point. The rainfall data from the Real-

Time Tropical Rainfall Measuring Mission's (TRMM) Multi-Satellite Precipitation Analysis (TMPA-RT) 3B41RT v7 data set, which has hourly temporal resolution and 0.25° × 0.25° spatial resolution [20].

3. Result and Discussion

3.1. MCC contribution to total rainfall during 15-years

A total of 1028 MCCs were identified and tracked during 15-years for a period of 2001 – 2015 as shown in figure 1 (a). The MCCs developed in several concentrated areas; some of them were over the ocean and coastal areas, and others were over inland and high elevation areas. By examining the location of the interior cloud centroid for all of the MCCs, one can see that MCCs are mostly frequent occur in nine regions as shown in red dotted line in figure 1 (a), i.e., 1). Indian Ocean (hereafter IO), 2) along the western coast of Sumatra (hereafter WS), 3) South China Sea (hereafter SCS) near of northern coastal of the Borneo Island, 4) the central Kalimantan (hereafter CK) around 3°S-1°N; 111° -115°E, 5) the east Kalimantan (hereafter EK) around 1° - 4°N; 116° - 120°E, 6) Makassar Strait (hereafter MS), 7) the central Sulawesi (hereafter CS) around 4° - 2°S; 120° - 123°E, 8) Merauke (hereafter MR) in location around of 8° - 5°S and 135.5° - 140°E and 9) Cendrawasih Bay (hereafter CB) in location around of 132.5° - 136.5°E and 4° - 1°S.

Most of the MCCs in the IMC occur on the continent than the ocean where this research found 31.23% MCCs over the IMC occurred over the ocean, 26.46% occurred over the coastal and most of the theMCCs occurred in the continent around 42.32%. It is similar with the characteristics of MCC globally by Laing and Fritsch [21] that found most of the MCCs is located in continents almost of 91% though they can occur over oceans. However, MCC has a large scale are more prevalent over the oceans than on land especially over IO, and MCCs originating over the oceans are longer lasting originating over land. The continental MCC are mostly concentrated in near the mountainous and high elevation areas. A mountain breeze and a valley breeze are two related, localized winds that occur one after the other on a daily cycle which also given a contribution to convective activity over the IMC due to IMC is a unique geographical region that composed of a complex system of mountainous islands. If compared to the global population of MCCs, it is similar to Ashley et al. [7] and Laing and Fritsch [22] that found the mean location for MCC genesis and initiation for the Americas are located on the lee side of major mountain ranges, such as the Andes in South America or the Rockies in the United States.

The contribution of the MCCs to total rainfall over IMC is already observed in this research. Figure 1 (b) shows that frequency distribution of MCCs contribution to the total rainfall is very similar and consistent with the geographic distribution of the MCCs over IMC during 15-years. It is indicated that the greatest contribution of the MCCs is most frequently occur in the concentrated area of the MCC occurrences. The greater contribution of the MCCs found in several concentrated areas; some of them were over the ocean and coastal areas, and others were over inland and high elevation areas. One can see that maximum contribution of the MCCs in globally which around 18-20% are mostly frequent occur in the WS, SCS, CK, CS, MR, and IO. The greatest contribution from of them is found in the CK that reaches up to 20%. The amount of MCC is happening in this area, one of which is caused by the Borneo vortex that also related with mesoscale convective rainfall as described by Koseki et al. [23].

The greater contribution of the oceanic MCCs is just found in the IO that reaches up to 14-18%. In this area, the greater contribution of MCC occurred along the WS than on the IO because most of MCC that occurred in the IO will move to the top of the Sumatra Island (hereafter SI) when MCC extinct, so there are the effect of MCC over IO to rainfall over the SI. The oceanic MCC also give the great contribution over SCS in around 13-17%. In detail, MCCs give a contribution to the total rainfall over Papua Island (hereafter PI) reaches 14 - 18% in MR and CB. The great contribution around 13-17% also occurred off the coast of West Kalimantan (hereafter WK) and on two locations over Sulawesi, i.e., in South Sulawesi (hereafter SS) and South East Sulawesi. MCC contributed is very small in along the Java Island (hereafter JI) until East Nusa Tenggara due to MCCs was rarely found in this location.

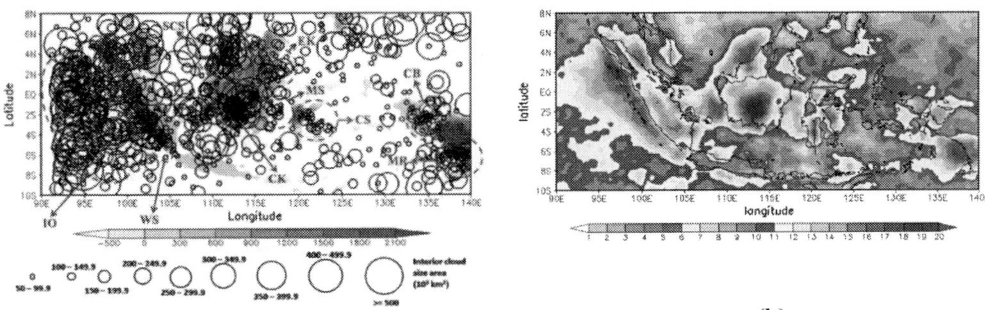

(a) (b)

Figure 1. (a) Geographical distribution of MCCs in IMC during 15-years (2001-2015). Locations are for the MCC at the time of maximum extent of the interior cloud size area when the mature stage. The circles representing the interior cloud size area (10^3 km^2). The legend representing the elevation with the unit in meters (m). (b)The frequency distribution of the MCCs contribution to total rainfall during 2001-2015, the legend representing the percentage of MCC contribution.

3.2. MCC contribution to seasonal rainfall

The Season in IMC is divided into four seasons, namely in December, January and February (DJF) as the wet season, March, April and May (MAM) as a transitional season from wet season to the dry season. June, July, and August (JJA) as dry season, and September, October and November (SON) is a season of transition from dry season to the wet season. In seasonally, the greatest contribution of the MCCs to the DJF total rainfall is concentrated in the coastal area near the northern part of SI, which reached 18-22% as shown in figure 2 (a). However, a considerable contribution ranges from 14-20% also found in several locations such as CK, SCS near Sarawak, SS, on the coast of CS and MR. During DJF season, the MCC also contributed significantly to the several regions over the Pacific Ocean (hereafter PO) and the IO up to 18%. Several of previous work has reported that the characteristics of rainfall over JI are heavy rainfall during DJF season, but the few systems of MCC that occur in near the JI are just contributing up to 6% - 8%, while the contribution of the MCCs over East Nusa Tenggara is less than 6 %.

The regions that have great contribution in DJF seasons are increased and expanded during MAM season as shown in figure 2 (b). The greatest contribution up to 24% found in SCS near the Brunei Darussalam and CK. The great contribution that occurs on the continent also found over Southern Sumatra and Sulawesi in several locations up to 16-20%. The contributions of MCC in several coastal regions were also great around 16-20% is in the area near the coast of North Sumatra and Malaysia, in the coastal area near WK, in the Java Sea near the coastal areas of JI and in the coastal area near MR. The great contribution also scattered in several locations in the IO and the PO. MCC also influencing the rainfall along from the JI until East Nusa Tenggara to remain weak in the MAM season, its contribution is very small along the land in this area, but the large contribution is seen in some coastal areas along this area.

During JJA, the impact of MCC to JJA rainfall is slightly less over continent where the MCC contribution just found up to 10% as shown in figure 2 (c). The MCC has a strong influence only at a few locations in the IO, in coastal areas near WK and coastal areas near MR. This was shown by the contribution is quite large compared with other regions which are about 16-22%. Figure 2 (c) also shows that the greatest contribution is just given by the oceanic MCC. The continental MCC is not too influencing to the rainfall during JJA. Figure 2 (d) shows the greatest contribution of the MCC during SON that around of more than 24% is concentrated on the Arafura Sea, Banda Sea, and the Flores Sea. The great contribution of the continental MCCs to SON total rainfall are mostly located over the CK and southern Sumatra, while the contribution of the coastal MCCs almost exists along the WS, SCS near Brunei Darussalam, MS, and CB.

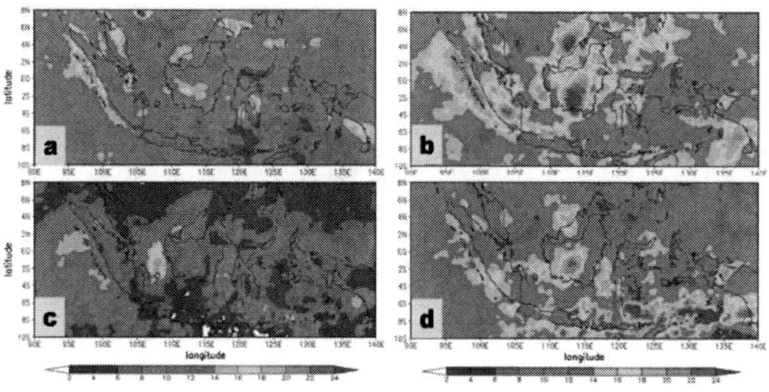

Figure 2. Frequency distribution of the MCC contribution to total rainfall during 2001-2015 in each season; (a) DJF, (b) MAM, (c) JJA and (d) SON. The legend represents the percentage of the MCC contribution.

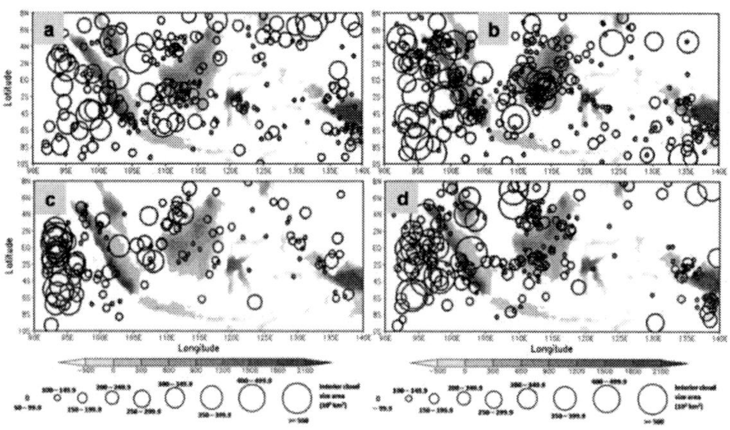

Figure 3. Same with Figure 1 (a) but for each season, (a) DJF, (b) MAM, (c) JJA and (d) SON.

By examining the location of the interior cloud centroid for each season, In DJF, one can see MCCs are most frequent concentrated in the continent at CK, SCS near northern Kalimantan, CS and MR as shown in figure 3 (a). However, that continental MCC has an interior cloud in small size around 50,000 - 149,999 km^2. In contrast, the large-size of MCC most frequently found over IO near SI. Most of them have an interior cloud in size more than 300,000 km^2. The MCC in medium size is also concentrated in PO. In MAM as shown in figure 3 (b), the frequency of MCCs was almost similar with DJF, but MCCs over CK became more widespread throughout to Kalimantan, and several MCCs with large size is indicated. MCCs over MR also became more widespread until Arafura Sea. Several MCCs with medium size around 200,000 - 300,000 km^2 was found over the SI and western coastal of the SI. The concentration of the oceanic MCC with large size still found over IO. Several MCC in small-size also occurs over JI. During JJA as shown in figure 3 (c), the continental MCC are rare occur, just found in several areas, among others; CK and northern Sumatra in small frequency. The MCCs are more concentrated over the ocean with the greatest concentration are over IO. MCC is never occurred in along JI until East Nusa Tenggara during JJA. There are five concentration regions of MCC events during SON as shown in figure 3 (d). i.e., over IO, SCS near northern Kalimantan, CK, CB, and MR. The MCC with large size is still concentrated over IO, while small size is concentrated over Kalimantan, especially CK. Overall, the frequency distribution of the MCCs contribution to the total rainfall in each season during the 15-years is very similar and consistent with the geographic distribution of the MCCs over IMC in each season during 15-years.

3.3. MCC contribution to monthly rainfall

Figure 4 shows the frequency distribution of the monthly contribution and location of the MCCs. In January, the continental MCC are concentrated in CK and CS but in small size less than 100,000 km^2. The MCCs with large size are predominantly the oceanic MCC that most frequently found over IO. The small size of MCC is also concentrated over the coastal of MR and the coastal of northern Kalimantan over SCS. The frequency of MCCs over the CK during February, but MCCs became widespread until to WK. MCCs is also concentrated in Lampung, southern Sumatra, and several MCC is found in the coastal of north Sumatra. The frequency of the MCC over the IO and the other ocean decreased, just occur in a few events. During March, MCCs just concentrated over CK with the frequency of MCCs which greater than February. There is large size of MCC occurs in CK. The small size of MCC is also occurred in SI, Sulawesi, and PI but in small frequency. A few of MCC with large size is also found over IO. The MCC with small size increased over CK in April, most of the MCC are also located in along western of SI. Most of them have medium size around 200,000 km^2. The MCC with medium size also occurs in IO near the coast of northern Sumatra. Several MCCs with small size is also found over PI and JI. Most of the MCCs that occur in May has a large size and located in IO. MCC is rarely found over Sulawesi, Papua, and Java. In June, most of the MCC concentrated over the ocean with the greatest frequency in IO. Only a few found the MCC on the continent, such as in CK and north Sumatra. Similar to May, MCC does not occur over Sulawesi, Java, and Papua. The maximum contribution of the MCCs to January total rainfall is mostly concentrated on the western near the northern Sumatra up to 30%. The MCC contributed 22-28% of the rainfall to the same area during February until May, while in June the contribution in this area less than 10%. The greatest contribution was also found in the CK for each month except June, even the contribution in this area reached more than 30% during March and April. The great contribution in the SCS near Brunei Darussalam occurs in March and April. The great contribution over Sulawesi is just found in April. The great contributions also found in the PO near Maluku which occurred in January and May. Several locations in the IO have great contributions almost on every month from January to June. The great contributions mostly occur on MR. MCCs contributed is very small over the JI.

The greatest contribution found in November in several areas that even reach more than 30% in Arafura Sea, Banda Sea, and the Flores Sea, this is possible because of there is an effect from MCC that occur over MR which moving toward that three sea areas after MCCs dissipated. The great contributions also found in several regions during November which located in the IO near Sumatra, on SCS near Borneo Island, on CK, on the coast of MR and CB.

Similar condition with November is also found in October but with the percentage less than November. During July until September and December, the contribution of the MCCs has considerably weak, where the maximum contribution just found in two small locations, like in July, the maximum contribution just found in the IO and the coastal of WK up to 18%. Even the contribution of MCCs along JI until East Nusa Tenggara is less than 4%, the same condition with August, but the contribution in the coastal of WK in August has the percentage value more than July up to 28%. Slight different to September, the maximum contribution up to 28% found in the coastal east on southern Sumatra, in the Java Sea near the CK and on Eastern Java. MCC is not too influencing the monthly total rainfall in December, where one can see that on average the contribution is less than 8%. The frequency of MCC was similar with June that only found over IO in longitude 91° - 97°E. The continental MCC is rare occur during this month. The frequency of oceanic MCC with large size over IO is increased in October and November, while in August is slightly same with July. However, this frequency decreased in September and December. In August and September, the frequency of continental MCC is slightly same with July, only a few found in the large island of the MCC. In fact, almost never occur along the JI to East Nusa Tenggara. The frequency of continental MCC over CK increased in October and November, However, this frequency decreased in August, September, and December. The coastal MCC is more concentrated in coastal of MR during October until December. Overall, similar with seasonal, the MCCs with large size more than 300,000 km^2 are most frequently occur over IO each month with the greatest frequency in October and November. CK is one favored region for MCC occurrences each month except

June until September, but most of the MCCs in this area have a small and medium size less than 300,000 km^2. Along of JI to East Nusa Tenggara is the region that rarely found of MCC events. Overall, the frequency distribution of MCCs contribution to the total rainfall in each month during the 15-years is very similar and consistent with the geographic distribution of the MCCs over the IMC in each season during 15-years.

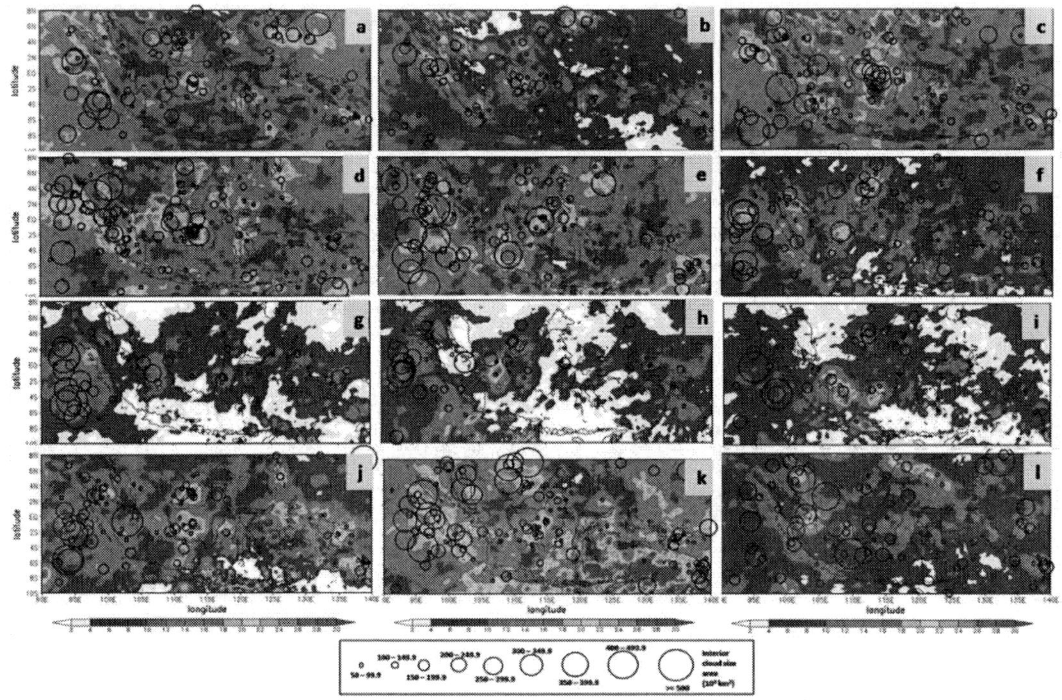

Figure 4. The frequency distribution of the MCCs contribution to monthly rainfall during 2001-2015. The legend represents the percentage of the MCC contribution. The circle representing to monthly distribution of the MCCs location in IMC during 15-years (2001-2015); (a) January, (b) February, (c) March, (d) April, (e) May, (f) June, (g) July, (h) August, (i) September, (j) October, (k) November, (l) December. Locations are for the MCC at the time of maximum extent of the interior cloud size area when the mature stage. The circles representing the interior cloud size area (10^3 km^2).

4. Conclusion

The contribution of the MCC to the total rainfall over the IMC has been determined in this study. One of the result shows that the MCC not only contribute to rainfall in the MCC area but the MCC give a contribution to its surrounding area. Most of the MCCs over IMC in this study found at mainly continental, strongly related with orographic. Overall, the contribution of the MCCs to total rainfall over IMC just up to 20% because the rainfall in the IMC is influenced by various weather systems occurring on a range of temporal and spatial scales, from individual cumulus clouds, through cloud superclusters and tropical storms, right up to planetary scale interannual variations which constitute the ENSO. These values are less than the result by Nesbitt et al. [12], but for MCS contribution, they stated that for the global tropics, MCSs produce more than 50% of the rainfall in heavily raining areas during 5-years. The greater contribution is mostly found in the areas which a lot of the large volumetric rainfall [24] and higher annual rainfall amounts [25]. Frequency distribution of the MCCs was found in this study have an almost similar pattern with the distribution of the MCSs for 4-years in IMC [26]. The greatest contribution of the MCC in this research is mostly concentrated in the continent and coastal than in the ocean. It is possibly related to the previous paper that stated most of the rainfall over IMC is concentrated in the land [27] and the high intensity of rainfall frequently occurs in the coastal region over the IMC [25][28].

The great contribution of the MCC over the CK and SCS near of the Kalimantan were found in each season, it is consistent with the frequency distribution of the MCC that found in each season. It is also possibly related to the cold surges and the Borneo vortex where the cold surge that also induces heavy rainfalls along the western coast of the SCS (Vietnam and Malaysia) near of the IMC, in particular when synoptic-scale cyclonic disturbances are formed near Kalimantan and the southern Philippines [29][30]. The greatest contribution of the MCC is more existence over continent than the ocean, where found almost in each season in several areas over the continent, except in July. While the greatest contribution of MCC oceanic just concentrated in IO near Sumatra in each season. The MCCs contribution over JI is almost small for each season. The result also shows that the greatest contribution of the MCC to total rainfall for DJF season are in the coastal area near the northern part of Sumatra, which reached 18-22%. The greatest contribution during MAM is relatively similar with DJF but the percentage value is a slight increase around of more than 24%. During JJA, the impact of MCC to JJA rainfall is slightly less over continent where the MCC contribution just found up to 10%. The greatest contribution of the MCC during SON that around of more than 24% is concentrated on the Arafura Sea, Banda Sea, and the Flores Sea

In the monthly analysis, the result shows the contribution of MCCs to monthly total rainfall up to 30% in several areas. During January, the maximum contribution of MCCs to total rainfall monthly was found on the west coast near the northern Sumatra up to 30%. The greatest contribution of the continental MCCs was also found on CK for each month except June, even a percentage value in this area reached more than 30% during the month of March and April. The greatest contribution of the coastal MCCs in the SCS near Brunei Darussalam most frequently occurred in March and April. The greatest contribution of the oceanic MCC in the IO most frequently occurred in November. The great contribution of the MCCs to November rainfall is also found in some area, i.e., on the IO, SCS, CK, MR, and CB. The contribution of the MCCs along JI until East Nusa Tenggara is less than 4% in each month.

Acknowledgments
Professor Y. Kodama, who is currently experiencing a serious health problem, contributed substantially to this paper to such an extent that he should be considered a co-author. The authors are grateful to the National Institute of Aeronautics and Space (LAPAN), which is the institution where the first author works. The Ministry of Research and Technology of the Government of Indonesia provided a scholarship to the first author.

References

[1] Maddox R A 1980 Mesoscale Convective Complexes *Bull. Amer. Meteor. Soc* **61** 1374-1387
[2] Mohr K I, Famiglietti J S and Zipser J E 1999 The Contribution to Tropical Rainfall with respect to Convective System Type, Size, and Intensity Estimated from the 85-GHz Ice-Scattering Signature *Journal of Applied Meteorology* **38** 596-606
[3] McAnelly R L and Cotton W R 1989 The Precipitation Life Cycle of Mesoscale Convective Complexes over the Central United States *Mon.* Wea. Rev **117** 784 - 808
[4] Houze R A Jr 1977 Structure and Dynamics of a Tropical Squall Line System *Mon. Wea. Rev* **105** 1540 - 1567
[5] Chong M P, Amayenc G S, and Testud J 1987 A Tropical Squall Line Observed during the COPT 81 Experiment in West Africa. Part I: Kinematic Structure Inferred from dual-Doppler Radar Data *Mon. Weather Rev* **115** 670–694
[6] Williams M and Houze R A Jr 1987 Satellite-observed Characteristics of Winter Monsoon Cloud Clusters *Mon. Weather Rev* 115 505–519
[7] Ashley W S 2003 A Distribution of Mesoscale Convective Complexe Rainfall in the United State *American Meteorology Society* **131** 3003 – 301

[8] Fritsch J M, Kane R J and Chelius C R 1986 The Contribution of Mesoscale Convective Weather System to the Warm Season Precipitation in the United States *J. Climate Appl. Meteor* **25** 1333 - 1345
[9] Laing A G 1999 Contribution of Mesoscale Convective Complexes to Rainfall in Sahelian Africa: estimates from Geostationary Infrared and Passive Microwave Data *Journal of Applied Meteorology* **38** 957 – 964
[10] Durkee J D, Mote T L and Shepherd M J 2009 The Contribution of Mesoscale Convective Complexes to Rainfall across, Subtropical South America *Int J. Clim.* **22**
[11] Blamey R C and Reason C J C 2012 The role of Mesoscale convective complexes in Southern Africa Summer Rainfall *J. Clim* **26**
[12] Nesbitt S W, Cifelli R, and Rutledge S A 2006 Storm Morphology and Rainfall Characteristics of TRMM Precipitation Features *Mon. Wea. Rev* **134**
[13] Yuan J, and Houze R A Jr 2010 Global variability of mesoscale convective system anvil structure from A-Train satellite data *J. Climate*, **23** 5864-5888
[14] Houze R A Jr 2004 Mesoscale Convective Systems *Rev. Geophys* **42**
[15] Ismanto H 2011 Characteristics of Mesoscale Convective Complexes over Maritime Continent *Magister Thesis* FITB ITB.
[16] Trismidianto 2012 Study of Mesoscale Convective Complexes in the Indian Ocean to Weather Pattern above Sumatra *Magister Thesis* FITB ITB
[17] Carvalho L M V and Jones C 2001 A Satellite Method to Identify Structural Properties of Mesoscale Convective Systems Based on the Maximum Spatial Correlation Tracking Technique (MASCOTTE) *J. of Applied Meteorology. American Meteorology Society* **40** 1683-1701
[18] Laing A G and Fritsch J M 1993 Mesoscale Convective Complexes Over the Indian Monsoon Region *American Meteorology Society* **121** 2254 – 2263
[19] Trismidianto, Hadi T W, Ishida S, Moteki Q, Mand A and Iizuka S 2016 Development Processes of Oceanic Convective Systems Inducing the Heavy Rainfall over the Western Coast of Sumatra on 28 October 2007 *SOLA* **12** 6−11
[20] Huffman G J and Bolvin D T 2013 TRMM and other data precipitation data set documentation Available online ftp://meso-a.gsfc.nasa.gov/pub/trmmdocs/3B42_3B43_doc.pdf (accessed 8 June 2014)
[21] Laing A G and Fritsch J M 1997 The Global Population of Mesoscale Convective Complexes *Q. J. R. Meteorol. Soc* **123** 389-405
[22] Laing A G and Fritsch J M 2000 The Large Scale environment of the Global Population of Mesoscale Convective Complexes. *Monthly Weather Review* **128** 2756 - 2776
[23] Koseki S, Koh T Y and Teo C K 2014 Borneo vortex and mesoscale convective rainfall *Atmos. Chem. Phys* 14 4539–4562
[24] Hirose M, Oki R, Shor D A and Nakamura K 2009 Regional characteristics of scale-based precipitation systems from ten years of TRMM PR data *Journal of the Meteorological Society of Japan* 87A 353-368
[25] Mori S, Hamada J-I, Tauhid Y I, Yamanaka M D, Okamoto N, Murata F, Sakurai N, Hashiguchi H and Sribimawati T 2004 Diurnal land-sea rainfall peak migration over Sumatra Island, Indonesian Maritime Continent, observed by TRMM satellite and intensive rawinsonde soundings *Mon. Wea. Rev* **132** 2021-2039
[26] Virts K S and Houze R A Jr 2015 Variation of lightning and convective rain fraction in mesoscale convective systems of the MJO *J. Atmos. Sci* **72** 1932-1944
[27] Qian J H 2008 Why precipitation is mostly concentrated over islands in the MaritimeContinent *J. Atmos. Sci* **65** 1428–1441
[28] Ogino S Y, Yamanaka M D, Mori S and Matsumoto J 2016 How much is the precipitation amount over the tropical coastal region? *J. Clim.* **29** 1231–1236

[29] Wu P-M, Fukutomi Y and Matsumoto J 2011 An observational study of the extremely heavy rain event in northern Vietnam during 30 October–1 November 2008 J. Meteorol. Soc Jpn **89A** 331–344
[30] Chen T C, Tsay J-D, Yen M-C and Matsumoto J 2013 The winter rainfall of Malaysia *J. Clim* **26** 936–958

Propagation of convective complex systems triggering potential flooding rainfall of Greater Jakarta using satellite data

Danang Eko Nuryanto[1,2*], Hidayat Pawitan[1], Rahmat Hidayat[1], and Edvin Aldrian[3]
[1]Departement of Geophysics and Meteorology, FMIPA Bogor Agricultural University, Kampus IPB Darmaga, Bogor 16680, West Java, Indonesia
[2]Research and Development Center, The Indonesian Agency for Meteorology Climatology and Geophysics, Jl Angkasa I No 2 Kemayoran Jakarta Pusat 10720, Indonesia
[3]Agency for The Assessment and Application of Technology, Jl MH Thamrin No 8 Jakarta 10340, Indonesia

E-mail : danang.eko@bmkg.go.id

Abstract. A convective cloud system that large, long lived, and exhibits a quasi-circular cloud shield could be called a mesoscale convective complexes (MCC) system. These systems produce a wide variety of severe convective weather such as heavy rainfall. On January 15th, 2013, Jakarta experienced an extraordinary heavy rainfall event. In this study, we examined the propagation of the convective complex system on that date that was triggering heavy rainfall occurred in the Greater Jakarta area, using observations from the Multi-functional Transport Satellite (MTSAT)-IR and the synoptic data. The convective complex system developed from midnight of January 14th, 2013 until the morning of January 15th, 2013 and it was intensified by the influence of low-level westerly winds. The convective complex systems were generated during the nighttime of January 14th, 2013 of over Sumatera, and propagate to the Southeast through Java Sea until the northern coast of the Great Jakarta in the morning of January 15th, 2013. These convective complex systems give heavy rainfall up to 63 mm/h on average over the Greater Jakarta area at 04.00 LT. This heavy rainfall had triggered flash flood in Jakarta January 2013.

1. Introduction

An extraordinary heavy rainfall/flood event had occurred in the Greater Jakarta (GJ) on January 15th, 2013. This event was one of a sequences of rainfall/flood events dated January 15 – 18th, 2013, with at least 20 people lost their lives and thousands of houses were inundated as noted by Wu et al. [1]. This event is one of the worst floods in recorded history over GJ.

The previous studies, an infrequently long duration of heavy rainfall implying the severe flood event that occurred in Jakarta during late January into early February 2007 was reported by Wu et al. [2] using observational data and by Trilaksono et al. [3, 4] using a numerical non-hydrostatic model. They show a constant strong trans-equatorial Asian winter monsoonal flow from Northern Hemisphere plays an important role in the formation of heavy rainfall in Jakarta during late January into early February 2007. While an extraordinary heavy rainfall event over Jakarta in the middle of January 2013, according to Wu et al. [1], was exerted by the eastward propagation of an active phase of the Madden-Julian Oscillation (MJO).

A heavy rainfall as one of wide variety of severe convective weather could be produced by Mesoscale Convective Complexes (MCC) [5-7]. There were studies on the contribution of MCC substantially to the total rainfall across the various region [8-11]. The MCC is a subclass of mesoscale convective systems (MCSs) defined as a convective cloud system that is large, long lived, and exhibits a quasi-circular cloud shield [5] for the important criteria.

We have identify the convective complex system as a convective cloud system that suited similar criteria with MCC but not showing a quasi-circular cloud shield [12]. In the present study we investigated the propagation of the convective complex system that triggered potential flooding rainfall occurred in the Greater Jakarta area, using observations analysis from the Multi-functional Transport Satellite (MTSAT)-IR and the rainfall data on January 14 – 15th, 2013.

2. Data and Methods

The 3B42 real-time TRMM Multi-Satellite Precipitation Analysis (TMPA-RT) version 6 data of 3 hourly precipitation from January 14 -15th, 2013 were used in this study. The 3 hourly dataset is high-frequency intervals temporal resolution with 0.25° x 0.25° spatial resolution where area study focus in the around GJ (figure 1). For detail information about this data can be found at the web page of Goddard Space Flight Center of National Aeronautics and Space Administration (https://pmm.nasa.gov/data-access/downloads/trmm). We examined heavy rainfall over GJ by TRMM dataset confirming the convective complex system. The equivalent temperature black body (TBB) that derived the hourly infrared data of Multi-functional Transport Satellite (MTSAT)-1R imagery from http://database.rish.kyoto-u.ac.jp/arch/ctop/index_e.html [13]. The satellite imagery was used to identify an objective of a convective complex system and their physical characteristics based on the parameters given by Maddox [5] and adopted by Blamey and Reason [14].

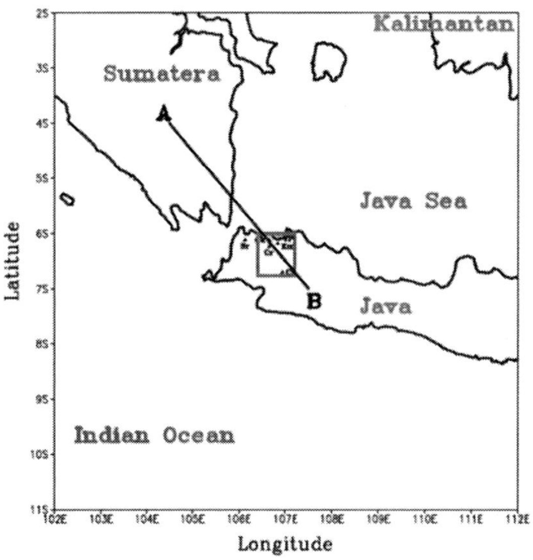

Figure 1. Area study in Greater Jakarta was indicated by a red box. Line AB is indicated for hovmoller plot in the next discussion.

To identify convective complex systems we implemented the "Grab 'em, Tag 'em, Graph 'em" (GTG) algorithm [15] as a tools a cloud-tracking and convective systems feature tracking algorithm based on graph theory in gridded infrared dataset. The graph theory implementation requires using graphs as objects along with some graph traversal methods. The criteria to applying GTG algorithm is based on table 1.

Table 1. Physical characteristics based on the parameters given by Maddox [5] and adopted by Blamey and Reason [14] to identify a convective complexes system in this study.

Physical characteristics	Blamey and Reason [14] Criteria
Cloud Top Temperature	≤ 221 K
Size	≥ 50000 km^2
Shape-eccentricity	≥ 0.7 at maximum extent
Duration	Size and temperature definition must be met for a period ≥ 6 hours
Initiation	Size and temperature thresholds are first met
Termination	Size definition are no longer satisfied

We plot the trajectories of the convective complex system center hour by hour from the first development until dissipation phase of the system for gain the description of propagation the convective complex system. To assess the relative position and strength of the convective cloud system, we apply time-crossection plot of TBB and overlying with rainfall in January 14 – 15th, 2013.

3. Results and Discussions

Figure 2 shows the three hourly temporal variation of the rainfall intensity over GJ indicated by red box (figure 1) during flood event (January 15th, 2013) in GJ area (about 7,392.6 km^2 of large). The peak of rainfall occured in the early morning of 15 January 2013. While the others times show less than 40 mm of rainfall.

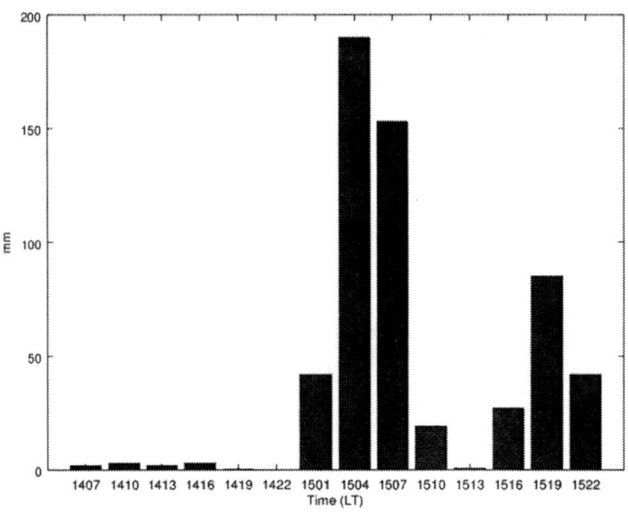

Figure 2. The accumulation diurnal variation of rainfall distribution (mm/3 hour) during January 14 - 15th, 2013 at around GJ (red box Fig. 1) about 7,392.6 km^2 of large. X-label is date and time, for example 1407 = 07.00 LT January 14th.

Figure 3 presents satellite images that depict the evolution of a convective complex cloud cluster near GJ on January 14 - 15th, 2013. The three stages in the life cycle of these convective system are development stage at 1600 LT January 14th, 2013 (figure 3a), mature stage at 0300 LT January 15th, 2013 (figure 3b) and dissipation stage at 0700 LT January 15th, 2013 (figure 3c). These three stages definition is refered to Maddox's definition [5]. The development stage have cloud cluster with first met criteria of temperature ≤ 221 K and size $\geq 50,000$ km^2. The mature stage have cloud cluster with maximum size of temperature ≤ 221 K. The dissipation stage have cloud cluster with last met criteria of temperature ≤ 221 K and size ≥ 50,000 km^2.

The eccentricity at development stages of convective system was 0.56 (< 0.7 as defined by Maddox [5]. The eccentricity at mature stages of convective system was 0.49 (< 0.7). The eccentricity at dissipation stages of convective system was 0.46 (< 0.7). The shape of these convective system was not match with Maddox's criteria of MCC. So these cloud cluster was not a MCC, although these convective system bringing heavy rainfall over GJ on January 15th, 2013. For a moment we call it as a convectice complex system which defined as a convective cloud system that large, long lived, and very cold cloud shield.

During the development stage in the afternoon (1600 LT) on January 14th, 2013, the group of clouds grew rapidly and propagated southeastward until around early morning (0300 LT) on January 15th, 2013, which marked as the mature stage. During the mature stage, the convectice complex system had a cloud shield with an area of around 249,731 km^2 at ≤ 221 K of cloud temperature. The center of the convectice complex system during this stage was around 5.6°S, 106.64°E with minimum temperature is 199 K. During the dissipation stage in the moring (0700 LT) on January 15th, 2015, the convectice complex system had split into small-scale clouds that propagated southwestward and northeastward toward the Indian Ocean and Java Sea respectively.

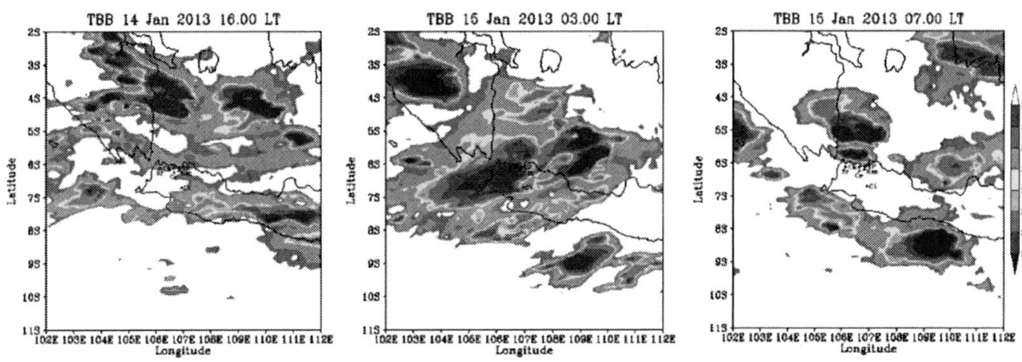

Figure 3. Horizontal distribution of black body temperature from infrared data obtained by MTSAT-IR over the Greater Jakarta area on January 14 - 15th, 2013, showing (a) development stage (1600 LT January 14th, 2013), (b) mature stage (0300 LT 15 January 2013) and (c) dissipation stage (0700 LT January 15th, 2013).

The life duration of the convective complex system are about 16 hour from 1600 LT January 14th, 2013 until 0700 LT January 15th, 2013. Figure 4 shows the trajectories of the convective complex system center that moved out of the Sumatera Island with southeastward, southwestward and southeastward propagations through northern coast of GJ and finally eastward propagation till Java Sea. The mature stage of the convective complex system activity occur in Java Sea that close to GJ. It is evident that these convective complex system propagation will generally impact heavy rainfall around of GJ during January 15th, 2013.

IOP Conf. Series: Earth and Environmental Science **54** (2017) 012028 doi:10.1088/1755-1315/54/1/012028

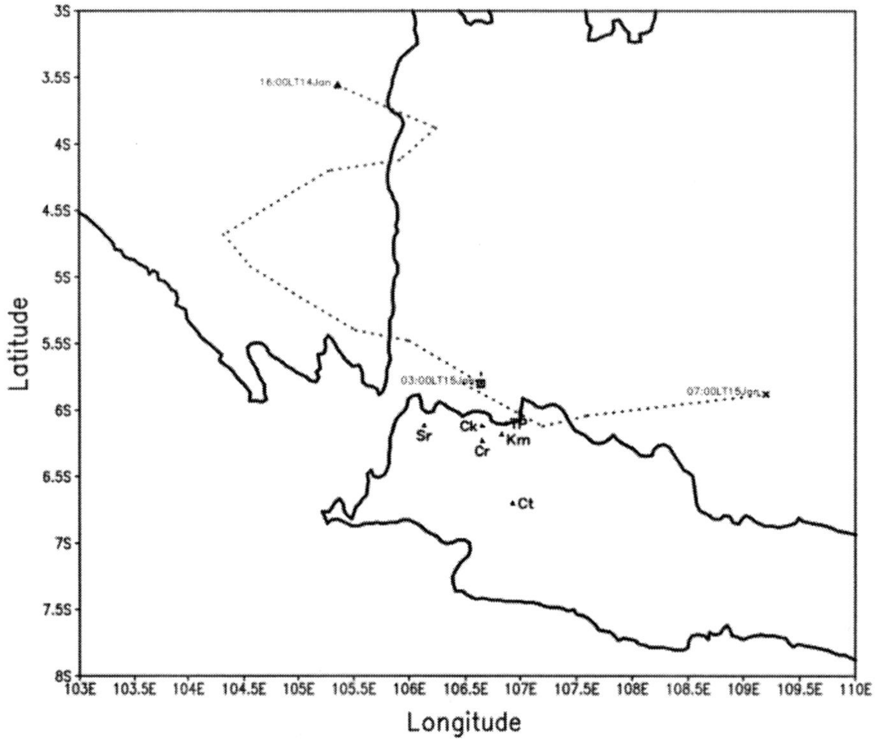

Figure 4. Trajectory of the convectice complex system center on January 14 - 15th, 2013 (dashed line). Cross, filled circle and triangle mark is development, mature and dissipation stages respectively of the convectice complex system. Sr, Ck, Cr, TP, Km and Ct denotes Serang, Cengkareng, Curug, Tanjung Priok, Kemayoran and Citeko respectively.

To assess or diagnose the behavior of the convective complex system over a span of A-B line (figure 1) through time, we showed time-crosssection plots of TBB and rainfall of A-B line from 1300 LT January 14th, 2013 until 1000 LT January 15th, 2013 (figure 5). Throughout in January 14 – 15th, 2013 the convective complex system, defined by ≤ 221 K of TBB, develop over the Sumatera Island with an amplitude of ≤ 221 K. The GJ area exists as is indicated by thick vertical dashed lines. Part of the convective complex system propagate over the GJ from 2100 LT January 14th, 2013 until 0800 LT January 15th, 2013. While the heavy rainfall (> 10 mm/hr) over the GJ from 0100 LT January 15th, 2013 until 0700 LT January 15th, 2013. This result has agreement with previous study that maximum rainfall was found at 0000 – 1100 LT over around the GJ [16]. The convective complex system has a southeastward propagation speed equal to 18.7 m/s. According to Wu et al. [1] with the eastward propagation of the active phase of the MJO, moderate westerly winds occurred over Java Island during the heavy rain event.

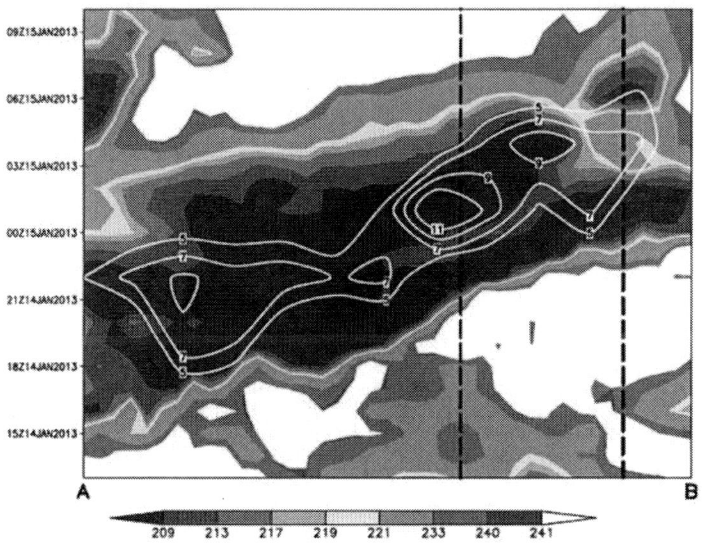

Figure 5. Time-crossection plot of TBB (shaded; K) and rainfall (contour; mm/hr) over AB line in January 14 – 15th, 2013. Thick vertical dashed lines indicate the GJ area

The convective complexes system developed from midnight on January 14th, 2013 until the morning of January 15th, 2013 and it was intensified by the influence of low-level westerly winds. The convective complex systems were generated during the nighttime of January 14th, 2013 of over Sumatera, and they propagate to the Southeast through Java Sea until northern coast of Great Jakarta the morning of January 15th, 2013. These convective complex systems give heavy rainfall up to 63 mm/h average over the Greater Jakarta area at 04.00 LT. The 63 mm/h of rainfall may triggered flash flood in Jakarta January 2013.

4. Conclusions

On January 15th, 2013, Jakarta experienced an extraordinary heavy rainfall that triggered flood event. The convective complex system developed from midnight on January 14th, 2013 until the morning of January 15th, 2013 and it was intensified by the influence of low-level westerly winds. The combined satellite and synoptic data analysis has successfully identified the convective complex systems that were generated during the nighttime of January 14th, 2013 over Sumatera, and propagated to the Southeast through Java Sea until northern coast of Great Jakarta the morning of January 15th, 2013. These convective complex systems generated heavy rainfall up to 63 mm/h on average over the Greater Jakarta area at 04.00 LT. This heavy rainfall had triggered flash flood in the Greater Jakarta basins.

Acknowledgments

This research was supported by Department of Geophysics and Meteorology, Bogor Agricultural University and the Indonesian Agency for Meteorology Climatology and Geophysics through the Graduate Study Program in Applied Climatology.

5. References

[1] Wu P, Arbain AA, Mori S, Hamada J, Hattori M, Syamsudin F and Yamanaka MD 2013 The Effects of an Active Phase of the Madden-Julian Oscillation on the Extreme Precipitation Event over Western Java Island in January 2013 *SOLA* **9** 79 - 83

[2] Wu P, Hara M, Fudeyasu H, Yamanaka MD, Matsumoto J, Syamsudin F, Sulistyowati R and Djajadihardja YS 2007 The impact of trans-equatorial monsoon flow on the formation of repeated torrential rains over Java Island *SOLA* **3** 93–96

[3] Trilaksono NJ, Otsuka S and Yoden S 2011 Dependence of model-simulated heavy rainfall on the horizontal resolution during the Jakarta flood event in January–February 2007 *SOLA* **7** 193–196

[4] Trilaksono NJ, Otsuka S and Yoden S 2012 A time-lagged ensemble simulation on the modulation of precipitation over West Java in January–February 2007 *Mon. Wea. Rev.* **140** 601– 616

[5] Maddox RA 1980 Mesoscale Convective Complexes *Bull. A. Met. Soc.* **61** 1374 – 1387

[6] Fritsch JM, Kane RJ and Chelius CR 1986 The Contribution of Mesoscale Convective Weather Systems to the Warm-Season Precipitation in the United States *J. Climate Appl. Meteor.* **25** 1333 – 1345

[7] McAnelly RL and Cotton WR 1989 The Precipitation Life Cycle of Mesoscale Convective Complexes over the Central United States *Mon. Wea. Rev.* **117** 784 – 808

[8] Laing AG, Fritsch JM and Negri AJ 1999 Contribution of mesoscale convective complexes to rainfall in Sahelian Africa: Estimates from geostationary infrared and passive microwave data *J. Appl. Meteor.* **38** 957–964

[9] Ashley WS, Mote TL, Dixon PG, Trotter SL, Durkee JD, Powell EJ and Grundstein AJ 2003 Effects of mesoscale convective complex rainfall on the distribution of precipitation in the United States *Mon. Wea. Rev.* **131** 3003–3017

[10] Durkee JD, Mote TL and Shepherd JM 2009 The Contribution of Mesoscale Convective Complexes to Rainfall across Subtropical South America *J. Climate.* **22** 4590 – 4605

[11] Blamey RC and Reason CJC 2013 The Role of Mesoscale Convective Complexes in Southern Africa Summer Rainfall. *J. Climate.* **26** 1654 – 1668

[12] Nuryanto DE, Aldrian E, Pawitan H and Hidayat R 2016 Application of Graph-Theory Based Algorithm for Identifying Convective Complex Systems over Greater Jakarta basins (*Unpublished*)

[13] Hamada A and Nishi N 2010 Development of a Cloud-Top Height Estimation Method by Geostationary Satellite Split-Window Measurements Trained with CloudSat Data *J. Appl. Meteor. Climatol.* **49** 2035 – 2049

[14] Blamey RC and Reason CJC 2012 Mesoscale Convective Complexes over Southern Africa. *J. Climate* **25** 753 – 766

[15] Whitehall K, Mattmann C, Jenkins G, Rwebangira M, Demoz B, Waliser D, Kim J, Goodale C, Hart A, Ramirez P, Joyce M, Boustani M, Zimdars P, Loikith P and Lee H 2014 Exploring a graph theory based algorithm for automated identification and characterization of large mesoscale convective systems in satellite datasets *Earth Sci. Inform.*

[16] Nuryanto DE, Pawitan H, Hidayat R and Aldrian E 2016 Heavy Rainfall Distributions Over Java Sea in Wet Season *Procedia Environ. Sci.* **33** 178 – 186

MJO modulation on diurnal rainfall over West Java during pre-monsoon and strong El Niño periods

E Yulihastin[1,2], Trismidianto[1,3], H Satyawardhana[2], G A Nugroho[2]
[1]Earth Science Department, Bandung Institute of Technology, Bandung, Indonesia
[2]Center for Science and Technology of Atmosphere, National Institute of Aeronautics and Space, Bandung, Indonesia
[3]Meteorology Department, Hirosaki University, Hirosaki, Japan

Email: erma.yulihastin@lapan.go.id

Abstract. This study was conducted to determine the MJO modulation ondiurnal rainfall in West Java during the pre-monsoon and strong El Niño periods in 2015 over West Java. The data used is a combination of satellite data, reanalysis data, radar stations data, and numerical weather prediction of Weather Research and Forecasting (WRF) data with spatial resolution of 5 km. The results confirmed that the strong MJO in 4 and 5 phases has modulated the amplitude of diurnal rainfall increase significantly over West Java in phase of lag+1. Modulation on diurnal cycle of rainfall was also indicated by the persistence of the rainfall and the formation of two peaks of maximum rainfall in the afternoon and early morning. In addition, modulation of rainfall for the southern part of Java was 50% greater than the north. Moreover, the MJO modulation mechanism was characterized by the formation of an active and extending of Meso-scale Convective System (MCS) which has a cycle of up to 12 hours and persistent from November 7-9 over West Java.

1. Introduction

Madden Julian Oscillation (MJO) is an atmospheric equatorial waves propagating from the western (Indian Ocean) to the eastern (Pacific Ocean) with a period of 30-90 days or 20-50 days at a speed of about 4-5 m/s [1-4]. Activities of MJO broad impact on a wide range of global weather and climate phenomena, one of which is relate to rainfall variability [5-7].

Previous research suggests that MJO modulated to the diurnal cycle of tropical Deep Convective Cloud (DCC). The most significant of DCC occurred in the Indian Ocean (50%) compared to the Pacific Ocean and IMC (20%) [8]. In the IMC, diurnal rainfall has a high variability of response to the MJO, depend on the season and local conditions of geographical features [9-11].

For the northern part of the IMC, the influencing of MJO on diurnal rainfall occurred in Sarawak, Borneo [9] which are identified by enhanced of convergence, land-sea breeze circulation, and maximum of the rainfall occurred on the beach at midnight before the active phase of the MJO. They also found a dynamically difference of the modulation rainfall for coastal and inland regions in Borneo. For coastal areas, there is frequency increased of diurnal rainfall, while the increased in the frequency and intensity of diurnal rainfall occurs in interior regions. For the eastern part of IMC, in New Guinea, the most attractive feature of the diurnal rainfall placed in the central part of mountain which is identified through the propagation of the diurnal rainfall in the afternoon which spread to the southeast and northwest associated with the easterly-westerly flow at the low level at speeds of 2-3 m/s [10].

Meanwhile, during the December-January-February (DJF) period in the western part of IMC (Sumatra and Java), convective activity increases when the MJO is in the active phase, otherwise it decreased when the MJO in an inactive state [11]. While MJO reach a mature stage, large variation of the diurnal rainfall occurs between the mainland (minimum) and sea (maximum). Moreover, it appears that rainfall more intensify in the morning over the Java Sea, while minimum rainfall occurs in Sumatra in the afternoon. Maximum rainfall which occurred in the morning over Java Sea is influenced by interaction of land-breeze and low level westerly generated by MJO which is reinforced by the Asian winter monsoon over the Java Sea, and is affected by the blocking effect of the mountains. On the other hand, the nocturnal diurnal rainfall over Java mainland generally occurs from the afternoon to midnight. It probably occurred because westerly and seasonal flow has hampered the convergence so that reducing rainfall in the mainland.

On the other hand, diurnal rainfall over Java Island has different characteristics between the northern and southern parts associated with the different topography of the region. The mountains in southern Java impact on rainfall diurnal became more intensive in the southern than the northern part. Interestingly, the diurnal rainfall in the southern part has increased in the event of El Niño during the period of DJF [12]. However, no studies have revealed the relationship between the MJO and strong El Niño in the transitional period before the monsoon due to rainfall season in Java, especially associated with the diurnal cycle response in the region. In this case, West Java was chosen as the study area because it's geographically closer to the Indian Ocean, an area where MJO undergoing a process of initiation. In addition, West Java geographically is also adjacent to the island of Sumatra where MJO influence in the region proved to be a strong and significant in influencing the diurnal cycle of rainfall. Therefore, this study aimed to explain the influence of the MJO in modulating diurnal rainfall that occurred in West Java during the period of transition to Asian winter monsoon due to rainfall season during a strong El Niño period in October-November 2015.

2. Data and Method

The research methodology developed in order to prove the MJO events in associated with the increased rainfall in West Java, with a case study during the strong El Niño period from October to November 2015.

2.1. Data

To prove the examination of the MJO, we investigated MJO indices data available in http://cawcr.gov.au/staff/mwheeler/maproom/RMM/RMM1RMM2.74toRealtime.txt. Meanwhile, evidence of the occurrence of El Niño by Nino 3.4 index data derived from: http://www.cpc.ncep.noaa.gov/products/precip/CWlink/MJO/enso.shtml.

Hourly rainfall data of 3B41RT derived from Tropical Rainfall Measuring Mission (TRMM) satellite data and WRF models. Data from the model used to supplement rainfall data from satellites, based on verification of hourly rainfall data from TRMM satellite which under-estimated compare to station rainfall data. Meanwhile, to determine meso-scale processes that affect the increase in rainfall, an analysis using the reanalysis data of Era-INTERIM of 6-hourly data of wind and rainfall, also clouds data of IR1 from Himawari satellite. All of data in this study had a range of periods from October to November, 2015.

2.2. Method

The method used to examine the increase in rainfall and meso-scale processes that causes it, consists of several stages. Firstly, examine the MJO index and ENSO index of Nino 34. Secondly, to categorize the MJO activity based on movement and strength. Thirdly, we used the hovmöller diagram analysis to examine the evolution of convective activity and rainfall that occurred in West Java during the period from October to November, 2015. Fourthly, perform spatial analysis on rainfall patterns in

West Java, and examine the difference between northern and southern regions in response to modulation of the MJO. The last, performing a meso-scale analysis to draw the mechanism of rainfall during the period of MJO active and strong El Niño period. The MCS calculated adapted from Trismidianto et al. [13].

3. Results and Discussion

Table 1 shows the MJO propagation of phase 9 (Africa) towards phase 8 (the Pacific Ocean), which lasted from October 5 to November 4, 2015. As for the range of 5 to October 20, 2015, the MJO is in a period of weak or inactive over the IMC (blue box). Furthermore, on October 25, 2015 to November 4, 2015 the MJO is active and stronger (yellow color in the red box) over the IMC, then pass to the Pacific Ocean.

Table 1. MJO activity over IMC (phase 4 and 5) during periods of weak, begin to active, mature, and dissipate.

Time\Fase	9	10	1	2	3	4	5	6	7	8
20151005	1.27	1.87	1.86	1.36	0.54	-0.60	-1.71	-1.75	-1.14	-0.12
20151010	1.58	2.31	2.32	1.70	0.68	-0.75	-2.13	-2.18	-1.41	-0.14
20151015	1.03	2.08	2.27	1.92	1.12	-0.20	-1.73	-2.14	-1.68	-0.64
20151020	0.39	1.56	1.86	1.85	1.37	0.36	-1.11	-1.81	-1.69	-1.01
20151025	-1.12	-0.08	0.14	0.98	1.45	1.51	0.71	-0.42	-1.15	-1.48
20151030	-1.56	-1.06	-0.83	0.03	0.80	1.45	1.44	0.60	-0.28	-1.09
20151104	-0.94	-0.97	-0.88	-0.44	0.09	0.70	1.06	0.78	0.28	-0.32
20151109	0.09	-0.42	-0.47	-0.59	-0.54	-0.28	0.20	0.54	0.61	0.51

Figure 1 describes the evolution of rainfall spatially in the area of West Java during the period of weak MJO (October 17-22, 2015), begin to initiate (October 23-31, 2015), and active (November 4-9, 2015) over the IMC. Seen over a period of weak MJO over IMC, but active in the Indian Ocean, based on rainfall data based on WRF model (0.1-10 mm/h) begin to occur in the southern region of West Java (Bogor and vicinity). This is supported by monsoon circulation at an altitude of 850 mb showing convergence over the area. Convergence was formed because there was a meeting of wind easterlies from the Java Sea and southeasterly from the Indian Ocean south of West Java. However, data from GSMaP not indicate formation of rainfall, as captured by the model WRF (Figure 2a and d).The inability of GSMaP data in capturing surface of rainfall has been studied previously for West Java, Jakarta and Bogor using surface station data, which shows the correlation <0.2 [14]. While the value of the model WRF rainfall is consistently showed values exceeding 3B42 TRMM satellite data (by a margin of 3 mm/3-time) for diurnal rain in parts of Indonesia [15].

Furthermore, when the MJO became active over IMC (October 26-31, 2015), rainfall began to increase and spread in the southern part of West Java with maximum rainfall occurs around the Bogor area (Figure 2b). The cause of the increase in rainfall is the same as the previous period, namely because there is convergence over the southern West Java. The maximum rainfall in the south around Bogor is confirmed by the GSMaP data (0.1-0.5 mm/h) despite the difference in intensity is very low when compared with data from the WRF model (25-30 mm h) (Figure 2e).

While MJO is an active period over IMC, rainfall in the southern West Java increased compared to the previous period. This is shown by the data of rainfall from WRF models (<20 mm/h) and GSMaP (<2 mm/h). This proves that the MJO activity is positively correlated with the formation of rainfall in the area of West Java, although it appears in the spatial pattern is not uniform between the intensity of rainfall in the northern and southern part of West Java.

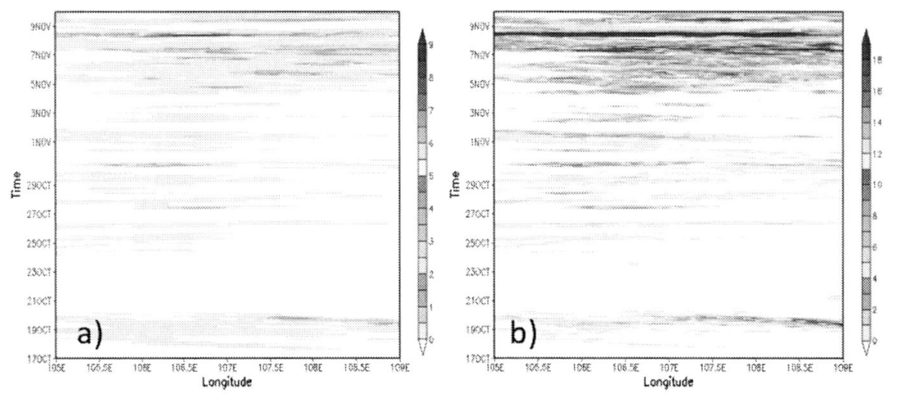

Figure 1. Hovmöller of Time-Longitude cross section during October 17 - November 9, 2015 based on TRMM 3B42RT hourly data (a) and convective activity based on Himawari satellite IR1 data (b).

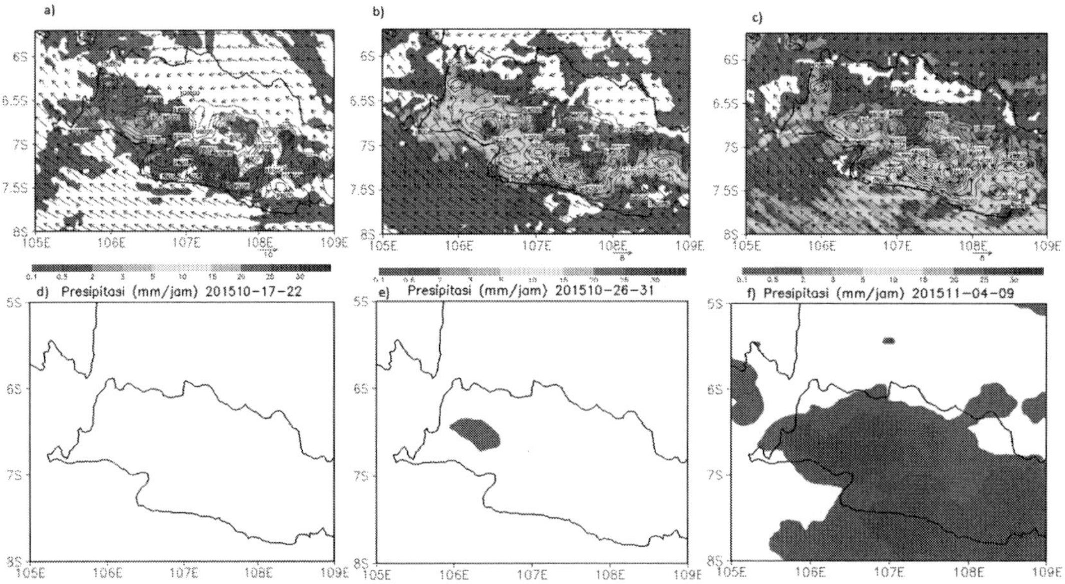

Figure 2. Rainfall at October 17-22 during weak MJO based on WRF model (a) and GSMaP (d), October 26-31 during MJO begin to active (phase 2 and 3) based on WRF model (b) and GSMaP (e), November 4-9 2015 during MJO active in phase 4 and 5 based on WRF model (c) and GSMaP (f).

Figure 3 draws the distribution of northern and southern regions in West Java where research was conducted. This is done because based on the results of the model WRF ((a) - (c)), there is a clear difference in rainfall between the northern and southern part during the pre-monsoon and MJO started and is active periods. Differences in rainfall between the northern and southern part in Java during El Niño was explained by previous investigation [12], which concluded that El Niño strengthen the diurnal cycle of rainfall in the southern part of Java so that the diurnal rainfall diurnal in southern Java climatologically increased during El Niño.

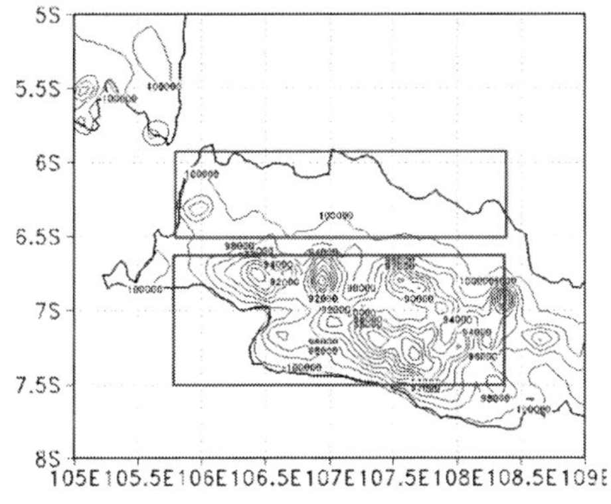

Figure 3. The division of the northern West Java region (blue) and southern (red) (a), the rainfall diurnal in West Java by WRF models for the northern region (blue) and south (red) on October 17-22 for weak MJO (b) and October 26-31, 2015during the active phase of the MJO started 4 and 5 (c), November 4-9, 2015 for the active MJO phase 4 and 5 (d).

Figure 4 shows rainfall based on GSMaP data occurred in the third period (November 4-9, 2015), when the MJO is strong and intensive over IMC. In general, in the north and the south, rainfall is formed throughout the day with increased rainfall began to occur at 13:00 pm and experiencing the peak at 17.00 pm for the next downhill at 18.00 pm. In addition, rainfall in the southern part of the higher of 50 percent compared to the northern region of West Java.

Rainfall data over the surface in Bandung, which represents the southern region of West Java using AWS and radar, are shown in Figure 5 (a) and (b). AWS data shows that the maximum rainfall occurred at 13:00 pm. In addition, at midnight the rain resumed and reaches a maximum value at 24.00 pm. The intensity of the rainfall that occurred during the night is lower than in the day based on AWS data. While over Bandung, radar data showed a pattern similar to GSMaP, rainfall began to occur around 10:00 am and reaches a maximum at 17.00 pm. Persistent of rainfall seemed occurred in middle night with a maximum value.

After the diurnal rainfall modulation in West Java evidenced by satellite data, models, and radar, further we investigated the mechanism that generates it. Hourly data from the Himawari satellite showed the development of MCS intensify during three days on November 7-9, 2015. Meanwhile, during the month of October there was no event of MCS observed from these data.

Figure 6-8 exhibits spatial evolution of the MCS life cycle from November 7-9, 2015. Figure 7 shows the MCS spread from north to the south, and MCS which are formed over Java appears related to initiation of MCS in Java Sea. Moreover, it describes that the MCS in the Java Sea near coastal is start growing from 9 am and began to propagate and extends towards the mainland at 11:00 LT. It causes maximum rainfall in the morning in the northern part of Banten and the surrounding areas (Figure 9). Maximum rainfall that occurred in the Java Sea and the northern coast of Java in the morning in the rainy season when the MJO is active in accordance with the opinion of previous research [11].

Figure 7 and 8 explain the MCS to develop after 01:00 pm in the afternoon and reached the mature conditions in the evening. This is evidenced by the occurrence of rain in the afternoon until the evening on November 8-9, 2015 (Figure 9). Figures 7 and 8 also show that MCS is growing on the mainland come from local convective cells joined reinforce each other and then turned into MCS. Merging several local convective cells that arise in coastal regions and mountains of the western part of the southeastern part of West Java seems massive and widespread cover the entire region of West Java (Figure 7).

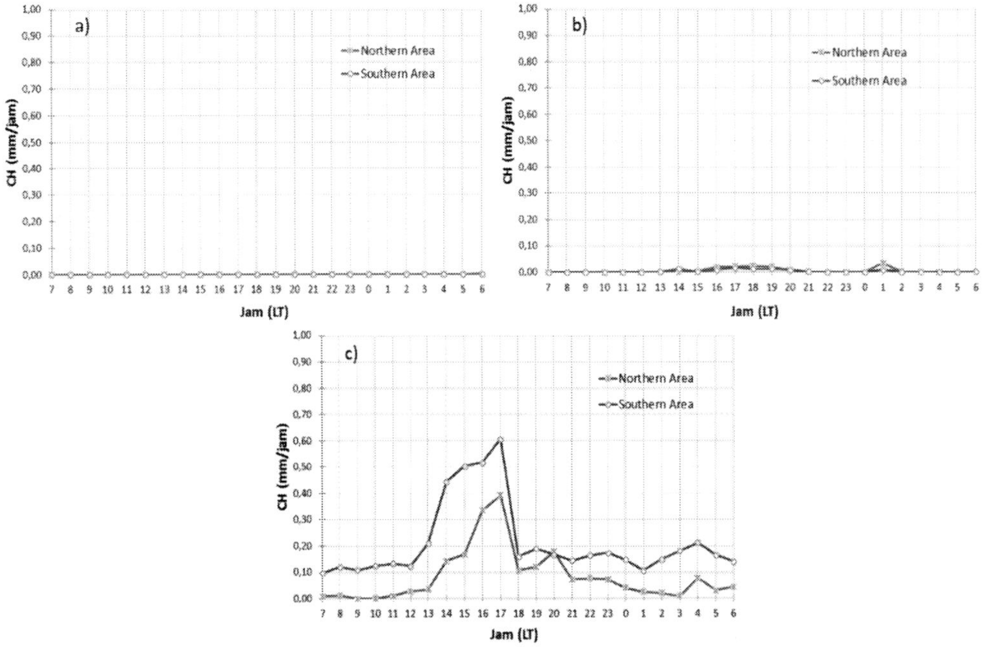

Figure 4. Diurnal rainfall in West Java GSMaP based on data for the northern region (blue) and south (red) on October 17-22 for weak MJO (a), October 26-31, 2015 during the active phase of the MJO started 4 and 5 (b), November 4-9, 2015 during an active MJO phase 4 and 5 (c).

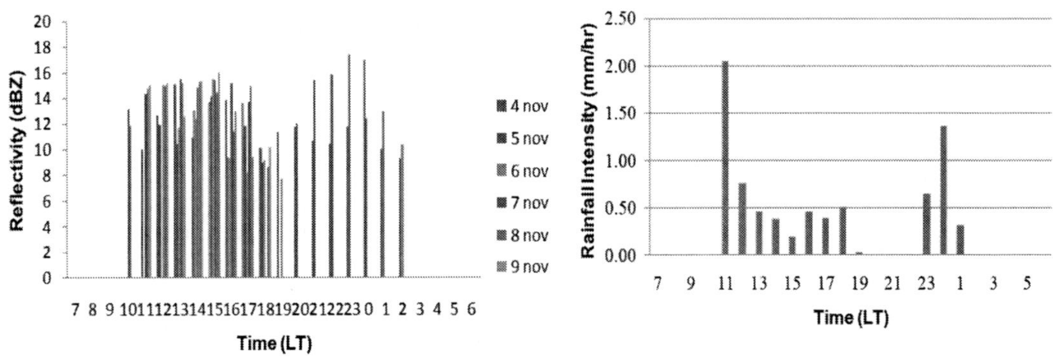

Figure 5. Diurnal rainfall over Bandung Space agency locations, Pasteur based on data from Furuno Radar (left-side) and the location of ITB Bandung, Dago (right-side).

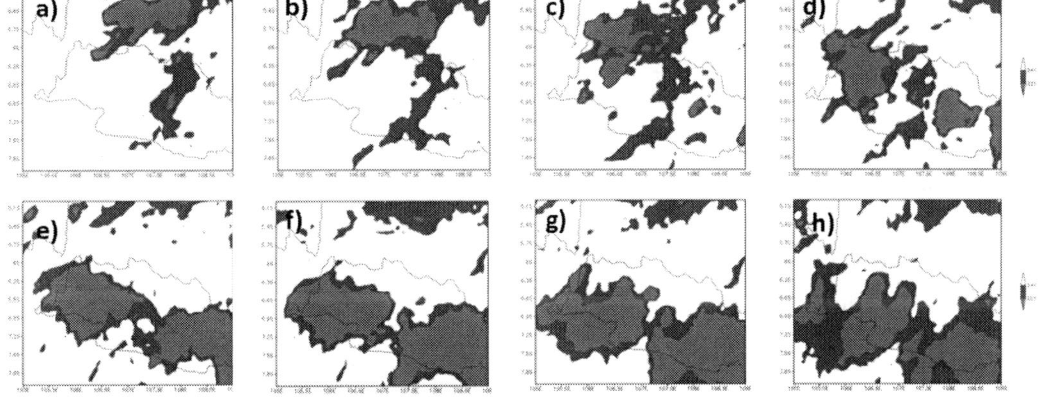

Figure 6. Evolution of MCS based on Himawari satellite at November 7, 2015 from 9.00 -16.00 LT (a-h).

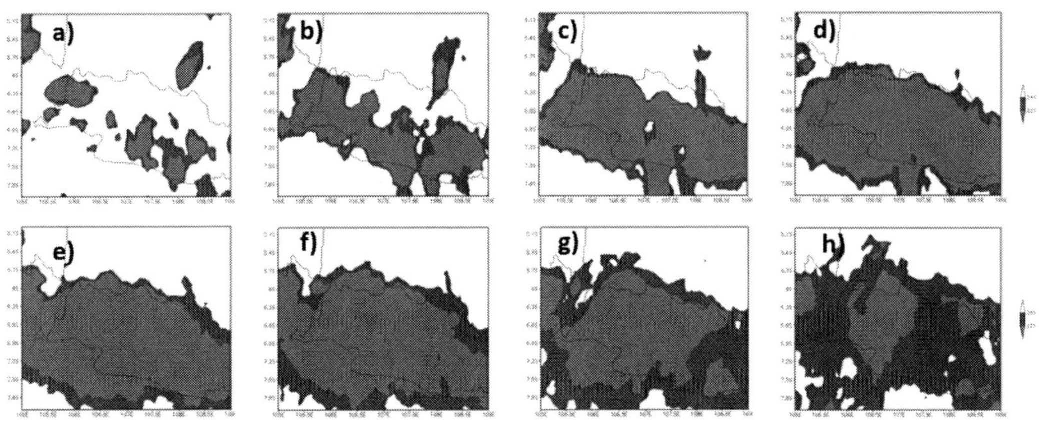

Figure 7. Same as Figure 6, but for 8 November 2015 from 01:00 -08:00 PM LT (a-h).

Meanwhile in figure 7 looks local convective cells initially developed over the mountainous area in West Java are joining the middle so that MCS widespread and extends only in the southern part of West Java. Figures 7 and 8 show that the diurnal convective activity generally occurs after the late afternoon, when the sun was warming in the atmosphere reached a maximum.

Having regard to the spatial patterns and the life cycle of MCS in figures 6-8, it appears that the diurnal aberration occurred on November 7, as evidenced by the occurrence of rain in the morning peak on land. This happens because of the influence of land-sea breeze circulation is increased, which looks very strong of sea breeze occurred during the day. Land-sea breeze circulation which appears clearly indicated that the influence of strong westerly circulation of MJO is omitted by a strong El Niño event so that local circulation factors emerge clearly and unambiguously. It also could suggested that strengthening land-sea breeze and this mountain-valley breeze circulation that have been modulate meso-scale rainfall increase over western Java on November 7-9, 2015.

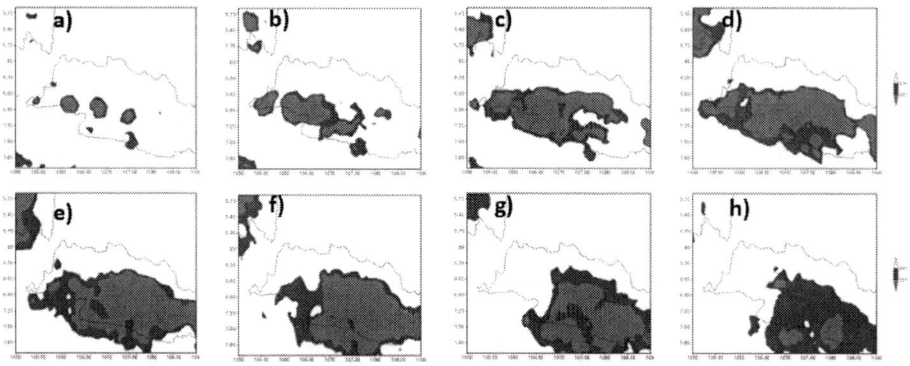

Figure 8. Same as Figure 7, but for 9 November 2015.

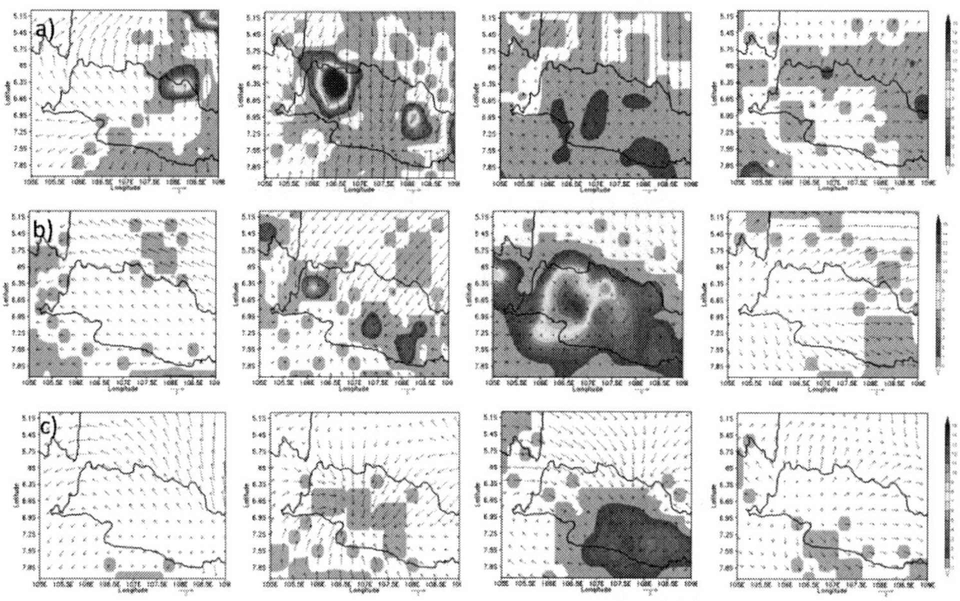

Figure 9. Rainfall and wind of 6-hourly reanalysis ERA-Interim data from left to right side of 07:00 AM, 01:00 PM, 07:00 PM, 01:00AM LT, 7 (a), 8 (b), 9 (c) November 2015.

4. Conclusion

Strong MJO from October 25 to November 4 in Indonesian Maritime Continent (IMC) has modulated the diurnal rainfall in West Java on November 7-9, which was affected by the formation of MCS and has a delay phase of two couple days with strong MJO of phase 5 occurrence. Northern and southern parts of West Java have different response to the diurnal pattern of diurnal rainfall, where the southern part has greater in rainfall amplitude than it's northern part. The occurrence of strong MJO which coincides with the strong El Niño have damp of each other so that the local circulation of land-sea breeze and mountain-valley breeze tend to enhance and interact to produce MCS, which are raised by local convective cells on the surface level and extended by meso-scale convective activity in the Java Sea.

Acknowledgments

We would like to thank the Head of Center for Science and Technology of Atmosphere, National Institute of Aeronautics and Space, which has given impetus and motivation to the authors to improve

IOP Conf. Series: Earth and Environmental Science **54** (2017) 012029 doi:10.1088/1755-1315/54/1/012029

the quality of this paper. This research was funded by the Ministry of Research Technology and Higher Education of the Republic of Indonesia through the Sistem Inovasi Nasional (SINAS) intensive research program of 2015. This research is part of a doctoral dissertation with the development of case studies on the incidence of MJO and strong El Niño periods.

5. References

[1] Hendon H H 2000 Impact of air-sea coupling on the Madden-Julian oscillation in a general circulation model *J. of the Atm. Sci.* **57** 3939-3952

[2] Waliser D, Stern W, Lau K M and Jones C 2003 Potential predictability of the Madden–Julian oscillation - *Bull. Amer. Meteor. Soc.* **84** 33-50

[3] Madden R A and Julian P R 1971 Detection of a 40-50 day oscillation in the zonal wind in the tropical Pacific *J. of the Atm. Sci.* **28** 702-708

[4] Madden R A and Julian P R 1972 Description of global-scale circulation cells in the tropics with a 40-50 day period *J. of the Atm. Sci.* **29** 1109-1123

[7] Seo H, Subramanian A J, Miller A J and Cavanaugh N R 2014 Coupled Impacts of the Diurnal Cycle of SPL on the MJO *J. of Clim.* **27** 8422-8433

[8] Tian B, Duane E W and Eric J F 2006 Modulation of the Diurnal Cycle of Tropical Deep Concevtive Clouds by the MJO*G.R.L.* **33** 1-6

[9] Kanamori H and Tetsuo Y 2013 Modulation of the Diurnal Cycle of Rainfall Associated with the MJO Observed by a Dense Hourly Rain Gauge Network at Sarawak, Borneo *J. of Clim.* **26** 4858-4875

[10] Ichikawa H and Tetsuo Y 2008 Intraseasonal Variability in Diurnal Rainfall over New Guinea and the Surrounding Oceans during Austral Summer *J. of Clim.* **21** 2852-2868

[11] Oh J-H, Kwang-Yul K and Gyu-Ho L 2012 Impact of MJO on the Diurnal Cycle of Rainfall over the Western IMC in the Austral Summer *Clim. Dyn.* **38** 1167-1180

[12] Qian J-H 2008 Why Precipitation Is Mostly Concentrated over Islands in IMC *J. of the Atm. Sci.* **65** 1428-1441

[13] Trismidianto, Hadi T W, Ishida S, Moteki Q, Manda A and Iizuka S 2016 Development Processes of Oceanic Convective Systems Inducing the Heavy Rain over the Western Coast of Sumatra on 28 October 2007 *SOLA.* **12** 6–11

[14] Wibowo, Y A 2010 Evaluation of GSMaP and TRMM TMPA Rainfall with Surface Rainfall on Jakarta-Bogor (Thesis) (Bogor: IPB)

[15] Teo C-K, Koh T-Y, Lo J C-F and Bhatt BC 2011 Principal Component Analysis of Observed and Modeled Diurnal Rainfall in the IMC *Int. J. of Clim.* **24** 4662–4675

Water allocation for agriculture complex terrain under changing climate

I Putu Santikayasa*, **Perdinan and Rizki Abdul Basit**
Department Geophysics and Meteorology, Bogor Agricultural University, Indonesia

E-mail: psantika@gmail.com

Abstract. The current water resources management in Indonesia requires the government to pay more attention on sustainable water management. Agriculture as the highest water demand in the country need better water management as the impact of future changing climate. Furthermore, the water managers as well as policy makers may require integrating the climate change assessment into water resources allocation policy and management. Agropolitan in Malang district, East java – Indonesia is an agriculture which is characterized by complex agricultural system and was assigned as a case study. The supply-demand water allocation approach was applied on allocating water to different water users under current and future climatic condition. Both climate and the changing nature of water demand have affected the development and evolution of water allocation. The result shows that the water supply is expected to decrease under future climate comparing with the current condition. Furthermore, it is required to incorporate the future climate information on design the future water policy and management to reduce the adverse impact of changing climate. This study also suggested policy actions as recommendation to better manage current climate variability as well as future uncertainty from climate change impacts on water allocation and resources management.

1. Introduction

Water resources sustainability means using the natural resource wisely and protecting the complex ecosystems with future generations in mind. But sustainability will not be achieved with current patterns of resource consumption and use [1]. It is therefore of paramount importance to rational planning and decision making in equitable water management. This must be undertaken within the widely accepted integrated approach at all levels of the society.

Population growth and the intensification of irrigation on agricultural lands have increased water demand over the past decade. As a result, water abstraction for irrigation, livestock and domestic use have severely stressed the water resources, particularly during dry seasons causing conflicts between upstream and downstream water users. There is therefore a need to understand the spatial and temporal water availability and to formulate a tool for planning and decision making in prioritisation of water allocation in the area or basin. However, given the complexity of the system and the interactions between water supply and demand, a large-scale water supply management tool would be useful for decision makers when formulating water management strategies for coping with future changes in water demands [2].

Poncokusumo is the area which developed from the agriculture area to be the agropolitan area. The development status of the area has increasing the water demand especially for the domestic and industry. The development of the infrastructure, to support the agropolitan area has increasing the water demand. Furthermore, the changing climate in which is expected to decrease the precipitation increases the water stress mostly during the dry season. Therefore, it is required to assess the impact of infrastructure development and water demand on the water allocation in the current and future period.

To contribute towards this goal, the current study adopted and applied Water Evaluation and Planning (WEAP) model as a decision support system (DSS) to assess water availability and investigating the impacts of different water allocation scenarios (water demand management strategies) aimed on assessing the impact of infrastructure development and future climatic change on water allocation in study area.

2. Data and Methodology

2.1. Study area

Poncokusumo is located about 112.42° - 122.54° E and 8.68° - 7.58° S, eastern part of Malang District, Java island (figure 1). The area is dominated by lowland with the altitude less than 1,000 meter above the sea level (masl). The area is hilly area which is mountainous around the area. The precipitation is dominated by the monsoonal pattern which is peak precipitation occurs during December to February. The dry season occurs during the period of June to August. The climate is identified based on the temperature and precipitation and is classified as the wet tropic based on Koppen climate classification.

The precipitation is about 2300 mm in average with the temperature is about 21.7° C in average annually. The 68 mm on monthly average of precipitation was observed during dry month in August and the maximum precipitation was observed about 423 mm during December. The area was also classified as humid area which is the relative humidity (RH) about 82%.

The Poncokusumo was selected as the study area of the impact assessment of the water resources under changing climate. The area is about 103 km^2 (3.46% from the Malang district area). The area is grouped into three area based on the topography. The flat area covers the area of nine sub districts namely Desa Karanganyar, Jambesari, Pajaran, Argosuko, Ngebruk, Karangnongko, Wonomulyo dan Belung. The hillside area which are dominated by agriculture covers about eight district namely, Desa Dawuhan, Sumberejo, Pandansari, Ngadireso, Poncokusumo, Wringinanom, Gubugklakah, dan Ngadas. The rest area is hilly area which is elevation more than 1200 masl.

2.2. WEAP for Water Allocation

WEAP, which is an object-oriented computer modelling package is an Integrated Water Resources Management (IWRM) tool designed for simulation of water resources systems and trade-off analysis. The tool considering that water supply is defined by the amount of precipitation that falls on a watershed or a series of watersheds, with the supply progressively becoming depleted through natural watershed processes, human demands and interventions, or enhanced through watershed accretions. These processes are governed by a water balance model concept that defines watershed scale evaporative demands, rainfall–runoff processes, groundwater recharge, and irrigation demands [3][4].

The model simulates water system operations within a river system with basic principles of water accounting on a user-defined time step, usually a month. Simulation allows the prediction and evaluation of "what if" scenarios and water policies such as water conservation programs, demand projections, hydrologic changes, new infrastructure and changes in allocations or operations [3][4][5][7]. Thus WEAP is considered as an integrated water management tool for evaluating water use and allocation with a greater focus on balancing supply and demand in a swift and transparent way.

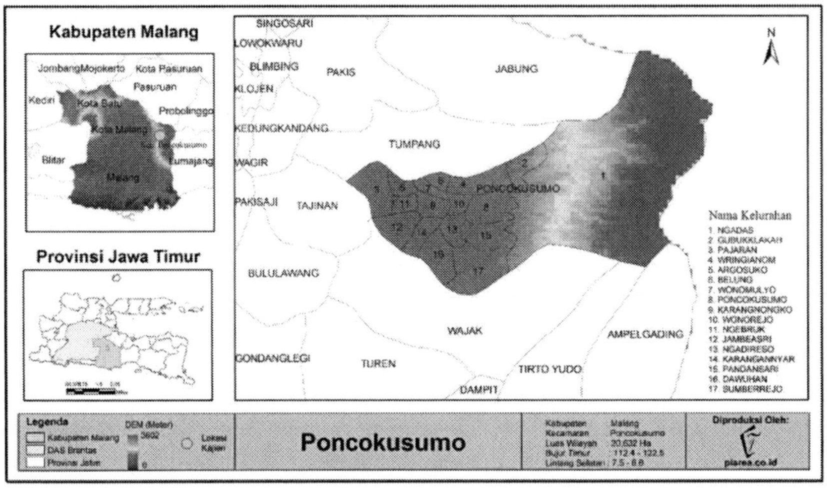

Figure 1. Map of the study area

WEAP model was applied by simulating recent base year or *'business as usual'* account, for which water availability and demand was determined. This information was obtained from different water users/stakeholders in the basin on year 2015 as the baseline. The application was defined by time frame, spatial boundaries, system components and configuration of the problem. The current account, which is the calibration step of the model, provided the actual water demand, resources and supplies for the system. Scenarios built on the current account enabled the exploration of the impact of alternative policies on future water availability and use. Construction of different scenarios was based on alternative sets of policies, which were evaluated with regard to water sufficiency, costs and benefits [6]. During this study, WEAP model was developed on the monthly time step.

2.3. Water Supply and Demand Assessment

The water availability was assessed based on the water discharge to the area. The discharge was modelled using hydrological rainfall – runoff model HEC HMS model. HEC HMS model is a lumped model which is simulates the hydrological processes in the area based on physical characteristic of the catchment. The water demand consists of three group namely domestic, industry and agriculture. The use of the water requirements of three user is based on field data indicating that the third is the dominant users of water in the study area. Water usage of those three group based on Indonesian National Standard (SNI) water use (SNI 19-6728.1-2002). In general SNI describe the amount of water use is approached by the population. Estimated water use for the domestic sector is done by using the approach of the number of people multiplied by water use national standard (SNI for domestic water demand). Uses of water for industry use a similar approach to the calculation of water demand for the domestic sector. The use of water for agriculture in this study is limited of water uses for irrigation. The water requirement for agriculture in units of cubic meters per hectare, therefore, the amount mentioned previously is multiplied by the area.

2.4. Scenario development

The scenario was developed to assess the impact of increasing water demand to the allocation as well as the decreasing of water supply. To assess those impacts, in this study, the water demand were simulated to increase by 10, 20, 30, 40 and 50% based on the baseline scenario. On the supply side, the water supply were simulated to decrease by 10, 20, 30, 40 and 50% based on the baseline scenario.

3. Result and Discussion

Estimated water demand for three water users were applied on the reference scenario. The water demand for the period of year 2010 – 2015 was used to estimate the water demand for three water users. In this study, the only water supply is from the surface water. The estimated water demand is based on the population and industrial group multiplied by the SNI of each water demand for domestic and industrial. The SNI of water demand per hectare was also applied to estimate irrigation water demand for agriculture. The total water demand was estimated by multiplying the water demand per hectare by total irrigation area on each district.

3.1. Water Allocation: Reference Scenario

Water allocation models must able to represent the model component of water resources in the model. The ideal water allocation model should simulate the water demand and supply. Priorities for different water demand and supply in this study was set as similar for each water users and supply. This priorities were able to describe the water condition in the field which is all of the water user has similar priorities to be fulfilled.

In this study, the water supply in the region comes from two main rivers. The water sources are Lesti and Amprong River. Lesti River located in the southern region of the area while Amprong River located on the northern region. The study on water availability approximated by calculating the value of the discharge of the two major river is. Based on the water supply analysis output, the availability of water in the area Poncokusumo is shown in figure 2.

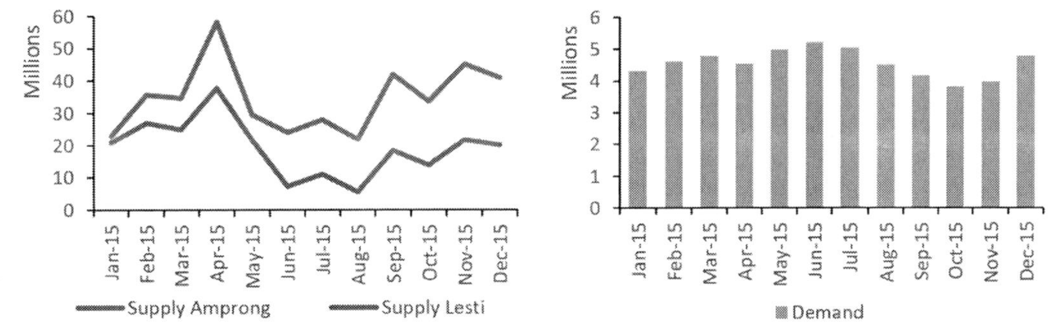

(a) (b)

Figure 2. Graph Poncokusumo water availability in the region, based on the flow of rivers and streams Amprong Lesti (a) and total water demand (industrial, domestic and agricultural) in the region Poncokusumo (b).

Based on figure 2(a), the highest and lowest water availability in the area is occurred in April and August, respectively. While compared with the precipitation, it is high confidence that the water availability pattern follows the precipitation pattern. It shows that as the region is influenced by monsoon, the precipitation in the area also follows monsoonal pattern. Furthermore, the changing of precipitation pattern on each month is directly affected the pattern of the water availability.

The water demand as shown in figure 2 (b) is the amount of water is required to meet the water requirements of domestic, industrial and agricultural. Water demands vary on each months, it is influenced by the number of days each month. It mainly affects the amount of water demands for domestic and industrial. Unlike the water needs for domestic and industrial, the water demands for agriculture are influenced by cropping patterns in respective fields. As shown in figure 2 (b), the high water demand in the dry season, it is because the water requirements of agriculture through irrigation during the dry season is increasing, as additional water for agricultural needs of decreasing precipitation. Based on the supply and demand (figure 2), it shows that for all water demands can be met by the availability of water in the area.

3.2. Scenario Analysis

In this section, the scenario of the increasing water demand and decreasing water supply were assessed. As there are unavailable data for the future water demand, the water demand was assumed to increase about 10, 20, 30 40 or 50% based on the reference scenario. On the other hand, the supply scenario for the future are based on the projection of future precipitation in year 2030 and decreasing precipitation of 10, 20, 30, 40 and 50%. Future precipitation is based on two different Global Circulation Model (GCM) namely CSIRO and MIROC.

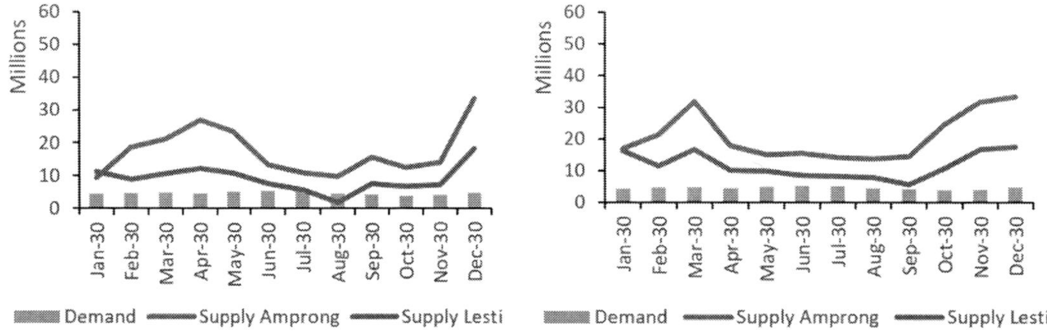

(a) (b)

Figure 3. Graph projections of water availability in 2030 using the CSIRO model (a) and projected water availability MIROC 2030 model (b) as well as the total water demand (industrial, domestic and agricultural) in the region Poncokusumo.

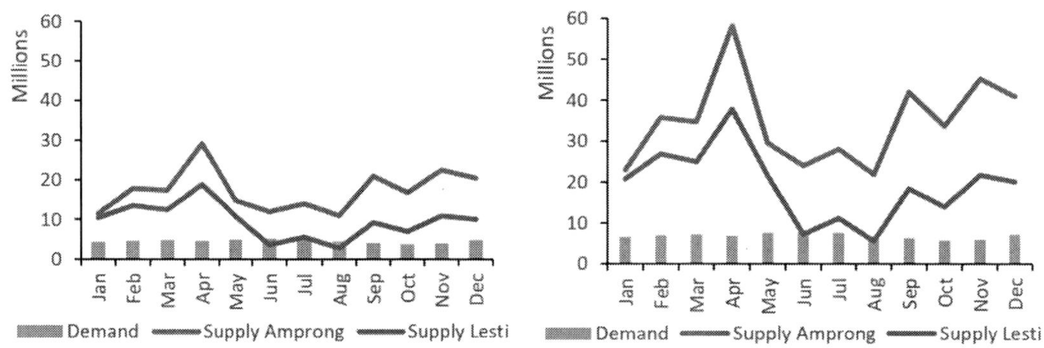

(a) (b)

Figure 4. Water resource allocation graph sensitivity to changes in water availability reduction by 50% from normal (a) and changes in water demand increase by 50% from normal (b).

Comparing with the reference scenario, the water availability is expected to decrease under CSIRO and MIROM model projection (figure 3). Under CSIRO model, the precipitation is expected to have a peak on April and December, with the lowest precipitation is expected to occur in August. However, the decreasing water availability in the future is able to fulfil the water demand if assumed that the future water demand are similar as the reference scenario. There are no water shortage in the area in the future.

Similarly, the MIROC model is also projecting the decreasing water availability in the future. The model projects that the precipitation is expected to be maximum in March and December, and achieved its minimum in September. The water availability under MIROC model is still able for fulfilling the water demand under reference scenario. The water availability in the area of Poncokusumo come from two sources as mentioned above, therefore, by this condition there are sufficient water for the future.

On this study, the sensitivity analysis were also assessed. The sensitivity of the decreasing water availability and also increasing water demand. The first scenario is decreasing water availability by 10,

20, 30, 40 and 50% comparing with the references scenario. The second scenario is increasing water demand about 10, 20, 30, 40 and 50% comparing with the water demand on references scenario. Based on those two scenario, the water availability in the Poncokusumo is still able to fulfil water demand of three water users; domestic, industry and agriculture. Moreover, the water is still available under dry season in which the water availability is expected to decrease up to 50% in this study (figure 4). On this research, all of the water availability is allocated only to those three water users. Furthermore, the efficiency of the water distribution is not considered and all of the water needs assumed fully utilized (100% consumed) with no backflow.

4. Conclusion and Recommendations

The water allocation is comparing the water demand and supply for certain area. Assessment of the water allocation in the future requires the future projection for the water demand and supply and also the allocation policy. The projection of the water supply can be expected using the precipitation projection based on the emission scenario under GCM model. On the other hand, the projection of the future water demand needs more parameters especially socio and economic parameters. On this research the future water demand projection using the assumption of increasing water demand of 10, 20, 30, 40 and 50%.

The result of this study shows that the Poncokusumo area is supplied by two sources water from Lesti and Amprong River. Those water sources is able to fulfil the water demand of domestic, industry and agriculture in the area. The increasing water demand is still able to be fulfilled in the future by up to 50% increasing water demand comparing to the reference period. Furthermore, it was expected that there will no water shortage in the future for the decreasing water availability up to 50% comparing the reference period. Furthermore, it is required to incorporate the future climate information on design the future water policy and management to reduce the adverse impact of changing climate. It was also suggested policy actions as recommendation to better manage current climate variability as well as future uncertainty from climate change impacts on water allocation and resources management.

References

[1] UN 2005 *The millennium development goals report 2005* (New York: USA)

[2] Chung G, Lansey K, Blowers P, Brooks P, Ela W, Stewart S, Wilson P 2008 A general water supply planning model: evaluation of decentralized treatment *Environ. Model Softw.* **23** 893–905

[3] Yates D, Sieber J, Purkey D, Huber-Lee A, Galbraith H 2005 WEAP21: A demand, priority, and preference driven water planning model: part 2, aiding freshwater ecosystem service evaluation *Water Int.* **30** 487–500

[4] Purkey D R, Hubber-Lee A, Yates D N, Henamann M, Herrod-Julius S 2007 Integrating a climate change assessment tool into stakeholder-driven water management decision-making processes in California *Water Resour. Manag.* **21** 315–329

[5] Raskin P, Hansen E, Zhu Z 1992 Simulation of water supply and demand in the Aral Sea *Region* **17** 55–67

[6] SEI 2008 *WEAP: Water Evaluation and Planning System, tutorial* (Stockholm Environment Institute: Boston Center)

[7] Santikayasa I P, Babel M S, Shrestha S, Jourdain D, Clemente R S 2014 Evaluation of water use sustainability under future climate and irrigation management scenarios in Citarum River basin, Indonesia *International Journal of Sustainable Development & World Ecology* **21** 181-194

Climate regionalization for main production areas of Indonesia: Case study of West Java

Perdinan[1,2,3*], Ryco Farysca Adi[2,3], Yon Sugiarto[1,3], Annisa Arifah[1], Enggar Yustisi Arini[3], and Tri Atmaja[2,3]

[1] Department of Geophysics and Meteorology, Bogor Agricultural University, Jalan Meranti Wing 19 Level 4 Gedung FMIPA Kampus IPB, Babakan, Dramaga, Bogor, Jawa Barat 16680

[2] PIAREA, Perumahan Griya Melati Blok C3 No.21/22 Bubulak Arah Cifor, Bogor Barat, Kota Bogor

[3] Generasi Hijau Indonesia, Desa Tegalwaru No.13, Ciampea, Kabupaten Bogor

E-mail: perdinan@gmail.com

Abstract. Spatially, climate condition is vary within a region and considered as essential information for planning activities such as agro-climate zonation. An approach to understand the spatial climate variability is the utilization of climate regionalization that is applied to rainfall data to distinguish differences in the pattern and magnitude (characteristics) of spatial rainfall variability over a region. Unfortunately, the application of climate regionalization poses a challenging issue in Indonesia, considering the availability of climate data. Recent advances in satellite and reanalysis data measuring climate variability over a large area provided an opportunity for the application of climate regionalization in the country. Using the West Java, one of main crop production regions in Indonesia, climate regionalization techniques were applied to map spatial variability of climate types based on rainfall data recorded by climate stations (point based analysis) and estimated by modeled/reanalysis data and satellite observations (gridded data). The regionalization derived from gridded rainfall data have reasonably better in capturing the zonal pattern of differences in climate types within the study region than the regionalization applied to insufficient numbers of site-based rainfall observation. This indicates that the gridded data offers an alternative for climate regionalization, when site-based observations are unavailable or limited.

1. Introduction

Climate is simply well-known as the average weather condition of a region in the long term period. Climate condition varies spatially over a region determined by topographical, geographical, and local condition. Climate classification is a method to identify and describe the climatic differences on a regional scale. Climate classification can be derived based on a number of factors such as rainfall, air temperature, humidity, geographical condition, and crop diversification [1]. As a tool to define differences in climate types over a region, climate classification is an essential element of regional climate information that can be used for various purposes such as agriculture, transport, and tourism.

It has been generally acknowledged that there are three climate classifications frequently employed in Indonesia, namely: Koppen, Oldeman, and Schmidth – Ferguson. Each classification technique has its own characteristics in assessment method and interpretation, which had been formulated in 1901 for Koppen [2], in 1975 for Oldeman and in 1951 for the Schmidt-Ferguson [3]. These three classification techniques required input of an essential climate-surfaced variable, i.e. rainfall, whose observations are coordinated by Badan Meteorologi, Klimatologi, dan Geofisika (Meteorology, Climatology, and Geophysics Agency) in Indonesia. The observations of rainfall are collected from rainfall or climate station over a region. Unfortunately, the limited number of stations is often being a constraint for Indonesia. Currently, available observational stations may not be sufficient to describe climate classifications for a region of interest. In this case, alternative data source to define climate characteristics of a region such as rainfall data derived from satellite, re-analysis, and models can be a good data source for employing climate classification techniques to define spatial variability of climate types over a region.

This research provides climate classification for crop production growing regions of West Java – Indonesia. The case study is selected considering West Java is one of primary crop growing regions or "food basket" in Indonesia. Defining climate types for this region can provide information on the contribution of climate types to spatial differences in crop production over the region. Concerning the limited records of site-observations, gridded climate data of WorldClim [4] and CHIRPS [5] are employed to define climate types over West Java.

2. Related works

Climate variables aproximate the patterns of climate variability over a region as the basis for climate classification. Rainfall is an essential climate variable that is often used for climate classification such as those applied by Aldrian [1] to characterize rainfall type in Indonesia. The climate classification is of interest considering it can be used for many purposes such as understanding agroclimate suitability. Previous study [6] applied climate classification to understand the suitability of climatic conditions for coconut plants, with the high production rate (~80% production target) for the region whose rainfall is about 2500-3600 mm/year. Ramala [7] defined regional suitability on the basis of irrigation, planting period, and agro-climatic zones for Cilacap. Another study, Fadholi and Supriatin [8] showed Priangan (the districs of Sukabumi, Cianjur, Sumedang, Garut, Ciamis, Tasikmalaya, and the city of Bandung) can be classified into five climate types according to Oldeman classification applied to rainfall data recorded by 41 site-observations, namely: B1, B2, C1, C2, C3. The above works demonstrated that climate classification is an essential information that also can be applied for evaluating crop-land-climate suitability over a region, understanding climate affects crop growth and development.

3. Methodology

The three climate classification method, namely: Koppen, Oldeman, and Schmidt-Ferguson, were employed to define spatial variability of climate types over West Java based on rainfall data recorded by 19-site observation stations (figure 1). In addition, climate regionalization technique developed using hierarchical clustering analysis was also applied to the site-observed monthly rainfall data to classify site-observations on the basis of statistical similatities, which is the monthly rainfall characteristic. The climate regionalization apporach was performed in several steps. First, analysis of principal component (PCA) was performed to avoid potential multicolleniarities among monthly rainfall data recorded by the site stations. The hierarchical clustering techniques were then applied to the results of PCA to group site stations with statistically similar rainfall characteristics. Several numbers of climate groups, i.e., five to nine, were used as input when perfomed the hierarchical clustering whose grouped observations based on a variety of linkage methods as discussed by Perdinan and Winkler [9]. The climate classification was also performed using gridded climate data of WorldClim for the climatologically period of 1971-2000 [4]. The main purpose is to evaluate the potential use of WorldClim as an alternative data source to define climate classification, understanding the limited availability of observation data (i.e., incomplete or unavailability of site-observations). Additionally, we also employed gridded rainfall data of CHRIPS [5], whose grid resolution is different from WorldClim to evaluate the effects of different grid resolution to regional climate classification for the study region.

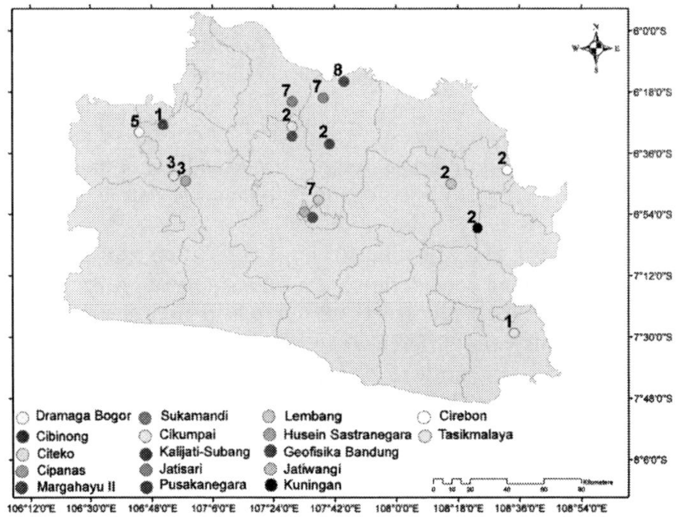

Figure 1. Location of the 19-site stations in West Java employed for the study

4. Results and discussion

The West Java Province is located between 5°50 '- 7°50' LS and 104°48 '- 104°48' E, with an area of 44354.61 km^2. West Java is directly adjacent to the Java Sea in the north, the Indian Ocean in the south, Central Java in the east, and DKI Jakarta and Banten in the west. The West Java has a monsoonal rainfall pattern. Climatologically, monthly-rainfall data of the 19-site stations (figure 1) for the period of 1950 to 2000 showed the highest rainfall amount is in January (350 mm) and the lowest rainfall is in July (97 mm), with the rainy season of November-April, and the dry season of May-October. The climate fluctuations affect the availability of water supply. For example, when the dry season is longer than normal then the needs for irrigation water also increased, which demanded for better water management strategies to storage water during the rainy season. The climate fluctuation is also influenced by global climate phenomena. Climate variability of West Java, its fluctuation and intensity, is affected by El Niño Southern Oscillation (ENSO) and Indian Ocean Dipole (IOD). These climate phenomena cause changes in the patterns and the intensity of rainfall, such as delaying the beginning of rainy season or declining the rainfall intensity such as those reported by Apriyana and Kailaku [10] for the period of September to November in the northern West Java.

The results of climate classification derived from different classification techniques shows a little bit of variation. Based on Koppen classification, West Java had two climate types, namely: Af and Aw. The area with Af climate type is in the western part of West Java, started from Bekasi to Garut and most part of Purwakarta, Subang, Ciamis, Tasikmalaya and Depok. This climate type has considerably higher rainfall amount over the year (above 60 mm for each month). The areas with Aw are considerably drier as this climate type is indicated by longer dry season. Oldeman classified West Java into 8 climate types, namely: A, B1, B2, C2, C3, D2, D3 and D4 (figure 2). The suitable areas for paddy production are areas with A and B1 climate type. The A climate type allows for paddy to continuously grow over the year; whereas, the B1 type suggests the needs for irrigation system to continuously grow paddy particularly over the dry season. The areas with the A climate type include Bogor, Garut, and Bandung; while, the areas with the B1 climate type encompass Sukabumi, Tasikmalaya, most areas of Ciamis, Cianjur, Bekasi and Depok. Furthermore, Smith – Fergusson classifies the study region into four climate types, namely: A, B, C and D. Most areas of West Java are classified into the A climate type, which means very wet. The B, C and D climate types are shown in some parts of the north-central West Java, and covered most areas in the northern Subang and Indramayu.

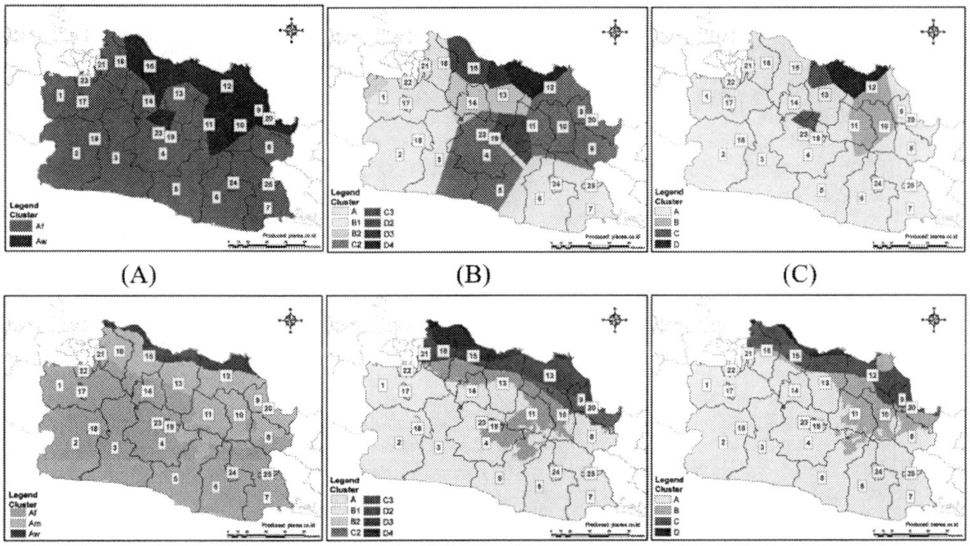

Figure 2. Climate classification for West Java developed using Koppen (a), Oldeman (b) and Smith Fergusson (c) derived from observed rainfall data (above) and WorldClim (below) of 1971-2000. The number indicates the name of district. (1) Kab. Bogor, (2) Kab. Sukabumi, (3) Kab. Cianjur, (4) Kab. Bandung, (5)Kab. Garut, (6) Kab. Tasikmalaya, (7) Kab. Ciamis, (8) Kab. Kuningan, (9) Kab. Cirebon, (10) Kab. Majalengka, (11) Kab. Sumedang, (12) Kab. Indramayu, (13) Kab. Subang, (14) Kab. Purwakarta, (15) Kab. Karawang, (16) Kab. Bekasi, (17) Kota Bogor, (18) Kota Sukabumi, (19) Kota Bandung, (20) Kota Cirebon, (21) Kota Bekasi, (22) Kota Depok, (23) Kota Cimahi, (24) Kota Tasikmalaya, (25) Kota Banjar

In addition to classifying climate types for the study region using site-observation data, the classification was also performed using grided climate data of WorldClim as an alternative of climate data source, providing insufficient availability of site-observations. The pattern of climate groups resulted from the grided climate data exhibits somewhat similar general pattern for each classification method (figure 2). The classification applied to site-observations showed more 'rough' climate cluster spatially, which may be the results of using insufficient numbers of site-observed stations. Meanwhile, climate classification derived from WorldClim resulted 'smoother' classifcation map, understanding high spatial resolution of WorldClim which gridded the study region into 43411 grids provide more detailed information for the study region. Eventually, the climate classification shows more detailed variation of climate classification map particularly in the central area of the study region, where topographical characteristics are more hilly than the nothern part of the study region. Site-observations are also very limited for the southern areas of West Java, which offer an opportunity for the utilization of gridded climate data such as WorldClim. Apart from the detailed spatial map, latitudinal pattern or zonation of rainfall classification is shown for the study region, regardless of site-observations or gridded data employed for running the classification methods.

For another comparison, climate information from BMKG which informed rainfall characteristics and the begining of rainy season supplied to distinguished climate zonation called topo (stands for Zona Musim - Seasonal Zonation) was evaluated. The ZOM supplies climate prediction for rainfall specifically for regions classified into a group, indicated by the same colors in figure 3. There are variations of rainfall conditions in the study area. The northern area includes Bekasi, Subang, Indramayu, Ciamis and Tasikmalaya, grouped into below normal rainfall. Meanwhile, the rest grouped into normal rainfall figure 3 (left).

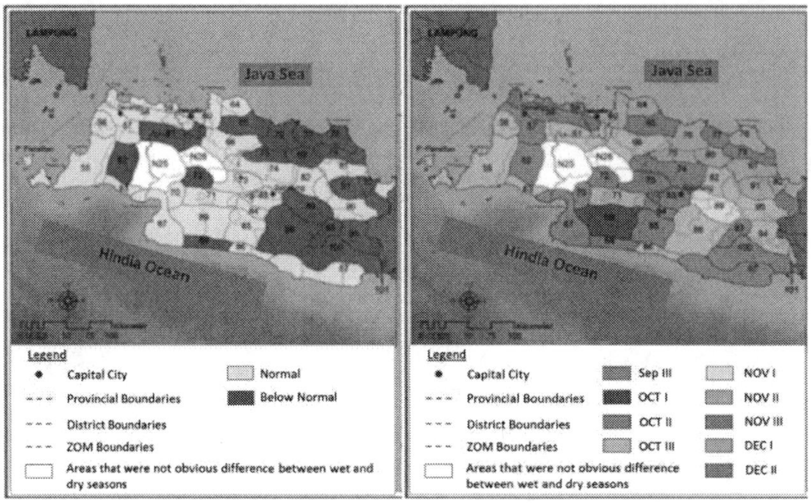

Figure 3. Climate prediction of rainfall characteristics (left) and the beginning of rainy season (right) in 2015/2016 in Banten, Jakarta and West Java (BMKG 2016)

Figure 3 (right) provides information on the climate prediction for the begining of rainy season. The district of Ciamis and Tasikmalaya were predicted to have the begining of rainy season in September; whereas, Sukabumi, Cianjur, Bogor, Cianjur and Bandung were predicted to be in October. Pangadaran, Garut, Sumedang and Majalengka would be in November. The other northern areas of West Java would have the beginning of rainy season in December. This climate information to some extend provides alternative to learn spatial variability of rainfall over the study region indicated by different pattern of the beginning of rainy season and charaterization of rainfall amount into above and below normal. However, it should be noted that the climate prediction is merely for a certain period of time, which may be different with the period of climate classification applied to climatological statistics of climate data such as rainfall over 30-year period as employed in this study.

Further analysis was conducted to classify climate types of the study region using the climate regionalization technique named hierarchical clustering analysis applied to rainfall data obtained from site-observations, WorldClim, and CHRIPS. The results showed that zonal pattern of rainfall are captured by climate regional derived from WorldClim and CHRIPS, although, WorldClim provide more detailed variation considering the higher spatial resolution of the data than the CHRIPS. The classification derived from rainfall data of site-observations provide ‘rough’ grouping, with unclear zonal pattern. Insufficient numbers of site-obsevation stations may contribute to the ‘rough’ patterns of the climate classification. Interestingly, the climate regionalization derived using gridded datasets (figure 4-middle and below) can almost imitate the zonal pattern of Oldeman classification (figure 2-middle). Evaluation of the zonal pattern explains that the southern and central parts of the study region exhibit high spatial variation for the climate groups derived from WorldClim and CHRIPS. This indicates that the gridded datasets have reasonably captured the variation of topography in the study region.

Further analysis was applied to evaluate rainfall characteristics of climate groups derived from the hierarchical clustering technique using rainfall data of site-observations. The evaluation was performed based on monthly rainfall fluctuation over a year for each station grouped into a specific climate type as presented in table 1 and figure 5.

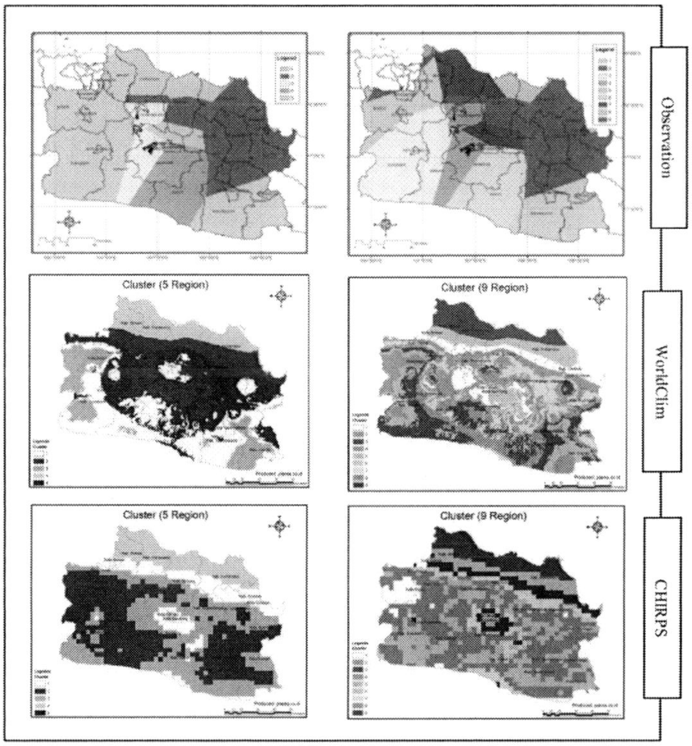

Figure 4. Climate regionalization for different data sources

Table 1. Climate characteristics of each climate group derived using hierarchical clustering technique applied to rainfall data of the 19-site observations

Group	Station	Characteristics
1	Atang Sanjaya Bogor, Cibinong, Tasikmalaya	High rainfall amount are in DJF and lower rainfall amount are in JJA whose monthly rainfall differences are small.
2	Cikumpai, Cirebon, Jatiwangi, Kalijati, Kuningan	Rainfall for the rainy season is similar to Group 1, but monthly rainfall for dry season is lower than Group 1.
3	Cipanas, Citeko	The wet season is in JFM and monthly rainfall is higher than that for Group 1 and Group 2. Dry season is in JJA.
4	Husein Sastranegara, Jatiluhur, Margahayu 2	Relatively lower monthly rainfall over a year than Group 1, 2, and 3. Dry season is in JJA.
5	Dramaga	Temporally, monthly rainfall is high throughout the year.
6	Geofisika Bandung	Higher monthly rainfall occurs in February, November and December.
7	Jatisari, Lembang, Sukamandi	The average monthly rainfall is considerably the lowest, compared to the other climate groups.
8	Pusakanegara	The longest period of dry season, compared to the other groups. Relatively high rainfall occurs in the beginning year.

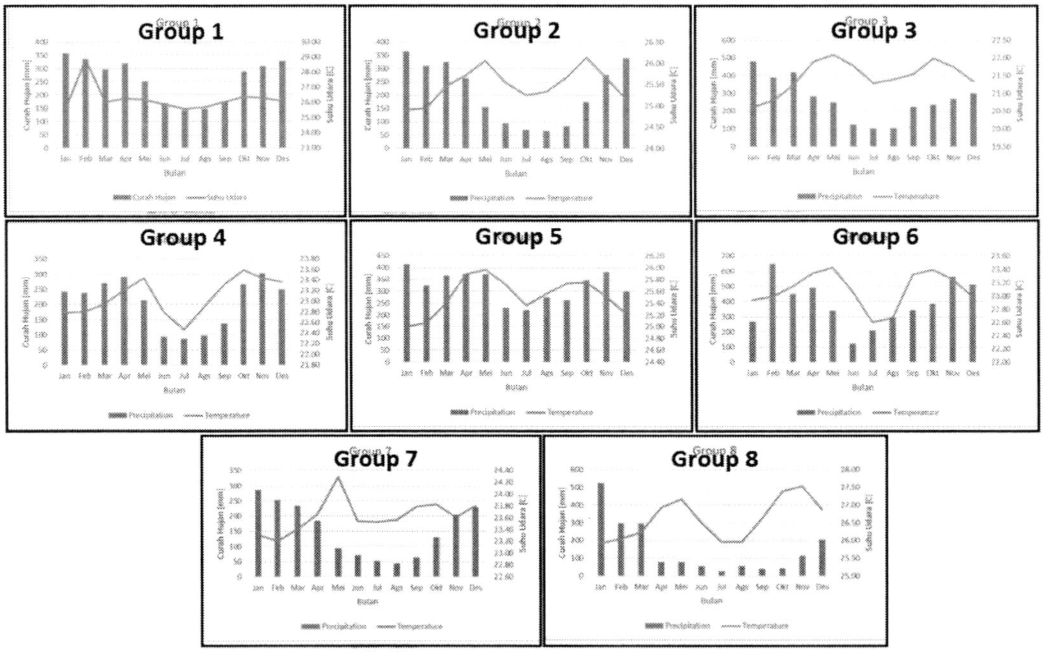

Figure 5. Monthly rainfall fluctuation for each climate group in the study region

5. Conclusion

Climate classification is one of essential tools in understanding spatial climate variation over a region, which can be further employed for planning decision on specific economic sector such as land-suitability for growing commodities. This research defines climate classification for West Java Indonesia using the three commonly used classification techniques (i.e, Koppen, Oldeman, and Schmidt Ferguson) and the climate regionalization method. The regionalization employed principal component analysis in combination with hierarchical clustering technique applied to monthly rainfall of site-observations and gridded climate data available for the study region, i.e., WorldClim and CHRIPS. The analysis revealed that the gridded climate data provide an alternative to perform climate classification when the site-observations are often limited. The climate classification applied to the gridded climate data exhibits a latitudinal variation of climate characteristics over the study region. The climate regionalization displays that the northern and southern areas of the region show distinct climate types; whereas, the central areas of the region show quite complex spatial variation, indicated the differences in climate types. The spatial variation of climate types may also in line with topographical condition of the region. Future works can further employ the regional climate classification to study land-climate suitable areas for many purposes such as distribution of agricultural commodities over crop production regions.

Acknowledgment

This research is supported by the grant of Research Applicative Strategy No. 079/SP2H/LT/DRPM/II/2016.

6. References

[1] Aldrian E 2001 Pembagian iklim Indonesia berdasarkan pola curah hujan dengan metoda "Double Correcaltion" *Jurnal Sains dan Teknologi Modifikasi Cuaca* **2** 11-18
[2] Handoko 1993 *Klimatologi dasar (landasan pemahaman fisika atmosfer dan unsur-unsur iklim)* (Bogor: Jurusan geofisika dan meteorologi FMIPA-IPB)
[3] Tjasyono B 2004 *Klimatologi* (Bandung: Bandung Press)
[4] Hijmans R J, Cameron S E, Parra J L, Jones P G and Jarvis A 2005 Very high interpolated climate surfaces for global land area *International Journal of Climatologi* **25** 1965-1978
[5] Funk C, Peterson P, Landsfeld M, Pedreros D, Verdin J, Shulka S, Husak G, Rowland J, Harrison L, Hoell A and Michaelsen J 2015 The climate hazard infred precipitation with stations- a new environmental record for monitoring extremes. *Scientific data*
[6] Dewi N K 2005 Kesesuaian iklim terhadap pertumbuhan tanaman *Mediagro* **1** 1-15
[7] Kamala R 2015 *Analisis agihan iklim klasifikasi oldeman menggunakan sistem informasi geografis di Kabupaten Cilacap* in Pulikasi Ilmiah (Surakarta: Universitas Muhammadiyah Surakarta)
[8] Fadholi A and D Supriatin 2012 Sistem pola tanam di wilayah priangan berdasarakan klasifikasi iklim Oldeman *Jurnal GEA* **12** 2
[9] Perdinan and Winkler 2014 Selection of climate information for regional climate change assessments using regionalization techniques: an example for the Upper Great Lake Region. International *Journal OF Climatology*
[10] Apriyana Y and T E Kailaku 2015 Varibilitas iklim dan dinamika waktu tanam padi di wilayah pola hujan monsunal dan equatorial *Pros Sem Nas Masy Biodiv Indon* Vol **1** (Bandung)

Characteristics of Kelvin waves and Mixed Rossby-Gravity waves in opposite QBO phases

Nur Zaman Fathullah[1*], Sandro W. Lubis[2] and Sonni Setiawan[1]
[1] Department of Geophysics and Meteorology, Bogor Agricultural University (IPB), Indonesia
[2] Physics of the Atmoshpere, GEOMAR Helmholtz Centre for Ocean Research Kiel, Germany

E-mail: zamanchaniago@gmail.com

Abstract. A 35-year ERA-Interim dataset from the European Center for Medium-Range Weather Forecasts (ECMWF) was used to study the characteristics of Kelvin waves and Mixed Rossby-gravity waves based on a Space-Time Spectral Analysis (STSA). The results show that Kelvin wave activity is stronger during easterly QBO phases, while Mixed Rossby-gravity waves are stronger during westerly QBO phases. Analysis on seasonal variations indicates that the Kelvin waves and Mixed Rossby-Gravity wave activities increase in JJA and SON, respectively. This is associated with a variation of basic mean flow in the lower stratosphere. In addition, the variations of Kelvin and Mixed Rossby-Gravity waves in the troposphere are not significantly affected by the QBO phases. In the troposphere, both Kelvin waves and Mixed Rossby-Gravity waves propagate with a lower phase speed compared to those observed in the stratosphere. This behavior is to be likely due to large.

1. Introduction

Equatorial Planetary Waves (EPWs) is an important class of tropical atmospheric disturbances. EPWs are generated by diabatic heating due to organized tropical large-scale convective heating in the equatorial belt, where their signatures are clearly evident between 20° N and 20° S [1]. Two examples of prominent EPWs are Kelvin waves and Mixed Rossby-Gravity (MRG) waves. These waves are generated in the troposphere by large-scale convective system in the equator [2]. Kelvin waves have a period of approximately 12 to 20 days, and propagate eastward with a vertical wavelength of about 6 - 10 km. On the other hand, MRG waves have a period of about 4 to 5 days, and propagate westward with a vertical wavelength of about 4 - 8 km [2]. These waves play pivotal role in controlling disturbances in the equatorial stratospheric (e.g., QBO) and tropospheric circulation and precipitation [3, 4].

The quasi-biennial oscillation (QBO) is the primary mode of variability of the equatorial mean wind in the stratosphere (~16-50 km), which is characterized by downward propagating easterly and westerly wind regimes, with a variable period averaging approximately 28 months [5]. The maximum amplitude is observed in the middle and lower tropical stratosphere, with the easterly phase having larger amplitudes compared to the westerly phase. The twofold-structure of QBO descends downward at rate of ~1 km/month. QBO is forced by the interaction of long-period vertically propagating tropical atmospheric waves (e.g., inertia gravity waves, Kelvin waves and mixed Rossby-gravity waves) with the zonal mean flow [1].

The focus of this study is to examine the characteristics of Kelvin and MRG waves in different QBO phases, including their horizontal structures in the stratosphere and troposphere, by using the Space-Time Spectral Analysis (STSA) technique. Since the Kelvin waves and MRG waves effectively modulate the temperature anomalies in the tropics [2, 3], we therefore used temperature fields as a proxy to isolate the Kelvin wave and MRG wave activities.

2. Methods

We used both daily and monthly averaged temperature, zonal wind, and sea surface temperature datasets obtained from the ERA-Interim reanalysis from January 1979 to December 2013 (35 years). Our study focused on the latitudinal bands between 20° N - 20° S (with a spatial resolution 2.5°x 2.5°) and altitudes between 100 hPa and 1 hPa. In addition, we also used outgoing long wave radiation (OLR) from NOAA with the same spatial and temporal resolutions.

2.1. Composite Analysis

Composite analysis is used to determine the distribution of atmospheric basic state at the same conditions. In this study, analysis of composites is used to analyze the variations of zonal mean wind in the stratosphere in opposite QBO phases.

2.2. Space-Time Spectral Analysis (STSA)

STSA is used to analyze the propagation properties of atmospheric disturbances in the zonal direction. This method decomposes the anomalous field in space and time as a function of wavenumber (k) and frequency (ω) [6]. A brief formulation can be expressed as follow:

$$x(\lambda, t) = \textstyle\sum_k x\, k(\lambda, t) \tag{1}$$

$$x(\lambda, t) = Re \textstyle\sum_{k,\omega}[W_{k,\omega}\, e^{i(kx+\omega)} + W_{k,-\omega} e^{i(kx+\omega)}] \tag{2}$$

$$x_k(\lambda, t) = \sum_k \sum_{\pm\omega} W_{k,\pm\omega} \cos(k\lambda \pm \omega t + \varphi_{k,\pm\omega}) \tag{3}$$

The sign (-) and (+) correspond to retrogressive and progressive components refer to eastward and westward mowing waves. $W_{k\pm\omega}$ and $\varphi_{k\pm\omega}$ are amplitude and phase angle of progressive and retrogressive waves with various wavenumber and frequency. The decomposition is conducted based on 2D Fourier transformation (in space and time domains). STSA procedure was performed with NCAR Command Language (NCL). The output is the power spectrum atmospheric disturbances as a function of frequency and zonal in the symmetric and asymmetric components.

3. Results

3.1. Equatorial QBO Profile

Figure 1 shows a time-height section of monthly mean zonal mean wind as a function of height (100 to 1 hPa) and time. We show that the signature of QBO charactherized by a reversal of the westerly to easterly wind in time is dominant in the lower-stratosphere at ~100 - 5 hPa (~16 - 35 km). Above 5 hPa (> 35 km), Semiannual Oscillation (SAO) dominates the variability of the tropical zonal wind. The oscillation peaks around the stratopause with amplitudes of around 45 m/s in the zonal wind speed. In general, the magnitude of easterly phase of QBO is stronger than the magnitude of westerly phase of QBO. The signature of QBO is not visible below 100 hPa and above a height of 5 hPa. Below 100 hPa, the amplitude of QBO decays as a result of fewer waves breaking that drives such oscillation, whereas above 1 hPa the oscillation was strongly influenced by the SAO in upper stratosphere [7].

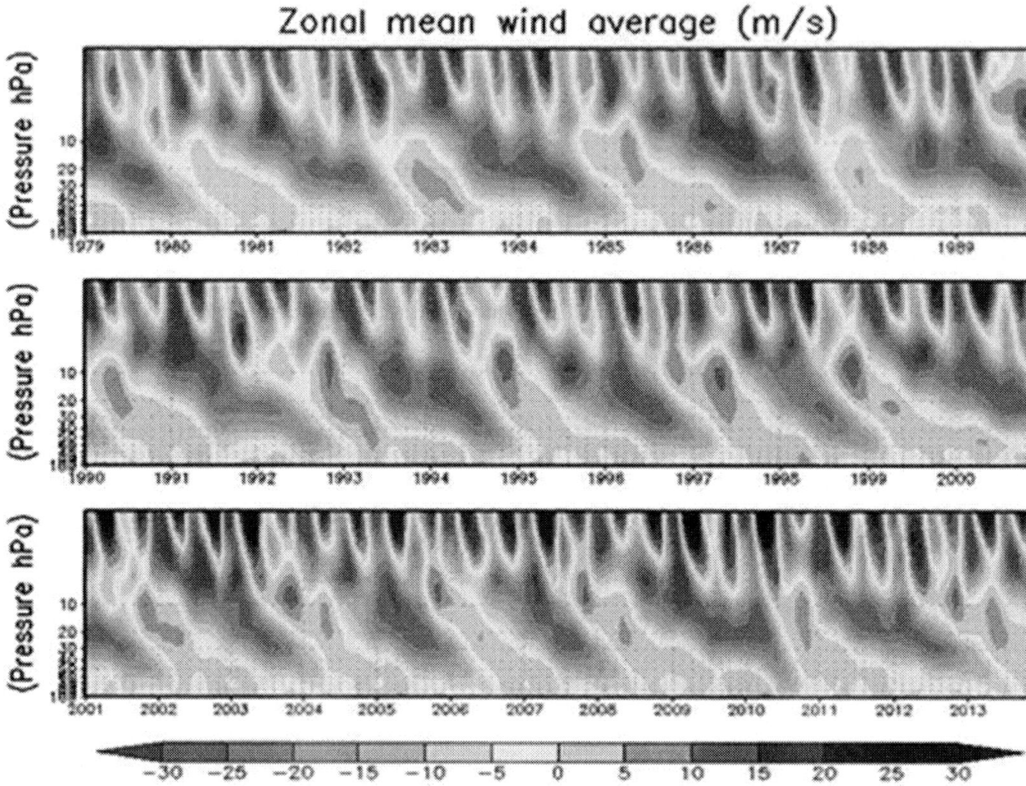

Figure 1. Monthly mean zonal mean wind between 100 and 1 hPa from 1979 to 2013 obtained from ERA-Interim dataset. Blue (green) color indicates easterly (westerly) wind.

3.2. *Definition of QBO Phases*

Different studies use different definitions for the QBO. Baldwin and Dunkerton [8] defined the QBO index as a zonal mean wind averaged over 40 hPa and between 5°N to 5°S. Gray [9] used QBO index as a zonal mean wind averaged at 44 hPa, while Yamashita [10] used QBO index as zonal wind oscillation averaged at 50 hPa between 10°N and 10°S. In this study, we used a similar definition of QBO as in Yamashita [10]. We define a QBO time series as the zonal mean zonal wind averaged between 10°N and 10°S at 50 hPa. We label a time step as QBO west phase (QBOW) when the average exceeds 5 m/s and QBO east phase (QBOE) when it falls below -5 m/s (figure 2). Using this definition, we obtained in total 9 years for each QBO phase. In particular, the WQBO occurred in 1981, 1983, 1986, 1988, 1993, 1995, 2000, 2009, 2011 and the EQBO occurred in 1982, 1984, 1987, 1992, 1994, 1996, 1998, 2001, 2010.

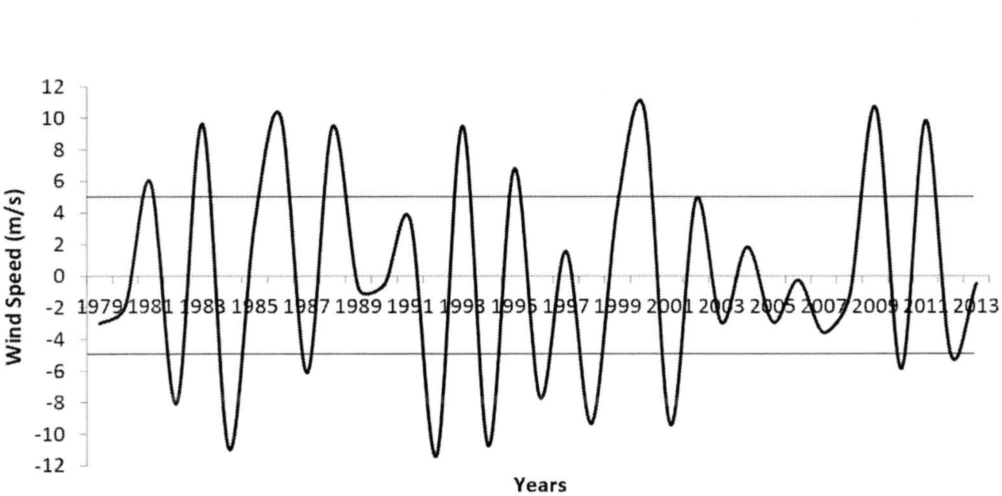

Figure 2. Time series of zonal mean zonal wind averaged between 10°N and 10°S at 50 hPa for 35 years (1979-2013).

3.3. *STSA of stratospheric temperature field*

As we mentioned earlier, the STSA technique is used to analyze the signals of planetary scale equatorial wave based on 2D Fast Fourier Transformation (FFT) into wave activity (spectrum) as a function of wavenumber and frequency. We used daily temperature anomalies by removing seasonal climatology in the data. To avoid any aliasing in the data, the linear trend has been removed by using linear regression curve fit. The results of the analysis in principle indicate the wave activity as a function of their propagation structure (symmetric and asymmetric component).

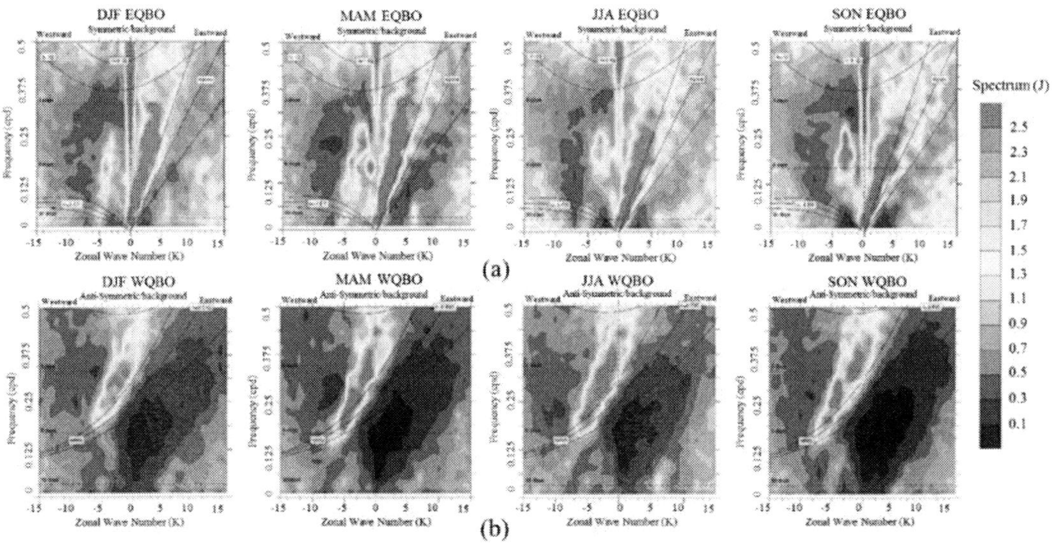

Figure 3. (a) Symmetric component of STSA temperature for EQBO at 50 hPa and (b) asymmetric component of STSA temperature for WQBO at 50 hPa for (left to the right) DJF, MAM, JJA, and SON.

Figure 3 shows a symmetric (asymmetric) component of STSA temperature for EQBO (WQBO) composite. In general, we can see that Kelvin waves (symmetric component) are stronger during the easterly phase of the QBO with a frequency of 0.063 to 0.313 cycles/day (3-16 days) and the zonal wavenumber 1 to 6, whereas MRG waves (asymmetric component) are stronger during the westerly phase of the QBO with a frequency of 0.188 to 0.313 cycles/day (3-5 days) and the zonal wavenumber -1 to -7. These results are consistent with propagation theory and the zonal wind distribution [11, 12], that in the easterly phase year (easterly basic flow) there will be more eastward-moving Kelvin waves propagating upward into the lower stratosphere and in the westerly phase year (westerly basic flow) there will be more westward-moving waves propagating upward. We also note that there is a seasonal change in the magnitude of the power spectrum of tropical wave for EQBO and WQBO, indicating that Kelvin waves and MRG wave activities are varied in season for different phase of the QBO. We thus proceed to examine this behavior in the following section.

3.4. Seasonal variation of Kelvin waves and MRG waves

Figure 4 shows differences in wave power spectrum between EQBO and WQBO for both symmetric and asymmetric components in each season. The positive values in symmetric component indicates that Kelvin waves are stronger in EQBO than in QBO, on the other way around the negative values in asymmetric component means the MRG waves are stronger in WQBO than in EQBO.

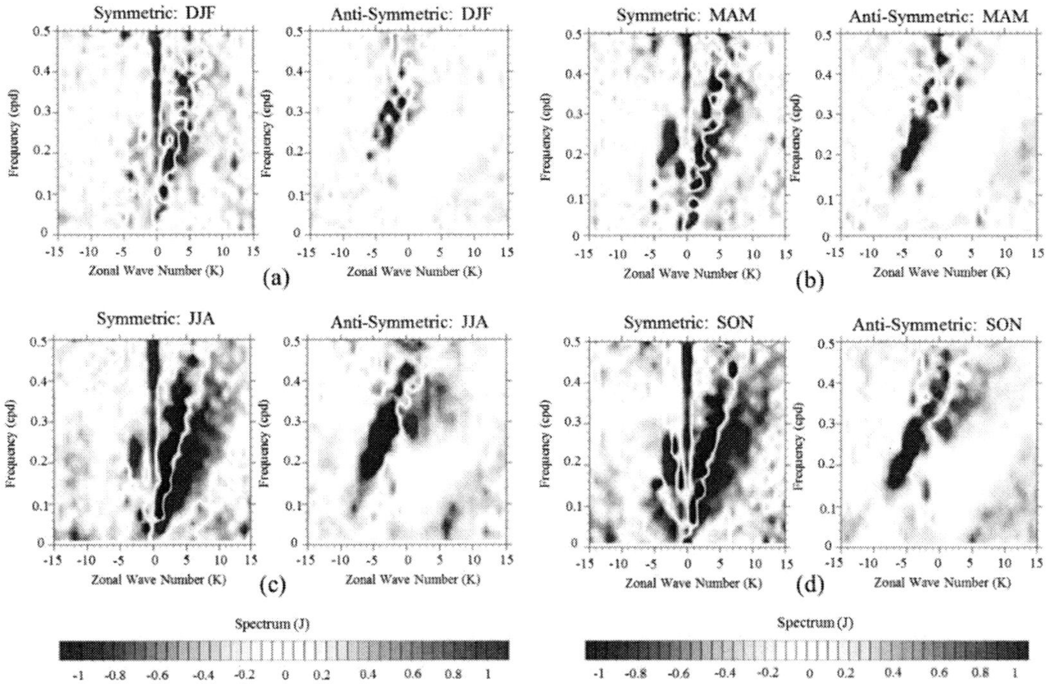

Figure 4. Differences in wave power spectrum between EQBO and WQBO for (a) DJF, (b) MAM, (c) JJA, and (d) SON at 50 hPa.

The results show that there are prominent differences in wave power spectrum in JJA and SON for both symmetric and asymmetric components at 50 hPa. In particular, positive spectrum anomalies in symmetric component are observed in JJA and SON for zonal wavenumber 1 to 10. This indicates that Kelvin wave activity is strongly observed in JJA and SON during EQBO. On the other hand, we also see that there is large negative value of wave spectrum in the asymmetric component for the same season. This indicates that MRG wave activity is enhanced in JJA and SON during WQBO.

3.5. Mechanisms for changes in the seasonality of Kelvin waves and MRG waves

To understand what causes increased Kelvin wave and MRG wave activities during particular season, we first analyze the zonal mean wind profile at 50 hPa for both EQBO and WQBO (figure 5). We show that enhanced Kelvin wave activity in JJA and SON during EQBO is consistent with the basic-state flow in the stratosphere. In particular, the basic mean flow in the stratosphere during this season is strongly easterly. Thus, this provides a favorable condition for more upward propagating Kelvin waves to the stratosphere. This is consistent with a upward propagating wave theory associated with the QBO from Plumb [12], showing that easterlies basic-state flow strengthens the Kelvin wave activity and allows these waves to propagate and penetrate at a higher level in the stratosphere.

Increased MRG wave activity in JJA and SON during WQBO is also consistent with the zonal mean flow in the stratosphere. In particular, in JJA the zonal mean flow is strongly easterly during EQBO, while during the WQBO the zonal mean flow is westerly. The easterlies prevent upward propagation of MRG wave to the mid/upper stratosphere, resulting weaker MRG wave activity in JJA during EQBO. Likewise, in SON the zonal mean flow during WQBO is more favorable for upward propagating MRG waves.

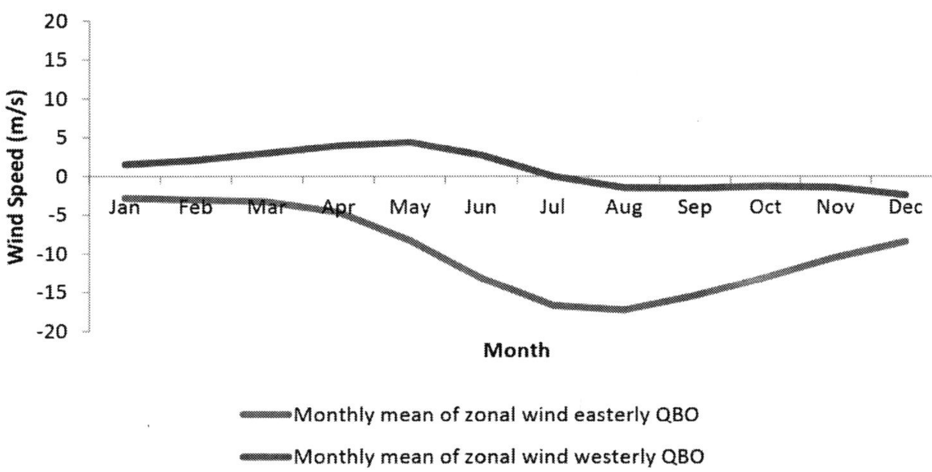

Figure 5. Zonal mean wind profile at 50 hPa during WQBO (blue) and EQBO (green).

Another possible mechanism leading to increased upward Kelvin waves and MRG waves in opposite QBO phase is the variation in lower tropospheric wave forcing. To investigate this, we composite sea surface temperature (SST) anomalies over the tropics for WQBO and EQBO phase as shown in Fig. 6. Positive sea surface temperature anomalies are indicated by red shading, while blue shading indicates the negative values. We see that there is increased SST variation over the tropical eastern Pacific in JJA and SON during both EQBO and WQBO phases. Increased SST leads to more evaporation and thus formation of convective clouds. This tropospheric forcing may cause an increased upward propagating Kelvin waves and MRG waves in the lower stratosphere.

These results confirm a previous study conducted by Yang [11], showing a possible influence of area of warm pool in the eastern Pacific on the upward propagation of Kelvin waves and MRG waves in the lower stratosphere during QBO event. Further studies are required to understand this mechanism in more detail.

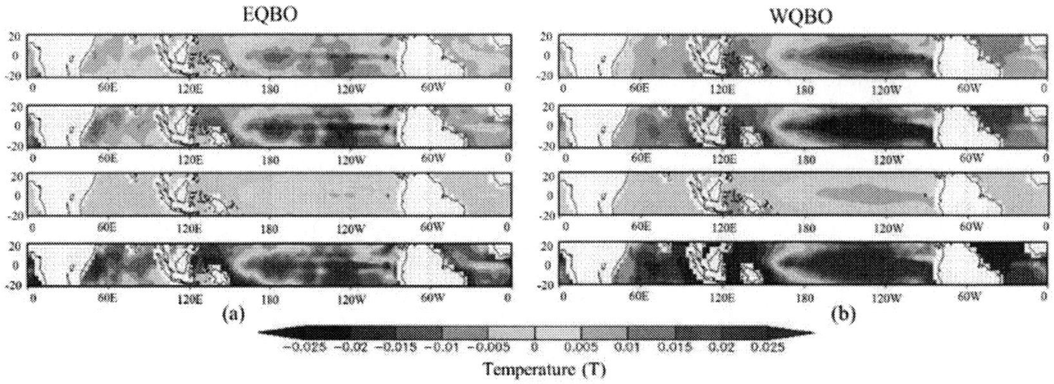

Figure 6. Composite of SST in (top to bottom) DJF, MAM, JJA, and SON for (a) WQBO and (b) EQBO.

3.6. Impact of QBO on Kelvin waves and MRG waves in the troposphere

To investigate if the opposite QBO phase has an implication for Kelvin wave and MRG wave activities in the troposphere, we repeat the STSA diagnostic for analyzing tropical wave in the troposphere that is coupled with convection. We used outgoing longwave radiation (OLR) as a proxy of convectively coupled equatorial wave (CCEWs) as in Lubis and Jacobi [4].

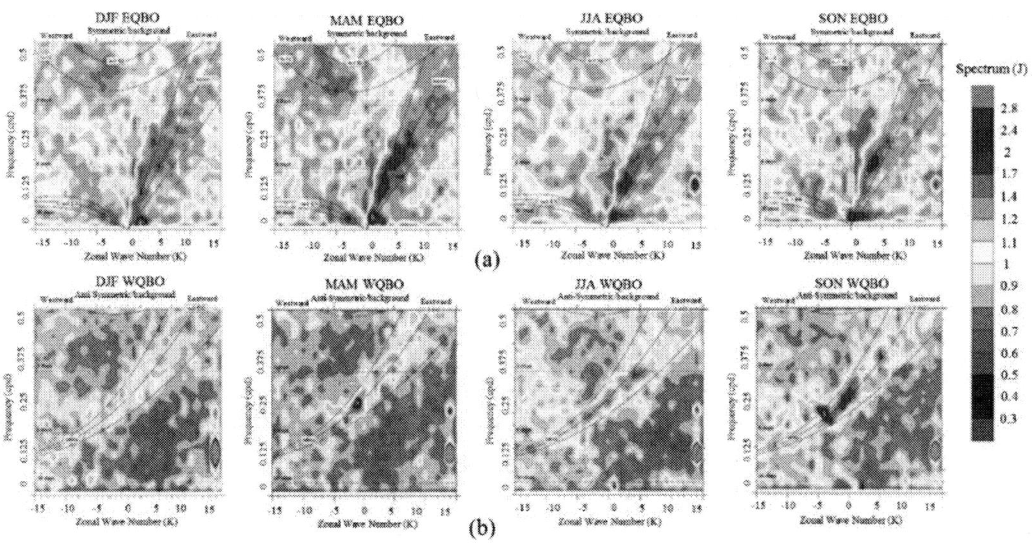

Figure 7. (a) Symmetric component of STSA OLR for EQBO and (b) asymmetric component of STSA OLR for WQBO in (left to the right) DJF, MAM, JJA, and SON.

Figure 7 shows seasonal variation of wave spectrum diagram obtained from OLR data. In general, the phase speed of Kelvin and MRG waves in troposphere is lower than in the stratosphere, because they are coupled with the convection [4]. General characteristics of the Kelvin wave in the troposphere have a frequency of approximately 0.0625 to 0.4375 cycles per day (a period of about 2 – 16 days) and zonal wavenumber approximately 1 - 15. On the other hand, MRG wave in the troposphere have a frequency of approximately 0.1875 to 0.3125 cycles per day (a period of about 3 - 5) and zonal wavenumber approximately -1 - (-5).

To understand if the seasonal variation of Kelvin wave and MRG wave activities in the troposphere is affected by different QBO phases, we calculated the differences (EQBO-WQBO) in wave spectrum diagram for both symmetric and asymmetric components (Fig. 8). We can see that there is no pronounced difference in the wave spectrum for both components. The results indicate that seasonal variation in Kelvin waves and MRG waves is not significantly influence by the QBO.

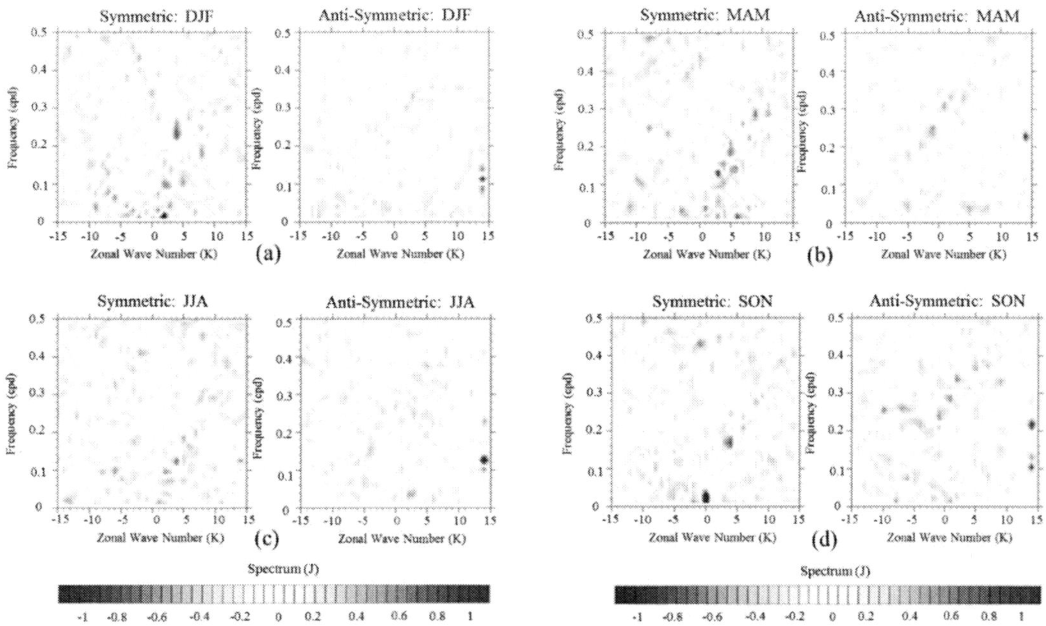

Figure 8. Differences in wave power spectrum in OLR between EQBO and WQBO for (a) DJF, (b) MAM, (c) JJA, and (d) SON. (a) Symmetric component of STSA OLR for EQBO and (b) asymmetric component of STSA OLR for WQBO in (left to the right) DJF, MAM, JJA, and SON.

4. Summary

We have studied the characteristics of Kelvin waves and Mixed Rossby-gravity waves in opposite QBO phases using the ERA-Interim reanalysis dataset from 1979-2013. The wave properties were analyzed by using space-time spectral analysis (STSA). The key results of the current study are summarized as follow:

(1) Kelvin waves are stronger during the easterly phase of the QBO with a frequency of 0.063 to 0.313 cycles/day (3-16 days) and the zonal wavenumber 1 to 6, whereas MRG waves are stronger during the westerly phase of the QBO with a frequency of 0.188 to 0.313 cycles/day (3-5 days) and the zonal wavenumber -1 to -7.

(2) Seasonal variations of Kelvin waves are significantly influenced by the QBO phases. Kelvin waves are stronger in JJA and SON during EQBO, while the MRG waves are stronger in JJA and SON during WQBO. This variation is significantly controlled by the variation in zonal mean flow in the lower stratosphere.

(3) Seasonal variations of Kelvin waves and MRG waves in the troposphere are not significantly affected by the QBO variations, which are to be likely due to a weaker variation in basic mean flow.

We also showed that Kelvin waves and MRG waves in the troposphere have a lower phase speed than in the stratosphere, which is consistent with equatorial waves in the troposphere being coupled with large-scale convective processes that dampen the phase velocity of the waves. Further studies are required to understand variations of Kelvin and MRG waves during other tropical atmospheric phenomena such as MJO, SAO etc.

Acknowledgment
We thank ECMWF for providing the ERA-Interim reanalysis dataset. The authors are grateful to Givo Alsepan from the laboratory of the Physical Oceanography and Climate, Hokkaido University, Japan, for his helpful comments and discussion.

5. References

[1] Holton JR and Lindzen RS 1968 A note on Kelvin Waves in the atmosphere *Monthly Weather Rev.* **96** 385-386
[2] Holton JR 2004 *An Introduction to Dynamics Meteorology* 4th Ed (Burlington: Elsevier)
[3] Setiawan S 2010 *Perumusan Struktur Horizontal Gelombang Atmosfer Ekuatorial* Departemen Geofisika Meteorologi IPB (Bogor: IPB)
[4] Lubis W S and Jacobi C 2015 The modulating influence of Convectively Coupled Equatorial Waves (CCEWs) on the variability of tropical precipitation. *Int. J. Climatol.*
[5] Baldwin MP 2001 The Quasi-biennial Oscillation *Geophys. Rev.* **39** 179 – 229
[6] Wheeler M and Kiladis GN 1999 Convectively coupled equatorial waves: Analysis of clouds and temperature in the wavenumber-frequency domain *J Atmos Sci.* **56** 374-39
[7] Pascoe C L 2005 The Quassi-biennial Oscillation: Analysis using ERA-40 data J. *Geophys. Res.* **110**
[8] Baldwin MP and Dunkerton TJ 1998 Biennial, Quasi-biennial, and Decadal Oscillations of potential vorticity in the Northern Stratosphere *J. Geophysics.* **103** 3919 – 3928
[9] Gray 2004 Solar and QBO influences on the timing of Stratospheric sudden warnings *J. Atmos. Sci.* **61** 2777 – 2796
[10] Yamashita 2011 Dynamical response in the Northern Hemisphere Midlatitude and High-Latitude Winter to the QBO simulated by CCSR/NIES CCM *J. Geophysics.* **116**
[11] Yang 2011 Equatorial waves in opposite QBO phases *J. Atmos. Sci.* **68** 839–862
[12] Plumb RA 1997 The interaction of two internal waves with the mean flow: Implications for the theory of the Quasi-Biennial Oscillation *J. Atmos. Sci.* **35** 1827-1839

Intraseasonal rainfall variability in North Sumatra and its relationship with Boreal Summer Intraseasonal Oscillation (BSISO)

Akhmad Faqih* **and Desiana Nurussyifa**
Department of Geophysics and Meteorology, Faculty of Mathematics and Natural Sciences, Bogor Agricultural University (IPB)
Kampus IPB Dramaga, FMIPA Building, Wing 19, Level 4, Jalan Raya Dramaga, Bogor, Jawa Barat, Indonesia 16680

E-mail: akhmadfa@apps.ipb.ac.id

Abstract. The Boreal Summer Intraseasonal Oscillation (BSISO) is a mode of climate variability on intraseasonal timescales that dominantly occurs during Northern Hemisphere summer monsoon. Its propagation and activity can be monitored by using indices namely as BSISO1 and BSISO2. The rainfall anomalies and extremes in North Sumatra may be potentially affected by BSISO due to its location within the path of their propagations. This study analyzes the relationship of BSISO with daily rainfall anomalies and extremes during different phases and types of BSISO. The results of spectral analysis show some agreements in the periodicity of BSISO1 and BSISO2 with daily rainfall anomaly. This study also identifies extreme rainfall during different phases of BSISO, showing the increase in the number of extreme rainfall events. High number of extreme rainfall events beyond thresholds at 90th and 95th percentiles are mostly found during phase 1, 2 and 3 of BSISO1 and phase 1 and 2 of BSISO2. Composite analysis conducted separately for spatial standard deviations of daily rainfall anomalies are also performed in this study. Similar results are found, where BSISO influences on rainfall variability especially during phase 1, 2 and 3 of BSISO1, and phase 1 and 2 of BSISO2.

1. Introduction

Recent studies indicated the identification of new intraseasonal phenomena namely as Boreal Summer Intraseasonal Oscillation (BSISO). BSISO is a mode of climate variability resulted from the sea-atmosphere interactions with the direction of movement from the Indian Ocean to the Western North Pacific Ocean [1]. The phenomenon has similarities with the Madden Julian Oscillation (MJO), a phenomenon around the equatorial climate characterized by the west to east propagation of convection processes from the Indian Ocean to the Pacific Ocean [2-4]. The main difference between BSISO and MJO is detected from the time of their occurrence; MJO dominates during the boreal winter while BSISO dominates during boreal summer [5]. Other differences were also found in their circulation period, MJO has a periods of 40-50 days [2], while BSISO consists of two types, i.e. BSISO1 that has 30-60 days oscillation periods and BSISO2 that has 10-20 days oscillation periods [1][6][7]. BSISO1 and BSISO2 also traverse differently according to their phases.

BSISO indices were constructed by using Principal Components (PC) of daily anomalies of outgoing longwave radiation (OLR) and zonal wind at 850 hPa (U850) in the region 10°S–40°N, 40°–160°E [1]. BSISO1 index uses the first two PCs, i.e. PC1 and PC2, while BSISO2 index utilizes the next two PCs, i.e. PC3 and PC4. Similar with the real time multivariate MJO (RMM) index [3], both BSISO1 and BSISO2 each consisting of eight phases, but with different area of propagations. BSISO1 has the first phase in the equatorial of Indian Ocean, phases 2-3 in the Indian Ocean and East Asia, phases 4-5 in India and maritime continent, phases 6-7 in the Bay of Benggala and the South China Sea, and phase 8 in the Western North Pacific Ocean. While BSISO2 has phase 1 and 2 in the Indian Ocean and the Philippine Sea, the phase 3 in India and South China Sea, phases 4-5 in the Bay of Benggala, phase 6 in Southeast Asia, phase 7 in Northeast Asia, and the phase 8 Western North Pacific Ocean [1].

BSISO phenomenon will affect rainfall variability in their propagation areas during their active phases. This includes the areas in the northern part of Indonesia, such as over North Sumatra province. North Sumatra is a region with the highest percentage of meteorological disaster on the island of Sumatra, accounting for 22% of flood events recorded from 1965 to 2007. The highest peak of rainfall and flood disasters is mostly found in December-January followed by the secondary peak in May [8]. The secondary peak coincides with the active period of BSISO during boreal summer season between May to October. Therefore, the rainfall during this period maybe related with the BSISO propagations. This study is conducted to identify the relation between rainfall variability and extreme rainfall during May-October periods in North Sumatra with the BSISO phenomenon.

2. Data and Methods

This study uses daily rainfall data over North Sumatra and BSISO indices (BSISO1 and BSISO2). The daily rainfall data is extracted from the Climate Hazards Group InfraRed Precipitation with Station data (CHIRPS) version 2.0 dataset [9][10] downloaded from the IRI Data Library (http://iridl.ldeo.columbia.edu/SOURCES/.UCSB/.CHIRPS/.v2p0/.daily/.global/.0p05/.prcp/). This is in addition to the daily rainfall data obtained from the record of several weather stations collected by Badan Meteorologi, Klimatologi, dan Geofisika (BMKG). BSISO indices are also downloaded from the IRI data library (http://iridl.ldeo.columbia.edu/home/mbell/.APCC/.BSISO/).

Several methods, such as spectral analysis, probability analysis and composite analysis, are used in this study. Spectral analysis is conducted to investigate and compare dominant oscillation of BSISO indices and rainfall. Probability analysis is performed to determine the threshold of extreme rainfall and investigate the probability of exceedance beyond the identified thresholds. The period of dataset used for this analysis is from May-October during 2008 to 2010 periods. Composite analysis is used to identify the rainfall characteristics in the region during different BSISO phases. Rainfall data used in the probability analysis and composite analysis is selected only for the periods between May to October during 1991-2010 periods. The criteria of the BSISO indices selected for the analysis is when the value more than 1 in order to focus only during active BSISO phenomenon.

3. Results and Discussions

3.1 Oscillation Periods of BSISO and daily rainfall in North Sumatra

Spectral analysis can be used to identify dominant oscillations of a time series data. We compare Power Spectral Density (PSD) graphs resulted from both BSISO indices (BSISO1 and BSISO2) and rainfall anomalies in the studied region. From this approach we can identify the influence of BSISO on rainfall variability by identifying whether the dominant spectrum found in PSD graph from the rainfall similar to the results found in both BSISO indices. The result of the analysis is shown in figure 1.

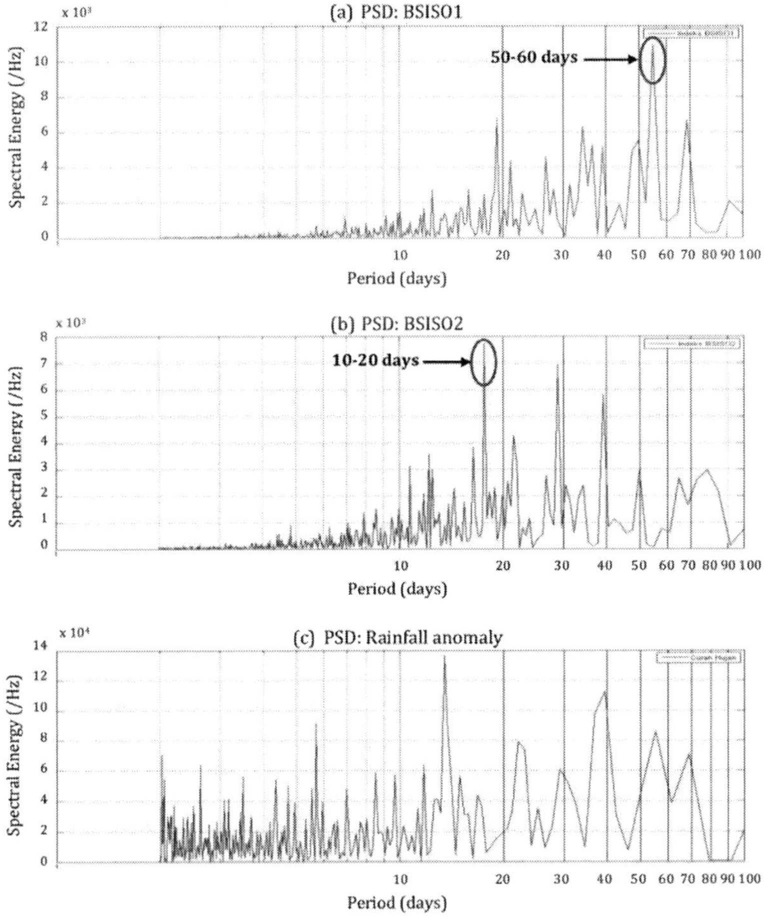

Figure 1. PSD of (a) BSISO1 index, (b) BSISO2 index, and (c) Rainfall anomalies in North Sumatra during May- October in 2008-2010 periods.

Figure 1a shows the PSD result calculated from BSISO1 index, showing several dominant power spectrals within the range of 10 to 70 days cycle with mostly dominant cycle found on 50-60 days oscillation. In figure 1b, there are three dominant power spectral from PSD of BSISO2 index, i.e. within the range of 10-20 days, 20-30 and 30-40 days oscillations. However, the strongest spectral is found in 10-20 days cycle. Both strong spectrals found in BSISO1 with 50-60 days oscillation and in BSISO2 with 10-20 days oscillation are the main distinguished characteristics as described by Lee et al. [1]. The results of PSD graphs in figure 1a and 1b are compared with PSD in figure 1c obtained from the rainfall data in North Sumatra. It is found that the rainfall have several dominant power spectrals that similarly occur within the time cycles where the BSISO are also dominant, especially within 50-60 days and 10-20 days oscilations, although the dominant cycle for 10-20 days between BSISO2 and rainfall did not perfectly match. Similar oscillations found in rainfall data indicate possible strong influence of BSISO phenomenon on intraseasonal rainfall variability in North Sumatra. Nevertheless, BSISO seems not the only phenomenon affecting the intraseasonal rainfall variability in the region since the PSD graph (figure 1c) also show strong spectral in 30-40 days oscillation. This strong 30-40 days cycle maybe related to the MJO phenomenon, which is more dominant in affecting the rainfall variability instead of the cycle within 50-60 days oscillation that is more representing BSISO.

3.2 Rainfall Extremes during active Phases of BSISO

Previous study used BSISO value more than 1.5 as threshold to define active strong BSISO indices [1]. However, as mentioned earlier, this study selects the BSISO index with value more than 1 in order to represent only active BSISO conditions. By referring to the BSISO occurrence with the value beyond this threshold, this study select and classify the rainfall data that occur at the same time with each BSISO phases. This approach is conducted separately for BSISO1 and BSISO2 indices and the results are shown in figure 2. From this figure, we can identify the phase(s) where the occurrence of extreme rainfall are dominant. The extreme rainfall is defined by the value beyond certain thresholds, i.e. at 90th, 95th and 99th percentiles of daily rainfall data based on Gamma distribution. Respectively, the identified extreme rainfall thresholds for the region are 16.91 mm/day for the 90th percentile, 21.62 mm/day for the 95th percentile, and 32.44 mm/day for the 99th percentile.

From the results show in figure 2a and 2b, we calculate the number of rainfall events occured beyond those three thresholds (Table 1). High number of extreme rainfall events beyond thresholds are mostly found during phase 1, 2 and 3 of BSISO1. This is possible since during those three phases the BSISO1 propagations are mainly located in the equatorial of Indian Ocean for phase 1 and in the Indian Ocean and East Asia in phase 2 and 3, which is relatively near to North Sumatra and could affect the increase of frequency and intensity of extreme rainfall in the region. Between those three phases of BSISO1, the highest number of rainfall event for the selected 20 year periods (1991-2010) is found when BSISO1 mainly located in the Indian Ocean during phase 2, with 59 rainfall events exceeding 90th percentile threshold and 17 rainfall events exceeding 95th percentile threshold. In addition, the extreme rainfall events are also found during other phases of BSISO1, but the number of events are not as many as in phase 1, 2 and 3.

During active BSISO2, the number of extreme rainfall events in North Sumatra are mostly high during phase 1 and 2 (Table 1), where the phenomenon is located in the Indian Ocean and the Philippine Sea, respectively. Similar with the result found in BSISO1, the highest number of extreme rainfall during the 20 year periods is also found during phase 2 of BSISO2 with the number of events is 54 for the rainfall exceeding 90th percentile and 16 for the rainfall exceeding 95th percentile. In addition, a number of extreme rainfall events is also found during phase 3 where the event is mainly located in India and South China Sea. The result also found with a relatively high number of event in phase 8, with 31 events of rainfall exceeding 90th percentile and 10 events for rainfall exceeding 95th percentile, but this is maybe considered irrelevant for directly affecting the extreme rainfall in the region since the main location during this phase is mostly in Western North Pacific Ocean. It is important to consider that other climate phenomena, such as MJO could also contribute to the increase in the frequency and intensity of extreme rainfall in North Sumatra.

3.3 BSISO-related Rainfall Variability in North Sumatra

Further analysis is performed in this study by selecting the rainfall data during the active phase of BSISO1 and BSISO2, separately. Instead of calculating the average values from the selected and classified rainfall data, we calculate the standar deviation of those rainfall data within each phase. The standard deviation is a statistical parameter used to determine the distance distribution of the data against the average value. By using this metric, we aim to investigate and compare the deviation of rainfall during different phases of BSISO. From this approach, it is expected that the spatial characteristics of the intraseasonal rainfall variability within North Sumatra influenced by BSISO can be identified.

Figure 3a shows the composite map of standard deviations of rainfall during different active phases of BSISO1. From the figure, it is clearly shown that the areas within North Sumatra dominantly have strong standard deviations during phase 1, 2 and 3. This is consistent with the results of previous analysis related to the number of extreme rainfall occurred during active phase of BSISO1. In figure 3a, the standard deviations are found to be weaker or smaller during other phases of BSISO1 (phase 4-8). Surprisingly, the area within Central Tapanuly Regency adjacent to Sibolga City is identified to have very strong standard deviations compared to the other regions in North Sumatra. This could indicates that BSISO1 event could strongly affect the intraseasonal rainfall variability in this specific region and it may be associated to the extreme rainfall conditions associated with climate-related disasters such as floods and landslides ever recorded in the region. For example, a flood event was recorded in Sibolga City on Wednesday, August 19, 2009 [11]. The flood consisting mud occurred due to heavy rainfall causing landslide and damaging hundreds of houses. Based on our analysis on the BSISO index during the day of the flood event, the BSISO1 index is located on phase 2 with the value of the index is 2.87. This value is far beyond 1 as the threshold of active BSISO or 1.5 as the threshold for strong BSISO index. In other word, the BSISO1 index during the flood event is very strong.

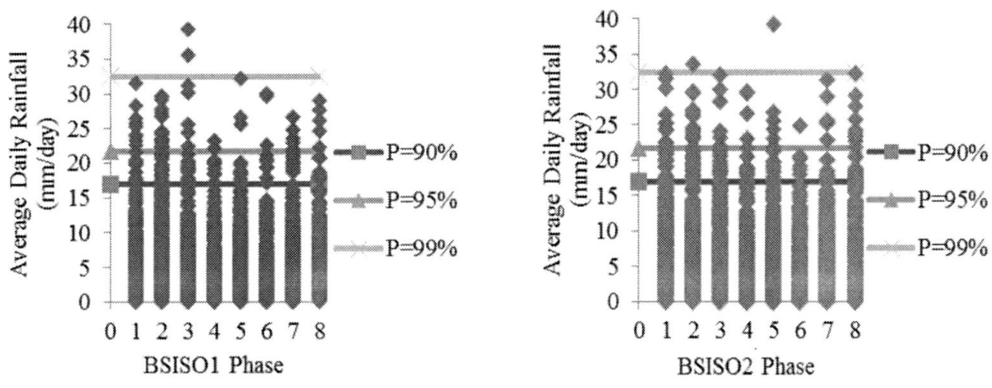

(a) (b)

Figure 2. Daily rainfall in North Sumatra during active phase of a) BSISO1 and b) BSISO2. The thresholds used are of 90th percentile (pink line), 95th percentile (green line) and 99th percentile (orange line).

Table 1. The number of extreme rainfall events that exceed the defined thresholds

Threshold beyond	BSISO1 Phase							
	1	2	3	4	5	6	7	8
90th Percentile	33	59	34	13	14	13	25	17
95th Percentile	11	17	9	2	3	3	8	5
99th Percentile	0	0	2	0	0	0	0	0
Threshold beyond	BSISO2 Phase							
	1	2	3	4	5	6	7	8
90th Percentile	40	54	28	19	25	7	16	31
95th Percentile	13	16	6	5	7	1	5	10
99th Percentile	0	1	0	0	1	0	0	0

The composite maps of standard deviation of rainfall during active phase of BSISO2 are shown in figure 3b. Similar to the previous results, the BSISO2 is identified to have strong influence on rainfall variability in North Sumatra especially during phase 1 and 2. This is shown by the composite maps of standard deviations that show strong value during those phases 1 and 2 of BSISO2.

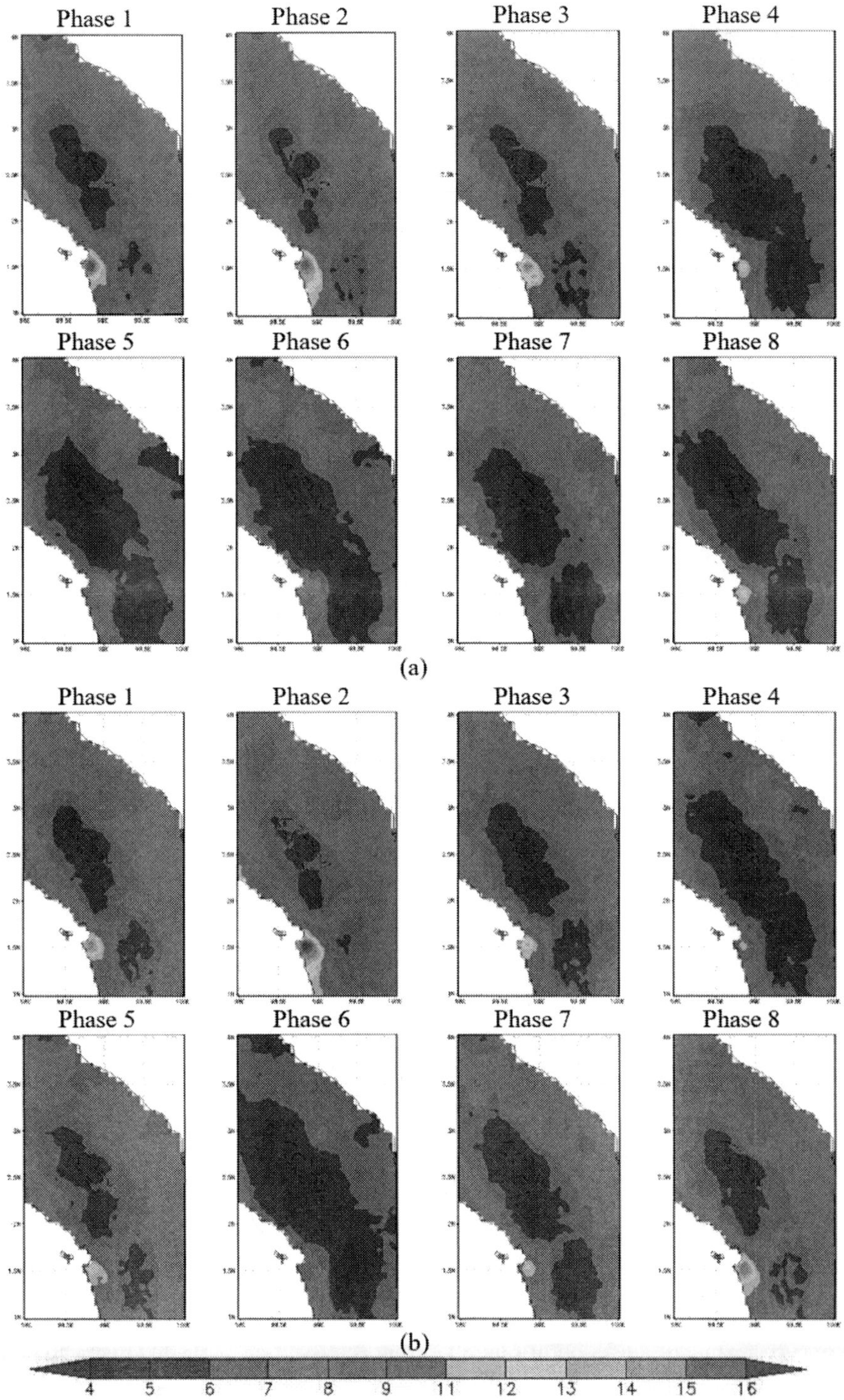

Figure 3. Composite maps of standar deviations of rainfall anomalies in North Sumatra on each phase of a) BSISO1 and b) BSISO2.

In general, from figure 3a and 3b, the influence of active phases of BSISO (BSISO1 and BSISO2) on rainfall variability is not similar across different region in North Sumatra. The strongest sign of BSISO influence is mostly found on the area adjacent to Sibolga City and then followed by the areas near the coastal areas surrounding the central areas within North Sumatra. The central areas mostly show weaker standard deviations, indicating less influence of BSISO.

4. Conclusions

This study identifies that the intraseasonal rainfall variability and extreme rainfall events in North Sumatra during May-October are strongly affected by BSISO phenomenon. These are shown by the results of spectral analysis comparing PSD graphs between BSISO indices and rainfall anomalies during 2008-2010 periods, the analysis of extreme rainfall during active phases of BSISO1 and BSISO2, and the composite maps of standard deviations of rainfall during active phases of BSISO 1 and BSISO2 taken from the 20 year datasets during 1991-2010 periods. The study found that the extreme rainfall could increased during both BSISO1 and BSISO2. High number of extreme rainfall events beyond thresholds at 90th and 95th percentiles are mostly found during phase 1, 2 and 3 of BSISO1 and phase 1 and 2 of BSISO2. Similar results are found from composite analysis of spatial standard deviations of daily rainfall anomalies, where BSISO influences rainfall variability in North Sumatra especially during the same phase 1, 2 and 3 of BSISO1, and phase 1 and 2 of BSISO2.

References

[1] Lee J Y, Wang B, Wheeler M C, Fu X, Waliser D E and Kang I S 2013 Real-time multivariate indices for the boreal summer intraseasonal oscillation over the Asian summer monsoon region *Clim Dyn* **40** 493–509
[2] Madden R A and Julian P R 1972 Description of global-scale circulation cells in the tropics with a 40-50 day period *J. Atmos. Sci.* **29** 1109-23
[3] Wheeler M C and Hendon H H 2004 An All-Season Real-Time Multivariate MJO Index: Development of an Index for Monitoring and Prediction *Monthly Weather Review* **132** 1917-32
[4] Donald A, Meinke H, Power B, Maia A d H N, Wheeler M C, White N, Stone R C and Ribbe J 2006 Near-global impact of the Madden-Julian Oscillation on rainfall *Geophysical Research Letters* **33**
[5] Kikuchi K, Wang B and Kajikawa Y 2012 Bimodal representation of the tropical intraseasonal oscillation *Climate Dynamics* **38** 1989-2000
[6] Annamalai H and Sperber K R 2005 Regional Heat Sources and the Active and Break Phases of Boreal Summer Intraseasonal (30–50 Day) Variability *Journal of the Atmospheric Sciences* **62** 2726-48
[7] Kikuchi K and Wang B 2010 Formation of Tropical Cyclones in the Northern Indian Ocean Associated with Two Types of Tropical Intraseasonal Oscillation Modes *Journal of the Meteorological Society of Japan. Ser. II* **88** 475-96
[8] Handayani A S 2010 Analisis daerah endemik bencana akibat cuaca ekstrim di Sumatra Utara *Jurnal Meteorologi dan Geofisika* **11** 54 - 8
[9] Funk C C, Peterson P J, Landsfeld M F, Pedreros D H, Verdin J P, Rowland J D, Romero B E, Husak G J, Michaelsen J C and Verdin A P 2014 A quasi-global precipitation time series for drought monitoring *U.S. Geological Survey Data Series* **832, 4 p**
[10] Funk C, Peterson P, Landsfeld M, Pedreros D, Verdin J, Shukla S, Husak G, Rowland J, Harrison L, Hoell A and Michaelsen J 2015 The climate hazards infrared precipitation with stations-a new environmental record for monitoring extremes *Scientific Data* **2**
[11] FPBI 2009 Banjir Sibolga merendam ratusan rumah. (http://fpbi-siaga.blogspot.co.id/2009/08/banjir-sibolga-merendam-ratusan-rumah.html

Impact of Madden-Julian Oscillation (MJO) on global distribution of total water vapor and column ozone

Irvan Fathurochman[1]*, Sandro W Lubis[2] and Sonni Setiawan[1]
[1] Department of Geophysics and Meteorology, Bogor Agricultural University (IPB), Indonesia
[2] Physics of the Atmoshpere, GEOMAR Helmholtz Centre for Ocean Research Kiel, Germany

E-mail: irvan.fathur@gmail.com

Abstract. The Madden-Julian Oscillation (MJO) is the leading mode of intra-seasonal variability in the tropical troposphere, characterized by an eastward moving 'pulse' of cloud and rainfall near the equator. In this study, total precipitable water (TPW) and total column ozone (TCO) datasets from ECMWF ERA-Interim reanalysis were used to analyse the impact of the MJO on the distribution of water vapor and column ozone in the tropics from 1979 to 2013. The results show that seasonal variations of TPW modulated by the MJO are maximized in the tropics of about 10°S-10°N during boreal winter, while the variation in TCO is maximized in the mid-latitudes of about 30°S - 40°N in the same season. The composite analysis shows that MJO modulates TPW and TCO anomalies eastward across the globe. The underlying mechanism of the MJO's impact on TPW is mainly associated with variation of tropical convection modulated by the MJO, while the underlying mechanism of the MJO's impact on TCO is mainly associated with an intra-seasonal variability of tropopause height modulated by the MJO activity. This knowledge helps to improve the prediction skill of the intra-seasonal variation of water vapor and column ozone in the tropics during boreal winter.

1. Introduction

Tropical atmospheric circulations are controlled by the large-scale convective system varying in space and time, which includes the equatorial planetary waves [1][2][3], intra-seasonal oscillation (MJO) [4], inter-annual oscillation ENSO [5], and decadal to multidecadal oscillation. On sub-monthly to intra-seasonal time scales, interaction between tropical waves and convection may result in unique spatial distributions and frequency characteristics of tropical precipitation [2].

The Madden-Julian Oscillation (MJO) is the leading mode of tropical climate variability on intra-seasonal timescale. MJO is characterized by an eastward moving 'pulse' of large-scale convection system elongated from the Indian Ocean to the Pacific Ocean [6]. This oscillation has an average phase speed of about 5 m s^{-1} and a period of about 30-90 days [7]. Previous research showed that the MJO can be identified by parameters of convection and other variables in the atmosphere, such as outgoing longwave radiation (OLR), precipitation, sea surface temperature, variations in wind, water vapor, and ozone [8][9][10].

This study examined the impact of the MJO on the distribution of total precipitable water (TPW) and total column ozone (TCO) in tropics, comprehensively. TPW is defined as the amount of water vapor contained in a vertical column of the atmosphere as measured from the surface to the top of the atmosphere [11]. Zhang [7] further explained that the intra-seasonal variation of water vapor in the troposphere modulated by the MJO could influence the impact of MJO on rainfall variability. In addition, TCO is defined as the total amount of ozone contained in a vertical column in the atmosphere from the surface to the top of the atmosphere [12]. More than 90% of the ozone in the atmosphere is in the stratosphere and the remaining of about 10% is in the troposphere. The ozone concentrations in the stratosphere are varying in latitude and height. Ozone is produced predominantly in tropical latitudes through a photochemical process [13].

To date, studies related to the influence of MJO on the distribution of water vapor and column ozone in the tropics has been very limited. Therefore, this study aims to investigate a possible impact of MJO on intra-seasonal variation of TPW and TCO in the tropics, and the associated underlying mechanisms.

2. Methods

The datasets used in this study are total precipitable water (TPW), total column ozone (TCO), zonal wind and meridional wind at 850 mb, and geopotential height at 150 mb obtained from ECMWF Reanalysis (ERA-Interim) [14]. Outgoing longwave radiation (OLR) dataset from NOAA [15] was used as the main proxy of convection. MJO events was identified using Real Time Multivariate (RMM) indices derived from the BOM Australia [16]. The data is the daily data with a 35-year period (January 1979 to December 2013) at 30°N - 30°S and at 40°N - 40°S (for TCO and geopotential height) in a band of tropics with spatial resolution of 2.5° x 2.5°.

2.1. Bandpass Filter

Intra-seasonal signal of MJO was isolated from daily data by applying a Butterworth bandpass-filter technique with period of (20-90) days. Filtering procedure is performed on a daily anomaly data over the period of 1979-2013.The anomalies were detrended by removing a linear trend in the data.

2.2. Seasonal Variance Analysis

Seasonal variance analysis is used to examine the spatial distribution intra-seasonal MJO oscillation. This analysis was performed by calculating the variance of each season of filtered anomaly data (20-90 days). The filtered anomalies were grouped into different season i.e., summer (NDJFMA) and winter (MJJASO).

2.3. Space-Time Spectral Analysis

Space-time spectral analysis (STSA) is used to analyze the wavy-like disturbances in the tropics that either propagates eastward or westward. This method decomposes the anomalous field in space and time as a function of wavenumber (k) and frequency (ω) [17][18]. The decomposition is conducted based on 2D Fourier transformation (in space and time domains).

2.4. Composite Analysis

Composite analysis is used to see the distribution of the data (TPW, OLR, and TCO) filtered anomaly (20-90 days) in each MJO events (8 phases). This composite technique is an ensemble average, the arithmetic average value of a set of data but in the same conditions.

3. Results

3.1. Intraseasonal Variance Related to MJO

Figure1 shows seasonal variance of MJO in (a,d) OLR, (b,e) TPW, and (c,f) TCO for both boreal summer and winter. Consistent with previous studies, MJO signals are generally stronger in the equator, from the Indian Ocean to the Western Pacific Ocean [8]. In particular, during the boreal summer (figure 1a), the MJO signal in OLR is strongly observed from the northern Indian Ocean to the western Pacific Ocean, while during boreal winter (figure 1b) the higher variance of OLR is dominant in the Southern Hemisphere and covering wide area, from the Indian Ocean (50°E) to the

IOP Conf. Series: Earth and Environmental Science **54** (2017) 012034 doi:10.1088/1755-1315/54/1/012034

eastern Pacific Ocean (about 120°E). The differences in MJO signal between boreal winter and summer is mainly due to the seasonal movement of the sun and the associated convective systems. Nevertheless, the stronger MJO activity is observed during boreal winter, consistent with previous studies [6][7].

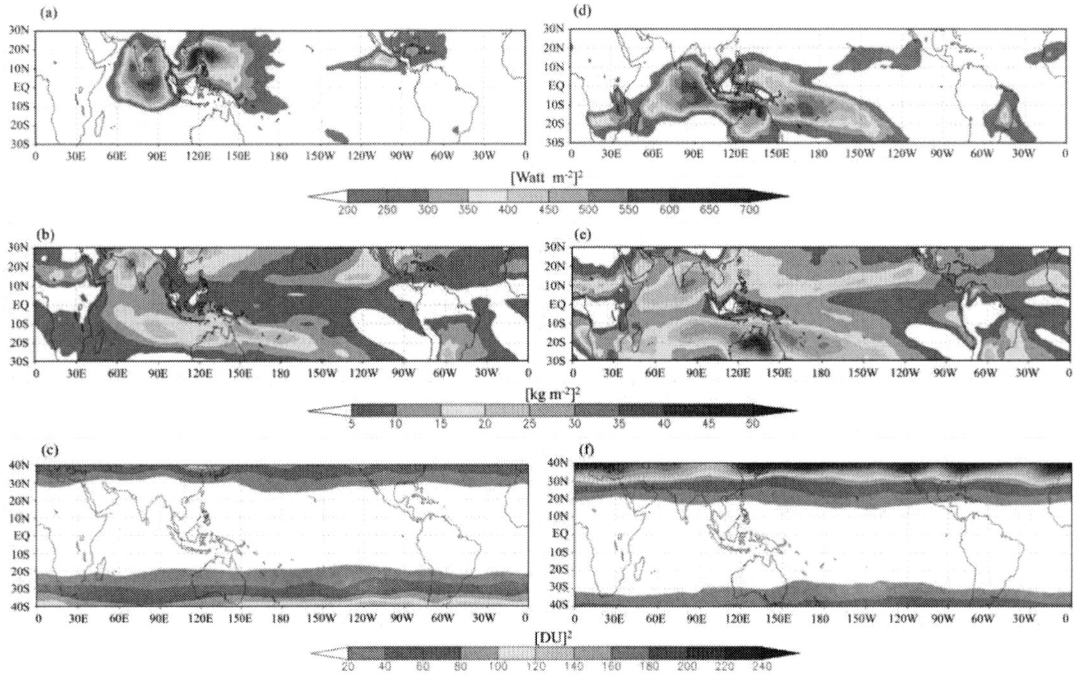

Figure 1. Intraseasonal variance associated with MJO (filtered (20-90 days)) for OLR, TPW, and TCO during (a-c) boreal summer and (d-f) boreal winter.

3.1.1. Intraseasonal Variance of TPW. The intra-seasonal variance of TPW exhibits different patterns compared to OLR variance. Figure 1(b,e) can be seen as the regions where the TPW are strongly affected by MJO activity. This result indicates that the effect of intraseasonal oscillation on TPW is stronger over the off equatorial region. According to Schreck *et al* [9], the intra-seasonal of water vapor is more sensitive to air mass subsidence of latitudes around the equator, but the higher variance can be seen over the Indian Ocean to the Western Pacific. During summer (figure 1b), a higher variance of TPW exist near the Northern Indian Ocean (10°N) and covering wide area in 15°S. During winter (figure 1e), the intraseasonal variance of TPW becomes stronger and maximizes between 10°N and 15°S over 50°E-120°W.

3.1.2. Intraseasonal Variance of TCO. Figure 1(c,f) shows the intraseasonal variance of TCO anomaly. In general, higher variance of TCO exist in mid-latitudes about 30°-40° both the Northern and the Southern Hemisphere. This is associated with the mean condition of TCO that is higher in mid-latitudes than in the tropics [10]. The seasonal distribution of ozone tends to be increased during the winter, and on the contrary, it is relatively decreased during summer in the mid-latitudes [13]. During summer (figure 1c), higher variance of TCO exists in the subtropical Southern Hemisphere. On the other hand, during winter (figure 1f) the higher variance of TCO maximizes in the subtropical Northern Hemisphere. In general, it suggests that the effect of intraseasonal oscillations on TCO is stronger in the mid-latitudes and during the winter, while the effect over the tropics is relatively weaker.

3.2. Wavenumber-Frequency Spectraof Seasonal MJO
Figure 2 shows the power spectrum of OLR, TPW and TCO anomalies for both winter and summer. The positive (negative) frequency indicates an eastward (westward) propagation. MJO signals in OLR anomalies (figure 2a,d) are dominant on intraseasonal period (frequency). In particular, the power spectra pattern of OLR anomalies tends to be stronger in boreal winter (figure 2d) than boreal summer (figure 2a). This pattern consistent with previous studies showing that the MJO signal on OLR anomalies was detected in intraseasonal period and with a zonal wavenumber 1-3 [8][18].

3.2.1. Wavenumber-Frequency Spectra of TPW. MJO signal on TPW anomalies shows a similar pattern with OLR anomalies (figure 2b,e). As in OLR, the power spectra of TPW anomalies associated with MJO seems to be stronger in boreal winter than boreal summer. The power spectra of TPW anomalies are stronger on intraseasonal scale (20-90 days), with zonal wavenumber (*k*) about +1 to +4, indicating an eastward propagation [9]. In boreal summer (figure 2b), power spectra of TPW anomalies peak on a frequency of 0.025-0.0125 day^{-1} or about 40-80 days. On the other hand, during winter, the power spectra are stronger and maximize in a frequency of about 0.025 – 0.011days^{-1} or 40-90 days.

3.2.2. Wavenumber-Frequency Spectra of TCO. Figure 2 (c,f) shows the space-time spectra analysis result for TCO anomalies. In general, MJO signal in TCO is relatively weaker on intraseasonal timescale (20-90 days). The prominent power spectra of TCO are detected in the period of about larger than 30 days. This indicates that the TCO anomalies in the tropical regions are relatively unaffected or weakly affected by intraseasonal phenomenon such as the MJO. Nevertheless, consistent with the finding of Tian *et al* [10], MJO activity has a higher relation with the variation of TCO in the subtropics.

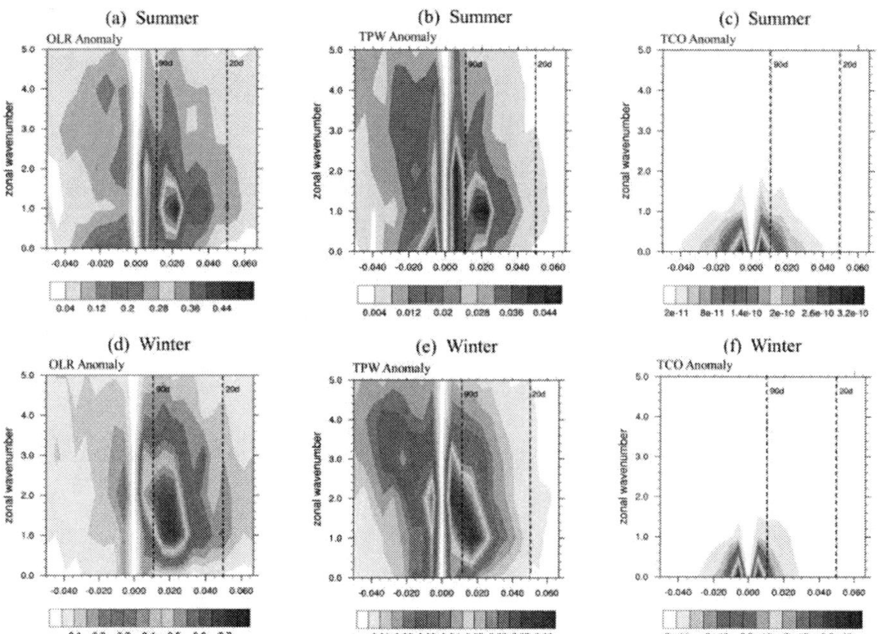

Figure 2. Space-time spectral analysis of OLR, TPW, and TCO during (a-c) boreal summer and (d-f) boreal winter.

3.3 Impact of MJO on TPW Distribution
The impact of MJO on TPW anomalies is shown in figure 3 for (a) boreal summer and (b) boreal winter. The positive (negative) TPW anomalies are shown in blue (red) shading, while the 850 mb

wind anomalies is indicated by arrows. It is shown MJO lead to eastward propagation of TPW anomalies in the tropics. The impact of the MJO on TPW reaches maximum during phases 4 and 5, when the convection is stronger over the maritime continent due to the strengthening of westerlies and easterlies wind from the Indian Ocean and the Western Pacific ocean. The stronger MJO convection causes the stronger TPW distribution, and vice versa. The maximum impact of MJO on TWP is stronger during boreal winter, consistent with previous STSA analysis (figure 2e).

It is well acknowledged that MJO modulates the tropical convection, as shown in OLR anomalies (figure 4). Comparing figure 3 and figure 4, it can be seen that the TPW-MJO pattern is anti-correlated with OLR-MJO pattern. In particular, positive TPW anomalies coincide with area of enhanced MJO convection, while negative anomalies coincide with suppressed MJO convection (figure 5). This indicates that MJO convection significantly affects the distribution of TPW anomalies (figure 5). Based on this composite results, we conclude that the underlying mechanism of MJO's impact on TPW is mainly associated with tropical convection modulated by the MJO.

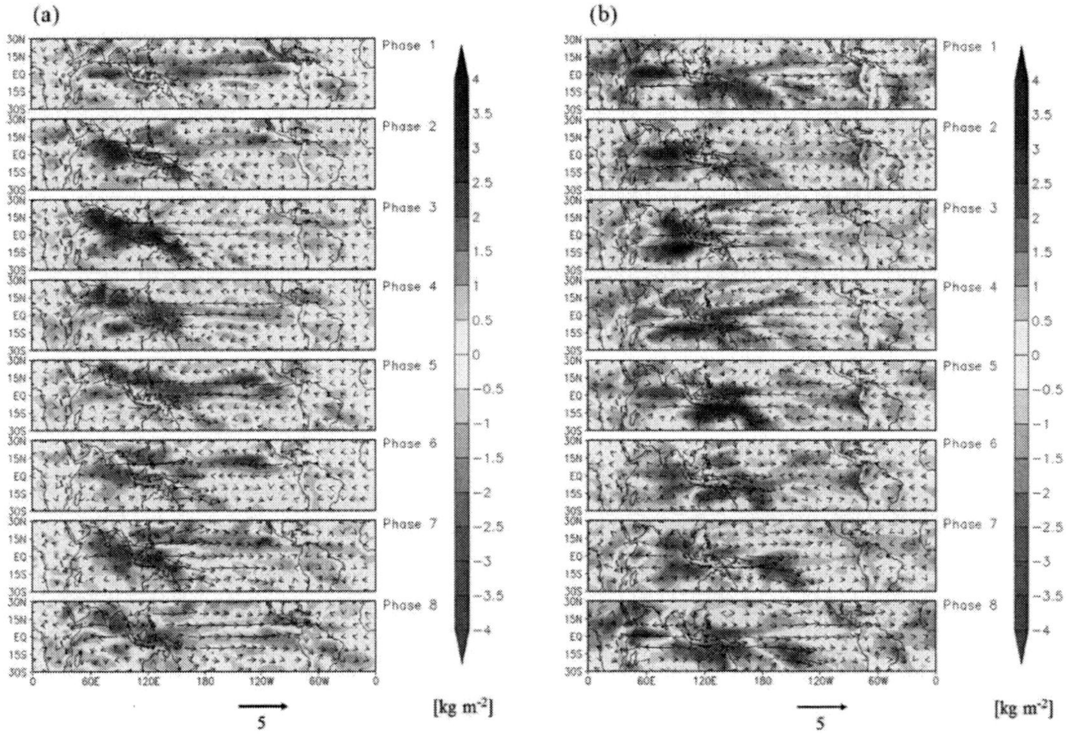

Figure 3. Composite of filtered TPW anomalies (shading) and 850 mb wind anomalies (vector) in (a) Boreal summer (b) Boreal winter.

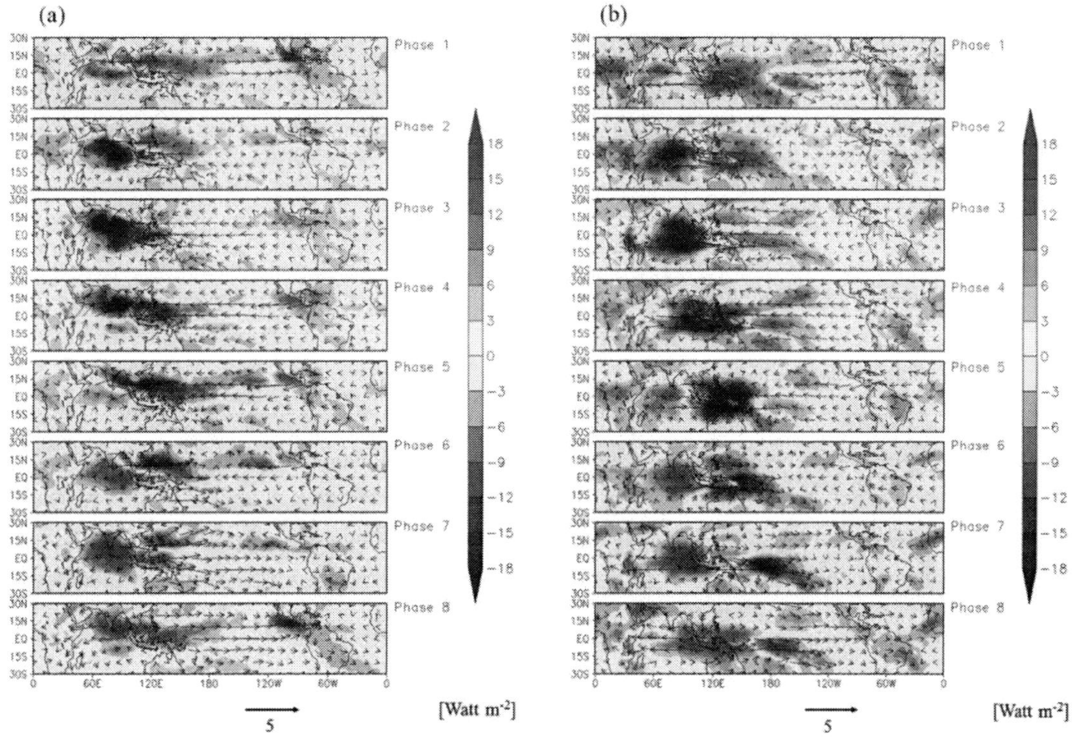

Figure 4. As in Fig. 3, but shading indicates filtered OLR anomalies.

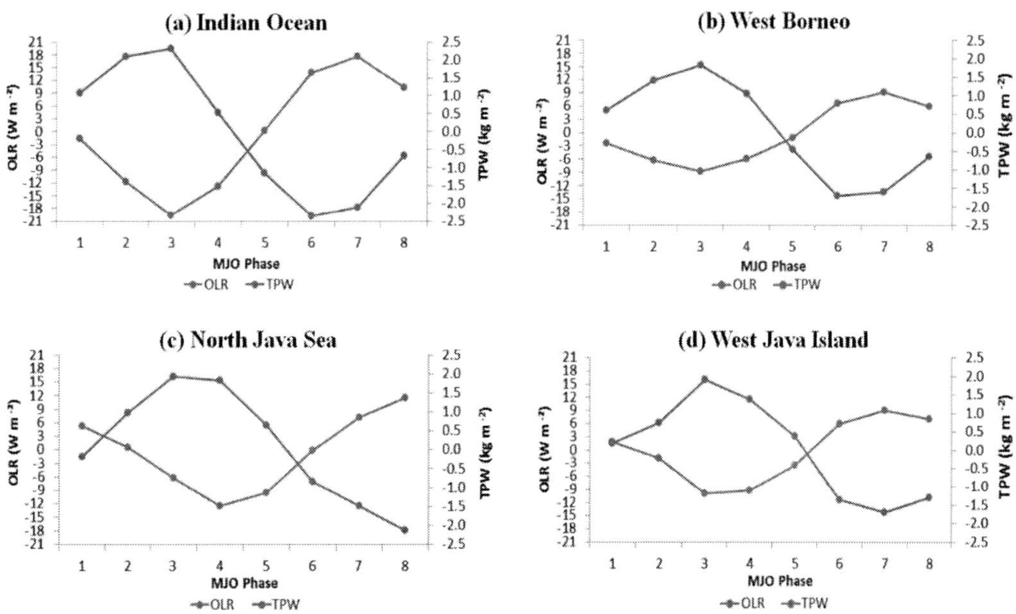

Figure 5. Composite of OLR and TPW anomalies for areal average of (a) Indian Ocean (2°S-1°S and 94°E-97°E), (b) West Borneo (1°S-1°N and 109.5°E-112°E), (c) Java Sea (6°LS-4.5°S and 113°E-118°E), (d) North Java Island (7.5°S-6.5°S and 107°E-108.5 E).

3.3. Impact of MJO on TCO Distribution

Variation of TCO in the atmosphere is controlled by processes that happen in the lower-stratosphere [19]. In this study, the geopotential height anomalies at 150 mb were used to examine the relationship between the fluctuations in the tropopause height and transport of ozone modulated by the MJO activity. Figure 6 shows the composite of TCO and geopotential height anomalies following the MJO phases. It can be seen that MJO lead to eastward movement of TCO and geopotential height in the tropics. The MJO impact on TCO is stronger and maximize during boreal winter (figure 6b) consistent with previous STSA results. In addition, there is a significant anti-correlation relationship between TCO and geopotential height anomalies during MJO phases (figure 7). In particular, positive (negative) TCO anomalies coincide with negative (positive) geopotential height anomalies. This suggests that fluctuations of tropopause height may potentially affect the distribution of ozone in the lower-stratosphere. Our results are in agreement with previous satellite measurement analysis from Tian *et al* [10], showing a strong relationship between the MJO, TCO and tropopause heights.

A possible mechanism to explain the impact of MJO impact on TCO is as follow: (1) reduction in height of the tropopause layer (characterized by negative anomaly of geopotential height) tends to increase the TCO by replacing the air masses of less ozone in the upper-troposphere with the air of rich ozone in the stratosphere. (2) Conversely, an increasing in height of the tropopause layer (characterized by positive anomalies of geopotential altitude) resulting in decreased of ozone transport from stratosphere to troposphere. This effect is really visible in off equatorial region, between 20°N-40°N. Moreover, the impact of MJO on TCO in tropical regions is relatively small with the magnitude is only about ± 1 DU, consistent with previous study of Tian *et al* [10].

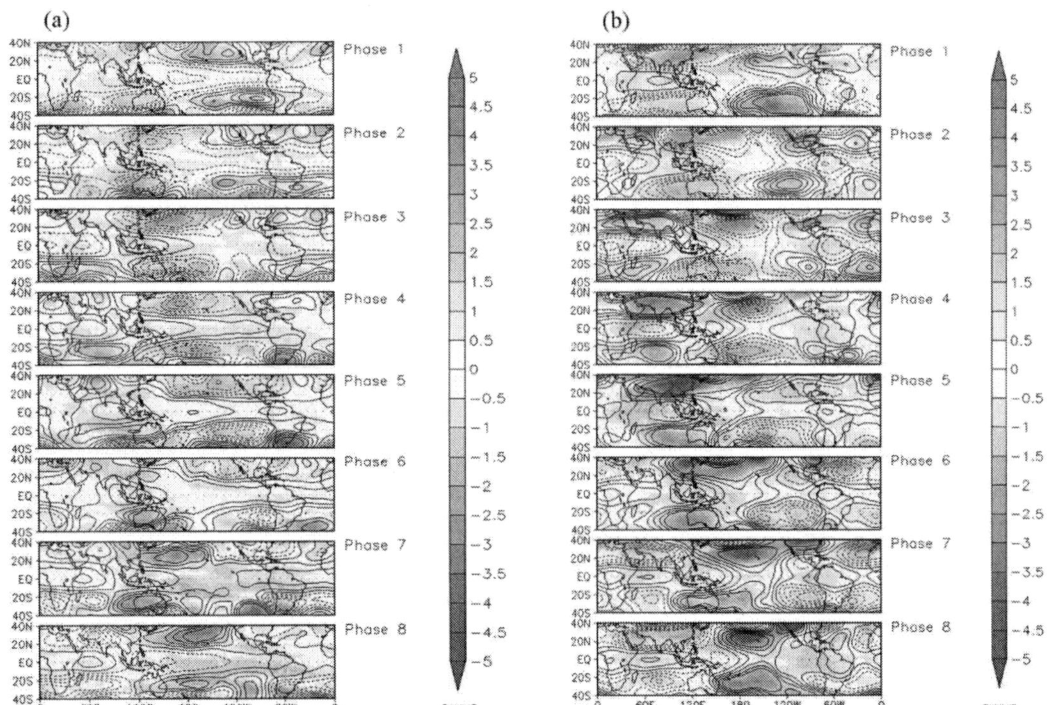

Figure 6. Composite of filtered TCO (shading) and 150 mb geopotential height (contour: positive (solid line), negative (dash line)) in (a) Boreal summer (b) Boreal winter.

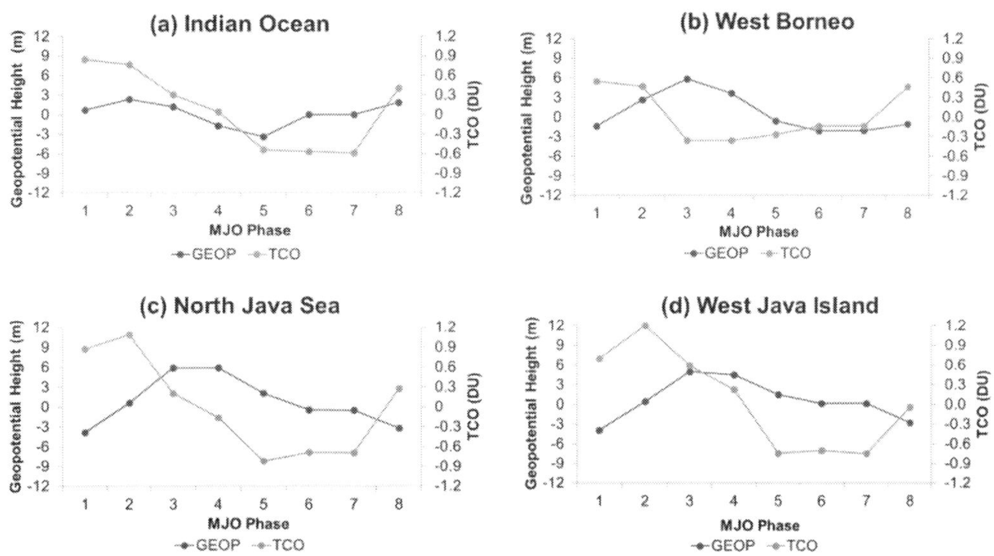

Figure 7. Composite of TCO and 150 mb geopotential height anomalies for areal average of (a) Indian Ocean (2°S-1°S and 94°E-97°E), (b) West Borneo (1°S-1°N and 109.5°E-112°E), (c) Java Sea (6°LS-4.5°S and 113°E-118°E), (d) North Java Island (7.5°S-6.5°S and 107°E).

4. Conclusion

We have examined the impact of MJO on TPW and TCO in the tropics using the ERA-Interim reanalysis dataset from 1979-2013. The MJO signal was isolated by using space-time spectral analysis (STSA) and composite analysis. The key results are summarized as follow:

(1) MJO influences zonal distribution of TPW and TCO anomalies.

(2) The seasonal variations of TPW modulated by the MJO are maximized in the tropics of about 10°-20° in latitude during boreal winter, while the variation in TCO is maximized in the mid-latitudes of about 30°-40° at the same season.

(3) The underlying mechanism of the MJO's impact on TPW is associated with tropical convection modulated by the MJO, as indicated by anti-correlation relationship between TPW and OLR.

(4) The underlying mechanism of the MJO's impact on TCO is mainly associated with an intra-seasonal variability of tropopause height modulated by the MJO activity. The positive (negative) anomalies of geopotential height modulated by the MJO are associated with low (high) ozone transport from the stratosphere to the troposphere. In addition, the MJO impact on TCO is more noticeable in subtropical (consistent with the results of STSA).

The results of this research generally indicate that the MJO may affect intra-seasonal variability of TPW and TCO in the tropical region, both in global and regional scales. The results of this study are expected to improve our understanding on the impact of MJO on water vapor and column ozone in both the global and regional scales. This study will hopefully helps to improve the prediction skill of the intra-seasonal variation of water vapor and column ozone in the tropics during boreal winter.

Acknowledgment

We thank ECMWF for providing the ERA-Interim reanalysis dataset. The authors are grateful to Givo Alsepan from the Laboratory of Physical Oceanography and Climate at Hokkaido University, Japan, for his helpful comments and discussion.

References

[1] Matsuno T 1966 Quasi-geostrophic motions in equatorial area *J. Meteor. Soc. Japan* **44** 25–43
[2] Lubis S W and Jacobi C 2015 The modulating influence of convectively coupled equatorial waves (CCEWs) on the variability of tropical precipitation *Int. J. Climatol.* **35** 1465–1483
[3] Lubis S W and Setiawan S 2010 Identifikasi Gelombang Kelvin Atmosfer Ekuatorial di Indonesia Berbasis Data NCEP/NCAR Reanalysis I *J. Fisika* **10** 71–82
[4] Madden R A and Julian P R 1971 Detection of a 40 –50 day oscillation in the zonal wind in the tropical pacific *J. Atmos. Sci.* **28** 702–708
[5] Shienbaum J 2003 Current theories on el nino-southern oscillation: a review *Geof. Int.* **42** 291–305
[6] Madden R A and Jullian P R 1994 Obsevations of the 40-50-day tropical oscillation-a review *Mon. Wea. Rev.* **122** 814–837
[7] Zhang C 2005 Madden-Julian Oscillation *Reviews of Geophysics* **43** RG2003
[8] Waliser D *et al* 2009 MJO simulation diagnostics *J. Clim.* **22** 3006–3030
[9] Schreck C J, Shi L, Kossin J P and Bates J J 2013 Identifying the MJO, equatorial waves, and their impacts using 32 years of HIRS upper tropospheric water vapor *J. Clim.* **26** 1418–1431
[10] Tian B, Yung Y L, Waliser D E, Tyranowski T, Kuai L, Fetzer E J and Irion F W 2007 Intraseasonal variations of the tropical total ozone and their connection to the Madden-Julian Oscillation *J. Geophys. Res.* **34** 1–7
[11] Stull R B 2000 *Meteorology for Scientist and Engineer 2nd ed* (California: Brooks/Cole)
[12] WMO 2008 *Guide to Meteorological Instruments and Methods of Observation 7th ed* (Geneva: World Meteorological Organization)
[13] Mohanakumar K 2008 *Stratosphere Troposhpere Interaction: An Introduction* (Netherlands: Springer)
[14] Liebmann B and Smith C A 1996 Description of a complete (interpolated) outgoing longwave radiation dataset *Bull. Amer. Meteor. Soc.* **77** 1275–1277
[15] Dee D P *et al* 2011 The ERA-Interim Reanalysis: Configuration and Performance of the Data Assimilation System *Q. J. R. Meteorol. Soc.* **137** 553-597
[16] Wheeler M C and Hendon H H 2004 An all-season real-time multivariate MJO index: Development of an index for monitoring and prediction *Mon. Wea. Rev.* **132** 1917–1932
[17] Hayashi Y 1982 Space-time spectral analysis and its applications to atmospheric waves. *J. Meteor. Soc. Japan* **60** 156–171
[18] Wheeler M and Kiladis GN 1999 Convectively coupled equatoril waves: analysis of clouds and temperature in the wavenumber-frequency domain *J. Atmos. Sci.* **56** 374–399
[19] NASA 2000 Studying Earth's Environment From Space [cited 2016 May 23]; Available from: http://www.ccpo.odu.edu/SEES/index.html

IOP Conf. Series: Earth and Environmental Science **54** (2017) 012035 doi:10.1088/1755-1315/54/1/012035

Impact of ENSO on seasonal variations of Kelvin Waves and mixed Rossby-Gravity Waves

Saeful Rakhman[1]*, Sandro W Lubis[2] and Sonni Setiawan[1]
[1]Department of Geophysics and Meteorology, Bogor Agricultural University (IPB), Indonesia
[2]Physics of the Atmoshpere, GEOMAR Helmholtz Centre for Ocean Research Kiel, Germany

E-mail: saefulrakhman88@gmail.com

Abstract. Characteristics of atmospheric equatorial Kelvin waves and mixed Rossby-Gravity (MRG) waves as well as their relationship with tropical convective activity associated with El Niño-Southern Oscillation (ENSO) were analyzed. Kelvin waves and MRG waves were identified by using a Space-Time Spectral Analysis (STSA) technique, where the differences in the strength of both waves were quantified by taking the wave spectrum differences for each ENSO phase. Our result showed that Kelvin wave activity is stronger during an El Nino years, whereas the MRG wave activity is stronger during the La Nina years. Seasonal variations of Kelvin wave activity occurs predominantly in MAM over the central to the east Pacific in the El Nino years, while the strongest seasonal variation of MRG wave activity occus in MAM and SON over the northern and southern Pacific during La Nina years. The local variation of Kelvin wave and MRG wave activities are found to be controlled by variation in lower level atmospheric convection induced by sea surface temperature in the tropical Pacific Ocean.

1. Introduction

Equatorial planetary waves (EPW) are one of the dominant modes of synoptic-to-subseasonal variability in the tropics. They are generated by diabatic heating due to organized tropical large-scale convective heating in the equatorial belt [1][2][3]. The amplitude is captured clearly in the equatorial belt between 20°N and 20°S [1]. EPW causes predominant disturbances in the equatorial atmosphere such as inducing mean-meridional circulation that is important for the heat balance of the equatorial belt [4][5], affecting the patterns of low level moisture convergence, and controling the distribution of tropical convective heating and storms in large longitudinal distances [4]. In addition, EPW also plays an important role in modulating rainfall in the tropics [3], circulation Walker [6], Madden and Julian oscillation (MJO) [7] and El Nino-Southern Oscillation (ENSO) [5]. Two dominant atmospheric equatorial waves has an important role in the region is equatorial Kelvin waves and Mixed Rossby-Gravity (MRG) waves. In this study our analysis will focus on understanding the nature of Kelvin waves and MRG waves during ENSO events.

El Niño Southern Oscillation (ENSO) is a coupled atmosphere-ocean phenomenon characterized by anomalous ocean warming over the eastern equatorial Pacific ocean [8]. ENSO can be classified into normal, El Niño and La Niña phases. The term El Niño refers to a large-scale periodic warming in sea surface temperatures across the central and east-central Equatorial Pacific. On the other hand, La Niña represent periods of below-average sea surface temperatures across the east-central Equatorial

Pacific. This phenomenon affects the climate over sizable portions of the globe, including some regions far removed from the tropical Pacific Ocean [9].

Warming in the tropical ocean (i.e., increased sea surface temperature – SST) will increase the evaporation and hence, the formation of large scale convective clouds. The process causes latent heating in the upper atmosphere which is very important ingredient to generate more EPW activity [10]. Kelvin waves and MRG waves are well observed within the anomalies of SST, precipitation, and outgoing longwave radiation (OLR) in the tropics [3]. Therefore, in this study we used both SST and OLR datasets as a proxy to isolate the Kelvin wave and MRG wave activity. The goal of the current study is to investigate the impact of ENSO on seasonal variation of Kelvin wave and MRG waves and the involved mechanisms.

2. Methods

ENSO phases were identified by using Running 3-month ONI Mean Values data. Sea surface temperature profile data is displayed using the NCEP Optimum Interpolation Sea Surface Temperature (OISST) in the period from 1982 to 2013 and a spatial resolution of 1° x 1°. In addition, OLR profile displayed using daily interpolated of outgoing longwave radiation (OLR) data in the period from 1974 to 2013 from National Oceanic and Atmospheric Administration (NOAA) Global CPC (except 1978). Details interpolation technique can be seen in the OLR data [11]. Horizontal wind datasets (u,v) at an altitude of 1000 hPa and 850 hPa and vertical velocity (ω) in the troposphere (1000 hPa-100 hPa) from National Centre for Environmental Prediction (NCEP/NCAR reanalysis 1: Pressure) each at 20°N-20°S in the equatorial bands with a spatial resolution of 2.5° x 2.5°.

2.1. Composite Analysis

Composite analysis is a method used to determine the field distribution of certain atmospheric conditions. In this study, composite analysis is used to analyze the distribution of sea surface temperature, outgoing longwave radiation and vertical velocity for each phase of ENSO. Composite technique is nothing but the ensemble average, the average value is calculated from a set of data, but in the same condition.

2.2. Space-Time Spectral Analysis (STSA)

STSA is one of the methodology that been used in this study. STSA is a method used to analyze the wave propagation in the zonal field. This method decomposing the physical data field in the domain of space and time into a data field in the domain of wavenumber and frequency of the waves that propagate to westward and eastward [12]. In principle, this decomposition process is the Fourier transform of the data in space and time domain into data in frequency (ω) and wavenumber (k) domain. STSA procedure contained in NCL software (NCAR Command Language) with SpaceTime function. The output of this method is the contours of the power spectrum that contained in the dispersion curve as a function of frequency and zonal wavenumber that classified into symmetric and asymmetric components.

3. **Result**

3.1. Identification Phase of ENSO

ENSO occurrences in this study were identified using Nino 3.4 index. The index is an index based on the average of sea surface temperature anomalies (SST) in 5°S-5°N, 120°-170°W. Nino 3.4 index 3-month running average of the period 1974 to 2013 (figure 1). The number of events each ENSO phase are listed in Table 1.

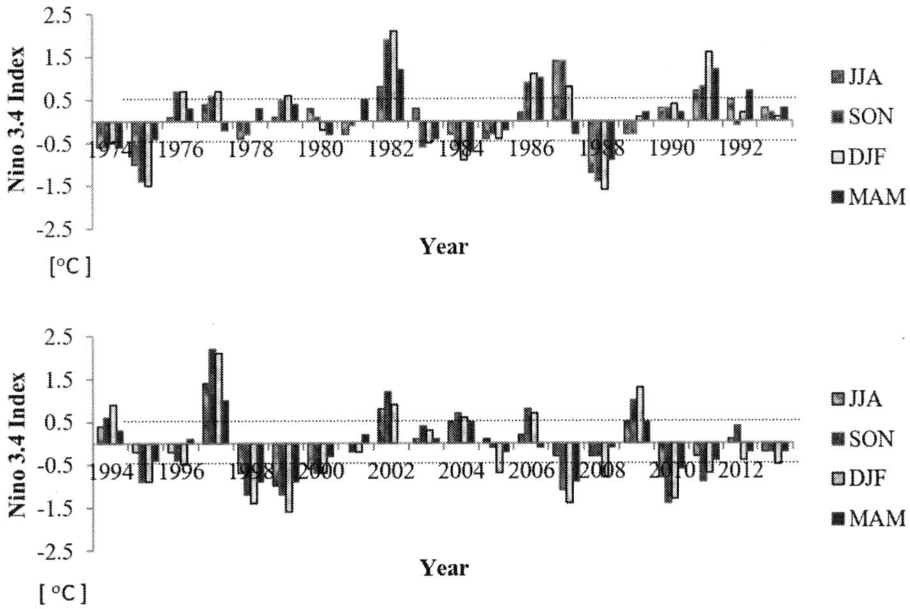

Figure 1. Average of 3-month run of sea surface temperature anomalies (SST,°C) Nino 3.4 index from 1974 to 2013.

Table 1. Season Number of events at each phase of ENSO (1974-2013)

Condition	**DJF**	**MAM**	**JJA**	**SON**
Normal	12	25	25	15
El Nino	13	8	8	13
La Nina	15	7	7	12

3.2. Seasonal Variations of Kelvin Wave and Mixed Rossby-Gravity (MRG) Wave at each Phase of ENSO

Identification of the equatorial planetary wave activity (EPW) were calculated using STSA [12]. The data used is the average of the daily OLR anomalies for ~ 40 years (1974-2013) on 15°LU-15°LS. Previously, the data is grouped into seasons for each phase of ENSO. Results of STSA OLR with an effective depth of 12-50 m are selected is shown in figure 2 (Symetric Component) and figure 3 (Antisymetric Component). The black box shows the waves become the focus of study in this research.

IOP Conf. Series: Earth and Environmental Science 54 (2017) 012035 doi:10.1088/1755-1315/54/1/012035

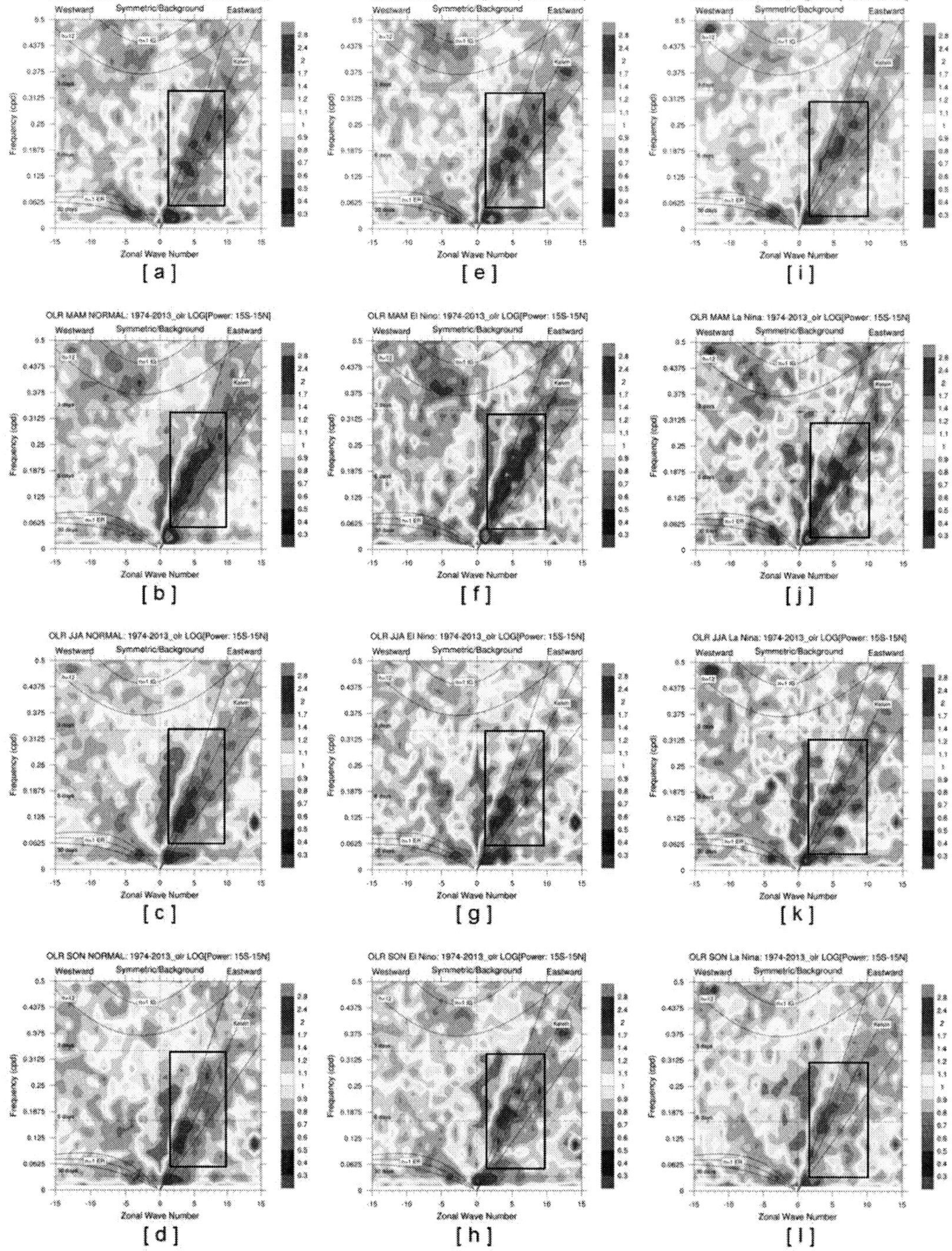

Figure 2. Symmetric STSA of outgoing longwave radiation in the normal phase (DJF (a), MAM (b), JJA (c), SON (d)), El Nino phase (DJF (e), MAM (f), JJA (g), SON (h)), La Nina phase (DJF (i), MAM (j), JJA (k), SON (l)).

IOP Conf. Series: Earth and Environmental Science **54** (2017) 012035 doi:10.1088/1755-1315/54/1/012035

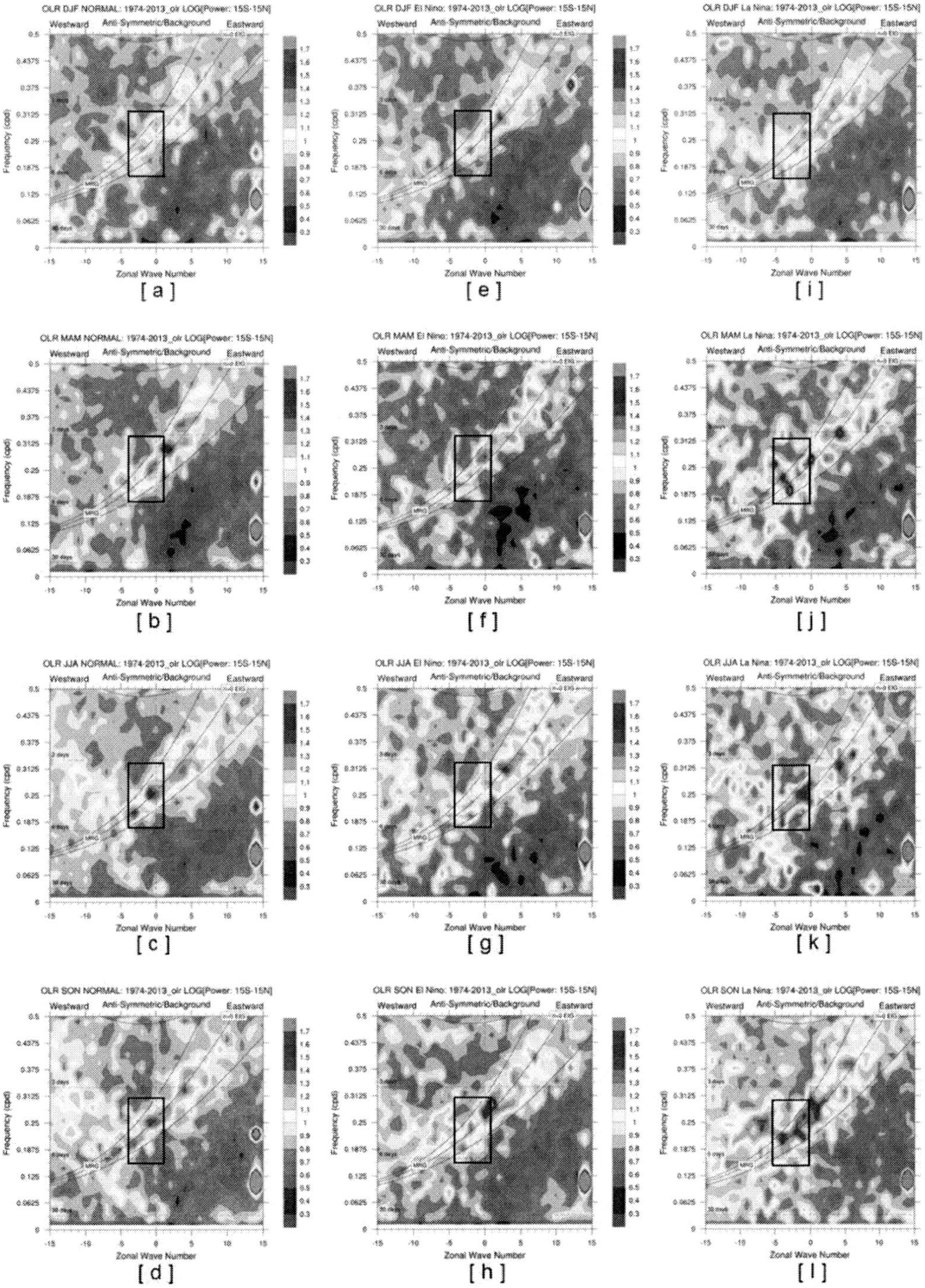

Figure 3. Anti-Symmetric STSA of outgoing longwave radiation in the normal phase (DJF (a), MAM (b), JJA (c), SON (d)), El Nino phase (DJF (e), MAM (f), JJA (g), SON (h)), La Nina phase (DJF (i), MAM (j), JJA (k), SON (l)).

Based on figure 2 and figure 3, shows that there are many waves associated with ENSO phenomenon. This wave is the result of the classification of basic types of interference waves around the equator by n integer associated with the number called nodal meridional wave mode [13]. For n = -1, commonly known as the Kelvin wave, only affects the zonal oscillations [14][4]. For n = 0, there are two groups of waves have propagation to the east and the west. The wave that moves to the east known as Inertio-waves or Gravity-wave. While the wave that spreading to the west is called Mixed Rossby-Gravity waves. This wave modes associated with antisimetry perturbation of meridional wind and equatorial relatively perturbation of meridional wind [13]. In general, for n ≥ 1 there are three groups, namely the Rossby for the wave that move to the west and Gravity waves for the wave that move to the east and to the west [13].

Overall for each phase of ENSO, at a equivalent depth of 12-50 m to a Kelvin wave signal (n = -1) was detected in symmetrical components. Results STSA OLR at latitude 15°N-15°S shows that the Kelvin wave is detected at a frequency of approximately 0.3125 to 0.0625 cycles per day or a period of about 3.2 to 16 days by the zonal wave number 1-10 and speed phase from 10.8 to 22.1 m s^{-1}. These results are consistent with research conducted by [12][10][15]. In figure 2 symmetrical components, looks a Kelvin wave rose during ENSO phenomenon. With Kelvin wave activity in El Nino conditions are stronger than La Nina.

Kelvin waves are detected in symmetrical components for each season and for all conditions (shown in figure 2). Strong signal the presence of Kelvin wave activity occurred during DJF to MAM, were MAM as the peak. It describes the results of research conducted by [12] using the same data (OLR, years 1979-1996). Wheeler and Kiladis (1999) [12] identify EPW wave of cases in the southern hemisphere summer (November-April) and the Northern Hemisphere summer (May to October). The results showed that the Kelvin wave activity more powerful in the summer in the southern hemisphere. In that case, the results of the study of the influence of ENSO phenomenon is often stronger in the summer in the southern hemisphere (SON, DJF), at specific intervals followed by the increasing of Kelvin wave activity in the DJF to MAM period were MAM is the the strongest period of wave activity to every condition. The results according to research conducted by Huang and Huang (2011), Lubis and Jacobi (2015) 15][3], which states that the strongest Kelvin wave activity occurs during MAM period. The results further showed that compared with normal conditions, the strongest Kelvin wave activity occurs in El Nino conditions.

On the other hand, Mixed Rossby-Gravity wave (MRG, n = 0) was detected in antisymmetric components for each phase of ENSO on the selected depth equivalent 12-50 m (figure 3). Results of STSA OLR at latitude 15°N-15°S shows that the MRG wave detected at a frequency of approximately from 0.3125 to 0.1875 cycles per day, or about 3.2 to 5.3 daily periods by zonal wave numbers is 1-5 and phase speed of 10.8 to 22.1 m s^{-1}. These results are consistent with research conducted [16][12][3]. In figure 3 antisymmetric component, MRG visible wavelength obtained during La Nina.

Overall, MRG waves detected at antisymmetric component for each season and for all conditions. Indicates the presence of a strong MRG wave activity in the MAM and SON period with SON period as the peak. It describes the results of research conducted by the MRG. WK99 show that wave activity more powerful in the summer in the northern hemisphere. The results according to research conducted by [15][3] which states that the strongest activity of MRG waves occur during the SON. The results further showed that compared with normal conditions, the strongest MRG wave activity occurred in La Nina conditions.

3.3. Difference of STSA (outgoing longwave radiation field)

Differences in the power spectrum of Kelvin waves and MRG wave on El Nino and La Nina conditions for the each season is shown in figure 4. Dominance of wave spectrum on the symmetric component represented by the red color were MAM period is the strongest event. That means the power of Kelvin wave in the MAM period in El Nino conditions is stronger than other periods. While the antisymmetric component of the wave spectrum dominance shown by the blue color and the strongest in the period MAM which means that MRG wave at MAM period of La Nina conditions are stronger than other periods.

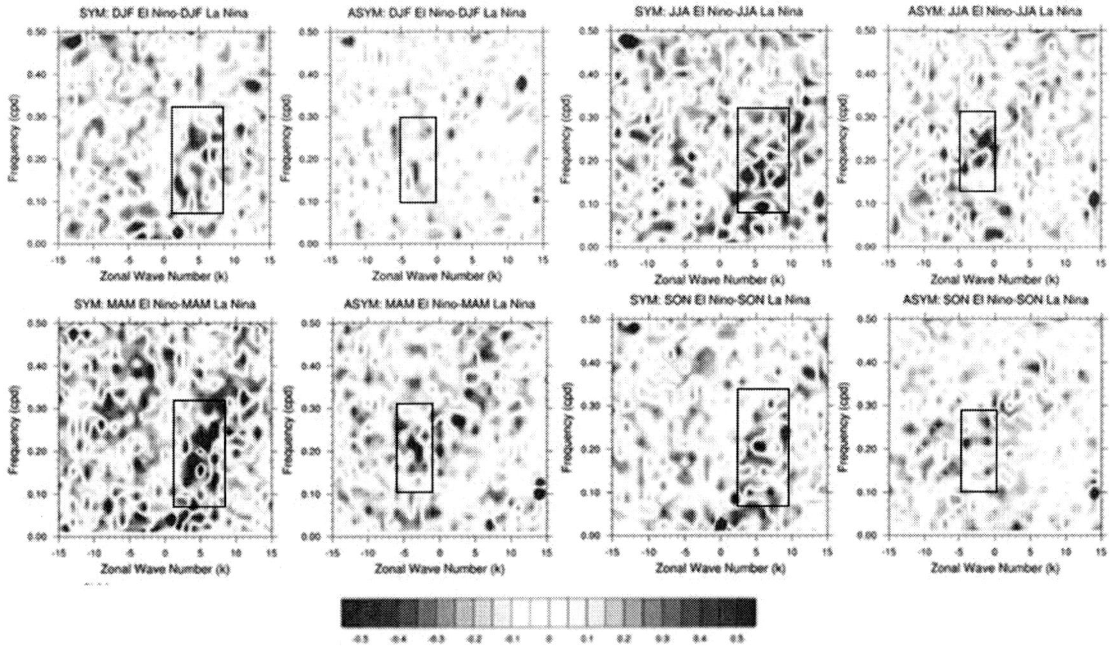

Figure 4. Difference of power spectrum in outgoing longwave radiation between El Nino and La Nina phases.

3.4. Spatial Distribution of Kelvin Wave and Mixed Rossby-Gravity (MRG) at each phase of ENSO
Spatial distribution profile of EPW, in this case the wave Kelvin and MRG not afford shown by STSA OLR. The profile is obtained by calculating the value of the variance results of each wave filter according to the method performed by [3]. Large variance stating how large wave activity in a region. Spatial distribution of Kelvin and MRG wave at each phase of ENSO is shown in figure 5. Kelvin wave activity more powerful than MRG wave. This is indicated by the value of the variant Kelvin wave is larger than the MRG wave (Varian Kelvin wave reached 20.5 (Watt $m^{-2})^2$, while MRG wave only 10 (Watt $m^{-2})^2$).

Kelvin wave spatial distribution is shown in figure 5a. These results have compatibility with research conducted by [12][15][3]. Kelvin wave spatial distribution of the strongest normal conditions occur in three regions covering the Indian Ocean (90°E), the Pacific ITCZ (5-12°N), and the Atlantic Ocean to Africa, South America (0-5°N). In the period from DJF to MAM. In El Nino conditions, the activity is centered in the middle of the Pacific Ocean to the east (90-150°W). At the La Nina conditions the strongest wave activity in the Atlantic Ocean south America to Africa (0-5°N) occurred in the MAM period.

MRG wave spatial distribution is shown in figure 5b. In this figure shows that the spatial distribution of MRG wave is divided into north and south equatorial Pacific around latitude 8°, with activity in the northern hemisphere is stronger than in southern [12][3]. In normal conditions, these waves are very weak activity with the distribution pattern in the north and south equatorial Pacific (120°E-90°W). In El Nino conditions, the activity is slightly increased compared to normal conditions and centered in the northern equatorial Pacific Ocean (120°E-90°E) in the JJA up to SON period. At

IOP Conf. Series: Earth and Environmental Science **54** (2017) 012035 doi:10.1088/1755-1315/54/1/012035

the La Nina conditions are very strong wave activity in the north and south of the equator of the Pacific in the period MAM and SON.

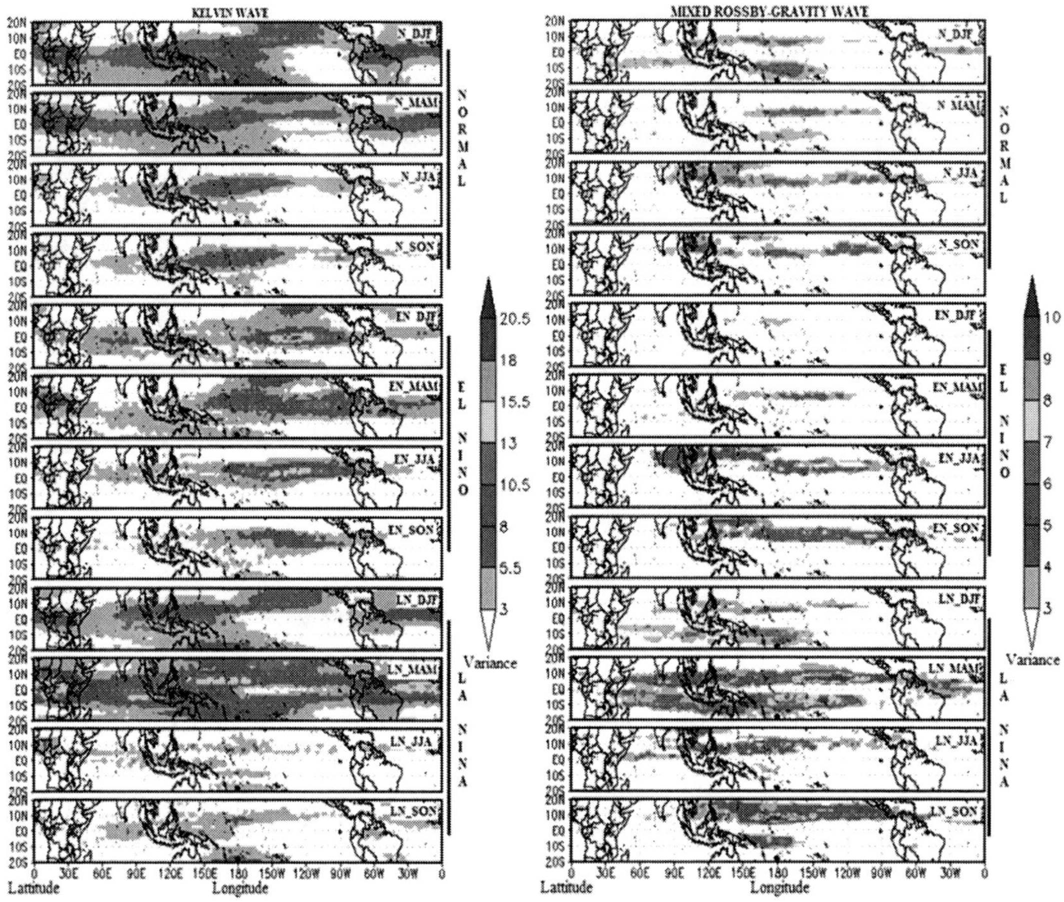

(a) (b)

Figure 5. Spatial distribution of Kelvin wave (a) and Mixed Rossby-Gravity (b) wave on Normal, El Nino and La Nina condition.

According to Holton (2004)[4], is one of convective heating plant wave activity. In connection with this, the ENSO phenomenon is a phenomenon that is closely associated with changes in regional centers convection due to warming sea surface temperatures will be analyzed for compliance with regional centers of activity EPW wave height. The suitability of the spatial distribution patterns Kelvin and MRG waves with convection activity described in the following sections.

3.5. Factors affecting Seasonal Variations Kelvin wave and Mixed Rossby-Gravity (MRG) wave at each phase of ENSO

ENSO phenomenon is closely related to a shift in the central area of convection activity and subsidence as a result of the warming and cooling of the Pacific sea surface temperature [17]. The event signal profile illustrated with OLR anomalies [2][3]. The focus of study in this research is convection. This relates to the statement [4], the formation of the cloud will generate waves.

Seasonal variations Kelvin waves and Mixed Rossby-Gravity (MRG) waves at every phase of ENSO suspected to be affected by seasonal variations in convection activity shown in the Walker circulation (figure 6a) and the profile of OLR anomalies (figure 6b) due to seasonal variations in heating and cooling of the sea surface temperature (figure 6c) in each ENSO phase. Convection activity is indicated by negative anomalies of vertical velocity (blue in figure 6a), as proof of the existence of convection signal (negative anomalies of OLR blue in figure 6b), due to the warming of the sea surface temperature (red in figure 6c).

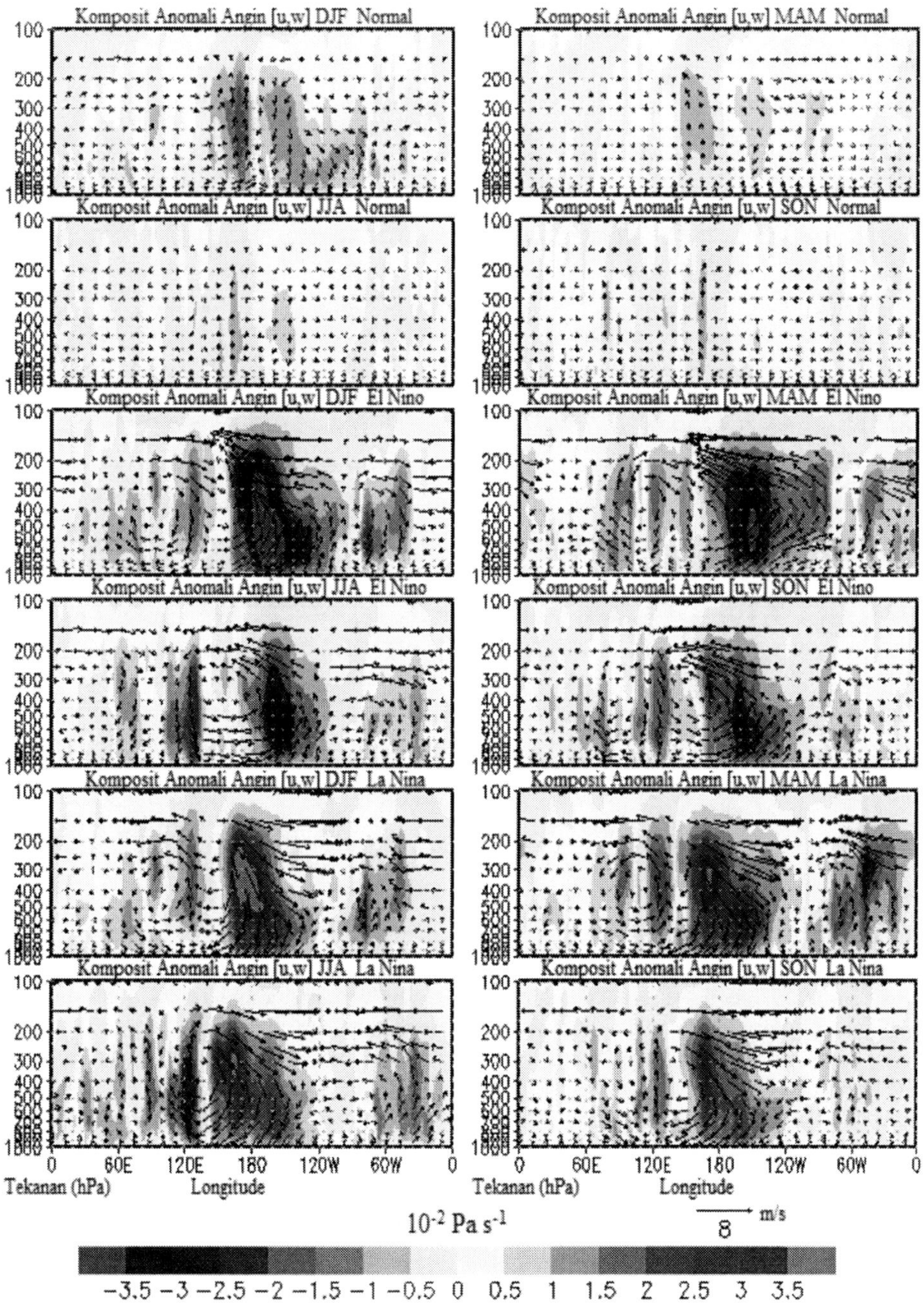

(a)

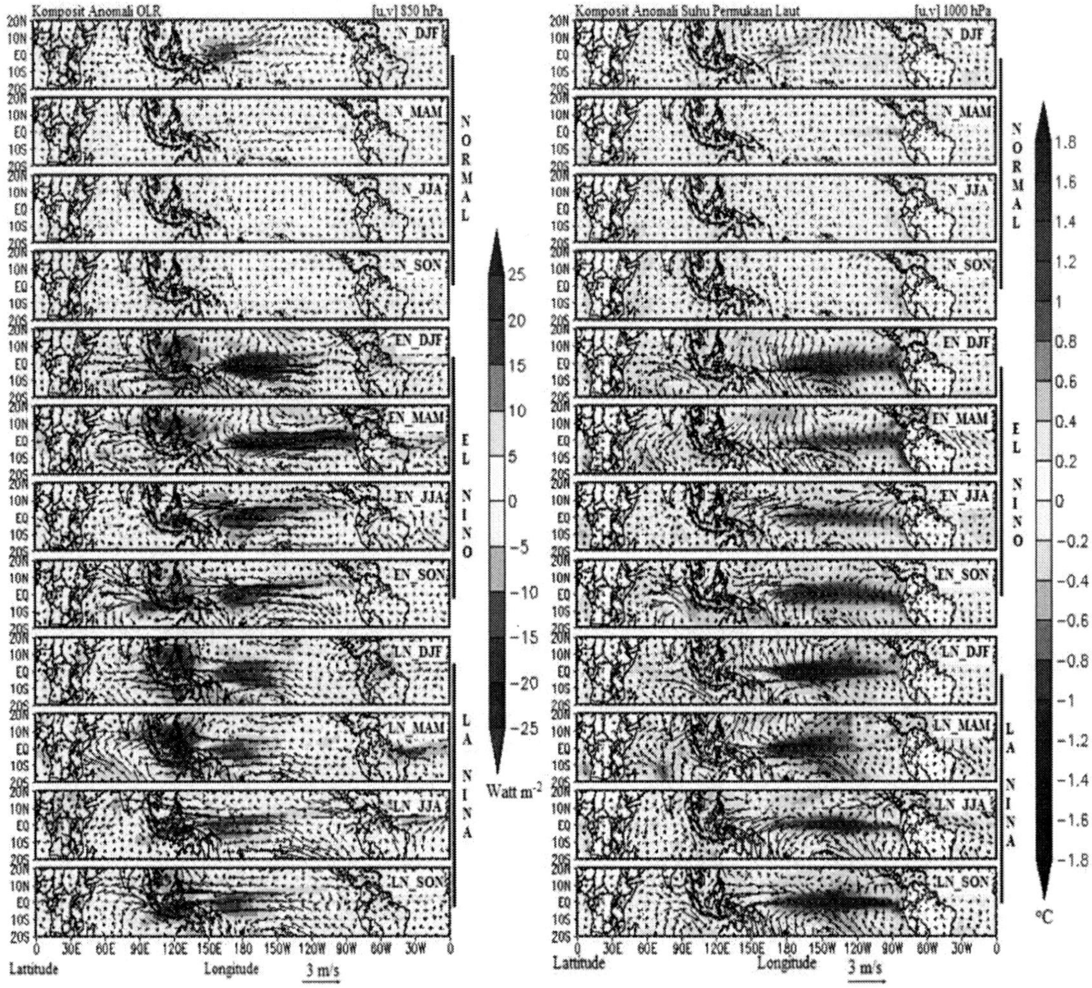

(b) (c)

Figure 6. Seasonal Composite of vertical velocity (Pa s^{-1}), OLR (Watt m^{-2}), and SST (°C) Normal condition, El Nino and La Nina. Anomalies calculated from the monthly average of the year 1982 to 2013.

Overall, convection activity for each season in each phase of ENSO indicated by Walker circulation anomaly profile (Figure 6a) have compatibility with convection signal shown by OLR profile (figure 6b). Likewise with patterns of warming and cooling of the sea surface temperature (figure 6c) behind the convection activity. There is a lag between SST anomalies seen with OLR anomalies and Walker circulation. It is seen from the incident phase of the strongest El Nino conditions and the La Nina that occurs in the SON and DJF period followed by convective signal and Walker circulation strong at DJF and MAM period. Similarly, the pattern of warming and cooling of sea surface temperatures MAM and JJA period will be followed by convection and circulation signal corresponding Walker circulation. As for normal conditions, do not see the lag time between the heating and cooling of the sea surface temperature OLR anomaly and the Walker circulation.

Based on the above, the mechanism of wave generation Kelvin and MRG can simply be explained as follows. The existence of the warming and cooling of the sea surface temperature (phases of ENSO, figure 6c) at specified intervals will increase the convection activity indicated by the presence of convective signal amplification. Convective signal profile shown by OLR anomalies (figure 6b) and is evidenced by convection shown into the Walker circulation anomaly profile (figure 6a). Activities convection underlying wave generation Kelvin and MRG indicated by symmetric and antisymmetric components shown by STSA OLR (figure 2 and 3) and their spatial distribution is obtained by displaying the results of calculation of variance. The higher the value of the variant in a region showing increasingly strong wave activity in the region (figure 5a and 5b).

4. Conclusion

We have examined the impact of ENSO on seasonal variation of Kelvin waves and MRG waves by using NOAA outgoing longwave radiation (OLR) dataset and SST. The Kelvin waves ad MRG waves have been isolated by using space-time spectral analysis (STSA) and composite analysis. The key results are summarized as follow:

(1) ENSO phenomenon significantly influences seasonal variation of Kelvin and MRG wave activity.
(2) Kelvin waves are stronger during El Nino years with frequency of about 0.3125-0.0625 cpd (T: 3.2-16 day), zonal wave number k=1-10, and phase speed of about 10.8-22.1 m/s.
(3) MRG waves are stronger during La Nina years with frequency of about 0.3125-0.1875 cpd (T: 3.2-6 day), zonal wave number k=(-1)-(-5), and phase speed of about 10.8-22.1 m/s.
(4) The strongest Kelvin wave activity during El Nino years occurs in MAM periods, while MRG wave during La Nina years occurrs in MAM and SON periods.
(5) Maximum activity of Kelvin wave activity is observed over the central to the east Pacific in the El Nino years, while MRG wave activity are observed over the northern and southern Pacific during La Nina years.

The local variation in Kelvin waves and MRG waves is to be likely controlled by the variation in the lower level convection due to warming in sea surface temperatures over these regions. It is indicated by a shift in the center of convection activity towards the central Pacific to the east following the shift in Kelvin wave acitivity. While increase in wave activity MRG are more related to the convection over the the northern and southern equatorial Pacific ocean. The results of this study is expected to improve the prediction skill of the intra-seasonal climate in the tropicas based on the knowledge of atmosphere-ocean coupled mechanism. Further studies are required to investigate in more detail how tropical ocean influence the generation of Kelvin waves and MRG waves in opposite ENSO phase and their associated implication on surface climate.

Acknowledgment

We thank NOAA for providing the OLR, OISST, zonal and meridional wind (u,v) and vertical velocity (ω) dataset that used in this study.

References

[1] Holton JR. and RS Lindzen 1968 *Monthly Weather Review* **96** 385
[2] Kiladis G, Wheeler M, Haertel P, Straub K, Roundy P 2009 Convectively coupled equatorial waves. *Review Geophys* **47**
[3] Lubis SW and Jacobi C 2015 The modulating influence of convectively coupled equatorial waves (CCEWs) on the variability of tropical precipitation *Int. J. Climatol.* **35** 1465–1483
[4] Holton JR 2004 *An Introduction to Dynamics Meteorology* Ed ke-4. Burlington: Elsevier
[5] Dima IM and Wallace JM 2006 Structure of annual mean equatorial planetary waves in the ERA-40 Reanalyses *J.Atmos.Sci.* **64** 2862-2880
[6] Gill A E 1980 Some simple solutions for heat-induced tropical circulations *Quart. J. Roy. Meteor.Soc.* **106** 447-462
[7] Lau KM and L Peng 1987 Origin of low-frequency (intra-seasonal) oscillation in the tropical atmosphere. Part I: Basic theory *J. Atmos. Sci.* **44** 950-972

[8] Bjerknes J 1969 Atmospheric Teleconnections from the Equatorial Pacific *Mon. Weather Rev.* **97** 163 172
[9] Mc Bride J. Haylock M.R and Nicholls N 2003 Relationships between the Maritime Continent Heat Source and the El Nino-Southern Oscillation Phenomenon *J. Climate.* **16** 2905-2914
[10] Yang Gui-Ying and Hoskin B 2013 Enso Impact on Kelvin Waves and Assosiated Tropical Convection *J. Atmos Sci.* **70** 3513-3532
[11] Liebmann B. and C.A. Smith 1996 Description of a Complete (Interpolated) Outgoing Longwave Radiation Dataset *Bulletin of the American Meteorological Society* **77** 1275-1277
[12] Wheeler M and Kiladis GN 1999 Convectively coupled equatorial waves; analysis of clouds and temperature in the wave-number-frequency domain *J Atmos Sci* **56** 374-399
[13] Yanai M and Marukami M 1970 Spectrum analysis of symmetric and antisymmetric equatorial waves *J. Atmos. Sci.* **48** 331-347
[14] Matsuno T. 1966 Quasi-geostrophic motions in equatorial area *J Meteor Soc Japan.* **44** 25-43
[15] Huang P and Huang R 2011 Climatology and interannual variability of convectively coupled equatorial waves activity *J. Climate.* **24** 4451-4465
[16] Yanai M and T Maruyama 1966 Stratospheric wave disturbances propagating over the equatorial Pacific J. Meteor.Soc.Japan **44** 291-294
[17] Chiodi A M and Harrison D E 2015 Global Seasonal Precipitation Anomalies Robustly Associated with El Nino and La Nina Events-An OLR Perspective *J. Climate.* **28** 6133-6159

A composite analysis of the Mesoscale Convective Complexes (MCCs) development over the Central Kalimantan and its relation with the propagation of the rainfall systems

Trismidianto[1,2*], E Yulihastin[2], H Satyawardhana[2] and S Ishida[1]
[1]Meteorological Laboratory, Graduate School of Science and Technology, Hirosaki University, Hirosaki, Japan
[2]Center for Atmospheric Science and Technology, National Institute of Aeronautics and Space, Jakarta, Indonesia

E-mail: trismidianto@lapan.go.id

Abstract. The composite analysis for 45-cases of the MCC which identified by using infrared satellite imagery over the Central Kalimantan (110° - 116°E, 4°S - 1°N) has been observed. The data used is a combination of satellite data and reanalysis data. This study reported that the MCCs develops triggered by the orographic convective that helped by the convergent surface wind flow through interaction with the sea breeze in the afternoon until midnight and dissipated in the morning. The new convective systems are generated by the divergent outflow of the cold pool, in conjunction with the morning land breeze during MCCs mature. After dissipated, the new convective systems induce the land convection over the Java Island that became heavy rainfall. The initial and mature region are characterized by weak low-level convergence and upper-level divergence, but the low-level divergence begin appear during mature. The MCC develops largely driven by MCC-scale moisture convergence in the lower troposphere and cold core structure in the lower level. The weak surface divergence and upper-level divergence, warm advection in the lower atmosphere are dissipation characteristics. MCCs develop due to low-level cold advection and temperature and separated when dissipated that indicate the existence of the new convective systems propagation.

1. Introduction

Prior to 1980, the study of convective activity at the mesoscale level had been limited to tropical phenomena like cloud clusters, squalls and hurricanes, midlatitude squall lines and land-sea breeze systems. In 1980, Robert A. Maddox introduced the concept of the MCC (hereafter MCCs) after a careful study of satellite IR images over the central United States during 1978. MCCs are a special case of mesoscale convective systems (hereafter MCSs), where MCSs are generally defined as an organized ensemble of convective elements, whose lifecycle is longer than that of the individual convective elements, and the largest of the convective storms. Maddox [1] original definition stated that MCCs have cloud shield with continuously brightness temperature or black body temperature (hereafter T_{BB}) $\leq$ -32°C or 241 K must have an area $\geq$ 100,000 km^2 and the interior cold cloud $T_{BB} \leq$ -52°C or 221 K must have an area $\geq$ 50,000 km^{2}for over six hours, although the cloud shield does not have to maintain an eccentricity of $\geq$ 0.7 for its entire life cycle. The eccentricity is also important to distinguish MCC from the other MCS types, for example; squall line, bow echo, cloud cluster, and the others. The areal extent and eccentricity values were used to determine the duration and life cycle of the MCCs, while the centroid position indicated the path of propagation of the systems.

The large-scale environment is very effect to the development of MCCs. Laing and Fritsch [2] is also found that MCCs around the globe typically form near a maximum in low-level shear, lending support to the idea that low-level shear may be important for the development of long-lived MCSs. And also globally, MCCs tend to initiate in locations where there is: (1) a low-level jet (LLJ); (2) a low-level convergence; (3) upper-level divergence; (4) an approaching mid-level vorticity maximum (associated with a weak short wave trough).The early stage of the MCC lifecycle is characterized by convergence, vertical motion, and heating being centered in the lower troposphere. The systems achieve and maintain its maximum divergence, upward motion, and anticyclonic vorticity in the upper troposphere during the latter half of the life cycle [4]. The MCC typically forms in association with a weak mid-tropospheric short-wave trough and a weak surface front or outflow boundary. Its environment often exhibits pronounced low-level temperature and moisture advection in association with a well-defined low-level jet [3][4][5].

In generally, the previous researcher found that some area has environment are very similar and exhibit many of the same dynamic and thermodynamic structure that is present with systems but several areas has different environmental condition during MCCs event. It is interesting to analyze the environmental condition during MCC that occur over IMC and compare its result with the previous work in several regions due to the analyses about that is not well understood. A Recent study by Trismidianto et al. [6]also reported that several convective systems were generated during the decay stage of the MCC over the Indian Ocean (hereafter IO) because of convergence between the land breeze and westerly wind. The new convective systems around the decaying MCC were generated during MCC event, and they propagated to the surrounding area because of the divergent outflow from the cold pool (hereafter CP). The combination of the land breeze and the CP outflow from the decaying MCC was a significant factor in the formation of the convective system that caused the heavy rainfall. Therefore, the goal of this paper is to analyze the development and movement of MCC using composite analysis with taking the 45-cases of MCC over Central Kalimantan (hereafter CK).

2. Data and Method

To identify the MCC in this study, we use $T_{BB}$from the Himawari geostationary satellites, operated by the Japan Meteorological Agency (JMA) that consists of Himawari-5/GMS-5 for data from January 2001- April 2003, Pacific GOES/GOES-9 for data from May 2003- June 2005, Himawari-6/MTSAT-1R for data from July 2005 - June 2010, Himawari-7/MTSAT-2 for data from July 2010 - June 2015, and Himawari-8 for data from July 2015 - December 2015. That satellite data has spatial and time resolutions 0.05°×0.05° and one hour, respectively. This paper adapts the method from Ismanto [7] and Trismidianto [8] that based on the maximum spatial correlation tracking technique (MASCOTTE) method by Carvalho and Jones [9] and almost similar way with Laing and Fritsch [10] to identify MCC with input the temperature, latitude, and longitude values for each cloud shield pixel that obtained from IR to a modified version of a computerized MCC program using MATLAB for measuring the satellite-observed characteristics of the MCCs by Maddox [1] as described in introduction section.

To explain the evolution and propagation of the MCCs in this study is utilized composite analysis that adapted from McAnelly and Cotton [11][12]. The focus of this study is to the analysis of 45-cases of the MCCwhich occurred in the same area center over CK (110° - 116°E, 4°S - 1°N), in the next will be mentioned as MCCs composite.In the composite category, MCCs composites are considered to have common composite times of initiation, mature, decay and dissipation. Besides the similarities in genesis region, maximum area, and mature duration, the MCCs in each category were reasonably consistent in term of other T_{BB} characteristics. These include; the duration of initiation to mature stage, mature to decay and decay to dissipation or post-MCC stage. The composite for initiation, mature, decay and dissipation are determined by the average the hour of initiation, mature, decay and dissipation for each MCC, around to the nearest hour. In the first, we determined the average the hour of the initiation of all case in each region, and then plus 1, 2 or 3 hours until reaches the average the hour of mature, the same way until reaches the average the hours of dissipation or post-MCC. So that, the average from *initial-to-maximum, maximum-to-dissipation* with 1, 2, or 3 duration time.

The estimated rainfall data, corresponding to the MCCs, were obtained from the Real Time Tropical Rainfall Measuring Mission's (TRMM) Multi-Satellite Precipitation Analysis (TMPA-RT) 3B41RT v7 data set, which has hourly temporal resolution and 0.25° × 0.25° spatial resolution [13]. To analyses the surface wind is utilized several data from Cross-Calibrated Multi-Platform (CCMP), which are available at 6-hourly intervals with 0.25° horizontal resolution and the environments condition during MCC event is utilized several data from the European Centre for Medium-Range Weather Forecasts (ECMWF) ERA-Interim analysis fields, which are available at 6-hourly intervals with 1° x 1°° horizontal resolution [14]. Maddox [3] presents a summary of environmental conditions for MCC generation and development in the USA using a composite of the wind and thermodynamic fields. Cotton et al. [4] also reported the environmental condition during MCC over the USA using composite model. This study adapts the method like them to analysis the MCC development. The composite analysis of the environmental condition is just analyzed during the critical stage of MCCs. i.e., initial, mature and dissipation stage which following the data available in 6-hourly at 0000, 0600, 1200 and 1800 UTC. In this study, we consider the data on time in 2300, 0000 and 0100 UTC as the time on 0000 UTC, and the data on time in 0500, 0600 and 0700 UTC as the time on 0600 UTC, while the data on time in 1100, 1200 and 1300 UTC as the time on 1200 UTC, and the data on time in 1700, 1800 and 1900 UTC as the time on 1800 UTC.

3. Result and Discussion

3.1 Composite analysis of the MCC evolution.

The life-cycle of the MCCs composite has been quantified by the composite analysis as shown in figure 1. On average, the evolution of the development of the MCCs composite began in the late afternoon at around 1900 LT. The MCC that generated by the orography along the mountain in the northern of the MCC area is one of the contributors to the development of the MCC as shown in figure 1(a) and (b). The system reaches the maximum size around 4 hours from the initiation. The mature stage of the MCC occurs in midnight around 0100 LT as shown in figure 1 c) and (d). This system might be able to survive for 4 hours in maximum size and will begin decayed at around 0400 LT - 0700 LT as shown in figure 1(e) and (f). MCCs became dissipated in the early morning at around 1000 LT - 1100 LT (figure 1(g) and (h)). However interesting for the analyzed from this MCCs development is the existence of convective clouds over the Java Island and southern Sumatra after a few hours of the MCC dissipated, as shown in figure 1(i), (j), (k) and (l).

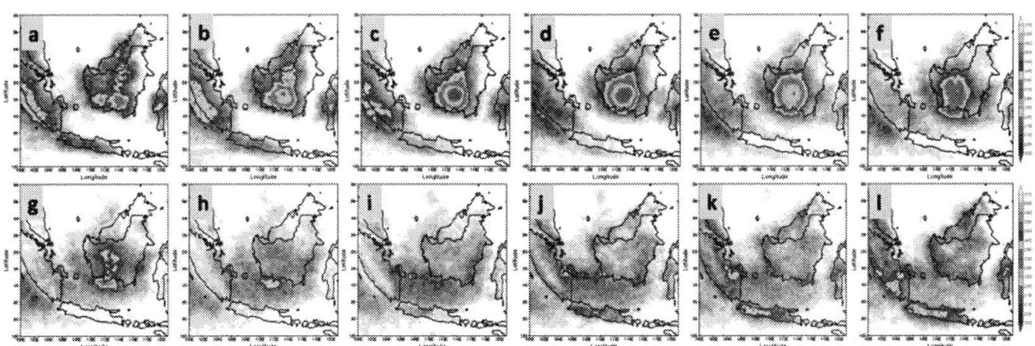

Figure 1. Horizontal distribution of composite of the T_{BB} for MCCs composites: (a), (b) Initial stage (around 1800 LT and 2000 LT)), (c),(d) mature stage (around 2300 LT to 0100 LT), (e),(f) decay stage (around 0400 LT and 0700 LT), (g),(h) dissipation stage (around 1000 LT and 1100 LT), (i) post-MCC stage (1400 LT) and (j), (k), (l) a few hours after post-MCC stage. The time illustrates the average

timing of the hours of initiation, mature, decay and dissipation for the 45-case composite in region 3 in term of the normalized life cycle using composite analyses. Unit for T_{BB} is Kelvin.

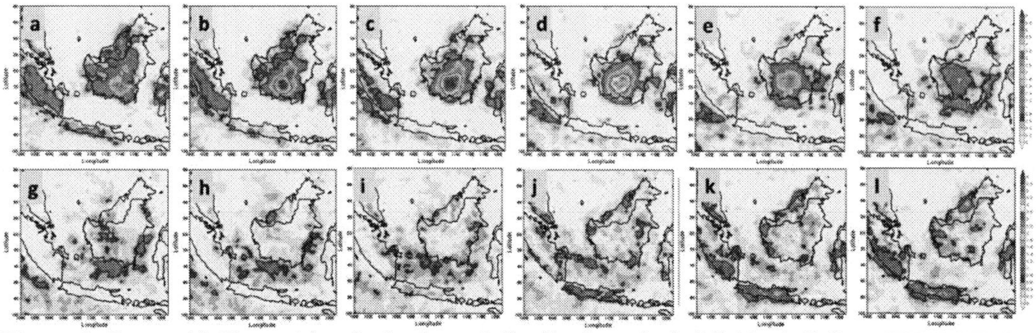

Figure 2. Same with Figure 1 but for horizontal distribution of rainfall (shaded) from TRMM TMPA-RT 3B41RT v7 data. Unit for rainfall is mm/hr.

figure 2 shows that most of the MCCscomposites were developed by the orographic rainfall as shown in figure 2(a) and (b). During the MCC reach maximum size, the rainfall system also reaches the peak value, on average around more than 6 mm/hour as shown in figure 2(c) and (d). The intensity of the rainfall began decreased during the MCC start decayed (figure 2(e) and (f)), but new rainfall clouds begin generated during the MCC dissipated as shown in figure 2(g) and (h). Even the new rainfall clouds were propagated until to Java Island and make the rainfall over Java Island after a few hours from MCC dissipated. The comparison figure 1 and 2 shows that there is relationship about the MCC to the rainfall system over the CK, similar with Trismidianto et al. [6] that have been found the MCC over the IO influencing the rainfall over Sumatra Island.

3.2 The propagation of rainfall systems during MCC events

Figure 3 shows the horizontal distribution of rainfall, wind surface vector anomaly, and surface potential temperature (theta) during the initial, mature, decay and post-MCC stage of the MCC composites. Figure 3(a) shows that during the initial stage in the evening around 1900 LT, the convergent surface wind flow that indicated as the westerly wind from Sumatra, northerly wind from South China Sea and the southerly wind from Java triggered convection through interaction with the strong sea breeze circulation and through the topography lifting effect as a mountain breeze from the north. This consistent with several previous papers that stated the complex topography of the IMC leads to strong land–sea breeze circulations, with integral convection and precipitation over Borneo/Kalimantan Island [15]. During the mature stage of the MCC in midnight around 0100 - 0200 LT based on the composite analysis (figure 3(b)), the convergent wind begin weakened and seen the divergent wind from the leading edge of MCC system. Perhaps that is due to by the CP where the surface theta at the center of the MCC was also relatively low (at around 297.5 K) compared with the surrounding area (298–300 K). According to Engerer et al. [16], this area is the so-called CP, which is an area of downdraft air cooled by evaporation that spreads out horizontally beneath a precipitating cloud. In addition, the CPs were associated with theta decreases [17] and generated by these individual convective cells in an MCS typically spread out over the surface and combine to form a large mesoscale CP covering a contiguous area on the scale of the entire MCS [18]. Trismidianto et al. [6] also reported that the existence of new convective systems generated by the CP during MCC system over the IO.

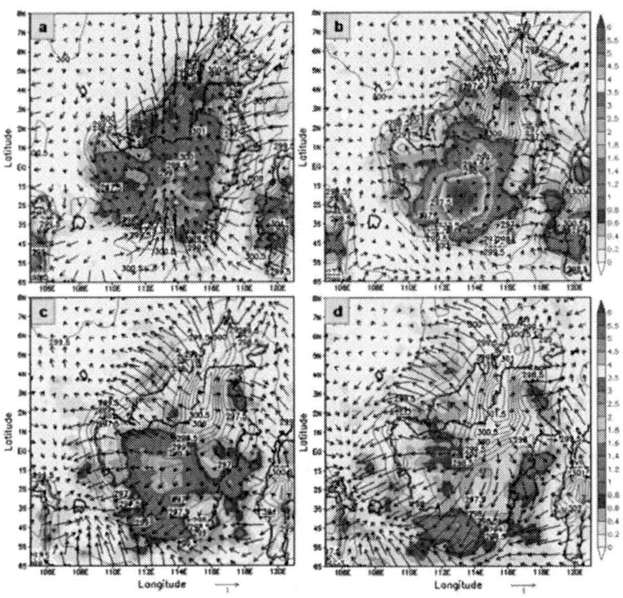

Figure 3. Horizontal distribution of rainfall (shaded) from TRMM TMPA-RT 3B41RT v7 data, wind surface vector anomaly (vector) from CCMP data, and surface potential temperature (contour) from the ECMWF ERA-Interim data during MCCs composite. where; (a) composite in initial time with the dominant time are 1800 and 1900 LT, (b) composite in mature time with the dominant time are 0100 and 0200 LT, (c) composite in decay time with the dominant time are 0700 LT, and (d) composite in post-MCC time with the dominant time are 1200 and 1300 LT.

The CP still see clearly during the MCC decayed which indicated by the existence of the theta decreases in the leading edge of the MCCs system as shown in figure 3(c). The difference in surface theta could have acted as a trigger for the development of new convective systems to form along the leading edge of the CP, which is consistent with the findings of Wilson and Schreber [19]. The several of new rainfall clouds begin developed on the leading edge of the MCCs, which are generated by the CP in the below of MCC system. The new rainfall cloud begins propagated to the surrounding area of the MCCs system due to of the divergent outflow of the CP, in conjunction with the morning land breeze. The divergent wind becomes stronger after the MCC dissipated due to the existence of the mountain breeze as shown in figure 3(d). The rainfall system propagates toward to the Java Sea, this is possible make the rainfall system over Java are more extensive during the evening (1900 LT) after a few hours from MCC dissipated due to the new convective systems induce the land convection over the Java Island as shown in figure 2. Moncrieff and Miller [20] that factors other than the strength of the shear and the CP affect the structure, organization, and longevity of the MCSs or/and MCCs. Therefore, this study gave a deeper insight into the mechanism of the propagation of rainfall system during MCC event. The structure and evolution of the MCC composites are related to the rainfall system.

3.3 Analysis of the environmental condition during MCC

3.3.1 Conditions over the initial region

During the MCC initial stage, the initial region is characterized by weak low-level convergence (surface to ~750 hPa) with a deep layer of a weak divergence (figure 4(a), (b), (c) and (d)). The strong wind convergence in the surface level is very clearly from the north. This is similar with MCC in USA that has been found by Maddox [3] and Cotton et al.[4], they stated that the early stage of the MCC life cycle is characterized by convergence, vertical motion, and heating being centered in the lower troposphere. An important area of convergence can be seen over the center of the MCC area in the black box but that is weak convergences. The initial region is characterized by a short-wave trough pattern of the geopotential height in surface and the low-level (figure 4(a) and (c)). The short-wave trough is also very clearly in temperature field at 850 hPa (figure 4(b)). Figure 5(b) also shows that the relative humidity is higher than the surrounding area and MCC became an alow-temperature center in the level of the 850 hPa which consistent with the Maddox [3].

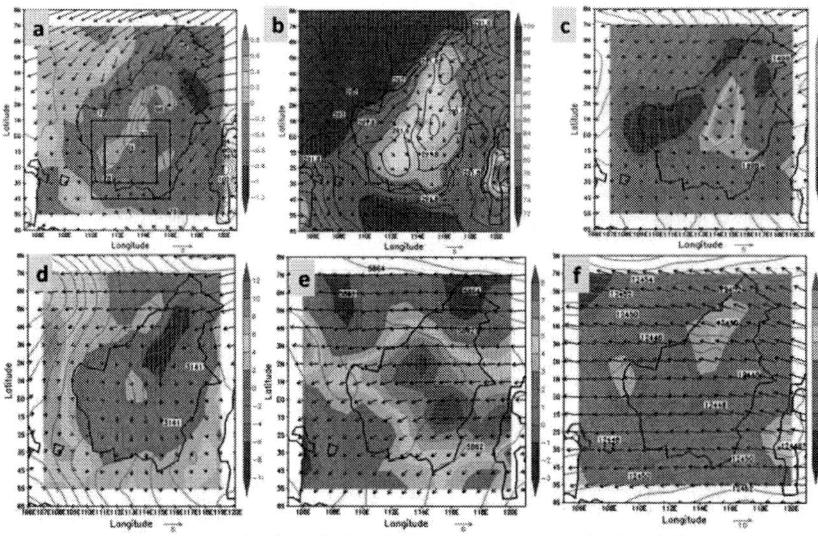

Figure 4. Composite analysis of the MCC composites during initial stage using ECMWF ERA-Interim data, showing (a) 1000-hPa divergence (10^{-5} s^{-1}, shaded), 1000-hPa geopotential height (m, contour) and 1000-hPa wind vector (m/s, vector); (b) 850-hPa relative humidity (%, shaded), 850-hPa temperature (K, contour) and 850-hPa wind vector (m/s, vector); (c) 850-hPa geopotential height (m, contour), 850-hPa moisture flux convergence (10^{-3} g kg^{-1} s^{-1}, shaded) and 850-hPa wind vector (m/s, vector); (d) 700-hPa moisture flux convergence (10^{-3} g kg^{-1} s^{-1}, shaded), 700-hPa geopotential height (m, contour) and 700-hPa wind vector (m/s, vector); (e) 500-hPa geopotential height (m, contour) and 500-hPa vorticity (10^{-6} s^{-1}, shaded) and 500-hPa wind vector (m/s, vector); (f) 200-hPa divergence (10^{-5} s^{-1}, shaded), 200-hPa geopotential height (m, contour) and 200-hPa wind vector (m/s, vector).

The 850 hPa geopotential field and moisture flux convergence (hereafter MFC) shown in figure 4 (c). The initiation of the MCCs is largely driven by MCC-scale moisture convergence in the lower troposphere which is then transported to the midlevel, even up to upper levels through deep convection as shown in figure 4(c) and (d), but the moisture convergence in upper levels is not shown in this study. The initial region is characterized by the region between the positive/negative patterns of the MFC. This region seen clearly at the level of 700 (figure 4(d)), this condition shows that possible to generates the deep convection in this area as initial development for MCCs. MFC is high at a surface level over the mountains that contribute to the development of the MCCs in the CK. The short-wave trough is very clearly in the initial region at 700 hPa as shown in figure 4(d). The southwesterly jet is not clearly seen at 850 hPa but the mid-level jet to the south from the north is clearer (figure 4(c) and (d)). This is a contrast with Cotton et al. [4] and Maddox [3] that shows the strong low-level jet over the initial region. The strong positive vorticity at 500 hPa over the MCCs area indicates counterclockwise rotation of the wind and it is associated with cyclones or storm at upper levels as shown in figure 4(e). This is similar with Maddox [3] that stated the composite MCC initially exhibits strengthening cyclonic vorticity confined within the surface to 700 hPa layer and anticyclonic vorticity located exclusively in the upper troposphere and Laing and Fritsch [10] that stated the MCC in the USA develops just to the northeast of the weak cyclone at 1000 hPa. The divergence on the upper level is weak over the initial region of MCC as shown in figure 4(f). The presence of a low-level convergence and upper-level divergence couplet helps in maintaining the inflow and outflow of mass necessary for long periods of sustained deep convection, while the attendant release of latent heat further enhances the low- to mid-level convergence, thus continuously providing fuel for the MCC systems, But the latent heating is not shown in this study.

3.3.2 Condition over the mature region

Figure 5(a) shows the weak convergence at surface level is still presented during mature stage, but seen clearly the weak divergence in the western and eastern leading edge in the mature region, this is indicates that possibly due to the CP exist in the surface below of MCC system that make the surface wind divergence in the leading edge. However, Maddox [3] stated that the low-level divergence seen in the mature stage show the low-level downdraft on mature stage. The MCC has left the region of surface convergence and the mature region is characterized by weak divergence. The short-wave trough is still clearly seen in the mature region at surface level. The relative humidity is still high more than 88% during at this stage (figure 5(b)), by the mature stage nocturnal cooling, and evaporatively chilled mesohigh at the surface effectively decouple the MCC from the boundary layer, and the system is sustained by convectively unstable air from 900-700 hPa. The MCC exhibits a cold-core structure at surface level (figure 5(b)). This is similar with MCC in the USA by Maddox [3]. The tropopause cooling and rising reach their maximum intensity at the mature stage, reflecting the contribution of mesoscale ascent and longwave radiative cooling by the upper-level cloud shield.

Figure 5(c) and (d) shows the geopotential height and MFC over 850 hPa and 700 hPa, respectively. The region of the positive/negative pattern of the MFC is very clearly at 850 hPa and become stronger in the middle level, according to Maddox [3], The MCC is maintained by intensified mid-level convergence in association with developing mid-level cyclonic shear. The negative of MFC in surrounding area of the mature region shows that the downdraft motion start occurs below 700 hPa. The short-wave trough of geopotential height is still clearly seen at 850 hPa and 700 hPa. The strong wind convergence flows from the northeastern toward the mature area, and there are change the motion of the wind significantly from the north to the south cross the mature region at mid-level 700 hPa. figure 5(e) show the weak positive vorticity during the mature stage, but the vorticity is located in between the two area of negative vorticity, this condition indicates the existence of cyclones or storm at the upper level. The upper-level jet stream and strong divergence are presented during the mature stage that produced by storm's divergent outflow as shown in figure 5(f).

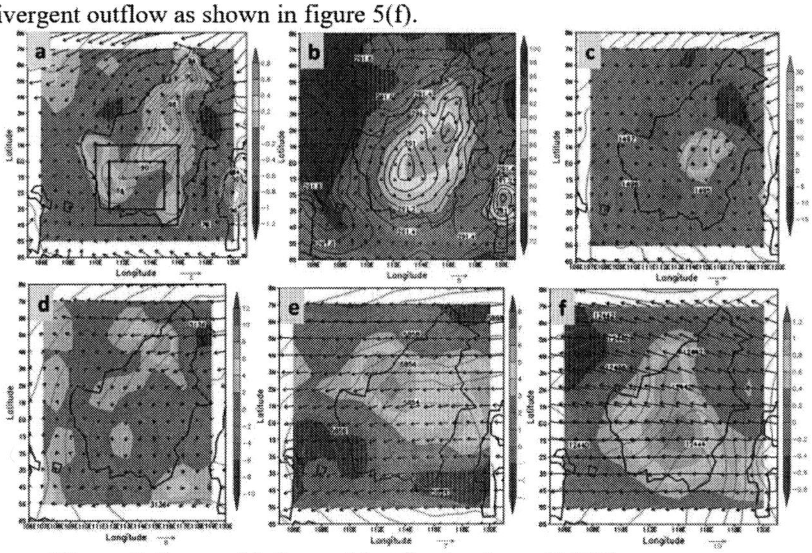

Figure 5. Same with figure 4 but for the time of MCC mature stage.

3.3.3 Condition over the dissipation region

The dissipation region is characterized by the strong surface divergence as shown in figure 6(a). The convergence is start left the MCC area, is similar to Cotton et al. [4]. The strong convergence is presented in the south of dissipation region (on Java Island) so that the strong convergent wind toward to Java Island. At 850 hPa as shown in figure 6(b) and (c), the strongest temperature gradients remain meridional and a cold trough is evident at the western edge of the dissipation region that content the cold advection, this result is very similar to the result by Maddox [3].

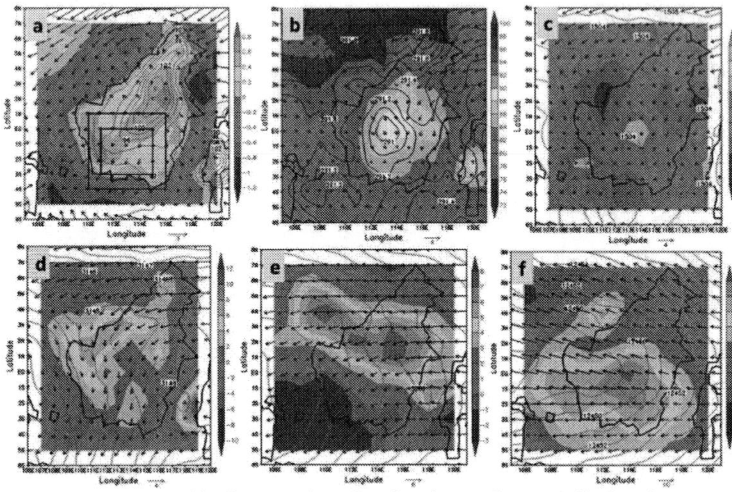

Figure 6. Same with figure 4 but for the time of MCC dissipation stage.

The cold trough makes the MCC as cold-core in dissipation region. The relative humidity at 850 hPa decreased in dissipation region. The MFC is begin left the MCC area to the western south and eastern north leading edge at 850 hPa but the high MFC (negative value) in the south and north of leading edge of dissipation region. In the mid-level the weak MFC is presented in the dissipation region, this indicates that downward motion still occurs in the dissipation region. The short-wave trough remain meridional of geopotential height still found in dissipation region still presents from low-level to the upper level (figure 6(c), (d), (e) and (f)), is similar with Maddox [3] that stated the short-wave trough is very pronounced in the height and temperature field. figure 6(e) shows that the cyclonic shear in the upper level in dissipation region is left from the MCC area. The weak divergence with the strong short-wave trough of the geopotential height still occurs on the upper level, that indicate the jet streak still present in dissipation region as shown in figure 6(f).

3.4 Vertical cross section analysis

The vertical distribution of divergence and vertical velocity during a critical stage of the MCC composites is presented in figure 7. The MCC composites are also triggered by the strong surface and low-level convergence at the initial stage. Surface convergence is confined to the lower half of the troposphere (from surface until 850 hPa), at the same time, strong divergence prevails above 700 hPa until 400 hPa as shown in figure 7 (a). The vertical velocity with the short-wave trough/ridge/trough pattern in the strong low-level convergence area is seen clearly. The negative value of vertical velocity indicates ascending air. The ascending motion is associated with cloudiness and rain. The wind convergences from the upper level to the initial region indicate there is moisture support from the upper-level by the wind. The presence of a low-level convergence and upper-level divergence couplet helps in maintaining the inflow and outflow of mass necessary for long periods of sustained deep convection, while the attendant release of latent heat further enhances the low- to mid-level convergence, thus continuously providing fuel for the MCC systems.

By the mature stage, as shown in figure 7 (b), the development of the MCC in this stage is triggered by the westerly and southerly wind that interacted the mountain breeze which makes merging the clouds system that shown in the initial stage. The upward process in the growth of the cloud has been started in this stage with averaged vertical velocity around 12 x 10^{-2} Pa s^{-1} in the upper level. The convergence during mature became weak in surface until low-level and only in the western side of the system. In fact, the strong divergence began to appear on the surface of the system, it is due to the effect from midnight land breeze. Upper-level strong divergence has developed a sharp maximum at 200 hPa. The updraft also becomes stronger with the averaged vertical velocity around -12 x 10^{-2} Pa s^{-1}. At the dissipation stage, as shown in figure 7 (c), the convergence is becoming dissipated and changed to strong surface divergence. The divergence also expands in upper-level above 300 hPa. However, the strong divergence also appears on the upper level. This condition indicates that the downdraft process already occurs during decay in the system. This convergence is southward widespread, it indicates the new convective system began generated in the southern and northwestern side of the MCCs system. The new convective systems indicated by the propagation of the convergence zone from the system.

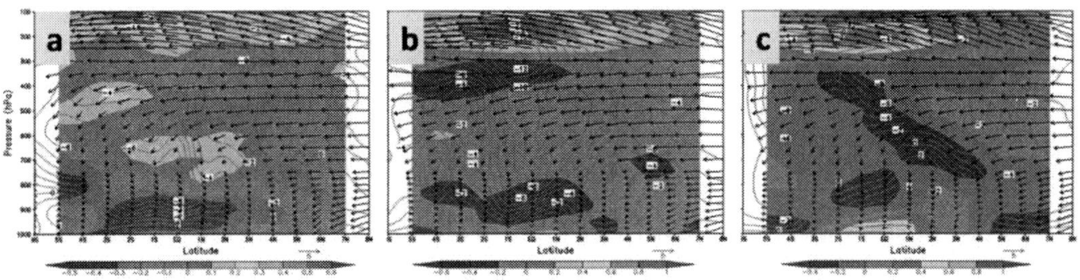

Figure 7. Pressure-latitude cross section (averaged for longitude 110° - 116°E) of divergence (10^{-5} s^{-1}, shaded) and vertical velocity (10^{-2} Pa s^{-1}, contour) for MCC composites during (a) initiation, (b) mature, (c) dissipation.

The vertical distribution of temperature advection and vorticity advection during critical stages of the MCC composite is presented in figure 8. The initial region of the MCC composites is characterized by the strong cold advection in the surface of the initial region. However, the strong warm advection is presented in the level of 900 - 700 hPa, while the weak cold advection appearance in the midlevel until upper level. This condition just followed the positive of vorticity advection in surface until midlevel as shown in figure 8(a). The upward motion also occurs in the initial region by the strong surface cold advection. During mature stage as shown in figure 8(b), cool advection is dominant on the southern edge of the system, possible due to the land breeze. However, the strong warm advection still appears in the lower level of the system. It indicates the existence of the process of updraft and downdraft motion. The surface cold advection is dissipated and widespread throughout the surrounding area of the MCC system during dissipation stage as shown in figure 8(c). The cold advection migrates eastward and westward by helped the westerly wind and sea breeze.

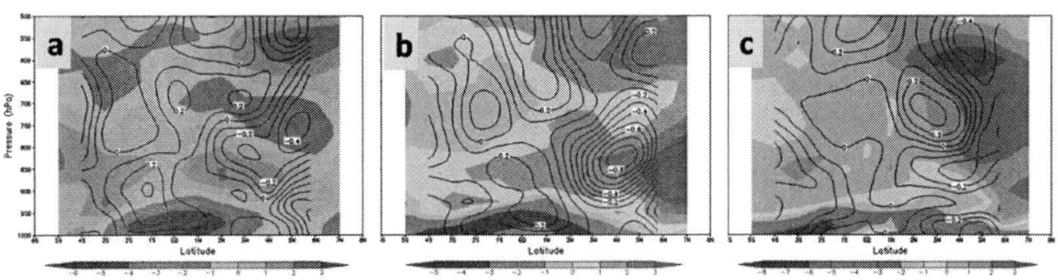

Figure 8. Same with figure 7 but for temperature advection (shaded: 10^{-5} Ks^{-1}) and vorticity advection (contour: 10^{-5} s^{-1}).

4. Conclusion

The composite analysis of the evolution and propagation for 45-cases of the MCC which occurred in the same area center over CK (110°E - 116°E, 4°S - 1°N) has been observed. On average, MCCs began develop in the late afternoon which generated by the orographic convective that interaction with the other convective cloud in the surrounding of the MCC area. The convergent surface wind flow that indicated as the westerly wind from Sumatra, northerly wind from the South China Sea and the southerly wind from Java through interaction with the strong sea breeze circulation and through the topography lifting effect as a mountain breeze from the north is helping the MCC development. The MCCs system reaches the maximum size in midnight and will begin decayed and dissipated in the morning. During the mature stage, the convergent wind begins weakened and seen the divergent wind from the leading edge of the MCC system due to by the CPs which generate the new convective systems when the MCCs decayed. The new convective systems southward propagated toward Java Island due to of the divergent outflow of the CPs, in conjunction with the morning land breeze. The rainfall systems over the Java Island are more extensive after a few hours of the MCCs dissipated due to the new convective systems induce the land convection over the Java Island.This study also shows that there is relationship about the MCC to the rainfall system over the CK.

The composite analysis of the environmental condition during critical stages of the MCC composites has been presented. The initial and mature regions are characterized by weak low-level convergence and upper-level divergence, but during mature, the weak divergence in the western and eastern leading edge in the mature region. The convergence is start left the MCC area when the MCC dissipated, while the weak divergence with still occurs at the upper level. This convergence is southward widespread, it indicates the new convective system began generated in the southern and northwestern side of the MCC system. The new convective systems indicated by the propagation of the convergence zone from the MCC systems. The wind convergences from the upper level to the initial region indicate there is moisture support from the upper-level by the wind. The presence of a low-level convergence and upper-level divergence couplet helps in maintaining the inflow and outflow of mass necessary for long periods of sustained deep convection, while the attendant release of latent heat further enhances the low- to mid-level convergence, thus continuously providing fuel for the MCC systems. The favorite areas for the MCC development are indicated by the short-wave of the geopotential height in each level from initial until MCC dissipated. The MCC composite development is largely driven by MCC-scale moisture convergence in the lower troposphere which is then transported to the midlevel, even up to upper levels through deep convection. MCCs become the center of cold advection and temperature at more a low level which followed the higher relative humidity during develop and then cold core separated when the MCC dissipated that indicate the propagation of the new convective systems. The strong midlevel positive vorticity that indicates counterclockwise rotation of the wind and it is associated with cyclones or storm at upper levels during the initial and mature stage. The upper-level jet stream and strong divergence are presented during the mature stage that produced by storm's divergent outflow.

Acknowledgments
Professor Y. Kodama, who is currently experiencing a serious health problem, contributed substantially to this paper to such an extent that he should be considered a co-author. The authors are grateful to the National Institute of Aeronautics and Space (LAPAN), which is the institution where the first author works. The Ministry of Research and Technology of the Government of Indonesia provided a scholarship to the first author.

References

[1] Maddox R A 1980 Mesoscale convective complexes *Bull. Amer. Meteor. Soc.* **61** 1374-1387
[2] Laing A G and Fritsch J M 2000 The large scale environment of the global population of mesoscale convective complexes. *Monthly Weather Review* **128** 2756 - 2776
[3] Maddox R A 1983 Large Scale Meteorological Conditions Associated with Midlatitude Mesoscale Convective Complexes *Mon. Wea. Rev.* **111** 1475 - 1493
[4] Cotton W R, Lin M S, McAnelly R L, Tremback C J 1989 A composite model of mesoscale convective complexes *Mon. Wea.. Rev.* **116** 939-949

[5] Augustine J A and Howard K W 1991 Mesoscale convective complexes over the United States during 1986 and 1987 *Mon. Wea. Rev.* **119** 1575–1589
[6] Trismidianto, Hadi T W, Ishida S, Moteki Q, Mand A, Iizuka S 2016 Development processes of oceanic convective systems inducing the heavy rainfall over the western coast of Sumatra on 28 October 2007 *SOLA* **12** 6−11
[7] Ismanto H 2011 *Characteristics of mesoscale convective complexes over maritime continent* Magister Thesis (Bandung: Bandung Institute of Technology)
[8] Trismidianto 2012 *Study of mesoscale convective complexes in the Indian Ocean to weather pattern above Sumatra* Magister Thesis (Bandung: Bandung Institute of Technology)
[9] Carvalho L M V and Jones C 2001 A satellite method to identify structural properties of mesoscale convective systems based on the maximum spatial correlation tracking technique (MASCOTTE) *J. of Applied Meteorology. American Meteorology Society* **40** 1683-1701
[10] Laing A G and Fritsch J M 1993 Mesoscale convective complexes over the indian monsoon region *American Meteorology Society* **121** 2254 – 2263
[11] McAnelly R L and Cotton W R 1986 Meso-β-scale characteristics of an episode of meso-α-scale convective complexes *Mon. Wea. Rev.* **114** 1740-1770
[12] McAnelly R L and Cotton W R 1989 The precipitation life cycle of mesoscale convective complexes over the central United States *Mon. Wea. Rev.* **117** 784 - 808
[13] Huffman G J and Bolvin D T 2013 TRMM and other data precipitation data set documentation available online ftp://meso-a.gsfc.nasa.gov/pub/trmmdocs/ 3B42_3B43_doc.pdf (accessed on 8 June 2014).
[14] Dee D P, Uppala S M, Simmons A J, Berrisford P, Poli S, Kobayashi S, Andrae U, Balmaseda M A, Balsamo G, Bauer D, Bechtold P, Beljaars A C M, Van de Berg L, Bidlot J, Bormann N, Delsol C, Dragani R, Fuentes M, Geer A J, Haimberger L, Healy S B, Hersbach H, Holm E V, Isaksen L, Kallberg K, Kohler M, Matricardi M, McNally A P, Monge-Sanz B M, Morcrette J J, Park B K, Peubey C, De Rosnay p, Tavolato C, Thepaut J N, Vitart F 2011 The ERA-Interim reanalysis: configuration and performance of the data assimilation system *Quart. J. Roy. Meteor. Soc.* **137** 553-597
[15] Ichikawa H, Yasunari T 2007 Propagating diurnal disturbances embedded in the madden-julian oscillation *Geophys Res. Lett.* **34** L18811
[16] Engerer N A, Stensrud D J and Coniglio M C 2008 Surface characteristics of observed cold pools. *Mon. Wea. Rev* **136** 4839-4849
[17] Tompkins A M 2001 Organization of tropical convection in low vertical wind shears: the role of cold pools *J. Atmos. Sci.* **58** 1650-1672
[18] Houze R A Jr 2004 Mesoscale convective systems *Rev. Geophys.* **42** 43
[19] Wilson J W and Schreiber W E 1986 Initiation of convective storms at radar-observed boundary-layer convergence lines *Mon. Wea. Rev* **114** 2516-2536
[20] Moncrieff M W and Miller M J 1976 The dynamics and simulation of tropical cumulonimbus and squall lines *Quart. J. Roy. Meteor. Soc* **102** 373–394

Satellite radiance data assimilation for rainfall prediction in Java Region

Novvria Sagita[1,2*], Rini Hidayati[2], Rahmat Hidayat[2] and Indra Gustari[1]
[1]Indonesian Agency for Meteorology Climatology and Geophysics (BMKG)
Jl Angkasa I No 2 Kemayoran Jakarta Pusat 10720, Indonesia.
[2]Department of Geophysics and Meteorology, Bogor Agricultural University (IPB)
Jl. Meranti Wing 19 Level 4 Gedung FMIPA Kampus IPB, Jawa Barat 16680

E-mail: novvria.sagita@bmkg.go.id

Abstract. This study examined the influence of satellite radiance data assimilation for predicting two days of heavy rainfall in the Java region. The first case occurred from 22 to 23 on January 2015 while the second case occurred from 1 to 2 on February 2015. The analysis examined before and after data assimilation in the two cases study. The Global Forecast System (GFS) data were used as initial condition which was assimilated with several data such as surface observation data, radiance data from AMSUA sensor, radiance data from HIRS sensor, and radiance data from MHS sensor. Weather Research and Forecasting Data Assimilation (WRFDA) is a tool which is used in this study for assimilating process with Three Dimensional Variation (3D-Var) method. The Quantitative Precipitation Forecast (QPF) skill was used to evaluate influence data assimilation for rainfall prediction. The result of the study obtained different rainfall prediction with different data assimilation. In general, the surface observation data assimilation has lower QPF skill than the satellite radiance data assimilation. Even thought radiance data assimilation has slightly contribution on rainfall prediction, but it gave better accuracy on rainfall prediction for two heavy rainfall cases.

1. Introduction
One of problem on weather modeling is inaccuracy in initial condition data. According to some previous studies, the solution of inaccuracy in initial condition can be solved by data assimilation [1]. Data assimilation is a combination of various data in order to determine as accurately as possible state of nature [2]. This study combined Global Forecast System (GFS) model data with observation data for rainfall estimation. In this study, data assimilation was used in numerical weather prediction with observational data for increasing accuracy of rainfall prediction [3][4]. Many sources of observational data for data assimilation, e.g. surface observation data, radiance satellite data, and radar data. Combining observation data from many sources with the GFS model use the data assimilation technique. There are two approaches of data assimilation namely sequential assimilation and non-sequential assimilation. Sequential assimilation is data assimilation consider observation data from past until time analysis. Non-sequential assimilation is data assimilation considers observation data from past until future time analysis [5].

Nowadays, improving accuracy of numerical weather prediction for heavy rainfall event is widely required for warning information for operational purposes. The heavy rainfall condition is rainfall which has intensity 50 millimeter/day or 10-20 millimeter/hour [6]. This study examines two cases of heavy rainfall events in Java i.e. on 22–23 January 2015 (first case) and 1-2 February 2015 (second case). Those heavy rainfall events were identified as severe flood in Jakarta and surrounding area. The early analysis used wind analysis, to know characteristic of rainfall event and potential location of cloud growth. Figures 1 show the wind vector at 850 hPa at 00.00 UTC of 22 January 2015 and 00.00 UTC of 01 February 2015 derived from JRA-55 reanalysis. Figure 1a shows wind vector and convergence pattern over the Java Sea. This condition influence strong convective process along the Java Region.

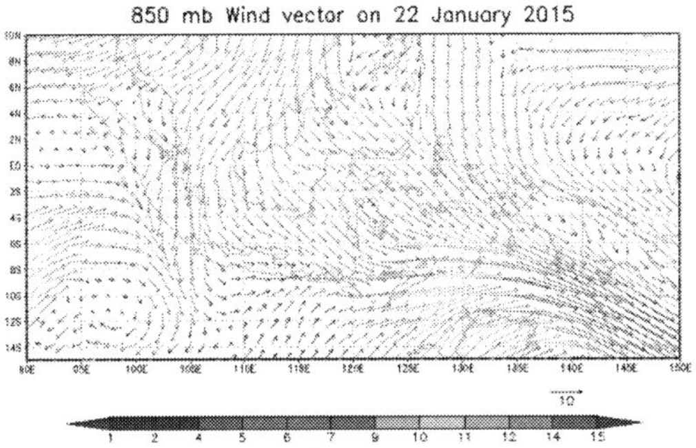

(a)

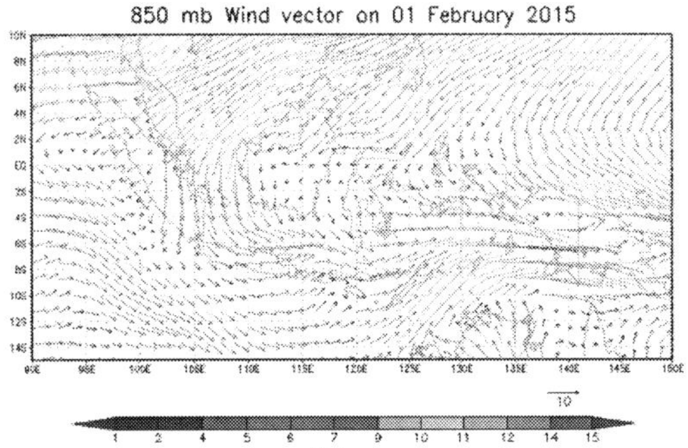

(b)

Figure 1. Gradient Wind Analysis on 22th January 2015 (a) and 01th February 2015 (b) [7].

Figure 1b shows wind vector on Java Sea, so it can generate highly cloud growth in the along Southern Sumatra until Western Java. The extreme rainfall event occurred 22-23 January 2015 and 1-2 February 2015 at Java region especially Jakarta area. The daily rainfall reached 134 mm on 22-23 January 2015 and 75 mm on 1–2 February 2015. The object of this study examined influence surface observation data and satellite radiance data assimilation for prediction of rainfall.

2. Data and Methodology

Weather Research and Forecast Data Assimilation (WRFDA) [8][9] is a tool for data assimilation. This study uses Global Forecast System (GFS) data with resolution 0.5° x 0.5° [10], surface observation data in the BUFR format [11], satellite radiance data from many sensors (AMSUA, HIRS, MHS) [11]. This study used rain rate data from GSMAP product for verifying the output of the models. The domain of this study is a Java region with resolution 9 kilometers (figure 2). Red circles on domain map are location of surface observational stations which data are used for data assimilation and data verification. The rainfall physical parameterization schemes were used that is the Thompson graupel microphysical parameterization [12], and the Betts-Miller-Janjic scheme cumulus parameterization [13].

Satellite radiance data were used from NOAA, METOP, and EOS satellite. This study used three kinds sensor of radiance data: AMSU-A, HIRS, MHS.

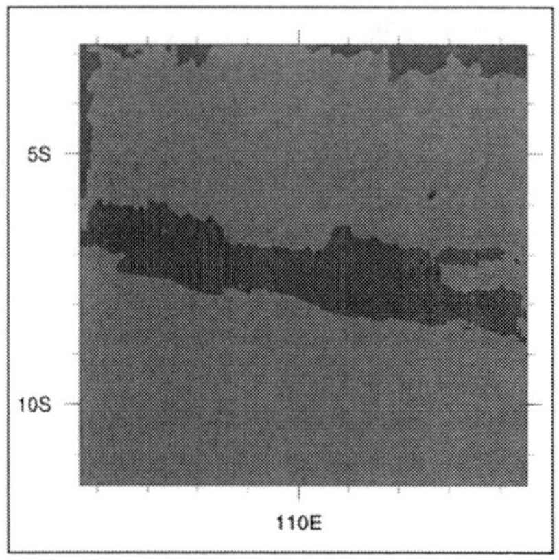

Figure 2. The domain used to model simulation

AMSU-A is The Advanced Microwave Sounding Unit-A. HIRS is a 20-channel infrared scanning radiometer that performs operational atmospheric sounding. Microwave Humidity Sounder (MHS) is a self-calibrating, cross-track scanning, five-channel microwave, full-power radiometer, operating in the 89 to 190 GHz region [14].

The method that was used for the data assimilation is a three-dimensional variation analysis (3D-Var). Formula of 3D-Var is [9]:

$$J(x)=\frac{1}{2}(x-x_b)^T B^{-1}(x-x_b)+\frac{1}{2}[y-H(x)]^T R^{-1}[y-H(x)] \quad (1)$$

J (x) is a function that calculates the cost function between models with observational data surface. The calculation of the cost function J (x) describes in equations 1 where x and xb is an analysis of expected data and the background data.

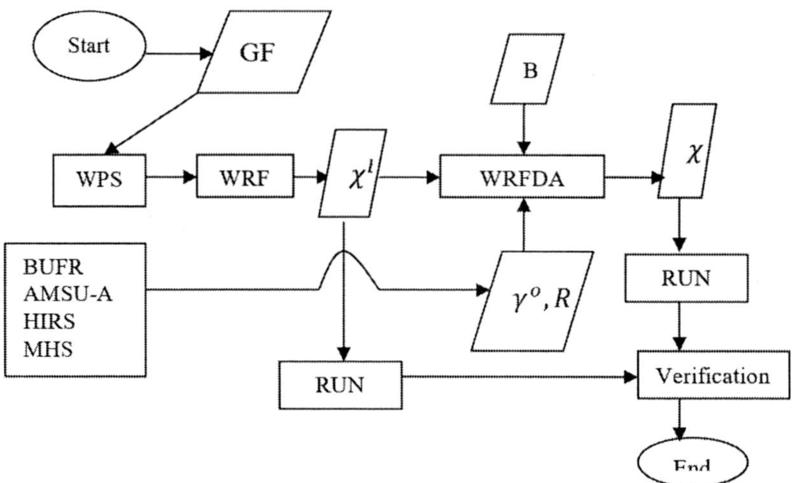

Figure 3. Flowchart.

H is the observation operators include Radiative Transfer Model, R is the observation error covariance operator, and y^o is an observation data in the model grid.

The flowchart of this study can be seen in figure 3. Earlier, this study sets the domain of research on WRF Processing System (WPS) which used GFS as input data. To get initial condition, this study run

output WPS on WRF. Three components are required on WRFDA such as initial condition χ^b, observation data γ^o, R, and background data (B). This study used the data assimilation with the background of global error that has been provided in the application WRFDA. An observation data will be assimilated. Observation data has an error value that was shown by the parameter R. The output of WRFDA is second guess of initial condition χ^a. The first guess of initial condition and second guess of initial condition were used in the forecast process which resulted without data assimilation and data assimilation. The last is verify process by comparing QPF skill between without data assimilation, surface observation data assimilation, AMSU-A radiance data assimilation, HIRS radiance data assimilation, and MHS radiance data assimilation. QPF method used threat score (TS) and false alarm ratio (FAR).
The TS was defined in equation 2 as:

$$\text{Threat Score (TS)} = \frac{\text{Hits}}{\text{Hits+False Alarm+Misses}} \quad (2)$$

The FAR score defined in equation 3 as :

$$\text{False Alarm Ratio (FAR)} = \frac{\text{False Alarm}}{\text{Hits+False Alarm}} \quad (3)$$

Hits are amounts of the event that are detected by observation and forecasting, false alarm is the amount of no event detected by observation, but event detected by forecasting, and misses is the amount of event detected by observation, but no event detected by forecasting.

3. Result

Figure 4 here shows six maps of 24 hours accumulated rainfall in Java region on 22 – 23 January 2015. Figure 4 (a) is 24 hours accumulated rain rate from GSMAP data. GSMAP data is accumulated rainfall that is measured by many satellites. Five maps others are accumulated rainfall prediction from (a) without data assimilation, (b) surface observation data assimilation, AMSU-A data assimilation, HIRS data assimilation, and MHS data assimilation. The severe weather happened in Northern West Java on 22 – 23 January 2015. This was indicated by heavy rainfall more than 80 mm. Comparing rainfall predicted from WRF model, WRF model without data assimilation (b) result heavy rainfall smaller area than others (c, d, e, f). Without data assimilation analysis gave result rain in most of area domain, but observation data from GSMAP gave information there was no rain in many Indian Ocean areas. Improvement result has been produced from surface observation data assimilation and MHS data assimilation, many areas in the Indian Ocean were no rain. This was better than without the data assimilation scheme. Whereares forecasting rainfall by data assimilation still gave overestimate rainfall like without data assimilation, but data assimilation reduced overestimate rainfall.

Figure 5 presents 24 hours accumulated rainfall between 00.00 and 24.00 UTC on 22 January 2015 from both observation data from GSMAP product (a) and scheme of experiments: without data assimilation (b) surface observation data assimilation (c), AMSU-A data assimilation (d), HIRS data assimilation (e), and MHS data assimilation (f). According observation data from GSMAP product (a), Northern West Java was an area where was main heavy rainfall happened (Figure 5a). Compared with observation, the main heavy rainfall region in most areas in Java sea generated from all experiments (Figure 5b, 5c, 5d, 5e, 5f). All experiments have resulted that main heavy rainfall prediction in land is smaller region than the sea. HIRS data assimilation (Figure 5e) reduced overestimation rainfall in Java Sea. Evenly, all experiments have resulted overestimate rainfall.

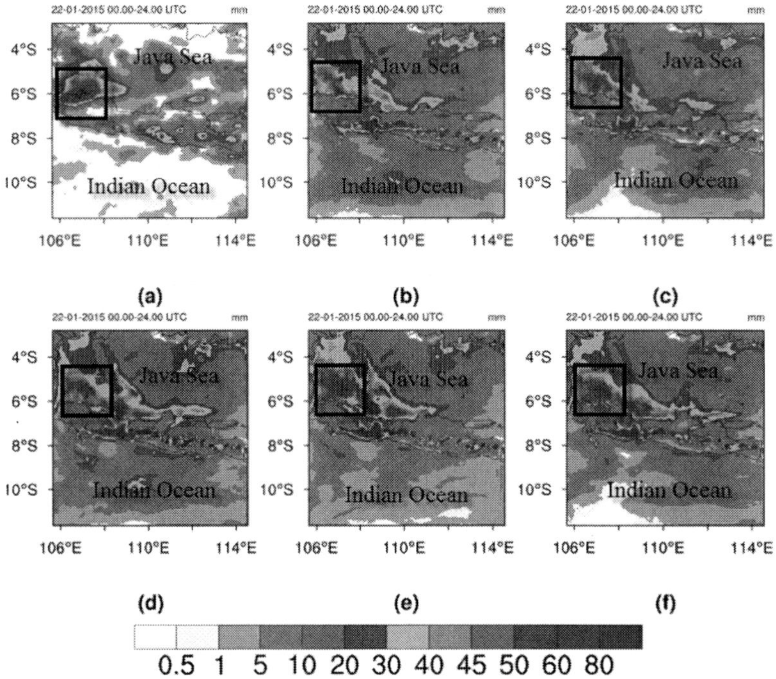

Figure 4. Map of 24 hours accumulated rainfall on 22 – 23 January 2015: (a) GSMAP, (b) without data assimilation, (c) surface observation data assimilation, (d) AMSU-A data assimilation, (d) HRS data assimilation, and (e) MHS data assimilation.

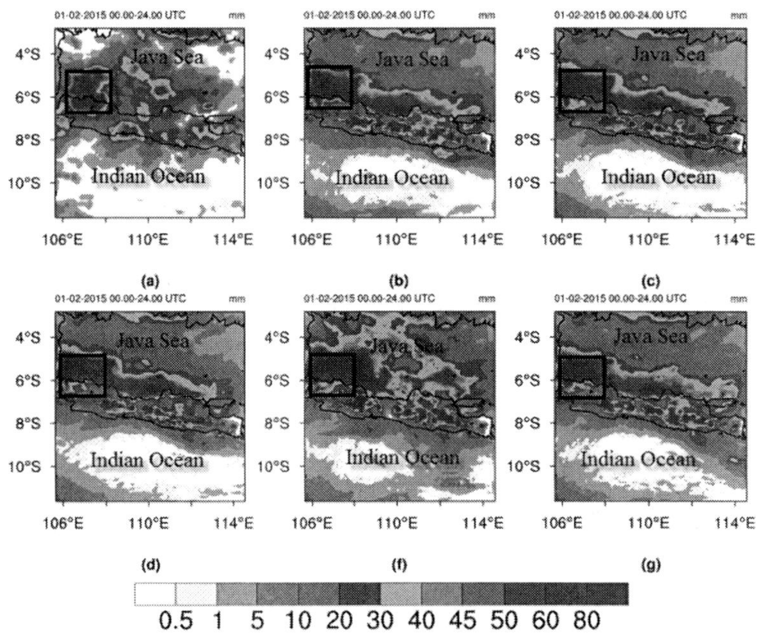

Figure 5. Map of 24 hours accumulated rainfall on 1-2 February 2015: (a) GSMAP, (b) without data assimilation, (c) surface observation data assimilation, (d) AMSU-A data assimilation, (d) HIRS data assimilation, and (e) MHS data assimilation.

Quantitative Precipitation Forecast (QPF) skill used threat score (TS) and false alarm ratio (FAR) score for evaluating all experiments result on rainfall prediction. A higher TS indicates a more skillful rainfall prediction. A lower FAR score indicates a few error rainfall predictions. The range value of TS and FAR score are between 0 and 1. The maximum value of TS and FAR are 1.

Figure 6 shows threat score (TS) (figure 6a) and false alarm ratio (FAR) (figure 7.a) for 24-hourly accumulated rainfall with threshold of 1 mm, 5 mm, and 10 mm from all experiments for case 22 – 23 January 2015. For the verifications in all thresholds , the TS in MHS data assimilation experiment are bigger than other experiment (figure 6a). Verification in 1 mm threshold, the TS in AMSU-A data assimilation is smaller than others experiment. Verification in 5 mm threshold, the TS in HIRS data assimilation is smaller than others experiment. Verification in 10 mm threshold, the TS in AMSU-A and HIRS data assimilation are smaller than others experiment. For the verifications in all thresholds, the FAR score in MHS data assimilation is smaller than another experiment. It was indicated MHS data assimilation has the smallest error. Verification in 1 mm threshold, the FAR score in AMSU-A data assimilation is higher than others experiment. Verification in 5 mm threshold, the FAR score in AMSU-A and HIRS data assimilation is higher than others experiment. Verification in 10 mm threshold, the FAR score in AMSU-A and HIRS data assimilation is higher than others experiment. MHS data assimilation for case 22 – 23 January 2015 was the best rainfall prediction that was indicated by the highest TS and smallest FAR score.

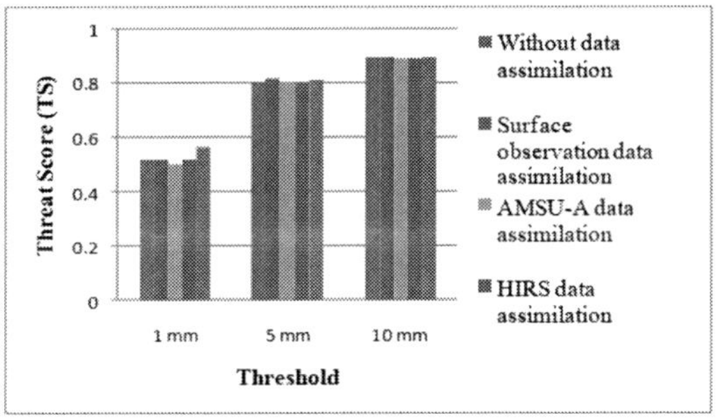

(a)

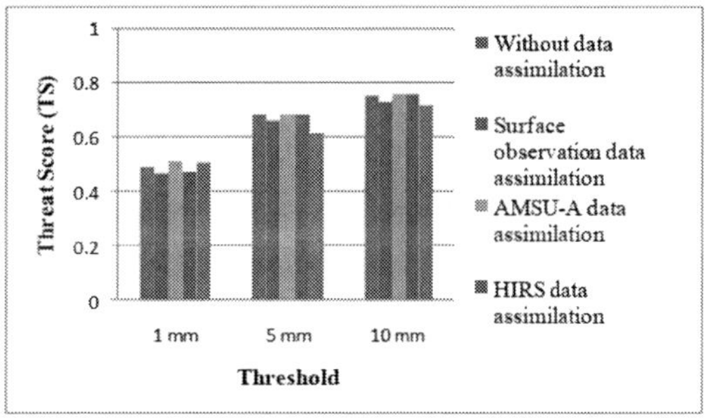

(b)

Figure 6. Threat Score (TS) for 3-hourly accumulated rainfall with threshold of 1 mm, 5 mm, and 10 mm from five experiments for case study 22 – 23 January 2015 (a) and case study 1-2 February 2015.

Figure 7 shows threat score (TS) (figure 6b) and false alarm ratio (FAR) (figure 7b) for 3-hourly accumulated rainfall with a threshold of 1 mm, 5 mm, and 10 mm from all experiments for case study 1–2 February 2015. Verification in 1 mm threshold, the TS in AMSU-A and MHS data assimilation is higher than others experiment. Verification in 5 mm threshold, the TS in AMSU-A and HIRS data assimilation is higher than others experiment. Verification in 10 mm threshold, the TS in AMSU-A and HIRS data assimilation are smaller than others experiment. For the verifications in all thresholds, the FAR score in MHS data assimilation is smaller than others experiment. It was indicated MHS data assimilation have smallest error. Verification in 1 mm threshold, the FAR score in AMSU-A data assimilation is smaller than others experiment. Verification in 5 mm and 10 mm threshold, the FAR score in AMSU-A and HIRS data assimilation is smaller than others experiment 11 August 2016. Generally, all radiance data assimilation experiments have given improving rainfall prediction in all rain threshold for case 1-2 February 2015.

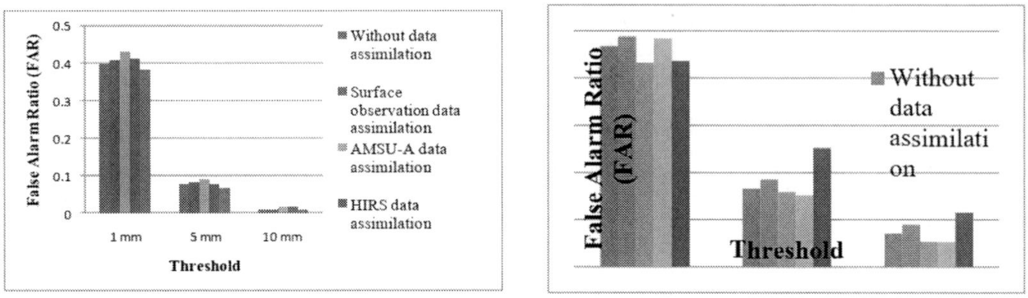

(a) (b)

Figure 7. False Alarm Ratio (FAR) (a) score for 3-hourly accumulated rainfall with threshold of 1 mm, 5 mm, and 10 mm from five experiments for case study 22–23 January 2015 (a) case study 1-2 February 2015 (b).

4. Summary and Conclusions

This study was one of the effort to improve rainfall prediction of Weather Research and Forecasting (WRF) model. Data assimilation experiments for 22 – 23 January 2015 and 1 – 2 February 2015 heavy rainfall event over Northern West Java was performed. This study was emphasized in evaluating the additional influence of radiance data assimilation on rainfall prediction. Data assimilation for all experiments still indicated overestimate rainfall prediction. Radiance data assimilation generated slight influence, but generally positive influence on region main heavy rainfall area. On 22 – 23 January 2015 MHS data assimilation was better than other experiment data assimilation showed by the smallest FAR score and highest TS. On 1 – 2 February 2015 AMSU-A data assimilation better than others experiment data assimilation showed by higher TS than others and smaller HIRS than others. The different of the best radiance data assimilation was caused different amount daily radiance data which was collected by satellite.

Acknowledgments

The author would like to express many thanks to Japan Aerospace Exploration Agency, Earth Observation Research Center (JAXA / EORC) for providing GSMAP data and to Indonesia agency for Meteorology, Climatology, And Geophysical Agency (BMKG) for providing surface observation data.

References

[1] Kalnay E 2003 *Atmospheric modeling, data assimilation, and predictability* (New York: Cambridge University Press) p 18

[2] Talagrand O 1997 Assimilation of observation, an introduction *J. Met.Soc Japan Special* **75** 191-209

[3] Xue M, Wang D, Gao J, Brewster K, Droegemeier et al The Advanced Regional Prediction System (ARPS), storm-scale numerical weather prediction and data assimilation *Meteorol. and Atmos. Phys.* **82** 139-170

[4] Wang X et al. 2008 A hybrid ETKF-3DVAR data assimilation scheme for the WRF model Part I: Observing system simulation experiment *Monthly Weather Review* **136** 5116-5131

[5] Bouttier F, and Courtier P 2002 *Data assimilation concepts and methods* Meteorological Training Course Lecture Series ECMWF

[6] BMKG 2010 *Prosedur standar operasional pelaksanaan peringatan dini, pelaporan & diseminasi informasi cuaca ekstrim* Badan Meteorologi, Kimatologi, dan Geofisika.

[7] Okamoto K, Iguchi T, Takahashi N, Iwanami K and Ushio T 2005 The Global Satellite Mapping of Precipitation (GSMaP) project *25th IGARSS Proceedings* 3414-3416

[8] Skamarock W C, Klemp J B, Dudhia J, Gill D O , Barker D M, Wang W and Powers J G 2005 *A description of the advanced research WRF version 2 (No. NCAR/TN-468+ STR)*

National Center For Atmospheric Research Boulder Co Mesoscale and Microscale Meteorology Div.
[9] Liu Z 2006 Radiance Data Assimilation in WRFDA 2011 Summer WRFDA tutorial.
[10] Rutledge G K, Alpert J, Ebuisaki W 2006 NOMADS: A Climate and Weather Model Archive at the National Oceanic and Atmospheric Administration *Bull. Amer. Meteor. Soc.* **87** 327-341
[11] National Centers for Environmental Prediction/NCEP Department of Commerce 2008 NCEP ADP Global Upper Air, GDAS Satellite Data and Surface Weather Observations May 1997 - Continuing Research Data Archive at the National Center for Atmospheric Research, Computational and Information Systems Laboratory Boulder CO [Available online at http://rda.ucar.edu/datasets/ds337.0/.] Accessed 11August 2016
[12] Thompson G, Rasmussen R M, Manning K 2004 Explicit Forecasts of Winter Precipitation Using an Improved Bulk Microphysics Scheme Part I: Description and Sensitivity Analysis *Monthly Weather Review* **132** 519–542
[13] Betts A K and Miller M J 1986 A new convective adjustment scheme Part II: Single column tests using GATE wave, BOMEX, ATEX, and Arctic Airmass data sets QQ *J. R. Meteorol. Soc.* **112** 693-710.
[14] Goodrum G, Kidwell K B, Winston W 1999 NOAA KLM user's guide (US: US Department of Commerce, National Oceanic and Atmospheric Administration, Satellite Services Branch)

Estimation of river based transportable volcanic material distribution using satellite DEM and precipitation data

A Wisoyo[1*] and S S Putra[2]

[1] Barunadri Engineering Consultant, JL RS. Fatmawati, No. 50 A-9, Commercial Area Fatmawati Festival, Jakarta 12430, Indonesia
[2] Balai Sabo, Ministry of Public Works & Housing, Sopalan, Maguwoharjo, Yogyakarta 55282, Indonesia

E-mail: wisoyoandre@gmail.com

Abstract. The transportable volcanic materials of Mt. Marapi, which are deposited around the caldera, as a result from last eruption in November 2015, must be estimated as a first step to handle the lahar/ debris flow disaster. In this research, the method used to determine the amount of river-based transportable volcanic material distribution was offered. The LIDAR Satellite DEM and TRMM data from Global Precipitation Climatology Project (GPCP) have been used to estimate the deposited volcanic materials. Based on the GPCP data analysis, it was found that the rainfall pattern distribute into two area, which are 0.52 mm/ hr on west side and 0.61 mm/ hr on east side relative to the mountain summit. The deposited materials from Mt. Marapi 14 November 2015 eruption (volcanic boulders and lava) were located in the upstream of six prioritized watershed. The transportable volcanic material will predominantly flows to the South and North West direction. The potentially transported boulder and lava are around 80.33 % of the total erupted material that are deposited in the river upstream. Batang Kadurang Watershed has the highest transportable material of 1,905.18 m^3. The results of study can be used as a rapid disaster countermeasure for lahar disaster mitigation.

1. Introduction

Mt. Marapi, located close to Padang city in West Sumatera, Indonesia, has long history of eruptions that were ever recorded [1]. Instead of all these worrisome evidence, many people still choose to live in the mountain slope. It was noticed that there are about 57,453 inhabitants live within 10 km range from the mountain top in the recent period [2]. The eruption of Mt. Marapi volcano on November 14, 2015 had urge the river authority to prepare the lahar flow disaster mitigation strategy and planning in the region. In order to attain that objective, the river authority needs the estimation of transportable volcanic material distribution around the river on the volcano slope. The site visit to the volcanic material deposited area of Mt. Marapi on May 8, 2016 had shown that the material is very exposed to the rain triggered erosion [3]. The erupted materials that are mainly formed of sand, gravel, and boulders were depicted in Figure 1. The large amount of deposited materials on Mt. Marapi raises our research question on how the materials will be distributed into the watersheds within the mountain slope. There is no measurement at the top of Mt. Marapi. Therefore the method used in this paper aims to be an optimum solution when no erupted material deposit measurement data is available. This research proposes a new method on using the global free availability satellite data for river-based transportable volcanic material distribution analysis.

Figure 1. Sediment deposited in the top of Mt. Marapi volcano, West Sumatera, Indonesia.

2. Related works

In the recent years, there are several data that are used for volcano erupted material deposition study. The common ground data collection is Thermal Infrared (TIR) sensor collected with airborne LiDAR (Light Detection and Ranging). The mentioned method will show the hot erupted material from the crater. The latest research demonstrates the use of cosmic-ray muon radiography (muography) that will produce a quantitative mass loss inside the crater during the eruption [4]. The two method utilize some sophisticated and exorbitant ground instrument risky operation within a volcanic eruption event. An approach in using the satellite imagery for estimating the volcanic erupted material volume had also been initiated. A recent paper mentions the use of ALOS Interferometric Synthetic Aperture Radar (INSAR) as an alternate for Advanced Spaceborne Thermal Emission and Reflection Radiometer (ASTER), Spinning Enhanced Visible and InfraRed Imager (SEVIRI), and Moderate Resolution Imaging Spectroradiometer (MODIS) [5]. However, these methods are highly dependent to the data availability within a certain spatial and temporal variability of the eruption event. In the case of 14 November 2015 Mt. Marapi eruption event, the mentioned instruments and data are hardly available for the river authority. The most reasonable method that is practical to be used by the local river authority is the Volcanic Explosivity Index (VEI) approach. The VEI analysis then usually combined with the lahar source zone digitation in order to map the material dispersion [6]. The calculation of deposit areas and volumes in June 2011 from satellite imagery and the average deposits thicknesses estimations from the field for each affected river can be used to estimate the material distribution within the adjacent volcanic river [7]. This research is trying to involve the rainfall pattern in the volcanic area as a governing parameter for the river based transportable volcanic material distribution analysis.

3. Methodology

The LIDAR Satellite Digital Elevation Model (DEM) and Tropical Rainfall Measuring Mission (TRMM) data from Global Precipitation Climatology Project (GPCP) are used to estimate the transportable volcanic material into the watershed within a single phase of volcano activity. All the dataset from Mt. Marapi case study are recommended to have EPSG: 32747 - WGS 84/ UTM Zone 47S coordinate system projection, as previewed in Figure 2. In order to achieve the goal, the erupted volcanic material needs to be estimated using Volcanic Explosivity Index (VEI) [8] and visual eruption plume height measurement (simple classical method that is used by The Mt. Marapi Monitoring Station Personnel). The technique is explained in equation 1, as follows:

$$EPh = (D_{CoE} \times \tan(\theta_{ToP})) - (Z_{lip} - Z_{sta}) \tag{1}$$

The eruption plume height (EPh) is calculated by subtracting the visual plume height with the crater lip reference. The visual plume height can be calculated by multiplying the horizontal distance from the centre of eruption point to the monitoring station (D_{CoE}) with tangent of top of plume viewpoint (θ_{ToP}). The D_{CoE} can be quantified from the google earth measurement by knowing the centre of eruption coordinate from the seismic analysis. On the other hand, the θ_{ToP} was measured by the volcano monitoring person by using inclinometer. The elevation of the crater lip (Z_{lip}) and the monitoring station (Z_{sta}) is usually had been recorded before the eruption event. All of these variables were in SI Unit. On 14 November 2015, the centre of eruption coordinate was projected at 0°23'28.44"S and 100°27'25.04"E with the crater lip elevation of + 2,716 m a.s.l. (metres above sea level). The θ_{ToP} is in the range from 7.55° to 7.9°, while the D_{CoE} was measured ± 17,270 m. Meanwhile, the monitoring station is located at 0°27'20.3"S and 100°35'53.9"E with the elevation of + 472 m a.s.l. Therefore, the calculated EPh value is in the range 50 m to 150 m. The symbol's indexes are available at https://goo.gl/PzOdSU.

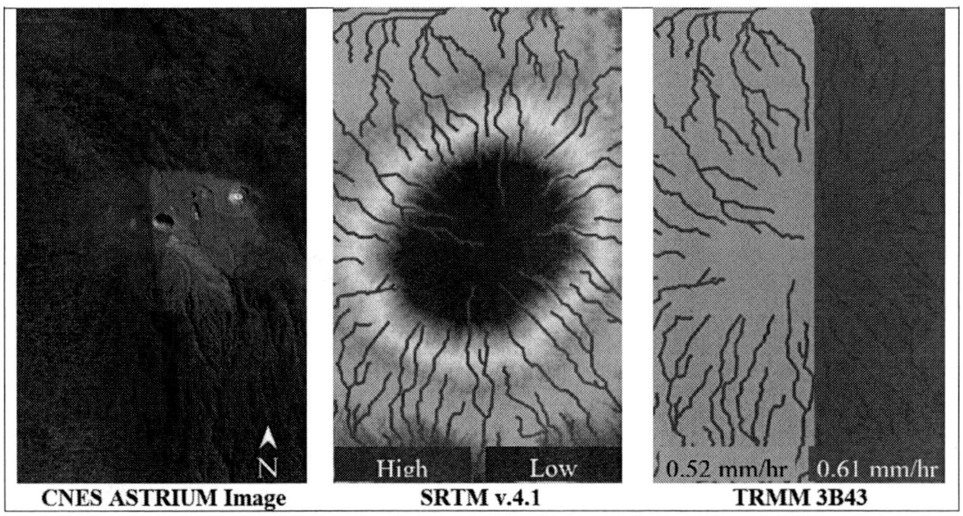

Figure 2. The open access satellite data that are available around the Mt. Marapi.

3.1. Satellite DEM analysis

The DEM that is used in this research is CIGAR-CSI SRTM v.4.1 data. This Shuttle Radar Topographic Mission DEM version 4.1 is an improved version, where the no data grid had been filled with a certain interpolation algorithm [9]. The data can be freely downloaded from the official CGIAR CSI website (http://srtm.csi.cgiar.org/SELECTION/inputCoord.asp). Another alternative DEM data that has potential to be used is ASTER GDEM v.2 data [10].

The elevation data that is encapsulated in the DEM and was analysed within the Geographical Information System Software (GIS). The input for the analysis is the watershed outlets within the radial range of 7 km from the Mt. Marapi crater. The DEM analysis outputs will be a sink filled raster, an AGREE surface reconditioned raster, a flow direction raster, a flow accumulation raster, a stream definition raster, a watershed definition raster, and a watershed boundary shapefile. In case of no watershed boundary was created, the selected watershed outlets must be shifted to exact overlaid the stream definition grid. The watershed boundary was used to calculate the watershed coverage area, to crop the GPCP data, and to determine the volcanic deposit watershed intersection area.

3.2. GPCP data analysis

The precipitation pattern analysis elaborate the monthly TRMM data, which is also freely downloadable from NASA MIRADOR Data Access Website (http://mirador.gsfc.nasa.gov). In concurrence with the aim of only identifying the rainfall pattern, the 3B43: Monthly 0.25 x 0.25 degree merged TRMM data is considered to be chosen. The 3B43 TRMM data was merged from 3B-42 TRMM data with some rain gauge estimates improvements [11].

The rainfall data were correlated and accumulated to each watershed. The rainfall data of November 2015 up to March 2016, which are 152 days in duration, in Mt. Marapi area were analysed based on the finding that hot pyroclastic material will not directly transported by rainfall occurrence until a certain period of time [12]. The first step before the rainfall accumulation step is to correlate the rainfall data with each watershed. This technique will result a rainfall intensity value that corresponds with each definite watershed. If there are more than one rainfall intensity value within a single watershed boundary, the final rainfall intensity value was weighted average rainfall data based on the rainfall value coverage area. The final step is to accumulate the rainfall within the selected period. It is necessary to remember that the rainfall unit in TRMM data is in mm/hr. Consequently, the accumulated rainfall within the month can be ensued by multiplying the rainfall intensity value with the number of days of the month and a constant of 24 (twenty four, which representing the number of hour in a day).

3.3. Transportable material estimation

The transportable volcanic material estimation into each watershed was calculated based on equation 2. The area of the watershed has 20 % weight and the watershed deposit intersection was attributed with 40 % weight. The reason is that the expected proportion of the material is about 59 % of boulders and 41 % of lapilli and ash, which is a typical for a Strombolian eruption [13,14]. The rest of 40% weight was appointed to the rainfall distribution variable due to the concern that rainfall is the main triggering factor for transporting the deposited material. The watershed rainfall data was considered based on its deviation divided by sum of deviation, as stated on equation 2 and 3.

$$V_i = \left(\left(A_i . A_{tot}{}^{-1}\right) \times 20\% + \left(D_i . D_{tot}{}^{-1}\right) \times 40\% + I_i \times 40\%\right) \times V_{tot} \quad (2)$$

$$I_i = (|x_n - m(X)|) \times (\textstyle\sum_1^n |x_n - m(X)|)^{-1} \quad (3)$$

The mentioned variables are watershed index (i), rainfall period index (n), transportable volcanic material in each watershed in m^3 (V_i), total volume of the erupted material in m^3 (V_{tot}), a certain watershed area in m^2 (A_i), sum of all watershed area in m^2 (A_{tot}), sum of all deposit area in (D_{tot}), rainfall deviation proportion for a certain watershed (I_i), rainfall in a certain period in mm (x_n), and rainfall central tendency or rainfall data mean in mm ($m(X)$). The watershed deposited intersection (D_i) is the intersection of watershed boundary and the deposited area boundary, measured in m^2. The deposited area boundary is digitized based on Landsat 7 ETM+ SLC-OFF that was acquired on 23 June 2015, Landsat 8 OLI that was acquired on 14 April 2016, and Google Earth CNES Astrium that was taken on 15 October 2015 from + 8,310 m a.s.l. The Landsat Satellite imagery is freely available from The USGS webpage (http://landsatlook.usgs.gov/viewer.html).

4. Result and discussion

The deposited materials from Mt. Marapi 14 November 2015 (volcanic boulders and lava) were identified in the upstream of 6 (six) prioritized watershed. The boulders and lava covered ± 2.36 km^2 area on the mountain top. The pyroclastic materials (lapilli, sand, and ash) were spread onto a total of 28 lahar-risked watersheds around the volcano slope. The eruption has volcanic explosivity index of 1, with the plume height of ± 150 m [15,16]. These facts are used for estimating the deposited material in the mountaintop, which is resulting a maximum number ± 10,000 m^3 total volcanic materials. The deposited material site within the volcano vicinity is appropriate presented on Figure 3. The thickness of the sediment deposition cannot be presented here due to the lack of sediment boring log data in the location. The Figure 3 also previews the separation of pyroclastic and lahar zone within the radial boundary of 3 km, 5 km, and 7 km from the mountain top. This boundary were generated based on the latest measured existing data in Mt. Marapi area [17].

The 6 (six) prioritized rivers that have volcanic boulder and lava deposition in the upstream acquire larger portion of estimated transportable volcanic material compared to the rest 22 (twenty two) rivers that have no volcanic boulder and lava deposition in their watersheds. The boulder and lava deposited at the upstream area of the rivers potentially transport 80.33 % of the total erupted materials as summarized at Table 1. Interestingly, Batang Kadurang Watershed has the highest transportable materials compared to the other prioritized rivers, followed with Batang Katik in the second rank. It means that the transportable volcanic material will predominantly flows to the South and North West direction relative to the mountain summit. The materials which have high probability flowing into the rivers in each watershed will not completely all transported, some of them remain at the top due to self-consolidation burden [18].

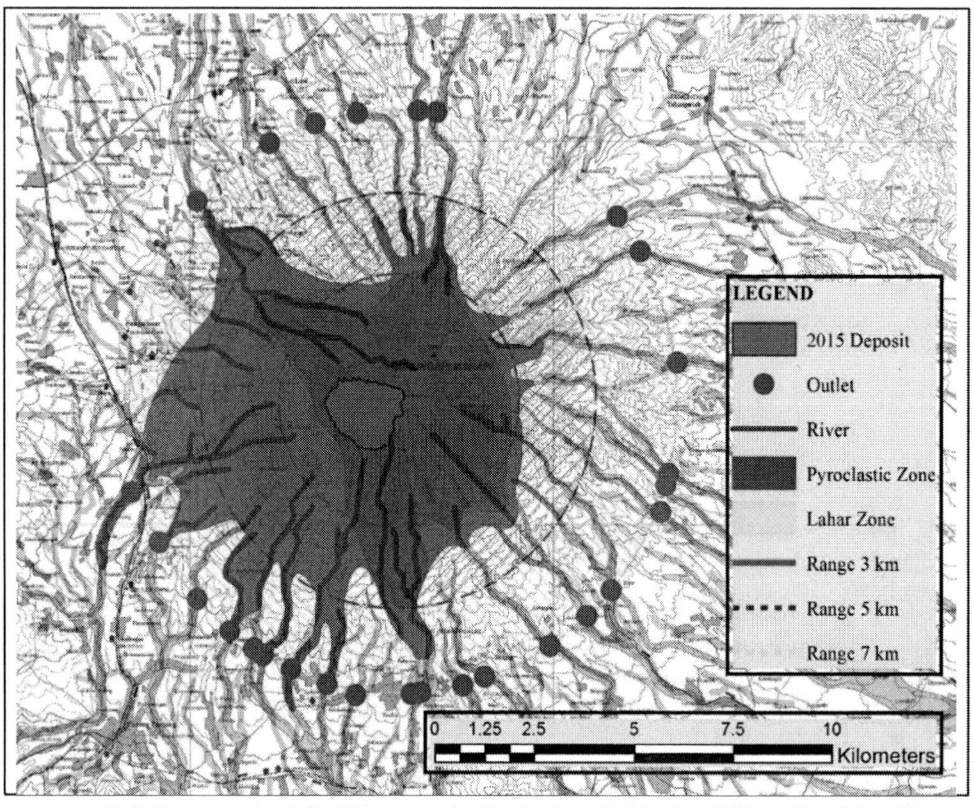

Figure 3. Volcano risk map overlaid with prioritized watershed outlet.

Table 1. Transportable materials estimation based on Mt. Marapi 14 November 2015 eruption.

Prioritized Watershed	Outlet Coordinate	Area	Deposit intersection	Rainfall accumulation	Transportable material estimation
		[km^2]	[km^2]	[mm in 152 days]	[m^3]
BA. Katik	100.418568°E, 0.346353°S	13.14	0.69	1,026.2	1,796.58
B. Malana	100.507400°E, 0.436777°S	6.96	0.04	1,280.2	1,530.47
B. Kadurang	100.455062°E, 0.454103°S	5.10	0.83	1,023.2	1,905.18
B. Arau	100.440153°E, 0.448458°S	4.19	0.41	1,023.2	1,176.48
BA. Siririt	100.431505°E, 0.443938°S	2.63	0.11	1,023.2	639.13
B. Gantung	100.403345°E, 0.409848°S	5.72	0.28	1,023.2	984.45
Others	-	70.33	0	1,181.5	1,967.70
Overall	-	108.07	2.36	1,143.1	10,000.00
Weight	-	20 %	40 %	40 %	-

The transportable volcanic materials estimation that is developed in this paper has high correlation with material deposit extent situation. Therefore, the watershed that has no deposit intersection is presumed to have smaller amount that is transportable into its outlet. Most of the materials transported in the watershed that has no deposit intersection, are predicted to be lapilli and volcanic ashes. The lava and volcanic boulders may be transported in the watershed that has intersection with the deposit area. The rainfall distribution around Mt. Marapi area also gives some significant effect to the final predicted volcanic material volume. The use of TRMM 0.25° x 0.25° gridded precipitation data is quite enough in representing the rainfall distribution in Mt. Marapi into the east side and the west side rainfall distribution pattern. Finally, the proposed transportable volcanic material method still have uncertainties, which mainly caused by the total erupted material prediction. In case of disaster preparedness task force, it is suggested to use maximum possible erupted material for each class of Volcanic Explosivity Index.

5. Conclusion

The river based transportable volcanic materials can be estimated by considering the area of deposited materials, watershed terrain profile, and rainfall pattern of the studied area. The web based free available satellite imagery, digital elevation model, and rainfall distribution data will be an alternative solution when there are no ground measurement data of the required input parameter, especially within the rapid disaster countermeasure program. This method can be used to identify the prioritized watershed that will transport the erupted materials in the watershed outlet, including the maximum estimated material. In the near future, it is recommended to calibrate the model by watershed outlet sediment transport measurement data or lahar flood occurrence record to enhance the model for wider application.

Acknowledgements
Authors appreciate the great research collaboration by the experts from The Experimental Station for Sabo and Barunadri Engineering Consultant. We would also thank Mr. Edy Sulistyono and Dr. Samuel J. Sutanto for providing perception and expertise that sharpening the research findings.

References

[1] Salisbury M J, Patton J R, Kent A J R, Goldfinger C, Djadjadihardja Y and Hanifa U 2012 Deep-sea ash layers reveal evidence for large, late Pleistocene and Holocene explosive activity from Sumatra, Indonesia *J. Volcanol. Geotherm. Res.* **231-232** pp 61–71
[2] Program G V 2013 Volcanoes of the World, v. 4.5.0. Venzke, E (ed.) *Smithson. Inst.*
[3] Hardian Y 2016 Merpati Peak Path, Mt. Marapi - Mei8, 2016 *maps.google.com*
[4] Tanaka H K M, Uchida T, Tanaka M, Takeo M, Oikawa J, Ohminato T, Aoki Y, Koyama E and Tsuji H 2009 Detecting a mass change inside a volcano by cosmic-ray muon radiography: First results from measurements at Asama volcano, Japan *Geophys. Res. Lett.* **36** L17302
[5] Chaussard E and Amelung F 2012 Precursory inflation of shallow magma reservoirs at west Sunda volcanoes detected by InSAR *Geophys. Res. Lett.* **39**
[6] Carranza E J M and Castro O T 2006 Predicting Lahar-Inundation Zones: Case Study in West Mount Pinatubo, Philippines *Nat. Hazards* **37** 331–72
[7] de Bélizal E, Lavigne F, Hadmoko D S, Degeai J P, Dipayana G A, Mutaqin B W, Marfai M A, Coquet M, Mauff B L, Robin A K, Vidal C, Cholik N and Aisyah N 2013 Rain-triggered lahars following the 2010 eruption of Merapi volcano, Indonesia: A major risk *J. Volcanol. Geotherm. Res.* **261** pp 330–47
[8] Newhall C G and Self S 1982 The volcanic explosivity index (VEI) an estimate of explosive magnitude for historical volcanism *J. Geophys. Res.* **87** p 1231
[9] Jarvis A, Reuter H I, Nelson A and Guevara E 2008 Hole-filled SRTM for the globe Version 4 *available from CGIAR-CSI SRTM 90m Database (http//srtm. csi. cgiar. org)*
[10] Putra S S and Neilzon B 2016 River bed stabilization structures placement determination based on satelite data *The 5th HATHI International Seminar on Water Resilience in a Changing World* ed S H Brotowiryatmo, N Anwar, M Amron, Suripin, D Yudianto and I G B S Dharma (Denpasar, Bali: Indonesian Association of Hydraulic Engineer) p 894
[11] Bolvin D T and Huffman G J 2015 *Transition of 3B42 / 3B43 Research product from monthly to climatological calibration / adjustment* (Greenbelt: MD)
[12] Putra S S, Hassan C and Hariyadi S 2012 Hot pyroclastic deposit as lahar resistor: a case study of Gendol River after the Mt. Merapi 2010 eruption *Monitoring, Simulation, Prevention and Remediation of Dense and Debris Flows IV* ed D de Wrachiend, C A Brebbia and S Mambretti (Southampton: WIT Press) pp 97–109
[13] Alvarado G E and Schmincke H 2013 The 1723 A.D. Violent strombolian & phreato-magmatic eruption at Irazú Volcano, Costa Rica *Rev. Geológica América Cent.* **48** pp 41–61
[14] van Otterloo J, Cas R A F and Sheard M J 2013 Eruption processes and deposit characteristics at the monogenetic Mt. Gambier volcanic complex, SE Australia: Implications for alternating magmatic and phreatomagmatic activity *Bull. Volcanol.* **75** pp 1–21
[15] Global Volcanism Program 2016 *Report on Marapi (Indonesia). In: Sennert, S K (ed.), Weekly Volcanic Activity Report, 20 January-26 January 2016* (Washington, DC)
[16] Center of Volcanology and Geological Hazard Mitigation 2016 *Evaluasi tingkat aktivitas level ii (waspada) G. Marapi pada 20 November 2015* (Bandung)
[17] Dana L N, Santoso M S, Karim A and Riyadi 2006 Volcanic Hazard Map of Marapi Volcano, West Sumatera Province, Indonesia, 2006
[18] Lipman P W 2007 Incremental assembly and prolonged consolidation of Cordilleran magma chambers: Evidence from the Southern Rocky Mountain volcanic field *Geosphere* **3** pp 42–70

Evaluation of CMIP5 models on sea surface salinity in the Indian Ocean

Ibnu Fathrio[1,2*], Atsuyoshi Manda[3], Satoshi Iizuka[4], Yasu-Masa Kodama[1] and Sachinobu Ishida[1]
[1]Meteorological Laboratory, Graduate School of Science and Technology, Hirosaki University, Hirosaki, Aomori, Japan.
[2]National Institute of Aeronautics and Space of Indonesia, Bandung, West Java, Indonesia.
[3]Climate and Ecosystems Dynamics Division, Graduate School of Bioresources, Mie University, Tsu, Mie, Japan.
[4]Storm, Flood, and Landslide Division, National Research Institute for Earth Science and Disaster Resilience, Tsukuba, Ibaraki, Japan.

E-mail: ibnu.fathrio@lapan.go.id

Abstract. Prior to future climate assessment of the 5th Coupled Model Intercomparison (CMIP5) experiments, how well CMIP5 models simulates present climate should be examined. Sea surface salinity (sss) play important role in ocean stratification and indirectly affects air sea interaction. However, few studies have been carried out to evaluate sss in CMIP5 models. In this study, performance of CMIP5 models in simulating sss in Indian Ocean was examined with respect to the observation. Our results showed that multi model ensemble (MME) mean of CMIP5 models displayed annual and seasonal salinity bias in three regions i.e. Western Indian Ocean (WIO), Bay of Bengal (BOB) and Southeastern Indian Ocean (SEIO). CMIP5 models overestimate sss in BOB about 1.5 psu and underestimated sss in WIO and SEIO about 0.4 psu. Biases in WIO and BOB were mainly attributed to bias in precipitation. CMIP5 models overestimated (underestimated) precipitation in WIO (BOB) with greater bias found during Boreal summer to winter. Meanwhile, advection process was responsible for negative SSS bias in SEIO.

1. Introduction

Influence of seas surface salinity (sss), as proxy of upper ocean salinity, on ocean upper ocean stratification has attracted many studies. Low salinity water influences the mixed layer depth in several places in the tropical ocean [1,2] by increasing stratification in ocean mixed layer that leads to barrier layer (BL) formation in Indian Ocean, BL exist in Bay of Bengal (BOB) [3], Eastern Indian Ocean (EIO) [4,5] and Southeastern Arabian Sea [6]. Recent studies also discussed the mechanism of how low salinity changes mixed layer density and forming BL formation. Presence of BL formation induced by low salinity water may prevent entrainment of cool water and maintains mixed layer warms [7,1,2]. As consequence, changes in mixed layer may further influence air-sea interaction and local-global weather/climate.

Low salinity water in Bay of Bengal is attributed to large excess of freshwater from precipitation and river discharge during summer-autumn. Ocean dynamics plays important role in salinity distribution through advection process. Monsoonal wind affects seasonal variation of current that transport salinity water. In the North of Indian Ocean, monsoon currents facilitate the water exchange between the Arabian Sea and the Bay of Bengal. High-salinity water is advected towards Bay of Bengal during summer while low-salinity water is advected towards southeast Arabian Sea in winter [8,9]. Near the equator of Indian Ocean, seasonal variation of salinity is affected by Yoshida-Wyrtki Jet [10,11], which carries high salinity water eastward in spring and autumn.

Modeling studies of salinity effect by [12,13] discussed the impact of precipitation/freshwater flux on the ocean stratification and equatorial circulation in the TIO which remarked that salinity stratification greatly affect the ocean dynamic and air-sea interaction [14]. The simulations with and without salinity effect in the coupled model reveal a highly seasonal influence of salinity and the barrier layer (BL) on oceanic vertical density stratification, which is in turn linked to changes in sea surface temperature (SST), surface winds, and precipitation [15].

The Coupled Model Intercomparison Project phase 5 (CMIP5) models provide the opportunity to investigate role of salinity on past and projected future climates [16,17]. Previous studies showed that CMIP5 models have common error in equatorial Indian Ocean which simulate positive IOD-like pattern as shown in figure 1 [18] during summer and fall indicated by warm SST bias in WIO and cool SST bias in Eastern Indian Ocean (EIO). This is accompanied by bias and strong easterly wind bias in the Eastern Indian Ocean (EIO) [19,20]. The thermocline is also tilted toward EIO [21]. [22] and [19] also remarked that SST biases over WIO are related to a weaker southwesterly summer monsoon.

Despite evaluation of CMIP5 models in representing SST variations, only few studies has been conducted to evaluate how sea surface salinity (sss) is being represented by CMIP5 in the present climate. Therefore, this study will investigate how CMIP5 models simulate seasonal variation of sss. The purpose of this study is to investigate the underlying cause in salinity bias by considering systematic bias of CMIP5 models in Indian Ocean.

2. Methods

We examined the dataset of CMIP5 that is referred to historical runs covering much of the industrial period from the mid-nineteenth century to near present [17]. Table 1 shows a list of 21 CMIP5 models used in this study. Further information on each model is available online at http://www-pcmdi.llnl.gov. In this study, monthly-mean outputs from 20-year CMIP5 simulations (1985–2004) were examined, including ocean salinity and currents, precipitation and evaporation. The similar averaging period was often used in the previous studies for mixed layer depth (MLD) (e.g., [23]), documenting that the simulated MLD shows weak decadal and longer variability, which was sufficient to analyze the mean seasonal cycle of sss in the tropical Indian Ocean. In order to calculate the multi-model ensemble-mean (MME), outputs of CMIP5 models were horizontally interpolated onto a 1° x 1° uniform horizontal grid and vertically interpolated onto a 5 m uniform vertical grid from ocean surface down to 100 m and 25 m uniform vertical grid below 100m, using linear interpolation. We noted that salinity at 5 m was used as sss in this study because the first ocean vertical level is 5 m in most of the models.

In order to analyze biases of sss in the models, we used the monthly mean ocean salinity provided by the In Situ Data Analysis System (ISAS) 13 [24, 25], which is an averaged over the period 2004–2014 (11 years) with horizontal grid resolution of 0.5° and vertical grid resolution of 152 levels from 0 to 2000m. The levels above 100m have higher vertical resolution with 5m thicknesses. Near-surface winds at the 1000-hPa level were obtained from the European Center for Medium-Range Weather Forecasts (ECMWF) Re-Analysis (ERA) Interim product [26], averaged over the period 1985–2004. The horizontal grid for near-surface wind data is 0.75°. Precipitation data were obtained from the Global Precipitation Climatology Project (GPCP) [27]. Evaporation data were provided by the Woods Hole Oceanographic Institute OAFlux project at http://oaflux.whoi.edu [28]. The horizontal grid resolution for precipitation and evaporation are 2.5° and 1° respectively. For ocean current comparison, we used monthly mean ECMWF Ocean Reanalysis System 4 (ORAS4) [29] with 1° horizontal grid resolution. ECMWF ORAS4 data were obtained from http://www.ecmwf.int/en/research/climate-reanalysis/ocean-reanalysis. All observation data were averaged over the period 1985-2004. Linear interpolation was also applied onto horizontal and vertical grid of the observation for calculating the bias in MME of CMIP5 models.

Table 1. List of the 21 CMIP5 models used in this study.

Model Name	Institution	Label
MRI-CGCM3	Meteorological Research Institute (Japan)	M01
CMCC-CM	The Centro Euro-Mediterraneo sui Cambiamenti Climatici Climate Model (Italy)	M02
MRI-ESM1	Meteorological Research Institute (Japan)	M03
BCC-CSM1-1	The Beijing Climate Center Climate Model (China)	M04
HADGEM2-ES	Met Office Hadley Centre (UK)	M05
IPSL-CM5B	Institut Pierre-Simon Laplace (France)	M06
CNRM-CM5-2	Centre National de Recherches Meteorologiques/Centre Europeen de Recherche et Formation Avancees en Calcul Scientifique (France)	M07
ACCESS1.3	CSIRO-BOM (Australia)	M08
CNRM-CM5	Centre National de Recherches Meteorologiques/Centre Europeen de Recherche et Formation Avancees en Calcul Scientifique (France)	M09
NorESM1-M	Norwegian Climate Centre (Norway)	M10
GFDL-CM3	Geophysical Fluid Dynamic Laboratory (USA)	M11
IPSL-CM5A-LR	Institut Pierre-Simon Laplace (France)	M12
MIROC-ESM	Atmosphere and Ocean Research Institute (The University of Tokyo), National Institute for Environmental Studies, and Japan Agency for Marine-Earth Science and Technology	M13
GISS-E2-R	NASA Goddard Institute for Space Studies (USA)	M14
CanESM2	Canadian Centre for Climate Modeling and Analysis (CCCMA)	M15
MPI-ESM-LR	Max Planck Institute for Meteorology (Germany)	M16
GFDL-CM 2.1	Geophysical Fluid Dynamic Laboratory (USA)	M17
MPI-ESM-P	Max Planck Institute for Meteorology (Germany)	M18
CSIRO	CSIRO Atmospheric Research (Australia)	M19
GFDL-ESM2G	Geophysical Fluid Dynamic Laboratory (USA)	M20
MIROC5	Atmosphere and Ocean Research Institute (The University of Tokyo), National Institute for Environmental Studies, and Japan Agency for Marine-Earth Science and Technology (Japan)	M21

We examined the possible cause of sss bias by using salt budget analysis. Salt budget analysis had been used in similar manner to heat budget analysis to study salinity variability in Northeast Pacific Ocean [30]. In this study, we used fixed depth (50m depth) salt following [31] and [32].

$$\frac{\partial s}{\partial t} = -u\frac{\partial s}{\partial x} - v\frac{\partial s}{\partial y} - w\frac{\partial s}{\partial z} + \frac{So(E-P)}{h} + Residual, \quad (1)$$

where S, u, v, and w represent climatological monthly means of ocean salinity, zonal, meridional, and vertical ocean current velocities respectively, and h is a thickness of the water column (= 50 m). All the terms were averaged over the top 50 m. Using fixed depth of 50 m avoids complications associated with a multi-model ensemble such as varying mixed layer depth amongst models [32]. *So* was defined as vertical average of salinity over 50 m. *E* and *P* were defined as evaporation and precipitation rate respectively.

3. Results

3.1. SST and surface wind bias

In this study, we investigated whether the sss bias was influenced by CMIP5 common bias. Bias in precipitation and evaporation may increase/decrease the ocean salinity. Bias in surface wind may induce changes in evaporation and also affects salt advection through ocean current bias. Multi model ensemble (MME) mean of CMIP5 models used in this study displayed common bias as shown in figure 1 that is consistent with previous CMIP5 studies. Warm SST bias was present during summer (June to August, called JJA) and autumn (September to November, called SON) in the WIO. At the same time, strong easterly bias was pronounced in Eastern Indian Ocean accompanied by cold SST bias. These warm-cold SST bias over India Ocean exhibited similar pattern like positive Indian Ocean Dipole (IOD) SST anomalies. [33] reported that significant biases in annual zonal wind at the equator (figure 1a) resulting too small of an east-west gradient of sea level. Here, our result (figure 1) reaffirmed previous studies in CMIP5 bias.

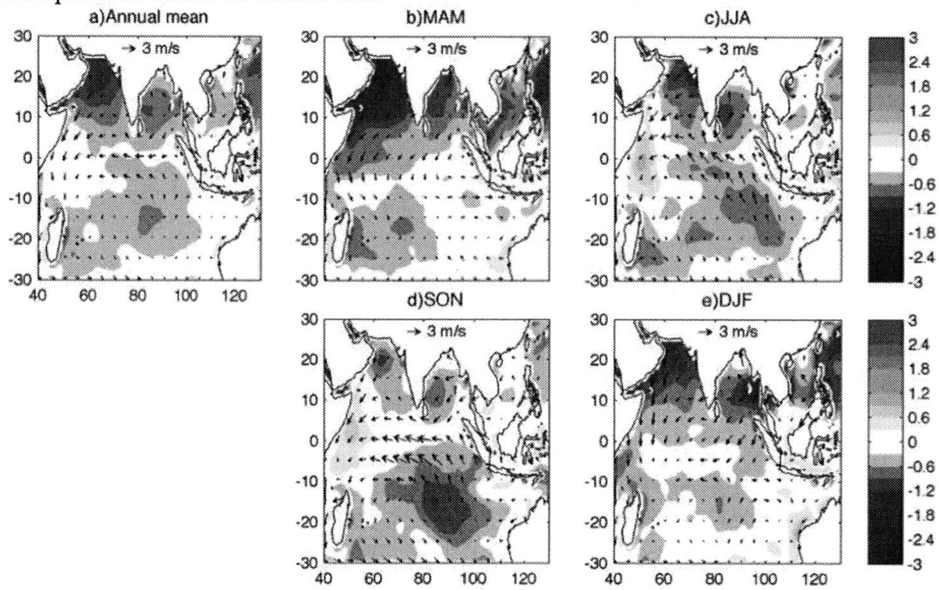

Figure 1. Seasonal SST bias overlied by surface wind bias. Here a) MAM (March to May, Boreal spring), b) JJA (June to July, Boreal summer), c) SON (September to November, Boreal autumn), and DJF (December to February, Boreal winter).

3.2. Sea surface salinity bias

Annual mean of sss over Indian Ocean is shown in figure 2a, indicating high sss presents in Arabian Sea and WIO and low sss presents in BOB. Low sss was present in BOB due to large amount of precipitation and low salinity water input from river discharge. Low sss along the coast of Java-Sumatra was attributed to annual precipitation in maritime continent throughout the year. MME CMIP5 models successfully simulated annual mean SSS in Arabian Sea, yet they underestimated annual sss in WIO and SEIO and overestimated sss in BOB (figure 2b-c). We defined box 1 (40-70E; 10S-10N), box 2 (80-100E; 8-20N) and box 3 (90-120E; 20-10S) as illustrated in figure 2c, which correspond to the area where annual sss bias was present, i.e. WIO, BOB and SEIO, respectively. We excluded the maritime continent in this study. More detail discussion for bias in maritime continent must consider influence from Western Pacific Ocean, which is beyond our discussion.

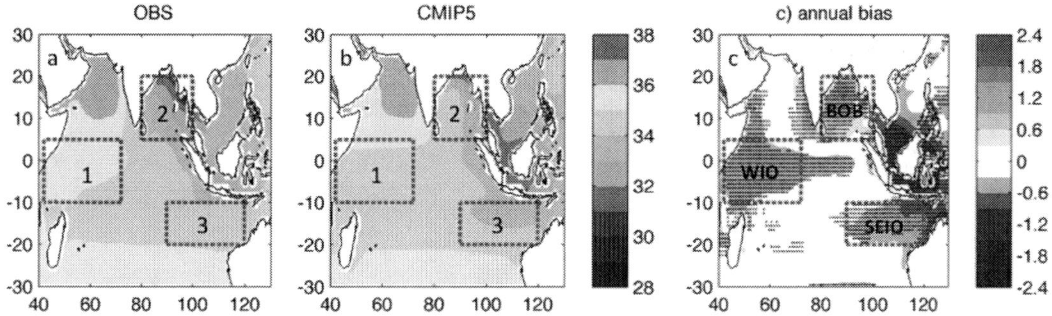

Figure 2. Annual mean of sea surface salinity for a) the observation, b) MME of CMIP5 models and c) MME annual mean bias. The dots in figure 2c shows that the bias is represented by 75% of models. The area of study are denoted by the box i.e. Western Indian Ocean (WIO, 40-70E; 10S-10N), Bay of Bengal (BOB, 80-100E; 8-20N) and Southeastern Indian Ocean (SEIO, 90-120E; 20-10S).

In general, MME mean of CMIP5 models overestimated seasonal variation of sss in BOB while CMIP5 underestimated sss in WIO and SEIO and in the maritime continent. Our study regions (figure 2) displayed significant seasonal sss bias. It is interesting to note that sss biases in the study area are greater in Boreal autumn to winter (December to February, called DJF) season as shown in figure 3 and 4. In particular the bias in the BOB (figure 3b) where most of models failed to simulate realistic sss in SON. MME of CMIP5 models overestimated the sss with bias about 1.5 psu greater than the observation. Meanwhile, MME of CMIP5 models simulated sss lower than the observation in WIO and SEIO with bias about 0.4 psu in SON-DJF. Although there was inter-model variation among CMIP5 models, we confirmed that the bias in three regions was significant which was represented by at least 15 of 21 models (75%) denoted by the dots in figure 4.

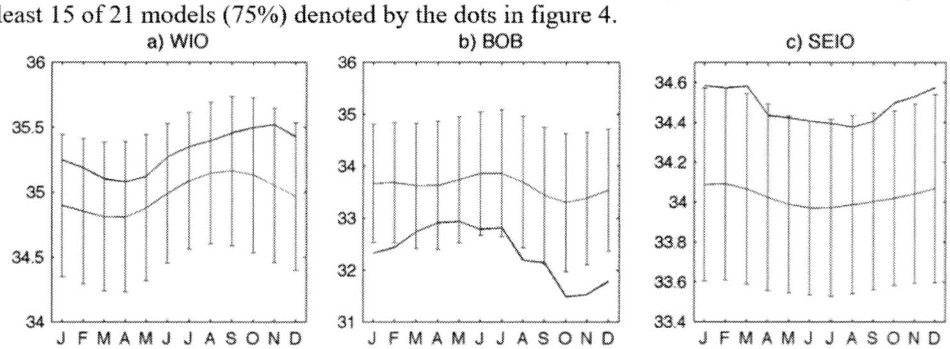

Figure 3. Seasonal time series of SSS in a) WIO, b) BOB and c) SEIO are indicated by red solid line. Standard deviation is indicated by error bar. Solid black line denotes the observation. The x and y axis denote time (months) and sss value (psu).

3.3. Precipitation bias

Precipitation is among the key factors controlling the sss. Realistic reproduction of precipitation is challenging task in state-of-the-art GCMs simulation. Positive precipitation bias in WIO existed in all seasons with greater bias presented during autumn and winter (figure 5a-d). Positive precipitation bias had appeared after warm SST bias developed in summer (figure 1c). In BOB, strong negative precipitation bias occured during summer when Indian summer monsoon precipitation emerged. In Eastern Indian Ocean, positive precipitation bias was present in maritime continent throughout the year. In the west of Sumatra, near the equator, negative precipitation bias was present during autumn. Positive bias of precipitation in WIO and negative bias of precipitation in the west of Sumatra exhibited positive IOD-like pattern of precipitation, which was mentioned earlier as common bias in

CMIP5 models. Precipitation bias in SEIO (box 3) was well simulated, except in spring (March to May, called MAM), where positive precipitation bias appeared. Horizontal distribution of evaporation minus precipitation rate (E-P) as proxy of freshwater flux (figure 5e-h) displayed similar pattern as precipitation bias. Positive bias indicated that evaporation exceeded precipitation and vice versa.

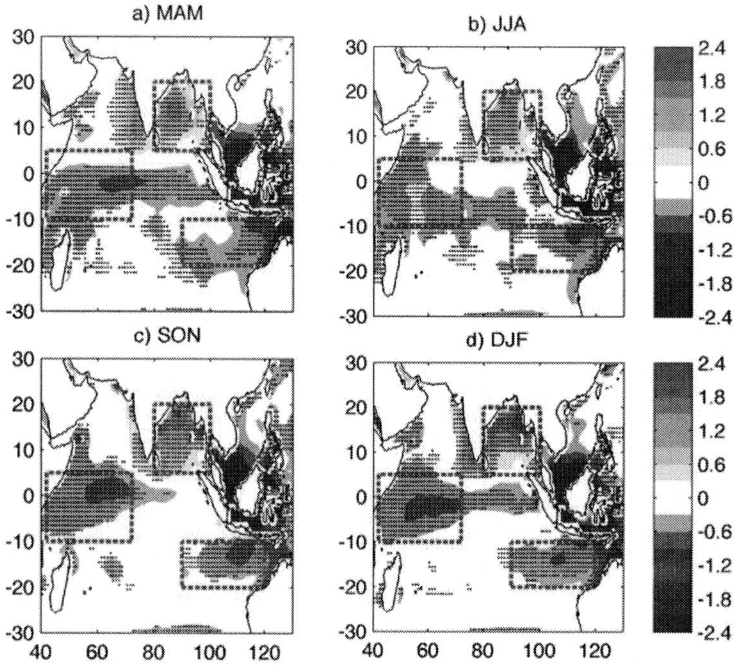

Figure 4. Horizontal distribution of seasonal sea surface salinity bias for a) MAM, b) JJA, c) SON and d) DJF. The boxes denote area of study as defined in figure 2. The dots show that the bias is represented by 75% of models

3.4. Fixed-depth salt budget analysis

In order to examine the possible cause of bias, we used fixed depth salt budget equations. Time series of fixed-depth salt budget calculation in WIO is shown in figure 6. Salt tendency and advection terms were well simulated by CMIP5 models. CMIP5 models simulated more positive contribution from residual terms, representing diffusion and mixing process. Freshwater flux during summer to winter (red solid line) was more negative than the observation, which was consistent with positive precipitation bias shown in figure 5. Apparently, unrealistic high precipitation was the main cause for negative sss bias in WIO. We obtained similar pattern of horizontal bias distribution between precipitation and sss bias in WIO by comparing figure 4 and 5.

In BOB, CMIP5 models reasonably simulated seasonal variations of all budget terms. CMIP5 simulated more positive freshwater flux and vertical advection during summer and winter respectively. The former one was mainly caused by negative precipitation bias during summer-autumn season. Unfortunately, river discharge was not available in all CMIP5 models outputs. Nevertheless, negative bias of precipitation in India continent (figure 3) may leads to decrease of water discharge from the river to the Bay. As a result, less precipitation and water discharge from the river maintained high sss water in the Bay. Saline waters were then distributed to the entire of the Bay by advection process. Figure 8 shows scatterplot between sss bias in DJF and precipitation. Strong correlation (-0.75) indicated that sss bias in winter (DJF) was mainly influenced by simulated precipitation in JJA (summer). Models that underestimated (overestimated) precipitation in JJA tend to simulate positive (negative) sss bias in BOB.

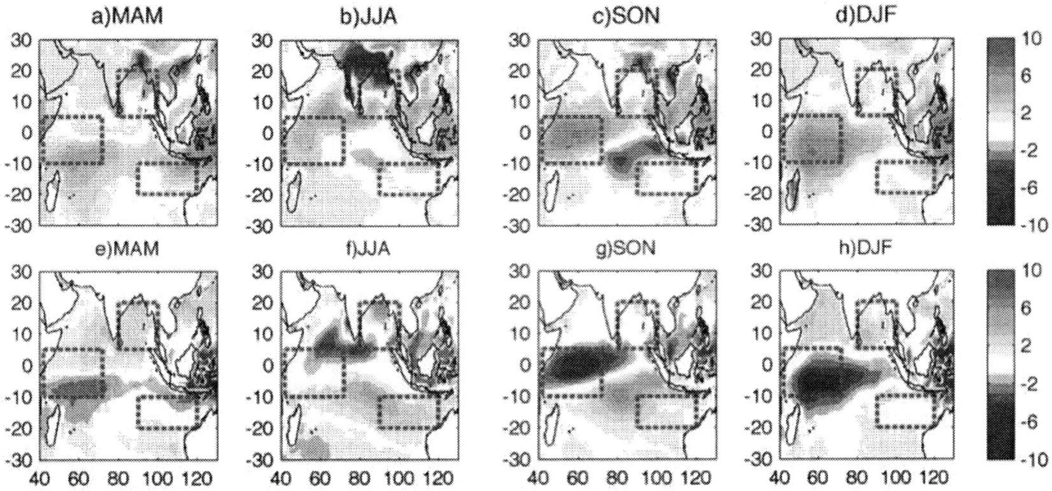

Figure 5. Horizontal distribution for precipitation bias (a-d) and E-P bias (evaporation minus precipitation) bias (e-h). The boxes denote area of study as defined in figure 2.

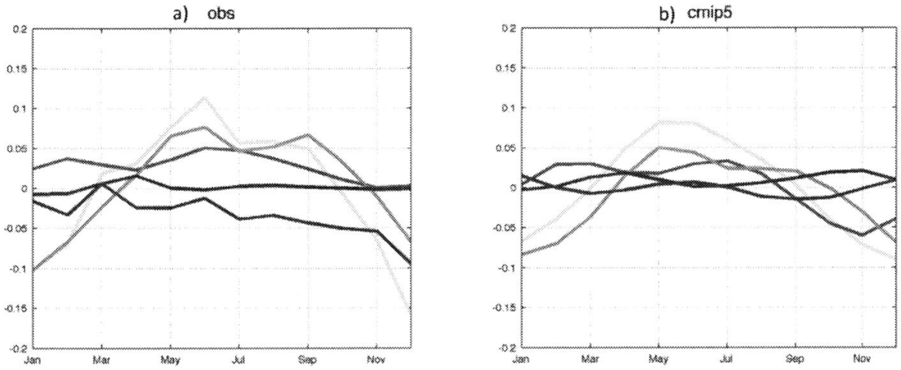

Figure 6. Time series of fixed depth salt budget equation for the a) observation and b) CMIP5 models (solid line) in WIO. Black, red, green, blue and purple indicate salinity tendency, freshwater flux, horizontal advection, vertical advection, and residual terms respectively. The x and y axis denote time (months) and sss tendency (psu month^{-1}).

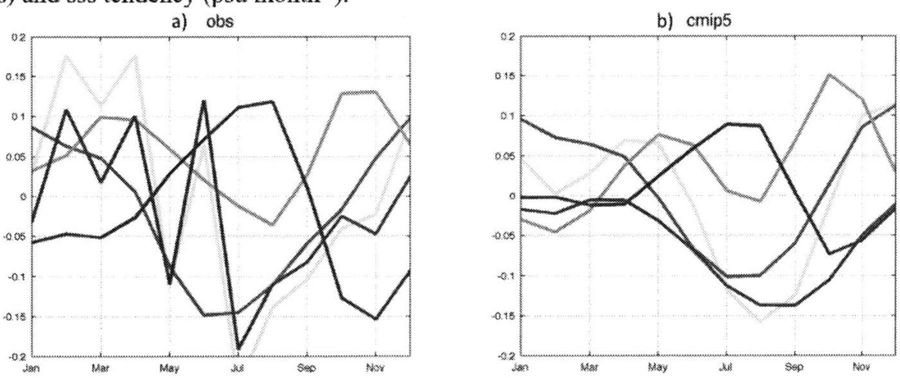

Figure 7. Same as figure 6 but for BOB

IOP Conf. Series: Earth and Environmental Science 54 (2017) 012039 doi:10.1088/1755-1315/54/1/012039

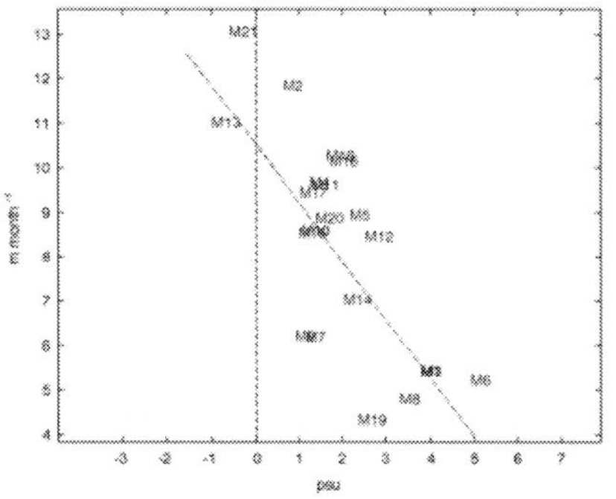

Figure 8. Scatterplot between sss bias (psu) in DJF (x-axis) and precipitation (mm month^{-1}) in JJA (y-axis).

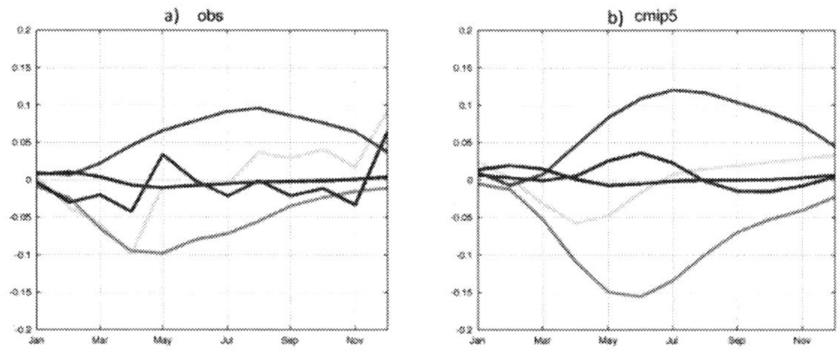

Figure 9. Same as figure 6 but for SEIO

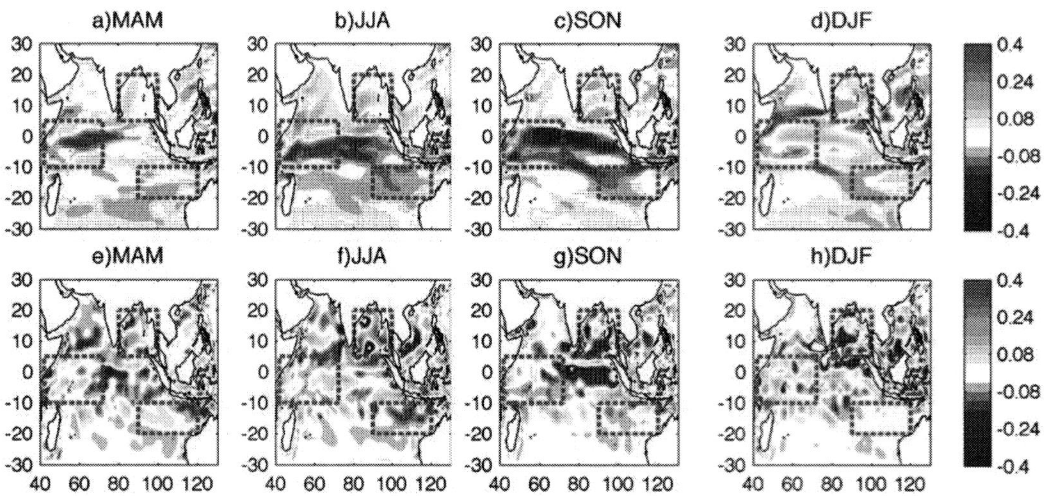

Figure 10. Horizontal map of seasonal distribution for eastward surface current bias (a-d) and horizontal advection bias (e-h). The boxes denote area of study as defined in figure 2.

Precipitation was not the only cause of negative sss bias since evaporation exceeded precipitation in SEIO as displayed in figure 5. This was shown by salt budget analysis where CMIP5 models simulated more positive freshwater flux compared to the observation throughout the year (figure 9). Figure 5 shows that significant positive bias of precipitation only presents in spring (MAM), which partly contributed to sss bias in this season. Meanwhile, sss bias continued to exist in the rest of season although E-P and precipitation are well simulated by CMIP5 models. These suggested that ocean process was more responsible to cause the bias than precipitation/freshwater flux.

figure 4 shows that negative sss bias are present in south of Java coast, which is still probably related to positive bias of precipitation in maritime continent. However, the negative sss bias extended further to the SEIO (box 3). Seasonal variation of salt budget terms were well simulated, however horizontal salt advection in CMIP5 models was more negative than the observation during spring to winter (figure 9). Too strong eastward current in SEIO possibly induced this bias during summer-autumn (figure 10b-c). Consequently, negative advection bias (figure 8) distributed low sss from the south of Java coast further to the west (box 3, SEIO).

4. Conclusion

This study examined the sss biases over Indian Ocean in the CMIP5 models. It was evident that CMIP5 models underestimated sss in WIO and SEIO and overestimated sss in BOB. Precipitation bias was found as the primary cause for sss bias in WIO and BOB, while horizontal advection and precipitation were responsible for negative sss bias in SEIO. Salt budget analysis indicated that CMIP5 overestimated precipitation shown by negative bias of freshwater flux in the WIO. Negative bias of precipitation/freshwater flux in BOB during summer monsoon was responsible for positive sss bias during winter. In SEIO, CMIP5 simulated strong eastward current bias in SEIO, transporting low salinity water during summer-autumn which caused negative sss bias.

Our analysis revealed the cause of common bias in CMIP5 on sss in Indian Ocean was closely related to systematic bias in CMIP5 models. Precipitation bias in BOB was indirectly caused by weak southwesterly summer monsoon wind in CMIP5 models, which considerably decreased moisture transport towards Indian continent. Bias of sss in WIO was attributed to unrealistic simulation of precipitation that very related to positive IOD-like bias where positive precipitation bias developed during autumn. Meanwhile, strong eastward current bias in SEIO was induced by strong easterly wind considered as systematic bias in CMIP5 that emerged in summer to autumn near the equator. Bias in SEIO was related to systematic bias where most of models tend to simulate too strong easterly wind near equator.

This study illustrated that most of CMIP5 models fail to simulate the seasonal variations of sss over the tropical Indian Ocean, which has not been fully examined in previous studies. [34] showed that better freshwater forcing and model physics in state-of-the-art GCMs might reduce the sss bias. Better ocean model and the use of high resolution might contribute to improve ocean dynamic biases in climate models. It was documented that the CMIP5 models showed a future IOD-like climate change [35]. Investigation of the process causing the future SSS changes over the Indian Ocean will be performed in future work.

References

[1] Sprintall J, Tomczak M 1992 Evidence of the barrier layer in the surface layer of the Tropics *J. Geophys. Res.* **97** 7305–7316

[2] Murtugudde R, Busalacchi AJ 1999 Interannual variability of the dynamics and thermodynamics of the tropical Indian Ocean *J.Clim* **12** 2300–2326

[3] Vinayachandran P N, Murty V S N, Ramesh Babu V 2002 Observations of barrier layer formation in the Bay of Bengal during summer monsoon *J. Geophys. Res.* **107** 8018

[4] Eriksen C C 1979 An equatorial transect of the Indian Ocean *J. Mar.Res.* **37** 215–232

[5] Godfrey J S, Bradley E F, Coppin P A, Pender L F, McDougall T J, Schulz E W, Helmond I, 1999 Measurements of upper ocean heat and freshwater budgets near a drifting buoy in the equatorial Indian Ocean, *J. Geophys. Res.* **104** 13269–13302

[6] Shankar D, Gopalakrishna V, Shenoi S, Durand F, Shetye S, Rajan C, Johnson Z, Araligidad N, Michael G 2004 Observational evidence for westward propagation of temperature inversions in the southeastern Arabian Sea *Geophys. Res. Lett.* **31** L08305

[7] Lukas R, Lindstrom E 1991 The mixed layer of the western equatorial Pacific Ocean *J. Geophys. Res.* **96** 3343–3357

[8] McCreary J P, Han W, Shankar D, Shetye S R 1996 Dynamics of the East India Coastal

Current: 2. Numerical solutions *J. Geophys.Res.* **101** 13993–14010
[9] Jensen T G 2003 Cross-equatorial pathways of salt and tracers from the northern Indian Ocean: Modelling results *Deep Sea Res.* **50** 2111–2127
[10] Zhang Y, Du Y 2012 Seasonal variability of salinity budget andwater exchange in the northern Indian Ocean from HYCOM assimilation *Chin. J. Oceanol* Limnol **30** 1082–1092
[11] Yoshida K 1959 A theory of the Cromwell current and of the equatorial upwelling - An interpretation in a similarity to a coastal circulation *J. Ocea-nogr. Soc. Japan* **15** 159–170
[12] Wyrtki K 1973 An equatorial jet in the Indian Ocean *Science* **181** 262–264
[13] Masson S, Delecluse P, Boulanger J P, Menkes C 2002 A model study of the seasonal variations and formation mechanisms of the barrier layer in the eastern equatorial Indian Ocean *J. Geophys. Res.* **107** 8017
[14] Masson S, Menkes C, Delecluse P, Boulanger J P 2003 Impacts of salinity on the eastern Indian Ocean during the termination of the fall Wyrtki Jet *J. Geophys. Res.* **108** 3067
[15] Masson S, Boulanger J P, Menkes C, Delecluse P, Yamagata T 2004 Impact of salinity on the 1997 Indian Ocean dipole event in a numerical experiment *J. Geophys. Res.* **109**
[16] Seo H, Xie S P, Murtugudde R, Jochum M, Miller A J 2009 Seasonal effects of Indian Ocean freshwater forcing in a regional coupled model *J. Climate* **22** 6577–6596
[17] Meehl G A, Covey C, Taylor K E, Delworth T, Stouffer R J, Latif M, McAvaney B, Mitchell J F 2007 The WCRP CMIP3 multimodel dataset: A new era in climate change research *Bull. Am. Meteorol. Soc.* **88** 1383–1394
[18] Taylor K E, Ronald J S, Meehl J S 2012 An overview of CMIP5 and the experiment design *Bull. Am. Meteorol. Soc.* **93** 485–498
[19] Saji N H, Goswami B N, Vinayachandran P N, Yamagata T 1999 A dipole mode in the tropical Indian Ocean *Nature* **401** 360–363
[20] Li G, Xie S P, Du Y 2015 Monsoon-induced biases of climate models over the tropical Indian Ocean *J. Climate* **28** 3058–3072
[21] Nagura M, Sasaki W, Tozuka T, Luo J J, Behera S K, Yamagata T 2013 Longitudinal biased in the Seychelles dome simulated by 35 ocean-atmosphere coupled general circulation models *J. Geophys. Res.* **118** 831-846
[22] Cai W, Cowan T 2013 Why is the amplitude of the Indian Ocean Dipole overly large in CMIP3 and CMIP5 climate models? *Geophys. Res. Lett.* **40** 1200–1205
[23] Levine R C, Turner A G, Marathayil D, Martin G M 2013 The role of northern Arabian Sea surface temperature biases in CMIP5 model simulations and future projections of Indian summer monsoon rainfall *Clm. Dyn* **41** 155–172
[24] Huang C J, Qiao F, Dai D 2014 Evaluating CMIP5 simulations of mixed layer depth during summer *J. Geophys. Res.* **119** 2568–2582
[25] Gaillard F, Reynaud T, Thierry V 2016 In situ-based reanalysis of the global ocean temperature and salinity with ISAS: Variability of the heat content and steric height *J. Climate* **29** 1305–1323
[26] Gaillard F, Autret E, Thierry V, Galaup P, Coatanoan C, Loubrieu T 2009 Quality control of large Argo datasets *J. Atmos. Oceanic Tech.* **26** 337-351
[27] Dee D P, and co-authors 2011 The ERA-Interim reanalysis: Configuration and performance of the data assimilation system *Quart. J. R. Meteorol. Soc.* **137** 553–597
[28] Adler R F, Huffman G J, Chang A, Ferraro R, Xie P P, Janowiak J, Rudolf B, Schneider U, Curtis S, Bolvin D, Gruber A, Susskind J, Arkin P, Nelkin E 2003 The version-2 global precipitation climatology project (GPCP) monthly precipitation analysis (1979–present) *J. Hydrometeor.* **4** 1147–1167
[29] Yu L 2008 Sea surface exchanges of momentum, heat, and freshwater determined by satellite remote sensing. In: Steele J, Thorpe S, a nd K. Turekian (eds.), Encyclopedia of Ocean Sciences, First Online Update, (London: Academic Press)
[30] Balmaseda M A, Mogensen K, Weaver A T 2013 Evaluation of the ECMWF ocean reanalysis

system ORAS4 *Q.J.R. Meteorol. Soc.* **139** 1132–1161
[31] Ren L Riser S C 2009 Seasonal salt budget in the northeast Pacific Ocean *J. Geophys. Res.* **114**
[32] Wells N C, Josey S A, Hadfield R E 2009 Towards closure of regional heat budgets in the North Atlantic using Argo floats and surface flux datasets *Ocean Science* **5** 59-72
[33] Ng B, et al Nonlinear processes reinforce extreme Indian Ocean Dipole events *Sci. Rep.* **5** 11697
[34] Lee T, Waliser D E, Li J L F, Landerer F W, Gierach M M 2013 Evaluation of CMIP3 and CMIP5 wind stress climatology using satellite measurements and atmospheric reanalysis products *J. Climate* 26 5810–5826
[35] Parekh A, Chowdary J S, Sayantani O, *et al* 2016 Tropical Indian Ocean surface salinity bias in Climate Forecasting System coupled models and the role of upper ocean processes *Clim Dyn* **46** 2403

Observed and blended gauge-satellite precipitation estimates perspective on meteorological drought intensity over South Sulawesi, Indonesia

A M Setiawan [1, 2], Y Koesmaryono[1], A Faqih[1*] and D Gunawan[2]
[1] Department of Geophysics and Meteorology, Bogor Agricultural University (IPB), Bogor , Indonesia
[2] Centre for Climate Change Information, Indonesia Agency for Meteorology Climatology and Geophysics (BMKG), Jakarta, Indonesia

E-mail: akhmadfa@apps.ipb.ac.id

Abstract. South Sulawesi province as one of the rice production center for national food security are highly influenced by climate phenomenon that lead to drought condition. This paper quantifies meteorological drought based on Standardized Precipitation Index (SPI) recommended by the World Meteorological Organization (WMO) and Consecutive Dry Days (CDD) as one of the extreme indices recommended by the Expert Team on Climate Change Detection and Indices (ETCCDI). The indices were calculated by using (i) quality controlled daily and monthly observational precipitation data from 23 weather stations of various record lengths within 1967-2015 periods, and (ii) 0.05° x 0.05° blended gauge-satellite of daily and monthly precipitation estimates of the Climate Hazards Group InfraRed Precipitation with Stations (CHIRPS) dataset. Meteorological drought intensity represented by Average Duration of Drought Intensity (ADI) from three-monthly SPI (SPI3) show spatial differences characteristic between eastern and western region. Observed and CHIRPS have relatively similar perspective on meteorological drought intensity over South Sulawesi. Relatively high values of ADI and longest CDD observed mainly over south western part of study area.

1. Introduction

Government and policies makers are pay attention about drought, especially when it's occurs during long time period and impact on sensitive sector, such as agriculture that act as thread for food security issue. National food security takes higher priority especially when drought occurred in food production center areas. Drought over Indonesia is commonly associated with warm phase of El Niño Southern Oscillation (ENSO), known as El Niño[1–4].

South Sulawesi as one of the primary national rice production centers also can be affected by El Niño indicated by wider drought coverage area [4,5]. The potential impact of drought due to the weak and moderate El Niño occurrences in Indonesia is such that yields are reduced by about 40 % in average. The most drought- prone areas are located in South Sulawesi for August–October [5].

Climate Hazards Group Infrared Precipitation with Station data (CHIRPS) is a new land-only gridded database of precipitation. It blend three different types of information: global climatologies, satellite estimates and in situ observations. This dataset incorporates monthly precipitation climatology CHP Clim (Climate Hazards Group Precipitation Climatology), quasi-global geostationary thermal infrared satellite observations, Tropical Rainfall Measuring Mission's (TRMM) 3B42 product, atmospheric model rainfall fields from NOAA CFS (Climate Forecast System), and precipitation observations from various sources, including national or regional Meteorological Services [6]. CHIRPS can be promising to be used as alternative meteorological drought monitoring tools because its showing good performance during evaluation with rainfall observation data [7]

CHIRPS dataset already used to monitor meteorological drought all over the world [8–10] and also applied for seasonal drought forecast in Africa [11]. Nevertheless, evaluation when CHIRPS used as drought monitoring tools in South Sulawesi region are still required. The objectives of this study was to carry out the perspective agreement on meteorological drought intensity monitoring between CHIRPS and observation.

2. Data and Methodology

2.1. Data

This paper quantifies meteorological drought based on Standardized Precipitation Index (SPI) recommended by the World Meteorological Organization (WMO) [12] and Consecutive Dry Days (CDD) as one of the extreme indices recommended by the Expert Team on Climate Change Detection and Indices (ETCCDI) [13]. The SPI and CDD were calculated by using (i) quality controlled daily and monthly observational precipitation data from 23 weather stations of various record lengths within 1971-2015 periods, and (ii) 0.05° x 0.05° blended gauge-satellite of daily and monthly precipitation estimates of the Climate Hazards Group Infra Red Precipitation with Stations (CHIRPS) dataset. This gridded dataset were obtained from Climate Hazards Group/ The Department of Geography, University of California Santa Barbara (ftp://ftp.chg.ucsb.edu/pub/org/chg/products/CHIRPS-2.0/) for the 35-year period from 1981 – 2015.

Table 1. List of observational station used in this paper from various altitude and type such as rain gauge (Obs Gauge), Agricultural Meteorological station (AgriMet), and BMKG weather station (BMKG). Various precipitation record lengths within 1967-2015 periods and well spatially distributed over South Sulawesi.

Station Name	District	Altitude (m.a.s.l)	Station Type	Data Range	Period (years)	Quality (%)
ANABANUA	Wajo	15	Obs Gauge	1977 - 2015	39	94.87
BATUBASSI	Maros	13	Obs Gauge	1976 - 2015	40	97.92
BATUKAROPA	Bulukumba	81	AgriMet	1967 - 2015	49	99.83
BIRINGROMANG	Makassar	8	BMKG	1972 - 2015	44	96.21
BPPBENGO	Bone	77	Obs Gauge	1971 - 2015	45	99.63
BPPDOPING	Wajo	4	Obs Gauge	1976 - 2015	40	89.38
BPPKGALESONG	Takalar	15	Obs Gauge	1980 - 2015	36	97.69
BPPMALAKAJI	Gowa	750	Obs Gauge	1976 - 2015	40	96.25
BUKITHARAPAN	Pare-pare	80	Obs Gauge	1976 - 2015	40	97.92
ENREKANG	Enrekang	65	Obs Gauge	1976 - 2015	40	90.00
MACOPE	Bone	9	Obs Gauge	1970 - 2015	46	99.82
MENGE	Wajo	38	Obs Gauge	1977 - 2015	39	94.02
PGCAMMING	Bone	132	AgriMet	1983 - 2015	33	95.71
PGTAKALAR	Takalar	15	AgriMet	1983 - 2015	33	90.91
SIWA	Wajo	25	Obs Gauge	1977 - 2015	39	90.38
STAKLIMMAROS	Maros	13	BMKG	1985 - 2015	31	90.86
STAMARPAOTERE	Makassar	2	BMKG	1986 - 2015	30	99.44
STAMETHASANUDDIN	Maros	14	BMKG	1972 - 2015	44	96.78
STAMETMASAMBA	Luwu utara	50	BMKG	1983 - 2015	33	95.71
STAMETPONGTIKU	Tana toraja	829	BMKG	1998 - 2015	18	98.15
STASIUNGEOFISIKA	Gowa	28	Obs Gauge	1978 - 2015	38	92.11
SUMPANGBINANGAE	Barru	7	Obs Gauge	1977 - 2015	39	98.93
TODOKKONG	Pinrang	18	Obs Gauge	1985 - 2015	31	98.66

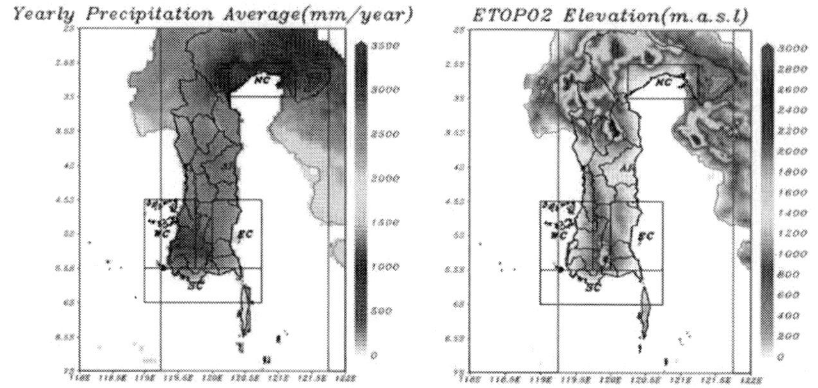

Figure 1. Yearly total precipitation average from monthly CHIRPS dataset (left) and elevation characteristic (right) over study area (meter above sea level/ m. a. s. l) with regional boxes chosen for Western Coast (WC), Southern Coast (SC), Eastern Coast (EC), Northern Coast (NC) and All Region (AR).

2.2. Methodology

Daily precipitation from observation used to calculate maximum number of consecutive days with daily precipitation amount < 1 mm (CDD) [13]. CDD values were also calculated from CHIRPS daily precipitation, for each grid in the study area with a 0.05 x 0.05° horizontal resolution.

SPI was developed in Colorado by McKee et al [14], is based on the probability distribution of precipitation. SPI can be used as valuable estimator of drought severity which requires less input data and efforts then another meteorological drought indices [15]. Furthermore, SPI was reported to be able to identify emerging droughts sooner than Palmer Index [16]. The basis of SPI approach is the calculating probabilities of precipitation for each time scale [14]. SPI values at 3-month time scales (SPI3) were calculated from observation and the CHIRPS monthly precipitation, for each grid in the study area. SPI3 were selected because its represents the typical time scale for precipitation deficits to affect usable water sources and soil moisture important for agriculture [14]. Based on paddy growth period, SPI3 would be able to capture the presence and severity of drought during its growth period [5]. Furthermore, agricultural drought (with soil moisture content as proxy) could be best replicated by SPI on a scale of 2–3months [17], SPI3 also suggested for seasonal drought studies because the 3 months droughts are having a drastic impact on the agriculture [18].

A drought event onset defined when SPI value are less than -1.0 and terminates when SPI value becomes positive again [14,19]. The positive sum of the SPI for all the months within a drought event is referred as drought magnitude (DM) [14,20]. The average drought intensity (ADI) was calculated based on drought magnitude (DM) divided by number or duration of consecutive months during drought event [14,21,22], using observation and CHIRPS.

Level of agreement between the observed and CHIRPS based meteorological drought assessment for each referred station grid value was conducted using temporal Pearson correlation coefficient (R) [23]. Spatial agreement in CDD and SPI pattern between observations and the CHIRPS is analyzed using pattern correlations. This analysis were widely used previously on several references [23–26].

3. Result and Discussion

3.1. Consecutive Dry Days

Maximum consecutive dry days and correlation coefficient for 35 years period (1981 – 2015) can be seen in Figure 2. South western region experience relatively longer dry spell period compared to eastern and northern region. Observational based CDD value also agreed with gridded data when explain spatial distribution of maximum CDD. Temporal agreement between CDD CHIRPS and observation were showed by temporal correlation coefficient values 0.3< R≤ 0.8 in Figure 2 (right). Furthermore, there are spatial agreement between CHIRPS and observational based maximum consecutive dry days during analysis period could be explained by pattern correlation value (R = 0.631).

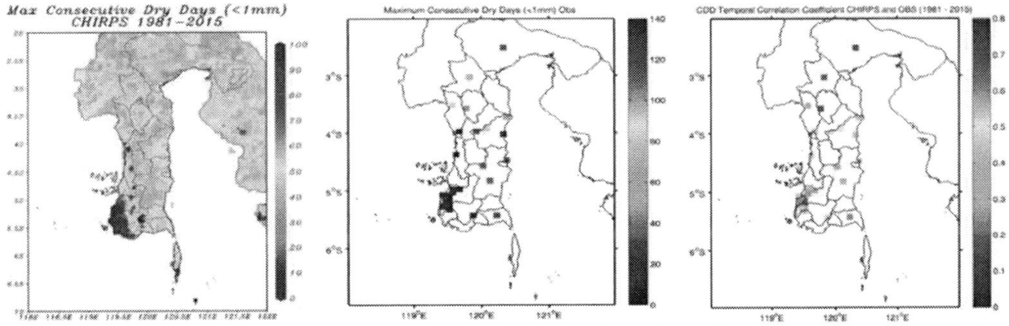

Figure 2. Maximum Consecutive Dry Days (CDD) comparison for each year based on CHIRPS data (left), rain gauge observation (middle) and correlation coefficient between CDD CHIRPS versus observation (right) during 35 years period (1981 – 2015). Longest consecutive number of days with total precipitation amount less than 1 mm/day period and correlation coefficient explained by color bar.

Average consecutive dry days for all time period can be seen in Figure 3. These results are relatively similar with maximum CDD. High pattern correlation value (R = 0.813) between CHIRPS and observation average CDD also show spatial agreement of these two dataset. South western region experience relatively longer dry spell period compared to eastern and northern region. Observational based CDD value also agreed with gridded data when explain spatial distribution of average CDD.

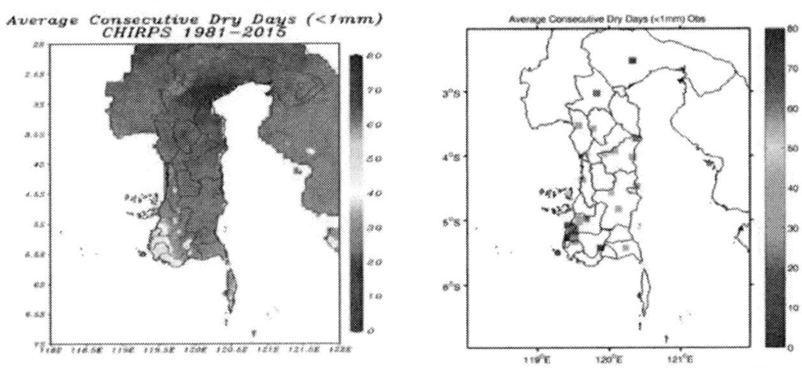

Figure 3. Same as Figure 1 except for Average Consecutive Dry Days (CDD) comparison for each year based on CHIRPS data (left) and rain gauge observation (right).

3.2. Standardized Precipitation Index (SPI)

Average duration of drought intensity (ADI) based on SPI3 for 35 years period (1981 – 2015) can be seen in Figure 4 (left and middle). Temporal agreement between CDD CHIRPS and observation were showed by temporal correlation coefficient values 0.4< R≤ 0.8 in Figure 4 (right). Different ADI value between gridded data and observation mainly occurs in south western region. Nevertheless, spatial distribution of ADI based on CHIRPS generally similar with observational based.

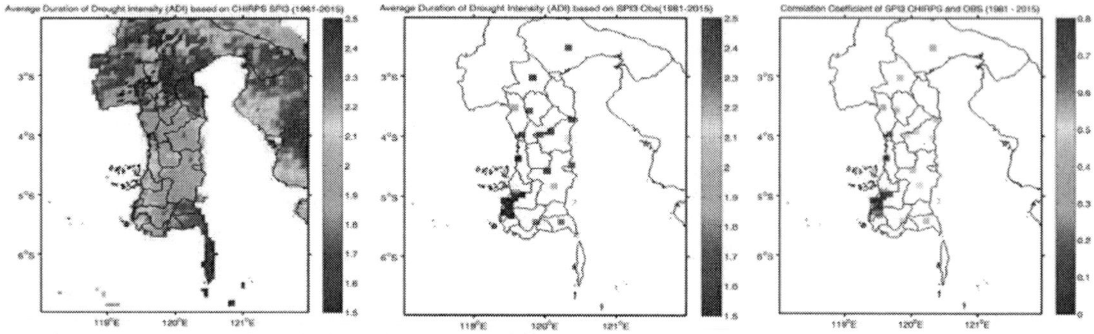

Figure 4. Average drought intensity (ADI) of 3-monthly Standardized Precipitation Index (SPI3) based on CHIRPS data (left), rain gauge observation (middle) and correlation coefficient between SPI3 CHIRPS versus SPI3 observation (right) during 35 years period (1981 – 2015).

Maximum Drought Duration (DD) for all time period can be seen in Figure 5. Different DD value between gridded data and observation generally occurs in all of region. Gridded data slightly over estimate the DD value compare to observation. Nevertheless, spatial distribution of ADI based on CHIRPS generally similar with observational based.

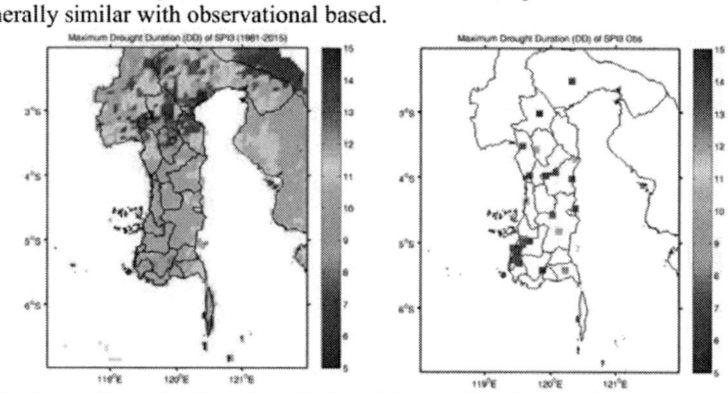

Figure 5. Maximum Drought Duration (DD) of 3-monthly Standardized Precipitation Index (SPI3) based on CHIRPS data (left) and rain gauge observation (right) during 35 years period (1981 – 2015). Longest consecutive number of month with SPI3 values less than -1.0 period explained by color bar.

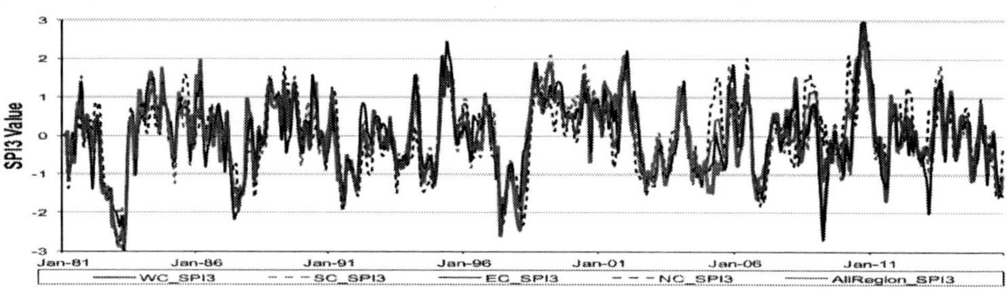

Figure 6. Time series of 3-monthly Standardized Precipitation Index (SPI3) area averaged over five different region i. e. West Coast (WC_SPI3) South Coast (SC_SPI3), East Coast (EC_SPI3), North Coast (NC_SPI3), and All South Sulawesi region (AllRegion_SPI3). More detailed descriptions about region boundary explained in text.

Time series of 3-monthly Standardized Precipitation Index (SPI3) area averaged over five different region i. e. West Coast (WC_SPI3), South Coast (SC_SPI3), East Coast (EC_SPI3), North Coast (NC_SPI3), and All South Sulawesi region (AllRegion_SPI3) can be seen in Figure 6. Strong El Niño mpact on SPI3 are clearly shown, especially during 1982/1983 and 1997/1998 (Figure 7). West coast region are highly effected by El Niño compared to another region.

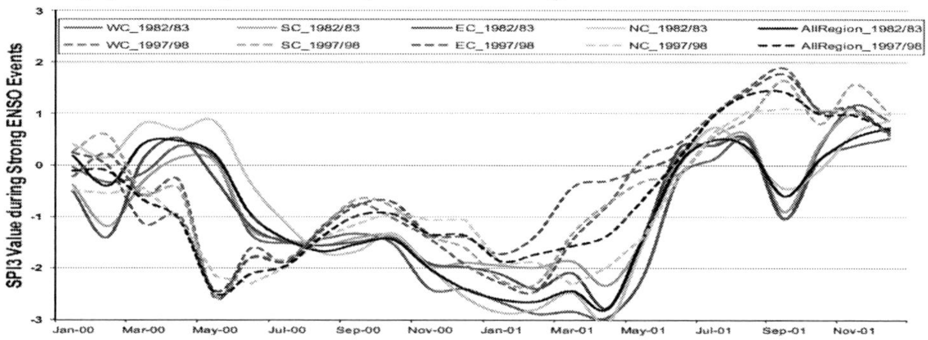

Figure 7. Same as Figure 6, except for two selected strong ENSO event : 1982/1983 (solid lines) and 1997/1998 (dashed lines). Horizontal axis describe ENSO years month (Jan-00 to Dec-00) and following ENSO years (Jan-01 to Dec-01)

4. Conclusion

Observed and blended gauge-satellite precipitation estimates (CHIRPS) have relatively similar perspective on meteorological drought intensity over South Sulawesi. Temporal agreement between CHIRPS and observation meteorological drought were showed by CDD (SPI3) temporal correlation coefficient values 0.3< R≤ 0.8 (0.4< R≤ 0.8). Spatial agreement between CHIRPS and observational based maximum consecutive dry days during analysis period also found and could be explained by CDD spatial pattern correlation value (R = 0.631). Meteorological drought intensity represented by Average Drought Intensity (ADI) from 3-monthly SPI (SPI3) show spatial differences characteristic between west and east coast South Sulawesi region. Lowest SPI3 value occurred especially during strong El Niño years.

References

[1] Nicholls N 1981 Air-sea interaction and the possibility of long-range weather prediction in the Indonesian archipelago *Mon. Weather Rev.* **109** 2435–43

[2] D'Arrigo R, Allan R, Wilson R, Palmer J, Sakulich J, Smerdon J E and Ngkoimani L O 2008 Pacific and Indian Ocean climate signals in a tree-ring record of Java monsoon drought *Int. J. Climatol.* **1901** 1889–901

[3] Quinn W H, Zopf D O, Short K S and Kuo Yang R T W 1978 Historical trends and statistics of the Southern Oscillation, El Nino, and Indonesian droughts (Peru). *Fish. Bull.* **76** 663–78

[4] Setiawan A M 2011 *Determination of Reference ENSO Index for Indonesian Region Based on Correlation Analysis of Spatial and Temporal Pattern with Standardized Precipitation Index (SPI)* (Bandung: Institut Teknologi Bandung (ITB)

[5] Surmaini E, Hadi T W, Subagyono K and Puspito N T 2015 Early detection of drought impact on rice paddies in Indonesia by means of Niño 3.4 index *Theor. Appl. Climatol.* **121** 669–84

[6] Funk C C, Peterson P J, Landsfeld M F, Pedreros D H, Verdin J P, Rowland J D, Romero B E, Husak G J, Michaelsen J C and Verdin a. P 2014 A quasi-global precipitation time series for drought monitoring *U.S. Geol. Surv. Data Ser.* **832** 4

[7] Duan Z, Liu J, Tuo Y, Chiogna G and Disse M 2016 Evaluation of eight high spatial

resolution gridded precipitation products in Adige Basin (Italy) at multiple temporal and spatial scales *Sci. Total Environ.* **573** 1536–53

[8] Funk C C, Peterson P J, Landsfeld M F, Pedreros D H, Verdin J P, Rowland J D, Romero B E, Husak G J, Michaelsen J C and Verdin A P 2014 A Quasi-Global Precipitation Time Series for Drought Monitoring *U.S. Geol. Surv. Data Ser.* **832** 4

[9] Toté C, Patricio D, Boogaard H, van der Wijngaart R, Tarnavsky E and Funk C 2015 Evaluation of satellite rainfall estimates for drought and flood monitoring in Mozambique *Remote Sens.* **7** 1758–76

[10] Zhan W, Guan K, Sheffield J and Wood E F 2016 Depiction of drought over sub-Saharan Africa using reanalyses precipitation data sets *J. Geophys. Res. Atmos.* 3405–20

[11] Shukla S, McNally A, Husak G and Funk C 2014 A seasonal agricultural drought forecast system for food-insecure regions of East Africa *Hydrol. Earth Syst. Sci.* **18** 3907–21

[12] Hayes M, Svoboda M, Wall N and Widhalm M 2010 The Lincoln Declaration on Drought Indices: Universal Meteorological Drought Index Recommended *Bull. Am. Meteorol. Soc.* **92** 485–8

[13] Klein Tank A M G, Zwiers F W and Zhang X 2009 *Guidelines on Analysis of extremes in a changing climate in support of informed decisions for adaptation* ed A M G K (Geneva, Switzerland: World Meteorological Organization)

[14] McKee T B, Doesken N J and Kleist J 1993 The Relationship of Drought Frequency and Duration to Time Scales *Proceedings of the 8th AMS Conference on Applied Climatology* (Anaheim, California)

[15] Keyantash J and Dracup J A 2002 The Quantification of Drought: An Evaluation of Drought Indices *Bull. Am. Meteorol. Soc.* **83** 1167–80

[16] Hayes M J, Svoboda M D, Wilhite D A and Vanyarkho O V. 1999 Monitoring the 1996 Drought Using the Standardized Precipitation Index *Bull. Am. Meteorol. Soc.* **80** 429–38

[17] Szalai S and Szinell C S 2000 Comparison of two drought indices for drought monitoring in Hungary—a case study (Springer) pp 161–166

[18] Edwards D C 1997 *Characteristics of 20th Century drought in the United States at multiple time scales.*

[19] Guttman N B 1999 Accepting The Standardized Precipitation Index: A Calculation Algorithm *J. Am. Water Resour. Assos.* **35** 311–22

[20] Morid S, Smakhtin V and Moghaddasi M 2006 Comparison of seven meteorological indices for drought monitoring in Iran *Int. J. Climatol.* **26** 971–985

[21] McKee T B, Doesken N J and Kleist J 1995 Drought Monitoring with Multiple Time Scales *Proceedings of the 9th AMS Conference on Applied Climatology* (Dallas, Texas: AMS) pp 233–6

[22] Nosrati K and Zareiee A R 2011 Assessment of meteorological drought using SPI in West Azarbaijan Province , Iran

[23] Sohn S J, Tam C Y and Ahn J B 2013 Development of a multimodel-based seasonal prediction system for extreme droughts and floods : a case study for South Korea *Int. J. Climatol.* **805** 793–805

[24] Ganguli P and Ganguly A R 2016 Robustness of Meteorological Droughts in Dynamically Downscaled Climate Simulations *J. Am. Water Resour. Assoc.* **52** 138–67

[25] Compo G P and Sardeshmukh P D 2009 Oceanic influences on recent continental warming *Clim. Dyn.* **32** 333–42

[26] Sohn S J, Tam C Y, Ashok K and Ahn J B 2012 Quantifying the reliability of precipitation datasets for monitoring large-scale East Asian precipitation variations *Int. J. Climatol.* **32** 1520–6

IOP Conf. Series: Earth and Environmental Science **54** (2017) 012041 doi:10.1088/1755-1315/54/1/012041

Anomaly of surface circulation and Ekman transport in Banda Sea during 'Normal' and ENSO episode (2008-2011)

Selfrida M. Horhoruw[1], Agus S. Atmadipoera[1]*, Pieldrie Nanlohy[2] and I Wayan Nurjaya[1]
[1] Department of Marine Science and Technology, FPIK IPB Bogor
[2] Department of Physics, FMIPA UNPATTI Ambon

E-mail: atmadipoera_itk@ipb.ac.id, idahorhoruw@gmail.com

Abstract. Banda Sea (BS) is a deep and large ocean basin in interior Indonesia Seas which may play a significant role on regulating ocean dynamics and regional climate variability such as El Nino Southern Oscillation (ENSO). It also provides a 'temporary reservoir' for Indonesian Throughflow (ITF), and as a region of seasonal alternating upwelling/downwelling. The objective of present study is to investigate changes of surface ocean circulation and associated upwelling anomaly during 'normal' (2008/2009), El Nino (2009/2010) and La Nina (2010/2011) years in BS, using data sets from validated INDESO model outputs. The results show that surface circulation in BS is regulated by both local forcing of monsoon winds and inflow of ITF via Flores Sea (FS), Lifamatola Strait, and Halmahera/Seram Seas. Inflow from FS is drastically increased during northern winter and 'normal' year. The upwelling event occurs along the eastern edge of BS with southwestward Ekman transport and its kinetic energy of above 10 Joule. The Ekman layer depth in BS is about 60 m. During La Nina event surface circulation is much stronger with kinetic energy of about 700 J, causing Ekman layer shallower due to increased Ekman transport and upwelling intensity.

1. Introduction

Banda Sea (BS) is one of important fishing ground areas for pelagic fisheries in Indonesia, due to the upwelling processes that impact positively on high primary productivity [1]. Furthermore, BS is a unique confluence region of different water masses, originated from North and South Pacific Ocean [e.g., 3, 9].

The seasonal reversal monsoonal winds are fully developed here during peaks of the Northwest monsoon (December-February) and Southeast monsoon (July-September) periods, causing seasonal reversal of upper ocean circulation, known as the Indonesian monsoon current (or Armondo). During the Northwest monsoon (NWM) period, Armondo transports water masses from the Java and Flores Seas into BS, resulting in accumulation of water masses and causing downwelling. During the Southeast monsoon (SEM) period, Armondo brings water masses westward from Banda, so there is an imbalance of water mass volume which generate a vertical movement of water masses from deeper to surface layers (or upwelling) [1, 2]. Circulation in BS is also affected by interannual climatic variability such as El Nino Southern Oscillation (ENSO). It is found that during the El Nino phase, seawater temperature is much cooler related to weakening Ekman pumping than that during the La Nina event [3].

Theoretically, upwelling event is associated with Ekman pumping and Ekman transport of water mass. Those are generated by the steady and strong winds fields. Ekman transport plays an important role on transporting water masses away from the center of upwelling region. Due to the Earth rotation, wind-driven surface currents are deflected about 45° to the right (left) to the winds direction in the northern (southern) hemispheres [4]. In BS, peak of Ekman upwelling occurs in May and June with upwelling transport volume of 2.5= Sv (1 Sv=10^6 m^3s^{-1}), but peak of Ekman downwelling occurs in February with transport volume of 1.0 Sv [3].

This paper aims to investigate changes of surface circulation patterns during 'normal' 2008/2009, El Nino 2009/2010 and La Nina 2010/2011 events (hereinafter referred to as 'normal', El Nino and La Nina), and its implication to Ekman layer and upwelling intensity in BS. Since anomaly of interannual climatic variability such as ENSO impacts significantly on ocean-atmosphere interaction and marine ecosystem in Indonesian Seas, particularly in Banda Sea, so this study may provide the first insights into a better understanding of circulation anomaly in BS during ENSO year 2009/2011.

2. Data and data analysis

2.1. Study area

This study was conducted from May to August 2016 in the Physical Oceanography Laboratory, Department of Marine Science and Technology (ITK) - FPIK IPB Bogor. The study area is in deep BS basin (figure 1). Lines that parallel and perpendicular to the eastern outer islands arcs of BS denote sections for Ekman transport and upwelling calculations.

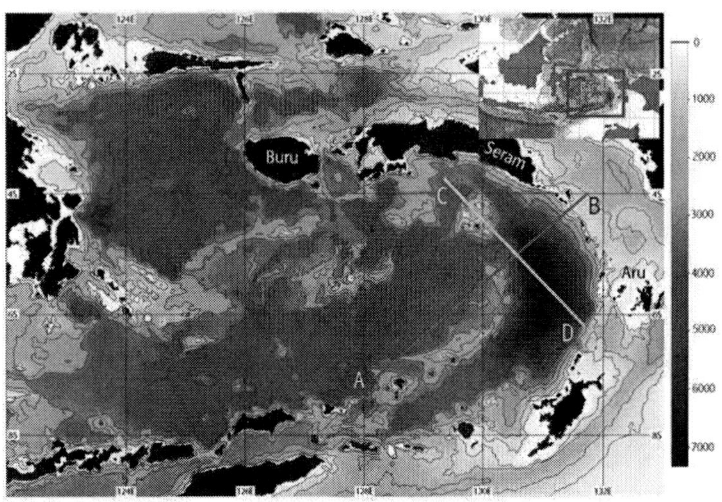

Figure 1. Bathymetry in study area in BS. Lines of A-B and C-D denote for cross-coast velocity and along-coast velocity for further calculations of Ekman upwelling with a reference to BS outer-islands arcs between Seram and Aru Islands. Interval of contour is every 1000 m.

2.2. The data

2.2.1. INDESO model outputs

Model outputs from a series of simulation of NEMO ocean general circulation model of INDESO configuration (INDESO: Infrastructure Development of Space Oceanography), performed by the CLS France & BPOL Bali, were used in this study. The outputs of data-series span from June 1st 2008 to May 31 2011, covering 'normal' and ENSO years. INDESO is a scientific cooperation program at the Ministry of Marine Affairs and Fisheries with a consortium of joint European Marine Research Institute

(MERCATOR-Ocean and *Collecte Localisation Satelittes* - CLS) France. A brief description of INDESO configuration is illustrated here. The model used 3-dimensions NEMO modeling system, developed by [5]; Atmospheric forcing is derived from the ECMWF reanalysis data with temporal resolution of 6-hours; Both barotropic and baroclinic tidal forcing was applied explicitly to the model; Lateral boundary forcing was extracted into the model from the 1/4° global ocean circulation of Mercator-Ocean; Horizontal grid resolution is 1/12° (approximately 9.25 km) with 50 depth levels; bathymetry of the model is derived from ETOPO02 data [6]. Model simulation was covered from 2008 to 2015. Daily averaged of 3-dimensions seawater temperature, salinity and ocean current fields were used for this study.

2.3. Model validation

To assess a performance and reliability of the model results, time-series data of sea surface temperature anomaly (SSTA) and sea surface height anomaly (SSHA) from the model outputs were compared with those observed from the NOAA AVHRR high-resolution SST data and AVISO satellite altimetry data, in the center basin of BS (figure 2; inset). A statistical correlation coefficient between SSTA and SSHA model output and the data are reasonably high of 0.99 and 0.94, for SSTA and SSHA, respectively. It is shown clearly that fluctuations of SSTA and SSHA both in the model and data are in excellent agreement. The SSTA time-series data reveal strong seasonal variations, which are minima during the SEM period and double maxima during the monsoon transition and peak of the NWM periods (figure 2A). Furthermore, SSHA minima occur during the peak of SEM period and their maxima are during the peak of NWM (figure 2B). These results are consistent with [3]. Hence, model outputs of SST and SSH from INDESO model reproduced well the observed satellite data and can be used for further analysis in this study.

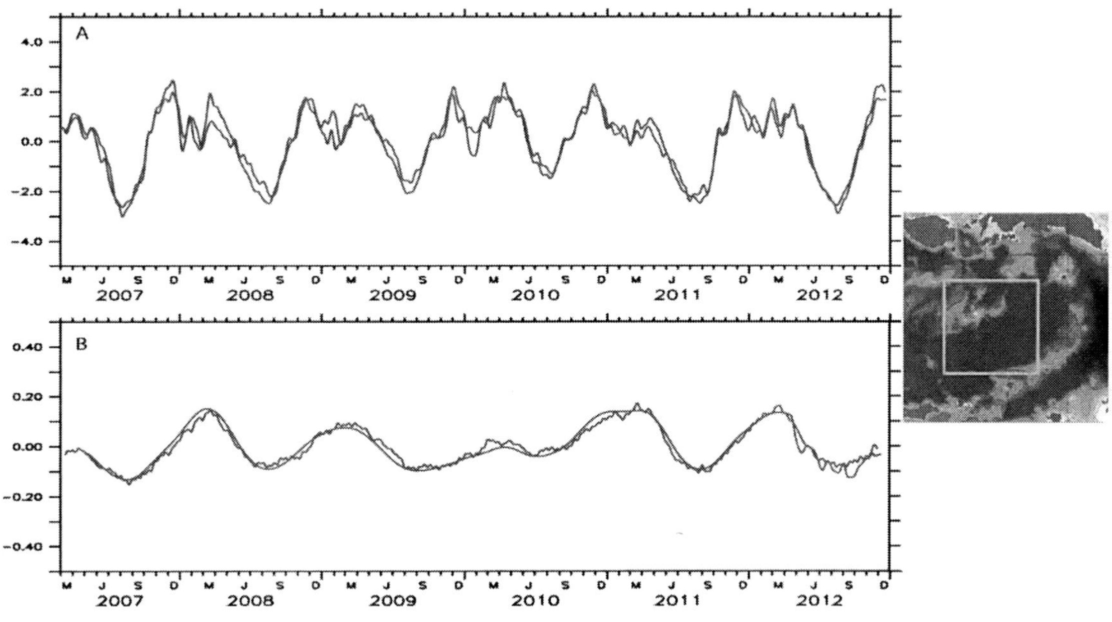

Figure 2. Time-series of sea surface temperature anomaly (A) and sea surface height anomaly (B) from the INDESO model output (red) and satellite data (blue). The data were extracted from a sampling box at the center of BS (see inset map). Statistical correlation coefficient for SSTA and SSHA are 0.94 and 0.96, respectively. Plot of time-series data are smoothed by 21-day Hanning low-pass filtered.

2.3. Data analysis

The 3-dimensions time-series data of INDESO model of seawater temperature, salinity and zonal & meridional current components, and time-series data of El Nino Southern Oscillation (ENSO) indices (NINO3.4 and SOI) between 2008 and 2011 were analyzed to understand monthly evolution of oceanic conditions (circulation and seawater properties distribution) during 'normal' year, El Nino and La Nina years.

Surface kinetic energy (KE) was calculated from ocean current fields and was analyzed to understand circulation pattern and upwelling dynamic in BS, using the equation as follows:

$$KE = \frac{1}{2}\rho V^2 \quad (1)$$

where V is total current vector of zonal and meridional components, and ρ is seawater density.

To obtain cross-coast velocity that will be calculated for Ekman transport and vertical cross-section of upwelling evolution, zonal (u) and meridional (v) current components were rotated anticlockwise by 45° from a reference of the BS islands arc between Seram and Aru Islands with the northwest-southeast orientation (see figure 1). According to [7], rotated current component (u and v) from the North reference into local reference with θ is presented in equation 2 and 3, as follows [7]:

$$U' = u\ \cos\theta + v\ \sin\theta \quad (2)$$

$$V' = -u\ \sin\theta + v\ \cos\theta \quad (3)$$

where, u and v are zonal and meridional current components; θ is degree of rotation with local reference; U' and V' are rotated cross-coast velocity (CCV) and along-coast velocity (ACV). Illustrations of CCV and ACV are shown in Fig. 1, as A-B line and C-D line, respectively.

Analysis of monthly averaged current fluctuation was focused in the CCV transect (A-B line, in Fig. 1). Furthermore, Ekman transport was calculated across the ACV transect (C-D line, in Fig. 1). Based on this calculation, magnitude and direction of Ekman transport volume away from the reference (islands) is considered to be upwelling parameter. Ekman transport volume estimate (Q) was calculated across C-D section (Fig.1) using the equation below:

$$Q = \sum_{k=1}^{n} V'A \quad (4)$$

where V' is the cross-coast velocity (CCV) perpendicular to the BS islands arc that obtained from equation (3), and A is cross-sectional area for i[th] grid, and k=1 to n is number of grids.

3. Results and discussion

3.1. Surface circulation during 'normal', El Nino, and La Nina years

Monthly mean of kinetic energy and current vectors during the SEM and NWM periods in three different years are presented in figure 3. Evolution of monthly surface circulation during 'normal' and ENSO events (2008-2011) showed that the surface flow in BS is driven by monsoonal winds and is also influenced by upper component of Indonesian Throughflow (ITF) that is derived from three main inflows via Flores Sea, Lifamatola Strait and Halmahera Sea. Previous studies showed that the main transport volume of ITF pass by the western route via Sulawesi Sea-Makassar Strait-Flores Sea [8-11]. Large seasonal reversal flows are related to seasonal changes of monsoonal winds over the study area [3, 12].

In general, the KE and current vectors showed high amplitude of surface currents in Flores Sea during the NWM period, which is coherent with strong surface winds over BS. High kinetic energy in the entrance of Flores Sea apparently brings surface water mass derived from western part of interior seas, such as Makassar Strait, Java Sea and southern South China Sea [2, 13, 26]. From December to January, surface current in eastern Flores Sea turned to the North along the western edge of Buru Island and joined surface current from Lifamatola Strait. So that, circulation in Seram Sea is indicated by eastward flows that may be a confluence region of different water masses from from Flores Sea and Maluku Sea/Lifamatola Strait. This is consistent with [2], during the NWM period, surface water flows eastward and enters Halmahera and Maluku Seas. From February to March, counter-clockwise circulation (eddy) was found in northwestern BS.

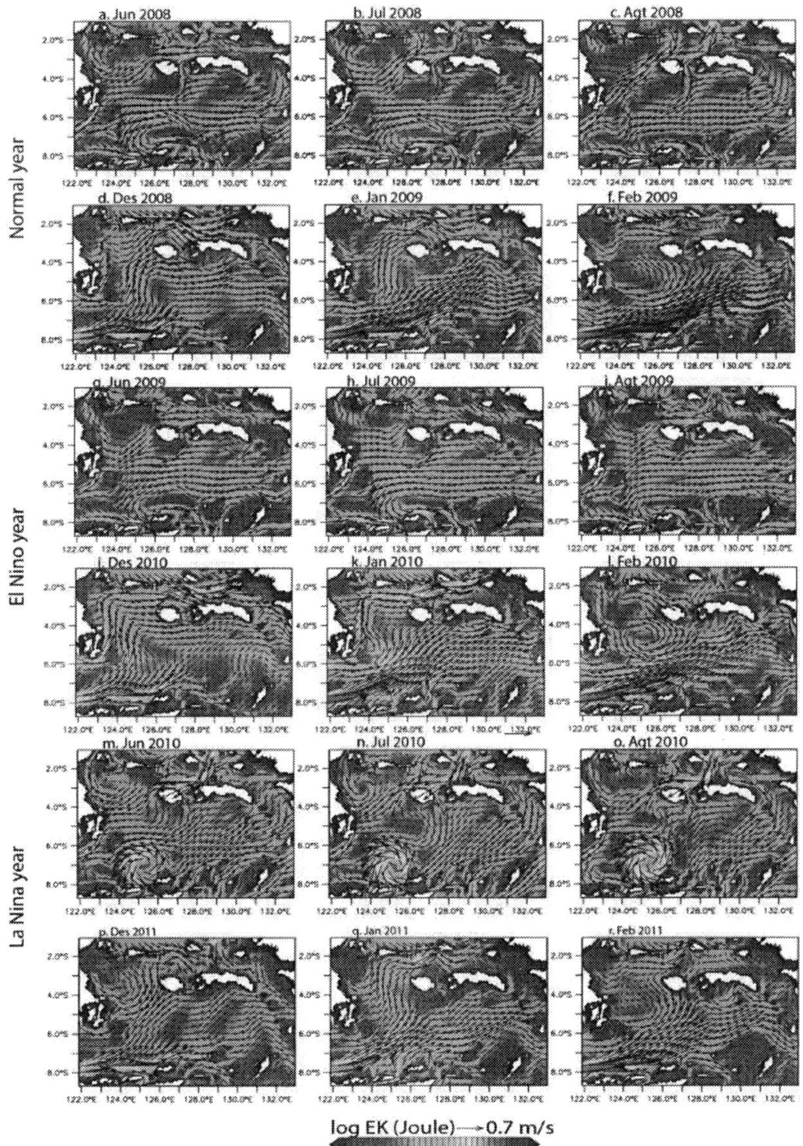

Figure 3. Monthly surface kinetic energy, overlaid with current vectors during the Southeast monsoon (SEM) and Northwest monsoon (NWM) periods in 'normal', El Nino and La Nina in BS.

The fully developed SEM winds cause reversal of surface circulation patterns, where surface current in Halmahera Sea becomes much stronger. Surface current in Halmahera flows westward along the northern edge of Seram Island and small branch of these currents enter Buru Strait where the other current down the western edge of Buru Island. During this period, a high kinetic energy anti-clockwise eddy was formed in the Southwest Maluku Island. The existence of this eddy was relatively permanent during the SEM period. Its size and kinetic energy varied during different ENSO period. Radius of eddy formation was about 300 km. According to [2, 13], the wind-driven currents in BS pushes back surface water from eastern BS westward into Flores Sea, thus weakening the flow in Flores Sea. Formation of anti-clockwise eddies in southern hemisphere causes a convergent circulation that leads to water mass sinking (downwelling), while clockwise eddies, enhanced by the SEM winds, cause a divergent circulation in eastern Banda region, transporting surface water mass away from the coastal area between Seram and Aru Islands. This offshore transport is compensated by upwelled water from deeper layer to surface layer to maintain continuity of seawater volume. [3] revealed that surface divergent circulation influences significantly on surface component of ITF in BS.

Variation of surface circulation in BS is also affected by the ENSO phenomena. During the SEM period in 'normal' year 2008 (figure 3), the flow via Halmahera Sea branched, where one part flows down along northern Obi Island and the rest turns into Seram Sea through Lifamatola Strait. Surface circulation is predominantly westward to western Seram Sea and merges with the flow from Lifamatola Strait. In addition, there a northward flow along the eastern boundary of Papua Island near 1°S latitude.

Hence, during 'normal' year 2008 the main inflow for circulation in BS was via Halmahera Sea and Lifamatola Strait through Buru Strait (western part of Seram Island) and western part of Buru Island. Furthermore, a divergent circulation occurs between eastern part of Seram and western side of Papua (near ~1°S), where the northward flow was predominant. During this period, upwelling also occurs in western part of Papua. Previous studies reported the occurrence of less salty water from eastern Papua [2, 14, 15]. The westward flow into BS from eastern side brings surface water away from eastern Banda Islands arcs, which extends from southern Buru Island to northern Southwest Maluku. This implies a generation of upwelling event during the SEM period.

During the SEM period in the El Nino event 2009, the main inflow from Halmahera and Lifamatola that supply surface flow in BS was reduced, but the divergent circulation in western Papua (~ 1°S) strengthened, where one branch of the current flows northward up to northern Buru which push back surface water into Halmahera Sea and Lifamatola Strait, and part of current branch enters BS via its eastern side (figure 3).

During the SEM period in La Nina event 2010, the trade winds in the equatorial Pacific were much stronger [16, 17], which increased an accumulation of 'warm pool' in western equatorial Pacific. Furthermore, a 'leakage' of ITF inflow via Halmahera Sea and the Lifamatola Strait also strengthened during this period, and divergent circulation in western Papua decreased (figure 3 m-o). Sea surface height anomaly in BS during La Nina event was much higher, compared to 'normal' and El Nino years (figure 2b) [3]. During this event, eddies formation in northern Southwest BS was developed much stronger with kinetic energy of about 400 Joule, and eddies diameter of about 290 km (Fig. 3o). During late of the SEM and transition monsoon periods the divergent circulation in BS was weakened (not shown).

During the NWM period in 'normal' year 2008, surface circulation in Flores Sea was much stronger with the highest kinetic energy was found in its main axis of about 100-750 Joules (figure 3 d-f). The maximum kinetic energy was also found in northern Alor (7-9°S and 123°E). However, during the NWM period in El Nino year 2009, surface circulation in Flores Sea was reasonably reduced in February with averaged kinetic energy in its main axis between 100-680 Joule.

Surface flows in eastern Flores Sea weakened during the NWM period in La Nina event 2010 with kinetic energy between 50-600 Joules, but kinetic energy of the circulation increased above 700 Joule in March (not shown). Inflow into BS was also increased (figure 3 p-r). This suggests that there was about a month phase delay of circulation peak during La Nina, compared to El Nino and 'normal' years. It is considered that much stronger kinetic energy of surface circulation via Flores Sea during La Nina event was associated with an increased flow of ITF along its western pathway [11, 18-20].

3.2 Surface temperature and salinity anomaly during 'normal' and ENSO years.

It is interesting to note that during the SEM 'normal' year 2008, SST anomalies in BS were much lower (negative) (figure 4a). However, in the El-Nino event 2009/2010, positive SST anomalies were seen in western BS, while negative anomalies occurred in eastern and northern region (figure 4b). During La Nina year 2010/2011 much higher positive anomalies of SST were evident in BS with weak negative anomalies occured in eastern Flores Sea (figure 4c).

It is expected that SST variation during El Nino event is much lower than that during 'normal' year [3]. It is suggested that during El Nino year SST anomaly is much cooler related to weakening Ekman pumping, in contrast to those observed during La Nina year [3]. In contrary, here, during 'normal' year 2008 mean SST distribution varied between 26.7 and 29.1°C, which was much cooler than that during El Nino year (28.8 and 29.5°C). This is consistent with observed time-series SSTA data in Fig. 2, showing that SSTA minima during 2008 ('normal' year) was much lower than that during 2009 (El Nino year). It is likely that weak El Nino 2009/2010 may be a possible explanation to this anomalous warmer SST during El Nino year. But, further study is needed to understand its mechanism. During strong La Nina 2010/2011 SST in BS was much warmer than 'normal' year, which is in good agreement with [21].

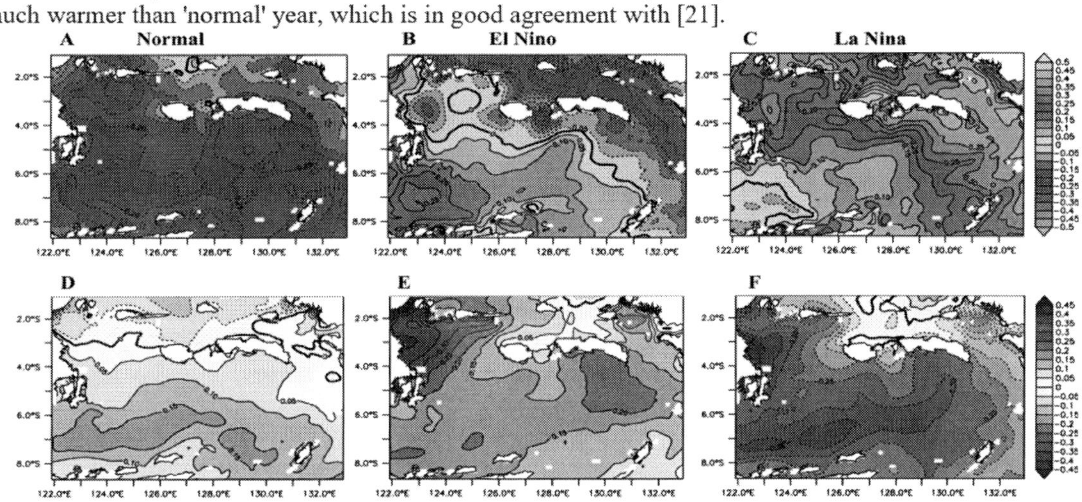

Figure 4. Anomaly of sea surface temperature (A-C) and salinity (D-F) in BS during 'normal' year 2008 (A and D), El Nino year 2009/2010 (B and E), and La Nina year 2010/2011 (C and F).

Furthermore, cross-section of seawater temperature along the C-D transect (not shown) showed that during the SEM in 'normal' year 2008, SST minima varied between 26.0 and 27.5°C with the Ekman depth of about 70 m in August 2008. During El Nino 2009/2010 surface temperature was about 0.5°C higher than that during 'normal' year 2008 of about 27-28.5°C and Ekman layer is shallower to about 60 m. During the SEM in La Nina year 2010/2011, surface temperature varied from 27 to 28.5°C and Ekman layer was much shallower (about 40 m).

Strong positive salinity anomalies were identified during El Nino year (figure 4e), while low salinity anomalies were found during La Nina year (figure 4f). Positive salinity anomalies were increased by about 0.15 psu during El Nino year, compared to salinity in 'normal' year and also higher than La Nina year with different salinity anomalies about 0.4 psu. Cooler surface temperature and saltier water masses in BS during El Nino create much denser water that decrease sea surface height [3,21]. Much denser surface water masses are associated with shallower mixed layer. Much saltier salinity during El Nino (Fig. 4e) may be associated with a loss of direct precipitation, since convective clouds shifted eastward from western equatorial Pacific. In contrast, during La Nina year (figure 4f), high precipitation rate was impact directly to a decrease of surface salinity.

High surface temperature variation was found in eastern BS (not shown), associated with upwelling events during the SEM period, and weak surface temperature variation occurs during La Nina 2010. However, surface salinity variation was high in northeastern BS and in eastern Flores Sea. Surface monsoon current flows eastward during the NWM period caused high surface salinity variation in western boundary of BS [15, 23].

3.3 Variation of Ekman transport during 'normal', El Nino and La Nina years

During the SEM period both in 'normal' and ENSO years, Ekman transport in the Ekman layer (55.76 m) was negative which means that transport direction was away from the coastal region in the eastern BS between Seram and Aru Islands. This is indicated by negative values and blue color (figure 5a). Consequently, upwelled water mass from deeper layer into the sea surface layer must be taken place to compensate transported water volume offshore. In contrary, during the NWM period in 'normal' and ENSO period, the Ekman transport direction was toward the coastal region (positive values; red color), where accumulation of surface warm and les salty water develop downwelling process. Those dynamics of divergence and convergence appear seasonally as a response of surface layer in BS to the monsoonal winds [3, 14, 22].

In 'normal' year 2008, the southwestward (negative) Ekman transport maximum was about 1.84 Sv (1Sv = 10^6 m^3s^{-1}). During the SEM period in El Nino year the Ekman transport was slightly decreased by 1.58 Sv with one-month lead in June, while in La Nina year the offshore Ekman transport maximum was about 1.24 Sv in June (figure 5b).

In contrary, the convergent (positive) Ekman transport towards the coastal waters occur during the peak of NWM period. In 'normal' year 2008, the peak of positive Ekman transport revealed two maxima in December 2008 and February 2009 with averaged transport of 1.4 and 1.61 Sv. During El Nino year 2009/2010, the divergent (positive) transport maximum was 2.6 Sv. However, in La Nina year 2010/2011 the divergent transport toward the coastal region was decreased to 2.12 Sv. Interestingly, the divergent (positive) transport maximum occurred during El Nino year 2009/2010 which was relatively short duration, from December 2009 to April 2010. However, during La Nina year 2010/2011 the divergent transport was much longer in duration from September 2010 to April 2011 (figure 5).

4. Conclusion

Surface circulation in BS is mainly driven by local reversal monsoonal winds, and influenced by upper component of the ITF. Its magnitude and spatial variation of surface circulation are significantly modified during the ENSO year. Model suggested that during the SEM period in El Nino year 2009/2010 surface inflow from Lifamatola and Halmahera was weakened and in western Papua waters a divergent circulation was established. In La Nina year 2010/2011 the inflow from northern Banda was increased drastically, which diminished divergent circulation in western Papua and increased sea surface height in BS. Transient eddies with high kinetic energy were formed in southwest BS only during La Nina event. During the NWM period, the inflow from Flores Sea into BS was increased.

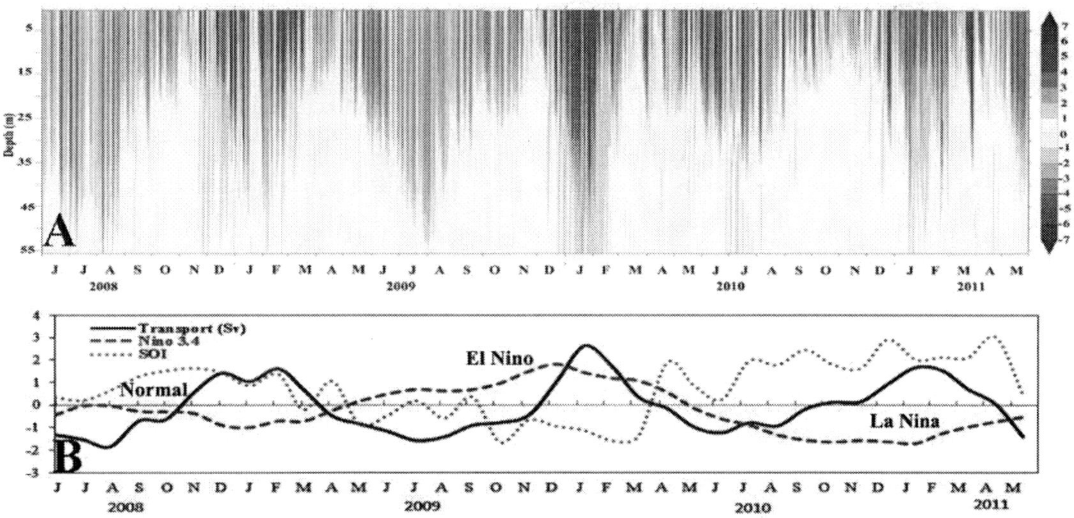

Figure 5. (A) The average of Ekman transport between surface and 56 m depth along the C-D transect (see figure 1); and (B) time-series data of total Ekman transport, Nino3.4 and SOI indices from June 2008 to May 2011. Positive/red (negative/blue) transports mean downwelling (upwelling) episode.

Much higher kinetic energy of circulation in Flores Sea was found during 'normal' year, weakened during El Nino year, and increased during La Nina year with one-month delay phase of inflow peaks. Seawater temperature within Ekman layer during the SEM period in El Nino 2009/2010 was anomalously much warmer than that during 'normal' year 2008. However, the thickness of Ekman layer in 'normal' year was much lower than that in El Nino year, and the shallowest Ekman layer occurred during La Nina year 2010/2011. The maximum divergent (negative) transport water mass that generates upwelling was revealed during 'normal' year, but minimum during La Nina year, due to positive transport (or downwelling) was maximum during La Nina year with short time duration, compared to 'normal' year.

Acknowledgments

The INDESO model outputs were provided by CLS France. The Authors would like to thank INDESO project office Jakarta, BPOL Bali, CLS and MERCATOR-Ocean Toulouse France, particularly to Dr. M. Eko Rudianto, Dr. Philippe Gaspar and Dr. Benoit Tranchant of CLS Toulouse. This study was partly financed by *KEMENRISTEK DIKTI* research grants 2016. Comments and suggestions from the Reviewers to improve this manuscript are highly appreciated.

References

[1] Ilahude A G and Nontji A 1993 Oseanografi Indonesia dan Perubahan Iklim Global (El Nina dan La Nina). *Lokakarya Kita dan Perubahan Iklim Global: kasus El Nino-La Nina. Akademi Ilmu Pengetahuan. Jakarta. 18-19 Mei.*

[2] Wyrtki K 1961 The physical oceanography of South East Asian Waters *Naga Report* 2 (California: University California Press) 195p.

[3] Gordon A L and Susanto R D 2001 Banda Sea surface-layer divergence *Ocean Dyn.* **52** 2-10. Doi: 10.1007/s10236-001-8172-6.

[4] Ekman V W 1905 On the influence of the earth's rotation on ocean currents. *Arkiv. For Matematik, Astronomi och Fysik,* **2** 52

[5] Madec G and Nemo Team 2008. *NEMO Ocean Engine.*
[6] Tranchant B, Reffray G, Greiner E, Nugroho D, Koch-Larrouy A and Gaspar P 2015 Evaluation of an operational ocean modal configuration at 1/12° spatial resolution for the Indonesian seas - Part 1 *Oce Phys Geosci Model Dev. Disc.* **8** 6611-6668
[7] Emery W J and Thomson R E 2004 *Data Analysis Methods in Physical Oceanography second and Revised Edition.* (Boulder: Colorado (USA))
[8] Bingham F and Lukas R 1994 The Southward Intrusion of North Pacific Intermediate Water along the Mindanao Coast *J Phys Oceanogr.* **24** 141-154.
[9] Gordon A L and Fine R A 1996 Pathways of water between the Pacific and Indian oceans in the Indonesian seas. *Nature.* **379** 146–149.doi:10.1038/379146a0.
[10] Ilahude A G and Gordon A L 1996 Thermocline stratification within the Indonesian Seas. *J. Geophys. Res.* 101 12401-12409. doi: 10.1029/95JC03798.
[11] Sprintall J, Gordon A L, Koch-Larrouy A, Lee T, Potemra J T, Pujiana K and Wijffels S E 2014 The Indonesian seas and their role in the coupled ocean-climate system *Nature Geoscience.***7** 487-492. doi:10.1038/NGEO2188.
[12] Chang C P, Harr P A and Chen H J 2005 Synoptic disturbances over the equatorial South China Sea and western Maritime Continent during boreal winter *Monthly Weather Review* **133** 489-503.
[13] Gordon A L, Susanto R D and Vranes K 2003 Cool Indonesian throughflow as a consequence of restricted surface layer flow *Nature.* **425** 824–828. doi:10.1038/nature02038.
[14] Waworuntu J M, Fine R A, Olson D B and Gordon A L 2000 Recipe for Banda Sea water *J Mar Res.* **58** 547-569.
[15] Zijlstr J J, Baars M A, Tijssen S B, Wetsteyn F J, Witte J I J, Ilahude A G and Hadikusumah 1990 Monsoonal effects on the hydrography of the upper waters (300 m) of the eastern Banda Sea and northern Arafura Sea, with special reference to vertical transport processes. Neth. *J. Sea Res.* **25** 431–447.
[16] Cane M A and Zebiak S E 1985 A theory for El Nino and the southern oscillation *Science.*
[17] Bradley R S and Lan W 2001 Climate change and society (Cheltenham Stanley: Thornes)
[18] Horhoruw S M 2016 Struktur dan Variabilitas Arlindo di Selat Makassar [Tesis]. (Bogor (ID): Institut Pertanian Bogor)
[19] Atmadipoera A S, Horhoruw S M, Purba M and Nugroho D Y 2016 Variasi spasial dan temporal Arlindo di Selat Makassar *ITKT.* **8** 299-320
[20] Ffield A, Vranes K, Gordon A L and Susanto RD 2000 Temperature variability within Makassar Strait *J Geophys Res Lett.* **27** 237-240
[21] Kadmaer E M Y 2013 Variabilitas klorofil-a dan beberapa parameter oseanografi, serta hubungannya dengan Monsoon, ENSO dan IOD di Laut Banda [Tesis]. (Bogor (ID): Institut Pertanian Bogor)
[22] Moore T S, Marra J and Alkatiri A 2003 Response of the Banda Sea to the southeast monsoon *Mar. Ecol. Prog. Ser.* **261** 41-49
[23] Atmadipoera A, Molcard R, Madec G, Wijffels, Sprintal J, Larrouy A K, Jaya I and Supangat A 2009 Characteristics and Variability of the Indonesian Throughflow Water at the Outflow Straits. *Deep Sea Research 1.* **56** 1942-1954

IOP Conf. Series: Earth and Environmental Science 54 (2017) 012042 doi:10.1088/1755-1315/54/1/012042

Prototype of the Mexican spatial data infrastructure for climate raster models and satellite imagery ("VISTA-C")

S Couturier*, J Osorno Covarrubias, V Magaña Rueda, I Martínez Zazueta and G Vázquez Cruz

Laboratory for Geospatial Analysis (LAGE), Geography Institute,
National Autonomous University of Mexico (UNAM)
Circuito Exterior s/n, Del. Coyoacán, CP: 04510, Mexico City, Mexico

E-mail: stephcamelo@mailcity.com

Abstract. In the face of climatic uncertainty and its impacts on agriculture yields, there is a growing need for public institutions of subtropical countries to access as reliable as possible meteorological models and transmit a representation of their results in an effective way to stakeholders in agriculture. In many of these countries however, broad climatic regions and point-based statistics remain the core of these representations. The use of satellite imagery is largely limited to visual assessment, although it could serve as complementary data to meteorological raster models and the basis for spatially consistent quantitative impact assessments of meteorological events. In view of this situation in Mexico, a project developed by the Institute of Geography at UNAM university, and promoted by the National Institute of Geography and Statistics, consisted in the development of a climate monitoring system, which includes three main features: 1) a modular array storage system containing NOAA and GOES satellite imagery acquired though a receiving station (ERISA), 2) a climate modeling squeme based on successive error corrections of climate raster maps and associated models using the above mentioned imagery, and 3) an online, dynamic geovisualization of the results of the models. We discuss the implemented technologies and illustrate the VISTA-C prototype which has been released.

1. Introduction

Climatic variability and environmental pollution impact multiple aspects of the survival strategy of human societies. Especially, food security, public health, disaster mitigation, are aspects which appear as a priority in the national agenda of many sub-tropical countries [1, 2]. For example, strong elements of evidence support that climatic change induces the occurrence of more frequent extreme hydrometeorological events causing droughts or floods [3].

Government institutions acknowledge important limitations in their capability to respond the associated crisis [4]. Consequently, there is a growing need for these institutions to access as reliable as possible meteorological and impact models and propose a representation of their results in an effective way to a range of stakeholders [2]. In the last decade, with the purpose of making available key information to a growing audience, geoportals with cartographic representation displays have been developed for government agencies. In Mexico, variables such as environmental, demographic, economical, have been released by multiple agencies e.g. [5-8]. However, these agencies recognize several limitations in the technology of these platforms, which may still hinder an optimal impact on the public and on targeted stakeholders.

At the international level, the GEOSS infrastructure has been a decentralized effort for international collaboration (Global Earth Observation System of Systems; [9]) in the climate, meteorology and agriculture topics, with the development of web-based geoportals such a EUMETSAT (Europe), CMA (China), and GEONETCast (United States of America: [10]). In Central America, the regional www.servir.net geospatial data portal is part of this GEOSS international network [10].

However, by its decentralized design, this international effort largely depends on the capacity of the participant countries. In many of the sub-tropical countries, broad climatic regions and point-based statistics remain the core of climate representations. The use of satellite imagery is largely limited to visual assessment, although it could serve as a valuable complement to meteorological raster models and the basis for a spatially consistent tool for risk management.

In this research, promoted by the National Institute of Geography and Statistics in Mexico (INEGI) and developed at the Geography Institute of the National Autonomous University of Mexico (GI-UNAM), we propose to enhance the participation of Mexico in the global network and create the Mexican spatial data infrastructure for climate raster models and satellite imagery (VISTA-C platform), with the generation of robust climatic raster models, the storage of these models and associated satellite imagery in a spatial data infrastructure, and the advanced geovisualization of the meteorological / climatic time sequence outputs of these models through the development of online cartographic applications.

2. Data and methodology

To illustrate the establishment of the VISTA-C platform, the following developments are presented: 1) the technological architecture of the platform supporting the acquired satellite imagery, 2) the construction of robust climate raster maps at two scales (at the national level and at regional scale in the Mexico Central Valley) and the incorporation of satellite imagery in the models for useful applications, and 3) the geovisualization of time sequences of the results of these models.

2.1. Data and technological architecture of VISTA-C

The acquisition of satellite imagery by the ERISA receiving station at the GI-UNAM allowed the collection and archiving, since 1996, of most variables registered as Essential Climatic Variables (ECV) by GCOS [11], over the Mexican terrestrial and maritime territory. Most of these are derived from AVHRR and GVAR sensors, from the NOAA and GOES satellite constellations (see table 6 in [12]). The acquisition of a GOES-R Nextgen ground station is planned at the end of 2016, and a GEONETCast America antenna (part of the GEOSS network) has just been acquired.

Apart from satellite data, point-based climatic data was fundamental for the construction of climatic raster maps; information from the national meteorological station network was stored at national level (daily data from 1979 to present). Additionally, in the Mexico Central Valley, time series of autonomous meteorological stations were collected since 2014 for this project.

The technological architecture that has been applied for the VISTA-C platform includes a satellite receiving station, a modular array storage system (24TB), a spatial database, a geoportal and a visualizer (see figure 1). Several components of this architecture are common to the platform of the Mexican National Climate Atlas [13] named UNIATMOS, although the focus of that platform is not on raster climate models. The metadata of both infrastructures are handled with Geoserver, OpenLayers and Geonetwork. Features of our data handler and visualizer are further explained in section 2.3.

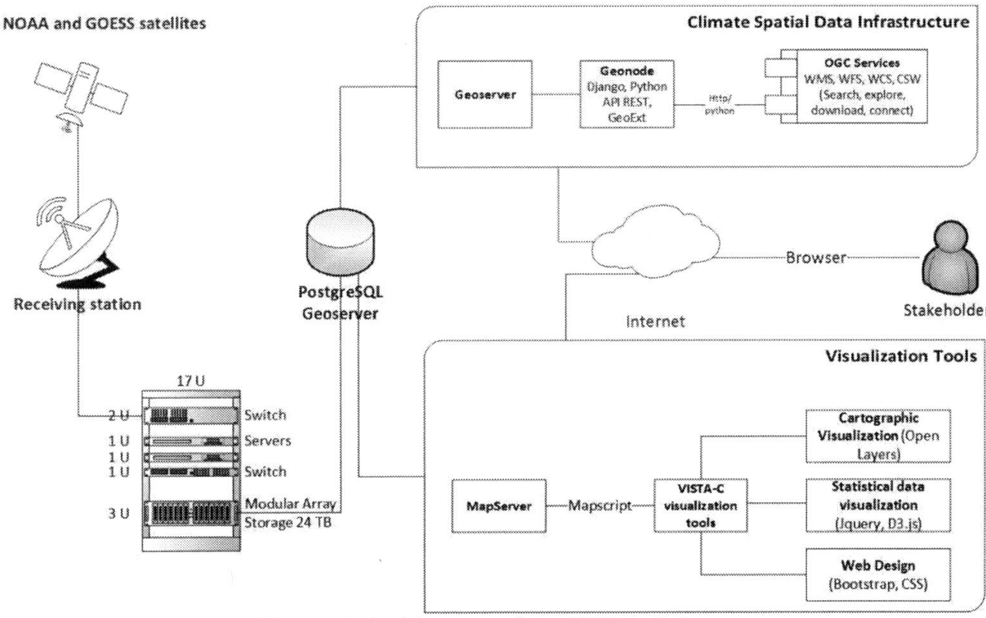

Figure 1. Architecture of the VISTA-C System

2.2. Construction of climate raster maps

The generation of robust climate raster maps allows the use of archive satellite imagery to build prospective models of biophysical variables for agroclimatic applications [14]. A range of initiatives has focused on the generation of climatic information in Mexico; however, quality criteria in terms of processes have often been neglected. For example, simple interpolation schemes involving static parameters such as topography do not take into account the coherence of the relationships between climate variables within a dynamic physical framework.

Two strategies have been developed in the Geography Institute in order to establish robust climatic raster models in Mexico. Both consist in down-scaling a well established numerical climatic model to a higher spatial resolution. In the case of the national level, a Global model (IPCC AR4 climate change scenarios) was used; the temperature and precipitation simulations of the Atmosphere Ocean General Circulation Models (AOGCMs) were corrected for systematic errors via the Climate Predictability Tool (CPT: [15]) and their representation downscaled to a 50x50km grid (see [14]). In the case of the Mexico Central Valley, the precipitation variable from the MM5 model was downscaled using the objective analysis by successive corrections ([16], cited by [17]). The variable was first assessed for coherence and consistency and then downscaled to a 4x4km grid, the interpolation process benefitting in this case of the high spatial frequency of meteorological stations in the Mexico Central Valley.

Next, variables derived from the acquired and stored satellite imagery were selected for associated applications of the mentioned climate rasters: NOAA-derived NDVI images (a proxy for soil moisture conditions) in the semi-arid northern regions of the national grid, and Landsat - derived impervious surface ([18]) in the urbanized Mexico Central Valley.

2.3. Geovisualizer of time sequences of the raster maps

The visualizer of atmospheric tendencies and climatic scenarios in VISTA-C was based on Mapserver, Openlayers and other applications/tools. Three major differences with the Mexican National Climate Atlas [13] that impact on the visualization are 1) our system includes a satellite imagery ground station for near real time modeling and visualization, 2) we developed an application that takes advantage of the OpenLayer capabilities to handle multi-date raster datasets for dynamic visualization, and 3) Geonode was implemented as an additional collaborative tool in order to scale up the use and reuse of

IOP Conf. Series: Earth and Environmental Science **54** (2017) 012042 doi:10.1088/1755-1315/54/1/012042

spatial data among the scientific community in and other stakeholders in Central America. The two first features of the visualizer are similar to the global NOAA geovisualizer [19].

The Central American SERVIR platform [10] also offers a dynamic visualization although it is built on the proprietary Google Earth API and on the kml format. That technology has very attractive assets for visualization but on the other hand is deficient in terms of metadata handling, a highly sensitive point for scientific data.

3. Results

The VISTA-C geovisualizer, set on the maritime, satellite – derived variable option (in this case, AVHRR-derived Superficial Sea Temperature), is illustrated in figure 2.

Figure 2: The VISTA-C Visualizer (June 2005 AVHRR-derived Sea Surface Temperature).

Climate raster maps were generated for the Mexico Central Valley region (figure 3a: precipitation). Urban impervious surface of the Mexico megalopolis (see figure 3b) extracted using the GI-UNAM methodology of Couturier *et al.* [18] from Landsat imagery (stored in the VISTA-C platform), is an input for a series of sensitivity experiments for precipitation broadcast with the MM5 model, in order to try and solve for discrepancies between predicted and observed precipitation patterns (see [17]). In the future, vertical profiles derived from GOES-R, acquired from our ground station, may also enhance the comprehension of the MM5 model in the Central Mexico Valley [17].

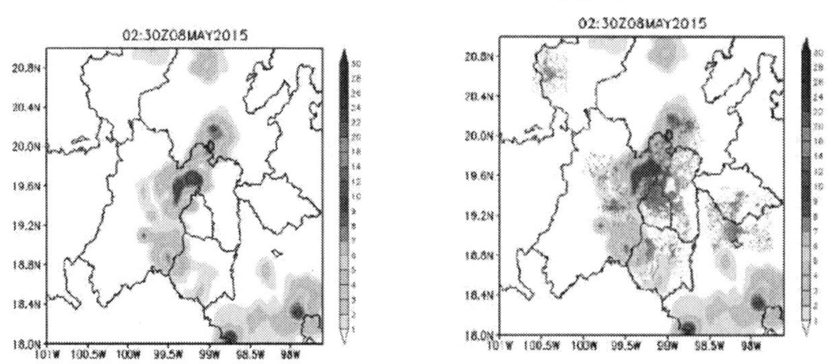

(a) (b)

Figure 3: (a) Precipitation raster map in Mexico Central Valley (mm/h) as a result of our MM5 downscaling method (b) Precipitation distribution and urban impervious surface derived from Landsat imagery in Mexico Central Valley (pink tones).

Finally, figure 4 illustrates multiple-year cumulative precipitation raster maps derived from the AOGCM model corrections, with the modality of time sequence visualization at national level. In this case, AVHRR – derived NDVI imagery was used for spatial drought representations associated to climate change in the region, modeled by Magaña *et al.* [14]. NDVI images are potential predictors of hydrological drought and changes in soil moisture, using surface temperature projections as predictors [14].

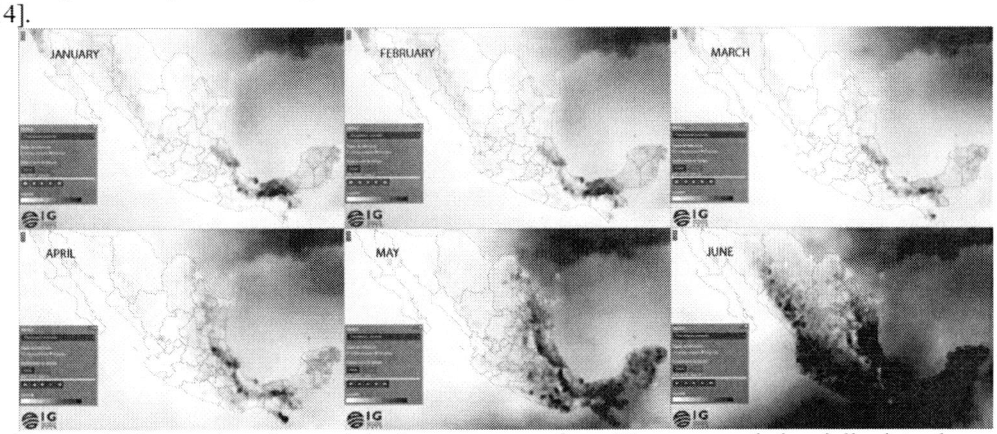

Figure 4. Time series of a precipitation raster map from 33 year cumulative daily data along dry season months in Mexico, derived from global climate models down-scaling with objective successive corrections.

In both case studies, satellite imagery derived variables were used to extend the climate raster models' explicative power and are susceptible of generating improved climatic spatial representations.

4. Conclusion

International collaborative efforts such as the GOESS network need substantial improvement in the participation of subtropical countries in terms of robust regional climate modeling and cartographic online applications. In this paper, we present the first spatial data infrastructure in Mexico dedicated to the online representation of climatic raster models and associated satellite imagery for climate applications. Two of these applications are exemplified at regional and national levels, for which the online platform may be of utility in climate change mitigation (case of the raster at national level for soil moisture prediction in the Sonora Northwestern Mexican region) or risk management (case of the regional raster for precipitation extreme events prediction using complementary impervious surface information). Also, an extension of Openlayer in Geoserver was an effective way to obtain an attractive design for time sequence displays on the platform. At the international level, the NOAA platform also makes use of OpenLayers with a remarkable design which offers an outstanding user-friendly interface to handle multi-temporal spatial data. Some characteristics of this platform may stir future designs for our VISTA-C visualizer. The use of free software enables public universities of sub-tropical countries, which are confronted with repeated budget constraints due to volatile global finances, to build and maintain useful interfaces for national priority issues in a sustainable, reproducible manner.

Acknowledgments
This research was developed in the context of two academic projects financed by public institutions in Mexico: the INEGI - CONACYT financed Project number 208637, entitled "Visualizer of Satellite Information for Environmental and Climate Tendencies (VISTA-C)", and the PAPIIT (DGAPA – UNAM financed) project IA300515, entitled "A multiscale territorial characterization of the contribution of the Mexican society to global environmental change". Thanks are due to Olivia Salmerón García for helping in the storage of maritime AVHRR – derived datasets.

References

[1] Delgado, Jazcilevich J A, Cram S, Siebe C, Ruiz N, Ángeles G and Hernández M 2006 *The environment or how social issues affect the commitment of environmental tasks,* in Randall L. (ed.), Changing structure of Mexico. 297-329.
[2] Landa R., Magaña V and Neri C 2009 *Clima y agua: Elementos para la Adaptación al Cambio Climático*. Ed. SEMARNAT., p 133.
[3] UNO 2009 *Report of Global Assessment on Disaster Risk Reduction* Genova, Switzerland, 218 p.
[4] Galindo L-M 2009 *La Economía del Cambio en México (Sintesis)* (Mexico: Gobierno Federal Government) p 35
[5] CONABIO 2015 *National Information System on Biodiversity,* Mexico, accessed in August 2015 at http://www.conabio.gob.mx/informacion/gis/
[6] CONAGUA 2014 *National Information System on Water*, Mexico, accessed in November 2014 at http://www.cna.gob.mx/Contenido.aspx?n1=3&n2=60&n3=60
[7] INECC 2015 *DGIOECE online cartography,* accessed in February 2015 at http://mapas.inecc.gob.mx/
[8] INEGI 2015 *National System on Geographic and Statistics Information,* Mexico, accessed in November 2015 at http://www.snieg.mx/
[9] GEO 2016a *GEOSS architecture,* accessed at http://www.earthobservations.org/ geoss_cl_tar.shtml
[10] GEO 2016b *GEONETCast Geoportal*, accessed at http://earthobservations.org/ geonetcast.shtml
[11] WMO 2013 *Essential Climate Variables*, accessed at http://www.wmo.int/pages/prog/gcos/ index.php?name=EssentialClimateVariables
[12] López Vega M A, Couturier S and Barrera González KY 2015 Design scheme for a spatial database of climatic and environmental variables in Mexico, integrating Big Data Technology *Procedia Computer Science* **55C** 503-513.
[13] CCA UNAM (Centro de Ciencias de la Atmósfera UNAM) 2016 *Unidad de Informática para las Ciencias Atmosféricas y Ambientales (UNIATMOS)*, accessed at http://uniatmos. atmosfera.unam.mx/ACDM/
[14] Magaña, Zermeño V D and Neri C 2012 Climate change scenarios and potential impacts on water availability in northern Mexico *Climatic Research*. **51** 171-184
[15] IRI (International Research Institute for Climate and Society) 2009,*CPT tutorial.* Accessed at: http://portal.iri.columbia.edu/portal/server.pt
[16] Cressman G P 1959 An operational objective analysis system *Mon. Wea. Rev.* **87** 367-374.
[17] Magaña V, López LC and Vázquez G 2013 El pronóstico de lluvias intensas para la Ciudad de México *TIP: Revista Especializada en Ciencias Químico-Biológicas.* **16** 18-25.
[18] Couturier S, Ricárdez M., Osorno J and López-Martínez R 2011 Morpho-spatial extraction of urban nuclei in diffusely urbanized metropolitan areas *Landscape and Urban Planning* **101** 338-348.
[19] NOAA 2016 *NOAA View Global Data Explorer,* accessed at http://www.nnvl.noaa.gov/view/ globaldata.html

Future rainfall variability in Indonesia under different ENSO and IOD composites based on decadal predictions of CMIP5 datasets

Harisa Bilhaqqi Qalbi, Akhmad Faqih*and Rahmat Hidayat
Department of Geophysics and Meteorology, Faculty of Mathematics and Natural Sciences, Bogor Agricultural University (IPB), Kampus IPB Dramaga, FMIPA Building, Wing 19, Level 4, Jalan Raya Dramaga, Bogor, Jawa Barat, Indonesia 16680

E-mail: akhmadfa@apps.ipb.ac.id

Abstract. El Niño-Southern Oscillation (ENSO) and Indian Ocean Dipole (IOD) are amongst important climate drivers that play a significant role in driving rainfall variability in Indonesia, especially on inter-annual timescales. The phenomena are suggested to have an association with interdecadal climate variability through the modulation of their oscillations. This study aims to analyse the characteristics of future rainfall variability in Indonesia during different condition of ENSO and IOD events based on decadal predictions of near-term climate change CMIP5 GCM data outputs up to year 2035. Monthly data of global rainfall data with 5x5 km grid resolutions of CHIRPS dataset is used in this study to represent historical rainfall variability as well to serve as a reference for future rainfall predictions. The current and future rainfall and sea surface temperature data have been bias corrected before performing the analysis. Given the comparison between rainfall composites during El-Nino and positive IOD events, the study showed that the future rainfall conditions in Indonesia will become drier than the historical condition resulted from the same composite approach. In general, this study showed the Indonesian rainfall variability in the future is expected to respond differently to a different combination of ENSO and IOD conditions.

1. Introduction

El Niño-Southern Oscillation (ENSO) is a natural climate variability in Pacific Ocean, namely Niño 3.4 (5^0N-5^0S, 120^0W-170^0W) [1] and Indian Ocean Dipole (IOD) in Indian Ocean which defined from SST anomalies difference between the tropical western Indian Ocean (50^0E-70^0E, 10^0S-10^0S) and the tropical south-eastern Indian Ocean (90^0E-110^0E, 10^0S-equator) [2] are the most important forcings of the Indonesia rainfall variability [3].

The influence of ENSO and IOD phenomenon to Indonesia rainfall variability are on inter-annual timescales. The phenomena are also related to inter-decadal climate variability through the modulation of their oscillation [4] [5]. Therefore, some knowledge from ENSO and IOD forecasting that can be applied to decadal climate prediction [6]. Decadal prediction model is skill in near-term climate simulation out to 10-30 years, which is a time period of interest for informing adaptation strategies as climate varies and changes. This study focus on the influence of ENSO and IOD phenomenon to Indonesia rainfall variability for several decades until 2035, based on decadal prediction of GCM CMIP5 datasets.

2. Methods

The identification of ENSO and IOD years were done by analysing the sea surface temperature (SSTs) anomalies in their respective territories, based on Extended Reconstructed Sea Surface Temperature version 4 (ERSST.v4) dataset [7] for periods 1981-2010 and Coupled Model Intercomparison Project phase 5 (CMIP5) dataset with variable temperature ocean surface (Tos) for periods 2006-2010. Find periods of ENSO (IOD) during which 5(3)-month running means of the monthly SST anomalies in the above $\pm 0.4^{0}$C or more for at least six consecutive months [1][8]. The 5 or 3-month running mean of SST anomalies are made in order to smooth out the possible intraseasonal variation in the tropical ocean.

Analysis of ENSO and IOD influence phenomenon to Indonesia rainfall variability are used composite analysis. Data for composite analysis is from ENSO and IOD years, during September-October-November (SON) season Indonesia rainfall anomalies are strongest correlated with SST [3].

The current and future rainfall and sea surface temperature data model CMIP5 have been bias corrected before performed. The bias correction is calculated using statistical bias correction method [9]. Four models (MPI-ESM-LR, CMCC-CM, MIROC5, and IPSL-CM5A-LR) are representative of GCM CMIP5 dataset. In this study is based on the initial of the CMIP5 data model baseline have the same statistical intensity distribution as the observation. The bias of SST data is corrected by function from comparison of cumulative density function (CDF) SST anomalies between ERSST.v4 and CMIP5 baseline model are assumed a normal distribution. Monthly data of global rainfall data with 5 km x 5 km grid resolutions of the Climate Hazards Group Infra-Red Precipitation with Stations (CHIRPS) dataset [10] is used in this study to represent historical rainfall variability as well as for corrected rainfall bias in CMIP5 dataset. For rainfall data, the process is similar to SST data, but the statistical distribution approximated by gamma distribution.

3. Result

3.1. Identification of ENSO and IOD years Period 1981-2010

ENSO and IOD phenomenon is indicated by SST anomalies in their respective area, for current time (1981-2010) based on ERSST.v4, the phenomena show have different condition of each other (figure 1). That explained the ENSO phenomenon is independent of the Dipole Mode event in the Indian Ocean [2]. The single event of strongest ENSO was seen in 1987, 2003 and 2010. In the other side, the single event of strongest IOD is seen in 1992 and 1996. In addition, there are several years of ENSO and IOD which they are concurrently occurred, such as 1982, 1994, 1997 and 1998. The years of ENSO, IOD and the both combinations summarised in the table 1.

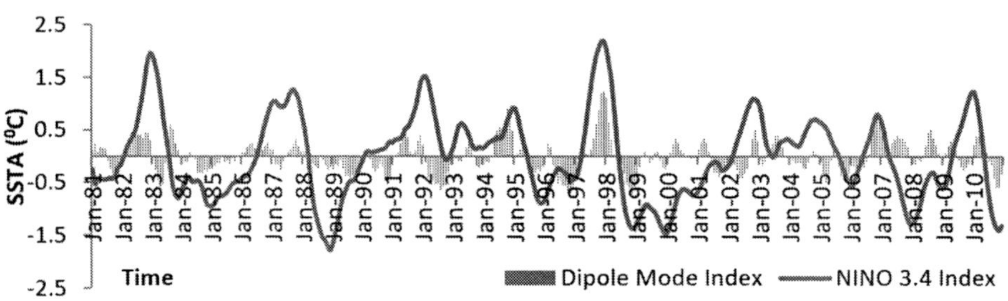

Figure 1. Index of Nino 3.4 and IOD during January 1981 to December 2010, Positive (Negative) SST anomaly showed El Niño (La Niña) and Positive (Negative) IOD.

Table 1. The years of ENSO, IOD and the both combination.

	IOD Positive	Normal	IOD Negative
El Niño	1982, 1991, 1994, 1997, 2006	1986, 1987,1993, 2002, 2004, 2009	–
Normal	–	1990, 2001, 2003, 2005,	1992, 1996,
La Niña	–	1981, 1983, 1984, 1985, 1988, 1989, 1995, 1999, 2000, 2007, 2008	1998, 2010

Table 1 show that ENSO is more intensive than IOD during 1981-2010. Data in table 1 is used for composite analysis to investigating Indonesia rainfall variability when the phenomena of ENSO, IOD and combination of both are occurred.

3.2. Indonesia Rainfall Variability when the Phenomena ENSO, IOD and the Both Combination During 1981-2010

3.2.1. El Niño and El Niño with Positive IOD

Indonesia rainfall in single El Niño event (1988, 1999, and 2007) is decreased up to 150 mm/month and El Niño occurred with positive IOD event (1982, 1997 and 2006) is extremely decreased up to 200 mm/month. It is indicated by negative anomalies seen in figure 2a and figure 2d. The highest decreased of Indonesia rainfall when El Niño phenomena in West Papua province. In the other side, rainfall in Indonesia also increased is seen in West Sumatera, West Java, and Papua province. When a El Niño and positive IOD are concurrently occurred, Indonesia is more significantly decreased than in the single El Niño event. Overall, Indonesia received a negative anomaly of rainfall is around 50-200 mm/month.

3.2.2. La Niña and La Niña with Negative IOD

Indonesia rainfall anomaly when the single La Niña event (1986, 2002 and 2000) is increased around 50-150 mm/month and the highest increased in Papua region (figure 2b). When La Niña and negative IOD (1998 and 2010) concurrently occurred is more significant increased Indonesia rainfall than the single event of La Niña and negative IOD up to 200 mm/month, except some region in Indonesia such as North Sumatera and Papua province (figure 2e).

3.2.3. Negative IOD

Anomaly of Indonesia rainfall when the single event of negative IOD (1992, and 1996) is indicated positively increased up to 200 mm/month (figure 2c). Meanwhile, some region of Indonesia is decreased rainfall up to 50 mm/month such as Papua, NTB/NTT, and North Sumatera province.

3.2.4. Normal (No ENSO and IOD)

Normal condition during SON showed that Indonesia rainfall in general is not drastically increase of about 50-100 mm/month. The highest rainfall increased in South Sumatera and West Java. Meanwhile, some region in Indonesia is decreased rainfall up to 50 mm/month such as East Kalimantan, Papua, Bali and NTT province.

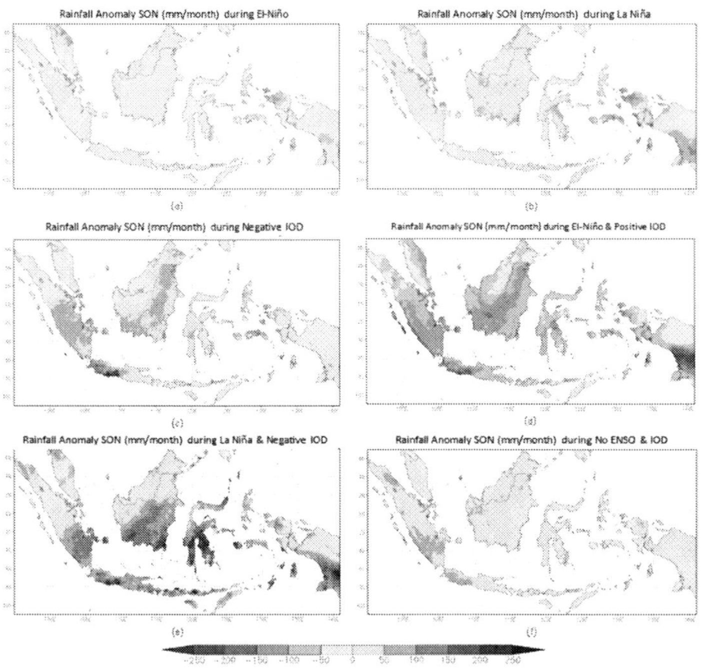

Figure 2. Composite of Indonesia rainfall anomaly during SON on (a) El Niño, (b) La Niña, (c) negative IOD, (d) El Niño and positive IOD, (e) La Niña and negative IOD, and (f) Normal (no ENSO and IOD)

3.3. Index of ENSO and IOD Period 2006-2035

ENSO and IOD index in period 2006-2035 are shown by the four model GCM CMIP5 (MPI-ESM-LR, CMCC-CM, MIROC5 and IPSL-CM5A-LR), indicates that it has different condition of SST anomaly each other (figure 3). In generally, the phenomena of ENSO in during 2006-2035 is more intensive and strong up to 4^{0}C (figure 3b) than IOD.

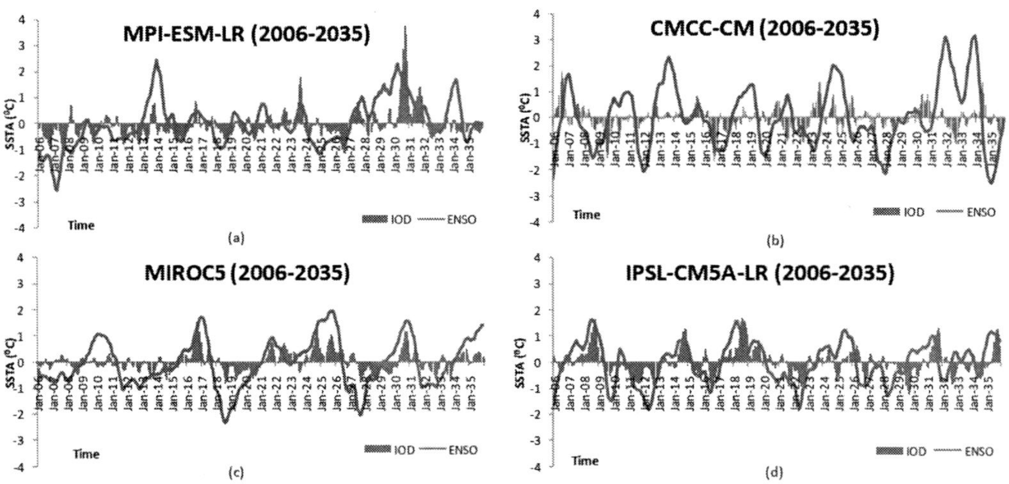

Figure 3. The index of ENSO and IOD during 2006-2035

3.4. Indonesia Rainfall Variability when the Phenomena ENSO, IOD and the Both Combination During 2006-2035

3.4.1. El Niño and El Niño with Positive IOD

Indonesia rainfall variability during 2006-2035 at each model CMIP5 in single El Niño (figure 4) and El Niño occurred with positive IOD event (figure 5) are more significant decreased than Indonesia rainfall variability in period 1981-2010 is over to 250 mm/month. It is shown in Java on model CMCC-CM (figure 4b and figure 5b) and Kalimantan on model MIROC5 (figure 5c).

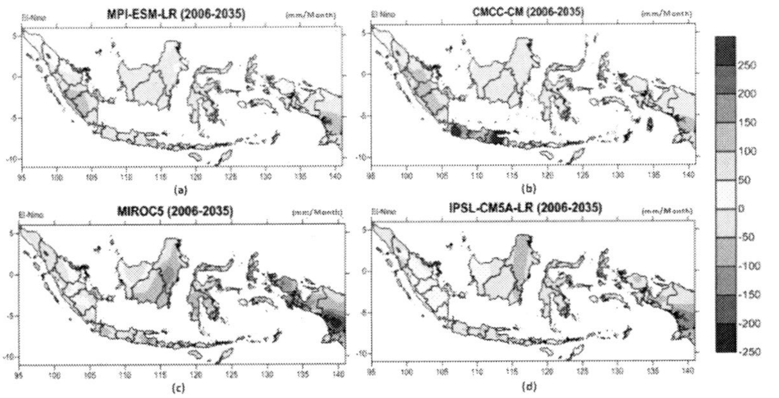

Figure 4. Composite of Indonesia rainfall anomaly during SON when El Nino on (a) MPI-ESM-LR, (b) CMCC-CM, (c) MIROC5, and (d)IPSL-CM5A-LR

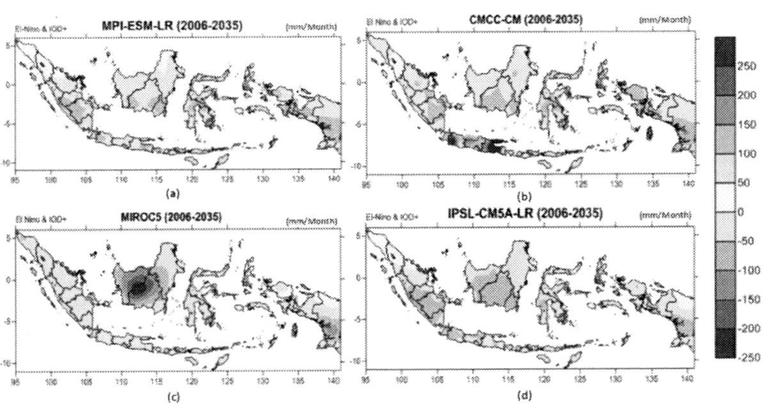

Figure 5. Composite of Indonesia rainfall anomaly during SON when El Nino occurred with positive IOD on (a)MPI-ESM-LR, (b)CMCC-CM, (c)MIROC5, and (d)IPSL-CM5A-LR

3.4.2. La Niña and La Niña with Negative IOD

Increased of Indonesia rainfall anomaly in single La Niña event during SON 2006-2035 is equal to Indonesia rainfall anomaly period 1981-2010 around 50-150 mm/month (figure 6). When La Niña and positive IOD concurrently occurred is extremely increased Indonesia rainfall variability up to over 250 mm/month in Java as in the CMCC-CM model (figure 7b). But, the Indonesia region has decreased rainfall up to 150 mm/month and extend to some region in Sumatera, Java and Papua is showed in MIROC5 (figure 6c and 7c) and IPSL-CM5A-LR (figure 6d and 7d).

IOP Conf. Series: Earth and Environmental Science **54** (2017) 012043 doi:10.1088/1755-1315/54/1/012043

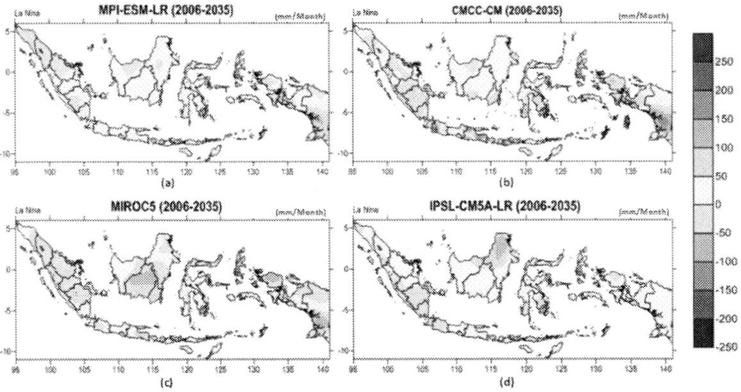

Figure 6. Composite of Indonesia rainfall anomaly during SON when La Niña event on (a)MPI-ESM-LR, (b)CMCC-CM, (c)MIROC5, and (d)IPSL-CM5A-LR

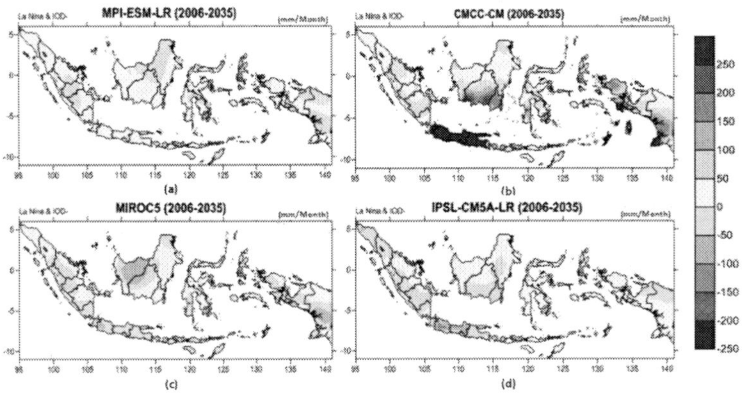

Figure 7. Composite of Indonesia rainfall anomaly during SON when La Niña occurred with Negative IOD on (a)MPI-ESM-LR, (b)CMCC-CM, (c)MIROC5, and (d)IPSL-CM5A-LR

3.4.3. *Negative IOD*

Rainfall anomalies of Indonesia on each model when the phenomena negative IOD is not drastically increased in the west Indonesia region of about 50-200 mm/month such as in Sumatera, Kalimantan and Java (figure 8). While, in some region in Indonesia has decreased rainfall especially in eastern region such as Papua, Sulawesi and Kalimantan as well as the highest rainfall decreased up to 200 mm/month in Java is showed by CMCC-CM model (figure 8b).

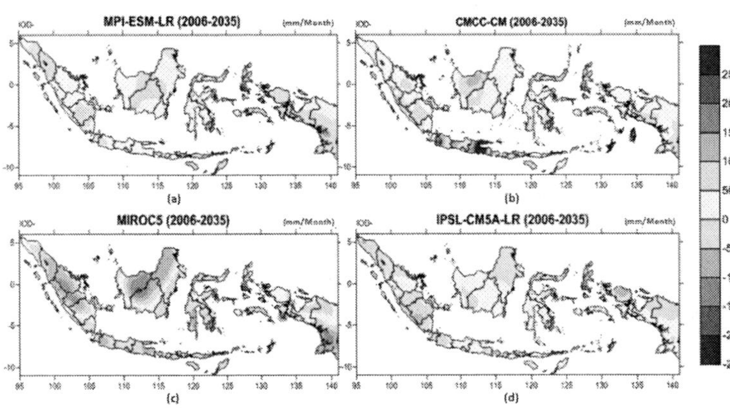

Figure 8. Composite of Indonesia rainfall anomaly during SON when Negative IOD on (a)MPI-ESM-LR, (b)CMCC-CM, (c)MIROC5, and (d)IPSL-CM5A-LR

3.4.4. Normal (No ENSO and IOD)

Normal condition when no ENSO and IOD during SON 2006-2035 showed that Indonesia rainfall variability is more intensive decreased of about 50-250 mm/month in several region in Indonesia (figure 9) and the highest rainfall decreased in Java is seen CMCC-CM model (figure 9b). Meanwhile, rainfall of some region in Indonesia is also increased up to 50 mm/month.

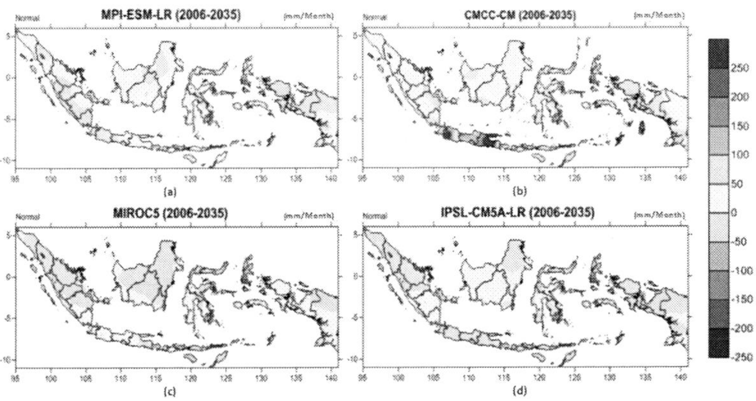

Figure 9. Composite of Indonesia rainfall anomaly during SON when Normal (No ENSO and IOD) on (a) MPI-ESM-LR, (b) CMCC-CM, (c) MIROC5, and (d)IPSL-CM5A-LR

4. Conclusion

The phenomena of ENSO and IOD during historical condition (1981-2010) showed that ENSO is more intensive than IOD and it will very intensive and strong up to 4^0C during future condition (2006-2035). The composite of Indonesia rainfall anomalies when ENSO, IOD and the both combination has different response. The impact of El Nino and positive IOD are concurrently occurred in Indonesia rainfall variability during historical condition is more significant decreased than in the single event of El Nino or positive IOD, and in the future rainfall will become drier than historical condition. In the other side, the phenomena La Nina concurrently occurred with negative IOD in Indonesia rainfall during historical condition is more significantly increased than the single event of La Nina or negative IOD. The future Indonesia rainfall when La Nina concurrently occurred with negative IOD will become wetter than historical condition up to over 250 mm/month, but the Indonesia region has decreased rainfall up to 150 mm/month and extend to some region in Sumatera, Java and Papua. The future rainfall when normal condition will become drier than historical condition.

References

[1] Trenberth K E 1997 The Definition of El Niño *Bulletin of the American Meteorological Society* 2771-2777

[2] Saji N H, Goswami B N, Vinayachandran P N, and Yamagata T 1999 A dipole mode in the tropical Indian Ocean. *Nature* **401** 360-363

[3] Hendon H H 2003 Indonesian Rainfall Variability: Impacts of ENSO and Local Air–Sea Interaction *American Meteorological Society* 1775-1790

[4] Ashok K, Chan W L, Motoi T, Yamagata T 2004 Decadal variability of the Indian Ocean dipole *Geophysical Research Letters* 1-4

[5] Trenberth K E, Jones P D, Ambenje P, Bojariu R, Easterling D, Tank A K, Parker D, Rahimzadeh F, Renwick J A, Rusticucci M, Soden B, and Zhai P 2007 Observations: surface and atmospheric climate change *Climate Change 2007: The Physical Science Basis* 235-336

[6] Meehl G A, Goddard L, Murphy J, Stouffer R J, Boer G, Danabasoglu G, Dixon K, Giorgetta M A, Greene A M, Hawkins E, Hegerl G, Karoly D, Keenlyside N, Kimoto M, and Kirtman B 2009 DECADAL PREDICTION : Can It Be Skillful? *American Meteorological Society* 1467-1485

[7] Huang B, Banzon V F, Freeman E, Lawrimore J, Liu W, Peterson T C, Smith T M, Thorne P W, Woodruff S D, Zhang H M 2015 Extended Reconstructed Sea Surface Temperature Version 4 (ERSST.v4) Part I: Upgrades and Intercomparisons *American Meteorology Society* **28** 911-930

[8] Saji N H and Yamagata T 2003 Possible impacts of Indian Ocean Dipole mode events on global climate *Climate research* **25** 151-169

[9] Piani C, Haerter J O, Coppola E 2010 Statistical bias correction for daily precipitation in regional climate models over Europe *Theor Appl Climatol* 187-192

[10] Funk C, Peterson P, Landsfeld M, Pedreros D, Verdin J, Shukla S, Husak G, Rowland J, Harrison l, Hoell A, Michaelsen J 2015 The climate hazards infrared precipitation with stations-a new environmental record for monitoring extremes *Nature* 1-21

Forest cover dynamics analysis and prediction modelling using logistic regression model (case study: forest cover at Indragiri Hulu Regency, Riau Province)

Irmadi Nahib* and Jaka Suryanta
Geospatial Information Agency
Jl Raya Jakarta Bogor KM 46 Cibinong, Jawa Barat, 16911, Indonesia

E-mail:irmnahib@gmail.com, irmadi.nahib@big.go.id

Abstract. Forest destruction, climate change and global warming could reduce an indirect forest benefit because forest is the largest carbon sink and it plays a very important role in global carbon cycle. To support Reducing Emissions from Deforestation and Forest Degradation (REDD +) program, people pay attention of forest cover changes as the basis for calculating carbon stock changes. This study try to explore the forest cover dynamics as well as the prediction model of forest cover in Indragiri Hulu Regency, Riau Province Indonesia. The study aims to analyse some various explanatory variables associated with forest conversion processes and predict forest cover change using logistic regression model (LRM). The main data used in this study is Land use /cover map (1990 – 2011). Performance of developed model was assessed through a comparison of the predicted model of forest cover change and the actual forest cover in 2011. The analysis result showed that forest cover has decreased continuously between 1990 and 2011, up to the loss of 165,284.82 ha (35.19 %) of forest area. The LRM successfully predicted the forest cover for the period 2010 with reasonably high accuracy (ROC = 92.97 % and 70.26 %).

1. Introduction

Forest is a natural resource that is very important and useful for life and living either directly or indirectly. Direct benefits from the existence of the forest are timber, non-timber products and wildlife. While the indirect benefits are environmental services, such as watersheds, aesthetic function, an oxygen supplier and carbon sink [1].

Indonesia has the highest deforestation rates in the world, exceeding even Brazil while having only a quarter of Brazil's forest area [2]. The average annual deforestation for the period 2000-2012 was 671,420 hectares, accounts for 525,516 ha of deforestation in mineral land 600 and 145,904 ha of deforestation in peat land. During this period, more than 80 percent of deforestation occurred in Kalimantan and Sumatra, while Sulawesi and Papua follow with 9 percent and 6 percent, respectively [3]. Meanwhile, refers to [4] the rate of deforestation in the Riau Province for the period 2000-2009 was 200,290.8 hectares.

Forest destruction, climate change and global warming reduce the indirect benefit of forest because forest is the largest carbon sink and play a very important role in global carbon cycle and can hold carbon at least 10 times greater than other vegetations prairie grass, crops and tundra [5]. Assessing the conversion of a forested landscape may help us to understand the way of natural resources extraction occurs, and consequently the human influences on the forest ecosystem services. For better understanding of the impact of forest cover change, factors affecting it must be fully studied [6].

Forest cover dynamics is rate, pattern, spatial distribution, and quantity of change from forest cover to other land coverages due to either natural condition or human induced causes. The constant interplay of various human induced disturbances along with topographic and climatic factors can gradually degrade a healthy forest cover or change it to other land use/land cover category. Understanding of the causes of land use change has moved from simplistic representations of two or three driving forces too much more profound understanding that involves situation-specific interactions among a large number of factors at different spatial and temporal scales. Studies related to forest cover change using satellite-derived information help in understanding the phenomena like carbon dynamics, climate change and threat to biodiversity [6].

Recent advancement in remote sensing and GIS methods also enable researchers to model and predict land use/land cover more efficiently than ever-before. Several approaches have also been developed to model and predict the dynamics of land use / land cover [7].

The present study aims to analyze explanatory variables associated with forest conversion process and to model the forest cover change using logistic regression model.

2. Methods

2.1. Study area and data

The present study was carried out in Indragiri Hulu Regency, Riau Province, Indonesia (figure 1).

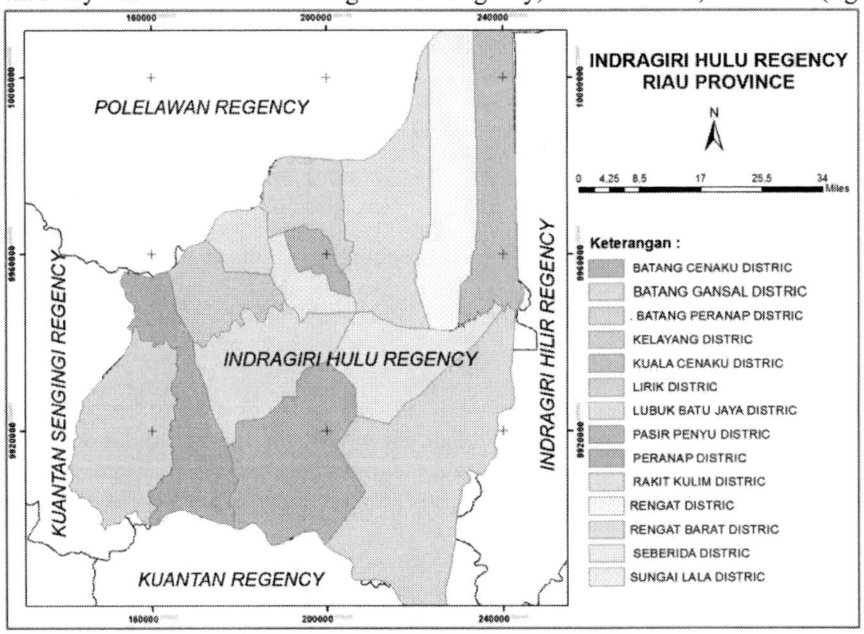

Figure 1. The map of study area

Indragiri Hulu is a regency of Riau, Indonesia. It has an area of 819,826 ha consists of Forest zone 189,368 ha and Forest for ohers (APL) 630,458 ha. The regency is divided into 14 districts, the seat of the regency is located at Rengat. The Forest zone of 189,368 Ha, consists of 84,753 National Park of Bukit Tigapuluh (NP BT), protected forest 34,758 and 69,856 wildlife reserve. Data used in the assessment:

- Indonesian Topographic Map, Scale 1: 50,000, Geospatial Information Agency (BIG)

- Forest Area, Inland Water, Coastal and Marine Ecosystem Map of Indragiri Hulu Regency, Riau Province, Scale 1: 250,000 Directorate General of Forestry Planning Ministry of Forestry.
- Land Use/Cover Map of Indragiri Hulu Regency, Riau Province, Scale 1: 250,000 (1990, 2000, 206 and 2011), Directorate General of Forestry Planning, Ministry of Environment and Forestry.
- Indragiri Hulu Regency Administration Map, Planning and Regional Development Agency (*Bappeda*) of Indragiri Hulu Regency, Riau Province.

Landcover change analysis was performed by a comparison method of landcover map series produced by Directorate General of Forestry Planning, Ministry of Environment and Forestry. This land cover map was produced through a visual interpretation of the 324 Landsat image. The determination of land cover area used the spatial analysis which is done by Land Change Modeler in Indrisi TerrSet process of the these map of Indragiri Hulu Regency in 1990, 2000, 2006 and 2011. Flowchart stage research activities are presented figure 2.

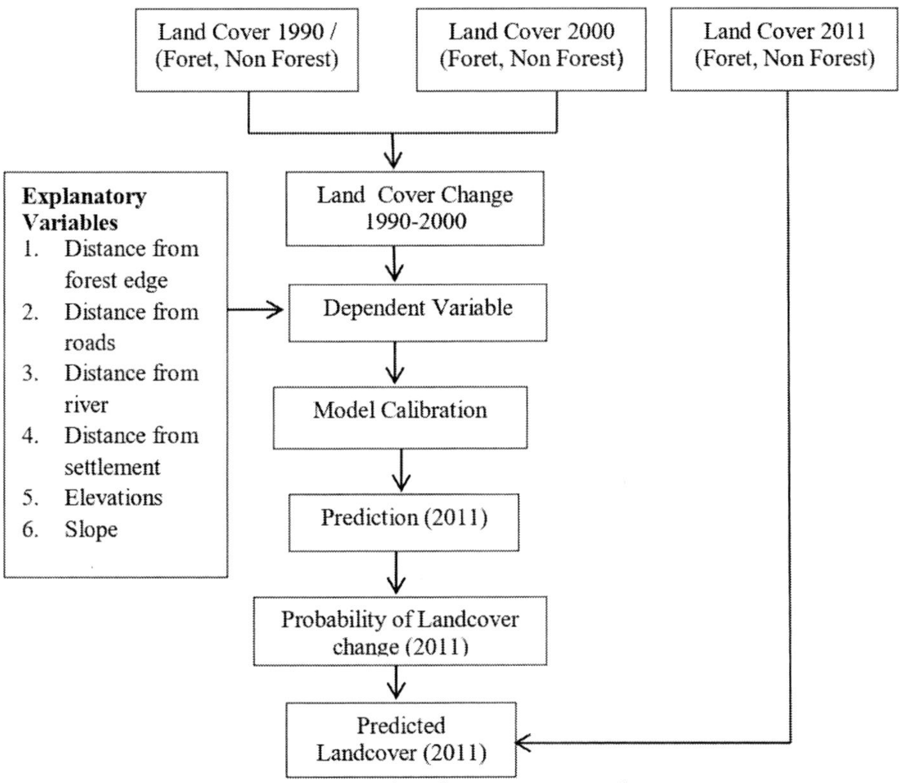

Figure 2. Flowchart stage research

The row data (Land Use/Cover Map) within 23 landcover catagories. For simplify of comparison, the catagories have been reduced to two categories of forest and non-forest. Forest is consist of the dryland forest, swamp forest, mangrove forest and timber forest plantation. The other catagories are included in the non-forest. Finally a boolean map of forest and non-forest was generated for all the four periods (figure 3).

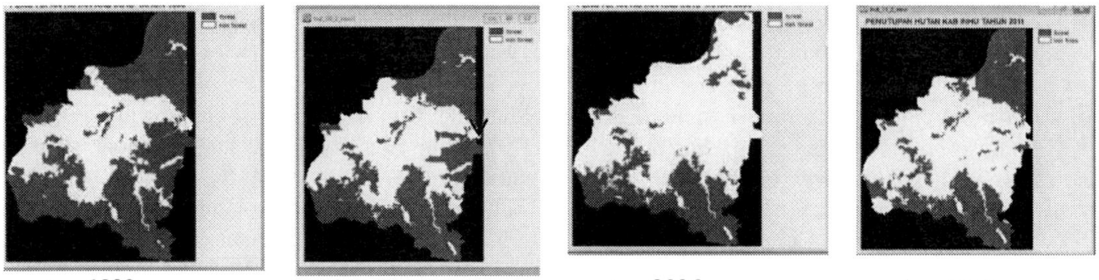

a. 1990 b. 2000 c. 2006 d. 2011

Figure 3. Forest cover during (a) 1990, (b) 2000, (c) 2006 and (c) 2011.

Based on Indonesian Topographic Map which collected from Geospatial Information Agency, we obtained information of road, river, settlement, and elevation layer. Morever, we considered four distance parameter as the independent variables: distance from forest edge, roads, river, settlements, and slope; and forest cover change as the independent variable (figure 4).

2.2. Logistic regression model

A Boolean image with the categories 'forest change' (forest to non-forest) and 'no change' (forest remained unchanged) were generated for the period 1990–2000 and 2000–2011. Distance measures the euclidean distance between each cell and the nearest of a set of target features. Deforestation tends to start from the edge of existing forest [8]. Hence, Distance from forest edge was considered as one of the explanatory variables of forest cover change. The distance from forest edge for 1990 were generated from the forest boundary of the respective years.

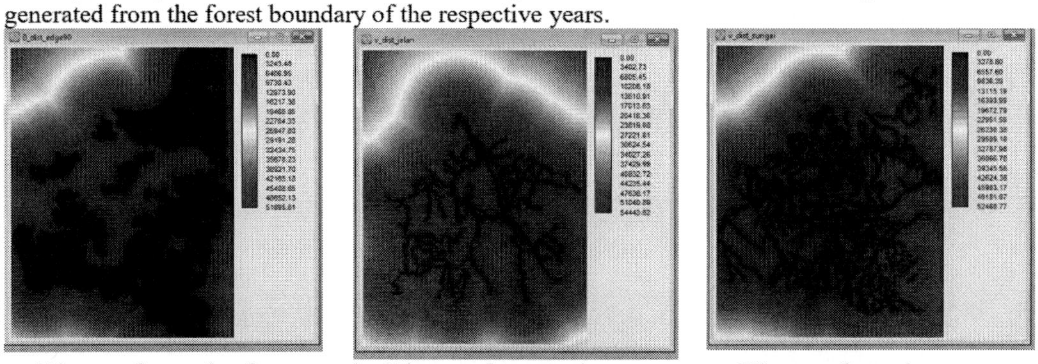

a. Distance from edge forest b. Distance from roads c. Distance from river

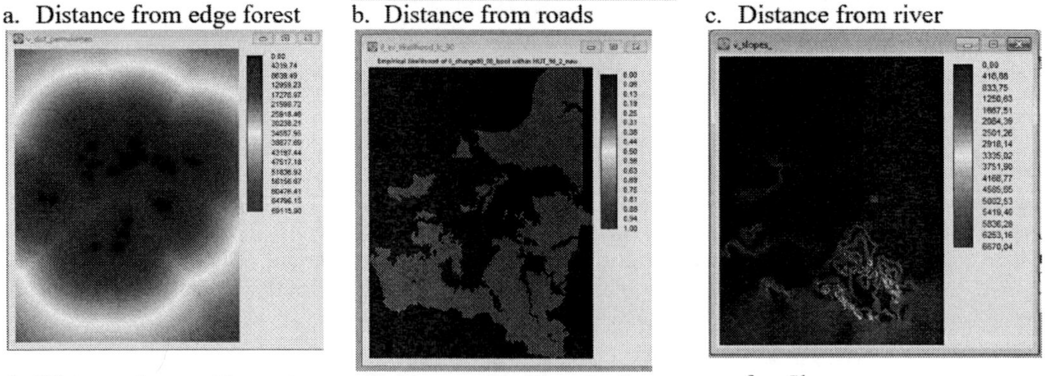

d. Distance from settlement e. Evidence likelihood forest changes f. Slopes

Figure 4. Independent variable (environmental variable)

LRM was used to model and analyze the forest change in IDRISI TerrSet. The objective of the present study was to assess the importance of the explanatory variables on forest change from 1990 to 2000 and predicting the probability of change by 2011.

The explanatory variable test procedure is based on Cramer's V contingency table analysis which can test the strength of the association between the dependent variable. The binary presence or absence is the dependent variable (1 = forest changeand 0= no change) for the periods 1990–2000.

The probability of forest change is considered to be a function of the explanatory variables. It is a monotonic curvi linear response bounded between 0 and 1 [10] and defined by the logistic function:

$$p = E(Y)\frac{e^{\beta_0+\beta_1\chi_1+\beta_2\chi_2+\beta_3\chi_3+\beta_4\chi_4}}{1+e^{\beta_0+\beta_1\chi_1+\beta_2\chi_2+\beta_3\chi_3+\beta_4\chi_4}} \tag{1}$$

where

p = the probability of forest change,

$E(Y)$ = the expected value of the dependent variable Y,

β_0 = a constant to be estimated,

β_i = the coefficient to be estimated for each explanatory variable χ_i.

This logistic function (equation1) can be transformed (equation2) into linear function (equation 3) which is called a logit or logistic transfor-mation

$$logit\ (p) = log_e\left(\frac{p}{1-p}\right) \tag{2}$$

$$logit\ (p) = \beta_0 + \beta_1\chi_1 + \beta_2\chi_2 + \beta_3\chi_3 + \beta_4\chi_4 \tag{3}$$

2.3. Model calibration and prediction

All the independent variables were normalized between 0.1 and 0.9 before introducing them in the model. The natural log transformation was done for the continuous variables (distances). For the categorical explanatory variable (slope position class), the evidence of likelihood transformation was applied. The regression equation of the best-fitted predictor set and the probability of forest change for 2000 were generated.The probability of forest change for 2000 was used to predict the change in forest between 2000 and 2010. The regression equation of the best-fitted predictor set and the probability of forest change for 2011 were generated. The threshold value was selected from the residual file of probability of forest change of 2010 and it was used to generate change - no change binary image (0 = forest to forest, non forest to non forest) and 1 = forest to non-forest). The resulted image is the predicted forest cover for the period 2011.

2.4. Validation of prediction

The predicted forest cover of 2010 was validated using Relative Operating Characteristic (ROC) /Area Under Curve (AUC) module of IDRISI TerrSet. The ROC module is comparing a suitability image depicting the likelihood of that class occurring (the input image) and a boolean image showing where that class actually exists (the reference image).

The ROC curve is the true positive fraction vs false positive fraction and the AUC is a measure of overall performance. The predicted forest cover map of 2011 was compared with actual forest cover mapof 2011 using 100 sampling points. A ROC/AUC curve was generated between the false positive (%) and true positive (%).

3. Results

Figure 5 shows the relationship between dependent variable and the explanatory variables. This relationship was useful for selection of the explanatory variables for forest change in the study area.

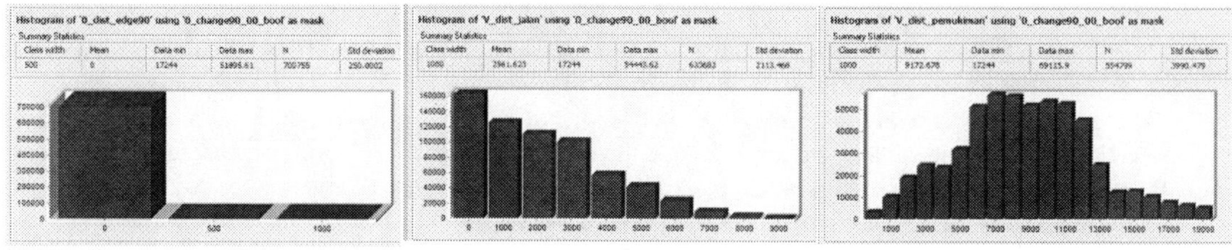

(a) vs. distance from forest edge, (b) vs. distance from roads (c) vs. distance from settlements

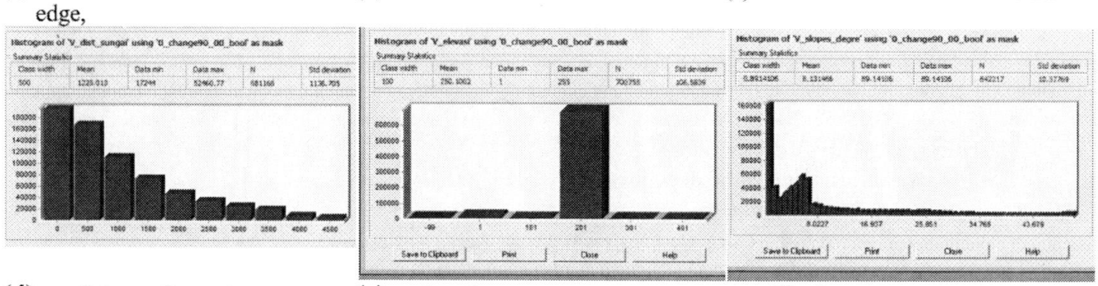

(d)vs. distance from streams (e)vs elevation (f) slopes

Figure5. Relationship between dependent and explanatory variables Forest cover change (a) vs. distance from forest edge, (b) vs. distance from roads, (c) vs. distance from settlements, (d) vs. slope position classes, (e) vs elevation, (f) slopes

Most of the forest change has taken place near the forest edges. Occurrences of forest change within 500 m of forest edge were noticed to be highly distinct but forest change dropped off to virtually nil beyond that, indicating a non-linear relationship between forest edge and forest change. The incidence of forest change near roads was found to be noticeable but dropped off to zero after 7 km from the roads. Influence the distribution of the settlement to the forest changes occurred in around 1 km upt o 13 km from the settlements.

Therefore, these variables were tested with Cramer's V and then included in the model.

Table 1. Association between dependent variable (forest cover change) and explanatory variables using Cramer's V

Explanatory variable	Forest changes 1990-2000		Forest changes 2000-2006	
	Cramer's V	P value	Cramer's V	P value
Dist from forest edge	0.5549	0.0000	0.4035	0.0000
Dist from roads	0.3796	0.0000	0.2126	0.0000
Dist from settlements	0.4412	0.0000	0.3261	0.0000
Distance from streams	0.2046	0.0000	0.1193	0.0000
Slope position	0.11682	0.0000	0.3284	0.0000
Evidence likelihood forest changes	0.5981			

V values for variables : distance from roads, settlements and forest edge, werefound to be between 0.1 and 0.5 with a p value of 0.0, indicating a poor association, there is no strong relationship with forest change.Logistic regression modelingrefers to equation 3,

$$Y_1 = -0.1406 - 0.1661\, x_1 - 0.0001068\, x_2 - 0.00001650 x_4 - 0.0005861 x_5 - 0.0024 x_6 \quad (4)$$

$$Y_2 = -1.0715 - 0.7204 x_1 + 0.0001487 x_2 - 0.0001362 x_3 + 0.00008544 x_4 + 0.0077 x_5 \quad (5)$$

where :x_1 = distance from edge forest,x_2 = distance from roads, x_3 = distance from settlement

x_4 = distance from streams, x_5 = elevation, x_6= slope

Based on table 2, The predictor based on forest changes 1990-2000, more hight (of model chi-square) than forest changes 1990-2000. A high value indicates that the forest changewas less expected under the null hypothesis than the full regression model. Pseudo R^2 value greater (0.5946 and 0.3750) than 0.2 indicates that the model is a relatively good fit for thedata.

Table2. Statistic of logistic regression

	Based on forest changes 1990-2000	Based on forest changes 2000-2006
Number of total observations	15,113,120	15,113,120
-2logL0	1,111,834.2390	54,2066.4826
-2log(likelihood)	450,748.1071	338,814.2830
Goodness of fit	395,226.7384	572,979.8677
Pseudo R^2	0.5946	0.3750
Chi-square (*df=5*)	661,086.1318	203,252.1996
ROC	0.9614	0.9287
AUC	0.9044	0.7026

The area under forest and nonforest during 1990, 2000,2006 and 2010 (actual and predicted) is shown in table 3.

Table 3. Areas under forest cover during 1990, 2000, 2006, and 2010 (actual and predicted).

Area (ha)	Area (Percentage)				
	1990	2000	2006	2011	2011 (*predicted*)
Forest	60.93	52.75	30.09	39.48	44.92
Non Forest	39.07	47.25	69.91	60.52	55.08
Total	100.00	100.00	100.00	100.00	100.00

It was observed that the study area experienced a continuous loss of forest cover, leading to the loss of 165,284.82 ha (35.19 %) of forest during 1990–2011. Average the rate of deforestation in Indragiri Hulu Regency for the period 1990-201 was 7,870.71 ha hectares.

The accuracy estimate of the forest cover maps of 1990, 2000, 2006 and 2010 is shown in table 3. All the maps were having accuracy of more than 85%. ROC/AUC graph generated between model predicted forest cover and the actual forest cover of 2011 is shown in figure 6. The area under ROC curve are 92.97 % and 70.26 % (based on forest changed 1990-2000 and 2000 – 2006) which gives an accuracy of 92.97 % % and 70.26 % for the predicted forest cover of 2011.

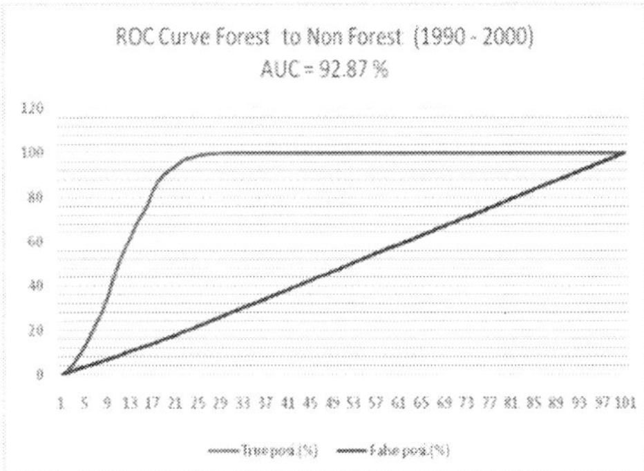

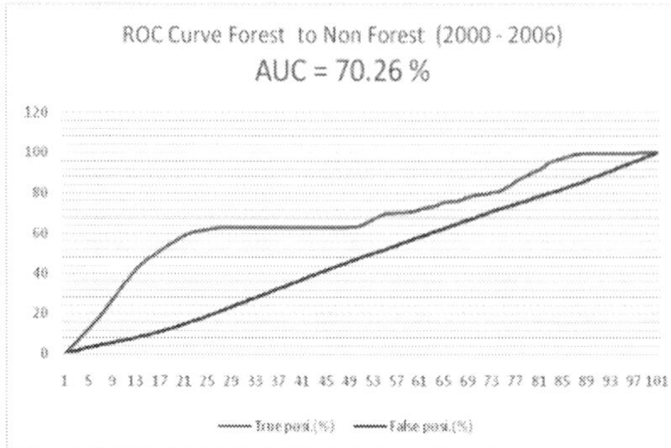

a. Based on forest changes 1990-2000 b. Based on forest changes 2000-2006

Figure 6. Validation of LRM prediction (AUC/ROC)

The driving factors of forest cover change may vary from place to place. In the present study, the selected explanatory variables encompass a substantial share of the factors driving forest cover changes. Specifically, the accessibility variables seem to be more important than the topographical ones. Many studies have indi-cated that most of these factors were also found to be important in other areas. Proximity to road, town and forest edge were found to be the important factors of forest change in southern Cameroon [10]

Refer to [11] modeled distance from roads, settlements and forest edge, and topography for tropical forest conversion and found that all these explanatory variables were highly significant predictor of forest conversion (ROC = 87%). Meanwhile [7] studied the effects of six factors : distance from roads and settlements, forest fragmentation index, elevation, slope and distance from the forest edge, on deforestation. An inverse relationship of forest change with distance from roads, settlements, forest edge and elevation was observed. The ROC of prediction of forest cover change was found to be 96%.

4. Conclusions

The analysis of forest cover change revealed that the forest area has undergone continuous change leading to the loss of 165,284.82 ha (35.19 %) forest in the past 1990–2011. The explanatory variables, i.e., distance from forest edge, roads and settlements, slope position classes were significantly associated with the forest cover change in the study area.

The LRM efficiently modelled all the explanatory variables associated with forest change and helped in analyzing them for their relative significance on forest change process in the area. The model for predicting the forest cover change :

$Y_1 = -0.1406 - 0.1661\,x_1 - 0.0001068\,x_2 - 0.00001650x_4 - 0.0005861x_5 - 0.0024x_6$ and

$Y_2 = -1.0715 - 0.7204x_1 + 0.0001487x_2 - 0.0001362x_3 + 0.00008544x_4 + 0.0077x_5$.

It predicted the forest–non forest cover of the area for 2011 with an accuracy of 92.87 % and 70.26 %.

Acknowledgments

The authors are thankful to the Centers for Research, Promotion and Cooperation, Geospatial Information Agency (BIG) for the data and financial support.

References

[1] Nahib I and Widjojo S 2016 *The Impact of Landcover Changes on Carbon Stock :A Study Case In Central Kalimantan Forest* (ACRRS)

[2] Margono B A 2014 Primary forest cover loss in Indonesia over 2000-2012 *Nature Climate Change* **4** 730-735

[3] Government of Indonesia 2014 National Forest Reference Emission Level for Deforestation and Forest Degradation in the Context of the Activities Referred to Decision 1/CP.16. Paragraph 70 (REDD+) Under the UNFCCC. Submission by Indonesia

[4] Sumargo W, Nanggara S G, Nainggolan F A and Apriani I 2011 *Potret Keadaan Hutan Indonesia 2000-2009 Edisi I* (Forest Watch Indonesia)

[5] Adiriono T 2009 Method of Measuring Carbon (Carbon Stock) on Industrial Plantation Forest Crassicarpa Acasia Type. *Thesis*. Faculty of Forestry Programs Graduate. University of Gadjah Mada. Yogyakarta. not published

[6] Kumar S, Nandy S, Agarwal R and Kushwaha S P S 2014 Forest cover dynamics analysis and prediction modeling using logistic regression model *Ecological Indicators* **45** 444–455

[7] Arekhi S 2011 Modeling spatial pattern of deforestation using GIS and logistic regression: a case study of northern Ilam forests, Ilam province, Iran *Afr. J.Biotechnol* **10** 16236–16249

[8] Eastman J R 2006 *IDRISI 15 Andes.Guide to GIS and Image Processing* (Worcester, MA: Clark University)

[9] Pontius Jr R G and Schneider L C 2001 Land-cover change model validation by an ROC method for the Ipswich watershed, Massachusetts, USA *Agric. Ecosyst. Environ.* **85** 239–248.

[10] Mertens B and Lambin E F 2000 Land-cover-change trajectories in southern Cameroon *Annals of the association of American Geographers* **90** 467-494

[11] Siles N J S 2009 Spatial Modelling and prediction of tropical forest conversion in the Isiboro Secure National Park and Indigenous Territory (TIPNIS), Bolivia. *Graduate Thesis*. International Institute for Geoinformation Science and Earth Observation, Enschede, The Netherlands.

IOP Conf. Series: Earth and Environmental Science **54** (2017) 012045 doi:10.1088/1755-1315/54/1/012045

Mapping *Shorea* natural distribution in the last remaining forests in Riau as a baseline information for conservation strategy

Atok Subiakto[1*], Henti Hendalastuti Rachmat[1], and Kesuma Wijaya[2]

[1] Center for Forestry Research and Development, Forestry Research, Development, and Innovation Agency (FORDIA), Ministry of Enviromental and Forestry, Jalan Raya Gunung Batu No. 5, Bogor 16118, Indonesia

[2] District Secretaries of Lingga Islands Regency, Jl. Istana Robat, Daik Lingga, Riau Islands Province

E-mail: atoksubiakto@gmail.com

Abstract. *Shorea* divided into four groups: *Shorea Red Meranti, White Meranti, Yellow Meranti* and *Balau*. In the past, conservation was not an important issue as this groups were common and abundant. However, Sumatran rain forest were cleared and converted at annual rate of 500.000 Ha with the most extensive in the Province of Riau where had lost 63% between 1985 and 2009. This study conducted to determine the potency and availability of most valuable tropical *Shorea* dipterocarps in Riau Province. Lines transect and point count methods used to determine the presence any of the *Shorea* species in the designated remnants area. Leaves samples for morphological identification collected for further taxonomic identification. Data analysis for mapping conducted by overlying the secondary and primary data source. The result showed that the patchy remnants forests in Riau still conserve at least of around 22 *Shorea* species, included 14 species those of *Shorea Red Meranti*, 1species *Shorea White Meranti*, 4 species *Shorea Yellow Meranti*, and 3 species of *Shorea Balau*. However, in average the numbers of individual found for each of the species was low with uncomplete occurrence of the life stage in several spots, showing the fragility of local species loss and extinction.

1. Introduction

Dipterocarpaceae is the dominant timber trees family of tropical rain forest in Malesia region, including Indonesia, Malaysia, Brunei, the Philippines and eastward to New Guinea. Dipterocarps is economically important where its timber (round wood logs) counted for 25% of total global consumption of tropical hardwood [1]. Ecologically, dipterocarps is significantly important in tropical primary forest where its wood volume was estimated 211.75 m3 ha^{-1} which means it counts for 86,9% of total volume [2].

Shorea is the biggest genus emergent dipterocarps which comprise of more than 190 species. They naturally grow from sea level up to 1750 m altitude [3]. The majority of *Shorea* species are listed as being critically endangered due to a population reduction over 80% in the past three generations, caused by a decline in its natural range and exploitation. In global market, the four group of *Shorea* are well known for their common identification, those represented by *Shorea Red Meranti, Shorea White Meranti, Shorea Yellow Meranti,* and *Shorea Balau*. The grouping of *Shorea* is mainly based on their wood colour appearance and grain. Based on their timber characteristic on water, the three group of

Shorea Red, White and Yellow Meranti are categorized as light hardwood while Shorea Balau is categorized as medium to heavy hardwood.

In the past, conservation of the dipterocarp family was not that important as the family was still very common and abundant. However, forest conversion and fragmentation together with habitat deterioration are happening progressively. In the case of Sumatran rain forest, the loss of forest area mainly because of the combination from followed factors: 1) industrial plantations; 2) small-scale forest clearing; 3) roads development; 4) and semi-wildfires [4]. Only in less than two decades, Sumatra has lost its 12.5 million ha of natural forest and this amount is nearly half of its natural cover forest covers in 1985 [5]. The annual conversion rate of Sumatran tropical forest itself is said to be at 500,000 hectares, or 2.56 % yr^{-1}. Even though there are significant numbers of protected areas available in Sumatera, however it is well understood that these protected areas have generally been established in remote highland and other agriculturally unsuitable areas where its biological diversity is the lowest [4]. Contrast to that situation, more than 80 percent of forest loss occurred in the most accessible lowland areas, where the most biodiversity and carbon-dense ecosystems are found. We can see clearly that fact from the largest Sumatran national park of Kerinci Seblat and Gunung Leuser that covers large pieces of inaccessible forests in the mountains [4,6] while all oil palm and timber concessions are granted in lowland forests.

Among all provinces in Sumatra, the most extensive of forest loss was happened in Riau Province. Riau Province lost 63% or 4.4 million hectares of its 6.9 million hectares of forest cover between 1985 and 2009 and this was accounted for nearly half of Sumatra's total forest loss between 2000 and 2009 [5]. The remnants forest in Riau now is very fragmented and scattered into pieces surrounded either by oil palm plantation and or exotic timber plantation, e.g *Acacia* and *Eucalyptus*. This study was carried out to determine the potency and availability of dipterocarps *Shorea* species in the remnant forests of Riau Province. The result will be very important to describe current condition of the species and can be put as baseline information to propose appropriate conservation strategy.

2. Methods

2.1 Primary Data

The presence of any *Shorea* species were obtained by conducting several field surveys using methods including; purposive sampling by line transect and point count. The surveys were conducted in several remnant forested areas in Riau Province, covered Tesso Nilo National Park (3 spots), Bukit Tiga Puluh National Park (4 spots), Bukit Rimbang Bukit Baling (4 spots), Ampang Delapan, Rumbio Indigeneous Forest (2 spots), Hapanasan Pasir Pangaraian (2 spots), Sutan Syarif Qasim Minas, Riparian RAPP Kuansing, and Siak river watershed. In most of the case, it took around 3-4 days to conduct survey at each of the sites. The length of survey mainly depends on the difficulties or accessibility of the area. Interviews were also conducted with local people as well as forest practitioners, (e.g. forest ranger, staff and or head of governmental conservation body, non-governmental conservationist, environment division of forest plantation companies) to get information on the presence of *Shorea* species in non-direct surveyed areas of study interest. This non-direct survey was essential to do a complete *Shorea* distribution checklist in Riau considering the constraint of our limitation in budget and time. Primary data were collected in form of GPS positioning for each tree individuals during field exploration trips using Garmin Dakota 20.

2.2 Secondary data, analysis and mapping

Data used in this study were mainly secondary data such as model position, digital elevation model, administrative map, and land typology. In this system, we turned these maps into each tematic map. Then, the species distribution data were overlaid with these tematic maps. We did not execute spatial data because we only showed the position of the object (trees) into the background data. The species distribution map was modify based on the latest information.

3. Results and Discussion

The four groups of *Shorea* are well known in global timber market based on their timber color appearance. Classification using vernacular names is divided *Shorea* into four groups, namely *Balau (Selangan Batu), White Meranti (Meranti Pa'ang), Yellow Meranti (Meranti Damar Hitam)* and *Red Meranti* [7]. All four group of *Shorea* were found during our field surveys. Among 22 species, *Shorea Red Meranti* are most dominant and consist of 14 species, followed by 4 species of *Shorea Yellow Meranti*, 3 species of *Shorea Balau* and only 1 species of *Shorea White Meranti.*

3.1 Shorea Red Meranti

Shorea Red *Meranti* is usually large tree with cylindrical bole. The timber is generally used for sliding, interior joinery, flooring, furniture and weatherboarding. In particular market such as in Europe, it is preferred for making window frames and sills. Morphologically, this *Meranti* group is easily recognized by their drooping twigs and branchlet.

Among 23 species of *Shorea Red Meranti* reported to grow in Sumatera [3], there were 14 species within the group that still can be found in remnants forest. This numbers is the most abundant among the four groups to be discovered in Riau. As seen in distribution map of *Shorea Red Meranti* showed in Figure 1, it is known that this group naturally grows both at peat swamp and also mineral soil type, from lowland to the lower montane zone. Figure 1 is also helping us to recognize that the species diversity in one particular place located at mineral soils is richer than those at peat swamp area. Tesso Nilo National Park, Bukit Tigapuluh National Park, Rumbio, Bukit Rimbang Bukit Baling (BRBB), Hapanasan, and Bukit Bungkuk are among the sites that preserved 8 or more *Shorea Red Meranti* species inside their areas. Figure 1 is also helping us to determine that *S. leprosula* and *S. acuminata* have a wide distribution in Riau Province that could occupy both mineral and rather swampy area.

The high diversity at both national parks is mainly due to the forest status where all extraction activities are prohibited and counted as illegal and against the law. Meanwhile, the high of species richness of Bukit Rimbang Bukit Baling was attributed to the site condition where BRBB is well known for its extreme contour that is hilly with frequent extreme slope. It is actually one of Bukit Barisan mountain ranges located in Riau Province. This naturally gifted condition makes the area less attractive to be converted to agricultural areas.

At peat swamp areas, Sinepis forest holds the richest biodiversity compare to others. This mostly because this area has not been logged so much. The high biodiversity in this swampy area is in accordance with mammals condition where often reported this site as one of the limited valuable pocket of the natural habitat for Sumatran tiger.

3.2 Shorea Yellow Meranti

Shorea Yellow Meranti is also known as *Meranti Damar Hitam* because the exudation of black or brown damar on bark. The timber colour is yellow to yellow-brown and will get darken along the age. The timber is generally less preferred compare to those of *Shorea Red Meranti*. The timber is also reported as non-durable to rot attacks. In workability, this group of *Meranti* is typically rather easy to work with. However, some species may have a slight blunting effect on tools due to small levels of silica present in the wood.

Among 8 species of *Shorea Yellow Meranti* reported to grow in Sumatera [3], there were found 4 species within the group that still can be found in remnants forest in Riau. Figure 2 showed us that the species within the groups, including *Shorea balanocarpoides*, *S. conica*, *S. multiflora*, and *S. peltata* always grows naturally at mineral soils type habitat. None of the *Shorea Yellow Meranti* species found to grow at peat swamp area in our field surveys. Figure 2 is also helping us to recognize that Bukit Tigapuluh National Park own all 4 *Shorea Yellow Meranti* inside its area. It also determined that *S. balanocarpoides* and *S. conica* are spread widely while *S. peltata* showed high level of habitat specificity that it confined to only one area of Bukit Tigapuluh National Park.

3.3 Shorea Balau

Shorea Balau is known for its heavy timber. The heartwood is very hard and logs are drowning into the water. Species within this group are usually have straight cylindrical bole with concave buttresses.

The timber colour can be highly variable depending on the species: ranging from a pale straw colour to a darker reddish brown. Even though the timber is hard, but in most of the case they are considered as non-durable in regard to decay resistance, and are also susceptible to insect attack. As the timber is hard, it is also typically hard to work on, due to its high density. The practical use of the wood is usually for plywood, veneer, general construction, flooring, as well as a number of other general utility purposes.

Among 11 species of *Shorea Balau* reported to grow in Sumatera [3], there were only 4 species found in remnants forest in Riau during our field surveys, those including *S. atrinervosa, S. falcifera and S. sumatrana*. Figure 3 showed us that the species within the groups are mainly grow in mineral soil type. Through Figure 3 it is also known that *S. sumatrana* found not only can grow in dry mineral soil but also can grow crossing into wetter area along Siak River. This distribution map also showed that Tesso Nilo National Park own more species than that of Bukit Tigapuluh National Park. In general all three species seems to be distributed more evenly where one species can appear in 4-5 areas in Riau. None showed the very high distribution ability or habitat specificity.

3.4 Shorea White Meranti

Shorea White Meranti is less popular for sawmilling and joinery because of the abundant silica present in ray cells that dulls saw and tools. The colour of its heartwood is a pale yellowish-orange when freshly cut, aging to a golden yellow-brown. Because of that, species within this group is preferred to be peeled for plywood and has bright white colour that make the species attractive for face veneer.

Among 9 species of *Shorea White Meranti* reported to grow in Sumatera [3], only 1 species was found during our field survey, namely *S. bracteolata*. Figure 4 showed the distribution of *S. bracteolata* in Riau Province. From this figure it was clearly recognizable that *S. bracteolate* is mineral soil type Meranti, that means its natural distribution is only confined to mineral soil habitat and never be found in peat swamp areas. It is also found from lowland forest until lower montane. Based on our direct field survey and non-direct survey, it is determined that the species distributed at least in 13 areas of remaining forested land in Riau.

3.5 Why mapping natural distribution of Shorea is necessary for conservation strategy

Manifesting direct data survey into distribution map of the species is necessary to do. It is beneficial to describe the distribution pattern of the species along with their brief description of the habitat. Rather than reading detailed information in form of descriptive paragraph, showing the result in form of distribution map will deliver the message easier and faster. However, detailed information in form of text is still crucial to those who want to know more detailed information. Overlying the primary data of species position into the background data (e.g elevation, habitat type, administrative area, etc) will give clearer description on how and where certain species located in one compact landscapes. In case of *Shorea*, the distribution map of each figure (Figure 1 – Figure 4) clearly described on how one species distribute in Riau Province, where it is distributed (mineral soil vs peat swamp), how many species own by each certain area (species abundance), and how the sites can be accessed (far vs near the road, etc). The benefit of displaying species distribution into a map is its congeniality of the exact position. Certain species both flora and fauna has a very high economic or ecological value. Publishing their position clearly into exact coordinate for public consumption can be harmful for their conservation status because it can trigger massive exploitation or hunting. Presenting their distribution and existence into distribution map only give rough location without discarding important information of its location in whole.

4. Conclusion

Mapping species distribution in forestry is important. It can show clearly how the species distribute together with their habitat condition. Natural distribution of *Shorea* (Dipterocarpaceae) in Riau showed that the species within the Genus is currently still available in small pockets of the remnants forest. Both mineral soil and peat swamp area still preserved several common and or endangered ones. Remnants forest is characterized by its mineral soil and tend to have higher diversity than those of peat swamp, while forest status with high and intense prevention for extraction (e.g National Park) together with difficulties in accessibility were known to be effective in maintaining high diversity of the *Shorea* species. Based on distribution map resulted in this study, concern on conservation strategy of the species can be designed both by species based and or area based.

Acknowledgements
This exploration was funded by Project Collaboration between Advance Research Department of KOMATSU with Centre for Forestry Research and Development. The authors are also grateful to Mr. Saiful Ikhsan, Forest Plantation Fiber Technology Research and Development Institute, for his kind help in plotting and mapping of primary survey data.

References

[1] ITTO 2008 *Annual review and assessment of the world's timber situation*. Yokohama
[2] Bischoff W, Newbery D M, Lingenfelder M, Schnaeckel R, Petol G H, Madani L and Ridsdale CE 2005 Secondary succession and dipterocarp recruitment in Bornean rain forest after logging *For. Ecol. Manage* **218** 174-192
[3] Newman M F, Burges P F and Whitmore TC 1996 *Manual of dipterocarps for forester: Borneo island light hardwoods* Jakarta: Royal Botanic Garden & CIFOR
[4] Gaveau D L A, Linkie M, Suyadi, Levang P and Wiliams N 2009 Three decades of deforestation in southwest Sumatra: Effect of coffee prices, law enforcement and rural poverty *Biol. Conserv* **142** 597-605
[5] Uryu Y, Purastuti E, Laumonier Y, Sunarto S B Y S A, Hadian O, Kosasih D A and Stüwe M 2010 Sumatra's Forests, their Wildlife and the Climate, windows in Time: 1985, 1990, 2000, and 2009. WWF-Indonesia Technical Report Jakarta Indonesia
[6] Curran L, Trigg S N, McDonald A K, Astiani D, Hardiono M, Siregar P, Caniago I and Kasischke E 2004 Lowland forest loss in protected areas of Indonesian Borneo, *Science* **303** 1000– 1003
[7] Symington C F 1943 *Foresters Manual Dipterocarps*, Mal. For. Rec. 16

The effect of image radiometric correction on the accuracy of vegetation canopy density estimate using several Landsat-8 OLI's vegetation indices: A case study of Wonosari area, Indonesia

R P Dewa[1] and Danoedoro P[1]*
Faculty of Geography, Universitas Gadjah Mada

E-mail: projo.danoedoro@geo.ugm.ac.id

Abstract. Recent studies on the use of spectral indices have involved radiometric correction as a prerequisite. However, study on the effect of radiometric correction level on the accuracy of biophysical parameters' estimate is still rare in Indonesia. This study tried to investigate the influence of various radiometric correction levels and the number of vegetation strata on the accuracy of vegetation density estimates using NDVI, MSAVI2 and GEMI of Landsat 8 OLI. In this study, the dataset covering vegetated area in Wonosari, Gunung Kidul Regency, Indonesia was processed radiometrically using eight different methods, i.e. spectral radiance, at sensor reflectance, sun elevation correction, histogram adjustments using original DN, spectal radiance, at sensor reflectance, and sun position correction respectively, as well as dark object subtraction (DOS). Every image with specific correction level was then transformed using the aforementioned indices, in order correlate with the field-measured canopy density. The analysis were carried out by considering the number of canopy layers. This found that different radiometric correction methods resulted canopy density estimates with different accuracies. The number of canopy strata also played an important role. Every vegetation index transformation performed its best accuracy by using different radiometric correction method and different number of canopy layers.

1. Introduction

Vegetation mapping is one of the most prominent applications of remote sensing. When a quantitative assessment of biophysical parameters of vegetation is required, vegetation index transformation is a common approach to this activity [1]. On the other hand, many studies also showed that the use of spectral transformations might lead to different results when the spectral bands are neither calibrated nor radiometrically corrected [2-4]. Therefore, several authors also accentuated the importance of radiometric corrections in vegetation spectral analyses [5-7]. Accuracy of vegetation biophysical parameters' estimates have been studied by several authors [8][4]. Although the vegetation indices generally require radiometric preprocessing, there were only limited number of papers explicitly reported that those treatment had been carried out.

Many authors also claimed that there are several factors related to biophysical parameters of vegetation causing variations in their spectral responses, e.g. vegetation types/species, soil background, number of canopy layers or strata [8-9,1]. During the past 20 years, many researchers processed multispectral image up to at-surface reflectance as the input to the vegetation index transformation, assuming that the parameter is the best measurable factor to represent the biophysical phenomena.

Based on the aforementioned background, it is important to assess the influence of radiometric correction and calibration on the vegetation canopy density estimate. Therefore, the objectives of this study were twofold, *i.e.* (a) to assess the importance of radiometric correction/calibration on the use of three vegetation indices (NDVI, MSAVI2, and GEMI) for estimating vegetation density in relatively flat area, (b) to evaluate the effect of radiometric correction/calibration on the accuracy of vegetation index-based canopy density estimate by considering the number of strata.

2. Previous Works

Vegetation index is a group of image transformation that uses of two or more vegetation-sensitive bands to generate new images that are more representative to show density-related variation. [3]explained that there are two groups of vegetation index, i.e. generic (*e.g.* SR, NDVI) and empirical ones. The empirical vegetation indices can be grouped into (a) the ones that can suppress the soil background influence (*e.g.* SAVI, TSAVI, MSAVI, MSAVI2), and (b) the ones than can suppress the atmospheric effects (*e.g.* ARVI and GEMI). Studies carried out by [11] showed that many applications in Indonesia simply used a vegetation index formula without a deep consideration on its development background.

In terms of vegetation composition[12], and [13] explained that vegetation can be described with respect to either its floristic or species composition or its structural composition. Any species may differ from others as viewed from its spectral response in particular range of wavelengths [10]. On the other hand, different vegetation structural compositions also give significant difference in spectral responses, since different canopy geometry and number of leaves' layer absorb, reflect and transmit electromagnetic radiation differently [14]. In short, the number of canopy layer (strata) plays an important role in absorbing and reflecting the EMR, particularly when near infrared spectrum is involved.

Discussed the effect of radiometric correction on SPOT HRV and Landsat TM NDVI data [2]. Their results show that: (a) the conversion of DN into apparent reflectance is the most important step for NDVI correction. Without this correction, a relatively constant error affects NDVI depending on the sensor considered; (b) MTF correction did not affect the average NDVI value; its interest is to restore the radiometric level of individual pixels that have a large contrast with their surroundings; and (c) the atmospheric effects were similar in the homologous spectral bands of SPOT-HRV and Landsat TM. Their correction increased the dynamic range of NDVI variation (around 24% in the example presented) and consequently the contrast between different targets.

The effect of radiometric corrections on carbon stock estimate using vegetation index approach [5]. However, their study only took into account four correction levels, *i.e.* histogram adjustment, shadow calibration, at-sensor reflectance, and at-surface reflectance. The study was carried out in urbanised area with heterogeneous terrain, and it mostly covered heterogeneous vegetation. Six vegetation indices were involved, *i.e.* EVI2, NDVI, TVI, ARVI, SAVI, MSARVI. They found that the at-sensor and at-surface reflectance calibrations were the most effective and stable corrections for estimating biomass and carbon stock at the study area. In addition, MSARVI was the most accurate vegetation index based on the at-sensor and at-surface reflectance images. The study did not include MSAVI2 and GEMI.

3. Study Objectives

This study tried to investigate the influence of various radiometric correction methods/levels and the number of vegetation strata on the accuracy of vegetation density estimates using NDVI, MSAVI2 and GEMI of Landsat 8 OLI. In this study, the dataset covering vegetated area in Wonosari, Gunung Kidul Regency, Indonesia was processed radiometrically using eight different methods, i.e. spectral radiance, at sensor reflectance, sun elevation correction, histogram adjustments using original DN, spectal radiance, at sensor reflectance, and sun position correction respectively, as well as dark object subtraction (DOS).

4. Materials and Methods

4.1 Materials and methode

This study made use of Landsat-8 OLI image covering Wonosari area, Gunung Kidul Regency (path 120 row 065, see Figure 1). The image was recorded on 24 June 2013. Preprocessing stages comprised geometric and radiometric corrections. The geometric correction referred to RBI topographic map at scale of 1:25.000 using third order transformation, since the coverage is mountainous in some parts and flat until rolling in other parts –although the study area is situated in relatively flat up to gently sloping.

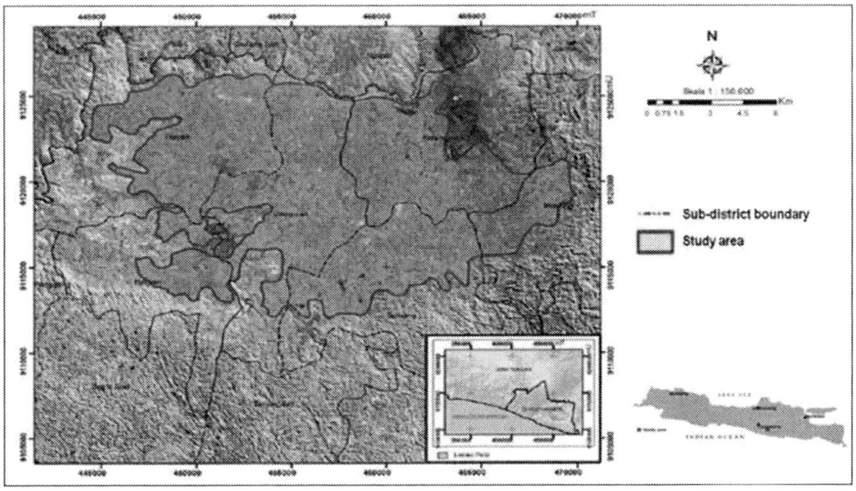

Figure 1. Study area

4.1.1. Radiometric corrections

The radiometric correction comprised eight different types or level of corrections, including (a) calibration from DN to spectral radiance, (b) calibration from DN to histogram, (b) adjustment of original DN to reflectance, (c) sun position adjustment, (d) histogram adjustment of original DN, (e) histogram adjustment of the spectral radiance, (f) histogram, adjustment of the reflectance, (g) histogram adjustment of the sun position, and (h) dark object subtraction of the reflectance. Topographic correction was not carried out since the study area is relatively flat or at least the topographic variation is minimum.

The spectral radiance $L\lambda$ is defined as $L\lambda = MLQcal + AL$, where $L\lambda$ = TOA spectral radiance *(Watts/(m2 * srad * µm))*, *ML* = Band-specific multiplicative rescaling factor from the image header (Radiance Multiplicative Band_*x*, where *x* is the band number to be corrected), *AL* = Band-specific additive rescaling factor from the image header (Radiance Additive_Band_*x*, where *x* is the band number to be corrected), *Qcal* = Digital Number of the band to be corrected.

The top of atmosphere planetary (TOA) reflectance $\rho\lambda'$ is defined as $\rho\lambda' = M\rho Qcal + A\rho$, where $\rho\lambda'$ = TOA planetary reflectance, without sun angle correction, $M\rho$ = Band-specific multiplicative rescaling factor from the image header (Reflectance Multiplicative Band_*x*, where *x* is band number to be corrected), $A\rho$ = Band-specific additive rescaling factor from the image (Reflectance Additive Band_*x*, where *x* is band number to be corrected), and *Qcal* = Digital Number (DN).

The sun position correction generate $\rho\lambda$, where $\rho\lambda = \rho\lambda'/cos(\theta_{SZ}) = \rho\lambda'/sin(\theta_{SE})$, where $\rho\lambda'$ = TOA planetary reflectance, without sun angle correction, θ_{SE} = local sun elevation angle. The scene center sun elevation angle in degrees is provided in the metadata (SUN_ELEVATION)., and θ_{SZ} = Local solar zenith angle; $\theta_{SZ} = 90° - \theta_{SE}$.

The dark object subtraction method produced corrected image with new value $Rc = Rs - Rsi$, where Rc = Object reflectance value wihch is atmospherically corrected, R= Object reflectance value before correction, $Rsi = (Mean\ Rw-(2*Standard\ Deviation\ Rw)$, and Rw = spectral value which is considered as an offset.

In the next stage, all datasets with different level of processing were then transformed using three different vegetation indices, *i.e.* NDVI, MSAVI2, and GEMI, resulting 24 different images containing vegetation index information. The indices involve brightness values (BV) in various expressions, including the original digital number (DN), spectral radiance, and reflectance. The formula for NDVI is $NDVI= (BVI_R - BV_R) /(BVI_R + BV_R)$, where R and IR are red and near-infrared bands respectively. The formula for $MSAVI2$ is $MSAVI2 = (0.5)*((2*(BV_{IR}+1))-(\sqrt{(((2*BVI_R)2)-(8*(BV_{IR}-BV_R))}))))$, while the formula for GEMI is $GEMI = eta*(1-(0,25*eta)) - ((BV_R -0,125)/(1-BV_{IR}))$, where $eta = ((2(BV_{IR2} - BV_{R2})+(1,5*BV_{IR})+(0,5*BV_R)))/(BV_{IR}+BV_R+0,5))$.

4.1.2. *Field Measurement and Data Analysis*

Vegetation canopy density measurement was carried out one year after, at the same date. The field data was grouped into two separate subsets, *i.e.* field data for developing regression model, and field data for accuracy assessment. Measurement of canopy density was undertaken using circular fish eye camera (5.6 mm focal length) at 30x30 m2, and by taking into account the number of strata (layer), ranging from one up to four, and combinations between them, *e.g.* two and three strata, two, three and four strata. Analysis of canopy density was done using image processing software to measure the vegetation percentage.

Statistical analysis was carried out to correlate the canopy density and the vegetation index value of each image, at various strata. Based on this analysis, the best radiometric correction in relation with the best vegetation index for estimating the canopy density could be found. Moreover, advantages and disadvantages of each correction and index could also be discussed.

5. Resuts and Discussion

5.1. *Radiometric Corrections*

An RMS error of 0.56 was obtained using a third order transformation, indicating that the image is suitable enough for field orientation, vegetation canopy density measurement and mapping. In this geometric correction, a nearest neighbor interpolation was chosen, instead of bilinear or cubic convolution resampling, in order to maintain the original pixel values. The original DNs range from 0 to 65,535 (16 bit coding), and they were calibrated/corrected using eight different methods as shown in Figure 1, which depict spectral pattern of vegetation in the study area.

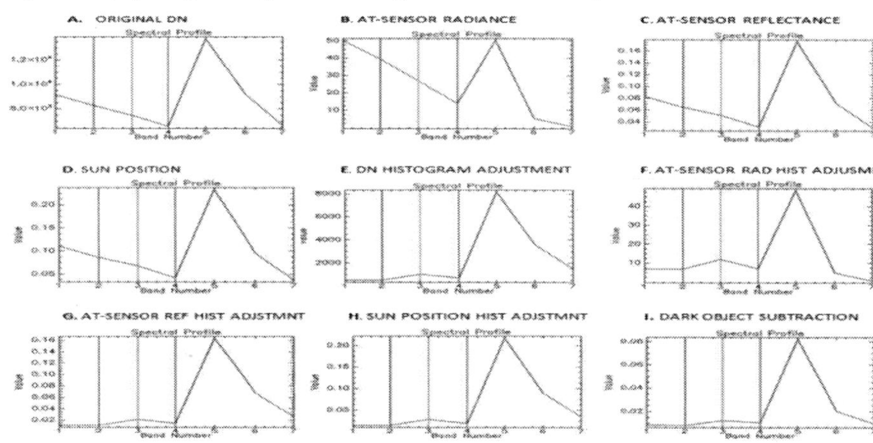

Figure 2. The graphical results of radiometric corrections (B up to I) as compared to original DN (A).

The figures show that only some of the radiometric correction methods (E – I) could derive spectral patterns similar to theoretical or ideal vegetation spectral curve [14]. It is understandable since the corrections/calibrations in Figure 2 B-D only either shifted the DN to 0 or converted the original DN to spectral radiances. The corrections shown in E-I combined them, which means that they also reduced the atmospheric effects. The most affected bands in these stages were visible ones. The Dark Object Subtraction method could also reduced the intensity of SWIR1 band values so that they dropped into much lower values than those of NIR band..

5.2. Vegetation Canopy Density Measurement

Vegetation types and maturity in the study area varied. The variation was related to the number of strata. Teak (*Tectona grandis*) and understory vegetation such as shrub and herb dominated two strata category. Vegetation with three strata was dominated by herbs and a mixture between teak stands and cahscrops (palawija), including ground nuts, cassava and sweet potato. Vegetation with four strata was dominated by various types of trees, shrub, and herbs. Vegetation with one or five strata was also found, but in general they only occupied areas smaller than the minimum area of observation (less than 64x64 m2)

Vegetation canopy measurement was carried out by taking into account the pixel size and RMS error. The width of the sample's rectangle A is defined as A= P(1+2*RMSe), where P is the spatial resolution (30 m), while RMSe is 0.56 according to the geometric correction results. Area of 64x64 m2 size for each sample was then set to represent the canopy density of the homogeneous pixels.

In the each area sample, nine points were used as the basis for making circular fisheye photographs of the canopy. The photos were taken upward from the ground. The averaged value of every nine photos was used to represent the canopy density. This study separated samples into two groups, *i.e.* the one for building up regression model and the other one for accuracy assessment. The measured canopy densities were then correlated with the vegetation indices.

5.3. Correlation between the canopy density and the vegetation indices

Correlation and regression analyses of the canopy density and vegetation index for different radiometric correction methods and different number of strata were carried out, resulting 96 variations of correlation coefficients and regression equations, which were then compared and evaluated. It is interesting to note that the simplest radiometric correction, *i.e.* histogram adjustment of the original DN did not perform badly, except for the GEMI vegetation index. The use of NDVI and MSAVI 2 of the corrected image gave a relatively high correlation coefficient, at the same values, i.e. 0.817 for four strata. Correlation between the corrected image and the canopy density at three strata gave slightly lower r, i.e. 0.792 and 0.799 respectively. With the GEMI, the correlation coefficients between canopy density and the vegetation index for various canopy strata were too low to be proceeded to regression-based estimation model.

For the NDVI, the most consistent radiometric correction method in terms of correlation coefficients with the canopy density (for various numbers of strata) was the dark object subtraction. The dark object subtraction method performed good r for all number of strata, with the lowest value of 0.81. The lowest correlation coefficient with the canopy density was obtained using at-sensor reflectance method, which showed relatively inconsistent r values for different strata. The MSAVI2 showed inconsistent r values when the histogram adjustment of at-sensor reflectance and dark subtract correction methods were used. Although the r values were relatively high, the discrepancy between them was also wide. A better performance in terms of high and consistent r values was found with the at-sensor radiance correction method. When the correlation coefficients of all combination have been found high enough, and they were statistically significant, this study continued with the regression-based equations to estimate the canopy density. Several results obtained using GEMI with different radiometric correction methods were not used due to their insignificant correlation coefficients.

5.4. Estimated Canopy Density Accuracies

This study found that different radiometric correction methods generated canopy density models with different accuracy levels. Besides the radiometric corrections, variation in the number of strata also played an important role. Generally speaking, there was no single radiometric correction that can be used as the best method for all numbers of canopy strata and all vegetation indices involved.

When the regression models were used to estimate the canopy density of the study area using NDVI, this simplest correction method also showed a higher accuracy level than other correction methods except dark subtract. When the MSAVI2 was used, the accuracy level achieved for four strata was still higher than the majority of radiometric corrections, except at-sensor radiance, at-sensor reflectance, and histogram adjustment of the TOA. Table 1 also shows that with NDVI, different radiometric corrections methods gave different canopy density estimate accuracies, when the number of strata was taken into account. MSAVI2 had the same tendency. On the other hand, GEMI only worked well for estimating the canopy density when the correction methods were applied using at-sensor reflectance, sun position, and histogram adjustment of those two resultants.

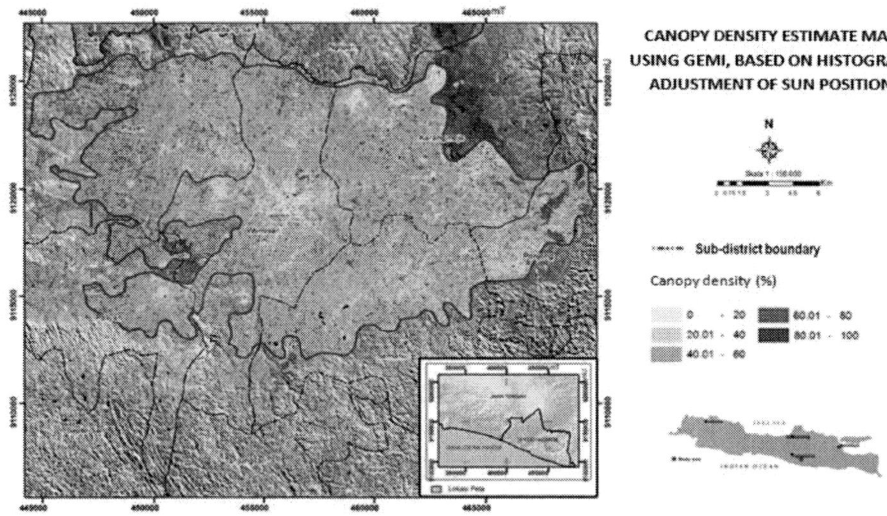

Figure 3. Example of the canopy density estimation map using GEMI with histogram adjustment of sun-position correction method.

Table 1. Summary of the result

Accuracy of vegetation canopy density estimate using original DN or corrected images:

Correction method / Vegetation Index	DN	At-Sensor Radiance	At-Sensor Reflectance	Sun Position correction	Histogram justment of original DN	Histogram adjustmt of At-sensor Radiance	Histogram adjustmt of At-Sensor Reflectance	Histogram adjustment of Sun Position	Dark Object Subtraction
NDVI	High, for 4 strata	High for 3 strata and combined 2,3 and 4 strata	High for 4 strata	High for 3 strata, 4 strata and combined 3 and 4 strata	High for 2 strata	High for 2 strata	High for 2 strata	High for 2 strata	High for 2 strata and 4 strata and cmbined 2&3 strata
MSAVI2	High for combined 2&3 strata	High for 3 strata, combined 2 and 3, combined 3 and 4, and combined 2, 3 and 4 strata	High for 4 strata	High for combined 2 and 3 strata and combined 3 and 4 strata	High for 2 strata, 3 strata and combined 2, 3 and 4 strata	High for 2 strata	High for 4 strata	High for 4 strata	High for 4 strata an combined 2, 3 and 4 strata
GEMI	X Correlation coefficient is too low	X High correlation coefficient, but inversed	High for combined 3 and 4 strata	High for 3 strata, combined 2 and 3, and combined 2, 3 and 4 strata	X High correlation coefficient, but inversed	X High correlation coefficient, but inversed	High for all variations of the number of strata	High for 2 strata, and 4 strata	High for 4 strata

* categorised as 'high' if the accuracy is the highest in comparison with the other strata variation under the same variable

6. Conclusions

This study concluded that radiometric corrections could increase the accuracy of vegetation canopy density maps, which were generated from Landsat-8 OLI using vegetation indices, in comparison with the original image. Different radiometric correction methods with different vegetation index formulas delivered different accuracy levels. The best canopy density map estimation accuracy was achieved using NDVI image that has been radiometrically corrected using Dark Object Subtraction, or using MSAVI2 with the at-sensor radiance image, or using GEMI with the sun position correction image. If the image data does not have any non-image information (i.e. image header) for the basis of radiometric correction, the original DN could still be used to deliver relatively accurate estimates for all number of strata, particularly using DNVI and MSAVI2. The number of strata gave significant effect on the accuracy of vegetation canopy density. By using NDVI, MSAVI2 or GEMI, it showed that the higher the number of strata, the higher the accuracy

References

[1] Danoedoro P, Phinn S and McDonald G 2004 Developing A Versatile land-use information system based on satellite imagery for local planning in Indonesia Phase I: Establishment of classification scheme. *7th International Seminar on GIS for Developing Countries (GISDECO-7)" GIS Capacity Building and Infrastructure"* (Universiti Teknologi Malaysia, Johor)

[2] Guyot G and Gu X 1994 Effect of radiometric corrections on NDVI-determined from SPOT-HRV and Landsat-TM data *Remote Sensing of Environment* **49** 3p 169-180

[3] Jensen J R 2005 Introductory Digital Image Processing Englewood Cliff (NJ: Prentice Hall)

[4] Danoedoro P, Kristian G and Rahmi K N I 2015 Pengaruh metode koreksi radiometrik citra ALOS AVNIR-2 terhadap akurasi hasil estimasi karbon vegetasi tegakan di wilayah kota Semarang bagian timur *Prosiding PIT MAPIN XX* (Bogor: IPB)

[5] Myneni R B, Hoffman S, Knyazikhin Y, Privette J L, Glassy J, Tian Y. 2002. Global products of vegetation leaf area and fraction absorbed PAR from year one of MODIS data. *Remote Sensing of Environment* **83** p 214–231

[6] Akkartal A, Turudu O and Erbek F S 2004 Analyis of changes in vegetation biomass using multisensor satellite data *Proceedings of the ISPRS Congress-Youth Forum* p 181–185

[7] Hadjimitsis D G, Papadavid G, Agapiou A, Themistocleous K, Hadjimitsis M G, Retalis A, Michaelides A, Chrysoulakis N, Toulios L and Clayton C R I 2010 Atmospheric correction for satellite remotely sensed data intended for agricultural applications: impact on vegetation indices Nat. *Hazards Earth Syst. Sci.* **10** p 89–95

[8] Huete A R 1988 A soil-adjusted vegetation index (SAVI) *Remote Sensing Of Environment* **25** p 295-309

[9] Danoedoro, P. 1989. Hubungan antara Konsentras Klorofil dengan Informasi Spektral Vegetasi pada Data Digital SPOT Multispektral Wilayah Lereng Gunungapi Merapi Bagian Selatan. Skripsi. Fakultas Geografi UGM

[10] Howard, J. A. (1990). Remote Sensing of Forest Resources: Theory and Applications. Melbourne: Chapman and Hall.

[11] Danoedoro P, Phinn S and McDonald G 2004 Developing a versatile land-use information System Based on Satellite Imagery for Local Planning in Indonesia Phase I: establishment of classification scheme. 7th International Seminar on GIS for Developing Countries (GISDECO-7) GIS Capacity Building and Infrastructure" (Universiti Teknologi Malaysia, Johor)

[12] Kuchler W, and Zonneveld I E 1988 *Vegetation Mapping* (Dordrecht: Kluwer Academic Publisher)

[13] van Gis H, Zonneveld I S, Van der Zee D and Huizing H 1990 *Land Ecology and Land-use Survey* (Enschede: ITC)

[14] Hoffer R M 1978 Biological and physical considerations in applying computer-aided analysis techniques to remote sensor data In: Swain P H and Davis *S M Remote Sensing The Quantitative Approach* (New York: Prentice Hall)

Soil erosion and its correlation with vegetation cover: An assesment using multispectral imagery and pixel-based geographic information system in Gesing Sub-Watershed, Central Java, Indonesia

Diwyacitta Dirda Gupita[1]* and Sigit Heru Murti B S[1,2]
[1] Department of Remote Sensing, Faculty of Geography, Gadjah Mada University, Sekip Utara, Sleman, Yogyakarta 55281
[2] PUSPICS Faculty of Geography, Gadjah Mada University, Sekip Utara, Sleman, Yogyakarta 55281

E-mail: diwyacitta.dirda.g@mail.ugm.ac.id

Abstract. Soil erosion in caused by five factors: rainfall erosivity, soil erodibility, slope and slope length, crop management, and land conservation practices. In theory, vegetation as one of the affecting factors has inversed correlation with soil erosion. This research is aimed to: (1) model RUSLE using pixel-based GIS, and (2) prove whether or not vegetation really has the said correlation with the soil erosion that occurs in Gesing Watershed. The method used in this research is divided into two: the use of RUSLE to estimate the soil erosion rate; and the use of fractional vegetation cover (FVC) formula to estimate the vegetation density in the area. Both methods used Landsat-8 OLI imagery, which is used to extract the RUSLE parameters as well as to derive the vegetation density through NDVI, and pixel-based GIS. The mapping of soil erosion rate distribution done in this research demonstrated that pixel-based modeling is able to represent a much more detailed and logical distribution of a phenomenon. The distribution of soil erosion rate in Gesing Watershed showed that the erosion rate in this area is relatively minor. About 1425.99 hectares and 1587.57 hectares of the total area have erosion rate of 0 – 15 tons/ha/yr (very mild) and 15 – 60 tons/ha/yr (mild) respectively.

1. Introduction

Soil erosion is one of many environmental problems that has to be managed. This natural phenomenon is the main cause of many cases of land degradation, regardless of the environmental condition whether it is arid or wet. Furthermore, the high intensity of rainfall may cause soil erosion to rapidly transport soil from its original place through overland flow [1]. Soil erosion which occur in a watershed boundary is basically caused by four factors: climate, the characteristics of soil, topograhy of the area, and landcover vegetation [2]. Vegetation effect on soil erosion can be identified from two conditions, canopy condition and basal area condition [3]. Tree canopy condition is related to the ability of trees to intercept rainfall and reduce the destructive energy to create raindrop erosion or splash erosion. Basal area vegetation, such as the presence of dry leaves, grass and other plant residue, can reduce overland flow which is the cause of sheet erosion. Moreover, basal plant roots provide open pores on the ground's surface, where water can seep through and increase rainfall infiltration. The roots interweave with each other and create a more solid mass of soil which is more resistant to erosion [4]. Therefore, vegetation cover gives a very big influence in the soil erosion rate that occur in a certain watershed or area. This research is aimed to show the correlation between soil erosion, which is obtained by using RUSLE formula, and vegetation cover density, which is extracted from fractional vegetation cover (FVC).

2. Area of study

This research was held in Gesing Sub-Watershed, that is located in the eastern District of Purworejo, Central Java. This sub-watershed covers the area of 47.09 km^2 and most of its area is located in western part of Menoreh Hills, which is shown in figure 1.

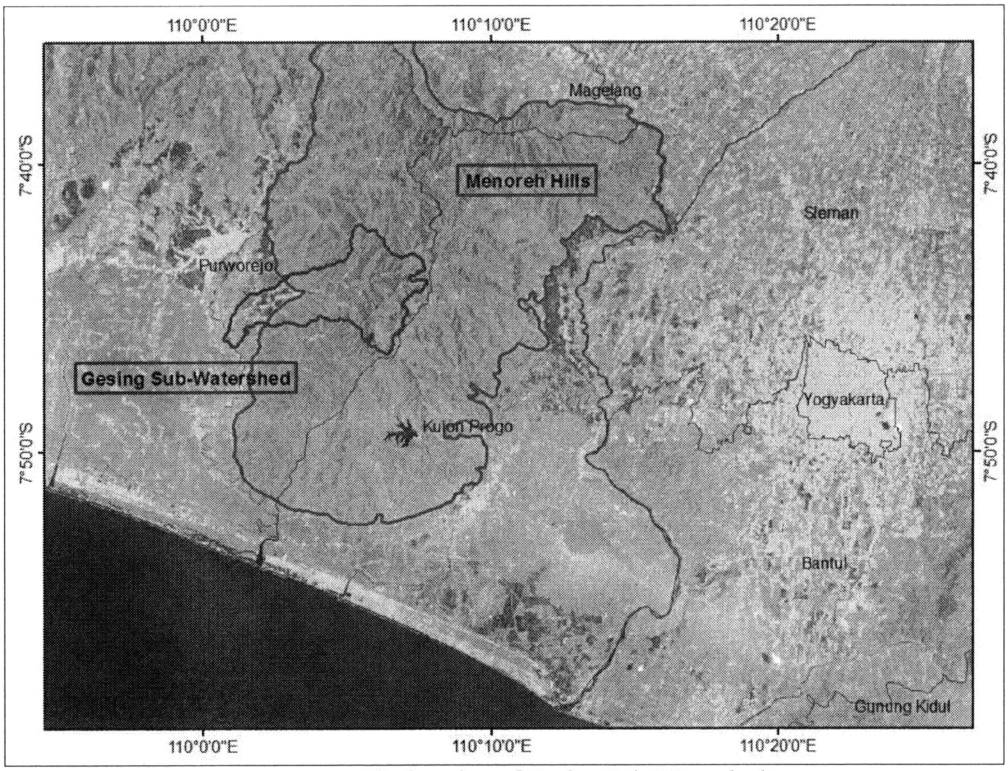

Figure 1. The location of Gesing Sub-Watershed

3. Data and methods

3.1. Data

The remotely sensed data used in this study in Landsat-8 OLI imagery (path 120 row 60) which was obtained in February 22nd 2015. Not all bands are used in this study, only band 2, band 3, band 4, and band 5 were used to extract landuse information as well as land unit information, with the help of a geologic map in a 1:50000 scale.

Peta Rupabumi Indonesia (Indonesian Base Map) was also utilized to get the contour data which was used to delineate the watershed boundary (in the form of triangulated irregular network), also as an input in the calculation of slope and slope length indices (SL in RUSLE, in the form of *flow accumulation).*

The secondary data of monthly rainfall from the year of 2004 until 2013 was obtained from the Central Java Meteorology and Climatology Agency.

3.2. *Methods*

3.2.1. *RUSLE Model*

In RUSLE model, there are five parameters affecting the rate of soil erosion: rainfall erosivity (R), soil erodibility (K), slope and slope length (LS), crop/ cover management (C), and land conservation practices (P). All of those are combined to form this equation:

$$A = R.K.L.S.C.P \tag{1}$$

3.2.2. *Rainfall Erosivity Index (R)*

Soil erosion, is basically the loss of soil caused by the detachment from the ground's surface by raindrops. The heavier the rain, the bigger the raindrops, which will cause more soil particles to be detached from the surface. Using the rainfall data in 10 years span, rainfall erosivity is calculated using the equation formulated by Mahmud and Utomo [5].

$$Rb = 10{,}80 + 4{,}15(Hb) \tag{2}$$

Where Rb is the monthly erosivity index, and Hb represent the average rainfall per month in centimeter unit. The use of this equation is primarily caused by the unavailability of daily rainfall data provided by the Meteorology and Climatology Agency.

3.2.3. *Soil Erodibility Index (K)*

Erodibility is defined as the soil vulnerability to erosion [2]. It is mostly dependant on the texture of the soil, the stability of soil aggregation, infiltration capacity, and also the the contents of organic and chemical materials [6]. The equation for K index used in this study is formulated as follows [6]:

$$\mathrm{K} = 1{,}292\begin{bmatrix} 2{,}1 \cdot 10^{-6}\left(\%silt \times (100 - \%clay)\right)^{1.14}(12 - \%OM) + 0{,}0325\,(Ss - 2) + \\ 0{,}025\,(fpm - 3) \end{bmatrix} \tag{3}$$

Where Ss is the soil structure with values of: 1 for very fine granular, 2 for fine granular, 3 for medium or coarse granular, and 4 for blocky, platy, or massive; *fpm* stands for soil permeability class with the values of: 1 for very slow infiltration, 2 for slow infiltration, 3 for moderately slow infiltration, 4 for moderate infiltration, 5 for moderately rapid infiltration, and 6 for rapid infiltration.

3.2.4. *Slope and Slope Length Index (LS)*

Area with higher slope and longer slope length tend to have higher soil erosion rate. In flat surface, raindrop splash will only cause the soil particles to detach in various directions, whereas in a sloped surface, those particles will detach only in the direction of the slope [7]. How much soil detached from ground surface, in this case, depends on the degree of the slope; the greater the slope, the larger the amount of soil lost because of detachment. The LS index is extracted using *flow accumulation* which can be formulated using the Moore and Burch equation [8]:

$$LS = \left(\frac{[flow\ accumulation \times cellsize]}{22{,}13}\right)^{m} \times \left(\frac{\sin slope}{0{,}0896}\right)^{1{,}3} \tag{4}$$

3.2.5. *Crop Management Index (C)*

Crop and management index can be obtained by using multispectral classified landuse information and a C-index table from literatures, but this method can be a little difficult to work with sometimes, especially if the C-index table cannot represent the landuse present in the study area. In this case, this study utilizes the C-factor from normalized vegetation index (NDVI) as suggested by De Jong [9]. The NDVI indices of vegetated area (forest) and open area (bare soil) present in Gesing Watershed are then analyzed through a regression analysis as dependent variable, while the independent variable is the sample index of 1 for forest (or dense vegetation) and 0 for bare soil.

3.2.6. Conservation Practice Index (P)
Multiple informations, such as administration boundary, landuse, and slope, were used to obtain the P-factor. Slope can sometimes give variations to the conservation practice of one type of landuse, certain landuse in steep slope are sometimes left as is it with no conservation applied. On the other hand, administration boundary affect the conservation practice through how agricultural instructions from agricultural/ forestry instituions are delivered. The value of each P-factor is obtained through field observation and matching the practices found in field with literary P-factor table by Abdurakhman [10].

3.2.7. Fractional Vegetation Cover (FVC)
Fractional vegetation cover refers to percentage of vegetation cover in one pixel [11]. FVC can be considered as a derivative of Normalized Difference Vegetation Index since it is formulated using NDVI values.

$$FVC = \left(\left(\frac{1}{NDVI_v - NDVI_s}\right) \times NDVI + b\right) \times 100 \qquad (5)$$

$NDVI_v$ refers to the NDVI value for forest or dense vegetation (NDVI = 1 or $NDVI_{max}$), while $NDVI_s$ refers to NDVI value of bare soil (NDVI = 0 or $NDVI_{min}$). Field observation was done in the study area to retrieve 25 point samples in total of places with dense vegetation cover and places with bare soil. The vegetation density of the 25 sample points extracted from both FVC and field observation are then analyzed using regression to obtain the actual vegetation density.

4. Results and discussion

4.1. Soil erosion in Gesing Sub-Watershed
The soil erosion in Gesing Sub-Watershed, form the appearance of it in figure2, can be considered not very serious. The areas with varying rates are mostly located in the mid-watershed section. Table 1 shows that erosion rate of 15-60 t/ha/yr is dominating the sub-watershed with the area of 1587.57 ha, while the more severe rate of 60-180 and 180-480 t/ha/yr are only distributed only in some areas.

Table 1. The area of each soil erosion rate class in Gesing Sub-Watershed (per section)

Watershed Section	Soil Erosion Rate (ton/hectare/year)					Area (ha)
	0 – 15	15 - 60	60 - 180	180 – 480	>480	
Lower	594.82	109.70	13.30			717.82
Mid	455.98	1080.23	1198.53	267.55	7.83	3010.12
Upper	375.19	397.64	154.39	14.49		941.71
Area (ha)	1425.99	1587.57	1366.22	282.04	7.83	4669.65

4.2. Vegetation cover of Gesing Sub-Watershed
The information of vegetation cover and density are retrieved through field observation in 25 points all over Gesing Watershed. From the regression analysis with the FVC vegetation density, the correlation coefficient of 0.615 shows the great correlation between the field vegetation density and the FVC density, and by using the regression formula, the actual vegetation density is extracted and is shown in figure 3.

Table 2. The area of each vegetation density class in Gesing Sub-Watershed

No.	Vegetation Density	Area (ha)
1.	0 – 20%	227.96
2.	20 – 40%	116.12
3.	40 – 60%	473.95
4.	60 – 80%	2017.9
5.	80 – 100%	1872.88

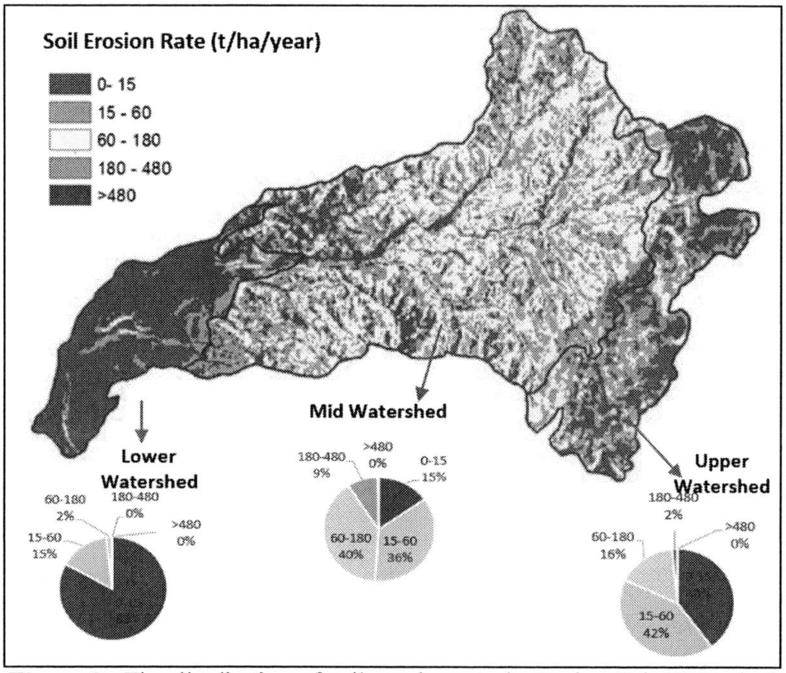

Figure 2. The distribution of soil erosion rate in Gesing Sub-Watershed based on each watershed section (the percentage shown is based on each section's total area)

The classification for the vegetation density is divided into 5 classes standardized by the Department of Forestry in 2004 [12]. From figure 3, it can be seen that the vegetation density of Gesing Sub-Watershed is somewhat dense, because most of its area is dominated by forests, as seen in Figure 4. It is visible that the mid section and the upper section of the watershed has fairly high vegetation density, but compared to the upper section, the mid section has more areas with very high vegetation density (80-100% coverage). The mid section of the watershed is dominated by forests; natural forests in the mid-lower section and production forest in the mid-upper section. Natural forest consists of mainly wild bamboos and shrubs grown in sloped areas with no conservation practice, while the producion forest consists of allelopathic trees such as teak (*Tectona grandis*) and mahogany (*Swietenia mahogani*) which are able to retards the growth of the basal vegetation.

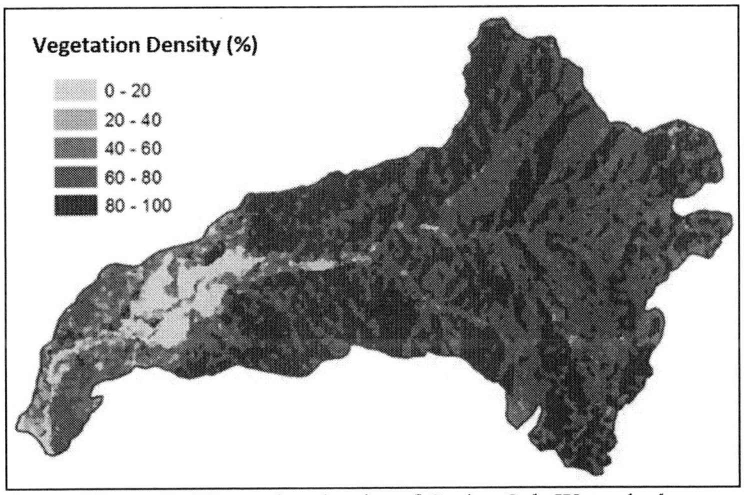

Figure 3. Vegetation density of Gesing Sub-Watershed

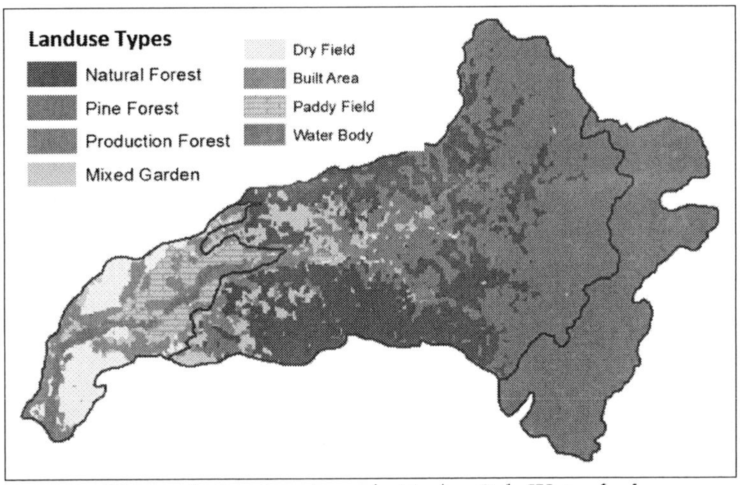

Figure 4. Landuse types in Gesing Sub-Watershed

4.2.1. Correlation between vegetation cover density and soil erosion rate

Table 3 shows that soil erosion is still inevitable in areas with high vegetation density (60-100% coverage). The theory says that the higher the vegetation coverage/ density, the lower the erosion rate; but this is not the case in this study. From the condition of the sub-watershed itself, the areas with high vegetation density are located mostly in the mid-watershed section. Moreover, the areas with high erosion rate are also located in the mid-watershed section. In this case, the high vegetation density in Gesing sub-watershed does little effect in minimizing soil erosion rate. The cause may come from the vegetation types existing within the mid-watershed section that are mostly allelopathic or grown in areas without conservation practices. Allelopathic vegetation negatively affects the growth of other plants, especially basal vegetation. It had been said before that basal cover helps reduce soil erosion by encouraging rainfall infiltration, but because of the lack of basal cover present, water tends to run on the ground surface in the form of overland flow. On the other hand, natural forest with high density of shrubs and wild bamboos in the areas without any kinds of conservation practice makes it easier for the soil to detach and move down the slope, hence the occuring soil erosion.

Table 3. Area of Gesing Watershed based on the vegetation density and soil erosion

Soil Erosion Rate	Vegetation Cover Density					Area (ha)
	0 - 20%	20 - 40%	40 - 60%	60 - 80%	80 - 100%	
0 - 15 ton/ha/yr	146.97	67.80	202.51	370.54	638.17	1425.99
15 - 60 ton/ha/yr	64.08	28.52	108.91	609.48	776.57	1587.57
60 - 180 ton/ha/yr	13.14	10.83	101.62	819.83	420.80	1366.22
180 - 480 ton/ha/yr	2.13	6.22	50.71	197.10	25.88	282.04
>480 ton/ha/yr	0.81	0.64	2.84	3.30	0.24	7.83
Area (ha)	227.14	114.01	466.59	2000.25	1861.67	4669.65

5. Conclusion

Gesing Sub-Watershed, although most of its area is covered by high density vegetation, still has many areas with severe soil erosion rate. The vegetation cover is supposed to have an inversed correlation with the rate of soil erosion, but that does not seem to be case in this study. Based on the result of this research, the internal and external conditions of the vegetation in the area give an important influence on the vegetation-erosion relationship. The allelopathic characteristic of the planted vegetation present in the study area turns out to be giving a significant effect on how the presence of basal cover affects the rainfall infiltration. Conservation practice also gives a crucial influence on soil erosion, especially in sloped areas.

References

[1] Sulistyo B 2011 *Penginderaan Jauh Digital: Terapannya dalam Pemodelan Berbasis Raster* (Yogyakarta: Penerbit Lokus)

[2] Blanco-Canqui H and Lal R 2008 *Principles of Soil Conservation and Management* (New York: Springer Dordrecht Heidelberg)

[3] Morrow S, Smolen M, Stiegler J and Cole J 2014 Using Vegetation for Erosion Control on Construction Sites," 2014. [Online]. Available: http://pods.dasnr.okstate.edu/docushare/ds web/ Get/Document-2264/BAE-1514web.pdf..

[4] Styczen M and Morgan R 1995 *Engineering Properties of Vegetation in Slope Stabilization and Erosion Control: A Bioengineering Approach*, R. Morgan and R. Rickson, Eds., E and F.N. Spon pp. 5-58.

[5] Utomo W 1994 *Erosi dan Konservasi Tanah* (Malang: Penerbit IKIP Malang)

[6] Morgan R 1995 *Soil Erosion and Conservation, 2nd ed* (Essex: Addison Wesley Longman Limited)

[7] Goldman S, Jackson K and Bursztynsky T 1986 *Erosion and Sediment Control Handbook* (New York: McGraw Hill Book Co.)

[8] As-Syakur A 2008 Prediksi Erosi dengan Menggunakan Metode USLE dan Sistem Informasi Geografis (SIG) Berbasis Piksel di Daerah Tangkapan Air Danau Buyan in *PIT MAPIN XVII*, Bandung

[9] Erencin Z C 2000 Factor Mapping Using Remote Sensing and GIS: A Case Study of Lom Sak / Lom Kao, Thailand (Enschede: International Institute for Aerospace Survey and Earth Sciences)

[10] Purnama N 2011 *Pendugaan Erosi dengan Metode USLE (Universal Soil Loss Equation) di Situ Bojongsari, Depok* Bogor

[11] Zhang X, Liao C, Li J and Quan S 2012 Fractional Vegetation Cover Estimation in Arid and Semi-Arid Environment Using HJ-1 Satellite Hyperspectral Data *International Journal of Applied Earth Observation and Geoinformation,* **21** 506-512

[12] Lathifah D 2012 *Hubungan antara Fungsi Tutupan Vegetasi dan Tingkat Erosi DAS Secang Kabupaten Kulonprogo* Yogyakarta

Age-surface temperature estimation model: When will oil palm plantation reach the same surface temperature as natural forest?

S B Rushayati[1*], R Hermawan[1] and R Meilani[1]
[1]Department of Forest Resources Conservation and Ecotourism Faculty of Forestry Bogor Agricultural University (IPB) PO Box 168 Bogor 16001

E-mail: rus_badriyah@yahoo.co.id

Abstract. Oil palm plantation has often been accused as the cause of global warming. However, along with its growth, it would be able to decrease surface temperature. The question is 'when will the plantation be able to reach the same surface temperature as natural forest'. This research aimed to estimate the age of oil palm plantation that create similar surface temperature to those in natural forest (land cover before the opening and planting of oil palm). The method used in this research was spatial analysis of land cover and surface temperature distribution. Based on the spatial analysis of surface temperature, five points was randomly taken from each planting age (age 1 15 years). Linear regression was then employed in the analysis. The linear regression formula between surface temperature and age of oil palm plantation was Y = 26.002 – 0.1237X. Surface temperature will decrease as much as 0.1237 ^{0} C with one year age growth oil palm. Surface temperature that was similar to the initial temperature, when the land cover was natural forest (23.04 ^{0}C), was estimated to occur when the oil palm plantation reach the age 24 year.

1. Introduction

1.1. Background

The world is currently facing global warming issue. Air temperature of the earth's surface has increased between 0.3°–0.6°C at the end of 1800 and will keep increasing up to 1°–3.5°C on the upcoming 2100 [1]. The raise of earth's surface temperature was triggered by the increase of greenhouse gasses effect intensity due to greenhouse gasses [2]. Global emission of greenhouse gasses has increased about 45% since 1990, and about 30% since 2000 [3]. The increase of greenhouse gasses was caused by two factors, i.e.: (1) the increase of CO_2 emission from various activities; (2) decrease of CO_2 absorbent agents. Oil palm plantation is amongst the activities that cause greenhouse gasses emission, and therefore contribute to the increase of earth's surface temperature that led to global warming. To cope with the increase of earth's surface temperature, all parties and sectors, including oil palm plantation sector, have to take part in global warming adaptation and mitigation effort [4, 5].

Oil palm plantations were developed on an area with Other Uses Area (Known as APL) status with various land cover condition, such as shrubs, imperata, and might be remnants of secondary forest. The accusation that oil palm plantation is one of causes of global warming is not always true. Oil palm plantation will provide a temperature decreasing effect through its planting activities and the increasing age (growth) of the plants. Vegetation can decrease temperature through direct shade and evapotranspiration [6]. Canopy of vegetation, which provide shading, will reduce the amount of solar radiation that reach earth's surface and reflects the radiation back into the atmosphere. Surface temperature of vegetation is lower than temperature of open land and other man-made objects due to relatively high albedo [7].

As the oil palm plants grow, the canopy will get denser and the area covered by the plants' canopy will be wider, and therefore solar radiation reaching the earth's surface will decrease; thus the surface temperature will decrease. As a reference, the ideal surface temperature is one found in natural forest ecosystem that has crucial role in controlling the climate [8, 9].

Oil palm plantation with single canopy layer and only one plant species is predicted could not generate surface temperature similar to natural forest with various canopy layers and plants composition [10]. However, oil palm plantation could be expected to have surface temperature close to natural forest. Therefore, this research was carried out to examine the extent to which the growth of oil palm trees can reduce the surface temperature so that the surface temperature approaching natural forests.

1.2. Objective
The objective of this research was to identify the age of oil palm plantation that can generate a closer surface temperature to natural forest.

2. Method
The research was carried out at Oil palm Plantation in Kubu Raya District. Land cover and surface temperature analysis for 1989, 1997, and 2014 was carried out. Materials used in the analysis were Landsat 5 TM imageries for 1989 and 1997, and Landsat 8 OLI/TIRS imageries for 2014 (table 1). Erdas software was used for the analysis, while ArcGIS software was used for sample point generation for statistical analysis and map layout.

Focusing on calculating surface temperature for oil palm, the land cover classes was generated based on imagery interpretation and field observation. There were only 5 classes, which included water body (natural water body, such as river, and man-made water body, such as pool), tree vegetation, non-tree vegetation (included oil palm vegetation, and imperata, grasses, and shrubs in 1989), built land, and open land.

There were 20 training areas for 1989 imagery, 39 training areas for 1997 imagery, and 100 training areas for 2014. Ground check was employed to identify oil palm age and land cover, while surface temperature was calculated using formula based on Landsat.

Table 1. Types and source of study materials.

No.	Types of study materials	Source
1.	Landsat 8 OLI/TIRS *path/row* 121/60 imagery captured on 09:52:23 acquired date on 21 August 2014	http://glovis.usgs.gov
2.	Landsat 5 TM *path/row* 121/60 imagery captured on 09:20:16 acquired date on 13 June 1989	http://glovis.usgs.gov
3.	Landsat 5 TM *path/row* 121/60 imagery captured on 09:21:52 acquired date on 19 June 1997	http://glovis.usgs.gov

Pre-processing of the imageries was carried out before conducting the analysis. Pre-processing include channel composite, image projection change, and image cut. Channel composite used in land cover analysis was channel 1–7, while the channels used in temperature analysis was channel 10 for Landsat 8 and channel 6 for Landsat 5. The projection used was Universal Transverse Mercator (UTM) 49 S zone. Analysis phases is presented in figure 1.

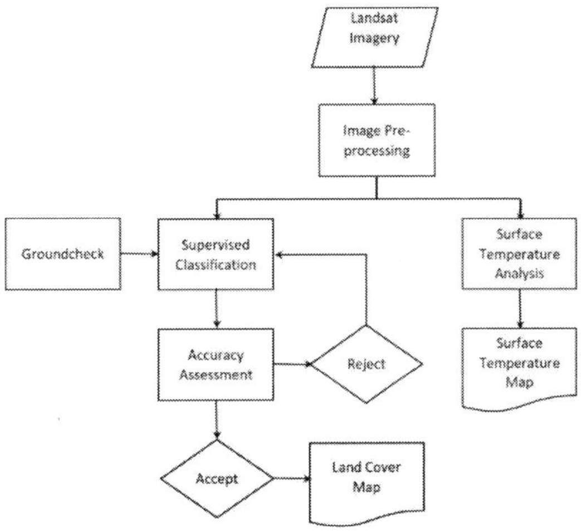

Figure 1. Flow chart of analysis phases.

2.1. Land cover analysis

Supervised classification with maximum likelihood classifier method was employed in land cover analysis. Land cover was classified into 5 classes, i.e. water body, tree vegetation, non-tree vegetation, build land, and open land. There were unclassified pixels due to image cutting with administrative boundary and cloud and cloud's shadow, which thereafter was presented as no data available. Accuracy assessment was conducted after classification. Training area for planting year was determined based on planting year block obtained from map using the tools create random points in ArcGIS application. Training area was determined based on imagery interpretation by user, aided by ground check data.

2.2. Surface temperature analysis

Calculation of surface temperature value was carried out in stages [11-13]. First, calculation was conducted from temperature channel to obtain spectral radian value using equation (1). Spectral radian value was then converted into temperature in Kelvin (K) unit using equation (2). After surface temperature value in K unit was obtained, it was then converted into Celcius (°C) unit using equation (3).

$$L_\lambda = ML \cdot Qcal + AL \tag{1}$$

Explanation:

L_λ = Spectral radian value (Watt/m^{2}s rad µm)
ML = Specific multiplier factor of thermal band
Qcal = Specific digital imagery value of thermal band
AL = Specific additional factor of thermal band

$$Ts = K2 / \ln[(K1/L_\lambda)+1] \tag{2}$$

Explanation:

Ts = Surface temperature (K)
$L\lambda$ = Spectral radian value (Watt/m^{2}s rad µm)
K1 = 1 constant conversion (774.89 for Landsat 8 imagery and 607.760 for Landsat 5 imagery)
K2 = 2 constant conversion (1,321.08 for Landsat 8 imagery and 1,260.56 for Landsat 5 imagery)

$$Ts = Ts\ (K) - 273 \tag{3}$$

Explanation:

Ts = Surface temperature (°C)

2.3. *Correlation analysis between variables of surface temperature and oil palm age*

Correlation analysis between age of oil palm and surface temperature was carried out by determining sample points on oil palm plot from 1–15 year (15 age classes) with 5 replications for each plot that the total points being analyzed was 75 points. The oil palm plot data were obtained from field observation. The map of surface temperature sample points' distribution is presented in figure 2.

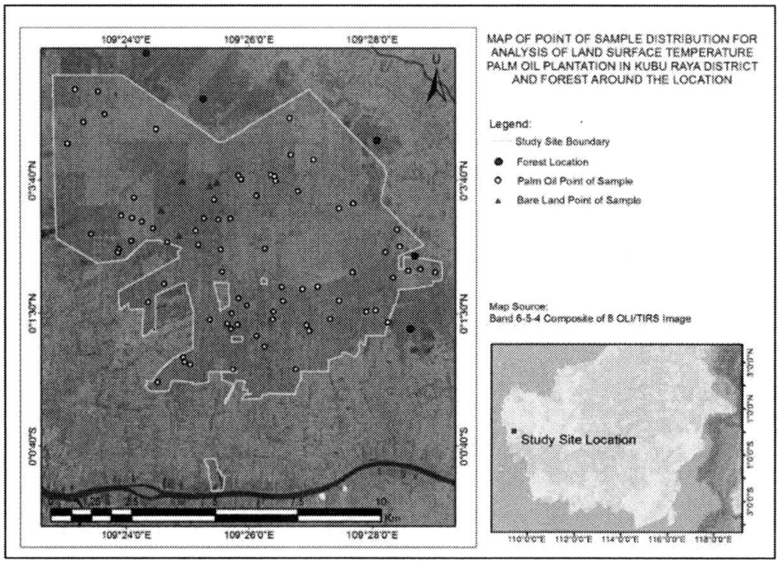

Figure 2. Map of surface temperature sample points' distribution.

Correlation analysis was used to measure the strength of relation (linear relation) between two variables. After correlation analysis, a test of correlation hypothesis was also carried out based on the following conditions:

1. $H_0 : \rho = 0$ (there is not any correlation between X and Y)
   $H_1 : \rho \neq 0$ (there is correlation between X and Y)
2. Level of significance $\alpha = 0.05$
3. Reject H_0 if $t < -t_{n-2,\ \alpha/2}$ or $t > t_{n-2,\ \alpha/2}$
   and using P-Value, if P-Value $< \alpha$ then H_0 is rejected.
4. Statistic test:

$$t_{0.05} = \frac{r - \rho}{\sqrt{\dfrac{1 + r^2}{n - 2}}}\ t_{tabled} \tag{4}$$

When the $t_{0.05} > t_{tabled}$, then the H_0 is rejected, or using the P-Value, when P-Value < 0.05 then H_0 is rejected. Thus the conclusion would be that there is correlation between X and Y.

3. Result and Discussion

3.1. *Land cover changes*

Evaluation of classification result was carried out through accuracy assessment using GPS coordinate data. A good overall accuracy based on the criteria from United States Geological Survey (USGS) is above 85% [14]. The overall accuracy value resulted from land cover analysis as much as 80%, not much different from the USGS's criteria that the accuracy can be stated as good. This is further supported by kappa accuracy as much as 0.6827. The closer kappa accuracy value to 1 (the higher kappa accuracy value), means that the classification is better.

Analysis of land cover in 1989, 1997 and 2014 showed a decrease of tree vegetation. In 1989, tree vegetation covered 4,261 ha, which decreased to 1,523 ha in 1997 and 33 ha in 2014. On the contrary, non-tree vegetation (oil palm) had been increasing from year to year. In 1989, it covered an area of 1,998 ha, which increased to 1,131 ha in 1997, and kept increasing to 5,777 ha in 2014. Land cover changes is presented in figure 3 and figure 4.

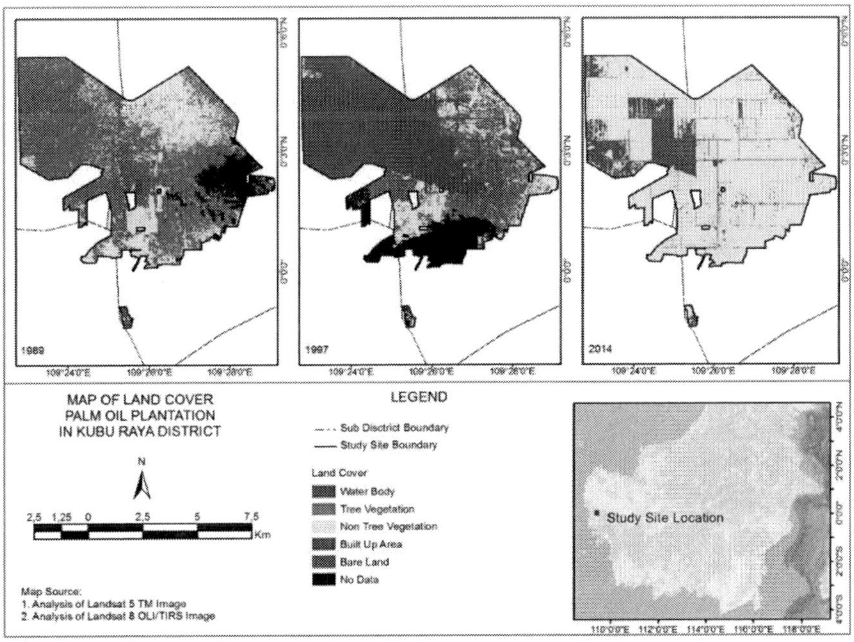

Figure 3. Map of land cover in 1989, 1997, and 2014.

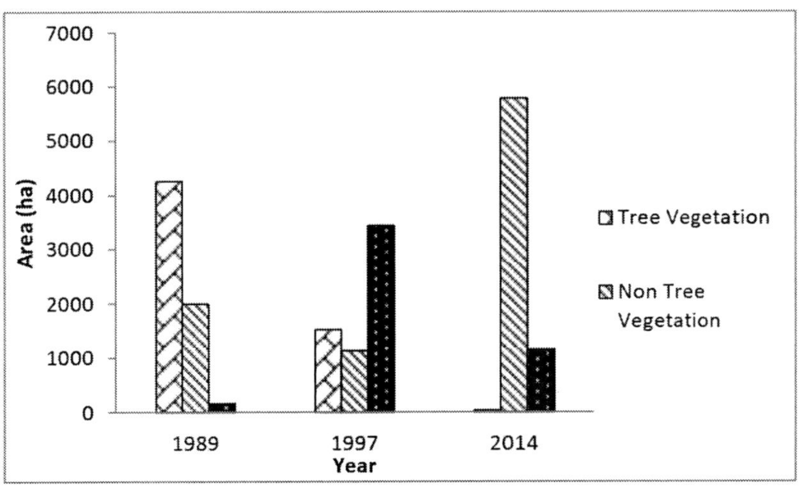

Figure 4. Land cover changes

3.2. Surface temperature changes

In 1989, the highest percentage (89%) was those of surface temperature ranging from 21–22 °C. At the time, land cover was dominated by tree vegetation. Land opening and oil palm planting has been started in 1997. Area with surface temperature ranging from 21–22 °C decreased to 14% in 1997. In that year, 38% of the area has surface temperature on 23–24 °C, and 34% 20–21 °C. In 2014, the coverage of non-tree vegetation increased. The surface temperature on 23–24 °C increased to 43%. From 1989 to 2014, there had been a shift of surface temperature. The highest range of surface temperature in 1989 was 23–24 °C, while in 1997 the highest range increased to a range of 25–26 °C. In 2014, the highest range of surface temperature rose to 28–29 °C. Apparently, the highest surface temperature will increase with the increase of changes from tree vegetation to non-tree vegetation. This condition due to a decrease of percentage of radiation interception by tree canopy.

There is a relation between vegetation density and climate, particularly air temperature [15]. The increase of area stress occurred in forest lands had caused the increase of air temperature. However, though the land opening from tree-vegetated area causes an increase of air temperature, the growth of oil palm plants (non-tree vegetation) can increase carbon absorption, proven in research by [16], which stated that biomass carbon stock (excluding fresh fruit bunches) of oil palm agroecosystem on the age of 1 to 18 year ranged from 0.7–16.43 ton/ha. Based on this research, a prediction can be made that oil palm absorbs CO_2 as much as 2.57–60.30 ton/ha. Oil palm growth can increase carbon stock and decrease greenhouse gasses, particularly CO_2.

A study in Beijing, China using a variety of spatial resolution of the imagery a consistent result that the higher coverage of green open space will decrease land surface temperature [17]. Similar condition applies in oil palm plantation, where we found an increase of surface temperature at the beginning of land opening. However, along with the increasing age of the oil palm (and the increase of vegetated area), carbon stick will also increase and therefore reducing surface temperature. Thus, surface temperature of oil palm plantation is greatly influenced by land opening activities that can increase surface temperature, but also by the growth of oil palm that can reduce surface temperature through radiation interception by oil palm canopy and the increase of greenhouse gasses (CO_2) absorption in its photosynthesis process. The changes of surface temperature due to land cover changes, particularly in oil palm plantation area from 1989 to 1997 and 2014 is presented in figure 5.

3.3. Relation between oil palm age and surface temperature

A comparison of oil palm plantation's surface temperature and tree vegetation's surface temperature was also employed in order to estimate the age of oil palm plantation that generates surface temperature close to tree vegetation. Therefore, spatial analysis was also carried out on area surrounding the oil palm plantation, resulted in map of surface temperature distribution (figure 6).

As the age of oil palm plantation increase, the solar radiation interception will also increase, and therefore reducing radiation reaching to land surface. The condition will cause a decrease of surface temperature. Statistical analysis showed that $t_{0.05} > t_{tabled}$, and the P-value < 0.05, therefore H0 is rejected. It can be concluded that there is correlation between X (oil palm age) and Y (surface temperature) with an equation of Y = 26.002 - 0.1237 X. With an addition of one year of X (oil palm age), the Y (surface temperature) will decrease as much as 0.1237 °C. In other words, an addition of oil palm age will caused lower surface temperature. The older oil palm age, the lower surface temperature would be, and this condition proved that oil palm also has a role in absorbing CO_2 and decreasing surface temperature. This finding is in line with [18], which stated that one of mitigation options in Agriculture, Forestry, and Other Land Use (AFOLU) is reducing greenhouse gasses by enhancing the uptake of carbon in terrestrial, and thereby removing CO_2 from the atmosphere. Oil palm plantation growth is enhancing the uptake of carbon in terrestrial, which can reduce the concentration of greenhouse gasses and decrease the surface temperature. The relation between surface temperature and oil palm age is presented in figure 7.

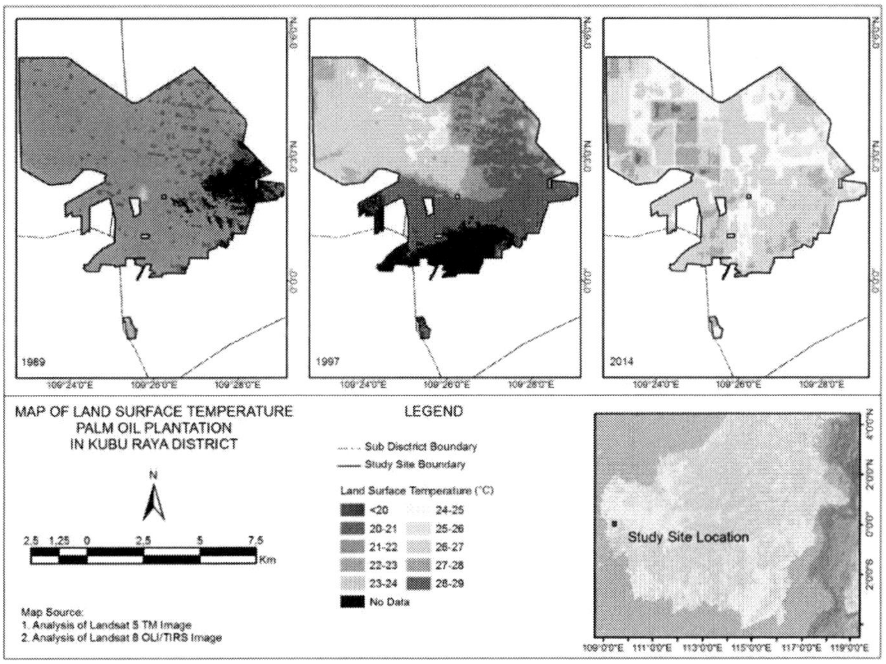

Figure 5. Map of surface temperature distribution changes.

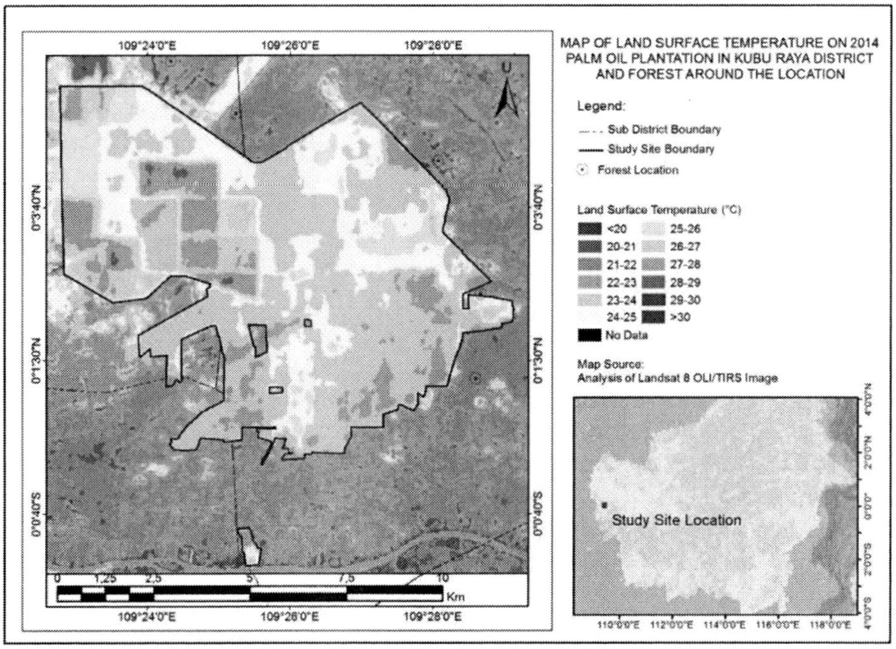

Figure 6. Map of surface temperature distribution in 2014.

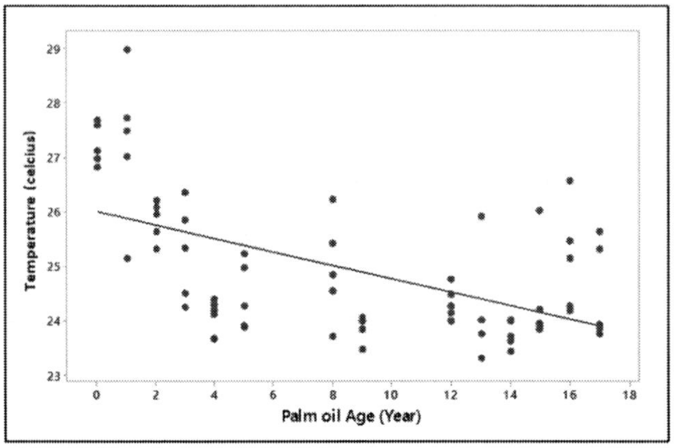

Figure 7. Graph on the relation between oil palm age and surface temperature.

The statistical equation enables us to predict that oil palm plantation's surface temperature close to the surface temperature of tree vegetation (land cover before conversion from tree vegetation to oil palm; the surface temperature was 23.04 ^{0} C) will be reached when the oil palm plantation has reached the age of 24 year.

4. Conclusion

Linear regression analysis between oil palm plantation age and surface temperature enables us to conclude that the surface temperature will decrease as much as 0.1237 ^{o}C with one year age growth oil palm. The decrease of surface temperature proved that solar radiation interception and CO_2 absorption occur in the process of oil palm growth, which causes a decrease of surface temperature along with oil palm age increase. Surface temperature that was similar to the initial temperature, when the land cover was natural forest, was estimated to occur when the oil palm plantation reach the age 24 year.

5. References

[1] Hamburg S P, Harris N, Jaeger J, Karl T R, McFarland M, Mitchell J F B, Oppenheimer M, Santer S, Trenberth K E and Wigley T M L 1997 *Common questions about climate change* (United Nation Environment Programme: World Meteorology Organization)

[2] Tjasyono B 2004 *Klimatologi* (Bandung: Penerbit ITB)

[3] Olivier J G J, Janssens-Maenhout G, Peters J A H W and Wilson J 2011 *Long-term trend in global CO2 emissions 2011 report* (The Hague: PBL/JRC)

[4] Gabungan Pengusaha Kelapa Sawit Indonesia 2013 *Indonesia dan Perkebunan Kelapa Sawit dalam Isu Lingkungan Global* (Jakarta: GAPKI)

[5] Murdiyarso D 2007 *CDM: Mekanisme Pembangunan Bersih (Seri Perubahan Iklim)* (Jakarta: Penerbit Buku Kompas)

[6] Yang J J, McBride, Zho J and Sun Z 2005 The Urban Forest in Beijing and its Role in Air Pollution Reduction *Urban Forestry & Urban Regreening* **3** 65–78

[7] Gao W, Sugiyama H and Ojima T 1995 Field study of effect of street and its trees on thermal environment of sidewalks. *Journal of Architecture and Planning Environment Engineering* **469** 53

[8] RECOFIC 2012 *Kita, Hutan dan Perubahan Iklim: Peningkatan Kapasitas Akar Rumput untuk REDD+ di Kawasan Asia Pasifik* (Bangkok: RECOFIC)

[9] Freer-Smith P H, Broadmeadow M S J and Lynch J M 2007 Forest and climate change: the knowledge-base for action. In Eds. Freer-Smith P H, Broadmeadow M S J, Lynch J

M *Forestry and Climate Change* (London: CABI International)

[10] Kusmana C and Istomo 1995 *Ekologi Hutan* (Bogor: Laboratorium Ekologi Hutan, Fakultas Kehutanan, Institut Pertanian Bogor)

[11] Chander G, Markham BL, and Helder DL 2009 Summary of current radiometric calibration coefficients for Landsat MSS, TM, ETM+ and EO-1 ALI sensors *Remote Sensing of Environment* **113** 893-903

[12] United States Geological Survey 2015a Landsat 7 Data User Handbook [Online] Available: http://landsathandbook.gsfc.nasa.gov/data_prod/prog_sect11_3.html [Accessed 9 September 2015]

[13] United States Geological Survey 2015b Using The USGS Landsat 8 Product [Online] Available: http://landsat.usgs.gov/Landsat8_Using_Product.php. [Accessed 9 September 2015]

[14] Lillesand TM and Kiefer RW 1997 *Penginderaan Jauh dan Interpretasi Citra* Ed ke-3 Sutanto, penerjemah; Dulbahri, Suharsono P, Hartono, Suharyadi, editor (Yogyakarta: Gadjah Mada University Press.) Translated from: Remote Sensing and Image Interpretation

[15] Bachelet D, Neilson R P, James M, Lenihan J M and Drapek R J 2001 Forest Biodiversity under Global Change, Climate Change Effects on Vegetation Distribution and Carbon Budget in the United States *Ecosystems* **4** 164–85

[16] Yulianti N, Winarna and Murtilaksono K 2009 Perhitungan biomassa dan cadangan karbon dari agroekosistem kelapa sawit, studi kasus di Kebun Meranti Paham dan Panai Jaya, Sumatera Utara. *J. Pen. Kelapa Sawit* **17** 108–13

[17] Li X, Zhou W and Ouyang Z 2013 Relationship between land surface temperature and spatial pattern of greenspace: What are the effects of spatial resolution? *Landscape and Urban Planning* **114** 1–8

[18] [IPCC] Intergovernmental Panel on Climate Change 2014 *Mitigation of Climate Change* (New York: Cambridge University Press.) p 818

Geospatial approach in determining anthropogenic factors contributed to deforestation of mangrove: A case study in Konawe Selatan, Southeast Sulawesi

Dewi Nurhayati Yusuf [1*], L B Prasetyo[2], C Kusmana[3] and Machfud[4]
[1]The Study Program of Natural Resources and Environmental Management, Graduate School of Bogor Agricultural University, Graduate School Building 2nd Floor, Campus of Baranangsiang Bogor 1614, Indonesia
[2]Departement of Forest Resources Conservation and Ecotourism, Faculty of Forestry, Bogor Agricultural University, Jl. Meranti, Dramaga, Bogor 16680, Indonesia
[3]Departement of Silviculture, Faculty of Forestry, Bogor Agricultural University, Jl. Meranti, Dramaga, Bogor 16680, Indonesia
[4]Departement of Agroindustrial Technology, Faculty of Agricultural Technology, Bogor Agricultural University, Jl. Meranti, Dramaga, Bogor 16680, Indonesia

E-mail: wafa.yusuf88@yahoo.com

Abstract. Human activities play an important role in ecosystem damage occurrence, particularly located in rapid developing city. Coastal area is one that has high resource potential, but also susceptible to disturbance since most of the population live in these areas. The development of coastal areas is also influenced by rapid development activities on land and sea. Mangrove is one of the dominant coastal ecosystems, highly productive, complex and high-value benefits. Therefore it is important to learn how anthropogenic factors affect the level of disruption to mangrove ecosystem. In this paper, several factors which considered as dominant have been evaluated through a case study in Konawe Selatan, Southeast Sulawesi. Geospatial model was used to mapping zone with disturbance level based on the anthropogenic factors. The result showed areas with highest disturbance level identified in some areas of District Tinangge; Palangga Selatan; Laeya; Lainea; middle of Kolono Bay; along coast area of Moramo and Moramo Utara. It can be concluded that mangrove ecosystem disruption level due model to anthropogenic factors may determine closely to the factual situation involving socio-economic data. This study provide a guide for future studies on mangrove susceptibility mapping and references in determining the strategy of sustainability mangrove management.

1. Introduction

Human activities play an important role in the occurrence of damage to the ecosystem, especially those located in cities with rapid development areas. Human influence on the ecosystem has been increasing worldwide due to rapid population growth, technology and lifestyle, which have resulted a global biodiversity crisis that threatens the world's species and ecosystems. The coastal area is one that has a very high resource potential, but it also susceptible to damage since most of the population live in there.

The development of coastal area is also greatly influenced by the rapid development activities on land and sea.

Mangrove is one of the dominant coastal ecosystem, highly productive, complex and high-value benefits. It is one of the coastal resources of Indonesia which has huge potential and renewable resources. The mangrove ecosystem has several important functions, such as the ecological, social-economic [1], and physical functions [2]. Mangrove ecosystem also has several benefits, both direct and indirect. Direct benefits of mangrove ecosystem are those which directly perceived its usefulness for human life and measurable, either production or services. Indirect benefits are often difficult to feel and measure, but these benefits actually have a strategic value in supporting human life, such benefits in the areas of research and education, germplasm resources, maintain the climate and hydrology [1]. Direct benefits for human are what generally cause high pressure on the mangrove ecosystem. Unfocused utilization and development not consider to the ecological and physical functions causes high levels of mangrove degradation around the world.

Indonesia is country with the largest mangrove ecosystem in the world [3]. Indonesia is also the world's center of diversity of mangroves [4]. Based on FAO publication [3], in 2003 the estimated area of mangrove in Indonesia about 3.0623 million ha. It covers about 19% of the total area of mangrove in the world. However, it has been significantly degraded. In 1980 the mangrove forests in Indonesia reached 4.2 million ha. There was a decrease of 1.8% in 1990 to 3.5 million ha. Then it continues decreased by 1% in 2000 to 3.15 million ha. In 2005, FAO reported that mangrove forests in Indonesia have only 2.9 million ha remain.

Degradation of mangrove in Konawe Selatan is currently very concerning. In 2009, mangrove area reached 23.195 ha in Distric of Konawe Selatan and approximately 2.940 ha in critical condition [5]. In 2011, Statistical Agency of Konawe Selatan [6] reported that mangrove area is about 15.522 ha, consist of 4.735 ha of forest cultivation area and 10.787 ha of protected forest area. It shows that about 33% mangroves have lost in 2 years.

Mangrove ecosystems are extremely important, but it has been degraded as a result of natural factors and human activities such as urbanization, industrial development, agriculture and aquaculture [7]. Anthropogenic factors identified as the main cause of mangrove degradation in Indonesia. The pressures of human activity on coastal ecosystems are often very high through the competition of land use for aquaculture, agriculture, infrastructure construction and tourism [3]. Mangrove conversions to other uses have recently become very concerning. Rakotomavo and Fromard [8] stated that the mangrove destruction is caused mainly by human activities, therefor it must be controlled to reduce the damages to the environment. Understanding the relationship between human activity and the ecological impact on the mangrove ecosystem, and mapping these impacts could be an important step in the management of mangrove ecosystems. The aims of this study are to identify what are the human activities that most threatening the existence of mangrove ecosystems and develop the spatial model of mangrove disruption level due to anthropogenic factors. Here, we assessed direct human impacts on the mangrove areas using the Human Influence Index (HII) as a proxy [9]. Several factors that are considered as dominant such as access to mangrove area, port activities, aquaculture ponds activities, mining activities, mangrove cover (canopy) density, designation types of forest areas as well as the capacity of local communities have been evaluated through a case study in Coastal Area of Konawe Selatan, Southeast Sulawesi Province.

2. Material and Method

2.1. Study Area

This research was conducted in Coastal Area of Konawe Selatan Region (4° – 4°31' S and 122° – 122°55' E), Southeast Sulawesi Province, on March 2015 to July 2016. This area consists of 8 districts, namely District Tinanggea, Palangga Selatan, Laeya, Lainea, Kolono, Laonti, Moramo and Moramo Utara. Figure 1 shows the location and distribution of ground control points.

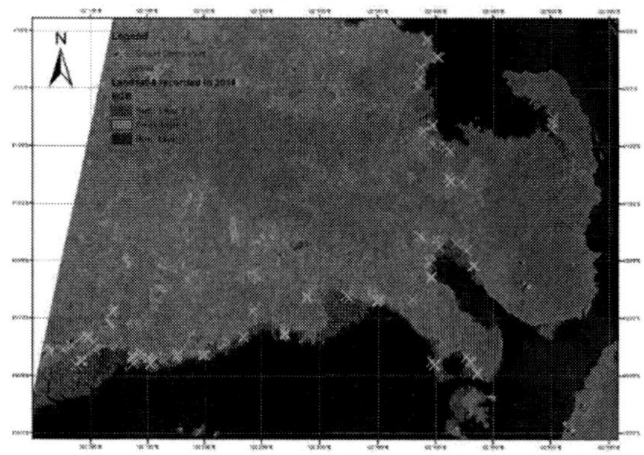

Figure 1. Study area and ground control point.

2.2. Tools and Materials

The tools used in this research include NIKON Digital Camera, Garmin GPS V+, Global Mapper, ESRI ArcGIS 10.1 and ERDAS ER Mapper 9.1 and a set of HP laptop Core i3 RAM 2 GB. The materials used in this research include interview and focus group discussion (FGD) guideline; Ground Control Point (GCP); Landsat-8 raw data full band 15m resolution in panchromatic and 30m resolution in multispectral, path 112 row 63, recorded on October 9th, 2014 in GeoTIFF format; thematic data such as road, river and coastline year 2009 in shape file format obtained from Geospatial Information Agency of Indonesia; port, mining and designation types of forest areas year 2013 in shape file format obtained from Planning and Regional Development Agency of Konawe Selatan Region; population growth and majority type of worker per district year 2014 in tabular data obtained from Statistical Agency of Konawe Selatan Region; and literature.

2.3. Research Procedure

2.3.1. Literature study, FGD and interview

Literature study was obtained from various sources, such as book reports, theses, journals, internet and other documents related to the research subject. Literature study aims to obtain secondary data that can be used in completing the research data. Data and information related to the condition of the coastal environment and public perception obtained through FGD method. Discussions were held by stakeholders, consists of community groups, fishing ponds, entrepreneurs, researchers, government and social/environmental organizations. Interview with the expert aims to get more specific and in-depth information.

2.3.2. Initial processing of satellite imagery
In this step include radiometric correction, geometric correction, composites, fusion and sharpening image contrast. Histogram adjustment method used in radiometric correction, based on several considerations: (1) The process does not use a complicated algorithms and processes can be performed quickly and easily; and (2) The method does not change the value of imagery data pixels, because the authenticity of the pixel values is very important to produce a high accuracy in the spatial information extraction process based on spectral approach, such as digital classification and image transformation process conducted in this research. The nearest neighbour method used in rectification process based on considerations: (1) The process does not use complex mathematical models; (2) Computational process is faster; and (3) Does not change the value of pixel image produced [10].

The colour composite used in image analysis was the original true colour composite RGB channel 321 and false color composite RGB channel 432. Pan sharpened image process conducted on satellite images Landsat-8 multispectral channel which have a 30m spatial resolution with panchromatic channel which has a 15m spatial resolution. The fusion process carried out with the aim to improve the quality of visual display of imagery for the purposes of interpretation and digital classification. Contrast enhancement used to increase the contrast difference between objects on the mainland, coastal and waters. Linear Transformation Method with auto clip of 99% used for contrast enhancement process.

2.3.3. Land cover mapping
Land cover data generated through the processing of Landsat-8 satellite imagery and field survey. At first, supervised classification digital feature extraction method used for land cover classification mapping. These procedure based on the spatial pattern recognition. Based on field knowledge, reclassify the results of the initial land cover classes were not as expected. These procedure also using land use map year 2009 from Geospatial Information Agency as the class reference. Classes of land cover mapped in this research were forest, crop, mangrove, settlement, bush, aquaculture ponds, open land and farm-field. Furthermore, the class of mangrove, settlement and aquaculture ponds thematic layer were used as input parameters in the preparation of the spatial model of mangrove disruption level due to anthropogenic factors.

2.3.4. Survey
Field verification aims to correct the results of remote sensing data processing carried out in the laboratory. Field verification done by the number of survey points (GCP) that correspond to the field conditions using purposive sampling method.

2.3.5. Accuracy assessment
Accuracy assessment aims to measure the validity of the thematic layer data to be accepted as an input parameters in the preparation of model, which were delivered from land cover map. We tested the accuracy of the land cover map resulted from image processing compared with the factual data in the field, through a set of GCP. Accuracy assessment was done using both Confusion Matrix [11] and Kappa Index Method [10].

2.3.6. Mangrove cover (canopy) density
One of the input parameter is the mangrove cover (canopy) density. Layer of canopy density obtained from the polynomial regression correlation between NDVI values with the calculated mangrove canopy density (%) in the field using a measuring instrument canopy density [12]. The calculation formula is as follows (equation 1):

$$Mangrove\ cover\ density\ (\%) = 99.739x2 + 44.942x + 7.4204\ ;\ Where\ x = NDVI \quad (1)$$

2.3.7. Spatial model of mangrove disruption level due to anthropogenic factor.

Spatial model of mangrove disruption level due to anthropogenic factor structured through 7 (seven) input parameters with different weight capacities based on the level of influence as the main driver of mangrove disruption. The process of identification and weighting the parameters based on the study of literature, focus group discussion (FGD), expert's comparative judgement and factual condition in the field. First, literature study listed any of parameters as a driven of disturbance on mangrove ecosystem through human activities. FGD aims to assess what parameters are happening on the site and to get any information from stakeholders about the condition of mangrove ecosystem on the site. Expert's comparative judgement aims to get the information of which parameters considered the most influential among all others, then weighting of each parameter based on the level of disruption through pairwise comparation method [13]. List of 7 parameters representing human disturbance on site and result of weighting of each parameter by the experts presented in table 1.

Table 1. Weight and value of variables in mangrove disruption level due to anthropogenic factor model.

No.		Parameters / Variables	Weight	Value		
				1	2	3
1		Access to the mangrove area (*Lgi.Ak*)	**0,2500**			
	A	Distance from settlement (*Lgi.Akm*)	*0,1125*	4 - 19.9 km	2 - 4 km	0 - 2 km
	B	Distance from main road (*Lgi.Akj*)	*0,0625*	2 - 10.5 km	1 - 2 km	0 - 1 km
	C	Distance from river (*Lgi.Aks*)	*0,0375*	500m - 2.5 km	250 - 500 m	0 - 250 m
	D	Distance from coastline (*Lgi.Akp*)	*0,0375*	500m - 2.5 km	250 - 500 m	0 - 250 m
2		Port activities (*Lgi.Plb*)				
		Distance from port	**0,1000**	2 - 15.5 km	1 - 2 km	0 - 1 km
3		Aquaculture ponds activities (*Lgi.Ptb*)				
		Distance from aquaculture ponds	**0,2000**	4 - 26.9 km	2 - 4 km	0 - 2 km
4		Mining activities (*Lgi.Apt*)				
		Distance from mining	**0,1000**	> 2 km	1 - 2 km	0 - 1 km
5		Mangrove cover (canopy) density (*Lgi.Tjm*)				
		Percentage of mangrove cover (canopy) density	**0,1000**	> 60%	30 – 60%	0 – 30%
6		Conservation Area (*Lgi.Kkv*)				
		Types of designation forest areas	**0,1500**	HSA* & HL**	-	APL***
7		The capacity of local communities (*Lgi.Km*)	**0,1000**			
	a	Population growth (*Lgi.Kml*)	*0,0500*	0 - 3 %	3 - 6 %	> 6 %
	b	Community work in sector: farming, fishing, hunting, forestry, plantation, mining and quarrying (*Lgi.Kmu*)	*0,0500*	0 - 30 %	30 - 60 %	> 60 %
		Cumulative	1,0000			

*HSA = preserves forest; **HL = protected forest; ***APL= another area of use

There were 3 different units in the parameters; distance, percentage and class type. Each layer of input parameter reclassified based on the range value of parameter. The range of each value of parameter defined through study of literature and factual condition on the site. The aims of these step was to get the same unit of each parameter. We were working on cell based to process the spatial data. Geospatial model approach was used to map the zone with the various level of disturbance (high, medium and low), based on the anthropogenic factors. It aims to represent direct human influence over the mangrove area as a gradient ranging from 0 (safest area) to the highest value (highest disturbance area). We used the 30x30m pixel raster map based on the source of the land cover thematic map, Landsat-8, which has 30m spatial resolution. Mangrove disruption level due to anthropogenic factors defined as the total cumulative of value multiplied by weight from each parameter. Based on the table 1, the spatial model of mangrove disruption level due to anthropogenic factor compiled through the following equation 2:

$$\textit{Mangrove disruption level (Lgi)} = (Lgi.Ak \times 0,25) + (Lgi.Plb \times 0,1) + (Lgi.Ptb \times 0,2) + (Lgi.Apt \times 0,1) + (Lgi.Tjm \times 0,1) + (Lgi.Kkv \times 0,15) + (Lgi.Km \times 0,1) \quad (2)$$

3. Result and Discussion

3.1. Land Cover Mapping and Accuracy

Based on classification of Landsat satellite imagery-8 multiband, the total forest area is about 131.695,449 ha, crop 75.322,070 ha, mangrove 11.849,992 ha, settlement 5.979,207 ha, bush 25.074,413 ha, aquaculture ponds 4.202,614 ha, open land 49.536,326 ha and farm-field 22.978,604 ha. The result of land cover mapping in 8 districts of coastal area in Konawe Selatan presented in table 2.

Table 2. Land cover mapping area, producer accuracy, user accuracy, overall accuracy and Kappa Index.

No.	Land cover Year 2014	Area (ha)	Producer Accuracy (%)	User Accuracy (%)
1	Forest	131.695,449	100,00	100,00
2	Crop	75.322,070	81,82	90,00
3	Mangrove	11.849,992	95,83	95,83
4	Settlement	5.979,207	95,00	82,61
5	Bush	25.074,413	85,71	85,71
6	Aquaculture Ponds	4.202,614	87,50	93,33
7	Open Land	49.536,326	80,00	80,00
8	farm-field	22.978,604	80,00	88,89
	Amount	326.638,675		
	Overall Accuracy (%)		89,22	
	Kappa Index		0,87	

Accuracy assessment conducted on land cover map shows that the value of overall accuracy about 89.22%, that's above the minimum tolerance value of 80%. It means the land cover map in accordance with the existing conditions in the field and acceptable validity. In user accuracy assessment of each classes of land cover show that the value range between 80 to 100%, correspond the minimum acceptable value of 80%. This indicates that each layer of land cover assessed has a good accuracy and can be used for analysis. It's very important because some layers such as mangrove, settlement and aquaculture ponds used as an input parameters for the preparation of mangrove disruption level due to anthropogenic factor model. In Kappa Index assessment, the accuracy value is 0.87. Kappa accuracy value is in the range 0-1. Closer to 1 means the accuracy value occurred by chance is 0, or means that the accuracy resulted not by accident.

3.2. Mangrove Cover (Canopy) Density

NDVI is an index measuring the level of greenness of a plant, ranging from -1 (which is generally water) to +1 (heavy vegetation). The NDVI value is defining characteristics of the vegetation, but it does not show the detail part of the vegetation. To get the value of mangrove cover (canopy) density, we proceed the NDVI formula to get the detail value of canopy density. Canopy is the cover of branches and foliage formed by the crown of trees. The percentage area of land covered by the canopy of trees is called canopy density. Vegetation density shows the health status of the forest which depends on the percentage of the canopy cover of the mangrove forest in the area. Dense mangrove shows good health status of mangroves which indicates tall trees, fine distribution, diversity extent and excellent local habitat for the mangrove growth [14]. NDVI was performed for estimating the vegetation cover of the mangrove canopy, this indicates biomass of forest. The NDVI data derived from satellite images was used to compute percentage of canopy closure from 0% to100% [15]. The distribution of mangrove cover (canopy) density can be seen in figure 2.

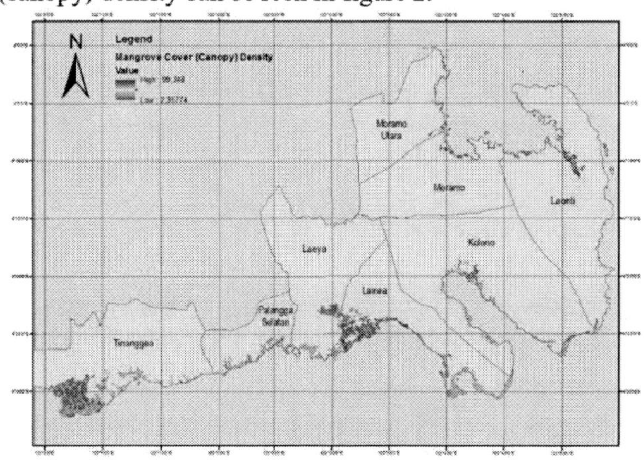

Figure 2. Mangrove cover (canopy) density. Green colours indicating the high density of mangrove, yellow colours indicating the middle density and red colours indicating the low density of mangrove.

3.3. Ecologycal Function of Mangrove

Mangrove usually located in the estuarine areas, which as the final destination of organic particles or silt washed from upstream areas due to erosion. Thus, the mangrove area is a fertile area, both land and waters, caused by the transport of nutrients due to tides [16]. Mangrove forests are communities of tropical and subtropical coastal vegetation consisting of several types of species that can grow and thrive in the muddy coastal tidal areas [17]. Based on the types of trees that making up the formation of mangrove forests from the sea to the mainland, then mangrove can be divided into several zones, namely: Zone *Avicennia sp.,* the area closest to the sea, with a little sandy substrate, often covered by *Avicennia sp.*, the usual association with *Sonneratia sp.* which is dominant to grow on mud rich in organic material; Zone *Rhisophora sp.*, more landward, mangrove forests are generally dominated by *Rhizophora sp.* In this zone also encountered *Bruguiera sp.* and *Xylocarpus sp.*; Zone *Bruguiera sp.*, dominated by *Bruguiera sp.* and Zone *Nypah sp.*, the transition zone between mangrove forest with the usual lowland forest overgrown by Nypah fructicans and some other palm species [17].

Mangrove ecosystem support the conservation of biodiversity, by providing shelter, breeding grounds, nursery and feeding sites of various animal species. Including several endangered animal groups, ranging from the class of reptiles, amphibians, aves and mammals. Mangrove ecosystems can also protect coral reef and sea grass. Various types of fish whether commercial or non-commercial also depend on the existence of mangrove ecosystems [3]. It has been estimated that nearly 90% of all marine animals spend some part of their life cycle within a mangrove ecosystem [18]. Mangroves play an important role within these areas as they guard the low-lying coastal land by forming a protective barrier. This biological barrier reduces damaged caused by storms by limiting wave energy and preventing the land from being flooded. This has become even more apparent after the 2004 Asian tsunami [19]. Mangrove also provide various ecological and economical ecosystem services contributing to coastal erosion protection, provision of building material and medicinal ingredients, and the attraction of tourists [20].

3.4. Human Activities
The influences of human activities are the main drivers of global ecosystem damage [21]. Changes in the size, composition, and distribution of human populations affect coastal regions by changing land use and land cover. Fishing or harvesting, the destruction of mangroves, and pollution and sedimentation from human activities all can affect the coastal environment [22]. Aquaculture, the world's fastest growing food production activity, can lead to the destruction of mangroves and may lead to irreversible damage to both estuarine and offshore fisheries and by modifying habitats [22]. Unconventional mining activity damages of the mangrove forest. Mining activity causes loss and reduces the mangrove area. This condition makes the mangrove lost its function, in ecological, physical, and economical benefit [14]. In Southeast Sulawesi, mangrove mostly converted into fishponds and settlements, the use of firewood, industrial, raw materials household furniture, as well as roads and ports [23].

3.5. Mangrove Disruption Level Due to Anthropogenic Factor
From Landsat satellite year 2014, BPDAS Sampara [5] and Statisitcal Agency of Konawe Selatan Region [6], show that a degradation of mangrove in Konawe Selatan Region around 48% from 2009 to 2014. Access to the mangrove area indicated as the main factor with highest weight (0.25), especially main road. Infrastructure building is one of the important things in development, but also become a bridge that could threaten the natural ecosystems. Aquaculture ponds activities also indicate as the important factor causing the damage of the mangrove with weight of 0.20. Mostly, aquaculture ponds were opened by illegal-cutting of mangrove forest. Once this area cleared, then mangrove area around it becomes more vulnerable to damage. Port activities, mining activities, mangrove cover (canopy) density and the capacity of local communities also contributing as the factor causing the damage of mangrove ecosystems with weight of 0.1. Conservation area is supposed to be a barrier of mangroves destruction, but in this case, designation as protected forest or preserves forest is not help a lot. This shows the weaknesses of the monitoring system and existing of law enforcement. Spatial model of mangrove disruption level due to anthropogenic factor is presented in figure 3.

The result shows the distribution of most and less effected level of disruption. The red colour showing the area with high level of disruption, the middle and low level of disruption showing in yellow to green colours. Areas in circle from figure 3 showing the highest to middle level disruption of mangrove. From the study, the high level identified in some areas, such as most of District Tinangge to the areas bordering the Palangga Selatan District; District Palangga Selatan and Laeya that abut each other; in the middle of coast area of District Lainea; in the middle of Kolono Bay; and along coast area of District Moramo and Moramo Utara.

The shape and nature of the human activities, the distance range to the location and ecological impacts from human activities are important information for the preparation of strategic management and oversight priorities for ecosystem management [24]. This study can be used to meet the priority area of working for the management programmes purpose.

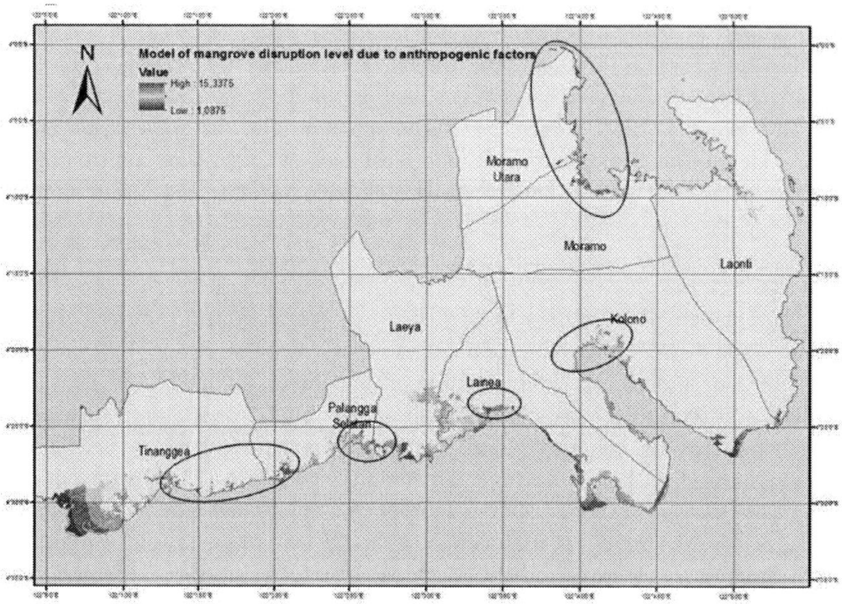

Figure 3. Spatial model of mangrove disruption level due to anthropogenic factor.

4. Conclusion

From the study, it can be concluded that:

1. Human activities that most threatening the existence of mangrove ecosystems are access to the mangrove, specially distance from the settlement and main road; aquaculture ponds activities; and types of designation forest areas.
2. The high level of mangrove disruption identified mostly in District Tinangge to the areas bordering the Palangga Selatan District; District Palangga Selatan and Laeya that abut each other; in the middle of coast area of District Lainea; in the middle of Kolono Bay; and along coast area of District Moramo and Moramo Utara.

5. Suggestion

From the study, we suggest that:

1. The spatial model with cell-based data analysis gives result for preparing zoning system of the mangrove area for management purposes.
2. Spatial model of mangrove ecosystem disruption level due to anthropogenic factors may be determined more closely to the factual situation by involving more recently socio-economic data such as the educational degree of local communities and other human activities spatial layer such as fishing ground and agricultural area.

References

[1] Kustanti A 2011 *Manajemen Hutan Mangrove* (Bogor: IPB Press)
[2] Kusmana C, Wilarso S, Hilwan I, Pamoengkas P, Wibowo C, Tiryana T, Triswanto A, Yunasfi, and Hamzah 2003 *Teknik Rehabilitasi Mangrove* (Bogor: Fakultas Kehutanan IPB)
[3] FAO 2007 *The World's Mangroves 1980 – 2005, A Thematic Study Prepared in the Framework*

of the Global Forest Resources Assessment 2005 FAO Forestry Paper 153 (Rome Italy)
[4] Spalding M D, Blasco F and Field C D 1997 *World Mangrove Atlas* (Okinawa: The International Society for Mangrove Ecosystems)
[5] Badan Pengelola Daerah Aliran Sungai Sampara 2009 *Inventarisasi dan Identifikasi Lahan Kritis Mangrove di Sulawesi Tenggara* Laporan Akhir Kegiatan BPDAS Sampara (Kendari)
[6] Badan Pusat Statistik Kabupaten Konawe Selatan 2013 *Kabupaten Konawe Selatan Dalam Angka* (Konawe Selatan)
[7] Manson F J, Loneragan N R, McLeod I M and Kenyon R A 2001 Assessing techniques for estimating the extent of mangroves: topographical maps, aerial photographs and Landsat TM imagery *Marine and Freshwater Research* **52** 787-792
[8] Rakotomavo A and Fromard F 2010 Dynamics of mangrove forests in the Mangoky River delta, Madagascar, under the influence of natural and human factors *Forest Ecology and Management* **259** 1161–1169
[9] SEDAC 2005 *Last of the Wild Data Version 2 (LWP-2): Global Human Influence Index (HII)* (New York: WCS and CIESIN)
[10] Lilliesand T M, Kiefer R W and Chipman J 2007 *Remote Sensing and Image Interpretation* 6th Edition (New York: John Wiley and Sons)
[11] Sutanto 2013 *Metode Penelitian Penginderaan Jauh* (Yogyakarta: Geography Faculty (BPFG) of Gadjah Mada University)
[12] TOTAL E&P Indonesie 2014 *Mahakam Satellite Imagery Services* Research Report (Balikpapan: PT TOTAL E&P Indonesie)
[13] Saaty T L 1993 *Pengambilan Keputusan Bagi Para Pemimpin, Proses Hirarki Analitik Untuk Pengambilan Keputusan Dalam Situasi Yang Kompleks* (Jakarta: Pustaka Binaman Pressindo)
[14] Sari S P and Rosalina D 2016 Mapping and monitoring of mangrove density changes on tin mining area *Procedia Environmental Sciences* **33** 436–442
[15] Yuvaraj E, Dharanirajan K, Saravanan, Karpoorasundarapandian N 2014 Evaluation of vegetation density of the mangrove forest in South Andaman Island using Remote Sensing and GIS techniques *Int. Res. J. Environment Sci.* **3** (8) 19-25
[16] Gunarto 2004 Konservasi mangrove sebagai pendukung sumber hayati perikanan pantai *Jurnal Litbang Pertanian* **23** (1) 15-21
[17] Bengen D G 2002 *Ekosistem dan Sumberdaya Alam Pesisir* (Bogor : Centre of Coastal and Marine Resources Assessment Bogor Agricultural University)
[18] Benfield S 2002 *An Assessment of Mangrove Cover and Forest Structure in Punta Mala Bay, Panama City, Panama by Means of Field Study and GIS Analysis of Aerial Photographs* (Scotland: Heriot Watt University Edinburgh)
[19] Betts T 2006 *An Assessment of Mangrove Cover and Forest Structure in Las Perlas, Panama* (Scotland : Heriot Watt University Edinburgh)
[20] Kuenzer C, Bluemel A, Gebhardt S, Quoc T V and Dech S 2011 Remote sensing of mangrove ecosystems: a review *Remote Sensing Journal* **3** 878-928
[21] Sanderson E W, Jaiteh M, Levy M A, Redford K H, Wannebo A V and Woolmer G 2002 The human footprint and the last of the wild *BioScience* **52** (10) 891-904
[22] Creel L 2003 *Rippel Effects : Population and Coastal Regions* (Washington DC: Population Reference Bureau – Measure Communication)
[23] Pemerintah Daerah Provinsi Sulawesi Tenggara 2014 *Rencana Strategis Wilayah Pesisir dan Pulau-Pulau Kecil Provinsi Sulawesi Tenggara Tahun 2014 – 2034* (Kendari)
[24] Selkoe K A, Halpern B S, Ebert C M, Franklin E C, Selig E R, Casey K S, Bruno J and Toonen R J 2009 A map of human impacts to a ''pristine'' coral reef ecosystem, The Papahanaumokuakea Marine National Monument *Coral Reefs International Journal* **28** 635-650

Leaf Area Index (LAI) in different type of agroforestry systems based on hemispherical photographs in Cidanau Watershed

Rahmi Nur Khairiah[1,2]*, Yudi Setiawan[1,2], Lilik Budi Prasetyo[1] and Prita Ayu Permatasari[2]

[1] Department of Forest Resource Conservation and Ecotourism, Faculty of Forestry, Bogor Agricultural University, Kampus IPB Dramaga, Bogor, Indonesia

[2] Center for Environmental Research, Bogor Agricultural University, PPLH Building 2nd-4th Floor, Jl. Lingkar Akademik, Kampus IPB Dramaga, Bogor, Indonesia

E-mail: rahmi.khairiah@apps.ipb.ac.id

Abstract. Ecological functions of agroforestry systems have perceived benefit to people around Cidanau Watershed, especially in the protection of water quality. The main causes of the problems encountered in the Cidanau Watershed are associated with the human factors, especially encroachment and conversion of forest into farmland. The encroachment has made most forest in Cidanau Watershed become bare land. To preserve the ecological function of agroforestry systems in Cidanau Watershed, monitoring of the condition of the vegetation canopy in agroforestry systems is really needed. High intensity thinning of crown density due to deforestation can change stand leaf area index dramatically. By knowing LAI, we can assess the condition of the vegetation canopy in agroforestry systems. LAI in this research was obtained from Hemispherical Photographs analysis using the threshold method in HemiView Canopy Analysis Software. Our research results indicate that there are six types of agroforestry in Cidanau Watershed i.e. Sengon Agroforestry, Clove Agroforestry, Melinjo Agroforestry, Chocolate Agroforestry, Coffee Agroforestry, and Complex Agroforestry. Several factors potentially contribute to variations in the value of LAI in different types of agroforestry. The simple assumptions about differences ranges of LAI values on six types of agroforestry is closely related to leaf area and plant population density.

1. Introduction

Agroforestry system is an intensified land use system for swidden agriculture in Indonesia. It is adaptation of shifting cultivation due to high pressure of population in Java. Java is the most populous island in Indonesia. The population density of Java Island is expected to increase from 1,068 per sq. km. in 2010 to 1,304 per sq. km. in 2035 [1]. Kosuke *et. al.,* (2013) [2] explain that increasing population density necessitates the shortening of fallow period in a shifting cultivation cycle. If the fallow period is too short, the forest will end up becoming degraded lands. Classical literature on forest ecology suggests that human intervention threatens the balance of nature of forest ecology. For example, Mikesell, (1960) [3] suggests that the major causes of deforestation have been and continue to be: domestic and industrial consumption of wood, burning to clear land for cultivation, and destruction of palatable plants by livestock. Even in modern literature, major causes of deforestation are believed to include population pressure inducing forest conversion into agricultural land and the demand for fuelwood, development project, logging and forest concession and fire loss [4, 5]

Cidanau Watershed is the only water resources for nearly 100 industries which are operating in Cilegon [6]. Ecological functions of agroforestry systems have perceived benefit to people around Cidanau Watershed, especially in the protection of water quality. The main causes of the problems encountered in the Cidanau Watershed are associated with the human factors, especially encroachment and conversion of forest into farmland. However, opportunities felling of trees in Cidanau Watershed very high. The encroachment has made most forest in Cidanau Watershed become bare land. However, the society as a whole bears the cost of lost biodiversity, global warming, smoke pollution and the degradation of water resources. The condition of seriously damaged watersheds will be a challenge to conserve. To preserve the ecological function of agroforestry systems in Cidanau Watershed, monitoring of the condition of the vegetation canopy in agroforestry systems is really needed.

LAI is defined as one sided is the total area of leaf tissue per unit ground surface area [7]. LAI has emerged as an important quantity for determining the impacts of global environmental change on forests [8] and highly correlated with forest growth and yield [9-11]. LAI is also commonly used as a measure of crop and forest growth and productivity at spatial scales ranging from the plot to the globe [12]. Canopy size determines the productivity of forest stands through its role in radiation interception. Canopy size is often measured as leaf area index, which plays a key role in interception models [13, 14]. High intensity thinning of crown density due to deforestation can change stand leaf area index dramatically. It is also likely that the discontinuous canopy resulting from thinning affects the radiation available to individual trees and may alter the structure of tree crowns and the stand as a whole [15]. By knowing LAI, we can assess the condition of the vegetation canopy in agroforestry systems. The condition of the vegetation canopy LAI would be considered good if the value doesn't decline drastically. If there is a drastic reduction in LAI, then we can know that a decline in leaf area. To measure the LAI of different type of agroforestry Systems over large areas would require measurements at many different locations. Leaf area estimation can be performed by direct or indirect methods. Direct methods are expensive and complex, since the sampling restricts the amount of measurements to be performed. On the other hand, indirect methods provide non-destructive and reliable results [16]. Hemispherical Photographs was one of the first methods used to indirectly estimate LAI and well established method to optically assess ecological parameters related to plant canopies [17]. The acquisition of high definition photos under the canopy is quickly analyzed by proprietary software, taking as a basis algorithms with the zenith angle, the light attenuation and the contrast between plants and sky. The advantages of using Hemispherical Photographs, including high-resolution images to LAI estimation are discussed by [18-20]. The objective of this research was to describe the characteristics of LAI in Different Type of agroforestry systems based on Hemispherical Photographs.

2. Data and methods

2.1. Study area

Cidanau Watershed is located in Serang District and Pandeglang District, Banten Province, Indonesia (06°07'30"-06°18'00"S and 105°49'00"-106°04'00"E). Cidanau Watershed cover an area of 22,620 Ha, which encompasses 999.29 Ha area of Pandeglang District and 21,620.71 Ha area of Serang District. Cidanau Watershed has the various topographical condition, from 0% up to 135%.

2.2. Data collection

Field observation was carried out to collect The Hemispherical Photographs data, names of plant species (ranging from samplings, pole, and tree) and ground control points (GCP). Sample points were then placed based on the 10 classes of stand density types, under the stratified random sampling method. In total, 140 plots were used for sampling, distribution of sample points can be seen in Figure 2.

Each observation plots represented one measurement plot of which Huang *et. al.*, (2006) [40] suggested that measurement plot which used a 30 meter x 30-meter resolution image should take a dimension of 50 meters x 50 meters. Going forward, estimating LAI will be done using Landsat Imagine, so as to the size of the sample plot, adapted to the size of the pixels on the Landsat data. The data collection was conducted on the structure of tree species growing stakes, poles, and trees with a trunk diameter ≥ 1.5 cm [41]. Tree type identification was conducted by observing the characteristics of each species through generative organs such as fruit and flower seeds or vegetative parts such as stems and leaves.

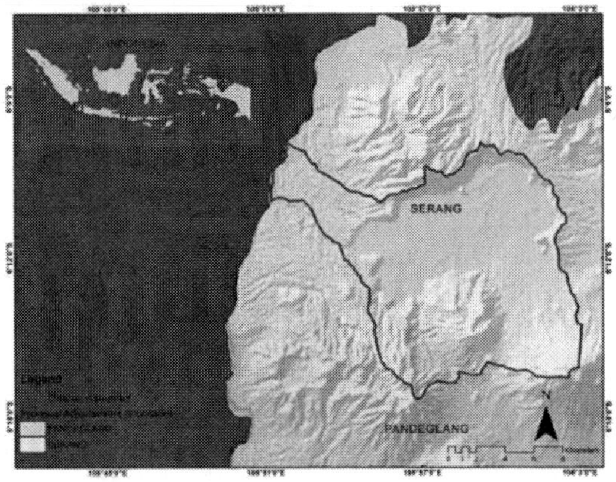

Figure 1. The map of study area

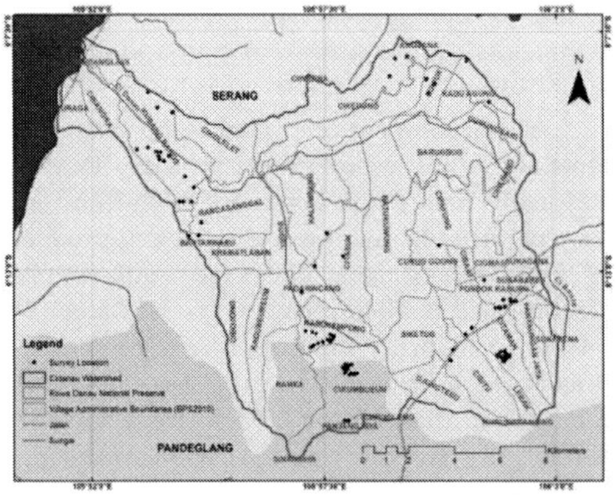

Figure 2. The map of distribution of sample plots

Hemispherical Photographs data collection was conducted by using a DSLR camera with a fisheye lens. Hemispherical Photographs was conducted at the midpoint of the plot with the camera facing upwards on a tripod with a height of ± 1.5 meters to avoid shrubs blocking the view and monitor the camera facing towards the north compass [21]. The camera can be raised or lowered to the proper height. In the case of the camera is very close to the ground, it is possible to squat or even lie to get out of the field of view lens [22]. Then the camera is positioned correctly and leveled before the photo was taken. The proper position for taking photos that include location horizontally and has a flat field, while the camera orientation refers to the rotation of the camera relative to the north, it is adjusted so that magnetic north is straight to the top of the image [23].

2.3. LAI calculation

LAI in this research was obtained from Hemispherical Photographs (figure 3) analysis using the threshold method in HemiView Canopy Analysis Software [6]. The threshold was increased or decreased until a match of classification result figure and the original figure was found, and so a clear boundary of the canopy covered area and open area was obtained. High quality figures with a clear boundary between canopy covered area and open area are needed to minimize subjectivity of this method [22]. The LAI estimated by the inversion process may not be an exact measure of the LAI of the real canopy, even if that could be measured accurately. However, an ideal canopy with the LAI estimated will behave in a very similar way to the real canopy as far as radiation interception is concerned.

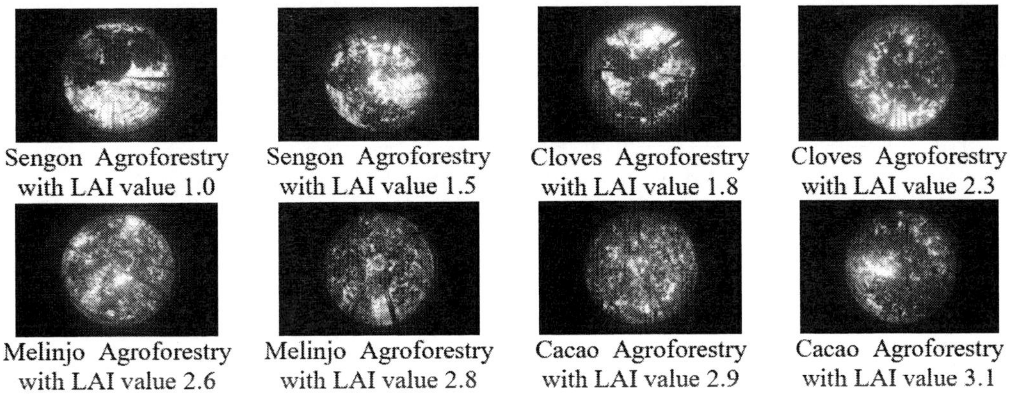

Figure 3. Hemispherical photographs in different type of agroforestry systems

Agroforestry is planted with a mixture of many species of trees but may be dominated by one species, e.g. bamboo, in which case it is named after this species, so the name is bamboo agroforestry [2]. Analysis of the type of agroforestry was conducted by quantifying a number of major crops in each plot. Each type of agroforestry reflects the dominant plant that assembles which in this study accounted more than 80% stands on the land. If there is no major crops then it is will described as a Complex Agroforestry. Complex Agroforestry is defined as forest structures planted and managed by farmers for the production of various forest and agricultural products on the same piece of land [24]. Complex Agroforestry Systems in intimately association with a high number of components (trees, lianas, herbs), and either primary or secondary forests are Able to sustain both biodiversity conservation and farmers' economic needs [25].

3. Result and discussion

Agroforestry system combining forestry component (woody plants) to the agricultural component (or non-wood plant). Woody plants intended long rotation (tree crops) and non-timber plants of this type of seasonal crops (annual crops). The advantage of agroforestry systems that was felt by the farmers is agroforestry systems have composition of diverse plants and can be used by the owner (farmer) at any time in accordance with the wishes of the owner.

Cidanau Watershed is the only water resources for nearly 100 industries which are operating in Cilegon [6]. Ecological functions of agroforestry systems have perceived benefit to people around Cidanau Watershed, especially in the protection of water quality. The role of the community in the upstream felt very important to communities downstream. Water quality protection requires a multi sectoral approach are integral in the management of Cidanau Watershed. Rivers through agricultural land is often not overgrown vegetation on the banks of rivers and streams zone of surface water containing excessive fertilizer and pesticide. Agroforestry system, can be said to be like a forest buffer on the river bank, it is very effective in reducing water pollution from agricultural activities. One of the efforts for soil preservation or soil erosion control is by means of vegetative, i.e. planting forestry plants, planting a cover cropland and planting crops is parallel to contour lines, planting crops in the strip, planting crops in rotation, and mulching or plant biomass utilization [26].

A total of 20,091 trees from 79 species were recorded in the 140 survey plots, there are: Akasia (*Acacia auriculiformis* A. Cunn. ex Benth.), Angsana (*Pterocarpus indicus* Willd.), Asam Jawa (*Tamarindus indica* L.), Asam Kranji (*Dialium indum* L.), Bacang (*Mangifera foetida* Lour), Balsa (*Ochroma pyramidale*), Bayur (*Pterospermum javanicum* Jungh.), Belimbing Wuluh (*Averrhoa bilimbi* L), Bisbul (*Diospyros blancoi* A. DC.), Bungur (*Lagerstroemia loudonii* T.& B), Cempedak (*Artocarpus integer*), Clove (*Syzygium aromaticum* (L.) Merr. & L. M. Perry), Dadap Merah (*Erythrina cristagalli* L.), Duku (*Lansium domesticum* Corr), Durian (*Durio zibethinus* Murr), Flamboyan (*Delonix regia*), Gamal (*Gliricidia maculata*), Gandaria (*Bouea macrophylla* Griff.), Huru

(*Macaranga rhizinoides* (Blume) Muell Arg.), Jabon (*Anthocephalus cinensis*), Jambu Air (*Eugenia aquea* Burm.F), Jambu Batu (*Psidium guajava* L.), Jarak Pagar (*Jatropha curcas* L.), Jati (*Tectona grandis* L.f.), Jati Putih (*Gmelina arborea* Roxb.), Jengkol (*Archidendron jiringa (Jack) I.C.Nielsen.*), Jeruk lemo (*Citrus amblycarpa* (Hassk.) Ochse), Cacao (*Theobroma cacao* L.), Kamboja (*Plumeria acuminata* Ait), Karet (*Hevea brasiliensis* Muell. Arg), Karet Kebo (*Ficus elastica* Roxb. ex Hornem), Kayu Afrika (*Maesopsis eminii* Engl.), Kayu Manis (*Cinnamomum burmannii* (Nees &Th. Nees)), Kayu Pedang (*Oroxylum indicum* (L.) Vent.), Kecapi (*Sandoricum koetjape* (Burm.F.) Merr), Kedondong (*Spondias dulcis*), Kedoya (*Dysoxylum gaudichaudianum* (A.Juss.) Miq.), Kelapa (*Cocos nucifera* L.), Kembang Sepatu (*Hibiscus rosa*-sinensis L.), Kemiri (*Aleurites moluccana* (L.) Willd), Kenanga (*Cananga odorata* (Lamk.) Hook.), Kenari (*Canarium indicum*), Kepuh (*Sterculia foetida* L.), Kersen (*Muntingia calabura* L.), Keruing (*Dipterocarpus cornutus* Dyer), Coffee (*Coffea arabica* L.), Laban (*Vitex pubescens* Vahl.), Mahoni Daun Besar (*Swietenia macrophylla* King.), Mahoni Daun Kecil (*Swietenia mahagoni* (L.) Jacq.), Maja (*Aegle marmelos* (L.) Corr), Mangga (*Mangifera indica* L.), Manggis (*Garcinia mangostana* L.), Manglid (*Magnolia Blumei* Prantl.), Mareme (*Glochidion borneense* Boerl.), Melinjo (*Gnetum gnemon* L.), Mengkudu (*Morinda citrifolia* L.), Meranti Tembaga (*Shorea leprosula* Miq.), Mindi (*Melia azedarach* L.), Nangka (*Artocarpus heterophyllus* Lam), Pachira (*Pachira affinis* Decne), Pala (*Myristica fragrans* Houtt), Palem Jawa (*Ceratolobus glaucescens*), Pepaya (*Carica papaya* L.), Petai (*Parkia speciosa* Hassk), Pinang (*Areca catechu* L.), Pisang (*Musa paradisiaca*), Pulai (*Alstonia scholaris* R. Br.), Pulus (*Laportea stimulans* (Lf) Gaud), Rambutan (*Nephelium lappaceum* L.), Randu (*Ceiba pentandra* L. Gaertn), Sagu (Metroxylon sagu Rottb.), Salam (Syzygium polyanthum Wigh Walp), Sawo Duren (Chrysophyllum cainito L.), Sawo Kecik (*Manilkara kauki* Dub), Sengon (*Paraserianthes falcataria* (L.) Nielsen), Singkong (*Manihot esculenta* Crantz), Sukun (*Artocarpus communis* Forst), Tanjung (*Mimusops elengi* L.), and Waru Gunung (*Hibiscus macrophyllus* Roxb. ex Hornem). These plants are grown there and there are growing naturally in farmers' fields. Clove (*Syzygium aromaticum* (L.), Cacao (*Theobroma cacao* L.), Melinjo (*Gnetum gnemon* L.) dan Sengon (*Paraserianthes falcataria* (L.) Nielsen)) are booming because of their good price in the markets. Even on smaller farms with a priority on crop production, planting tree with high economic value on boundaries away from crops or at low densities seems to be an attractive option to enhance income.

Turner *et. al.*, (1999) [27] has conducted research on various types of tropical forest vegetation and crop areas (gardens) with a calculation method directly in the field. The study involved three National Science Foundation Long Term Ecological Research sites differing widely in vegetation physiognomy, climate, and topography. Vegetation types at the three sites included grassland, shrub-land, hardwood forest, and conifer forest. At each site, measurements of LAI were made at a number of plots with locations determined by the Global Positioning System (GPS). Furthermore, another study shows that the in situ measurements for LAI in evergreen broadleaf forest in East Africa vary between 3.95 and 7.4 [28]. Global averages from ground measurements of tropical evergreen broadleaf forests around 4.9 [29] and Aragao *et. al.,* (2005) [30] measured slightly lower LAI values for the Amazonian evergreen broadleaf forest between 3.25 and 5.1.

Table 1 Value ranges in tropical forests [27]

Number	Vegetation Types	Number of Samples	Estimation Method of LAI	Ranges of LAI
1	Burned grassland	6	Clipping	2.5-6.3
2	Unburned grassland	7	Clipping	2.5-3.2
3	Broadleaf forest	14	Litterfall	4.4-8.4
4	Conifer forest - Young	10	Litterfall	1.4-3.9
5	Shrub-land	3	Plant Cover	1.0-3.3
6	Conifer Forests - Young	6	Sapwood Area	5.3-9.6
7	Conifer Forests - Old	7	Sapwood Area	7.9-13.0

LAI in our research was obtained from Hemispherical Photographs by using the threshold method in HemiView Canopy Analysis Software. Each one plot represented by a Hemispherical Photographs which is then analyzed into one LAI value. Thus, in this study we have 140 LAI values are grouped into several types of agroforestry systems based on the calculation of plant components in the plot. Once grouped, we describe a range of values LAI on any type of agroforestry systems.

Our research results indicate that there are six types of agroforestry in Cidanau Watershed i.e. Sengon Agroforestry, Clove Agroforestry, Melinjo Agroforestry, Chocolate Agroforestry, Coffee Agroforestry, and Complex Agroforestry. Complex agroforestry is a dominant type of agroforestry in Cidanau Watershed, this is because the types of crops in agroforestry is very diverse. Complex Agroforestry consists of 11-36 plant species, so the ranges bound of the LAI value is very difficult to observe. LAI Ranges values in Coffee Agroforestry can not be known, this is because only one number of samples have been found. As can be seen in Table 2, LAI interval of Sengon Agroforestry (1.0 -1.5), Cloves Agroforestry (1.8-2.3), Melinjo Agroforestry (2.6-2.8), Cacao Agroforestry (2.9-3.1), and the Complex Agroforestry (1.0-5.3). This range value showed minimum and maximum range of LAI values forming fairly open to very dense canopy in six types of agroforestry in Cidanau Watershed. These results can be used to monitor vegetation conditions based on the value of LAI in Cidanau Watershed.

Table 2. Ranges of LAI in different type of agroforestry

Number	Agroforestry Type	Number of Samples	Estimation Method of LAI	Ranges of LAI
1	Sengon Agroforestry	4	Hemispherical Photographs Analysis	1.0-1.5
2	Cloves Agroforestry	5	Hemispherical Photographs Analysis	1.8-2.3
3	Melinjo Agroforestry	6	Hemispherical Photographs Analysis	2.6-2.8
4	Cacao Agroforestry	5	Hemispherical Photographs Analysis	2.9-3.1
5	Coffee Agroforestry	1	Hemispherical Photographs Analysis	2.5
6	Complex Agroforestry	119	Hemispherical Photographs Analysis	1.0-5.3

Several factors potentially contribute to variations in the value of LAI in different types of agroforestry. The simple assumptions about differences ranges of LAI values on six types of agroforestry is closely related to leaf area and plant population density. This can be explained by more extensive the leaf area, leaf density or plant population density, the higher the value of LAI. This can happen due to the more extensive and dense, the more leaves shade each other, so that the area of leaves will accumulate and become larger. This is consistent with the statement Koesmaryono, (1996) [31], which explained that the increase in plant populations will increase the value of LAI. However, additional research needs to to prove the relationship between LAI values in any type of agroforestry system with leaf area and plant population density. At the leaf level, these include influences associated with variation in leaf pigments, leaf internal structure, and the orientation of the leaf relative to solar radiation [32-35]. At the multi tree scale, heterogeneity in tree height and the number and size of tree gaps influences reflectance [36-39].

4. Conclusion

The results of this study showed that six types of agroforestry system applied in Cidanau Watershed by farmers. LAI ranges value of five agroforestry types has been obtained, namely Sengon Agroforestry (1.0-1.5), Cloves Agroforestry (1.8-2.3), Melinjo Agroforestry (2.6-2.8), Cacao Agroforestry (2.9-3.1), and Complex Agroforestry (1.0-5.3). Several factors potentially contribute to variations in the value of LAI in different types of agroforestry. The simple assumptions about differences ranges of LAI values on six types of agroforestry is closely related to leaf area and plant population density. However, additional research needs to to prove the relationship between LAI values in any type of agroforestry system with leaf area and plant population density.

Acknowledgments

We would like to thank the Center for Environmental Research, Bogor Agricultural University (PPLH-IPB), Indonesia, for giving us the opportunity to obtain funding research for publication.

References

[1] Ferraris J, Makalew R J, Posselt H and Harahap D P 2015 The 2010–2035 Indonesian Population Projection: Understanding the Causes, Consequences and Policy Options for Population and Development. UNFPA or the Government of Indonesia.
[2] Kosuke M, Mugniesyah S S, Herianto A S and Hiroshi T 2013 Talun-huma, swidden agriculture, and rural economy in West Java, Indonesia *Southeast Asian Studies*. **2** 351–381
[3] Mikesell and Mike W 1960 Deforestation in Northern Morocco *Science* **132** 441-445
[4] World Bank 1990 *Indonesia: Sustainable Development of Forests, Land and Water* (Washington, D.C.: The World Bank)
[5] World Resources Institute 1991 *World Resources 1990-1991* (New York: Oxford University Press)
[6] Khairiah R N, Prasetyo L B, Setiawan Y and Kosmaryandi N 2016 Monitoring model of payment for environmental service (PES) implementation in Cidanau Watershed with stands density approach *Procedia Environmental Sciences*. **33** 269-278
[7] Chen J M and Black T A 1992 Defining leaf area index for non-flat leaves plant *Cell Environ*. **15** 421–429.
[8] Thomas, Sean C and Winner William E 2000 Leaf area index of an old-growth Douglas-fir forest estimated from direct structural measurements in the canopy *Canadian Journal of Forest Research*. **30** 1922-1930
[9] Waring R H 1983 Estimating forest growth and efficiency in relation to canopy leaf area *Adv. Ecol. Res.* **13** 327-354.
[10] Fownes J H and Harringron R A 1990 Modelling growth and optimal rotations of tropical multipurpose trees using unit leaf rate and leaf area index *J. Appl. Ecol.* **27** 888-896.
[11] Harringron R A, Fownes J H, Meinzer F C and Scowcroft P G 1995 Forest growth along a rainfall gradient in Hawaii: Acacia koa stand structure, productivity foliar nutriens, and water and nutrient use efficiencies *Oecologia*. **102** 227-284.
[12] Anonymous 2014 *LAI Theory and Practice* (New York: Decagon Devices, Inc)
[13] Jarvis P G and Leverenz J 1983 Productivity of temperate, deciduous and evergreen forest. In *Encyclopedia of Plant Physiology. New Series,* Vol. 12D. Eds. O.L. Lange, P.S. Nobel, C.B. Osmond and H. Ziegler. (Berlin: Springer-Verlag) pp 233–80.
[14] Landsberg J J and Hingston F J 1996 Evaluating a simple radiation/dry matter conversion model using data from Eucalyptus globulus plantations in Western Australia *Tree Physiol.* **16** 801–808.
[15] Medhurst J L and Beadle C L 2001 Crown structure and leaf area index development in thinned and unthinned *Eucalyptus nitens* plantations *Tree Physiology* **21** 989–999
[16] Brandão and Zonta 2016 Hemispherical photography to estimate biophysical variables of cotton. *Rev. bras. eng. agric. ambient*. **20** 789-794.
[17] Glatthorn J and Philip B 2014 Standardizing the protocol for hemispherical photographs: accuracy assessment of binarization Algorithms *PLoS One*. **9** e111924.
[18] Jonckheere I, Fleck S, Nackaerts K, Muys B, Coppin P, Weiss M and Baret F 2004 Review of methods for in situ leaf area index determination. Part I. Theories, sensors and hemispherical photography *Agricultural and Forest Meteorology*. **121** 19-35.
[19] Chianucci F and Cutini A 2012 Digital hemispherical photography for estimating forest canopy properties: Current controversies and opportunities *Forest Biogeosciences and Forestry*. **5** 290-295.
[20] Zhao, F., Strahler, A. H., Schaaf, C. L., Yao, T., Yang, X., Wang, Z., Schull, M. A., Román, M. O., Woodcock, C. E., Olofsson, P., Ni- Meister, W., Jupp, D. L. B., Lovell, J. L., Culvenor, D. S.,
[21] Djumaher M 2003 Pendugaan leaf area individual dan stands basal area menggunakan Landsat 7 ETM+. *Bachelor Thesis* (Bogor: Institut Pertanian Bogor)

[22] Rich P M, Wood J, Vieglais D A, Burek K. and Webb N 1999 Hemiview User Manual. United of Kingdom: Delta-T Devices LTD. **106** 285–304.
[23] Rich P M 1990 Characterizing plant canopies with hemispherical photographs *Remote Sensing Reviews* **5** 13-29.
[24] De Foresta,H and Michon G 1996 Tree improvement research for agroforestry: A note of caution *Agroforestry Forum.* **7** 8-10.
[25] De Foresta H and Michon G 1994 Agroforest: an original agro-forestry model from smallholder farmers for environmental conservation and sustainable development. In *Traditional Technology for Environmental Conservation and Sustainable Development in the Asian-Pacific Region. Japan* pp 52-58.
[26] Senoaji G 2012 Pengelolahan lahan dengan sistem agroforestri oleh masyarakat baduy di banten selatan *Bumi lestari.* **12** 283-293.
[27] Turner D P, Cohen W B, Kennedy R E, Faanacht K S and Brings J M 1999 *Relationship Between Leaf Area Individual and Landsat TM Spectral Vegetation Individuals across Three Temperate Zone Sites* (New York: Elsevier Science Inc.)
[28] Kraus T 2008 Ground-Based Validation of the MODIS Leaf Area Index Product for East African Rain Forest Ecosystems. *Ph.D. Thesis*, Friedrich-Alexander (Erlangen: Universität Erlangen-Nürnberg Germany)
[29] Wang Y, ian Y, Zhang Y, El-Saleous N, Knyazikhin Y, Vermote E and Myneni R B 2001 Investigation of product accuracy as a function of input and model uncertainties: Case study with SeaWiFS and MODIS LAI/FPAR algorithm *Remote Sens. Environ.* **78** 299–313.
[30] Aragao L E O C, Shimabukuro Y E, Espirito-Santo F D B and Williams M 2005 Spatial validation of the collection 4 MODIS LAI product in eastern Amazonia. IEEE Trans. *Geosci. Remote Sens.* **43** 2526–2534.
[31] Koesmaryono Y, Haseba T, Sugimoto H and Ito D 1996 Growth and yield of soybean cultivated under different population density with special reference to unusual summer eather in 1994. In Crop research in Asia: Achievements and perspective. *Proc. 2nd Asian Crop Sci. Conf.* 478-479.
[32] Baret F and Guyot G 1991 Potentials and limits of vegetation indices for LAI and APAR assessment. *Remote Sens. Environ.* **35** 161–173.
[33] Williams D L 1991 A comparison of spectral reflectance properties at the needle, branch, and canopy level for selected conifer species *Remote Sens. Environ.* **35** 79–93.
[34] Bouman B A 1992 Accuracy of estimating the leaf area index from vegetation indices derived from crop reflectance characteristics, a simulation study. *Int. J. Remote Sens.* **13** 3069–3084
[35] Yoder B J and Waring R H 1994 The normalized difference vegetation index of small Douglas-fir canopies with varying chlorophyll concentrations. *Remote Sens. Environ.* **49** 81–91
[36] Guyot G, Guyon D and Riom J 1989 Factors affecting the spectral response of forest canopies: a review. *Geocarta Int.* **4** 3–18.
[37] Cohen W B and Spies T A 1992 Estimating structural attributes of Douglas-fir/western hemlock forest stands Landsat and SPOT imagery. *Remote Sens. Environ.* **41** 1–17.
[38] Cohen W B, Spies T A and Bradshaw G A 1990 Semivariograms of digital imagery for analysis of conifer canopy structure *Remote Sens. Environ.* **34** 167–178.
[39] Leblon B, Gallant L and Grandberg H 1996 Effects of shadowing types on ground-measured visible and near-infra- red shadow reflectances *Remote Sens. Environ.* **58** 322–328.
[40] Huang D, Yang W, Tan B, Rautiainen M and Zhang P 2006 The importance of measurement errors for deriving accurate reference leaf area index maps for validation of moderate resolution satellite LAI products. J. IEEE Transactions *Geoscience and Remote Sensing.* **44** 1866-1871
[41] Standar Nasional Indonesia 2011 *Pengukuran dan Perhitungan Cadangan Karbon-Pengukuran Lapangan untuk Penaksiran Cadangan Karbon Hutan.* (Jakarta: Badan Standarisasi Nasional) **25** 320–333

Use of Geographical Information System (GIS) and remote sensing in development of urban forest types and shapes in Tangerang Selatan City

Gumilar Adam, Rachmad Hermawan* **and Lilik Budi Prasetyo**
Urban Forest and Environmental Services Division, Forest Resources Conservation and Ecotourism, Forestry Faculty Bogor Agricultural University Darmaga Campus, Bogor, 16680, Indonesia

E-mail: racher67@gmail.com

Abstract. The development of a city could create adverse effects, such as increase in air pollutant, and decrease in amenity. One of the ways to overcome these adverse effects is developing urban forest. For maximizing the function of urban forest, the appropriate types of urban forest to be developed, should be determined first. The aims of this study were to determine the appropriate types of urban forest and to identify the shape of urban forest in Tangerang Selatan City by using GIS and remote sensing. Urban forest shape was identified on the basis of shape and distribution of land unit. The steps of the study comprised data collection, map preparation and data analysis. Landsat 8 satellite imagery was interpreted for land use/cover classification. Scoring based on air temperature, land slope, and soil types was used to determine priority of the urban forest locations. Besides that, land use planning was considered to determine the appropriate urban forest type. The results of the study show that the appropriate urban forest types are residential area urban forest, industry urban forest, and recreation urban forest. On the other hand, the appropriate urban forest shapes are strip, scattered, and clustered pattern.

1. Introduction

City is an area which constitutes a center of activities with main activities in the form of service and trade. Consequence of being the center of activities is that the city area exhibits rapid development in economic sector and often be accompanied with problems of environmental quality deterioration, such as increasing in air pollution, occurrence of urban heat island (UHI), decreasing in convenience and decreasing in biodiversity [1, 2, 3]. Such city or urban problem should be controlled so that the area does not go toward necropolis condition [4].

Tangerang Selatan town is a satellite city situated around Jakarta city. The city exhibits rapid development as shown by development of shopping centers and residential areas, traffic jam prevalence, decreasing in biodiversity and others, so that the city possesses problems similar with those of other big cities.

One of the efforts to overcome problems in urban areas is development of urban forest. According to RI Government Regulation Number 63 Year 2002 [5], urban forest was defined as land covered with tree growth in public or private land, which is formally established as urban forest by authorized official, with minimum area size of 0.25 ha. Furthermore, Irwan [6] defined urban forest more completely as vegetation community in the form of trees and their associates which grow in urban land or around urban area, and could be in the form of strips, scattered spots, or clumps with structure resembling natural forest, creating habitat which facilitate life for wild animals and create aesthetics and healthy and convenient environment. Urban forest exhibits social benefits, aesthetics, recreation, and phenomena of engineering, economy and ecology.

For creating maximum function, development of urban forest should be preceded by determining the type of the urban forest. Determination of urban forest type is based on the objective of urban forest development and objects which will be protected, while considering problems and potential problems existing in the area and functional potency of the area. According to RI Government Regulation Number 63 Year 2002 [5], urban forest types comprise residential area type, industrial area type, recreation type, germplasm conservation type, protection type, and security type. Shapes of urban forest to be developed is based on forms of land unit and distribution of land unit in the urban areas [5,6], whereas Fahutan IPB [7] distinguished shapes of urban forest on the basis of location, important function, vegetation, and management intensity.

GIS and remotes sensing are valuable tools in planning urban forest. Rushayati et al. [1] used these tools in determining location priority of green open spaces based on surface temperature in Bandung Regency. These tools also were used by Humaida et al. [8] in determining location priority of green open spaces based on vegetation density (NDVI), temperature humidity index (THI), population density, and land price in Banjarbaru City.

The objectives of this study were determining the appropriate types of urban forest and identifying shapes of urban forest in Tangerang Selatan City by using GIS and remote sensing. The benefit of this study is to provide an input for the local government in determining types and shapes of urban forest to be developed.

2. Method

2.1. Location and time of study

This study was conducted in Tangerang Selatan City, the province of Banten (figure 1) from August through December 2015. Data processing was conducted in Laboratory of Environmental Analysis and Spatial Modeling, Department of Forest Resources Conservation and Ecotourism, Faculty of Forestry, Bogor Agricultural University (IPB).

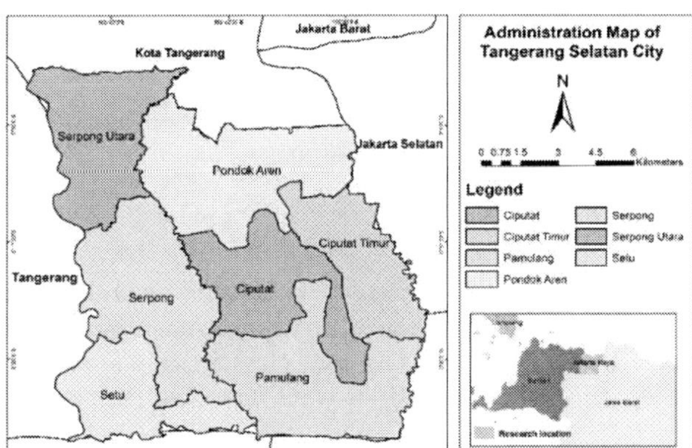

Figure 1. Map of Tangerang Selatan City

2.2. Materials and equipments

Equipment being used in this study were pro-summer camera for activities documentation, GPS (Global Positioning System), Microsoft Office software, Arc GIS 10.2 software, Erdas Imagine 9.1 software. Materials being used were: (1) from BAPPEDA (Municipal Development Planning Agency) of Tangerang Selatan City namely Administration Map of Tangerang Selatan City, Land use Map of Tangerang Selatan City of Year 2011, Soil Types Map of Tangerang Selatan City of Year 2011;

(2) Landsat 8 satellite imagery with acquisition date 13th September 2014 Path/Row 122/064; and (3) Digital Elevation Model (DEM) produced by Shuttle Radar Topography Mission (SRTM).

2.3. Stages of Study

2.3.1. Data collection

Data collection was conducted by literature review, observation and ground check. The types of data collected were air temperature, rainfall, soil types, land use, topography, and land cover.

2.3.2. Preparation of maps

2.3.2.1. Map of air temperature

Air temperature is classified into three classes, namely below convenient level (< 24 °C), convenient level (24°C - 28°C) and above convenient level (> 28 °C) [9]. Determination of surface temperature value used Erdas Imagine 9.1 software. Afterwards, a model was constructed in the available model maker to convert pixel values at landsat 8 and band 10. DN (Digital Number) value was converted to radiation value [10]. The next stages were as follows: calculation of albedo, calculation of net radiation, calculation of reflected short wave radiation, calculation of incoming short wave radiation, calculation of reflected long wave radiation, estimation of Normalized Differential Vegetation Index (NDVI), calculation of soil heat flux, calculation of sensible heat flux [11], and determination of values of temperature [12].

2.3.2.2. Maps of rainfall, soil types and land use.

Maps of rainfall, soil types and land use were obtained from BAPPEDA (Municipal Development Planning Agency) of Tangerang Selatan City.

2.3.2.3. Topographic/slope map

Topographic map was processed from DEM produced by SRTM. Slope classes were categorized into seven classes [13], namely: (a) nearly level: 0 - 3%; (b) gently sloping/undulating: 3 - 8%; (c) rolling: 8 - 15%; (d) hilly: 15 - 30%; (e) moderately steep: 30 - 45%; (f) steep: 45 - 65%; (g) very steep: > 65%. The creation of such map was conducted by processing DEM produced by SRTM, into slope map.

2.3.2.4. Land use/cover

This map was obtained by interpreting landsat 8 imagery year 2014 with acquisition date 13th September 2014 Path/Row 122/064 with Supervised Classification Method. Land cover were classified into water bodies, tree vegetation, non tree vegetation, and built-up.

2.4 Data analysis

2.4.1 Determination of priority location of urban forest.

Determination of priority location of urban forest was conducted by scoring techniques [8]. For each criteria being used (temperature map, soil type map, and slope classes map) there was classification and labeling of different values. Labeling of values for each class referred to research by Humaida et al. [8] and Kridalaksna [14]. Stages in determination of types and shapes of urban forest are presented in figure 2.

Combination of the three criteria would produce maximal score 9 and minimal score 3. Maximal and minimal values (scores) were divided into 3 ranges, namely score ≥3 - <5, ≥5 - <7 and ≥7 - 9.

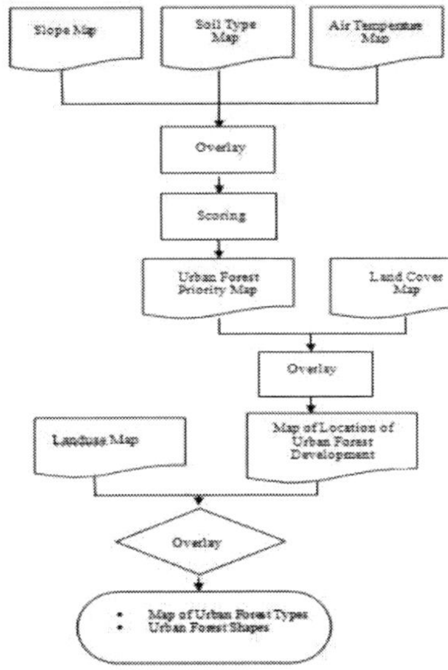

Figure 2. Stages in determination of types and shapes of urban forest

Those scores will be used as reference for determining location of urban forest in Tangerang Selatan city. Those ranges will serve as reference in determination of location priority, where the first priority was for area which have score ≥7 – 9; second priority for area with score ≥5 - <7, and third priority for area with score ≥3 - <5 as shown table 1.

Table 1. Value for each criteria for determination of urban forest location

No.	Criteria	Classes	Score
1	Temperature	> 28 °C	3
		< 24 °C	2
		24 °C - 28 °C	1
2	Land slope	> 15 %	3
		8 - 15 %	2
		0 - 8 %	1
3	Soil types*)	Regosol	3
		Latosol	2
		Aluvial	1

*) [15, 16]

2.4.2 Determination of types and shapes of urban forest in Tangerang Selatan city

After the urban forest location priority had been obtained, the next step was determining the types and shapes of urban forest. Determination of urban forest types was based on data of field observation, and land use map which referred to RI Government Regulation Number 63 Year 2002 [5] and Irwan [6], and the types were as follows (1) Residential area urban forest, (2) Industry urban forest, (3) Recreation urban forest, (4) Conservation urban forest, (5) Protection urban forest, and (6) Security urban forest.

Afterwards, on the basis of the dominant land form, the shapes of urban forest were determined by referring to RI Government Regulation Number 63 Year 2002 [5] and Irwan [6] and were categorized into 3 shapes (forms), namely: (1) Clumped, which is a category of urban forest whose vegetation community is concentrated in a particular area with vegetation abundance of minimally 100 trees with close spacing and irregular arrangement; (2) Scattered, which is a category of urban forest which does not have particular pattern, whose vegetation community grows in scattered pattern in the form of clusters or small clusters; (3) Strip, which is a category of urban forest whose vegetation community grow in land in the form of straight or curved strip following the path of river, road, canal and others.

Besides that, shapes of urban forest were also identified by referring to Fahutan IPB [7] with the following categories: (1) City park; this shape of urban forest is allocated for areas which have industrial urban forest type, residential areas and center of activities; (2) Plantation /home garden; this shape of urban forest is suitable for residential area urban forest type occurring in fertile area; (3) Green belt; this shape of urban forest is developed in road area and conservation area; (4) Forest; this shape of urban forest is suitable to be established in protected area.

3. Results and Discussion

3.1. Physic environment condition

3.1.1. Air temperature

One of the problems which is often faced by urban community recently is decreasing of convenience due to temperature increase. Temperature of > 28 °C covered area of 5,066.19 ha or 36.36 % of the area size of Tangerang Selatan city as shown on figure 3.

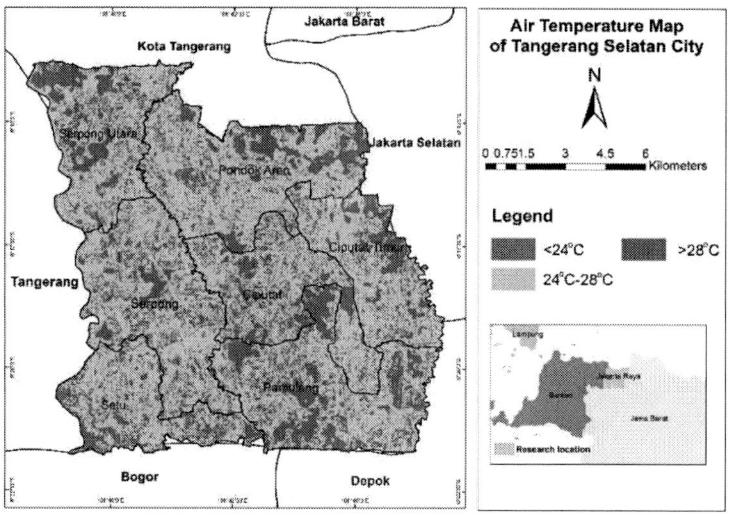

Figure 3. Air Temperature Map in Tangerang Selatan City

3.1.2. Soil types and land slope

Soil types in Tangerang Selatan is dominated by soil association of Alluvial, Latosol, and Regosol. Soil type which exhibited the largest area was Reddish Brown Latosol which covered area of 8,403.30 ha or 50.80 % of the total area of Tangerang Selatan City. Area in Tangerang Selatan City is dominated by slope of 3 - 8 % with area size of 88,041.43 ha or 53.21 % of the area size, whereas the smallest area was for slope class > 65 % (steep). Maps of soil types and land slope can be seen on figure 4.

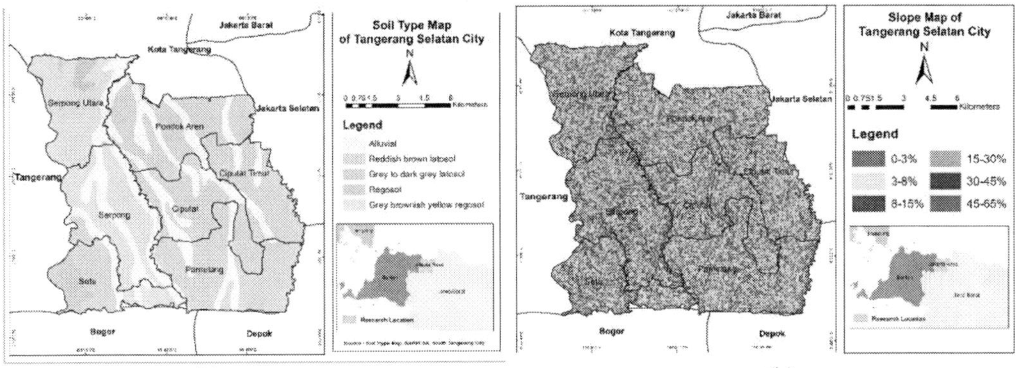

(a) (b)

Figure 4. Thematic maps in Tangerang Selatan City: (a) soil type map; (b) land slope map

3.1.3. Land cover dan land use

Land cover constitutes the physical manifestation of an area surface without considering the allocated use of that area [17]. Land cover in Tangerang Selatan city was dominated by built-up with area size of 10,525.75 ha or 63.65% of the total area size. Land use in Tangerang Selatan City are categorized into 15 types with domination by moderate density residential area, with area size of 6,751.70 ha or 38.17% of the total area. Open space such as plantation/farm land, rice field, or bushes occupied only 20.59% of the total area. Maps of land cover and land use are shown on figure 5.

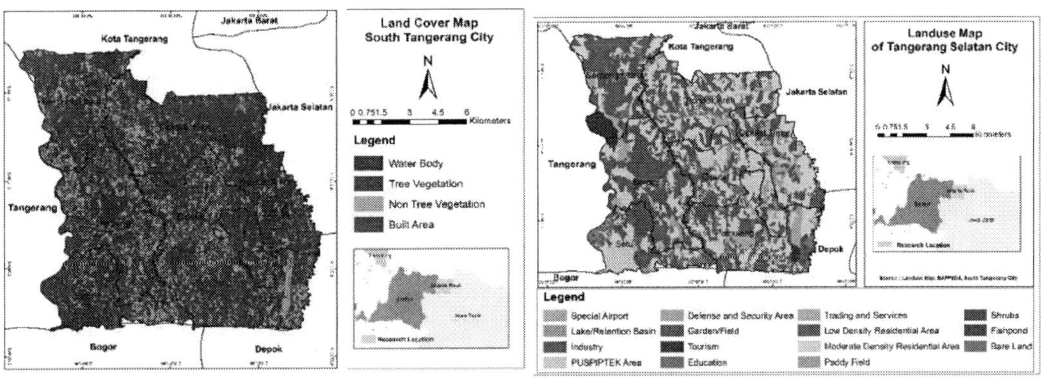

(a) (b)

Figure 5. Thematic maps in Tangerang Selatan City: (a) land cover map; (b) land use map

3.2. Priority location of urban forest development

From results of overlay process for air temperature map, slope and soil types map, and of the scoring process, there were obtained priority locations for urban forest development. Area size of 1st priority was only 1,521.33 ha or 9.22 % of the city area size. More complete information of each forest location priority is shown in table 2, with distribution as shown in figure 6.

Table 2. Area size of each priority classes of urban forest development for each sub district in Tangerang Selatan city

No	Sub district	Area size (ha)		
		1st Priority	2nd Priority	3rd Priority
1	Serpong Utara	637.47	1,247.99	339.70
2	PondokAren	42.52	1,054.17	1,896.03
3	Ciputat	60.56	797.67	1,255.78
4	Pamulang	85.50	1,342.51	1,446.44
5	Serpong	435.55	1,490.40	909.17
6	CiputatTimur	53.45	656.73	1,067.94
7	Setu	206.26	841.14	626.13
	Total	1,521.33	7,430.64	7,541.26

On the basis of Table 2, it could be seen that sub district which possessed the largest area for 1st priority was sub district of Serpong Utara with area size of 637.47 ha, followed with sub district of Serpong as large as 435.55 ha, while that of the smallest area size was Ciputat Timur with area size of 53.45 ha. Afterwards, 1st priority area was further categorized into area which became location of urban forest priority as strategy of urban forest development acceleration, namely area which had possessed vegetation, dominated either by tree or non tree vegetation, as shown in figure 6.

After the location of urban forest was known, the next step was determination of urban forest types which constituted an initial step and reference for future development of urban forest. Determination of urban forest type was based on objectives and objects being protected, which was adjusted with present condition of land use and plan for the future. Afterwards, on the basis of overlay between location map urban forest development and land use, there could be determination of types of urban forest, and therefore there were obtained urban forest types of residential area, industry and recreation, with area size of each type shown in table 3, while the distribution is shown in figure 6.

Table 3. Area size of development of urban forest type in each sub district in Tangerang Selatan City

No.	Sub district	Area size (Ha)		
		Residential area	Recreation	Industry
1	Pamulang	12.47	0.00	0.00
2	CiputatTimur	6.22	0.00	1.14
3	Serpong	166.34	2.63	0.00
4	Serpong Utara	132.04	22.82	9.67
5	Setu	38.50	0.00	0.00
	Total	355.57	25.45	10.81

IOP Conf. Series: Earth and Environmental Science **54** (2017) 012051 doi:10.1088/1755-1315/54/1/012051

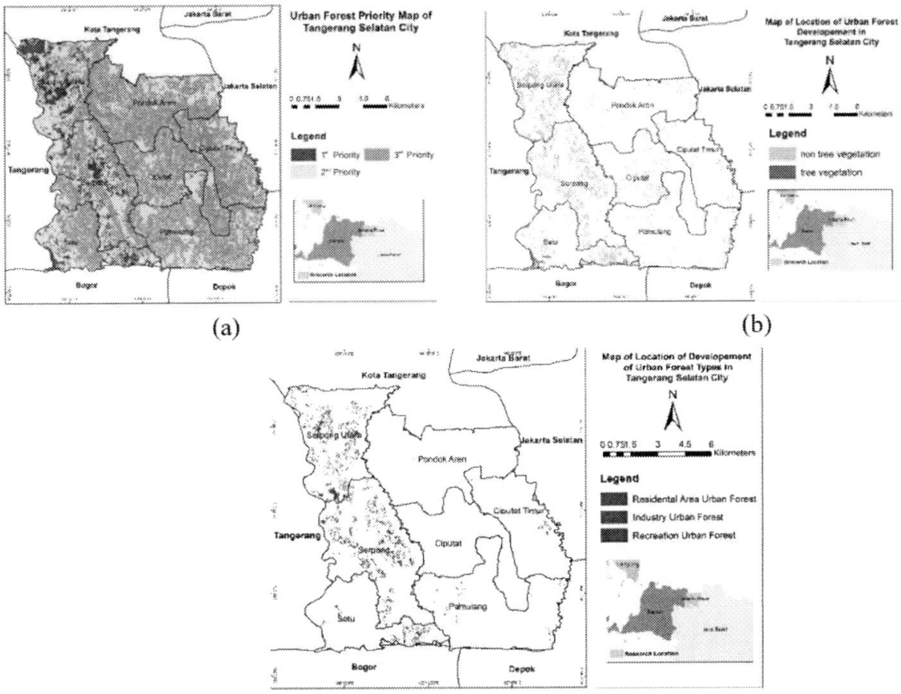

Figure 6. Maps used to determination of urban forest type in Tangerang Selatan City: (a) urban forest priority map; (b) map of location of urban forest development; (c) map of location of urban forest types

Identification of urban forest shapes was conducted by reference to two approaches, namely Fahutan IPB [7] and RI Government Regulation Number 63 Year 2002 [5] and Irwan [6]. Determination of urban forest shapes by Fahutan IPB [7] was based on location, important function, vegetation condition and management intensity, whereas that of RI Government Regulation Number 63 Year 2002 [5] and Irwan [6] were based on land shapes and their distribution.

Results of identification show that shapes of urban forest according to Fahutan IPB [7] which are suitable for: (1) Residential area urban forest type are city park, home garden and green belt along road sides in residential area; (2) Industry urban forest type are city park and green belt surrounding industrial area. This city park functions as absorbent and adsorbent of pollutant, and as resting places for workers. Green belts function as buffer for the surrounding areas which are affected by air pollution; (3) Recreation forest type are city park which is dominated by scenic tree species and managed with high intensity level of management.

Identification of urban forest shapes by RI Government Regulation Number 63 Year 2002 [5] and Irwan [6] shows that in sub district of Pamulang and sub district of Ciputat Timur, the suitable shape of urban forest is that which has scattered pattern, whereas for those of sub districts of Serpong, Serpong Utara and Setu, the shape of urban forests are those with clumped/clustered pattern. More complete description of urban forest shapes of each sub district of Tangerang Selatan City is presented in table 4.

Table 4. Development of urban forest shapes for each sub district in Tangerang Selatan City

Subdistrict	Types and shapes of urban forest					
	Residential area		Industry		Recreation	
	A*)	B**)	A*)	B**)	A*)	B**)
Pamulang	City park, Home garden,	Scattered				
Ciputat Timur	City park, Home garden	Scattered	City park, Green belt	Scattered, clustered, strip		
Serpong	City park, Home garden, Green belt	Clustered, Strip			City park	Clustered
Serpong Utara	City park, Home garden, Green belt	Clustered, Strip	City park	Clustered	City park	Clustered
Setu	City park, Home garden, Green belt	Clustered, Strip				

*) [5] **[6,7]

3.3 Implication of development of types and shapes of urban forest.

The priority locations of urban forest are those which have high air temperature [3, 18], steep slope and high susceptibility to erosion [19]. Therefore, as an initial step, the arrangement and choice of planted tree species are directed toward overcoming problem of inconvenience and soil erosion [20, 21]. Such directives should be further specified on the basis of urban forest types to be developed. Analysis results show that in Tangerang Selatan city, urban forest can be developed with three types as follows:

(1) *Residential area urban forest type.* Urban forests which are developed in residential area have the main function in microclimate amelioration, noise reduction, oxygen producer, carbon dioxide absorbent and wind break [5, 7, 22, 23, 24], so the residents could enjoy tranquility, convenience and health. This type of urban forest possessed the largest area size, namely 355.57 ha. This could be understood because 61.79 % of the territory of Tangerang Selatan City was in the form of residential area. To create maximum benefits, the tree species to be developed should have strong root system, strong branch and twigs which do not break easily, leaves which are not easily shed, and produce flower/fuit/ seeds which have economic value [5, 7, 25].

(2) *Industry urban forest.* Urban forests which are developed in industrial area have the main function to reduce air pollution and noise created by industrial activities [5,7, 26, 27, 28]. Development of this urban forest type was in accordance with policy of Pemkot Tangsel [21] concerning the control of pollution and environmental destruction; tree species which could be developed should have leaves which are wide and shady, hairy or pubescent, have rough/lobed surface; thick crown, and high ability to absorb and adsorb air pollutant [7, 25, 29, 30, 31, 32, 33].

(3) *Recreation urban forest type*; this type of urban forest is developed with the main function to fulfill the need for recreation by urban people community [5, 7], while also could serve as facilities for environmental education [7, 34]. Tree species being planted for this type possesses high aesthetic value, either in terms of crown shape, leaf color, flower color; does not shed much leaves, are not thorny, and does not produce dangerous fruits [15, 25, 26, 35]. This type of urban forest is developed by considering the presence of blocking allocation, namely intensive block and non intensive block, circulation path, facility types, and activities [35].

As one of the efforts to conserve biodiversity, there can be developed species which are endemic/ rare/protected [36, 37, 38], especially for urban forest with residential area type and recreation type, while the species should remain fulfilling the required characteristics for those types. Species which could be developed are among others *Stelechocarpus burahol, Diospyros celebica, Sterculia foetida,* and *Dalbergia latifolia.*

In accordance with RI Government Regulation Number 63 Year 2002 [5], urban forest should have minimal area size of 0.25 ha. Therefore for each classification, urban forests in the shape of strip or clusters should remain to have minimal area of 0.25 ha. Urban forest with scattered form constitutes one management unit comprising separate strip shape and clustered shape, and each of them should be minimally 0.25 ha. Therefore, as a follow up action, there is a need to identify management units of urban forests in accordance with shapes which will be developed.

4. Conclusion

Location and priority for development of each type of urban forest are as follows: (a) residential area urban forest type with area size of 355.57 ha was distributed in sub districts of Pamulang, CiputatTimur, Serpong, Serpong Utara, and Setu; (b) Industry urban forest type with area size of 10.81 ha was distributed in sub district of Ciputat Timur and Serpong Utara; (c) recreation urban forest type with area size of 25.45 ha was distributed in sub districts of Serpong and Serpong Utara. On the basis of land forms and their distribution, the shapes of urban forest that could be developed in: (a) residential urban forest type are clustered/clump, scattered, and strip; (b) industrial urban forest type are clustered/clump, scattered, and strip; (c) recreation forest type are clustered/clump. On the other hand, on the basis of location, important function, and management intensity, the forms of urban forest which are developed in: (a) residential area urban forest type are city parks, home garden, and green belt; (b) industry urban forest type are city park, and green belt; (c) recreation urban forest type are city park.

References

[1] Rushayati S B, Alikodra H S, Dahlan E N and Purnomo H 2011 Green open space development based on distribution of surface temperature in Bandung regency *Forum Geografi* Vol. **25** 17 - 26
[2] [DKP-DKI Jakarta] Marine and Agricultural Services of DKI Jakarta 2012 *Master plan of urban forest in DKI Jakarta* (Jakarta: Marine and Agricultural Services of DKI Jakarta)
[3] Rushayati S B and Hermawan R 2013 Characteristics of urban heat island condition in DKI Jakarta. *Media Konservasi* Vol. **18** 96-100
[4] Marbun B N 1990 *Indonesia city future: problem and prospect* (Jakarta: Penerbit Erlangga)
[5] [RI Government] Government of Republic of Indonesia 2002 Government regulation number 63 year 2002 concerning urban forest (Jakarta: State Secretariat of RI)
[6] Irwan Z J 1997 *Environmental challenge and landscape of urban forest* (Jakarta: PT Pustaka CIDESINDO)
[7] [Fahutan IPB] Faculty of Forestry, Bogor Agricultural University 1987 *Concept of urban forest development* (Bogor: Faculty of Forestry, IPB)
[8] Humaida N, Prasetyo L B and Rushayati S B 2016 Priority assessment method of green open space (case study:Banjarbaru City) *Procedia Environmental Sciences* **33** 354 – 364
[9] Karyono T H 2001 Research on thermal comfort in Jakarta as reference for convenient temperature for Indonesian people *Dimension Architectural Technique* Vol **29** 24 – 33
[10] [USGS] United States Geological Survey 2013 *Landsat 8 science data user handbook (*America: USGS)
[11] Khomarudin M R 2005 *Estimation of regional scale evapotranspiration using remote sensing satellite data* [thesis] (Bogor: Bogor Agricultural University)
[12] Monteith J L and Unsworth M H 1990 *Principles of environmental physics.* 2nd ed. (London: Edward Arnold)
[13] [Deptan] Department of Agriculture 2008 *Technical guidelines of upstream watershed conservation* (Jakarta: Directorate of Land Management)
[14] Kridalaksana A 2014 Application of *geographic Information System (GIS) for determining location of urban forest and example of pre-design of urban forest in subdistric* of Banyuwangi, District of Banyuwangi [undergraduate thesis]. (Bogor: Faculty of Forestry Bogor Agricultural University)

[15] [Dep-PU] Department of Public Works 2007 Regulation of Public Works Minister Number: 41/PRT/M/2007 concerning *Guidelines on Technical Criteria of Cultivation Area* (Jakarta: Department of Public Works)
[16] Rachim D A and Suwardi 2002 Morphology and classification of soils (Bogor: Faculty of Agriculture, Bogor Agricultural University)
[17] Lillesand T M and Kiefer R W 1997 *Remote sensing and image interpretation* (Third Ed. Sutanto) (Yogyakarta: Gadjah Mada University Press)
[18] Rushayati S B, Prasetyo L B, Puspaningish N and Rahmawati E 2016 Adaptation strategy toward urban heat island at tropical urban area *Procedia Environmental Sciences* **33** 221 – 229
[19] Mahon J R and Miller R W 2003 Identifying high-value greenspace prior to land development *Journal of Arboriculture* **29** 25-33
[20] Armson D, Stringer D and Ennos A R 2013 The effect of street trees and amenity grass on urban surface water runoff in Manchester, UK *Urban Forestry and Urban Greening* **12** 282-286
[21] [PemkotTangsel] Muncipal Government of Tangerang Selatan City 2011 Local Regulation of Tangerang Selatan City Number 15 year Tahun 2011 concerning *Spatial plan of Tangerang Selatan City area year 2011 – 2031* Government of Tangerang Selatan City
[22] Cho S H, Poudyal N C and Roberts R K 2008 Spatial analysis of the amenity value of green open space. *Ecological Economics* **66** 403-416
[23] Nasir R A, Ahmad S S, Ahmed A Z and Ibrahim N 2015 Adapting human comfort in an urban area: the role of tree shades towards urban regeneration *Procedia - Social and Behavioral Sciences* **170** 369 – 380
[24] Moufida B and Jamel A 2012 Impact of vegetation on thermal conditions outside, thermal modeling of urban microclimate, case study: the street of the republic, Biskra *Energy Procedia;* 73 – 84
[25] Dahlan E N 2013 *Green city of urban forest* (Bogor: Department of Forest Resources Conservation and Ecotourism, Faculty of Forestry, Bogor Agricultural University)
[26] Grey G W and Deneke F J 1986 *Urban forestry (second edition)* (New York: J. Wiley & Sons, Inc)
[27] Miller R W 1988 *Urban forestry: planning and managing urban green spaces* (New Jersey: Prentice Hall)
[28] Tyrväinen L, Pauleit S, Seeland K and de Vries S 2005 *Benefits and uses of urban forests and trees. In*: Urban forest and trees (Eds: Konijnendijk CC, Nilsson K, Randrup TB, Schiperrijn J. (Springer: Berlin Heidelberg New York)
[29] Cavanagh J E, Reza P Z and Wilson G J 2009 Spatial attenuation of ambient particulate matter air pollution with an urbanized native forest patch Urban Forestry & Urban Regreening **8** 21-30
[30] Hermawan R 1997 *Effects of simulated acid rain on the growth and leaf anatomical structure of Mimusops elengi L. Seedlings* (Goettingen: Georg-August University)
[31] Hermawan R, Kusmana C, Nasrullah N and Prasetyo L B 2011 The effects of the plant row number of roadside vegetation in reducing lead (Pb) particles emitted by motor vehicle (Case study of Acacia mangium green belt, Jagorawi highway) *Media Konservasi* Vol. **16** 55-64
[32] Lei W, Lian-you L, Shang-yu G and Zhi W 2006 Physical characteristics of ambient particle setting upon leaf surfaces of urban plan in Beijing *J Environ Sci* **18** 921-926
[33] Purnomohadi S 1995 *Role of green open space in control of air quality in DKI Jakarta* [dissertation] (Bogor: Graduate Program, Bogor Agricultural University)
[34] Hermawan R, Kosmaryandi N and Ontarjo O 2008 Study on type and shape of urban forest in Danau Raja area, Rengat City, Indragiri Hulu Regency, Riau Province *Media Konservasi* Vol. **13** 71-78
[35] [Fahutan IPB] Faculty of Forestry Bogor Agricultural University 2008 *Compilation of master plan for urban forest Bundayati Tanjung Selor, District of Bulungan.* Cooperation between Faculty of Forestry, IPB with Forestry Service Agency, Bulungan District
[36] [IUCN] International Union for Conservation of Nature and Natural Resources 2014 *IUCN Red List of Threatened Species.* [Internet]. [downloaded 2016 April 22]. Available at: http://www.iucnredlist.org/

[37] [LIPI] Indonesian Scientific Institute 2001 Indonesian rare plants. Bogor (ID): Puslitbang Biologi-LIPI
[38] Yuniastuti E, Handayani T and Djoar D W 2009 *Identification and selection of pranajaya (Sterculiafoetida Linn.) plant diversity and technology of in vitro plant propagation for raw material supply of biofuel*. [Internet].[downloaded 2016 April 21]. Available at: http://lppm.uns.ac.id/identifikasi-dan-seleksi-keragaman-tanaman-pranajaya-sterculia-foutida-linn-serta-tek-nologi-perbanyakan-tanaman-secara-in-vitro-untuk-penyediaan-bahan-baku-biofuel-universitas-sebelas-maret.html

Monitoring tropical peatland ecosystem in regional scale using multi-temporal MODIS data: Present possibilities and future challenges

Y Setiawan[1,3]*, H Pawitan[2], L B Prasetyo[1] and P A Permatasari[3]
[1] Department of Forest Conservation and Ecotourism, Faculty of Forestry, Bogor Agricultural University, Kampus IPB Darmaga, Bogor 16680, Indonesia
[2] Department of Geophysics and Meteorology, Faculty of Mathematics and Natural Sciences, Bogor Agricultural University, Kampus IPB Darmaga, Bogor 16680, Indonesia
[3] Center for Environmental Research, Bogor Agricultural University, Kampus IPB Darmaga, Bogor 16680, Indonesia

E-mail: setiawan.yudi@apps.ipb.ac.id

Abstract. Many studies on peatland ecosystem have been focused on forest conversion or forest degradation as a single pathway, meanwhile; in the context of peatland ecosystem, the change in land surface is more complicated since it can be categorized into two types and mechanisms: 1) gradual change, caused by interannual climate variability and forestland degradation, and 2) abrupt change, caused by disturbances such as deforestation and wildfires. Understanding this change types is needed for conservation and management, particularly to improve understanding of terrestrial environmental change in peatland ecosystem. In such situation, simultaneous analysis of land surface attributes from long-term datasets and seasonal variation seems to be a way to monitor the tropical peatland ecosystem. This analysis provides information about how the changes occurred accurately as well as how big are these affected areas. In this article, the feasibility of using long-term MODIS data for monitoring the dynamics change in peatland ecosystem is examined. The temporal vegetation dynamics of long-term MODIS datasets offer great promise for characterizing gradual change as well as abrupt change at large scale, however, the mixed pixel issue and some residual noises in temporal sequences are quite problematic when using MODIS data.

1. Introduction

Importance of peatland in the tropical environment has been studied by many scientists, such as estimation of CO_2 emission released by forest fire [1], variation in its ecological function [2], and loss of biodiversity through land conversion [3]. Consequently, the study of sustainable peatland management has been highlighted as a priority study in mitigation of climate change in agriculture and forestry sector [4]. These studies indicated the need to construct an update and accurate database concerning tropical peatland environment, their meaning, their pace and the explanatory factors prompting their dynamic changes.

As an essential ecosystem, the peatland ecosystem is vulnerable to be changed in their water supply [5], and it is expected that climate change will have a pronounced effect on peatlands through alterations in hydrological regimes with climate variability [6]. Therefore, available information on characteristics of peatland ecosystem and its dynamics change is necessary for many aspects of peatland conservation and management.

Peatlands are complex systems that reflect their geographical and ecological setting. Their characteristics and its changes are the results of interactions between socio-economic and cultural conditions, biophysical constraints and land use history [7]. Development and stability of peatlands are determined as interactions between different land use/cover types, hydrological setting and spatial elements of the landscape [8].

This paper evaluates the feasibility of using time series satellite datasets to consider the peatland characteristics and its dynamics changes occurred in this ecosystem. In some earlier works, monitoring approach to recognizing the peatland characteristics and its dynamic changes on a regional scale is due to simultaneous analysis of land surface attributes from long-term data sets and seasonal variation. Monitoring of land surface continuously can provide information on the ecosystem characteristics, including how the change occurred accurately as well as how big are these affected areas [9].

2. Remote sensing approach for identifying the ecosystem characteristics

Advances in remote sensing technology enable land scientists to characterize some terrestrial ecosystems using satellite data sets. Monitoring of land surface continuously in space and time, at both seasonal and inter-annual scales allows characterization of the temporal vegetation dynamics of terrestrial ecosystem, including peatlands [10].

Characterization of vegetation dynamics has often been made by using vegetation index values [11]. The temporal dynamics of those index values are useful for distinguishing land surface conditions by differentiating among vegetation types and their distributions [12]. The vegetation indices relate to the amount of green leaf biomass. They are computed by algebraic combination of two or more spectral bands and are designed to enhance the contribution of vegetation properties [13].

The normalized difference vegetation index (NDVI) and enhanced vegetation index (EVI) are commonly used to measure reliable spatial and temporal inter-comparisons of terrestrial photosynthetic activity and canopy structural variations [14]. The EVI is more responsive to canopy structural variations, including leaf area index (LAI) and canopy type, rather than the NDVI, as it was developed to optimize the vegetation signal with improved sensitivity in high biomass regions through a de-coupling of the canopy background signal and a reduction in atmosphere influences [14].

The uses of multi-year datasets are necessary to develop methodologies that utilize information on inter-annual variations to increase the accuracy of the land surface characterization. The Moderate Resolution Imaging Spectroradiometer (MODIS) satellite datasets provide basic information related to vegetation dynamics in peatland ecosystem. These datasets are provided by the U.S. Geological Survey - Land Processes Distributed Active Archive Center (USGS LPDAAC). The MODIS Vegetation Index composited 16-day product (MODIS MOD13Q1) is available for packaged MODIS that should be useful for a specific need for land surface monitoring and historical trend analysis. Such kind of the high temporal datasets is required to characterize the vegetation dynamics in peatland ecosystem, including detecting the event of forest cover change.

In the previous work, the 25 types of EVI pattern, which represent a specific class of land use in regional scale of Java Island, was determined [9]. The accuracy of this classification revealed the overall accuracy and the overall k-statistics to be 80.16% and 70.94%, respectively. As pointed out previously, locations displaying similar temporal patterns inferred to have a similar ecosystem type, not limited to land use type, which are have relatively homogeneous characteristics such as fractions of vegetation covers and environmental conditions. The global ecosystem characteristics was determined based on their seasonal land cover, floristic properties, climate, and physiognomy [15]. At a continent scale, monitoring the reliable spatial and temporal characteristics of vegetation dynamics allows the classification of African ecosystem at 1 km resolution [16].

Distinguishing among temporal patterns of EVI pattern accomplished using the K-means clustering method based on Euclidean distance in an EVI-space, in which each EVI images provides one dimension of feature space, analogous to spectral clustering. The clustering method yields sets of clusters, which each cluster represents a significant different EVI pattern of detailed information of characteristics/type of peatlands. As shown by figure 1, the method distinguished temporal vegetation dynamics of the peatlands into different land cover properties such as secondary swamp forest, bush/swamp bush, agricultural lands/plantation and swamp forest mixed bush.

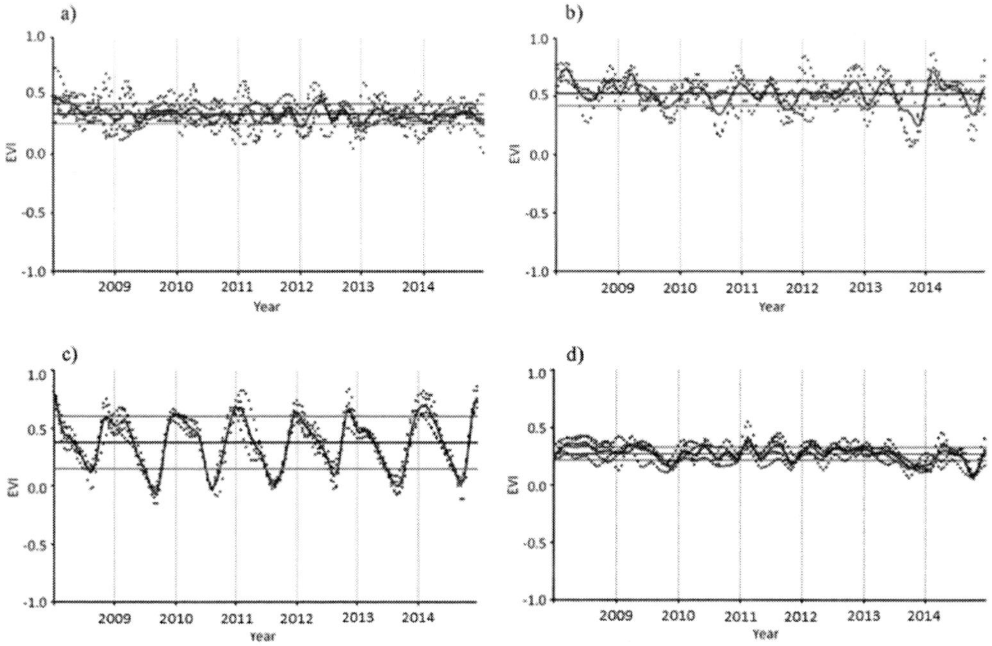

Figure 1. Temporal pattern of EVI accomplished based on K-mean clustering in peatland ecosystem following different land use/cover types: a) swamp forest mixed bush, b) bush/swamp bush, c) agricultural lands, and d) swamp forest

3. Characterization of tropical peatland ecosystem

The peatland ecosystem of Sumatra was defined by differentiating the long time-series of typical EVI patterns, which distinguished the specific type of peatland and/or different response of this ecosystem to the environment changes in these given land areas [10]. The study performed initial segmentation of the long time-series of EVI composites by the clustering method and differentiated into 23 clusters. This number of cluster seems sufficient to represent various patterns of vegetation dynamics in Sumatran peatland ecosystems. Each class pattern represents the specific peatland ecosystem type; in turn inferred to have relatively homogeneous environmental characteristics such as peat depth, specific land cover type, and other environmental parameters (table 1).

Most of class patterns in the peatlands were associated mainly with six land cover types; namely secondary swamp forest, swamp bush, timber forest, plantation, agricultural lands and bare lands. Consequently, more than one typical pattern might be characterized by one land cover type, which represents different environmental characteristics of a specific ecosystem. The distribution of each typical pattern in Sumatra is shown in figure 2.

Table 1. Summary of typical pattern characteristics in Sumatran peatland regarding to EVI-based class pattern

No	Land cover type	Peat depth (cm)	Performance of typical vegetation pattern
1	Water features including lakes and reservoirs	< 50	
2	Water features including lakes and reservoirs; mixed vegetated areas	< 50	
3	Swamp bush	100-200	Pattern change occurred in 2002 (Jan–Jun)
4	Swamp bush, paddy field/intensive agricultural lands	50-100	Pattern change occurred in 2002 (Apr-Sep)
5	Swamp bush	> 400	Pattern change occurred in 2006 (Apr-Dec)
6	Industrial forest plantation, swamp bush	50-100; 100-200,	Pattern change occurred in 2005(Apr-Sep)
7	Swamp bush	50-100	Pattern change occurred in 2013 (Apr-Sep)
8	Industrial forest plantation	> 400	Pattern change occurred in 2002 (Jul-Des), 2008 (Jan-Jun), 2013 (Apr-Sep)
9	Bare land, swamp bush	300-400	Pattern change occurred in 2011 (Apr-Sep), 2013 (Apr-Sep)
10	Secondary swamp forest	300-400, > 400	Not significantly changed
11	Industrial forest plantation	> 400	Pattern change occurred in 2006 (Jul-Des), 2011 (Apr-Sep)
12	Secondary swamp forest	> 400	Not significantly changed
13	Industrial forest plantation	> 400	Pattern change occurred in 2002 (Jul-Des), 2009 (Jul-Des)
14	Secondary swamp forest	> 400	Pattern change occurred in 2013 (Apr-Sep)
15	Industrial forest plantation, swamp bush	50-100	Pattern change occurred in 2002 (Jul-Des), 2006 (Apr-Des)
16	Secondary swamp forest, swamp bush	100-200	Pattern change occurred in 2002 (Jul-Des)
17	Swamp bush	100-200	Not significantly changed
18	Plantation, swamp bush	100-200	Pattern change occurred in 2005 (Jul-Des)
19	Plantation, swamp bush	50-100	Pattern change occurred in 2006 (Jul-Des)
20	Plantation	50-100	Pattern change occurred in 2006 (Jul-Des)
21	Secondary swamp forest	50-100	Not significantly changed
22	Secondary swamp forest, swamp bush	50-100	Pattern change occurred in 2005 (Jul-Des)
23	Secondary swamp forest	50-100	Not significantly changed

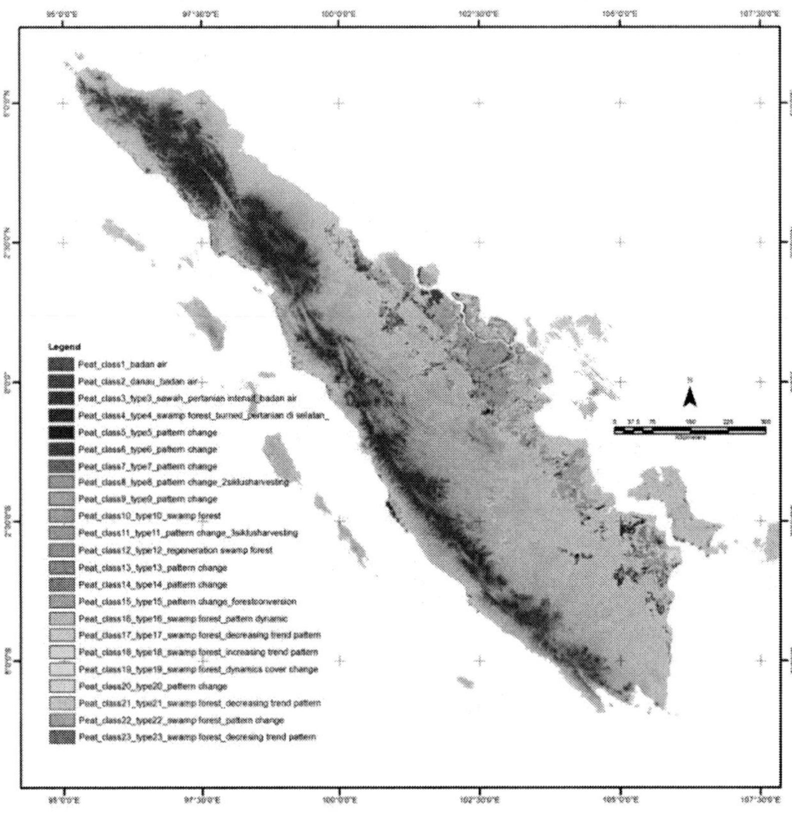

Figure 2. Distribution of each typical pattern in peatlands of Sumatra

Due to the temporal vegetation pattern, dynamics changes and their properties could be detected by the distance comparison of EVI. This detection method has been discussed in [17]. Regarding the identification of the pattern change, many significant patterns indicated specific changes of land use/cover in peatland ecosystem. Figure 3 indicates some examples of change trajectories, as follows: a) forest mixed bush converted to barren land through burning at the end of October 2009, b) agricultural development due to crop planting in barren land, and then developed into intensive agricultural lands from July 2011, c) .re-growth/re-vegetation processes occurred in a land affected by forest fire/wildfires, where vegetated lands drastically changed to open land by fire on October 2009, and then the recovery process occurs gradually over long periods of time.

Moreover, due to overlay with hotspot data (https://firms.modaps.eosdis.nasa.gov/), the results indicated that some areas of specific peatlands had never been burned from 2000 to 2014. These peatland ecosystems support a natural tree cover and important ecosystem to be preserved. Some of abrupt pattern changes occurred in swamp bush and located at a peat depth of more than 400 cm, which is widely distributed in the central raised part of the peat swamp area. These typical conditions have numerous consequences relevant to the environment as well as changes in carbon storage, land degradation and loss of biodiversity.

Indeed, characterization of typical peatland ecosystem based on temporal pattern analysis would provide useful information regarding the change dynamics in the peatland ecosystems.

IOP Conf. Series: Earth and Environmental Science 54 (2017) 012052 doi:10.1088/1755-1315/54/1/012052

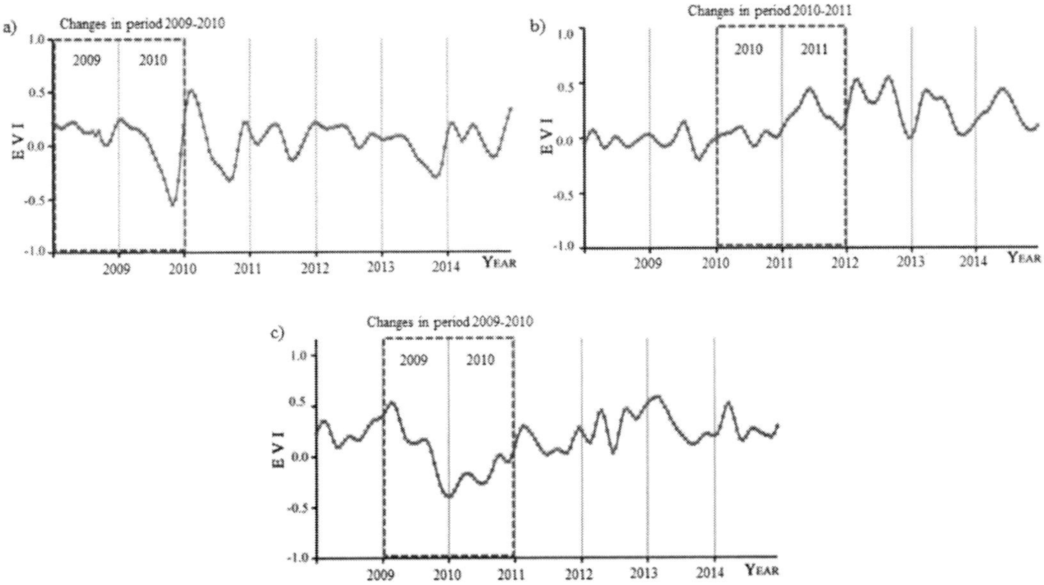

Figure 3. Some of significant pattern change in peatlands ecosystem could be recognized by the change pattern detection method [10].

Although MODIS data have some advantages in providing basic information related to vegetation dynamics pattern, time-series of these data inevitably contain disturbances caused by atmospheric variability and aerosol scattering [18]. Such noise degrades the data quality and introduces considerable uncertainty in temporal sequences, confusing the analysis of temporal images sequences by introducing significant variations in the EVI time series data. Therefore, noise reduction (de-noising) or fitting a model to observe data is necessary before vegetation dynamics pattern can be determined.

Various strategies for image preprocessing including smoothers have been applied, such as: median filters [19], a moving window to select the local maximum VI [20], principal component analysis (PCA) [21], and the Savitzky-Golay filter approach [22]. Several spectral-frequency techniques have also been used, including Fourier-based fitting methods for separating the high-frequency components of noise and the low-frequency components of seasonal changes of VI [23] and, most recently, wavelet decomposition, which has been used to de-noise and to build a vegetation signatures through feature extraction of MODIS multi-temporal [9].

Moreover, the mixed pixel issue associated with the 250 m grid data is another issue for using the MODIS data. Overlaying the specific coverage type in peatlands ecosystem in Java Island over the image classification result using Landsat ETM+ revealed that in general, approximately only 20% of MODIS pixels contained homogeneous cover types [9]. This result shows that most image pixels contain a mixture of land use/cover classes, which is referred to as sub-pixel mixing. Such issue may lead to the difficulty of obtaining the actual number of clusters.

Regarding the accuracy assessment issue, the creation of a reference data set in order to assess the accuracy of characterization result of the peatland ecosystem is one important step of the study. A complicating factor in assessing the accuracy derived from 250 m image data was the disparity between the image pixel size and the average patch size of the landscape. The same issue was mentioned by [24], that one problem of moderate spatial resolution imagery is that the intrinsic scale of spatial variation in many land covers is finer than the scale of sampling imposed by the image pixels.

Although the mixed pixel issue is quite problematic when using MODIS data, the results indicated that MODIS data offer great promise for monitoring the dynamics change processes of the peatland ecosystem, including how and when the change occurred accurately. Furthermore, the understanding of the terrestrial environmental changes such as carbon storage change and sequestration by terrestrial plants will be improved. The research of temporal vegetation dynamics could be carried out under the framework of regional analysis, so that better understanding of a typical ecosystem can be achieved in a regional scale.

4. Conclusions

Monitoring land surfaces continuously based on temporal vegetation dynamics allows characterization of the typical ecosystem in peat swamp area. This study is based on the approach that each pattern of vegetation dynamics represents the specific peat swamp ecosystem type; in turn inferred to have relatively homogeneous environmental conditions such as land cover type and peat depth.

Considering the results, the peat swamp ecosystems can be categorized into two types as follows: 1) developed land as the forest plantation and agricultural lands, and 2) secondary swamp forest and swamp bush. Different typical pattern of these ecosystems represented different environmental characteristics such as peat depth. Moreover, the abrupt changes in the secondary swamp forest and swamp bush in peatlands ecosystem, by either peat-fire or human factors, especially located at very deep peat soil, have numerous consequences relevant to the environmental changes.

Regional shifts in temporal vegetation dynamics, including the actual changes in the peatland and temporary changes of land cover, have numerous consequences relevant to the environment as well as changes in carbon and nitrogen storage, land degradation and loss of biodiversity. Determining the change of temporal vegetation dynamics is the first step in understanding their implications, for example, long-term crop production, and environmental, agricultural and economic sustainability. An understanding of temporal vegetation dynamics to explain the mechanisms and pathways of dynamics change of peatland ecosystem is important because of the relationship between ecosystem characteristics and socio-economic attributes of the land.

Acknowledgement

We would like to thank the Center for Environmental Research, Bogor Agricultural University (PPLH-IPB) for giving us opportunity to get the research funding. This work was supported by Indonesian Directorate General of Higher Education (DIKTI) for fiscal year 2016 under the scheme of Universities Excellent Research (PUPT) Letter No. 244/IT3.11/LT/2016.

References

[1] Page S E, Rieley J O, and Banks C J 2010 Global and regional importance of the tropical peatland carbon pool *Glob Change Biol* **17** 798818
[2] Achard F, Eva H D, Stibig H J, Mayaux P, Gallego J, Richards T, and Malingreau J P 2002 Determination of deforestation rates of the world's humid tropical forests *Science* **297** 999-1002
[3] Myers N, Mittermeier R A, Mittermeier C G, da Fonseca G A B, and Kent J 2000 Biodiversity hotspots for conservation priorities *Nature* **403** 853–8
[4] Rieley J 2014 Utilization of peatlands and peat *Towards climate-responsible peatlands management Mitigation of Climate Change in Agriculture* (Series 9) ed R Biancalani and A Avagyan (Rome: FAO) p 117
[5] Whittle A and Gallego-Sala A V 2016 Vulnerability of the peatland carbon sink to sea-level rise *Scientific Report* **6** 28758
[6] Erwin K L 2009 Wetlands and global climate change: the role of wetland restoration in a changing world *Wetlands Ecol Manage* **17** 71
[7] Cole L E S, Bhagwat S A and Willis K J 2015 Long-term disturbance dynamics and resilience of tropical peat swamp forests *J. Ecol* **103** 16–30
[8] Evans T P and Moran E F 2002 Spatial integration of social and biophysical factors related to landcover change *Popul Dev Rev* **28** 165–186

[9] Setiawan Y, Yoshino K and Philpot W D 2013 Characterizing temporal vegetation dynamics of land use in regional scale of Java Island, Indonesia *J. Land Use Sci.* **8** 1–30

[10] Setiawan Y, Pawitan H, Prasetyo L B, Lubis M I, Parlindungan M and Nurdiana A 2016 Characterizing spatial distribution and environments of sumatran peat swamp area using 250 M multi-temporal MODIS data *Proc Env Sci* **33** 117–127

[11] Knight J F, Lunetta R S, Ediriwickrema J and Khorram S 2006 Regional scale land cover characterization using MODIS-NDVI 250 m multi-temporal imagery: A phenology-based approach *GISci.Remote Sens* **43** 1–23

[12] Hansen M, DeFries RS, Townshend J and Sohlberg R 2000 Global land cover classification at 1 km spatial resolution using a classification tree approach *Int J. Remote Sens* **21** 1331–1364

[13] Huete A R, Liu H Q, Batchily K and Van Leeuwen W 1997 A Comparison of vegetation indices over a global set of TM images for EOS-MODIS *Remote Sens. Env* **59** 440–451

[14] Huete A, Didan K, Miura T, Rodriguez E P, Gao X and Ferreira L G 2002 Overview of the radiometric and biophysical performance of the MODIS vegetation indices *Remote Sens. Env* **83** 195–213

[15] Muchoney D, Strahler A, Hodges J and Locastro J 1999 The IGBP discover confidence sites and the system for terrestrial ecosystem parameterization: Tools for validating global land cover data. *Photogram Eng Remote Sens* **65** 1061-67

[16] Kaptue Tchuente A T, Roujean J L and Faroux S 2010 ECOCLIMAP-II: An ecosystem classification and land surface parameters database ofWestern Africa at 1 km resolution for the African Monsoon Multidisciplinary Analysis (AMMA) project *Remote Sens Env* **114** 961–976

[17] Setiawan Y, Yoshino K and Prasetyo L B 2014 Characterizing the dynamics change of vegetation cover on tropical forestlands using 250m multi-temporal MODIS EVI *Int J. Appl Earth Obs Geo* **26** 132–144

[18] Xiao X M, Braswell B, Zhang Q Y, Boles S and Frolking S 2003 Sensitivity of vegetation indices to atmospheric aerosols: Continental-scale observations in Northern Asia. *Remote Sens Env* **84** 385–392

[19] Van Dijk A, Callis S L, Sakamoto C M and Decker W L 1987 Smoothing vegetation index profiles: An alternative method for reducing radiometric disturbance in NOAA/AVHRR data. *Photogram Eng Remote Sens* **53** 1059–1067

[20] Viovy N, Arino O and Belward A S 1992 The Best Index Slope Extraction (BISE) : A method for reducing noise in NDVI time-series *Int J. Remote Sens* **13** 1585– 1590

[21] Li Z and Kafatos M 2000 Interannual Variability of Vegetation in the United States and Its Relation to El-Nino Southern Oscillation *Remote Sens Env* **71** 239–247

[22] Chen J, Jonsson P, Tamura M., Gu Z, Matsushita B and Eklundh L 2004 A simple method for reconstructing a high-quality NDVI time-series dataset based on the Savitzky-Golay filter *Remote Sens Env* **91** 332–344

[23] Lunetta R S, Knight J F, Ediriwickrema J, Lyon J G and Worthy L D 2006 Land-cover change detection using multi-temporal MODIS NDVI data *Remote Sens Env* **105** 142–154

[24] Atkinson P M, Cutler M E J and Lewis H 1997 Mapping sub-pixel proportional land cover with AVHRR imagery *Inter J. Remote Sens* **18** 917–935

Forest fires detection in Indonesia using satellite Himawari-8 (case study: Sumatera and Kalimantan on august-october 2015)

Fatkhuroyan*, TrinahWati and Andersen Panjaitan
Indonesia Agency for Meteorology Climatology and Geophysics
BMKG, Kemayoran, Central Jakarta, Indonesia

E-mail: fatkhuroyan@bmkg.go.id

Abstract. Forest fires in Indonesia are serious problem affecting widely in material losses, health and environment. Himawari-8 as one of meteorological satellites with high resolution 0,5 km x 0,5 km can be used for forest fire monitoring and detection. Combination between 3, 4 and 6 channels using Sataid (Satellite Animation and Interactive Diagnosis) software will visualize forest fire in the study site. Monitoring which used Himawari-8 data on August, September and October 2015 can detect the distribution of smoke and the extents of forest fire in Sumatera and Kalimantan. The result showed the extent of forest fire can be identified for anticipation in the next step.

1. Introduction

Every year forest fire inflicts smoke problem in most of province in Sumatera and Kalimantan on August to November.The impacts are seriously on environment, health and socio-economic problems [1]. The loss of forest fire in 2015 was around 16.09 million US dollars and Kalimantan Island was the most severe area affected by forest fire and smoke haze [2]. The impact of smokes were not only happened in Indonesia region, but also in neighbouring countries such as Malaysia and Singapore and this condition affects the relations between those countries. Some studies use weather and environment satellite data for smoke monitoring with polar orbital which is good in spatial resolution and have various wavelength, yet weak in temporal resolution [3,4], while the others used geostationary orbital types such as MTSAT (Meteorological Satellite) that can produce image every hour and has 5 channels [5].

Himawari-8 is a new Japanese geostationary meteorological satellite with optical sensors significantly higher in radiometric, spectral and spatial resolution than previous geostationary orbit such as MTSAT. There are 16 observation bands with 0.5 or 1 km spatial resolution for visible and near-infrared bands and 2 km for infrared bands. The shortened revisit times around 10 minutes for full disk provide new levels of capacity for identification and tracking of rapidly changing weather phenomena and for the derivation of quantitative products [6,7]. Himawari-8 was launched from Japan's Tanegashima Space Centre using an H-IIA rocket on & October 2014 and settled in geostationary orbit on 16 October. JMA (Japan Meteorological Agency) has operated Himawari-8 since 7 July 2015 [7]. Comparison the result of RGB (Red Green Blue) false colour image (1 visible channel and 2 near infra-red channels) between Himawari-8 and Terra-Aqua MODIS satellites with other supporting data (weather report and hotspot data) showed a similar pattern image between them which consist of channels combination of 3, 4 and 6 and also had corresponded to the report of smoke from weather station and hotspot data [8].

The aim of this paper is to investigate the smoke detection using Himawari-8 satellite with RGB false colour technical method which located in Sumatera and Kalimantan regions on August, September and October 2015.

2. Data and methods

Data that were used in this study are:

a. Images of Satellite Himawari-8, channel 3,4 and 6 on August 25th, September 10th and October 19th 2015.
b. Synoptic reports from meteorological stations
c. Hotspot data in Sumatera and Kalimantan based on Terra-Aqua Modis satellite

The Sataid software was used to manage RGB image combination of Himawari-8 from channel 3, 4 and 6. Sataid is software developed by JMA as an application to display satellite imagery and conduct daily weather analysis including tropical cyclone monitoring. One of the core of Sataid system is GMSLPD with specialized functions for Dvorak analysis, display and overlay between satellite imagery and numerical model, use many functions such as vertical cross-sectional chart and time-series chart [9]. The method of Dvorak has been applied to Himawari-8 data set from the start of its operation. The Objective Dvorak Method is being updated to obtain maximal use from the data of new-generation geostationary satellites such as Himawari-8/9 and GOES-R [10]. RGB composite imagery is a technique to display a colour using the property of the three primary colours of light. In satellite imagery processing, RGB technique is used to combine some different channels to make an image which better result rather than only one channel [11]. The locations of the study are Sumatra Island and Kalimantan Island.

3. Results and Discussion

3.1 August 25th 2015

The observation of Himawari-8 imagery has revisit time every 10 minutes. However, in this paper the imagery was taken only at 0600 UTC as samples to monitor the spread of smokes from forest fire. The observation using Sataid GMSLPD as seen on Fig 1.a and Fig 1.b on August 25th 2015 0600 UTC showed smoke of forest fire area with brown colour [12][13] in Sumatera (Riau and South Sumatera provinces) and Kalimantan (Central Kalimantan Province).

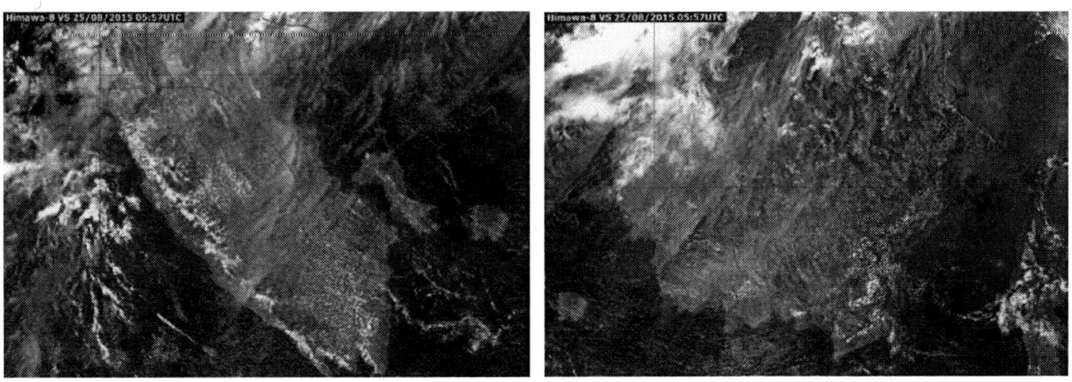

(a) (b)

Figure 1. Himawari-8 Imagery at August 25th 0600 UTC for forest fire area in (a) Sumatra (b) Kalimantan

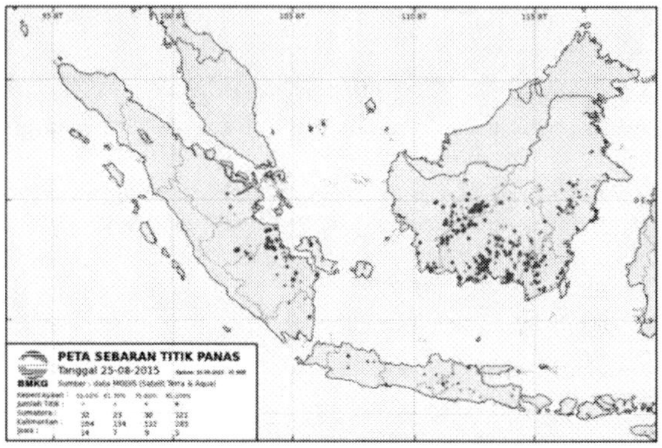

Figure 2 : Hotspot area, August 25th
source : BMKG

In order to endorse the result from RGB false colour analysis, supporting data from hotspot data of Sumatera and Kalimantan on august 25th also provided at Fig 2, the figure showed that most of area which the same area as brown colour in Fig 1.a and Fig 1.b have many hotspots as the signals of forest fire.

The hotspot data showed that the burning area are not exactly the same with the brown colour on the imagery, wind direction played a role in the spread of smoke (Table 1). Wind direction from weather stations in Pekanbaru, Rengat and Palembang for Sumatera were dominated from south and southeast, so that the spread of smoke tend toward northwest of Sumatera. While wind direction in Kalimantan was dominated from southeast and east (Palangkaraya, Pangkalan Bun and Sampit), so the smokes spread toward central and west of Kalimantan. Smoke visibility also was reported from weather stations in Sumatera and Kalimantan as seen on Table 1.

Table 1. Weather Condition on August 25th 2015

Station Name	Weather Condition	Wind Direction
	25-Aug	25-Aug
PekanBaru	Smoke	South
Rengat	Smoke	Southeast
Palembang	Smoke	Southeast
Palangkaraya	Smoke	East
Pangkalan Bun	Smoke	Southeast
MuaraTeweh	Smoke	Calm
Sampit	Smoke	Southeast

3.2 September 10th 2015

The smoke on September 10th as seen at Fig 3.a and Fig 3.b showed that at 0600 UTC, the locations of hotspot in Sumatera were the same as the smoke and wind direction dominated calm (Table 2), while in Kalimantan the hotspots mostly located in all region (West, Central and South of Kalimantan Provinces) but the brown colour intensively located in central Kalimantan. Wind direction in Kalimantan was dominated from east. Weather stations reported the smoke visibility as seen on Table 2.

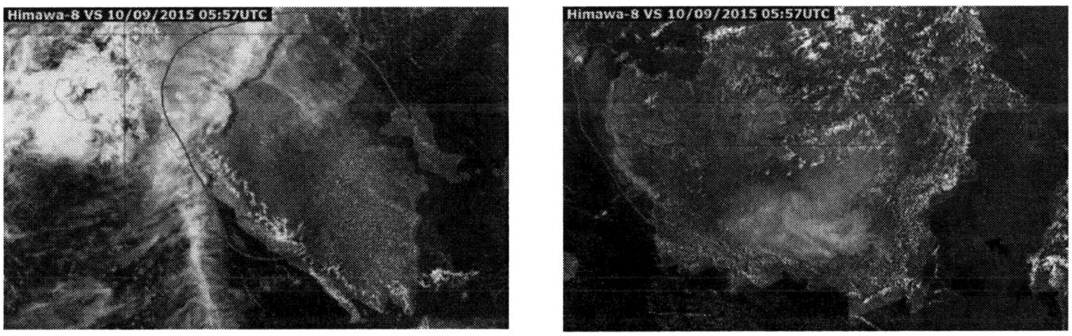

Figure 3. Himawari-8 Imagery at September 10th 06 00 UTC for forest fire area in (a) Sumatra (b) Kalimantan

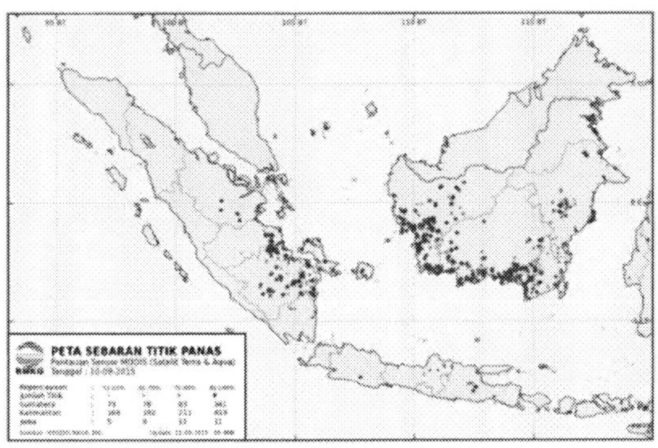

Figure 4 : Hotspot area, September 10th
source : BMKG

In order to endorse the result from RGB false colour analysis, supporting data from hotspot data of Sumatera and Kalimantan on September 10th also provided at Fig 2, the figure showed that most of area which the same area as brown colour in Fig 3.a and Fig 3.b have many hotspots as the signals of forest fire.

The hotspot data showed that the burning area are not exactly the same with the brown colour on the imagery, wind direction played a role in the spread of smoke (Table 2). Wind direction from weather stations in Pekanbaru, Rengat and Palembang for Sumatera were dominated by calm, so that the spread of smoke tend to stay calm. While wind direction in Kalimantan was dominated from east (Palangkaraya, Pangkalan Bun and Sampit), so that the spread of smoke was toward west of Kalimantan. Smoke visibility also was reported from weather stations in Sumatera and Kalimantan as seen on Table 2.

Table 2. Weather Condition on September 10th 2015

Station Name	Weather Condition	Wind Direction
	10-Sep	10-Sep
Pekan Baru	Smoke	Calm
Rengat	Smoke	Calm
Palembang	Smoke	East
Palangkaraya	Smoke	East
pangkalan Bun	Smoke	East
MuaraTeweh	Smoke	Calm
Sampit	Smoke	East

3.3 October 19th 2015

The spread of smoke on October 19th 0530 UTC as seen on Fig 5.a and Fig 5.b showed that the location of smoke in Sumatera were the same as the hotspots because wind direction mostly calm (Pekanbaru and Rengat on Table 3). The spread of smoke in Kalimantan mostly in central and eastern part of Kalimantan, while hotspots were located in Central, South and East Kalimantan, mostly wind direction were calm.

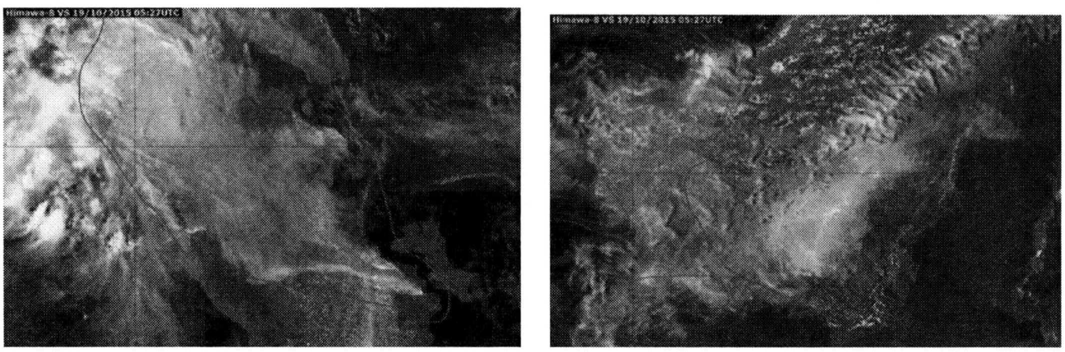

(a) (b)

Figure 5. Himawari-8 Imagery at October 19th 0530 UTC for forest fire area in (a) Sumatra (b) Kalimantan

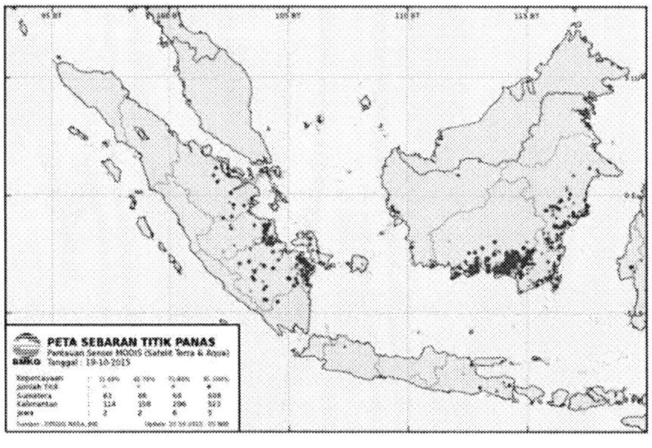

Figure 6 : Hotspot area, Oct 19th
source : BMKG

In order to endorse the result from RGB false colour analysis, supporting data from hotspot data of Sumatera and Kalimantan on October 19th also provided at Fig 6, the figure showed that most of area which the same area as brown colour in Fig 5.a and Fig 5.b have many hotspots as the signals of forest fire.

The hotspot data showed that the burning area are not exactly the same with the brown colour on the imagery, wind direction played a role in the spread of smoke (Table 3). Wind direction from weather stations in Jambi and Palembang for Sumatera were dominated from southeast, so that the spread of smoke tend toward northwest of Sumatera. While wind direction in Kalimantan dominated from east (Palangkaraya and Sampit), so that the spread of smoke toward central and west of Kalimantan. Smoke visibility also were reported from weather stations in Sumatera and Kalimantan as seen on Table 3.

Table 3. Weather Condition on October 19th 2015

Station Name	Weather Condition	Wind Direction
	19-Oct	19-Oct
PekanBaru	Smoke	Calm
Rengat	Smoke	Calm
Palembang	Smoke	Southeast
Jambi	Smoke	Southeast
Palangkaraya	Smoke	East
pangkalan Bun	Smoke	Calm
MuaraTeweh	Smoke	Calm
Sampit	Smoke	East
pontianak	Smoke	Calm

The smokes from forest fire on August, September and October 2015 using RGB technical method could be displayed properly with Himawari-8 especially for areas that were not covered with cloud. The brown colour pattern on RGB false colour imagery of Himawari-8 indicated the smoke from forest fire and land [12][13]. The analysis also supported by the smoke visibility reported from weather stations around the location and hotspots location map resulted from MODIS satellites. However, the location of smoke did not always the same as hotspots because the influence of wind direction. The brown colour as indicator of smoke is the combination of three channels namely channel 3 with the wavelength of 0.64μm (interval between 0.2 – 0.55 μm) as red component, channel 4 with the wavelength of 0.86 μm (interval between 0.3 – 0.55 μm) as green component and channel 6 with the wavelength of 2.3 μm (interval between 0.04 – 0.21 μm) as blue component.

4. Conclusions

It can be concluded that Himawari-8 has advantages in temporal resolution which can revisit every 10 minutes, spatial resolution until 0.5 km and spectral resolution with 16 channels. This indicates that RGB technical method can be applied for forest fire smoke detection. Then the spread of forest fire smoke can be monitoring for anticipation purposes. The limitations of Himawari-8 satellite in smoke detection according to this study that it has passive sensor that very dependent on the reflection of solar radiance. So, it can only monitor the forest fire during day-time. The more brown colour detected the thicker of the smoke, and vice versa. Therefore, supporting instrumentation observation such as LIDAR (Light Detection and Ranging) observation can be obtained in order to complete the smoke condition monitoring.

Acknowledgements
The authors are grateful to BMKG as Indonesia agency for Meteorology Climatology and Geophysic (satellite images data management sub division) for Himawari-8 satellite images, weather report and hotspot data referred to in this study.

References
[1] Tacconi L 2003 Fires in Indonesia: causes, costs and policy implications *CIFOR Occasional paper no. 38* (Bogor:CIFOR)
[2] World bank 2016 The Cost of fire. An economic analysis of Indonesia's 2015 *Fire crisis indonesia sustainable landscapes knowledge note: 1* (Jakarta :TheWorld bank)
[3] Tjahjaningsih A, Sambodo A K and Prasasti I 2005 Sensivity analysis of Modis channels for hotspots and fire smoke detection (in Indonesian) *Pertemuan Ilmiah Tahunan MAPIN XIV Pemanfaatan efektif penginderaan jauh untuk peningkatan kesejahteraan bangsa* September 14 – 15, 2005 Surabaya
[4] Xie Y 2009 Detection of smoke and dust aerosols using multi-sensor satellite remote sensing measurements *Dissertation George Mason University*
[5] Stewart N R 2015 Exploration of the MTSAT2 satellite capabilities for real time detection and characterization of volcanic emissions *Master's Thesis Michigan Technological University*
[6] Kushardono D 2012 *Study of new generation remote sensing weather satelite himawari 8 and 9* (In Indonesian) *Jurnal Inderaja* **03** December 2012
[7] Bessho K *et al* 2016 An introduction to Himawari-8/9-Japan's new-generation geostationary meteorological satellites *Japan J. Meteor Soc* **94** 151
[8] Panjaitan B S and Panjaitan A L 2015 Data utilization of new generation of weather satellite himawari-8 for forest and land fire smoke detection in Indonesia (Case study: forest and land fire in Sumatera and Kalimantan Island on September 2015) (In Indonesian) *Seminar Nasional Penginderaan Jauh* 2015
[9] Tanaka Y 2009 SATAID-Powerful tool for satellite analysis RSMC Tokyo-typhoon center Japan Meteorology Agency (JMA)
[10] Goodman S J *et al* 2012 The GOES-R proving ground: accelerating user readiness for the next-generation geostationary environmental satellite system *Bull. Amer Meteor Soc* **93** 1029–1040
[11] Shimizu A 2015 Outline of RGB Composite Imagery. Meteorological Satellite Center, Japan Meteorology Agency (JMA)
[12] Melbourne Vlab CoE 2016 [cited 2016 6 June]; Available from: www.virtuallab.bom.gov.au/training/hw-8-training/introduction-resources-and-case-studies/
[13] Kerkmann J 2012 Effective training and use of RGB satellite products for forecasters. *Eumetsat RGB Satellite Products Workshop 2012*

Clustering of CO and CO_2 concentration from Sumatra peat fire haze using HYSPLIT and K-means algorithm

M Ni'am[1*], I S Sitanggang[1] and D E Nuryanto[2]
[1] Computer Science Department, Bogor Agricultural University, Bogor, Indonesia
[2] Indonesian Agency for Meteorological, Climatological and Geophysics, Jakarta, Indonesia

E-mail: muslichatunniam49@gmail.com

Abstract. Peat fires in Indonesia could have a negative impact for human life such as the emergence of haze. Therefore, this research aims to analyze concentration of CO and CO_2 due to peat fires in Sumatra in 2015 using the Hybrid Single-Particle Lagrangian Integrated Trajectory (HYSPLIT) and K-Means algorithm. The results of this study indicate that HYSPLIT can be used to obtain the concentration of CO and CO_2 from the haze. Clustering using K-Means algorithm produces an average of the highest concentration of CO is 11.1471 $\mu g/m^3$ and CO_2 is 88.5882 $\mu g/m^3$. Generally, pollutants have an average concentration of 0.0487 $\mu g/m^3$ for CO and 0.3687 $\mu g/m^3$ for CO_2. The pollutant concentration is contained in 45 525 (95%) positions in haze trajectory and the haze spread starting from Riau to Nanggroe Aceh Darussalam.

1. Introduction

Indonesia has abundant natural resources, one of which is peat that has area more than 10% of Indonesia's land area. Peatland is one of the natural resources that has an important role in the economy of the country is as a provider of various timber and non-timber products. Additionally, peatlands also provide a range of benefits that are important to people's lives, including water supply, flood control, carbon storage and protection for biodiversity that live in it [1]. Given the importance of the peatlands role, peatlands should be prevented from any destruction including peat fires.

Peat fires is more difficult to handle than the fires in the highlands because the fire spread under the surface [2]. Global Forest Watch analysis shows 75% warning of hotspots occur in peatlands which is mostly composed of decomposed organic material [3]. Meanwhile, in this decade area of peat fires reached an average of 32.1% in Sumatra and 25.1% in Kalimantan [4].

Peat fires can cause a very broad impact for human life, not only nationally but also globally. One of the impacts of peat fires is the emergence of haze. Haze due to forest fires may bother some neighbouring countries. That is because the haze is an object that the direction of movement influenced by other factors such as weather, directions and speed of wind which making it difficult to predict which areas will be affected by the fire. Negative impact of fire smoke can trigger respiratory disease, decline in crop production and fisheries, disruption of transportation services, and tourism industries. The worst impacts experienced in Indonesia is deteriorating levels of health, loss of standing timber, a plane crash, and the possible extinction of some species [5].

The forest and peat fires that were occurred in Indonesia left a devastating trail of destruction, one of which produces harmful haze in large numbers. This resulted in the closure of hundreds of schools and several local airports, as well as respiratory disorders that affect more than 50,000 people [3]. Anticipating the negative impact of peat fires can be done by analyzing the trajectory patterns and the distribution of pollutant concentrations of haze of fires. By knowing the trajectory and the concentration distribution, information regarding to which areas, how far haze can spread and how much concentrations of pollutants can be obtained.

One indicator of peat fires is a hotspot. Various satellites such as National Oceanic and Atmospheric Administration (NOAA) and MODIS monitor hotspots periodically. The location of hotspots occurred sequentially can be used as the initial location of the movement of haze due to peat fires. Negative impact of haze due to peat fires can be anticipated based on the haze trajectory information, pollutant concentration and location of the hotspots. For example, by knowing the direction of haze movements related parties could inform early warning to the community about the haze and pollutant contained. In addition, this information is also important in evacuation activities.

Research on trajectory pattern has been done before by Anugrah [6] using The Air Pollution Model (TAPM). In addition to TAPM, the Hybrid Single-Particle Lagrangian Integrated Trajectory (HYSPLIT) can be used to identify trajectory patterns. HYSPLIT model has been used to simulate the spread of pollutants NO_2 and SO_2 in the study by Fairuzi [7]. The study by Wang *et. al* [8] identified the transportation path of PM_{10} in Beijing. In addition, the work by Luo & Chen [9] studied the potential sources and transportation path of $PM_{2.5}$ in Shanghai using HYSPLIT modelling, K-Means clustering algorithm, and methods of analysis Concentration-Weighted Trajectory (CWT). This study aims to obtain haze distribution due to peat fires in Sumatra in 2015 using HYSPLIT model and to cluster CO and CO_2 concentrations using K-Means clustering algorithm. It is expected that the results can be used in anticipating the negative impact of fires such as health problems by providing the information about CO and CO_2 concentrations of haze from forest and land fires.

2. Data and methods

2.1. Dataset

The data used in this research are sequence hotspots in peatland of Riau in Sumatra in 2015 and the meteorological data in 2015. Hotspot sequences were obtained from the previous work by Abriantini [10] in which the duration of occurrence is two days. The meteorological data were collected from the Global Data Assimilation Systems with a resolution of one degree (GDAS1). The data are available on the website of Air Resources Laboratory (ARL) NOAA. Each grid data on the GDAS1 has size of 111.1984 km for every longitude and latitude. Meteorological data GDAS1 have several attributes including complex surface pressure (hPa), speed of vertical pressure (hPa/s), height of geopotential (m), temperature (°C), the components of wind U and V (m/s), relative humidity (%), wind direction (degrees) and wind speed (m/s). Wind direction and wind speed are two attributes that have much influence in haze trajectory pattern of peatland fires.

In addition to hotspot sequence data and meteorological data, this study also used emissions data and characteristics of each pollutant as input parameters for determining the concentration of pollutants. Data of CO and CO_2 emissions were obtained from the Global Fire Emissions Database version 4.1 (GFED4.1s) [11]. Emissions data are available on GFED4.1s which are annual emission data for different types of fires distinguished by each region in the world. The data used in this study are peat fire emissions for the region of Indonesia that has the name of Equatorial Asia (EQAS) on GFED4.1s. Pollutant characteristics required in submenu of pollutants deposition in HYSPLIT include molecular weight (g/mol), velocity (m/s), and Henry's constant (mol/L.atm). This study uses the following values of pollutant characteristics. The molecular weight of CO is 28.01 g/mol and 44.01 g/mol for CO_2. Average speed for CO and CO_2 are 454 m/s and 362 m/s respectively [12], Henry's constant is 9.7×10^{-6} mol/L.atm for CO and 3.3×10^{-4} mol/L.atm for CO_2 [13].

2.2. Trajectory pattern mining

Trajectory pattern mining is an extension of the paradigm of sequential pattern mining which analyze the trajectory of moving objects. Trajectory pattern is a pair of dimensional space (areas visited during the movement) and time (duration of the movement). Trajectory pattern mining evolves studies of the spatio-temporal sequential pattern and temporally annotates sequences. In trajectory pattern mining, the Region of Interest (RoI) is associated with the movement area of an object and typical travel time is associated with the movement of an object from one region to another [14].

2.3. Hybrid Single-Particle Lagrangian Integrated Trajectory (HYSPLIT)

HYSPLIT is one of the models used to predict the trajectory, dispersion and concentration of pollutants. This model explains the relationship between the dispersion of pollutants with meteorological components in a certain area. In addition, this model can be applied to identify sources and pathways of pollutant transport. HYSPLIT was developed by the National Oceanic and Atmospheric Administration (NOAA). According [15] HYSPLIT applies two approaches that are Eularian and Lagrangian. Lagrangian approach is based on a parcel of air flowing in the trajectory which is affected by meteorological factors. Change movement of the pollutants from the initial location is included in the Lagrangian approach. Eularian approach is based on the use of grid in the model. Changes in the concentration of pollutants are involved in this approach. Transport processes and chemical reactions are influenced by meteorological factors occurred in the grid. The process causes change of the pollutant concentration at any time, therefore the concentration of pollutants act as a function of time.

2.4. Clustering

Clustering is a technique to divide objects into groups or clusters so that objects that are in the same cluster have similar characteristics to each other and different from the objects that are in another cluster [16]. The K-Means algorithm is a partitional clustering algorithm which is often used because it is simple and efficient [17]. K-Means algorithm is summarized as follows [16]:

1. Select k initial objects as the centroid of the cluster,
2. Insert the objects into clusters whose object is most similar based on the average value of objects that are in a cluster,
3. Renew the centroid of cluster by calculating the average value of objects for each cluster,
4. Repeat step two and three until the objects in the cluster have not changed.

3. Result and discussion

Stages of this research include pre-processing data, determining the location of the initial movement of haze, determining the concentration of pollutant using HYSPLIT, and clustering of CO and CO_2 concentration using K-Means algorithm.

3.1. Pre-processing data

In the pre-processing step, we selected hotspot sequences in the range of July until October 2015 as shown in table 1. Meteorological data were selected based on the hotspots sequence. Meteorological data for each month are divided into five sections namely w1 containing data on 1st to 7th day, w2 containing data on 8th to the 14th day, w3 containing data on the 15th to 21st day, w4 containing data on the 22th to 28st day, and w5 containing data on 29th to 31st day. Table 2 provides the periods of meteorological data that were determined based on hotspot sequences.

Table 1. Location of hotspot sequences

Period of sequence (2015)	First location (*longitude, latitude*)	Last location (*longitude, latitude*)
July 09 – July 11	100.457, 2.083 ; 100.455, 2.082	100.460, 2.077 ; 100.463, 2.078
	100.459, 2.082	100.456, 2.079 ; 100.459, 2.082
July 22 – July 23	101.610, 1.727 ; 101.613, 1.733	101.608, 1.728 ; 101.605, 1.726
July 26 – July 29	102.707, -0.407	102.712, -0.413
August 30 – September 01	102.272, 0.262 ; 102.269, 0.262	102.268, 0.257
September 01 – September 02	101.354, 0.449	101.348, 0.448
	101.347, 0.447 ; 101.348, 0.452	-
October 21 – October 22	102.501, -0.297 ; 102.496, -0.302	102.497, -0.298 ; 102.495, -0.295

Table 2. Meteorological data selection based on hotspots sequence

Date of the first hotspot occurrence	Date of the last hotspot occurrence	Meteorological data
July 09, 2015	July 11, 2015	gdas1.jul15.w2
July 22, 2015	July 23, 2015	gdas1.jul15.w4
July 26, 2015	July 29, 2015	gdas1.jul15.w4 and gdas1.jul15.w5
August 30, 2015	September 01, 2015	gdas1.aug15.w5 and gdas1.sep15.w1
September 01, 2015	September 02, 2015	gdas1.sep15.w1
October 21, 2015	October 22, 2015	gdas1.oct15.w3 and gdas1.oct15.w4

3.2. Determining initial point of trajectory
According to table 1, there are several locations for each period of hotspot sequence. By applying spatial operation, a representative point among the hotspot locations is selected for each period of hotspot sequence. The representative point located in the centre of all hotspots is used as an initial point for haze trajectory. Table 3 presents the selected initial point of haze trajectory.

Table 3. Selected initial point of trajectory for each sequence

Period of sequence (2015)	Longitude	Latitude
July 09 – July 11	100.445	2.080
July 22 – July 23	100.550	1.845
July 26 – July 29	102.662	-0.583
August 30 – September 01	102.103	0.660
September 01 – September 02	101.281	0.432
October 21 – October 22	102.493	-0.279

3.3. Determining concentration of pollutant using HYSPLIT
HYSPLIT results CO and CO_2 concentrations based on the characteristics of the pollutant emissions. Parameters required in determining concentration are emissions (per hour), the duration of emission (h), date and time of starting and ending the simulation, and chemical characteristics of each pollutant, such as molecular weight (g/mol), density (kg/m^3), velocity (m/s) and Henry's constant (mol/L.atm).

CO and CO_2 emissions data until 2014 are available in GFED4.1s. This study approximate emissions in Sumatra in 2015 based on the emissions in 2014 and area of forest fires in Indonesia in 2014 and 2015 that are obtained from the Sipongi application developed by Ministry of Environment and Forestry of the Republic [18]. The estimation values of emissions are 778 423.973 kg/h of CO and 6 260 780.82 kg/h of CO_2. The duration of pollutants emission is per three hours.

Several parameters are assigned to obtain a concentration using HYSPLIT. The parameters include initial location of trajectory, pollutant source height, date, duration and simulation maximum height, as well as the trajectory approach used. The initial location, date and duration are determined based on hotspots sequence. Pollutant source height used is 10 m with maximum height of 10 km. Both values of the parameter are used based on the assumption that all vegetation burned is trees and the maximum of source height is the limit of earth's troposphere. Trajectory approach applied is the Forward Trajectory because the objective of this study is to explore the direction of the haze from the source. HYSPLIT model results as many 47 718 data of pollutant concentration. Figure 1 shows a plot of the concentration of CO and CO_2 which occurred on September 1 at 00:00 until 03:00 UTC or at 07.00 until 10.00 WIB. The concentration plot shows a difference between the CO and CO_2. The concentration of CO at that time have a maximum value of 7.7 μg/m^3 and a minimum value of 6.5 × 10-6 μg/m^3. Meanwhile, at the same time and the same location, CO_2 have a maximum value of 60.0 μg/m^3 and a minimum value of 9.4 × 10-5 μg/m^3. In the first 3 hours on September 1, 2015, the result shows that CO_2 concentration is higher than those of CO.

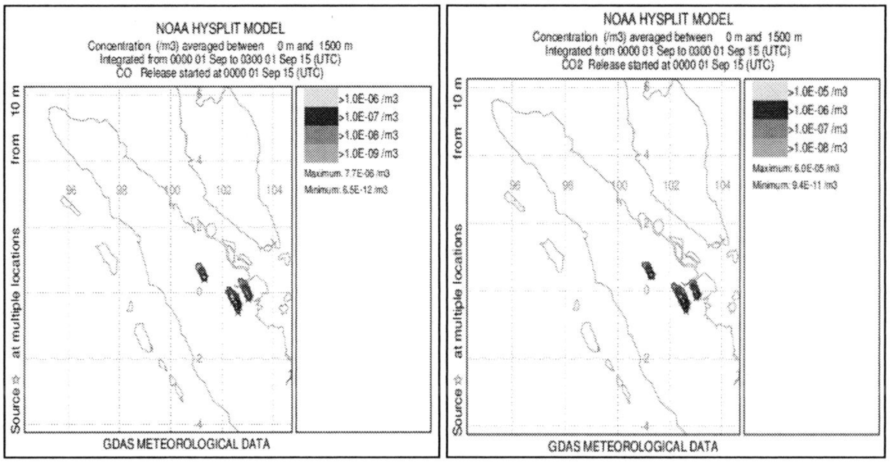

Figure 1. Plot of CO and CO_2 concentration on September 1, 2015

3.4. Clustering of CO and CO_2 concentration

Pollutant concentration data as output of HYSPLIT were converted from kg/m^3 to a common unit, namely μg/m^3 before clustering. Clustering was performed using the K-Means module on Flexible Procedures for Clustering (FPC) package that is available in R. Variables included in clustering are concentrations of CO and CO_2, whereas time (date of occurrence) and the location (longitude and latitude) are used as identifier of pollutant concentration data. Sum Square Error (SSE) of K-means clustering on the concentration of CO and CO_2 on July 9, 2015 until October 22, 2015 showed that the best clustering was obtained at number of cluster (k) of 5. The results of clustering using the K-Means algorithm are given in table 4.

Table 4. K-Means clustering result with 5 clusters

Cluster	Haze locations	Average concentration of pollutants (μg/m^3)		Percentage of cluster members
		CO	CO_2	
1	369	1.7245799	14.5298103	0.773
2	17	11.1470588	88.5882353	0.036
3	1,714	0.4561499	4.1485998	3.592
4	93	4.3258645	35.1397849	0.195
5	45,525	0.0487478	0.3687339	95.404

According to table 4, high concentrations of CO and CO_2 on July 9, 2015 until October 22, 2015 are found in the cluster 2 whereas low concentrations are found in the cluster 5. As many 17 haze locations are members of cluster 2. Most of haze locations, which are 45,525 data, are members of cluster 5. Figure 2 shows haze locations in cluster 2 in which the average concentration of pollutants in these locations is 11.1471 μg/m^3 for CO and 88.5882 μg/m^3 for CO_2. Pollutants concentrations in cluster 2 spread in several sub districts in Riau province which are Kubu, Tanah Putih, Bangko Pusako, Rimba Melintang, Tanah Putih Tanjung Melawan, Rupat, Bukit Batu, Putri Puyu, Kuala Cenaku, Bagan Sinembah, Siak, Rengat, Batang Gansal, Kempas, Dayun, Bandar Petalangan, Gaung, Rengat Barat, Gaung Anak Serka, and Teluk Meranti, converging toward the northwest to the District of Siak, Rupat, Kuala Cenaku, Rengat, Rengat Barat, and Teluk Meranti. These sub districts are located in Rokan Hilir, Bengkalis, Meranti island, Indragiri Hulu, and Siak district. Time periods of the initial movement of haze are July 9, July 22, July 26, August 30 and October 21, 2015 in the first 3 hours. Table 5 provides concentration value at each period.

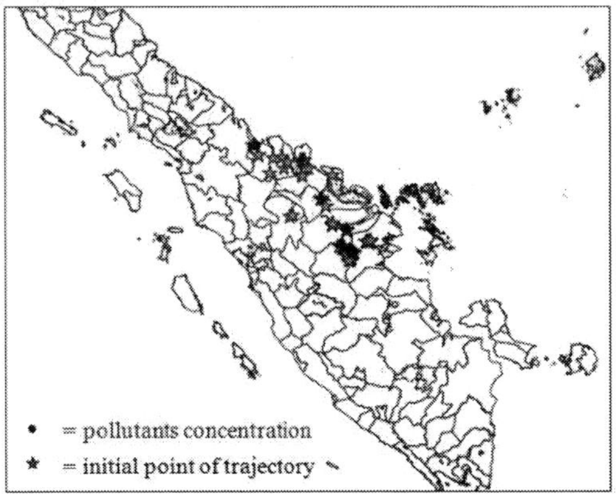

Figure 2. Distribution of the pollutants concentration in cluster 2

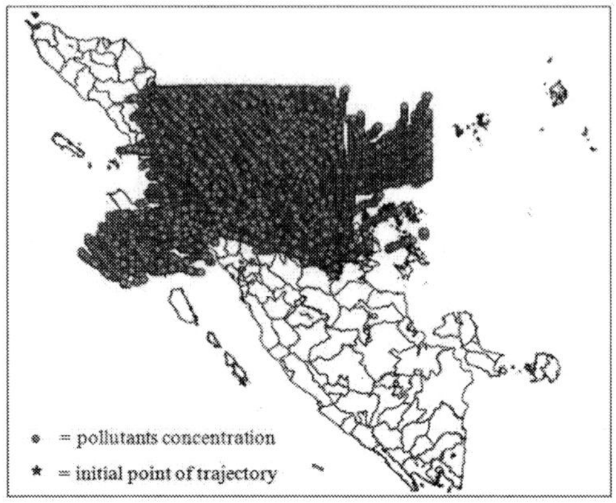

Figure 3. Distribution of pollutants concentration in cluster 5

Table 5. Maximum pollutant concentrations for each period of haze movement in cluster 2

Date	Maximum CO concentration (μg/m³)	Maximum CO_2 concentration (μg/m³)
July 09, 2015	18.0	130.0
July 22, 2015	11.0	96.0
July 26, 2015	20.0	160.0
August 30, 2015	10.0	77.0
September 01, 2015	7.7	60.0
October 21, 2015	17.0	140.0

Pollutants concentration in cluster 5 has an average value of 0.0487 μg/m³ for CO and 0.3687 μg/m³ for CO_2. Plot of pollutants concentration in cluster 5 is given in figure 3. Number of haze movement locations in cluster 5 are 45,525 locations in which only 22,363 locations are found in Sumatra island. As many 23,162 movement are located outside Sumatra island. In general haze movements in cluster 5 occurred on all dates of hotspot sequences namely July 9 until October 22, 2015. Pollutants concentration in cluster 5 spread from Riau Province to the west, northwest, east, northeast, north and to Nanggroe Aceh Darussalam.

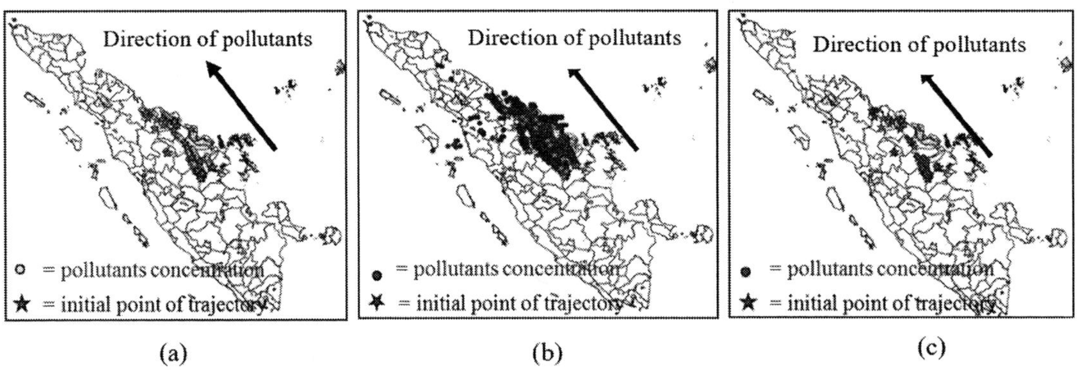

Figure 4. Distribution of pollutants concentration in (a) cluster 1, (b) cluster 3, and (c) cluster 4

Figure 4 shows plot of haze movement locations in cluster 1, cluster 3, and cluster 4. Summary of pollutants distribution in each cluster is stated in table 6.

Table 6. Summary of pollutants distribution in cluster 1, cluster 3, and cluster 4

Cluster	Average concentration (µg/m³)		Number of haze movement	Location of movement	Period of movement (in 2015)
	CO	CO_2			
Cluster 1	1.7246	14.5298	369	Outside Sumatra island as many as 17	July 9, July 22, July 26, August 30, September 1 and October 21
Cluster 3	0.4561	4.1486	1,714	1566 locations in Sumatra island, 148 locations outside Sumatra island, spread toward Riau, Nanggroe Aceh Darussalam and North Sumatra	July 9, July 22, July 23, July 26, July 27, August 30, September 1, October 21, October 22
Cluster 4	4.3259	35.1398	93	Riau	July 9, July 22, July 26, August 30, September 1 and October 21

Distribution of concentration of pollutants in all clusters is shown in figure 7. According to figure 7 and table 6, our study conclude that high pollutants concentration is found in small clusters, cluster 2 and cluster 4, with few number of haze movement locations. These locations are found around the initial points of trajectory that were identified based on hotspot sequences. However, low pollutant concentrations are found in large clusters such as cluster 5 and cluster 3. Haze movement locations in these clusters spread away from the initial location of trajectory. Distribution of concentration of pollutants in each cluster can be seen in figure 7.

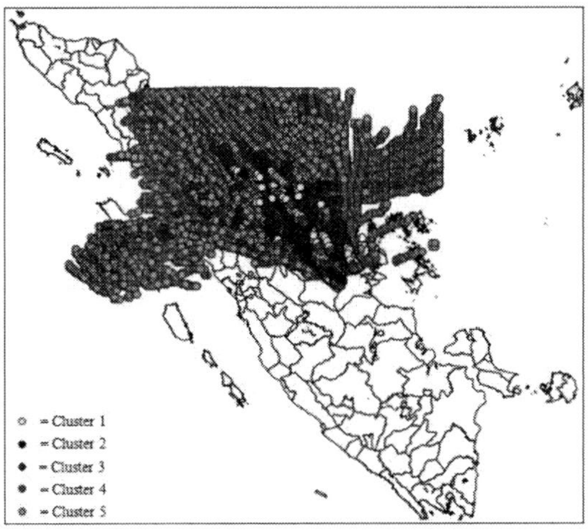

Figure 7. Distribution of the pollutants concentration in all clusters

4. Conclusion

This study found that CO and CO_2 pollutants from peat fires in Riau Province in 2015 spread to the west, northwest, east, northeast, and north of the Riau Province to the province of West Sumatra, North Sumatra, Aceh and Riau Islands. On July 9, July 22, July 26, August 30 and October 21, 2015, high pollutants concentration spread in some districts in Riau province including Rokan Hilir, Bengkalis, Meranti island, Indragiri Hulu, and Siak. The average concentration of pollutants in these locations is 11.1471 µg/m³ for CO and 88.5882 µg/m³ for CO_2.

Distribution pattern of haze and CO and CO_2 concentration from forest and land fires are important information to anticipate negative impacts of forest and land fires especially in health problems of community and global warming. Based on the information of haze trajectory and pollutant concentration, authorized parties can perform warning and early evacuation for people who live around the area affected by haze with high pollutant concentration.

References

[1] Y R Noor and J Heyde 2007 *Community-Based Management of Peatlands in Indonesia. Climate Change, Forests and Peatlands in Indonesia Project* (Bogor: Wetlands International-Indonesia Programme and Wildlife Habitat Canada) pp 2-4

[2] W C Adinugroho, I N N Suryadiputra, B H Saharjo and L Siboro 2005 *Manual for the Control of Fire in Peatlands and Peatland Forest. Climate Change, Forests and Peatlands in Indonesia Project* (Bogor: Wetlands International-Indonesia Programme and Wildlife Habitat Canada) pp 8-9

[3] N Sizer, A Leach, S Minnemeyer, M Higgins, F Stolle, J Anderson and J Lawalata 2014 Preventing forest fires in Indonesia: Focus on Riau Province, peatlands, as well as illegal burning (Online) Available: http://www.wri.org/blog/2014/06/mencegah-kebakaran-hutan-di-indonesia-fokus-pada-provinsi-riau-lahan-gambut-serta

[4] [WWF] World Wide Fund for Nature 2015 Could the decline in hotspots is reached? (Online) Available:http://www.wwf.or.id/tentangwwf/upayakami/iklimdan_nergi/solusikami/ adaptasi/fores fire.cfm.

[5] L Syaufina 2008 *Land and Forest Fires in Indonesia: Fire Behavior, Causes and Impact of Fire* (Malang : Bayu Media)

[6] D F Anugrah 2008 Analysis of haze trajectories from forest fire using The Air Pollution Model (Case Study Kalimantan Forest Fire 2006) Thesis (Bogor: Institut Pertanian Bogor)

[7] F Fairuzi 2012 Use of Hybrid Single-Particle Langrangian Integrated Trajectory (HYSPLIT) for simulating the distribution of pollutants NO_2 and SO_2 Thesis (Bogor: Institut Pertanian Bogor)

[8] F Wang, D S Chen, S Y Cheng, J B Li, M J Li and Z H Ren 2010 Identification of regional atmospheric PM_{10} transport pathways using HYSPLIT, MM5-CMAQ and synopticpressure pattern analysis *J. Environmental Modelling & Software* **25** p 927-934

[9] M Luo and C Chen 2015 *Potential sources and transport pathways of $PM_{2.5}$ in Shanghai, China* 978-1-4799-7749-9 /15/$31.00

[10] G Abriantini, I S Sitanggang, R Trisminingsih. Hotspot Sequential Pattern Visualization in Peatland of Sumatra and Kalimantan using Shiny Framework. *The 3rd International Symposium of LISAT on Food Security and Environmental Monitoring on October 25-26, 2016 in Bogor*

[11] G R Van Der Wer 2016 Global Fire Emission Database (Online) Available: http://www.falw.vu/~gwerf/GFED/GFED4/tables/GFED4.1s

[12] [NPL] National Physical Laboratory 2015 Mean Velocity, Free Path, and Size of Moleculer (Online) Available: www.kayelaby.npl.co.uk/general_physics/2_2/2_2_4.html

[13] R Sander 2015 *Compilation of Henry's law xonstants (version 4.0) for water as solvent J. Atmospheric Chemistry Department* **15** 4399-4981

[14] F Giannotti, M Nanni, D Pedreschi and F Pinelli 2007 Trajectory pattern mining *Proc. International Conference on Knowledge Discovery and Data Mining (KDD'07)* (USA: San Jose, California) p 330-339

[15] R R Draxler and G D Hess 1998 An overview of the HYSPLIT_4 modeling system for trajectories, dispertion, and deposition *J. Australian Metereological Magazine* **47** 295-308

[16] J Han, M Kamber and J Pei 2011 *Data Mining: Concepts and Techniques 3th Ed* (San Francisco: Morgan Kaufmann) p 361-373

[17] B Santoso 2007 *Data Mining:* Data Utilization Techniques for Business Purposes (Yogyakarta: Graha Ilmu)

[18] [Kemenlhk RI] Kementerian Lingkungan Hidup dan Kehutanan Republik Indonesia 2016 Luas Kebakaran Hutan di Indonesia (Online) Available: http://sipongi.menlhk.go.id/hotspot/ luas_kebakaran

Visualization of spatial decision tree for predicting hotspot occurrence in land and forest in Rokan Hilir District Riau

Aji Primajaya[1*], Imas Sukaesih Sitanggang[2], Lailan Syaufina[3]

[1] Department of Informatics Engeneering, Faculty of Computer Science, University of Singaperbangsa Karawang, Karawang 41361, Indonesia
[2] Department of Computer Science, Faculty of Natural Science and Mathematics, Bogor Agricultural University, Bogor 16680, Indonesia
[3] Department of Silviculture, Faculty of Forestry, Bogor Agricultural University, Bogor 16680, Indonesia.

E-mail: aji.primajaya@staff.unsika.ac.id

Abstract. Visualization is an important issue in datamining to easy understand patterns extracted from dataset. This research applied the Bottom-Up Approach method to develop a visualization module for a spatial decision tree in a geographic information system. Spatial data used in this work consists of nine explanatory layers and one target layers. Explanatory layers include maximum daily temperature, daily precipitation, wind of speed, distance of nearest river, distance of nearest road, land cover, peatland type, peatland depth, income source. The target layer contains hotspot and non-hotspot points that occurred in 2008. The result is the visualization module of spatial decision tree that has three main features including mapping window, interactive window, tree node and tabular visualization for predicting hotspot occurrence.

1. Introduction

Predicting hotspot occurrences is important to prevent area from fire risk. Knowledge discovery in database (KDD) is an approach used to search knowledge in large size dataset [1]. Spatial decision tree as one of methods for KDD can be applied to extract knowledge on spatial database that has high accuracy on predicting hotspot occurrence [2]. Output of spatial decision tree is rule based knowledge. On the other hand, the output is not easy to understand by users.

Visualizing output of spatial decision tree algorithm in geographical information system is essential so that users can easily use the system for predicting hotspots occurrence in a certain location. Visualization is the last step in data mining process. Visualization provides several facilities to analyze patterns and spatial relationship in spatial data distribution [3]. A Web-GIS based visualization as a combination of the web technology and the GIS technology [4] can be applied to visualize output of spatial decision trees. The system can retrieve, store, request, analyze, display of spatial data [5]. Developing the system as web engineering needs multidiscipline collaboration [6].

Several studies have been conducted on visualization of knowledge discovery in database. A spatial decision tree developed from the Iterative Dicotomizer 3 (ID3) algorithm was applied in [2] to predict hotspot occurrences in District of Rokan Hilir Riau Indonesia in which the result has higher accuracy than ID3. In addition, a spatial decision tree algorithm was applied on the forest fire dataset that result the classifier with the higher accuracy than C4.5 [7]. The Spatial decision tree also has higher accuracy than logistic regression to predict hotspot occurrences [8]. Visualization method was introduced in [9] based on multiple trees. Tree map method was proposed to visualize spatio-temporal data in [10]. Reviewing classification of visualization method was done in [11]. The interactive map was proposed to visualize non-spatial data in [3]. The interactive map provides several facilities that support users to input non-spatial parameters and to interpret the output of C4.5 algorithm.

The bottom-up approach was applied in [12] to develop a web-GIS system. This approach consisted of four steps that was used to develop a geographic information system for health exposure assessment in Europe area [12]. This research results a system that analyze real spatial data uploaded by experience users or non-experience users. The system was developed using open source software. Therefore the cost of system development can be minimized. The system was tested on arsenic concentration of Europe area for impact analysis. A web-GIS was applied in [6] to visualize the pest life cycle simulation using the Software as Service Approaches (SaaS) method. The research developed a system that capable to analyze impact of olive fruit fly to agricultural goods damage. It predicts olive fruit fly life cycle. The system used PHP CRUD framework and Google Map API. A web-GIS was adopted to visualize analysis of land suitability in the study area Yogyakarta and Central Java in [11]. The research provides spatial information for supporting agricultural businesses. A web-GIS to visualize agricultural characteristics and the ecological system was applied in [13]. The system has four main features including thematic map, spatial data, data analysis and decision support facilities. Research on the development of geographic information system to help mitigate natural hazard in Aceh has been done on [14].

This work developed a visualization module in a form of a web-GIS using the Bottom-Up Approach method that adopted from [12] to visualize output of the spatial decision tree algorithm that was developed in the previous work [7]. The method is composed by four main stages namely system requirements, basic system design, software selection and system development [12]. The system predicts hotspot occurrences in land and forest for the study area of Rokan Hilir District, Riau Indonesia. The system was developed by open source software therefore the development cost is low. The spatial data in the system consisted of nine explanatory layers and one target layer. Explanatory layers consist of maximum daily temperature, daily precipitation, wind of speed, distance of nearest river, distance of nearest road, land cover, peatland type, peatland depth, income source and the target laye is the hotspot layer.

2. Research method

2.1. Study area and forest fire data

The research developed a visualization module of spatial decision tree for predict hotspot occurrences in the study area Rokan Hilir district. The district was located in Riau Province where the coordinate is between 100°16' – 101°21' East Longitude and 1°14' – 2°30' North Latitude [7].

Explanatory layers include physical layers, socio-economic layers, weather layers, peatland type layers and peatland depth layers. Physical layers consist of river, road and land cover. Weather layers are precipitation layer, screen temperature layer and wind speed layer. Socio-economic data is income source. The target layer is hotspot data on the year of 2008. The data sources are provided on Table 1.

Table 1. Data and its source.

Data	Source
Spread and coordinates of hotspot 2008.	FIRMS MODIS Fire/Hotspot, NASA/ University of Maryland
Weather data 2008 : maximum daily temperature, daily precipitation, wind of speed	Meteorological Climatologically and Geophysical Agency (BMKG)
Physical data 2008: distance of nearest river, distance of nearest road, land cover	National Coordinating Agency for Survey and Mapping (BAKOSURTANAL)
Peatland type layer, peatland depth layer	Wetland International
Income Source layer	BPS -Statistic Indonesia

2.2. System development method

This study uses the Bottom-Up Approach to develop the system. The method is divided into four main steps. The steps are system requirements, basics system design, software selection and system development.

- System requirement
  The system requirement includes two tasks namely essential requirement and additional requirement.
- System design
  The system must have user friendly interface where divided into three main parts, namely spatial database, map sever, web viewer. Geodatabase is a part of the system that store spatial database. Map server connects between database and user interface or web viewer. Web viewer is a part of the system that interacts with users. The design of the system was recommended to use open source software and users do not need additional software to interact with the system.
- Software selection
  The selection of software was done based on the system design. Geodatabase needs open source software that can conduct spatial query and store the spatial data.
- Implementation and testing
  The implementation was done based on the system design and system requirements. System testing was done on this stage to evaluate the system performance to visualize of rule based knowledge.

2.3. Spatial Decision Tree Algorithm

The spatial decision tree algorithm in [2] was developed from the ID3 algorithm. It was one of several data mining algorithms to create classification models on spatial data [2]. This algorithm has higher accuracy than several methods such as non-spatial ID3 [2], logistic regression [8], C4.5 [7]. Spatial dataset for creating spatial decision tree includes explanatory layers and target layer.

There are several measures to be calculated in the spatial decision tree algorithm such as Spatial Join Relation, spatial entropy, spatial informational gain. Spatial Join Relation (SJR) is developed from Spatial Join Index (SJI). SJR is the result of features calculation on spatial relation between two layers [2]. Spatial entropy is entropy applied for spatial dataset. This algorithm needs several inputs including spatial dataset, explanatory layers, target layer and SJR [7]. This algorithm creates spatial decision tree as the output. From the tree several rules can be generated. There are 131 rules from spatial decision tree that was visualized by the system. Five examples of rules are as follows :

- Rule 1

IF Income source = Forestry AND Land Cover = Bare Land AND Wind Speed = 0 m/s – 1 m/s AND Screen Temperature = 297°K – 298°K AND Peatland Depth = very deep/ very thick > 400 cm THAN Hotspot Occurrence = False.

- Rule 2
  IF Income source = Forestry AND Land Cover = Bare Land AND Wind Speed = 0 m/s – 1 m/s AND Screen Temperature = 297°K – 298°K AND Peatland Depth = Moderate 100 – 200 cm THAN Hotspot Occurrence = True.
- Rule 3
  IF Income source = Forestry AND Land Cover = Bare Land AND Wind Speed = 1 m/s – 2 m/s THAN Hotspot Occurrence = True.
- Rule 4
  IF Income source = Forestry AND Land Cover = Shrubs AND Wind Speed = 1 m/s – 2 m/s AND Screen Temperature = 297°K – 298°K THAN Hotspot Occurrence = False.
- Rule 5
  IF Income source = Forestry AND Land Cover = Shrubs AND Wind Speed = 1 m/s – 2 m/s AND Screen Temperature = 298°K – 299°K THAN Hotspot Occurrence = True.

3. Development of spatial decision tree visualization

3.1. Concept of Spatial Decision Tree Visualization

This visualization was made in the form of Web GIS that represents the knowledge as output of the spatial decision tree algorithm. It can predict potential forest and land fires based on hotspot occurrences in the district of Rokan Hilir. This ability can support decision making for preventing and detecting land and forest fire. In order to use the system, users are not required to have background of spatial analysis.

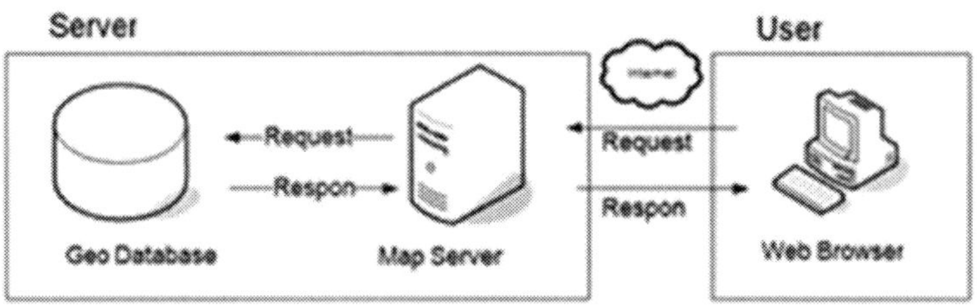

Figure 1. Main components of Web GIS

Figure 1 illustrates the main components of web-based GIS for spatial decision tree visualization. Target data and the explanatory data are stored on the spatial database. Users interact with the system using web browser and spatial data can be retrieved through map server to predict hotspot occurrences using interactive facilities in the system.

3.2. Spasial Visualization of Spatial Decision Tree

3.2.1. System requirement. Basic requirements of the system are visualize spatial data and presenting model. Additional requirements are tree node facility and tables for displaying query results of spatial data.

3.2.2. System design. The base map layer of Rokan Hilir is stored in the spatial database . The Google map that shows Rokan Hilir is displayed on the main page. The explanatory layers and the target layers can be displayed using visualization tool in the web page. The basic system design is showed on Figure 2. In general, the overall system design has two parts, namely client and server. There are several facilities on client side that allow users to interact with the system. Users can query layers through a browser and the associated query will be forwarded to the server side. There are map server and database server in the server side. Map server will change spatial data that are stored in the database to be layers so it can be queried by the client side. The database server stores spatial data that are needed by the system. Spatial data are vector format in which the target data are point features and the base maps consist of polygon and line features. In addition, the district border layer is stored in the database. The Google base map layer is not stored on the database server but it is queried directly in the form of vector.

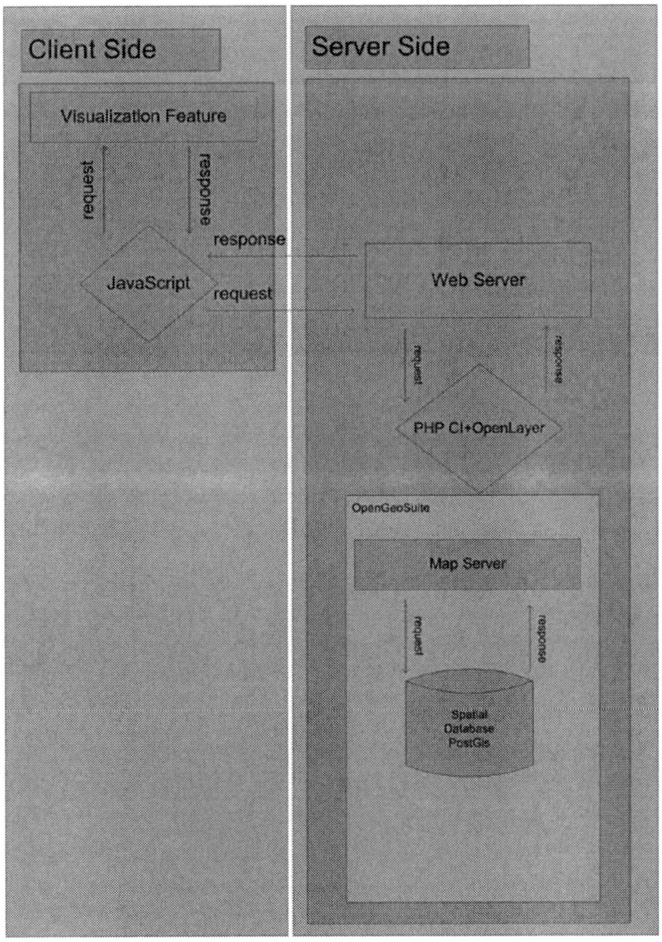

Figure 2. Architecture of the system

3.3. *Software selection*

This work utilizes OpenGeoSuite 3.0, OpenLayer, PHP CI, Proj4Js. OpenGeoSuite 3.0 which consists of PosgreSQL, PostGIS, Geoserver and GeoWebChace. There are several open source software that can be applied to develop the geodatabase such as PostgreSQL and MySQL. PostgreSQL is the open source software that can facilitate requirement of database system. PostgreSQL is combined with PostGIS to perform spatial data processing. The software was selected because it has larger range of spatial data processing than MySQL [12]. The software is able to perform spatial queries because it has spatial analysis function for developing GIS. The software can store spatial data in tabular form and it can manage geometry data using spatial queries.

Map server can be used to connect spatial data to the internet in which it can be requested to be displayed on a Web GIS page. Geoserver is the alternate map server software that can be adopted to the system. The spatial data is connected to a map server by creating workspace and then it is uploaded to the spatial map server with certain name. The name becomes the layer name and it can be requested to display spatial data or to visualize results of spatial data analysis using a certain algorithm.

The web viewer needs software to visualize of rule based knowledge from the spatial decision tree. Several frameworks can be used to develop web viewer. PHP CI is the framework to develop an user interface. It is combined with several frameworks to make the web-GIS viewer. The other frameworks that were used are proj4js, openlayer. Proj4js is the javascript software to convert the coordinate system of layer. The layer that has certain coordinate system can be easy converted to other coordinate system by this framework. OpenLayer is the framework to construct web-GIS with several libraries. OpenLayer can be embedded to PHP CI by requesting libraries. OpenLayer provides several functions to request layers. It is collaborated with Google API.

3.4. Implementation and Testing

Implementation of the system produces three types of visualization facilities. The facilities are the map window, an interactive window, tree node and tabular visualization. Black box was used for system testing to ensure that the system is running well. Figure 3 shows the main web page of the system.

3.4.1. Map Window. The spatial data requested by users will be visualized on several parts of webpage including the map window. The map window shows the base map layer and the marker as an indicator of selected target (hotspot). The marker gives information about selected hotspot location and prediction results of the spatial decision tree. The red marker indicates true hotspot occurrence and the blue for false hotspot occurrence. Zooming scale allows user to enlarge an important area.

3.4.2. Interactive Window. The interactive window provides several features that is composed by scroll down feature, prediction button, location button, and prediction result window. Scroll down feature facilitates users to choose explanatory attributes. Prediction button is the prediction trigger. The prediction result of hotspot occurrence is showed in the prediction result window. Location button is the trigger of marker on mapping window.

3.4.3. Tree Node and Tabular Visualization. Tree node facilitates user to query all explanatory layers and the target layer. It displays query results in the tabular form and maps. Tabular visualization provides information about prediction of hotspot occurrences.

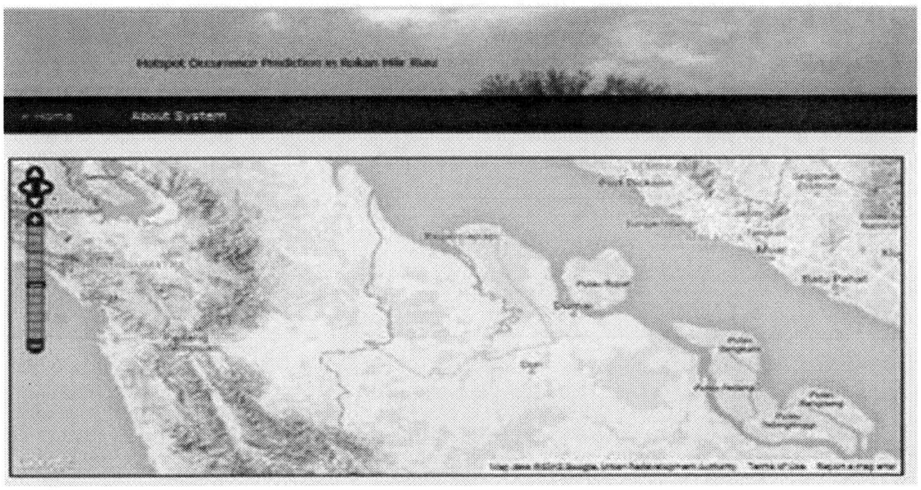

Figure 5. Map window

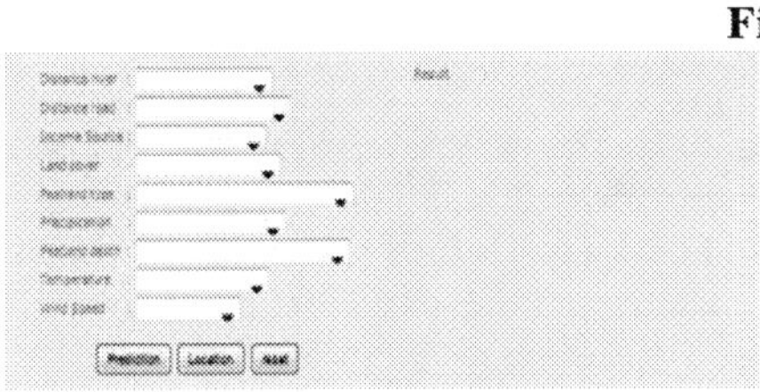

Figure 6. Input parameter

Figure 7. Tree node

4. Results and discussion

The visualization of spatial decision tree has facilities to provide prediction of hotspot occurrences so we can determine potential high risk of forest and land fire in the Rokan Hilir. Figure 4 the visualization result of rule 117 (if distance river is low and distance road is low and income source is plantation than false) on spatial ID3 models for predicting hotspot where the visualization shows that the predicted locations are not potential hotspot.

The system development method has the similarities with [12] but there are novelty. The first is on classification method used to create the model in which the method in this study is spatial ID3 whereas in [12] 4 types of classification include equal range, equal count, natural breaks, and standard deviation. Equal method divides the data into several class with the same interval range for each class. Equal count method is a method of classification that divides the data into several classes which have the same proportion to the sample. Natural breaks method is a method where the classification is based on the grouping of making the interval on real data. Standard deviation method is a method in which the making class based on the value of the standard deviation. The second is in terms of the cases studied. In this study, statistical analysis was not performed but only testing that is done at the final stage with black box method.

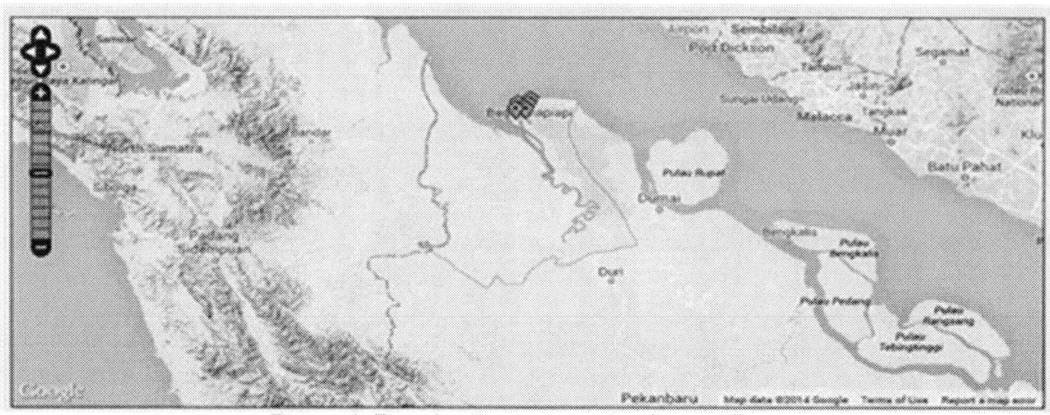

Figure 4. Visualization result of a classification rule

5. Conclusion

The study produces a visualization system which is able to provide several facilities that can make visualization of hotspot occurrences prediction on land and forest in Rokan Hilir district Riau Province. The system provides output of spatial decision tree algorithm in the form of mapping window, interactive window, tree node and tabular forms. The visualization allows users to easy understand the spatial decision tree to predict hotspot occurrences based on spatial data that are stored in the database. In further research, a facility to accommodate users to input new data will be developed.

References

[1] Han J, Kamber M and Pei J 2011 *Data mining : Concept and Technique 3nd edn* (US: Morgan Kauffman)
[2] Sitanggang I S, Yaakob R, Mustapha N and Ainuddin A N 2013 *Journal of computer science* **9** 244-251
[3] Andrienko G L and Andrienko N V 1999 *Proceeding of the 3rd International Symposium IDA-99* vol **1642** (Amsterdam) p 747-755
[4] Zhishana Y, Run'e L, Yanjiang W and Xiaoling S 2012 *International Conference on Future Energy, Environment, and Materials Energy Procedia* vol **16** (Hong Kong) p 1199 – 1205
[5] Chang K T 2008 *Introduction to Geographic Information System 4th edn* (New York: Mcgraw-hill)
[6] Karydis I, Gratsanis P, Semertzidis C and Avlonitis M 2013 *Proceeding of the 6th International Conference on Information and Communication Technologies in Agricultural, Food and Environment (HAICTA)* vol **8** (Corfu Island) p 526-529
[7] Sitanggang I S, Yaakob R, Mustapha N and Ainuddin A N 2013 *TELKOMNIKA* **12** p 511-518
[8] Sitanggang I S, Yaakob R, Mustapha N and Ainuddin A N 2013 *Journal of applied sciences* **13** p 252-261.
[9] Graham M and Kennedy J 2010 *Information Visualization* **9** p 235-252
[10] Slingsby A, Dykes J and Wood J 2008 *Information Visualization* **7** p 210-224
[11] Hazain F A, Harisno and Legowo N 2013 *International Conference of Advances Science and Contemporary Engineer* vol **50** (Jakarta) p 532-543
[12] Evans B and Sabel C E 2012 *International Journal of Health Geographics* **11** p 2
[13] Zhang H, Yi S and Wu Y 2011 22 International Conference in Energy Engineering (Bangkok) **14** p 382-386.
[14] Nasaruddin, Munadi K, Dirhamsyah M and Yuliansyah D 2011 *TELKOMNIKA* **9** p 89-98

Spatial temporal clustering for hotspot using kulldorff scan statistic method (KSS): A case in Riau Province

S A Hudjimartsu[1*], T Djatna[2], A Ambarwari[1] and Apriliantono[1]

[1] Department of Computer Science, Faculty of Math and Natural Sciences, Jl. Meranti Wing 20 Level 5 Kampus IPB, Babakan, Dramaga, Bogor, Jawa Barat
[2] Department Agroindustrial Technology, Faculty of Agricultural Technology, Gedung Fateta lt.2, Kampus IPB, Dramaga, Bogor, Jawa Barat

E-mail: shudjimartsu@gmail.com

Abstract. The forest fires in Indonesia occurs frequently in the dry season. Almost all the causes of forest fires are caused by the human activity itself. The impact of forest fires is the loss of biodiversity, pollution hazard and harm the economy of surrounding communities. To prevent fires required the method, one of them with spatial temporal clustering. Spatial temporal clustering formed grouping data so that the results of these groupings can be used as initial information on fire prevention. To analyze the fires, used hotspot data as early indicator of fire spot. Hotspot data consists of spatial and temporal dimensions can be processed using the Spatial Temporal Clustering with Kulldorff Scan Statistic (KSS). The result of this research is to the effectiveness of KSS method to cluster spatial hotspot in a case within Riau Province and produces two types of clusters, most cluster and secondary cluster. This cluster can be used as an early fire warning information.

1. Introduction

Recently the forest fires in Indonesia is an annual catastrophe that occurs when the dry season. About 61% of forest fires in Southeast Asia occurred in Indonesia in the period 1990-2010 [1]. Almost all of the causes of forest fires are caused by the human activity itself, among others forest fires and land clearing for plantations, sparks from plantation or forest and sabotage [2]. The impact of forest fires is the loss of biodiversity, pollution hazard and harm the economy of surrounding communities. Given the impact of the fires are very harmful and the factors causing the fire complex, it is important to develop early warning systems for the prevention of forest fires. Indication of land and forest fires can be known through the hotspots detected in a location at a certain time.

The hotspots area data derived from remote sensing satellite imagery. Hotspots data provide information such as location and time of occurrence, and as an early indicator of land and forest fires. To provide better information to assess that hotspots are a fire, there should be a method in processing. One of method is by clustering, grouping spatio temporal data to produce clustering that can be judged from grouping, the clustering can be regarded as a fire or not. Clustering method for spatio temporal data frequently used and adopted with KSS (Kulldorff Scan Statistic) [3]. KSS method was first introduce by Kulldorff et al. (1997) which is the result of the development of scan statistics method [4]. Other than that KSS method has been able to detect clustering of point in 3 dimensions, the dimension spatial, the dimension of time (temporal) and as well as dimension of spatial and time (spatio temporal) [5].

2. Previous Work

Research on data processing hotspot has been done to prevent forest fires through the development of early warning systems. Sitanggang and Ismail (2010) observed hotspots data processing with classifying by decision tree method, using on algorithm j48 or C4.5 algorithms, the resulting in a ruleset that is used to predict the occurrence of hotspots based on parameters such as type of land, roads and river network [6]. In 2014 Sitanggang and Khoiriyah develop research in previously (2010), which connects between the spatial data in the data processing so as to generate spatial decision tree [7].

In addition to classification, data processing hotspot as an early warning system can also be done using clustering techniques. The study was conducted by Wulandari (2012), which applies the Dynamic Density Based Clustering Algorithm (DDBC) are known to be able to handle spatial and temporal data proposed [8]. Usman (2015) observed the clustering based on density for the distribution of hotspots as indicators of forest fires and peatland in Sumatra in 2002 and 2013. The method used for clustering is DBSCAN-in this study found 53 clusters in 2002 and 42 clusters in 2013 [9].

Annisa (2015) using Kuldorff Scan Statistics (KSS) the great method for detecting hotspots grouping based on spatial and temporal aspects. Clustering of hotspot using KSS method can detect where or when the occurrence grouping of hotspots and areas of distribution [10].

3. Methodology

This research was divide into three parts: area, tools and stages of research. Area of research in Riau Province with approximation of area 89,150.16 km^2 and has an area of peat land 40,087.76 km^2. The data that used are hotspots of the year 2010 – 2014, sourced from FIRM MODIS fire can downloaded at website http://earthdata.nasa.gov/data/nrt-data/firms. For peat land data sourced from ministry of environment and for administrative boundaries Riau Province sourced from geospatial information agency (BIG). Tools that used in this research are: R application, "SpatialEpi" and Quantum GIS. The stages are doing in this research consisted of six stages: collection and analysis data, preprocessing of data, implementation KSS method with Poisson model for determining likelihood, determine hotspots clustering with KSS method, cluster validation and visualization.

3.1. Collection Data

Hotspots data sourced from FIRM MODIS fire, which is data spatio temporal consist of location and time dimension. Dimensional location of the hotspot data is the latitude and longitude, and the time dimension is the date of occurrence of hotspots.

3.2. Preprocessing of Data

3.2.1. Preprocessing of hotspots data

Preprocessing on the hotspot data is divide into three phases:

a) Data Extraction
At this phase, selected attribute from hotspot data which consist of latitude, longitude, and date only, the other attribute will be eliminated and divide data into two types (overlay/clip)
b) Selection of hotspots based on peatland
This following result overlay (clip) with peatland.

Table 1. Number of hotspots in 2010-2014

Year	Before overlay (clip)	After overlay (clip)
2010	4,148	2,988
2011	6,864	4,827
2012	7,873	4,688
2013	15,103	10,948
2014	21,591	19,332

c) Selection distribution of hotspot locations by district
In the area of Riau Province grouping based on subdistrict boundary, so that the occurrence of hotspots can be found in each subdistrict.

3.3. Implementation of KSS method
The implementation of KSS method is based on the determination of the likelihood value on the scanning results, shape scanning of KSS method is a circular window. According to Wen (2009) that the radius of the circular window size varies from zero to a predetermined upper on any given centroid. Each circular window has a likelihood ratio λ (Z) which of each was calculate based on observed data and the probability model that has been chosen, in this research were taking Poisson models. Further, look for the maximum likelihood ratio and the candidate most likely cluster obtained, then the next stage is to test the significance of using a Monte Carlo approach.

3.4. Scanning Window
Clustering in KSS method based on scanning window with the phases as follows:
a) Random centroid/cell in research area.
b) Make a circle centered at centroid/cell of the closest neighborhood to the farthest distance and not exceed 50% of the population (Kulldorff et.al 1997).
c) Count the number of incident cases (the number of hotspot in an area) and population (the entire area of peatland) for each centroid neighborhood.
d) Calculating of the likelihood ratio of each pairwise.
e) Repeat steps 1 – 5 for all centroid in research area.

3.5. Determining of Hypothesis
Scanning window is used to detect hotspots, then grouped based on the average value of the hotspots in the scanning window. Let p be the average value of the occurrence of hotspots in the scanning window and q is the average value of the occurrence of hotspots outside the scanning window.
H0 : There is no clustering of hotspots in peatland (p = q).
H1 : There is clustering of hotspots in peatland (p > q) in circular scanning window (Z).
p = q can be interpreted by the number of case is same in every area/spread evenly. Meanwhile, p > q is defined as occurring grouping of hotspots in the scanning window.

3.6. Determining of Likelihood Ratio
Probability function, f(x) which represented the probability of the number of occurrences within a cell x described in equation 1 as follows (Kulldorff et.al 1997):

$$(x)=\begin{cases}\dfrac{e^{-p\mu(x)}(p\mu(x))^{n_x}}{n_{x!}}, x\in Z\\[2ex] \dfrac{e^{-p\mu(x)}(p\mu(x))^{n_x}}{n_{x!}}, x\notin Z\end{cases} \quad (1)$$

Where,
f(x) = Occurrence probability in cell x.
p = Average value of occurrence of hotspots in the scanning window.
q = Average value of occurrence of hotspots outside the scanning window.
μ(x) = Number of population in the cell x.
n_x = Number of cases in the cell x.
Z = circular scanning window.
e = 2.71828.

KSS method and Poisson models were used to compare the number of cases that are inside and outside the scanning window, which is used to find the hotspots. To calculate the likelihood ratio function λ (Z) with Poisson models for each circular scanning window Z (Kulldorff 2014) can be seen in the following equation 2.

$$\lambda(Z)= \begin{cases} \left(\frac{n_z}{e_z}\right)^{n_z} \cdot \left(\frac{n_G-n_z}{n_G-e_z}\right)^{n_G-n_z}, & \text{if } n_z>e_z \\ 1, & \text{other} \end{cases} \quad (2)$$

Where,
n_z = Number of cases in the scanning window Z.
e_z = Expected of cases in the scanning window Z.
n_G = Number of cases in research area G.

In which expected value of cases e_z obtained from following equation 3.

$$e_z=\mu(Z)\left(\frac{n_G}{\mu_G}\right) \quad (3)$$

Where,
μ(Z) = Number of population in the scanning window Z.
n_G = Number of cases in the research area G.
μ(G) = Total of population in the research area G.

3.7. Cluster Validation
The results of the validation can be determined by calculating the statistical significance p-value or calculate using a Monte Carlo approach (Kulldorff et.al 1997). To calculate the p-value can use the following equation 4.

$$p=\frac{(T(x))\geq t_0}{m+1} \quad (4)$$

Where,
T(x) = The highest likelihood ratio value of data replication.
t_0 = The highest likelihood ratio value of real data.
m = Many simulations.

Monte Carlo calculations used to find the value of p that good, then the number of simulations is limited to values ending in 99, such as 1999, 2999, or 9999. To accept or reject H 0 is typically used alpha level value (α) by (α) = 0.1, (α) = 0.05, (α) = 0.01 and (α) = 0.001 (Kulldorff et.al 1997).

3.8. Cluster Analysis
At this stage, a cluster analysis is done based on physical characteristics of peat in Riau province, to get information on the peatland hotspot. The physical characteristics of peatland including the type and thickness of the peat land and land cover.

4. Results and Discussions

The results of the study the distribution of hotspots in the province of Riau, especially in peatland can be seen in the following figure.

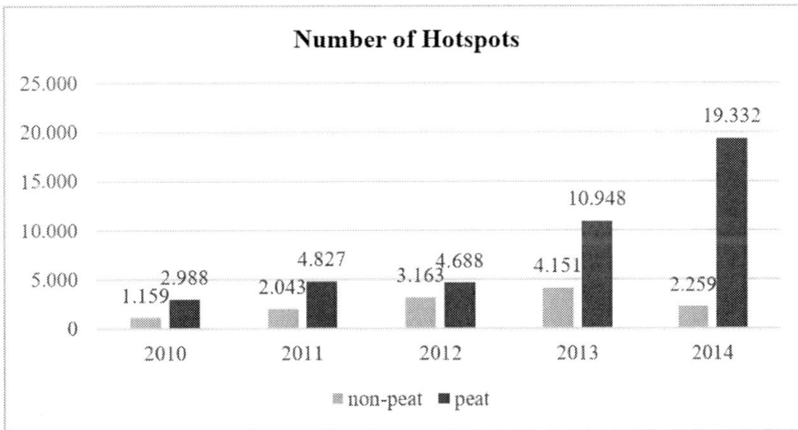

Figure 1. Number of hotspots in peat and non-peat Riau Province

Based on Figure 1, shows that the frequency of occurrence of hotspots in Riau particularly on peatland in the period 2010-2014 has increased.

Peatland is distinguished based on the type of maturity and thickness. Type maturity consists of Fibrists (immoldy), Hemists (half moldy), Saprists (molded) and the combination of these three types of maturity. Meanwhile, peatland based on the thickness divided into three: moderate depth, deep and very deep. The distribution of hot spots in the type of maturity peatlands can be seen in table 2.

Table 2. Distribution hotspots in type of maturity peatlands in 2010-2014

Type of maturity peatlands	2010	2011	2012	2013	2014	Average/year
Fibrists/Saprists (60/40),moderate depth	1	0	0	3	2	1.2
Hemists/min (30/70),shallow	11	22	52	95	33	42.6
Hemists/min (30/70), moderate depth	2	0	7	9	23	8.2
Hemists/min (90/10), moderate depth	0	1	2	0	1	0.8
Hemists/Saprists (60/40),deep	297	355	324	675	906	511.4
Hemists/Saprists (60/40),very deep	1,095	1,094	1,103	1,096	1,097	1,097
Hemists/Saprists (60/40), moderate depth	394	510	309	1,007	3,598	1,163.6
Saprists (100),deep	0	0	0	0	0	0
Saprists (100),moderate depth	68	178	121	131	303	160.2
Saprists/Hemists (60/40),deep	239	576	453	519	2894	936.2
Saprists/Hemists (60/40),very deep	620	900	1,011	1,546	3,315	1,478.4
Saprists/Hemists (60/40), moderate depth	121	125	75	153	413	177.4
Saprists/min (50/50),shallow	2	15	17	78	42	30.8
Saprists/min (50/50), moderate depth	30	77	104	470	143	164.8
Saprists/min (90/10), moderate depth	109	418	336	380	1106	469.8

From the type of peat maturity, which occurs on the type Hemists hotspot / Saprists (60/40) is very deep, Saprists / Hemists (60/40) is very deep, and Hemists / Saprists (60/40) deep enough. While the thickness of the peat is divided into five, which consists of:

1) Peaty soil, < 50cm (D0)
2) Shallow, 50-100cm (D1)
3) Moderate, 100-200cm (D2)
4) Deep, 200-400cm (D3)
5) Very deep, >400cm (D4)

Table 3. Distribution of hotspots in thickness of peat in 2010-2014

Thickness of peat	2010	2011	2012	2013	2014	Average/year
D0	2	22	24	9	9	13.2
D1	134	213	167	376	709	319.8
D2	796	1398	1087	2363	5933	2,315.4
D3	701	1440	1706	3345	4449	2,328.2
D4	1356	1767	1706	4863	8243	3,587

From Table 3, hotspot often occurs at the level of D4 (very deep, >400cm), D3 (in, 200-400cm) and D2 (Moderate, 100-200cm). This indicates that the thicker or more in peat, then the frequency of occurrence of a more frequent hotspot. Aside from factors peat, hotspots can occur because of land cover in Riau Province.

Table 4. Distribution of hotspots in cover in 2010-2014

Land Cover	2010	2011	2012	2013	2014	Average/year
Thicket swamp	296	396	327	932	2407	871.6
Swamp forest	2,091	3,444	3,653	8,530	13,718	6,287.2
Rubber plantation	22	24	20	66	74	41.2
Coconut on a former swamp forest> 5 years	199	334	260	416	1,553	552.4
Palm on former swamp forest <5 years	148	174	203	203	496	244.8
Palm on former swamp forest >5 years	77	158	69	323	204	166.2
Forest land concessions	36	164	59	199	191	129.8
Land cultivation of industrial plants	1	0	9	3	3	3.2
Plantation preparation	10	16	18	13	239	59.2
Rice fields and coconut	103	108	51	250	425	187.4
Intensive rice fields	0	0	0	0	0	0
Shrubs and marsh grasses and former fire	5	17	12	16	29	15.8
Shrubs, grass on a former rice field	1	5	9	5	4	4.8

Distribution of hot spots based on land cover with high frequency occurs in swamp forest in the period 2010 to 2014, with an annual average of about 6,287.2 hotspots.

4.1. Spatial Clustering with KSS Method

Results of clustering the hotspot data using KSS method within period from 2010 to 2014, obtained cluster consisting of two classes cluster composed of most cluster and the secondary cluster.

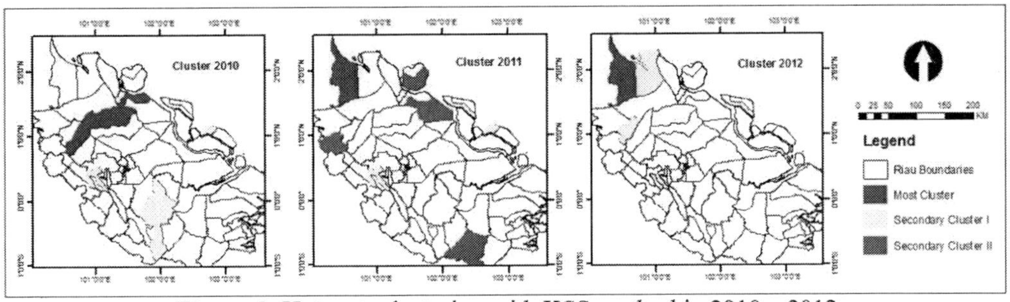

Figure 2. Hotspots clustering with KSS method in 2010 – 2012

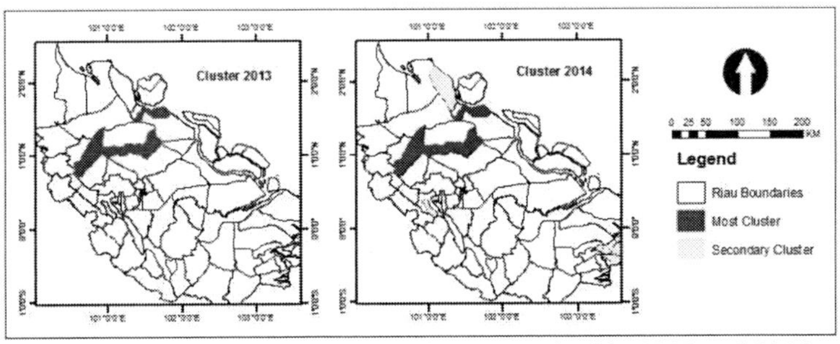

Figure 3. Hotspots clustering with KSS method in 2013 and 2014

From Figure 2 in 2010 occurred grouping with the largest density value is 0.498 hotspot/km^2 in the Bukit Kapu subdistrict. Whereas in 2010 and 2011, the grouping also occurs in the same district, started Sine districts with a density of 0,265 hotspot/km^2 and 0.185 hotspot/km^2. In 2013 the grouping occurred in the district of Pinggir with a density of 0.767 hotspot/km^2 and 2014 occurred in the district of Dumai west with a density value of 0.943 hotspot/km^2.

Referring to Usman (2014), that using DBSCAN in 2013 the density of hotspots highest on peatlands of Sumatra was found in the province of Riau, with a density value of 0.056 km^2, where most hotspots are in Rokan Hilir, Bengkalis, Dumai and Siak District. Compared with KSS method, the results are almost the same, because some districts which formed part of the subdistrict.

5. Conclusions

Based on the results obtained, it can be concluded:

1. Kulldorf Scan Statistic (KSS) is the good methods to analyze data with the spatial and temporal dimensions, such as the hotspot data.
2. In our case, the subdistrict that has the highest frequency of occurrence of hotspots is a subdistrict of Bukit Kapu, Bagan Sine and Dumai.
3. Based on the type of the maturity and thickness of the peat, the emergence of hotspots dominated by Hemists/Saprists (60/40), with the thickness of the peat is deep and very deep. While on land cover, hotspot dominated by peat swamp forests.

References

[1] Stibig H J, Achard F, Carboni S and Raši R and Miettinen R 2013 Change in tropical forest cover of southeast asia from 1990 to 2010 *Biogeosciences* **11** 247–258

[2] Syaufina L 2008 Kebakaran hutan dan lahan di Indonesia *Malang Bayumedia Publishing*

[3] Wen S and Kedem B 2009 A semiparametric cluster detection method a comprehensive power comparison with kulldorff âA˘Zs method *International Journal of Health Geographics BioMed Central USA*

[4] Naus J I 1965 Clustering of random points in two dimensions *Biometrika 52 263-267*

[5] Kulldorff M 1997 A spatial scan statistic *Communications in Statistics: Theory and Methods* **26** 1481–1496
[6] Sitanggang L S and Ismail M H 2010 Hotspot occurrences classification using decision tree method: Case study in the rokan hilir, Riau province, Indonesia In *Proc. Eighth Int. Conf. ICT and Knowledge Engineering* 46–50
[7] Khoiriyah Y M and Sitanggang I S 2014 A spatial decision tree based on topological relationships for classifying hotspot occurences in bengkalis riau Indonesia In *Proc. Int Advanced Computer Science and Information Systems (ICACSIS) Conf* 268–272
[8] Wulandari F 2012 Penerapan dynamic density based clustering pada data kebakaran hutan [Undergraduate Thesis] Bogor (ID): Bogor Agricultural University
[9] Usman M, Sitanggang I S and Syaufina L 2015 Hotspot distribution analyses based on peat characteristics using density-based spatial clustering *The 1st International Symposium on LAPAN IPB Satellite for Food Security and Environmental Monitoring, Procedia Environmental Sciences*
[10] Kirana, Annisa P, Sitanggang I S and Syaufina L 2016 Hotspot pattern distribution in peat land area in Sumatera based on spatio temporal clustering *Procedia Environmental Sciences* **33** 635-645

Hotspot sequential pattern visualization in peatland of Sumatera and Kalimantan using shiny framework

G Abriantini[1], I S Sitanggang[1*] and R Trisminingsih[1]
[1]Computer Science Department, Bogor Agricultural University, Bogor, Indonesia

E-mail: gema.abriantini@gmail.com

Abstract. Fires on peatland frequently occurred in Sumatra and Kalimantan. Fires on peatland can be identified by hotspot sequential patterns. Sequential pattern mining is one of data mining techniques that can be used to analyse hotspot sequential patterns. Sequential pattern discovery equivalent classes (SPADE) algorithm can be applied to extract hotspot sequential patterns. The objectives of this work are: 1) to obtain hotspot sequential pattern in Sumatra and Kalimantan in 2014 and 2015, and 2) to develop a web based application using Shiny framework that is available in R package for hotspot sequential pattern visualization in peatland of Sumatra and Kalimantan. Hotspot sequential patterns were obtained using minimum support of 0.01 with the focus of analysis is the hotspot sequences with length two or more events. This work generated as many 89 sequences with length 2 or more in Sumatra in 2014, 147 sequences in Sumatra in 2015, 48 sequences in Kalimantan in 2014, and 51 sequences in Kalimantan in 2015. Hotspot sequential patterns are visualized based on peatland's characteristics, weather, and social economy. The features in this web based application have been tested and the results show that all features work properly according to the test scenario.

1. Introduction

Peatland in Indonesia has area around 20 million hectares [1]. The majority of peatland in Indonesia in located in Sumatra and Kalimantan. Ironically, peatland in Kalimantan have been burned for around 2.66 million hectares (3.58% from total). In Sumatra, around 1 million hectares' peatlands were burned. This will reduce the capability of peatland absorbing water during rainy season and this will prevent the flood occurrences.

Fires in peatland has been an annual disaster in Indonesia. Fires in peatland are usually influenced by several factors including human and weather factors. Human factors are difficult to remove because of the increasing of human needs that lead to vegetation burning for land conversion. In addition, drought is one of the weather factors that influence fires in peatlands. Fires in peatland can be detected using hotspot data. Hotspot data are collected coordinate points of area that have high temperature around those areas [1]. Hotspots that appear in sequence of at least two times in the same location are a strong indicator for peatland fires.

The emergence of hotspot that causes fires can be identified using data mining techniques. Hotspot data could be processed using sequential pattern mining which is one of the methods in data mining. There are several algorithms in sequential pattern mining, namely, generalized sequential pattern (GSP), sequential pattern discovery using equivalent classes (SPADE), frequent-pattern projected sequential pattern mining (FreeSpan), prefix projected sequential pattern mining (PrefixSpan), and mining closed frequent sequential pattern mining (CloSpan) [2].

Several researches in the references [3-7] have been doing researches about the processing of the hotspot dataset and those implementations. Hotspot sequence patterns were integrated with weather data in Riau to observe weather conditions for each sequence of hotspot using CloSpan algorithm [3]. There was also a research that found the patterns of hotspot sequences in Riau using PrefixSpan algorithm [4]. There was a research that detected outliers of daily hotspot data using K-Means clustering method on a web based application which was built using the Shiny framework [5]. A web based application was also developed using the Shiny framework for DBSCAN clustering of data hotspot and visualized plot of hotspot clusters in Sumatra [6]. A web based application was also built using the Shiny framework for hotspot data classification using the C5.0 algorithm [7].

The objectives of this research are to obtain hotspot sequential patterns in Peatland of Sumatra and Kalimantan and to visualize those sequential patterns in a web based application. The sequential patternswere observed based on the location of physical peatland fires. The algorithm which was applied in this research is a sequential pattern discovery using equivalent classes (SPADE). The SPADE algorithm finds patterns of sequential data quickly [8]. The SPADE algorithm already exists in R package so the sequences can be easily visualized using Shiny framework.

2. Data and methods

2.1. Dataset

This study used hotspot dataset from FIRMS MODIS NASA (http://earthdata.nasa.gov). The hotspot data in Sumatra and Kalimantan used in this study startedfrom January 2014 to December 2015. The attributes of the hotspot dataset are latitude, longitude, brightness temperature, scan, track, acq_date, acq_time, satellite, confidence, version, bright_t31, and fire radiative power. The main attributes that used in this study are latitude, longitude, acq_date, and confidence. Latitude and longitude are the attributes ofhotspots location. Acq_date is the day that those hotpots were acquired by the NASA satellite. Confidence is confidence degree of hotspot.

In addition to hotspot data, this work utilizes Sumatra and Kalimantan peatland maps in shapefile format (.shp) that were obtained from Wetland International. Those maps are used to select hotspots that are located only in peatland, and to determine the physical characteristics of peatland where hotspot regularly occurred. QuantumGIS was used in this study to select hotspots that are only located in peatland. Characteristics of the peatland in Sumatra and Kalimantan were obtained from the peatland layer in shapefile format. There are three physical characteristics of the peatland namelypeatlanddepth, peatland type, and landuse.

2.2. Sequential pattern mining

Hotspot data will continue to grow in size over time. A database that consists of a group of sequences of events recorded based on the valid time is called a time-series database [8]. Hotspot dataset which have time attribute is an example of time-series databases. There are four patterns that can be obtained by processing the time-series data, i.e. trend analysis, similarity search, sequential pattern, and periodical pattern [9].

Sequential pattern is a sequence of consecutive item-set appearance and all of the items occurred in almost the same time [8]. A sequence $\langle a_1, a_2, \ldots, a_n \rangle$ is also included into the sequence $\langle b_1, b_2, \ldots, b_m \rangle$, if $i_1 < i_2 < \ldots < i_n$ is on $\langle a_1\ b_{i1}, a_2\ b_{i2}, \ldots, a_n\ b_{in} \rangle$. For example, the sequence ‹(3)(6, 7, 9)(7, 9)› can be called as part of the sequence ‹(2)(3)(6, 7, 8, 9)(7, 9)› in which (3) ⊆ (3), (6, 7, 9) ⊆ (6, 7, 8, 9), (7, 9) ⊆ (7, 9). However, ‹(2) (3)› is not part of the ‹(2, 3)› because ‹(2) (3)› means that 3 occurs after 2 and ‹(2, 3)› means 2 coincides with 3. A sequence is called to be maximal sequence if it is not included in any sequence [9]. As an illustration in the hotspot dataset, ‹(2) (3)› indicates the occurrence of hotspots in the time period 2 followed by the occurrence in the time period 3. Moreover, the sequence ‹(2, 3)› denotes occurrences of hotspot in the time period 2 and 3.

Sequential pattern mining is a process of extracting sequential pattern that its support value exceeds the specified minimum support [2]. Sequential data typically include large dataset, so it would be more efficient if pruning their sequences are performed by support values. Minimum support value is specified by user according to the needs of each user. Support is the percentage of transaction that contains a certain sub-sequences of total transactions.

2.3. SPADE algorithm

Sequential Pattern Discovery Algorithm using Equivalence classes (SPADE) is an algorithm to find patterns that use sequential equivalence classes to be divided the main problem into sub-problems that can be solved separately using join operations [8]. The input of SPADE algorithm is a sequential dataset. An example of sequential data is given in table 1.

Table 1. Example of sequential data.

Sequence ID	Event ID	Size	Items
1	10	2	{C,D}
1	15	3	{A,B,C}
1	20	3	{A,B,F}
1	25	4	{A,C,D,F}
2	15	3	{A,B,F}
2	20	1	{E}
3	10	3	{A,B,F}
4	10	3	{D,G,H}
4	20	2	{B,F}
4	25	3	{A,G,H}

Steps of the SPADE algorithm are as follows [8]:

1. Calculate the support value of all the items.
2. Calculate all items up to a maximum length sequences until no more sequences exceeding the specified minimum support.
3. Conduct the decomposition of a class of all sequences based on the length of each sequences.
4. Enumerate the entire sequence to generate a new sequence.

2.4. Research steps

Stages of this research are data pre-processing, generating hotspot sequential pattern, selecting hotspot sequential pattern, visualizing hotspot sequential pattern, and testing the application. Data pre-processing consists of several steps including data selection, data transformation, and sequential data generation. Data selection was performed to select the attributes used for mining process and to select hotspots in the study area namely peatland of Sumatra and Kalimantan. Data transformation was done to prepare sequential data of hotpots whereas sequential data generation was conducted to prepare sequential data of hotspot that its format meets the SPADE algorithm that is available on the package 'arulesSequence' in R software.

2.5. Data pre-processing

In this research, there are 3 stages on data pre-processing:

- Data selection, this stage was done to select all hotspots that occurred in peatland of Sumatra and Kalimantan. Longitude and latitude were rounded to two decimal digits in order to get several hotspots in the area with radius about 1 km. In addition, hotspots were selected only those with confidence value greater than or equal to 70% because according to the forest expert, hotspots at this confidence level have high possibility to be real fires.
- Data transformation, in this stages we transform the dataset into data format properly in the next stage. In this step, we create new attribute size and sequence ID (SID). In hotspot

dataset, size specifies the number of hotspots that occur at the same location and date. SID attribute represents the code for each different location.

- Creating sequential data, we prepare input data for the SPADE algorithm to generate the sequential pattern as shown in figure 1. In the figure 1, the columns from left to right are the SID, date_code, size, and items. SID is sequence ID. Column items will appear as many the value in the column size. Date_code represents code of the date when the hotspot occurred. The code is represented in integer. The column items represent one or more date_code indicating the sequences of hotspot occurrences.

```
1 1 1 1
2 8 1 8
2 62 1 62
2 65 2 65 65
3 19 1 19
4 19 1 19
4 26 1 26
4 28 1 28
4 30 1 30
5 20 1 20
5 31 1 31
5 32 1 32
5 34 2 34 34
5 36 1 36
5 40 1 40
```

Figure 1. Sequential dataset of hotspots, from left to right are SID, date_code, size, and items

2.6. Hotspot sequential pattern generation

Hotspot data as the results of pre-processing stage were processed using R by applying the SPADE algorithm. In this research, the minimum support value of 0.01 was used to obtain sequential patterns at least two hotspot occurrences. Number of sequential patterns that were generated using the SPADE algorithm is provided in table 2.

Table 2. Number of sequential pattern.

Dataset	1-Sequence	2-Sequence	3-Sequence	4-Sequence
Sumatra 2014	61	24	4	-
Sumatra 2015	62	70	14	1
Kalimantan 2014	41	7	-	-
Kalimantan 2015	44	7	-	-

2.7. Hotspot sequential pattern selection

Our analysis focuses on the sequential patterns that have length of 2 or more events. This is based on consideration that hotspots which occur sequentially in at least two days are considered as strong indicator for burning peatland. Some examples of 2-sequences of hotspot sequential pattern are shown in figure 2.

```
          sequence      support
62          <{81},{83}> 0.01431493
63          <{79},{81}> 0.01003904
64          <{67},{70}> 0.01505856
65          <{69},{70}> 0.01543038
66 <{67},{69},{70}> 0.01003904
67          <{65},{69}> 0.01338539
68          <{67},{69}> 0.03811117
69 <{65},{67},{69}> 0.01134040
70          <{67},{68}> 0.01189812
```

Figure 2. Hotspot sequential pattern.

In Fig. 2, the 2-sequence <{67},{69}> has the highest support value of 0.03811117. The date code 67 means 11 March 2014, whereas the date code 69 means 13 March 2014. The 2-sequence <{67},{69}> states that several hotspots were occurred on 11 March 2014 and then the occurrences were followed by other hotspots on 13 March 2014. The pattern is supported by about 3.81117 % sequence data of hotspot.

2.8. Hotspot sequential pattern visualization

A web based application for visualization of hotspot sequence patterns was built using Shiny framework. The application consists of two main files, namely Server.R as a server file and UI.R as an interface of the application. Data input of this application is the dataset in csv format that includes longitude and latitude of hotspot in the sequence patterns in peatland of Sumatra and Kalimantan.

There are four main features in the application, 1) plotting sequence patterns based on peatland's characteristics in Sumatra, 2) plotting sequence patterns based on peatland's characteristics in Kalimantan, 3) plotting sequence pattern based on weather data, and 4) plotting sequence pattern based on socio-economic data. There are additional features in the application that display a summary table for easily interpretation of sequence patterns. As shown in figure 3, the application has several features including selecting datasets, main menu, sub-menu, and legend of each plot. Each colour in peatland map represents the peatland's characteristics, such as type of peatland, peatland depth, and landuse type where the hotspot sequence occurred. Figure 4 shows a plot of hotspot sequences based on the peatland depth. Another feature in the application is summary table of the dataset that is provided in figure 5.

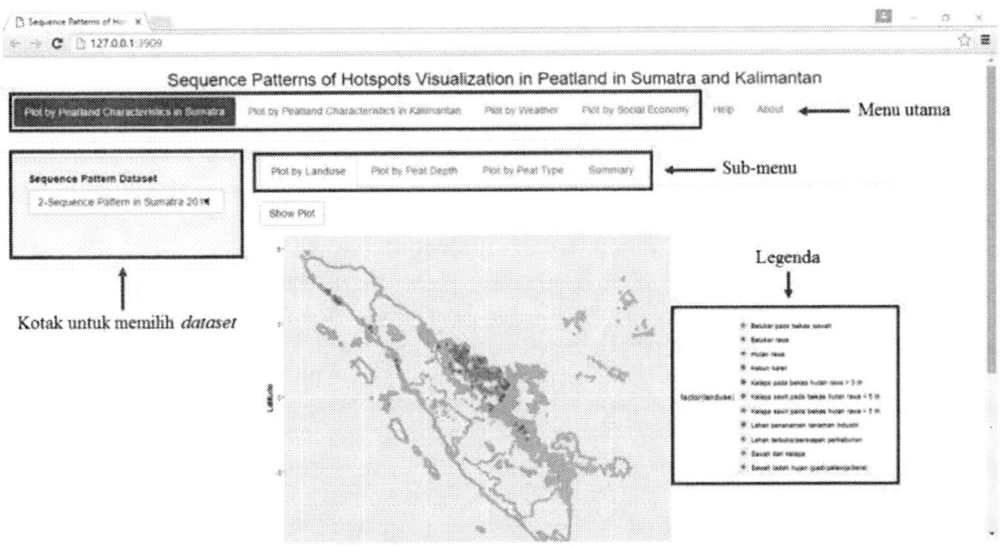

Figure 3. Plotting sequence patterns based on peatland types.

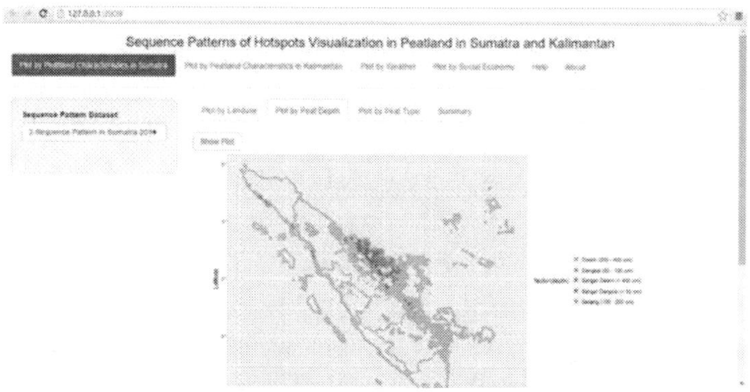

Figure 4. Plotting sequence patterns based on peatland depth.

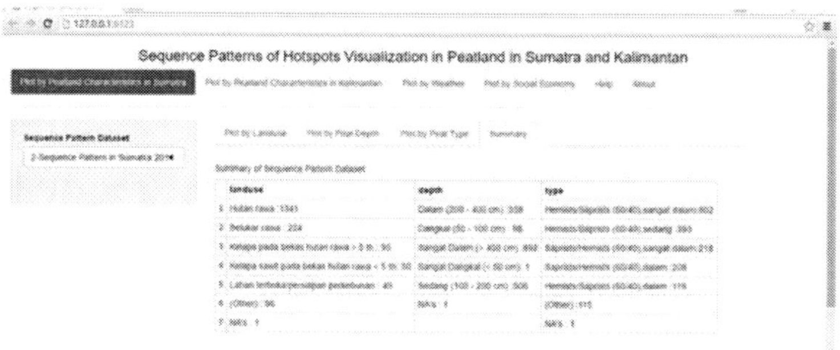

Figure 5. Summary of the dataset.

2.9. Application testing

The testing phase was carried out with several test scenarios and compared results with dataset visualization in Quantum GIS. The main features and additional features are worked properly and its outputs met the test scenario. The testing results are shown in table 3.

Table 3. Test scenarios.

Features	Test Scenarios	Expected Output	Work Properly?
Plotting Peatland Characteristics in Sumatra	Users select the "Show Plot"menu	Peatland map of Sumatra with a plot based on landuse, the type of peat, and peat depth is appeared	Yes
Plotting Peatland Characteristics Kalimantan	Users select the "Show Plot" menu	Peatland map of Kalimantan with a plot based on landuse, the type of peat, and peat depth is appeared	Yes
Plotting sequences by Weather	Users select the "Show Plot"menu	Peatland map of Sumatra with a plot based on average temperature, average humidity, and precipitation is appeared	Yes
Plotting sequences by Social Economy	Users select the "Show Plot"menu	Peatland map of Sumatra with a plot based on fire on land, income source, and number of school is appeared	Yes
Summary	Users select the Summary tab	Summary table of each dataset is displayed	Yes

3. Conclusion

This work results a web-based application for hotspot sequences visualization using the framework Shiny. Main features in the application includes plotting sequence patterns of hotspots in form of map based on weather, socio-economics and the characteristics of the peatland in Sumatra and Kalimantan including peat types and peat depth. Additional feature of the application is the summary table for displaying the brief description of each dataset. All features have been tested and the testing results based on the test scenarios shows that those features can work properly. This application has advantages in providing the summary of hotspot sequences and visualization of hotspot distribution in sequences. The information is important for identifying hotspots that are considered as strong indicators of peatland fires. A challenge for further study is how to improve the application by developing an additional feature that can automatically generate sequences from hotspot datasets.

Acknowledgments

The authors would like to thank to Directorate of Research and Community Service, Ministry of Research, Technology and Higher Education, Republic of Indonesia for the research grant (Contract No. 601/IT3.11/PN/2016).

References

[1] [KLH] Kementrian Lingkungan Hidup 2006 Koordinasi kelembagaan pengelolaan lahan gambut di Indonesia (Online) Available: http://www.menlh.go.id/koordinasi-kelembagaan-pengelolaan-lahan-gambut-di-indonesia
[2] Han J, Pei J and Yan X 2005 Sequential pattern mining by pattern-growth: principles and extensions *J. StudFuzz* **180** 183–220
[3] Agustina T and Sitanggang I S 2015 Sequential patterns for hotspot occurences based weather data using Clospan algorithm*Proc. 3rdInternational Conference on Adaptive and Intelligent Agroindustry* (Indonesia: Bogor) p 301–305
[4] Nurulhaq N Z and Sitanggang I S 2015 Sequential Pattern Mining on hotspot data in Riau province using the Prefix pan algorithm *Proc.3rdInternational Conference on Adaptive and Intelligent Agroindustry* (Indonesia: Bogor) p 257–260
[5] Suci A M Y A and Sitanggang I S 2016 Web-based application for outliers detection on hotspot data using K-Means clustering algorithm and Shiny framework *Conf. IOP Conference Series: Earth and Environmental Science 31*
[6] Hermawati R and Sitanggang I S 2016 Web-based clustering application using Shiny framework and DBSCAN algorithm for hotspots data in peatland in Sumatra *Proc.Procedia Environmental Sciences 33* (Indonesia) p 317–323
[7] Siknun G P and Sitanggang I S 2016 Web-based classification application for forest fire data using the shiny framework and the C5.0 algorithm *Proc.Procedia Environmental Sciences 33* (Indonesia) p 332–339
[8] Zaki M J 2001 SPADE: An efficient algorithm for mining frequent sequences *J. Machine Learning* **42** p 31–60
[9] Zhao Q and Bhowmick S S 2003 Sequential pattern mining: A survey *J. Technical Report CAIS* 118

Typology of land and forest fire in South Sumatra, Indonesia Based on Assessment of MODIS Data

M Ardiansyah[1,2*], R Boer[1,3] and A P Situmorang[1]
[1]Center for Climate Risk and Opportunity Management in Southeast Asia and Pacific, Bogor Agricultural University, Kampus-IPB Baranangsiang, Jl. Pajajaran No. 1, Bogor 16145
[2]Department of Soil Science and Natural Resources, Faculty of Agriculture Bogor Agricultural University, Jl. MerantiKampus IPB Darmaga, Bogor 16680
[3]Department of Geophysics and Meteorology, Faculty of Mathematic and Natural Sciences Bogor Agricultural University, Jl. MerantiKampus IPB Darmaga, Bogor 16680

E-mail: ardysaja@gmail.com

Abstract. In 2015, Sumatera and Kalimantan, in particular, has undergone dramatic fires. The fires were particularly bad in 2015 because of a prolonged dry season caused by the El Nino weather pattern and creating a lot of greenhouse gas emissions. Between about July and December, more than a million hectares of forest were burned. South Sumatra is one of the provinces with the highest of hotspots number and of fire area on this period. The aim of the study was to find burned area that caused by fire activity in 2015 and to identify a typology of land and forest fire the South Sumatera. In our study showed that between July and December 2015 the estimated burned area during El Nino in South Sumatra was 422,718 ha, of which 163,143 ha in mineral soil and 260,575 ha in peat soil. The majority of burned area occurred outside concession and inside concession with following typology: the fire activity in the HTI on non-forested land (26%), in the HTI on forested land (24%), in oil palm on non-forested land (17%), and in oil palm on forested land (2%).

1. Introduction

Land and forest fires have become an increasing problem in Indonesia in the past decades and occurred almost every year with varying intensity and burned area. These fires have many adverse environmental, social and economic effects. They affect global carbon dynamics and the haze from peat fires in this region has serious negative impacts on the ecosystem, economy and human health. The intensity of land and forest fires tends to rise in El Niño years. In 1997–98, fires associated with an exceptional drought caused by El Niño/Southern Oscillation (ENSO) devastated large areas of Indonesian tropical rain forests [1]. Indeed, several authors have reported strong correlations between large scale fire activity in Indonesia and El Niño, in particular for fire episodes in 1982–83, 1997–1998, 2002 and 2006 [1-4]. The last drought in 2015 associated with ENSO creating conditions that affected extensive fires in Indonesia's tropical forests and peatland. The fire last year was one of the worst after the 1997-1998 event, where between June and October 2015, an estimated 2.6 million hectares burned [5]. These fires were equivalent to four times the island of Bali [6].

In 2015, Sumatera and Kalimantan have undergone the most extensive fires. The fires were particularly bad in 2015 because of a prolonged dry season caused by the El Nino weather pattern and creating a lot of greenhouse gas emissions. According to Ministry of Environment and Forestry,

Indonesian National Institute of Aeronautics and Space, Global Fire Emission Database, and former REDD+ Agency,whichhave been using different types of data source of fires activity and different technique, the estimated emissionsfrom land and forest fires Indonesia ranged from 0.8 to 1.9 Gt CO2-e [7]. The Indonesian Government estimated that between February and October, fires affected more than 2.6 million hectares [5,7], which spread-out all over Indonesia with mostly concentrated in the two provinces: South Sumatra and Central Kalimantan, where peatlands may have accounted for more that 600 thousand hectares [7]. The South Sumatra is one of the provinces with the highest of burned area on this period.

The aim of the study was to map burned area that caused by fire activity in 2015 and to identify a typology of land and forest fire the South Sumatera. The results of this study are expected to evaluate the characteristics of fires that occur in South Sumatra Province and eventually to be used for fire prevention and mitigation.

2. Material and Method

2.1. Study Area

The study was conducted in the South Sumatra Province (Figure 1) by observation fire activity from July to December 2015. The Ministry of Environment and Forestry estimated that during drought session of July and October 2015, Sumatra Selatan was one of provinces with the most extensive forest and land fire, in addition to Central Kalimantan, of which the fires mostly may have occurred in mineral soil [7]. Administratively, South Sumatra Province consists of 13 Districts and 14 Municipal Government, with Palembang as its capital. The District of OganKomeringIlir is the largest area and followed by Musi Banyuasin. Rubber and Oil palm is the main commodity of agriculture sector, while forest plantation estate in forestry sectors [8].

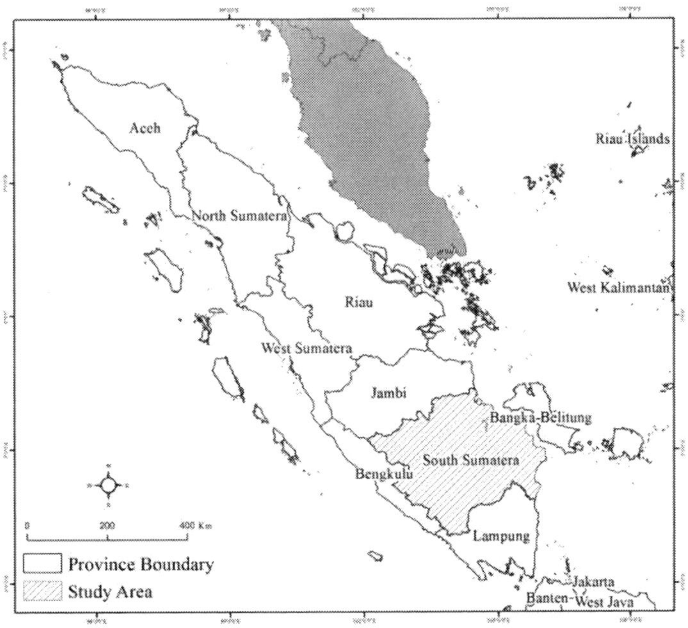

Figure 1. Map of study area

2.2. Data
Data on fire activity was derived from Moderate Resolution Imaging Spectrora-diometer (MODIS) covering the period July – December 2015. To understand the typology of fire in South Sumatera, we studied the relationship between burned area and several biophysical data. These were concession map (2011), forest area map (2010) and land cover map (2015) producedbyMinistry of Forestry as well as peat soil map sourced from Ministry of Agriculture (2014).

2.3. Method
Fire activity on the ground was not collected due to the large surface areas to be surveyed. Identification of fire activity relies only from satellite data. Using satellite data of MODIS (Terra and Aqua), the identification of fire activity was performed by using hotspot detection and detection of burnt area.

Hotspots detected by both MODIS (Terra and Aqua) sensors were used to identify fire activity. For the purpose of this study, the hotspot number for South Sumatra was computed for each month between July and December 2015. A simple method developed REDD+ Agency was used to identify fire activity. The burned area was estimated using MODIS hotspot data with confidence level ≥80%, which overlaid with a 1×1 km grid map. Grids without hotspots were considered as areas that were not burned and excluded in the estimation of burned area. Each grid pixel with at least one hotspot is considered to be burned with the burned area of 75 ha (75% of the grid area). This rule applies for each grid regardless of the number of the hotspots within a particular grid [9].

Since the burned area have different thematic areas and therefore different typology of fire, we also computed the typology of fire identified in South Sumatra province by overlaying the estimated burned area by the concession map, forest area map, land cover map, and peat soil map.

3. Results and Discussions
The temporal analysis of the hotspot number detected in South Sumatra province (Figure 2) shows interesting features: the fire activity season occurred in 2015 with peaks between June and December and there was fire activity in rainy season namely in February. The fires activity was particularly worst from September through November 2015 because of a prolonged dry season caused by the El Nino weather pattern. The highest hotspot number occurred in September, while the lowest hotspot number in December. The decreasing of fire activity in South Sumatra after November 2015 may due to success of firefighting by the government, private and community as well as a lot of rain fell in some area [6, 10].

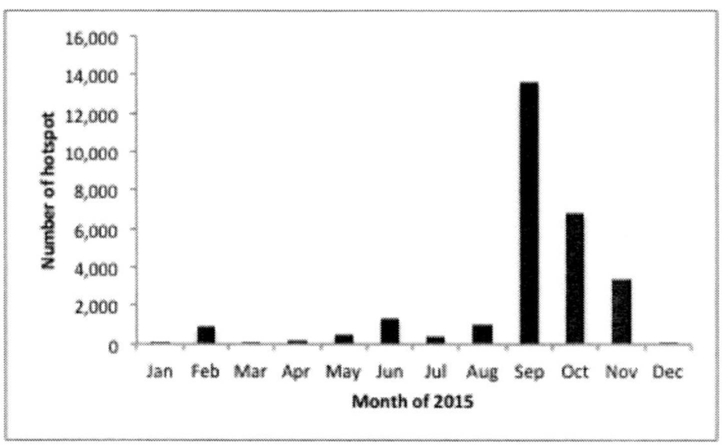

Figure 2. Number of hotspots detected in South Sumatra in 2015

Between July and December 2015, an estimated 422,718 ha burned, where 163,143 ha occurred in mineral soil and 260,575 in peat soil. The peatland area covers 14% of total area and mostly located in eastern part of South Sumatra. The rainfall influences the moisture content of the peat soil. Along with increasing rainfall, peatlands store large amount of water so that the peat water content increases and it is difficult to burn. When the rainfall decreases, the peat moisture content is to be on the wane and peat is to be dry. In the seasons with lower rainfall, water level drops to a critical threshold that causes the peat is highly flammable [11]. The dry peat with very low moisture in the last extreme drought season will become a fuel that is very combustible. This is also confirmed by [12] that the dried peat soil cannot absorb any water when it was watered and it is inflammable.

Spatially, the fire activity is occurred and correlated in a certain land cover and land use since land cover and land use changes are driven by fire. However changes in land cover and land use will ultimately have an impact on fire regimes through changes in the ignition frequency, fuel load, and land management. The overall impact of land cover land use change on fires often is specific to location, ecosystems, land use, and climate that underlie a particular place [13]. Based on land cover, the fire activity is mostly occurredin forested land (36%), grassland (36%), cropland (18%), and other land (10%) as depicted in Table 1. The grassland is a result of land use changes from one dominant vegetation type to grassland, and it may change the ignition or land management. The study result in South Sumatra showed that the grassland burn more frequently than active cropland. The grassland is an area that is prone to burning because it provides abundant fuel and flammable. Likewise the grassland fire may larger than other in agricultural lands, since the fire still used as mechanism for opening area suitable for agriculture [13-15].

Table 1. Area burnt by land use category

Land Use Category	Area (ha)	Percentage (%)
Forestland	152,221	36
Cropland	77,858	18
Grassland	165,232	36
Otherland	27,407	10
Total area burnt	422,718	100

Area burnt by land use type during drought session in 2015 is presented in Table 2. More than 165 thousand hectares of grassland and 152 thousand of forestland in this period was burnt due to fire exposure (Figure 3). The area burnt of forest plantation was larger than other forest cover, followed by secondary swamp forest as the second largest area burnt, where the fire in peatlands may be larger than in mineral soil.

Data analysis by overlapping burned area with licensed/ land concession reveals the burned area occurred in palm oil plantation (19%) was less than in industrial forest plantation (50%) as presented in Figure 4. In addition, there wasburned area occurred outside thelicensed/ concession (31%).Based on licensed/ land concession, fires in the HTI (forest plantation) occurred innon-forested land was 26% and inlands that are still covered by forest (forested land) 24%, while fires in the oil palm occurredin non-forested land was 17%, and in forested land 2%. This data indicated that some of licensed and concession for the palm oil plantation and the HTI was not yet planted. Fires in this region may be suspected since some the community/farmers or companiesopen the land for cropland using slash and burn technique that is still widely used in the study area.

Table 2. Area burnt by land use category during the 2015 drought session

Land use category	Land use type	Area (ha)	Percentage (%)
Grassland	Grass	13,079	3.09%
	Shrub	41,098	9.72%
	Swamp Shrub	111,055	26.27%
Otherland	Bare land	26,820	6.34%
	Mining	587	0.14%
Foresland	Primary dryland forest	10,958	2.59%
	Secondary dryland forest	4,979	1.18%
	Primary mangrove forest	317	0.07%
	Secondary mangrove forest	267	0.06%
	Secondary swamp forest	24,856	5.88%
	Forest plantation	110,844	26.22%
Cropland	Plantation	22,976	5.44%
	Dryland agriculture	14,906	3.53%
	Mix drayland agriculture	37,801	8.94%
	Paddy field	2,141	0.51%
	Transmigration	34	0.01%
Total area burnt		422,718	100.00%

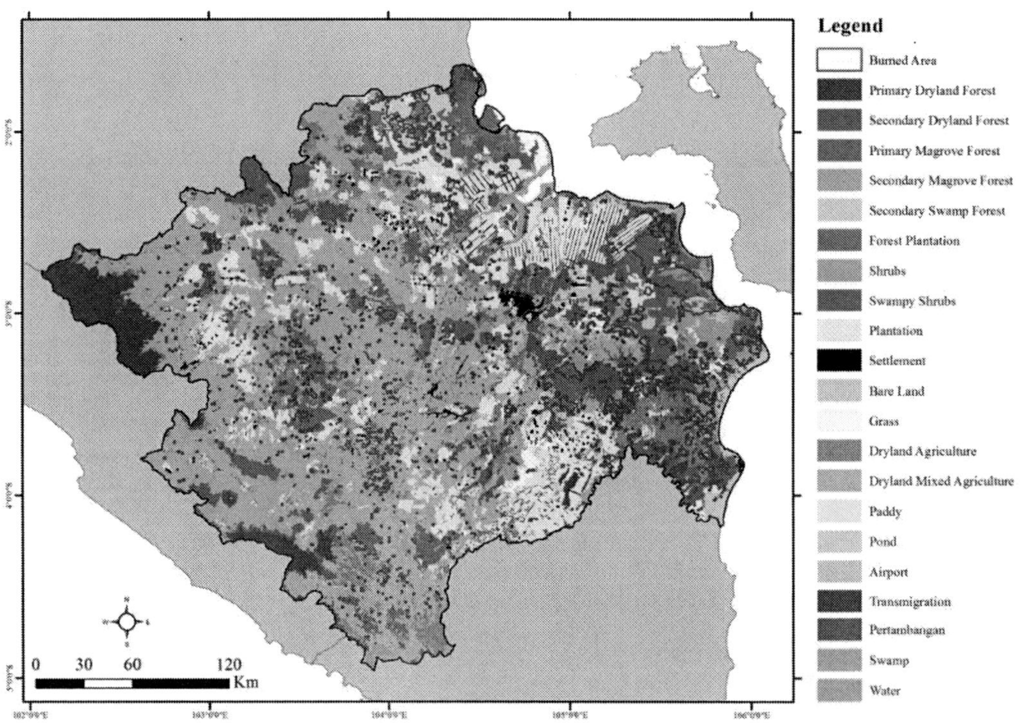

Figure 3. Distribution of estimated area burnt by land cover type of South Sumatera

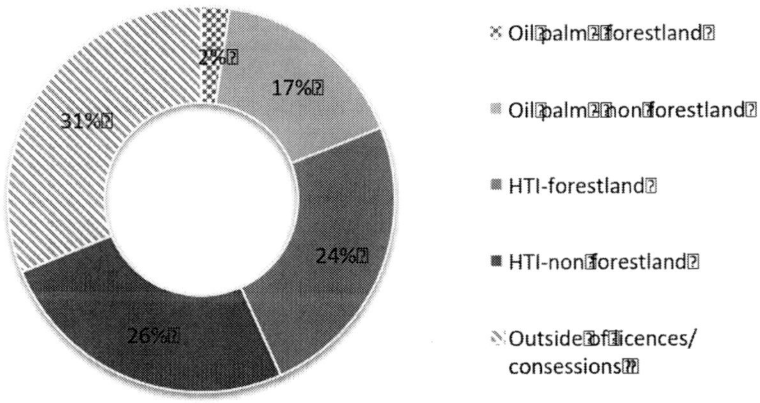

Figure 4. Percentage of area burn by licensed/concessions

4. Conclusion

In our study showed that between July and December 2015 the estimated burned area during El Nino in South Sumatra was 422,718 ha, of which 163,143 ha in mineral soil and 260,575 ha in peat soil. The grassland fire was larger than other land use category as the grassland is prone to burning because of abundant fuel and flammable.

The majority of burned area occurred outside licensed/concession (31%) and followed in concession with following typology: the fire activity in the HTI on non-forested land (26%), in the HTI on forested land (24%), in oil palm on non-forested land (17%), and in oil palm on forested land (2%).

Acknowledgements

This research was funded by the Ministry of Environment and Forestry of Indonesia and the Center Climate Risk and Opportunity Management in Southeast Asia and Pacific, Bogor Agricultural University, under USAID-UNOPS Grant No. 96764/2016/LPPM-IPB/01. The authors would like to convey gratitude for continuous support.

References

[1] Siegert F, Ruecker G, Hinrichs A, Hoffmann A A 2001 Increased damage from fires in logged forests during droughts caused by El Niño. *Nature* **414** p 437-440

[2] Siegert F, Hoffmann A A 2000 The 1998 Forest fires in East Kalimantan (Indonesia): A quantitative evaluation using high resolution, multitemporal ERS-2 SAR images and NOAA-AVHRR hotspot data *Remote Sensing of Environment* **72** p 64-77

[3] Murdiyarso D, Adiningsih E S 2007 Climate anomalies, Indonesia vegetation fires and terrestrial carbon emissions. In Murdiyarso D, Lebel L (Ed.) *Mitigation and Adaptation Strategies for Global Change* **12** 1 (The Netherlands: Springer) p 101-112

[4] Van der Werf G R, Dempewolf J, Trigg S N, Randerson J T, Kasibhatla P S, Giglio L, Murdiyarso D, Peters W, Morton D C, Collatz G J, Dolman A J, DeFries R S 2008 Climate regulation of fire emissions and deforestation in equatorial Asia *Proceedings of the National Academy of Sciences of the United States of America* **105** 51 p 20350-20355

[5] The World Bank e Cost of Fire: An Economic Analysis of *Indonesia's 2015 Fire Crisis* (Indonesia Sustainable Landscape Knowledge, 2016)

[6] Nugraha I, In three months, the forest and land area burnt equals four times the area of Bali in Indonesian (online) Available: http://go.nature.com/ITwzW7 [31 October 2015]

[7] [MoEF] The Ministry of Environment and Forestry 2016 Understanding Estimation of Emission From Land and Forest Fires in Indonesia 2015. Directorate General of Climate Change The Ministry of Environment and Forestry Indonesia

[8] [BPS] Badan Pusat Statistik 2015 Statistics Indonesia: Sumatra Selatan in Figure 2015 Statitics of Sumatra Selatan

[9] [MoEF] The Ministry of Environment and Forestry 2015 National Forest Reference Emission Level for REDD+ In the Context of Decision 1/CP.16 Paragraph 70. Directorate General of Climate Change The Ministry of Environment and Forestry Indonesia

[10] Abdullah Y, 2015 Hotspots of Fire In South Sumatera Already Lost. Antara Megapolitan (online) Available: http://megapolitan.antaranews.com/berita/17028/titik-panas-api-di-sumsel-sudah-hilang [18 November 2015]

[11] Wosten J H M, Clymans E, Page S E, Rileyand J O, Limin S H 2008 Peat–water interrelationships in a tropical peatland ecosystem in Southeast Asia. *Catena* **73** p 212–224

[12] Agus F, Subiksa I G M 2008 *Peatland: Potency for agriculture and its environmental aspects* (Bogor: Soil Research Centre and World Agroforestry Centre (ICRAF))

[13] Van Butsic, Kelly M and Moritz M A 2015 Land Use and Wildfire: A Review of Local Interactions and Teleconnections. *Land* **4** p 140-156 doi:10.3390/land4010140

[14] Tacconi L 2003 Fire in Indonesia: Causes, Cost and Policy Implications. Center for International Forestry Research

[15] Adinugroho W C, Suryadiputra I N N, Saharjo B H, Shibora L 2011 Manual for control of Fire in Peatlands and peatland forest (Bogor: Wetland International-Indonesia Programme)

Classification Model for Forest Fire Hotspot Occurrences Prediction Using ANFIS Algorithm

A K Wijayanto[1], O Sani[2], N D Kartika[2,3], Y Herdiyeni[4]

[1]Center for Environmental Research, Bogor Agricultural University, Indonesia, Gedung PPLH lt 3, Jl Lingkar Akademik, Kampus IPB Darmaga, Bogor, Indonesia, 16680

[2]Master of Science in Information Technology for Natural Resources Management, Bogor Agricultural University SEAMEO BIOTROP Campus, km 6 Tajur, Bogor, West Java, Indonesia, 16721

[3]The Agency of The Assessment and Application of Technology (BPPT), Jakarta, Indonesia

[4]Computational Intelligence Lab, Department of Computer Science, Bogor Agricultural University, Darmaga Campus, Wing 20 Level V-VI, Bogor, West Java, Indonesia, 16680

E-mail: akwijayanto@apps.ipb.ac.id

Abstract. This study proposed the application of data mining technique namely Adaptive Neuro-Fuzzy inference system (ANFIS) on forest fires hotspot data to develop classification models for hotspots occurrence in Central Kalimantan. Hotspot is a point that is indicated as the location of fires. In this study, hotspot distribution is categorized as true alarm and false alarm. ANFIS is a soft computing method in which a given inputoutput data set is expressed in a fuzzy inference system (FIS). The FIS implements a nonlinear mapping from its input space to the output space. The method of this study classified hotspots as target objects by correlating spatial attributes data using three folds in ANFIS algorithm to obtain the best model. The best result obtained from the 3rd fold provided low error for training (error = 0.0093676) and also low error testing result (error = 0.0093676). Attribute of distance to road is the most determining factor that influences the probability of true and false alarm where the level of human activities in this attribute is higher. This classification model can be used to develop early warning system of forest fire.

1. Introduction

The factors that influence fire occurrence in the boreal forest include the properties of the forest vegetation, weather, and ignition agents [1]. Forest fire mostly occurs in Central Kalimantan Province that has large area of peatland and this type of land makes is more dangerous than other forest fire types because fire in peat penetrates the bottom layer of the soil to form a funnel hole, and then the fire spreads horizontally beneath the surface, with addition that climate variations play an important role in influencing peatland fire vulnerability [2].

In Kalimantan, the number of hotspot during the year of 2002-2006 was in Central Kalimantan Province [3]. The impact of forest fire is not only experienced by Indonesian people but also the people who live in neighbour countries near Indonesia. It can affect health, social, politic and economic sector.

The government-led fire management practice in Central Kalimantan has been focused on short-term fire suppression during the May to September dry season. It depends upon a fire monitoring and warning system, using weather and environmental data from agencies such as the Indonesian National Institute of Space and Aeronautics (LAPAN), Ministry of Environment, ASEAN Specialized Meteorological Centre (ASMC), and the Indonesian Bureau of Meteorology, Climatology and Geophysics (referred to as the Indonesian meteorological service or BMKG). At the provincial level, the Natural Resource Conservation Agency (Badan Konservasi Sumberdaya Alam or BKSDA) uses hot spot data and the Fire Spread Risk Index Map (FSRIM), combined with on-the-ground observations, to assess current fire risk levels across the province. [4]

Previous researches developing forest fire risk model have been conducted by using data mining technique as C4.5 algorithm on Weka toolkit: J48, SimpleCart and NaveBayes. Classical data mining methods can extract patterns easily from spatial datasets which is more difficult by other methods, numeric and categorical spatial data types are complex but classical data mining methods do not support locations of objects or relationships between objects that implicitly exist in a spatial dataset [5]. Therefore the methods cannot be utilized to discover knowledge from spatial datasets. Locations of objects determine relations of the objects to its neighbours. So by generating data, classification can be done by data mining algorithm.

Vega-Garcia et al [6] adopted Neural Networks (NN) to predict human-caused wildre occurrence. The data were analyzed using logistic regression analysis (binary logit model), which served as the domain expert to identify the important input variables. The resultant model had four input nodes and two output nodes, and correctly predicted 85 percents of the no-fire observations and 78 percents of the fire observations. Cortez [7] also done a similar research by using data mining approach to predict forest fires using meteorological data. The Data Mining (DM) approach were explored to predict the burned area of forest res. Five different DM techniques, e.g. Support Vector Machines (SVM) and Random Forests, and four distinct feature selection setups (using spatial, temporal, FWI components and weather attributes), were tested on recent real-world data collected from the northeast region of Portugal.

In this study, Adaptive Neuro-Fuzzy Inference System (ANFIS) which is classification algorithms of data mining in Matlab software is used to determine classifier rule. It will show that hotspots occurrences can be predicted by using nearest distance relation between target (hotspot data) and factor (spatial data).

The objective of this project is to develop a decision support system using Adaptive Neuro-Fuzzy Inference System (ANFIS) algorithm for predicting hotspot occurrence in Central Kalimantan Province, Indonesia. The result will be based on two classes: true alarm and false alarm.

2. Methodology

2.1. Materials and Tools

The dataset for this project are physical and hotspot data taken from project research of Spatial Model of Land and Forest Fire Risk Index [3]. Explanatory objects are physical data including land cover, roads, rivers, city centre, village and settlement that will classify the target objects into false or true alarm.

The tools used in this study were ArcGIS and Matlab. ArcGIS used to calculate distance of attributes to hotspot data as target object. Then, ANFIS tool in Matlab used to generate rules in model building.

2.2. Study Area

The study area is in Central Kalimantan Province. The chosen spatial data of the boundary area was based on hotspot occurrence in Central Kalimantan province with coordinate on longitude 113°30'E latitude 02°00'S.

2.3. Overall Step Flow

This study conducted through several steps as shown by Figure 1. Data collected in data collecting step is spatial and hotspot data from a research conducted by Samsuri in 2008 [?]. The detail of each next step explained in the next sub-chapter.

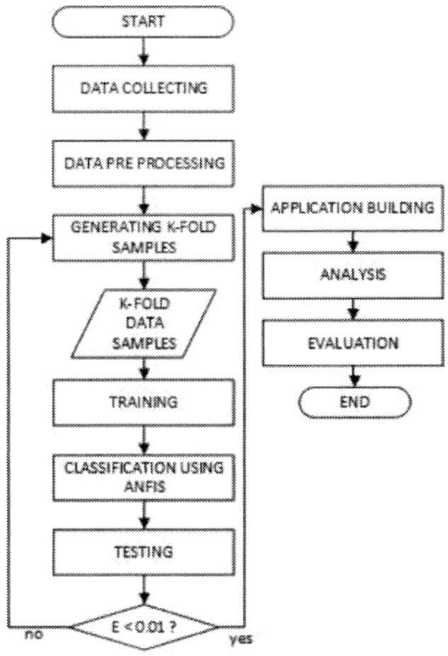

Figure 1. Overall stepflow of research.

2.4. Data Pre-processing

Two main tasks conducted in constructing a forest fires dataset are: 1) creating the target attribute and populating its value from the target objects, and 2) creating explanatory attributes. Target objects are true and false alarm data. True alarm data (positive examples) are hotspots that spread in Central Kalimantan. False alarm data (negative examples) were randomly generated and they are located within the areas at least 1 km away from any true alarm data.

Preprocessing steps were performed to relate the explanatory attributes to target objects by applying topological and metric operations. The relations between target objects and spatial objects which are settlement, roads, rivers, village and city centroids defined by calculating distance from target objects to nearest spatial data by using ArcGIS.

2.5. Generating Rule

Method used to generate rules is hybridization of ANN (Artificial Neural Networks) and FIS (Fuzzy Inference System) which called ANFIS model. The Adaptive Neuro Fuzzy Inference System (ANFIS), was first introduced by Jang, in 1993. The ANFIS is a neural network that is functionally the same as a TakagiSugeno type inference model [8]. The architecture of an ANFIS has introduced the first order Sugeno-style FIS at first. A first-order Sugeno-style FIS model is a system that manages the process of mapping from a given crisp input to a crisp output, using fuzzy set theory [9]. Considering a Sugeno type of fuzzy system having the rule base:

(i) If x is A1 and y is B1, then f1 = c11x+c12y+c10

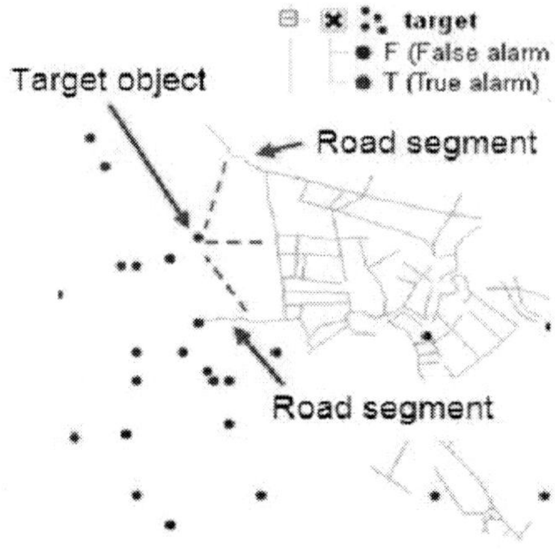

Figure 2. Distance near neighbour between Target object and Road Segment

(ii) If x is A2 and y is B2, then f2 = c21x+c22y+c20

All computations can be presented in a diagram form. ANFIS normally has 5 layers of neurons of which neurons in the same layer are of the same function family (Figure 3)

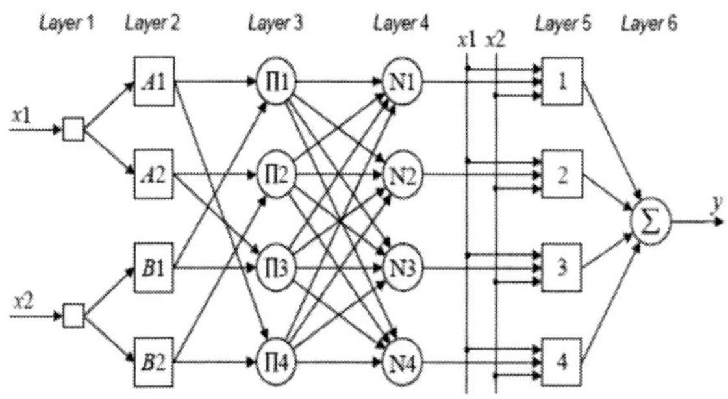

Figure 3. Structure of the ANFIS network

From Figure 3, attribute data (x1) in Layer 1 is divided into three membership functions in Layer 2: near (A1), medium (A2), far (A3). While the next layers become the hidden layers which are structured by ANN algorithm. Layer 3 nodes calculate the ratios of the rules firing strength to the sum of all the rules firing strength. While layer 4 nodes compute a parameter function on the layer 3 output. Parameters in this layer are called consequent parameters. The result is a normalised firing strength. Layer 5 normally a single node that aggregates the overall output as the summation of all incoming signals. The output layer (Layer 6) is the summation layer.

To choose the best combination of training and testing data, cross-validation is done. The basic idea of cross-validation is to split the training set into two disjoint sets, one which is actually used for training, and the other, the validation set, which is used to monitor performance. An extreme case of k-fold cross-validation is obtained for k = n, the number of training cases, also known as leave-one-out cross-validation (LOO-CV). Of- leave-one-out ten the computational cost of LOO-CV (training n models) is prohibitive, but in certain cases, such as Gaussian process regression, there are computational shortcuts [10]

In this study, the data is divided into three k-fold data samples. From the 100 samples data, three samples obtained by combining training and testing data, as shown by Table 1.

Table 1. K-Fold data samples

Fold (disjoint set)	Training-Testing combination
Fold 1	60 tr 40 ts
Fold 2	40 ts 60 tr
Fold 3	20 ts 60 tr - 20 ts

To do the training process, hybrid method is used instead of the backpropagation method by using 100 epochs. In this study, different number (200, 500, and 1000) of epoch are already tried. However, since the error value is already stable on 3rd epoch, so it is not necessary to do too much epoch.

3. Result and Discussion

3.1. Preprocessing Result

The rules were developed from a dataset contained of physical data and target data. The number of training data was 100 and being set randomly. The training data were consisted of 50 true hotspot alarms and 50 false hotspot alarms. The dataset had 6 attributes and each attribute represented:

(a) Distance from target data to nearest city in meter (NEAR_CITY).
(b) Distance from target data to nearest village meter (NEAR_VILLAGE).
(c) Distance from target data to nearest road in meter (NEAR_ROAD).
(d) Distance from target data to nearest river in meter (NEAR_RIVER).
(e) Distance from target data to nearest settlement in meter (NEAR_SETTLEMENT).
(f) Target attribute containing true alarm and false alarm (TARGET)

The distance is calculated by using ArcGIS and stored as a .dat file. This file then being divided into three k-fold samples and used in the training stage. The training stage result is explained in the next sub-chapter.

3.2. Rules Generation

The dataset was split into two disjoint sets consisted of 60 percent of training data and 40 percent of testing data. In this project, the dataset was divided into three samples with different k-fold. All the samples were trained using Matlab anfisedit as a preloaded feature to generate rules for determining true alarm and false alarm and also to get error for each training data.

From Table 2 and Figure 4, the highest error is obtained from fold 1 by combination of first 60 data for training and other 40 data for testing. The lowest score obtained from fold 2 by combination of first 40 data for testing and 60 other data for training. The best fold is not the highest nor the lowest error, but the average. The 3rd fold gives error which is closest to the average. Thus the 3rd fold is chosen to generate rules to determine true and false alarm.

Table 2. Error comparison of 3 folds

Fold (disjoint set)	Training error	Accuracy (a)	a-x̄
Fold 1 (60 tr 40 ts)	0.014777	99.9852	0.0057
Fold 2 (40 ts 60 tr)	0.0030858	99.9969	0.0060
Fold 3 (20 ts 60 tr 20 ts)	0.0093676	99.9906	0.0003
	Average (x̄)	99.9909	

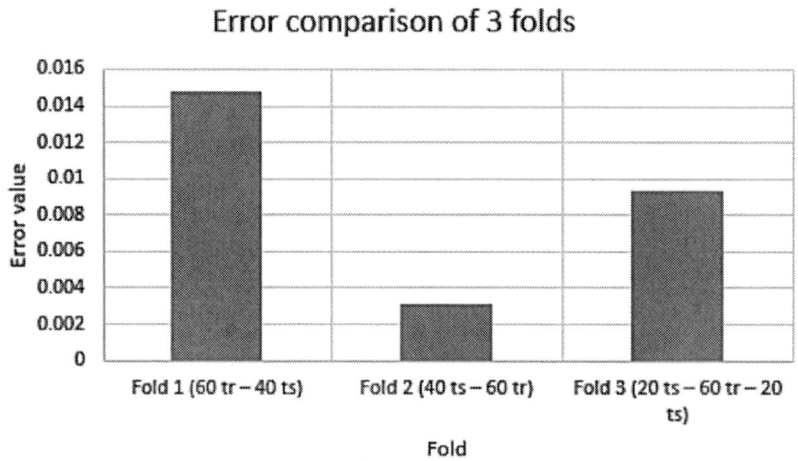

Figure 4. Chart of error comparison of 3 folds

Each input data for ANFIS were divided into three membership functions (near, middle, far) that represented the distance from hotspot to any place where humans did their activities (cities, villages, roads, rivers, settlements). Total node of membership function is fifteen and each node of membership function was generated to be the rule base that represented Sugeno type of fuzzy system function where:

If x is A and B and C and D and E, then Output Parameter

(A,B,C,D, and E is Attribute Input)

The output parameter that is influenced by attribute NEAR_ROAD is the most determined class of true alarm or false alarm. The attribute NEAR_ROAD is the determining factor because the level of human activities that influence forest fire occurrences was high and it means that the probability of hotspot occurrences is also high. This result can be said as match to the result of research conducted by Sitanggang [11] who did a research about hotspot occurrences prediction in Riau Province Indonesia, by using C4.5 algorithm. The research also used attribute of distance to determine the occurrences of forest fire. From the result, it is proven that attribute of distance can be used as determining attribute to forest fire occurrences.

The rule of training result then being tested by using testing data. Testing process provided low error by 0.0093676. Because of the low error, the rule can be used for further utilization.

4. Conclusion and Future Work

ANFIS algorithm can be used to generate expert system for predicting hotspot occurrences. The proposed method provided low error for training result (error = 0.0093676) and also low error testing result (error = 0.0093676). Attribute NEAR_ROAD is the most determine factor that influences the probability of true and false alarm where the level of human activities in this attribute is high. The higher level of human activities, the higher probability of hotspot to be forest fire. This classification model can be used to build early warning system for forest fire.

Another attribute that might determine probability of forest fire occurrences as like land use type, land cover type, and soil type, might be considered in the future work. The more attribute expected to make higher accuracy of prediction.

References

[1] Anderson K, Martell D, Flannigan M and Wang D 2000 *Climate Change, and Carbon Cycling in the Boreal Forest* (New York: Springer)
[2] Syaufina L 2002 *The Effect of Climate Variation on Peat Swamp Forest Condition and Peat Combustibility* Ph.D. thesis University Putra Malaysia
[3] Samsuri 2008 *Spatial Model of Land and Forest Fire Risk Index, Case Study in Central Kalimantan Indonesia* Ph.D. thesis Bogor Agricultural University
[4] Ceccato P, Jaya I, Qian J, Tippet K, Robertson W and Someshwar S 2009 Early warning and response to fires in kalimantan, indonesia *Advances in Operational Weather Systems for Fire Danger Rating* ed Sivakumar M (New York: Springer)
[5] Zeitouni K 2000 *Proc. Information Resources Management Association International Conference on Challenge of Information Technology Management in the 21st century* (Palo Alto)
[6] Vega-Garcia C, Lee B, Woodard P and Titus S 1996 *AI Application* **10** 9
[7] Cortez P and Morasis A 2007 *Proc. the 13th Portuguese Conference on Artificial Intelligence (EPIA 2007), New Trends in Artificial Intelligence* (Guimares)
[8] Jang J 1993 *IEEE Trans On Systems Man and Cybernetics* **19** 665
[9] Loganathan C and Girija K 2013 *International Journal of Engineering and Science* **2** 06
[10] Rasmussen C and Williams C 2006 *Gaussian Processes for Machine Learning* (Massachusetts: The MIT Press)
[11] Sitanggang I, Yakoob R, Mustapha M and Ainuddin A 2012 *Journal of Theoretical and Applied Information Technology* **43** 214

Greenhouse gas and air pollutant emissions from land and forest fire in Indonesia during 2015 based on satellite data

A Pribadi[1,2*] and G Kurata[2]
[1]Department of Civil and Environmental Engineering, Bogor Agricultural University IPB Darmaga Campus, PO-BOX 220, Bogor 16002, Indonesia
[2]Department of Environmental Engineering, Kyoto University C-cluster, Kyoto-Daigaku-Katsura, Kyoto 615-8540, Japan

E-mail: andik.pribadi.23s@st.kyoto-u.ac.jp

Abstract. Land and forest fire still become a major problem in environmental management in Indonesia. In this study, we conducted quantitatively assessment of land and forest fire emissions in Indonesia during 2015. We applied methodology of emission inventory based on burned area, biomass density, combustion factor and emission factor for each land cover type using several satellite data such as MODIS burned area, Pantropical National Level Carbon Stock Dataset, as well as Vegetation Condition Index. The greenhouse gases emissions from land and forest fire in Indonesia during 2015 were (in Gg) 806,406 CO_2, 8,002 CH_4, 96 N_2O, while pollutants emissions were (in Gg) 85,268 CO, 1,168 NOx, 340 SO_2, 3,093 NMVOC, 1,041 NH_3, 259 BC, 1,957 OC, 4,118 $PM_{2.5}$ and 5,468 PM_{10}. September was the peak of fire season that generate 58% (species average) of total emissions for this year. The largest contribution was from shrubland/savanna burning which account for 66% (species average) of the total emissions, while about 81% of the total emissions were generated from peatland fire. The results of this study emphasizethe importance of proper peatland management in Indonesia as land and forest fire countermeasures strategy.

1. Introduction

Land and forest fire is a major cause of land surface changes with impacts on the climate system, vegetation composition and the chemical composition of the atmosphere. It is the largest source of primary fine carbonaceous particles and the second largest source of trace gases in the global atmosphere [1]. Global impact of forest fire can be caused by greenhouse gases (GHGs) emissions which lead to global warming, while regional to local impact can be affected by pollutant gases and particles emissions which lead to air pollution.

Severe land and forest fire frequently occurred in Indonesia that caused negative impacts on atmospheric environment, not only in Indonesia it self but also to surrounding countries such as Malaysia and Singapore. For example, during the dry season of 1997, 2006 and 2015, severe forest fire occurred in Sumatra and Kalimantan, Indonesia, and it resulted in various types of air pollution over large area by emitting aerosol particles and photochemical precursors [2–4]. On a recently study, Huijnen revealed that with a mean emission rate of 11.3 Tg CO_2 per day during Sept-Oct 2015, emissions from Indonesia forest and peatland fires exceeded the fossil fuel CO_2 release rate of the European Union (EU28) (8.9 Tg CO_2 per day) [4].

Many studies on emission inventory from forest fire or generally biomass burning in Indonesia have been conducted by some researchers, either in global scale, regional scale, or specifically for

Indonesia region [4–7]. However, these studies investigated thefire emissions in Indonesia during the last decade, or had a relatively coarse spatial resolution. Moreover, some previous studies used land type-based aboveground biomass density[5], or measurement data in a typical study plot or statistical data as available fuel with country or province-based above ground biomass density [6,8], which means the available fuels in an area or land type were usually used to represent the whole area or land type. Such representations usually do not reflect the spatial variations and heterogeneities of biomass within the area or land type because the biomass varies significantly with the type of vegetation and geographical location.

Therefore, this study was conducted to quantify emissions of land and forest fire in Indonesia during the last year 2015, which was the largest fire year since 1997 [4]. This study assessed the generated emissions from land and forest fire, including shrubland, plantation and agricultural land fire, using the latest available data, which are expected to give more accurate results. Several important species of fire emissions were quantified, consisted of GHGs (CO_2, CH_4, N_2O) and air pollutants (CO, NOx, SO_2, NH_3, NMVOC, OC, BC, $PM_{2.5}$, PM_{10}). Updated land and forest fire emissions inventory will provide a general assessment of the potential effects of forest fire, as well as important information, which can be used for formulation of appropriate mitigation measures and evaluation of emissions reduction policy in Indonesia.

2. Methods and Data

Emissions from land and forest fire, or commonly open burning biomass, are generally calculated using the burned area, available fuels, combustion factor, and emission factor with the following equation [9]:

$$Es = \sum_{i=1}^{n} (B \times F \times CF \times EF_s)_i \quad (1)$$

where Es is emissions for each species (g), B is burned area (m^2), F is available fuels for combustion (kg/m^2), CF is combustion factor, defined as the fraction of combusted fuel to the total amount (unitless), EFs is emissions factors for species, defined as the mass of species per mass of dry matter burned (g/kg), and i is land cover types. We applied Equation 1 for land and forest fire emissions calculation on monthly basis, i.e. the calculation was performed using monthly data for the year 2015.

2.1. Burned Area

Burned area data were derived from the MODIS direct broadcast burned area product (MCD64A1), which was used to estimate the entire area that have been burned in Indonesia during 2015. MCD64A1 is a MODIS satellite-derived product that uses surface reflectance, daily active fire, and land cover products to delineate burned areas and burn cells which are tagged with approximate burn date[10]. MCD64A1 has a spatial resolution of 500 m and temporal resolution of a month, and the day of burning and a temporal uncertainty range of the burn date are noted for every pixel.

The MCD64A1 are available online at ftp://fuoco.geog.umd.edu/db/MCD64A1/, and provided in Hierarchical Data Format (HDF), 10° tile size with SIN projection. In this study, we created mosaic of the data tiles that cover the study area (Indonesia region) using Modis Reprojection Tool (MRT). Then we extracted the Julian date of burning from the data, converted the raster data to polygon and overlaid with national land cover and peatland distribution map using ArcGIS.

Figure 1 shows the spatial variation of burned area average (ha) in Indonesia during 2015. In this map, the burned area were aggregated to 0.25° resolution. However during our emissions calculation, we preserved the original resolution of the data (500 m). The area in Indonesia that have relatively large annual burned area is the southern part of Central Kalimantan, which have burned area exceeded 45,000 ha inside one pixel area (0.25°). Another area that also have large burned area is the eastern part of South Sumatera, which approximately have burned area up to 30,000 ha in one pixel area.

2.2. *Fuels*

In this study, we took into account both above ground biomass (AGB) and below ground biomass (BGB) as available fuels for combustion. We used satellite derived biomass density data to determine the AGB, namely the Pantropical National Level Carbon Stock Dataset, which was published by Woods Hole Research Center. This dataset was generated using three-stage data collection and modelling strategy that is specifically designed to allow measurements collected on the ground to be up-scaled to the 500m resolution of MODIS imagery using GLAS (Geoscience Laser Altimeter System) LiDAR data for the period circa 2007-2008[11].

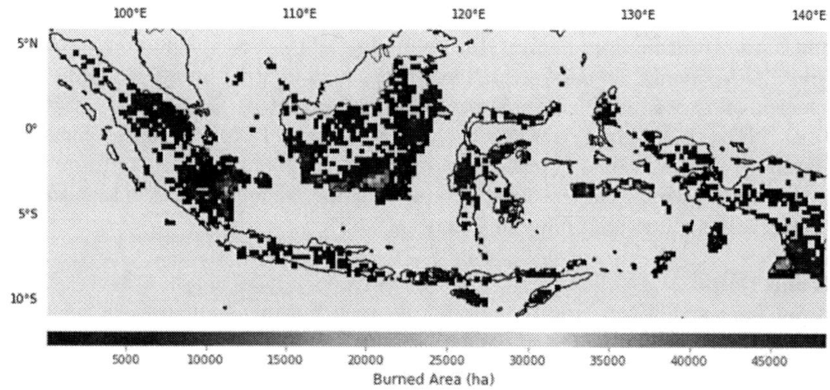

Figure 1. Burned area in Indonesia during 2015 derived from MODIS MCD64A1 (aggregated to 0.25° resolution).

The Pantropical National Level Carbon Stock Dataset are available online at http://whrc.org/publications-data/datasets/, and provided in GeoTIFF format with 500 m resolution. In this study, we converted the raster data to polygon and overlaid with burned area to obtain the biomass density in each burned area. Based on this dataset, the area that have relatively high biomass density are the western part of Sumatera, the middle part of Kalimantan and Papua Islands, where the biomass density in these area reached up to 464 Mg/ha.

Meanwhile, the below ground biomass (BGB) were derived from the Indonesia peatland distribution data, which was provided by Wetlands International. The data was distributed online at data.globalforestwatch.org, and provided in ArcGIS layer format. In this study, we converted the data to polygon and overlaid with burned area to obtain the peat availability in each burned area. This data provide peat depth distribution that classified into 6 categories, as shown on Figure 2. This figure describe that peatland area in Indonesia mostly distributed in eastern part of Sumatera, southern part of Kalimantan and southern part of Papua. The thickest layer of peat soil were located at the southern part of Central Kalimantan where the peat soil layer exceeded 10 m depth.

Previous studies did not consider peatland fire [6] or assumed constant depth of burned peat soil [12]. Here, we assumed the depth of burned peat soil based on the peat depth category. On the first category (peat depth <50 cm), we took the middle value between 0-50 cm i.e. 25 cm as the burned peat depth. While on the other category (peat depth > 50 cm), we assumed 51 cm as the burned peat depth. The peat dry bulk density was assumed 100 kg/m^3. These assumptions referred to some previous studies that reported that the burned peat depth were varied between 20-150 cm [13]and 25-85cm [14] based on Indonesian field measurements.

2.3. *Combustion Factor*

The combustion factor (CF) is the fraction of biomass actually consumed in a fire as determined by fuel type and moisture conditions. It varies greatly among the different biomass of different fuel types.

We determined fuel moisture conditions according to the time series of MODIS Vegetation Condition Index (VCI), which were derived from Normalized Difference Vegetation Index (NDVI) [15]. These data were available online at ftp://ftp.star.nesdis.noaa.gov/ and provided in weekly temporal resolution with 1/112° spatial resolution.

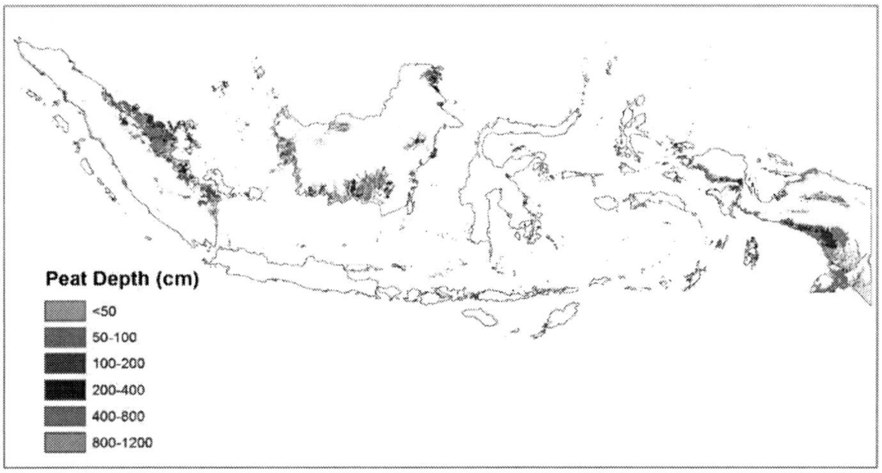

Figure 2. Peat depth distribution that represent below ground biomass in Indonesia.

In order to calculate CF, we equally divided VCI values, which range from 1 to 100, into six different categories, representing fuel moisture conditions of very dry, dry, moderate, moist, wet, and very wet, respectively[16]. The monthly variation in moisture conditions were used to associate with fuel moisture category factors (mcf), as shown on Table 1.

Table 1. Moisture category factor (mcf) (from [17]).

Moisture condition	Canopy	Shrub
Very dry	0.33	0.25
Dry	0.5	0.33
Moderate	1	0.5
Moist	2	1
Wet	4	2
Very wet	5	4

The CF is a function of fuel type and moisture category and was calculated from the following model [17]:

$$CF_i = (1 - e^{-1})^{mcf}, \qquad (2)$$

whereCFi denotes the fraction of fuel loading consumed for fuel type i, which is forest (canopy) and shrub; and mcf is the moisture category factor. The mcf value for each fuel type was determined using the fuel moisture condition derived from MODIS VCI data (Table 1), which increases from dry to wet fuel conditions.

2.4. Land Cover and Emission Factors

The land cover data was derived from the Indonesian Land Cover Map 2011 from the Ministry of Forestry, Indonesia. The land cover map were available online at http://webgis.dephut.go.id:8080/kemenhut/index.php/id/fitur/unduhan, which provided in kml format for each province. The land covers in this data were classified into 21 categories, plus 2 additional categories for cloud and no data. The dataset with the original land cover classification system were then re-classified into six broad categories of vegetation susceptible to fires in Indonesia (tropical forest, woodland, shrubland, cropland, paddy field and peatland). The re-classification was performed in order to match with the available emission factor (EF) for each land cover type.

The EFs for this study were selected from several available publications. We referred to some previous EFs compilation studies to determine EFs for tropical forest and shrubland [6,18], woodland [18,19], cropland [19,20], paddy field [12,20] and peatland [12,19]. The selected EFs for all species from different biomes used in this study are summarized in Table 2.

Table 2. Selected emission factors for each land cover (re-classified) in Indonesia.

No	Land cover (re-classified)	Emission Factors (g/kg)											
		CO_2	CH_4	N_2O	SO_2	NOx	CO	NMVOC	NH_3	BC	OC	$PM_{2.5}$	PM_{10}
1	Tropical Evergreen Forest	1601	6.44	0.2	0.43	1.44	106	8.1	1.1	0.64	6.79	14.8	18.5
2	Woodland	1636	4.4	0.21	0.54	2.19	81	5.21	1.44	0.52	3.76	7	10.2
3	Shrubland/Savanna	1685.8	2	0.21	0.9	3.9	63	3.4	0.56	0.37	2.62	5.4	8.3
4	Farmland (Combined Crops)	1130	4.56	0.1	0.216	0.7	86.3	7	1.3	0.48	0.7	3.9	8.05
5	Paddy field	1177	9.59	0.07	0.18	2.28	93	7	4.1	0.52	2.99	8.3	9.1
	Peatland	1703	20.8	0.2	0.71	2.26	210	7	2.55	0.57	4.3	9.05	11.8

3. Results and Discussion

3.1. Annual Emissions

The annual emissions of land and forest fire in Indonesia for each species (CO_2, CH_4, N_2O, CO, NO_x, SO_2, NMVOC, NH_3, BC, OC, $PM_{2.5}$, and PM_{10}) for the year 2015 are presented in Table 3. Throughout this year, emissions of the three main GHGs from land and forest fire in Indonesia were 806,406 Gg/year, 8002 Gg/year and 96 Gg/year of CO_2, CH_4 and N_2O respectively. These amounts are equal to 1004 TgCO_2-e/year based on Global Warming Potential (GWP) values of IPCC 4th Assessment Report for 100 years time horizon. As a comparison, according to Indonesia 1st Biennial Update Report (2016), GHGs emissions in Indonesia during 2000-2012 period (exclude LULUCF) was 612 Tg CO_2-e/year. This comparison could illustrate the large contribution of land and forest fire to national GHGs emissions in Indonesia, which is almost twice than GHGs emissions from all sector (excluding LULUCF) based on 2000-2012 average.

Table 3. The annual emissions of land and forest fire in Indonesia for the year 2015.

Species	Emissions (Gg)	Species	Emissions (Gg)
CO_2	806,406	NMVOC	3,093
CH_4	8,002	NH_3	1,041
N_2O	96	BC	259
CO	85,268	OC	1,957
NO_x	1,168	$PM_{2.5}$	4,118
SO_2	340	PM_{10}	5,468

3.2. Spatial Distribution

In this study, we found that the emissions of all species presented similar spatial distribution even though their magnitudes differed greatly. This is reasonable since all species would be emitted from the existed burned area with different magnitude depend on the EF of each species and the availability of burnable peat soil. Figure 3 depicts the spatial distribution of annual average of CO_2 emissions during 2015, which is also similar to the other species. This spatial pattern is also comparable to the distribution of burned area as shown on Figure 1. In general, the emissions could be observed throughout most of Indonesia area with high spatial variability. The western part of Indonesia generally generated higher emissions than the eastern area. The area that has relatively large annual emissions is the southern part of Central Kalimantan, which generate CO_2 emissions more than 25 Tgin one pixel area (0.25°).

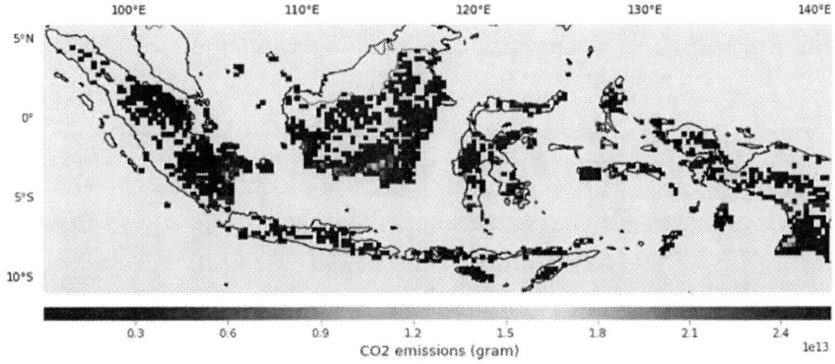

Figure 3.Spatial distribution of CO_2 emissions.

3.3. Spatio-Temporal Variation

Assessment of monthly variation in CO_2 emissions revealed that the highest emissions were generated during September-October period, which is correspond to peak of dry season in most part of Indonesia. The dry season in Indonesia generally lasts from July through October when there is relatively little precipitation. Local farmers usually prepare fields by cutting down the accessible forest and short vegetation, which they then allow to dry until immediately before the rainiest part of the year to ensure an effective burn. Consequently, values in burned area and CO_2 emissions are typically high during September-October period.

To investigate the spatio-temporal variations, we plot the peak month of CO_2 emissions for each pixel area, as shown on Figure 4. Here we can observed that the middle and northern part of Sumateraand the northern part of Kalimantan have earlier peak of fire season which are occurred during May-July period, while the other area (mostly the middle to eastern part) have peak of fire season during Sep-Nov period. This spatio-temporal variability were likely due to variability on annual rainfall pattern. Aldrian and Susanto identified that there are three rainfall regions in Indonesia which have distinct annual rainfall pattern, namely region A, B and C [21]. By referring the rainfall region map in those study, it can be seen that the monthly peak of emissions were associated with the dry season in particular region.

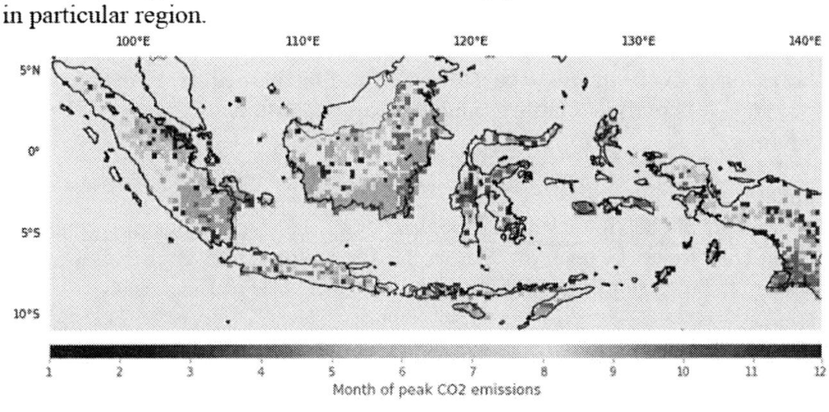

Figure 4. Spatio-temporal variation of CO_2 emissions.

3.4. Land Cover and Peatland Contributions

The contributions of each land cover type to the total emissions were graphed for all trace gases and aerosols (Figure 5). Here we also separated the emissions that originated from peatland area and non-peatland area for each land cover type. The results showed that for different emission species, the contribution of each land cover type to the total amount varied moderately.However, on average, emissions from shrublands/savanna burning are the largest contributor to the total emissions (66%; 56% on peatland + 10% on non-peatland), followed by burning of woodlands (19%; 15%+4%), farmland (9%; 8%+1%), and tropical evergreen forest (6%; 2%+4%).

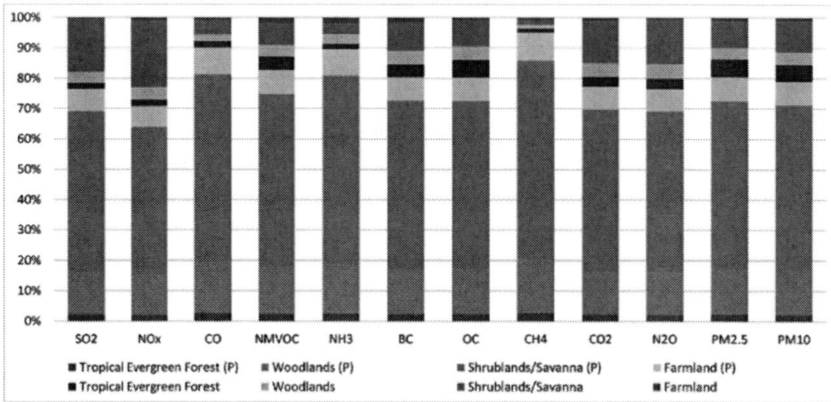

Figure 5. Contributions of each land cover type to the total emissions.

We found that peatlands fire had a very significant contribution to the total emissions. Proportion of burned area on peatlands was 69% of the total burned area during this year (1,598,756 ha of the total 2,321,141 ha). However, peatlands fire contributedas much as 81% (on species average) to the total emissions. This was due to peatlands having a relatively higher density of biomass that consist of above and below ground biomass.

4. Conclusions

Emissions of GHGs and pollutants from land and forest fire in Indonesia during 2015 have been assessed in this study. We found that the most severe forest fire event during this period was occurred on the southern part of Central Kalimantan. The peak of emissions was occurred during September which account for 58% of the total annual emissions. Peak season of land and forest fire occurred earlier in western part of Indonesia (May-Jul) than the eastern area (Sep-Nov). The largest contribution was from shrubland/savanna burning which account for 66% (on average) of the total emissions. Burned area on peatlands was 69% of the total burned area in this country, whereas peatlands fire (comprise above and below ground biomass) contributedas much as 81% (on average) to the total emissions.

Acknowledgments

This research was part of the first author's doctoral study that was financially supported by DIKTI scholarship from Directorate General of Research, Technology and Higher Education Resources, Ministry of Research, Technology and Higher Education, Republic of Indonesia.

References

[1] Andreae M O and Merlet P 2001 Emission of trace gases and aerosols from biomass burning *Global Biogeochem. Cycles* **15** 955–66
[2] Tsutsumi Y, Sawa Y, Makino Y, Jensen J B, Gras J L, Ryan B F, Diharto S and Harjanto H

1999 Aircraft measurements of ozone, NOx, CO, and aerosol concentrations in biomass burning smoke over Indonesia and Australia in October 1997: Depleted ozone layer at low altitude over Indonesia *Geophys. Res. Lett.* **26** 595–8
[3] Zhang L, Li Q B, Jin J, Liu H, Livesey N, Jiang J H, Mao Y, Chen D, Luo M and Chen Y 2011 Impacts of 2006 Indonesian fires and dynamics on tropical upper tropospheric carbon monoxide and ozone *Atmos. Chem. Phys.* **11** 10929–46
[4] Huijnen V, Wooster M J, Kaiser J W, Gaveau D L A, Flemming J, Parrington M, Inness A, Murdiyarso D, Main B and van Weele M 2016 Fire carbon emissions over maritime southeast Asia in 2015 largest since 1997 *Sci. Rep.* **6** 26886
[5] Chang D and Song Y 2010 Estimates of biomass burning emissions in tropical Asia based on satellite-derived data *Atmos. Chem. Phys.* **10** 2335–51
[6] Permadi D A and Kim Oanh N T 2013 Assessment of biomass open burning emissions in Indonesia and potential climate forcing impact *Atmos. Environ.* **78** 250–8
[7] Gaveau D L a, Salim M A, Hergoualc'h K, Locatelli B, Sloan S, Wooster M, Marlier M E, Molidena E, Yaen H, DeFries R, Verchot L, Murdiyarso D, Nasi R, Holmgren P and Sheil D 2014 Major atmospheric emissions from peat fires in Southeast Asia during non-drought years: evidence from the 2013 Sumatran fires. *Sci. Rep.* **4** 6112
[8] Song Y, Chang D, Liu B, Miao W, Zhu L and Zhang Y 2010 A new emission inventory for non agricultural open fires in Asia from 2000 to 2009 *Environ. Res. Lett.* **5** 14014
[9] Seiler W and Crutzen P 1980 Estimates of gross and net fluxes of carbon between the biosphere and the atmosphere from biomass burning *Clim. Change* **2** 207–47
[10] Giglio L, Loboda T, Roy D P, Quayle B and Justice C O 2009 An active-fire based burned area mapping algorithm for the MODIS sensor *Remote Sens. Environ.* **113** 408–20
[11] Baccini A, Goetz S J, Walker W S, Laporte N T, Sun M, Sulla-Menashe D, Hackler J, Beck P S A, Dubayah R, Friedl M A, Samanta S and Houghton R A 2012 Estimated carbon dioxide emissions from tropical deforestation improved by carbon-density maps *Nat. Clim. Chang.* **2** 182–5
[12] Shi Y and Yamaguchi Y 2014 A high-resolution and multi-year emissions inventory for biomass burning in Southeast Asia during 2001–2010 *Atmos. Environ.* **98** 8–16
[13] Boehm H-D V, Siegert F, Rieley J O, Page S E, Jauhiainen J, Vasander H and Jaya A 2001 Fire impacts and carbon release on tropical peatlands in Central Kalimantan, Indonesia *22nd Asian Conference on Remote Sensing* pp 5–9
[14] Page S E, Siegert F, Rieley J O, Boehm H-D V, Jaya A and Limin S 2002 The amount of carbon released from peat and forest fires in Indonesia during 1997 *Nature* **420** 61–5
[15] Kogan F N 1997 Global Drought Watch from Space *Bull. Am. Meteorol. Soc.* **78** 621–36
[16] Zhang X, Kondragunta S, Schmidt C and Kogan F 2008 Near real time monitoring of biomass burning particulate emissions (PM2.5) across contiguous United States using multiple satellite instruments *Atmos. Environ.* **42** 6959–72
[17] Anderson G K, Sandberg D V. and Norheim R A 2004 *Fire Emission Production Simulator (FEPS) User's Guide Version 1.0*
[18] Akagi S K, Yokelson R J, Wiedinmyer C, Alvarado M J, Reid J S, Karl T, Crounse J D and Wennberg P O 2011 Emission factors for open and domestic biomass burning for use in atmospheric models *Atmos. Chem. Phys.* **11** 4039–72
[19] Van der Werf G R, Randerson J T, Giglio L, Collatz G J, Mu M, Kasibhatla P S, Morton D C, DeFries R S, Jin Y and van Leeuwen T T 2010 Global fire emissions and the contribution of deforestation, savanna, forest, agricultural, and peat fires (1997–2009) *Atmos. Chem. Phys.* **10** 11707–35
[20] Kanabkaew T and Kim Oanh N 2011 Development of Spatial and Temporal Emission Inventory for Crop Residue Field Burning *Environ. Model. Assess.* **16** 453–64
[21] Aldrian E and Dwi Susanto R 2003 Identification of three dominant rainfall regions within Indonesia and their relationship to sea surface temperature *Int. J. Climatol.* **23** 1435–52

Habitat Suitability Index (HIS) of Surili (*Presbytis comata* Desmarest, 1822) in mixed forest of Kuningan District, West Java-Indonesia

Lilik Budi Prasetyo[1*], Toto Supartono[2], Agus Priyono Kartono[1], Agus Hikmat[1], Syahru Ramdhoni[1]
[1]Forest Resources Conservation Department, Forestry Faculty, Bogor Agriculture University
[2]University of Kuningan

E-mail : lbprastdp@ipb.ac.id

Abstract. Java has been experiencing deforestation due to high population pressure. A lot of natural forests which play an important role as wildlife habitat are loss. The remaining natural forest distribute in mountainous areas in the form natural conservation area, meanwhile the others have been converted into settlement andinfrastructure, food crops, cash crops plantation, estate and private forest plantation. Javan langur (*Presbytiscomata*) is an endemic species of Java and used to utilize natural forest as their habitat. However, in a recent observation the species isfound inside plantation forest in Kuningan district, West Java. This is a unique finding, due to the fact that a plantation forest is not suitable habitat for Javan langur. The research is aimed to develop Habitat Suitability for this species based on physical, biological and human disturbance factors. Data on Javan langur presence and its habitat component were derived from field observation and secondary data/map. Result showed HSI could be developed based on 4 PC and showed that the study area mostly is occupied by low HIS Index or not suitable area for surili.

1. Introduction

Surili is an endemic species of Java Island and found only in West Java and Central Java[1] [2]. This is one of endangered species [3]and listed in the Appendix II of CITES and protected by Indonesian government through Government Regulation No. 7/1999. Indonesian Ministry of Forestry incorporates this species into itslist of conservation priority species (Ministry Regulation No.P.57/Menhut-II/2008).Based on Hoogerwerf (1970) [4], this species is mainly foraging insidenatural forest. In addition,Suriliwas found in highland natural forest of Mount Halimun [5] [6] [7], Mount Salak [8], Mount Gedeand Pangrango [2], Mount Ciremai[9] [10][11], Nature Reserve of Mount Tukung Gede [12] and lowland natural forest in Ujung Kulon [13]. However, recent records show that suriliis also found in mixed forest, which is a mixture of secondary forest, pine plantation and agroforestry adjacent to agricultural land and human settlement in Regency of Kuningan [14]. This discovery is very interesting to be studied in-depth because the fact of the distribution mentioned aboveis different from the nature of suriliwhich is very sensitive to human disturbance as well asto feedand habitat preferences in natural forests. Perhaps this is one of adaptation strategies applied by the surili in a state of urgency, due to the existence of natural forests that are drastically reduced, degraded and fragmented. There is a necessity to investigaterelationshipsbetween the existence of the species on the one hand and various variables of components of habitat that affect the distribution of surili on the other hand, both abiotic (elevation, slope) and biotic (feed, cover) components and disturbance from human activity, so that one can predict or map levelsof habitat suitability of surili in form of index on a landscape scale. This is important as a basis for conservation of species, habitat and risk management.

In spatial term, thedetermination of habitat suitability index (HSI) can be done by overlaying a variety of thematic maps, and spatial analysis can be done either in subjective or objective manners. Subjective spatial analysis can be done by simple weighting in which the amount of weight is determined based on the justification of researchers (expert judgment) or by an expert panel with Analytical Hierarchy Process approach.Objective analysis can be done by statistical analysis, especially if the relationship between the presence of this species andthe habitat component is unknown yet. Information on the ecology of suriliis very limited compared to the other primates species.

Among other statistical analysis are linear regression, general linear model/GLM/Logistic regression, and Multivariate. Recently, there is a developmentof an analysiswith approach of Machine Learning Method including ANN, Genetic Algorithm and Maximum Entropy. The chosentechnique of spatial analysis depends on the availability and types of data, both independent and dependent variable data.In cases of studies of wildlife that are always moving, in many cases one reliesmerely on the data of the species presence, thus it limits the choosing of methods of analysis. In addition to using Maxent, analytical technique that can be used isPrinciple Component Analysis/PCA. PCA is a technique to reduce and create new variable (Principle Component) based on original variables. The study aims to determine factors that influence the distribution of surili and to establishHSI models by using PCA.

2. Methodology

2.1. Study area

The study was conducted in the southern part of KuninganRegencythat bordering with CiamisRegency, CilacapRegency and BrebesRegency in the south (Figure 1). Land cover is dominated by pine plantations, secondary natural forests, agroforestry and agricultural fields (dry fields and paddy rice fields). The pine plantation forest and the natural forest are managed by Perhutani, while the agroforestry gardens are a form of community forests owned by local community. The community forest is planted with various species of trees either timber or fruit such as: sengon (*Paraserianthesfalcataria*), mahogany (*Swieteniamahagoni*), teak (*Tectonagrandis*), bitter bean (*Parkiaspeciosa*), and jackfruit (*Artocarpusheterophyllus*) [15].

2.2. Design of HSI prediction model

HSI prediction model is done by using an approach of Principle Component Analysis (PCA). The geo-location where surilifound were recorded. Independent variables (factors) thatmight play important roles in determining the existence of surili at the study area were extracted. The data of suriliwere divided into two groups. The first group is used to develop the model, while the second group is used to validate the model.There are 13 factors used in this study categorized as physical factors (elevation, slope, distance from river), biological factors (Normalized Difference Vegetation Index/NDVI, Bare soil index/BSI, tree density, density of feed trees, Basal Area/BA of feed and all trees, and distance from forest patch), and human disturbance (distance from settlement, distance from farms and distance from settlement). The value of each factor/variable for each observation was extracted using spatial analysis tool in ArcGIS software.

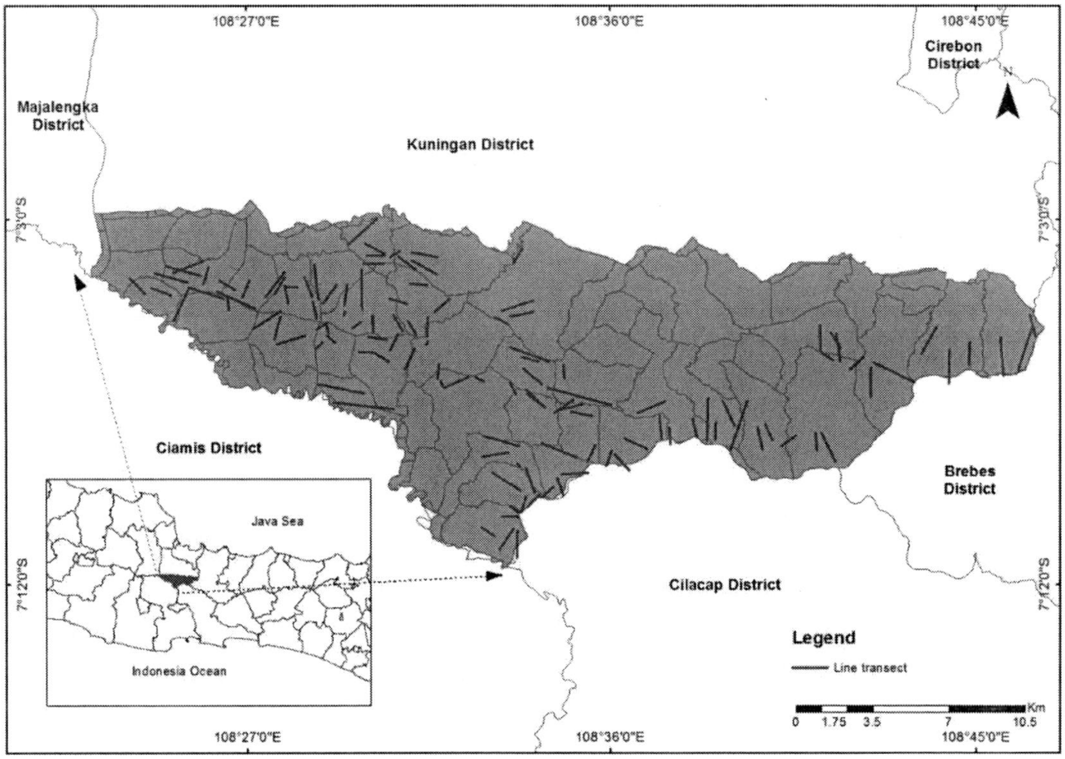

Figure 1. Study Location

PCA calculation results are considered appropriate if they meet several criteria, as presented in Table 1.

Table 1. Criteria and minimal requirement of PCA

Criteria	Minimal requirement
Ratio of observation number to variables	5 : 1[16]
Presence of substantial correlation	Some correlation greater than 0.30[17]
Sampling adequacy for individual variables	Acceptable if greater then 0.5[17]
Sampling adequacy for set variables	Acceptable if greater than 0.5[17]
Barlet test of sphericity	Less than 0.001[17]
Communalities	Greater than 0.5[18]
Total variance of the original variables	Between 70 – 90% [19]

PCA operation is repeated until all of the minimum requirements in Table 1 are met. After these minimum requirements arefulfilled, it will calculate number of newly formed variables (Principle component/PC) based on eigen value of bigger than 1. Total score is obtained by summing the PC multiplied by each of loading factor, with a general formula as follows:

$$Score = \sum_{i=1}^{n}(LF_i * PC_i) \quad (1)$$

Where:
LF_i: Loading factor of i^{th}PC
PCi: i^{th} PC
Afterward, result of the calculation in the form of floating number is spatialized and converted into HIS index scale of 0-1by using ERDAS Imagine software.

2.3. Surili existence survey

The purpose of the survey is to obtain the informationof suriliand its geo-locations. Initial information of surili presence is obtained from local community members. Subsequently, the survey is designed by using line transect method [20] [21]. The survey is started in the morning (around 6:00 am) until noon (around 12:00 pm), an investigator is walking slowly on existed paths or paths that are made available for observation [22]. Total length of transect in each forest is 5-6 km and the length of each track is 0.5-3 km. Length of each tractis adjusted to the width and the vast of the area of the forest block. Placement of transect in each forest depends on the distribution of forest and cover types as well as accessibility. This arrangement is made because the forest areawhere the research is located has a very steep topography with a lot of ravines (Figure 2).

2.4. Physical variable (elevation and slope)

2.4.1. Elevation and slope

Elevation and slope variables are generated from Digital Elevation Model (Aster G DEM) with a spatial resolution of 30 m (http://earthexplorer.usgs.gov/) by using Erdas Imagine software (Figure 2a and 2b).

2.4.2. Distance from rivers

River data of both temporal and permanent rivers is digitized based on topographical map based on a scale of 1: 25 000. Distance of existence of surilifrom riveris calculated spatially by using Euclidean Distance (ED) on ArcGIS software with spatial resolution of 30 meter (Figure 2c).

2.5. Biological variables

2.5.1. Tree density and tree basal area

Density and basal area of trees is calculated based on vegetational inventory using transect method [23]. Sample plots were established at a distance of every 100 meters (Figure 2) along surili observation transect, which is placed on each type of forest cover. The size of each sample plot is 20m x 20m [24]. Vegetation data collected includes name of tree species and tree diameter at breast height for each individual trees with diameter $\geq$ 10 cm [25].

Considering that the suriliis arboreal species [26] [27], data of undergrowth and trees with diameters of less than 10 cm were not collected. Information on the types of trees that have potential as food sources were obtained in sample plots by direct observation in the field and by studyingresults of previous studies elsewhere [26].After data of tree density and basal area of each track is obtained, then it is converted into spatial data by using interpolation technique of Inverse Distance Weight/IDW on ArcGIS software (Figure 3a, 3b, 3c and 3d).

2.5.2. Vegetation index (normalized different vegetation index/NDVI)

NDVI indicates level of greenness of earth's surface, which has continue value from -1 upto +1. Vegetation index is calculated by using data from Landsat satellite imagery 8 OLI (row/path: 121/065), which was acquired on 11 October 2015. The higher the NDVI value indicatedthe faster process of photosynthesis of those areas (Figure 3e). NDVI is calculated by using the following formula:

$$NDVI = (\text{band NIR-Red})/ (\text{band NIR} + \text{Red}) \qquad (2)$$

2.5.3. Bare soil Indeks (BSI)

This index is calculated by using the image that is used to calculate NDVI. Contrary to NDVI, Bare soil index indicates the openness of the land, which has continues value from -1 to +1, where thehigher BSI, the more open the area [28] (Figure 3f). BSI formula as follows:

$$BSI = (Band\ NIR + Band\ Red - Band\ Green) / (Band\ NIR + Band\ Red + Band\ Green) \quad (3)$$

2.5.4. Distance from forest patch

Forests provide foodsources, protectionfrom disturbance and predators, and facility for movement. Thus, the distance between where the suriliwere recorded and the forests is considered to be a potential variable to understand the distribution of surili in the study area. Forest cover is interpreted by using satellite images of Landsat 8 OLI (path/row: 121/065), which was acquired on 11 October 2015. Landsat interpretation was done by supervised classification technique on ERDAS Imagine software (Figure 3g).

2.6. Human activity or disturbance factors

Human disturbance factor is represented by the distance from roads, residence and farms. Road data is digitized from basic/Topographic maps with a scale of 1:25 000, while the distance from farms is based on land cover maps that were constructed from Landsat satellite image data. Distance of the presence of suriliis calculated spatially by Euclidean Distance technique in ArcGIS software (Figure 4a,4b and 4c).

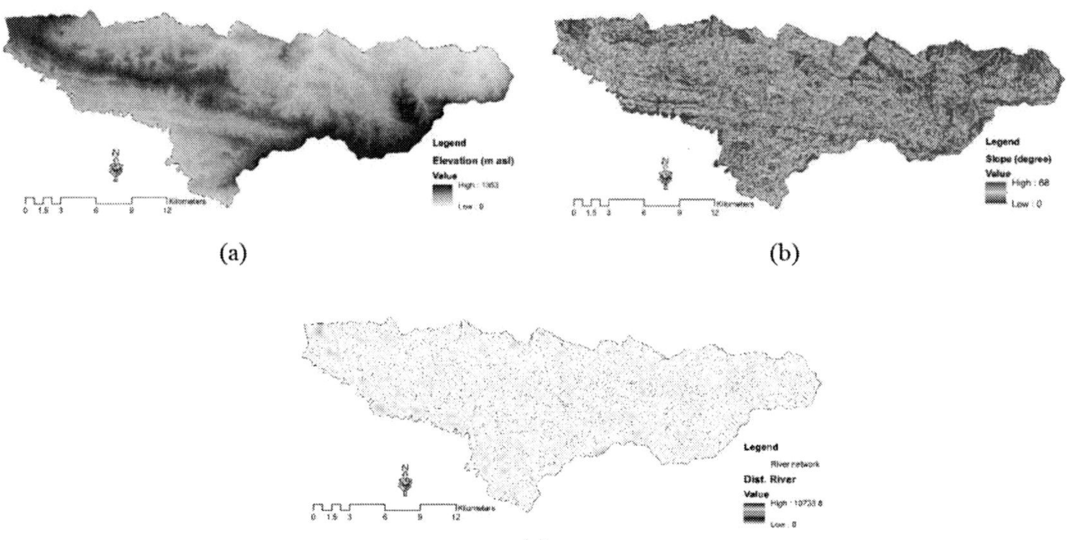

(a) (b) (c)

Figure2. Physical Variables: (a) Elevation, (b) Slope, (c) Distance from river

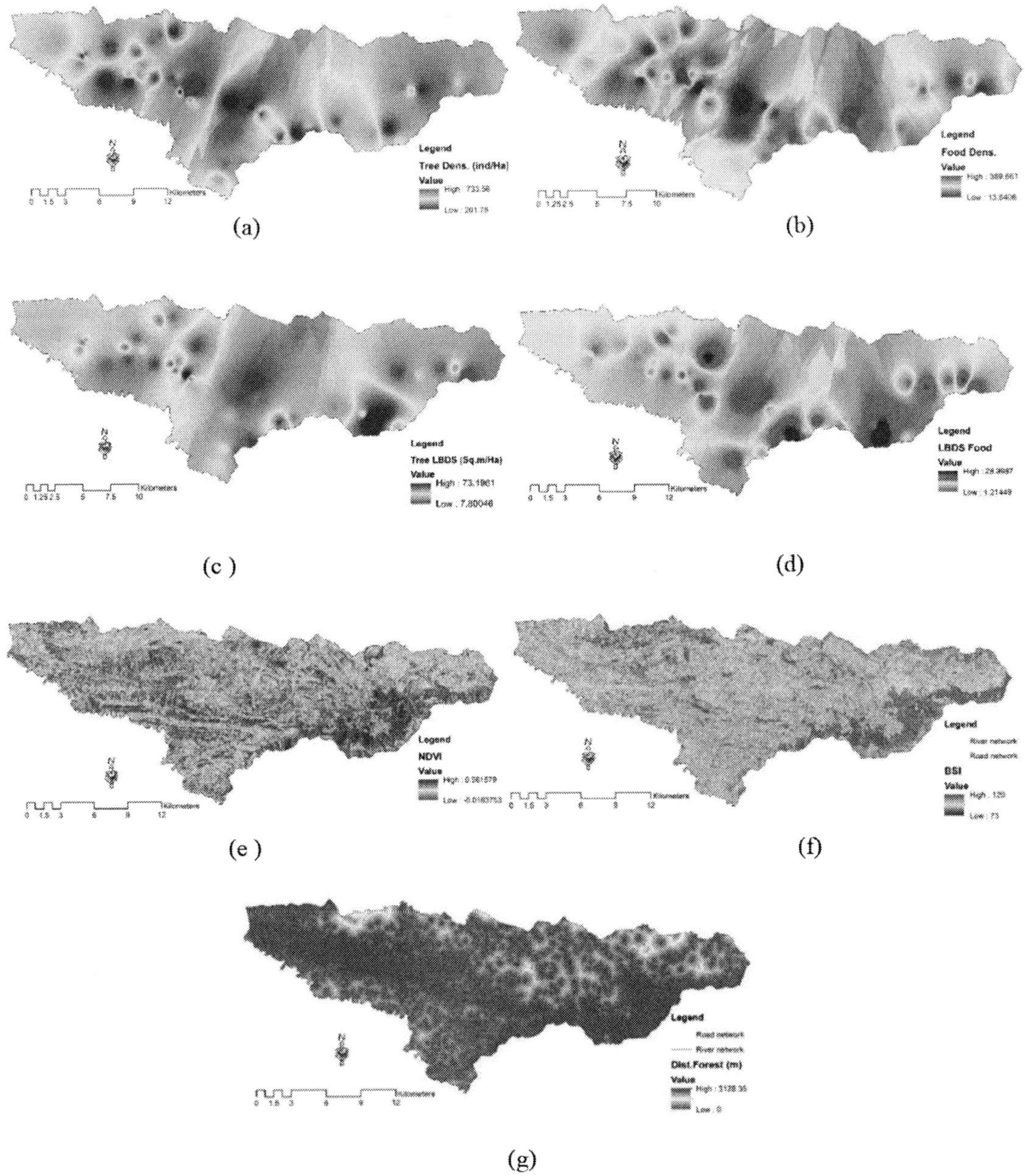

Figure 3. Biological variables: (a) Tree density (b) Feed tree density (c) Basal area of tree, (d) Basal area of feed tree, (e) NDVI, (f) BSI, (g) Distance from forest patch

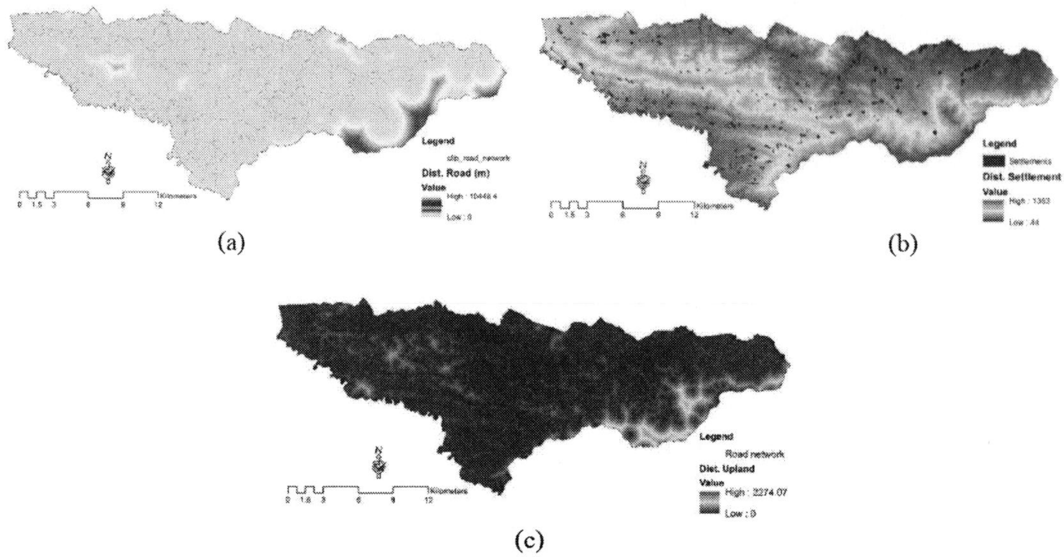

Figure 4. Disturbance Variables: (a) Distance from road, (b) Distance from settlement (c) Distance from uplands

3. Result &Discussion

3.1. Distribution of Surili

There were 156 locations of surili recorded during the survey period.When those positions are overlaid with each variable, it can be recognized the way surili respondsto each of these variables. However, these correlations cannot explain properlythe why surili occurred in a certain area. It is entirely possible that the existence of suriliis caused by many factors which are interrelated and influencial to each other (multicollinearity). Thus, the influence of each factor can be best understood by looking at its contribution to the formation of a new variable or Principal Components/PCs.

(a) Physical Factors

Factors that are grouped into physical factorsare distance from rivers, elevation and slope. The minimum distance from rivers, elevation and slope is 0 meter, 236 meters and 2 degrees,respectively. While the maximum valuesare 361.25 meters, 1271 meters and 48 degrees. From the distribution of data, it can be seen that the closer to rivers there is a tendency of more surilioccurence.This may be due to the river condition at the study site is located in a steep valley, with good forest cover.

In connection with the elevation and slope, surilidid not provide a consistent response to the increase in slope and elevation.This may be related to the fact that suriliis an arboreal animal and still be able to adapt to the height range of the study sites. At a certain slope and elevation, the occurrences of surili seem to decrease (Figure 5).

(b) Biological Factors

Minimum values of bare soil Index and NDVI at the point of surili encounteredwere77 and 0.155, respectively. Meanwhile maximum valueswere98 and 0.47 respectively.Response ofsurili to Bare soil Index and Normalized vegetation index is inversely proportional. The higher the bare soil, the less of numbers ofsurili found, while the higher the NDVI, the moresurili found (Figure 6a, b).Distance from the nearest forest is 0 meters (in the forest) and the farthest is 573.15 meters. Distribution of the data shows that the farther from the forest the less number of pointsof surili presence (Figure 6g).

Minimum values of density of all kinds of trees and feed treesare, respectively, 202and 14individuals per hectare, while the maximum valuesare, respectively, 734 and 389 trees per hectare. Pattern of distribution of the data shows that the higher the density of trees,the less the point where surili presents. Conversely the higher the density of food trees, the more points of surili encountered (Figure 6c, 6d).

Minimum basal areas of trees and feed treesare, respectively, 7.8 square meters and 1.3 square meters. Whereasmaximum valuesare, respectively, 73.1 square meters and 28.37 square meters. Surili's responses to the basal areas of trees and to feed trees are the same, i.e. the higher the basal area the less the point of encounter of surili (Figure 6e, f). This fact shows that the actual preference of surilitowards the basal area is not linear. Probably, the preference towards the basal area decreases at a certain value.

(c) Distance from human activity

Factors related to human activity is represented by distance from roads, residential and farmland. Although suriliis known as shy animal and tend to avoid humans, our observation data shows that this animal did not avoid human activities. The minimum distance of surilipresence from roads, residence and farmlandis, respectively, 0 meter, 30 meters and 0 meter. While the maximum distance of surili presence from roads, residential and farmland is 1776.09 meters, 3552.41 meters and 982.7 meters,respectively. From the data, there is a trend that the closer to the roads, residence and farms, the more the points of surili encounter. Many things that might explain this, and among them is the habit of local people who do not bother or expel the surili compared with that of to macaques. In addition, the roads in the research location are small roads whichare located onsteep slopeswherecanopies of trees are still interconnected. Similarly, the location of the community residence and farms areat the edge of mixed forests (Figure7).

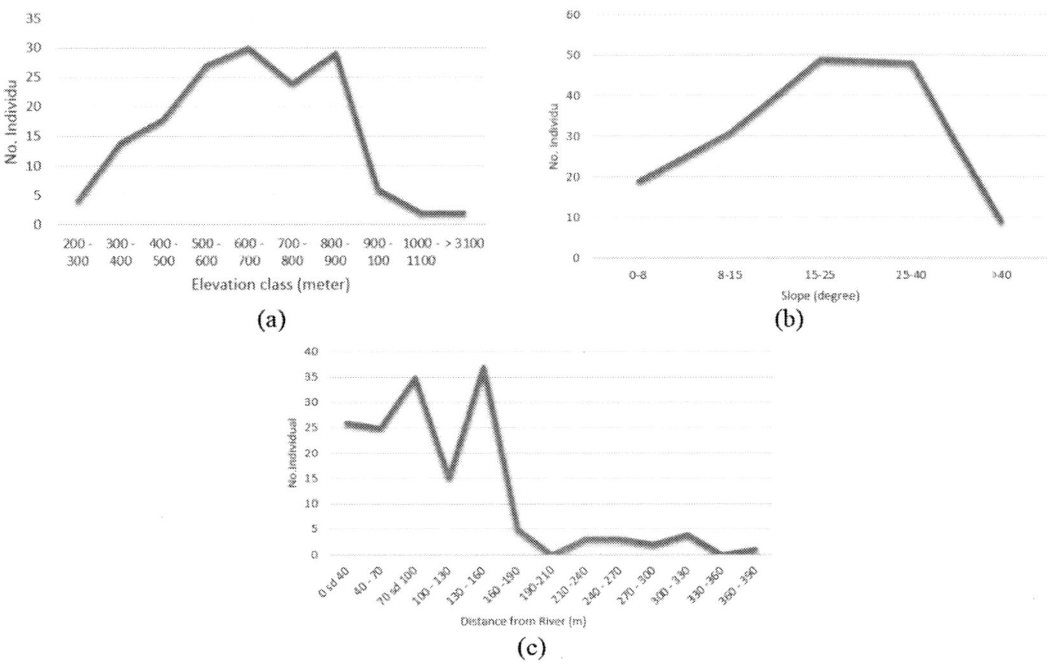

Figure 5. Number of Surili with regard to Physical variables (a) Elevation, (b) Slope and (c) Distance to River

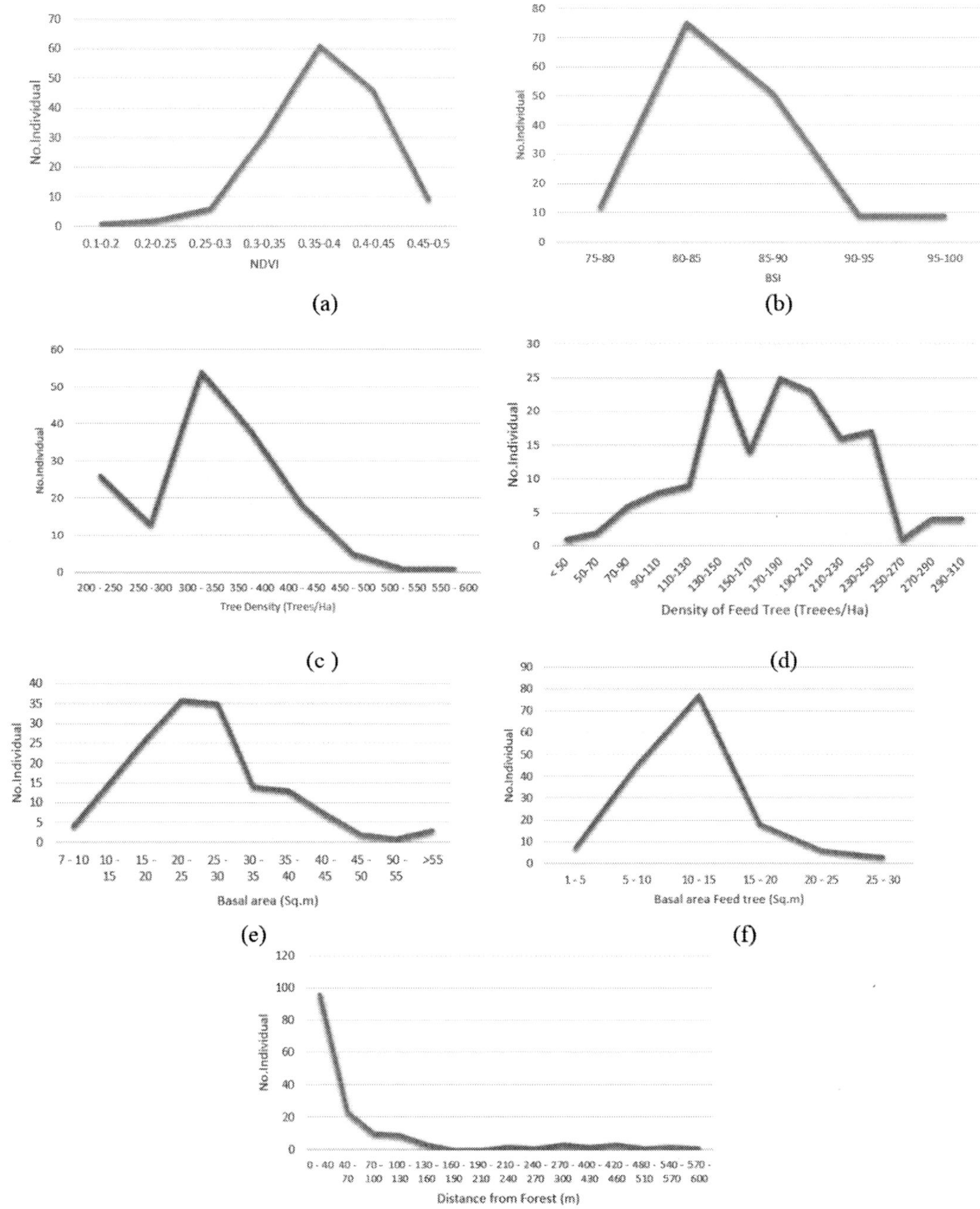

Figure 6. Number of Surili on various Biological factor variables (a) NDVI, ((b) BSI, (c) Tree density, (d) Feed tree density, (e) Basal area of trees, (f) Basal area of feed trees, (g) Distance from Forest

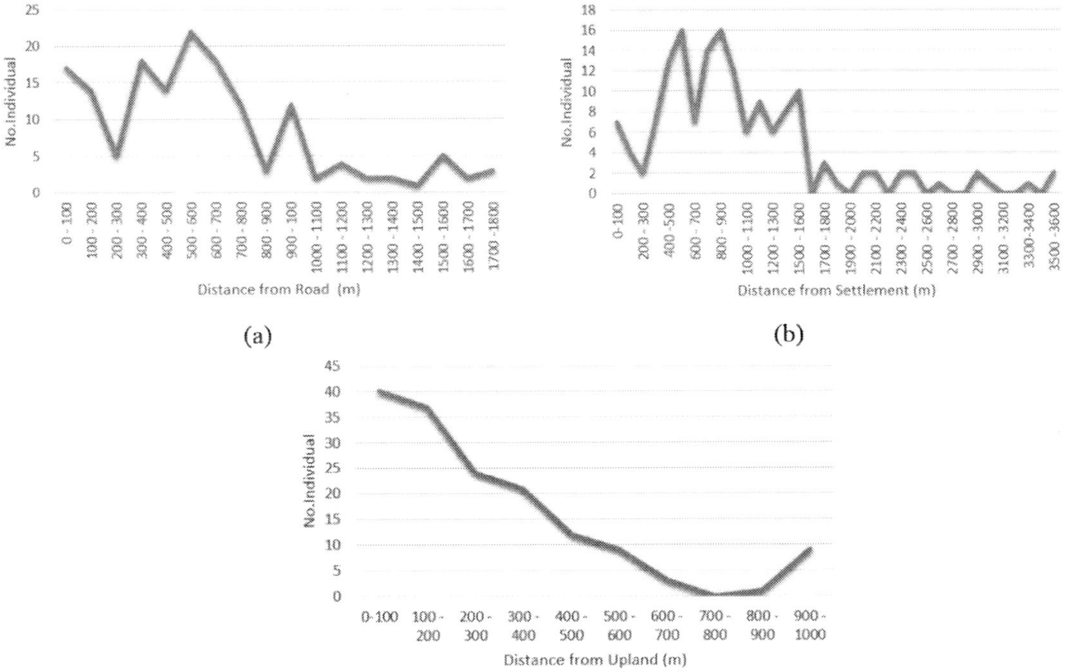

(c)

Figue7. Number of surili on human disturbance factor (a), distance from road, (b) distance from settlement and (c) distance from farms/upland

3.2. *PCA Calculation Results*

A total of 152 points of surili were recorded,98 points from this data were randomly selected to build the model, whilst 54 points used to validate the model. The random selection of the points is conducted by using ArcGIS software. Operation of PCA is performed 3 times of analysis to get an acceptable value in accordance with Table 1, so that the analysis can be continued.

Number of factors that have communalities value larger than 0.5 and having correlation greater than 0.5 are 10 factors, namely BSI, elevation, distance from roads, distance from farms, distance from forest, distance fromresidence, distance from rivers, NDVI, basal area of trees and basal area of feed trees. Thus, the ratio between the number of observation and the variables is 9.8: 1. In addition, KMO value as much as 0.655 and Bartlett's test of Sphericity with a significance level of 0.00.

There were4 PC selected, and each of them has an Eigenvalue of greater than 1. In cumulative PC 1, PC 2, PC 3 and PC4 contribute as much as, respectively, 33.512%, 19.677%, 14.88% and 11.367%, and the total four PCs can explain 79.44% of the data variance. PC1 through PC4 haveloading factors of3.351, 1.968, 1.489 and 1.137, respectively (Table 2).

When we examine further it appears that each PC represents different factors. From the contribution of each factor of its original variables (Table 3), the PC 1 explains the role of human disturbance factors which positively correlated with the presence of surili, that is the farther from the road and settlement the more suitable the habitat for the surili. This new variables explain that surili is still very sensitive to human activities. This fact is in contrast to when we see only partial human disturbance factors in the above description.

Table2. Variance contribution

Component	Initial Eigenvalues		
	Total	% of Variance	Cumulative %
1	3.351	33.512	33.512
2	1.968	19.677	53.189
3	1.489	14.888	68.077
4	1.137	11.367	79.444
5	.798	7.982	87.426
6	.369	3.694	91.119
7	.293	2.928	94.047
8	.271	2.710	96.757
9	.195	1.949	98.706
10	.129	1.294	100.000

Table 3. Rotated Component Matrix

Variable	Component			
	1	2	3	4
BSI	-.231	-.860	-.206	-.066
NDVI	.166	.908	-.004	.109
Distand from road	.887	.166	.049	.175
Dist_settlmtn	.919	.119	-.181	-.034
Distance from upland	.566	.402	.470	.261
Elevation	.069	.208	.830	.178
Distance from Forest	.254	-.381	-.735	.164
Dist_river	.008	-.363	.623	-.049
Lbds_Trees	.054	.112	.066	.897
Lbds_Food	.115	.039	-.029	.906

On the PC 2 there are two original influential variablesnamely the BSI which is correlated negatively, and the NDVI factorwhich is positively correlated. This means that the lower the Bare Soil Index and the higher the NDVI, the more suitable the habitat for surili. Both of these factors are consistent with the initial analysis.

On the PC 3, factors that have high contribution are elevation, distance from rivers and forest. The distance from rivers and the elevation appear to be more in explaining that a place that is relatively far from human activities are more suitable. Likewise, the distance from the forest which negatively correlated, suggesting the role of forests as protection. The farther a spot from forest, the more unsuitable it is for habitat of surili.

PC 4 reflects the need of surili on feed and cover which is represented by the values of basal area of feed trees and basal area of all species of trees (Table 3).

3.3. Habitat Suitability Index and Its Accuracy

Result of HSI calculations indicates that the location is dominated by areas with low HSI, which are located in areas that are relatively lower area relatively flat which dominated by upland agriculture with high disturbance from road and settlement. While high HSI are located at locations with high elevation and slope, good forest cover, and relatively far from residence, roads and upland agriculture/farms (Figure 8). The location is spatially fragmented, thus maintaining forest connectivity betweenthose locations becomes very important.

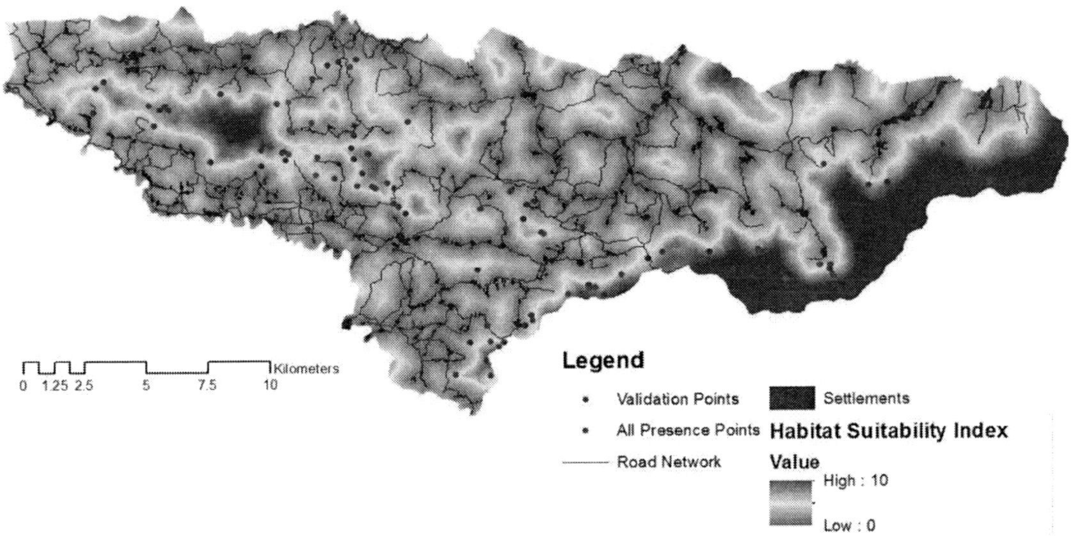

Figure 8. HSI of surili

If the HSI is divided into 5 classes, then most of the area study is unsuitable and poorly suitable, meanwhile, only small suitable area situated in forest area on hilly area (Figure 9 and Table 4). The overlay between point for validation (54 points) showed that 18 points (33.33 %) located on not suitable, 28 points (51.85 %) on Poorly suitable area, 7 points (12.96%) located on Moderately suitable and 1.85 % on suitable areas.

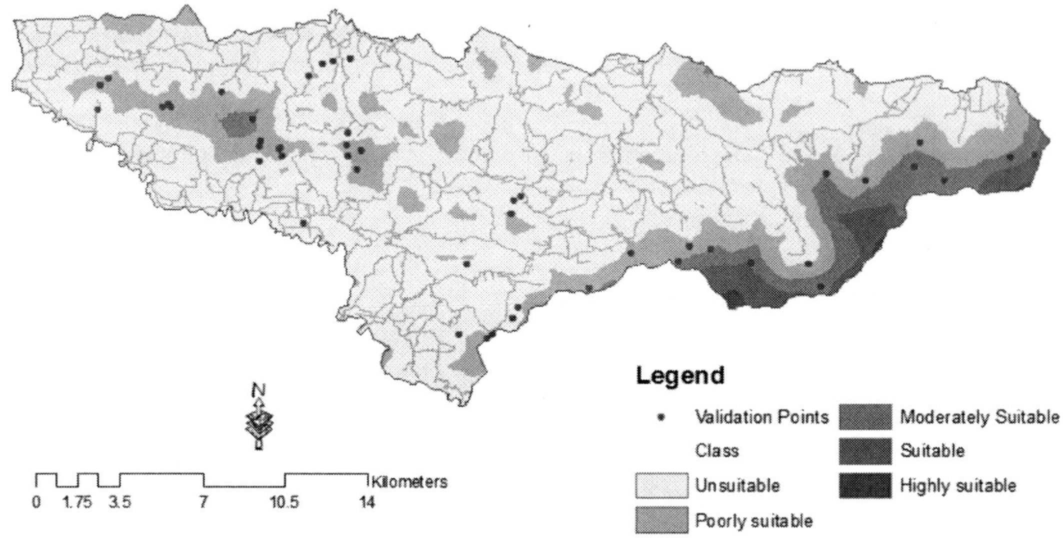

Figure 9. Suitability habitat class of surili

Table 4. Suitability class of surili

Class	Area (Ha)	Percentage (%)
Unsuitable	27,292.84	75.05
Poorly suitable	6,075.19	16.71
Moderately Suitable	1,926.53	5.30
Suitable	996.63	2.74
Highly suitable	73.61	0.20
Total	36,364.81	100.00

4. Conclusion

Surili occurrence is influenced by disturbance factor (distance from road, settlement and upland), followed by biological factor (BSI & NDVI), Physical factors (elevation, forest and distance from river) and cover and feed. The HSI could be developed based on 4 PC and showed that the study area mostly is occupied by low HIS Index or not suitable area for surili.

5. References

[1] Nijman V and van Balen S (Bas) 1998 A faunal survey of the Dieng Mountains, Central Java, Indonesia: distribution and conservation of endemic primate taxa. *Oryx* **32** 145

[2] Nijman V 1997 On the occurrence and distribution of Presbytis comata (Desmarest, 1822) (Mammalia: Primates: Cercopithecidae) in Java, Indonesia *Contrib Zool* **66** 247–256

[3] Eudey AA 1987 Priorities in Asian primate conservation *Primate conservation* **8** 172-174

[4] Hoogerworf AA 197) *Udjung Kulon: The land of the last Javan rhinoceros*. (Brill, Netherlands)

[5] Kool KM 1992 Food selection by the silver leaf monkey, Trachypithecus auratus sondaicus, in relation to plant chemistry *Oecologia* **90** 527–533

[6] Tobing ISL 1999 *Pengaruh perbedaan kualitas habitat terhadap perilaku dan populasi primata di Kawasan Cikaniki, Taman Nasional Gunung Halimun, Jawa Barat* (Bogor: Institut Pertanian Bogor)

[7] Farida WR and Harun 2000 The diversity of plants as feed resources for the java gibbon (Hylobates moloch), grizzled langur (*Presbytis comata*), and silver langur (*Trachypithecus auratus*) in Gunung Halimun National Park *J Primatol Indonesia* **3** 55–61
[8] Siahaan AD 200) Pendugaan parameter demografi populasi surili (*Presbytis aygula* Linnaeus 1758) di Kawasan Unocal Geothermal Indonesia, Gunung Salak (Bogor: Institut Pertanian Bogor)
[9] Gunawan, H dan M. Bismarck 2007 Status konservasi dan populasi satwaliar mamalia di taman Nasional Gunung Ciremai *Jurnal Penelitian Hutan dan Konservasi Alam* **8** 117-128
[10] Kartono AP, Gunawan, Maryanto I dan Suharjono 2009 Hubungan mamalia dengan jenis vegetasi di Taman Nasional Gunung Ciremai *J Biol Indonesia* **5** 279–294
[11] Supartono T 2010 *Karakteristik habitat dan distribusi surili (Presbytis comata) di Taman Nasional Gunung Ciremai* (Thesis)
[12] Melisch R dan Dirgayusa IWA 1996 Notes on the grizzled leaf monkey (*Presbytis comata*) from two nature reserves in the West Java, Indonesia *Asian Primates* **6** 5-11
[13] Heriyanto NM dan Iskandar S 2004 The population status and habitat of grizzled-leaf monkey *Presbytis comata* Desmarest in Kalajeten-Karangranjang forest complex, Ujung Kulon National Park *J Penelitian Hutan dan Konservasi Alam* **1** 89-98
[14] Supartono T 2016 *Distribusi dan habitat surili Presbytis comata di hutan campuran di luar kawasan konservasi* (Bogor: Institut Pertanian Bogor)
[15] Rikimaru A, Roy P S and Miyatake S 2002 Tropical forest cover density mapping *Tropical Ecology* **43** 39–47
[16] Bryant F B and Yarnold P R 1995 Principal components analysis and exploratory and confirmatory factor analysis In: Grimm, L.G. and Yarnold P.R., Eds., *Reading and Understanding Multivariate Statistics* (Washington DC: American Psychological Association) p 99-136
[17] Beaumont R 2012 An introduction to Principal Component Analysis & Factor Analysis Using SPSS 19 and R (psych package)
[18] MacCallum R C, Widaman K F, Zhang S and Hong S 1999 Sample size in factor analysis *Psychological Methods* **4** 84-99
[19] Landau S and B S Everitt 2014 *A hand book of Statistical analysis using SPSS* (Washington: Chapman & Hall/CRC) 335 p
[20] Greenwood J J D and Robinson R A 2006 General census methods In: Sutherland WJ, editor. *Ecological Census Techniques* Second Edition (New York: Cambridge University Press)
[21] Martins M M 2005 Density of primates in four semi-deciduous forest fragments of Sao Paulo, Brazil *Biodiversity and Conservation* **14** 2321-2329
[22] Estrada A and Coates-Estrada R 1996 Tropical rain forest fragmentation and wild population of primates at Los Tuxtlas, Mexico *International Journal of Primatology* **17** 759-783
[25] Soerianegara I dan Indrawan A 2005 *Ekologi Hutan Indonesia* (Bogor: Institut Pertanian Bogor)
[24] Kusmana C dan Istomo 1995 *Ekologi Hutan* (Bogor, Institut Pertanian Bogor)
[25] Onderdonk D A and Chapman C A 2000 Coping with forest fragment: the primates of Kibale National Park, Uganda *International Journal of Primatology* **21** 587-611
[26] Ruhiyat Y 1983 Socio-ecological study of *Presbytis aygula* in West Java *Primates* **24** 344-359
[27] Gunawan, Kartono A P dan Maryanto I 2008 Keanekaragaman mamalia besar berdasarkan ketinggian tempat di Taman Nasional Gunung Ciremai *Jurnal Biologi Indonesia* **4** 321-324
[28] Rikimaru A, Roy P S and Miyatake S 2002 Tropical forest cover density mapping *Tropical Ecology* **43** 39–47

Agarwood-planted tree inventory in Indonesia

Maman Turjaman[1*], and Asep Hidayat[2]
[1] Forest Microbiology Research Group, Forest Research and Development Centre, Environment and Forestry Research, Development, and Innovation Agency (FORDA), Ministry of Enviromental and Forestry, Jalan Raya Gunung Batu No. 5, Bogor 16610, Indonesia
[2] Forest Plantation Fiber Technology Research and Development Unit, Environment and Forestry Research, Development, and Innovation Agency (FORDA), Ministry of Environmental and Forestry, Jalan Raya Kuok Bangkinang Km. 9 Kotak Pos 4/BKN, Bangkinang, Riau 28401, Indonesia

E-mail: turjaman@gmail.com

Abstract. Indonesia has as a country that has a high diversity of agarwood-producing trees (APT) species compared to other countries in Asia. Unfortunately, the populations of APT species have declined significantly. The purpose of this study was to record and maps the agarwood-planted trees in Indonesia as a baseline for future management of this species. The questioners were distributed to 31 of provinces in Indonesia. The feedback came from 21 prefectures (67.7%), consisting from 121 regencies (36.6%) those in detail came from 579 district, 1,257 villages and 4,757 farmers group. The major of APT species planted by farmer groups are *Aquilaria malaccensis, A. microcarpa,* and *Gyrinops versteegii*. The potency of APT in Indonesia is 3.4 million trees, consisting from 0.2 million tree with DBH > 20 cm and 3.2 million tree with DBH < 20 cm. The highest APT in Indonesia is located in Central Kalimantan (24.7%) followed by North Sumatera (17.9%). The prediction of agarwood products and its derivate will be obtained in 2020 with economic value might be reached 1.6 trillion rupiahs if the inoculation technique used the standard procedure recommended by FORDA. These results showed how huge the potential of APT will be developed in the future.

1. Introduction

In accordance to the very high demand of agarwood product in the global market, *Aquilaria* and *Gyrinops* as agarwood-producing tree (APT) have been listed in Appendix II of the Conservation on International Trade in Endangered Species of Wild Fauna and Flora (CITES) since 2005 [1]. The demand for agarwood is believe to keep increasing year by year because agarwood-based products have a high economic important to be transformed to many products such as digestive, sedative and antiemetic drug, and also as incense and perfume [2]. However, the production of wild agarwood has been facing to serious depletion by illegal APT harvesting in forest and also the rapid loss of tropical rain forest as their natural habitats [3]. Similar trend have been also happening in Indonesia and as consequences, will never meet the agarwood demand if it is still depend on the product from natural habitat.

High demand of agarwood product in the market is not followed by comprehensive understanding of it formation. The formation of natural agarwood is a mystic, complex and slower time process [4]. The formations are usually trigered by wounding of twigs or branches caused by wind, thunder, mechanical hacking, slashing, cork boring, or wild animals [5]. Borer insects and caterpillars could also make a hole in the tree and trigger the agarwood formation. The condition then supported by the present of infectious fungi which are abundant in nature in the form of spores or hyphae, which is dispersed through water, wind and soil. Based on those hypotheses, we have been developed a bio-induction technology by using endophytic fungi and promoted agarwood formation in Indonesia [5] supported by FORDA, the Ministry of Environment and Forestry since 1984. Five years after that first induction, various quality of agarwood could be found into market.

The way to make meet between demand and production of agarwood, the sustainable mass planting of APT and bioinduction to obtained agarwood formation should be regulated well. Current condition showed that APT plantation in Indonesia are established in small-scale and run traditionally by farmers, hence they are still sporadic in its pattern of distribution and most are unrecorded. This situation is rather bit different compared to other Asian country [6]. The aim of this study was to record and maps the agarwood-planted trees in Indonesia as a baseline for future management of this species.

2. Methods

2.1 Data

Data were obtained from questioners, open discussion with agarwood farmers and direct field surveys conducted from April to October 2014. Questioners were distributed to 31 of provinces covering 329 of regencies and 25 natures reserve institution all over Indonesia archipelago. Other data were collected from discussion with key-person through direct interview. The key-persons were selected based on their intensive communication in attending technology-transfer of bioinduction organized by FORDA. Meanwhile, other data related to APT activities (cultivation method, APT species, etc) were collected by mean of field surveys in four representative location; West Kalimantan, Central Kalimantan, North Sumatra, and Riau Islands according to map in previous technical report [7].

2.2 Data Analysis and mapping

MS Excel were used to compile data based on APT species, diameter breast high (dbh), the age of the tree, and bioinduction treatment. The APT distribution, based on the number of its dbh classes > 20 cm and < 20 cm was plotted by using ArcView. The Indonesian spatial data map was downloaded from geospatial data gateway site (http://www.bakosurtanal.go.id/download-peta-rupabumi-indonesia-skala-1-1-000-00/, last accessed Feb, 30, 2014). The prediction of agarwood products was calculated by method describe by Turjaman et al. (2014) [7], after two years of inoculation to those of individuals with dbh > 20, agarwood will be formed and yield 2 kg of gubal A, 5 kg of kemedangan and 5 mL of agarwood oil.

3. Results and Discussion

APT in Indonesia may take from various ways, those: pure stand (monoculture), agroforestry (mixed with cash crops such as citronella, banana, lemon, guava, soursop, durian, etc). According to data collected in this study, there are 4,757 farmers groups have been establishing APT, which is distributed in 1,259 villages, 597 district, 121 regency and 21 provinces. The biggest APT plantation owned by farmer's group was found in Central Kalimantan which area covered of 140 ha; while the smallest was 0.1 ha. It is also common that APTs were planted as border trees delineating certain area of ownership. Contrasting with those of APT characteristics in other countries such such as Laos, Vietnam, Cambodia, Thailand, and Malaysia [8], APT in Indonesia are still purely traditional and those they were not integrated in its management.

3.1 APT distribution

Total population of APT in Indonesia is 3.4 million trees (figure 1). The highest potency of APT was recorded from Central Kalimantan (0.8 million trees), while the lowest was Jambi (818 tress).

Sumatera and Kalimantan Island contribute more than 85% of APT plantation in Indonesia and both islands are known as the main producer of wild agarwood, which called "malaccensis" as its trading name. People in both islands have been very familiar with the product becauses of its high economic values.

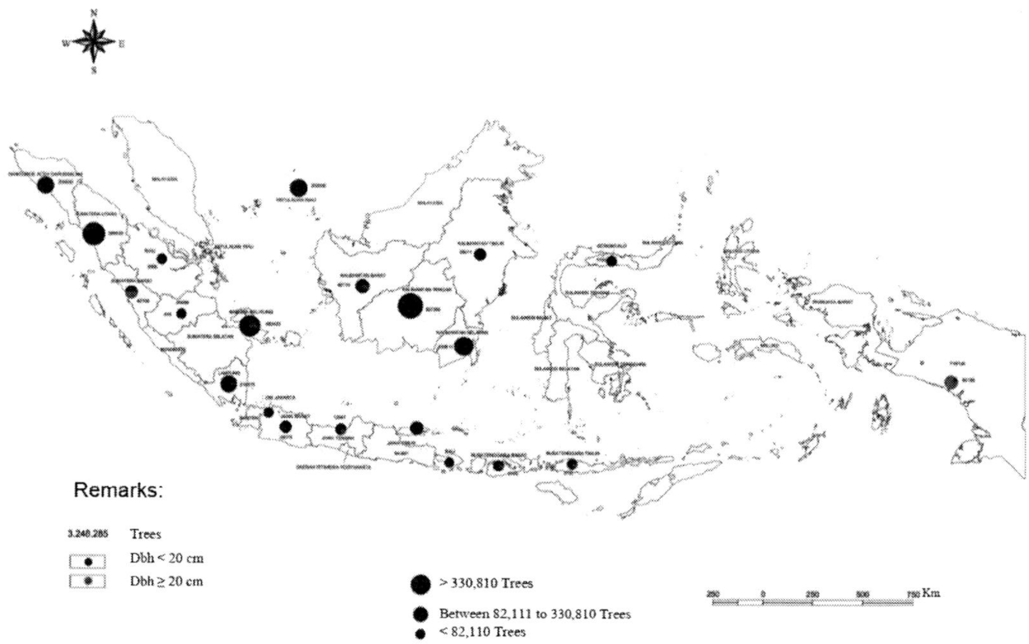

Figure 1. The potency of APT cultivation in Indonesia

The population of APT for each province was showed in the pie pattern; the bigger population of APT was indicated with the bigger of the pie pattern compared to each other. The size of dbh were categorized to be < 20 cm (black color), and > 20 cm (red color). The APT population distribution for each province is displayed in figure 1. The population of APT dominated by smaller one (dbh < 20 cm, 97%), totally. These indicated that farmers started to intensively plant APT for since 10 years ago.

Among those of APT species, *A. malaccensis* gains its popularity to be the APTS chosen in western Indonesia, followed by *A. microcarpa*, *A. beccariana* and *A. hirta*. In eastern Indonesia, *Gyrinops versteegii*, *A. filaria* and *A. cumingiana* are APT species which is having interest to be planted by farmers. Locally the species is known also as "*filaria*". The most familiar "*filaria*" species is *G. versteegii* which distributed in the island of Lombok, Flores, Sumba, Sumbawa, Celebes, Moluccas, and Papua [9]. Table 1 shows several species, which is commonly planted as APT with the high preference of the farmers were *A. malaccensis* and *G. versteegii* [10,11]. Furthermore, table 1 also shows the natural distribution and cultivation APT species in Indonesia. There are also exotic species were planted in very small scale by farmers groups in Deli Serdang – North Sumatera, which is include the species of *A. subintegra* and *A. crassna*. From the Table 1, we also can distinguish that APT of species *G. versteegii* have been widely planted to the western part of Indonesia.

Table 1. Distribution of APT species cultivation in Indonesia

No.	Scientific Name	Distribution	
		Cultivation	Natural
1.	*A. malaccensis*	West Java, Central Java, East Java, Bali, Bangka Belitung, Riau, Aceh, West Sumatera, North Sumatera, West Kalimantan, Central Kalimantan, East Kalimantan, and South Kalimantan	Bangka, Jambi, Riau and South Sumatra, Kalimantan, Sulawesi, Moluccas and Papua *
2.	*A. microcarpa*	North Sumatera, and South Kalimantan	Malay Peninsula, Sumatra (Sijunjung, Palembang and Lampung), Belitung, Bangka and throughout Borneo *
3.	*A. subintegra*	North Sumatera	Thailand *
4.	*A. crassna*	North Sumatera	Indochina and Thailand *
5.	*A. filaria*	Papua	Moluccas, Nusa Tenggara and Papua **
6.	*G. versteegii*	West Java, Central Java, East Java, Bali, West Kalimantan, East Kalimantan, West Nusa Tenggara, East Nusa Tenggara, Gorontalo, and Papua	Lesser Sunda Islands (Lombok, Sumbawa, Flores, Sumba) and North Celebes (Minahasa) *

Sources : *) [10]; **) [11]

3.2 *Prediction of Agarwood and its derivate products*

From all current potency of APT plantation, 50% (0.09 million trees) of the 0.19 million trees (dbh $\geq$ 20 cm) were inoculated by bio-induction techniques proposed by FORDA. Bali become the area with highest application of FORDA bio-induction technique followed by Bangka, Belitung, Aceh and Central Kalimantan. Although West Sumatera, East Nusa Tenggara, and Papua own more than 0.02 million trees with dbh $\geq$ 20 cm, but is no record that farmer do FORDA's bio-inoculation. in Bali the technique have been intensively introduced and disseminated since 2011. This program seemed to attack farmers' interest and started to practise the techniques to their APTs.

Current condition shows that bio-inoculation technique was quite successful to accelerate agarwood formation, hence its products were marketable. However, establishment of APT will require long time of investment. So APT cultivation need more diverse of product, not only rely on its agarwood product but also other product that can be optimized. Based on our previous serial research and field experience, the whole product APT (Glubal A, Kemedangan, and agarwood oil) can be predicted and calculated.

With the increasing demand of the agarwood products and its derivate, it is automatically will increase the farmers interest in cultivating APT. This will triggered the development of the APT in many provinces. Based on our finding in the abundances of APT (figure 1), other provinces in Indonesia will turn to the new APT centre area such as North Sumatra, Riau, Bangka-Belitung, Lampung, South Kalimantan and Central Kalimantan. Some problems could influence quantity and quality of cultivated agarwood products, i.e. pest and disease, fake inoculant, and agarwood marketing. APT cultivated in a large scale is prone to pest and disease attack. Recently, the most important pest found is *Heortia vitessoides* Moore (Odontiinae, Crambidae) that has caused severe damage at several APT in Indonesia [9]. Many fake inoculants sell to agarwood farmers, but the result of inoculation was not good. Furthermore, there is an urgent task for the Indonesia government to be present as both regulator and also facilitator to support this huge opportunity of community economic growth. Agarwood cultivated need special rule and regulation to transport, market, and export non-quota, it should be difference with wild agarwood regulation which is under control of CITES management [12].

4. Conclusion

As a non-timber forest product, wild agarwood had an impact to social, cultural and economic to society and environment. Now, it had significantly declined and the effort to replace its should be developed by agarwood cultivation. In this study, APT plantations were found in 21 provinces distributed in 121 district with more than 4,757 farmer both individuals or groups. The total area of APT varied from 0.1 to 140 ha through monoculture, and agroforestry plantation technique. The total population of APT was recorded 3.4 million trees approximately, including *A. malaccensis, A. microcarpa, A. subintegra, A. crassna, A. filaria*, and *G. versteegii* tree species. In 2020, the economic value for plantation was predicted reaches about 1.6 trillion rupiah. Based on this finding, Indonesian governments should be supposed the agarwood community to overcome the problem and create a regulations to guarantee and ensure that all agarwood cultivation will protect the farmers and profitable to society.

Acknowledgements

Some part of this research was sponsored by the International Tropical Timber Organization, and the Ministry of Forestry for Indonesia. The authors are grateful to Mr. Saiful Ikhsan, Forest Plantation Fiber Technology Research and Development Unit, for cooperation in plotting and mapping of data. We thanks to our colleagues for help and support to did this research and critical comments in writing of manuscript.

5. References

[1] CITES 2004 *Consideration of Proposals for Amendment of Appendices I and II- Aquilaria spp. and Gyrinops spp.* (Bangkok: 13th *meeting* conf.) CoP17 Prop: XXX
[2] Yang L Qiao C Xie D Gong N-B Lu Y Zhang J Dai J and Guo S 2013 Antidepressant abietane diterpenoids from chinese eaglewood *J. Nat. Prod* **76** 216-222
[3] Lee S H and Mohamed R 2016 *The Origin and Domestication of Aquilaria, an Important Agarwood-Producing Genus* ed R Mohamed (Singapore: Springer) chapter 1 pp 1-20
[4] Santoso E 2013 *Agarwood formation by fungal bioinduction technology* ed A Susmianto M. Turjaman and P Setio Track Record : Inoculation agarwood (Indonesia-Bogor: FORDA Press) chapter 3 pp 22-66
[5] Lee S Y, Mahat Ng W L, Nazre M N and M Mohamed R 2016 DNA barcoding of the endangered *Aquilaria* (Thymelaeaceae) and its application in species authentication of agarwood products traded in the market *Plos One* 1-21
[6] Turjaman M, Hidayat A and Santoso E 2016 *Development of Agarwood Induction Technology Using Endophytic Fungi* ed R Mohamed (Singapore: Springer) chapter 4 pp 57-71
[7] Turjaman M, Hidayat A, Rosita T and Santoso E 2014 Estimation of annual agarwood production and their quality Technical Report (Indonesia-Jakarta: PHKA-ITTO) p 70
[8] Turjaman M 2013 Agarwood upstream-downstream industry ed A Susmianto M Turjaman P Setio Track Record: Inoculation agarwood (Indonesia-Bogor : FORDA Press), Bogor, chapter 12 pp 185-216
[9] Sitepu I R, Santoso E, Siran S A and Turjaman M 2011 Fragrant wood gaharu: when the wild can no longer provide. (Indonesia-Bogor: ITTO PD425/06 Rev.1.; R&D Center for Forest Conservation and Rehabilitation) p 67
[10] Traffic 2004 Recommendations on the proposals to amend the CITES appendices at the 13th meeting on the conference of the parties Bangkok (Thailand) 2-14 October 2004 p 33

[11] Susilo A, Kalima T and Santoso E 2014 Field guide: identification of *Gyrinops* agarwood producing trees International Tropical Timber Organization (Indonesia-Bogor: ITTO) CITES Phase II Project p 34
[12] Santoso E 2015 Valuation of agarwood cultivated technology (Indonesia-Bogor: FORDA Press) p 167

Spatial analysis on school environment characteristics in mangrove management based on local wisdom (Case study at Lhokseumawe, Aceh)

Dewi Susiloningtyas*, Tuty Handayani, Naila Amalia, and Arum Ira Nadhira
Department of Geography, Faculty Mathematics and Natural Science, University of Indonesia, 16424 Depok, West Java. Indonesia

E-mail: dewi.susiloningtyas@ui.ac.id

Abstract. After 2004 tsunami, lots of efforts have been made, such as building school and distributing mangrove forests. This study examines the perception of teachers and students about mangrove management which spread in the administrative area of Lhokseumawe to become a reference then applied as local education regarding mangrove after tsunami disaster. This paper was based on primary data taken using questionnaire with a predetermined analysis unit to interview teachers and students in the study area. The result presented with quantitative and descriptive analysis. The result is of the total number of junior high schools in the city of Lhokseumawe as many as 41 Public Schools, Private and Religious School, there are 31 schools with priority for local wisdom education implemented mangrove. The result is classified with 3 class.The school's first priority is schools with a melee, with mangroves mangrove poor condition.Educational priority 2 is schools with close proximity to the mangrove and mangrove condition with moderate levels of damage. Schools with third priority are school with a close range, and mangrove good condition. Priority I as many as 18 schools, 10 schools priority II and 3 school for priority with learning competency standards that differ from each other.

1. Introductions

The earthquake and tsunami that hit Aceh on the 26 of December 2004 had created a terrible disaster of a scale that has never been faced by Indonesia. Mangrove forests were among the worst damaged wetland ecosystems in Aceh, but sources vary widely as the extant of mangrove forest damaged by the tsunami, Unfortunately, area of mangrove forest damaged by tsunami in Lhokseumawe, Aceh is 308.6 ha [1]. A number of organizations, both governmental and non-governmental, initiated a variety of environmental restoration activities, in particular the planting of mangrove. It takes a conservation program that is continuous and educative through the education, in this case the application of local wisdom education of mangrove. But in this case, teachers and students who are on the coastal environments abrasion and damage to mangrove forests on the environmental conditions should know about local wisdom based education mangrove. So, perception from them was needed to assessment how far they understand about local wisdom based education mangrove.

Disaster prone areas in the city of Lhokseumawe, covers regions prone to earthquakes, tsunami disaster prone areas, flood prone areas, areas prone to erosion and tidal wave. These disaster prone areas are in protected areas and some cultivated areas, necessitating intensive management primarily located in cultivated area. With the risk of vulnerability to disasters in the cultivation area this does not mean that the region cannot be built, but its use must be accompanied by efforts to anticipate / reduce the possibility of the impact of natural disasters (mitigation)

Earthquake prone areas scattered throughout the region in the city of Lhokseumawe. Utilization of cultivated area is located in disaster-prone areas should consider the level of the earthquake, especially to the toughness of the structure. As a reference in the toughness of the structure is to consider the level of power of the earthquake that occurred in 2004. In addition, a disaster prone area will abrasion, tidal waves, is Ujong Blang beach, Rancung, Meuraxa. While the area prone to floods are in the District of Banda Sakti, covering the village of Java, Java village of old, Lancang Garam and Tumpok Teungoh . Utilization of cultivated area is located in disaster prone areas should also pay attention to the potential for abrasion and tidal wave, and the potential and magnitude of the flooding.

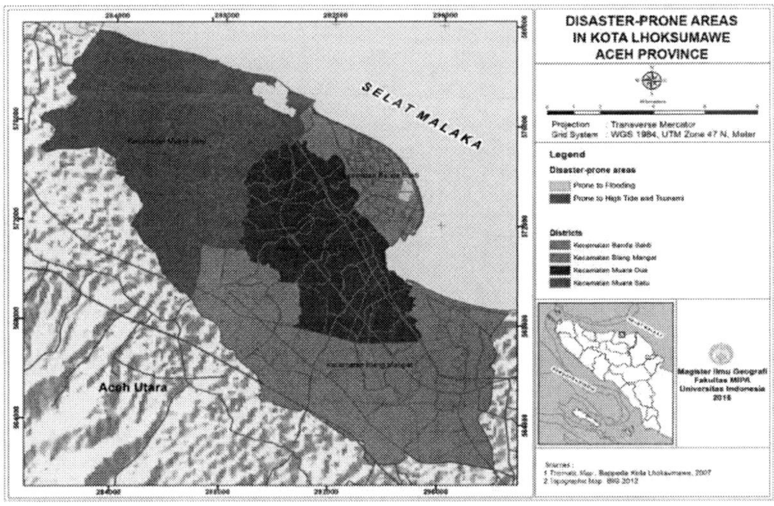

Figure 1. Disaster-Prone areas in Lhokseumawe City

Lhokseumawe (5°11'17" N - 97°8'25"E) is the second largest city in Aceh Special District, in the north of Sumatra, Indonesia (figure 1). The city covers an area of 181.06 square kilometers, and had a population of 171.163 at 2010 Census. Being between Banda Aceh and the large southern city of Medan, the town is a key regional center important for the economy of Aceh. Lhokseumawe set the status of a city based on Law Number 2 of 2001, dated June 21, 2001. Lhokseumawe City is divided into 4 districts, 9 residents, 68 villages, and 259 sub-villages with the boundaries of the area: the north from Lhokseumawe is Malacca Strait; the south from Lhokseumawe is Kecamatan Kuta Makmur; the west from Lhokseumawe is Kecamatan Dewantara; and the East from Lhokseumawe is Kecamatan Syamtalira Bayu.

According to Ghuffran (2012) [2], mangrove forest referred to as mangrove swamp forest is an ecosystem that is constantly undergoing development pressure. Mangroves grow optimally in coastal areas that have a large river estuaries and deltas that flow of water contains a lot of mud. While in coastal areas that are not contained estuaries, mangrove forest growth was not optimal. Mangrove is difficult to grow in coastal areas are steep and large choppy with strong tidal currents, because these conditions do not allow the deposition of silt that is a substrate required for the growth of mangroves [3]. To maintain the sustainability of mangrove forest functions, integrated management effort required. Stages of conservation efforts through mangrove management can be done early, namely through the understanding of children and young people to know the various aspects related to the existence of mangrove forest as an ecosystem are important, which are in the neighborhood of their lives. According Saraswati, *et al* (2010) [4] basic education development of local content greatly affect the quality of schools and students in the region. Implementation of local content are integrated in environmental education curriculum has also been done by Susiloningtyas and Handayani (2007) [5]. The results showed that school activities based on local wisdom turns perceived positively and contribute to improving the perception of the high importance of education in small island. Educational institutions are not only limited to fulfilling the education and research (research), but also plays an important role in supporting regional development through the development of the region in economic, social and cultural [6]. In this study, an assessment of the perception of the environment around the respondent in cases the teachers and students who are on the coastal environments and damage to mangrove forests on the environmental conditions.

In conducting this study, the authors use this line of thinking described in the figure below:

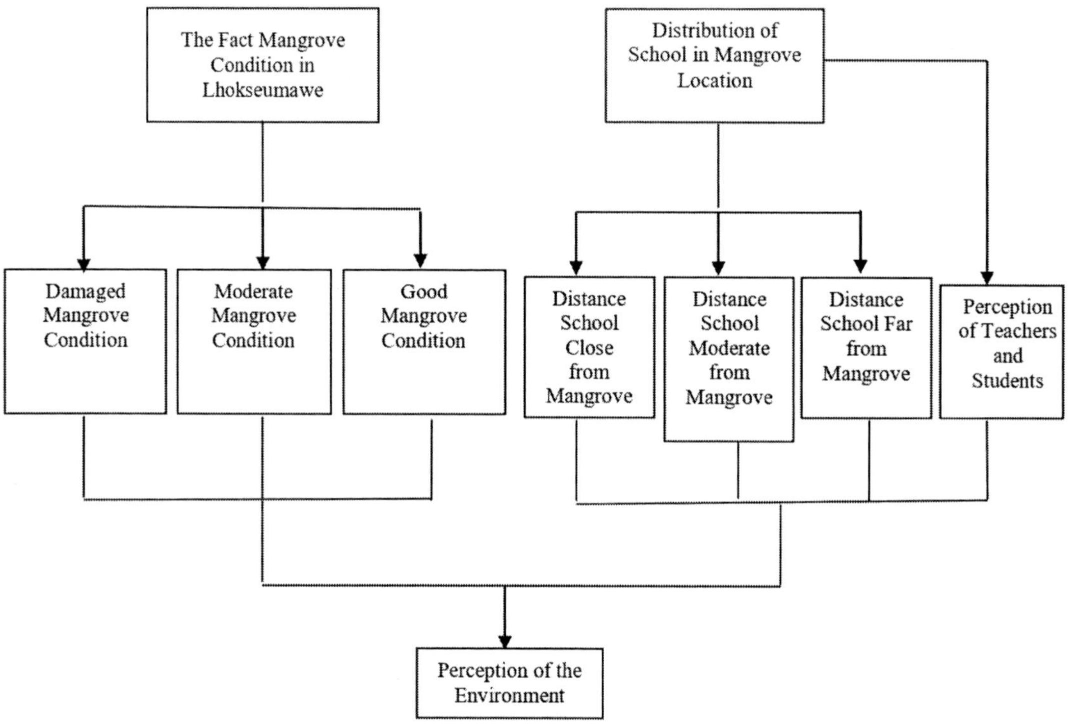

Figure 2. The decision how to improve the perception of teachers and students

Lhokseumawe city lies between 04°54'N - 05°18'N and 96°20'E - 97°21'E with an average altitude of 13 meters above sea level. Since forming in 2001, until now Lhokseumawe city is divided into four districts, 9 settlement, 68 villages and 259 hamlets. Lhokseumawe City is as follows:

- Northern Strait of Malacca.
- West of the slams Kecamatan Dewantara North Aceh.
- South side with Kuta Makmur sub-district, North Aceh regency.
- East side with Syamtalira Bayu Sub-district, North Aceh regency

The city covers an area of 181.06 square kilometers, and had a population of 171,163 at the 2010 Census; the latest official estimate (as at 2014) is 180,200. Being between Banda Aceh and the large southern city of Medan, the town is a key regional center important for the economy of Aceh.

Lhokseumawe City is an urban expansion of Aceh Utara Regency and is located on the east coast of Sumatra. Being between Banda Aceh and Medan, Lhokseumawe City holds a strategic position as a distribution channel for trading traffic in Aceh Special region.

When formed in 2001, the city was dividing administratively into three districts (*kecamatan*), since then, the creation of Muara Satu District (formerly part of Muara Dua District) has increased the number to four:

Table 1. Vast districts and the number of population

District	Area (km^2)	Population at 2010 Census
Banda Sakti	11.24	79,336
Blang Mangat	56.12	22,850
Muara Dua	57.80	46,646
Muara Satu	55.90	32,975

Source: Lhokseumawe Dalam Angka 2013/ Lhokseumawe in Figures 2013

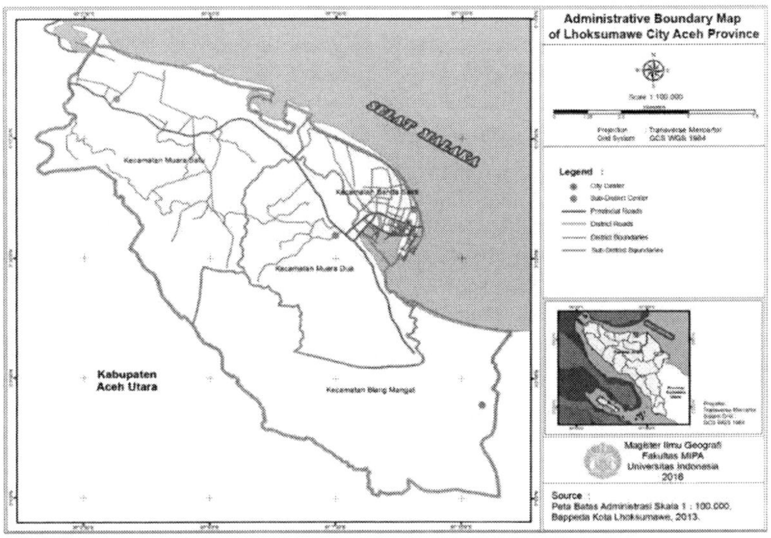

Figure 2. Administrative boundary map of Lhokseumawe City Aceh Province

2. Methods

The research location is in the province of Aceh, Lhokseumawe, located between 04°54'N - 05°18'N and 96°20'E - 97°21'E. This research is done in April and Mei 2016. Sampling method for this study is purposive sampling. Purposive sampling is samples are carefully selected by taking the person or object research was selective and has specific characteristics. Samples taken have special characteristics of the population that is considered quite representative [7]. The samples in this study were selected to support the strengthening of the data or to clarify the data. In the study, emphisize on the perception of teachers and students to Education Local Wisdom Mangrove.

The sampling technique used purposive sampling with the following criteria:

1. Students and teachers at the Junior High School who were in school in the study area in accordance with a predetermined analysis unit.
2. Students and teachers who have a residence in accordance with a predetermined analysis unit.

Data needed in this research consisted of primary data and secondary data. Secondary data used in this study was obtained through the relevant agencies such as BPS, Bappeda Lhokseumawe City, the Department of Education and Culture in Lhokseumawe, Regional Disaster Management Agency Lhokseumawe and SPOT 5 satellite imagery acquired from Bappeda Aceh province. While at the stage of primary data collection, researchers use interviewing techniques and questionnaire, the questionnaire technique using a list of questions. In this study used a questionnaire with the aim that the respondent can answer the appropriate answer options are available, but some of the questions answered in accordance free respondents thought.

The processing data method for this study is quantitative descriptive analysis. Descriptive analysis is a behavioral sensory evaluation approach that uses descriptive panels to measure a product's sensory characteristics. Panel members use their senses to identify perceived similarities and differences in products, and articulate those perceptions in their own words. Sensory evaluation is a science that measures, analyzes, and interprets the reactions of the senses of sight, smell, sound, taste, and texture (or kinesthesis) to products. It is a people science; i.e., people are essential to obtain information about products.

Descriptive research in this study is intended to get an overview and particulars of school conditions and quality of junior high school in the city of Lhokseumawe linked to local knowledge - based mangrove education and respondents' perceptions of local knowledge based education mangrove. While the quantitative data was processed by SPSS (Statistical Product and Service Solutions).

Methods of data processing are spatial and tabular data processing with the Geographic Information System (GIS). The process is done through a phase of spatial data entry, editing, labeling, annotation, joint map, overlay, classification, data entry attribute to the output / output. Then performed to classify the degree of damage to the mangrove and the distance of the school. Steps being taken are as follows: Classification of mangrove destruction is done by looking at the average area is obtained from the Spot 5 satellite imagery analysis and observations from researchers in the field on the existing condition of mangrove. Then, Distance to school with mangrove is used to determine the spatial pattern analysis indigenous mangrove education, by looking at the distribution of schools and mangrove. The distance of the school and are grouped into three classes mangrove distance by using the same method with the classification of mangrove destruction, with the results of three classes: near, medium and far. After that, the quality of the school is the result of the calculation of scores of the quality of teachers, students and school infrastructure quality result school quality are also grouped in three classes, namely schools with high quality, medium and low.

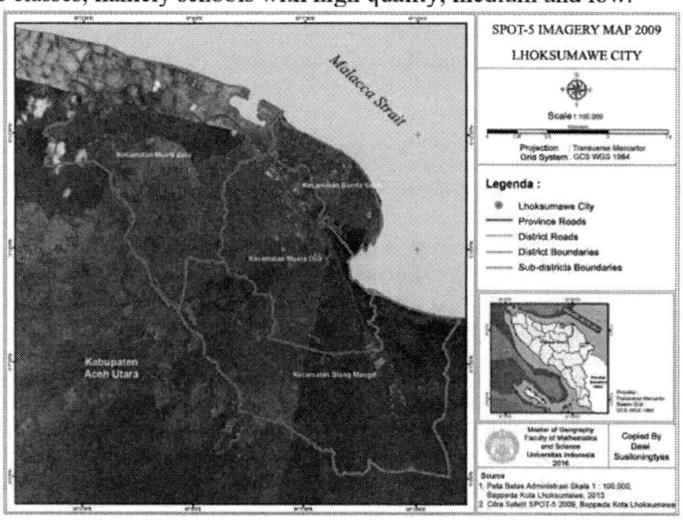

Figure 3. Satellite image Lhokseumawe City, Aceh Province

Then, Distance to school with mangrove is used to determine the spatial pattern analysis indigenous mangrove education, by looking at the distribution of schools and mangrove. The distance of the school and are grouped into three classes mangrove distance by using the same method with the classification of mangrove destruction, with the results of three classes: near, medium and far. After that, the quality of the school is the result of the calculation of scores of the quality of teachers, students and school infrastructure quality result school quality are also grouped in three classes, namely schools with high quality, medium and low.

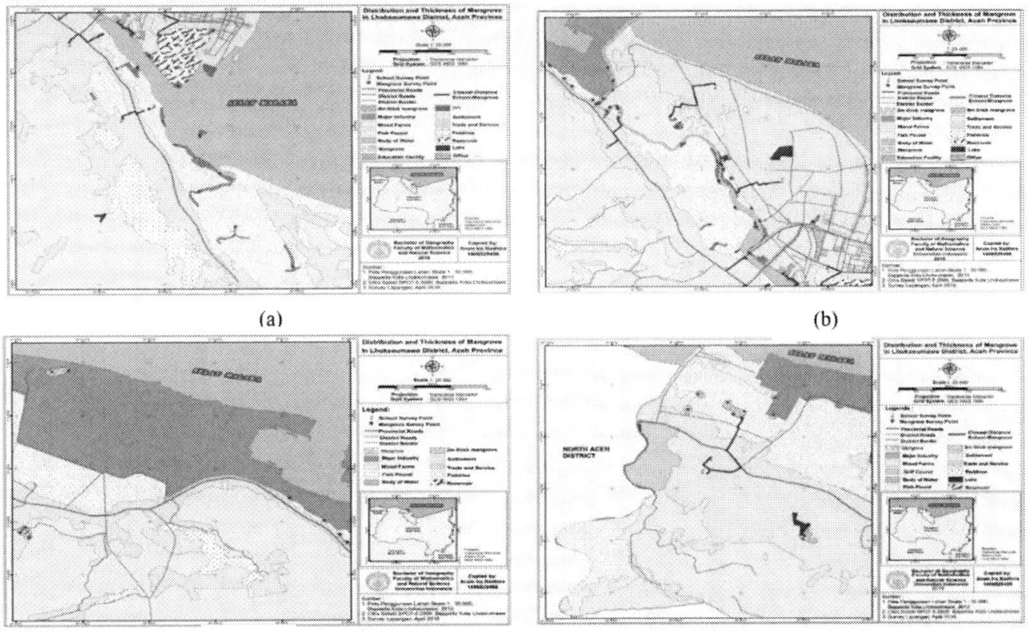

(a) (b) (c) (d)

Figure 4. Distribution and tickness of mangrove in Lhokseumawe District, Aceh Province

Then, Distance to school with mangrove is used to determine the spatial pattern analysis indigenous mangrove education, by looking at the distribution of schools and mangrove. The distance of the school and are grouped into three classes mangrove distance by using the same method with the classification of mangrove destruction, with the results of three classes: near, medium, and far. After that, the quality of the school is the result of the calculation of scores of the quality of teachers, students and school infrastructure quality result school quality are also grouped in three classes, namely schools with high quality, medium and low. Most of the junior high school level is close to the location of the mangrove, i.e. with a percentage of 78% of the total. Districts with the highest population are Kecamatan Banda Sakti and Muara Dua is the center of the city of Lhokseumawe. Both districts are also the districts with the largest mangrove are poor to moderate anyway. While schools with medium and far distances percentage of each of the 7 and 15% presented in picture below:

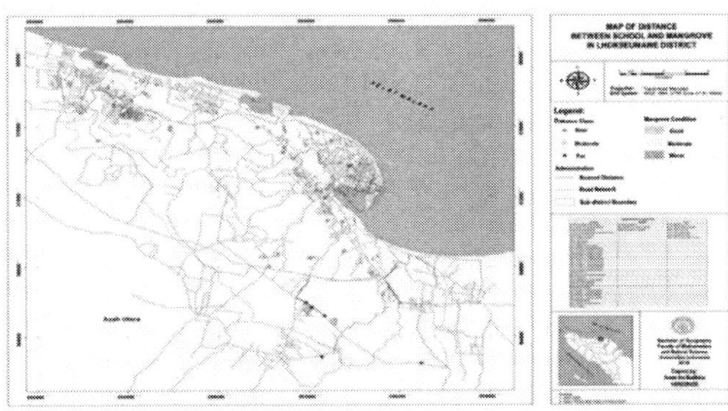

Figure 5. Map of distance between school and mangrove in Lhokseumawe District

3. Result and discussion

In this study, an assessment of the perception of the environment around the respondent in this cases the teachers and students who are on the coastal environments abrasion and damage to mangrove forests on the environmental conditions. There are 4 questions in the questionnaire which are about Insights and Knowledge about Mangrove, Perception of Teachers about Mangrove, Teacher Participation on Mangrove Local Wisdom and Learning Methods Local Content in Schools and Management.

For in this case, taken 3 teachers from 41 schools to be measured insight, perception and participation on the coastal environment and mangroves as well as to know the opinion of the teachers education based on local wisdom mangrove. And this is the result from the questionnaire.

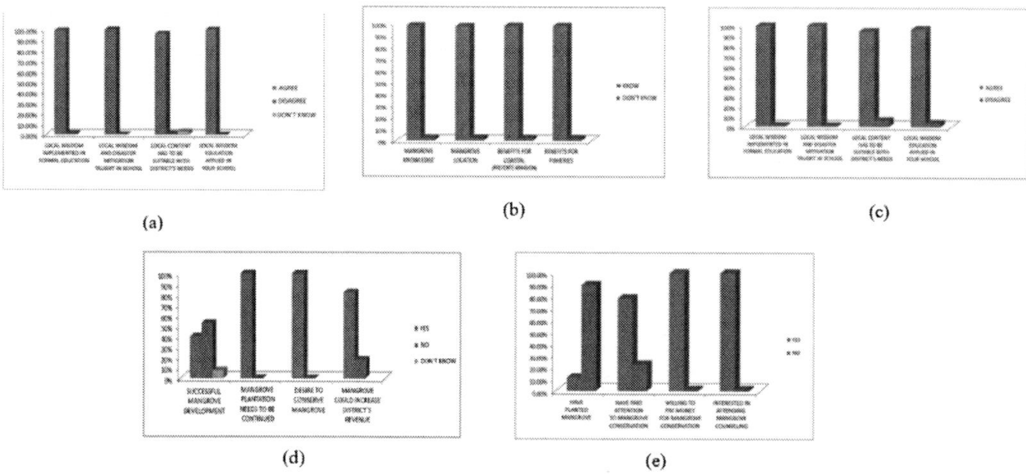

Figure 6. (a) Insight and knowledge about mangrove from teachers. (b) Perception of teacher about mangrove management. (c) Teacher participation on mangrove management. (d) Local wisdom and learning methods local content in schools from teachers. (e) Local wisdom and learning methods local content in schools from teachers.

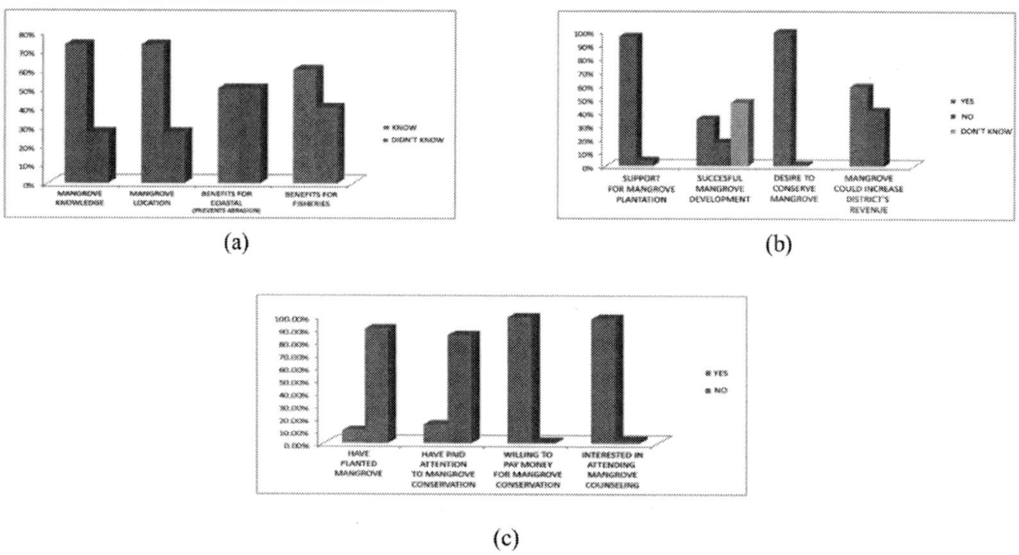

Figure 7. (a) Insight and knowledge about mangrove from students. (b) Perception of students about mangrove (c) Student participation on mangrove management

After that, we discussed about the relationship between qualities of School with Student Perceptions. The analysis used to look at the relationship the perception of respondents (teachers and students) with distance and condition of mangrove is a simple correlation analysis. Simple correlation analysis is used to describe the strength and direction of the relationship between the two variables. The method used is the Pearson product moment. Figures correlation ranged from -1 s / d +1. The closer to 1, the closer the correlation is perfect. While the negative and positive values indicate the direction of the relationship.

Hypothesis is:
H0: There is no relationship between the qualities of schools with student's perceptions
H1: There is a relationship between the qualities of schools with students's perceptions

Table 2. Correlation between the qualites of schools with student's perceptions

Correlations

		Quality of School	Students Perception
Standard of School	Pearson Correlation	1	.352**
	Sig. (2-tailed)		.000
	N	410	410
Students Perception	Pearson Correlation	.352**	1
	Sig. (2-tailed)	.000	
	N	410	410

**. Correlation is significant at the 0.01 level (2-tailed).

From the analysis above correlation (r) obtained correlation between the qualities of schools with Student Perceptions (r) is 0.352. This shows that the relationship between the quality of schools with Student Perceptions is low because it is located between the range of 0, 20 to .399. While the direction of the relationship is positive, this means that if X goes up, then Y rises or in other words the higher the value the better the quality of schools Perception of Students. Correlation is significant at 0.01 which means that H0 is rejected or there is a relationship between the qualities of schools with student perceptions.

Then, we discussed about the relationship between qualities of School with Teachers perceptions. The method used is the Pearson product moment too.

Hypothesis is:
H0: There is no relationship between the qualities of schools with teachers' perceptions.
H1: There is a relationship between the qualities of schools with teachers' perceptions.

Table 3. Correlation between the qualites of schools with teachers's perceptions

Correlations

		Quality of School	Teachers Perception
Standard of School	Pearson Correlation	1	.209*
	Sig. (2-tailed)		.021
	N	123	123
Teachers perception	Pearson Correlation	.209*	1
	Sig. (2-tailed)	.021	
	N	123	123

*. Correlation is significant at the 0.05 level (2-tailed).

From the analysis above correlation (r) obtained correlation between the qualities of schools Perception of Teachers (r) is 0.209. This shows that the relationship between the quality of schools with teachers Perception is low because it is located between the range of 0, 20 to .399. While the direction of the relationship is positive, this means that if X goes up, then Y rises or in other words the higher the value the better the quality of schools Perception Teachers. A significant association was 0.05, which means that H0 is rejected or there is a relationship between the qualities of schools with teachers' perceptions.

Schools are prioritized for implementation of education based wisdom mangrove is a school with a close-range mangrove bad condition , then melee mangrove condition medium and close range mangrove good condition are divided into three classes of priorities, namely :

1. The school's first priority is schools with a melee, with mangroves poor condition. There are 18 schools in Lhokseumawe City.
2. Educational priority 2 is schools with close proximity to the mangrove and mangrove condition with moderate levels of damage. There are 10 schools in Lhokseumawe City.
3. Schools with third priority are school with a close range, and mangrove good condition. There are 3 Schools in Lhokseumawe.

About the matrix of the infrastructure quality schools, the condition of mangrove and quality of schools, teachers and students can be seen in table below:

Table 4. Matrix of research results

Number	School's Name	Quality of Teachers	Quality of Students	Infrastructure Quality Schools	Quality of School	Mangrove's Condition
1	SMP NEGERI 1 LHOKSEUMAWE	High	High	Good	High	bad
2	SMP NEGERI 2 LHOKSEUMAWE	Moderate	High	Good	High	bad
3	SMP NEGERI 3 LHOKSEUMAWE	High	Low	Good	moderate	good
4	SMP NEGERI 4 LHOKSEUMAWE	Moderate	Low	Good	moderate	good
5	SMP NEGERI 5 LHOKSEUMAWE	High	High	Good	High	bad
6	SMP NEGERI 6 LHOKSEUMAWE	High	Low	Good	moderate	moderate
7	SMP NEGERI 7 LHOKSEUMAWE	Moderate	Low	Good	moderate	bad
8	SMP NEGERI 8 LHOKSEUMAWE	High	Low	Good	moderate	moderate
9	SMP NEGERI 9 LHOKSEUMAWE	Low	Low	Good	moderate	bad
10	SMP NEGERI 10 LHOKSEUMAWE	High	Low	Good	moderate	bad
11	SMP NEGERI 11 LHOKSEUMAWE	High	low	Good	moderate	bad
12	SMP NEGERI 12 LHOKSEUMAWE	Moderate	low	Good	moderate	moderate
13	SMP NEGERI 13 LHOKSEUMAWE	High	low	Good	moderate	good
14	SMP NEGERI 14 LHOKSEUMAWE	Low	low	Low	Low	moderate
15	SMP NEGERI 15 LHOKSEUMAWE	Low	low	Good	moderate	bad
16	SMP NEGERI UJUNG PANCU	Low	low	Low	Low	bad
17	MTSN LHOKSEUMAWE	Moderate	moderate	Good	moderate	bad
18	SMP NEGERI ARUN	High	high	Good	High	moderate
19	SMP SWASTA MUHAMMADIYAH 6	Low	low	Good	moderate	bad
20	MTSS AL-WASHLIYAH CUNDA	Low	low	Low	Low	bad
21	MTSS BATUPHAT	Moderate	low	Low	Low	moderate
22	MTSS IHYAAUSSUNNAH	Low	low	Low	Low	bad
23	MTSS MUHAMMADIYAH	Low	low	Moderate	Low	bad
24	MTSS ULUMUDDIN	Low	high	Good	moderate	bad
25	MTSS BABUL HIDAYAH	Low	low	Low	Low	moderate
26	MTS SWASTA HIDAYATULLAH	Low	low	Low	Low	bad
27	SMPS SUKMA BANGSA	High	high	Good	High	moderate
28	MTs SWASTA AL – MUARIF	Moderate	low	Moderate	moderate	moderate

Number	School's Name	Quality of Teachers	Quality of Students	Infrastructure Quality Schools	Quality of School	Mangrove's Condition
29	MTs SWASTA BANDA MASEN	Low	low	Low	Low	bad
30	MTs SWASTA YAPENA	High	high	Good	High	moderate
31	MTSS DARUL FAIZIN	Low	low	Low	Low	bad
32	MTSS SAFINATUSSALAMAH	Low	high	Moderate	moderate	moderate
33	MTSS MISBAHUL ULUM PALOH	Low	high	Good	moderate	moderate
34	MTSS BABURRAHMAH	Low	high	Moderate	moderate	bad
35	MTSN PUNTEUT	Moderate	high	Moderate	moderate	bad
36	MTSS RAUDHATUL MUTA'ALAMIN	Low	high	Low	moderate	bad
37	MTSS DARUL ULUM	Low	high	Moderate	moderate	good
38	MTSS RAUDHATUL FATA	Low	high	Low	moderate	bad
39	MTSS RAUDHATUL JANNAH	Moderate	high	Moderate	moderate	bad
40	SMP SWASTA 1 SERAMBI MEKKAH	Low	high	Moderate	moderate	bad
41	MTsS AL-MUNAWWARAH	Moderate	high	Low	moderate	bad

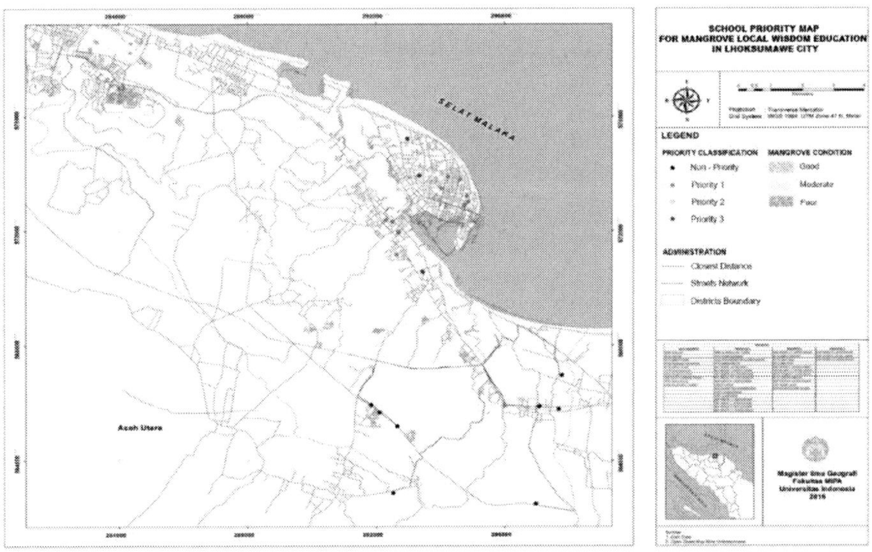

Figure 8. School priority map for mangrove local wisdom education in Lhokseumawe City

4. Conclusion

The distribution of mangrove in Lhokseumawe can be analyzed from mangrove location, and land use classification of mangrove destruction is done by looking at the average area is obtained from the SPOT 5 satellite imagery analysis and observations from research in this field about mangrove existing condition. The destruction level of mangrove measured from mangrove thickness and size of mangrove. The destruction level of mangrove divided into 3 classes such as the good condition, the moderate condition and bed condition.

Of the total number of junior secondary schools in the city of Lhokseumawe as many as 41 Public Schools, Private and Religious School, there are 31 schools with priority for local knowledge-based education implemented mangrove. Priority I as many as 18 schools, as many as 10 schools priority II and III priority as much as 3 school. The main determining factor is the distance between the school and mangrove, it is because the local knowledge-based education involves students mangrove planting mangrove in mangrove nearest location of the school.

There are differences in standards of competence and basic competences indigenous mangrove education that will be imposed on each of the patterns formed. At the first priority school learning activities focused on mangrove planting activities of the students, as well as to the school with the second priority, but the priority II school students will also study the utilization and planning of mangrove. On the third priority school students focused on maintaining the sustainability of mangrove activities and get to know other economic benefits of mangrove.

References

[1] Suryadiputra 2006 Coastal Ecosystem Restoration: Lessons Learned in Aceh since the Tsunami. [Unpublished]
[2] Ghufran, H K K M 2012 *Ekosistem Mangrove: Potensi, Fungsi dan Pengelolaan* (Jakarta: Bineka Cipta)
[3] Kordi M G 2010 *Panduan Lengkap Memelihara Ikan Tawar di Kolam Terpal* (Yogyakarta: ANDI)
[4] Saraswati R, Susilowati D and Susiloningtyas D 2010 Penerapan Muatan Lokal dan Kajian Kurikulum, serta Model Wilayah Prioritas Pengembangan Pendidikan Dasar Sesuai dengan Struktur Ruang Kota Serang Provinsi Banten. [Unpublished].
[5] Susiloningtyas D and Handayani T 2007 Penerapan muatan Lokal dan Kajian Kurikulum serta Persepsi Ibu Rumah Tangga terhadap Pendidikan di Pulau Kecil (Studi Kasus Pulau Pramuka dan Pulau Panggang Kepulauan Seribu) *Proceeding PIT IGI and National Seminar IGI. Medan.*
[6] Pabundu Tika M 2005 *Metode Penelitian Geografi* (Jakarta: PT. Bumi Aksara)
[7] Arbo P and Benneworth P 2007 Understanding the Regional Contribution of Higher Education Institutions: A Literature Review. [Unpublished]

IOP Conf. Series: Earth and Environmental Science **54** (2017) 012064 doi:10.1088/1755-1315/54/1/012064

Potential habitat of Javan Hawk-Eagle based on multi-scale approach and its implication for conservation

C Nurfatimah[1], Syartinilia[1]*, Y A Mulyani[2]

[1] Landscape Architecture Department, Faculty of Agriculture, Bogor Agricultural University (IPB), Jl. Meranti, Kampus IPB, Darmaga - Bogor 16680, Indonesia.

[2] Department of Forest Resource Conservation and Ecotourism, Faculty of Forestry, Bogor Agricultural University (IPB), Jl. Meranti, Kampus IPB, Darmaga – Bogor 16680, Indonesia

Email: syartinilia@yahoo.com

Abstract. In Indonesia the Javan Hawk-Eagle has been designated as one of the 25 top priority protected species to be increased by 10% of current population number. Lack of suitable habitat is most likely the reason for the decline of the species in landscapes subject to major human modification. Central part of Java Island has suffered the most severe forest damage and fragmentation compared to the western part and eastern part of the island. This study presents the number of predicted suitable habitats for Javan Hawk-Eagle in the central part of Java Island based on habitat probability model. Multi-scale approach was being used to determine the accuracy level of patches reading between different image resolutions. 38 patches were detected at 30 m^2, 28 patches at 90 m^2, and 19 patches were detected at 250 m^2 images resolutions. Higher reading implied more landscape structures within different regions should be considered during management of habitat conservation. Therefore, larger scale of conservation management application should be conducted as well.

1. Introduction

Javan Hawk-Eagle (*Nisaetus bartelsi*) is an endemic species raptor from Java Island. Its habitats characterized by tropical forest either primary or secondary, and located on steep slopes [1-3]. The home range of this species can reach the area as wide as 3000 ha within fragmented landscape [4]. With its vast home range Javan Hawk-Eagle (JHE) could be an umbrella species which could save the other species within its reach if we apply the right conservation management system [5].

The Indonesian government had been officially announced JHE as one of 25 top priority species to be increased in population by 10% in the period of 2015 until 2019 [6]. Predicted population number of JHE by 2002 was 325 pairs all over Java Island [2]. The rate of land use change and fragmentation in Java Island was so high [7]. The fragmented landscape reduced the ability of animal to disperse and build up new colonies and most likely induced the inbreeding potential [8, 9]. Inbreeding among species will lead to genetic depletion and increase the extinction risk [10-12]. Up until this study started in 2015 there were very limited new information about current JHE's habitat condition or population. Therefore, this study tried to evaluate the latest condition of JHE's habitat and determined whether there is any change that happened in the landscape.

Landscape-scale conservation based on remote-sensing sometimes is facing several issues regarding technical obstacles [13]. Due to the limitation on satellite image availability and software'sability the previous study was only able to use the coarse resolution in the process of predicting habitat suitability [2]. In regional management planning the coarse resolution might be enough to monitor any ecological change, but not very useful in habitat mapping and change detection for sensitive species or structure at site level [14, 15]. Different resolution will affects the management it needs for different ecosystem services [16]. In this study we try to analyze the result of the use of different image resolutions in data processing and its accuracy in habitat suitability reading. Then we suggested the suitable management conservation program by the result.

2. Method of analysis

2.1. The study area

This study focused on the central part of Java Island. The boundary of Central Java Province including Special Region of Yogyakarta is used as our study area (figure 1). It covers an area of 37,300 km^2. Central part of Java Island had the least habitat quality compared to eastern part and western part [2]. More than 80% of natural dry land forest had lost due to extensive human activity [7]. There were three protected areas in the central part of Java. Mount Merbabu National Park and Mount Merapi National Park are located inside the study area. Meanwhile, Karimunjawa National Park was located in the Karimunjawa archipelago of Java Sea and excluded from this study.

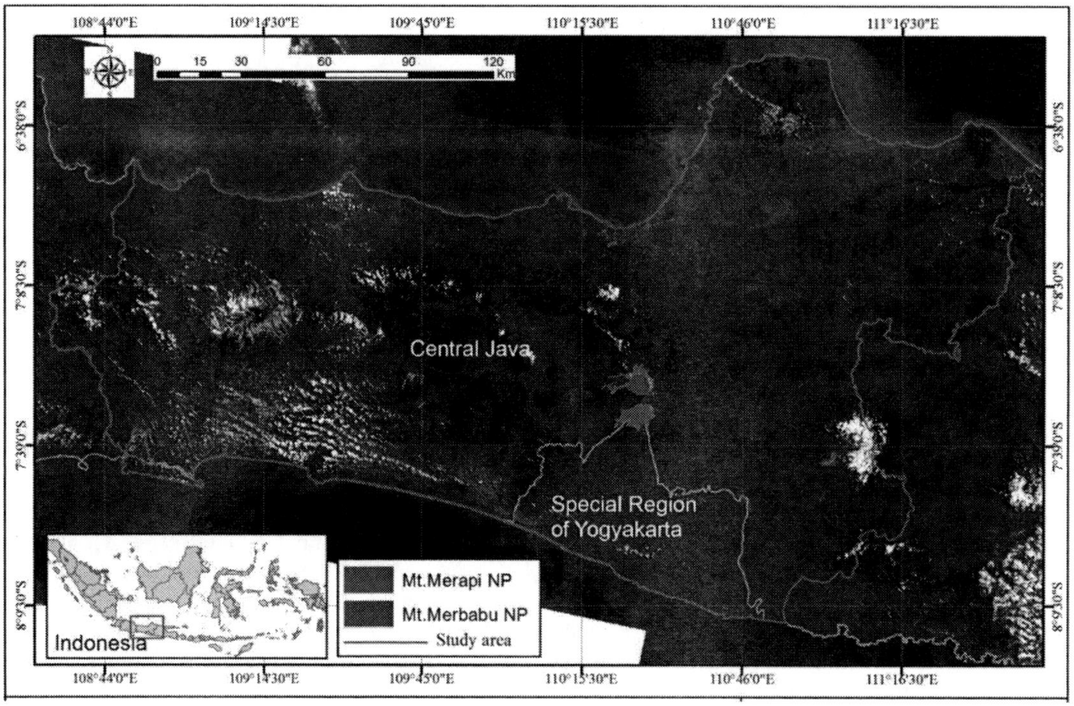

Figure 1. Study area: Central part of Java Island (Source: Landsat 8 OLI/TIRS)

2.2. Model update

Logistic regression was used to determine the suitable habitat for JHE on Java Island [16]. The model was built based on presence and pseudo-absence nests located around Gunung Gede-Pangarango National Park in West Java. The environmental variables that being input into regression were terrain factors, vegetation cover, proximity to water source and proximity to human activity. Thus, produced the environmental variables preferred by JHE as its habitat which are slope, elevation, and vegetation indices. The model equation (1) is as below:

$$Pi = \frac{1}{1+\exp[-(0.3804SLP+0.0148ELV-0.1229NDVI-19.6518)]} \quad (1)$$

SLP and ELV derived from the Shuttle Radar Topography Mission (SRTM) 1 Arc-Second Global digital elevation data of 2014. Landsat 8 OLI/TIRS scene (path/row) 119/66, 119/65, 120/65, 121/65 and scene (path/row) 120/64 were used to generate the NDVI value. All environmental variables were on spatial resolution of 30 m^2. Those 30 m^2 then resampled with Resample tool to 90 x 90 meter and 250 x 250 meter resolution to analyze the multi-scale images (table 1).

Table 1. Multi-scale image resolution visual comparison

Map	Resolution		
	30m	90m	250m
Elevation	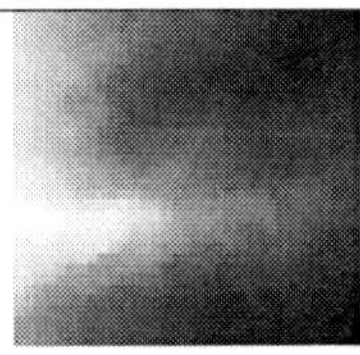	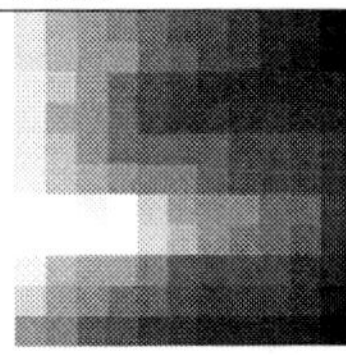	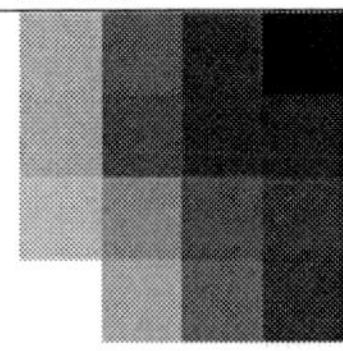
Slope	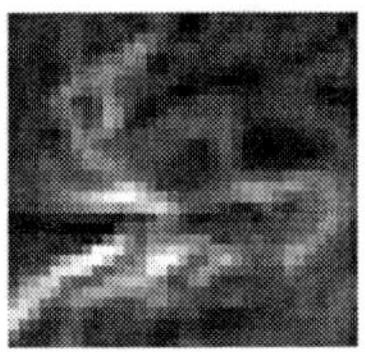	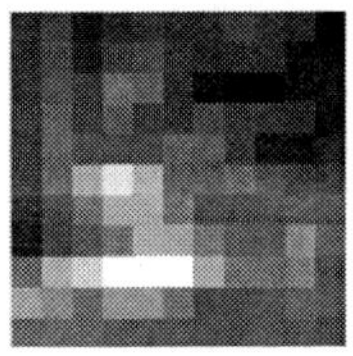	
NDVI	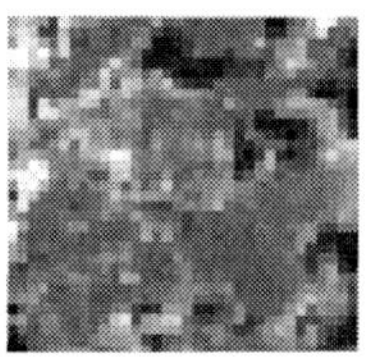	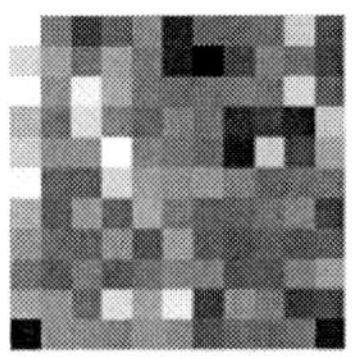	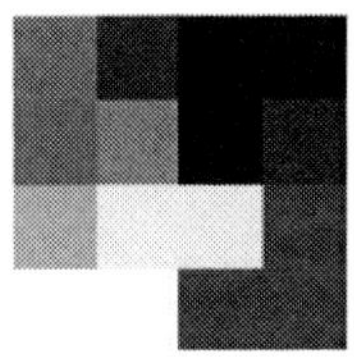
Patch habitat Blue=habitat Grey=non-habitat	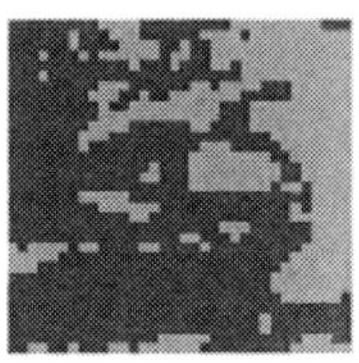	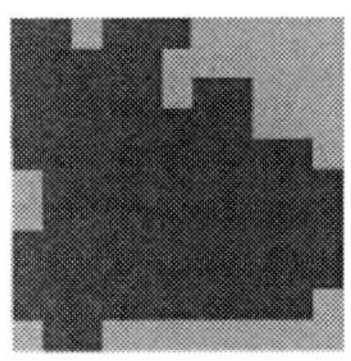	

Multi-scales of environmental variables were then being processed in RAMAS GIS using the probability of a habitat suitability model. The threshold value was 0.5 and the result was ranged from Unsuitable to Suitable area (figure 2).

IOP Conf. Series: Earth and Environmental Science **54** (2017) 012064 doi:10.1088/1755-1315/54/1/012064

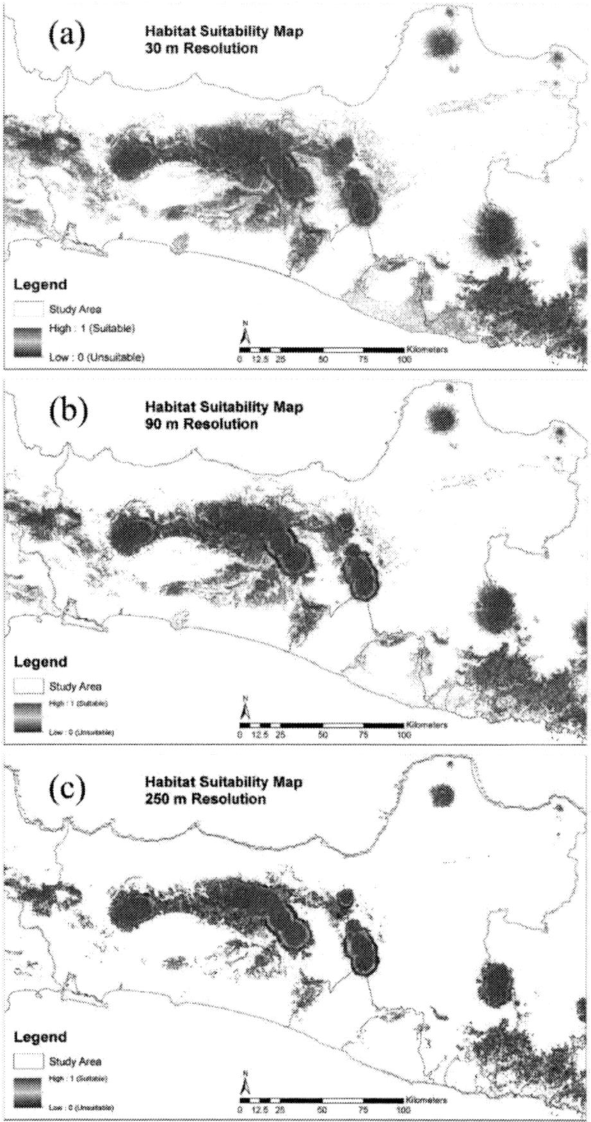

Figure 2. Multi-scale habitat suitability map with probability threshold 0.5 based on JHE probability of habitat suitability model: (a) 30 x30 meter resolution; (b) 90 x 90 meter resolution; (c) 250 x 250 meter resolution. White: unsuitable landscape; green: partly suitable landscape; yellow: suitable landscape; red: highly suitable landscape. The study site boundary shown with red polyline.

3. Result

3.1. Updated JHE habitat distribution

Habitat patch distribution was obtained by selecting the most suitable habitat with minimum size of 4 km^2 as the minimum home range of JHE [1]. Multi-scale image resolution had a different ability in detecting habitat patches (figure 3). Thirty-eight patches were detected on 30 m^2 resolution, 28 patches were detected on 90 m^2 resolution and 19 patches were detected on 250 m^2 resolution (figure 4).

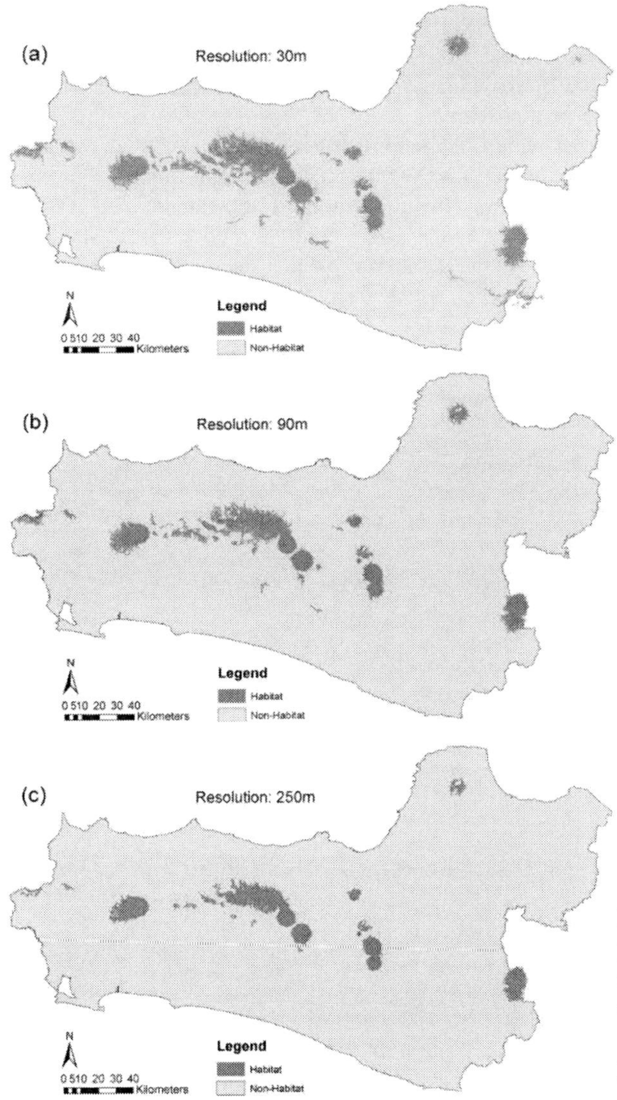

Figure 3. 2014 multi-scale JHE habitat patches distribution with minimum patch size 4 km^2: (a) 30 x 30 meter resolution, (b) 90 x 90 meter resolution and (c) 250 x 250 meter resolution. Green: habitat patches; Grey: non-habitat patches.

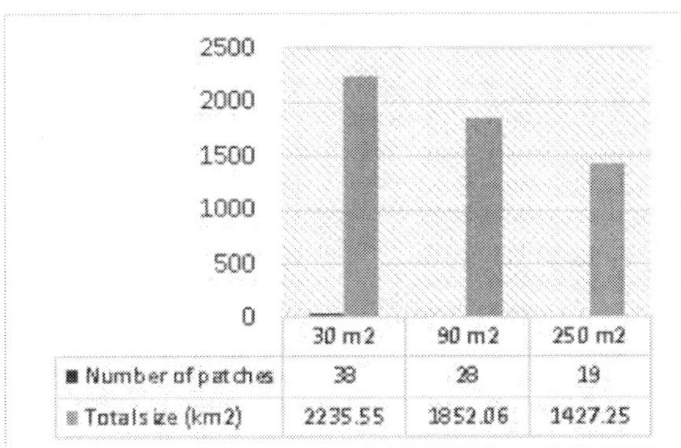

Figure 4. Number of predicted suitable habitat patches using different image resolutions of 2014.

On 250 m^2 resolution there were 19 habitat patches detected all over the central part of Java with total area of 1427.25 km^2 (figure 5). The 8 historical locality records were detected among those patches. The largest patch size was 115.2 km^2 and the smallest patch size was 4.5 km^2 (table 2). Patch 19 (Mount Lawu) was still considered as one patch and not divided administratively even though it was located between the Central Java province and East Java province.

Evaluation of the suitable habitat changes which happened between 2002 and 2014 was done by comparing between models. We use the 2002 MODIS NDVI 250 m^2 resolution model from previous study to compare with the 2014 LANDSAT 8 250 m^2 resolution model of this study. The 2002 model predicted 22 patches with a total patch size of 1477.51 km^2. Between 2002 model and 2014 model, the patch size only decreased by 3.4% or 50.01 km^2, but the number of patches decreased by 13.63%. There were only two protected areas in the study area which located close to each other: Mount Merbabu National Park (57.3 km^2) and Mount Merapi National Park (64.1 km^2). These protected areas only covers 8.5% of the total patch size.

Table 2. 250 m^2 resolution suitable habitat patches distribution and size in 2014

Patch number	Location	Patch size (km^2)
Patch 1	Mt. Muria*	46.36
Patch 2	Mt. Sagara*	9.00
Patch 3	Mt. Palasari	6.61
Patch 4	Mt. Subang	23.8
Patch 5	Mt. Ungaran*	37.16
Patch 6	Mt. Imus, Mt. Beser, Mt. Langit, Mt. Igirtipis, Mt. Bongkok,	22.25
Patch 7	Mt. Pesangon	5.19
Patch 8	Mt. Cupu*, Mt. Kraton, Mt. Jaha	13.25
Patch 9	Mt. Beser, Mt. Butak, Mt. Malang, Mt. Tumbang, Mt. Bopong	6.13
Patch 10	Mt. Pogong	5.09
Patch 11	Mt. Slamet*	238.2
Patch 12	Mt. Pandan	4.50
Patch 13	Mt. Lumping*, Mt. Lebakbarang*, Mt. Kemulan, Mt. Sindoro	511.2
Patch 14	Mt. Telomoyo	26.3
Patch 15	Mt. Andong	5.64
Patch 16	Mt. Sumbing	107.6
Patch 17	Southern hill of Mt. Sumbing	4.43
Patch 18	**Mt. Merapi dan Mt. Merbabu***	160.7
Patch 19	Mt. Lawu*	193.8

Note:
- Bold names are protected areas
- (*) historical locality record of JHE occurrences [2, 17]

 IOP Publishing
doi:10.1088/1755-1315/54/1/012064

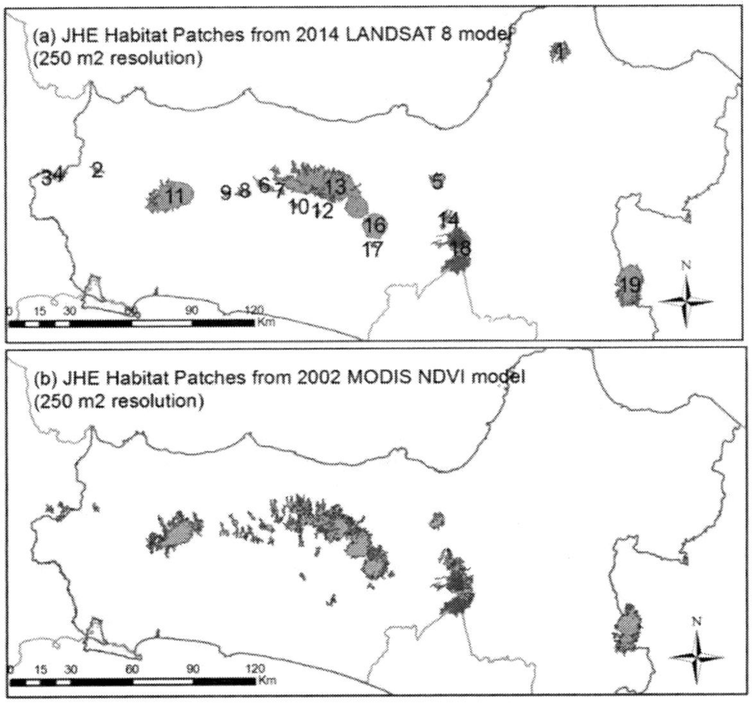

Figure 5. Javan Hawk-Eagle habitat patches distribution comparison. (a) JHE habitat patches derived from 2014 LANDSAT 8 model; (b) JHE habitat patches derived from 2002 MODIS NDVI model; Blue patch is Mount Merbabu National Park; Purple patch is Mount Merapi National Park.

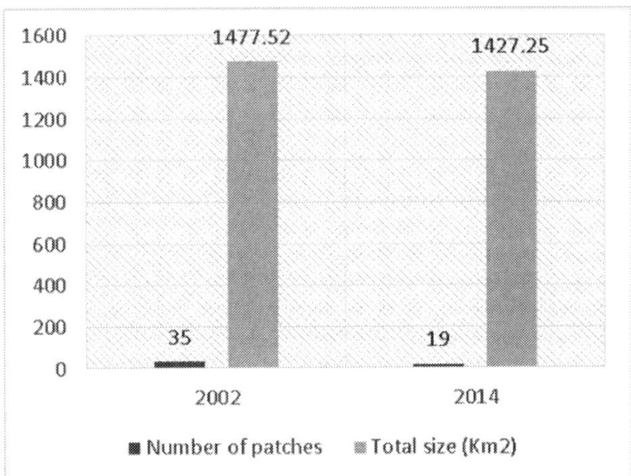

Figure 6. Number of detected patches and total size comparison between 2002 MODIS NDVI model (250 m^2 resolution) and 2014 LANDSAT 8 model (250 m^2 resolution).

The suitable habitat of JHE in the central part of Java Island had been experienced loss, shrank and fragmented [18]. There were 2 small patches southern to Patch 13 detected on 2002 model but undetected on 2014 model. On 2014 model, Patch 13 was separated from Patch 16. Meanwhile, on the 2002 model Patch 13 and Patch 16 were on one single patch and not fragmented. Aside from habitat loss and fragmentation there was new suitable habitat detected which is Patch Mount Muria (Patch 1). This patch was detected on the 2014 model, but was not detected on the 2002 model (figure 6).

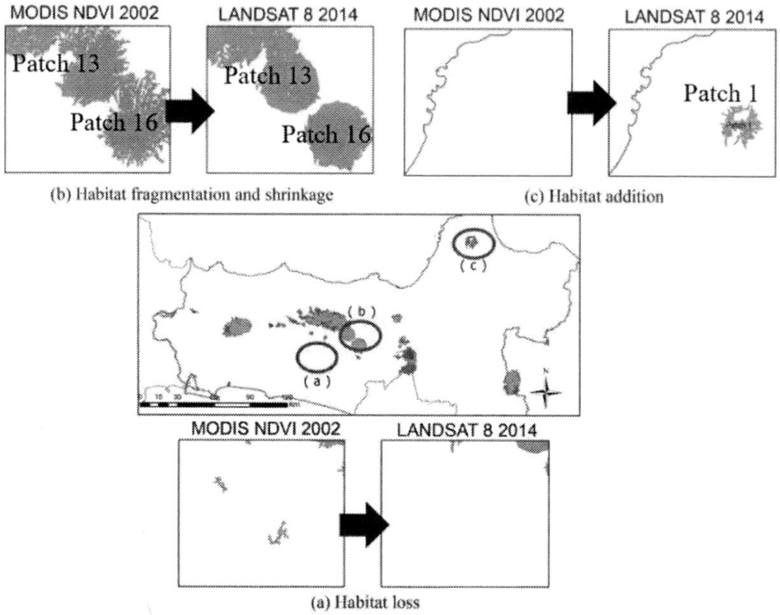

Figure 7. Landscape changing process of JHE habitat in the central part of Java between 2002 MODIS NDVI (250 m^2 resolution) model and 2014 LANDSAT 8 (250 m^2 resolution) model.

Habitat loss or expansion could happen because of land conversion which reduce or add the overall size of native habitat [18]. Habitat fragmentation is happened when a habitat patch is divided into smaller fraction [18, 19]. Habitat fragmentation can also happened as the after effect of human disturbance at larger scale: landuse change from natural forest to production forest, extensive farming practice, illegal logging, edge-effect, settlement development and infrastructure expansion [20-22].

4. Discussion

In understanding the environmental problem and providing adequate conservation measure we should optimize the function of scale [23, 24]. Implementing different scale in habitat monitoring allowed us to configure different variables in modeling habitat suitability [25, 26]. Landscape configuration and composition which determined the predicted distribution pattern of a species may vary between local-scale, patch-scale and landscape-scale [27, 28]. The multi-scale model could provide guidance in conservation management ranging from the territorial area up to national level [29, 30].

This study showed that the finer-resolution of satellite images were able to detect more predicted suitable habitat for the sustainability of Javan Hawk-Eagle than coarse-resolution images (figure 4). The coarse-resolution (Landsat 8 at 250 m^2 and 90 m^2) resulted by resampling the finer-resolution pixel (Landsat 8 at 30 m^2 resolution). This could happen because when resampling the fine-scale image to coarse-scale image there were some information which lost during process [33]. Thus, eliminating potential variable smaller than the resampled pixel size and fewer patch detected (table 1). By analyzing the potential habitat on a multiple scale we could identify the true extent of suitable habitat [24].

Coarse-scale and fine-scale were both important for habitat management conservation. By looking at the coarse-scale model the managers and conservationists will be able to identify all of the hotspots of predictive suitable habitat in general [2], especially the locations where management decisions should take into account the jurisdictional boundaries. Also, the coarse-scale model is able to help in assessing the need of ecological networks through valuing the connectivity between widely scattered fragmented patches [31, 32]. Meanwhile, more specific management conservation guideline can be produced better through the use of finer-scale model. By using the fine-scale model we can specify the detailed habitat preferences, ecological needs and even foraging ground location for JHE [15, 30]. With multi-year coarse-scale model we can identify the landscape change in regional scale (figure 5). But, to figure out what causing those changes, the fine-scale is better to determine the landscape configuration [27].

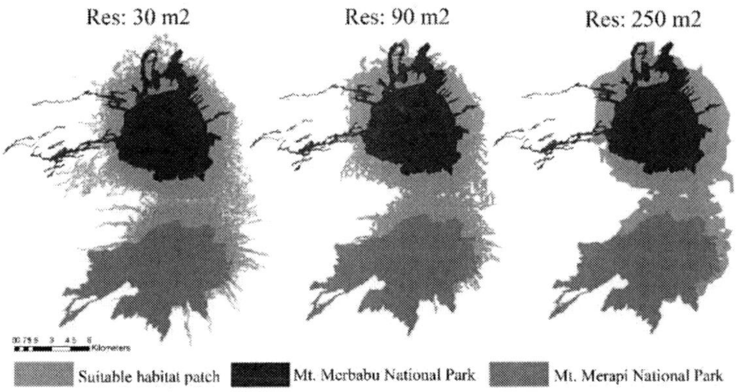

Figure 8. Multi-scale comparison of suitable habitat patches on protected areas. Each predicted habitat has different size: 191.9 km^2 at 30 m^2 resolution; 173.5 km^2 at 90 m^2 resolution; 160.7 km^2 at 250 m^2.

Our study predicted the suitable area for Javan Hawk-Eagle habitat larger than the designated park within a single patch (Figure 6). At coarse-scale, the model was good enough to predict the extended habitat outside the protected area. To apply the management conservation species specific the fine-scale was better in determined the habitat edge delineation.

In conclusion, the logistic-regression model to predict suitable habitat for JHE shows different results for different image resolution. Finer-resolution method and coarse-resolution method were both important for wildlife management depending on the purpose and goal of conservation. Finer-resolution can provide higher accuracy in mapping the forest and predicting nesting site [29, 34]. But to provide the finer-resolution data will result in higher cost in human resources and technology [35, 36]. Therefore, the managers and conservationists can choose the most suitable image scale for managing, planning and monitoring based on the scale of application—site-scale or regional scale.

Acknowledgments
The authors would like to thank the Ministry of Research, Technology and Higher Education of Republic Indonesia under "Penelitian Unggulan Perguruan Tinggi" grant in fiscal year of 2016.Also thanks to Raptor Indonesia especially Asman A Purwanto for providing us the local information and guiding us during field survey.

References

[1] Gjershaug J O, Rov N, Nygrd T, Prawiradilaga D M, Afiyanto M Y, Hapsoro and Supriatna A 2004 Home-range size of the Javan Hawk-Eagle (Spizaetus bartelsi) estimated from direct observation and radiotelemetry *Journal of Raptor Research* **3** 343-349
[2] Syartinilia, Tsuyuki S and Lee J S 2009 GIS-based habitat model of Javan Hawk-Eagle (Spizaetus bartelsi) using inductive approach in Java Island, Indonesia *Wildlife: Destruction, Conservation and Biodiversity*ed J D Harris and P L Brown (New York: Nova Science Publisher, Inc) pp 301-312
[3] Sozer R, Nijman V, Setiawan I and Rakhman Z 2012 *Panduan Inventarisasi Elang Jawa Nisaetus bartelsi* (Bogor: Raptor Indonesia) p 9
[4] van Balen S, Nijman V and Prins H H 2000 The Javan Hawk-Eagle: misconception about rareness and threat *Biological Conservation* **96** 297-304
[5] Simberloff D 1997 Flagship, umbrellas, and keystones: is single-species management passé in the landscape era? *Biological Conservation* **832** 47-257
[6] SK Dirjen Perlindungan Hutan dan Konservasi Alam No.200/IV/KKH/2015

[7] Gunawan H, Prasetyo L B, Mardiastuti A and Kartono A P 2010 Fragmentasi hutan alam lahan kering di Provinsi Jawa Tengah *Jurnal Penelitian Hutan dan Konservasi Alam* **7** 75-91
[8] Aurambout J P, Endress A G and Deal B M 2005 A spatial model to estimate habitat fragmentationandits consequences on long-term persistence of animal populations *Environmental Monitoring and Assessment* **109** 199-225
[9] Row J R, Blouin-Demers G and Lougheed S C 2010 Habitat distribution influences dispersal and fine-scale genetic population structure of eastern foxsnakes (Mintonius gloydi) across a fragmented landscape *Molecular Ecology* **19** 5157-5171
[10] Kempenaers B, Adriaensen F, Van Noordwijk A J and Dhondt A A 1996 Genetic similiarity, inbreeding and hatching failure in Blue Tits: are unhatched eggs infertile? *Proceedings: Biological Sciences* **263** 179-185
[11] Frankham R 1998 Inbreeding and extinction: island populations *Conservation Biology* **12** 665-675
[12] Szulkin M and Sheldon B C 2008 Correlates of the occurrence of inbreeding in a wildbird population *Behavioral Ecology* **19** 1200-1207
[13] Wang K, Franklin S E, Guo X, He Y and McDermid G J 2009 Problems in remote sensing of landscapes and habitats *Progress in Physical Geography* **33** 747-768
[14] Nagendra H, Lucas R, Honrado J P, Jongman R H G, Tarantino C, Adamo M and Mairota P 2013 Remote sensing for conservation monitoring: Assessing protected areas, habitat extent, habitat condition, species diversity, and threats *Ecological Indicators* **33** 45-59
[15] Brambilla M, Casale F, Bergero V, Crovetto G M, Falco R, Negri I, Siccardi P and Bogliani G 2009 GIS-models work well, but are not enough: Habitat preferences of Lanius collurio at multiple levels and conservation implications *Biological Conservation* **142** 2033-2042
[16] Syartinilia and Tsuyuki S 2008GIS-based modelling of Javan Hawk-Eagle distribution using logistic and autologistic regression *Biological Conservation*doi:10.1016/j.biocon.2007.12.030
[17] Setiadi A P, Rakhman Z, Nurwatha P F, Muchtar M and Raharjaningtrah W 2000 Status, distribution, population, ecology and conservation Javan Hawk-eagle Spizatus bartelsi Stresemann, 1924 on southern part of West JavaFinal Report BP/FFI/BirdLife International/YPAL-HIMIBIO (Bandung: UNPAD)
[18] Collinge S K 2009 *Ecology of Fragmented Landscape* (Maryland: The John Hopkins University Press)
[19] Goodwin B J and Fahrig L 2002 How does landscape structure influence landscape connectivity? *OKIOS* **99** 552-570
[20] Coulon A, Josson J F, Angibault M, Cargnelluti B, Galan M, Morellet N and Hewison J M 2004 Landscape connectivity influences gene flow in a roe deer population inhabiting a fragmented landscape: an individual-based approach *Molecular Ecology* **13** 2841-2850
[21] Crist M R, Wilmer B and Aplet G H 2005 Assessing the value of roadless areas in a conservation reserve strategy: biodiversity and landscape connectivity in the Northern Rockies *Journal of Applied Ecology* **42** 181-191
[22] Yin K, Xie Y and Wu N 2006 Corridor connecting giant panda habitats from north to south in the Min Mountains, Sichuan, China *Integrative Zoology* **1** 170-178
[23] Farina A 2006 *Principles and Methods in Landscape Ecology: Toward a Science of Landscape* (Netherland: Springer) pp 87-108
[24] Denoël M and Lehmann A 2006 Multi-scale effect of landscape process and habitat quality on newt abundance: Implication for conservation *Biological Conservation* **130** 495-504
[25] Martin R D, Brabyn L and Potter M A 2011 Sensitivity of GIS-derived terrain variables at multiple scales for modelling stoat (*Mustela erminea*) activity *Applied Geography* **31** 770-779
[26] Revermann R, Schmid H, Zbinden N, Spaar R and Schröder B 2012 Habitat at the mountain tops: how long can Rock Ptarmigan (Lagopus muta Helvetica) survive rapid climate change in the Swiss Alps? A multi-scale approach *J Ornithol* **153** 891-905

[27] Garden J G, McAlpine C A and Possingham H P 2010 Multi-scaled habitat considerations for conserving urban biodiversity: native reptiles and small mammals in Brisbane, Australia *Landscape Ecology* **25** 1013-1028

[28] Duren K R, Buler J J, Jones W and Williams C K 2011 An improved multi-scale approach to modeling habitat occupancy of Northern Bobwhite Journal of *Wildlife Management* **75** 1700-1709

[29] Illera J C, von Wehrden H and Wehner J 2010 Nest site selection and the effects of land use in a multi-scale approach on the distribution of a passerine in an island arid environment *Journal of Arid Environment* **74** 1408-1412

[30] Razgour O, Hanmer J and Jones G 2011 Using multi-scale modelling to predict habitat suitability for species of conservation concern: The grey long-eared bat as a case study *Biological Conservation* **144** 2922-2930

[31] Doko T, Fukui H, Kooiman A, Toxopeus A G, Ichinose T, Chen W and Skidmore A K 2011 Identifying habitat patches and potential corridors for remnant Asiatic black bear (*Ursus thibetanus japonicas*) populations in Japan *Ecological Modelling* **222** 748-761

[32] Baranyi G, Saura S, Podani J and Jordán F 2011 Contribution of habitat patches to network connectivity: Redundancy and uniqueness of topological indices *Ecological Indicators* **11** 1301-1310

[33] Colditz RR, Acosta-velázquez J, Gallegos JRD, Lule ADV, Rodríguez-zúñiga MT, Maeda P López MIC and Ressl R 2012 Potential effects in multi-resolution post-classification change detection *International Journal of Remote Sensing* **33** 6426-6445

[34] Gairola S, Proches S and Rocchini D 2013 High-resolution satellite remote sensing: a new frontier for biodiversity exploration in Indian Himalayan forest *International Journal of Remote Sensing* **34** 2006-2002

[35] Spanhove T, Borre JV, Delalieux S, Haest B and Paelinckx D 2012 Can remote sensing estimate fine-scale quality indicators of natural habitats? *Ecological Indicators* **18** 403-412

[36] Corbane C, Lang S, Pipkins K, Alleaume S, Deshayes M, Millan VEG, Strasser T, Borre JV, Toon S and Michael F 2015 Remote sensing for mapping natural habitats and their conservation status – New opportunites and challenges International Journal of Applied Earth Observation and Geoinformation **37** 7-16

Mangrove mapping and change detection in Sungai Asam Village, Indragiri Hilir Regency, Riau Province

Prita Ayu Permatasari[1]*, Yudi Setiawan[1,2], Rahmi Nur Khairiah[1] and Dadan Mulyana[3]

[1] Center for Environmental Research, Bogor Agricultural University, PPLH Building 2nd-4th Floor, Jl Lingkar Akademik, Kampus IPB Dramaga, Bogor, Indonesia

[2] Department of Forest Resource Conservation and Ecotourism, Faculty of Forestry, Bogor Agricultural University, Kampus IPB Dramaga, Bogor, Indonesia

[3] Department of Silviculture, Faculty of Forestry, Bogor Agricultural University, Kampus IPB Dramaga, Bogor, Indonesia

E-mail: pritapermatasari@gmail.com

Abstract. Indonesia is an archipelago in the tropical climate that has the largest mangrove forest in the world. Based on data from the Ministry of Environment and Forestry, mangrove forest area is estimated about 9 million hectares in 2000. However, this amount is reduced drastically. Approximately, sixty percent of mangrove forest area in 2006, has been lost, damaged, and changed into ponds, plantation, agriculture area, and settlement. Indragiri Hilir is one of regency in Riau Province which has the high potential diversity of mangrove. Based on interpretation of Landsat 8 satellite imagery in June - July 2013, mangrove forest in Indragiri Hilir reached 100,211.23 hectares. The area of mangrove forest is estimated to decrease, due to land use and land cover change. Based on land cover change map, mangrove forest area in Sungai Asam Village has increased in 20 years.

1. Introduction

Like many other countries in Southeast Asia, the mangrove areas in Indonesia have decreased markedly. Indonesia is an archipelago in the tropical climate that has the largest mangrove forest in the world. Based on data from the Ministry of Forestry, mangrove forest area is estimated at about 9 million hectares in 2000. However, this number is reduced drastically. Around 60% of mangrove forest area in 2006, has been lost, damaged, changed into ponds, plantation, agricultural area, and settlement. It is estimated that mangrove forests in Indonesia is currently the only remaining 3.6 million hectares [1]

Mangrove forest are among the most important and productive of ecosystems and provide habitat for wildlife [2]. Mangrove ecosystems occure world-wide on tropical and sub-tropical coastline [3]. The mangrove is a unique and complex ecosystem. Moreover, mangrove ecosystems stabilize coastlines, clean water, protect land from erosion, and in many cases promote coastal accretion, and provide a natural barrier against storms, cyclones, tidal bores and other potentially damaging natural forces [4].

Mangrove ecosystems are reported to disappear worldwide by 1 to 2% per year, which is the same or even worse as compared to the nearby coral reefs or tropical rainforests. Mangroves are species-poor ecosystems and the number of mangroves has been reported to be directly correlated with forest size [5, 6]. So, further decline in mangrove area is likely to be followed by accelerated functional losses in the future. The world-wide distribution of mangroves is a good thing, however, it remains uninvestigated how bad their decrease in functionality is.

For centuries, mangroves have contributed significantly to the socioeconomic lives of coastal dwellers as well as the people in Sungai Asam Village. Sungai Asam is one village in the Reteh District, Indragiri Hilir Regency, Riau Province. Since 2002, people in this village conducted community-based mangrove cultivation because of sea water intrusion in their coconut plantations that made their land unproductive. Mangrove reforestation is also conducted for the purpose of conservation and ecotourism development. Mangrove density in Sungai Asam expanding from year to year. In this village there are various types of mangrove species and wildlife that live inside.

Mangrove forest area changes in the last 20 years can be viewed using satellite imagery. Satellite remote sensing data can be used for large areas over time and thus represent an indispensable tool for mangrove forests monitoring. The objectives of this study are to map mangrove forests using LANDSAT imagery in Sungai Asam Village in 1996, 2006, and 2016.

2. Material and methods

2.1. Study area

Sungai Asam is a village located in the Reteh District, Indragiri Hilir, Riau Province. The village is situated in Indragiri River estuary. Indragiri river estuary waters have a number of alluvial inhabited and uninhabited islands (sediment originating from river mud). It has wet tropical climate with rainfall of 1300 mm throughout the year, the rains falling from October to March and the dry season lasts from November to February. The east coast of Indragiri Hilir delta is covered with massive mangroves forests. This forest is naturally plays into the ecological chain to keep the beaches from the waves and abrasion. Sungai Asam is surrounded by coconut plantation in the north and west, Asam River in the south, and Rawi Ditch in the east. Vegetation in the Reteh District dominated by mangrove species of *Rhizophora apiculata* and *Nypa fruticans*. Most of people in this village are working as a farmer in coconut plantation or rice field and the others are working as fishermen [7].

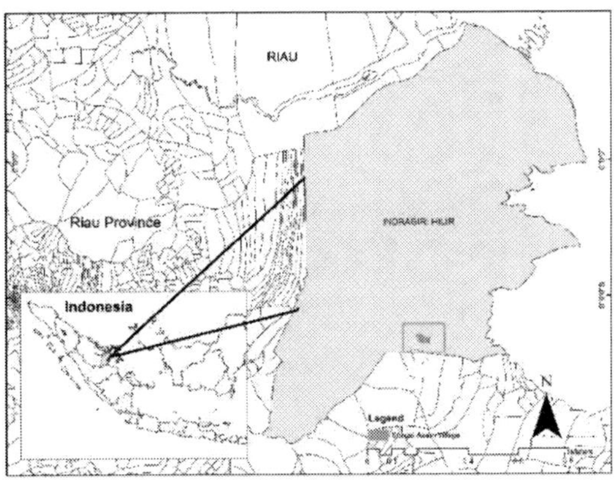

Figure 1. Study area

2.2. Satellite imagery

Multi-temporal medium resolution LANDSAT imagery was used to obtain comprehensive coverage and analysis of the current and historical mangrove conditions. The used imagery was acquired by several different satellite sensors including LANDSAT TM (March 8, 1996 and July 26, 2006) and LANDSAT 8 (June 3, 2016). The Landsat satellite data was georeferenced to UTM WGS 84 Zone 48 south. Spatial data were processed using ArcMap 10.3 software. Data processing using geographic information systems (GIS) consists of some steps such as data entry, data analysis, and data display.

Some important spatial data processing operations used in this research were the map digitation, map overlay, and area calculation (figure 2).

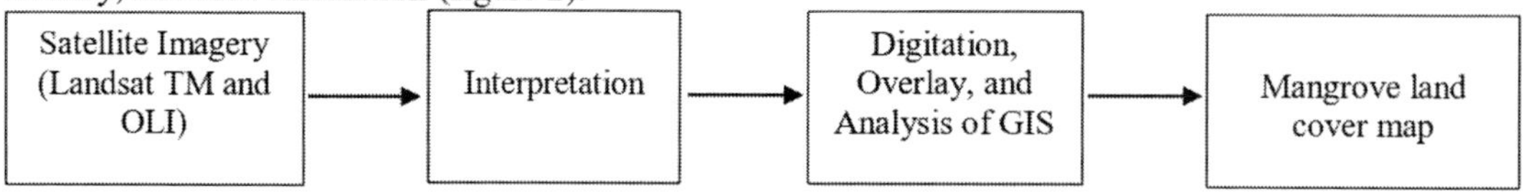

Figure 2. Flowchart of the image processing

2.3. Field survey data

The field data was collected during survey carried out during November 2015. The foundation for these field trips was the collection of land cover information in particular mangrove forests. Some points in Sungai Asam Village were visited and informations were collected ground-truth points using GPS (Global Positioning Systems) to create training data for supervised classification and generate accuracy assessment for the post classification of the land cover map in 2016.

3. Result and discussion

3.1. Mangrove forest land cover change from 1996-2016

Based on Landsat map obtained in 1996, 2006 and 2016, differences of mangrove forest land cover within 20 years can be seen. In 1996, the mangrove forest area was still dominated by swamp area (figure 3a). Before the mangrove reforestation in 2002, Sungai Asam Village was an unproductive coconut plantation due to sea water intrusion.

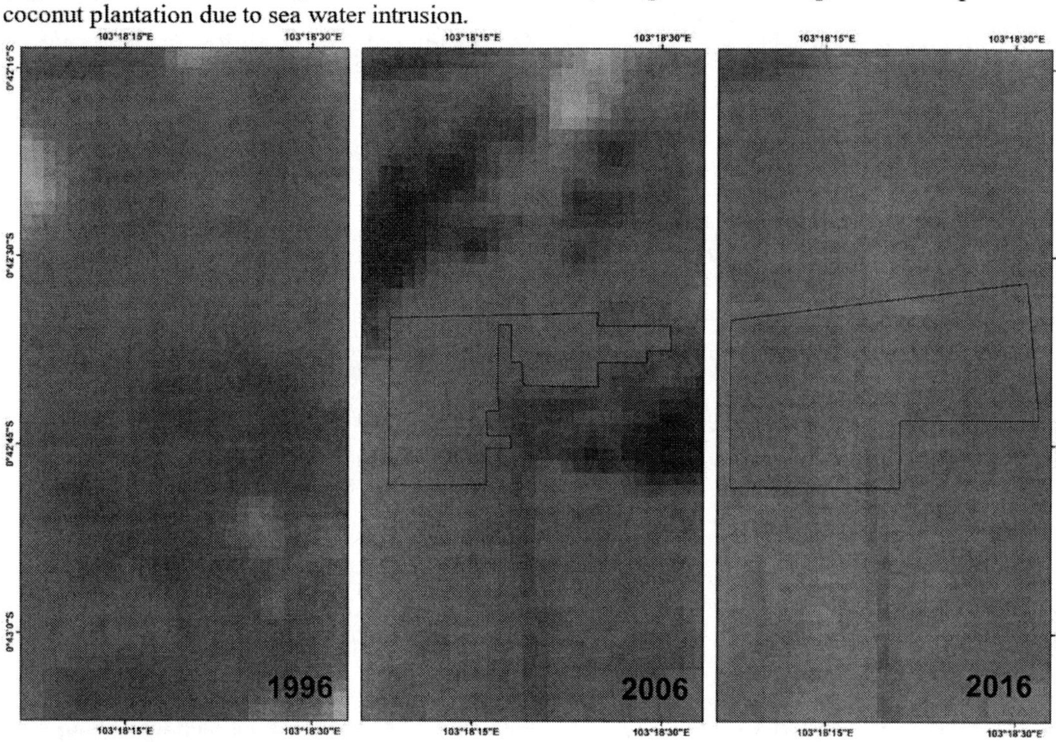

Figure 3. Land cover of mangrove forest in (a) 1996, (b) 2006, and (c) 2016

After the mangrove cultivation in 2002, swamp area was dominated by young Rhizophora sp. (figure 3b). In 2016, the mangrove forest area are expanded, denser, and dominated by mature stand of

Rhizophora sp. (figure 3c). Based on the results of extensive calculations, mangrove forest area from 2006 to 2016 increased significantly to about 25 times (table 1).

Table 1. Mangrove forest area in 2006 and 2016

Year	Total Area (hectare)
2006	1.37
2016	28.42

At first, the idea of mangrove reforestation (*Rhizophora apiculata*) comes from one of the village dweller after observing mangrove forests, especially in the Reteh District, those were decreasing and heading for extinction. Though mangrove forests have economical functions as a habitat for a variety of aquatic fauna (shrimp, crabs and fish), the idea was not initially supported by the other village dwellers, who mostly work as coconut farmers and fishermen, because they are less profitable.
After learning many positive impacts of mangrove planting, the community began to keen and interested in joining. In mid-2005, farmer groups formed "Forum for Sustainable Mangrove" with the aim of preserving the mangrove forests in the area of Sungai Asam. Now, the mangrove forest in Sungai Asam become green, a new sanctuary for many species of aquatic fauna and birds, and often used as a research location by universities [7].

3.2. Mangrove species
Based on survey results, seven mangrove species were found (table 2). The number is still less diverse thus the species enrichment is really needed.

Table 2. Plant species in mangrove forest

No.	Local name	Latin name	IUCN Red List Status
1	Api-api daun lebar	*Avicennia marina*	-
2	Bakau bandul	*Rhizophora mucronata*	Least Concern
3	Bakau kacang	*Rhizopora apiculata*	-
4	Pedada	*Sonneratia ovata*	-
5	Nipah	*Nypa fruticans*	Least Concern
6	Krakas	*Acrostichum aureum*	Least Concern
7	Drujon	*Acanthus ilicifolius*	Least Concern

Nypa fruticans is known to occur on soft, fine-grained substrata with a perennial (high) input of fresh water. *Rhizopora apiculata* grew at low to mid-tide level. *Sonneratia ovata* was usually planted in coastal areas to control erosion of tidal river banks. While, *Avicennia marina* and *Rhizophora mucronata* usually occur close to the sea, indicating their tolerance of high salinity [8-10]. Two shrubs species were found are *Acrostichum aureum* and *Acanthus ilicifolius*. Those species are growing along lakes, marshes, sea shores, and tolerant of raised salinity levels.

Table 2 shows that the majority of plant species in the mangrove forests is categorized as least concern and several other species aren't on the IUCN red list. Among all species encountered, *Rhizophora apiculata* is a species that dominates the mangrove forests. This species is cultivated by the community. Bakau kacang (*Rhizophora apiculata* Bl.) is the most economically important commercial species of mangrove trees, have been intensively exploited for their valuable wood for house construction [11]. The contribution this species towards the growth and primary production of mangrove ecosystem is remarkable. This is due to the fact that *R. apicutata* colonizes relatively faster; grows quicker and taller than other species [12, 3]. *R. apiculata* is the most dominant and productive species. It is also the most important species of commercial mangrove timber in the Asia-Pacific region.

4. Conclusion

Based on land cover map in 1996, 2006 and 2016, total area of mangrove forests in Sungai Asam Village has increased significantly. Village dwellers' awareness of the importance of mangrove functions for the environment make they conduct mangrove reforestation on abandoned land. Based on survey results, seven mangrove species were found. *Rhizhophora apiculata* is a species that dominated mangrove forest. Most of species is categorized as least concern. The amount of species is still too few for a mangrove forest biodiversity. Therefore, species enrichment is really needed to increase the value of biodiversity.

Acknowledgment

This research was financially funded by Government Environmental Agency, Indragiri Hilir Regency, Riau Province for fiscal year 2015.

References

[1] Noor Y R, Khazali M and Suryadiputra I N N 2006 *The introductory of mangrove in Indonesia* (Bogor: Wetland International Indonesia Programme) (in Indonesian)

[2] Wolanski E, Brinson M M, Cahoon D R and Perillo G M E 2009 *Coastal Wetlands: An Integrated Ecosystem Approach* (Amsterdam: Elsevier)

[3] Tomlinson P B 1986 *The Botany of Mangroves* (Cambridge: Cambridge University Press) pp 22-29

[4] Pham T D and Yoshino K 2015 Mangrove mapping and change detection using multi-temporal Landsat imagery in Hai Phong city, Vietnam. *The International Symposium on Cartography in Internet and Ubiquitous Environments 2015 17th - 19th March, Tokyo.*

[5] Duke N C, Ball M C and Ellison J C 1998 Factors influencing biodiversity and distributional gradients in mangroves *Global Ecology and Biogeography Letters* **7** 27-47

[6] Ellison A M 2002 Macroecology of mangroves: large-scale patterns and processes in tropical coastal forests *Trees - Structure and Function* **16** 181-194

[7] Suwignyo R A, Munandar, Sarno, Ulqodry T Z and Halimi E S 2011 Assistance experience in forest management mangrove community. *Workshop on Regional Mangrove Working Group Formation (KKMD) South Sumatra Province Palembang, 26 Mei 2011.*

[8] Wells A G 1982 Mangrove vegetation of Northern Australia In (Clough B F ed) *Mangrove ecosystem in Australia: function and management* (Canberra: Australian National University Press) pp 57-78

[9] Imbert D, Rousteau A and Scherrer P 2000 Ecology of mangrove growth and recovery in the Lesser Antilles: state of knowledge and basis for restoration projects *Restor. Ecol.* **8** 230 236

[10] Saenger P 2002 *Mangrove ecology, silviculture, and conservation* (London: Kluwer Academic Publishers) pp 360

[11] Sarno, Suwignyo R A, Munandar, Dahlan Z and Ridho M R 2010 The status of mangroves ecosystem reforested of *Rhizophora apiculata* Bl. and natural mangrove forest in Riau. *Proceeding of an International Seminar-Workshop on Integrated Lowland Development and Management Palembang March 18 - 20, 2010*

[12] Ong J E, Gong W K and Wong C H 2004.Allometry and partitioning of the mangrove, *Rhizophora apiculata. Forest Ecology and Management.* **188** 395-408.

Quadcopter applications for wildlife monitoring

S Radiansyah[*] **, M D Kusrini and L B Prasetyo**

Tropical Biodiversity Conservation Studies Program, Graduate School of IPB, Darmaga, Bogor, Indonesia

E-mail: rade03499007@gmail.com

Abstract. Recently, Unmanned Aerial Vehicle (UAV) had been use as an instrument for wildlife research. Most of that, using an airplane type which need space for runaway. Copter is UAV type that can fly at canopy space and do not need runaway. The research aims are to examine quadcopter application for wildlife monitoring, measure the accuracy of data generated and determine effective, efficient and appropriate technical recommendation in accordance with the ethics of wildlife photography. Flight trials with a camera 12 - 24 MP at altitude ranges from 50-200 m above ground level (agl), producing aerial photographs with spatial resolution of 0.85 – 4.79 cm/pixel. Aerial photos quality depends on the type and setting of camera, vibration damper system, flight altitude and punctuality of the shooting. For wildlife monitoring the copter is recommended to take off at least 300 m from the target, and flies at 50 - 100 m agl with flight speed of 5 - 7 m/sec on fine weather. Quadcopter presence with a distance more than 30 m from White-bellied Sea Eagles (*Haliaeetus leucogaster*) nest and Proboscis Monkey (*Nasalis larvatus*) did not cause negative response. Quadcopter application should pay attention to the behaviour and characteristic of wildlife.

1. Introduction

Unmanned Aerial Vehicles (UAV), often called the drone, is a flying robot with a remote control which is capable of carrying payloads in accordance with the purpose and designation. Some wildlife studies used airplane/fixed wing type of drone [1][2][3][4][6][9] and multirotor [1]
[5]. Small UAVs to be useful as management or research tools, they should be durable, modular, electric powered, launchable and recoverable in rugged terrain, autonomously controllable, operable with minimal training, and collect georeferenced imagery [2]. Multirotor drone such as quadcopter is a drone type which is appropriate for research in the forest with large canopy space and unavailable of runway [5]. Quadcopter can be operated manually and automatically with a programming of flight path using open-source software.

Drone applications for wildlife data collection depends on the purpose and environmental parameters which will be taken at each site. This requires UAV system settings that vary according to the capacity of UAV, the camera and the distance or scope of the study area [1][2][3][4][6][9]. Photographing using quadcopter in wildlife monitoring is part of the wildlife photography. Principle ethic of wildlife photography is doing a photo shoot by minimizing the impact / disturbance to wildlife and the environment. Information required in the application of this principle is about the wildlife objectives, locations, regulations and expertise photographer [7].

The fundamental questions about the use quadcopter as wildlife monitoring instrument that is how are the technical applications, how well the accuracy of the data generated and how is the response of wildlife to the application. To answer these questions then conducted research with the aim to examine quadcopter application and measuring accuracy of the data generated and the factors that influence it. From the results and based on wildlife target responses to the presence of quadcopter, then determined technical recommendation of quadcopter application which is effective, efficient and appropriate with the ethics of wildlife photography.

2. Method

The study was conducted over 15 months. Preparation and quadcopter development conducted in March-May 2015 (3 months). While field data collection held in June 2015 - May 2016 (12 months).

Quadcopter (multirotor drone with 4 arms, 4 rotors and 4 propellers) used in this research is a type of assembly results from multi-branded component. Dimensions of quadcopter are: width 700 mm, height 415 mm (landing skids installed), and 80 mm ground clearance. Empty weight without batteries and camera is 1800 g. Quadcopter equipped with a battery checker (10 g), LiPo iP3 8000 mAh main battery (850 g), or 6200 mAh (600 g) as reserve battery. Quadcopter flight endurance ± 25 minutes and capability of carrying payloads up to 1200 g.

The camera is mounted on the front and facing downward (camera angle is 90° to the quadcopter/flight direction). Quadcopter can be flown manually using a radio controller or automatically flight using autonomous mode. Quadcopter called "Aerialview-650", since assembled using Tarot 650 frame Sport with semi automatic landing skids (
Figure **1**).

2.1. Quadcopter Applications Trial

The quadcopter application trial conducted in Dramaga Campus, Bogor Regency, West Java Province, at 3 locations, namely:

1. LSI Lake - Rektorat IPB (S 6°33'35.98" – E 106°43'34.25");
2. Soccer field at Soka street (S 6°33'12.92" – E 106°43'22.34");
3. Horse ranch at Cendana street (S 6°33'14.45" – E 106°43'04.93").

Flight endurance test based on payload is done by a flight without a load (without camera mounting and cameras) and flight with load, with the addition of camera mounting and 5 types of camera (Table 1). Flight mode used is:

1. Hovering, done manually or semi-manual on flying altitude 2-30 meters above the ground level at the take off (m agl) within ± 1 ha area (visually controlled).
2. Flight path, flown automatically in accordance with flight path which has been set previously using mission planner program (open-source software).

Data analysis quality which generated from quadcopter application has been conducted through automatic flight path by 3 altitude levels (200, 100 and 50 m agl). Flight altitude of 200 m agl using action cam w9 sport, flight altitude of 50 and 100 m agl using sony α5100 single lens 16-50 mm mirrorles camera. While shooting a flight speed of 5 m/s, is set regardless height level. Photo shooting is done at various intervals, adjusting to the flight altitude and type of camera used. The flight time is done at various times, from 07:00 am to 18:00 pm, not in rainy day (quadcopter designed for non waterproof condition). One flight takes an average time of one hour.

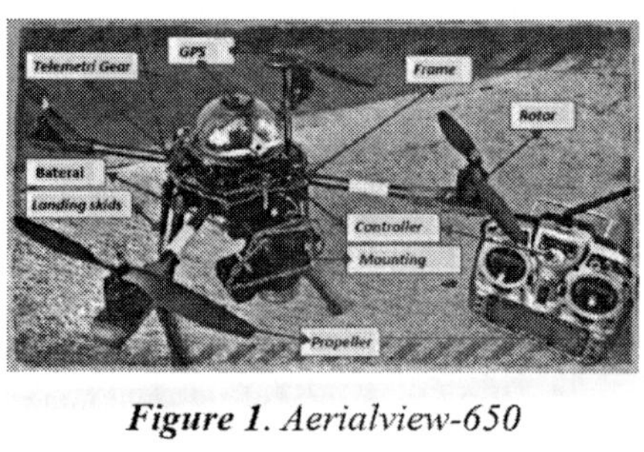

Figure 1*. Aerialview-650*

Table 1. Camera used for Quadcopter Applications Trial

No	Camera type	Brand	Resolution (Pixel)	Lens/ Focal length (mm)	Aperture	Weight (g)	Production year
1	Action cam	W9 Sports Camera	12 MP HD 1080P	Fisheye	f/2.8	150	2015
2	Action cam	GoPro HERO4 Silver	12 MP HD 1080P	Fisheye 17-34	f/2.8	150	2015
3	Digital compact	Canon power-shoot A1400	16 MP	28-140	f/2.8-6.9	300	2015
4	Mirrorles	Sony α5100	24.3 MP APS-CMOS	Pancake 16	f/2.8	370	2015
5	Mirrorles	Sony α5100	24.3 MP APS-CMOS	Single lens 16-50	f/3.5-5.6	400	2015

Noise level analys based on the sound of quadcopter that flown in hover. The shortest distance between quadcopter and recorder is 5 m (vertical and horizontal). The recordings were analyzed using praat sound analysis program to get the value of the noise intensity.

2.2. Quadcopter Application for Wildlife Monitoring

2.2.1. White-bellied Sea Eagles (Haliaeetus leucogaster)

White-bellied Sea Eagles nest monitoring carried out on the tower of 500 kV high-voltage wires in PT PJB UP Paiton, Probolinggo, East Java (S 7°42'55.96" - E 113°34'44.76"). Quadcopter flown across the eagle's nest tower with flying height 100, 80 and 70 m agl with 5 m/s flight speed when recording. Interlude between flight altitude level around 15-30 minutes (Figure 2a). Flights performed at 7:15 to 9:00 am. The timing was based on the eagle's daily behaviour which has been observed previously by researcher. At that time male eagle leaving the nest to the sea for foraging, while the female perched in another tower (adjacent tower to the nest). The female always keep an eye on the nest and territory while sunbathing. Eagle nest is empty, there are no eggs and chicks. The camera used in the first repeat is action cam GoPro HERO4 silver, while in the second repetition using the mirrorles camera sony α5100, pancake lens 16 mm. The image is taken automatically.

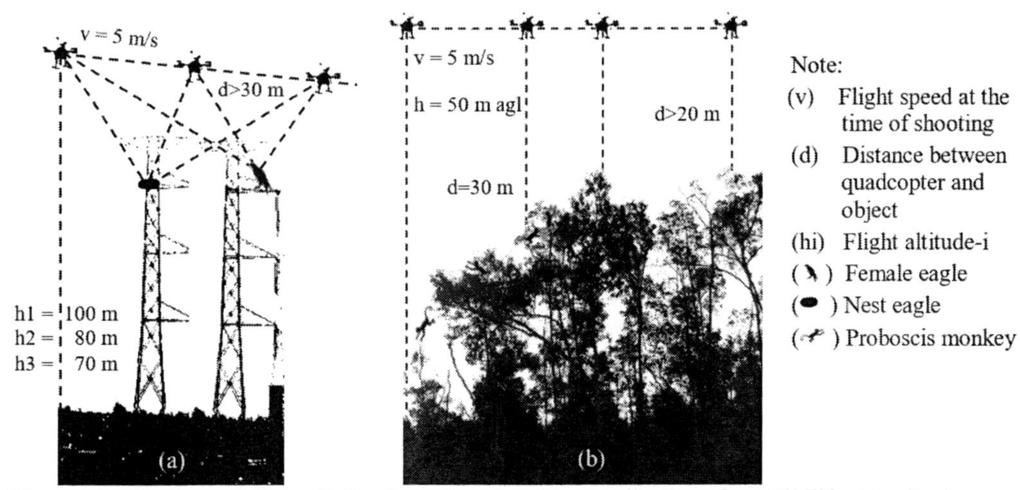

Figure 2. Design of Data Collection: Quadcopter Application for Wildlife Monitoring
(a) White-bellied Sea Eagles (*Haliaeetus leucogaster*) nest
(b) Proboscis Monkey (*Nasalis larvatus*)

2.2.2. Proboscis Monkey (Nasalis larvatus)

Quadcopter application for Proboscis monkey in riparian zone at High Conservation Value Areas (HCVA) of Palm Oil Plantations PT AMR, Kota Waringin Barat Regency, Central Kalimantan Province (S 2°19'14.88" – E 111°46'0.58") performed in the morning (7:00 to 08:00 am) and in the afternoon (15:00 to 16:00 pm). The timing is based on daily behaviour information of wildlife in HCVA by PT AMR. Flights carried out on flight path with altitude 50 m agl, and 5 m/s flight speed at the time of shooting (Figure 2b). Photo shoots using sony α5100 mirrorles camera, 16 mm pancake lens, shutter speed of 1/2000 second, 16 mm focal length, f/2.8 for aperture and 320 for ISO, is set manually.

2.3. Data Analysis

1. Spatial analysis for object identification on aerial photographs and orthophoto mosaic establishment (using agisoft photoscan software).
2. Descriptive statistical analysis to provide weights and flight duration relationship, and the parameters that affect the accuracy of aerial photographs.
3. Effectiveness and efficiency analysis for quadcopter technical applications and disturbance to the wildlife target using criteria and indicators in Table 2.

3. Result and Disscusion

3.1. Aerialview-650 Development and Application

Aerialview-650 is quadcopter assembly, which is developed as wildlife monitoring instrument. Considerations for the use of quadcopter assemblies are:

1. Designed for mapping and monitoring wildlife.
2. Quadcopter can load different types of cameras as needed.
3. Flight endurance can be adjusted with the weight of battery and type of camera system (payload).
4. Application software used open-source (free of charge).
5. Quadcopter can be partially repaired if damaged.
6. Easiness of obtaining replacement component.

Overall quadcopter application trial (28 flights) indicate an inreament both in the aviation system and aerial photographs produced. Extra vibrator damper and enhancement in the type of camera (camera resolution 12 to 24 MP) results in quality improvement of aerial photographs. Constraints experienced in the application of Aerialview-650 include:

1. Maneuver spontaneously due to the displacement of two GPS navigation systems.
2. Take off and landing on narrow areas manually.
3. Missing propeller can result quadcopter crash.
4. Running out of battery during the flight resulting in hard landing.
5. Bad heading estimate (navigation disturbance) resulting in quadcopter uncontrolled.

Table 2. Criteria and Indicators of Effectiveness and Efficiency Quadcopter Applications and Disturbance to Wildlife

	Criteria	Indicators
Effectiveness	The data generated can be used for wildlife identification (large mammals and birds)	Objects can be differentiated well Objects colors properly translated Wildlife identification to the species level Accuracy photo reaches 10 cm/pixel
Efficiency	Quadcopter applications with a combination of flight altitude and time to generated the most optimal data	The shortest flight time Fewest number of aerial photos Wide range of aerial photographs greatest
Disturbance	Quadcopter noise level	Noise intensity > 60 dB
	Wildlife response to the presence of quadcopter	View, voice, panic, avoidance and attack

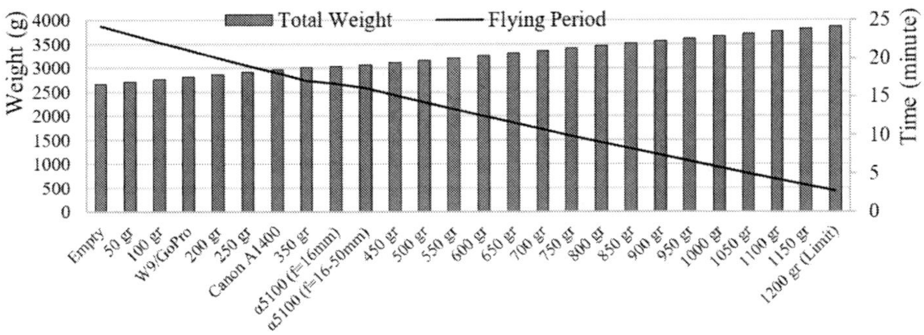

Figure 3. Relationship between Weight (Quadcopter and Camera) and Flight time

In broad outline, quadcopter development parameters are similar to the type of airplane/fixed wing. The addition of damping system, an increase in the type/quality of cameras used and the settings for the shooting is done to improve the performance and quality UAV aerial photographs produced [2][3]. The time limit quadcopter for safe flight is 15 minutes. To get these flight time, weight of camera that can loaded Aerialview-650 below than 500 g (Figure 3). Type of camera with weight less than 500 g has become alternate in quadcopter application.

3.2. *Data Quality Generated*

Spatial resolution of orthophoto mosaic generated from flight at altitude of 50-200 m agl is 0.85 to 4.79 cm/pixel (Table 3). Flight at altitude of 50 m agl generates fivefold greater detail image compared to 200 m agl. The quality of orthophoto mosaic between flights at an altitude of 50 and 100 m agl did not show significant difference. Both analysis in the difference of flight altitude emphasis on identification of objects from aerial images produced. Aerial photo at an altitude of 50-200 m agl can be used for identification of large wildlife (with a size more than 1 m). The results of aerial photography at flight altitude of 200 m agl using action cam w9 sports produce greater detail images than conservation drone (airplane type). Conservation drones using Canon or Pentax camera with a focal length of 5.7, 200 m agl altitude, produce 5.3 cm/pixel aerial photo resolution [3].

Application quadcopter on the flight 100 m agl produce aerial photographs that can be used to detect the presence of wildlife to the species level. This is similar to aerial images produced by FolBat (airplane UAV) an altitude of 100-150 m, which is able to distinguish some seabirds (*Egretta* sp. & *Mycteria americana*) as well as medium-sized vertebrates [2]. The resulting aerial photographs on a flying height of 50 m agl can be used for identification of wildlife species measuring up to 30 cm. The quality of the data generated proving that quadcopter applications is effective for medium-sized wildlife monitoring.

Trials with nearly the same flight time indicate the flight at an altitude of 200 m agl produce coverage area of twenty-five-fold compared to low flight of 50 m agl. Flight at 100 m agl require shorter shooting time by half compared to low flight of 50 m agl. The number of images produced on the flight of 100 m agl is 30% of the total number of photos on the flight of 50 m agl.

Table 3. Data Quality Based on Flight Altitude and Photo Shoot Parameter

No	Camera type/ brand	Flight altitude (m agl)	Shutter speed (s)	Focal length (mm); Aperture; ISO	Photo Overlap/ Sidelap (%)	Dimensional aerial images (pixel)	Orthophoto mosaic resolution (cm/pixel)
1	Action cam/ W9 Sports	200	1/30-1/60	- ; f/2.8; 200-400	80/60	4000 x 3000	4.79
2	Mirrorles/ Sony α5100	100	1/1250	16; f/5.6; 320-640	80/60	6000 x 4000	0.94
3	Mirrorles/ Sony α5100	50	1/2000	17-21; f/4-f/5; 320	60/40	6000 x 4000	0.85

Aerial photo which generated in sunny weather, whether it is taken in the morning or in the afternoon produce good quality. So did the shooting in the morning until noon on cloudy days. However, the shooting on the afternoon at 17:24 pm with cloudy weather, produced dark images. This indicates the quality of images strongly influenced by the lighting which has received by the camera lens. Despite the cloudy weather, when the lighting of camera is adequate, could produce good quality image. Dark aerial photos can not be combined into orthophoto mosaic.

The main factors affecting thoroughness of mosaic orthophoto are the quantity and quality of aerial photographs which has been combined. In the early stages of trials, there is an imperfect fusion in orthophoto mosaic. This is due to less good quality images which became constituent for mosaic (the image is blur). The high frequency of quadcopter vibration is the main factor. To overcome this requires appropriate camera and shooting settings. The next trial used a camera with better specifications and equipped with a vibrator damper as well as a mounting and vibration dampers. Vibrator dampers can minimize the vibration frequency at the same speed shooting, so that the camera sensor can capture images in a good and clear condition. Automatically camera settings to the lighting and focal length is recommended. To reduce blur, the shooting speed is set at 1/320 - 1/1000 and added damping system made from low density foam packaging [3].

The number of photos produced is not necessarily all of them can be used to build the mosaic. The more aerial photos that can be combined from the total of aerial photographs produced, the more perfect orthophoto mosaics that can be created. However, in case coordinate of aerial photographs produced has too large deviation, the result of orthophoto mosaics will be imperfect. (Table 4). This condition is a result of constantly changing from focal length determination. To overcome the changing, focal length is locked by fixed setting as well as can be done by replacing the lens using pancake lens [4][9].

Relationship among camera weight, quadcopter, flight time, camera type, flight altitude and accuracy of the data generated, can be used to formulate compatible quadcopter application with appropriate output data as needed. For example, wildlife monitoring that requires output spatial resolution of 2-5 cm/pixel, can be done at flight altitude of 100-200 m agl and using 12-16 MP camera (camera's weight less than 300 g). Whereas to generate aerial photos with a spatial resolution of 0.85 to 1 cm/pixel, the flight can be done at altitudes of 50-100 m agl using 24 MP camera with a weight of 350-500 g.

3.3. Effectiveness and Efficiency of Quadcopter

Based on effectiveness criteria which has been established, the result of aerial photographs from entire flight has met all indicators. Criteria and indicators of efficiency that has been set by the quadcopter application is fulfilled by the flight altitude of 100 m agl. Despite the quality of images is good enough, the identification of wildlife to the species level with a size range of 30-100 m still have to use images from flight altitude of 50 m agl.

Technical recommendations for wildlife monitoring which effective and efficient are:

1. Flight altitudes of 50-100 m agl. The distance between wildlife and the highest object with the quadcopter is as far as 20-30 m. Flight will adjust to the size and characteristics of the wildlife.
2. The use of light-weight type of camera (< 500 g) with high resolution (≥ 20 MP), vibration damper system, fixed or pancake lens, shutter speed between 1/1250 - 1/2000 and flight speed 5 m/s.
3. Images are taken in the morning until afternoon (7:00 am to 17:00 pm) in sunny weather so the lighting needed by the camera is fulfilled.

Table 4. Estimation of Camera Position Error Based on the Precision of Aerial Photos

No	Flight altitude (m agl)	Σ Total aerial photo (N)	Σ Appropriate aerial photo (n)	Proportion n/N (%)	X *error* (m)	Y *error* (m)	Z *error* (m)
1	200	53	53	100.00	57.48	12.93	1.42
2	100	23	23	100.00	28.46	13.27	0.22
3	50	72	61	84.72	40.67	29.07	10.54

3.4. Wildlife Response to the Quadcopter Application

3.4.1. Noise Level
Rotation of the rotor and propeller provide noise pollution. Results of identification issued quadcopter sound intensity at a distance of 5 m reach 26.5 dB. This value is relatively low and below the threshold of disturbance for wildlife in the vicinity. Noise levels ≤ 60 dB of quadcopter at a distance of 2 m is considered leave no impact for shorebirds [8][9][10].

3.4.2. Monitoring of White-bellied Sea Eagles (Haliaeetus leucogaster) nest
Quadcopter flown across the nest of White-bellied Sea Eagles near the breeding season. The identification of the eagle nests indicate that nest each empty, no eggs and chicks. Even the nest looks clean, no visible remains of food such as fish bones, skull of fish or small mammals and litter, just visible twigs with leaves that are still green (new branches). The dimensions of the nest by measurements on aerial photographs show nests longest diameter of 201 cm and 159 cm shortest (Figure 4a).

Overall, there was no negative response shown by the presence of quadcopter that flew over the eagle nest. This is consistent with the results of research on bird behavior to the presence of quadcopter with the closest distance of 4 m from the bird [8]. Quadcopter application for eagle nest monitoring should be performed once without a repetition in adjacent time. Repetion of data retrieval can be done with different intervals and taking into account the characteristics and daily behavior of the eagles.

3.4.3. Monitoring of Proboscis Monkey (Nasalis larvatus)
The result of aerial photographs taken in the morning at HCVA of PT AMR, has been identified some Proboscis Monkey above trees canopy on riparian. Overall, quadcopter application does not have any impact on the foraging activity of Proboscis Monkey at their habitat. Proboscis Monkey remain active as usual at the time of quadcopter passed over the trees. These suggest that the Proboscis Monkey is not responsive to the presence of quadcopter.

Proboscis Monkeys in this area has been accustomed with the presence of wildlife that pass in the air, since this area is also the habitat of Hornbill. Several Hornbills, alone in a group (with a number of 2-5 individu) often flew over the trees where Proboscis Monkeys foraging. The size of hornbill is larger than the dimensions of quadcopter and has a sound of wing stroke noisier than the sound of quadcopter. Morphometry of Proboscis Monkeys were captured based on spatial resolution aerial photo shows a body length is 31.4 cm and 35.2 cm, which is juvenile category (Figure 4b; Figure 4c).

Figure 4. Identification Result of Wildlife from Aerial Photo
(a) White-bellied Sea Eagles nest, aerial photo size of 768 x 768 pixels, photo insert of 200 x 200 pixels with a magnification of four-fold, spatial resolution of 1 cm/pixel, L1 (longest diameter), L2 (shortest diameter)
(b) (c) Proboscis Monkey, aerial photo size of 2000 x 2000 pixels, photo insert of 200 x 200 pixels with a magnification of twenty five-fold, spatial resolution of 1 cm/pixel, L (body length)

3.5. Quadcopter Application which in Accordance with the Wildlife Photography Ethics

Technical recommendations in accordance with the ethics of wildlife photography are:

1. Before flight, the daily behavior of wildlife target should be known in advance.
2. Take off at least 300 m of wildlife to avoid being detected.
3. Minimum distance between quadcopter and wildlife target or highest object is 30 m.
4. Flight speed at 5-7 m/s in accordance with quadcopter and camera capabilities.
5. Flight repetitions require interval (pending bit longer or different days)

4. Conclusion

Aerial photographs which generated from the application of quadcopter can be used to identify wildlife species up to 30 cm of size. Flight altitude at 50-100 m agl with a flight speed while shooting or recording 5 m/sec and using a camera weighing <500 g, is the technical recommendation which is effective and efficient of quadcopter application for wildlife monitoring. Flights abiding the ethics of wildlife photography do not cause negative response from wildlife target.

Acknowledgements

This project is partially supported by PT Meganesia Tirta Foresta (MeTTa), PT PJB UP Paiton & Astra Group. We are grateful to PT Riap Indonesia & MeTTa for master schollarship programme; Malang Regency Goverment for study permit; Bogor Agricultural University, PT PJB UP Paiton, PT AMR for research location and amenities. In particular we also would like to thank AN Putra, E Juarsa, LM Laban, A Suprabhana, IS Sugato for their material support; RW Subekti "Nano" as UAV coach, WN Akbar "Waladi" as Aerialvew-650 maker; Zulham, BA Yulianto, H Farmen, R Hardansyah, A Kurniawan, Rismunandar, FI Mansyur, A Chandra, A Herdiyanto for their comments on this manuscript; MeTTa Institute, Paiton Eagle Team, Kumaigreen HCV Team for their technical assistance.

References

[1] Hodgson J C, Baylis S, Mott R, Herrod A and Clarke R 2016 Precision wildlife monitoring using unmanned aerial vehicles *Scientific Report* **6** 22574

[2] Jones G P, Pearlstine L G and Percival H F 2006 An assessment of small unmanned aerial vehicles for wildlife research *Wildlife Society Bulletin* **34** 750-758

[3] Koh L P and Wich S A 2012 Dawn of drone ecology: low-cost autonomous aerial vehicles for conservation *Tropical Conservation Science* **5** 121-132

[4] Martin J, Edwards H H, Burgess M A, Percival H F, Fagan D E, Gardner B E, Ortega-Ortiz JG, Ifju P G, Evers B S and Rambo T J 2012 Estimating distribution of hidden objects with drone: from tennis balls to manatees *PLoS ONE* **7** e38882

[5] Paneque-Galvez J, McCall M K, Napoletano B M, Wich S A and Koh L P 2014 Small drone for community-based forest monitoring: an assessment of their feasibility and potential in tropical areas *Forests* **5** 1481-1507

[6] Pazmany M M, Stolper R, van Essen L D, Negro J J and Sassen T 2014 Remotely piloted aircraft systems as a rhinoceros anti-poaching tool in Africa *PLoS ONE* **9** e83873

[7] Podduwage D R 2016 An ethical model for the wildlife photography of Sri Lanka *Journal of Aesthetic and Fine Arts* **1** 98-129

[8] Vas E, Lescroel A, Duriez O, Boguszewski G and Gremillet D 2015 Approaching birds with drones: first experiments and ethical guidelines *Biology Letters* **11** 20140754

[9] Watts A C, Perry J H, Smith S E, Burgess M A, Wilkinson B E, Szantoi Z, Ifju P G and Percival H F 2010 Small unmanned aircraft systems for low-altitude aerial surveys *Journal of Wildlife Management* **74** 1614–1619

[10] Wright M D, Goodman P and Cameron T C 2010 exploring behavioural responses of shorebirds to impulsive noise *Wildfowl* **60** 150-167

Spectral radiation of tree leaves in Bogor Agricultural University campus

Deki Andika Purbaya[*], **Siti Badriyah Rushayati, Lilik Budi Prasetyo**
Department of Forest Resource Conservation and Ecotourism, Bogor Agricultural University

E-mail: andikadeki100@yahoo.com

Abstract. Every anthropogenic activities that use fossil fuels will produce pollutants such as greenhouse gases. CO_2 with other greenhouse gases increase urban air temperatures through the greenhouse effect. The aims of this study are to measure spectral radiation of several species of trees leaves in Bogor Agricultural University Campus and determine types of trees that are effective in absorbing CO_2. Data was statistically analyzed based on the order of spectral radiation value. Meanwhile, grouping the ability of species to absorb CO_2 was done based on normal curve distribution. Spectral radiation value is inversely proportional to the ability of plants to absorb CO_2. The tree species classified as having a high ability to absorb CO_2 is *Tamrindus indica*, *Adenanthera pavoniana*, *Samanea saman*, and *Ceiba pentandra* whereas the species classified as low capacity in absorbing CO_2 is *Annona murricata*, *Pterocarpus indicus*, *Acacia mangium*, and *Canangium odoratum*, the rest classified as having moderate capability.

1. Introduction

Every anthropogenic activities that use fossil fuels will produce pollutants such as greenhouse gases, one of that is carbon dioxide (CO_2). Anthropogenic activities will be increased to the center of the city. Many modern cities in the world and including Indonesia can not escape from those kind of problem. CO_2 with other greenhouse increase urban air temperatures through the greenhouse effect. CO_2 can increase the temperature of the air by absorbing near-infrared wave [1], so it becomes important to note as the greatest contribution in increasing air temperature [2].

One of the way to tackle the problem is developing urban forests and other forms of green open space. In urban forest and other forms of green open space generally planted by trees. Different types of tree has primary function of each, such as a shade, fruit producing, windbreaker, good-sided view, CO_2 absorber, reduce the noise, etc. Trees that have a primary function in sinking CO_2 is needed to reduce the greenhouse effect and decrease the air temperature. Therefore it is necessary to do research on plants (in this case is a tree) that are effective in absorbing CO_2. Several studies that had been conducted to estimate the ability to absorb CO_2 on some tree species are as follows, [3][4][5][6][7][8]. Most of these studies using carbohydrate analysis and light response curve to estimate the ability of trees to absorb CO_2 but very few of them use the information about spectral radiation of tree leaves.

Predicting the ability of plants to absorb CO_2 by using the value of its spectral radiation is a new approach. Spectral radiation is the amount of energy emitted per unit surface area of thermal emitting objects. These predictions are not directly related with the ability of photosynthesis of a plant species. If it compared with other methods, this method does not require a long time and the results can be quickly obtained

2. Objectives and Benefit

The purpose of this study is to measure the spectral radiation value on some types of trees leaves in Bogor Agricultural University campus based on its using that has been widely planted in Bogor city and determine the types of trees that are effective in absorbing CO_2 based on the spectral radiation. The results of this study can be used as the basis for choosing the type of tree that is effective in absorbi ng CO_2

3. Methods

3.1. Location and Time

Research and data collection was conducted in Bogor Agricultural University (IPB) campus and LAPAN Rancabungur, Bogor on June 16th, 2016.

3.2. Tools and Materials

Equipment used were Handheld 2 Spectroradiometer, software HH2 Sync, Icebox, labels, pole, styrofoam, and stationery. Styrofoam was formed in 45^0 of angle from research object. The materials used were ice cubes

3.3. Research Object

Research object were leaf samples from twenty five of tree species which exist in Bogor Agricultural University campus. Selection of tree species referred to [3][9]. The selected trees were the trees that had been widely planted in Bogor city, the location was not too far apart and the leaves and branches still reachable by a pole.

3.4. Type of Collected Data

Type of collected data in this study were only primary data. Primary data were obtained from value of spectral radiation from each leaf sample. Spectral radiation value is in range of 325-1075 nm.

3.5. Collecting Data

3.5.1. Literature study. Literature study is an activity related to any literature review, showing tree species that has been widely planted in Bogor city and those kind of trees were exist in IPB Darmaga campus.

3.5.2. Collecting Spectral Radiation Value

There are 3 leaves were taken from each tree species of twenty five selected trees. It were the fifth leaves from of the end of the branch. All leaves were taken in the morning and kept on ice box during the trip to LAPAN. Measurement of spectral radiation value by using Handheld 2 Spectroradiometer from 12:20 to 13:00 pm. Before measuring spectral radiation value on each leaf, Handheld 2 Spectroradiometer need to be calibrated by using white reference. Measurement of the spectral radiation value was done by placing the Handheld 2 Spectroradiometer on each leaf in 45^0 angle with the distance from the tool to the leaf as far as 4 cm. Collecting spectral radiation value of small leaf (length and width <2 cm) was to be stacked and for the leaf with the length and width more than 2 cm was not stacked. Handheld 2 Spectroradiometer was placed in side of styrofoam to remain static while capturing spectral radiation value. It would automatically capture leaf spectral radiation value. Spectral radiation value stored in a text document format (.txt). Before doing the method of collecting spectral radiation value as described above, it has been carried out trial and error in orde to get spectral radiation value accurately

3.6. Processing and Data Analysis

3.6.1. Selected Spectral Radiation Value

The value of spectral radiation was modified form .txt to Ms. Excel format. It were not all of wavelengths of spectral radiation values were used, only red and blue wavelength of sunlight as the most effective wavelength to perform photosynthesis [10][11]. Red wavelengths (450-520 nm) and blue (630-690 nm) based on by [12] (Table 1.)

Table 1. Characteristics of the sensor on Landsat TM, ETM +

	Band	*Wavelength (μm)*	*Resolution (m)*
Blue	1	0.45 - 0.52	30
Red	3	0.63 - 0.69	30

Spectral radiation value of tree leaf from one tree species were averaged. Spectral radiation value obtained from averaging the value of nine repetition. Leaf spectral radiation value of twenty-five species of trees are sorted from smallest to largest. Tree species that has the lowest spectral radiation value is a species of tree that has the highest CO_2 absorption ability due to spectral radiation value is inversely proportional to the ability of photosynthesis. Spectral radiation value is only inform in sorting the ability of photosynthesis and absorbing CO_2 of plant species, from the largest to the lowest level, but can not provide a picture of how the mass of CO_2 absorbed per time unit.

3.6.2. Grouping Spectral Radiation Value

Grouping spectral radiation values are divided into classifications of low, medium, and high. The ability of photosynthesis is inversely proportional to the value of the spectral radiation. Once all types of trees leafe has grouped its spectral radiation values and relate it with photosynthetic ability, the type of tree that is effective in absorbing CO_2 can be directly determined. Grouping spectral radiation values using a simple statistical calculation based on the normal distribution curve. Twenty-five species of trees that have been determined the spectral radiation value of its leaf were calculated its standard deviation (Sd), mean, mean-Sd, and mean+Sd. Furthermore, spectral radiation value that included in low classification is if X <Mean - Sd, high if X> Mean - Sd, and moderate if spectral radiation value is between X <Mean-Sd and X> Mean - Sd.

3.6.2.1. Calculation of mean

$$\mu = \frac{\Sigma Xi}{N} \tag{1}$$

3.6.2.2. Calculation of standard deviation (Sd)

$$\sigma \sqrt{\frac{\Sigma X_i^2 - \left(\frac{\Sigma Xi}{N}\right)^2}{N}} \tag{2}$$

3.6.2.3. Calculation of mean – Sd

$$\mu = \frac{\Sigma Xi}{N} - \sigma \sqrt{\frac{\Sigma X_i^2 - \left(\frac{\Sigma Xi}{N}\right)^2}{N}} \tag{3}$$

3.6.2.4. Calculation of mean + Sd

$$\mu = \frac{\Sigma Xi}{N} + \sigma \sqrt{\frac{\Sigma X_i^2 - \left(\frac{\Sigma Xi}{N}\right)^2}{N}} \tag{4}$$

note:

σ : Standard deviation; $\sum$: Total population; N : Population; x_i : Data ke-i; μ: average

4. Results and Discussion

4.1. Spectral Radiation Value

Spectral radiation values that can be captured by Handheld 2 Spectroradiometer has range of 325-1075 nm. It can not be directly processed by using Ms. Excel. HH 2 Sync software can convert the data spectral radiation value into .txt format and further processed by using Ms. Excel. Spectral radiation value of 25 types of tree leaves can be seen in Figure 1.

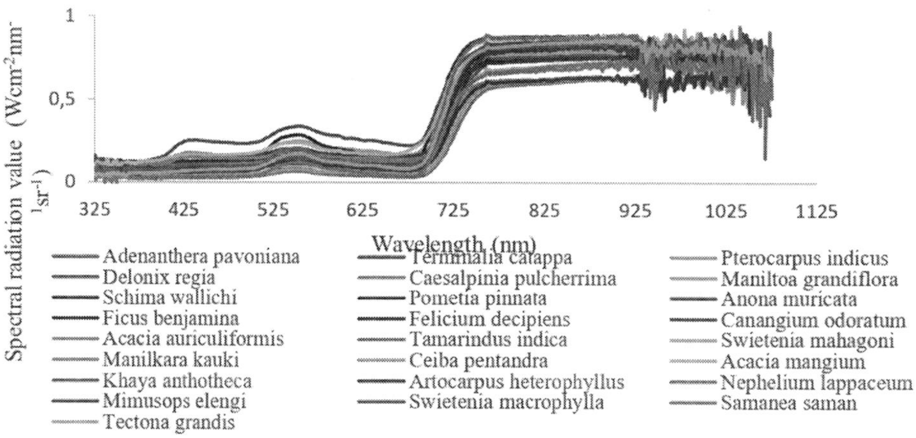

Figure 1. Spectral radiation value of 25 types of tree leaves

Image above only shows distribution of spectral radiation value of each type of leaf, but it has not shown how the pattern spectral radiation value itself clearly on blue and red wavelength. Although the data is only focused on those kind of wavelength, it does not mean a wavelength of 325-1075 nm has no value. Figure 2 shows an example pattern of spectral radiation value of saga leaf (*Adenanthera pavoniana*) on 325-1075 nm wavelength as well as Figure 3 and Figure 4 shows the pattern of spectral radiation value of light wavelength that is used for photosynthesis (400-700 nm) on trembesi (*Samanea saman*) and Angsana (*Pterocarpus indicus*).

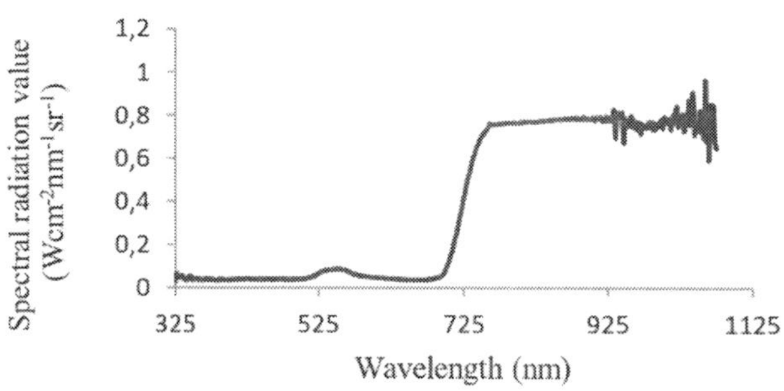

Figure 2. Spectral radiation value of saga leaf (*Adenanthera pavoniana*) on 325-1075 nm wave

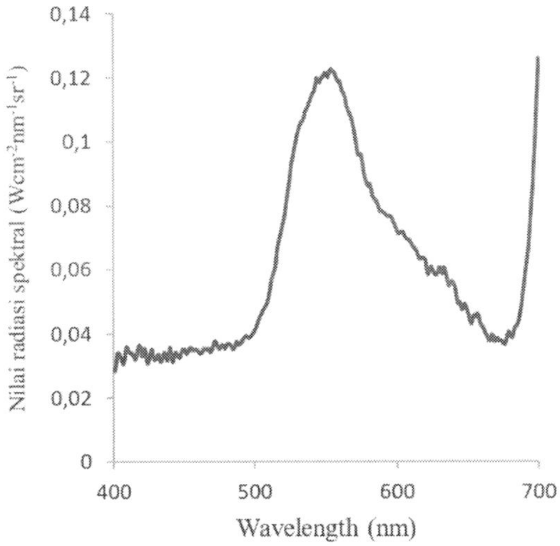

Figure 3. Spectral radiation value of trembesi leaf

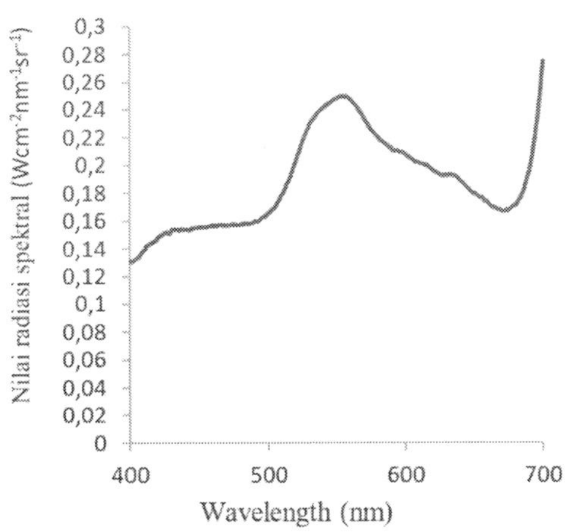

Figure 4. Spectral radiation value of angsana leaf

Related with spectral radiation value which is mentioned above will be further described by the following literature. Spectral radiation value is a reflection of the sun's radiation, so it applied the principle of light reflection. Solar radiation that comes and touches a surface can be reflected, absorbed or transmitted. Light reflected by an object called the reflectance / reflection, whereas the absorbance / absorption is light absorbed by an object located at a particular wavelength. Each different objects reflect and absorb different wavelengths.

Reflectance or reflection has no units or expressed in percent while spectral radiation value has range between 0 and 1, the unit is $Wcm^{-2}nm^{-1}sr^{-1}$. Spectral radiation value indicates how much power per square centimeter per steradian per namometer of sunlight reflected by the leaves, so based on the value, it can be seen how well the ability of type of tree leaves in using the energy of solar radiation for photosynthesis. The greater spectral radiation value leaves on blue and red wave the less solar radiation energy that is used for photosynthesis. Figure 5 shows the spectral reflectance characteristics of green plants leaf. Spectral radiation value in Figure 3 and Figure 4 are accordance with what is stated by [13]. Spectral reflectance of green leaf tend to be high on the wave of green (520-600 nm) and low in red and blue wave. It shows that the blue and red wave are light wave which is effectively used for photosynthesis.

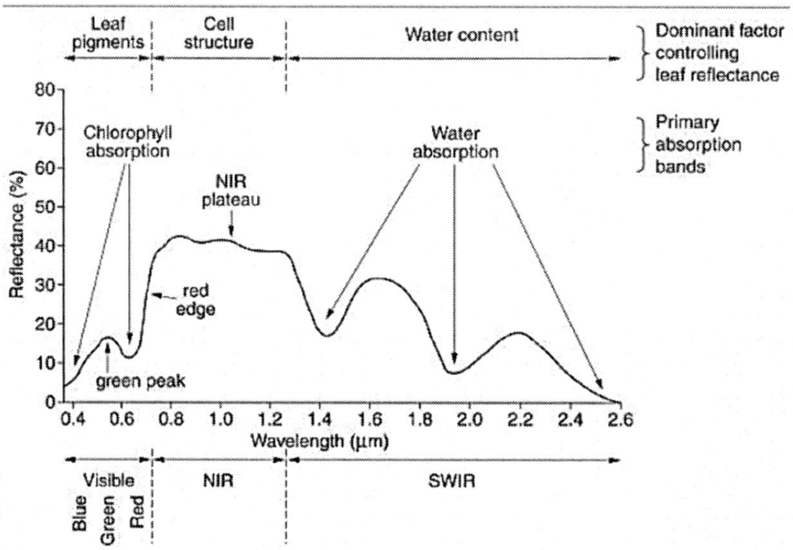

Figure 5. Characteristics of the spectral reflectance of green leaf [13]

Related to spectral radiation, solar light is a form of radiation that can be reflected, absorbed, or transmitted. Radiation can be defined as the energy emitted in the form of particles or wave. Radiation consists of several kinds, and every kind of radiation has a wavelength of each. In term of the mass, the radiation can be divided into electromagnetic radiation and particle radiation. Electromagnetic radiation is radiation that has no mass. This radiation consists of radio wave, microwave, infrared, visible light, X-rays, gamma rays and cosmic rays. Radiation particle is radiation in the form of particles that have a mass, such as beta particles, alpha and neutron [14]. In the process of photosynthesis, chlorophyll have very important role because it can absorb the sun light that is form of electromagnetic radiation in visible spectrum [10].

4.2. The Order of Spectral Radiaton Value

It is called as PAR (*photosyntetically active radiation*) in photosynthesis which is known as effective wavelengths for photosynthesis ranged between 400-700 nm. The wavelength range of 400-700 nm is called visible light [6][15][16][17]. As explained earlier the wave of blue and red is the most effective sunlight wavelenght in photosynthesis process. Thus the calculation of spectral radiation value focused on the visible light. Once spectral radiation value of all the leaves were known then the spectral radiation values are sorted from the smallest to the largest. The order of the spectral radiation values on blue and red wave are shown in Figure 6 and Figure 7.

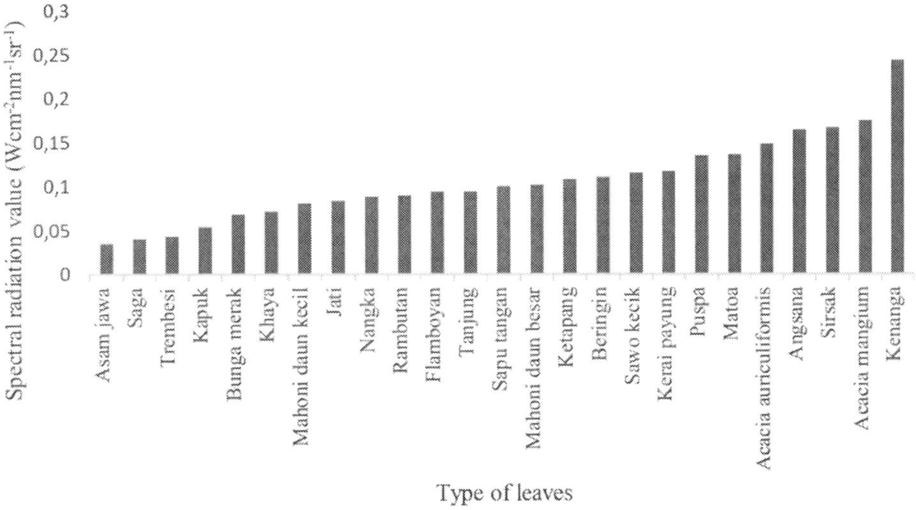

Figure 6. The order of radiation spectral value of blue wave (450-520 nm)

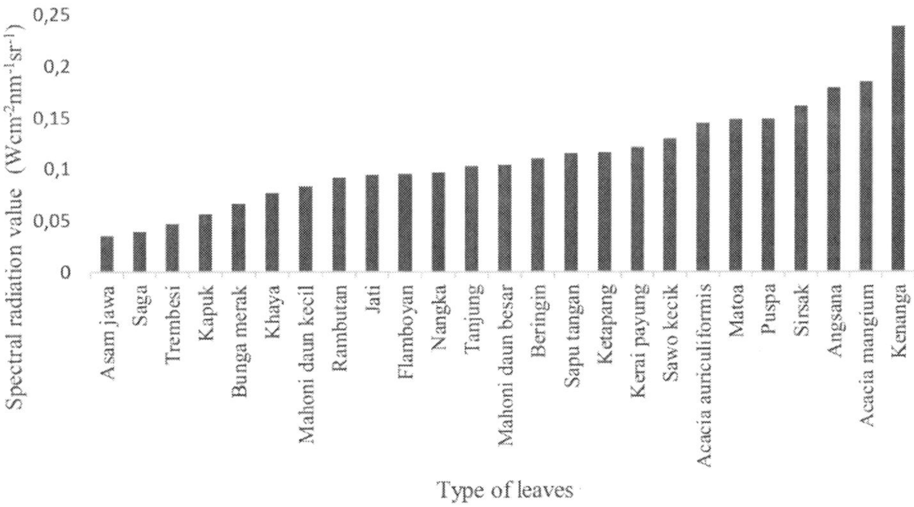

Figure 7. The order of radiation spectral value of red wave (630-690) nm

4.3. Grouping Spectral Radiation Value

Grouping spectral radiation value based on normal distribution curve. Normal curve can only be made if the calculation of standard deviation, mean, min, and max are known. Figure 8 and Figure 9 shows the normal curve value of spectral radiation blue and red wave. The next stage was spectral radiation value of blue and red wave were averaged for classifying overall of spectral radiation value as a reference in grouping capability of types of trees leaf in absorbing CO_2 and the result was made as normal curve (Figure 10).

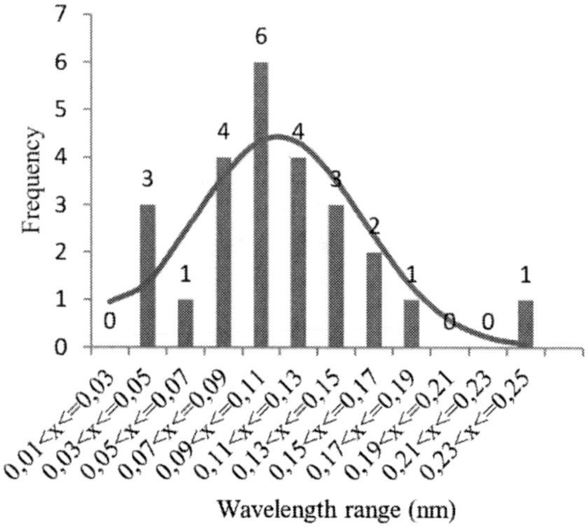

Figure 8. Normal curve of radiation spectral value of blue wave (450-520 nm)

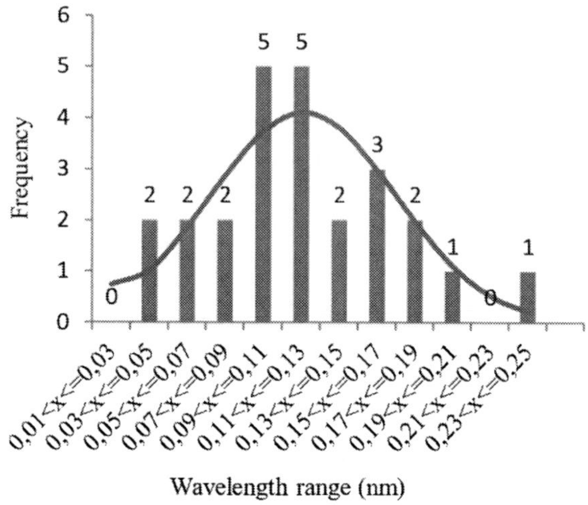

Figure 9. Normal curve of radiation spectral Value of red wave (630-690 nm)

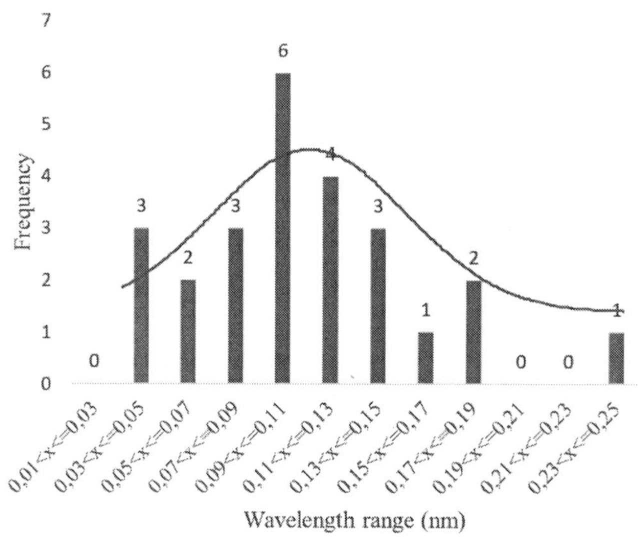

Figure 10. Averaged-normal curve of radiation spectral value of blue and red wave (450-520 nm)

4.4. Relation between Spectral Radiation Value and CO_2 Absorption

The leaf has a pigment molecule that has a light absorption characteristic of each. There are two major types of pigment contained in leaf, they are chlorophyll and carotenoids. Compounds of chlorophyll found in green plants is divided into two, they are chlorophyll-a and chlorophyll-b. Chlorophyll plays an important role in the process of photosynthesis and it is the main light-absorbing pigment found in thylakoid membranes [11]. Carotenoids are compounds that give red and orange on fruits, algae, microorganisms, vegetables and crustacean outer frame such as shrimp. Carotenoids that are found in nature are very numerous. Carotenoids also play an important role in photosynthesis and the absorption of oxygen for the survival of plants. Each spectrum shows the difference wavelength absorbed by any pigment. The light spectrum that can effectively absorbed by chlorophyll-a and chlorophyll-b are the red and blue.

Leaf as the main photosynthesis organ also functioned as the light receiver [18]. [19] suggested that photosynthesis does not depend on the total energy of light, but the number of photons or quanta absorbed. Photons are packets of energy that is fragmented, each having a specific wavelength. The characteristic of light as well as a wave have an effective wavelength for photosynthesis ranged between 400-700 nm. Therefore wavelengths between 400-700 nm is also called radiation for photosynthesis or photosyntetically active radiation (PAR). Campbell *et al.* [11] stated that the absorption spectrum of chloroplast pigments provide guidance on the relative effectiveness of various wavelengths to drive photosynthesis, because light can work only if it is absorbed in the chloroplast. Another fact mentioned that the speed of carbon change in the form of inorganic (CO_2) into organic matter (glucose) in net primary productivity (NPP) is influenced by the availability and intensity of photosyntetically active radiation (PAR) and the distribution of pigment photoactive and light efficiency absorbed [20].

In photosynthesis, plants utilize solar energy to oxidize water to release O_2, and reduce CO_2 to form larger carbon compounds, especially sugars [21]. Abiotic factors such as light, temperature, CO_2 concentration, moisture, presence of nutrients have a major effect on net photosynthesis, which in turn affects growth and productivity. The ability of photosynthesis is not only influenced by abiotic factor but also influenced by age and position of leaves in the canopy. Different types or genus has a different rate of photosynthesis. Dahlan [3] refers to various references explained that the rate of leaf photosynthesis of plants is influenced by several factors, namely: air pollutants, leaf health, the availability of ground water and humidity, concentration of CO_2, sunlight intensity, the growth phase, the location of the leaf, the age of leaves, and different types of plants. Another explanation said that the temperature of the air through a complex scheme also affects photosynthesis [21].

Based on information from the description above it can be concluded that solar radiation plays a role in the ability of leaf photosynthesis. Plants that have a lower spectral radiation value in blue and red light wave (more sunlight energy absorbed) can be interpreted that these plants use the sun's light wave effectively to carry out photosynthesis, so the ability of photosynthesis can is high. Related to that fact it is automatically concluded if the leaf have high photosynthetis ability, CO_2 absorption is also high. Thus the relation between spectral radiation value and the ability of plants to absorb CO_2 is inversely proportional. Table 2, Table 3, and Table 4 shows grouping of spectral radiation value and ability to absorb CO_2 from the 25 types of tree leaf.

Table 2. Grouping spectral radiation values and the ability to absorb CO_2 low and high classification

No	Scientific name	Classification of Spectral radiation value	Classification in absorbing CO_2
1	*Tamarindus indica*	Low	High
2	*Adenanthera pavoniana*	Low	High
3	*Samanea saman*	Low	High
4	*Ceiba pentandra*	Low	High

Table 3. Grouping spectral radiation values and the ability to absorb CO_2 intermediate classification

No	Scientific name	Classification of Spectral radiation value	Classification in absorbing CO_2
5	*Caesalpinia pulcherrima*	Intermediate	Intermediate
6	*Khaya anthotheca*	Intermediate	Intermediate
7	*Swietenia mahagoni*	Intermediate	Intermediate
8	*Nephelium lappaceum*	Intermediate	Intermediate
9	*Tectona grandis*	Intermediate	Intermediate
10	*Delonix regia*	Intermediate	Intermediate
11	*Artocarpus heterophyllus*	Intermediate	Intermediate
12	*Mimusops elengi*	Intermediate	Intermediate
13	*Swietenia macrophylla*	Intermediate	Intermediate
14	*Ficus benjamina*	Intermediate	Intermediate
15	*Maniltoa grandiflora*	Intermediate	Intermediate
16	*Terminalia catappa*	Intermediate	Intermediate
17	*Felicium decipiens*	Intermediate	Intermediate
18	*Manilkara kauki*	Intermediate	Intermediate
19	*Acacia auriculiformis*	Intermediate	Intermediate
20	*Pometia pinnata*	Intermediate	Intermediate
21	*Schima wallichi*	Intermediate	Intermediate

Table 4. Grouping spectral radiation values and the ability to absorb CO_2 high and low classification

No	Scientific name	Classification of Spectral radiation value	Classification in absorbing CO_2
22	*Anona muricata*	High	Low
23	*Pterocarpus indicus*	High	Low
24	*Acacia mangium*	High	Low
25	*Canangium odoratum*	High	Low

5. Conclusion

Twenty five types of tree leaves of the research object have a various spectral radiation value. The lowest spectral radiation value is tamarind and the highest is ylang on blue and red wave. Low classification of spectral radiation value is tamarind leaves, sage, tamarind, and cotton. Moderate spectral radiation value is peacock flower, khaya, small leaf mahogany, rambutan, teak, flamboyant, jackfruit, capes, big leaf mahogany, banyan, handkerchiefs, ketapan, sunshade umbrella, sapodilla kecik, Acacia auriculiformis, matoa, and flower. High spectral radiation value is soursop, Angsana, Acacia mangium, and ylang-ylang. Spectral radiation value is inversely proportional to the ability of plants to absorb CO_2. So the type of leaves that were classified as having high ability to absorb CO_2 is tamarind, saga, tamarind, and cotton while the type of tree leaves were classified as having low ability to absorb CO_2 is soursop, Angsana, Acacia mangium, and ylang, the rest classified as having moderate capability.

References

[1] Chou M D 1990 Parameterization for the absorbtion of solar radiation by O_2 and CO_2 with application to climate studies *J of Climate* **3** 209-217
[2] Henning R 1990 A Comparison of the contribution of various gases to the greenhouse effect *J of Science* **248** 1217-1219
[3] Dahlan E N 2008 Jumlah emisi gas CO_2 dan pemilihan jenis tanaman berdaya rosot sangat tinggi: studi kasus di Kota Bogor *J Media Konservasi* **13** 85-89
[4] Dahlan E N 2007 Analisis kebutuhan luasan hutan kota sebagai *sink* CO_2 antropogenik dari bahan bakar minyak dan gas di Kota Bogor dengan pendekatan sistem dinamik [Dissertation] (Bogor (ID): Bogor Agricultural University)
[5] Indriyana N D 2014 Analisis biomassa dan kandungan karbon pada jenis serasah daun tanaman hutan kota di arboretum arsitektur lanskap IPB Bogor [Undergraduate Thesis] (Bogor (ID): Bogor Agricultural University)
[6] Triono S 2004 Potensi penyerapan karbondioksida pada akasia (*Acasia crassicarpa*) dan gmelina (*Gmelina Arborea* Linn.) berdasarkan model pertumbuhan logistik dan kurva respon cahaya [Undergraduate Thesis] (Bogor (ID): Bogor Agricultural University)
[7] Malayanda Y 2007 Kajian daya rosot karbondioksida oleh jenis tanaman hutan kota di hutan penelitian darmaga [Undergraduate Thesis] (Bogor (ID): Bogor Agricultural University)
[8] Sinambela T S P 2006 Kemampuan serapan karbon dioksida 5 (lima) jenis tanaman hutan kota [Undergraduate Thesis] (Bogor (ID): Bogor Agricultural University)
[9] Wungkar M G 2005 Evaluasi aspek fungsi dan kualitas estetika arsitektural pohon lanskap jalan kota bogor [Thesis] (Bogor (ID) : Bogor Agricultural University)
[10] Handoko P and Fajariyanti Y 2013 Pengaruh spektrum cahaya tampak terhadap laju fotosintesis tanaman air *Hydrilla verticillata. Seminar Nasional X Pendidikan Biologi FKIP UNS*. Kediri (ID): Universitas Negeri Sebelas Maret. 1-9
[11] Campbell N A, Reece J B, Urry L A, Cain M L, Wasserman S A, Winickoff B and Jackson R B 2010 *Biologi*. Ed ke-8, Jilid 1. Hardani W, editor. Wulandari DT, penerjemah. Jakarta (ID): Penerbit Erlangga. Terjemahan dari: Biology
[12] [CRISP] Centre for Remote Imaging Sensing and Processing. 2001. Optical remote sensing [downloaded on 2016 March 26 at http://www.crisp.nus.edu.sg/~research/tutorial/optical.htm]
[13] Pinker R T and Laszlo I 1992 Global distribution of photosynthetically active radiation as observed from satelites *J of Climate* **5** 56-65
[14] [BATAN]. Badan Tenaga Nuklir Nasional. 2016. Radiasi dan dunia yang kita huni [downloaded 2016 March 25 at http://www.batan.go.id/pusdiklat/elearning/proteksiradiasi/pengenalan_radiasi/1-1.htm]
[15] Nyamsi W W, Espinar B, Blanc P and Wald L 2015 Estimating the photosynthetically active radiation under clear skies by means of a new approach *Adv Sci. Res* **12** 5-10
[16] Qin J, Yang K, Liang S and Tang W 2011 Estimation of daily mean photosynthetically active radiaton under all-sky conditions based on relative sunshine data. *J of Applied Meteorology and Climatology* **51** 150-160
[17] Pitman J I 2000 Absorption of photosynthetically active radiation, radiation use effeciency and spectral reflectance of bracken [*Pteridium aquilinum* (L) Kuhn] canopies. *J of Annals of Botany*. **85** 101-111
[18] Sitompul S M and Bambang G 1995 *Analisis Pertumbuhan Tanaman* (Yogyakarta (ID): Gajah Mada University Press)

[19] Salisbury F R and Ross C W 1992 Fisiologi Tumbuhan.Jilid ke-3. Terjemahan Diah RL (Bandung (ID): ITB Press)

[20] Pinker R T and Laszlo I 1992 Global distribution of photosynthetically active radiation as observed from satelites. *J of Climate* **5** 56-65

[21] Berry, Raison O L, Nobel P S, Osmond C B and Zieger H 1981 *Physiological Plant Ecology I.* (New York (US): Springer-Verlag Berlin Heidelberg)

Wildlife species composition in various forest types on Sebuku Island, South Kalimantan

C Kusmana, A Manshur, O Rusdian, H R Putro, F Hakim and M Ermyanyla
Center for Regonal Development and Planing Study, Bogor Agricultural University

E-mail: ckmangrove@gmail.com

Abstract. Sebuku is one of the small islands in South Kalimantan Provincehaving various forest types with high potential economic in mining sector. Based on it's business permit, the island has been divided up into several mining concessions. So that biological diversity studies in this island is an interesting in order to serve biological baseline data if someday this island to be extractedfor mining. This research was conducted on 28th November to 5th December 2015 aims to explore wildlife species inhabit mangrove forest, beach forest, and lowland forest usinga rectangle transect (40 x 1000 meter) in each forest type. The results show there are 90 wildlife species identified in Sebuku Island.The beach forest has the highest wildlife species richness (36 species), while the area having the highest protected wildlife species isthe lowland forest. Mangrove forests generally have a lower wildlife species richness. Nevertheless, in Sebuku Island, can be found mangrove forest that have a quite high wildlife species richness (28 species, 50% protected). It is due to silt sedimentation in the estuary area, so that this area become feeding ground for shore and migratory birds.

1. Introduction

Sebuku island is one of the small islands in South Kalimantan, having the area estimated at 245.5 km2. Geographically, the island located at between 4°09'27" – 4°05'35" North Latitude and 117°031'39" – 117°35'55" East Longitude. This island covers various vegetated land cover types which has important role for wildlife habitats. Furthermore, the island has abundance of coal and minerals. Based on it's business permit, the island has been divided up into several mining concessions.

A high potential of coal and minerals in Sebuku Island led to mineral exploration and mining activity. Mineral exploration and mining activity are contributed to increasing of foreign exchange of the country. However, the mining activities provide impact damage to wildlife habitat.

Continuously and uncontrolled mining activity could result in loss of wildlife habitat, that could cause further impact on the wildlife extinction. Hence the mineral potential of the abundant coal and minerals can be a threat to the existence of habitat and wildlife in Sebuku Island. So that biological diversity studies in this island is an interesting in order to serve biological baseline data if someday this island to be extracted for mining. The research was carried out to explore wildlife species inhabit three forest types (mangrove forest, beach forest, and lowland forest).

2. Methods

This research was conducted on 28th November to 5th December 2015 in three forest types (mangrove forest, beach forest, and lowland forest). In order to explore wildlife species, a rectangle transect (40 x 1000 meter) was constructed in each forest types.

3. Results

3.1 Various types of ecosystems on Sebuku Island

Based on Sebuku Island beach geomorphology, there are 3 Sebuku Island typologies: (1) mangrove area in the West and North of the island, (2) the coastal and the coral areain the East of the island, and (3) coral-mangrove area in the southern part of the island. If it is based on the existence of wildlife, the important ecosystems on the Sebuku island is divided into 6 ecosystem types (Figure 1). Thereare:

1. Primary mangrove ecosystem with mud substrate, this ecosystem is represented by a plot in the nature reserve of the Straits Sebuku precisely bordering SP 23B (village Serakaman). This Habitat is dominated by mangrove flora with high diameter and high canopyclosure. However, this region was once polluted by acid mine and in the central part of the region is used to be an embankment. The study also pointed out the existence of illegal logging activities in the area.
2. Primary mangrove ecosystem with coral substrate, this ecosystem is represented by a plot in Tanjung Karang. Habitat in this area is dominated by mangrove flora is a fairly tight on the edge but open on the middle section hence the canopy closure stillclose to each other. This area was polluted by waste oil that resulted in death of some plants in the central part. This occured because the observation location is near the coal transportatin port. In addition the presence of roads cutting mangrove forests lead to lack of tidal sea water reaching the middle part due to relying wall and embankment.
3. Secondarymangrove ecosystems represented by Selat Sebuku nature reserve in particular on the island of Sekapung, the village of Keluang and Sejakah Villages. Generally it is the area of mangrove forests on the outer side of the secondary especially on the north side of the island of Keluang. The presence of wood in order to capture the making of chart by people thought to be the cause of this ecosystem is degraded. The mangrove forest in the village of Sejakah is still experiencing growth characterised by a large number of still permudaan. The main thing that is causing the expansion of the vast area of mangroves in this area is the result of silting and sedimentation jedidah River which is a good media for mangrove plants. Unlike the acreage of Mangrove Forests in Sekapung where this area is located close to the settlement. Basically this area blends in with the acreage of mangrove forests in the Cape of the head, nevertheless in some locations in the forest can be found the former embankment by the community even though it is currently not in use, however the area still leftopen.
4. Coastal forest ecosystems are represented by ecosystems in eastern of SebukuIsland. Characteristic of this ecosystem is the ecotone zone that connected beach forest with secondary forest ecosystems. This area is stil in a good condition due to further distance community and mining activity. Presence of Clear water (on the Hill of Saung) and mystical area is one of the reasons why the area was protected by local community aside fromBeach forest role as coatal protectionyet on a beach part ,there is a sea sand mining used for the construction of the settlement by communities.
5. Lowland forest ecosystems generally merges with forest beach. This ecosystem is represented by the Bukit Saung. The area around West Bukit Saung was a HPH (logging) area. This area has also been occupiedby the community into a rubber plantations (Hevea sp.) without felling the tree. Currently, the rubber plantation is not well-managed, the community tend to exploit the land around their settlement becauseof the remote location and decreasing of rubber price.
6. Another ecosystem that is generally artificial ecosystems such as the plantation with a homogeneous rubber plant, agricultural fields and pastures with low diversity of fauna. In this region only viewed the portrait area but habitat analisys and Wildlife observation are not acomplished

3.2 Wildlife on Sebuku Island

In total 90 wildlife species belong to 40 families has been identified in Sebuku Island area, consist of 14 species of mammals (10 protected species), 67 species of birds (protected species) and 9 species of Reptiles (3 protected species). Therefore, it can be concluded that almost half of the species is a protected species (44 kind). The number of species found in each observation plot are presented in Table 1 while list of identified species in each plot presented in Table 2.

Table1. Number of wildlife species found in different plots in Sebuku Island

No	Location	Number of species			Protected			Total	
		Mammal	Birds	Reptile	Mammal	Birds	Reptile	Species	Protected
1	Lowland Forest Bukit Saung	9	19	2	6	10	0	30	16
2	Beach Forest Eastern part	5	27	4	3	10	1	36	14
3	Mangrove Pulau Keluang	0	8	2	0	3	0	10	3
4	Mangrove Sejakah	1	28	3	0	14	1	32	15
5	Mangrove SP. 23B	2	4	3	1	3	1	9	5
6	Mangrove Tanjung Kepala	2	14	1	1	3	1	17	5
7	Mangrove Desa Sekapung	1	14	0	0	5	0	15	5

Result shown that in Sebuku island area there was a suitable area for birds, reptile and some endanger and protected mammals. In some area of the island can be encountered forest area with high tree canopy for mammals and partly provide good nesting sites or foraging for birds. Based on the data on Table can be inferred that the birds group is the most widely observed while the reptile group is least identified. However, mammals are the most sensitive to changes in the environment and this indicates that these species have specific habitat requirements, unlike the reptile which tend to be unobserved because this group generally nocturnal.

In Sebuku area can be found mangrove and sediment, whose existence is very important for some of fish-eating birds species (Figure 1), such as Alcedinidae (Cekakak and Raja Udang), beach birds and Sea birds such as species includes in family Ardeidae, Anatidae, Burhinidae, Ciconiidae dan Scolopacidae. The existence of sediment are believed as effective natural traps because of the tides.

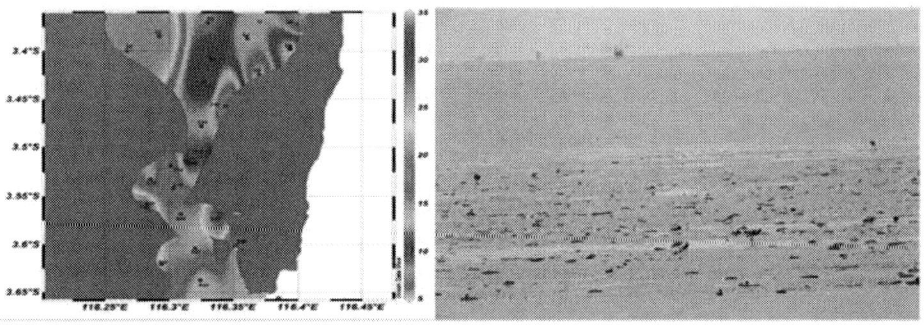

Figure 1. Distribution of the total surface of suspended solids in the Selat Sebuku (mg/l) and sample of feeding ground locations for Birds in Muara Sejakah

Field observasion result shown that Beach Forest area in East coast of the island and the Mangrove Forests Sebuku Sejakah has a high diversity than other observation plot. These condition rise to allegations that beach forest and mangroves are more stable than other observation plot. This is proved by the number of species that are able to adapt on the conditions and has interactions in the ecosystem.

3.3 Key Habitat in Sebuku Island

The "habitat "term is an axiomatic start from environment scope that its resources are really utilized by wildlife until the wide scale as a vegetation form in a landscape [1]. Tews et al. [2] defined the habitat as a division of a location that is generally based on the structure of vegetation formations. Different with [3] which defines the habitat as the all resources and environmental conditions at a location that is able to maintain the survival and reproduction of organism. The more obvious habitat concept put forward by [4] by defining the habitat at 3 scales: (1) the location of organism live, (2) physical and chemical aspects of environment, and (3) the concept of community in habitats.

Habitat was the basis of a functional ecosystem, therefore this information is important in the wildlife utilization directly [5] as well as arranging spatial plan area [6].[7] States that the successes of wildlife conservation efforts depend on collecting, analyzing, and interpreting reliable information about the wildlife habitat.

Assessment a habitat become important to be protected or not, generally use the biodiversity approach. The measurement of biodiversity is the simplest way to count the amount of species [8, 9]. Understanding on levels of species was the basis in understanding biodiversity due to the level of species, almost universally used as units where biological diversity is measured [10].

The diversity of wildlife can be used to estimate biology quality such as environmental and ecological pollution of an ecosystem through the community structure (1984) and its monitoring. However that biodiversity only can describe community structure and ecological quality of an ecosystem so that only simplify and describe the real biodiversity. One of the important aspects in the study of biodiversity is that not all species found are the same. The species vary in amount, form, abundance, distribution, trophic position, ecology function, and food selection. Therefore, similarity index of community that incorporates species in it is more suitable to use in the determination of the key habitats in Sebuku Island.

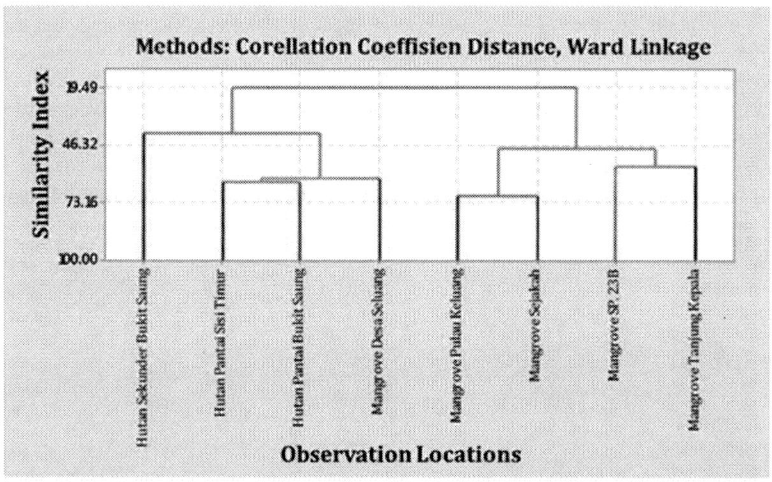

Figure 4. Similarity index of community at various locations on the Sebuku Island

Based on the level of community similarity showed that the each various observation location on the Sebuku Island have separate community characteristic and different with each other. The analysis result showed that the largest similarity index of community is only 70.04% namely Keluang Mangrove and Sejakah Mangrove Islands (Figure 2). This indicates that the entire ecosystem in the Sebuku island is a unique ecosystem and important so needs to be managed and maintained to keep it together. In this analysis can also note that factor of location proximity is at least gives implications on equal species though is not significant. It also showed that several species especially bird species using more than one type of location as Habitat.

Shorebirds and scaffolding birds use Keluang Mangrove as habitat for sleep while in the morning until in the afternoon these species foraging around the Mangrove Sejakah. The existence of embankment and fish trapped in the sediment makes it easy to find these species feed.

3.4 Potential Threats of Wildlife

Based on the definition of habitat on the previous section, each wildlife animal has its own characteristics; although perhaps on some species have similarities. Therefore, knowledge of the selected habitat characteristics can provide fundamental information about behavioral change as well as patterns of adaptation its physiology [11]. Wildlife needs in an environment can generally be divided into three, namely [12]:

1. Food and beverage to enhance nutrition/energy and
2. Shelter and protection to avoid themselves from predators and weather influences such as heat and rain
3. Space to get food, water and attract mates.

Several wildlife species in Sebuku Island is inseparable from the threat of its existence in nature. In the theory of biogeography, biodiversity size is equal with its wide island. In addition, over exploitation in the island caused the size of habitat become narrower due to conflict of utilizing interest in the small island. This cause wildlife habitat only concentrated on a very narrow area. This will cause the high intensity of threats to biodiversity, both natural and anthropogenic influences. Naturally the extinction can be caused by competition between individual and species in fighting over resources. There are various threats to the wildlife in the Sebuku Island, i.e.:

3.4.1 Coverage changes and Land Use

Land clearing activities undertaken by communities, usually use conventional pattern, where land burning is the most realistic and affordable because it is cheap and easily applied. This condition indicates that land clearing indirectly shows burning land activities. Many researchers believe that food quality is a limiting factor for many herbivores. Several species of primates give responses to the habitat changing due to forest fire, they found in large amount on the forest floor and choose certain food in primary forest or consume immature leaves and insect in large amount. This is due to losing of high tree as Enggang nest and a few fruit trees as food sources of Enggang.

The forest land clearing for development interests to increasing human living standard has led to wildlife population that used to be in its habitat becomes fragmented to find and occupy the remaining habitat. The remaining Habitat is typically relatively small area that is not supported by feed conditions. Based on the above conditions then it can be inferred that the existence of socio-economic dynamics of the society in General, will have an impact for wildlife, include:

a. The decline in biodiversity; Deforestation and forest degradation into gardens and fields will cause decreasing of the total of wildlife biodiversity. This is caused by two things, namely: (1) the inability of wildlife due to habitat changes became extinct in local and (2) natural habitat damage will force most some wildlife will migrate to the outside of its habitat to look for a better habitat.
b. Pest increasing; environment changes that are likely to lead to environmental damage due to environmental exploit that exceed environmental carrying capacity led to an imbalance of the ecosystem and resulted in a conflict of wildlife. Conflicts give rise to lack of human appreciation on the wildlife. Human activities in forest cause wildlife habitat be narrow and forcing wildlife to find new space to live and resulted in conflicts between the community and wildlife. The most frequent species of wildlife damage the fields and plant community is the long-tailed Macaca (*Macaca fascicularis*)and boar (*Sus scrofa*). Long-tailed Macaca (*Macaca fascicularis*) and boar included mammals species that easily adapt to its environment. Along with getting the narrowness of forest area, the role of this species is no longer as ecosystem stabilizers but as farmers enemy and can be source of pest in rice field and plantations. In communities field, both of them are damaging to the various species of commodities that became the object of economic society, including: corn (*Zea mays*), banana (*Musa paradisiaca*), cassava (*Manihot esculenta*), long beans (*Vigna sinensis*), palm oil (*Elaeis guinnensis*) and rubber (*Hevea braziliensis*). In addition, farm fields is also often used as a place for wildlife to play around that damage other crops despite the plants is not taken asfood.
c. Reducing wildlife populations; Forest fragmentation causes the size of the habitat and resources in it became limited. The limited food resources led to the low population density of animals especially

herbivores (First Consumers) who later also determine the existence of large carnivores (Second Consumers).In addition changes into the gardens and fields, the road network that was constructed to the gardens and fields also led to fragmentation on forest landscape. High levels of habitat fragmentation causes wildlife habitat become fragmented. This condition then makes the animals can't interact (social or reproductive) because each of them is isolated from other habitats. On the long term, at least showed decreasing in the wildlife population. The pattern of closed reproduction resulted in increasing inbreeding chances that result in decreasing in genetic variation and in some cases even gave rise to genetic disorders. While the limited amount of food will result in the competition process to obtain access to resources between individuals become very close. This means animals will often fight to gain access to resources compared to invest the rest of its energy for growth or breeding.

The vegetation is the main variable in quality habitat analysis and environment changes. McKinney M L [13] warns that the loss of habitat is the main cause of the alleged loss of biodiversity. As for the negative effects of habitat fragmentation include:

a. Abiotic component changes on fragmented habitats. This will result in mismatch environment for native wildlife. Further the fragmentation also results in modifications and changes to surrounding environment topography, increasing the noise intensity (degree of decibels increase), changes the dynamics and the water cycle as well as air [14].
b. The existence of a linear fragmentation (road construction) will have direct negative impacts to wildlife populations due to the increasing mortality rate for wildlife (accident), exotic species introduction (alien species) that will ultimately change the natural behavior fauna
c. The function of ecosystems may be disrupted due to habitat fragmentation due to the reduced capacity of the components in the ecosystem to sustain ecological processes that occur naturally, raises barriers to gene flow and the grouping of animals in a habitat.

The fragmentation in the short term will basically provide benefits for wildlife and plants or where there is a succession process and produce a fast crop especially for herbivores species. However, in the long term this condition will result in competition between wildlife and the individuals as well as increasing the chance of predation (natural as well as human). Based on a list of fauna species that can be found in plot observations on Sebuku Island, then adaptability of wildlife on cover changes and land used in Sebuku Island are as follows (Table 2).

Tabel 2. Wildlife Adaptability on Land Use

No	Wildlife Species	Land uses and changes						
		F	G	E	M	A	S	Sh
1	*Sus scrofa*	*****	****	*	**	****	***	***
2	*Callosciurus notatus*	*****	****	*	*	****	****	*****
3	*Lariscus insignis*	*****	***	*	*	*	*	**
4	*Nasalis larvatus*	*****	*	*	*	*	*	*
5	*Ratufa affinis*	*****	**	*	*	*	*	**
6	*Muntiacus muntjak*	*****	*	*	*	*	*	****
7	*Galeopterus variegates*	*****	**	*	*	*	*	**
8	*Nycticebus coucang*	*****	**	*	*	*	*	*
9	*Macaca fascicularis*	*****	****	*	***	***	***	****
10	*Tragulus kanchil*	*****	**	*	*	**	*	****
11	*Rusa unicolor*	*****	***	*	*	**	*	****
12	*Tarsius tarsier*	*****	**	*	*	*	*	*
13	*Tupaia glis*	*****	**	*	*	**	*	***
14	*Tupaia dorsalis*	*****	*	*	*	*	*	**

IOP Conf. Series: Earth and Environmental Science 54 (2017) 012068 doi:10.1088/1755-1315/54/1/012068

15	*Leptoptilos javanicus*	*****	*	****	*	*	*	***
16	*Psittacula alexandri*	*****	***	*	*	**	***	*
17	*Psittacula longicauda*	*****	****	*	*	**	*	*
18	*Gallinago sp.*	*****	*	**	*	*	*	**
19	*Ardeola speciosa*	*****	*	***	*	*	*	**
20	*Lonchura fuscans*	*****	****	***	****	****	****	*****
21	*Centropus bengalensis*	*****	****	*	*	**	*	****
22	*Passer montanus*	***	****	**	****	****	****	****
23	*Nectarinia calcostetha*	*****	****	*	*	**	***	***
24	*Nectarinia sperata*	*****	****	*	*	**	***	***
25	*Anthreptes malacensis*	*****	****	*	*	**	***	**
26	*Nectarinia jugularis*	*****	****	*	*	**	***	**
27	*Dicaeum trigonostigma*	*****	****	*	*	**	***	**
28	*Dicaeum ignipectus*	*****	****	*	*	**	***	**
29	*Hemicircus concretus*	*****	****	*	*	*	**	***
30	*Picoides moluccensis*	*****	****	*	*	*	**	***
31	*Ardea sumatrana*	*****	*	****	*	*	*	**
32	*Ardea purpurea*	*****	*	****	*	*	*	**
33	*Todirhamphus chloris*	*****	***	****	*	**	**	****
34	*Chloropsis sonnerati*	*****	****	*	*	**	***	*
35	*Chloropsis cyanopogon*	*****	****	*	*	**	***	*
36	*Orthotomus ruficeps*	*****	***	*	*	*	*	**
37	*Onychoprion fuscatus*	*****	*	***	*	*	*	*
38	*Sterna sumatrana*	*****	*	***	*	*	*	*
39	*Haliastur indus*	*****	****	*	**	***	***	***
40	*Icthyophaga ichthyaetus*	*****	*	***	*	*	**	**
41	*Haliaeetus leucogaster*	*****	*	***	*	*	**	**
42	*Numenius arquata*	*****	*	****	*	*	*	**
43	*Numenius phaeopus*	*****	*	****	*	*	*	*
44	*Lalage nigra*	*****	***	*	*	*	*	*
45	*Apus affinis*	***	****	**	*	***	*****	**
46	*Amaurornis phoenicurus*	*****	****	**	**	***	**	***
47	*Merops philippinus*	*****	*	***	*	*	*	*
48	*Butorides striatus*	*****	*	***	*	*	*	*
49	*Nycticorax caledonicus*	*****	*	***	*	*	*	*
50	*Copsychus saularis*	*****	***	*	*	*	*	**
51	*Egretta alba*	*****	*	****	*	*	*	**
52	*Egretta sacra*	*****	*	****	*	*	*	***
53	*Egretta garzetta*	*****	*	****	*	*	*	***
54	*Egretta intermedia*	*****	*	****	*	*	*	***
55	*Hirundo tahitica*	*****	****	***	*	***	***	***
56	*Pycnonotus goiavier*	*****	***	*	*	**	**	***
57	*Pelargopsis capensis*	*****	***	*	*	*	*	**
58	*Dryocopus javensis*	*****	**	*	*	*	*	**
59	*Ducula pickeringii*	*****	*	**	*	*	*	**

60	*Geopelia striata*	*****	***	**	**	**	***	**
61	*Aplonis panayensis*	*****	***	**	**	**	***	***
62	*Treron fulvicollis*	*****	*	**	*	*	*	**
63	*Treron vernans*	*****	***	**	*	*	***	**
64	*Ceyx erithaca*	*****	***	*	*	**	**	*
65	*Alcedo atthis*	*****	***	*	*	**	**	*
66	*Alcedo euryzona*	*****	*	*	*	*	*	*
67	*Alcedo meninting*	*****	***	*	*	**	**	*
68	*Buceros rhinoceros*	*****	**	*	*	*	*	*
69	*Loriculus galgulus*	*****	***	*	*	**	**	***
70	*Cyanoptila cyanomelana*	*****	***	*	*	**	**	*
71	*Cyornis turcosus*	*****	***	*	*	**	**	*
72	*Pernis ptilorhynchus*	*****	***	*	**	***	***	*
73	*Gracula religiosa*	*****	****	*	*	**	**	****
74	*Actitis hypoleucos*	****	*	***	*	*	**	**
75	*Tringa erythropus*	****	*	***	*	*	**	**
76	*Collocalia linchi*	***	*****	**	****	*****	*****	****
77	*Collocalia esculenta*	***	*****	**	****	*****	*****	***
78	*Aerodramus maximus*	***	*	***	*	*	****	**
79	*Esacus magnirostris*	*****	*	***	*	*	*	*
80	*Cacomantis merulinus*	****	***	*	*	**	**	**
81	*Cacomantis variolosus*	*****	***	*	*	**	**	**
82	*Varanus salvator*	*****	***	**	**	**	**	***
83	*Crocodylus porosus*	*****	*	*	*	*	*	*
84	*Eutropis multifasciata*	*****	*****	*	*****	*****	*****	*****
85	*Eutropis rudis*	*****	*	****	*	*	*	*****
86	*Orlitia borneensis*	*****	*	**	*	*	*	**
87	*Gecko gecko*	***	***	*	*	*	****	*
88	*Boiga wallachi*	*****	***	**	*	*	*	**
89	*Ahaetulla prasina*	*****	****	*	*	***	***	***
90	*Cerberus rynchops*	**	*	*****	*	*	*	*

Source : www.iucnredlist.org; www.animaldiversity.org; www.birdlife.org; www.reptile-database.org (processed).

Description: F = forest, G = gardens, E = embankment, M = mining A = agriculture, S = settlement, Sh = shrubs, * = not adaptable Animals; ** = low adaptability wildlife; = The ability of animals adaptation is quite good; = Adaptable Animals; = Highly adaptive Animals.

Based on the tables above, it can be noted that adaptability of wildlife to land use and cover changes are different from each other. In the Mammalia class, only boar (*Sus scrofa*), Palm squirrel (*Callosciurus notatus*) and long-tailed Macaca (*Macaca fascicularis*) that havehigh adaptability towards land use changes in Sebuku Island. Other mammals have low adaptability to land use changes, except on land use for plantation that have quite high density such as plantations and agroforestry. For example Bekantan (*Nasalis larvatus*) requires high density and high conopy coverage as well as sustainability as its habitat in order to make this species can move freely because it is arboreal species and also keep body temperature more stable. In addition, in the area should also contain a high diversity of plants because of high plant diversity is a representation of amount of insects in it, which is the main feed of insects for Bekantan.

Unlike the bird class that has the ability of adaptation which tends to be random depends on a way of life. The species of scaffolding and shorebirds tend to be in the same group of mammals in general, i.e. has low adaptability to environment changes. Species of land mammals tend to be more adaptive on environmental conditions despite a low capability. On the class of reptiles, only species of Kadal Kebun (*Eutropis multifasciata*) and Ular Pucuk (*Ahaetulla prasina*) that have a high power of adaptation, while on the other species have specific habitat. This shows that adaptability is very low even several species do not tolerate land use and changes.

3.4.2 Illegal Logging

One of the activities that can decrease biodiversity of wildlife is illegal logging (Figure 8). Ilegal logging that had been identified in the natural reserve area which is directly adjacent to the PT. BCS in the west side on and on Keluang Island. The need of community in wood mainly to the needs of wood as the material for home and constructions will reduce the existence of some tree species which are natural feed for wildlife. If this condition occurs massively, then the species of natural feed will be reduced in high amount and the animals will suffer hunger and death. In addition, some types of trees which have large diameter, high height and high density of canopy which can provide the hive for animals, especially primates, and shorebirds. This condition will decrease the chances of wildlife to reproduce because of habitat disturbance.

3.4.3 Illegal Hunting

Direct threat to the diversity of wildlife on Sebuku Island is illegal hunting. Illegal hunting eliminates individu and species from its natural habitat. Basically, there are several categories of illegal hunting in Sebuku Island: (1) The hunt for hobbies (2) the hunt for reducing the population of animals that are considered as pests, (3) the hunt for economic motives. Illegal hunting for the hobby generally done by the community with motives. To fulfill the desire of hunting or poaching. Generally the species that are hunted are deer (Indian muntjac), various species of squirrels (Tupaiidae) and (Sciuridae) and also various species of birds that have various types of voice, shape and color. The short term hunting will not have a massive impact on biodiversity, however in the long term, hunting will decrease the wildlife diversity.

Different with hunting that aimed to reduce animal population. Basically, the purpose of this type of hunting is to keep wild animals for not disturbing the cultivated area. In this context, based on the legal aspects, hunting is permitted. Ecological functions of a species will appear when the animal's prey or predators increased the population to destructive and detrimental to humans. However, historical records show that the methods used are sometimes not in accordance with the rules of environmental preservation. It identified have occurred on the Laut Timur Island. When the migration season of birds, many migratory birds (shore bird), shorebirds and scaffolding are looking for fish in the community's embankment. This caused some embankment deliberately added with toxic (heavy metals) in high concentrations so that the bird population can be decreased.

Hunting for the economic motive is usually done by community who did wildlife trading. Some records show that the species of potential warbler can be traded. Moreover, Pangolin (Manis javanica), had become the animal target hunted in the area of Pulau Laut island because of its high economic value. This hunting model is certainly very dangerous, given the number of hunted and hunter motivation is proportional to the economic value of the hunted animal target.

4. Conclusion

Sebuku Island is covered by three forest types, mainly mangrove forest, beach forest, and lowland forest in which 90 wildlife species live in. There are 9 – 32 species (3 – 15 protected) in mangrove forest, 36 species (14 protected) in beach forest, and 30 species (16 protected) in lowland forest.In all forest type, dominant wildlife groupisbirds.Furthermore various observation location on the Sebuku Island have separate community characteristic and different with each other.

References

[1] Tagliapetra, Sigovini 2010 Benthic fauna: collection and identification of macro benthic invertebrates *Terre et Environment* **88** p 253-261
[2] Tews et al. 2004 Animal species diversity driven by habitat heterogeneity/diversity: The importance of keystone structures *Journal of Biogeography* **31** p 79–92
[3] Johnson M D 2007 Measuring habitat quality: a review *Condor* **109** p 489–504
[4] Mitchell S C 2005 How useful is the concept of habitat? *Oikos* **110** 3 p 634-638
[5] Morris D 2003 Toward an Ecological Synthesis: A Case for Habitat Selection *Oecologia* **136** p 1-13
[6] Hostetler M E, Drake D 2009 Conservation subdivisions: a wildlife perspective *Landscape and Urban Planning* **90** 3-4 p 95-101.
[7] Morrison et al. 2006 Population estimates of North American shorebirds *Wader Study Group Bull* **111** p 67-85
[8] Poole R W 1974 *An Introductuon To Quantitative Ecology* (New York: McGraw-Hill)
[9] Krebs 1978 *Ecology: The Experimental Analysis of Distribution and Abundance* (New York: Harper and Row Distribution)
[10] [WCMC] World Conservation Monitoring Centre *1992 Global Biodiversity: status of the Earth's living resources* (London: WCMC)
[11] Horne J S, Garton EO, Rachlow J L 2008 A synoptic model of animal space use: simultaneous estimation of home range, habitat selection, and inter/intra-specific relationships. *Ecol. Modell.* **214** p 338–348
[12] Hall L S, Krausman P R, Morrison M L 1997 The habitat concept and a plea for standard terminology *Wildlife Soc. Bull.* **25** p173–182
[13] McKinney M L 2006 Urbanization as a major cause of biotic homogenization *Biological Conservation* **127** 3 p 247-260
[14] Coffin A W 2007 From roadkill to road ecology: A review of the ecologicaleffects of roads *Journal of Transport Geography* **15** p 396-406

Spatial distribution and biophysics chemistry characterization of Pearl Oyster farming in Semau Strait, East Nusa Tenggara

Enggar Yulia Wardani[1*], Yusli Wardiatno[2] and Syamsul Bahri Agus[3]
[1]Student of Coastal and Marine Resources Management Study Program, Department of Aquatic Resources Management, Bogor Agricultural University, Indonesia
[2]Lecturer of Aquatic Resources Management Department, Bogor Agricultural University, Indonesia
[3]Lecturer of Marine Science and Technology Department, Bogor Agricultural University, Indonesia

E-mail: enggaryulia20@gmail.com

Abstract. Pearl oyster is one of marine organisms which has high of economic value. One of Indonesia region that potential for pearl oyster farming activity is Semau Strait-East Nusa Tenggara. The purpose of this study is mapping and characterize spatial distribution based on biophysics chemistry condition of pearl oyster farming. Mapping using Geographic Information System (GIS) tools, while grouping and characterize using hierarchical clustering and Mann-Whitney method. Grouping based on character similarity of biophysics chemistry divided the stations distribution into three groups: Group I is excellent group, Group II is good group and Group III is bad group for pearl oyster farming. Group I is excellent area because it has the highest values of dissolved oxygen, chlorophyll-a, and phytoplankton abundance. Group III is bad area because it temperature is high relatively.

1. Introduction

Pearl oyster is one of fisheries resources which has high economic value in Indonesia. The harvest of oyster can be used as jewellery or cosmetic ingredients. Sustainable utilization of pearl oyster resource is very needed, because of pearl market demands are increase. However, the oyster in nature getting limited. Marine culture is one of way to preserve the nature conditions [1]. Technology of pearl oyster farming to produce pearl can take oyster with or within nucleus [2]. An oyster has a double function, they are as the mantle tissue and recipients of mantle donor [3]. A region of Indonesia that potential for pearl oysters farming is Semau Strait, which is that area connecting from Kupang-East Nusa Tenggara to Semau Island.

The condition of biophysics chemistry such as temperature, salinity, pH, dissolved oxygen, ammonia, nitrate, orthophosphate, silica, chlorophyll-a, and phytoplankton abundance are the important parameters to determine of water quality condition. Information about distribution and characterization of biophysics chemistry is needed for pearl oysters farming, thus to know the dynamics of aquatic environments and monitoring in the pearl oyster farming areas.

Distribution of water quality condition can be interpreted by mapping. Mapping is a communicative way to provide information related biophysics chemistry quality in the water. Mapping use Geographic Information System. Geographic Information Systems (GIS) is tools used to collect, store, manipulate, analyse, and display of spatial information in map. GIS is a program that converts data become an attractive information to communicate and support to decision-making [4].

GIS can be used to transform the information into maps through analysis tools. GIS can present the maps as like as display pictures [5]. GIS also can inform character of the resource, identify the potential, and analysis of area suitability for its purpose [6]. One of analysis tools in GIS is spatial analysis [7]. Spatial analysis are working through data interpolation principle and linked the score in each station [8], so the score in the field sample which cannot be taken can be predicted by interpolation in GIS. The purpose of this study are mapping and characterize the spatial distribution based on biophysics chemistry condition in the pearl oyster farming, Semau Strait-East Nusa Tenggara.

2. Methods

2.1 Time and Location

The research consist are observations and sampling in 20 points of pearl oyster farming, Semau Strait - East Nusa Tenggara (figure 1). Field research conducted in December 2015.

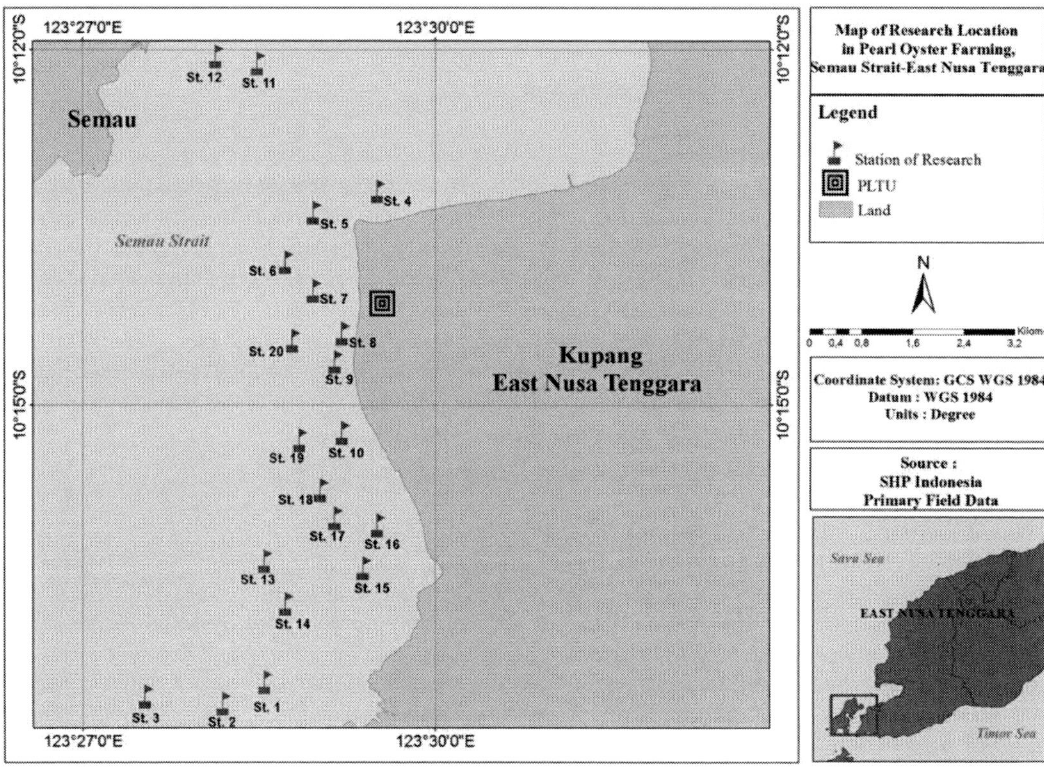

Figure 1. Research Location

Parameter analysis and data processing in Aquatic Productivity and Environment Laboratory, Micro Biology Laboratory and Mapping and Geospatial Modelling Laboratory, FPIK-IPB. The activity was conducted in January-March 2016.

2.2 Data Collecting

Total research points are 20, horizontally on the surface of the sea. The primary data consists are physics, chemistry, and biology parameters. Physic parameter is temperature. Chemistry parameters are salinity, pH, dissolved oxygen, nitrate, orthophosphate, ammonia and silica. Biology parameters are chlorophyll-a and phytoplankton. Research are in situ observed at the field and analysis laboratory. Biophysics chemistry parameters analysis refer to standard method of [9] that shown below on table 1.

Table 1. Collecting biophysics chemistry parameters data

Parameters	Units	Tools	More information
Physic			
Temperature	°C	Thermometer	*in situ*
Chemistry			
Salinity	‰	Refractometer	*in situ*
pH		pH paper	*in situ*
Dissolved oxygen	mg/L	DO meter	*in situ*
Nitrate	mg/L	Spectrophotometer	Laboratory
Silica	mg/L	Spectrophotometer	Laboratory
Orthophosphate	mg/L	Spectrophotometer	Laboratory
Ammonia	mg/L	Spectrophotometer	Laboratory
Biology			
Chlorophyll-a	mg/m^3	Spectrophotometer	Laboratory
Phytoplankton	$cell/m^3$	SRC (*Sedgwick Rafter Chamber*)	Laboratory

2.3 Data Analysis

Data were collected analysed by spatial and statistical analysis. Spatial analysis using Geographic Information Systems (GIS) which spline with barrier interpolation method. Meanwhile, statistical analysis use hierarchical clustering and Mann-Whitney as the methods. Spatial analysis have function to determine the distribution of parameters in study area, spatially. Statistical analysis used to determine the similarity based on the score range of biophysics chemistry parameters and significant differences in each groups. Group of station generated from spatial analysis based on distribution, hierarchical clustering, and further correction through Mann-Whitney. Then, characterized based on significant differences in each groups.

3. Result and Discussion

Human factor only affect about 10% influence the growth rate, such as determining the hue and size of pearls oyster that produced [10]. In the other hand, the most important factor for pearl oyster growth is environment factors. Thus, biophysics chemistry parameters is needed to observe. Table 2 shown on the result of biophysics parameters.

Observation of chlorophyll-a, phytoplankton abundance and temperature as biophysics parameters at 20 stations provide different results on each station observation. Based on the table known that the highest value of chlorophyll-a, phytoplankton and temperature are is 2.198 mg/m3, 14900 sel/m3 and 34oC. Moreover, the lowest are 0.308 mg/m3, 3898 sel/m3 and 28oC.

Chlorophyll-a is a pigment in phytoplankton to photosynthesis reactions and produce of organic compounds. Chlorophyll-a pigment is the most dominant and the largest in comparison with chlorophyll-b and chlorophyll-c [11]. However, phytoplankton are plankton species that have function to photosynthesis and primary productivity in the aquatic. Phytoplankton also have the important function in pearl oyster farming. Phytoplankton is the main food of pearl oysters, because they are filter feeders organism. Hence it, they need phytoplankton as their feeds. Phytoplankton and organic matter in the aquatic is obtained through gills absorption or membrane coat of pearl oyster [12].

Table 2. The observation result of pearl oyster farming biophysics parameters

Parameters Stations	Biology		Physic
	Chlorophyll-a (mg/m^3)	Phytoplankton (sel/m^3)	Temperature (°C)
1	0.663	14900	28
2	0.339	14277	29

Parameters Stations	Biology		Physic
	Chlorophyll-a (mg/m^3)	Phytoplankton (sel/m^3)	Temperature (°C)
3	0.576	12663	28
4	1.599	4687	31
5	0.866	5200	32
6	0.543	4398	33
7	0.308	3898	34
8	0.732	9429	34
9	0.663	5269	34
10	0.766	6199	32
11	0.486	5834	30
12	1.687	5805	32
13	2.198	8693	31
14	1.123	6604	31
15	0.952	4930	31
16	0.935	10840	30
17	1.258	5563	30
18	0.646	8700	30
19	0.647	10862	30
20	0.953	8439	30

Based on spatial analysis result known that the chlorophyll-a in the pearl oyster farming ranges from 0308-2198 mg / m3. The highest of chlorophyll-a found at Station 13 and the lowest of chlorophyll-a at Station 7. Chlorophyll-a in some stations have not linier relation with phytoplankton abundance. It can be caused by phytoplankton distribution and it batimetry. Phytoplankton abundance in pearl oyster farming ranges from 3898-14900 cells / m3. The highest of phytoplankton abundance found at Station 1 and the lowest of phytoplankton abundance at Station 7. Phytoplankton which have the highest abundance are *Chaetoceros* sp. *Chaetoceros* sp. dominan in the aquatic due to this genus have growth rates higher than dinoflagellates group. *Chaetoceros* sp belong to the Bacillariophyta genus is dominant in the aquatic because has the ability to synthesize of nutrients, highly. *Chaetoceros* sp. also can save the nitrate and phosphate compounds as feed reserves in their cells [13]. Research of pearl oyster feed habits conducted by [14] stated that the main feed of pearl oyster is *Chaetoceros* sp. which belong to the Bacillariophyta.

Sea surface temperature is one of essential factor for living organisms in the sea. Temperature can affect the metabolic activity and proliferation of aquatic organisms [15]. Temperature also can describing that aquatic still be inhabited by aquatic organisms [16]. Sea surface temperatures in the upper layer is relatively homogeneous. In tropic, the homogeneous layer (mixed layer) in the depth of 50 m to 100 m [17].

Based on the results of spatial analysis known that the temperature in the pearl oyster farming have ranges between 28-34oC. The highest temperature are located at Station 7, 8 and 9 while the lowest temperature are located at Station 1 and 3. Some station point in the pearl oyster farming have the temperature higher than other, because the stations close with power plant activity. The disposal of hot waste water discarded into aquatic system, while around the aquatic there are some pearl oyster farming activities. The aquatic temperature increased by 10°C causes increase 2-3 times in oxygen consumption by aquatic organisms. Increasing temperatures can also cause increasing in the organic matter decomposition by phytoplankton [18]. Extreme increased of temperature can cause disruption for aquatic organisms metabolism and physiology. Near of mainland has lower salinity than the salinity where near of open sea. The condition occurs because near of mainland, the water quality are still influenced by river runoff. Therefore, salinity distribution was influenced by rainfall seasons, topography, tidal and evaporation [19].

The study also got information on chemistry parameters such as salinity, pH, dissolved oxygen, nitrate, silica, orthophosphate and ammonia. The result of chemistry parameter shown on table 3.

Table 3. The observation result of pearl oyster farming chemistry parameters

Parameters Stations	Salinity (‰)	pH	Dissolved Oxygen (mg/L)	Nitrate (mg/L)	Silica (mg/L)	Orthophosphate (mg/L)	Ammonia (mg/L)
1	35	7	10.3	0.066	0.291	0.025	0.08
2	36	7	10.2	0.059	0.194	0.023	0.063
3	37	7	8	0.034	0.172	0.023	0.082
4	35	6.5	9.5	0.009	0.104	0.031	0.072
5	35	6	10.9	0.004	1.592	0.026	0.08
6	34	7	8.1	0.001	0.179	0.03	0.094
7	33	6.5	8.4	0.016	0.196	0.035	0.067
8	33	6.5	8.5	0.014	0.316	0.035	0.109
9	33	6.5	8.8	0.015	0.316	0.031	0.135
10	35	7	11.1	0.025	0.264	0.036	0.098
11	36	7.5	9.2	0.053	0.111	0.033	0.166
12	35	7.5	8.5	0.009	0.206	0.049	0.099
13	37	7.5	6.3	0.028	0.386	0.051	0.161
14	37	7.5	7.9	0.037	0.615	0.034	0.166
15	36	6.5	3.9	0.026	0.311	0.067	0.234
16	35	6	4	0.019	0.564	0.063	0.177
17	34	7	5.6	0.025	0.554	0.034	0.162
18	35	7	5.5	0.016	0.773	0.042	0.189
19	35	6.5	3.3	0.018	0.211	0.02	0.206
20	36	6.5	3.7	0.031	0.537	0.033	0.308

Observation of salinity, pH, dissolved oxygen, nitrate, silica, orthophosphate, ammonia as chemistry parameters at 20 stations gave different results on each station observation. Based on the table known that the highest value of salinity, pH, dissolved oxygen, nitrate, silica, orthophosphate, ammonia are 37‰, 7.5, 11.1 mg/L, 0.308 mg/L, 0.066 mg/L, 0.67 mg/L, 1.592 mg/L. Besides, the lowest are 33‰, 6, 3.3, 0.063 mg/L, 0.001 mg/L, 0.020 mg/L, 0.104 mg/L.

Toxicity of chemical compounds are affected by pH. Alkaline conditions (high pH) are commonly form non-ionized ammonia and toxic. Some stations have a low pH value ranged of 6-6.5. That assessment because the station near the coast allow for anthropogenic pollution and an acidic waters in the rainy season. Rainwater contains carbonic acid that may affect the acidity in the waters [20].

Based on results of spatial analysis known that salinity in the pearl oyster farming ranges 33-37 ‰. The highest salinity located at station 13 and 14, while the lowest salinity at Station 7. 8 and 9. Salinity differences affect to aquatic organisms especially in osmotic pressure. The balance of osmotic pressure in the aquatic organisms need stabilized to make balance between internal and external fluid. Large fluctuations can affect the performance of internal and external fluids of organisms and in the pearl oyster [11]. Besides, pH in pearl oysters farming ranges from 6 to 7.5. The pH value of this research still within the range allowed for marine culture [21]. The pH value still to be volatile until reaches ≥ 4 concentration in the maximum salinity up to 35 ppm [22]. The pH value still in the normal range if they have value ranges of 6-8 [16]. Otherwise, if pH value is less than 6 will affect the diversity and abundance of phytoplankton [18].

Dissolved oxygen are used by marine organisms for respiration. Sources of dissolved oxygen comes from atmosphere and photosynthesis activity by aquatic plants and phytoplankton [18]. Marine organisms need oxygen in sufficient concentration. The concentration of dissolved oxygen in the water less than 5.0 mg/L induce to decrease of marine organisms types even will die [18]. However, dissolved oxygen used by phytoplankton for photosynthesis.

Based on spatial analysis known that the dissolved oxygen in pearl oysters farming ranges from 3.3 to 11.1 mg/L. The highest of dissolved oxygen at station 10 and the lowest at station 19. The values of dissolved oxygen in the research has the lowest value in the coastal which near mainland. It related to the turbidity of seawater and caused by increasing microorganisms activity to break down organic matter into inorganic substances that use dissolved oxygen in these waters [23]. Horizontally, dissolved oxygen concentration will decrease if go to the onshore [24]. Some stations such as stations 19 and 20 have low dissolved oxygen. It caused that the station uses a raft system that has lack of covering the pearl oyster farming area to photosynthesis, consequently the photosynthesis process is inhibited. That is giving impact are declining the dissolved oxygen caused by respiration and decomposition (decomposition) process of organic matter and evaporation processes [25]. Declining of dissolved oxygen is inverse to the increasing levels of ammonia. It was shown at station 20. Station has low pH values induce the low of dissolved oxygen while increasing carbon dioxide concentrations [18]. Carbon dioxide is a form of gases from respiration result [16].

Some of ammonia source in the water are from breakdown of organic nitrogen (protein and urea) and contained both inorganic nitrogen in the soil and water from organic materials decomposition such as aquatic organisms had death [18]. The excessed of ammonia value will interfere the growth of marine culture organisms [26]. The un-ionised ammonia is more easily absorbed into the aquatic organisms as compared to ammonium [18]. Based on the results of spatial analysis are known that the ammonia in the pearl oyster farming ranges from 0.063 to 0.308 mg / L. The highest ammonia at station 20 and the lowest at station 2. According to data, ammonia value was high and exceeds the threshold, it can be an indicator of organic matter contamination derived from domestic sewage, industrial and river runoff. Some points of the research station has high ammonia value compared with other stations point. It was induced that some stations close to the mainland which there are settlements and rivers in the region. However, some of the station point applied raft system in the pearl oyster farming to give higher ammonia input compared with the floating system. This was caused, raft systems contain higher organic matter from wood or bamboo and there was activity and preparation work hut X-ray pearl oysters in the area around the station.

Nitrate (NO3) is the main form of nitrogen and a major nutrient for the growth of aquatic plants and phytoplankton. Nitrate is soluble in the water and stable [18]. Nitrates are not toxic for aquatic organisms and relatively stable to aquatic plants and phytoplankton in the quality standards of set limits. The high nitrate value in the waters near the river run of, that is suspected because of the organic material input from any settlement or other anthropogenic activities. Based on the results of spatial analysis known that nitrate in the pearl oysters farming ranges from 0.001 to 0.066 mg/L. The highest nitrate contained in Station 1 and the lowest nitrate contained in Station 6.

Orthophosphate is a form of soluble phosphate. Source orthophosphate is derived from the decomposition of organic materials. Industrial waste, domestic waste, and other anthropogenic activities, such as waste of detergent [18]. Orthophosphate acts as a nutrient in the water. The nutrients can be used to help the photosynthesis performed by aquatic plants and phytoplankton. Orthophosphate value is relatively high in the waters where near river. It is suspected because of the input of organic material from any settlement or other anthropogenic activities. Based on the results of spatial analysis known that orthophosphate in the pearl oyster farming ranges from 0.02 to 0.067 mg/L. The highest of orthophosphate at stations 7 and 8, and the lowest at station 19.

Generally, silica in the aquatic are soluble or bonded to the element oxide (SiO4). The existence of silica does not generated toxic. Silica is an essential element for living beings. Some algae, especially diatoms (Bacillariophyta) requires frustule or silica to form the cell wall [18]. Based on the results of spatial analysis known that the silica in the pearl oysters farming ranges from 0.104 to 1.592 mg/L. Source of silica in the coastal commonly from the weathering of soil minerals containing silica, then dissolved in streams [27]. The silica value at stations where close to the coast is relatively high compared with other station area. Station 5 has the highest concentration, because rocks from the beach morphology entered into the sea. Additionally, high silica can be caused by high mineral content in the aquatic. That is related to the condition of water bottom substrate and water depth

Grouping station by similarity of biophysics chemistry character in the pearl oyster farming Semau Strait-East Nusa Tenggara in December 2015 shown in figure 2.

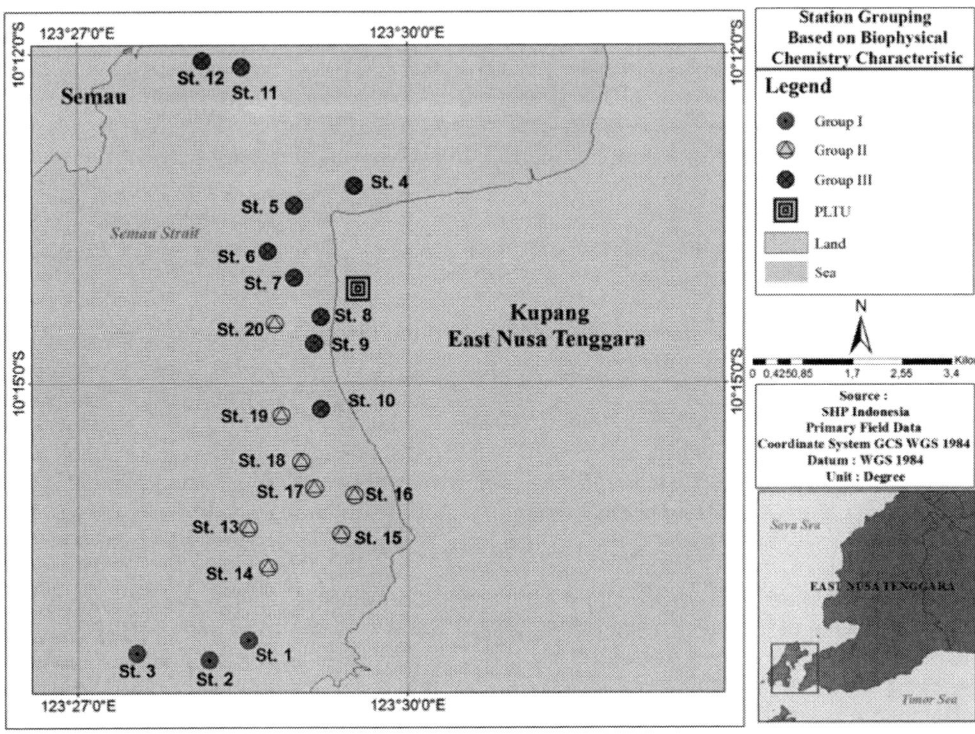

Figure 2. Station grouping by aquatic biophysics chemistry character similarity in pearl oyster farming

Based on the data, whole stations are dividing into four groups: Group I consists of Stations 1, 2 and 3; Group II consists of stations 13, 14, 15, 16, 17, 18, 19 and 20; Group III consists of Station 4, 6, 7, 8, 9, 10, 11 and 12; and Group IV is Station 5. Then the correction of the Mann-Whitney known there are no significant differences between group IV with three other groups. Furthermore. The identified within the nearest known that Group IV. Group IV has a closeness with Group III. Based on the analysis. grouping stations are divided into three groups: Group I consists of Stations 1, 2. and 3; Group II consists of stations 13, 14, 15, 16, 17, 18, 19 and 20; and group III consists of stations 4, 5, 6, 7, 8, 9, 10, 11 and 12.

Spatially, grouping station by similarity of aquatic biophysics chemistry character divided into three groups: Group I, II and III. Similarities of biophysics chemistry influenced by morphology area and the distance between stations. Morphology of area has same character and location of each station could be expected to provide the effect between the stations. Group I is the excellent group for pearl oyster farming, the located quiet far to the coastal area that has a lot of anthropogenic activities and power plant. The second group is the group that located in the centre of all area in pearl oyster farming and away from the power plant activities. That area is close to the coast and estuaries. Group II is good group. Group III is the closest group to power plant activities. Group III is bad group for pearl oyster farming. Group I is the excellent group for pearl oyster farming activities because it has the highest value of dissolved oxygen, chlorophyll-a and phytoplankton abundance compared to other groups. Group II is good group for pearl oyster farming activities, but need monitoring of increasing in nutrients and eutrophication in the area. Group III is bad group for pearl oyster farming activities because it has high temperature

Significant differences of parameters between Group I and Group II contained in ammonia, temperature, dissolved oxygen, nitrate, silica, chlorophyll-a and phytoplankton abundance parameters. Chlorophyll-a, dissolved oxygen and phytoplankton abundance in Group I are higher than Group II. Significant differences of parameters between Group I and III contained in temperature. Salinity, nitrate, orthophosphate and phytoplankton abundance. Temperature in Group III is higher than Group I. The significant difference of parameters between Group II and III contained in temperature, salinity, silica, dissolved oxygen, nitrate, ammonia and phytoplankton abundance. Temperatures in Group III is higher than Group II.

Further research is needed to determine the suitability, carrying capacity of area, monitoring of biophysics and chemistry parameters in pearl oysters farming-Semau Strait, temporally. The influence of marine environment and the study of pearl oysters on integrated coastal management of each stakeholder in the coastal area, Semau Strait-East Nusa Tenggara.

4. Conclusion

Spatial mapping by biophysics chemistry similarity dividing the stations into three groups. Group I consists are Stations 1, 2 and 3; Group II consists are stations 13, 14, 15, 16, 17, 18, 19 and 20; and group III consists are stations 4, 5, 6, 7, 8, 9, 10, 11, and 12. All groups have different characteristics and categories for pearl oyster farming activity. Group I is excellent group, because it group has the highest values of dissolved oxygen, chlorophyll-a, and phytoplankton abundance. Group II is good group, because it group has moderate biophysics chemistry quality in each parameters and Group III is bad groups, because it group has high temperature.

Acknowledgement

Authors wishing acknowledge, especially for financial support from Collaborate Research Project between IPB and Ehime University. Therefore, special for our research location in PT Timor Otsuki Mutiara Kupang, East Nusa Tenggara and those technical staff who help us. Besides, acknowledge to Laboratory of Aquatic Productivity and Environmental, Bogor Agricultural University, Laboratory of Micro Biology, Bogor Agricultural University and Laboratory of Mapping and Geospatial Modelling, Bogor Agricultural University

References

[1] Kordi K M G H 2011 Marikultur: Prinsip dan Praktik Budidaya Laut (Yogyakarta: Lily Publisher)
[2] Cartier L E and Krzemnicki M S 2013 New Developments in farming pearl production: Use of organic and Baroque Shell Nuclei *Journal of The Australian Gemmologist* **25** 6-13
[3] Hanni H A 2007 A Description of pearl farming with Pinctada Maxima in South Asia *Journal Gemmoligist* **30** 357-365
[4] Bunch M J 2012 Using geographic information systems (GIS) for spatial planning and environment management in India: Critical considerations *International Journal of Applied Science and Technology* **2** 1-15
[5] Wedding L M, Gibson B A, Walsh W J and Battista 2011 Integrating remote sensing products and GIS tools to support marine spatial management in West Hawai *Journal of Conservation Planning* **7** 60-73
[6] Yunandar 2007 *Analisis pemanfaatan ruang di kawasan pembangunan perikanan pesisir muara kintap Kabupaten Tanah Laut Propinsi Kalimantan Selatan* thesis (Semarang: Dipenogoro University)
[7] Daly C 2006 Guidelines for Assessing the Suitability of Spatial Climate Data Sets *International Journal of Climatology* **26** 707-721
[8] Chiu C A, Lin P H and Lu K C 2009 GIS-based tests for quality control of meteorological data and spatial interpolation of climate data *Journal of Mountain Research and Development* **29** 339-349
[9] APHA, AWWA, WEF 2005 *Standard Methods for examination of water and wastewater 22nd ed.* (Washington: American Public Health Association) pp 1360
[10] Harramain E H M 2008 *Kajian Faktor Lingkungan Habitat Kerang Mutiara (Stadia Spat) di Pulau Lombok, Nusa Tenggara Barat* undergraduate thesis (Bogor: Bogor Agricultural University)
[11] Kangkan A L 2006 *Studi Penentuan Lokasi untuk Pengembangan Budidaya Laut berdasarkan Parameter Fisika. Kimia dan Biologi di Teluk Kupang. Nusa Tenggara Timur* undergraduate thesis (Semarang: Dipenogoro University)
[12] Lucas J S dan Southgate P C 2003 *Aquafarming: Farming Aquatic Animals and Plants* (Australia: Blackwell Publishing)
[13] Sahubawa L 2001 Dampak pembuangan limbah terhadap perubahan kualitas oseanografi biofisik kimia dan produksi ikan teri (*Stolephorus spp.*) di Perairan Laut Teluk Ambon *Jurnal Manusia dan Lingkungan* **8** 15-29
[14] Anwar K, Toelihere M, Affandi R, Azwar N R dan Etty R 2004 Kebiasaan makan tiram Mutiara Pinctada maxima di Perairan Teluk Sekotong, Lombok *Jurnal Ilmu-Ilmu Perairan dan Perikanan Indonesia* **11** 73-79
[15] Limbong M 2008 *Pengaruh Suhu Permukaan Laut terhadap Jumlah dan Ukuran Hasil Tangkapan Ikan Cakalang di Perairan Teluk Palabuhanratu, Jawa Barat* undergraduate thesis (Bogor: Bogor Agricultural University)
[16] Purba N P and Khan A M A 2010 Karakteristik fisika-kimia perairan Pantai Dumai pada musim peralihan *Jurnal Akuatika* **1** 69-83
[17] Sediadi A 2004 Efek upwelling terhadap kelimpahan dan distribusi fitoplankton di perairan Laut Banda dan sekitarnya *Jurnal Makaira Sains* **8** 43-51
[18] Effendi H 2003 *Telaah Kualitas Air* (Yogyakarta: Kanisius)
[19] Nybakken J W 2000 *Biologi Laut: Suatu Pendekatan Ekologis* (Jakarta: Gramedia)
[20] Wardhani N K Ihwan A and Nurhasanah 2015 Studi tingkat keasaman air hujan berdasarkan kandungan gas CO_2, SO_2 dan NO_2 di Udara *Jurnal Prisma Fisika* **3** 9-14
[21] Keputusan Menteri Kependudukan dan Lingkungan Hidup No 2 Tahun 1988 tentang Pedoman Penetapan Baku Mutu Lingkungan diperuntukkan budidaya perikanan biota laut

[22] Lucas J S and Southgate P C 2003 *Aquaculture: Farming Aquatic Animals and Plants.* (Australia: Blackwell Publishing)
[23] Patty S I, Hairati A and Malik S 2013 Zat Hara (Fosfat, Nitrat), Oksigen Terlarut, dan pH Kaitannya dengan Kesuburan di Perairan Jikumerasa, Pulau Buru *Jurnal Pesisir dan Laut Tropis* **1** 43-50
[24] Nybakken J W 1988 *Biologi Laut: Suatu Pendekatan Ekologis* (Jakarta: Gramedia)
[25] Armita D 2011 *Analisis Perbandingan Kualitas Air di Daerah Budidaya Rumput Laut dengan Daerah Tidak Ada Budidaya Rumput Laut. di Dusun Malelaya. Desa Punaga, Kecamatan Mangarabombang, Kabupaten Takalar* undergraduate thesis (Makassar: Universitas Hasanuddin)
[26] Kordi K M G H and Tancung A B 2007 *Pengelolaan Kualitas Air dalam Budidaya Perairan* (Jakarta: PT Rineka Cipta)
[27] Lukman M, Nasir A, Amri K, Tambaru R, Hatta M, Nurfadilah and Rahmat J N 2014 Silikat Terlarut di Perairan Pesisir Sulawesi Selatan *Jurnal Ilmu dan Teknologi Kelautan Tropis* **6** 461-478

IOP Conf. Series: Earth and Environmental Science **54** (2017) 012070 doi:10.1088/1755-1315/54/1/012070

Evaluation of karst water quality as an early reference of land suitability mapping for vaname shrimp (*Litopenaeusvannamei*) culture media

D M Wildan[1*], R Affandi[1], N T M Pratiwi[1], M Krisanti[1], I P Ayu[1], and A Iswantari[1]
[1]Living Aquatic Resources Management, Faculty of Fisheries and Marine Science, Bogor Agricultural University. Jl. AgatisKampus, Dramaga, Bogor 16680

Email: dudi.mw@gmail.com

Abstract. Vaname shrimp (*Litopenaeusvannamei*) is one of the excellent fishery commodities in Indonesia. Vaname shrimp farming can be conducted in low salinity water. Low salinity water sources which could be used as culture media is karst water because it has a high mineral. The research was aimedto evaluate land suitability mapping for pond as the vaname shrimpculture mediaseen from the water quality. Research was conducted in May and August 2016. Water sampling was conducted in several locations; Ancol-Jakarta (seawater), Ciseeng-Bogor (karst water salinity), Ciampea-Bogor (karst freshwater), and Situ Gede Bogor (freshwater). Evaluating the suitability of karst water quality for vaname shrimp culture media, done by the results of karst water quality analysis compared with seawater and SNI 01-7246-2006 on shrimp vaname culture media. The results showed that Karst water of Ciseeng and Ciampea could not directly use as vaname shrimp culture media. It is needed water quality treatment of ozonation and aeration of karst water to improve water quality.Ozonation and aeration treatments were able to improve the quality of karst water up to approach the living quality standard of vaname shrimp media.

1. Introduction

Vaname shrimp aquaculture activities are generally carried out in high-salinity waters such as on coastal waters. Vaname shrimp which have euryhaline properties causing this organism, in addition to be able to be cultured in high-salinity waters, it can also be cultured in low salinity waters such as lakes, rivers, wetlands and reservoirs [1].

Vaname shrimp farming activities in the waters with low salinity is not only able to increase shrimp production but also provide other benefits such as employment, reducing transport costs due to the distance between farmers and consumers are relatively close, shrimp can be served in a fresh state, and shrimp farming activities in freshwater can prevent shrimp contracts a disease. However, when vaname shrimp is cultured in low salinity waters, it will cause vaname shrimp experience stress because it cannot maintain the osmotic pressure of the body [2,3]. This is because low salinity water has lower mineral content than the high salinity waters[4,5]. Therefore, it requires source of freshwater that has a high mineral content.

Low salinity water source that has the mineral potentials is karst water or limestone water [6]. Karst water contains minerals dominated by calcium, magnesium, and carbon. Mineral content in karst water is come from the activities of rocks dissolution by water (carstification) [7]. Abundant mineral content and availability throughout the year makes the karst water is a potensial ofvanameshrimpculture media. Limited information on this is an opportunity for the evaluation study on karst water quality conditions for vaname shrimp (*L. vannamei*) breeding media. The purpose of this study is to assessland suitability mapping for pond as the Vaname shrimp habitat seen from the water quality.

2. Material and Methods

2.1 Time and place

The study was conducted on May and August 2016. Water sampling was conducted in several locations, namely Ancol-Jakarta (sea water), Ciseeng-Bogor (karst water salinity), Ciampea-Bogor (fresh karst water), and Situ Gede Bogor (freshwater) (figure 1). sea water in the study was used as a positive control because it represents the habitat vaname shrimp, Ciseeng karst water in this study is used as a representative water karst having a value of high salinity, Ciampea karst water used in this study as a representative of the water karst not have salinits, while Situ Gedeis freshwater used as a negative control because it has a different character with shrimp habitat vaname.Water quality analysis was conducted at the Laboratory of Water Environment Productivity and the experiments were conducted in the laboratory of the Secondary Productivity, Ministry of Water Resource Management, Faculty of Fisheries and Marine Science, Bogor Agricultural University.

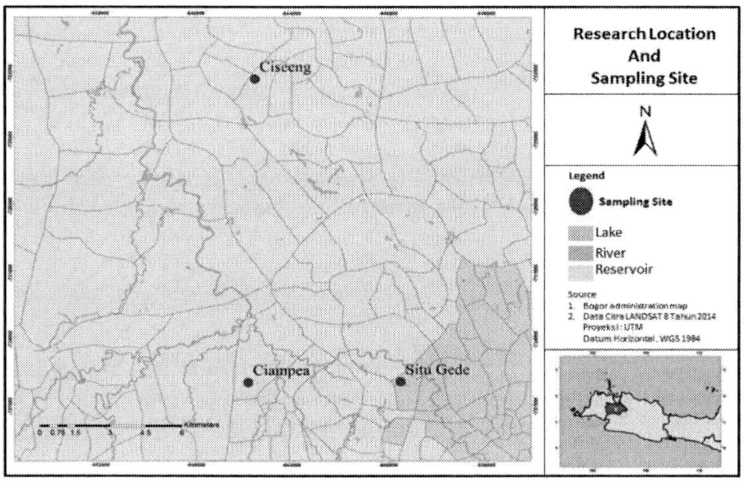

Figure 1 Research location and sampling site

2.2 Stages of Research

2.2.1 Karst water quality evaluation

Analysis was performed on each sample of the water source. Salinity and pH measurements were performed using a refractometer and pH meter. Furthermore, dissolved oxygen and temperature measurements were performed using a DO meter, electrical conductivity (EC), total hardness, SO_4^- (sulfate), alkalinity, NH_3^- (ammonia), Ca, Mg, K, and H_2S [8].

2.2.2 Karst water quality treatment
Treatmentwascarried out by ozonation and aeration. Prior to ozonation and aeration, the water was filtered using a sieve with a mesh size of 25 microns. The water is then stored in separate containers. Water was treated with ozonation and aeration for 24 hours and left a few hours until the odor generated by ozone treatment disappeared. Analysis of water quality on the same parameters as in the evaluation phase carried [8].

2.3 Data analysis
The evaluation on the suitability of water quality conditions for vaname shrimp culture media was carried out by comparing the overall results of the analysis of quality of karst water to sea water as its natural habitat, as well as to the SNI (Standar Nasional Indonesia) 01-7246-2006 for vanameshrimp culture media. Differences in karst water quality before and after the modification were analyzed using t-test.

3. Results and Discussion

3.1 Physical and chemical conditions of the water
Results of water quality measurements Ciseeng and Ciampea karst waters that includes physics and chemistry of water compared with the seawater and Situ Gede. This is done to determine the characteristics of karst water Ciseeng and Ciampea. The results of measurements seawater quality, Ciseeng and Ciampea karst waters and water Situ Gede is presented in table 1.

Table 1. The quality of sea water, karst Ciseeng and Ciampea and water Situ Gede

Parameter	Unit	Water Source			
		Seawater	Ciseeng	Ciampea	Situ Gede
Physics					
Temperature	°C	28,000±0,820	39,100±1,890	26,500±0,880	28,500±0,510
Conductivity	µS/cm	47740,00±1368,890	41972,500±2784,090	433,000±13,520	267,000±123,600
Chemistry					
Dissolved Oxygen	mg/l	4,400±0,320	0,275±0,150	4,500±0,260	4,050±0,240
Salinity	Ppt	33,250±0,960	33,250±0,960	0	0
pH	-	8,120±0,080	6,352±0,420	6,680±0,170	6,460±0,390
Hardness	mg/l $CaCO_3$	4103,900±622,580	8458,450±5262,32	84,080±30,840	62,560±18,590
Alkalinity	mg/l	186,00±5,160	1599,000±202,840	267,750±5,440	51,000±6,000
NH_3^-	mg/l	0,210±0,180	3,160±0,020	0,180±0,040	0,415±0,270
H_2S	mg/l	<0,001	<0,001	<0,001	<0,001
Ca	mg/l	652,460±99,470	2904,572±575,790	186,880±61,050	14,040±7,770
Mg	mg/l	907,460±78,400	354,181±9,440	3,870±1,280	2,670±0,220
K	mg$CaCO_3$/l	329,550±134,410	250,950±18,250	0,370±0,250	1,140±0,120
SO_4	mg/l	1234,620±132,920	8.360±0.26	5,270±0,760	8,470±0,520

Based on table 1 it can be seen that almost all parameters of Ciseeng karst water has higher value than the Ciampea karst water unless the value of dissolved oxygen and H_2S. Physical and chemical parameters of Ciseeng water tend to be similar to sea water, while Ciampea water tends to be similar to freshwater. This causes Ciseeng water is classified as water that has high salinity and Ciampea water is classified as freshwater or low salinity. The seawater has a salinity> 16 ppt, while freshwater has a salinity <0.5 ppt [9].

Ciampea and Ciseeng waters are potential for vanameshrimp living media. The Ciseeng water has salinity similar to salinity of vaname shrimp (*L. vannamei*) in their habitat, while Ciampea water has alkaline salinity similarity to the salinity in their habitat. However, to determine whether or the potential water of Ciseeng and Ciampea can be used as vanameshrimp (*L. vannamei*) living media, it is necessary to perform a comparison between physical and chemical water at Ciseeng and Ciampea with water quality for vanameshrimp (*L. vannamei*) by SNI 01-7246-2006.

3.2 Comparison of physical and chemical parameters of karst water with water quality for shrimp by SNI 01-7246-2006

The results of measurements of physical and chemical parameters of karst water at Ciseeng and Ciampea are compared to SNI-01-7246-2006 on quality standard of vaname shrimp living media. This is performed to determine the suitability of karst water quality at Ciseeng and Ciampea with tolerance range which is possessed by the vaname shrimp. Comparison results of karst water quality measurements of Ciseeng and Ciampea and SNI-01-7246-2006 were presented in table 2.

Table 2. Comparison of physical and chemical parameters of karstwater with water quality for shrimp by SNI 01-7246-2006

Parameter	Unit	Ciseeng karst water	Ciampea karst water	SNI 01-7246-2006
Temperature	°C	39,100±1,890	26,500±0,880	28,500-31,500
Dissolved Oxygen	mg/l	0,275±0,150	4,500±0,260	>3,500
Salinity	ppt	33,250±0,960	0.00	15,000-25,000
pH	-	6,353±0,420	6,680±0,170	7,500-8,500
Alkalinity	mg/l	1599,000±202,840	267,750±5,440	100,000-150,000
NH_3^-	mg/l	3,160±0,020	0,180±0,040	<0,010
H_2S	mg/l	<0,001	<0,001	<0,001

Based on Table 2, it can be seen that almost all physical and chemical parameters of Ciampea and Ciseeng karst water do not correspond to the water quality for vaname shrimp (*L. vannamei*) based on SNI 01-7246-2006. This is because the karst water at Ciseeng and Ciampea is relatively poor to be uses as vaname shrimp breeding media. Therefore, Ciampea and Ciseeng karst water requires water quality modification to be used as vaname shrimp breeding media. Water quality modifications that can be performed are ozonation and aeration treatments. Ozonation treatment can decrease the value of alkalinity[10]. Other than that ozonation lowers the alkalinity value by 21.5% and increases the pH value by 5.2% [11]. While the aeration treatment can boost dissolved oxygen and lower the value of ammonia by 98% and reduce the alkalinity value of 69% on a liquid fertilizer [12,13].

3.3 Karst water quality modification

Karst water quality modification aims in order that karst water can be used for vanameshrimp (*L. vannamei*) preserving media. Modification to lower the temperature of the water is performed by placing the karst water in a room so it has room temperature. Meanwhile, modification to increase oxygen and pH and lower alkalinity and NH_3^- were performed by providing aeration and ozone for 24 hours. Comparison Results of karst water quality after the modification is presented in Table 3.

Table 3. The quality of sea water, karst water of Ciseeng and Ciampea and water of Situ Gede after modification

Parameter	Unit	Ciseeng karst water	Ciampea karst water	SNI 01-7246-2006
Temperature	°C	27,15±0,129	27,250±0,058	28,500-31,500
Dissolved Oxygen	mg/l	5,15±0,129	5,200±0,183	>3,500
Salinity	Ppt	33,250±0,960	0.000	15,000-25,000
pH	-	8,0125±102	7,673±0,077	7,500-8,500
Alkalinity	mg/l	204,000±19,114	54,750±6,397	100,000-150,000
NH_3^-	mg/l	2,664±0,135	0,056±0,059	<0,010
H_2S	mg/l	<0,001	<0,001	<0,001

Karst water quality modification using ozonation and aeration in Table 3 indicates the change that approaches SNI 01-7246-2006 water quality standards for vanameshrimp.In karst water at Ciseeng, the parameters are changed significantly, those are temperature, dissolved oxygen salinity, pH, alkalinity, and NH 3 - ($p < 0.05$). Karst water at Ciampea experiences real changes including dissolved oxygen salinity, pH, and alkalinity ($p < 0.05$). However, the NH_3^- and temperature parameter decreases but is not significantly different ($p > 0.05$). Whereas H_2S parameter remains largely significantly better on Ciseeng karst water than on Ciampea karst water and still comply with the quality standards for vaname shrimp.
Value of salinity, alkalinity, and NH_3^- which is possessed in Ciseeng and Ciampea karst water despite undergoing changes it remains not meet the quality standards for vaname shrimp. Salinity value that does not comply will cause the growth of shrimp to be slow because of the energy required for growth will be exclusively used for osmoregulation [14]. Meanwhile, if the value of alkalinity does not comply, it will cause disruption of the process of cells osmoregulation in the body due to lack or excess of calcium[15]. NH_3^- value that is too high can inhibit the growth of shrimp, in addition to the direct influence of NH_3^- content is high but not damaging lethal gill tissue. The sheets will swell gills (hyperplasia) so that the gills as respiratory function will be impaired in terms of the binding of oxygen from the water. High ammonia levels in the water can also increase the concentration of ammonia in the blood, thereby reducing the activity of blood (hemocyanin) in binding oxygen. Additionally, high levels of ammonia can also increase susceptibility for shrimp to get disease[16].

The high value of salinity and alkalinity on Ciseeng karst water can be lowered by way of dilution of freshwater or distilled water in order to obtain the appropriate salinity and alkalinity. Meanwhile, to increase the value of salinity and alkalinity in the Ciampea karst water to conform to the quality standards can be done by adding seawater. Ammonia values that are too high in all karst water can be reduced by performing aeration treatment.

4. Conclusion

Karst watersresources of Ciseeng and Ciampea a decent location used as a source of water for aquaculture vaname shrimp (*Litopenaeusvannamei*)development in inland areas far from the beach.

Acknowledgements
Thanks to the Ministry of Research, Technology and Higher Education which has provided a research grantsas well as the BPPTN and LPPM as a facilitator so that the study can be conducted well. This research is part of a series of study activities entitled "Ecological Potential and Karst Water Utilization as Rich in Mineral Growing Media for Aquatic Micro Biota".

Reference

1. Bray WA, Lawrence A L, Leung T J R 1994 The effect of salinity on growth and survival of PenaeusVannamei, with observations on the interaction of IHHN virus and salinity. *Aquaculture*. 122 p 133-146
2. Mantel L H, Farmer L L 1983 Osmotic and Ionic Regulation. In: L. H. Mantel (ed), The biology of Crustacea, 5 internal anatomy and physiological regulation. (New York: Academic Press)
3. Pham D, Charmantier G, Wabete N, Boulo V, Broutoi F, Mailliez JR, Peignon JM, Charmantier DM. 2012 Salinity Tolerance, Ontogeny Of Osmoregulation And Zootechnical Improvement In The Larval Rearing Of The Caledonian Blue Shrimp, *LitopenaeusStylirostris*(Decapoda, Penaeidae). *Aquaculture* **362-363** p 10-17
4. Rump H H, Krist H 1992 *Laboratory Manual For The Examination Of Water, Wastewater And Soil*. (Germany: Weinheim)
5. Gunter G1977 Salinity In coastal ecosystem management. Ed. J Clark A -Ley *Interscience Publ.* New York
6. Hartmann A, Goldscheider N, Wagener T, Lange B J, Weiler M 2014 Karstwater resources In achanging world: Review of hydrological modeling approaches, *Rev. Geophys*. **52** 218–242
7. Ekmekci M 2005 Karst in turkish thrace: Compatibility between geological history and karst type *Turkish Journal of Earth Sciences* **14** p 73-90
8. APHA, AWWA, WEF 2012 *Standard Methods for examination of water andwastewater* 22nd ed. (Washington: American Public Health Association)
9. Mclusky, D S 1971 *Ecology Of Estuaries* The Scholarship (London: Hememann Education Books Ltd)
10. Summerfelt S T, Hochheimer J N 1997 Review of ozone processes and applications as an oxidizing agent in aquaculture *The Progressive Fish-Culturist* **59** p 94-105
11. Krisnawati A, Sururi M R, Ainun S 2014 Pengaruh karakteristik lindi terhadap ozonisasi konvensional dan advanced oxidation processes (Aop) *Jurnal Rekayasa Lingkungan* **2** p 2
12. García-Gonzalez MC, Vanotti MB, Szogi AA 2015 Recovery of ammonia from swine manure using gas-permeable membranes: Effect of aeration. *Journal Of Environmental Management.***152** 19-26
13. Sulihingtyas WD, Suyasa IWB, Wahyuni NMI 2010 Efektivitassistempengolahaninstalasipengolahan air limbahsuwung Denpasar terhadapkadar BOD, COD, danAmonia. *Jurnal Kimia*. **42** p 141-148
14. Davis D A, Saoud I P, Mcgraw W J, Rouse DB 2002 Consideration For *Litopenaeus Vannamei* Reared In Inland Low Salinity Waters. Di Dalam Cruz-Suarez Ie, Rieque-Marie D, Tapia-Salazar M, Gaxiola-Cortes Mg, Simoes N, Editor. *Avances En Nutricion Acuicola Vi Memories Del Vi Simposium Internacional De Nutricion Acuicola 3 Al 6 De September Del 2002*, Cancun, Quantana Roo p 73-90.
15. Araneda M, Pérez E P, Gasca-Leyva E 2008 White shrimp Penaeus vannamei culture in freshwater at three densities: Conditionstate based on length and weight. *Aquaculture.* **283** p 13–18
16. Lin Y C, Chen J C 2003Acute toxicity of nitrite on *Litopenaeus Vannamei* (Boone) Juveniles at different salinity levels *Aquaculture* **224** p 193–201

Determining Ecological Status of Two Coastal Waters in Western Java using Macrozoobenthic Community: A Comparison between North Part and South Part

Y Wardiatno, Y Qonita, Mursalin, R Zulmi, H Effendi, M Krisanti, A Mashar, S Hariyadi, AA Hakim, A Sahidin, B Widigdo and S Nursiyamah

Department of Aquatic Resources Management, Faculty of Fisheries and Marine Sciences, Bogor Agricultural University, Indonesia

E-mail: yusli@ipb.ac.id

Abstract. The main problem of coastal waters in Indonesia is generally caused by anthropogenic activities that promoting the accumulation of organic matters in sediments. It was well-known that organic matters in sediment might influence the macrozoobenthic communities. AMBI and M-AMBI are two recent developed indices which could be used to evaluate the health of soft bottom macrozoobenthic community and their ecological quality status. This present study was aimed to compare and evaluate the ecological status of macrozoobenthic communities in Tangerang coast (northern part of Java) and Palabuhanratu Bay (southern part of Java) based on AMBI and M-AMBI indices. This study was conducted in April-May 2013 in Tangerang coastal area, and in May 2016 in Palabuhanratu Bay. There are 18 sampling sites in the Tangerang coast, and 30 sampling sites in Palabuhanratu Bay. The results showed that the disturbance level of macrozoobenthos community in Palabuhanratu Bay based on AMBI analysis ranged from undisturbed to slightly disturbed, while in Tangerang coastal waters ranged from undisturbed to moderate disturbed. Ecological quality status based on the M-AMBI ranged from poor to high for both locations. The AMBI and M-AMBI indices showed that benthic environment in Palabuhanratu Bay was ecologically better than that inTangerang coastal waters.

1. Introduction

The main problems of coastal waters in Indonesia are generally caused by anthropogenic activities. Some anthropogenic activities (*e.g.* industrial wastewater and domestic sewage, shellfish aquaculture, oil and gas industry and marine transportation, etc.) are believed to promote the accumulation of organic matters in sediments. The organic matters in the sediments are well proved to play an important role in benthic communities affecting the trophic structure and biomass [1]. Many studies showed that benthic organisms were very powerful as bio-indicators of the environmental conditions, since benthos had a measurable response to natural environmental disturbances or human activity [2-5]. Several indices have been developed to evaluate the health of the aquatic environment based on benthic communities, such as the Norwegian Classification Tool [6], AMBI [7], and Bentix [8], etc. AZTI's Marine Biotic Index (AMBI), one of marine biotic indices, has been applied broadly across the globe. The index was successfully used in Spain [7, 9, 10], China [11-12], Southern California [13], Uruguay [14], and Slovenia [15].

Borja *et al.* [7] who developed the index, strongly recommended the application of AZTI Marine Biotic Index (AMBI) to evaluate benthic community response as a result of natural environment changes and anthropogenic pressures. This index can classify the level of disturbance or pollution in a region based on the level of health of the benthic community [16]. Assessment of AMBI index is based on the distribution and abundance of five ecological groups of macro-benthos [17] related to its sensitivity to the environmental pressure. Furthermore, evaluation by Borja *et al.* [9] showed the usefulness of the index for evaluating benthic community health due to different impact resources, for example the impact of organic enrichment, physical changes in habitat, the input of heavy metals, etc., along the coasts of Europe.

Although AMBI has been widely implemented in various coastal areas with different sources of pressures, yet it must be careful to use a single index for all systems due to the complexity of benthic communities and diversity of benthic environment gradients [18]. Therefore, Muxika *et al.* [19] suggested the use of M-AMBI to assess ecological status of benthic communities instead. M-AMBI is Multivariate- AZTI Marine Biotic Index, which is an integration of the Shannon diversity index, richness and AMBI into a multivariate approach.

Benthic ecological status assessment is essential considering the increase of anthropogenic activities over time as a result of the waste load into coastal areas. Coastal waters on the northern and southern part of Java have been affected by anthropogenic activities. Nevertheless, the south and north coastal of Java have different characteristics of bathymetry and oceanography as well as the impact resources. Palabuhanratu Bay, situated in south coast of Java is semi-closed waters with less sources of impact due to anthropogenic or industrial activities. Meanwhile, the northern part, particularly the Tangerang coastal waters is an open coastal area experiencing high anthropogenic pressures [20], from industrial activities, power plants, farms and residential of the mainland. The study was aimed to compare and evaluate the ecological status of macrobenthic communities in Tangerang coastal waters and Palabuhanratu Bay based on AMBI and M-AMBI indices.

2. Methods

2.1. Sampling

Tangerang Coastal is an open coastal waters located in the northern part of Java, precisely in the West Java area. Tangerang coastal waters is considered suffered from high pressure due to the increasing anthropogenic activities over years. The anthropogenic sources comes from industrial areas, residential, power plant, transportation, etc [20]. Palabuhanratu Bay is semi-enclosed waters located in the south of Java. The water body also has been affected by anthropogenic pressures due to tourism and residential activities. There is also a power plant constructed recently in the bay as a source of, at least, thermal polltion. The bay area is the largest bay on the south coast of Java and the wave conditioan is rough since it is directly facing Indian Ocean. This study was conducted in April-May 2013 in Tangerang coastal area, and in May 2016 in Palabuhanratu Bay. There are 18 sampling sites in the Tangerang coast, and 30 sampling sites in Palabuhanratu Bay. Macrozoobenthos was collected by a van Veen Grab (13x26 cm). The macrozoobenthos samples were then screened using 1 mm sieve and preserved in 10% formalin. Fig. 1 showed the map of study sites.

2.2. Data Analysis

Assessment of disturbance level of the macrozoobenthos community structure and determination of the ecological status was performed through a calculation using the AZTI Marine Biotic Index (AMBI) and M-AMBI. This biotic indices classified the type of macrozoobenthos on the sampling sites into five ecological groups (Ecological Group/EG) based on their sensitivity to pollutants [7]. Grall and Glemarec [17] described the five ecological groups of macrozoobenthos. Group I consists of species which is very sensitive to organic enrichment and present under unpolluted conditions (initial state). Group II consists of species which is indifferent to enrichment, always present in low densities with non-significant variations with time (from initial state to slight unbalance). Group III consists of species which is tolerant to excess organic matter enrichment. These species may occur under normal conditions, but their populations are stimulated by organic enrichment (slight unbalance situations). Group IV consists of species which is second-order opportunistic species (slight to pronounced unbalanced situations). Group V consists of species which is first-order opportunistic species (pronounced unbalanced situations). AMBI index is calculated with the formula below [7] [16].

$$AMBI = \frac{[(0)(\%EGI) + (1.5)(\%EGII) + (3)(\%EGIII) + (4.5)(\%EGIV) + (6)(\%EGV)}{100}$$

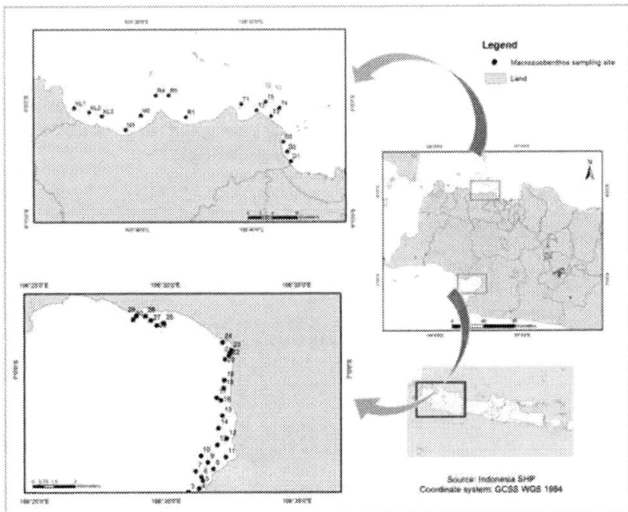

Figure 1. Study area of macrobenthos sampling sites in western Java, i.e. Tangerang Coastal waters (north part) and Palabuhanratu Bay (south part).

AMBI value indicate the level of environmental disruption of benthic habitat. After obtaining the value of AMBI, a multivariate analysis (M-AMBI) was required to measure the ecological status of the benthic environment [19]. M-AMBI calculation involves values of diversity index (Shannon-Wiener index), species richness, and AMBI index. The level of disturbance and ecological quality status were summarized in Table 1.

Table 1. AMBI value and M-AMBI value associate with disturbance classification and ecological status [7, 21, 22].

AMBI	Disturbance classification	M-AMBI	Ecological Quality Status
0.0<AMBI≤1.2	Undisturbed	>0.77	High
1.2<AMBI≤3.3	Slightly disturbed	0.53<AMBI<0.77	Good
3.3<AMBI≤ 5.0	Meanly disturbed	0.38< AMBI <0.53	Moderate
5.0<AMBI≤ 6.0	Heavily disturbed	0.20< AMBI <0.38	Poor
6.0<AMBI≤ 7.0	Extremely disturbed	<0.20	Bad

3. Results

Based on the analysis of 30 sampling sites in Palabuhanratu Bay, 65 taxa were obtained included in 6 groups consisting Echinodermata, Polychaeta, Crustaceans, Sipuncula, Nemertina, and Pelecypoda. Polychaeta is the group with the highest number of taxa, i.e. 30 taxa, followed by crustaceans (19 taxa), Pelecypoda (7 taxa), Echinodermata (5 taxa), Sipuncula (3 taxa), and Nemertina (1taxa) (Figure 2). The results of the analysis of macrozoobenthos community of 18 sampling sites in the coast of Tangerang showed that there were 81 taxa included in 15 groups (Polychaeta, Crustaceans, Bivalves, Anopla, Enopla, Clitellata, Gastropods, Eumidasanguinea, Sipunculidea, Ophiuroidea, Actinopterigii, Echiura, Anthozoa, Echinoidea, Ascidiaecae). Similar to Palabuhanratu Bay, Polychaeta also had the highest number of taxa in Tangerang coast, i.e. 24 taxa. Eumidasanguinea, Sipunculidea, Ophiuroidea, Actinopterigii, Echiura, Anthozoa, Ascidiaecae only had one taxa found in each group, thus they were grouped into miscellaneous group (Figure 3).

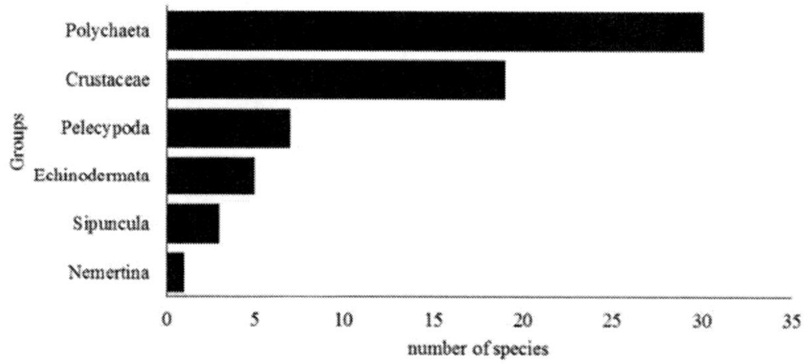

Figure 2. Macrozoobenthos composition from 30 sampling sites in Palabuhanratu Bay.

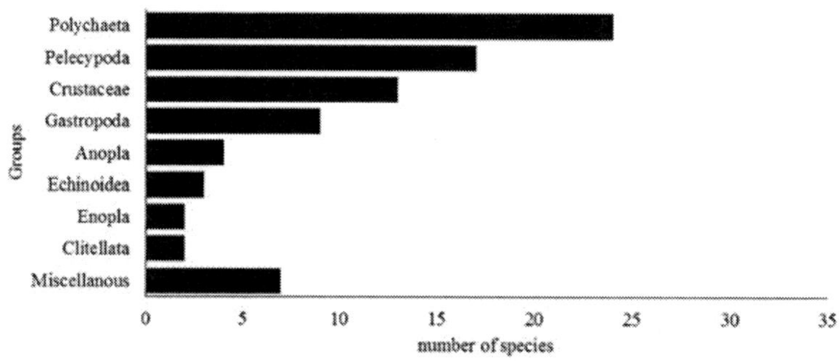

Figure 3. Macrozoobenthos composition from 18 sampling sites in Tangerang coastal waters.

Macrozoobenthos highest density in Palabuhanratu Bay was detected at station 22 (2021 ind/m^2), and the highest number of individual was observed in *Cirratulus grandis* belonging to Polychaeta. The lowest benthos density was observed at station 19 (44 ind/m^2) consisting of two species only, namely *Cirratulus* sp. and *Pontogeneia intermedia* which belong to Polychaeta and Crustacea, respectively. Total density of macrozoobenthos of each station in Palabuhanratu Bay was summarized in Figure 4. Macrozoobenthos density in Tangerang coast ranged from 0 to 1069 ind/m^2 (there was no macrozoobenthos samples found in R2 and R3 station). Unlike Palabuhanratu Bay, the highest density of macrozoobenthos in Tangerang coastal waters was observed at D3 station (Figure 5) whis was dominated by *Nuculana vernula* of Pelecypoda group.

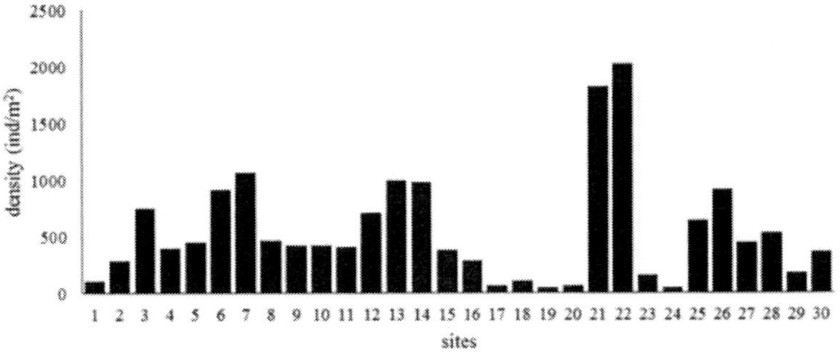

Figure 4 Macrozoobenthos at each sampling station density in Palabuahanratu Bay.

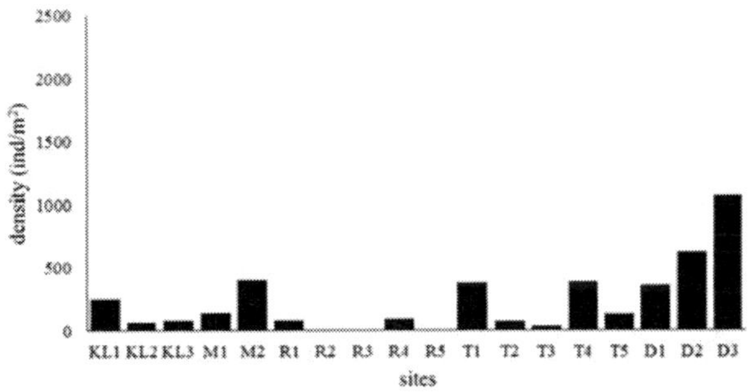

Figure 5. Macrozoobenthos density at each samplg station in Tangerang coastal waters. No fauna was found in stations R2 and R3.

From all of the macrozoobenthos community in Palabuhanratu Bay, no species belong to Ecological Group V of AMBI was observed. The mean value of AMBI obtained for each station ranged from 0.5 to 3.5, which implied the level of disturbance of macrozoobenthos communities classified as undisturbed until moderate disturbed (Table 2). Several stations were classified as undisturbed population, i.e. stations 8, 11, 15, 16, and 17; yet station 3 was classified as moderate disturbed. The rest is classified as slightly disturbed. This indicated that the condition of benthic communities in most of the sites in Palabuhanratu Bay was still in relatively good condition because it was in level of undisturbed and slightly disturbed. Macrozoobenthos composition based on the type of ecological group at each station in Palabuhanratu Bay is shown in Figure 6.

Although the condition of macrozoobenthos community in most of stations in Palabuhanratu Bay could be considered as relatively good, the number of taxa (S) was found in great variation (2-19 taxa). There were two stations which have the lowest number of taxa, i.e. station 19 and 24, and the highest number of taxa was observed at station 26. The value of diversity index (H') also varied from 0.95 to 3.90 and Simpson's dominance index (C) ranged from 0.08 to 0.72. These tell us the occurrence of disturbance in macrozoobenthos community at some sampling stations in Palabuhanratu Bay.

Table 2. AMBI and M-AMBI values of 30 sampling stations in Palabuhanratu Bay.

Sites	Ecological Group Presentation (%)						AMBI	Disturbance classification	M-AMBI	Ecological status
	I	II	III	IV	V	not assigned				
1	33.3	33.3	33.3	0.0	0	40	1.5	Slightly disturbed	0.52	Moderate
2	9.1	63.8	9.1	18.1	0	15.3	2.0	Slightly disturbed	0.64	Good
3	3.0	18.1	21.2	57.7	0	2.9	3.5	Moderate disturbed	0.65	Good
4	33.5	19.9	26.6	19.9	0	16.8	2.0	Slightly disturbed	0.73	Good
5	30.7	30.7	7.7	31.0	0	35.1	2.1	Slightly disturbed	0.76	Good
6	19.6	19.6	7.3	53.6	0	0	2.9	Slightly disturbed	0.59	Good
7	39.6	20.8	12.5	27.1	0	0	1.9	Slightly disturbed	0.85	High
8	66.7	28.5	0.0	4.7	0	0	0.6	Undisturbed	0.71	Good
9	23.5	29.4	41.3	5.8	0	10.4	1.9	Slightly disturbed	0.62	Good
10	37.6	6.2	31.1	25.1	0	15.9	2.2	Slightly disturbed	0.68	Good
11	64.2	7.1	28.7	0.0	0	22.3	1.0	Undisturbed	0.67	Good
12	28.0	60.1	7.9	4.0	0	21.9	1.3	Slightly disturbed	0.72	Good
13	11.1	84.5	2.2	2.2	0	0	1.4	Slightly disturbed	0.46	Moderate
14	42.5	20.0	5.0	32.5	0	9	1.9	Slightly disturbed	0.70	Good
15	64.8	35.2	0.0	0.0	0	0	0.5	Undisturbed	0.53	Moderate
16	54.7	27.2	18.1	0.0	0	15.3	1.0	Undisturbed	0.77	Good
17	66.7	33.3	0.0	0.0	0	0	0.5	Undisturbed	0.54	Good
18	40.0	40.0	0.0	20.0	0	0	1.5	Slightly disturbed	0.57	Good
19	0.0	50.0	0.0	50.0	0	0	3.0	Slightly disturbed	0.31	Poor
20	0.0	66.7	33.3	0.0	0	0	2.0	Slightly disturbed	0.44	Moderate
21	23.0	10.8	59.5	6.7	0	9.7	2.2	Slightly disturbed	0.66	Good
22	11.2	34.8	49.5	4.5	0	2.2	2.2	Slightly disturbed	0.66	Good
23	14.3	28.6	42.9	14.3	0	0	2.4	Slightly disturbed	0.56	Good
24	0.0	50.0	0.0	50.0	0	0	3.0	Slightly disturbed	0.31	Poor
25	13.2	47.9	21.6	17.3	0	20.7	2.1	Slightly disturbed	0.82	High
26	7.6	33.4	35.9	23.0	0	4.8	2.6	Slightly disturbed	0.87	High
27	11.0	5.5	55.6	27.8	0	9.9	3.0	Slightly disturbed	0.59	Good
28	41.7	33.4	12.4	12.6	0	0	1.4	Slightly disturbed	0.69	Good
29	28.6	28.6	14.3	28.6	0	12.5	2.1	Slightly disturbed	0.61	Good
30	23.6	32.9	14.5	29.0	0	18.6	2.2	Slightly disturbed	0.68	Good

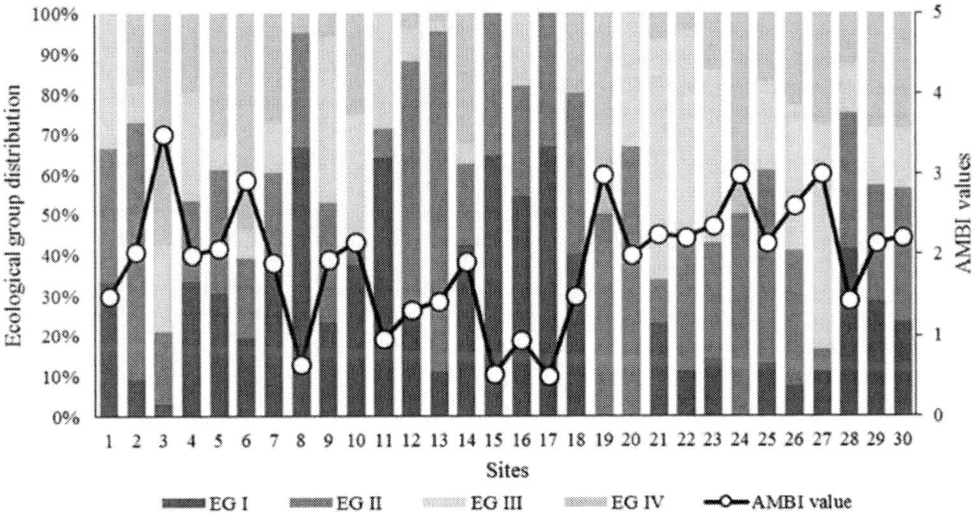

Figure 6. Ecological group composition and AMBI values in 30 sampling sites in Palabuhanratu Bay

Based on the M-AMBI value, the ecological status of macrozoobenthos community in Palabuhanratu ranged from Poor to High. There were 3 stations (10.00%) classified into high ecological quality status, 21 stations (70.00%) classified into good quality ecological status, 4 stations (13.33%) classified into moderate ecological quality status, and 2 stations (6.67%) classified into poor quality ecological status. Ecological status distribution based on the M-AMBI in Palabuhanratu Bay was shown in Figure 7.

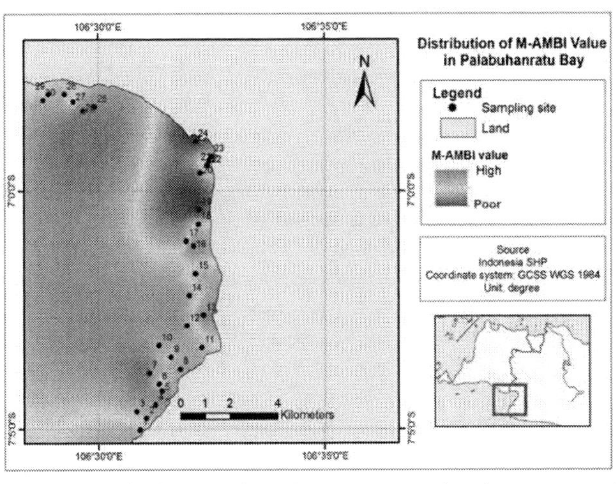

Figure 7. Distribution of ecological status based on macrozoobenthos community with M-AMBI value approach in Palabuhanratu Bay.

AMBI mean value in Tangerang coastal waters ranged from 0.075 to 4.682 (R2 and R3 stations were excluded from the analysis) (Table 3). Macrozoobenthos composition based on the type of ecological group at each station in Tangerang coastal waters is shown in Figure 8. The values indicated that the condition of macrozoobenthos community classified into undisturbed until moderate disturbed. Undisturbed population was examined at T5, D1, D2, D3 station, while the slightly disturbed population observed at T1, T2, T3, T4, KL1, KL2, KL3, M1, M2, and R1 station, however moderate disturbed population observed at R4 and R5 station.

Table 3. AMBI and M-AMBI values of 18 sampling sites in Tangerang coast

Sites	Ecological group presentation (%)						AMBI	Disturbance classification	M-AMBI	Ecological status
	I	II	III	IV	V	not assigned				
KL1	15.4	3.5	73.8	1.9	5.4	0.5	2.68	Slightly disturbed	0.49	Moderate
KL2	1.0	2.1	94.8	1.0	1.0	0	2.98	Slightly disturbed	0.27	Poor
KL3	2.5	0.0	87.4	2.5	7.6	1.7	3.19	Slightly disturbed	0.29	Poor
M1	21.3	1.0	64.9	7.9	5.0	0	2.61	Slightly disturbed	0.58	Good
M2	3.3	1.4	91.6	2.6	1.1	1.7	2.95	Slightly disturbed	0.45	Moderate
R1	45.9	14.4	21.6	2.7	15.3	5.9	1.91	Slightly disturbed	0.71	Good
R2	0.0	0.0	0.0	0.0	0.0	-	-	-	-	-
R3	0.0	0.0	0.0	0.0	0.0	-	-	-	-	-
R4	8.6	1.4	24.3	0.7	65.0	0	4.68	Moderate disturbed	0.32	Poor
R5	0.0	40.0	0.0	60.0	0.0	0	3.30	Moderate disturbed	0.25	Poor
T1	30.8	18.8	30.6	14.0	5.8	5.8	2.18	Slightly disturbed	0.82	High
T2	35.6	25.7	15.8	10.9	11.9	1	2.06	Slightly disturbed	0.85	High
T3	8.7	6.5	69.6	15.2	0.0	0	2.87	Slightly disturbed	0.40	Moderate
T4	48.9	23.4	9.7	6.6	11.4	6.5	1.62	Slightly disturbed	0.86	High
T5	75.5	1.6	17.9	4.9	0.0	0.5	0.78	Undisturbed	0.57	Good
D1	96.7	0.4	1.7	1.2	0.0	0	0.11	Undisturbed	0.47	Moderate
D2	97.5	0.0	2.5	0.0	0.0	0	0.08	Undisturbed	0.41	Moderate
D3	91.2	0.6	5.0	1.0	2.2	0	0.34	Undisturbed	0.47	Moderate

Number of taxa found in each station ranged from 0 to 26. The diversity index (H') of macrozoobenthos in Tangerang coastal waters ranged from 0.54 to 3.7 and Simpson's dominance index ranged from 0.11 to 0.86. The low value of the diversity index and high value of the dominance index at some stations indicated there was disturbance to benthic communities in the coast of Tangerang. With 2 stations without fauna (11.11%), the results of the analysis of M-AMBI in Tangerang coastal waters showed that 4 stations (22.22%) were identified as poor in ecological quality, 6 stations (33.33%) as moderate, 3 stations (16.67%) as good and 3 stations (16.67%) as high. Ecological status distribution based on the M-AMBI in Tangerang coastal waters was visualized in Figure 9.

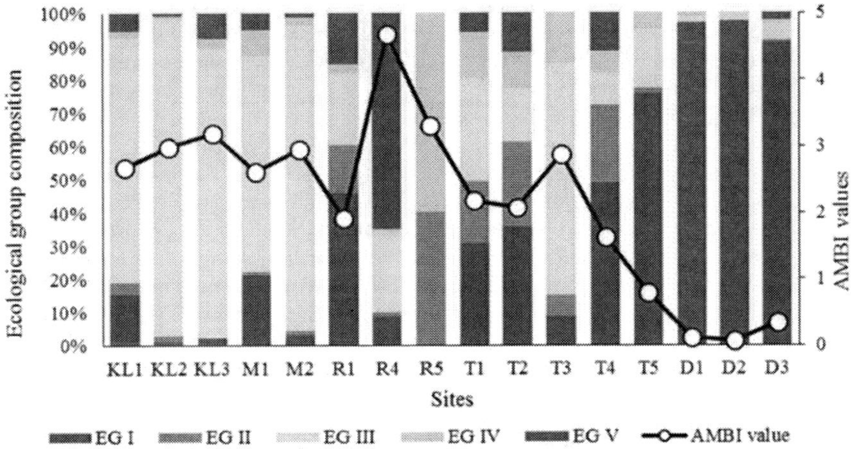

Figure 8. Ecological group composition in 16 sampling sites in Tangerang coast

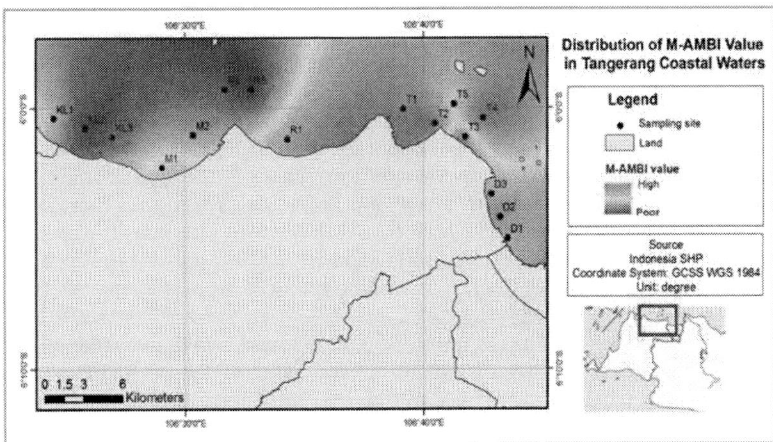

Figure 9. Distribution of ecological quality status based on macrozoobenthos community with M-AMBI value approach in Tangerang coastal waters.

4. Discussion

Analysis of benthic communities in Pelabuhanratu Bay (southern part of Java) and coastal Tangerang (northern part of Java) showed that there were 65 taxa and 81 taxa, respectively. The highest number of taxa in both waters come from the same group, i.e. Polychaeta. The similar results were also discoverd by Mavric *et al.* [15] in the Gulf of Trieste, Slovenian which was dominated by Polychaeta (65-84%), and by Zettler *et al.* [23] in the Baltic Sea. This phenomenon is not uncommon condition because polychaetes are usually the most abundant taxon in benthic communities of disturbed soft-bottom sediments, so that is why they are quite often utilized as indicator species of environmental conditions [24]. Morreover, since the taxon contains both sensitive and tolerant species, the polychaeta have been identified worldwide as group that respond quickly to environment disturbance [2, 25-27]. Polychaetes are an important component of macro-benthic community as they often dominate in term of abundance, biomass, and species or taxa. They play an important role in the stability and functioning of the benthic community and the ecological in general. They are extremely diverse taxa, with wide range of reproductive strategies and feeding guilds [27, 28].

The highest density of macrozoobenthos found at station 22 in Palabuhanratu Bay, and the most dominant species in this station was *Cirratulus grandis* which belong to the Ecological Group III. This group includes species tolerant to excess organic matter enrichment. *C. grandis* is a selective deposit feeder, consuming organic particles found within the sediments, so that *C. grandis* is commonly used as a test organism to study the concentrations and effects of various pollutants found within sediments in marine environments [29, 30]. *C. grandis* also helps to overturn and oxygenate sediment by mixing particles during burrowing and feeding activities [31]. The highest macrozoobenthos density in Tangerang coastal waters was found in station D3 dominated also by Pelecypoda group, *Nuculana vernula*. *N. vernula* is a species belong to family Nuculidae classified into EG I which including species that are very sensitive to organic matter enrichment. The species in this group is considered as an indication of the health environmental ecosystem or unpolluted conditions [7].

The analysis of AMBI showed that benthos community level of disturbance in the Palabuhanratu Bay ecosystem ranged from undisturbed until moderate disturbed, while the result of the M-AMBI analysis stated ecological status in these waters ranged from poor to high. There was a significant difference between the results of the AMBI and M-AMBI on several stations in Palabuhanratu Bay. Stations 19 and 24 were classified as slightly disturbed by AMBI, yet classified as poor of Ecological quality status based on the M-AMBI. This difference could be caused by low species richness (only 2 species) and low diversity index at both stations, therefore the value of M-AMBI become low. There were only two species found in station 19, i.e. *Cirratulus* sp. (EG IV) and *Pontogeneia intermedia* (EG II), while at the station 24, *Prionospio* sp. (EG IV) and *Pholoe* sp. (EG II) were found. According to Zettler *et al.* [23], the AMBI index considers only the abundance and species classification into its ecological group without considering the species richness in the area. Borja *et al.* [32] and Borja & Muxika [33] agreed with this disadvantage and limitation of using it. Therefore Muxika *et al.* [19] suggested the using of Multivariate-AMBI (M-AMBI) which integrates the species richness, diversity index, and the AMBI value. As a note, Borja and Muxika [33] stated if the species found in a location is less than equal to 3 species, then it should be removed from the analysis of AMBI, because the value of AMBI will be less accurate.

The same phenomenon occurred in the Tangerang coast, where the value of AMBI at station KL2 and KL3 belong to the slightly disturbed, but the value of the M-AMBI showed poor ecological quality status. Evaluating from the species richness, there were 6 taxa in both stations. These station also had a low diversity index. This indicated an unbalance in macrozoobenthos communities in both locations. This unbalanced can be caused by the location of the KL2 and KL3 station which located not far from the mouth of two major rivers, i.e. Kronjo and Cidurian Rivers. The two rivers discharged anthropogenic inputs of urban waste to coastal waters areas which might be affecting macrozoobenthos communities in the Tangerang coastal waters. In addition, the water currents of both river which met and bumped to coastal ocean current was indicated causing heavy turbulence which caused natural environmental pressure to the macrozoobenthos communities.

As AMBI was developed for assessing the health status of the macrozoobenthos community in European coastal areas, it is necessary to make adjustments when it is used in other regions, especially in the tropical regions, because the species composition of macrozoobenthos in each location are not the same and many local species is not included in the list. This caution opinion was also revealed by Liu *et al.* [12] who conducted a study in the Changjiang River, China and made some adjustments in classifying some local species to its ecological group by using the assignment of the same genus or expert judgment in accordance with Borja *et al.* [34]. In addition, according to Borja and Muxika [33], generally, not assigned taxa discovered in one location should be less than 15%. However, when the percentage of not assigned taxa is more than 20%, then the value of AMBI should be interpreted with care. Also, when not assigned taxa found more than 50%, then AMBI should not be used. In this study, there were no station had the percentage of not assigned taxa more than 50%, but there were several stations in Palabuhanratu Bay which had percentage of not assigned taxa more than 20%, i.e. station 1, 5, 11, 12, and 25, hence the AMBI value at these stations should be evaluated with care, especially in case of poor qualiity status.

Generally, marine environment ecosystem in Palabuhanratu Bay was relatively in better condition than the Tangerang coastal waters by seeing the values of AMBI and M-AMBI indices. In comparison of AMBI value, 96.67% of sampling stations in Palabuhanratu Bay falls into undisturbed and slightly disturbed condition, while in Tangerang coastal waters is only 77.78%. In terms of M-AMBI values, the ecological status with good to high quality in Tangerang coastal waters was only 33.34% sampling station with 11.11% was found witout fauna. While in Palabuhanratu Bay, the good to high ecological status was demonstrated by 80% of sampling stations with 0% stations without fauna.

The north coast of Java Island is more densely populated urban areas and has many industrial areas that contribute the large waste into the coastal waters of Tangerang. However, Palabuhanratu Bay was also not free from the anthropogenic activities but it was expected that anthropogenic pressures on the southern coast of Java was lower than that of the north coast of Java. In addition, the position of Palabuhanratu Bay which directly facing the Indian Ocean with huge wave would help the bay to accelerate the flushing of this semi-enclosed waters.

5. Conclusion

The disturbance level of macrozoobenthos community based on AMBI analysis was ranged from undisturbed to moderate disturbed in Palabuhanratu Bay and Tangerang coastal areas. Ecological quality status based on the M-AMBI was ranged from poor to high for both locations. These conditions indicated a disturbanbe in benthoc environment of both location due probably to anthropogenic activities discharging waste to the seas. By comparing the AMBI and M-MABI indices, it could be concluded that benthic environment in Palabuhanratu Bay (southern part of West Java) was less affected by anthropogenic activities than Tangerang coastal waters (northern part of West Java).

Acknowledgement

This study was supported by BOPTN budget of Higher Education Directorate (DIKTI) and PT Kapuk Naga Indah. We also wish to thank to research team of Department of Aquatic Resources Management for kindly helping on this research.

References

[1] Grebmeier JM, McRoy CP and Feder HM 1988 Pelagic-benthic coupling on the shelf of the northern Bering and Chukchi Seas. I. Food supply source and benthic biomass *Mar. Ecol. Prog. Ser.* **48** 57-67

[2] Pearson T and Rosenberg R 1978 Macrobenthic succession in relation to organic enrichment and pollution of the marine environment *Oceanogr. Mar. Biol. Annu. Rev.* **16** 229-311

[3] Thouzeau G, Robert G, and Ugarte R 1991 Faunal assemblages of benthic mega invertebrates inhabiting sea scallop grounds from eastern Georges Bank in relation to environmental factors *Mar. Ecol. Prog. Ser.* **74** 61-82

[4] Dauer DM 1993 Biological criteria, environmental health and estuarine macro benthic community structure. *Mar. Poll. Bull.* **26** 249-57

[5] Ritter C and Montagna PA 1999 Seasonal hypoxia and model of benthic response in a Texas bay. *Estuaries* **22** 7-20

[6] Molvaer J, Knutzen, J, Magnusson J, Rygg B, Skey JM and Sorensen J 1997 Classification of environmental quality in fjords and coastal waters. A guide. Veiledning 97:03. Norwegian Pollution Control Authority Oslo. 35 pp. [in Norwegian]

[7] Borja A, Franco J and Pérez V 2000 A marine biotic index to establish the ecological quality of soft-bottom benthos within European estuarine and coastal environments. *Mar. Poll. Bull.* **40** 1100-14
[8] Simboura N and Zenetos A 2002 Benthic indicators to use in Ecological Quality Classification of Mediterranean soft bottom marine ecosystems, including a new Biotic Index. *Mediterr. Mar. Sci.* **3**(2) 77-111
[9] Borja A, Muxika I and Franco F 2003 The application of a Marine Biotic Index to different impact sources affecting soft-bottom benthic communities along European coasts. *Mar. Poll. Bull.* **46** 835-45
[10] Muxika I, Borja A and Franco J 2003 The use of a biotic index (AMBI) to identify spatial and temporal impact gradients on benthic communities in an estuarine area. ICES CM2003/ Session J-01
[11] Cai W, Liu L, Meng W and Zheng B 2012 The suitability of AMBI to benthic quality assessment on the intertidal zones of Bohai Sea *Acta Scien. Circum.* **32** 992-1000
[12] Liu L, Li B, Lin K, Cai W and Wang Q 2014 Assessing benthic ecological status in coastal area near Changjiang River estuary using AMBI and M-AMBI *Chinese J. Oceanol. Limnol.* **32**(2) 290-305
[13] Teixeira H, Weisberg SB, Borja A, Ranasinghe JA, Cadien DB, Velarde RG, Lovell LL, Pasko D, Phillips CA, Montagne DE, Ritter KJ, Salas F and Marques JC 2012 Calibration and validation of the AZTI's Marine Biotic Index (AMBI) for Southern California marine bays *Ecol. Indic.* **12** 84-95
[14] Muniz P, Hutton M, Kandratavicius N, Lanfranconi A, Brugnoli E., Venturini N and Giménez L 2012 Performance of biotic indices in naturally stressed estuarine environments on the South western Atlantic coast (Uruguay): A multiple scale approach *Ecol. Indic.* **19** 89-97.
[15] Mavrič B, Orlando-Bonaca M, Bettoso N and Lipej L 2010 Soft-bottom macrozoobenthos of the southern part of the Gulf of Trieste: faunistic, biocoenotic and ecological survey *Acta Adriat.* **51**(2) 203-216
[16] Muxika I, Borja A and Bonne W 2005 The suitability of the marine biotic index (AMBI) to new impact sources along European coasts *Ecol. Indic.* **5** 19-31
[17] Grall J and Glemarec M 1997 Using biotic indices to estimate macro benthic community perturbations in the Bay of Brest *Estuar. Coast. Shelf Sci.* **44** (suppl. A) 43-53
[18] Engle VD and Summers JK 1999 Refinement, validation and application of a benthic condition index for Northern Gulf of Mexico estuaries *Estuaries* **22**(3A) 624-35.
[19] Muxika I, Borja A and Bald J 2007 Using historical data, expert judgement and multivariate analysis in assessing reference conditions and benthic ecological status, according to the European Water Framework Directive *Mar. Poll. Bull.* **55** 16-29.
[20] LPPM IPB & Kapuk Naga Indah 2013 Status terkini sumberdaya perikanan dan lingkungan di pesisir utara Kabupaten Tangerang. 259 p
[21] Borja A, Josefson AB, Miles A, Muxika I, Olsgard F, Phillips G, Rodríguez JG and Rygg B 2007 An approach to the intercalibration of benthic ecological status assessment in the North Atlantic ecoregion, according to the European Water Framework Directive *Mar. Poll. Bull.* **55** 42-52.
[22] Borja A, Miles A, Occhipinti-Ambrogi A and Berg T 2009 Current status of macroinvertebrate methods used for assessing the quality of European marine waters: implementing the Water Framework Directive *Hydrobiologia* **633**(1) 181-196.
[23] Zettler ML, Schiedek D and Bobertz B 2007 Benthic biodiversity indices versus salinity gradient in the southern Baltic Sea. *Mar. Poll. Bull.* **55** 258-70
[24] Dean HK 2008 The use of polychaetes (Annelida) as indicator species of marine pollution: a review. *Rev Biol Trop* (*Int J Trop Biol*) **56**(4) 11-38
[25] Pocklington P and Wells PG 1992 Polychaetes: key taxa for marine environmental quality monitoring. *Mar. Poll. Bull.* **24** 593-598

[26] Zajac RN and Whitlatch RB 2003 Community and population-level responses to disturbance in a sandflat community. *J. Exp. Mar. Biol. Ecol.* **294** 101-125
[27] Sivadas S, Ingole B and Nanajkar M 2010 Benthic polychaetes as good indicators of anthropogenic impact *Ind. J. of Mar. Sci.* **39**(2) 201-11
[28] Hutching PA 1998 Biodiversity and functioning of polychaetes in benthic sediments. *Biodivers. Conserv.* **7**(9) 1133-1145
[29] Rust AJ, Burgess RM, Brownawell BJ and McElroy AE 2004a Relationship between metabolism and bioaccumulation of benzo[*a*]pyrene in benthic invertebrates *Env. Toxic. Chem.* **23** 2587-93
[30] Rust AJ, Burgess RM, McElroy AE, Cantwell MG and Brownawell BJ 2004b Influence of soot carbon on the bioaccumulation of sediment-bound polycyclic aromatic hydrocarbons by marine benthic invertebrates: an interspecies comparison *Env. Tox. Chem.* **23** 2594-603
[31] Shull DH and Yasuda M 2001 Size-selective downward particle transport by cirratulid polychaetes *J. Mar. Res.* **59** 453-73
[32] Borja A, Franco J, Valencia V, Bald J, Muxika I, Belzunce MJ and Solaun O 2004 Implementation of the European Water Framework Directive from the Basque Country (northern Spain): a methodological approach *Mar. Poll. Bull.* **48** (3-4) 209-218
[33] Borja A and Muxika I. 2005. Guidelines for the use of AMBI (AZTI's Marine Biotic Index) in the assessment of the benthic ecological quality *Mar. Poll. Bull.* **50** 787-789
[34] Borja A, Dauer DM, Diaz R, Llanso R J, Rodriguez J G and Schaffner L 2008 Assessing estuarine benthic quality conditions in Chesapeake Bay: a comparison of three indices *Ecol. Indic.* **8** 395-403

The relationship between concentration of clorophyll-a with skipjack (*Katsuwonus pelamis, Linnaeus 1758*) production at West Sumatera waters, Indonesia

Usman*, T. Ersti Yulika Sari, Syaifuddin, Audina

Department of Aquatic Resources Utilization, Faculty of Fisheries and Marine Science, Riau University

E-mail: usman_psp@yahoo.co.id

Abstract. The regression and correlation technic was uses to evaluated the contribution of chlorophyll-a concentration on variation of longline skipjack tuna production. An analysis was performed by placing Chlorophyll-a as predictor and Skipjack (*Katsuwonus pelamis, Linnaeus 1758) p*roduction as dependent variable, using Chlorophyll-a derived from NPP VIIRS, and CPUE derived from longline fisherman log books for the year of 2013. Chlorophyll-a distribution which derived from NPP VIIRS between 0.13-0.26 mg/m3 whereas maximum CPUE as much as 0,1875 kg/trip in April. The regression equation obtained was *CPUE = -1.12 + 11.5 Chl-a.* Correlation between chlorophyll-a and CPUE have moderate relationship (r=0.51). From regression equation for those variables showed that the variation of chlorophyll-a had affected about 26% on variation of CPUE, only.

1. Introduction

Skipjack tuna (*Katsuwonus pelamis, Linnaeus 1758)* is one of the economically important marine fish resources resulting from marine waters of Indonesia, both as an export commodity as well as domestic consumption to meet national nutritional needs. Based on statistical data, the production of tuna fishery at West Sumatera since 2008 showed a very significant improvement. In 2008 the province recorded a production of 300 tons, which increased by more than double in 2009 which became 735 tons [1].

In 2006 the annual catch of skipjack in the Indian Ocean peaked at 620,000 tons and since then, catches have not exceeded 450,000 tons as reported by Indian Ocean Tuna Commission, 2010 *in* Dueri, Faugeras, and Maury, [2]. Commodity featured fish and potential for export are yellow fin tuna, bigeye tuna, skipjack, tuna and mackerel are mostly caught by fishermen using troll line, trolling and longline.

Fishing activities would be more efficient and effective when fishing areas can be predicted in advance, before the fishing fleet set out from the base. One step to determine potential fishing areas through the study of fishing grounds and its relation to oceanographic phenomenon in a sustainable manner [3].

Laevastu and Hela [4] stated that one of the properties tuna is highly migratory and hunting areas are plentiful source of food. As it is known that the chlorophyll-a can be used to determine the abundance of phytoplankton and can be used as indicators of marine productivities. Thus, indirectly, there is a relationship between phytoplankton to the presence of fish. The distribution of chlorophyll-a concentration can be detected by remote sensing, using satellite imagery MODIS aqua. On the other hand, some oceanographic parameters are not obtained directly from these sensors, but instead, they are estimated by the radiance reflected and/or emitted by the ocean surface. The conversion of this radiance into values of oceanographic parameters (e.g. sea surface temperature and chlorophyll-a concentration) estimate by using the specific algorithms. This allows the monitoring of surface oceanic parameters that influence the distribution of pelagic species, i.e. those living in surface waters.

Skipjack tuna, is a pelagic species adapted to perform large migrations that inhabit the three largest oceans of the world [5]. Skipjack tuna is the most important species targeting by local fishermen in the Indian Ocean, specialy long line and purse seine fisherman from West Sumatera. The distribution and migration of this species are strongly influenced by oceanographic factors such as distributions of SST and chlorophyll-a (hereafter Chl-a). SST and Chl-a derived from satellite data are determinant factors in predicting tuna forage and their habitat in Western North Pacific Ocean [3]. The objectives of this study are to explore the short term relationship between oceanographic factors and skipjack tuna abundance, and to predict potential fishing grounds using GIS techniques.

2. Material and methods

2.1. Study area

The study area covered skipjack tuna fishing ground at Indian Ocean, West Sumatera, Indonesia and located at 0°25'48"– 5°34'48" N and 95°25'23" - 100°57 ' E (Figure 1). Determining the location of the research is based on the logbook tuna fishing ground carried out by fishermen whose landing their catch in PPS Bungus

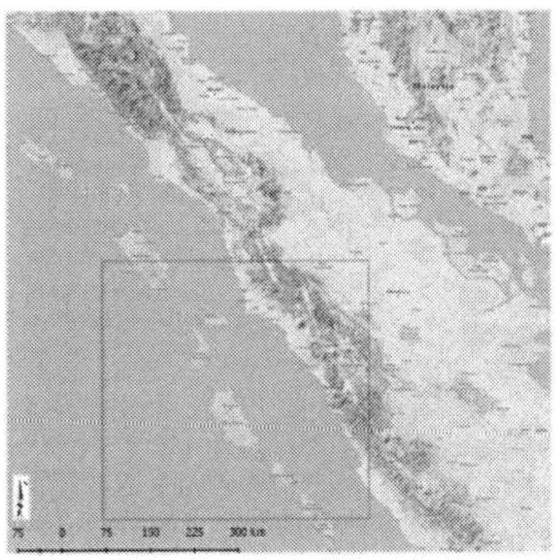

Figure 1. Study area

2.2. Fisheries data

Skipjack tuna daily catch data obtained from fishing logbook of long line fishery, for the period January to December 2013. These data comprised daily geo-referenced fishing positions (latitude and longitude), catch in kilogram and effort, and catch per unit effort (CPUE) was determined.

2.3. Remote sensed environmental data

By using the biological environmental data to describe the oceanographic conditions at the fishing ground were Chl-a, derived from SATELLITE IMAGE NPP VIIRS level 3 for the year of 2013.

2.4. Data analysis

In this study, firstly the fishing data were plotted on Chl-a image maps. Then the oceanographic data around the fishing positions were extracted to understand the spatial and temporal distribution pattern of skipjack fishing grounds relative to the oceanographic conditions. To describe the relationship between oceanographic conditions of chl-a, and skipjack production (CPUE), we were using regression and corellation analysis.

3. Result and Discussion

The potential areas of tuna fishing ground of West Sumatera waters in the position 0^0 25'48"N - 5^0 34' 48"S and 95^0 25'23"-$100^0$57'E. Fishing gears were standardized to purse seine since we found this gears commonly used by the fishermen. Total catching station on the east and west season are 21 and 24 stations, while the highest point in the monsoon II as many as 50 stations and in the monsoon II is 30 stations.

Determination of catching station based on the experience of the fishermen in the previous year. Natural conditions such as the rainy season and dry season which also affects the catch. The distribution of chlorophyll-a were analyzed by using satellite imagery NPP VIIRS level 3 in each fishing ground in 2013 had the highest level of chlorophyll-a at 1.22 mg m^{-3} and the lowest at 0,075 mg m^{-3}. Septiawan [6] stated that the highest level of chlorophyll-a in April included in the high category between 0.501 to 1.0 mg m^{-3} whereas chlorophyll-a has a value of 0.075 mg m^{-3} was the lowest. Based on the fertility level in the waters. Gunarso [7] explained that catching station which has a value of chlorophyll-a 1.22 mg m^{-3} included in the category of fertile i.e 0.4 to 2 mg m^{-3} while the chlorophyll-a-value of 0.075 mg m^{-3} fertility rate is very low at <0.2 mg m^{-3} (Figure 2).

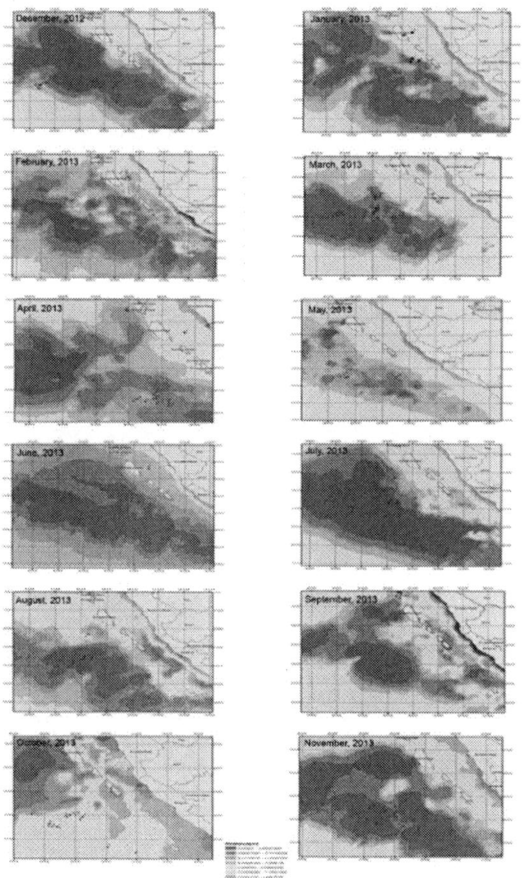

Figure 2. Monthly of Chl-a concentration distribution

Analysis of chlorophyll-a satellite image of NPP VIIRS show in 2013 recorded an average value of chlorophyll-a is 0.22 mg m^{-3}. Viewed from the fishing season the average value of chlorophyll-a increased. When the west season the average value of of chlorophyll-a is 0.19 mg m^{-3}, at monsoon 1 when the average value of chlorophyll-a is 0.20 mg / m3, at the east season 0.24 mg m^{-3} and 0.26 mg m^{-3} while monsoon II.

Based on the level of eutrophication of waters according to Bricker et al. in [8] average of chlorophyll-a in 2013 in the category low as 0-5 mg m^{3}. Nixon in [8] defined eutrophication as a process of increasing the rate of supply of organic material to the water body, whether originating from within the water itself (autochthonous) or from outside the waters (*allochthonous*). Enrichment of this water causes an increase in phytoplankton growth and turbidity. This condition causes a negative impact on aquatic ecosystems, such as reduced oxygen in the water column and in extreme conditions can cause death in fish and benthic organisms.

The regression equation obtained was CPUE = -1.12 + 11.5Chl-a. The result showed that increasing the chlorophyll-a had positive impact on the CPUE of skipjack (*Katsuwonus pelamis*). The highest CPUE occurred in November is 4.310 kg/settings as well as the concentration of chlorophyll-a in November is 0.287 mg m^{-3} (Table 1). This condition could be caused by the location of tuna fishing ground, environmental factors such as weather, wind, salinity and fishing season.

Table 1 Chl-a concentration and CPUE for skipjack tuna, 2013

Month	Chl-a	CPUE
January	0.128	0.631
February	0.169	1.000
March	0.174	1.000
April	0.207	1.494
May	0.206	1.579
June	0.239	0.554
July	0.241	0.608
August	0.254	1.000
October	0.229	1.774
November	0.287	4.310
December	0.260	1.180

The coefficient correlation between the concentration of chlorophyll-a and CPUE of skipjack (*Katsuwonus pelamis*) is 0.51. This means that the relationship between the concentration of chlorophyll-a and CPUE in tuna was moderate. In addition, the coefficient of determination between the two variables was only 26%, there was about 74% caused by others factors.

4. Conclusion

The regression equation obtained was CPUE = -1.12 + 11.5Chl-a and the increasing the chlorophyll-a had positive impact on the CPUE of skipjack. The coefficient of determination between the two variables was only 26%, there was about 74% caused by others factors

References

[1] Pillai G N P S 2012 *Biology, fishery, conservation and management of Indian Ocean Tuna Fisheries. Jurnal Ocean Sci.* J **47** 23

[2] Dueri, S, Faugeras B and Maury O 2012 Modelling the skipjack tuna dynamics in the Indian Ocean with APECOSM-E: Part 12 *Model formulation. Ecological Modelling* **245** 41–54.
[3] Dueri, S, Faugeras B and Maury O 2013 Modelling the effect of marine protected areas on the population of skipjack tuna in the Indian Ocean *Aquat. Living Resour.* **26** 171–178
[4] Laevastu T and Ilmo Hela 1970 *Fisheries oceanography and ecology* (London: Fishing News Book Ltd)
[5] ICCAT 2004 Report of the Standing Committee on Research and Satistics SCRS. International Commission of Atlantic Tunas. Madrid.PLE-025/2004.http://www.iccat.es
[6] Septiawan AW 2006 *Mapping the distribution of chlorophyll in the region strait Bali using remote sensing technology* (Surabaya)
[7] Gunarso W 1985 *Fish behavior: relationship with catching tools, methods and tactics* (Bogor: Bogor Agricultural University)
[8] CCME 2007 Canadian Guidance Framework for The Management of Nutrients in Nearshore Marine Systems. Scientific Supporting Document

Estimation on the concentration of total suspended matter in Lombok Coastal using Landsat 8 OLI, Indonesia

Emiyati*, **Anneke K S Manoppo and Syarif Budhiman**
Remote Sensing Application Center, Indonesian National Institute Aeronautics and Space (LAPAN) Jl. Kalisari No.8 Kelurahan Pekayon Kecamatan Pasar Rebo Jakarta Timur, Indonesia

E-mail: emiyati@lapan.go.id

Abstract. Total Suspended Matter (TSM) are fine materials which suspended and floated in water column. Water column could be turbid due to TSM that reduces the depth of light penetration and causes low productivity in coastal waters. The objective of this study was to estimate TSM concentration using Landsat 8 OLI data in Lombok coastal waters Indonesia by using empirical and analytic approach between three visible bands of Landsat 8 OLI subsurface reflectance (OLI 2, OLI 3 and OLI 4) and field data. The accuracy of model was tested using error estimation and statistical analysis. Colour of waters, transparency and reflectance values showed, the clear water has high transparency and low reflectance while the turbid waters have low transparency and high reflectance. The estimation of TSM concentrations in Lombok coastal waters are 0.39 to 20.7 mg/l. TSM concentrations becoming high when it is on coast and low when it is far from the coast. The statistical analysis showed that TSM model from Landsat 8 OLI data could describe TSM from field measurement with correlation 91.8% and RMSE value 0.52. The t-test and f-test showed that the TSM derived from Landsat 8 OLI and TSM measured in field were not significantly different.

1. Introduction

Sediment transport in the coastal waters influences not only the rates and types of biological and chemical processing, but also geomorphic processes constructing the landforms [1]. Land use intensification in coastal zone causes the increasing of sediment and nutrient to water bodies, which threatening water security for human and another waters organism [2]. Total Suspended Matter (TSM) is sediment particle (organic or inorganic) which suspended in water and originated from erosion and human activities. Higher concentration of sediment particles in water causes the increasing of TSM affecting water turbidity and reduce the ability of sunlight penetration into water. TSM is also one of water constituent which can be used as indicators for water quality. Monitoring of TSM concentration in Lombok coastal is necessary since its designation as tourism and business area for marine culture would affect their development [3].

Remote sensing images provide global or regional coverage of earth surface with potential monitoring platform from monthly to daily basis. Development of water quality monitoring method based on remote sensing data require information on apparent optical properties (AOP) and inherent optical properties (IOP) of water [4]. The interaction of light with sediment into water is complex because both materials scatter and absorb radiation [1][5]. The apparent upwelling radiance measured remotely a function of water and sediment properties and a function of initial solar input, atmospheric transmission of the radiation, the upwelling radiance at the water and specular reflection of water surface due to both sunlight and sky-light. The apparent upwelling is dependent on illumination geometry, atmospheric conditions and water surface parameters such as roughness. By counting all of these parameters can determine the relationship of remotely sensed reflectance and sediment [1][5][6]. Nevertheless, IOP are defined as the optical properties which determined only by the type and concentration of substance in water and it is independent with illumination geometry [4].

The wavelength between 400 to 1000 nm are the most effective for estimating the TSM concentration in water from reflectance because the low rate of absorption by the water high rate of scattering [1]. The previous studies interpretation quantitative and qualitative of grey level pattern on images described sediment concentration in water bodies. Many studies have been carried out for development of monitoring TSM method using remote sensing for coastal waters, among others which worked on multispectral satellite data, such (SeaWiFS) MODIS, OCM, Landsat and SPOT images [1][4][5][7][8][9][10][11][12].

The objective of this study is estimation of TSM concentrations using Landsat 8 OLI (Operational Land Imager). Landsat 8 OLI is multispectral satellite which divided 433 nm to 12.5 nm wavelength into eleven band [13]. Landsat 8 OLI have three spatial resolution which are 15 meter (band 8), 30 meter (band 1-7 and 9) and 100 meter (band 10 and 11). It also has temporal resolution 16 days to capture an image from certain location. This study will apply Tassan (1994) [7] algorithm which utilise three band of wavelength bands for estimating TSM (490 nm, 550 nm and 670 nm).

2. Material and Methods

The study area of this research is located at Lombok, Indonesia with latitude 8°2′39.68″ S to 9°10′6.27″ S and longitude 115°42′40.19″ E to 116°50′21.81″ E which is shown in Figure 1. This study use Landsat 8 OLI data on September 19, 2014 with synchronized field measurement. Field measurements were conducted under different tidal, sun angle and could cover conditions at three areas: Tanjung An, Gerupuk and Awang Bay areas as seen in figure 1. Chlorophyll-a (Chl-a) and TSM concentration were resulted from laboratory analysis of water sample collected from 15 observation stations which spread from Tanjung An, Gerupuk and Awang Bay. Coastal water analysis was carried out at certified laboratory in Bogor Agriculture University.

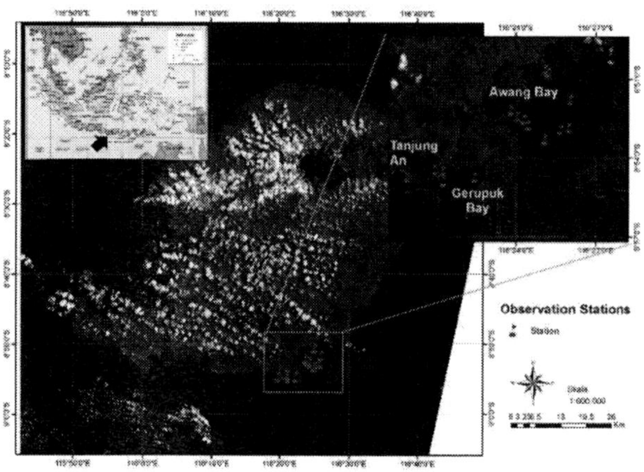

Figure 1. Location of study area and observation stations which indicated by red flag symbols.

In the laboratory, TSM concentration was determined using the gravimetric method based on Budhiman et al. 2012 [4] while Chl-a concentration was measured using spectrophotometer using filtered water samples referring to Budhiman et al. 2012 [4]. Coastal water transparency in field was measured by Secchi disc. Field observation sites were marked by GPS. The image processing and analysis were carried out using ER Mapper 2014, ENVI 5.0 and ArcGIS 10.1. ENVI 5.0 was used for atmospheric correction, while ER Mapper 2014 and ARGIS 10.1 was used for advanced image processing and spatial analysis. TriOS RAMSES as hyperspectral radiometer was used for reflectance measurement in the wavelength range of 400 - 700 nm with 10 nm interval. Processing for *in situ* reflectance from TriOS RAMSES was retrieved using MSDA_XE software and the statistical analysis was carried out using Microsoft Excel.

Empirical and analytical approach was used to estimation of TSM concentration. The empirical approach is based on AOP and the analytical approach is based on IOP [4][6]. AOP generate subsurface remote sensing reflectance based on water leaving radiance and irradiance, while IOP generate subsurface remote sensing reflectance based on absorption and scattering from each water constituent (Chl-a, TSM and CDOM). Optical measurements were conducted simultaneously with water sample collection using above water measurement technique performed with TriOS RAMSES hyperspectral radiometer.

AOP generate based on subsurface remote sensing reflectance TriOS RAMSES and compare with Landsat 8 OLI. TriOS RAMSES consists of three sensors which are irradiance sensor for Ed and two radiance sensors for Lsky (sky radiance) and Lu (upwelling radiance) measurement. The measurements were taken at 40 deg ($\theta = 40$ deg) relative to nadir and zenith [14]. Remote sensing reflectance from TriOS RAMSES got from water leaving radian divided by irradiance. Subsequently remote sensing reflectance (Rrs) of RAMSES is calculated as follows:

$$R_{rs}(\lambda,\theta) = \frac{L_w(\lambda,\theta)}{E_d(\lambda,\theta)} \quad (1)$$

According to Hommersom et al. (2012) [14], water leaving radiance $L_w(\lambda,\theta)$ is measured through formula:

$$L_w(\lambda,\theta) = L_u(\lambda,\theta) - 0.028 \times L_{sky}(\lambda,\theta) \quad (2)$$

Hommersom et al. (2012) [14] mentioned that $R_{rs}(\lambda,\theta)$ spectra were corrected for possible "white light" error ε to become $R_{rs}(\lambda,\theta)$ corrected. Thus, the corrected remote sensing reflectance is calculated as follows:

$$R_{rscor}(\lambda,\theta) = R_{rs}(\lambda,\theta) - \frac{2.35 \times Rrs(780) - Rrs(720)}{1.35} \quad (3)$$

Reflectance from Landsat 8 OLI was generated from Landsat 8 OLI level 1T on September 19th, 2014 which had been radiometric and atmospheric corrected previously. Radiometric and atmospheric corrected was conducted by using ENVI FIAASH radiometric calibration. Corrected radiometric and atmospheric image would generate water leaving reflectance data. According to Danbara 2014 [15], the water leaving reflectance $\rho_{w(\lambda)}$ retrieved from atmospheric correction can be converted to remote sensing reflectance [15] using formula:

$$R_{rs}(\lambda) = \frac{\rho_{w(\lambda)}}{\pi} \quad (4)$$

Remote sensing reflectance below the (subsurface remote sensing reflectance $r_{rs}(\lambda)$) is related to above surface remote sensing reflectance $R_{rs}(\lambda)$ [16], through:

$$r_{rs}(\lambda) = \frac{R_{rs(\lambda)}}{\left(0.52 + 1.7\, R_{rs}(\lambda)\right)} \quad (5)$$

Inherent optical properties (IOP) was generated from subsurface remote sensing reflectance $r_{rs}(\lambda)$ of water column [4][6] [17] using formula:

$$r_{rs}(\lambda) = 0.34 \frac{0.5b_w + Bb^*_{TSM}}{(a_w + 0.06a^*_{chl-a}C^{0.65})\left(1 + 0.2\exp(-0.014(\lambda - 440))\right) + (0.5b_w + Bb^*_{TSM})} \quad (6)$$

where bw= backscattering of water, B = ratio between backscatter and scatter of particle, b*TSM= backscattering of TSM, aw= absorption of water, a^*_{chl} = absorption of chl-a, C= Chl-a concentration, and f = 0.34. The value of absorption coefficient of water (aw) and backscattering of water (bw) were taken from Mobley (1994) [18]. The scattering coefficient of TSM (bTSM) was taken from Kopelevich (1983) [19] formula:

$$b_{TSM}(\lambda) = b^*_{TSM}(550)(\frac{550}{\lambda})^{\alpha} \quad (7)$$

where α is the spectral shape exponent and $b^*_{TSM}(550)$ is backscattering coefficient of TSM at reference wavelength of 550 nm [19].

In order to estimation of TSM concentration from Landsat 8 OLI for Lombok coastal waters effectively an attempt is made to generate site specific new coefficient for Tassan (1994) algorithms, as following formula:

$$\boldsymbol{TSM} = \boldsymbol{10}^{1.6+0.23\, log((R(550)-R(670))(\frac{R(550)}{R(490)})^{-0.1})} - \boldsymbol{10} \quad (8)$$

where TSM in mg/l. According to Landsat 8 OLI band specification R (490), R (550) and R (670) are corresponding to band of Landsat 8 OLI 2, OLI 3 and OLI 4 respectively. The statistical analysis has been carried out to compare the model output with in situ measurement for TSM concentration.

The accuracy of TSM algorithms was evaluated using best fit analysis by comparing TSM in situ measured versus TSM estimation [6]. Evaluation was carried out by using error estimation, t-test, f-test and regression analysis. According to the regression analysis, the model could be accepted if error estimation is low and t-test and f test is bigger than confident level (>0.05). The Root Mean Square error (RMSE) is absolute mean square error which describes the magnitude of difference between in situ measured and estimated concentrations. Root Mean Square Relative Error (RMSRE) and Mean Relative Error (MRE) are relative error in terms of error percentage relative to the measured concentrations. RMSRE and MRE are dimensionless. The RMSE, RMSRE and MRE are defined as formula bellow:

$$RMSE = \sqrt{\frac{1}{N}\sum_{i=1}^{N}(Ci_{estimated} - Ci_{measured})^2} \quad (9)$$

$$RMSRE = \sqrt{\frac{1}{N}\sum_{i=1}^{N}(\frac{Ci_{estimated}}{Ci_{measured}} - 1)^2} \quad (10)$$

$$MRE = \frac{1}{N}\sum_{i=1}^{N}\frac{|Ci^{estimated} - Ci^{measured}|}{Ci^{measured}} \quad (11)$$

where N is the number of measurement. TSM in Lombok coastal waters was mapped using Landsat 8 OLI based on site specific algorithm. This study was analysed the distribution of TSM concentration spatially based Landsat 8 OLI using density slicing classification.

3. Result and discussion

TSM concentrations in Lombok coastal waters were ranged between 10 - 21 mg/l with average concentration of 14.9 ± 3.4 mg/l. The TSM concertation was varied from off-shore to shore or coastal waters, where higher TSM concentration was located in the coastal areas near the shores. These values were similar with studies in Mahakam conducted by Budhiman et al. (2012) [4], in which TSM maximum was found in the coast and TSM minimum was in the off-shore. TSM concentrations in the coastal waters were generally influenced more by riverine inputs compare to tidal force [20]. Since TSM concentration measured in this study low then TSM dynamics in Lombok waters were influenced by tide. TSM concentrations in Lombok lower compare to Mahakam Delta (6-182 mg/l). According Budhiman (2004) [6], Lombok coastal water can be categorized as clear waters.

The maximum Chl-a concentration (2.547 μg/L) was found in the coastal area of Awang Bay and the minimum was in Tanjung An (0.086 μg/L) with average 1.255 μg/L. The concentration of Chl-a has decreasing trend from near river mouth to the sea. Chl-a in Lombok coastal water was lower than Mahakam delta which have Chl-a concentration 1.3 to 29.9 μg/L [6]. Water transparencies result from Secchi depth (Sd) at field measurement varied from 0.5 m to 7 m. the minimum Sd value located on

Awang Bay near port which actual depth was 3.9 m. The maximum value located on middle of Awang Bay which actual depth 31.9 m. Similar with Budhiman et al. (2012) [4] studied, the transparencies increased from the on-shore to the off-shore in each transect.

The AOP measurement in Lombok coastal water show in subsurface reflectance. Subsurface reflectance in situ measurement from TriOS RAMSES was obtained by using equation (1) to (5). The subsurface reflectance *in situ* and water colour are presented in Figure 2. This subsurface reflectance's were developed from water leaving radiance which dominated by absorption and backscattering from water molecule and particulate matter. Generally, the reflected energy in blue region is low. The reflectance spectra start increasing up to 490-570 nm and decreasing after 570 nm till 670 nm. Based on colour of waters, values of transparency and spectra reflectance showed that clear water which have high transparency and low subsurface reflectance value while the turbid waters have low transparency and high subsurface reflectance.

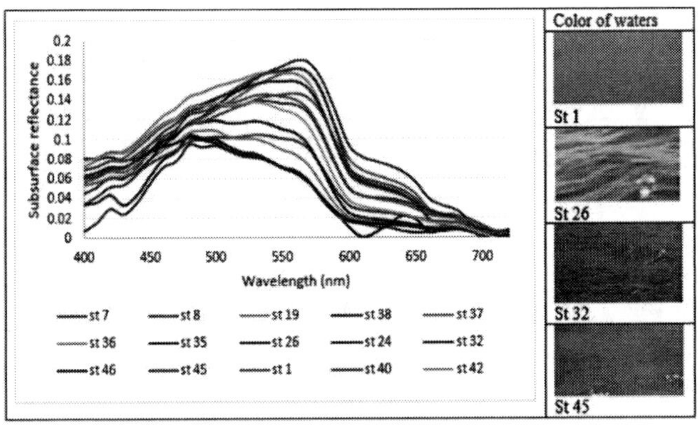

Figure 2. Subsurface reflectance in Lombok coastal waters from *in situ* measurement.

The reflectance of TriOS RAMSES measurements at 490 nm, 550 nm and 670 nm corresponds to Landsat 8 OLI 2, OLI 3, and OLI 4 respectively (Figure 3). Based on statistical analysis, the coefficient of determination (R^2) and RMSE are found to be 0.79 and 0.23 respectively. This analysis indicates that subsurface remote sensing reflectance of OLI 2, OLI 3 and OLI 4 are in close agreement with *in situ* observations and can be used to estimation the TSM concentrations effectively.

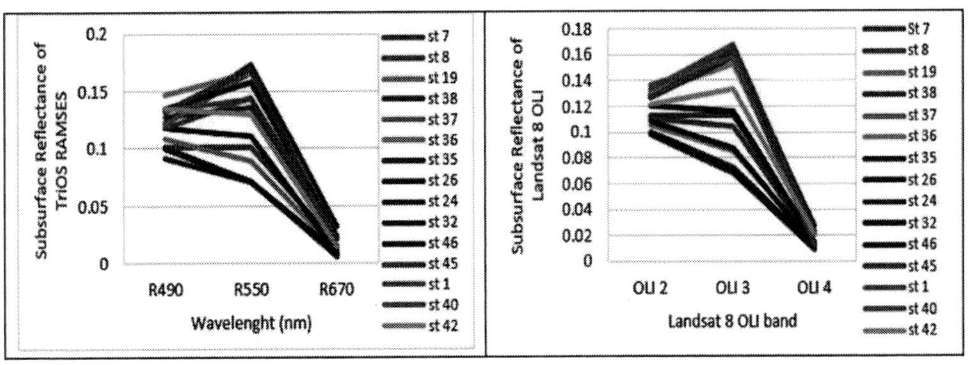

Figure 3. Subsurface remote sensing reflectance spectra of TriOS Ramses vs Landsat 8 OLI.

The IOP in Lombok coastal water described based on Chl-a absorption coefficient at 440 nm and backscattering coefficient of TSM at 50 nm. They retrieved from AOP measurement by inverting optical model using formula (6) and (7). Table 1 show summarizes of the absorption and backscattering result. The result indicates low variation of the IOP due to low standard deviations of all IOP. The low standard deviation related to the low variation of Chl-a and TSM concentration from 15 station observations. We used three sensors at the same time for AOP measurement so there is not time different between water, sky and irradiance measurement in one cloud conditions. It may have impact on the IOP result.

Table 1. IOP parameters values of Lombok coastal waters

IOP	a^*_{chl} (m^{-1})	bTSM (m^{-1})
Maximum	0.377	0.582
Minimum	0.07	0.214
Mean	0.18	0.352
Standard deviation	0.11	0.106

Figure 4 show the average of a^*_{chl} (m^{-1}), a_w (m^{-1}), bTSM (m^{-1}), which calculated using formula (6) and (7) respectively. It is seen that the absorption of Chl-a highly contributes in 400-570 nm. Absorption of water highly contributes in 610-720 nm. Backscatter of TSM value is bigger than water and Chl-a absorptions in 400-600 nm and lower than water absorption but bigger in Chl-a absorptions in 610-720 nm. The depression of TSM backscatter due to pigment of absorption can be seen, especially at high Chl-a concentration as shown in Figure 4. Compare with Mahakam delta turbid water had absorption of Chl-a 0.21 to 0.79 m^{-1} [6], Lombok coastal water have low absorption of Chl-a. the absorption of Chl-a in Lombok coastal water is near similar with Nha Phu Estuary, Vietnam [21].

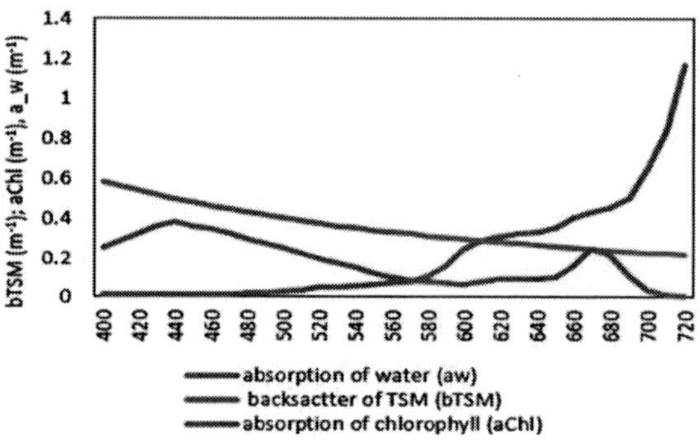

Figure 4. The average IOP of the optically active water constituent which consist of absorption of water, absorption of chl-a and backscattering of TSM in Lombok coastal waters.

The TSM concentrations were determined using the *in situ* observed subsurface reflectance values and modified algorithm presented in equation (8). The statistical analyses have been carried out to compare the estimated of TSM concentrations with the TSM measured (Figure 5). The coefficient of determination is found to be 0.918 with RMSE =0.52, RMSE=3.5 %, and MRE = 1.8%. Furthermore, p-value from t-test and f-test is greater than 0.05 (p(T<=t) =0.39 and p(F<=f) = 0.37. Thus, this modified algorithm resulted low error estimation and can describe TSM concentration in Lombok coastal waters very good. The f-test indicated the variances of TSM estimated and TSM measured were not significant different while t-test indicated the distribution of TSM estimated and TSM measured were not different either.

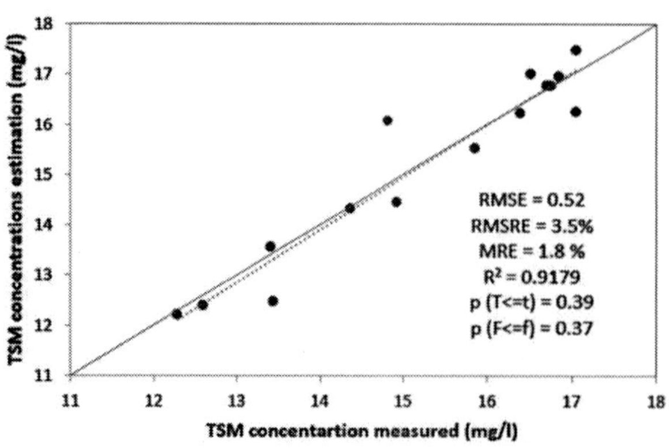

Figure 5. The comparison between estimated of TSM concentrations with TSM concentrations measured.

The TSM concentration map in Lombok coastal waters derived from Landsat 8 OLI data is presented in Figure 6. The minimum of TSM concentration is 0.39 mg/L and the maximum is 20.7 mg/L. The mean of TSM concentration is 11.78 mg/L with standard deviation 2.3. Furthermore, the high TSM concentration mostly the distributed along on-shore waters which the values are more than 15 mg/L especially in south part of Lombok such as Gerupuk and Awang Bay. The high value of standard deviation indicates that there is high variation of TSM concentration in Lombok coastal waters. However, the analysis indicates that Lombok coastal waters are clear water due to TSM concentrations are low. The map also showed that TSM concentrations becoming high when it is near coast (on-shore) and low when it is far from the coast (off-shore). The result is similar with field measurement and others studied which conducted by Budhiman et al. 2012 [4] and Dutrieux 1991 [20]. According to Dutrieux 1991 [20] the low value of TSM concentrations in Lombok coastal waters can be categorized as influenced by tide.

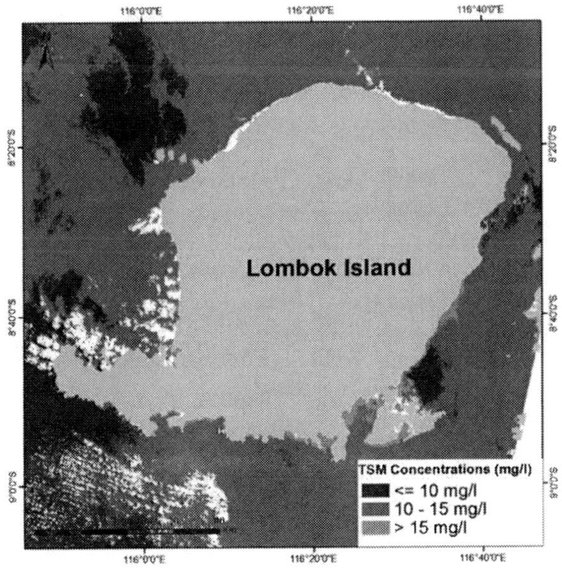

Figure 6. TSM concentration map in Lombok coastal waters derived from Landsat 8 OLI data.

4. Conclusion

The TSM concentration in Lombok coastal waters were analyzed using Landsat 8 OLI based empirical and analytical approach. The statistical analyses showed a coherent pattern between Landsat 8 OLI and in situ observation with R^2 and RMSE 0.79 and 0.23 respectively. It is suggested that Landsat 8 OLI can be used effectively to retrieved TSM concentrations of Lombok coastal waters. According to color of waters, values of transparency and spectra reflectance, the clear water have high transparency and low subsurface reflectance values while the turbid waters have low transparency and high subsurface reflectance values.

An attempt was made to estimated TSM concentrations in Lombok coastal waters and the site-specific algorithm resulted R^2 and RMSE 0.918 and 0.52 respectively with p-value from t-test and f-test is greater than 0.05. Thus, the modified algorithm was very well described TSM concentration in Lombok coastal water. The distribution of TSM concentration in Lombok coastal waters are between 0.39 to 20.7 mg/l with mean value 11.78 mg/L. The analysis indicates that TSM concentrations becoming high when it is on coast and low when it is far from the coast (off-shore). Since this research only used fifteen samples, there is a wide scope to improve the site-specific algorithm which statistically representative.

Acknowledgments
Authors would say thank you to remote sensing application center- LAPAN who's has facilitated this research. We also would say thank you to BPBL Lombok-Indonesian Ministry of Marine Affairs and Fisheries who have help us during field survey.

References

[1] Mertes L A K, Smith M O and Adams J B 1993 Estimating Suspended Sediment Concentration in Surface Waters of the Amazon River Wetlands from Landsat Images *Remote Sens. Environ.* **43** 281–301

[2] Lymburner L, Botha E, Hestir E, Anstee J, Sagar S, Dekker A and Malthus T 2015 Landsat 8: Providing continuity and increased precision for measuring multi-decadal time series of total suspended matter *Remote Sens. Environ*

[3] Pijar. 2012. Book: Evaluasi Kegitan Program 2011dan Rencana Kinerja Tahun 2012. Pemerintah Provinsi Nusa Tenggara

[4] Budhiman S, Suhyb Salama M, Vekerdy Z and Verhoef W 2012 Deriving optical properties of Mahakam Delta coastal waters, Indonesia using in situ measurements and ocean color model inversion *ISPRS J. Photogramm. Remote Sens.* **68** 157–69

[5] Kirk J T 1983 Light and photosynthesis in aquatic ecosystems (Cambridge Univeristy Press)

[6] Budhiman S 2004 *Mapping TSM Concretations from Multisensor Satellite Images in Turbid Tropical Coastal Waters of Mahakam Delta, Indonesia* Endesche Master Thesis (Netherlands, ITC Endesche) **33** 81–7

[7] Tassan S 1994 Local algorithms using SeaWiFS data for the retrieval of phytoplankton, pigments, suspended sediment, and yellow substance in coastal waters *Appl. Opt.* **33** 2369–78

[8] Parwati E, et al. 2008 The Study of Relationship Between TSS (Total Suspended Solid) and Land Used – Land Cover Using Remote Sensing Data in Berau Coastal Area, East Kalimantan *Proc. Int. Conf. Geomatics Fisheries and Marine Science for a Better Future and Prosperity* (Semarang)

[9] Parwati E 2014 Analisis Dinamika Fluktuasi TSS (Total Suspended Solid) Sepanjang DAS-Muara-Laut di Perairan Berau Kalimantan Timur *Proc. Seminar Penginderaan Jauh Nasional* (Bogor)

[10] Putra R M, et.al. 2014 Analisa Sedimen Tersuspensi (Total Suspended Matter) Di Perairan Timur Sidoarjo Menggunakan Citra Satelit Landsat Dan SPOT *Proc. Seminar Penginderaan Jauh Nasional* Bogor

[11] Qu L 2014 Remote Sensing Suspended Sediment Concentration in the Yellow River *Dr. Diss. Paper* 383

[12] Kumar A, Equeenuddin S M, Mishra D R and Acharya B C 2016 Remote monitoring of sediment dynamics in a coastal lagoon: Long-term spatio-temporal variability of suspended sediment in Chilika *Estuar. Coast. Shelf Sci.* **170** 155–72

[13] USGS 2016 Landsat 8 http://landsat.usgs.gov/landsat8.php (last accessed February 1st, 2016)

[14] Hommersom A, Kratzer S, Laanen M, Ansko I, Ligi M, Bresciani M, Giardino C, Beltrán-Abaunza J M, Moore G, Wernand M and Peters S 2012 Intercomparison in the field between the new WISP-3 and other radiometers (TriOS Ramses, ASD FieldSpec, and TACCS) *J. Appl. Remote Sens.* **6** 1-21

[15] Danbara T T 2014 Deriving water quality indicators of Lake Tana, Ethiopia, from Landsat-8 *Geo-Information Earth Obs.* 68 (Netherlands, Univ. Twente Fac.)

[16] Lee Z P, Darecki M, Carder K L, Davis C O, Stramski D and Rhea W J 2005 Diffuse attenuation coefficient of downwelling irradiance: An evaluation of remote sensing methods *J. Geophys. Res. C Ocean.* **110** 1–9

[17] Morel A 1991 Light and marine photosynthesis: a spectral model with geochemical and climatological implications *Prog. Oceanogr.* **26** 263

[18] Mobley C D 1994 Light and Water Radiative Transfer in Natural Waters *Academic Press*

[19] Kopelevich O V 1983 Small-parameter model of optical properties of sea waters In: Monin AS (Ed.) *Physical Ocean Optics* **1** 208–234

[20] Dutrieux E 1991 Study of the ecological functioning of the Mahakam Delta (East Kalimantan, Indonesia) *Estuarine, Coastal and Shelf Science* **32** 415–420

[21] Lund-Hansen L C, Hai D N, Lam N N, Nielsen M H 2010 Optical properties of a tropical estuary during wet and dry conditions in the Nha Phu estuary, Khanh Hoa province, south-east Vietnam *Hydrobiological* **644** 207–216

Detection of potential fishing zone for Pacific saury (*Cololabis saira*) using generalized additive model and remotely sensed data

Achmad Fachruddin Syah[1*], Sei-Ichi Saitoh[2], Irene D. Alabia[2], Toru Hirawake[3]
[1]Department of Marine Science, University of Trunojoyo Madura, Jalan Raya Telang PO BOX 2 Kamal, Bangkalan-Madura, Indonesia.
[2]Arctic Research Center, Hokkaido University, N21-W11 Kita-ku, Sapporo 001-0021, Japan
[3]Laboratory of Marine Environment and Resource Sensing, Faculty of Fisheries Sciences, Hokkaido University, 3-1-1 Minato-cho, Hakodate 041-8611, Japan

Email: fachrudinsyah@gmail.com

Abstract. Abstract. To evaluate the effects of oceanographic conditions on the formation of the potential fishing zones for Pacific saury in western North Pacific, fishing locations of Pacific saury from Defense Meteorological Satellite Program/Operating Linescan System (DMSP/OLS) and satellite-based oceanographic information were used to construct species habitat models. A 2-level slicing method was used to identify the bright regions as actual fishing areas from OLS images, collected during the peak fishing season of Pacific saury in the North Pacific. Statistical metrics, including the significance of model terms, and reduction in the Akaike's Information Criterion (AIC) were used as the bases for model selection. The selected model was then used to visualize the basin scale distributions of the Pacific saury habitat. The predicted potential fishing zones exhibited spatial correspondence with the fishing locations. The results from generalized additive model revealed that the Pacific saury habitat selection was significantly influenced by the SST ranges from 13-18°C, SSC ranges from 0.5-1.8 mg.m^{-3}, SSHA ranges from 5-17 cm and EKE ranges from 700-1200 cm^{2}s^{-2}. Moreover, among the set of oceanographic factors examined, SST explained the smallest AIC and is thus, considered to be the most significant variable in the geographic distribution of Pacific saury.

1. Introduction

Pacific saury (*Cololabis saira*) is one of the most commercially important fisheries, widely distributed in the North Pacific. It undergoes a seasonal migration, moving towards the northern waters of Japan in spring, staying in Oyashio region in summer, and migrating down south from autumn to winter season [1, 2]. Pacific saury was caught at night from August through December, with October as the peak fishing season [3]. Pacific saury is mostly found in the Oyashio areas, along oceanic fronts between Oyashio cold water and warm water from Kuroshio warm-core rings and Tsugaru warm waters [4, 5]. Therefore, oceanographic conditions are believed to largely influence its spatial distribution.

Due to high fluctuations in catches of Pacific saury [2, 6], prediction of potential fishing zones (PFZ) is essential to support fishery operations. Some models have been used to predict abundance and migration of Pacific saury, using single factor such as sea surface temperature [7-9]. However, we believed that the migration and distribution of Pacific saury could be potentially affected by other oceanographic factors. Therefore, this paper further explores a multi-parameter generalized additive model (GAM) to predict the potential fishing zones for Pacific saury and evaluates the effects of oceanographic conditions on its preferred habitat in the western North Pacific.

2. Material and methods

2.1. Study area and satellite-based data

This study was conducted in the western North Pacific (140°-150°E and 34°-46°N), a region where Pacific saury fishing vessels operate off the east coasts of Japan (figure. 1). This area is also influenced by dominant physical features such as the Oyashio Current, Kuroshio Current, eddies, and fronts [10, 11]. The presence of these oceanographic features creates an interesting region to study the relationships between oceanographic conditions and various fisheries resources.

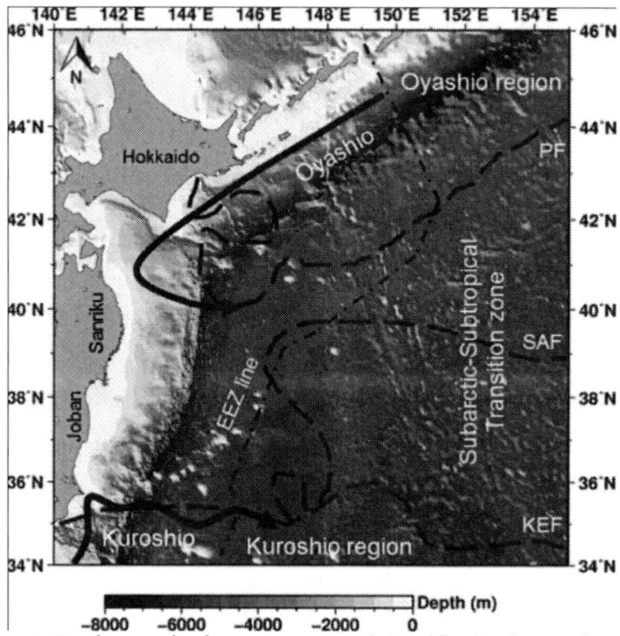

Figure 1. Study area in the western North Pacific (Redrawn from [3])

Fishing locations of Pacific saury, which are also assumed as the species occurrence points, were obtained from OLS nighttime images and were downloaded from the Satellite Image Database System of the Agriculture, Forestry and Fisheries Research Information Center of Japan Ministry of Agriculture, Forestry and Fisheries. We used the TeraScan system to analyze these images and process the nighttime lights into digital numbers within the range of 0-63. In addition, we utilized a 2-level slicing method [12] to detect fishing vessels. Since Pacific saury prefers cold areas as migration routes [4], SST was used to distinguish between the lights of vessels targeting Pacific saury and those of other fishing vessels [13].

The satellite-derived data of SST, SSC, EKE and SSHA from 2005 to 2013 were used as independent variables in GAMs. Daily SST and SSC were downloaded from NASA Goddard Space Flight Center (http://oceancolor.gsfc.nasa.gov/cms/data/aqua). We used the SeaDAS software to process these data and create maps with a 1-km resolution. The daily SSHA and geostrophic velocities (*u, v*) data from the Topex/Poseidon and ERS-1/2 altimeter were obtained from the Archiving Validation and Interpretation of Satellite Oceanographic Data (AVISO; http://www.aviso.oceanobs.com). The EKE was computed using the surface geostrophic velocities [14].

The monthly averages for each environmental variable from daily data sets were calculated with the grid function of the generic mapping tools (GMT 4.5.7) software. All the environmental variables were resampled to 1-km resolution prior to habitat model construction.

2.2. Generalized additive model

In this study, GAMs were constructed in R software (version 3.0.1) using the *gam* function of the *mgcv* package [15], with the Pacific saury presence as response variable and SSC, SST, EKE and SSHA as predictors. A model equation was given as follows:

$$g(u_i)=\alpha_o+s_1(x_{1i})+s_2(x_{2i})+s_3(x_{3i})+....s_n(x_{ni}) \quad (1)$$

where g is the link function, u_i is the expected value of dependent variable (presence or absence), α_o is the model constant, and s_n is a smoothing function with respect to the individual model covariates x_n.

Constructed GAMs can predict Pacific saury habitat suitability index (HSI) when a set of covariates used to build the model are provided. We constructed 15 models for the period between 2005 and 2010 and selected the best model based on Akaike's information criterion (AICs). The distribution of the GAM-predicted Pacific saury habitat were then compared with the monthly DMSP/OLS data.

3. Results and Discussion

We estimated distributions of Pacific saury in the study area, assuming that they were caught in areas where the fishing vessels were identified. Here, we focused on understanding the distributions of Pacific saury in October, corresponding to its peak fishing season [3].

Table 1 showed the statistical results for each of the 15 models earlier constructed using different oceanographic parameter combinations. Based on these results, all the predictor variables were highly significant ($p<0.01$). However, the single parameter models exhibited the lowest explained deviance for EKE and SSHA. Among the single parameter models, SST had the highest explained deviance followed by SSC. The deviance explained and AIC values varied, as predictor variables were added to subsequent models. For instance, a model constructed from EKE and SSHA showed the lowest explained deviance. However, a combination of two-parameter models (SST and SSHA or SST and SSC) recorded relatively higher and lower values for deviance explained and AIC, respectively. The combination of SSC, SST and SSHA showed that all predictor variables were highly significant. Overall, the four-parameter model was selected as the best model, with highest cumulative deviance explained (CDE) and lowest AIC values.

Individual effect of each predictor variable was shown in GAM plots (figure 2). The model results indicated that the favorable habitat for Pacific saury was within the SSC range from 0.5 to 1.8 $mg.m^{-3}$; SST from 14 to 16.5°C; EKE from 0 to 180 cm^2s^{-2} and less than 700 cm^2s^{-2}; SSHA less than -7 cm and from 5 to 16 cm. [6] reported that SST has high relation with medium and large-sized of Pacific saury. In addition, [16] pointed out the SST preference for Pacific saury in October is 15 to 16°C. Together with SST, SSC affects the growth, distribution and migration pattern of Pacific saury [17, 18]. To describe PFZ in relation to mesoscale oceanography variability, SSHA and EKE were used. Our results indicate that fishing activities occurred in areas with low both EKE and SSHA, reflecting the likely association of this species with eddies. [19] reported that eddies create good feeding opportunities through local enhancement of SSC and zooplankton abundance and through the aggregation of prey organisms.

IOP Conf. Series: Earth and Environmental Science **54** (2017) 012074 doi:10.1088/1755-1315/54/1/012074

Table 1. Results from presence-only GAMs developed for Pacific saury using multiple oceanographic factors. The best model was selected based on reduction of AIC and increase in CDE.

Model	Variable	*p*-value	AIC	CDE (%)
SSC	SSC	$<2.00 \times 10^{-16***}$	58410.9	13.4
SST	SST	$<2.00 \times 10^{-16***}$	47550.55	29.5
SSHA	SSHA	$<2.00 \times 10^{-16***}$	60237.73	10.7
EKE	EKE	$<2.00 \times 10^{-16***}$	61631.4	8.62
SSC + SST	SSC SST	$<2.00 \times 10^{-16***}$ $<2.00 \times 10^{-16***}$	46333.72	31.3
SSC + SSHA	SSC SSHA	$<2.00 \times 10^{-16***}$ $<2.00 \times 10^{-16***}$	55251.17	18.1
SSC+ EKE	SSC EKE	$<2.00 \times 10^{-16***}$ $<2.00 \times 10^{-16***}$	56609.64	16.1
SST+ SSHA	SST SSHA	$<2.00 \times 10^{-16***}$ $<2.00 \times 10^{-16***}$	45328.34	32.8
SST+ EKE	SST EKE	$<2.00 \times 10^{-16***}$ $<2.00 \times 10^{-16***}$	46465.89	31.1
SSHA+ EKE	SSHA EKE	$<2.00 \times 10^{-16***}$ $<2.00 \times 10^{-16***}$	57672.22	14.5
SSC+ SST+ SSHA	SSC SST SSHA	$<2.00 \times 10^{-16***}$ $<2.00 \times 10^{-16***}$ $<2.00 \times 10^{-16***}$	44014.9	34.8
SSC+ SST+ EKE	SSC SST EKE	$<2.00 \times 10^{-16***}$ $<2.00 \times 10^{-16***}$ $<2.00 \times 10^{-16***}$	45046.54	33.3
SST+ SSHA+ EKE	SST SSHA EKE	$<2.00 \times 10^{-16***}$ $<2.00 \times 10^{-16***}$ $<2.00 \times 10^{-16***}$	44551.71	34
SSC+ SST+ SSHA+ EKE	SSC SST SSHA EKE	$<2.00 \times 10^{-16***}$ $<2.00 \times 10^{-16***}$ $<2.00 \times 10^{-16***}$ $<2.00 \times 10^{-16***}$	43021.93	36.3

*** indicates statistical significance at the 0.001 level

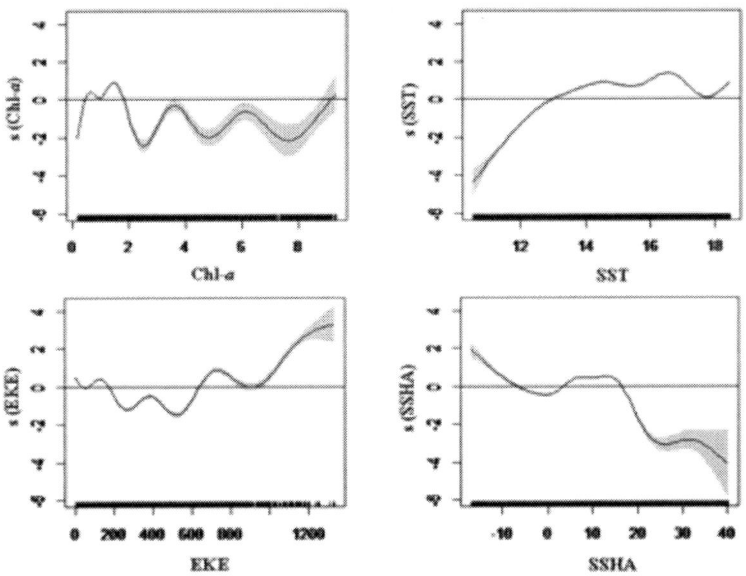

Figure 2. Effects of the 4 environmental variables –SSC, SST, EKE, SSHA, on Pacific saury presence derived from GAM for August 2005–2010. The x-axis shows the value of the explanatory variable, and the y-axis shows the contribution of the smoother to the fitted values. The tick marks on the horizontal axis represent the values of the observed data points; the thick line indicates the fitted function. Grey-shaded area represents the 95% confidence intervals. The horizontal line at zero indicates no effect. The percent of occurrence was higher for all values, for which the fitted GAM function was above the zero axis and lower for values less than 0.

3.1. Prediction and validation

Figure 3 showed the predicted HSI overlaid with actual presence of Pacific saury. During October, most of the fishing points appeared southeast Hokkaido and east of Sanriku (39°-42°N). The presence of fishing locations derived from the OLS images showed high especially in 2005-2007. Based on the predicted HSI maps, the probability of occurrence of Pacific saury in the east and southeast of Hokkaido and south of the Kuril Islands was high. Similarly, the results showed that high probability of occurrence were also found off the Kuroshio-Oyashio transition zone (40°N).

In October, most of the Pacific saury fishing vessels were located around the east coasts of Hokkaido and Sanriku. It has been believed that the observed southward movement was potentially driven by the southward extension of the Oyashio fronts [8, 16] and these features served as the preferred migratory routes of marine species including Pacific saury [4, 20]. In general, the predicted distributions of Pacific saury during October revealed areas of high probability of occurrence southeast off Hokkaido and east of Sanriku. The distribution and migration of Pacific saury in western North Pacific, during the fishing season from August to December, were explained in detail by [3].

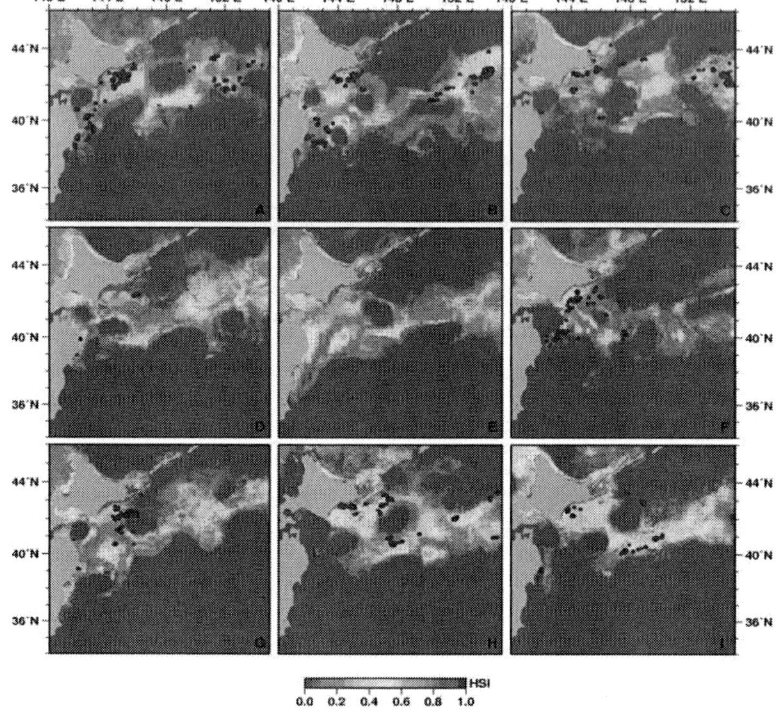

Figure 3. The spatial distributions of potential fishing locations of Pacific saury from DMSP/OLS (red dots) for (A) October 2005 to (I) 2013 overlain on habitat suitability maps. The suitability is depicted as the habitat suitability index (HSI) with values ranging from 0 to 1, representing "poor" to "good" habitat quality, respectively.

Among the set of oceanographic variables examined, SST showed the highest model contribution, thereby suggesting the sensitivity of Pacific saury distribution to temperature changes. For instance, increase in temperature could prevent or delay the southward migration of Pacific saury in winter [21]. The winter SST changes in Kuroshio–Oyashio Transition Zone and Kuroshio and Oyashio regions were also shown to impact the abundances of the large-sized (winter cohort) and medium-sized (spring cohort) groups of Pacific saury [6].

4. Conclusions

Pacific saury potential fishing zones mostly occurred around the east coast of Hokkaido and Sanriku in October. Generative additive models have shown the capability to predict the potential fishing zones for Pacific saury in the western North Pacific.

5. References

[1] Fukushima S 1979 Synoptic analysis of migration and fishing conditions of saury in the northwestern Pacific Ocean *Bulletin of Tohoku Regional Fisheries Research Institute* **41** 1–70 (in Japanese with English abstract)

[2] Kosaka S 2000 Life history of Pacific saury *Cololabis saira* in the Northwest Pacific and consideration of resource fluctuation based on it *Bulletin of Tohoku National Fisheries Research Institute* **63** 1–95 (in Japanese, English abstract)

[3] Syah AF, Saitoh SI, Alabia ID and Hirawake T 2016 Predicting potential fishing zones for Pacific saury (*Cololabis saira*) with maximum entropy models and remotely sensed data *Fisheries Bulletin* 330–42

[4] Saitoh S, Kosaka S and Iisaka J 1986 Satellite infrared observations of Kuroshio warm-core rings and their application to study of Pacific saury migration *Deep Sea Research Part A. Oceanographic Research Papers* **33** 1601–15
[5] Yasuda I and Watanabe Y 1994 On the relationship between the Oyashio front and saury fishing grounds in the north- western Pacific: A forecasting method for fishing ground locations *Fisheries Oceanography* **3** 172–81
[6] Tian Y, Akamine T and Suda M 2003 Variations in the abundance of Pacific saury (*Cololabis saira*) from the northwestern Pacific in relation to oceanic-climate changes *Fisheries Research* **60** 439–54
[7] Kosaka S and Tanno S 1984 On the annual fluctuation of the catch of saury *Cololabis saira* in the Sea of Kumano *Bulletin Tohoku National Fisheries Research Institute* **46** 21–26
[8] Watanabe K, Tanaka E, Yamada S and Kitakado T 2006 Spatial and temporal migration modeling for stock of Pacific saury *Cololabis saira* (Brevoort), incorporating effect of sea surface temperature *Fisheries Science* **72** 1153–65
[9] Tseng CT, Su NJ, Sun CL, Punt AE, Yeh SZ, Liu DC and Su WC 2013 Spatial and temporal variability of the Pacific saury (*Cololabis saira*) distribution in the northwestern Pacific Ocean. *ICES Journal of Marine Science* **70** 991–999
[10] Murase H, Hakamada T, Matsuoka K, Nishiwaki S, Inagake D, Okazaki M, Tojo N and Kitakado T 2014 Distribution of sei whales (*Balaenoptera borealis*) in the subarctic-subtropical transition area of the western North Pacific in relation to oceanic fronts *Deep-Sea Research* II **107** 22-28
[11] Shotwell SK, Hanselman DH and Belkin IM 2014 Toward biophysical synergy: Investigating advection along the Polar Front to identify factors influencing *Alaska sablefish* recruitment *Deep-Sea Research* Part II **107** 40-53
[12] Takagi M and Shimoda H 1991 *Handbook of image analysis* (Tokyo: University of Tokyo Press) (in Japanese)
[13] Mugo RM, Saitoh SI, Takahashi F, Nihira and Kuroyama T 2014 Evaluating the role of fronts in habitat overlaps between cold and warm water species in the western North Pacific: A proof of concept *Deep Sea Research Part II Topical Studies in Oceanography* **107** 29–39
[14] Steele JH, Thorpe SA and Turekian KK (eds.) 2010 *Elements of physical oceanography: A derivative of the Encyclopedia of Ocean Science* (London: Academic Press) 660 p
[15] Wood SM 2006 *Generalized Additive Models, An Introduction with R* (London: Chapman and Hall) 392pp.
[16] Tseng CTe, Sun CL, Yeh SZ, Chen SC, Su WC and Liu DC 2011 Influence of climate-driven sea surface temperature increase on potential habitats of the Pacific saury (*Cololabis saira*) *ICES Journal of Marine Science* **68**1105–13
[17] Oozeki Y, Watanabe Y and Kitagawa D 2004 Environmental factors affecting larval growth of Pacific saury, *Cololabis saira*, in the northwestern Pacific Ocean *Fisheries Oceanography* **13** 44–53
[18] Ito S, Sugisaki H, Tsuda A, Yamamura O and Okuda K 2004 Contributions of the VENFISH program: meso-zooplankton, Pacific saury (*Cololabis saira*) and Walleye pollock (*Theragra chalcogramma*) in the northwestern Pacific *Fisheries Oceanography* **13** 1–9
[19] Zhang TY, Shao QQ and Zhou CH 2001 Application of satellite altimeter data in fishery stock assessment *Fisheries Science* **20** 4–8
[20] Zainuddin M, Saitoh K and Saitoh SI 2008 Albacore (*Thunnus alalunga*) fishing ground in relation to oceanographic conditions in the western North Pacific Ocean using remotely sensed satellite data *Fisheries Oceanography* **17** 61–73
[21] Ito S, Okunishi T, Kishi MJ and Wang M 2013 Modelling ecological responses of Pacific saury to future climate change and its uncertainty *ICES Journal Marine Science* **70** 980–90

Coastal upwelling in Southern Coast of Sumbawa Island, Indonesia

Qamal Taufikurahman and Rahmat Hidayat*
Department of Geophysics and Meteorology, Faculty of Mathematics and Natural Science, Bogor Agricultural University, Bogor, Indonesia

E-mail: rahmath@apps.ipb.ac.id

Abstract. Circulation and water mass characteristics in southern coast of Sumbawa Island is strongly influenced by monsoon which change each season. This research aims to (1) identify the Ekman transport and Ekman pumping variations and (2) analyse the SST and chlorophyll-a variability. During Juni-July-August (JJA) season, the highest value of Ekman transport (-0.72 m^2 s^{-1}) moves the water mass towards Indian Ocean while Ekman pumping ($0.15x10^{-4}$ m^2 s^{-1}) rises the water mass to the surface (upwelling). Impact of coastal upwelling was identified by cooling SST and high chlorophyll-a. Coolest SST (26.58°C) associated with highest chlorophyll-a (0.6 mg m^{-3}) were found in JJA season. Inter-annual variation of SST and chlorophyll-a concentrations during 2003-2015 is driven by Ekman transport and Ekman pumping. Coolest SST and highest chlorophyll-a were found in JJA season in 2006 with an average value of 25.98°C and 0.78 mg m^{-3}, also in JJA season in 2008 with an average value of 26°C and 0.71 mg m^{-3}. Thus the process of Ekman transport and Ekman pumping has an important role in lifting the cold water high nutriens from the deeper layer to surface. The high concentration of chlorophyll-a in water is the impact of appointment nutrient from the deeper layer by the coastal upwelling.

1. Introduction

Circulation and water mass characteristics in southern coast of Sumbawa Island, located in the northeast of Indian Ocean, are affected by monsoon system resulting in seasonal variability. Non-local phenomena that interact with the system will affect monsoon semi-annual, annual and interannual variability [1][2]. Southern coast of Sumbawa Island has highest wind speed in JJA that is driven by southeast monsoon wind, where the wind move from southeast to northwest [1]. Southeast Monsoon wind caused Ekman transport which is directed away from south coastal of Java and Sumbawa. It cause vacuum in the water mass that result in increasing water mass (upwelling) from the deeper layer toward the surface layer [3][4].

Coastal upwelling is a type of upwelling, which occurred near the coast as a result of Ekman transport and –wind that blows uniformly along the shore which moving the water mass toward or away from the coast. Compensation of the water mass loss in the ocean moves the cooler water with rich of nutrients from the deeper layer to surface. The cooler water that lifted and rich of nutrients stimulates the growth of primary producers such as phytoplankton and fish production. Thus, coastal upwelling has a significant impact on ecosystem and fisheries [5].

Besides Ekman transport, Ekman pumping which driven by the wind convergence or divergence increase local upwelling significantly [6][7][8]. Ekman pumping process moves wind divergence in surface water and lifted the water mass rise to surface. Compared to Ekman transport, Ekman pumping

IOP Conf. Series: Earth and Environmental Science **54** (2017) 012075 doi:10.1088/1755-1315/54/1/012075

give the same or even greater contribution to the upwelling procces in many regions [7][8][9][10]. Driving factors of coastal upwelling is analyzed through Ekman transport and Ekman pumping values. Meanwhile, the impact of coastal upwelling is analyzed through the distribution of SST and chlorophyll-a concentration in surface water [11].

2. Methodology

2.1. Data

Data used for this research were monthly SST with spatial resolution 0.125° x 0.125° and monthly surface wind with spatial resolution 0.125° x 0.125°. Both dataset are provided by European Centre for Medium-Range Weather Forecasts http://www.ecmwf.int [12]. Monthly chlorophyll-a data with spatial resolution 0.036° x 0.036° is provided by Moderate Resolution Imaging Spectroradiometer (MODIS). Period of SST, wind, and chlorophyll-a data is from January 2003-December 2015 (13 years).

2.2. Methodology

2.2.1. Wind stress

The calculation of wind stress for each component of the wind using the assumption that the x-wind component in the direction parallel to the coast and the y-wind component in the direction of the shore. Calculation of the friction of the wind stress expressed in equation 1 [13] :

$$\tau_x = \rho_U C_D |\vec{v}_H| U_{10} \qquad \tau_y = \rho_U C_D |\vec{v}_H| V_{10} \tag{1}$$

Where :

τ_x = wind stress parallel to the coast (kg m^{-1} s^{-2})
τ_y = wind stress upright to the coast (kg m^{-1} s^{-2})
C_D = drag coefficient = 1.5 x 10^{-3}
ρ_u = air density = 1.3 kg m^{-3}
$|\vec{v}_H|$ = wind magnitude (m s^{-1})
U_{10} = zonal wind component (m s^{-1})
V_{10} = meridional wind component (m s^{-1})

2.2.2. Ekman transport and Ekman pumping Calculation

Ekman transport calculations in parallel and vertical components of the coast can be calculated using wind stress data. In addition, the calculation of Coriolis parameter can used the equation 2 :

$$f = 2\,\Omega \sin \Theta \tag{2}$$

where :

f = coriolis parameter (rad s^{-1})
Ω = rotation velocity = 7.29 x 10^{-5} rad s^{-1}
Θ = latitude

Ekman transport calculation using equation 3 [14] :

$$M_{xE} = \tau_y / \rho_a f \qquad M_{yE} = -\tau_x / \rho_a f \tag{3}$$

where :

M_{xE} = Ekman transport parallel to the coast (m^2 s^{-1})
M_{yE} = Ekman transport upright to the coast (m^2 s^{-1})
τ_x = wind stress parallel to the coast (kg m^{-1} s^{-2})
τ_y = wind stress upright to the coast (kg m^{-1} s^{-2})
ρ_a = water density = 1025 kg m^{-3}
f = coriolis parameter (rad s^{-1})

Ekman pumping calculations using Ekman transport data parallel and vertical components of the coast. Ekman pumping calculation using equation 4 :

$$-\mathbf{W}_E = \frac{\partial \mathbf{M}_{xE}}{\partial x} + \frac{\partial \mathbf{M}_{yE}}{\partial y} \tag{4}$$

where :
W_E = Ekman pumping ($m^2 s^{-1}$)
M_{xE} = Ekman transport parallel to the coast ($m^2 s^{-1}$)
M_{yE} = Ekman transport upright to the coast ($m^2 s^{-1}$)

W_E are the Ekman pumping events associated with convergence or divergence of Ekman transport. Positive Ekman pumping ($W_E > 0$) indicates upwelling and negative Ekman pumping ($W_E < 0$) indicates downwelling. The calculation of Ekman transport and Ekman pumping were done assess its contribution the coastal upwelling in southern coast of Sumbawa Island.

3. Results and Discussion

3.1. Ekman transport and Ekman pumping Calculation

Surface wind variations can moves the Ekman transport and Ekman pumping in Southern coast of Sumbawa Island (figure 1). Ekman transport dominantly moves the surface water mass to Indian Ocean. Ekman transport process occurs continuously may impacting water column and the pressure gradient along the inshore waters [14]. as a result of the pressure gradient then the water mass will balance, so that the water mass move towards the coast to follow the slope of the seafloor moving towards the surface and move the Ekman pumping process.

December-January-February (DJF) season has an average wind speed 3.19 m s^{-1} and dominantly moving from west to east. March-April-May (MAM) season has an average wind speed 3.19 m s^{-1} and changes in wind direction from southeast to southwest. It caused by the strengthening of the southeast monsoon winds over the water. JJA season has an average wind speed 4.44 m s^{-1} and Southeast monsoon wind strongly occurs in this season. September-October-November (SON) season has decreased in wind speed average from the previous season 3.44 m s^{-1} and changes in wind direction from southeast to the north. It caused by the influence of the southeast monsoon winds that has been decreased. Variations in wind speed and direction monsoon blows in Southern coast of Sumbawa Island will affect the variation of the Ekman transport and Ekman pumping.

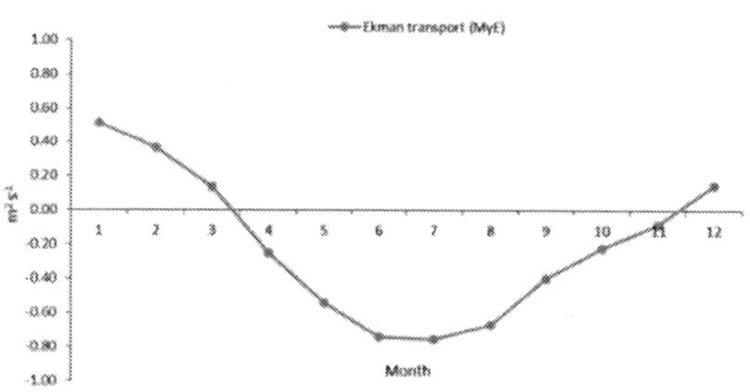

Figure 1. Monthly variation of Ekman transport in southern coast of Sumbawa Island during 2003-2015.

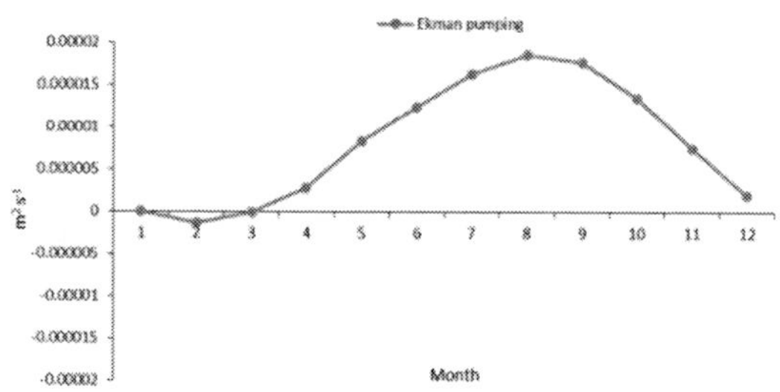

Figure 2. Monthly variation of Ekman pumping in southern coast of Sumbawa Island during 2003-2015.

Value Ekman transport components paralell to the coast (M_{yE}) was calculated using wind stress data component parallel to the coast (τ_x) (figure 1). The calculation is done to see the trend of the mass of water moving toward or away from the coast. Determining the direction Ekman transport leading to the coast is marked with a positive value (+) and towards the open sea is characterized by a negative value (-) [15].

Ekman transport in DJF season has an average value 0.32 $m^2 s^{-1}$ and showed an increasing trend of mass transport volume of water toward the coast. MAM season has an average value of -0.19 $m^2 s^{-1}$ and showed an increasing trend of mass transport volume of water towards the Indian Ocean. JJA season has an average value of -0.72 $m^2 s^{-1}$ and showed an increasing trend of volume water mass transport towards the Indian Ocean. SON season average value of -0.22 $m^2 s^{-1}$ and showed the downward trend in the volume of water mass transport towards the Indian Ocean. Wind stress component parallel to the coast (τ_x) dominantly move the water mass into the Indian Ocean. It proves by the trend of events in MAM, JJA and SON seasons.

Ekman pumping value is calculated using Ekman transport which paralell to the coast component (M_{xE}) and vertical to the coast (M_{yE}) (figure 2). Upwelling and downwelling phenomenon is illustrated by the analysis of vertical transport of water mass in Ekman layer. Positive value (+) indicates the lifting of the water mass (upwelling) and negative (-) indicates the sinking water mass (downwelling).

DJF season has an average value of 0.34×10^{-6} $m^2 s^{-1}$ and indicates the lifting of water mass (upwelling). MAM season has an average value -0.33×10^{-5} $m^2 s^{-1}$ and showed a tendency of sinking water mass (downwelling). JJA season has an average value of 0.15×10^{-4} $m^2 s^{-1}$ and showed a tendency of lifting of water mass (upwelling). SON season has an average value of 0.12×10^{-4} $m^2 s^{-1}$ and showed a tendency of lifting of water masses (upwelling). Ekman transport which paralell to the coast component (M_{xE}) and vertical to the coast (M_{yE}) variations dominantly move the water mass from bottom to surface along the coast. It proves by the trend of events in DJF, JJA and SON seasons.

Process Ekman transport and Ekman pumping values occur JJA season, caused by the influence of the movement of southeast monsoon winds which getting stronger within the season [16]. Ekman transport process has a greater contribution than the Ekman pumping on the incidence of coastal upwelling. It is caused by the loss of water mass on the surface that move towards the Indian Ocean. The loss of water mass will increase the occurrence of coastal upwelling, where there is appointment of water mass with high nutrient abundance from bottom water. Required lag time from appointment of nutrients to found an abundance of chlorophyll-a concentration on the surface of the south ocean of Sumbawa Island.

3.2. SST and Chlorophyll-a Variations

Variation value of Ekman transport and Ekman pumping has an impact on distribution of SST and chlorophyll-a in surface water (figure 3). The average value of SST from DJF to JJA season has decreased and increased in SON season (DJF:29.32°C; MAM:29°C; JJA:26.58°C; SON:27.85°C). Meanwhile, the average value of chlorophyll-a from DJF to JJA season has increased and decreased in SON season (DJF:0.13 mg m^{-3}; MAM:0.22 mg m^{-3}; JJA:0.6 mg m^{-3}; SON:0.37 mg m^{-3}). Another factor affecting the distribution of high concentration of chlorophyll-a in coastal area is nutrient supply which is product of run-off from the mainland and the stirring process from deeper layer (mixing) that intensive along the coast [17].

The strong lifting of water mass mostly found in cooler SST and high chlorophyll-a concentration [18]. The highest incidence of upwelling has SST range 25-26°C and chlorophyll-a range 0.6-0.7 mg m^{-3}. While upwelling occurs when SST range 29-30°C and chlorophyll-a concentration below 0.2 mg m^{-3}. SST distribution is inversely correlated with chlorophyll-a concentration. Lowest SST and highest chlorophyll-a found in JJA season, where the monthly variability, from June to August (JJA) was found the warming SST while chlorophyll-a is increase. Temperature affects the solubility of gases needed for photosynthesis such as CO_2 and O_2. These gases are easily dissolved at a lower temperature than high temperature, resulting in increased photosynthesis rate at a low water temperature.

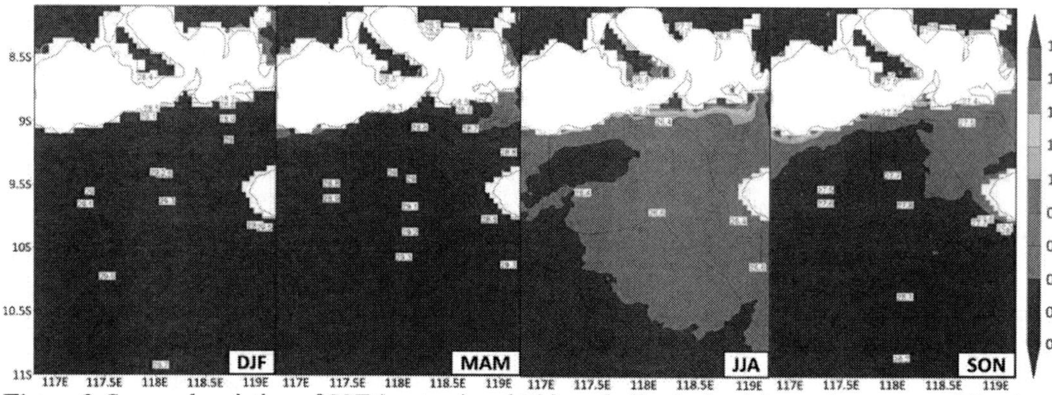

Figure 3. Seasonal variation of SST (contour) and Chlorophyll-a (colour) in southern coast of Sumbawa Island during 2003-2015.

3.3. *Inter-annual Variations of Coastal upwelling*

Inter-annual variation of Ekman transport and Ekman pumping during 2003-2015 is driven by the monsoon winds (figure 4). Variations in monsoon wind speed and direction will affect the fluctuating value of Ekman transport and Ekman pumping. The maximum value Ekman transport and Ekman pumping occurred in JJA season in 2006 with an average value of -0.85 $m^2 s^{-1}$ and $0.2x10^{-4}$ $m^2 s^{-1}$, also in JJA season in 2008 with an average value of -0.75 $m^2 s^{-1}$ and $0.21x10^{-4}$ $m^2 s^{-1}$.

Inter-annual variation of SST and chlorophyll-a concentrations during 2003-2015 is driven by the Ekman transport and Ekman pumping (figure 5). Coolest SST and highest chlorophyll-a were found in JJA season in 2006 with an average value of 25.98°C and 0.78 mg m^{-3}, also in JJA season in 2008 with an average value of 26°C and 0.71 mg m^{-3}.

Climate variability (ENSO and IOD) may affect the value of SST and chlorophyll-a. Upwelling intensity increases with cooling SST condition and high concentration of chlorophyll-a when IOD(+) and El Nino occur simultaneously in 2006 in the ocean of South Java [18]. Then in 2010, IOD(+) and La Nina occur, where IOD(+) has high impact on SST and chlorophyll-a value in ocean of South Java than La Nina [19].

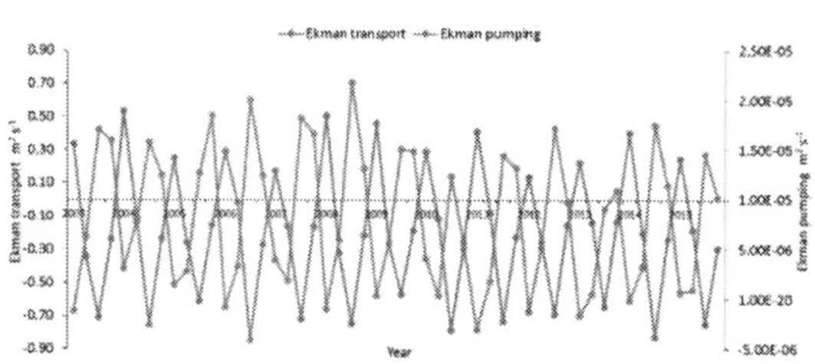

Figure 4. Inter-annual variation of Ekman transport and Ekman pumping in southern coast of Sumbawa Island during 2003-2015.

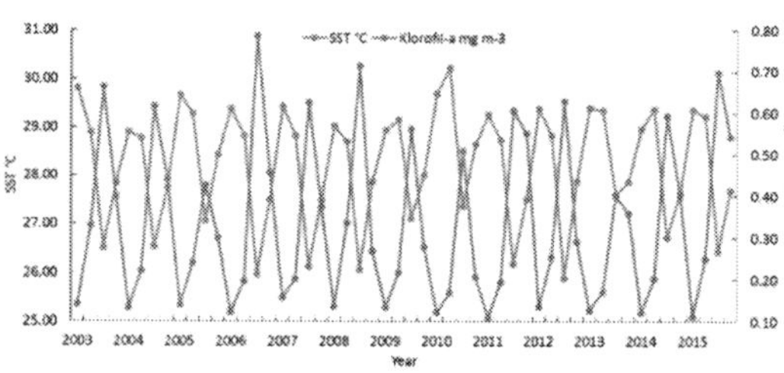

Figure 5. Inter-annual variation of SST and Chlorophyll-a in southern coast of Sumbawa Island during 2003-2015.

4. Conclusion

Coastal upwelling in Southern coast of Sumbawa Island is driven by Ekman transport and Ekman pumping. Ekman transport dominantly moves the water mass towards Indian Ocean and Ekman pumping dominantly lift the water mass from bottom to surface (upwelling). JJA season has the highest value Ekman transport (-0.72 $m^2 s^{-1}$) and Ekman pumping ($0.15x10^{-4}$ $m^2 s^{-1}$). Highest impact of coastal upwelling occurred in JJA season with the coolest SST (26.58°C) and the highest chlorophyll-a concentration (0.6 mg m^{-3}). Thus the process of Ekman transport and Ekman pumping has an important role in lifting the cold water high nutriens from the deeper layer to surface. The high concentration of chlorophyll-a in water is the impact of appointment nutrient from the deeper layer by the coastal upwelling.

References

[1] Wyrtki K 1961 *Physical Oceanography of the Southeast Asean Waters* (California: NAGA Rep 2 Scripps Inst of Oceanography La jolla)
[2] Fieux M, Andrie C, Charriaud E, Ilahude A G, Metzl N, Molcard R, Swallow J C 1996 Hydrological and chlorofluoromethane measurements of the Indonesian throughflow entering the Indian Ocean *Journal of Geophysical Research* **101** 12433-12454
[3] Wyrtki K 1962 The Upwelling in The Region Between Java and Australia During The South East Monsoon Aust *Java Mar Freshw Res* **17** 217-225
[4] Purba M, Nurjaya I W and Utaminingsih S 1992 *Variasi Suhu Permukaan Laut Yang Diukur Dengan Satelit NOAA dan Kaitannya Dengan Proses Upwelling di Perairan Selatan Jawa* (Bogor: Laporan Penelitian Fakultas Perikanan IPB)
[5] Barth J A, Menge B A, Lubchenco J, et al. 2007 Delayed upwelling alters nearshore coastal ocean ecosystems in the northern California current Proceedings of the National Academy of Sciences of the United States of America **104** 3719–3724
[6] Enriquez A G, Friehe C A 1995 Effects of wind stress and wind stress curl variability on coastal upwelling *J Phys Oceanogr* **25** 1651-1671
[7] Pickett M H, Paduan J D 2003 Ekman transport and pumping in the California Current based on the US Navy's high-resolution at-mospheric model (COAMPS) *Journal of Geophyscal Research: Oceans* **108**
[8] Castelao R M, Barth J A 2006 Upwelling around Cabo Frio Brazil: The importance of wind stress curl *Geophys Res Lett* **33**
[9] Croquette M, Eldin G, Grados C, et al. 2007 On differences in satellite wind products and their effects in estimating coastal upwelling processes in the south-east pacific *Geophys Res Lett* **34**
[10] Jing Z Y, Qi Y Q, Hua Z L, Zhang H 2009 Numerical study on the summer upwelling system in the northern continental shelf of the South China Sea *Continental Shelf Research* **29** 467-478
[11] Yunwei Y, Zheng L, Changlin C 2015 Winter coastal upwelling off northwest Borneo in the South China Sea *Acta Oceanol.Sin* **34** 3-10
[12] Dee D P, Uppala S M, Simmons A J, et al. 2011 The ERA-Interim re-analysis: configuration and perfomance of the data assimilation system *Quart. J. Roy Meteor. Soc.* **137** 553-597
[13] Kutsuwada K 1998 Impact of wind/wind-stress field in the North Pacific constructed by ADEOS/NSCAT data *J Oceanogr* **54** 443-456
[14] Pond S and Pickard G L 1983 *Introductory dynamical Oceanography Second edition* (New York: Pergamon Press)
[15] Tubalawony S, Kaswadji R F, Purba M, Wouthuyzen S and Soedharma D 2007 *Dampak Proses Fisik Terhadap Sebaran Klorofil* (Bogor: Bogor Agricultural University)
[16] Susanto R D, Gordon A L, and Zeng Q 2001 Upwelling Along the Coasts of Java and Sumatera and its Relation to ENSO *Geophysical Research Letters* **28** 1599 – 1602
[17] Valiela I 1984 *Marine Ecologycal Processes* (New York: Springer-Verlag)
[18] Kunarso, Hadi S, Ningsih N S and Baskoro M S 2011 Variabilitas Suhu dan Klorofil-a di Daerah Upwelling pada Variasi Kejadian ENSO dan IOD di Perairan Selatan Jawa sampai Timor *Jurnal Ilmu Kelautan* **16**171-180
[19] Kunarso, Hadi S, Ningsih N S, Baskoro M S 2012 Perubahan Kedalaman dan Ketebalan Termoklin pada Variasi Kejadian ENSO IOD dan Monsun di Perairan Selatan Jawa Hingga Pulau Timor *Jurnal Ilmu Kelautan* **17** 87-98

Biodiversity and Distribution of Horseshoe Crabs in Northern Coast of Java and Southern Coast of Madura

A Mashar*[1], NA Butet[1], B Juliandi[2], Y Qonita[1], AA Hakim[1], and Y Wardiatno[1]

[1] Department of Aquatic Resources Management, Faculty of Fisheries and Marine Sciences, Bogor Agricultural University, Indonesia
[2] Department of Biology, Faculty of Mathematics and Natural Sciences, Bogor Agricultural University, Indonesia

E-mail: alimashar75@gmail.com

Abstract. Horseshoe crab is an important component of macro-benthos communities in the fine sand or mud substrate in coastal waters, both in the tropical and temperate region. This primitive animal consists of four species in the world, and three species can be found di Asian region, including Indonesia, namely *Tachypleus tridentatus*, *T. gigas*, and *Carcinoscorpius rotundicauda*. Scientific information about species distribution of three Asian horseshoe crab in Indonesia is limited, also about morphometric characters. This study aims to determine the morphometric characters and species distribution of three Asian horseshoe crab in north coast of Java and south coast of Madura Island. This study was conducted on July-August 2016. The total number of three Asian horseshoe crab obtained in this study was 260 individuals, distributed along north coast of Java and south coast of Madura Island, respectively 176 individuals of *C. rotundicauda*, 35 individuals of *T. tridentatus*, and 49 individuals of *T. gigas*. *Tachypleus gigas* has the largest size and widest class interval among three Asian horseshoe crab species. Morphometric characters is differences among three Asian horseshoe crab species. Carapace width and telson length were not significantly different among sampling locations only in *T. tridentatus*.

1. Introduction

Horseshoe crab is an important component of the macro-benthos community in the coastal waters with fine sand or mud substrate, both in the tropics and in season four worldwide. Actually, horseshoe crabs are more closely related to spiders and scorpions than to crabs. There are four species of horseshoe crab found around the world. *Limulus polyphemus* (Linnaeus, 1758) occurs along the Atlantic coast of North America, while *Tachypleus tridentatus* (Leach, 1819), *Tachypleus gigas* (Müller, 1785) and *Carcinoscorpius rotundicauda* (Latreille, 1802) found in Asia, from India to Japan and south to Malaysia and Indonesia [1-12]. The tri-spine horseshoe crab *T. tridentatus* and the coastal horseshoe crab *T. gigas* live in sandy to muddy habitats [13-15], while the mangrove horseshoe crab *C. rotundicauda* inhabits muddy areas, commonly in brackish waters [13-14, 16]. In India and Indonesia, some *T. gigas* co-exist with *C. rotundicauda* [17-19]. In several area in Indonesia, such as Subang, Indramayu, and Surabaya coastal area, three Asian horseshoe crab species could be found at the same location [20]. Unlike the other Asian horseshoe crab species, *C. rotundicauda* does not move out to sea with the receding tide. Instead, adults can be found buried from 2 to 3 cm deep in the wet mud, while sub-adults and juveniles remain on the sediment surface.

IOP Conf. Series: Earth and Environmental Science **54** (2017) 012076 doi:10.1088/1755-1315/54/1/012076

Horseshoe crab can survive long time periods without showing any significant morphological changes, so that horseshoe crabs are often called as living fossil. Horseshoe crabs have changed very little over 500 million years and have survived from two mass extinctions [6, 21]. Status of all 4 species of horseshoe crab according to the IUCN Red Data Book are classified near threatened for *Limulus polyphemus* and deficient for the other three species [4].

Habitat throughout Asia has been diminishing [15, 18, 22-23], and evidence from the current study and information from local fishermen suggest that all species are rarely found and the population is probably declining. For example, in India, spawning activity has been adversely affected by the degradation and destruction of spawning beaches of both *Carcinoscorpius rotundicauda* and *Tachypleus gigas* [18]. In Hong Kong, juvenile densities have been declining for some years [22, 23], and juveniles are found in fewer locations [15]. *T. gigas* is now thought to be extirpated locally around Hong Kong [1]. In addition to loss of habitat, entrapment in fishing nets in coastal waters and commercial harvest for food remain threats for crab populations. In Hong Kong *T. tridentatus* is sold in restaurants [15], and in Thailand and Malaysia all 3 Asian horseshoe crabs species are consumed as an exotic delicacy [2].

Indonesian coastal waters is one of important distribution areas of horseshoe crabs in the world. Three of four species of horseshoe crabs in the world can be found in Indonesian coastal areas. In Indonesia, horseshoe crab is one of priority protected marine species [24]. Although Indonesian coastal area is primary distribution area of three Asian horseshoe crabs species, scientific information about horseshoe crab species and its distribution are limited. Even though, information about species, distribution, and quantitative data of horseshoe crabs are urgently needed to confirm the conservation status of these species, to monitor emerging threats to the remaining horseshoe crab population in Indonesia, and to justify actions for protecting these species. One of approaches to identify species of aquatic species, such as horseshoe crab, is by morphometric characters. Morphometric characters are effective tool to learn change and variation of organism body shape [25]. Morphometric study explained change and comparison of quantitative measurement, especially from the part of organism body. Significant variation of morphology characters of species can be guidance to identify population or genus of organism.This study was aimed to determine the morphometric characters and species distribution of three Asian horseshoe crabs in north coast of Java and south coast of Madura Island.

2. Method

2.1. Sampling

This study was carried out on July-August 2016, and sampling was conducted on 25 to 31 July 2016 on northern coast of Java, including north of Subang, Indramayu, Semarang, Demak, Rembang, Tuban, Lamongan, Surabaya; and south of Sumenep in Madura Island. Horseshoe crab specimens were captured using net by local fisherman. The specimens originating from these regions were measured in situ. Specimens of horseshoe crab were collected and then identified using the taxonomic books [6, 17, 21, 26]. Three morphometric characteristics were measured, including carapace length, carapace/prosomal width, and telson length (Figure 1). The measurrements were conducted using digital calipers to the nearest 0.01 mm.

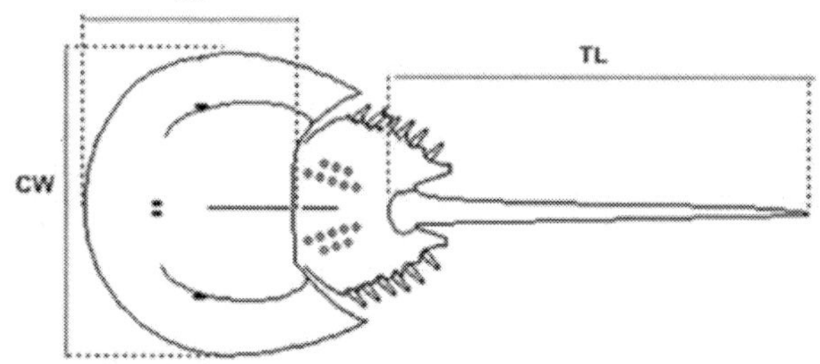

Figure 1. Measurements of morphological characteristics of horseshoe crab (CL: carapace length, CW: carapace width, TL: telson length)

2.2. Data Analysis

All morphology data were corrected for length to make comparisons among locations. Morphological data among locations were compared using Mann-Whitney.

3. Results and Discussion

3.1. Horseshoe crab diversity and distribution

Three species were found on the north coastal of Java, there are *Tachypleus gigas, T. tridentatus*, and *Carcinoscorpius rotundicauda*. The total number of samples were obtained from nine sampling locations is 260 individuals. Number of female horseshoe crab more than male for all three species at all sampling locations. Horseshoe crab distribution of collected samples shown in Figure 2.

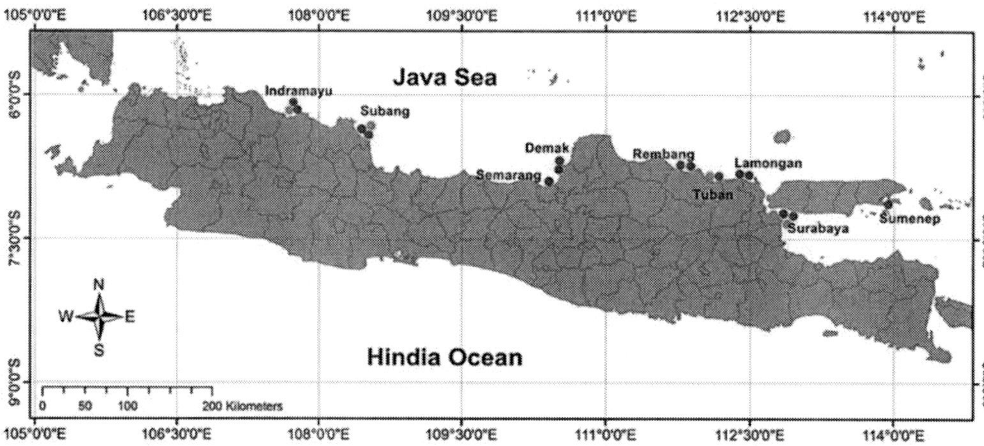

Figure 2. Distribution of horseshoe crabs on the north coast of Java (●: *Tachypleus gigas;* ● : *T. tridentatus*; ●: *Carcinoscorpius rotundicauda*)

Based of Figure 2 the distribution pattern of the horseshoe crabs is as follow: in north coast of Subang, Indramayu, and Surabaya three Asian horseshoe crab species could be found; in north coast of Demak, Rembang, and Lamongan the coastal horseshoe crab *Tachypleus gigas* and the mangrove horseshoe crab *Carcinoscorpius rotundicauda* existed; in north coast of Tuban *T. gigas* and *T.*

tridentatus occurred; and in north coast of Semarang and south coast of Sumenep only *C. rutundicauda* was presence. The co-existence of three Asian horseshoe crab species at the same location is the quite unique because they are ussualy rarely found co-existing. Meilana [20] also reported the same finding in Semarang coast. In most cases many researchers found one or two horseshoe crab species at the same location [6, 17, 19, 20, 27].

In this study, *C. rotundicauda* was found almost in all locations, except in Tuban coast. This is not an uncommon phenomenon. This species commonly lives in shallow waters with soft, sandy bottoms or extensive mud flats or the mangrove-mudflat ecosystem [18]. The mangrove horseshoe crab spends most of its life close to or at the bottom of a body of their brackish-swampy water habitat. The characteristics of research locations in north coast of Java are believed to be very suitable as habitat for *C. rotundicauda* [6-7, 19]. Actually, *T. tridentatus* and *T. gigas* also almost found as well in all location but in substantially less number. This condition occurred because *T. gigas* and *T. tridentatus* prefer different habitat with *C. rotundicauda*. All sampling locations allegedly not suitable or ideal habitat for horseshoe crab from genus *Tachypleus*. So, both species *T. gigas* and *T. tridentatus* were not found in all study locations and could only be found in small number. This finding strengthens the previous study where *T. tridentatus* and *T. gigas* were almost always found in small number and lesser than *C. rotundicauda* [20].

3.2. Length Frequency Distribution

The carapace length of 246 individuals of horseshoe crabs were measured. Figure 3 shows carapace length frequency distribution of the three Asian horseshoe crab species of north coast of Java.

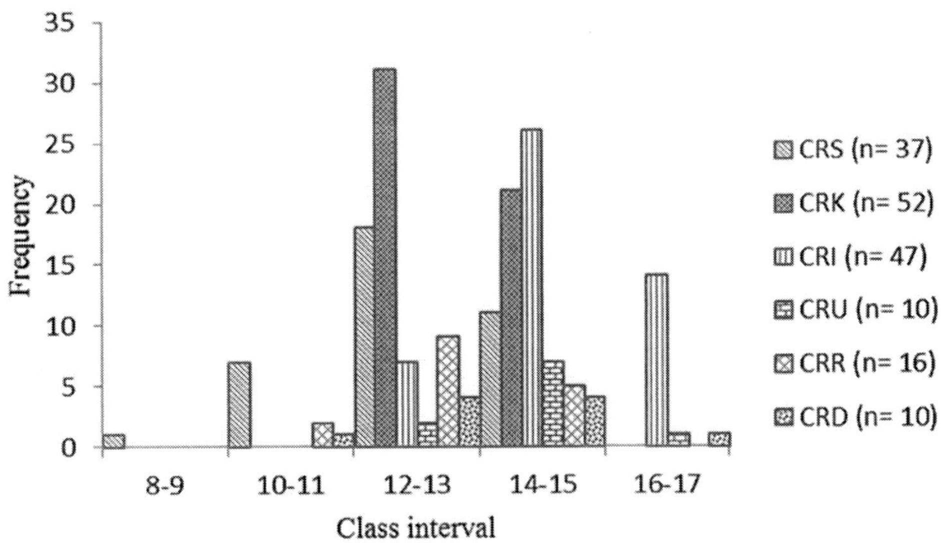

Figure 3. Distribution of frequency carapace length (a) *Carcinoscorpius rotundicauda*, (b) *Tachypleus tridentatus*, and (c) *T. gigas*. Note: Two letters at the beginning is the name of horseshoe crab species (CR= *Carcinoscorpius rotundicauda*; TT=*Tachypleus tridentatus*; TG=*T. gigas*) and One letter at the end is the name of sampling location (S=Semarang; K=Surabaya; I=Indramayu; U=Subang; R=Rembang; D=Demak)

Figure 3 showed that the cosmopolitant horseshoe ccarb species of along north coast of Java was *C. rotundicauda* with the highest number found in Surabaya coast. Overall, the number of *C. rotundicauda* were found in all study locations was 170 individuals, i.e. 9 individuals in Subang coast,

47 individuals in Indramayu coast, 37 individuals in Semarang coast, 9 individuals in Demak coast, 16 individuals in Rembang coast, and 52 individuals in Surabaya coast. The class interval of carapace length of *C. rotundicauda* ranged from 8-9 cm until 16-17 cm, with the highest number found in the class interval 14-15 cm (Figure 2a).

The total number of *T. tridentatus* was found from three locations during this study was 33 individuals, i.e. 16 individuals in Surabaya coast, 3 individuals in Indramayu coast, and 14 individuals in Subang coast. The tri-spine horseshoe crab from Surabaya and Subang tended to have the same carapace length, but was longer than those from Indramayu. The class interval of carapace length of *T. tridentatus* ranged from 9-10 cm until 17-18 cm, with the highest number found in the class interval 13-14 cm (Figure 2b).

The total number of *T. gigas* was found during this study is 43 individuals and they were collected from 5 study locations, i.e. 6 individuals from Surabaya coast, 12 individuals from Indramayu coast, 13 individuals from Subang coast, 2 individuals from Rembang coast, and 10 individuals from Demak coast. The coastal horseshoe crab from Subang had longer carapace length than other locations. The class interval of carapace length of *T. gigas* ranged from 15-16 cm until 25-26 cm, with the highest number found in the class interval 17-18 cm (Figure 2c).

In general, of the three species *T. gigas* had largest size and widest class interval. Size variation of horseshoe crab, including length and width of carapace, is probably influenced by the geographic location, population density, availability of food, and environmental condition [28, 29]. For example, carapace length of the mangrove horseshoe crab from Malaysia is longer than form India and Thailand [5].

3.3. *Morphometric Characters*

Comparison of morphometric characters of all three Asian horseshoe crab species in this study was performed by Mann-Whitney U-test ($p < 0.05$ and $p < 0.1$). The results are presented in Table 1, 2, and 3.

Table 1. Comparison of morphometric characters of *Carcinoscorpius rotundicauda* (mean±SE) among several sampling locations in this study

Character	Semarang	Surabaya	Indramayu	Subang	Demak	Rembang
Carapace width	1.38 ± 2.42^{ab}	0.97 ± 0.02^{abcd}	0.96 ± 0.04^{cde}	0.96 ± 0.98^{bcde}	0.95 ± 0.04^{bcdef}	0.93 ± 0.03^{ef}
Telson length	1.51 ± 2.79^{ab}	1.06 ± 0.12^{ab}	$1.06\pm0.07^{a*}$	1.08 ± 1.08^{ab}	$0.99\pm0.18^{b*}$	1.01 ± 0.25^{ab}

^{a,b,c,d,e,f} within the same row, values with different superscript are significantly different ($p < 0.05$ and $* = p < 0.1$).

Table 2. Comparison of morphometric characters of *Tachypleus gigas* (mean±SE) among several sampling locations in this study

Character	Subang	Indramayu	Surabaya	Demak	Rembang
Carapace width	1.87 ± 3.27^{ac}	$1.03\pm0.11^{b*}$	0.97 ± 0.04^{abc}	0.97 ± 0.04^{ac}	$0.95\pm0.01^{ac*}$
Telson length	0.92 ± 0.28^{ab}	1.06 ± 0.20^{a}	1.04 ± 0.09^{ab}	0.97 ± 0.15^{b}	1.04 ± 0.19^{ab}

^{a,b,c} within the same row, values with different superscript are significantly different ($p < 0.05$ and $* = p < 0.1$).

Table 3 Comparison of morphometric characters of *Tachypleus tridentatus* (mean±SE) among several sampling locations in this study

Character	Subang	Indramayu	Surabaya
Carapace width	0.99±0.03	0.98±0.03	0.98±0.03
Telson length	1.05±0.17	1.08±0.04	1.08±0.15

[a,b] within the same row, values with different superscript are significantly different ($p < 0.05$).

Table 1 shows that carapace width of *Carcinoscorpius rotundicauda* was significantly different among all study locations ($p < 0.05$). Telson length of *C. rotundicauda* was significantly different between Indramayu and Demak ($p < 0.1$). While in other sampling locations they were not significantly different, both carapace width and telson length ($p < 0.05$ and $p < 0.1$). Carapace width of *T. gigas* was significantly different between Indramayu and all sampling locations except Surabaya coast ($p < 0.05$ and $p < 0.1$). Telson length of *Tachypleus gigas* was significantly different between Indramayu and Demak ($p < 0.05$). In other locations it was not significantly different both carapace width and telson length ($p < 0.05$) (Table 2). Carapace width and telson length of *T. tridentatus* were not significantly different in all sampling locations ($p < 0.05$ and $p < 0.1$) (Table 3).

The variation and differences in morphometric characters of the horseshoe crabs might be affected by environmental and genetic factors. The significant difference in morphometric characters of the horseshoe crabs between two different habitats strongly suggests that these specimens might belong to two different independent stocks [5]. Krumholz and Cavanah [30] showed that differences in morphometric characters of *Carcinoscorpius rotundicauda* in different environments were due to environmental effects of the habitat or it might be related with genetic effect.

4. Conclusions

The three Asian horseshoe crabs species were found and distributed along north coast of Java and south coast of Madura Island; they are the tri-spine horseshoe crab *Tachypleus tridentatus*, the coastal horseshoe crab *Tachypleus gigas,* and the mangrove horseshoe crab *Carcinoscorpius rotundicauda*. Those three species were found co-existing in north coast of Subang, Indramayu, and Surabaya. In term of size, the coastal horseshoe crab, *T. gigas* showed largest size and widest class interval among the three Asian horseshoe crab species. Morphometric characters of the three species showed variations and differences with locations.

Acknowledgements
This study was supported by BOPTN program of Higher Education Directorate (DIKTI) fund. We wish to thank to Research Team of Department of Aquatic Resources Management for kindly helping on this research.

References

[1] Lee CN, and Morton B. 2005. Experimentally derived estimates of growht by juvenile *Tachypleus tridentatus* and *Carcinoscorpius rotundicauda* (Xiphosura) from nursery beaches in Hong Kong. *Jurnal Marine Biology Ecology* **318**:39-49

[2] Christianus A and Saad CR 2007 Horseshoe crabs in Malaysia and the world. *Fishery Mail* **16**:8-9

[3] Zaleha K, John BA, Atika HE, Kamaruzzaman BY, and Jalal KCA 2012 Spawning and nesting behaviour of *Tachypleus gigas* along the east coast of Peninsular Malaysia. *International Journal of Biology* **4**(2):102-111

[4] IUCN [International Union for Conservation of Nature] 2010 Red list of threatened species. www. Iuc-nredlist.org

[5] Srijaya TC, Pradeep PJ, Mitun S, Hasan A, Shaharom F, and Chatterji A 2010 A new record on the morphometric variation in the population of horseshoe crab (*Carcinoscorpius rotundicauda* Latreille) obtained from two different ecological habitats of Peninsular Malaysia. *Nature* **8** 204-211

[6] Cartwright-Taylor L, Lee J, and C.C. Hsu CC 2009 Population structure and breeding pattern of the mangrove horseshoe crab *Carcinoscorpius rotundicauda* in Singapore. *Aquatic Biology* **8**:61-69

[7] Cartwright-Taylor L, Bing YV, Chi HC, and Tee LS 2011 Distribution and abundance of horseshoe crabs *Tachypleus gigas* and *Carcinoscorpius rotundicauda* around the main island Singapore. *Aquatic Biology* **13**:127-136

[8] Ismail N, Taib M, Shamsuddin AA, and Shazani S 2011 Genetic variability of wild horseshoe crab, *Tachypleus gigas* (Müller) in Tanjung Dawai, Kedah and Cherating, Pahang of Peninsular Malaysia. *European Journal of Scientific Research* **60**(4):574-583

[9] Ismail N, Jolly JJ, Dzulkiply SK, Mustakim, MKM, Hafiz ANM, Huda AGI, Taib M, Shamsuddin AA, and Chatterji A 2012 Allometric variations of horseshoe crab (*Tachypleus gigas*) populations collected from Chendor and Cherating, Pahang, Peninsular Malaysia. *Journal of Sustainability Science and Management* **7**(2):164-169

[10] Chatterji A and Pati S 2014. Allometric relationship in the adult population of the Malaysian horseshoe crab (*Tachypleus tridentatus*; Leach). *International Journal of Research* **1**(11):1378-1385

[11] Robert R, Muhammad Ali SH, and Amelia-Ng PF 2014 Demographics of horseshoe crab populations in Kota Kinabalu, Sabah, Malaysia with emphasis on *Carcinoscorpius rotundicauda* and some aspects of its mating behaviour. *Pertanika J. Trop. Agric. Sci.* **37**(3):375-388

[12] Mohamad F, Manca A, Ahmad A, Sofa MFAM, Amirul Asyraf Alia'm AA, and Ismail N 2016 Width-weight and length-weight relationships ff the tri-spine horseshoe crab, *Tachypleus tridentatus* (Leach 1819) from two populations in Sabah, Malaysia: Implications for population management. *Journal of Sustainability Science and Management* **11**(1):1-13

[13] Chatterji A and Abidi SAHJ 1993 The Indian horseshoe crab: a living fossils. *Journal Indian Ocean Study* **1**:43-48

[14] Davidson GWH, Ng PKL, and Ho HC 2008 The Singapore Red Data Book: threatened plants and animals of Singapore. The Nature Society (Singapore), Singapore

[15] Shin P, Li HY, and Cheung SG 2009 Horseshoe crabs in Hong Kong: current population status and human exploitation. In: Tanacredi JT, Botton ML, Smith DR, Earle SA (eds) *Biology and conservation of horseshoe crabs*. Springer Science and Business Media, Heidelberg, p 347–360

[16] Ng PKL and Sivasothi N 2002 A guide to the mangroves of Singapore. I. The ecosystem and plant diversity, 2nd edn. BP Guide to Nature Series, Singapore Science Centre, Singapore

[17] Chatterji A 1999 New record of the sympatric distribution of two Asian species of the horseshoe crab. *Indian Journal of Marine Science* **77**(6):43-48

[18] Mishra JK 2009 Horseshoe crabs, their eco-biological status along the northeast coast of India and the necessity for ecological conservation. *In*: Tanacredi JT *et al.* (eds). 2009. *Biology and conservation of horseshoecrabs. Springer Science & Business Media* 89-96

[19] Rubiyanto E 2012 Studi Populasi Mimi (Xiphosura) di perairan Kuala Tungkal, Kabupaten Tanjung Jabung Barat, Jambi [Tesis]. Universitas Indonesia

[20] Meilana L 2015 Assessment of Horseshoe Crabs (Xiphosura, Limulidae) Morphology and Genetic as Basis for Conservation and Management in the Java Island Coast [Thesis]. Bogor (ID): Institut Pertanian Bogor

[21] Sekiguchi K1988 *Biology of horseshoe crabs*. Science House, Tokyo: ix + 428 pp.

[22] Morton B 1999 On turtles, dolphins and, now, Asia's horseshoe crabs. *Mar Pollut Bull* **38**:845-846

[23] Chiu HMC and Morton B 2003 The status of horseshoe crabs in Hong Kong. In: Morton B (ed) Perspectives on marine environmental change in Hong Kong and southern China. Proceedings of the Hong Kong workshops reunion conference on perspectives on marine environmental change in Hong Kong and southern China (1977–2001). Hong Kong University Press, Hong Kong, p 743-758

[24] KKP [Kementerian Kelautan dan Perikanan] 2013 Biota Perairan Terancam Punah di Indonesia. Direktorat Konservasi Kawasan dan Jenis Ikan Ditjen Kelautan, Pesisir, dan Pulau – Pulau Kecil Kementerian Kelautan dan Perikanan 257 pp.

[25] Webster M 2007 A cambrian peak in morphological variation within trilobite species. *Science* **317**(5837):499-502

[26] Sekiguchi K and Shuster CN 2009 Limits on the global distribution of horseshoe crabs (Limulacea): Lessons learned from two lifetimes of observations: Asia and America. In: Tanacredi JT, Botton ML, and Smith DR. (eds.) *Biology and Conservation of Horseshoe Crabs*. Heidelberg: Springer, 5-24

[27] Chiu HMC and Morton B 2004 The behaviour of juvenile horseshoe crabs, *Tachypleus tridentatus* (Xiphosura), on a nursery beach at Shui Hau Wan, Hong Kong. *Hydrobiologia* **523** 29–35

[28] Chatterji A, Parulekar AH, and Vijayakumar R 2000 Morphometric characteristics in the horseshoe crab *Tachypleus gigas* (Artropoda: Merostomata). *Indian Journal of Marine Science* **29** 333-335

[29] Zadeh SS, Christianus A, Saad CR, Hajeb P, and Kamarudin MS 2011 Comparations in prosomal width and body weight among early instar stages of Malaysian Horseshoe crabs, *Carcinoscorpius rotundicauda* and *Tachypleus gigas* in the laboratory. *In*: Tanacredi JT *et al.* (eds). 2011. *Biology and conservation of horseshoe crabs.* Springer Science & Business Media: 267-274

[30] Krumholz LA and Cavanah HS 1968 Comparative morphometry of freshwater drum from two Midwestern localities. *Trans Am Fish Soc* **97** 429-441

Assessment area development of sustainable shrimp culture ponds (case ctudy the gulf coast Banten)

M Farkan[1], D D Setiyanto[2], R S Widjaja[3], Kholil[4] and Widiatmaka[5]

[1] Jakarta Fisheries University, Ministry of Marine Affairs and Fisheries, Indonesia.
[2] Faculty of Fisheries and Marine, Bogor Agricultural University, Indonesia
[3] Ministry of Marine Affairs and Fisheries, Republic of Indonesia
[4] Environmental Engineering Department of Sahid University, Jakarta, Indonesia
[5] Department of Soil Sciences and Land Resources, Bogor Agricultural University, Indonesia.

Email: moch_farchan@yahoo.co.id

Abstract. Shrimp is a fishery commodity that has the economic value and important food provision, so that there is a need for increasing sustainability and continuity of the production. This research was conducted during March – December 2015 in Banten Bay, Indonesia. The objective of this research were: (1) to assess the land suitability for shrimp farming, (2) to analyze land carrying capacity for shrimp farming, (3) to establish the institutional model of shrimp farming management. The data used were primary data, collected from field survey and secondary data, collected from literature and research report which were done in the research area. The methods used to evaluate the land suitability were weighted spatial overlay. The carrying capacity were analyzed using two approaches: land suitability weight and water availability methods. The institutional model was established using *Interpretative Structural Modeling* (ISM). The results of the study showed that from a total area analyzed of 5.028.3 ha, it can be classified into two suitability classes: highly suitable (S1) area which is 141.7 ha (2,8 %) and suitable (S2) area which is 4.886.6 ha (97.2 %). In term of management, the area can be grouped as traditional farming area of 4.173.5 ha (83 %), semi-intensive farming area of 698.93 ha (13,9) and intensive farming area of 155.87 ha (3,1%). The institutional modelling shows that the most decisive institutions are universities and research institutions. The model designed showed an inter-related relationship between land suitability, carrying capacity, institutional, and social in order to increase the sustainability of shrimp farming management.

1. Introduction

Banten Bay coastal aquaculture area is ecologically a focus of activities on land and at sea. Carrying more than 351 industries that directly or indirectly impact on the Gulf coast of Banten. The area of aquaculture in coastal areas Banten Bay continues to decrease due to diversion into the industrial area, in 2015, all operating for shrimp farming only about 90 ha. The purpose of this research is to create a model of sustainable shrimp farming area management. As a case study is the Gulf coast of Banten.

2. Methodology

2.1 The location and time of the study

Aquaculture area of research was from the village of Banten with the coordinates of latitude 05 ° 57'13" 106 ° 6'6" BT until Ciujung river, village Tengkurak at coordinates 05 ° 57'48" LS 106 ° 21'26" BT. The field research was conducted in March-November, 2015.

2.2 Methods

Management model based on land suitability, carrying capacity and institutions that affect the sustainability of shrimp farming. The measured parameter is the quality of water, soil and infrastructure or supporting shrimp farming. Water quality parameters include: temperature, salinity, dissolved oxygen, pH, BOD_5, COD, TSS, ammonia, Fe and tides. The quality of the soil pH, KCl, soil texture (sand, silt, clay), redox, CEC, K, Ca, Mg, Fe, slope and elevation. While supporting parameters or infrastructure consists of the availability of infrastructure, distance from the river, the distance from the sea. Data is collected in 24-point spread in aquaculture area of the pond. Tools and materials used include test kits, pH paper, refractometer, sechi dish, thermometer, measuring tidal barrier and water and soil test equipment in the lab. Land suitability evaluation calculated using the weighting method to perform paired comparison is data obtained in the field was analyzed by comparison with the standard. To get the score level influence land suitability compared to the results of field measurements with standard shrimp farming in ponds. Figures obtained by expert opinion. The number of parameters of water and soil, and supporting shrimp farming are 25 parameters. The results of processing performed overlay with the application of land suitability map for shrimp farming.

As a basic method of determining the carrying capacity of the weighting method is very close association with qualitative land suitability later made quantitative. Rated capacity of the environment is a quantification of land suitability classes. The results declared by the pond intensive, semi-intensive and extensive. This method also uses calculations have been prepared by Prasita et al. [1] ie the assumption of maximum production and availability of sea water. Various studies have used the method to assess the carrying capacity of water supply [1, 4]. Although the development of the use of this method is different. The method used is a development method which is the adjustment of the method developed by Widigdo and Pariwono [4] based on the existing objective.

For the measurement of sea water availability as part of the calculation of the carrying capacity using a water supply line fields.

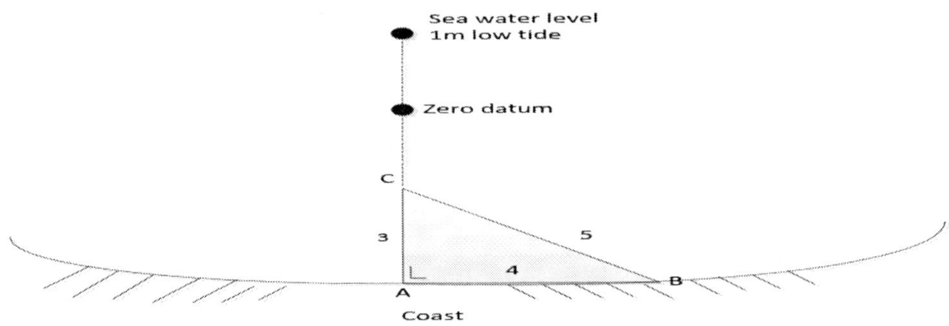

Figure 1. The straight line from the shore to the point of water height of 1 m at the time of lowest tide.

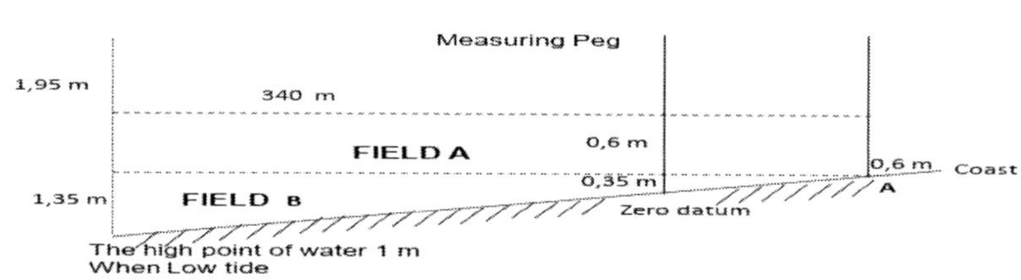

Figure 2. Field construction measurements at sea

Institutional data collection was done by direct surveys and interviews with stakeholders and experts. This institutional study using data analysis using Saxena [3]. The results of discussions with experts, analysis of secondary data from the field and 9 elements have five elements are used to assess the management of sustainable shrimp pond area that is 1) .Sektor influential community; 2) Requirement of the program; 3) The main trouble; 4) The purpose of the program; 5) Institutions involved in the implementation of the program.

3. Discussion

3.1 Water quality parameters

Based on the results of water quality measurements weight compared with a score of pairwise comparison results (pairwise comparisons) criteria is very appropriate, suitable and less suitable. These results are then carried analysis overlaying (over-lay) which resulted in land based water quality parameters as in Table 1 below.

Table 1. Total land based water quality

Criteria	Very Suitable		Suitable		Suitable Less		Total size of	
	Area (ha	%	Area (ha)	%	Area (ha)	%	Area (ha)	%
Temperature	2,564.2	51.0	2,098.2	41.7	366.0	7.3	5,028.3	100.0
Salinity	1,146.0	22.8	3,469.9	69.0	412.5	8.2	5,028.3	100.0
COD	1,606.0	31.9	2,713.6	54.0	708.8	14.1	5,028.3	100.0
TSS	5,015.0	99.7	-	-	13.3	0.3	5,028.3	100.0
pH	545.6	10.9	4,482.7	89.1	-	-	5,028.3	100.0
DO	-	-	4,899.5	97.4	128.8	2.6	5,028.3	100.0
Amonia	2,240.0	44.5	883.2	17.6	1,905.2	37.9	5,028.3	100.0
BOD_5	2,840.7	56.5	2,033.4	40.4	154.2	3.1	5,028.3	100.0
Fe	3,674.3	73.1	1,087.7	21.6	266.3	5.3	5,028.3	100.0
Tidal	-	-	-	-	5,028.3	100.0	5,028.3	100.0

3.2 Parameter soil quality

These results are then analyzed overlaying (over-lay) which resulted in a land area based on parameters such as soil quality in Table 2 below.

Table 2. Area of land based on soil quality

Kriteria	Very Suitable		Suitable		Suitable Less		Total size of	
	Area (ha	%	Area (ha)	%	Area (ha	%	Area (ha)	%
pH	1.955.3	38.9	2.558.6	50.9	514.4	10.2	5.028.3	100.0
Redox potential	5.028.3	100.0	-	-	-	-	5.028.3	100.0
KTK	561.3	11.2	3.091.8	61.5	1375.2	27.3	5.028.3	100.0
Texture	977.2	19.4	4.051.1	80.6	-	-	5.028.3	100.0
Magnesium	-	-	4.788.2	95.2	240.1	4.8	5.028.3	100.0
Potassium	1.457.4	29.0	3.486.8	69.3	84.1	1.7	5.028.3	100.0
Calcium -	-	-	-	-	5.028.3	100.0	5.028.3	100.0
Ironi	-	-	-	-	5.028.3	100.0	5.028.3	100.0
Slope	5.028.3	100.0	-	-	-	-	5.028.3	100.0
Elevation	5.028.3	100.0	-	-	-	-	5.028.3	100.0

3.3 Infrastructure

Results were analyzed overlaying (over lay) produce land based support such as in Table 3.

Table 3. Suitable criteria of infrastructure

Criteria	Very Suitable		Suitable		Suitable Less		Total size of	
	Area (ha	%	Area (ha)	%	Area (ha	%	Area (ha)	%
Distance from the road	1.795.7	35.7	1.693.4	33.7	1.539.3	30.6	5.028.3	100.0
The distance from the river	3.191.3	63.5	1.384.8	27.5	452.3	9.0	5.028.3	100.0
Distance from the beach	994.8	19.8	3.817.4	75.9	216.2	4.3	5.028.3	100.0
Rainfall	-	-	-	-	5.028.3	100.0	5.028.3	100.0

3.4 Comparison of pairs of parameters of water, soil and infrastructure

Analysis of land basetured on three parameters generated very suitable land, appropriate and less appropriate as in Table 4 below.

Table 4. The land area is based on a comparison of water quality, soil quality and support

Criteria	Very Suitable		Suitable		Suitable Less		Total size of	
	Area (ha	%	Area (ha)	%	Area (ha	%	Area (ha)	%
Water quality	655.1	13.0	4,351.0	86.5	22.2	0.4	5.028,3	100.0
The soil quality	4,358.9	86.7	669.4	13.3	-	-	5.028,3	100.0
Infrastructure	1,994.2	39.7	2.697.1	53.6	337.0	6.7	5.028,3	100.0

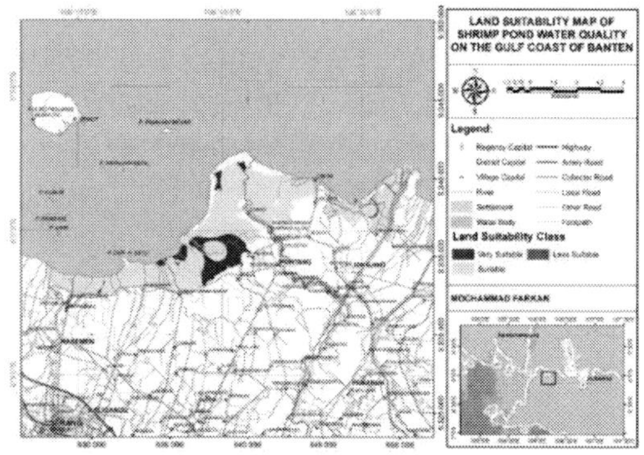

Figure 3. Map of the land suitability of water quality parameters

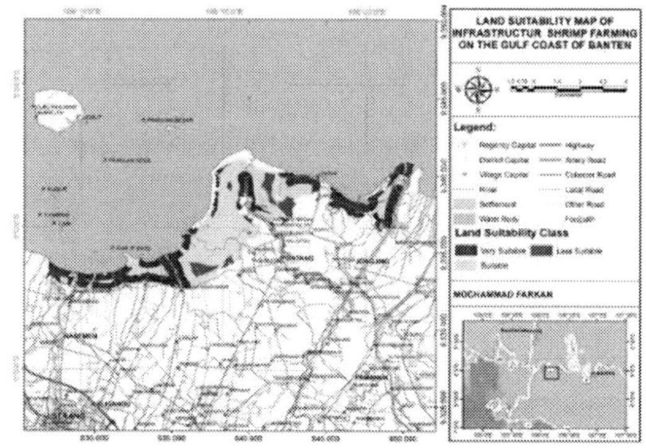

Figure 4. Map of the land suitability of infrastructure parameters

An overlay of the land suitability is based on the analysis of water quality, soil quality and land infrastructure can be seen very suitable and suitable as shown in Table 5.

Table 5. Summary of the extent of land suitability classes pond shrimp farming in the Gulf coast of Banten

Criteria	Extents	
	Ha	%
Very suitable	141.7	2.8
Suitable	4,886.6	97.2
Less suitable	-	-
Total	5,028.3	100.0

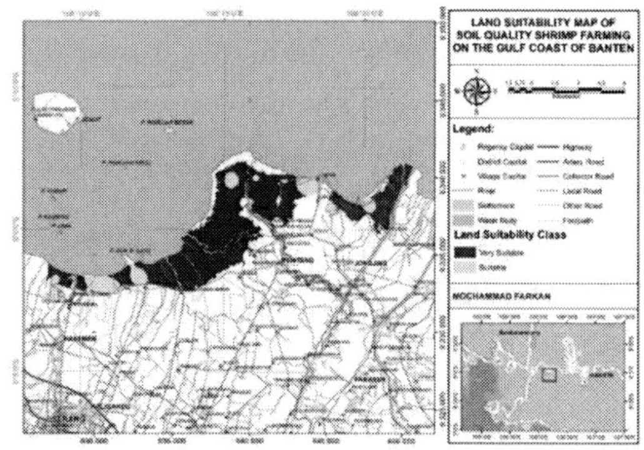

Figure 5. Land suitability maps of soil quality parameters.

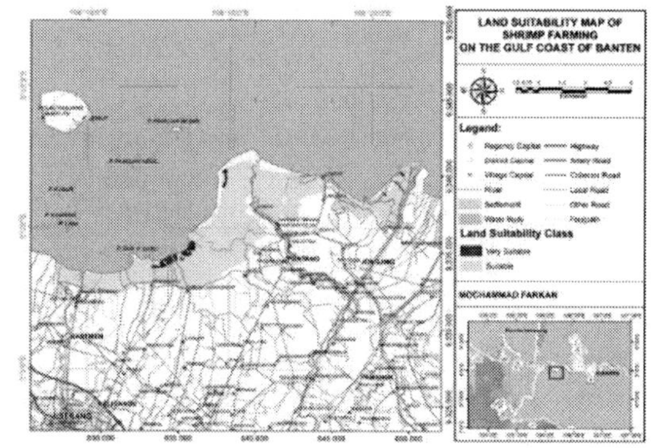

Figure 6. Land suitability maps for the Gulf Coast of pond in Banten

Measuring carrying capacity based on the availability of water. beach water volume available = 36 750 m x 433.5 m^2 = 15931.12 m^3. Teluk Banten occur post the mixture is lower so that the potential of the water volume 1.2 times tide. So the water volume Beach = 1.2 x 20.0655 million m = 19117.35 m^3. The volume provides a measure of the volume of waste ponds that can be supported at 19117.35 m^3 which also means the volume of sea water supply for cultivation in ponds 19117.35 m^3. Based Prasita et al. (2008), the width of intensive pond that can be supported by 191.17 ha or 3.8%. Commodity type of shrimp reared in the coastal bays offerings are tiger shrimp and shrimp vaname. Extensive maintenance tiger shrimp average production is 36: 2 = 18 tonnes / ha / year. So the maximum carrying capacity of these waters is 18 tonnes / ha x 191.17 ha = 3441.06 tons / yr. If the semi-intensive technology with a production capacity of semi-intensive maintenance tiger shrimp is 5.1 tonnes / ha / year, shrimp vaname 12 tonnes / ha / year, the average production is 17.1: 2 = 8.55 tonnes / ha / year , Comparison of intensive: semi-intensive = 36: 8.55 = 4.21. Thus the semi-intensive pond area becomes 804.82 ha or 16%. Extensive 4033.68 ha or by 80.1 coastal ponds Banten Bay area 5028.3 ha, the area can be mapped extensive technology 4033.68 ha of ponds; semi-intensive and intensive 804.82 ha of 300.98 ha or equal to (3.8%). This calculation is only relying on the availability of sea water and fresh water supplies have not counted provided by the river. Besides, the speed of decomposition of organic materials by microbes and oxygen availability. Production facilities and infrastructure that is easy to competent and utilization are balanced by the relevant sector affect other sub element in achieving sustainable shrimp farming.

Institutional influential in managing the sustainable shrimp aquaculture ponds has 11 sub-elements.

Table 6. Institutions that play a role in the area of sustainable shrimp farming in the Gulf persisir Banten.

Code	Elemen
L1	District Department of Agriculture and Fisheries of Serang
L2	Village Institution
L3	Department of Marine Fisheries, energy and mineral resources Serang District.
L4	Financial institutions
L5	Shrimp Farmers Group
L6	The Environment Agency and Spatial Serang District
L7	Marine and Fisheries Agency of Banten Province
L8	Extension and Food Security Agency Serang District
L9	Central Government (Ministry of Marine Affairs and Fisheries)
L10	Universities
L11	Research institutions

By using VAXO approach, the results can be known expert discussion final results of reachability matrix. The value then interpreted Driver Power and Dependence in Figure 7

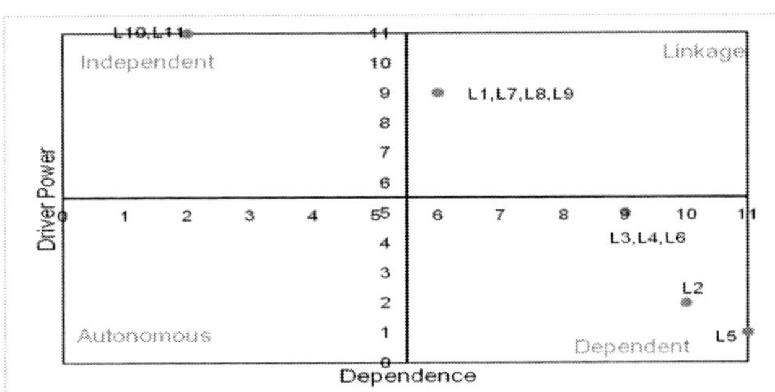

Figure 7. Matrix Driver Power and Dependence on elements of the agencies involved. In Figure 7 it appears that elements of the college (L10); research institutes (L11) be the independent sector (sector IV) which means sub-element has a very big role in the program.

That universities and research institutions have a balanced utilization by the sector can affect other sub elements in realizing sustainable shrimp farming. Based on the results of focus group discussion that using land suitability analysis, carrying capacity and institutional Banten Gulf coast region is divided into three, namely western region, central and east. Analysis result as in Table 6 and Figure 7.

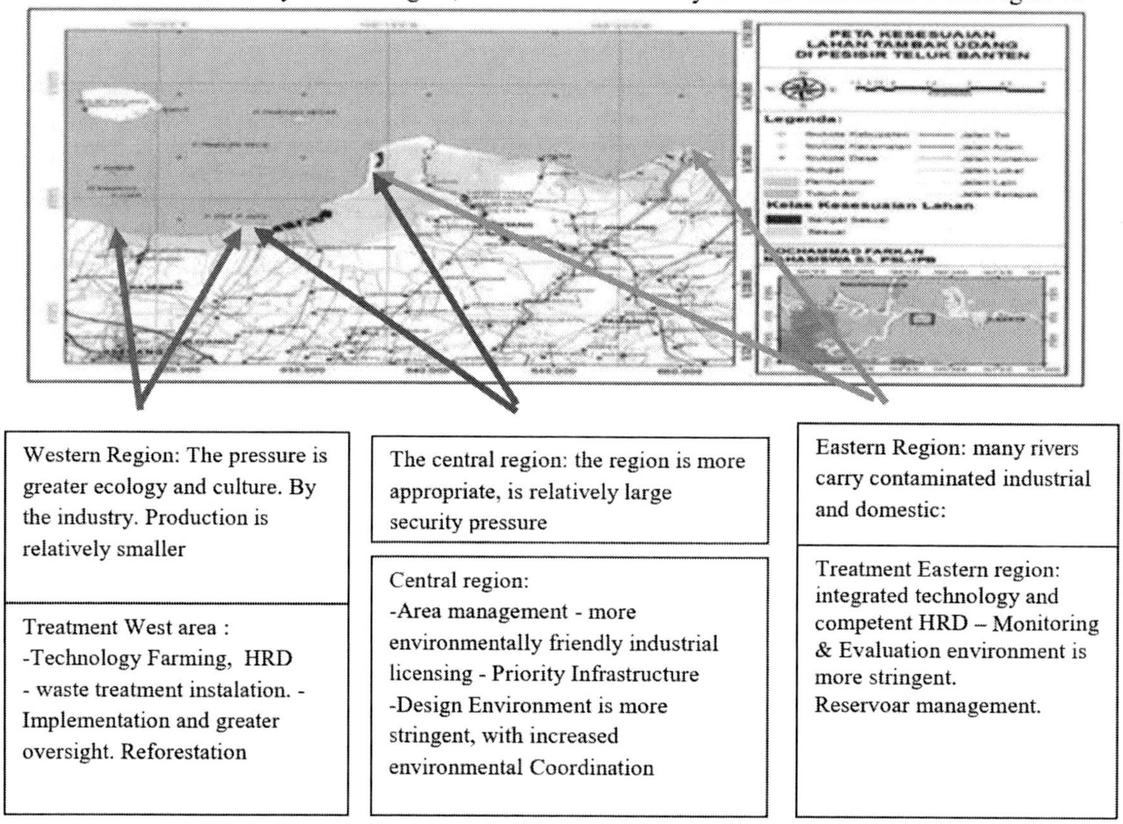

Figure 7. The district management area of shrimp farming in the Gulf coast of Banten

Table 6. Area Regional Management

No	Region	Suitability Class Size	Luas	
			Ha	%
1	West	Very suitable	0.3	0.0
		In suitable	651.6	13.0
2	Central	Very a suitable	129.7	2.6
		In suitable	1,984.6	39.5
3	East	Very suitable	10.8	0.2
		In suitable	2.251.3	44.8
		Total Area of	5,028.3	100.0

4. Conclusions

Land coast of Banten Bay area of 5028.3 ha based on the parameters of water quality, soil and infrastructure can be mapped into a highly suitable land 141.7 ha (2.8%) and suitable 4886.6 ha (97.2%). Carrying capacity pond Gulf coast region support offerings can be mapped to a traditional or extensive cultivation area of 4173.5 ha (83%), semi-intensive 698.93 ha (13.9) and intensive 155.87 ha (3.1%). To realize the management of shrimp farming area in the coastal bays sustainable offerings, the institution most responsible among other institutions are universities and research institutes.

References

[1] Prasita VD, Widigdo B, Hardjowigeno S, Budiharsono S 2008 Kajian Daya Dukung Lingkungan Kawasan Pertambakan Di Pantura Kabupaten Gresik Jawa Timur *Jurnal Ilmu-Ilmu Perairan dan Perikanan Indonesia* **2**: 95-102.

[2] Hossain MS, Das NG 2010 GIS – Based Multi – Criteria Evaluation To Land Suitability Modelling For Giant Prawn (Macrobrachium Rosenbergii) Farming In Companigons Upzila Of Nokhali, Bangladesh *Journal of Computers and Electronics in Agriculture* **70** 172-186.

[3] Saxena JJP, Sushil and Vrat P. 1992. Hierarchy and Classification of Program Plan Element Using Interpretive Structural Modelling *System Practice* **5** 651: 670.

[4] Widigdo B dan Pariwono J 2003 Daya Dukung Perairan di Pantai Utara Jawa Barat untuk Budidaya Udang (Studi Kasus di Kabupaten Subang, Teluk Jakarta dan Serang) *Jurnal Ilmu-ilmu Perairan dan Perikanan Indonesia*. **1** 10-17.

Delineation of maritime management boundary between Surabaya, Gresik and Bangkalan using SPOT 7

Khomsin and M Ayuningtyas
Geomatic Engineering Department – Institut Teknologi Sepuluh Nopember, Surabaya - Kampus ITS Sukolilo – Surabaya, East Java

Email: khomsin95@yahoo.com

Abstract Recently, a lot of marine resource management in this area such as shipping lane, Suramadu Bridge, Tanjung Perak Port, Teluk Lamong Port and other ports in Gresik and Bangkalan. According to the Law No. 32/2004 about local government stated that cities/regencies have full authorities to manage the sea untill 1/3 provincial authorities up to 12 nautial miles and then changed to the Law No. 24/2014, the authority of marine resource management is the provincial government up to 12 nautical miles from baseline. Regency and city governments just to get sharing of marine managed by province up to 4 nautical miles from baselines. The baseline as referenced in Law No. 32/2004 is low water line (LWL) and high water level (HWL) in Law No. 23/2014. The problem is maps (topographic map and nautical chart) which is used in this area were not referenced by HWL. Therefore, in this paper, we used SPOT 7 imagery to determine HWL in this area. After that, boundary of marine resource management is determined by median line and equidistance methods. Median line is used to determine maritime boundary between opposite cities/regencies (Surabaya - Gresik and Bangkalan) and equidistance is used to determine between adjacent cities/regencies (Surabaya and Gresik). The results of this study indicate that Surabaya, Gresik and Bangkalan have the right to earn sharing of marine resource management in the shipping lanes, Teluk Lamong Port and Suramadu Bridge. In this case, the change of Law No. 32/2004 to Law No. 23/2004 had an impact on the maritime boundary such as Surabaya getting loss of 70.844 ha, Bangkalan getting loss of 407.705 ha and Gresik benefit of 331.003 ha.

1. Introduction

Regional autonomy is the right, the authority, and the obligation of autonomous region to set up and manage their own affairs and interest local communities in the system of the Republic of Indonesia [1]. The implementation of regional autonomy is an opportunity for local government to improve people's welfare. Local governments free to be creative and express themselves in order to build their area according the potential and the peculiarities. Autonomous region is a legal community unit with region boundaries have authorized to organize and to manage governmental affairs and public interests at its own initiative based on the aspirations of the people in the system of the Republic of Indonesia [1]. To be able to implement regional autonomy, the region boundary should be clear where the location.

Region boundary has an important role in local governance and its impact would affect the policies that accompany it. For that, it needs to be organized well and responsibly. To be clear which region include his authority and to avoid disputes in the border region. Region boundary is dividing the territory of the one authority area to another [2]. Boundary area is divided into two, namely boundaries in the land and boundaries in the sea. Based on the authority of region government that regulated in Local Government Law, No. 23 / 2014, the regional maritime boundary consist of maritime management boundary for the province and maritime income sharing boundary for the regency/city.

Madura strait especially marine waters which is the border between Surabaya, Gresik and Bangkalan is an area of dense shipping lanes to Tanjung Perak Port. In addition, this location has also Suramadu bridges which connecting Surabaya and Bangkalan, Galang Island which is still disputed between

Surabaya and Gresik [3], and other areas that have potential to be developed and generate financial benefits. First, in the era of regional autonomy, law on maritime boundary discussed is Law No. 32/2004 about Local Government. In this law, regencies and cities have full authorities to manage the ocean territory up to 1/3 from provincial territory up to 12 nautcal miles [2]. Meanwhile, the Local Government Law No. 23/2014 state that provinces have the authority to manage marine areas up to 12 nautical miles from shoreline and regencies and cities just acquire a share of the marine management undertaken by the provincial government till 4 nautical miles [1]. Second, the reference used in the law No. 32/2004 and No. 23/2014 is different. In the law No. 32/2004 used low water level (LWL) as vertical reference [2] and the law No. 23/2014 used high water level [1].

The problem is that until now has not provided a topographical and hydrographical maps that refers to high water level. This is because Geospatial Information Agency according to Geospatial Information Law No. 4/2011 only provide Linkungan Pantai Indonesia (Indonesian Coastal Environment) Map which refers to low water level [4]. In this paper, we will discuss about how to construct marine boundaries between Surabaya, Gresik and Bangkalan using high resolution satellite imagery. In this case, we used SPOT 7 to generate coastline which refer to high water level. Then, this coastline to construct marine boundaries between Surabaya, Gresik and Bangkalan using equal distance and median line method [5].

2. Method

This study used High Resolution Satellite Image (HRSI) SPOT 7 in 2015 that have been corrected and obtained from the Nationa Institute of Aeronautics and Space (LAPAN) with .ers format. The location of this research is Madura Strait which is bounded by Surabaya, Gresik and Bangkalan (Figure 1). In the early stages of data processing is to perform the digitization of high water level shoreline on High Resolution Satellite Image (HRSI) SPOT 7 and digitizing of low water level shoreline on Lingkungan Pantai Indonesia (LPI) map. The next step is buffering to delineate as far as 4 nautical miles from the shoreline. For the overlapping sea territorial claims, it should be shared with equidistance method for adjacent regencies (Surabaya and Gresik) and median line for opposite regencies (Surabaya – Bangkalan and Gresik – Bangkalan) [5]. Equidistance and median line delineation should be done manually with a 3-points circle principle, three distance equal to the midpoint.

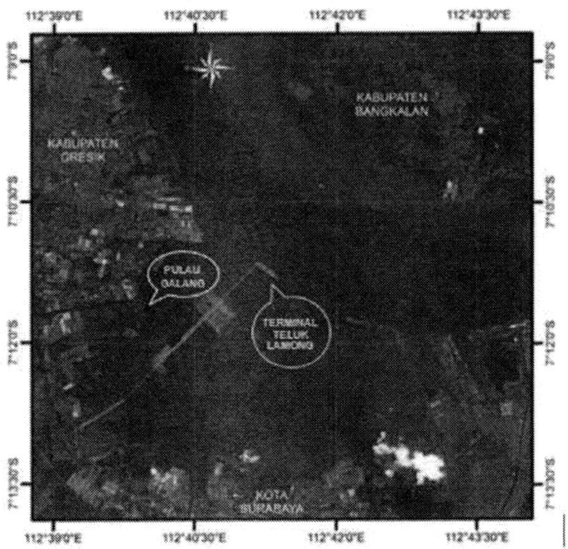

Figure 1. Research location

3. Results And Discussions

3.1 Shoreline delineation

Shoreline plays a very important role for the coastal researcher, coastal engineer and coastal manager to manage marine and coastal area. For practical purposes, shoreline is defined as the intersection of water and land surfaces and it depend on the spatial and temporal scale. It is being considered results in the use of a range of shoreline indicators, high water line, mean high water and low water line (Geospatial Information Law No.4/2011on Article 13 [4]. According to Local Government No. 32/2004 and Local Government Law No. 23/2014 Article 14 state that the determination of regency / city to calculate for marine outcome is the result of marine which is within the territory of four (4) miles measured from *shoreline* toward the open sea and/or toward the waters Island and its explanation, the shoreline in this article refers to low water level (Law No. 32/2004) and high water level (Law No. 23/2014).

Nautical charts in this area (and generally in Indonesia) which refer to LWL is available, but this chart to refer HWL are not available yet. Geospatial Information Agency (BIG) has duty to fulfill these nautical charts (Geospatial Information Law No. 4/2011 on Article 18:2) up to scale 1:10000 [4]. Because of financial and human resources limited, until now BIG have not yet fulfilled this nautical chart which refer to HWL shoreline. Therefore, in this study, to delineate LWL shoreline used LPI map (Figure 2 (right)) and to delineate HWL shoreline with the digitized High Resolution Satellite Image (HRSI) SPOT-7 (Figure 2 (left)).

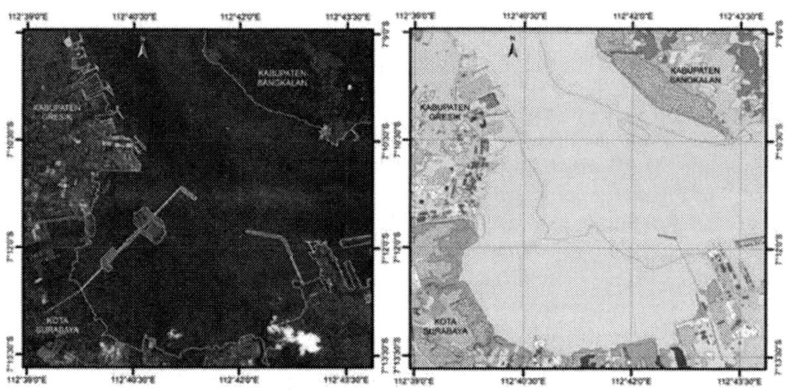

Figure 2. HWL (left-yellow line) and LWL (right-orange line) shoreline delineation

Shoreline changes detection can be approached and detected with multispectral remote sensing satellite provide digital imageries in infrared spectral bands where the land water interface is well defined [6]. Alesheikh, et al. [7] proposed combination procedure between histogram thresholding and band ratio techniques to detect coastline. The accuracy of the extracted coastline has been estimated as 1.3 pixels (pixel size=30 m). SPOT-7 satellite sensor built by AIRBUS Defence & Space was successfully launched on June 30, 2014. SPOT 7 will cover wider areas with a resolution of 1.5 m and genuine constellation, beyond ensuring service continuity from the SPOT family, will now bring new capabilities especially in terms of sharpness, responsiveness and collection capacity: 1.5m resolution suitable for 1:25.000 scale topographic mapping [8]. This digitization of SPOT 7 follow the rules as follows (Figure 3) [9]: the artificial beach, the shoreline is determined from the line of the outer limits of a permanent building on the beach such as port, bridge, sea wall, and others; the coastal areas of trees, shoreline vegetation represented the outer limits closed to the land; and the fish pond area, shoreline dijitized the outer embankment of fish pond close to the sea.

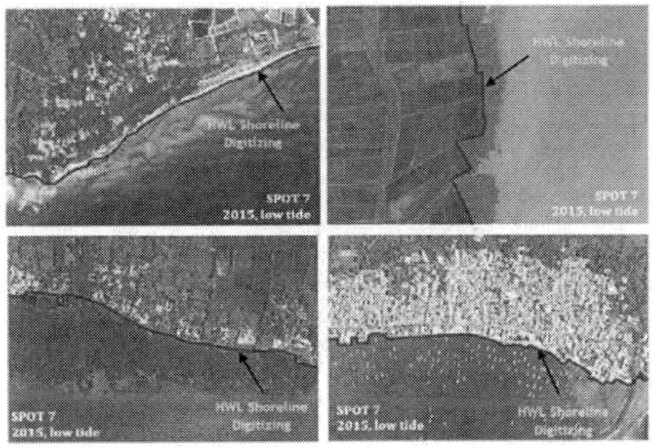

Figure 3. HWL shoreline digitizing on SPOT 7

3.2 Buffering to 4 Nautical Miles From Shoreline

Determination of regencies/cities sea boundaries for marine outcome based on four (4) miles measured from shoreline to the seaward (Local Government Law No. 23/2014 Article 14:6). Buffering far as 4 nautical miles from the Surabaya - Gresik shoreline and Bangkalan shoreline will produce overlapping area (Figure 4). This is because of a distance of Surabaya-Gresik and Bangkalan less than 8 nautical miles. Sea boundaries construction is created with the aim to address overlapping claims between Surabaya, Gresik and Bangkalan.

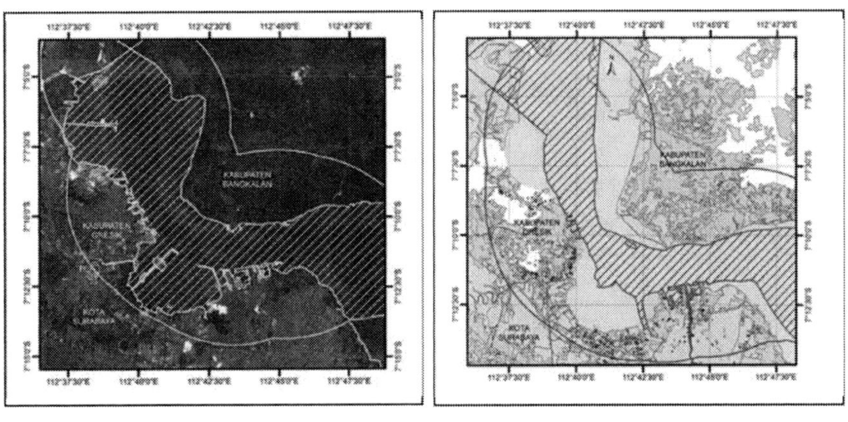

Figure 4. Buffering up to 4 nautical miles from HWL (left-magenta line) and LWL (right-red line) shoreline

The method used for the sea boundary construction refers to the annex regulation of minister of home affairs No 76/2012 those are equidistance method for the adjacent regency/city (in this case between Surabaya and Gresik) (Figure 5 (left)) and opposite regency/city (in this case between Surabaya-Gresik and Bangkalan) (Figure 6 (rigt)). The principle of delineation is by connecting three basic points in the first circle which touch three points. Determination of the base points made subjectively by a distance not far from each other and the outermost points of the coastline, so it does not cut off the coastline (land area) is too much.[10]

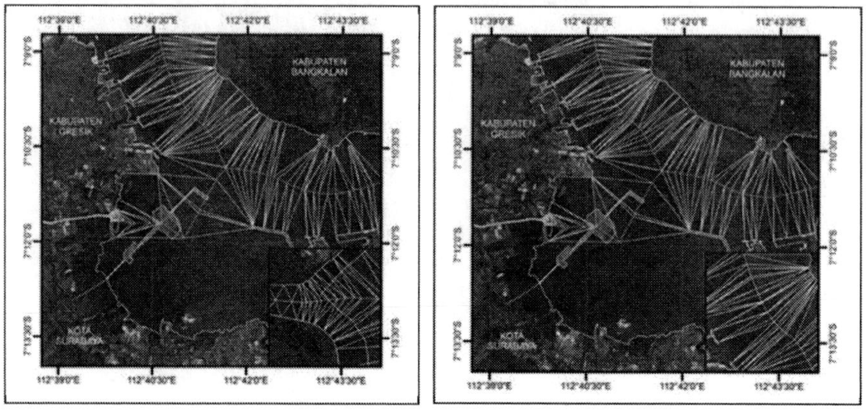

Figure 5. Equidistance (left) and median line (right) method

3.3 Maritime Boundaries Management Analysis

In this case, assumed that Galang Island is not exist, the delineation of maritime baoundaries management between Surabaya and Gresik and between Surabaya-Gresik and Bangkalan (UU No. 23/2014) compared with the delineation of maritime boundaries authorities (UU No. 32/2004) can be seen at Figure 7. The change in Local Government Law effected Surabaya and Bangkalan getting losses by decreasing the extent of 70.844 ha and 407.705 ha consecutively, meanwhile Gresik getting benefit by increasing the extent of 331.003 ha.

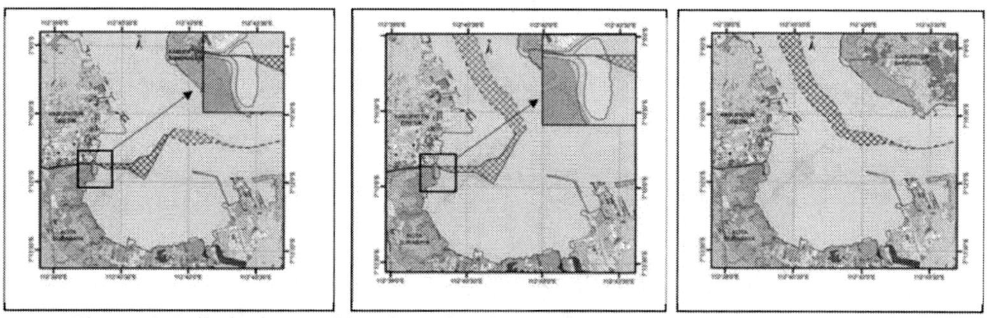

Figure 6. Delimitation maritime boundary between Surabaya, Gresik and Bangkalan

4. Conclusion

Changes in the Local Government Law from No. 32/2004 to No. 23/2014 changes the shoreline reference used is from low water level to high water level. Shoreline change used in the delimitation of the sea will change the sea boundary line. In this case, the change had an impact on the maritime boundary such as Surabaya getting loss of 70.844 ha, Bangkalan getting loss of 407.705 ha and Gresik benefit of 331.003 ha. However the change of the law has changed the regency's/cities' authority to sea before having full authority to manage their own the sea and now regencies/cities just earn a share of the provincial marine management

Acknowledgements
Thanks are going to Lembaga Penerbangan dan Antariksa Nasional (LAPAN) had provided SPOT 7 imagery and Badan Informasi Geospasial (BIG) had provided Lingkungan Pantai Indonesia Map in reseacrh area.

References

[1] Republik Indonesia 2014 Undang-undang Republik Indonesia nomor 23 tahun 2014 tentang Pemerintahan Daerah. Jakarta.
[2] Republik Indonesia 2004 Undang-undang Republik Indonesia nomor 32 tahun 2004 tentang Pemerintahan Daerah. Jakarta.
[3] Khomsin, Widyastuti R, Fayakun T, and Artanto E 2015 A determination analysis of regional maritime boundary based on regulation of Home Ministry Affair number 76 in 2012 (case study: dispute of Galang Island border between Surabaya and Gresik). Elsevier: Procedia Earth and Planetary Science.
[4] Republik Indonesia 2011 Undang-undang nomor 4 tahun 2011 tentang Informasi Geospasial. Jakarta
[5] Kementrian Dalam Negeri 2012 Lampiran peraturan Menteri Dalam Negeri Republik Indonesia nomor 76 tahun 2012 tentang pedoman penegasan batas daerah. Jakarta.
[6] Alesheikh A A, Ghorbanali and Nouri N 2007. Coastline change detection using remote sensing. Int. J. Environ. Sci. Tech., 4 (1): 61-66, 2007
[7] Van T T and Bin T T 2009 Application of remote sensing for shoreline change detection in Cuu Long Estuary. VNU Journal of Science, Earth Sciences 25. Page 217 – 222
[8] http://www.satimagingcorp.com/satellite-sensors/spot-7 : SPOT-7 Satellite Sensor.
[9] Khomsin 2016 Analysis of east Surabaya shoreline determination using tidal data, image satellite and RTK-GNSS. ISOCEEN. Surabaya
[10] Putra Y S 2016 Tutorial penarikan median line untuk pengelolaan laut menggunakan software ArcGIS. Pusat Pemetaan Batas Wilayah. Badan Informasi Geospasial. Cibinong.

Eddies spatial variability at Makassar Strait – Flores Sea

F Nuzula[1*], M L Syamsudin[2], L P S Yuliadi[2], N P Purba[2] and Martono[3]
[1]Bachelor Degree, Marine Science Program, Faculty of Fisheries and Marine Science, Padjadjaran University, Jl. Raya Bandung-Sumedang Km. 21, Jatinangor, West Java 45363, Indonesia
[2]Department of Marine Science, Padjadjaran University, Jl. Raya Bandung-Sumedang Km. 21, Jatinangor, West Java 45363, Indonesia
[3]Science Technology and Atmosphere Center, National Institute of Aeronautics and Space (LAPAN), Jl. Doktor Djunjunan No. 133, Padjadjaran, Cicendo, Bandung City, West Java 40173, Indonesia

E-mail: firdaoesnoezoela@gmail.com

Abstract. This study was aimed to get the distribution of eddies spatially and temporally from Makassar Waters (MW) to Flores Sea (FS), as well as its relations with the upwelling, the downwelling, and chlorophyll-a concentration. The study area extends from 115°–125° E to 2.5°–8° S. The datasets were consisted of monthly geostrophic currents, sea surface heights, sea surface temperatures, and chlorophyll-a from 2008 – 2012. The results showed that eddies which found at Makassar Strait (MS) has the highest diameter and speed of 255.3 km and 21.4 cm/s respectively, while at the southern MW has 266.4 km and 15.6 cm/s, and at FS has 182.04 km and 11.4 cm/s. From a total of 51 eddies found, the majority of eddies type was anticyclonic. At MS and FS, eddies formed along the year, whereas at southern MW were found missing in West Season. Moreover, the chlorophyll-a concentration was consistently higher at the *eddies* area. Even though, the correlation among eddies and the upwelling downwelling phenomena was not significantly as shown by sea surface temperatures value.

1. Introduction

Makassar-Flores Sea was the area with dynamic condition of oceanography. The water masses in this region are highly influenced by two systems of main current through it, the Indonesian through flow (ITF) and Indonesia monsoonal flow (IMF) [1]. Arlindo is a current flowing through Indonesian waters from the pacific ocean to the Indian ocean caused by differences in the sea level of height [2]. Through Arlindo, some waters get influenced by El Nino Southern Oscilation (ENSO), one of them is the southern ocean of Makassar – Flores Sea which located at the center of Indonesian waters. Furthermore, this phenomenon can affect the other oceanography events such as eddies and upwelling.

IMF in the west season is characterized by water mass that moves from west to east along with west monsoon wind which lead to the accumulation of water masses in Banda Sea and resulted the downwelling events. In the east season, water mass move in the reverse direction so that the water mass in Banda Sea is empty and the upwelling occurred [1]. Primer productivity in this region can be determined based on concentration of chlorophyll-a [3], because chlorophyll-a is an important pigment which is found in phytoplankton for photosynthesis, as well can be used to identifying upwelling and downwelling phenomenon [4].

Upwelling is a phenomenon when a cold water goes up from the deep layer to the surface caused by emptiness of water mass in the surface. Downwelling is a phenomenon where warm water goes down from the surface to the deep layer caused by an accumulation of water mass in a surface that must be distributed to the deep layer. Upwelling brings nutrients that required by phytoplankton, while downwelling brings a great oxygen saturation [5]. One of the cause of upwelling is eddies [6].

Eddies is a circular current that have scale of spatial ranged from dozens to hundreds of kilometers and temporal scale ranged from weekly to monthly. Eddies is one of the phenomena attracting a lot of attention for oceanography expertise because eddies have influences to biological aspect, physics and the dynamics of the atmosphere such as affecting the abundance of phytoplankton and the transfer of heat into the atmosphere [7].

Mc Gillicuddy et al. (1998) adduced that temperature, salinity, sea surface height, chlorophyll-a in eddies can changing based on the criteria of eddies itself [8]. On southern Makassar, water moving westward in May until September and moving eastward in November until March. In April and October the direction of water changed and eddies were occurred [9]. Then according to Nontji (2005), in April eddies were occurred from Karimata Strait until Flores Sea [10]. Martono (2010) has been proving by using hydrodynamics modeling of baroclinic three dimensions method and the results shown that eddies occured in April and October in Java Sea and Flores Sea [11]. Then Kartadikaria (2010) who used in situ data that has been through several processes of analyses has found three eddies in the north of Lombok Island [12].

2. Study area, material and method

2.1. Study area

This study was conducted in December 2014 to January 2015 with study area located in 2°30" – 8° S 115° – 125° E (Figure 1). Data processing carried out in National Institute of Aeronautics and Space (LAPAN) Bandung and computer laboratory at Fisheries and Marine Science Faculty, Padjadjaran University.

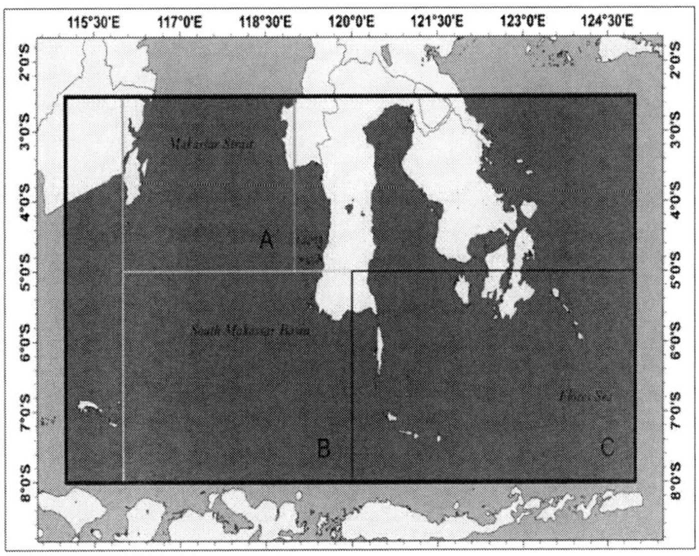

Figure 1. Study Area

2.2. Material and method

Software used in this research shown at table 1.

Table 1. Software

No	Software	Utility
1	Surfer 10	Create ocean current pattern map Create spatial map of SSH, SST, and Wind
2	ArcGis	Create study area map Create distribution of chlorophyll-a spatial map
3	ODV	Data conversing from .NC format to excel worksheet
4	SeaDas	Data extracting from .TIFF format
5	Microsoft Excel	Data control process
6	Microsoft Word	Writing paper

Materials used in this research shown at table 2.

Table 2. Materials

No	Dataset (2008-2012)	Data Resolution	Data Level	Source
1	Sea Surface Temperature	0.08° or ±9km	3	Terra/MODIS http.//coastwatch.pfeg.noaa.gov/erddap
2	Geostrophic Current	0.08° or ±9km	3	NOAA/AVHRR http.//atoll-motu.aviso.altimetry.fr/
3	Sea Surface Height	0.08° or ±9km	3	JASON2 http.//atoll-motu.aviso.altimetry.fr/
4	Chlorophyll-a	0.08° or ±9km	3	Aqua/MODIS http.//oceancolor.gsfc .nasa.gov

2.3. Eddies Spatial Distribution

Geostrophic current obtained from data processing and identified the existence of eddies by observing whether there are circular pattern on vector plot. It could be concluded as an eddies when circular pattern at vector plot is out or separate from the main flow. To strengthen the analysis, geostrophic current was compared with sea surface height in the same month and year.

Moreover, eddies were analyzed from its rotation direction, diameter, core, and geostrophic current velocity on each month and divided into three areas represent Makassar Strait (2.5° – 5° S and 116° – 119° E), Southern Makassar Waters (5° – 8° S and 116° – 120° E) and Flores Sea (5° – 8° S and 120° – 125° E). The direction of rotation was analyzed clockwise (cyclonic eddies) or counter clockwise (anticyclonic eddies). Eddies diameter could be known by counting longest distance between the circle [14]. The calculated longest distance could be reflected to the position of the latitude or longitude, simply written into the formula .

$$D = \frac{[|b| - |a|] x\ 111\ km}{1\ derajat}$$

Where:
D = Eddies diameter (km)
b = Easternmost edge of the circle of longitude or lower latitude edge of the circle (degree)
a = Westernmost edge of the circle of longitude or high latitudes edge of circle (degree)

Eddies core was determined by identifying the midpoint of the circular current visually and reflected on the position of latitude and longitude to obtain the eddies center point. Average geostrophic velocity in eddies could be calculated using the Pythagorean formula where in

$$c = (u^2+v^2)^{1/2}$$

Where:
c = The resultant speed of geostrophic current (cm/s)
u = Geostrophic current U component (cm/s)
v = Geostrophic current V component (cm/s)

2.4. Eddies relationship with upwelling and downwelling
Furthermore, in order to see eddies relationship with SSH, SST, and chlorophyll-a concentration as an indicator of upwelling and downwelling phenomenon. SSH, SST, and the concentration of chlorophyll-a in every month of which occurred in the formation of eddies analyzed with the surrounding area. At the time of eddies were formed or not formed was compared to the SSH, SST, and its concentration of chlorophyll-a. Also we analyzed the relationship between SSH, SST, and the concentration of chlorophyll-a with the direction of eddies.

3. Results and discussion

3.1. Eddies spatial distribution
Based on the results of data visualization, geostrophic currents and sea surface height on a monthly dataset, it is known that during the years 2008 - 2012 has happened 52 events of eddies in the study area that represent Makassar Strait, South Makassar Waters and Flores Sea. Within one month, eddies could formed up to four events, but the eddies is not always formed on each month.

3.1.1. Eddies Spatial Distribution at Makassar Strait
Eddies at Makassar Strait is formed due to the topography of the sea and coastline complexity. These waters have a depth of over 2000 meters [12] and is located between the two islands that flank. During the years 2008 - 2012 in Makassar Strait eddies were formed 29 events, which is 17 cyclonic eddies and 8 anticylonic eddies. In Makassar Strait, eddies was formed in June 2010 and has the longest diameter 255.3 km, which is included into the mesoscale eddies criteria [8]. Then the fastest flow velocity on the Makassar Strait was within the eddies formed in March, which is 21.4 cm/s.

In the Makassar Strait, eddies formed in January of 2009 have a central point around 117.88° E and 4.89° S. Eddies has a diameter along 132.09 km and geostrophic velocity average reached 10.4 cm/s, while in February 2011 had a central point around 118.74° E and 3.80° E with geostrophic velocity average of 5.2 cm/s with a diameter of 147.63 km.

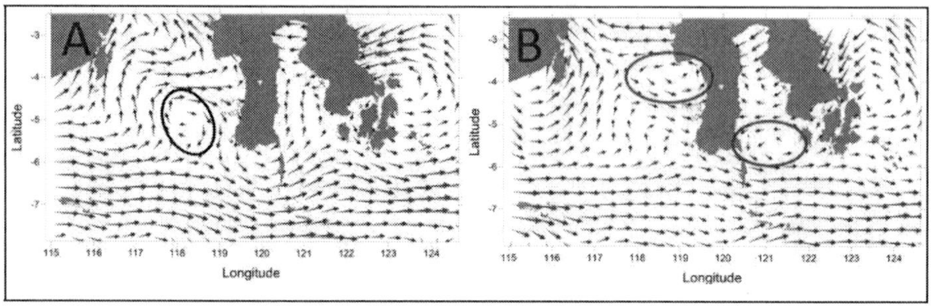

Figure 2. Geostrophic Current Pattern (A) January 2009 and (B) February 2011

Eddies formed in January 2009 was a cyclonic eddies (indicated by black circles) whereas eddies formed in February 2011 was an anticyclonic eddies (indicated by the red circle). It can be seen that in January 2009 in line with the direction of clockwise eddies (cyclonic) and in February 2011 rotates counter-clockwise (antisiklonik). Based on the map of the current pattern of geostrophic, the formation of cyclonic eddies in January 2009 were due to the dominance or strong currents forming a major which originated from the Java Sea, while anticyclonic eddies in February 2011 were due to the interaction with the mainland [14] as shown in (Figure 2).

In March of 2008, 2009 and 2012, there were formed eddies. Each point of the center of the eddies were 117.3° E and 4° S in 2008, 117.9° E and 4° S in 2009, 117.1° E and 4° S, 115.65° E and 4.8° S in 2012. Diameter of each eddies in March 2008 and 2009 were 177.6 km and 137.64 km, while the diameter in March 2012 was 164.28 km. The central point in 2009 is shifted, seen from the previous year which shifted about 0.6° eastward. This is presumably due to the mass of water that entered through the Makassar Strait in March 2008 stronger (8.5 cm /s) compared to the March 2009 (3 cm /s). Visually, it can be seen that the sea surface temperature of water masses that coming from the Pacific Ocean in March 2008 were colder than in March 2009. The differences could reach to 1.4° C, it could influence the position and size of the eddies [11].

Eddies that formed in April 2008 moved counter-clockwise with the center point that was around 117.6° E and 3.4° S. Diameter of the eddies in April of 2008 reached 139.86 km and geostrophic velocity average up to 10.6 cm/s. In 2009 and 2012 in the same month was also formed an eddies with a center point approximately 117° E and 3° S and diameter of 133.2 km, as well as 4,1°S and 117,1° E with a diameter of 124.32 km.

At Makassar Strait in May 2008 formed an eddies and its center in 117.9° E and 3.9° S with a diameter reached 155.4 km. In the same month in 2009, 2011 and 2012 were also formed an eddies in 117.2° E and 4° S with a diameter of 172.05 km, 117.1° E and 4.9° S with a diameter of 138.75 km, 118.3° E and 4.1° S diameter 142.08 km. Eddies in May 2008 and 2012 was a cyclonic eddies (indicated by black circles), while in May 2009 and 2011 was the anticyclonic eddies (marked by red circle). The cyclonic and anticyclonic types occurred because of differences in the density of the water masses that form the eddies. Eddies formation in May 2008, the water mass in these locations tend to be of low density, forming a cyclonic type of eddies, whereas in May 2009 tend to be high-density waters. Based on visualization, sea surface height at a central point eddies May 2008 is lower than the center point of eddies May 2009 (Figure 3).

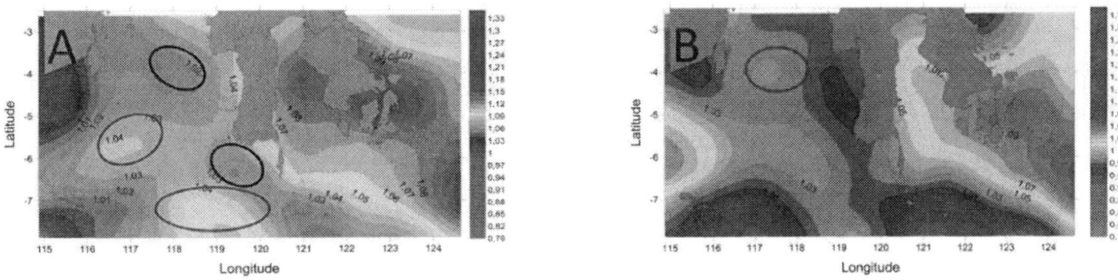

Figure 3. SSH Differences (A) May 2008 and (B) May 2009 That Shows Cyclonic and Anticyclonic Eddies

On June there were formed the eddies in 2008, 2009, 2010, and 2011. The focal point of each event is located in 117.9° E and 4° S, 117.8° E and 3.4°S, 118.4° E and 3.7° S, 118.3° E and 4.7° S with a diameter in the range of 179.8 - 255.3 km. The highest geostrophic velocity on this month was found in 2009 which is 23.5 cm/s.

Eddies formed in these waters in July of 2009 and 2010 have a central point in 117.9° E and 3.7° S, 118.6° E and 4°S. These both eddies diameter reached 155.4 km and 144.3 km. While the eddies in August 2008 and 2010 center point 117.9° E and 3.7° S, 4.5°S 118.1° E and with a diameter 187.59 and 213.2 km and geostrophic velocity average is 14.1 and 11.8 cm/s,

3.1.2. Eddies Spatial Distribution at South Makassar Waters

A total of 20 events eddies formed in the waters of South Makassar to limit 116° - 20° and 5° E - 8° S. A total of 15 events during the year 2008-2012 move anticlockwise and just five events that move clockwise. The most eddies found in one month occurred in May 2008 by three events with the center point 116.7° E and 5.7° S, and 6.3°S 119.4° E, 119.1° E and 7.6°S. With eddies diameter is 241.98 km, 166.5 km and 250.86 km.

In SMW, the center point is shifted in April and May 2008. Both of this eddies center point is 117.1° E and 5.4° S, 118.4° E and 5.6° S, diameter is 168.72 km and 210.9 km. Both eddies is moving anticyclonic (indicated by the red circle). According to Gordon et al. (2003), the waters of Indonesia got a very strong influence coming from NINO3 area in the Pacific Ocean on the west monsoon season and the transition period that goes through the Makassar Strait. This is evident through data visualization of geostrophic current, its shown that the center point shifted 0.4° southeastern (Figure 4).

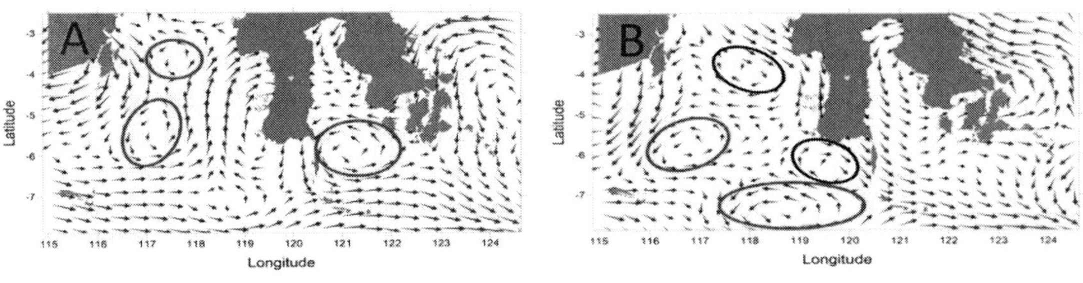

Figure 4. Geostrophic Current Pattern (A) April 2008 And (B) May 2008

SMW eddies formed in June of 2008, 2009, 2010, and 2012. These eddies central point is around 118.6° E and 6.6° S, 119.3° E and 7.7°S, 119.3° E and 6.3° S, 117.8° E and 6.4° S. In this month the highest diameter was found, its reached 228.66 km in 2009.

SMW in July formed an eddies in 2008 and 2010. The diameter of each eddies is 157.62 km and 166.5 km. Geostrophic velocity reach 9.1 cm/s and 4.6 cm/s. The speed differences of these eddies caused by its center point location. In July 2008 the central point of the eddies is in 118.4° E and 5.4° S whereas in July 2010 is 118.6° E and 7.1° S. The speed differences of these eddies caused by the proximity of eddies which formed closer to the major current from Arlindo and South China Sea. This is consisted with the statement from Mann and Lazier (2006), speed of geostrophic in eddies nearer the major current tends to be higher than the eddies that far from major current [14].

Two eddies found in August 2009 and 2012. Each eddies has a central point at 119.4° E and 7.7° S, 119.6° E and 6.4°S. Then in October formed an eddies in 2008 and 2012 that has a central point at 118.7° E and 6.3° S, 117.7° E and 5.6°S. Furthermore, in November eddies occurred in 2009 and 2011 with the center point at 118.6° E and 6.3° S, 119.4° E and 5.8° S.

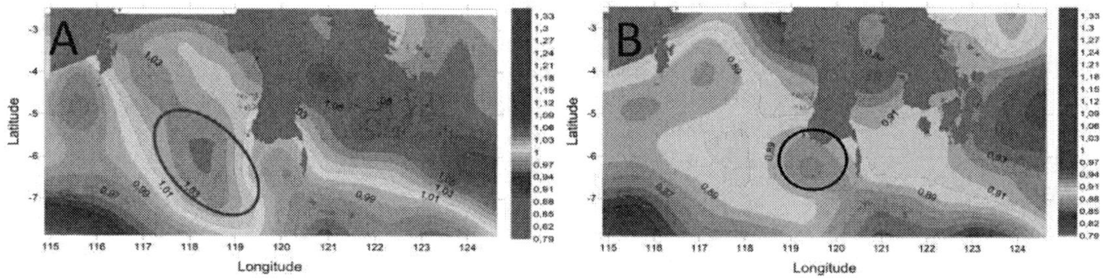

Figure 5. SSH Differences at Eddies Core (A) September 2010 And (B) September 2011

In September eddies was formed in 2010, 2011, and 2012. The center point of these eddies started at 118.2° E and 5.8° S, then shifted southeastern to 119.3° E and 6.3° S, then returns to 118.4° E but the latitude still at 6.9° S. Its happen because of the differences in the speed and direction of the major current of the years during 2008 - 2012. Then there are differences in sea level at the center of the eddies. The year 2011 had higher sea levels were lower compared by year 2010 and 2012 is 1.04 cm and 0.94 cm. Based on sea level at the center point eddies, eddies formed in September 2011 was a cyclonic eddies (indicated by black circles), while in September 2010 and 2012 is an anticyclonic eddies (marked by red circles) (Figure 5).

Eddies in these waters has a direction of rotation is counter clockwise except in June of 2011, which has a central point around 120.1° E and 7° S with a diameter of 129.87 km. Based on data processing of geostrophic currents and sea surface height on a monthly basis, the general direction of the rotation is counter clockwise (marked by red circle) which has a higher sea level at its center than the surrounding area. Otherwise when the direction of rotation is clockwise, the high sea level in its center point will be lower (indicated by the black circle) (Figure 6). This is consistent with the theory by Stewart (2008) that the level of sea surface height at the center point indicate the type which is cyclonic eddies or anticyclonic eddies [15].

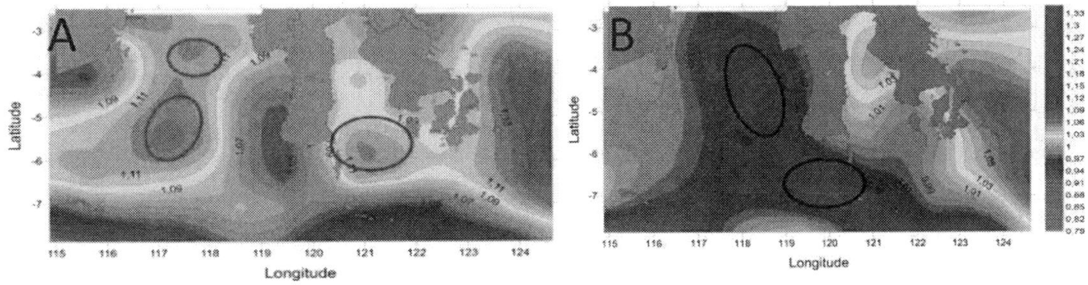

Figure 6. SSH Differences That Shows Anticyclonic eddies at (A) April 2008 And *Cyclonic* Eddies (B) June 2011

3.1.3. Eddies Spatial Distribution at Flores Sea

Eddies in the waters of the Flores Sea is the least formed. During the years 2008 - 2012 eddies formed only seven times, even in the year 2010 in these waters did not formed an eddies in throughout the year. It's possible because these waters are more open than in the other study area as well as away from the main water mass that can form eddies.

At Flores Sea in February 2011, March 2012 and April 2008 formed an eddies with a center point at 121.2° E and 5.4° S, 6.1° S and 124°E, 121° E and 5.4°S. The diameter of each eddies is 150.96 km, 182.04 km and 175.38 km with geostrophic speed average 4.1 cm/s and 11.4 cm/s and 10.6 cm/s. In December 2008 and October 2012 (Figure 7) also formed an eddies in Flores Sea this. Each of these events has a center point around 124.7° E and 5.7° S, 5.6° S and 123.8° E and with a diameter of 166.5 km and 157.62 km and geostrophic speed average is 8.9 cm/s and 8.5 cm/s.

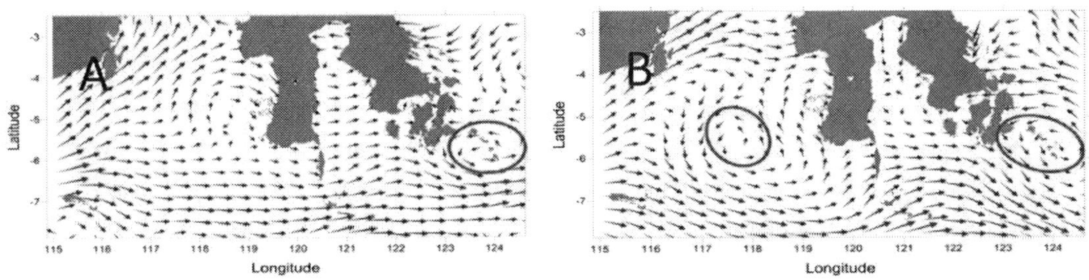

Figure 7. Geostrophic Current Pattern (A) December 2008 And (B) October 2012

Eddies formed in December 2008 and October 2012 is anticyclonic eddies where the direction of rotation counterclockwise (indicated by the red circle). It can be seen at the center of the eddies that has a higher sea level higher than the surrounding area.

4. Conclusion

Spatially, eddies formed at Makassar Waters – Flores Sea during 2008 – 2012 up to 52 events that spread to MS, SMW and FS. There was 29 events of eddies with 17 cyclonic eddies and 8 anticyclonic eddies. At SMW, there was 20 eddies formed with 15 cyclonic eddies and 5 anticylonnic eddies, whereas at FS there was 7 eddies with 5 cyclonic eddies and 2 anticyclonic eddies. Diameter of eddies can shrinking or developing by 5 until 100 km and the core of eddies can shifting by 0.1° until 1.0° to east-south-west in every month.

References

[1] Ilahude A G and Nontji A 1999 Oseanografi Indonesia dan perubahan iklim global (El Nino dan La Nina) *Lokakarya AIPI Puslitbang Oseanologi-LIPI Jakarta* p 12
[2] Hasanudin M 1998. Arus Lintas Indonesia (ARLINDO) *Oseana* **23** 2 p 1-9
[3] Kunarso S, Hadi N S, Ningsih M S, Baskoro 2011 Variabilitas suhu dan klorofil-a di Daerah *Upwelling* pada variasi kejadian ENSO dan IOD di perairan selatan Jawa sampai Timor *Jurnal Ilmu Kelautan* **16** 3 p 171-180
[4] Azis M F 2006 *Gerak* Air di Laut *Oseana* **31** 4 p 9-21
[5] Martono S, Hadi N S, Ningsih. 2008b. Studi *Eddies* Mindanao dan *Eddies* Halmahera *Jurnal Penelitian Perikanan* **11** 2 p 204-210
[6] Robinson A R 1983 Eddies in Marine Science (Berlin: Springer) p 609
[7] McGillicuddy D J, Robinson A R, Siegel DA, Jannasch H W, Johnson R, Dickey T D, McNeil J, Michaels A F, and Knap A H 1998 Influence of mesoscale eddies on new production in the sargasso sea *Nature* **394** p 263-265.
[8] Wyrtki K 1961 *Physical Oceanography of the South East Asian Waters* (San Diego: The Scripps Institution of Oceanography University of California)
[9] Nontji A 2005 Laut Nusantara (Jakarta: Djambatan) p 368
[10] Martono 2010 Pengaruh fenomena El Nino dan La Nina terhadap arus eddies di Laut Jawa dan Laut Flores
[11] Kartadikaria A R Y, Miyazawa K, Nadaoka A,Watanabe 2011 Existence of eddies at Crossroad of The Indonesian Seas *Springer-Verlag* in T Ezer (Ed.) Ocean Dynamics **62** 31-44.
[12] Mann and Lazier 2006 Dynamics of Marine Ecosystem *Blackwell Publishing Ltd* **3** 285-336
[13] Pranowo *et al.* 2005
[14] Vaillancourt R D *et al.* 2003 Impact of A cyclonic eddies on phytoplankton community structure and photosynthetic competency in subtropical North Pacific Ocean (Columbia: Pergamon)
[15] Aulia A 2013 Variabilitas arus eddies di perairan Selatan Jawa-Bali berdasarkan data Satelit *Jurnal Perikanan Kelautan* (Jatinangor: Universitas Padjadjaran)

Combining ocean numerical model and SAR imagery to investigate the occurrence of oil pollution, a case study for the Java Sea

A Setiawan[1*], M R Putri[2], M Gade[3], T Pohlmann[3] and B Mayer[3]
[1]Agency for Marine and Fisheries Research and Development, Ministry of Marine Affairs and Fisheries, Indonesia
[2]Research Group of Oceanography, Institute of Technology Bandung, Indonesia
[3]Institute of Oceanography, University of Hamburg, Germany

E-mail: setiawan.agus@gmail.com

Abstract. IndoNACE is an abbreviation of **Indo**nesian Seas **N**umerical **A**ssessment of the **C**oastal **E**nvironment, a pilot study between Indonesia and Germany that combining analysis of oil spills from SAR images and numerical tracer studies from 3-D numerical model. Aim of this study is to understand the observed seasonal variations in marine oil pollution. Within this study, a visual inspection of all available SAR images is performed in order to generate maps of oil pollution occurrence in the Java Sea. Afterward, a set of numerical models is applied to trace back the origin of oil pollution. Our results showed that by analysing 706 ENVISAT ASAR images, the highest number of oil spills occurrence in the Java Sea was found during the transition monsoons, i.e. March to May and September to November. Assuming Marine Protected Area (MPA) of Seribu Islands as the end position of oil trajectory, we found that the origins of oil pollutions in that area were mostly from north and east.

1. Introduction

The Java Sea is part of Sunda Shelf and has relatively shallow water with mean depth of 50 meters. The area is located between Kalimantan and Java Islands and has significant role on water mass exchange between the South China Sea and Indonesian seas. Moreover, Java Sea is also a major fishing ground within the Fishing Management Area (FMA) 712 that contributes to about 18% of total Indonesian capture fisheries production in 2014 [1]. Beside its potential in capture fisheries, some areas in the Java Sea are also potential for tourism and, moreover, have been declared as Marine Protected Area (MPA) such as Seribu Islands (approx. 1075 km^2) and Karimun Jawa (approx. 1100 km^2). There is also some oil and gas fields located in the Java Sea, particularly in the areas north of Cirebon in West Java to Seribu Islands in Jakarta (known as Offshore North West Java Block or ONWJ) and east of Sumatra Island (known as Offshore South East Sumatra Block), covered areas of approx. 8,284 km^2 and 5,852 km^2, respectively (see figure 1) [2]. In general, around 530 oil and gas platforms exist in the offshore Indonesian waters, where around 141 platforms are located in the ONWJ block and 82 of them still active. Pertamina Hulu Energy Offshore North West Java Ltd. (PHE) currently operates this block with average daily crude oil output reached 40,000 barrel per day in 2015 [3].

The Java Sea is prone to oil spills due to offshore oil exploration and production activities as well as ship traffic through the sea-lanes of the Indonesian archipelago (known as Alur Laut Kepulauan

Indonesia or ALKI). Based on historical data, about 2-4 oil spills occurred per year around Seribu Islands, where the main source of these spills is offshore oil production [4]. Moreover, these oil spills have also polluting the MPA in Seribu Islands. According to the local authority of Seribu Islands Regency, in 2003 and 2004 around 78 and 37 of 87 islands were polluted by the oil spills, respectively. Table 1 shows some occurrence of oil spills around Seribu Islands since 2003.

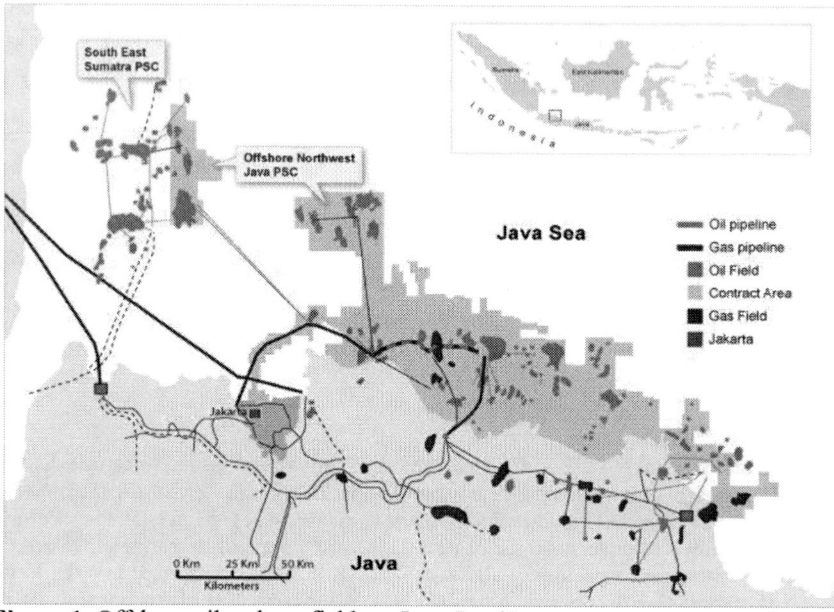

Figure 1. Offshore oil and gas fields at Java Sea (Source: www.energy-pedia.com)

Table 1. Oil spill occurrence around Seribu Islands

Time	Records of events
December 2003	Around 78 of 87 islands of Seribu Islands were polluted by crude oil spill, including islands in the Marine Protected Area (MPA)
April 2004	Around 37 islands were polluted by crude oil spill
May 2004	Oil spills were detected eastward of Pramuka Island until Peniki as well as Kelapa Islands
October 2004	Clumps of tar ball of 4-5 cm thick detected 1-2 km around Kotok Island
February 2006	Oil spill was detected in the MPA of Seribu Islands
October 2008	Waters in the southern part of Seribu Islands, i.e. around Pari, Tikus, Burung, and Payung Islands were covered by oil spill with thickness of 1-20 cm
February, March, and August 2011	Four islands, namely Putri, Payung, Tidung Kecil, and Tidung Besar were polluted by crude oil spill

Synthetic Aperture Radar (SAR) deployed in the satellite currently becomes an important tool that can be used to monitor the oil spill [5]. In the ocean oil spills appear as slicks that generally could be distinguished by SAR imagery as dark patches [5][6][7]. Even though SAR imagery does not have any capabilities on estimating oil spill thickness and recognizing oil types, its ability on detecting oil spills is still the most efficient and superior due to its wide area of coverage and capabilities to capture images in all weather condition day and night [5].

Since the MPA of Seribu Islands is located close to offshore oil exploration and production and for several times have already polluted by oil spills, continuous monitoring is a key of importance for this area. Based on this reason, a pilot study called Indonesian Seas Numerical Assessment of the Coastal Environment (IndoNACE) is proposed by combining the advantage of remote sensing technology and numerical model techniques to meet the needs. The objective of this joint Indonesia-German cooperation is to improve the information on the state of marine environment gained from satellite data, particularly SAR, in order to produce oil pollution density map and hydrodynamics and trajectory models in order to trace back the source of oil spills. This study has two regions of interest (ROIs) namely western Java Sea and Makassar Strait, and this paper will focus the discussion on results obtained for the western Java Sea region.

2. Materials and Methods

The IndoNACE pilot study will utilize historical and actual data from ESA's ENVISAT archive and Sentinel 1A Rolling Archive to generate map of occurrence of all detected oil pollution, binned on 0.05°x0.05° grids. Approximately, 706 ENVISAT ASAR images acquired over the study area were used in this study. Afterwards, in order to identify the source areas for oil, which was detected via satellites, as well as to confirm their pathways, a tracer model was applied in backward mode, making use of daily average surface current velocity obtained by the hydrodynamics model.

2.1. Regions of Interest

Based on its potential to oil spills occurrence, particularly due to offshore oil exploration and high density of MPAs as well as mangroves and coral reef ecosystem, western Java Sea (105°-111° E and 3°-7° S) has been chosen as region of interest (ROI) for this pilot study. The area is given in figure 2 as a hatched area.

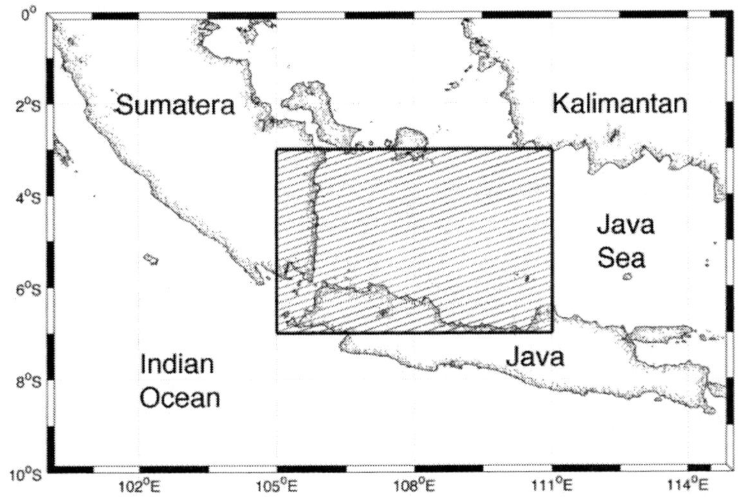

Figure 2. One of the regions of interest (ROI) for the pilot study of IndoNACE

2.2. Detection of Oil Spill from SAR Images

The detection of oil spills from SAR images was performed using visual inspection. All of SAR images that are available for the ROI were processed, along with manual registration, including geo-information (latitude, longitude, size, etc.) on the detected spills [6][7]. Special emphasis has been put on the discrimination between anthropogenic (mineral oil) spills and biogenic slicks, since both species tend to cause similar feature on SAR imagery [8][9]. Figure 3 shows the monthly coverage (October 2002 to April 2012) of ENVISAT ASAR of both IndoNACE ROI.

In general, the detection of oil spill consists of four operations, i.e. pre-processing (calibration and speckle filtering to source images), land-sea masking (in order to focus the detection only on area of interest), dark spot detection, and clustering and discrimination (clustering and then eliminating pixels that detected as part of dark spot based on dimension of the cluster and user selected minimum cluster size) [10]. All of these processes could be performed making use of Next ESA SAR Toolbox (NEST), an open source toolbox that available to read, post-process, analyse, and visualize the large archive data from ESA SAR missions.

2.3. Hydrodynamics and Tracer Models

Long-term simulation of meso-scale (approximately 11 km horizontal resolution) Hamburg Shelf Ocean Model (HAMSOM) covering the entire region of Indonesian waters has been carried out to produce velocity data used in tracer model. Results from Max Planck Institute-Ocean Model (MPI-OM) were used for open boundary values. Atmospheric data from National Centers for Environmental Prediction (NCEP) as well rivers discharge from Research Center for Water Resources (RCWR) was used as external forcing in order to obtain adequate results [11]. The tracer model used in this study was based on Lagrangian approach. Similar resolution was applied to the tracer model, where velocity data obtained by the HAMSOM were used to estimate the oil spills pathways in backward mode.

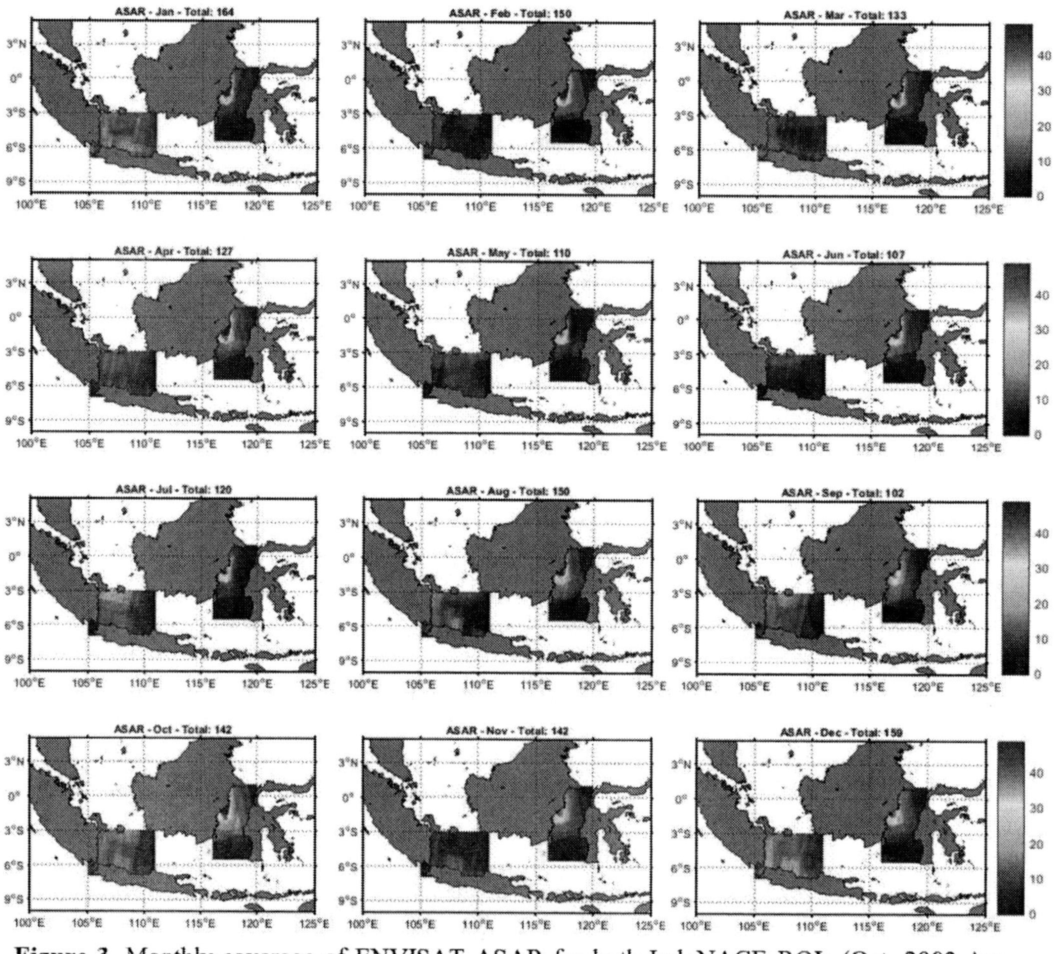

Figure 3. Monthly coverage of ENVISAT ASAR for both IndoNACE ROIs (Oct. 2002-Apr. 2012), unit for the color bar is number of coverage.

IOP Conf. Series: Earth and Environmental Science 54 (2017) 012080 doi:10.1088/1755-1315/54/1/012080

3. Result and Discussion

Figure 4 shows an example shows oil slicks around Seribu Islands after image processing. Meanwhile, figure 5 shows the map of pollution occurrence that includes all detected oil pollution generated from visual inspection over 706 ENVISAT ASAR images. From this figure, which is indicating the number of oil spills per 0.05°x0.05° cells, higher number of oil spills mostly detected in the area around offshore oil exploration, particularly Arjuna and Cinta fields, and Port of Tanjung Priok. These results clearly show that oil production and ship traffic putting the western Java Sea (including MPA of Seribu Islands) under severe threat. By calculating the ratio of the total number of oil spills detected in each month of the year and monthly distribution of the approximately 706 ENVISAT ASAR images, monthly distribution of oil spills was estimated and shown in figure 6. From this estimation, in general the highest number of oil spills was occurred during March to May and September to November, respectively.

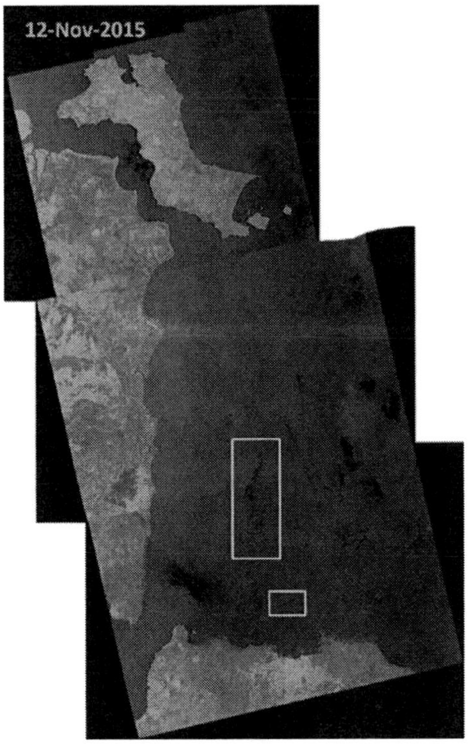

Figure 4. ENVISAT ASAR image for western Java Sea acquired on 12 November 2015 that show oil slicks close to Seribu Islands (indicated by yellow boxes).

Based on results of one-week simulation of backward tracer model, making use of daily average velocity data of years 2003-2011, during the northwest monsoon (represented by result in February) the MPA of Seribu Islands was most likely hit by oil spills originating in the area around Offshore Southeast Sumatera Block and ALKI I that passing through Sunda Strait-Karimata Strait-Natuna Sea-South China Sea (figure 7 top left panel, diamond marks indicate the position of oil spills a week before reach the MPA). In general, there is an eastward transport in the Java Sea during the northwest monsoon (December to February). Water mass from the South China Sea (SCS) enters the Karimata Strait and flows eastward to the Java Sea. However, water mass in the western Java Sea flows also south-westward to the Indian Ocean through Sunda Strait. On the other hand, during southeast monsoon (represented by result in August), the MPA of Seribu Islands mostly hit by oil spills originating in the area around ONWJ Block (figure 7 bottom left panel). During this period (southeast monsoon - June to August), main flow in the Java Sea is in opposite direction to northwest monsoon. Water mass from Makassar Strait flows westward into the Java Sea and then north-westward to the Karimata Strait. In the western Java Sea, water mass flows also south-westward to the Indian Ocean through Sunda Strait.

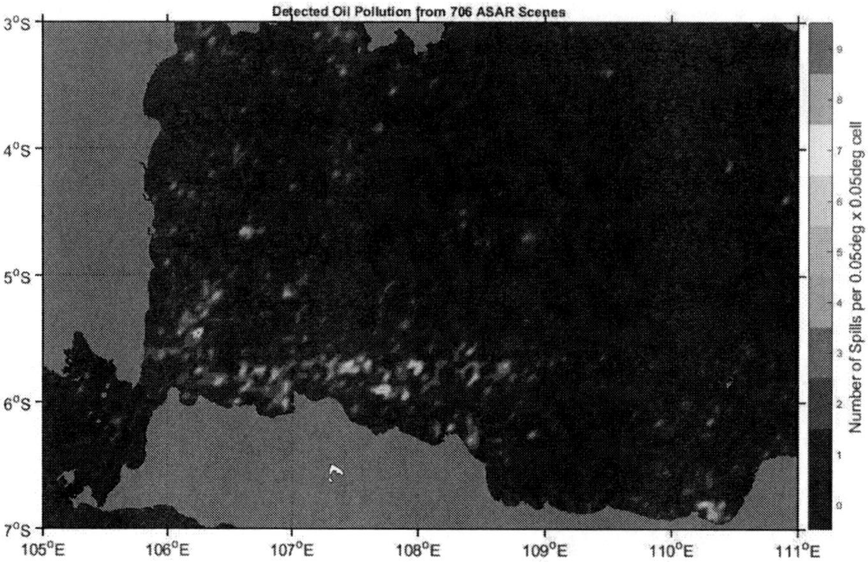

Figure 5. Number of oil spills per 0.05°x0.05° cells generated from visual inspection over 706 ENVISAT ASAR images

Interesting results were obtained during both monsoon transitions, where statistically the number of oil spills detected by satellite was higher than that during northwest and southeast monsoons (see figure 6). Comparing the condition with northwest and southeast monsoons periods, surface current velocity and residence time in the Java Sea during transition periods is weaker [4] and longer [11], while the sea surface current direction is varied without any dominant direction. Consequently, oil spills tends to accumulate in this region and this condition can be seen clearly in the results of backward trajectories in April and October as shown in figure 7 top and bottom left panels. Sources of oil spills reach the MPA of Seribu Islands during these periods were vary with year, but mostly dominated by spills originated from offshore oil exploration of ONWJ.

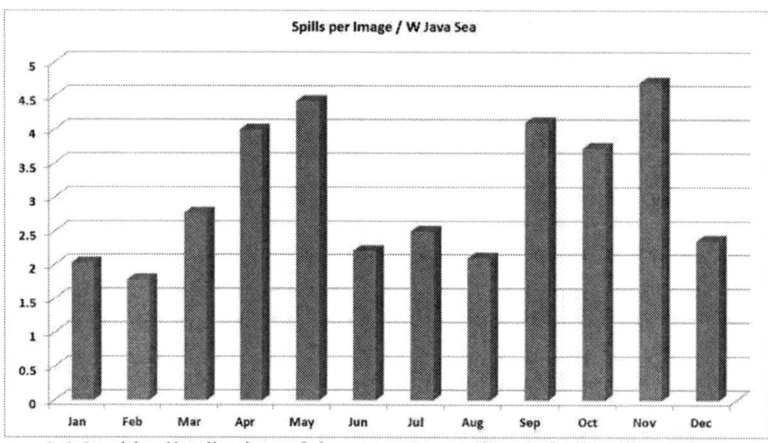

Figure 6. Monthly distribution of the average numbers of oil spills per SAR image

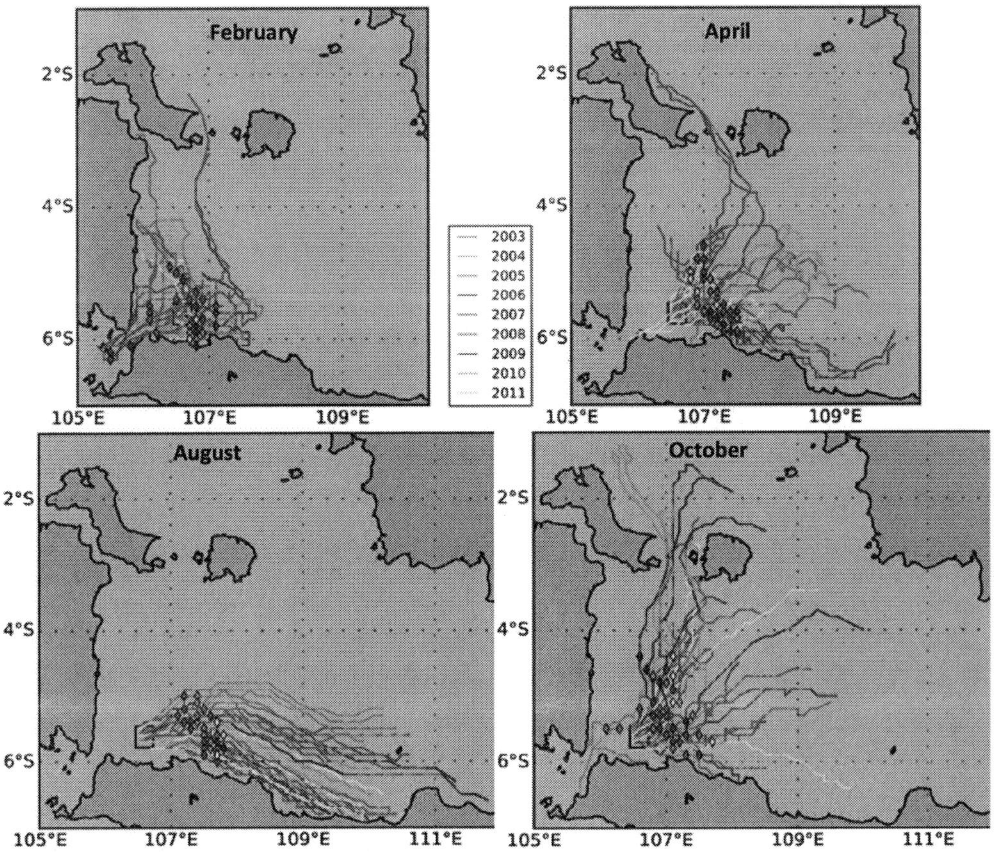

Figure 7. Backward trajectories of possible sources of oil spills reached the MPA of Seribu Islands for the years 2003 to 2011. The simulations were performed for 4 weeks and the diamonds mark the position a week before the oil spills reached the MPA.

4. Conclusion

In the case of oil spills in Java Sea, the number of spills during monsoon transition periods is higher than during northwest and southeast monsoons and the residence time in the Java Sea plays very important role to this condition. From the simulation of backward trajectories model, we found that the origin of spills is mostly from offshore oil exploration in ONWJ block. However, during northwest monsoon source from ALKI I have also detected.

In general, the pilot study of IndoNACE has successfully demonstrated how to manage data from SAR satellites and combine them with ocean numerical models in order to assess the incidence of oil spills and estimate the origin of spills that taking into account the seasonal aspect. However, since SAR imagery does not have any capabilities to recognize oil types, further laboratory analysis on oil spills sample is needed to improve the assessment, especially when dealing with trajectory model, since different type of oil needs different parameterization. Moreover, further information on weather conditions, particularly wind speed as the limiting factor for the visibility of oil pollution on SAR imagery, is also needed in order to improve the results of oil spills detection.

Acknowledgements
The pilot study **Indo**nesian seas Numerical Assessment of the Coastal Environment (IndoNACE) receives funding from the European Space Agency (ESA) under contract ITT AO 1-8176/14/F/MOS. ENVISAT ASAR imagery was kindly made available by ESA's EO Grid Processing On-Demand (EO0GPOD) Team.

References

[1] Ministry of Marine Affairs and Fisheries 2015 Keragaan Perikanan Tangkap
[2] Energy-pedia News Website 2010 http://www.energy-pedia.com/news/indonesia/inpex-sells-offshore-northwest-java-and-offshore-southeast-sumatra-interests-to-pertamina, accessed August 4th 2016
[3] Petro Energy Website 2016 http://www.petroenergy.id/article/---2015-phe-onwj-penuhi-target-produksi-minyak-40-ribu-bopd, acessed 4 August 2016
[4] Putri M R, Setiawan, T. Pohlmann, B. Mayer, M. Gade 2016 The assessment of oil pollution in Seribu Islands absed on remote sensing and numerical models *Proc. ESA Living Planet Symp* (Prague, Czech Republic)
[5] Brekke C and Solberg A H S 2005 Oil spill detection by satellite remote sensing *Remote Sensing of Environment* **95** 1-13
[6] Gade M, Mayer B, Pohlmann T, Putri M R, Setiawan A 2016 Using SAR data for a numerical assessment of the Indonesian coastal environment *Proc. ESA Living Planet Symp 2016* (Prague, Czech Republic)
[7] Gade M, Mayer B, Pohlmann T, Putri M R, and Setiawan A 2016 An assessment of the Indonesian coastal environment based on SAR imagery *Proc. Int. Geosci. Remote Sens. Symp. (IGARSS)* 2016 (Beijing, China)
[8] Gade M, Alpers W, Hühnerfuss H, Masuko H, Kobayashi T 1998 The imaging of biogenic and anthropogenic surface films by a multi-frequency multi-polarization synthetic aperture radar measured during the SIR-C/X-SAR missions *J. Geophys. Res.* **103** 18851-18866
[9] Gade M, Byfield V, Ermakov S, Lavrova O, Mitnik L 2013 Slicks as indicators for marine processes *Oceanography* **26** 138-149
[10] Solberg A S, Brekke C, Solberg R 2004 Algorithm for oil spill detection in Radarsat and ENVISAT SAR images *Geoscience and Remote Sensing Symposium IGARSS Proc. IEEE International* **7** p 4909-4912
[11] Mayer B, Stacke T, Stottmeister I, Pohlmann T 2015 Sunda shelf seas: flushing rates and residence times *Ocean Sci. Discuss.* **12** 863-895

Development of an automated processing system for potential fishing zone forecast

R Ardianto[1*], A Setiawan[1], J J Hidayat[1] and A R Zaky[1]
[1]Research and Development Centre for Marine and Coastal Resources
Agency for Marine and Fisheries Research and Development
Ministry of Marine Affairs and Fisheries

E-mail: romy.ard@gmail.com

Abstract. The Institute for Marine Research and Observation (IMRO) - Ministry of Marine Affairs and Fisheries Republic of Indonesia (MMAF) has developed a potential fishing zone (PFZ) forecast using satellite data, called *Peta Prakiraan Daerah Penangkapan Ikan* (PPDPI). Since 2005, IMRO disseminates everyday PPDPI maps for fisheries marine ports and 3 days average for national areas. The accuracy in determining the PFZ and processing time of maps depend much on the experience of the operators creating them. This paper presents our research in developing an automated processing system for PPDPI in order to increase the accuracy and shorten processing time. PFZ are identified by combining MODIS sea surface temperature (SST) and chlorophyll-a (CHL) data in order to detect the presence of upwelling, thermal fronts and biological productivity enhancement, where the integration of these phenomena generally representing the PFZ. The whole process involves data download, map geo-process as well as layout that are carried out automatically by Python and ArcPy. The results showed that the automated processing system could be used to reduce the operator's dependence on determining PFZ and speed up processing time.

1. Introduction

Since its first-established in 2005, the Institute for Marine Research and Observation (IMRO) has had routinely produce map of fishing ground for Indonesian waters based on sea surface temperature (SST) and chlorophyll-a (Chl-a) data from satellite. At the beginning, the map was produced manually and based only on very simple methods, i.e. identifying areas with low SST and high concentration of chlorophyll-a as a fishing zone. Afterwards, from time to time IMRO tried to improve its potential fishing zone (PFZ) map by implementing Python-based scripts that automatically estimate the potential area of fishing zone to reduce the operator's dependence and speed up the processing time.

The method applied to develop the automated processing system was edge-detection [1]. This method was used to identify the thermal fronts as an indication for some oceanographic phenomena such as upwelling fronts, frontal eddies, and fronts associated to geomorphic features [2] where some of those phenomena were occurred in the Indonesian waters. For example, upwelling along the Java-Sumatra Indian Ocean coasts that usually occurred during the southeast monsoon (June to October) [3--5], persistent upwelling and front over the Sulu Ridge [6], and front in the southern Banda Sea, shelf front north of Java in the Java Sea, shelf front west of Dolak Island, New Guinea in the northern Arafura Sea [7]. The utilization of satellite-retrieved SST data to estimate the ocean fronts have been started since 1970s, including some parts of Indonesian waters [7]. Since the upwelling regions usually nutrient-rich and the primary productivity tends to increase and can be represented by the increase of chlorophyll-a concentration, the utilization of remotely-sensed SST and concentration of chlorophyll-a were widely used to estimate the PFZ for pelagic fishes, making use of simple relationship of marine food webs [5, 8, 9].

2. Methods

2.1. Study Area

Currently, IMRO has produced several types of PFZ maps based on their spatial resolution (from national to fishing port scales) as well species of fishes (small and big pelagic fishes). In this paper, the automated system discussed was applied to only national scale fishing zone map, covering the whole area of Indonesian waters (figure 1). For our dissemination purposes, the PFZ maps were divided into 5 main regions afterward, i.e. Java-Bali-Nusa Tenggara, Kalimantan, Maluku-Papua, Sumatra, and Sulawesi (figure 2).

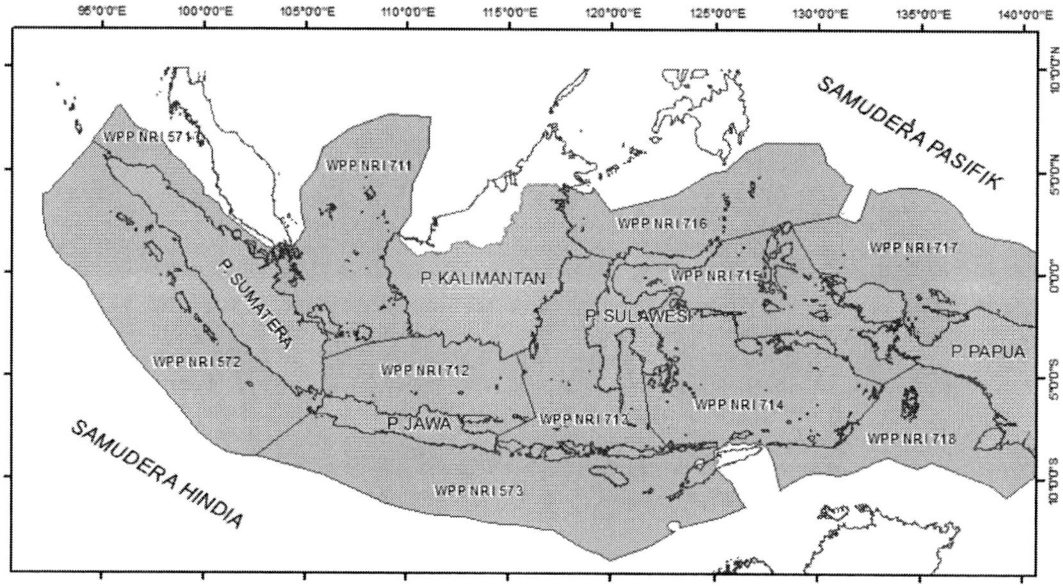

Figure 1. The study area

2.2. Satellite Data

Two types of dataset were used in this study, i.e. 3-days composite of Level 3 Terra and Aqua Moderate Resolution Imaging Spectroradiometer (MODIS) of sea surface temperature (SST) and chlorophyll-a concentration (Chl-a). These dataset are processed and distributed by the Ocean Biology Processing Group (OBPG) at http://oceancolor.gsfc.nasa.gov. These images were Level 3 Standard Mapped Image (SMI) products that are created from corresponding Level 3 binned products and have spatial resolution of 4.6 km at the equator [10].

2.3. Modules of Automated Processing System

The Python-based scripts developed in this study were divided into 3 modules, namely pre-processing, processing, and post-processing. In the pre-processing module, the script will handle automatically the process on cropping the global image into Indonesian region, masking the image with land and cloud, and compositing the raster. In the processing module, images obtained by the first module were processed automatically to detect the thermal fronts as well as regions with high primary productivity. Afterwards, results produced by second module were processed to produce a PFZ map.

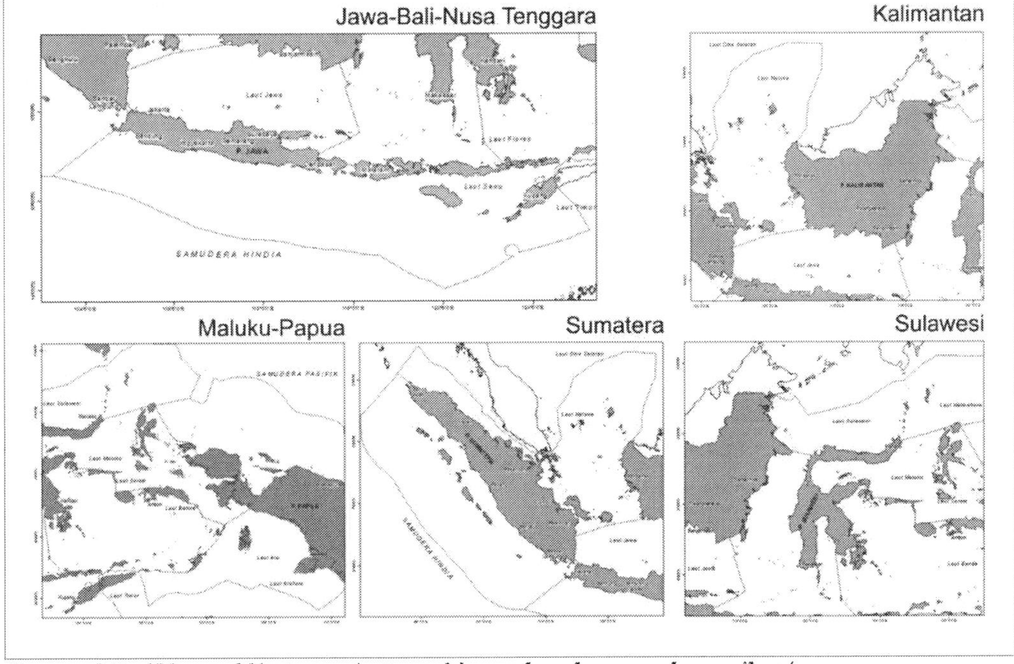

Source: http://kkp.go.id/category/peta-prakiraan-daerah-penangkapan-ikan/

Figure 2. Regions of national scale PFZ that disseminate to the public via official website of Ministry of Marine Affairs and Fisheries

2.3.1. Detection of Thermal Fronts
SST is usually used in predicting PFZ [5, 8, 9, 11]. Two water masses with temperature gradient known as thermal front were usually occur on sea boundary between two horizontal masses of seawater with different densities [12]. A thermal front zone usually nutrients rich and has an effect to stimulate primary production [13] indicated by the abundance of phytoplankton as producer in the marine food webs. Since in the thermal front zone, temperature gradient exists, an edge detection method could be used to identify this phenomenon [1]. As already mentioned previously, there are some areas within the Indonesian waters that have thermal fronts due to upwelling, eddies, and geomorphic features. Although in some cases the temperature gradient is not big enough, for example around 0.025°C/km over the Sulu Ridge [6], but the evidence showed that this phenomenon could increase the concentration of chlorophyll-a by 5 to 15 times greater [5, 6]. Therefore, thermal fronts are important and could be used as indicator to predict the PFZ. Previous study used temperature gradient of 0.5°C for every 3 pixels [9] while in this study, the value chosen for threshold is equal to 0.45°C. Before this value was chosen, an experiment making use of monthly SST data that shows upwelling in the Indian Ocean south of Java was done. Within this experiment several values of threshold were used in edge detection method in order to detect the thermal fronts and we found that 0.45°C gives reasonable result.

2.3.2. Identification of Chlorophyll-a for Ocean Productivity.
Chlorophyll-a is one of important oceanographic parameter to detect skipjack tuna habitat in tropical region [15]. The migration route and feeding habitat for albacore tuna in the central and eastern North Pacific Ocean have been detected using chlorophyll-a about > 0.2 mg/m^3 [16]. On the other hand, the distribution and migration of skipjack tuna are strongly influenced by oceanographic factors such as

distribution of SST and chlorophyll-a [14]. In the present study, it was observed that the favourable ranges of chlorophyll-a and SST for various pelagic species that can be used to find out the potential fishing ground are 0.2-0.5 mg/m^3 and 15-27°C, respectively [17]. But since for tropical region such as Indonesia the SST difference is not as high as sub-tropical regions, therefore, concentration of chlorophyll-a was used also as indicator to estimate the PFZ. Before the chlorophyll-a data were used, in order to reduce high and low pixel in the data, smoothing process using median and Wiener filters were applied [18].

2.3.3. PFZ Map Generation. After successfully identifying the thermal fronts as well as favourable range of chlorophyll-a concentration for various pelagic fishes species, the PFZ was generated in the post-processing step. There are two conditions for PFZ define in the map, i.e. based on thermal front and chlorophyll-a concentration, which is defined as PFZ with highest probability, and based only on thermal front, which is defined as PFZ with moderate to low probability (table 1). These conditions were defined in order to inform the fishers which areas should be chosen as priority.

Table 1. The PFZ conditions.

Dataset	PFZ Condition
Thermal front + favourable range of chlorophyll-a for various pelagic species	(Highest probabiity)
Thermal front only	(Moderate to low probability)

3. Results

Figures 3 and 4 show the example of 3-day composite of SST and concentration of chlorophyll-a data obtained by pre-processing module. These data are for 1-3 January 2016. By implementing this module, processing time to provide images to be used in the processing module was decreased by 30 minutes. Anyhow, from these figures we found that some area were covered by cloud and therefore cannot be included in the next process. Normally for tropical areas such as Indonesia, especially during the rainy season, cloud cover is one of crucial issue that sometime reduces the area can be analyzed. Method such as Data Interpolating Empirical Orthogonal Function (DINEOF) [19] could be a good alternative to be used in order to fill the area covered by cloud.

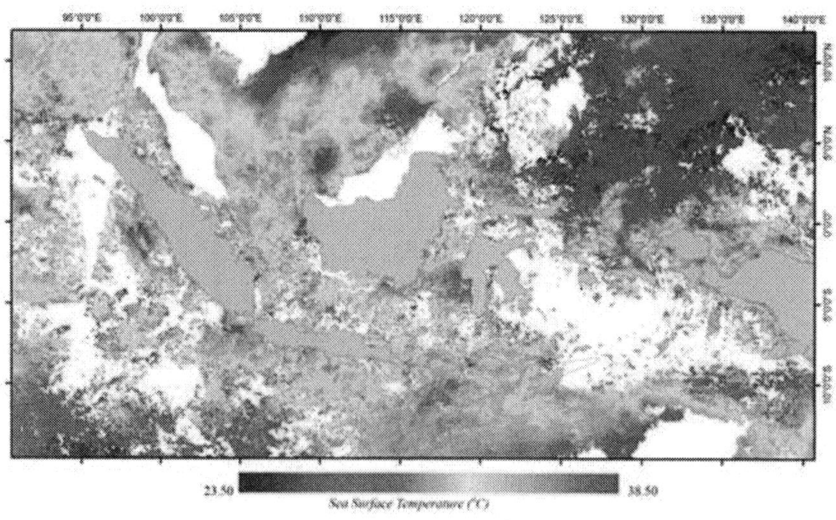

Figure 3. Sea surface temperature derived from MODIS Level 3.

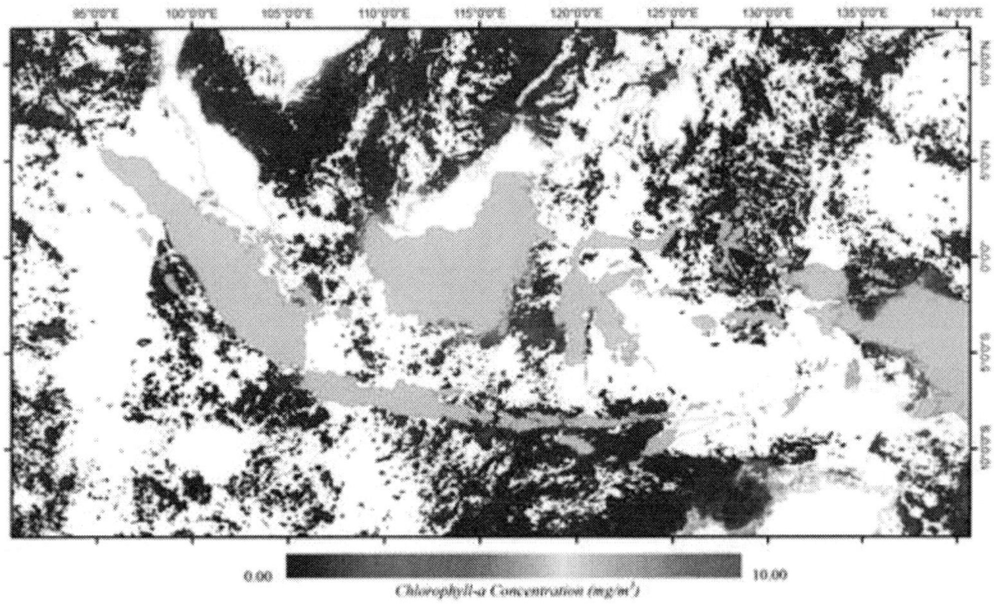

Figure 4. Chlorophyll-a concentration derived from MODIS Level 3

Figure 5 shows the example result after edge detection method was applied to the SST data where thermal fronts are found in some areas around the South China, Java, and Sulawesi Seas, Sulu Ridge, Makassar Strait, as well as the western part of Pacific Ocean.

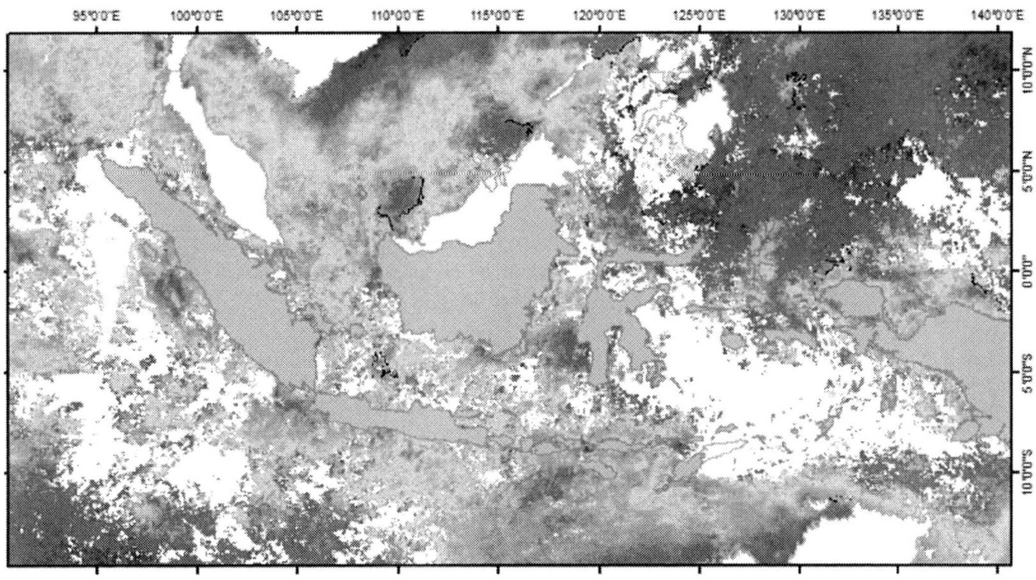

Figure 5. Thermal front detected by the edge detection method

Figure 6 shows favorable range of chlorophyll-a concentration after filtering. Overlaying this result with thermal front map, we found that in general, in the area where thermal front exists the concentration of chlorophyll-a is significantly high (see Figure 7).

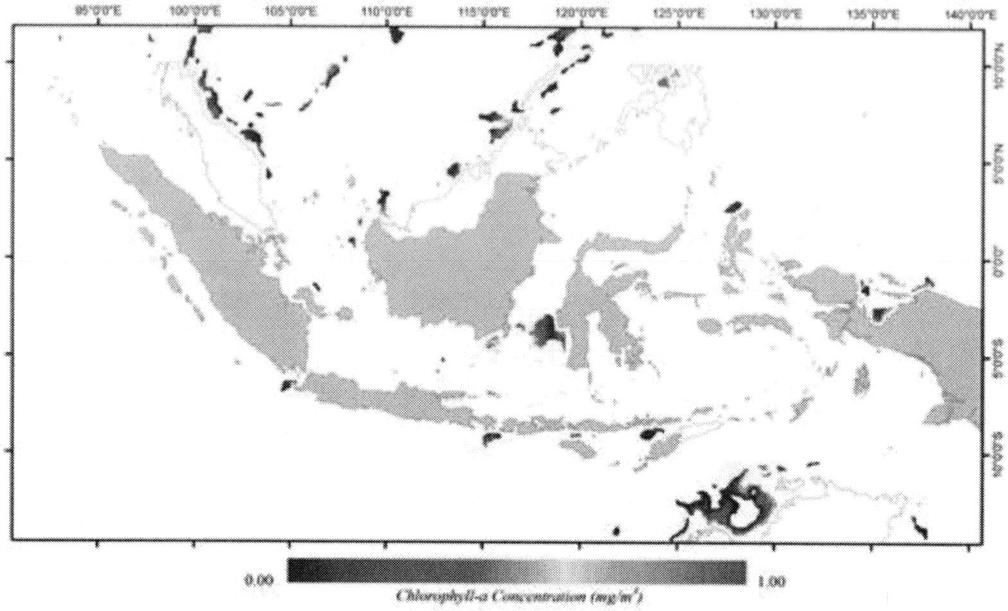

Figure 6. Favorable range of chlorophyll-a concentration for various pelagic species

After all processes to identify the thermal front and favorable range of chlorophyll-a concentration were finished, and based on classification given by table 1, the forecasting process to obtain PFZ was performed. The result obtained by this process was given in figure 8. In order to evaluate the overall improvement obtained by applying automated processing system, results obtained by this method were compared to results produced manually by operators. Two aspects were taken into account in the evaluation, i.e. the accuracy and processing time. It was found that by applying edge detection method on estimating the thermal front, the process was faster. Furthermore, since for tropical region such as Indonesia, thermal gradient is usually small and in some cases difficult to recognize manually, the edge detection significantly improve the accuracy.

In average, time needs to produce fishing zone map manually is approximately 1 to 2 hours, depend on the experience of operator. By applying the automation system, time needed to produce the fishing zone map is only 5 minutes, and this improvement of course very significant. The machine used to perform the process is personal computer Intel i5 with 8GB RAM.

IOP Conf. Series: Earth and Environmental Science **54** (2017) 012081 doi:10.1088/1755-1315/54/1/012081

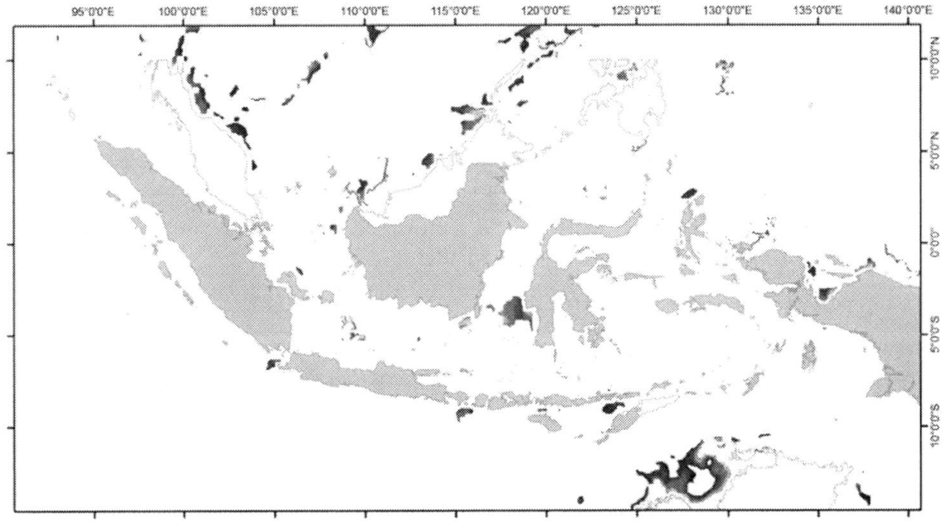

Figure 7. Overlay of thermal front and favorable range of chlorophyll-a for various pelagic species

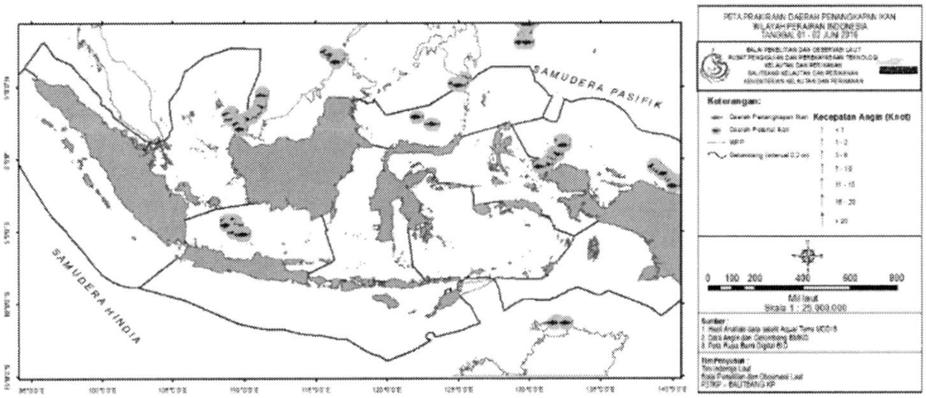

Figure 8. Map of PFZ obtained by the automated processing system

4. Conclusion

From this study, we found that the automation processes on producing fishing ground map have some advantages. In the pre-processing step, automation processes could decrease the processing time significantly, from 2 hours to only 5 minutes. In the processing step, besides decreasing the processing time, misinterpretation by operators could also be minimized, and therefore, results from this automation system could increase the accuracy. Similar advantages also found in the post-processing step, where the production of fishing ground could be done faster and easier.

However, in this study, issue related to cloud cover did not handle yet during pre-processing step, while this issue was very significant, particularly for tropical region such as Indonesia. We planned to apply method such as Data Interpolating Empirical Orthogonal Function (DINEOF) in our next generation of automated processing system.

Acknowledgements

The Authors would like to thank Ocean Color for data, Institute for Marine Research and Observation and all of PPDPI team for giving support.

5. References

[1] Cayula J F and Cornillon P 1992 Edge detection algorithm for SST images *Journal of Atmospheric and Oceanic Technology* **12** 821-829

[2] Acha E M, Piola A, Iribarne O and Mianzan H 2015 *Ecological processes at marine fronts: Oases in the ocean* (Berlin: Springer)

[3] Lumban-Gaol J, Leben R R, Vignudelli S, Mahapatra K, Okada Y, Nababan B, Mei-Ling M, Amri K, Arhatin R E, and Syahdan M 2015 Variability of satellite-derived sea surface height anomaly, and its relationship with Bigeye tuna (Thunnus obesus) catch in the eastern Indian Ocean *European Journal of Remote Sensing* **48** 465-477

[4] Susanto R D, Gordon A L and Zheng Q 2001 Upwelling along the coasts of Java and Sumatra and its relation to ENSO *Geophysical Research Letter* **28** 1599-1602

[5] Hendiarti N 2008 The existence of pelagic fish in relation to oceanographic phenomenon and seasonal climate change based on remote sensing data analysis (June 2008) *Proc. Globë* vol 10 pp 19-25

[6] Jing Z, Qi Y and Du Y 2012 Persistent upwelling and front over the Sulu Ridge and their variations *Journal of Geophysical Research* **117** C11011

[7] Belkin I and Cornillon P 2003 SST fronts of the Pacific coastal and marginal seas *Pacific Oceanography* vol. 1 (2) pp 90-113

[8] Sholva Y, Sitohang B and Wikantika K 2013 New approach to locate upwelling and thermal-front from satellite imagery data *Procedia Technology* vol 11 pp 317-326

[9] Hasyim B, Hartuti M and Sulma S 2009 Identification of fishery resources in Madura Straits based on the implementation of potential fishing zone information from remote sensing *International Journal of Remote Sensing and Earth Sciences* **6** 1-13

[10] NASA Goddard Space Flight Center, Ocean Biology Processing Group 2014 Algorithm descriptions for chlorophyll a. http://oceancolor.gsfc.nasa.gov/cms/atbd/chlor_a. Accessed 2016/12/05. Maintained by NASA Ocean Biology Distributed Active Archive Center (OB.DAAC), Goddard Space Flight Center, Greenbelt MD

[11] Kunarso S, Hadi N, Ningsih S and Baskoro M S 2011 Variabilitas suhu dan klorofil-a di daerah upwelling pada variasi Kejadian ENSO dan IOD di Perairan Selatan Jawa sampai Timor *J. Ilmu Kelautan* **16** 171-180

[12] Robinson and Ian S 2010 *Discovering ocean from space* (UK: Praxis Publishing)

[13] Li Q P, Franks, P J, Ohman M D and Landry M R 2012 Enhanced nitrate fluxes and biological processes at a frontal zone in the southern California current system *J. Plankton Res. 2012* **34** 790-801

[14] Zainuddin M 2011 Skipjack tuna in relation to oceanographic conditions of Bone bay using remotely sensed satellite data *Jurnal Ilmu dan Teknologi Kelautan Tropis* **3** 82-90

[15] Bertignac M, Lehodey P and Hampton J 1998 A spatial population dynamics simulation model of tropical tunas using a habitat index based on environmental parameters *J. of Fisheries Oceanography* **7** 326-334

[16] Polovina J J, Howell E A, Kobayashi D R and Seki M P 2015 The transition zone chlorophyll front updated: Advances from a decade of research *Progress in Oceanography* in press http://dx.doi.org/10.1016/j.pocean.2015.01.006

[17] Tanveer M, Mukherjee C K and Vinu Chandran R 2015 Application of ann to predict chlorophyll-a and sea surface temperature in demarcation of fishing zone *J. Water Technology* **5** 294-304

[18] Deshpande and Ashwini G 2000 Edge analysis of seasonal variability in chlorophyll maps of the black sea *Diss. Massachusetts Institute of Technology*

[19] Beckers J M and Rixen M 2003 EOF calculations and data filling from incomplete oceanographic datasets* *J. Atmospheric and Oceanic Technology* **20** 1839-1856

IOP Conf. Series: Earth and Environmental Science 54 (2017) 012082 doi:10.1088/1755-1315/54/1/012082

Estimation of suspended sediment concentration from Acoustic Doppler Current Profiler (ADCP) instrument: A case study of Lembeh Strait, North Sulawesi

Angga Dwinovantyo[1], Henry M Manik[2]*, Tri Prartono[2], Susilohadi[3] and Delyuzar Ilahude[3]

[1]Graduate School of Marine Technology, PMDSU Program Batch II, Bogor Agricultural University, Bogor 16680, Indonesia

[2]Department of Marine Science and Technology, Faculty of Fisheries and Marine Science, Bogor Agricultural University, Bogor 16680, Indonesia

[3]Marine Geological Institute, Bandung, Indonesia

E-mail: henrymanik@ipb.ac.id

Abstract. Measurement of suspended sediment concentration (SSC) is one of the parameters needed to determine the characteristics of sediment transport. However, the measurement of SSC nowadays still uses conventional technique and it has limitations; especially in temporal resolution. With advanced technology, the measurement can use hydroacoustic technology such as Acoustic Doppler Current Profiler (ADCP). ADCP measures the intensity of backscatter as echo intensity unit from sediment particles. The frequency of ADCP used in this study was 400 kHz. The samples were measured and collected from Lembeh Strait, North Sulawesi. The highest concentration of suspended sediment was 98.89 mg L^{-1} and the lowest was 45.20 mg L^{-1}. Time series data showed the tidal condition affected the SSC. From the research, we also made correction from sound signal losses effect such as spherical spreading and sound absorption to get more accurate results by eliminating these parameters in echo intensity data. Simple linear regression analysis at echo intensity measured from ADCP to direct measurement of SSC was performed to obtain the estimation of the SSC. The comparison result of estimation of SSC from ADCP measurements and SSC from laboratory analyses was insignificantly different based on t-test statistical analysis with 95% confidence interval percentage.

1. Introduction

One of the important parameters to understand the sediment transport is suspended sediments concentration (SSC). The measurement of SSC in the water column can be conducted in several ways, such as conventional, optical, and acoustic methods. Conventional method use gravimetric assessment in laboratory addressing some measurement issues and is relatively difficult to obtain because it need to be analyzed at laboratory [1] and require collection of numerous water samples in each point of research area; and lack of continuous data because it required continuous water sampling, where the necessary data within a particular period of observation. The other disadvantages of measurements by conventional method were slow and high cost [2].

One of the instruments that can estimate the concentration of suspended solids is Acoustic Doppler Current Profiler (ADCP). Through the acoustic technologies used, ADCP can measure various objects in the water column; one of them is the estimation of SSC [3]. The most important factor and advantage of using this instrument is to obtain as nearly as practicable the true concentration of SSC with long period of time [4]. The measurements result using this instrument have spatial and temporal resolution, the important factor what conventional measurements cannot provide [5][6].

Currently, the acoustic technology is growing and continues to be used in research in marine field. On the other hand, not many researchers use ADCP as a scientific echosounder for the detection and quantification of underwater targets. For more than two decades, ADCP have been used for estimate suspended sediments concentration [7], and is becoming common used for sediment transport studies in estuaries, bay system [8], lake [9], and mud-dominated environment [10]. The application of most of these studies used at high frequency ADCP (>100 kHz) to provide qualitative estimates of suspended sediment concentration [11]. Recent studies also found good correlation between acoustic backscatter from ADCP and measured concentration by direct measurement and showed a good correlation between these methods [12]. Many research also used numerical analysis to compare the measured SSC from ADCP and model [13]. But using this instrument in a strait with strong tidal effect is need to be observed, and this paper will discuss that ADCP can provide some advantages over conventional measurements for estimating time series analysis of SSC in that main observation.

The method of calculating conversion into SSC backscatter can be done based on the principle of sonar equation [14]. The direct measurement of SSC was then being correlated with echo intensity from ADCP using simple regression linear equations. After that, time series data of backscatter value from the ADCP can be converted into SSC in mg L^{-1}. This paper describes the results of a field experiment in which time series of SSC were estimated at multiple layer of depth from the ADCP. The aims of this study were as follows: (1) to estimate SSC and the temporal SSC data of Lembeh Strait, (2) to provide baseline information for sediment transport study, and (3) to assess acoustic technology using ADCP to monitor SSC and compare to in-situ SSC measurement.

2. Materials and methods

2.1. Research location

This research was conducted in April 2016 at Lembeh Strait, located at North East Coast of Sulawesi near the city of Bitung, North Sulawesi. ADCP instrument was deployed at 30 meters (visualized at figure 1) with geographical position 1°28'00.6" N - 125°14'07.7"E (figure 2). Lembeh Strait is a narrow strip of water between Sulawesi Island and Lembeh Island, stretching from Sulawesi Sea to Maluku Sea with depth ranging from 15 to 30 meters.

The oceanographic condition in Lembeh Strait is strongly influenced by the tidal currents [15], where the water moves from Sulawesi Sea to Maluku Sea and vice versa. Lembeh Strait has a reputation for its clear water visibility for diving activity, and it indicates that this site has low concentration of suspended sediments. The strait has a great biodiversity of colorful marine life, and also has transparency of 10-25 m, which is generally suitable for growth of benthic organism especially in coral reefs [15].

The main consideration of site selection is a large port called Port of Bitung there. This port is considered strategic for export-import activities, because it directly faces the Pacific Ocean region. The construction of the port activity such as dredging and disposal to the water column will affect the suspended sediment concentration.

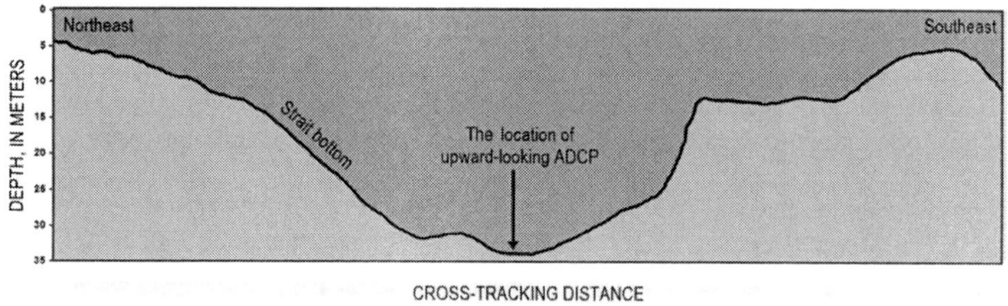

Figure 1. The location of moored ADCP in the bottom of the Lembeh Strait.

IOP Conf. Series: Earth and Environmental Science **54** (2017) 012082 doi:10.1088/1755-1315/54/1/012082

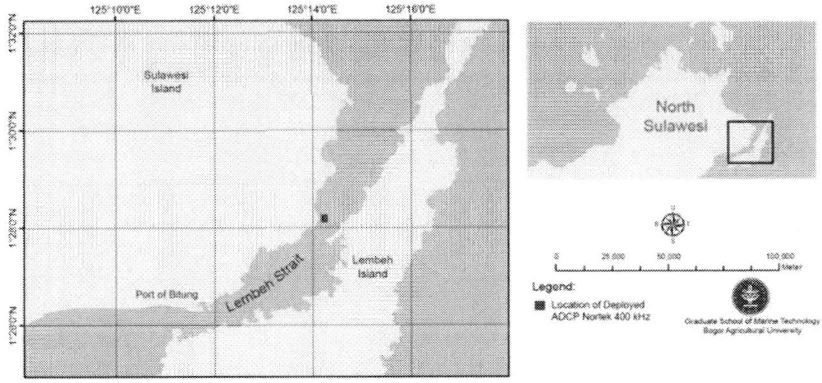

Figure 2. Study location in Lembeh Strait, North Sulawesi

2.2. Materials and equipment

Equipment used in this study was ADCP Nortek WAV 6579 400 kHz with upward-looking setting, bin size 1.5 m, and started to measure from 2.5 m above bed. The time series data was acquired from 4 April 2016 to 22 April 2016 with 20 minutes interval of data recording. Other equipment used was Horiba U-50 for temperature, salinity, and pH data loggers. Water samples were collected from 50 sampling depth. All water samples were preserved in pre-cleaned polyethylene bottles before filtered using Whatman filter paper (pore diameter: 47 mm). SSC direct sample determination referred to APHA gravimetric method [16]. Processing and data modelling were conducted using MATLAB 2016a (Mathwork, Inc.).

2.3. Data analysis

2.3.1. Counts to decibel (dB) conversion.

Echo intensity (EI) detected from this ADCP were converted to SSC through some processing. Conversion of EI to dB is stated on equation (1).

$$I_{dB} = K_c I_{counts} + TL - 10\log_{10}\left(\frac{L_{Xmit}}{\cos\theta}\right) \quad (1)$$

where K_c is the echo intensity scale, I_{counts} is echo intensity (EI) RSSI detected from ADCP, TL is two-way transmission losses, θ is the beam angle, and L_{Xmit} is the transmit length [7]. In the form of detection range, K_c is given as follows in equation (2):

$$K_c = \frac{127.3}{(Te+273)} \quad (2)$$

Te is the temperature measured by ADCP (°C). Based on the equation (1), echo intensity in dB is a function of distance, sound absorption, beam spreading, transmitted power, and the backscatter coefficient.

2.3.2. Transmission losses and absorption.

Some correction is required for sound propagation, for example geometrical spreading and attenuation beyond the near zone distance. Transmission losses were calculated by using the formula on eq. (3):

$$TL = 20\log_{10}R + 2\,\alpha R \quad (3)$$

Where α is sound absorption coefficient (dB m^{-1}) and R is distance between ADCP transducer to measured layer or bin (m). ADCP measured EI from each layer of depth bin with 1.5 meter measurement. The distance from ADCP to bin can be measured by equation (4):

$$R = r + \frac{D}{4} \quad (4)$$

Where r is distance from transducer to center of bin or a half of bin size, D is distance from each bin in meter. Absorption coefficient value was depending on environmental condition such as temperature, salinity, depth, pH, and material absorbing the sound wave. The materials were boric acid (f_1) and magnesium sulphate (f_2). Beside those factors, absorption coefficient also depends on frequency used in ADCP. Absorption coefficient can be measured using the formula in equation (5) [17]:

$$\alpha = 0.106 \frac{f_1 f^2}{f_1^2+f^2} e^{(pH-8)/0.56} + 0.52\left(1+\frac{T}{43}\right)\left(\frac{S}{35}\right)\frac{f_2 f^2}{f_2^2+f^2} e^{-D/6} + 0.00049 f^2 e^{-\left(\frac{T}{27}+\frac{D}{17}\right)} \quad (5)$$

Where f is the frequency of ADCP (400 kHz), T is the average water column temperature (°C), D is the maximum detected depth (m), pH is the average pH of water column, and S is the average salinity (psu). The f_1 and f_2 parameters are given as follows in equation (6) and (7):

$$f_1 = 0.78\left(\frac{S}{35}\right)^{1/2} e^{T/26} \quad (6)$$

$$f_2 = 42\, e^{T/17} \quad (7)$$

2.3.3. Estimation of suspended sediments concentration.
Calibration with known values of concentration by direct measurement of SSC is therefore required for ADCP-based estimation of SSC. Water sampling was conducted at some depth in the station along with its high and low water tidal condition. Water samples were taken at 10 different depths, started from 2 meters until 20 meter for each 2 meter, using van Dorn bottle. The amount of water sample was 1 liter and then stored in polyethylene bottles and place the bottles in the cooler box with ice cube (<4 °C) before analyzed in the laboratory. This process repeated for 5 times in a different water tidal condition (from high water tide, ebb tide, and then low water tide). The 200 mL of water (C) was then filtered using filter paper. Before the filtration process, the filter paper was dried at a temperature of 103-105 °C for 1 hour, and then the filter paper was cooled in a desiccator and weighed (B). After water filtration process, filter paper plus residue was dried again at least one hour at 103-105°C, cooled in desiccator and weighed (A). Direct SSC measurement (mg L^{-1}) can be calculated by equation (8):

$$SSC_{direct} = \frac{(A-B)\times 1000}{C} \quad (8)$$

Collecting data using ADCP at regular intervals can produce changes in I_{dB} value as movement of suspended particles. I_{db} value changes can be attributed to changes in the concentration of suspended particles (SSC). These relations are based on sonar equation by the equation (9):

$$SSC_{ADCP} = 10^{(Intercept\, +\, Slope\, \cdot\, I_{dB})} \quad (9)$$

Where *intercept* and *slope* can be calculated by empirical formula from the simple linear regression. Water sampling should be located at a depth equal to the depth of bin observed by ADCP [18]. The comparison between SSC measured by ADCP and SSC from laboratory analyses were then tested using an analysis of independent-samples t-test with 95% confidence interval percentage.

3. Results and discussion

3.1. Conversion of echo intensity in counts to dB
Above frequency of 100 kHz, absorption coefficient must be matched to local environmental condition [19]. By using equation (5) and substituting the environmental condition on research location with T = 27.5°C, S = 33.23 ppt, pH = 8.13, the sound absorption was 0.1618 dB m^{-1} and two-way transmission losses can be determined in each layer of depth. To produce an accurate measurement of echo intensity, determination of relative backscatter at each bin was conducted by removing RSSI reference level, correcting for transmission loss. Finally, for the determination of the echo intensity in dB, the following practical approaches were used, and signal coefficient factor as given on equation (2) was 0.42, and from the equation (1) can be calculated that the measured 1 count intensity equal to 0.4 dB.

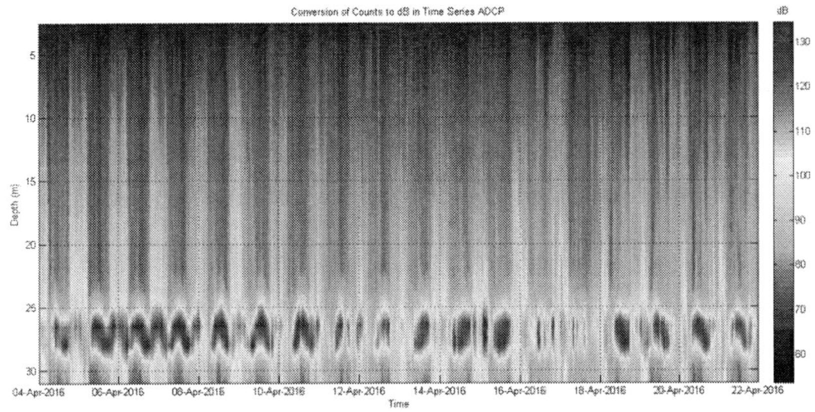

Figure 3. Echo intensity in dB after several processes of correction from transmission losses and sound absorption. The EI values showed that the higher acoustic signal on the bottom layers.

The acoustic signals in upper and lower layer were significantly different, while in bottom layer the echo intensity in dB was higher (figure 3). The acoustic backscatter value ranged from 53-89 dB in surface area, 68-107 dB in water column, and 72-134 dB in near bottom area. Until this stage, the transmission loss factors were no longer the issue at each depth. The sound attenuation due to prevailing sediments in suspension appeared to be calculated. Since acoustic signal was high, it indicated that the S would be high as well. But it still needs to be corrected and calibrated to actual SSC, in this case to direct measurement. Because the limitation using ADCP based on theoretical for the acoustical signal is Rayleigh parameters. The acoustic backscatter is very sensitive to the size of sediment particles. Most suspended sediments are small compared to the wavelength of the sound.

3.2. Calibration and validation

The echo intensity in dB and direct measurement SSC were used in the calibration and validation process (figure 4). From this analysis, the regression line correlated direct sample and echo intensity data from ADCP. Simple regression analysis was used to find the regression slope (A) and intercept (B). The range of direct measurement SSC was between 49 – 64 mg L^{-1} and EI in dB was between 64-90 dB.

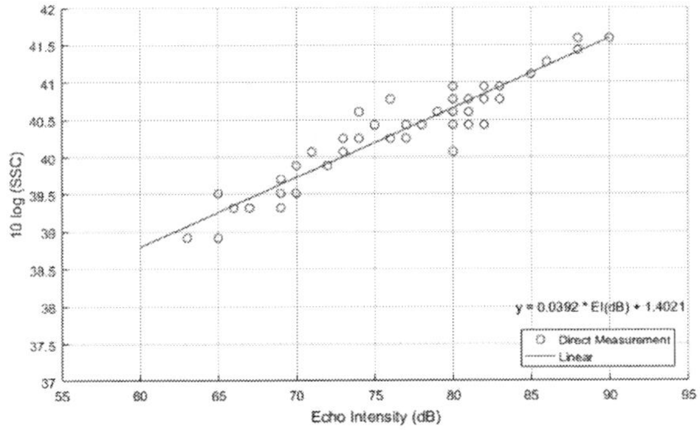

Figure 4. Calibration of echo intensity in dB using direct sample measurement expressed in mg L^{-1}. The direct suspended sediment concentrations were obtained in one hour period during high-to-low water tide (ebb tide) at near the deployed ADCP station.

The calibration show that a linear relation between echo intensity and SSC from direct measurement is generally valid, but the slope of the best-fit linear lines may be different. The constant of this proportionality needs to be assessed empirically for each site selection. This regression analysis employed 50 data pairs with result A and B were respectively 0.0392 and 1.4021. The determination coefficient (r^2) was found to be 0.8943.

3.3. Suspended sediment concentration

The time series from this ADCP exhibit a general agreement between the acoustic backscattering signals (dB) with the corresponding SSC direct measurement (figure 5). The concentration of SSC in near bottom area was higher than in the surface and water column. This indicates sediment resuspension in this site, where sediment settled to the bottom. The fluctuation of SSC between depths of 25-30 meters may occur due to tidal effect. The formation and destruction of flocculated sediment may result in an increase and decrease of the acoustic backscatter from ADCP, and it also will affect SSS in this case.

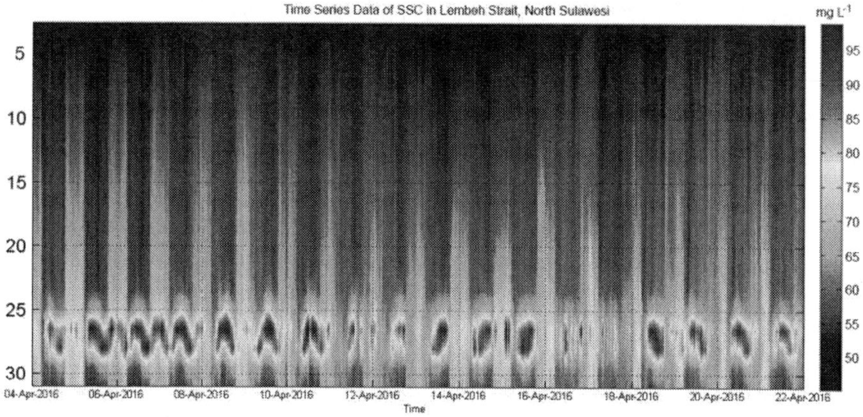

Figure 5. Time series of suspended sediment concentration in mg L^{-1}. From this data, it can be showed that at the upper layer of the water column has lower concentration compared to bottom layer.

The maximum SSC were evident during high water tides, the low water tides showed less concentration of suspended sediments (figure 6). The change of SSC in low and high water tides time also indicated there were effects of advection and tidal resuspension especially in near bottom area; although there was no evidence of any relation between the time series of the SSC and monthly tidal amplitudes. From this result, it proves that the tidal condition affected the SSC [7], especially tidal currents that provide the main energy source for particle resuspension and transport.

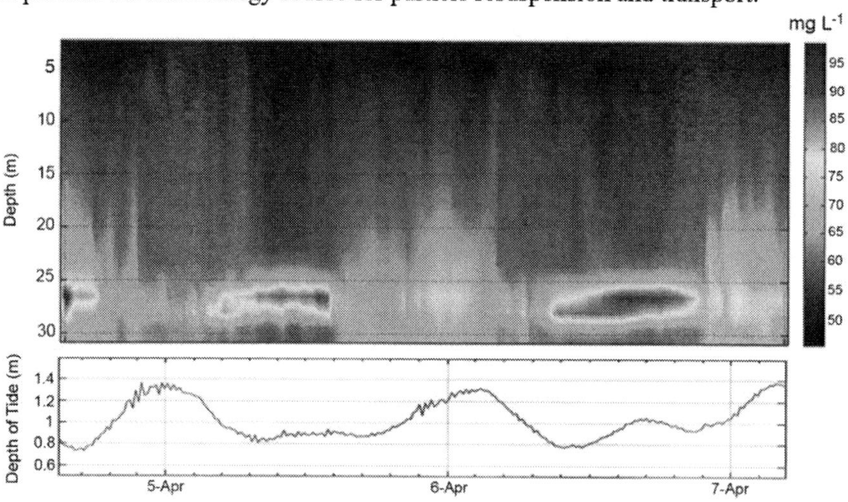

Figure 6. Estimated suspended sediment concentrations during the high and low water tide.

3.4. *Comparison results between ADCP and direct measurement*

In evaluating SSC measurement from ADCP results, it is important to understand that the results from ADCP have correlation with SSC from direct sampling. Since accuracy and precision are the advantages of using ADCP for SSC estimation, they still have calibration error. Results of the acoustic method compared to direct measurement show generally good qualitative agreement showed at figure 7 by R^2 = 0.7391. There are possible factors for this, for example different particle size distribution at different depth and differences in sample interval for the ADCP and direct measurement [20].

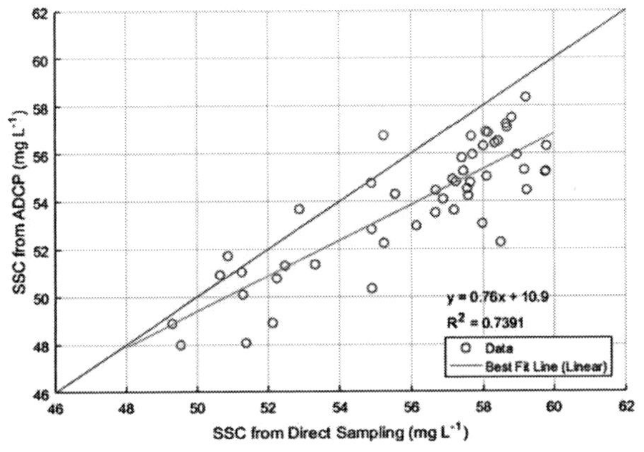

Figure 7. Validation of estimated SSC from ADCP compared to SSC from direct sampling. This analysis showed that a good correlation of direct sampling measurement and acoustic based with R_2 = 0.7391 although the measurement of ADCP was slightly lower than conventional analysis.

Comparing the 1:1 line with linear regression line, the ADCP-based SSC estimates were slightly lower than measured SSC. Despite the possible influence of flocculated sediment, large differences between the slopes in figure 5 may be at least attributed to grain size variation. This calculation has provided insight into potential of ADCP backscatter for the measurement of SSC [10].

Statistical analysis (t-test) was used to compare the results from ADCP and direct measurement. Based on comparison results (figure 7), with a 95% confidence interval statistical analysis, it was shown that the results from ADCP compared with direct measurement of SSC has similarities and insignificantly different. Averaged standard deviation from comparison of all measured data from ADCP and direct sampling was no more than 5 mg L^{-1}. This suggests that measurement of SSC from ADCP was as good as conventional method, and can be used as monitoring SSC in a long period time.

4. Conclusion

Suspended sediment concentration could be estimated using the acoustic backscatter intensity. The acoustic results confirmed a comparable estimation performance with respect to those given by direct measurement of SSC, although the result was slightly lower compared with direct measurement. Error of comparison between ADCP and direct measurement was not more than 5 mg L^{-1}. The monitoring of SSC with ADCP could be considered as a good method to observe time series data of SSC, where it relatively not possible with conventional measurement. Although ADCP show potential for estimating SSC quickly, several issues remain to be explored: Rayleigh parameter for different grain size of suspended sediment, a better understanding of instrument error, and still required a water sample analysis to calibrate the sonar equation.

Acknowledgements

The authors would like to thank Marine Geological Institute, Bandung, Indonesia for supporting this project. This work also supported by Directorate General of Higher Education (DIKTI), Indonesia for financially supporting this work through PMDSU Research Grant Scheme (No. 330/SP2H/LT/DRPM/IX/2016).

References

[1] Zhang T, Stansbury J, Branigan J 2013 Development of a field test method for Total Suspended Solids analysis *Final Reports & Technical Briefs from Mid-America Transportation Center* 52

[2] Holdaway G P, et al. 1999 Comparison between ADCP and transmissometer measurements of suspended sediment concentration *Continental Shelf Research* **19** 421-441

[3] Park H B, Lee G, 2016 Evaluation of ADCP backscatter inversion to suspended sediment concentration in estuarine environments *Ocean Science Journal* **51** 109-125

[4] Moore S A, et al. 2012 On the application of horizontal ADCPs to suspended sediment transport surveys in rivers. Continental Shelf Research **46** 50-63

[5] Gartner J W, Cheng R T 2001 The promises and pitfalls of estimating total suspended solids based on backscatter intensity from acoustic Doppler current profiler *Proc. of the Seventh Federal Interagency Sedimentation Conf. Reno* (Nevada) 119-126

[6] Tessier C, et al. 2008 Estimation of suspended sediment concentration from backscatter intensity of Acoustic Doppler Current Profiler Comptes Rendus Geoscience **340** 58-67

[7] Ghaffari P, et al. 2011 Estimating suspended sediment concentrations using a broadband ADCP in Mahshahr tidal channel *Ocean Science Discussion* **8** 1601-1630

[8] Gartner J W 2004 Estimating suspended solids concentrations from backscatter intensity measured by acoustic Doppler current profiler in San Francisco Bay, California *Marine Geology* **211** 169-187

[9] Wood T M, Gartner J W 2010 Use of acoustic backscatter and vertical velocity to estimate concentration and dynamics of suspended solids in upper Klamath Lake, South-Central Oregon: implications for Aphanizomenon flos-aquae *USGS Scientific Investigations Report* **5203** 1-20

[10] Hoitink A J F, Hoekstra P 2005 Observations of suspended sediment from ADCP and OBS measurements in a mud-dominated environment *Coastal Engineering* **52** 103-118

[11] De Stigter H C, et al. 2011 Recent sediment transport and deposition in the Lisbon–Setúbal and Cascais submarine canyons, Portuguese continental margin *Deep Sea Research Part II: Topical Studies in Oceanography* **58** 2321-2344

[12] Guerrero M, et al. 2016 The acoustic properties of suspended sediment in large rivers: consequences on ADCP methods applicability *Water* **8** 1-22

[13] Bayram A, et al. 2012 Estimation of suspended sediment concentration from turbidity measurements using artificial neural networks *Environmental Monitoring and Assessment* **184** 4355-4365.

[14] Wall G R, Nystrom E A, Litten S 2006 Use of an ADCP to compute suspended-sediment discharge in the tidal Hudson River, New York *Geological Survey Scientific Investigations Report* **5055** 16

[15] Baohong C, et al. 2016 A baseline study of coastal water quality in the Lembeh Strait of North Sulawesi, Indonesia, in 2013 *Marine Pollution Bulletin* **104** 364-370

[16] American Public Health Association 2012 Standard Methods for the Examination of Water and Wastewater 22nd Ed (Washington: APHA)

[17] Ainslie M A, McColm J G 1998 A simplified formula for viscous and chemical absorption in sea water *Journal Acoustic Society America* **103** 1671-1672

[18] Poerbandono, Suprijo T 2013 Modification of attenuation rate in range normalization of echo levels for obtaining frequency-dependent intensity data from 0.6 MHz and 1.0 MHz devices *Journal Engineering Technology Science* **45** 140-152

[19] Furusawa M 2015 Effect of noise and absorption on high frequency measurement of acoustic backscatter from fish *International Journal of Oceanography* **2015** 1-11

[20] Poerbandono Mayerle R 1998 Assesment of approaches for converting acoustic echo intensity into suspended sediment concentration *Proc. 3rd FIG Regional Conference* (Jakarta) 1-13

The utilization of Depth Invariant Index and Principle Component Analysis for mapping seagrass ecosystem of Kotok Island and Karang Bongkok, Indonesia

Agnestesya Manuputty, Jonson Lumban Gaol*, Syamsul Bahri Agus and I Wayan Nurjaya
Department of Marine Science and Technology, Faculty of Fisheries and Marine Science, Bogor Agricultural University, Jl Agatis, Kampus IPB Dramaga, Bogor 16680 Indonesia

E-mail: jonson_lumbangaol@yahoo.com

Abstract. Seagrass perform a variety of functions within ecosystems, and have both economic and ecological values, therefore it has to be kept sustainable. One of the stages to preserve seagrass ecosystems is monitoring by utilizing thespatial data accurately. The purpose of the study was to assess and compare the accuracy of DII and PCA transformationsfor mapping of seagrass ecosystems. Fieldstudy was carried out in Karang Bongkok and Kotok Island waters, in Agustus 2014 and in March 2015. A WorldView-2 image acquisition date of 5 October 2013 was used in the study. The transformations for image processing data were Depth Invariant Index (DII) and Principle Component Analysis (PCA) using Support Vector Machine (SVM) classification. The result shows that benthic habitat mapping of Karang Bongkok using DII and PCA transformations were 72%and 81% overall's accuracy respectively, whereas of Kotok Island were 83% and 84% overall's accuracy respectively.There were seven benthic habitat types found in karang Bongkok waters and in Kotok Island namely seagrass, sand, rubble, coral, logoon, sand mix seagrass, and sand mix rubble. PCA transformation was effectively to improve mapping accuracy of sea grass mapping in Kotok Island and Karang Bongkok.

1. Introduction

Seagrass beds are highly diverse and productive ecosystems and mostly covered by seagrass as their dominant vegetation [1]. Seagrasses form extensive beds or meadows, which can be either mono-specific (made up of a single species) or in mixed beds where more than one species coexist. According to [2], the seagrass ecosystem of Indonesia is 30,000 km^2. However, according to the last report it showed that 58% of world's seagrass meadows are currently decreasing, including Indonesia itself lost about 30-40% of its seagrass beds [3]. In Seribu Islands including Kotok Island and Karang Bongkok the coastal resources such seagrass ecosystem has been degradated caused by natural factors and human induced factors such as, unsustainable resource exploitation (destructive fishing) and overfishing [4].

Kotok Island and Karang Bongkok, have high potential of diversity coastal resources, one of which is seagrass ecosystems. Seagrass ecosystems should be preserved because they contribute to the improvement of the fishery and tourism sectors. Thus, seagrass is an important instrument for both ecologically and economically, they sequester large amounts of carbon within plants above and below sea-level as well as within soils [5, 6]. Although they are important, unfortunately, seagrass is seldom given the attention or protection to they deserves. Improved management and protection of seagrass are required to understand better the dynamic nature of these ecosystems.

Remote sensing data from satellite images had been utilized to generate objective information to monitor large of coastal areas [7]. Ecosystem mapping of shallow waters based on satellite imagery can be done using Depth Invariant Index (DII) transformation [8, 9]. DII transformation has also been utilized for mapping seagrass [10], but the accuration in Karang Bongkok and Kotok Island waters lower than 85% [11]. In addition to DII, Principle Component Analysis (PCA) was applied on satellite imagery hyperspectral for mapping seagrass. Mapping of *Posidonia oceanica* seagrass using PCA had successfully in the Mediterranean with more than 90% accuracy [12]. Accordingly, the authors conducted such study to assess whether high spatial resolution satellite images of PCA transformation can be used effectively for mapping diverse species compositions, cover, and above-ground biomass of seagrass in Kotok Island and Karang Bongkok.

Principle Component Analysis is a mathematical procedure that uses an orthogonal transformation to convert a set of observations of possibly correlated variables into a set of values of uncorrelated variables called principal components. The other main advantage of PCA isthat once wehave found these patterns in the data and we compress the data i.e. by reducing the number of dimensionswithout much loss of information[13]. This process is performed for the multispectral image, a particular class of images that require specialized coding transformations.In multispectral images, the same spatial region is captured multiple times usingdifferent imaging modalities, so that some objects in the image are hard to recognize [14]. Therefore, PCA is often viewed as an efficient compression technique for information. The purpose of thisresearch was to assess and compare the accuracy of DII and PCA transformationsfor mapping seagrass ecosystems.

2. Methods

2.1. Data collections

The study was carried out at two different locations namely Kotok Island and Karang Bongkok, in the Seribu Islands National Park, Jakarta (Indonesia) (figure 1). Field data collections were implemented in two steps, the first step was in August 2014 (represent East Monsoon) and the second step was in March 2015 (represent West Monsoon).

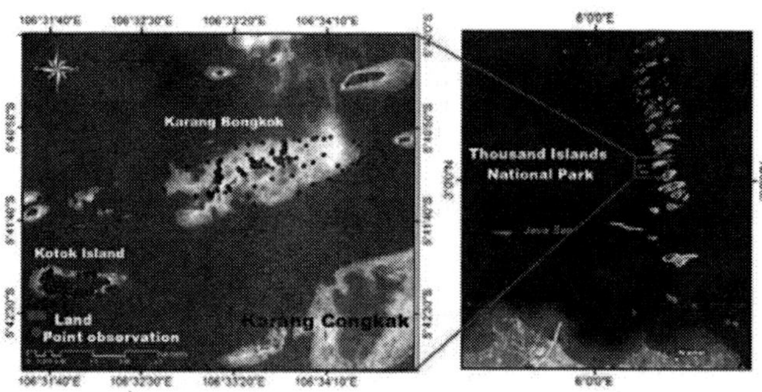

Figure 1. Study area

The tools used in this study were Global Positioning System (GPS) Garmin 76 CSX with 2m precision, transect quadrant, scale meter, stationery, The guidelines for seagrass identifier guideline (Seagrass Watch), underwater camera, snorkel and mask. The material used in this study was the satellite imagery of WorldView-2with the acquisition date was 5 October 2013. The WorldView-2 satellite imagery have 8 bands with high spatial resolutions, 1.84 m (multispectral) and 0.46 m(panchromatic).WorldView-2 images can register the energy reflected by the shallow water habitat different intervals of the electromagnetic spectrum with wavelengths ranging from the blue region to the mid-infrared (Coastal: 0.40–0.45μm; Blue: 0.52–0.51μm; Green:0.51-0.58μm; Yellow: 0.585-

0.625μm; Red: 0.63–0.69μm;Red-edge: 0.705–0.745μm;N-IR-1: 0.720–0.895μm; N-IR-2: 0.86–1.04μm).

Line transects and some spot check methods were used for collecting spatially referenced benthic cover of the seagrass beds [15, 16]. This transect was used again to conduct the substrate line transect. Transect lines used to collect the data of seagrass was 30 m perpendicular that parallel the coast. At both locations observations performed 3 times for each station

Quadrant size to collect seagrass is 1m x 1 m and divided into 4 segments which 25 cm size each segments placed along the transect line. Data observations were carried out using Seagrass watch monitoring methods [17].

2.2. *Image processing*

Image processing was performed with several stages such as geometric correction, radiometric conversion and atmospheric correction, image segmentation and enhancement. Image enhancement was conducted by correction of the water column with using the *DII* transformation. This method is based on the electromagnetic wave that are emitted by the sun will experience the gradual loss in intensity due to absorption and scattering by particles contained in water (attenuation). *DII* derivative of the transformation is as follows [18]:

$$DII_{ij} = \ln(L_i) - \lfloor (k_i / k_j)\ln(L_j) \rfloor$$

$$ki / kj = a + \sqrt{(a^2 + 1)}$$

$$a = \frac{\sigma ii - \sigma jj}{2\sigma ij} \quad (1)$$

Where, L_i and L_j = the reflectance values of the band all i and j, k_i / k_j= ratio attenuation coefficient of the band all i and j, σii =variance of the band i, σjj = variance of the band j and σij = covariance of the band.

PCA transformed value of digital data into the new smaller data.A satellite image can be expressed in matrix format in the following way [18]:

$$(C - \lambda_k I)w_k = 0, \quad (2)$$

Where, n represents the number of the pixels and b the number of bands. Considering each bandas a vector, the above matrix can be simplified as follows:

$$X_k = \begin{pmatrix} x_1 \\ x_2 \\ . \\ x_6 \end{pmatrix} \quad (3)$$

Where,k is the number of bands.

To reduce the dimensionality of the original bands the eigenvalues of the covariance matrix must be calculated. This matrix can be calculated as follows:

$$C_{b,b} = \begin{pmatrix} \sigma_{1,1} \cdots \sigma_{1,6} \\ . \\ \sigma_{6,1} \cdots \sigma_{6,6} \end{pmatrix} \quad (4)$$

where$_{i,j}$ is the covariance of each pair of different bands.

$$\sigma_{i,j} = \frac{1}{N-1} \sum_{p=1}^{N} (DN_{p,i} - \mu_i) - (DN_{pj} - \mu_j) \quad (5)$$

Where,$DN_{p,i}$is a digital number of a pixel p in the band i,DN_{pj} is a digital number of a pixel, p in the band j,μ_i and $\mu_j$are the averages of the *DN* for the bands i and j, respectively. From the variance-covariance matrix, the eigenvalues (λ) are calculated as the roots of the characteristic equation,

$$\det(C-\lambda I)=0,$$

Where, $C$is the covariance matrix of the bands and I is the diagonal identity matrix.
The PCA can be expressed in matrix form:

$$Y_6=\begin{pmatrix} y_1 \\ y_2 \\ . \\ y_6 \end{pmatrix}=\begin{pmatrix} w_{1,1}\ldots w_{1,6} \\ \ldots \\ w_{6,1}\ldots w_{6,6} \end{pmatrix}\begin{pmatrix} x_1 \\ x_2 \\ . \\ x_6 \end{pmatrix} \quad (6)$$

Where,Y is the vector of the principal components, W the transformation matrix, and X the vector of the original data. The eigenvectors can be calculated from the vector - matrix equation for each eigenvalue k,

$$(C-\lambda_k I)w_k=0, \quad (7)$$

Where, C is the covariance matrix, $k$is the k eigenvalues (six in our example), I is the diagonal density matrix, and $wk$is the k eigenvectors.

Image classification was performed using the Support Vector Machine (SVM). SVM have their roots in Statistical Learning Theory. SVM is a classification technique to find a vector or line that has functions as the separator between the classes [19] explained that the simple concept of SVM is the classification of the image pixel by finding the best hyperplane. Hyperplane serves as a divider of two classes in the input space. All classification transformations are based on the assumption that the image in question depicts one or more features and each of these features belongs to one of several distinct and exclusive classes. SVM classification is done on the image of the DII and PCA transformations.

The accuracy of satellite image classification was conducted by using the confusion matrix table. The confusion matrix is a rigorous statistical technique where number or value of pixels correctly assigned to each classification class and those missassigned to other classes are arranged in rows and columns relating alloted pixels in the classification image to a ground data. The percentage accuracy of classification is calculated from the ratio of the sample point calculation in the field (ground-truth) with image data classification results (the number of pixels). Test accuracies were calculated including User's Accuracy (UA), Producer's Accuracy (PA) and Overall's Accuracy (OA).User's accuracy occurs when pixels associated with a class are incorrectly identified as other classes, or from improperly separating a single class into two or more classes. Producer's accuracy is obtained by dividing the total pixels not correctly classified for each class in the reference data (column) by the total pixels for that class in the reference data/image (column total). Overall's accuracy is obtained by dividing the total number of correct pixels (diagonal) by the total number of pixels in the error matrix. Calculation accuracies of image classifications were performed with the following equation [20]:

$$\textit{Overall's accuracy:}(n_{11}+n_{22}+n_{kk})*100\% \quad (8)$$

$$\textit{Producer's accuracy}:\frac{n_{ii}}{n_{1+}}*100\%\% \quad (9)$$

$$\textit{User's accuracy}:\frac{n_{ii}}{n_{1+i}}*100\%n \quad (10)$$

		j = Columns (Reference)			Row Total
		1	2	k	n_{i+}
i = Rows (Classification)	1	n_{11}	n_{12}	n_{1k}	n_{1+}
	2	n_{21}	n_{22}	n_{2k}	n_{2+}
	k	n_{k1}	n_{k2}	n_{kk}	n_{k+}
Column Total n_{+j}		n_{+1}	n_{+2}	n_{+k}	n

3. Result and discussion

3.1. Seagrass distribution

Field observations revealed three types of seagrass in Kotok Island, namely *Thalassia hemprichii, Enhalus acroides, Halophila minor,* and two types in Karang Bongkok waters, namely *Halodule uninervis* and *Cymodocea rotundata* (table 1). The differences in the seagrass composition in Karang Bongkok and Kotok Island waters are due to their adaptability to changing environmental conditions [21]

Thalassia hemprichii species was distributed in both locations due to the facts that *Thallasia hemprichi* was a wide spread and common species. The population appears to be stable, despite considerable threats. This species grow in various types of substrates such as muddy sand, medium-sized and coarse sand, rubble until death [22].

Table 1. Composition of seagrass in Kotok Island and Karang Bongkok

Locations	Spesies				
	Thalassia hemprichii	*Cymodocea rotundata*	*Enhalus acroides*	*Halophila minor*	*Halodule uninervis*
Kotok Island	+	-	+	+	+
Karang Bongkok	+	+	-	-	-

Description: + = Found, - =unfound

3.2. Classification of benthic habitat

DII transformation uses a combination of band for the correction of the water column as coastal-blue bands (bands 1-2). The combinations of these three bands were used for the calculation attenuation coefficients (k_i/k_j). The k_i/k_j ranged between 0.59-0.89 (table 2) were the range of k_i/k_j values marine waters [8]. The attenuation coefficient varies depending on the specific type of waters or waters optical properties. The higher attenuation coefficient values in Karang Bongkok shows that Karang Bongkok was shallower than Kotok Island and the existence of mixed pixel between sand and sea grass patchy [23].

Table 2. Attenuation coefficients (k_j/k_i) determinant (r^2) on Karang Bongkok and Kotok Island

Pair bands	k_i/k_j		r^2	
	Karang Bongkok	Island Kotok	Karang Bongkok	Island Kotok
Coastal-Blue (1-2)	0.89	0.59	0.81	0.82

PCA transformation technique was conducted for bands-1, 2, 3, 4, 5 and 6 of WorldView-2 imagery. The sixth bands have important interconnected information [24], so the transformation of the PCA provided more complete informations. PCA transformationswere generated new bands, namely PC-1, PC-2 and PC-3, contains nearly all image information.

PC-1 costructed from band-1 up to band-6 contains 90% of the information on the image transformation [25, 26]. From band-1 up to band-4 provided more information on aquatic habitat, because the band 1 to band 4 penetrated the water column. PC-2 contains about 7% of the information, whereas the PC-3 contains about 2% the information from the transformation images [25]. PC-2 was constructed from band-1 up to band-4, while PC-3, the band-3 up to band-6.

DII and PCA transformations were performed on the World-View2 satellite imagery. There were seven benthic habitat types inkarang Bongkok waters and in Kotok Island namely seagrass, sand, rubble, coral, logoon, sand mix seagrass and sand mix rubble (figure 2). In general, the total area of each class object generated by PCA and DII transformations were relatively equal except for seagrass. The area of seagrass using PCA was larger than DII transformations (table 3). The use of PCA transformation could generated information completely from satellite imagery [13]. Meanhwhile,DII transformation could lead to the loss of some information recorded satellite sensors [25].

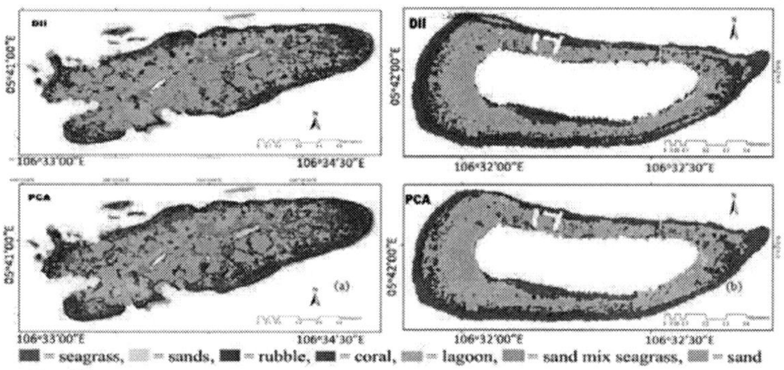

Figure 2. Map of benthic habitat using classification SVM transformation on the DII and PCA image transformation of (a) Karang Bongkok and (b) Kotok Island.

Table 3. The areal of seabed habitat classification

Class	Karang Bongkok (ha)		Island Kotok (ha)	
	DII	PCA	DII	PCA
Coral reefs	335.23	337.70	86.72	87.06
Seagrass	19.51	37.01	2.57	2.63
Lagoon	73.72	62.46	0.40	0.41
Sand	2.44	2.43	0.44	0.43
Rubble	66.94	75.80	7.90	5.96
Sand mix Seagrass	13.80	13.21	10.02	9.86
Sand mix Rubble	99.76	82.92	13.50	15.19

3.3. Classification accuracy of benthic habitat

Overall's accuracy values of the benthic habitat objects classification in Karang Bongkok and Kotok Island more than 70% have a high accuracy values (table 4). The OA value of objects classification in Karang Bongkok using images DII and PCA transformations were 72.38% and 81.00% respectively. Meanwhile, in Kotok Island were 83.00%and 84.29% respectively. An accuracy value of 65-70% can be considered good enough for the coastal habitat mapping using remote sensing [27]. PCA transformation for objects classification both of Karang Bongkok and Kotok Island was more accurate than DII transformation.

In general, the classification accuracy of each object benthic habitat with higher PCA transformation than DII transformation either in Karang Bongkok and Kotok Island (table 4). This means that the object classification of benthic habitat of the image of the PCA transformation is more accurate than the DII. Based on the type of object the object classification sand, and sand mixed seagrass, logoon have the highest accuracy values. In other words that the object is more easily distinguished class than other objects from satellite images. Rubble class is a class of objects most low-value accuracy it can happen because the object rubble covered with algae that can be classified into seagrass. Accuracy object class coral at Karang Bongkok was smaller than Kotok Island. One of the causes is because Karang Bongkok waters deeper than Kotok Island.

Table 4. Accuracy of benthic habitat classifications in Karang Bongkok and Kotok Island

Location	**PA**		**UA**		**OA**	
	DII	**PCA**	**DII**	**PCA**	**DII**	**PCA**
Karang Bongkok						
Coral reef	17%	50%	100%	100%		
Lagoon	100%	100%	63%	83%		
Seagrass	**73%**	**82%**	**96%**	**96%**	72.38%	81%
Sand	80%	80%	100%	100%		
Sand mix Seagrass	100%	100%	100%	83%		
Sand mix Rubble	100%	100%	50%	50%		
Rubble	57%	64%	35%	53%		
Island Kotok						
Coral reef	87%	80%	81%	80%		
Lagoon	100%	100%	100%	100%		
Segrass	**72%**	**82%**	**100%**	**98%**		
Sand	100%	100%	100%	100%	83%	84.29%
Sand mix Seagrass	95%	95%	66%	100%		
Sand mix Rubble	100%	88%	50%	44%		
Rubble	100%	83%	64%	75%		

3.4. Mapping of seagrass distribution

Seagrass distribution at Karang Bongkok and Kotok Island was sensed by satellite imagery with PCA transformation wider than the DII transformation (figure 3). Procedure accuracy test of seagrass classification has been carried out on 62 samples in Karang Bongkok and 76 samples in Kotok Island. The test results showed that of the 62 samples of DII classification of image transformation in Karang Bongkok, as many as 45 samples in accordance with field data, the other 5 samples are rubble and sand mix ruble. The test results showed accuracy in Kotok Island of 76 samples, 55 samples in accordance with field data, but samples are objects rubble, rubble and sand mix. Thus the value of procedure accuracy seagrass classification of the image in Karang Bongkok and Kotok was 72 % and 73 % respectively.

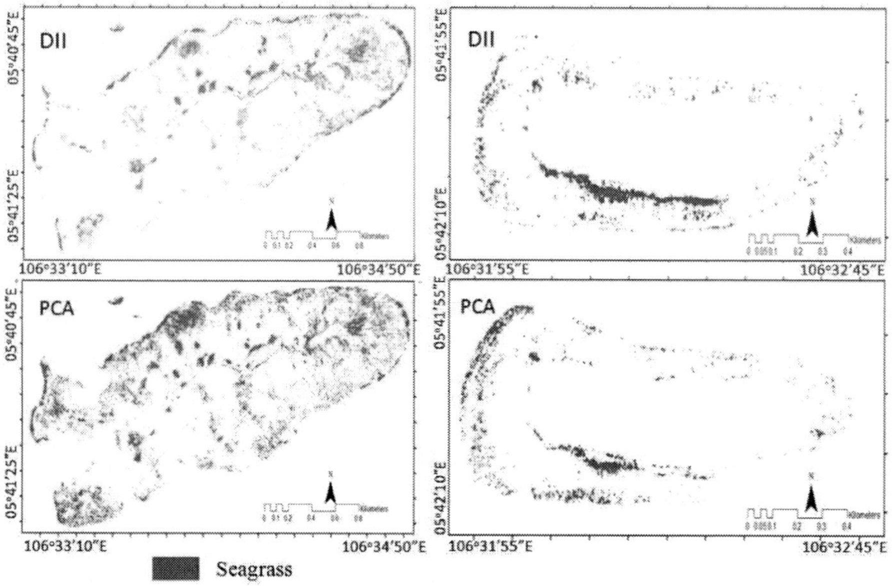

(a) (b)

Figure 3. Classification of seagrass utilizing SVM transformation by enhancement technique of DII and PCA in (a) Karang Bongkok and, (b) Kotok Island.

The test results obtained on the different accuracy classes of seagrass of image transformation and PCA in Karang Bongkok and Kotok Island. Of the 62 samples of seagrass object classification results of 51 samples in accordance with field data and 11 other samples are objects raubble and sand mixed rubble. From 76 samples of seagrass class in Kotok Island, 62 samples in accordance with field data, while 14 samples is the object of rubble and sand mixed rubble. Thus both the value of the procedure's accuracy of the classification seagrass PCA transformation in Karang Bongkok and Kotok Island were 82% (table 2). Value procedure's accuracy of the image classification seagrass PCA transformation is higher than DII eventhough in Karang Bongkok and Island Kotok (table 2).

The results of this study indicate that both the PCA and the DII transform a portion of the object classification seagrass in the field is the object rubble, sand mixed rubble. Bottom waters where covered by rubble generally overgrown by algae so that the spectral values similar to seagrass. Likewise, the sandy sea floor covered with sea grass and seaweed that is similar to the spectral values of seagrass.

User's accuracy values of seagrass both of in Kotok Island and Karang Bongkok were a range between 96-100%. These results were similar to previous studies using satellites SPOT with PCA transformation in Morenton Bay, Australia [12].

4. Conclusion

Utilization of satellite imagery Word-view-2 using PCA transformation for mapping benthic habitat of Kotok Island and Karang Bongkok was more accurate than DII transformation. There were seven benthic habitat types in Karang Bongkok waters and in Kotok Island namely seagrass, sand, rubble, coral, logoon, sand mix seagrass, and sand mix rubble.

Both of PCA and DII transformations were applicable for mapping seagrass ecosystems in Kotok Island and Karang Bongkok with very high accuracy. However, PCA has been applied effectively to improve mapping accuracy of sea grass mapping in Kotok Island and Karang Bongkok.

References

[1] Wimbaningrum R, Choesin D N and Nganro NN 2003 Komunitas lamun di rataanb terumbu Pantai Banna, Taman Nasional Baluuran, Jawa Timur *Ilmu Dasar*. 424: 31.
[2] Nontji A 2009 Rehabilitasi Ekosistem Lamun dalam Pengelolaan sumberdaya.
[3] UNEP 2004 Seagrass in the South China Sea. *UNEP/GEF/SCS Technical Publication No. 3. Bangkok, Thailand.*
[4] Kusumastanto T 2012 Economic Valuation of Coastal Resources Degradation in Seribu Islands: Is Blue Economy a Solution for Reversing the Degradation? Retrieved from http:// eascongress.pemsea.org/2012/sites/default/files/document-files/abstract-st12-kusumastanto.pdf.
[5] Duarte C M, Middelburg J J and Caraco N 2005 Major role of marine vegetation on the oceanic carbon cycle *Biogeosciences* **2** 1-8.
[6] Nellemann C 2009 Blue carbon: The role of healthy oceans in binding carbon. *A rapid response assessment* (Norway: United Nations Environment Programme)
[7] Mumby P J 2006 Connectivity of reef fish between mangroves and coral reefs: transformations for the design of marine reserves at seascape scales *Journal of Biological Conservation* **128** 215-222
[8] Lyzenga D R 198. Passive remote-sensing techniques for mapping water depth and bottom features *Applied Optics* **17** 379-383
[9] Siregar V 2010 Pemetaan Substrat Dasar Perairan Dangkal Karang Congkak dan Lebar Kepulauan Seribu Menggunakan Citra Satelit Quick Bird. *E-Jurnal Ilmu dan Teknologi Kelautan Tropis*. **2** 19-30
[10] Tamondong AM, Blanco AC, Fortes MD, Nadaoka K. 2013. Mapping of Seagrass and Other Bentic Habitat in Balinao, Pangasinan Using WorldView-2 Satellite Image. *IGARSS*. 1579-1582.
[11] Manuputty A, Lumban-Gaol J and Agus S B 2016 Seagrass Mapping Based on Satellite Image Worldview-2 by Using Depth Invariant Index Method *Indonesian Journal of Marine Science* **21** 37-44.
[12] Pasqualini V, Martini C P, Pergent G, Agreil M, Skoufas G, Sourbes L, Tsirika A 2005 Use of SPOT 5 for mapping seagrasses: An application to Posidonia oceanica *Remote Sensing of Environment*. **94** 39-45.
[13] Jensen J R 2005 *Introductory Digital Image Processing: A Remote Sensing Perspective. (3rd ed.)* Pearson Prentice Hall
[14] Richards J A 1999 *Remote Sensing Digital Image Analysis* (Berlin: Springer-Verlag) p. 240.
[15] Roelfsema C M, Phin S R, Udy N and Maxwell P 2009 An Interrated Field and Remote Sensing Approach for Mapping Seagrass Cover, Morenton Bay, Australia *Spatial Science*. **54** 45-62
[16] English S, Wilkinson C and Baker V 1997 *Survey Manual for Tropical Marine Resource* (Australia: Mc Graw Publication)
[17] McKenzie, Campbell S J and Roder C A 2003 *Seagrasswatch: Manual for mapping & monitring seagrass resources by community (citizen) volunteers 2sd edition* (Queensland: The state of Queensland, Department of Primary Industries, CRC Reef) pp 104
[18] Estornell J, Martı-Gavila J M, Sebastia M T and Mengua J 2013 Principal component analysis applied to remote sensing *Modelling in Science Education and Learning* **6** 83-89
[19] Vapnik V N and Kotz S 1982 *Estimation of dependences based on empirical data* (New York: Springer-Verlag)
[20] Congalton R G and Green K 2009 *Assesing the Accuracy of Remotely Sensed Data.Principles and Practices. 2nd Edition* (New York: CRC Press Taylor and Francis Group)
[21] Lefaan P T 2008 Kajian Komunitas Lamun di Perairan Pesisir Manokwari. *Graduate Thesis*. (Bogor (ID): Institut Pertanian Bogor)

[22] Takadengan K and Azkab M H 2010 Struktur komunitas lamun di Pulau Talise, Sulawesi Utara *Oseanologi dan Limnologi Indonesia*. **36** 85 - 95

[23] Syarif B and Parwati E 2012 The effect of the extent of coral reef area on uniform bottom reflectance determination for water column correction using Landsat-ETM. *International Journal of Remote Sensing and Earth Sciences*. **9** 88-99

[24] Chauvaud S, Bouchon C and Maniere R 2010 Remote sensing techniques adapted to high resolution mapping of tropical coastal marine ecosystems (coral reefs, seagrass beds and mangrove) *International Journal of Remote Sensing*. **19** 3625-3639

[25] Pasqualini V, Martini C P, Pergent, Vernandes C and Pergent G 1997 The use of airborne remote sensing for benthic cartography: Advantages and reliability *JRS*. **18** 1167-1177

[26] Danoedoro P 2012 *Pengantar Penginderaan Jauh Digital*. (Jakarta: C.V Andi Offset)

[27] Mumby P J, Green E P, Clark C D and Edwards A J 1998 Digital Analysis of Multispectral airbone imagery of coral reefs *Coral Reefs*. **17** 59-69

Seagrass species distribution, density and coverage at Panggang Island, Jakarta

Iswandi Wahab, Hawis Madduppa*, Mujizat Kawaroe*
Marine Science and Technology Dept. Bogor Agricultural University, Indonesia.

Email : hawis@ipb.ac.id/ mujizat@ipb.ac.id

Abstract. This study aimed to assess species distribution, density and coverage of seagrass in Panggang Island, within Kepulauan Seribu Marine National Park, northern Jakarta. Seagrass sampling was conducted between March to April 2016 at three observation stations in the West, East, and South of Panggang Island. A total of 6 seagrass species was recorded during sampling period, including *Cymodocea rotundata*, *C. serulata*, *Halodule uninervis*, *Syiringodium isoetifolium*, *Enhalus acoroides*, and *Thalassia hempricii*. All species were observed in the South station, while in the West and East station found only three species (*C. rotundata*, *E. acoroides*, and *T. hemprichii*). While, *C. rotundata* and *T. hemprichii* were observed at all station. The highest density was observed for *C. rotundata* (520 ind/m^2) and for *T. hempricii* (619 ind/m^2) in the West station and South Station, respectively. The lowest density was observed in South Station for *C. serulata* (18 ind/m^2), *Halodule uninervis* (20 ind/m^2), and *Syiringodium isoetifolium* (15 ind/m^2). Seagrass coverage of *Thalassia hempricii* was the highest (43.60%) and the lowest observed at *Syiringodium isoetifolium* (0.40%). This could be basic information for the management of seagrass ecosystem in the Kepulauan Seribu Marine National Park.

1. Introduction

Seagrass ecosystems have a high biodiversity and act as a contributor of nutrients for productivity of coastal waters. Coastal water get a nutrient supply from two places, land and sea, to form an organic ecosystem of high productivity and very supportive of seagrass plants can live in coastal areas and develop optimally. Seagrass is a flowering plant found only in coastal area that is able to live submerged in water [8,9]. The existence of seagrass in the world are so vast but the highest biodiversity found in the Indo-Pacific region including Indonesia [27].

Indonesia has 12 species of seagrass from 69 species in the world. Distribution and diversity of abundant seagrass found throughout the coastal areas of Indonesia. Seagrass is very important for a number of threatened organisms, including dugongs, turtles and sea horses. This association has a complex interaction with seagrass. Groups of fauna which are in seagrass dominated by benthic organisms such as sea cucumbers, crabs (Arthopoda), sea urchins, starfish, clams, brittle stars (Echinodermata), snails (mollusks), shrimp and sponge [2]. This is consistent with [24], the presence of seagrass in a body of water can attract different types of marine organisms to feed, spawn and settled.

Distribution for all types of seagrass is the result of a combination of several seagrass, sexual reproduction and growth of colonial and influenced by the spread and environmental limitations [21]. Many seagrass populations are largely rely on asexual reproduction for the maintenance of the population [16]. Seagrass produce a large number of sexual propagules with reproductive models varies and depend on the environmental conditions [6]. Seagrass is also the sexual reproduction that produce fruit and seeds (viviparous) [7]. A close relationship between the habitat and species of mussels obviously a lot happening in intertidal areas. All these systems exert a stabilizing effect on the environment, so that their physical and biological support is very important for other ecosystems [4].

Conditions seagrass ecosystems in some parts of Indonesia are under threat from human activities such as tourism, ports, aquaculture and sand mining. An estimated 58% of seagrass ecosystems in the world has decreased extents [27]. According to [24], the area of seagrass in Indonesia decreased by about 30-40%, and the greatest damage seagrass found in Java. Seagrass beds can be found in most of the waters of the island within the Thousand Islands Marine National Park, like Pramuka Island, Panggang island, Kelapa and Harapan Island. If islands change to tourist resorts and residential, so that damage seagrass easily occur on a large scale. However, this ecosystem might have experienced of the impact from antrophogenic activities of Jakarta, as coral reefs and its associated organism[12]. According to [1], there are some major factors that affect changes in seagrass beds in the region, that are development activities, daily activities on the islands, reclamation and sand backfilling.

According to Kepulauan Seribu Marine National Park Authority (BTNKpS), 12 different types of seagrass found in the waters of Indonesia, seven of which can be found in this marine park [1]. Changes that occur in seagrass can influence the presence of associated biota. The impact of human activities on seagrass influence on the distribution and condition of seagrass, both the density and the area of forest cover, particularly where seagrass close to the port, industry and coastal development [18, 19, 20]. The direct effect is dredging, reclamation beach, pier and aquaculture practices. Indirect impacts such as nutrient, suspended sediments, the removal of coastal vegetation and shoreline change resulted in reduced water clarity and the effect on the decline of seagrass.

Panggang Island that is the location of research is one of the islands located in the District of North Thousand Islands, with an area of approximately 9 hectares. Panggang Island geographically located on 5044'19,47 " S and 106036'01,43"E. and have sloping beach topography with a carpet of seagrass are quite extensive. Panggang island known have a number of very dense human population, it can be seen from the reclamation of the island, in order intended for residential development. Under these conditions, the need for protection of seagrass and can be started with in-depth research on the distribution and diversity of seagrass. This study aimed to analyze the diversity of seagrass in the waters of Panggang Island. The diversity of sea grass distribution illustrated by the percentage of coverage and seagrass density.

2. Materials and Method

2.1. Research Location

This research was conducted in March and April 2016 in seagrass beds ecosystem in Pangang Island, Thousand Islands, Jakarta. There are three observation stations were set based on the level of complexity and the different vegetation density and closing with a view seagrass species constituent (Figure 1).

2.2. Seagrass Sampling

The process of data collection was done systematically by using line transect method by placing quadrant 1m x 1m [3], which begins with a site survey research to look at the condition of vegetation seagrass along the seagrass beds.

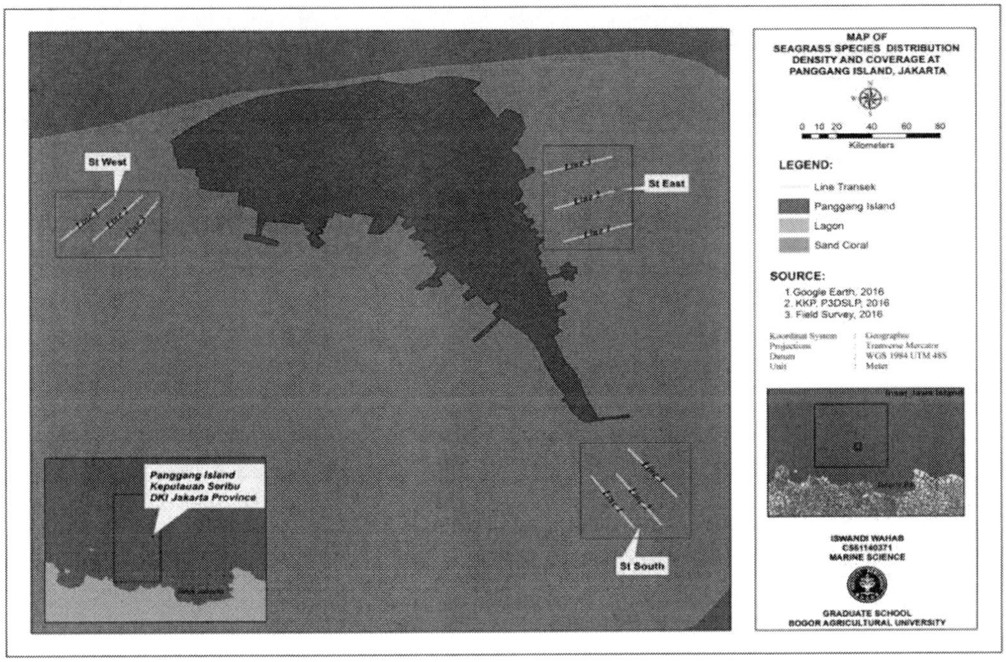

Figure 1. The location of sampling points at seagrass beds in Panggang Island, showing the placement of the line transects

Specify the sampling point on the third observation station that is in the western, eastern and southern of Panggang Island predetermined accompanied by a decision point using GPS coordinates. In every observation location, west, east and south, three replicates of lines transect consists at each station with a 25 m distance between one another. In a transect line consists of five quadrants with each quadrant distance of 20 m. Seagrass found in the square was calculated of the stand and cover by type seagrass.

2.3. Water Quality

Water quality parameters measured such as temperature, salinity, pH, flow, dissolved oxygen (DO), tidal and substrate fractions (sand, silt, clay). Data measurement of temperature, salinity, pH, current speed was done directly in the location, while the measurement of the fraction of the sediment carried in Environmental Laboratory of Aquaculture, Faculty of Fisheries and Marine Science IPB. All parameters were measured as much as 3 repetitions at each station. Analysis of the results of the substrate is inserted into the triangular miller to determine the composition of granular substrate [23].

3. RESULTS AND DISCUSSION

3.1. Water Quality Condition

Water quality conditions of the study sites showed they are in the optimum range for growth of seagrass (Table 1).

The water temperature on the island ranges between 30-31^{0}C. The value of this range is the optimum range in accordance with seagrass life. According [12] the temperature can affect photosynthesis for making nutrients depends on the water temperature. Events temperatures past the optimum value would mislead the seagrass is experiencing stress and eventually die [12].

Salinity waters in the study ranged from 31-32 ppt and is at the optimum value. According [17], the optimum salinity is at a value 31 ppt and 33 ppt. This is according to [15], changes in salinity may affect the spread of aquatic organisms and may indirectly alter the composition of organisms in a body of water.

Table 1. Water Quality Condition at research stations

Parameter	Station 1	Station 2	Station 3	Quality Standards
	Average	Average	Average	
Temperature (^{0}C)	30,93±0,55	30,97±0,49	31,27±0,67	≤ 32
Salinity (ppt)	31,17±0,06	31,5±0,17	31,43±0,12	≤ 34
pH	8,17±0,06	8,37±0,06	8,33±0,06	≤ 8,5
DO (mg/l)	7,7±0,36	8,1±0,10	8,37±0,50	> 5
Flow(m/det)	0,43±0,02	0,47±0,01	0,62±0,02	
Very coarse sand (%)	9,1±1,13	4,67±1,11	11,53±8,20	
Coarse (%)	24,73±4,21	24,53±1,14	24,07±4,80	
Medium (%)	48,37±6,23	57,03±2,11	46,3±17,34	
Fine sand (%)	5,43±1,03	3,97±0,83	4,58±1,02	
Very fine sand (%)	12,13±2,54	9,6±1,10	13,33±3,49	

Note: Quality Standard environment minister's decision No.51 of year 2004

3.2. Seagrass Density

The species of seagrass found in the waters of Panggang island bake consists of six types of seagrass that comes from the two tribes and the five clans. Cymodoceaceae tribe includes genera / species *Cymodocea rotundata*, *Cymodocea serulata*, *Halodule uninervis* and *Syringodium isoetifolium*. While tribal Hydrocharitaceae include; *Enhalus acoroides*, *Thalassia hemprichii*.

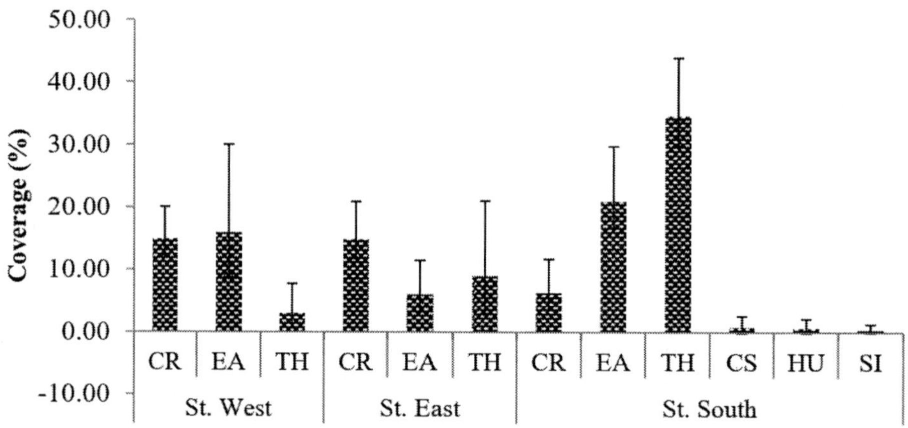

Figure 2. Density of seagrass spesies at each station in Panggang Island (CR = *Cymodocea rotundata*, HU = *Halodule uninervis*, EA = *Enhalus acoroides*, TH = *Thalassia hemprichii*, CS = *Cymodocea serulata*, SI = *Syiringodium isoetifolium*)

Seagrass density value in the western part of the observation station types *Cymodocea rotundata* namely 520 ind /m^2, *Enhalus acoroides* 167 ind /m^2, and *Thalassia hempricii* 30 ind /m^2. For observations in the eastern part of the station, the density of seagrass *Cymodocea rotundata* types namely 483 ind /m^2, *Enhalus acoroides* 36 ind /m^2 and *Thalassia hempricii* namely 95 ind/m^2. While at the station southern part of seagrass types *Cymodocea rotundata* 218 ind/m^2, Type *Enhalus acoroides* 162 ind/m^2, and the type of *Thalassia hempricii* 619 ind /m^2, type *Cymodocea serulata* namely 18 ind/ m^2, Type *Halodule uninervis* of 20 ind/m^2 and *Syiringodium isoetifolium* namely 15 ind/m^2 (Figure 2).

Seagrass density differences caused by several factors referred topography, physical condition, activity of coastal communities around the seagrass beds and seagrass adaptation. Other factors that influence the distribution of seagrass is predation and associated biota [5]. Additionally *Thalassia hempricii* and *Cymodocea rotundata* has a wide coverage area because it can tolerate a wide range of water conditions [22] The highest density of seagrass species for the entire observation stations found in the southern part of the observation station (Figure 2). This is presumably because the substrate in observation southern part is sandy, many types of seagrass *Thalassia hempricii* and away from the influence of local settlement activities. [18], states that the seagrass stand density is influenced by several factors such as the type of seagrass, substrate conditions, seasons, tides, wave energy strength, the content of organic matter in the sediment and other environmental factors.

The existence of seagrass *Cymodocea serulata, Halodule uninervis* and *Syiringodium isoetifolium* found very little and only in the south but the observation station density value is low, this is because all three of these species have the ability to adapt to different environmental conditions, and can not grow dominant. Seagrass adaptability to environmental conditions are very different from one species with other species [10]. In addition, the observations show that in general the seagrass species *Cymodocea rotundata, Thalassia hempricii* and *Enhalus acoroides* is the type commonly found in each observation station.

3.3. Seagrass Coverage

The highest value of seagrass in the western part of the station type *Enhalus acoroides* 15.87%. Seagrass type *Cymodocea rotundata* 14.87% and seagrass type *Thalassia hempricii* namely 3.00%. This is because the type of seagrass *Enhalus acoroides* have morphology great shape either stem, rhizome and leaves so dominant occupying more space than other type. Relating to the closure of seagrass habitat and morphological also the size of the species. One individual *Enhalus acoroides* will have higher closing value compared with one individual *Halodule uninervis* because its leaf size smaller than *Enhalus acoroides* [18].

For the coverage of seagrass in the eastern part of the observation stations are the type *Cymodocea rotundata* has a closing value namely 14.80%, followed by *Thalassia hemprichii* 9% and the type *Enhalus acoroides* namely 6%. The dominance of the seagrass *Cymodocea rotundata* due to a suitable substrate in the form of sand and rubble are peppered throughout these waters.

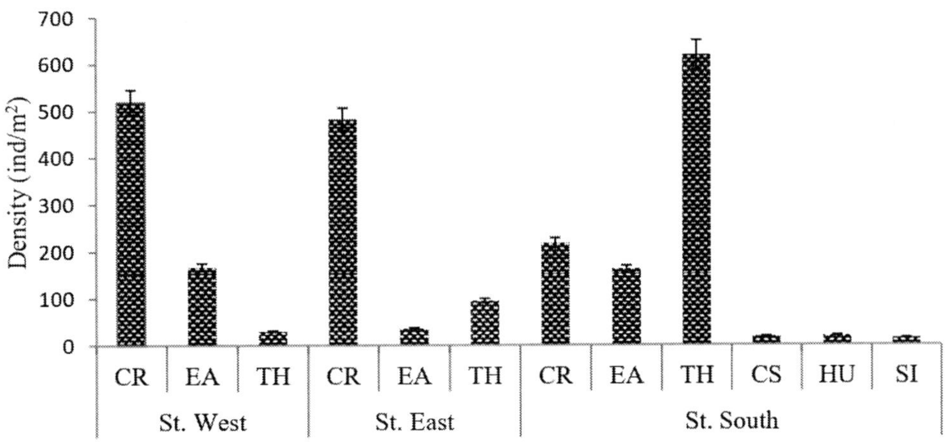

Figure 3. Coverage of seagrass spesies at each station (CR = *Cymodocea rotundata* HU = *Halodule uninervis* EA = *Enhalus acoroides* TH = *Thalassia hemprichii* CS = *Cymodocea serulata* SI = *Syiringodium isoetifolium*)

While *Thalassia hempricii* was observed high coverage (34.60%) at south station, followed by. *Enhalus acoroides* (21.00%), *Cymodocea rotundata* (6.27%), *Cymodocea serulata* (0.80%), *Halodule uninervis* (0.67%), and the lowest *Syringodium isoetifolium* 0.40%. Usually the species composition of seagrass consisting of 4-7 inclined *Cymodocea rotundata, Cymodocea serulata, Enhalus acoroides, Thalassia hemprichii, Halodule uninervis, Syringodium isoetifolium* [12]. Usually these seagrass dominated by *Enhalus acoroides* and *Thalassia hemprichii* [8, 9].

4. Conclusions

The highest number of seagrass species was observed in Panggang Island and they were distributed at all station with high percentage of coverage, with a dominant species *Thalassia hempricii*. This could be basic information for the management of seagrass ecosystem in the Kepulauan Seribu Marine National Park.

5. References

[1] BTNKpS Thousand Islands National Park Office 2005 Inventory of seagrass in the Thousand Islands National Park. Jakarta

[2] De Wilde PAWJ, Kastoro WW, Berghuis EM, Aswandy, Al Hakim, Kok A. 1989 Strucutre and Energy demand of benthic soft-bottom communities in the java sea and around the islands if Madura and Bali Indonesia. *Nether jour. Sea res.* **23** 449-461.

[3] English S, Wilkinson C, Baker V. *Survey Manual for Tropical Marine Resources* 1997 Australian Institute of Marine Science. Townsville.

[4] Gillander, B.M. 2006 Seagrasses. Fish, and fisheries. In: Larkum, A. W. D, Orth, R. J. Duarte, C.M. (Eds). Seagrass: Biology, Ecology and Conservation. Springer, The Netherlands 503-536.

[5] Heck KL., Valentine JF 2006 Plant-herbivore interactions in seagrass meadows. *J Exp Mar Biol Ecol* **330** 420-436.

[6] Kuo, J. Coles. R.G. Lee Long, W, J. Mellors, J.E 1991 Fruits and seeds of Thalassia hemprichii (Hydrocharitaceae) from Queensland, Australia. *Aquant Bot.* **40** 65-174.

[7] Kuo, J. Kirkman, H 1997 Floral and seedling morphology and anatomy of Thalassodendron pachyrhizum den Hartog (Cymodoceaceae) *Aquant Bot* 29, 1-7.

[8] Kawaroe M, Nugraha AH, Juraij. 2016 Seagrass Ecosystem. IPB Press, Bogor.

[9] Kawaroe M, Nugraha AH, Juraij, Tasabaramo I 2016 Seagrass biodiversity at three marine ecorgion of Indonesia, Sunda Shelf, Sulawesi ea and Banda Sea. Indonesia. *Bio Diversitas* 585-591

[10] Keough, M. J., and G. P. Jenkins 1995 *Seagrass meadows and their inhabitants in* A. J. Underwood and G. Chapman, editors. *Coastal marine ecology of temperate Australia*. University of New South Wales Press, Sydney

[11] Minister of State for the Environment. Decree of the Minister of Environment On Sea Water Quality Standard for Marine Life. Jakarta: KEP No 51 / MENLH / I / 2004.8 April 2004.

[12] Marsh JA, Dennison WC, Alberte RS 1986 Effects of irradiance, temperature, and nutrients on growth dynamics of seagrasses: a review. *J of Experiment Marine Biol and Ecol* **101** 257-267.

[13] Madduppa H., Subhan B, Suparyani E, Siregar AM, Arafat D, Tarigan SA, Alimuddin, Khairudi D, Rahmawati F, Bramandito A 2013. Dynamics of fish diversity across an environmental gradient in the Seribu Islands reefs off Jakarta. *Biodiversity,* **14** (1):17-24 DOI: 10.13057/biodiv/d140103.

[14] Nienhuis PH, Coosen J, Kiswara W 1989 Community Structure and Biomass Distribution of Seagrass and Macrofauna in The Flores Sea, Indonesia. *SeaResearch*. **23**:197-214.

[15] Odum EP. *Dasar-dasar Ekologi Edisi Ketiga*. Gadjah Mada University Press. Yogyakarta

[16] Rasheed, A.M. 1999 Recovery of experimentally created gaps within a tropical Zostera capricorni (Aschers). Seagrass meadow, Queensland, Australia. *J. Exp. Mar. Biol. Ecol.* **235** 183-200.

[17] Rattanachot E, Prathep A 2011 Temporal variation in growth and reproduction of *Enhalus acoroides* (LF) Royle at Haad Chao Mai National Park, Trang Province, Thailand. *Botan Mar* **54** 201-207.
[18] Short FT, Coles RG, (eds) 2003 *Global Seagrass Research Methods.* Amsterdam: Elsevier Science BV
[19] Short, F,T. Wyllie-Echeverria, S 1996 Natural and human-induced disturbances of seagrass. *Environ. Consery* **23** 17-27
[20] Short FT & Coles RG 2001 Global Seagrass research methods. Elsevier Science BV. Amsterdam
[21] Spalding, M., Taylor, M., Ravilious, C., Short, F, Green, E. 2003 Global overview; The Distribution and status of seagrasses. In: Green, E.P. Short. F.T. (Eds), World Atlas of seagrasses. University of California Press, Berkeley
[22] Tomasick T, Mah AJ, Nontji A, Moosa MK 1997 *The Ecology of the Indonesia Seas, Part One Periplus Edition*, Singapore.
[23] USDA United States Departement of Agricultural 2009 *Soil Survey Manual.* New York (US): Soils USDA Gov
[24] Vonk JA, Christianen MJA, Stapel J 2010 Abundance, edge effect, and seasonality of fauna in mixed-species seagras meadows in Sout-West Sulawesi, Indonesia. *Mar.Biol.Res.***6**: 282-291.
[25] Vo ST, Pernetta JC, Paterson CJ 2013 Status and trends in coastal habitats of the south China sea. Ocean Coast Manag **85** 153-163.
[26] Waycott M, Mahon KM, Mellors J, Calladine A, Kleine D 2004 *A Guide to Tropical Seagrass of The Indo-West Pacific*. Townsville-Queensland Australia; James Cook University
[27 Waycott M, Duarte CM, Carruthers TJB, Orth RJ, Dennison WC, Olyarnik S, Calladine A, Forqurean JW, Heck KL Jr, Hughes AR, Kendrick GA, Kenworthy Wj, Short FT, Williams SL 2009 Accelerating loss of seagrasses across the globe threatens coastal ecosystems. *Proc Natl Acad Sci USA* **106**;12377-12381.

IOP Conf. Series: Earth and Environmental Science **54** (2017) 012085 doi:10.1088/1755-1315/54/1/012085

Analysis of upwelling event in Southern Makassar Strait

F G Utama[1*], A S Atmadipoera[1], M Purba[1], E H Sudjono[2] and R Zuraida[2]
[1]Department of Marine Science and Technology, Bogor Agricultural University, Indonesia
[2]Marine Geological Institute, Bandung, Indonesia

E-mail: ferdy_37523@gmail.com

Abstract. The southeast monsoon (SEM) winds which blow in southern Makassar Strait, generate the coastal upwelling phenomenon. The wind data for one year, which is equipped with CTD data from MAJAFLOX cruise results, is used to analyze the phenomenon of upwelling in this region. During the SEM 2015 occurrence, the southeasterly winds speed were in average of 6 m/s, while the highest speed appeared in August and September. Using the Ekman theory's analysis of upwelling during this monsoon period, we could estimate the Ekman transport was about 8.50 m^2/s toward offshore (to the Southwest direction); the upwelled water, occurred from deeper layer, started from the coastal area with vertical velocity was about 6.87 x 10^{-5} m/s – 7.84 x 10^{-5} m/s; and The Ekman layer depth in the upwelling region was approximately 60 m and these were good agreement with CTD observation result.

1. Introduction

Upwelling is an oceanographic phenomenon that involves the rising of water mass at lower temperature which comes from the water column towards the surface layer to replace the surface water mass with higher temperature. This phenomenon will affect the water mass properties distribution; such as temperature, salinity, dissolve or undissolve compounds, and nutrient on the surface water column [1][2][3][4][5][6]. Previous studies have informed that there are some areas in Indonesia's seas that indicate the existance of upwelling phenomena; such as the southern waters of Java [7], the southern waters of the Strait of Makassar [2][4][8][9][10], Banda Sea waters, and the western Water of Sumatra [5].

Southern Makassar strait (SMS) is one of the region which its upwelling occurs annually during southeast monsoon period [2][4][8]. Upwelling phenomenon can be noticed from the decreasing of sea surface temperature (SST) in this region, especially in August – September with the range of temperature between 24 °C – 30.34 °C [4][10][11]. It will increase the amount of chlorophill-A in surface layer [1] also the productivity of this region. The number of chlorophill-A in SMS area when southeast monsoon occurs are 0.00 mg/m^3 – 1.144 mg/m^3 [1]. Meanwhile, the concentration of chlorophyll-A during southeast monsoon is in the range of 0.4 mg/m^3 - 0.7 mg/m^3 [12].

Makassar strait is also the main path of the ITF which passes through Indonesian waters. The ITF stream that passes through the Makassar strait is the stream that has followed the pattern of topographic depth. In southern part of the strait of Makassar (SMS region) show tthe distribution of flow. Most of the stream flow into the canal in the west, while some of it flow into the canal in the deeper eastern part. In southeast monsoon circumtances, the flow velocity in the area of SMS on surface layer is affected by the reversal of the direction of the wind [13]. This causes the surface current on this area strengthened this period.

Southeast monsoon winds that blow in the SMS area become the factor that evokes an upwelling. The wind blowing will generate a water mass transport known as Ekman transport. Generally, upwelling phenomenon in coastal area is formed by Ekman dynamic process in the surface layer [14]. The dynamics of the surface layer are due to the influence of the wind that will cause the displacement of water mass that influenced by the direction of Coriolis factor. While transport away from land boundary, according to the theory, there will be a balance from transport that leaves the boundary. This process will induce a lifting on water mass and it is known as upwelling. In the other hand, it will make a downwelling if the water mass is accumulated in land boundary which is caused by surface transport. In general, the dynamic caused by wind blowing in surface layer--will form a transport process known as Ekman transport and Ekman pumping [15].

Utilization of satellite imagery data, such as imagery SST and chlorophyll-A, is commonly occured in studying the phenomenon of upwelling in the SMS region. For complete information on upwelling in this region, the analysis of the Ekman dynamics that occur in the region needs to be done. This study proposes to complement the information about the phenomenon of upwelling by using the Ekman dynamic approach through transport value, Ekman depth, and the condition of the Ekman vertical speed when the occurrance take place.

2. Methods

Temperature data were collected from the field observation in August 2015 conducted by MAJAFLOX cruise. MAJAFLOX Cruise is a research activity conducted by multi institutionl i.e IPB, P3GL, and BPPL. Temperature data were recorded using CTD (Conductivity, Temperature, Depth) SBE V19plus version with 4Hz frequency rate. There were 27 stations fully recorded and these stations were divided into three main; i.e Java sea zone, southern Makassar Strait zone, and Flores Seas zone. This study used observation data station which located in southern Makassar strait (station 17 - station 23) (figure 1a). The data will show in cross section profile. Sea surface temperature (SST) is also provided to support the analysis, especially spatial distribution of surface temperature. This based on spesific time period downloaded from INDESO homepage (http://www.indeso.web.id/indeso_wp/index.php). Spatial resolution of data was about 0.02° x 0.02° from May to october 2015. The data was a composite result from four satellites that carry SST recording sensor. The wind data was downloaded from ECMWF homepage (http://apps.ecmwf.int/dataset/data/interim-full-daily) with spatial resolution 0.125° x 0.125°. Wind U10 component and wind V10 component were used in this studyas well as Wind data from May to october of 2015. The daily average data that were used are from six –hourly data. Both data (SST and Wind data) region are from 4S – 7S and 117E – 121E.

Surface layer dynamic analysis when upwelling occurred in the area of interest, is divided into three calculations. The first one is Ekman transport, the second one is vertical speed and the last one is Ekman depth. Ekman transport calculation is based on wind drag value (τ_n), homogenous water mass density (ρ), and coriolis parameter effect (f) [4][16] expressed in equation (1).

$$M_{yE} = -\frac{\tau_x}{\rho f};\ M_{xE} = \frac{\tau_y}{\rho f} \quad (1)$$

Where M_{yE} is Ekman transport perpendicular coastline (cross shore transport) and M_{xE} is Ekman transport alongshore coastline (alongshore transport). M_{xE} is the value that used to indicate transport magnitude while upwelling occurs. Ekman depth (D_E) indicate the depth below the surface layer which the wind is still influenced by Ekman dynamic. Ekman depth (D_E) can be presented in equation (2).

$$D_E = \frac{4.3\,W}{\sin|\theta|^{\frac{1}{2}}} \quad (2)$$

Where theta (θ) is position of latitude. The vertical speed (w_E) on Ekman layer was calculated by equation (3).

$$-\rho\, w_E = \frac{\delta M_{xE}}{\delta x} + \frac{\delta M_{yE}}{\delta y} \quad (3)$$

Vertical speed (w_E) indicates convergence (Ekman pumping) that is associated with downwelling if the value are plus and also indicates divergence (Ekman suction) that is associated with upwelling if the value are minus. New grid (1/3° x 1/3°) is used to calculate the number of average vertical speed by using an approach of vertical speed value inside the new grid (figure 1b).

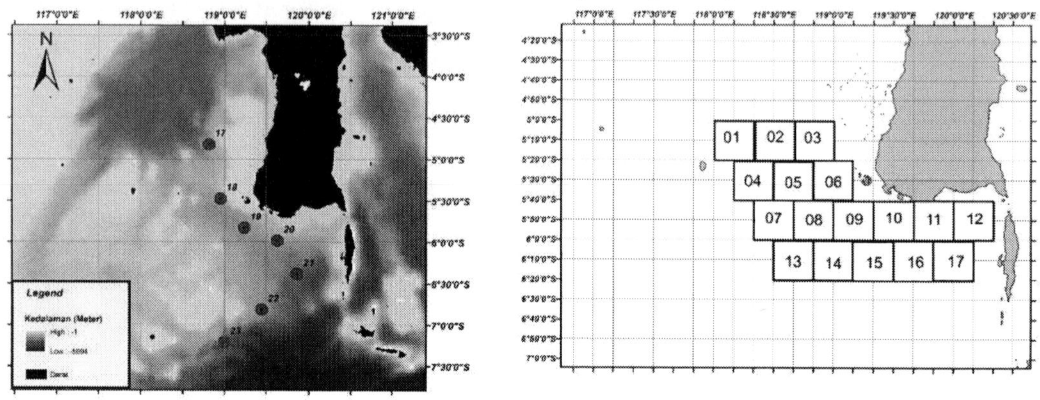

(a) (b)

Figure 1. (a) Station point of CTD Cast in MAJAFLOX Cruise 2015 over SMS area; (b) Area calculation of Ekman transport divided into 17 grid area. Resolution of grid area used 1/3° x 1/3°

Ekman transport estimation that describes the total number of transport average is calculated by integrating the dimension of length (distance) of the area of upwelling region from the distribution of SPL in August 2015. The size of the region of upwelling in question is the extent of the territory which has SST with 26°C isotherm. Determination of isotherms used in determining the area of upwelling is based on the distribution of temperature of the water column that looks raptured during the period of August, 2015 (figure 2).

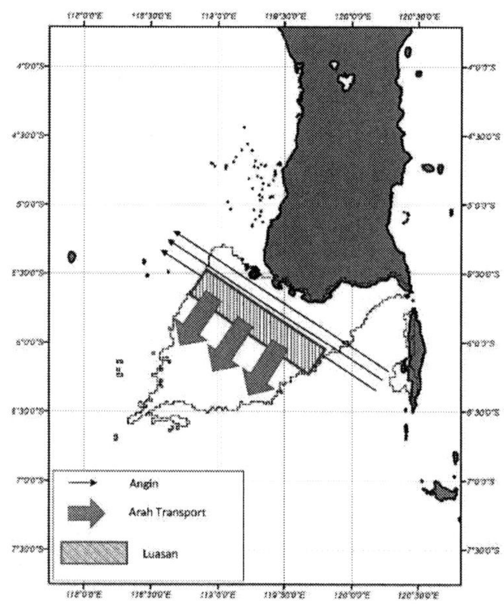

Figure 2. Calculation of transport based on isotherm 26. Integration of transport value along line A-B used to calculate total transport while upwelling occur

3. Results

SST distribution in the southeast monsoon period ranges in temperature between 24.22°C - 30.80°C. SST distribution in the southeast monsoon period from May to October shown the presence of SST low temperature (isotherm 24°C to isotherm 26°C) rather than the SST in the surrounding area. The existence of a low temperature SST looked expanding every month. The formation of the low SST detected came from the south coast of the mainland of southern Sulawesi and was expanding its range consistently until the period of September, which the beginning period was in the month of May.

The lowest surface temperature in May was 25.90°C. It was detected on southern region of Sulawesi (120.03 E, 5.55 S). The average of surface temperature for the entire water in research area was about 28.90°C. The temperature decreased on June up to 25.40°C with average surface was 28.16°C. On July, the surface temperature decreased continuously with minimum temperature was 24.32°C. Average surface temperature was 27.11°C. In August, the lowest surface temperature recorded was 24.43 °C and average of surface temperature was 26.70°C. Decreasing surface temperature still recorded in September. Lowest temperature in September was 24.22 °C with average surface temperature was 26.74°C. Temperature decreas stopped on October. Surface temperature increased 0.89 with average surface temperature was from 24.70°C - 30.26°C. The distribution of low surface temperature at southeast monsoon periode was oriented to southwest side of south Sulawesi.

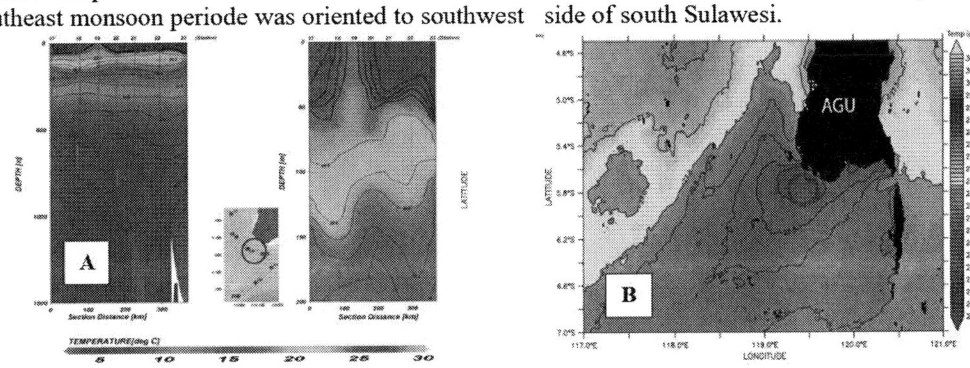

(a) (b)

Figure 3. (a) Cross section profile of temperature from MAJAFLOX station data (St 17 - St 23); (b) Red circle shows the area of low temperature develop in southeast monsoon periode. This is associated with water mass lifted in station 19 MAJAFLOX.

Cross section profile of temperature data from MAJAFLOX 2015 observation showed the presence of water mass lifting in low SST formation region. This is indicated by the increase of isotherm 24.5 to isotherm 26 toward surface layer (figure 3). Observation of the southern of Makassar Strait began at station 17 and it ended at station 23. The mixed layer of each station ranges from 30 meters to 50 meters below the surface waters with temperatures range from 24.00°C up to 27.50°C at the surface. Mixed layer in this area is strongly influenced by the pattern and strength of the wind that blows in the surface layer. Therefore, on the temperature transversed distribution results from stations 17 to 23, it showed the increase of isotherm 26.00 °C, isotherm 25.50°C, isotherm 25.00°C, and isotherm 24.5°C to the surface. This increase was clearly identified at station 17 to station 21, with the center of the appointment was in the station 19.

In the observation area, Southeast monsoon winds would form the dynamics of the surface layer commonly called Ekman dynamics. Ekman dynamics in the surface layer were composed of the surface Ekman transport and Ekman vertical velocity which could be associated with the occurrence of pumping Ekman (Ekman pumping/downwelling) and the increase of Ekman (Ekman suction/upwelling). For calculating the Ekman dynamics that affected the surface layer, Ekman depth were needed. The Ekman depth distribution in the southeast monsoon period is described as follows.

IOP Conf. Series: Earth and Environmental Science **54** (2017) 012085 doi:10.1088/1755-1315/54/1/012085

Southeast monsoon with a faster wind caused the formation of a deeper depth due to Ekman depth which is a function of latitude and wind speed. Areas on the eastern side of the Java Sea with the latitude of 5.4S to 6.4S is the region with the deepest Ekman layer compared to other water areas. Ekman depth in this region has reached 90 meters in depth with a pattern of widespread deployment towards the west side of the Flores Sea. Ekman depth of maximum territorial expansion in the southeast monsoon period continued along with the months and the maximum depth reached in August-September.

In the southern of Makassar Strait, Ekman layer depth in the southeast monsoon season ranged from 39.00 meters to 81.00 meters. In the southeast monsoon period, as well as in western monsoon period, it appeared that Ekman depth decreased as it approaches the mainland (coastal). In the waters near the southern of mainland southern Sulawesi, Ekman depth in this period are at a depth of 42.00 to 72.00 meters below the surface.

As in the period of upweling occurrence in southeast monsoon period, also the indication of the upweling formation region which was located in the waters near the southern mainland of Sulawesi (position 5.5S -5.7S 119.3E-119.6E), Ekman depth conditions in this area varies between a depth of 40.00 meters to 75.00 meters below sea level. This is linear with the detection of the mass removal of water from vertical profiles of temperature (figure 3) which indicates the existence of a mass removal of water from depths ranging from 45 meters to 60 meters to the surface. This information suggested that the existence of upwelling that occurs in the southern Strait of Makassar is caused by the wind which then form a pattern of Ekman transport to the surface waters.

Vertical speed at the observation area was calculated with the new grid area, as described in the method. Vertical speed when the monsoon period in the southeast region of observation is predetermined in the graph below.

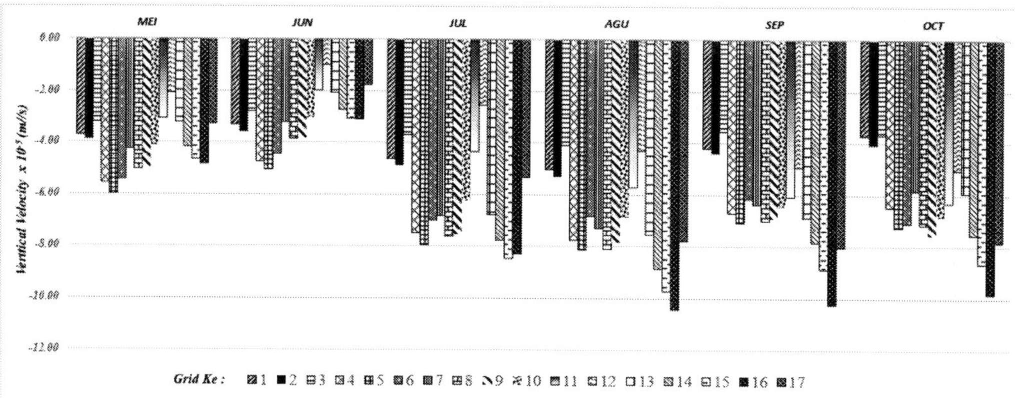

Figure 4. Vertical speed distribution while southeast monsoon. While southeast monsoon period, vertical speed gave negative value. That indicates the Ekman suction wasoCcur on this period. This phenomenon is associated with upwelling

The value of the vertical velocity of the grid element was calculated and shown a negative value. This indicates that during the period of monsoon southeast, the Ekman increase (Ekman suction) which are further associated with upwelling occurred throughout the period. Figure 6 indicates that in August and September, the average Ekman vertical velocity was increased. In May, the average vertical velocity of the Ekman grid observation reached 4.2 x 10^{-5} m/s. In August, the average vertical speed of Ekman was 7.15 x 10^{-5} m/s and it was maximum. Vertical speed of Ekman reached 6.56 x 10^{-5} m/s. In october, Ekman vertical velocity at observation grid average decreased in compare to september period. Ekman vertical velocity in october was at 6.51 x 10^{-5} m/s. On grid 09 (G09) as well as grid 10 (G10) as an area that represents the area of the formation of upwelling based on the distribution of surface temperature (SST) in the period of the monsoon southeast region observations which show the value of vertical speed during periods of maximum incidence of upwelling (august - september) that reached 6.87 x 10^{-5} m/s (G10) and 7.84 x 10^{-5} m/s (G09).

Ekman transport value that crossed coastline south Sulaweasi is the value of transport used in indicating and determining the upwelling region of the periodic monsoon southeast. The results of Ekman transport crosses the observation area in the southeast monsoon period indicates a minus value. This indicates Ekman cross transport to leave the beach during southeast monsoon period.

The distribution of Ekman transport which left shoreline in the area of observation during the southeast monsoon period indicated that the south area of the mainland of southern Sulawesi namely the position 6.4S - 6.6S and 118.0E - 120.0E is a region which has a maximum value of transport (figure 7). The maximum value of transport seem always to be in the region of the difference lies in the intensity of its transport. Intensity patterns of transport shows that during the period May to August, the intensity of transport has increased is continuous where in May, transport maximum detectable reached 8.60 m^3/s, and then continue to grow until the peak period in August the intensity of transport reaches 14.00 m^3/s. Entering the period september, the intensity of the maximum transport decreased slightly. Transport maximum value in this period in the range of 13 m^3/s to 13.5 m^3/s. Entering the month of October, the value of transport again decreased. Transport a maximum of 13 m^3/s with a distribution area that get maximum transport is not covering the previous period.

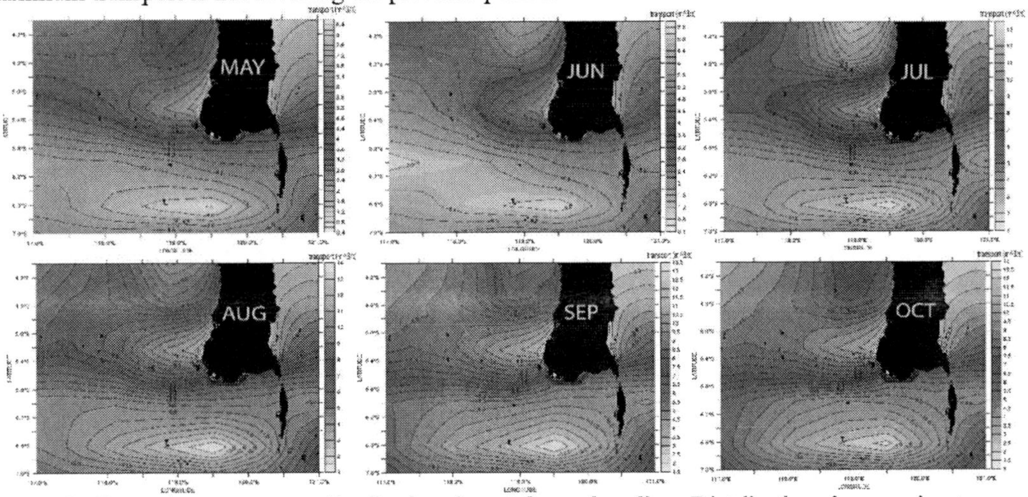

Figure 5. Cross shore transport distribution depart from shoreline. Distribution show on iso transport line contour that flow oriented more southwest.

In areas near the coast of southern Sulawesi which is based on the hue, SST suggest that the region as a region of mass removal of water which is marked by the decline in SST in the region, the transport value generated on the southeast monsoon period showed an increase as a common trait of the study area. The initial period of the southeast monsoon demonstrated in this area of transport ranging from 3.00 m^3/s – 4.50 m^3/s. Then the increas continues until in August the value of the transport reaches 8.50 m^3/s. The direction of movement of transport seemed to south - southwest of mainland southern Sulawesi.

4. Discussion and Conclusion

The SST distribution and vertical distribution conditions of variable temperatures in the period of August 2015 is the culmination of the southeast monsoon conditions in the region of observation, true upwelling events happening in the area. Decrease the temperature with a difference of nearly 3.00 °C be a strong indicator the mass removal of water from the water column to the surface. The most influential factor for this incident is the result of southeast monsoon winds that pass over the surface of the observation area thus causing the mass transport of the water leaving the beach.

Ekman depth value is based on the area around the site thought to be the point of appointment of the mass of water (stations 19 MAJAFLOX), Ekman depth value reaches 40 meters to 75 meters below the surface. This corresponds with the results of the cross section profile of temperature from stations 17 to station 23 indicating the mass removal of water from the water column at a depth of around 60 meters. This clearly illustrates that Ekman depth value which is a function of the wind blowing on the surface has a value conformity with the fact the mass removal of water through the results of cross section temperature profile.

Reviewing Ekman vertical velocity value associated with a value that indicates the speed of the moving mass of water column by depth changes due to events mass drownings or removal of the water, on the southeast monsoon period average vertical velocity shows a minus value. Minus value of the vertical velocity calculations indicates a phenomenon appointment Ekman Ekman upwelling is associated with the presence in the area. Ekman vertical velocity value in the region of the alleged establishment of upwelling area is at 6.87 x 10^{-5} m/s to 7.84 x 10^{-5} m/s.

Furthermore, based on transport leaving the beach caused by the Ekman process, transport distribution value indicates a negative value for the entire observation area. This indicates that the average transport that occurs is to keep off from the shore with the amount of different transport between regions. The closer to the land, the smaller transport value given. The total value of transport in the area bounded by isotherm 26 in SST were 175.5 m^3/s with the direction was to keep off the shoreline.

Acknowledgments
We would like to give thank to P3GL ESDM RI, IPB, and BPPL for the opportunity we could participate and to become part of the research team and a cruise MAJAFLOX 2015. Thanks also conveyed to the captain and all the crew and the crew of the research vessel GEOMARIN III for their help and dedication so that activities of MAJAFLOX could go according to plan. Agus Atmadipoera was funded by PUPTN research grant from Kemenristek Dikti (contract no. 714/IT3.11/PN/2016).

References

[1] Afdal, Riyono H 2004 Sebaran Klorofil-a Kaitannya dengan Kondisi Hidrologi Selat Makasssar *Oseanologi & Limnologi Indonesia* **36** 69-82
[2] Nababan B, Rosyadi N, Manurung D, Natih NM, Hakim R 2016 The seasonal variability of sea surface temperature and chlorophyll-a concentration in the south of Makassar Strait *Procedia Env. Sci.* **33** 583-599
[3] Dakui W, Hui W, Ming L, Guimei L, Xiangyu W 2013 Role of Ekman Transport Versus Ekman Pumping in Driving Summer Upwelling in South China Sea *J.Ocean Univ.China.* **12** 355-365
[4] Illahude A G 1970 On the occurence of Upwelling in the Southern Makassar Strait.
[5] Kemili P, Putri M R 2012 Pengaruh durasi dan intensitas upwelling berdasarkan anomali suhu permukaan laut terhadap variabilitas produktivitas primer di Indonesia *JITKT* 4 66-79
[6] Uibopin R, Laanemets J 2009 Upwelling Characteristic Derived from Satellite Sea Surface Temperature data in the Gulf of Finland, Baltic Sea *Boreal Envi. Research* **14** 297-304
[7] Purba M 2007 Dinamika perairan selatan Pulau Jawa – Pulau Sumbawa saat monsun tenggara *Torani* **17** 140-150
[8] Atmadipoera A S, Widyastuti P 2014 A Numerical Study of Upwelling Mechanism in Southern Makasssar Strait. *JITKT* **6** 355-371
[9] Inaku D F 2011 *Analisis Pola Sebaran dan Perkembangan Area Upwelling di bagian Selatan Perairan Selat Makassar* thesis (Bogor: Bogor Agricultural University)
[10] Wyrtki K 1961 Scientific Result of Marine Investigation of the South China Sea and the Gulf Of Thailand. La Jolla, California: Scripps Institution of oceanography *NAGA REPORT* **2**
[11] Gesteira M G, Moreira C, Alvarez I, deCastro M 2006 Ekman transport along galacian coast (Northwest Spain) calculated from forecasted wind *JGR* **11**
[12] Illahude A G, Gordon A L 1996 Thermocline stratification within the Indonesian seas *JGR* **101** 12401 – 12409
[13] Atmadipoera A S, Horhorouw S M, Purba M, Nugroho D Y 2016 Spatial and Temporal Variation of Indonesian Throughflow in the Makassar Strait *JITKT* **8** 299-320

[14] Akhir M F, Daryabor F, Husain M L, Tangang F, Qiao F 2015 Evidence of Upwelling along Panisular Malaysia During Southeast Monsoon *Open Journal of Marine Science* **5** 273-279
[15] Dakui W, Hui W, Ming L, Guimei L, Xiangyu W 2013 Role of Ekman Transport Versus Ekman Pumping in Driving Summer Upwelling in South China Sea *J.Ocean Univ.China.* **12** 355-365
[16] Smith R L Upwelling oceanography marine biology *Annual Rev.* **6** 11-46

Analysis of turbulent mixing in Dewakang Sill, Southern Makassar Strait

Risko[1*], A S Atmadipoera[2], I Jaya[2] and E H Sudjono[3]
[1] Graduate Program in Marine Science, Bogor Agricultural University, Indonesia
[2] Marine Science and Technology Dept. Bogor Agricultural University, Indonesia
[3] Marine Geological Institute, Bandung Indonesia

Email: risko.fisika4@gmail.com

Abstract. Dewakang Sill is located in southern Makassar Strait, conveying major path of Indonesian Throughflow (ITF), as a confluence region of different water masses, such as salty Pacific water and fresh Java Sea water. Its depth is about 680 m which blocks the ITF flow below this depth into Flores Sea. This research aimed to estimate turbulent mixing in the Dewakang Sill by applying Thorpe analysis using 24 hours "yoyo" CTD data sets, acquired from MAJAFLOX Cruise in August 2015. The results showed that stratification of water masses is dominated by Pacific water origin. Those are North Pacific Subtropical thermocline and Intermediate water masses. Mean dissipation of turbulent kinetic energy (ε) and turbulent vertical diffusivity (K_ρ) value in the Dewakang Sill are of $O(1.08 \times 10^{-6})\mathrm{Wkg}^{-1}$, and $O(2.84 \times 10^{-4})\ \mathrm{m^2s^{-1}}$ respectively. High correlation between epsilon and internal waves oscillation suggested that internal tidal waves activities are the major forcing for turbulent mixing in the study area.

1. Introduction

Makassar Strait conveys major Pacific-Indian transfer (ITF) with mean transport volume of about 9-11 Sv (1 Sv = $10^6\ \mathrm{m^3s^{-1}}$) [1, 2]. In southern part of Makassar Strait, there exists the Dewakang sill waters which is located in eastern channel of Southern Makassar Strait. This sill functions as a barrier of Indonesian Throughflow (ITF) water mass below the sill depth of about 680 m [3]. However, in the western chanel, part of the ITF flows into Lombok Strait and the rest continues flowing into Flores Sea [2, 4, 5].

Theoretically, an unstable stratification of water column leads fluids in the mixing process, and they are grouped into two parts i.e. static stability and double-diffusion [6]. This shows that the static stability density changes with depth while the dynamic stability changes with velocity shear. Furthermore, this double-diffusion is related to the salinity gradient and seawater temperature. Water mass movement which is caused by the variation of turbulent flow can form a fluid mixing with a very high fluctuation, and the process of coating or stratification of the water mass is influenced by differences in ocean temperature, salinity and density. The process of water mass stratification is caused by density differences which can create the mixing. This, according to Ffield [7], causes a change in the amount of heat, salinity, and water mass momentum.

Ffield and Gordon [8], estimated the value of the mixing in the thermocline layer in the Indonesia seas, and the value of $1.0 \times 10^{-4}\mathrm{m^2s^{-1}}$ was obtained. The value of the mixing coefficient is nearly the same as the result obtained by Koch-Larrouy [9] in which she conducted the research in the waters of Indonesia, i.e. $1.5 \times 10^{-4} m^2 s^{-1}$. Furthermore, the same research was also conducted by Hatayama [10] using a numerical modeling that produced a vertical diffusivity value in Dewakang sill, i.e. $6.0 \times 10^{-3} m^2 s^{-1}$. An approach with an indirect measurement was also carried out by Suteja [11] in Ombai Straits and Purwandana [12] in Alor Straits, where the avaraged turbulent mixing values obtained were $7.56 \times 10^{-3} m^2 s^{-1}$ and $1.0 \times 10^{-3} m^2 s^{-1}$, respectively.

Some of these approaches produced various turbulent mixing values. Therefore, it is necessary to apply another approach for calculating the turbulent mixing in Dewakang Sill that is by using Thorpe analysis method based on the vertical profile obtained from the hydrographic data (CTD). This research was carried out by field experiment in Dewakang Sill, southern Makassar Strait where the location is part of the Arlindo region, known to have strong internal tides and energetic mixing region [9, 13].

Turbulent mixing that occurs in Indonesia seas is primarily caused by the internal tidal, and one way to determine the internal tidal is by measuring the CTD data for 24 hours (one tidal period) or commonly called yoyo CTD [14, 15]. The aims of this study are to analyze stratification and characteristics of the Indonesian throughflow water masses; to estimate of turbulent mixing of water masses, and to investigate physical process influencing turbulent mixing, such as surface wind stress, internal tides and sea topographic slope in Dewakang Sill, that contributes to the transformation of water masses.

2. Method

2.1. Data aquisition

This research was conducted in Dewakang Sill, South Makassar Strait (figure 1). The study on turbulent mixing analysis is part of the Makassar - Java - Flores (MAJAFLOX) cruise which was held from August 4 to August 20, 2015 by using the Research Vessel of Geomarin III of the Marine Geological Institute (P3GL), Balitbang ESDM Bandung.

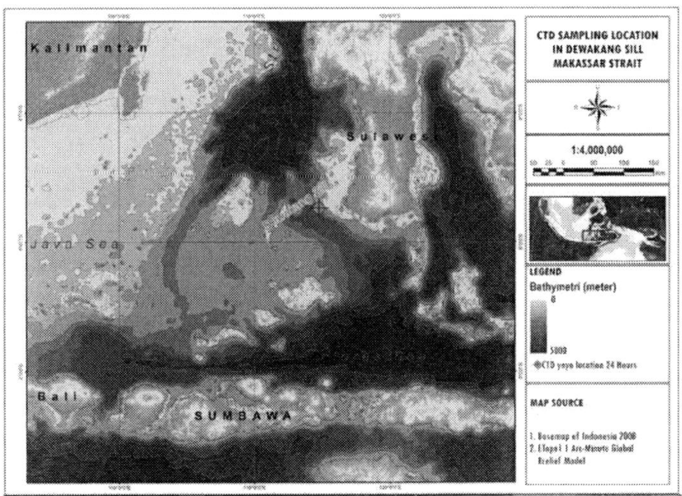

Figure 1. Bathymetry in the triangle seas of Java Sea – Makassar Strait – Flores Sea. Green dot denotes location of repeated CTD "yoyo" measurement for 24-hours in the Dewakang Sill Strait from August 10 to 11, 2015.

MAJAFLOX expedition activity is part of a research collaboration between the Department of Marine Science and Technology, Faculty of Fisheries and Marine Science of IPB and P3GL-ESDM.

The data of temperature, salinity and density were obtained by using the SBE (Sea Bird Electronic) CTD 19plus Version 2 for 11 casts repetition during 24-hours. The results of the data measurement were recorded on deck unit in the form of an analog signal which is then converted by CTD probe into a digital signal by connecting it directly to a computer with a data cable. The determination of the averaged depth of each parameter is 0.5 m obtained by using the software of SBE Data Processing 7.21e. Accuracy and resolution of the temperature sensor is 0.005°C and 0.0001°C, respectively, and for conductivity sensor is 0.0005 S/m and 0.00005 S/m, respectively [16].

Winds field data were obtained from the ECMWF (European Centre for Medium-Range Forecast) through the site of (www.ecmwf.int). The data are the result of repeated analysis and meteorological data interpolation which were obtained from various centers of observations and research on the world meteorological stations. The data used in this study include the wind vectors obtained from 10 to 11 August 2015 consisting of zonal (u) and meridional (v) components 10 m above sea level. Tidal data in Dewakang Sill from August 10 - 11, 2015 were obtained from website of Indonesian Agency for Geospatial Information (www.tide.big.id) at the point of the study area. The data which include the observations, predictions and model of the sea from the control link center of geodesy and geometrics are taken at the point of the research station.

2.2. Data analysis

The water mass stratification in the water column can be seen from the Brunt-Vaisala frequency (N^2) value. This is influenced by the gravitational force (g), which will cause fluids with a greater density to move downward and those with the smaller density to move upward. The calculation method of vertical mixing of water masses from the inner layer of the waters is a preliminary analysis used by calculating the Brunt-Vaisala frequency using the equation [17, 18] (1)

$$N^2 = -\frac{g}{\rho_0}\frac{d\rho}{dz}. \tag{1}$$

According to Ferron [19] the density value used in the Brunt-Vaisala frequency is derived from the density data compiled in a stable condition, where ρ_0 shows the average water density of the whole CTD data repetition (1.026.52 kg m^{-3}), $d\rho$ is the density change (gradient) density toward the depth change (1 m), and g is the earth gravitational acceleration (9.79423 m s^{-2}).

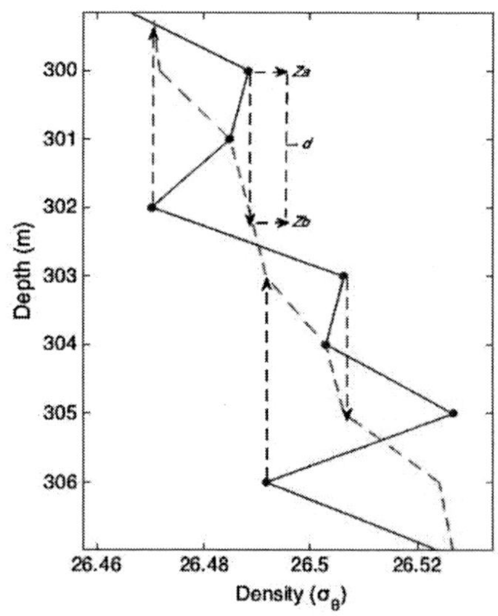

Figure 2. The schematic Thorpe displacement (d) calculation. The initial density profile (blue line) are reordered to seek for the density of the static stability condition (red dotted line). Distance from z_a to the depth of z_b is stated as a Thorpe displacement (d) value

The estimated value dissipation of turbulent kinetic energy (K_ρ) was conducted using the Thorpe scale analysis. Thorpe is determined from the density repetitions in the form of static stability in accordance with the initial density or depth (Figure 2). Thorpe displacement (*d*) can be calculated by the equation (2).

$$d = z_a - z_b \tag{2}$$

Where z_a and z_b are the initial pressure position and pressure after the repetitions [18, 20, 21]. The *d* positive value indicates that the water mass will move upward to search for the static stability in which this occurs if the water mass which has a low density is under the water mass with high density. Meanhile, the *d* negative value indicates that the water mass moves downward.

After calculating the *d* value, the minimum thickness displacement of CTD vertical resolution was estimated. The reason for this is that the *d* value becomes the real displacement value and does not come from the CTD noise. The principle of this estimation is based on the fact that CTD has a limited ability to detect inversion of water mass. This refers to the Nyquist sampling theory, where if a inversion occurring is two times lower than the vertical resolution, the inversion cannot be measured. Determination of a stronger reversal can be done if there are adequate samples in which based on the regulation, the number of samples must be at least five samples [22] or 7-8 samples [23].

Furthermore, the determination of the threshold of the reversal value of the data that have been replicated was carried out by using the GK method [23]. Based on this method, a minimum value (5 m) will be ignored and excluded for further calculations. Interval of the vertical depth (∂z) of CTD data is 1 meter to support the reversal of water mass threshold i.e. $L_z = 5\partial z$. The determination of the turbulent kinetic energy dissipation is used to describe the amount of kinetic energy that is lost or its form is changed in the waters. The calculation of the amount of kinetic energy experiencing a dissipation process based on the Ozmidov scale (1965) in Part [17] is

$$\varepsilon = L_O^2 N_i^3 \tag{3}$$

where L_o and ε denote the Ozmidov length scale and dissipation of turbulent kinetic energy. Furthermore, the determination of Thorpe scale was obtained by using an equation (4)

$$L_T = \left(\frac{1}{n}\sum_{i=1}^{n} d_i^2\right)^{1/2} \tag{4}$$

where d_i is the Thorpe displacement at the depth of *i* and *n* is the number of samples [17, 20, 21, 24]. Each L_T value was obtained from the average of *n* samples in the desired depth. The average L_T value of this research was obtained by dividing the depth of the waters into three layers including the mixed layer, thermocline layer, and homogeneous layer of the inner part. The depth of each layer varies depending on the vertical profile of water masses.

Thorpe scale was used in determining of the value of Ozmidov length scale at each layer and the calculation was carried by using an equation [20] (5)

$$L_o = 0.8 * L_\tau \tag{5}$$

The turbulent vertical diffusivity value at each depth was subsequently obtained by applying the equation (6)

$$K_{\rho i} = \frac{\Gamma \varepsilon_i}{N_i^2}, \tag{6}$$

where Γ states the mixing efficiency (0.2) [17, 24, 25].

The effects of wind stress on the vertical turbulent mixing in the mixed layer ware considered by calculating the wind stress at sea level with equation (7)

$$\tau_x = \rho_{air} C_D (U_{10})^2, \tag{7}$$

where ρ_{air} (the density of air) = 1.3 kgm^{-3}; C_D = 1.5x10 ms^{-1}; U = wind speed [26]. The relationship between the vertical turbulence and windstress was obtained by calculating their linear correlation coefficient.

3. Results and discussion

3.1. *Water mass static stability*

Static stability of the water masses is one of the basic variables in oceanography which define the increasing density with descending depth. Instability in the water column can lead to water mass mixing that can be identified by calculating the Brunt Vaisala frequency values (N^2). The calculation results of Brunt Vaisala frequency from the CTD yoyo deployment showed a relatively large value of N^2 was found in the thermocline layer $(6.39 \times 10^{-5} - 1.58 \times 10^{-4})$ cycl/s at depth of 100-200 m. Figure 3 shows a vertical profile of N^2 in Dewakang Sill waters. The similar profile to N^2 in this region was obtained by Park [17] and Suteja [11] in their research sites of Kerguelen Plateau and the Ombai Strait respectively. The high value of N^2 in the thermocline layer occurred because of the existence of the pycnocline layer that is a layer where the density gradient increases significantly with depth (pressure) [26]. The higher value of N^2 in a particular ocean layer, the larger the static stability of the layer. On the contrary, if the value of N^2 becomes more negative, the water column becomes more unstable, or in other words, it is in a state of static instability. In reference to all of the N^2 repetitions, it can be seen that most of them are relatively negative. This suggests that a highly instable water column condition is presumably linked to a high current condition as well as an interaction with the sill in the region of Dewakang Sill which triggers a turbulence to occur.

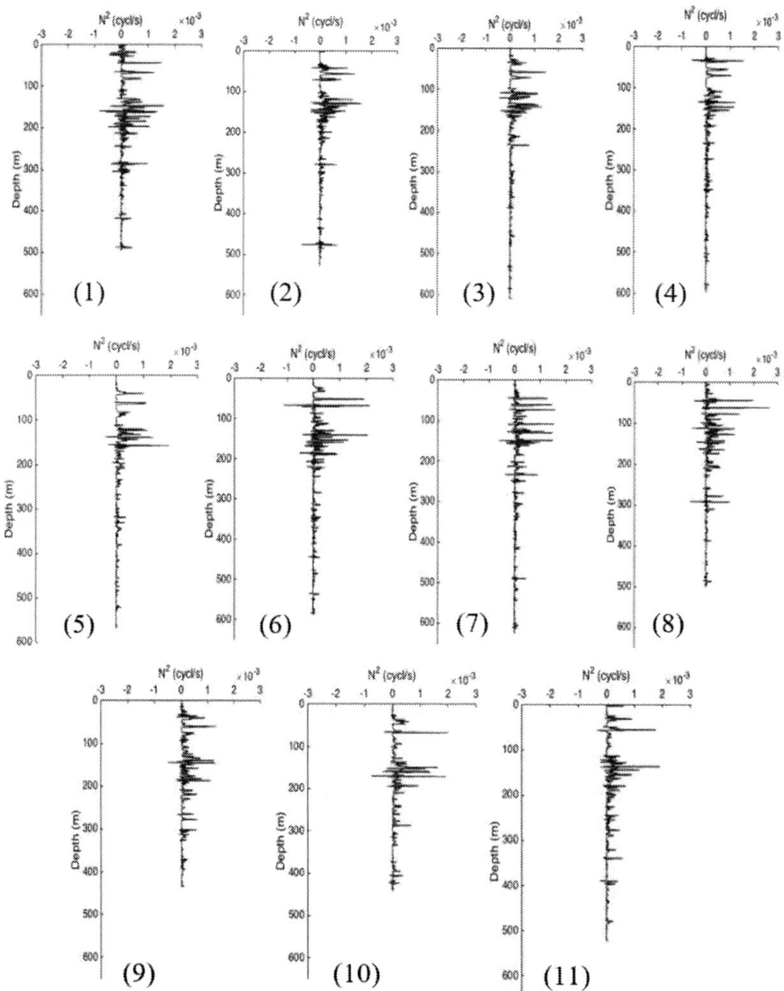

Figure 3. Brunt Vaisala frequency (N^2) profiles from 11 CTD repeated measurement in the Dewakang Sill.

3.2. Analysis of vertical turbulent mixing

3.2.1. Dissipation of turbulent kinetic energy

The estimated average value dissipation of turbulent kinetic energy (ε) obtained from Dewakang Sill waters at all levels is 1.08x10^{-6} W kg^{-1}. This result is similar to that obtained by Suteja [11] in Ombai Strait i.e. 10^{-6} Wkg^{-1}, however, it is larger than the outcome obtained by Purwandana [12] in Alor Strait i.e. around $O(1.08 \times 10^{-6})$Wkg^{-1} the Sotheast monsoon. The average value dissipation of turbulent kinetic energy (ε) at the average of every 50 m depth in Dewakang Sill is presented in Table 1. It can be seen that the bottom layer has the smallest value in comparison with the layers of turbulent mixing and thermocline. The low value of ε in deeper layer is suspected to result in smaller kinetic energy from the turbulent flow, which will break into another smaller form (dissipation) to transfer energy to another media.

Table 1. Tabulation of average values dissipation of turbulent kinetic energy (ε) on every 50 meter-depth in Dewakang Sill as resulted from 11 CTD profiles

Depth (m)	Dissipation of Turbulen Kinetic Energy (ε)(Wkg^{-1})	Standard Deviation
25.8	9.6×10^{-7}	$\pm13.58 \times 10^{-7}$
76	30.89×10^{-7}	$\pm67.74 \times 10^{-7}$
125.2	28.74×10^{-7}	$\pm31.82 \times 10^{-7}$
175.7	22.51×10^{-7}	$\pm28.11 \times 10^{-7}$
225.5	8.21×10^{-7}	$\pm9.48 \times 10^{-7}$
273.8	4.95×10^{-7}	$\pm11.08 \times 10^{-7}$
322.9	4.27×10^{-7}	$\pm4.67 \times 10^{-7}$
378.4	1.68×10^{-7}	$\pm2.34 \times 10^{-7}$
427.4	2.02×10^{-7}	$\pm3.01 \times 10^{-7}$
476.6	3.63×10^{-7}	$\pm7.59 \times 10^{-7}$
522.3	1.34×10^{-7}	$\pm1.88 \times 10^{-7}$
573.7	0.87×10^{-7}	$\pm0.98 \times 10^{-7}$

Dissipation of turbulent kinetic energy values, which are relatively large, are found in the thermocline layer i.e. $(8.35 \times 10^{-7} - 4.37 \times 10^{-6})$Wkg^{-1}. This indicates that the thermocline layer in Dewakang Sill waters is the layer where the kinetic energy experiences the highest breakage presumably happens due to the sill functioning as the barrier which causes the mixing process. The level of the values is presumably related to the internal wave activity and as a result of tidal phases on the location of the data collection, which then, will affect the range of values of displacement and dissipation of turbulent kinetic energy.

3.2.2. Turbulent vertical diffusivity

In general, the average value of turbulent vertical diffusivity (K_ρ) for the entire CTD measurements is 2.8×10^{-4} m^2 s^{-1}. This is considered to be larger than that from the research conducted by Koch-Larrouy [9] to obtain the average value of 1.5×10^{-4}m^{2}s^{-1} for the Indonesian waters and nearly the same as the research result obtained by Purwandana [12] in Alor Strait i.e. $O(10^{-4} - 10^{-3})$ m^{2}s^{-1} More intensively powerful turbulent mixing is found in the mixed layer surface which is presumably caused by relatively high wind sterss during the field experiment (the Southeast monsoon), as shown in Figure 4. This is marked by the considerably high correlation value between the dissipation of turbulen energy and wind stress (corr = 0.67). Therefore, it has more energy for the vertical turbulent to occur in the mixed layer. This reflects the major contribution of atmospheric cooling and air pressure to push the turbulent mixing in-to the surface layer.

Furthermore, stronger vertical turbulent mixing is also found in the thermocline layer and the layers near the bottom waters (Table 2). Turbulent vertical diffusivity (K_ρ) value in the thermocline layer is higher than that obtained by Ffield and Gordon [8] and Hatayama [10] at Dewakang Sill i.e. 1.0×10^{-4} m^2s^{-1} and 6.0×10^{-4} m^2s^{-1}, respectively. In this study, the range of K_ρ value is higher than the results obtained by Matsuno [27] that is equal to $O(10^{-6} - 10^{-4})$ m^2s^{-1} in the East China Sea at a depth of 200 m. The correlation value between dissipation of turbulent kinetic energy (ε) and internal wave ($A_{int} \times N^2$) is also reasonably high (corr=0.68). Subsequently, it is suggested that at this layer there are reasonably high influence of internal tidal waves. This condition indicates that Dewakang Sill is a region with reasonably high turbulent mixing. These results are consistent with Koch-Larrouy [9] that Dewakang Sill is an energetic turbulent mixing region.

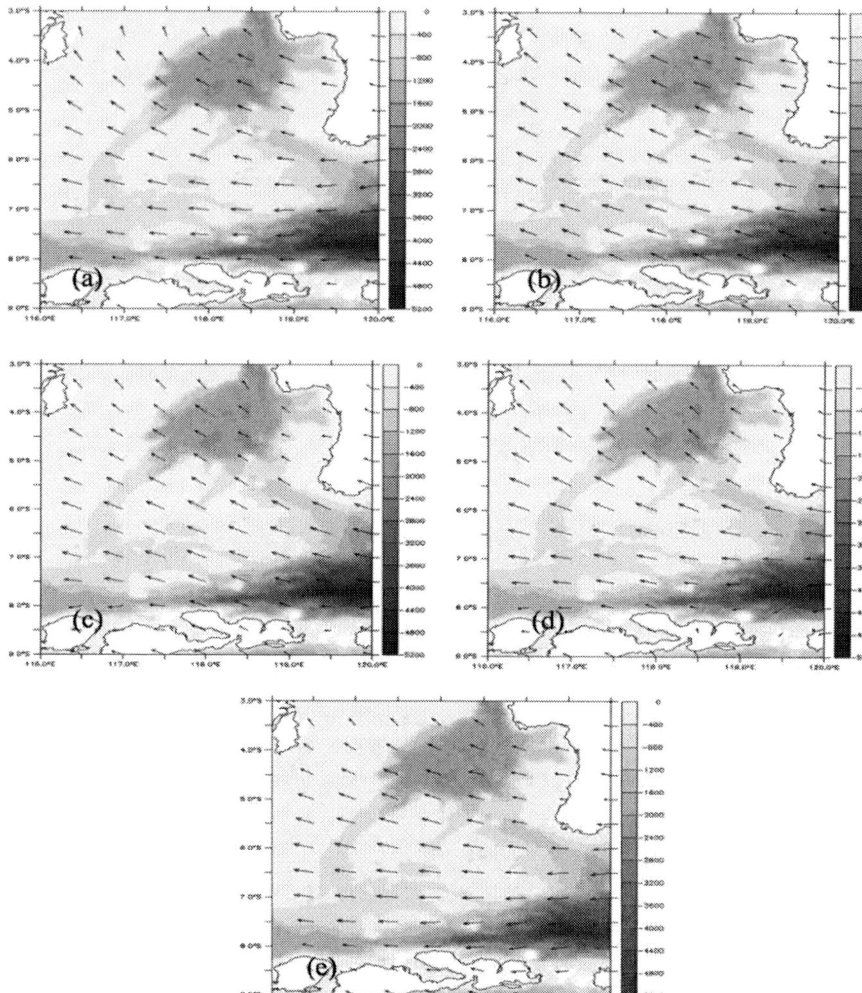

Figure 4. Snapshot of surface wind vector 10 m above sea level; (a) August 10, 2015 (18:00) (b) August 11, 2105 (00:00); (c) August 11, 2015 (06:00); (d) August 11, 2015 (12:00); (e) August 11, 2015 (18:00) during the data collection of "yoyo" CTD in Dewakang Sill and its adjacent waters

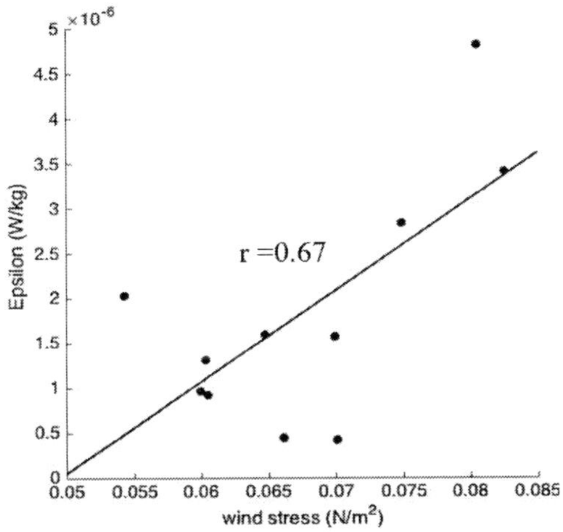

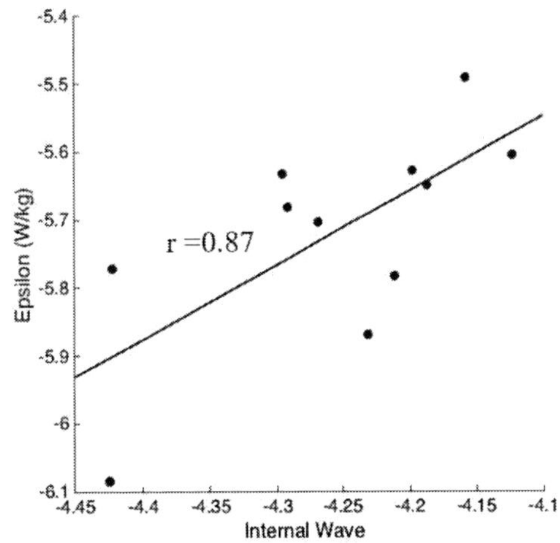

Figure 5. Correlation between the dissipation of turbulent kinetic energy and wind stress in the mixed surface layer (3-80 m) (a); (b) between the dissipation of turbulent kinetic energy and internal waves in the thermocline layer (80-250 m)

Table 2. Tabulation of averaged turbulent vertical diffusivity (K_ρ) on every 50 meter-depth in Dewakang Sill as resulted from 11 CTD profiles

Depth (m)	Vertical Eddy Diffusivity (K_ρ) (m^2/s)	Standard Deviation
25.8	4.74×10^{-3}	$\pm5.99 \times 10^{-3}$
76	4.91×10^{-3}	$\pm8.63 \times 10^{-3}$
125.2	3.60×10^{-3}	$\pm4.40 \times 10^{-3}$
175.7	2.96×10^{-3}	$\pm2.83 \times 10^{-3}$
225.5	2.68×10^{-3}	$\pm2.94 \times 10^{-3}$
273.8	2.00×10^{-3}	$\pm2.88 \times 10^{-3}$
322.9	2.64×10^{-3}	$\pm2.98 \times 10^{-3}$
378.4	1.36×10^{-3}	$\pm1.56 \times 10^{-3}$
427.4	1.99×10^{-3}	$\pm2.01 \times 10^{-3}$
476.6	2.91×10^{-3}	$\pm5.49 \times 10^{-3}$
522.3	2.07×10^{-3}	$\pm2.39 \times 10^{-3}$
573.7	2.20×10^{-3}	$\pm3.38 \times 10^{-3}$

Furthermore, the deeper layers are most likely related to the internal tidal influence and caused by the influence of the sill in this research area. Moum [28] reported that the sill is able to cause changes in the average flow velocity, modify the vertical gradient velocity, strengthen density gradient, and reduce the flow. In addition, Naulita and Kitade [29] also argued that the slope of the seabed topography corresponding to the tilt direction of propagation of internal tides can cause breaking internal tides, then,-resulting mixing in the water column.

4. Conclusion

Dewakang Sill is categorized as an area that has a strong turbulent mixing value. Thus, the highest values dissipation of turbulent kinetic energy (ε) and the turbulent vertical diffusivity were found in the mixed and thermocline layers with respective values of 1.08×10^{-6} Wkg^{-1} and 2.84×10^{-4} m^2 s^{-2}. The surface layer is most likely affected by wind stress, since high correlation is found between dissipation of turbulent kinetic energy (ε) and wind stress (corr = 0.67). Meanwhile, on the thermocline layer, dissipation of turbulent kinetic energy(ε) is highly correlated with internal tidal wave oscillation (corr=0.68). Those suggest that the strong Southeast monsoon wind stress and internal tidal waves activities are the major forcing for the energetic turbulent mixing within the mixed layer and thermocline layer in Dewakang Sill.

Acknowledgement

The MAJAFLOX cruise is a part of the collaborative research between Department of Marine Science and Technology FPIK-IPB and Marine Geological Intitute (P3GL) Bandung (Principal Investigator: A. Atmadipoera and R. Zuraida). A. Atmadipoera was supported by research grant from Kemenristek Dikti (PUPTN 714/IT3.11/PN/2016). We are very grateful to the crews of R.V. Geomarin III for their professional assistance during the field experiment. The first author would like to thank the LPDP Scholarship (Indonesia Endowment Fund for Education) for the master degree scholarship.

References

[1] Gordon A L, Orsi A H, Muench R, Huber B A, Zambianchi E and Visbeck M 2009 Western Ross Sea continental slope gravity currents *Deep-Sea Res. II.* **56** 796-817

[2] Susanto R D, Ffield A, Gordon A L and Adi T R 2012 Variability of Indonesian throughflow within Makassar Strait, 2004–2009. Lamont-Doherty Earth Observatory, Columbia University, Palisades, New York, USA. *J of Geophys Res*. **117**, doi :10.1029/2012JC008096

[3] Gordon A L 2001 Interocean exchange, in Ocean Circulation and Climate: Observing and Modeling the Global Ocean. Siedler G, Church J, and Gould J (Ed) *Int. Geophys*. **77** 303–314 (New York: Elsevier)

[4] Gordon A L, Giulivi C F and Ilahude 2003 A. Deep Topographic Barriers Within the Indonesian Seas *Deep-Sea Res*. **50** 2205-2228

[5] Atmadipoera A, Mocard R, Madec G, Wijffels S, Sprintall J, Koch-Larrouy A, Jaya I, Supangat A 2009 Characteristics and variability of the Indonesian Throughflow water at the outflow straits *Deep-Sea Res I*. **56** 1942-1954

[6] Stewart R H 2003 *Introduction to Phys Oceanogr*. Texas A and M University. Departement of Oceanography

[7] Ffield A 1994 Tidal Mixing in the Indonesia Seas. Paper presented at Internasional Scientific symposium at the IOC-WESTPAC, IOC Bali, Indonesia

[8] Ffield A and Gordon A L 1992 Vertical mixing in the Indonesian thermocline *J Phys Oceanogr*. **22** 184-195

[9] Koch-Larrouy A, Madec G, Bouruet-Aubertot P, Gerkema T, Bessières L and Molcard R 2007 On the transformation of Pacific Water into Indonesian Throughflow water by internal tidal mixing *Geophys Res Lett*. **34** 1-6

[10] Hatayama T 2004 Transformation of the Indonesian Throughflow Water by Vertical Mixing and Its Relation to Tidally Generated Internal Waves. Japan Agency for Marine-Earth Science and Technology, Yokohama Institute for Earth Sciences, Showa-machi, Kanazawa-ku, Yokohama, Kanagawa 236-0001, Japan *J of Oceanogr*. **60** 569-585

[11] Suteja Y 2015 Percampuran Turbulen Di Selat Ombai *J of Trop Mar Scien and Tech*. **7** 71-82.

[12] Purwandana A 2015 Distribusi Percampuran Turbulen di Perairan Selat Alor *J Mar Scien* **19** 43-54

[13] Atmadipoera A, Kusmanto E, Purwandana A, Nurjaya I 2015 Observation Of Coastal Front And Circulation In The Northeastern Java Sea, Indonesia *J of Trop Mar Scien and Tech* **7** 91-108

[14] Ffield A and Gordon A L 1996 Tidal mixing signature in the Indonesian Seas. *J Phys Oceanogr.* **26** 1924-1936

[15] Koch-Larrouy A, Atmadioera A, Van Beek P, Madec G, Aucan J, Lyard F, Grelet J and Souhaut M 2015 Estimates of tidal mixing in the Indonesian archipelago from multidisciplinary INDOMIX in-situ data *Deep-Sea Part I.* doi : 10.106/j.dsr.2015.09.007

[16] Backes J 2011 *SBE 19plus CTD Operating and Repair Manual* (Washington USA: Bellevue)

[17] Park Y H, Fuda J L, Durand I and A C N Garabato 2008 Internal tides and vertical mixing over the Kerguelen Plateau *Deep-Sea Res Pt II.* **55** 582-593

[18] Thompson A F, Gille S T, Mackinnon J A, Sprintall J 2007 Spatial and temporal patterns of small-scale mixing in Drake Passage *J Phys Oceanogr.* **37** 572-592

[19] Ferron B, Mercier H, Speer K, Gargett A, Polizin K 1998 Mixing in the Romanche Fracture Zone *J Phys Oceanogr.* **28** 1929-1945.

[20] Dillon T M 1982 Vertical overturns: a comparation of Thorpe and Ozmidov length scale *J Geophys Res.* **87** 9601-9613

[21] Finnigan T D, Luther DS and Lukas R 2002 Observation of enhanced diapycnal mixing near the Hawaiian Ridge *J Phys Oceanogr.* **32** 2988-3002

[22] Koch S E, Desjardins M and Kocin P J 1983 An interactive Barnes objective map analysis scheme for use with satellite and conventional data *J Climate Appl Meteor.* **22** 1487-1503

[23] Galbraith P S and Kelley E 1996 Identifying overturn in CTD profiles *J Atmos Ocean Tech.* **13** 688-702

[24] Cisewski B, Strass V H and Prandke H 2005 Upper-ocean vertical mixing in the Antarctic polar front zone *Deep-Sea Res II.* **52** 1087-1108

[25] Osborn T R 1980 Estimates of the local rate of vertical diffusion from dissipation measurements. *J. Phys. Oceanogr.* **10** 83–89

[26] Pond S and Pickard GL 1983 *Introductory Dynamical Oceanography Ed ke-2* (Oxford: Pergamon Press)

[27] Matsuno T, Shimizu M, Morii Y, Nishida H and Takaki Y 2005 Measurement of the Turbulent Energy Dissipation Rate around the Shelf Break in the East China Sea. *J of Ocean.* **61** 1029-1037

[28] Moum J N and Smyth W D 2001 *Upper Ocean Mixing Processes* (Oregon (US): Academic Press)

[29] Naulita Y and Kitade Y 2011 Observed Turbulence Properties over the Continental Shelf and Slope off Jogashima, Sagami Bay *Société franco-japonaise d'océanographie La mer.* **49** 1 – 15

Nutrient and chlorophyll-a distribution in Makassar Upwelling Region: From MAJAFLOX CRUISE 2015

A Rosdiana[1*], T Prartono[1], A S Atmadipoera[1] and R Zuraida[2]
[1]Department of Marine Science and Technology, Bogor Agricultural University, Indonesia
[2]Marine Geological Institute, Bandung, Indonesia

E-mail: annisyarosdiana@outlook.com

Abstract. Upwelling is an important mechanism on productivity enhancement in the Southern Makassar Strait (MAK). Previous studies found correlation between upwelling and chlorophyll-a blooms, which is perhaps caused by nutrient regeneration from deeper water. Through a multi-disciplinary study "MAJAFLOX Cruise" in August 2015, CTD casts and seawater samples have been collected from Southern Makassar Strait. The present study aims to investigate relationship between physical processes of upwelling and nutrient usage, especially for nitrate, phosphate and silicate from regenerated processes to be used for the primary productivity (chlorophyll-a). The results show that upwelling centre is found near Dewakang Sill around station 5, associated with the highest chlorophyll-a concentration (0.438 μgL-1). Relatively high nutrient is also revealed in other station of observation. In sea surface layer, nitrate is high in station 1 to 3, phosphate high in station 2 and silicate high in station 9. This shifting of high nutrient concentration may indicate spatial variation of biogeochemical processes in the surface layer associated with nutrient fluxes.

1. Introduction

Upwelling is an upward movement of deeper waters to the surface area. Upwelling generally bring materials from the bottom including nutrients [1,2], so that upwelling associated with the nutrient regeneration [3]. The nutrient regeneration may increase the rate of photosynthesis, thus potentially increasing marine productivity and fishery resources [4,5].

The Southern Makassar Strait (abbreviated MAK) lies between the southern part of Kalimantan and Sulawesi Island. Makassar Strait is the western path of Indonesian Throughflow (ITF), that carry the Pacific water into the Indian Ocean. Fluctuation of the throughflow is affected by the monsoonal wind [6], strong tidal current [7] and El Ninō-Southern Oscillation (ENSO) signals [8,9].

Former research found that MAK is an upwelling region with high fishery resources [10,11]. Winds that occur during Southeast Monsoon (May-October) are responsible for that upwelling formation. On the Southeast Monsoon, the surface wind blows nothwestward and led the Ekman current moves away from the coast. The empty space on surface area then filled by waters from deeper layer. The pattern of Southeasterly winds with a maximum speed occurs in the August [12]. Details of physical mechanism in the upwelling had been described by a numerical model approach Regional Ocean Modeling System (ROMS) [13].

Transport of deep and nutrient-rich waters to the surface are characterized by increase of chlorophyll-a concentration and decrease of sea surface temperature (SST) along the coast. Some evidences found phytoplankton blooms in MAK during Southeast Monsoon based on chlorophyll-a data from satellite imagery [12,14,15] and water sampling [16]. Indication of upweling variability is also found based on sea surface temperature from sattelite data [12,14,17].

In the tropical ocean, areas of upwelling have high nutrient concentrations during some periods of the year [18]. Nutrient supply (in particular N, P, and Si for diatoms) and its ratios have a decisive effect on the species composition of the phytoplankton [19], then change the primary production and sinking rate of organic matter [1]. This study is the first attempt to understand both physical and chemical interaction based on in-situ data from Makassar upwelling region. Two major objectives were established in this study : (1) to detect the center of upwelling and (2) to find possibility of nutrient regeneration.

2. Methods

This study was part of the cruise expedition "MAJAFLOX 2015". The main goal of this cruise is to understand the relation between water mass characteristics, biogeochemical distribution and aquatic ecosystem in the Java Sea, Makassar Strait and Flores Sea (JMF Triangle Seas). JMF triangle seas is an area crossed by ITF, where nearly 75% of total flow (15106 m^3s^{-1}) through the Makassar Strait [8] This area is also have potency on small pelagic fisheries such as mackerel and oil sardine [15,20].

This cruise is conducted on Research Vessel Geomarin III, which sailed from the north coast of Cirebon (Java Sea) to the southern Kangean Island (Flores Sea). Southern Makassar Strait (MAK) region represented by transect along 600 km, starting from Station 1 on the edge of the Java Sea to Station 9 in southwestern Selayar Island (figure 1).

This study is divided into three parts : (1) sampling and preservation, (2) sample analysis and (3) data processing. The sampling and preservation was conducted in August (year 2015) that represent the Southeast Monsoon in MAK. The instrument used for sampling is Rosette Sampler; consists of a Seabird CTD sensor V19 Plus and 12 Niskin Bottles (8 liters capacity). Rosette Sampler is operated on profile mode which CTD sensor continously record four data per second (4 Hz) of depth, temperature and salinity. At the same time, Niskin Bottle took seawater samples at nine depth points, i.e. 5, 20, 50, 100, 200, 300, 400, 500 and 1000 meters.

Seawater samples from Niskin Bottles preserved on board by filtration using Whatman membrane filter CNM 0:45 μm (Ø47 mm) and vacuum pump (with the suction force is less than 30 cmHg). Water samples then added by HNO_3 solution until pH = 2 then kept on HDPE bottle. The membrane filters also wrapped by aluminum foil. Water samples and membrane filters are stored in a cooler (with temperatures $\leq 4\ ^\circ$ C) until ready to be analyzed [21].

Nutrients in seawater samples analyzed in the laboratory using a spectrophotometer Shimadzu UV-1201V. Analysis of dissolved nitrate-N using cadmium reduction method, while the analysis of dissolved orthophosphate-P and silicate-Si using staining methods [22]. For chlorophyll-a analysis, membrane filters are destructed by Aseton solution (v/v 90%) then detected by a fluorometer Trilogy® [23].

Data processing is included CTD and seawater samples. Data records from CTD are converted and corrected using software SBE Data Processing series 7.21a. Data from the analysis of nutrients and chlorophyll-a tabulated using software Microsoft Excel. Data CTD, nutrient and chlorophyll-a then displayed using software Ocean Data View (ODV). Especially for the nutrient ratios shown in a diagram using software Microsoft Excel based on logaritmic scale [24,25].

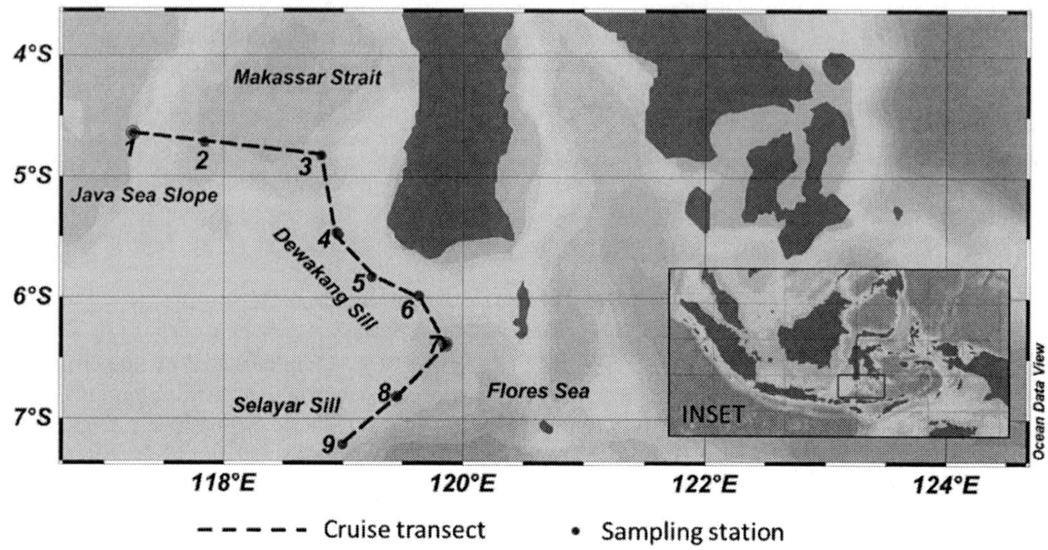

Figure 1. Map of study location in Southern Makassar Strait (MAK) region. The inset shows study location (red box) on the Indonesian map.

3. Results and Discussion

3.1 Water Mass Characteristics and Signals of Upwelling

Based on CTD record in this research, bathymetry between Station 1 to 3 forming a basin with maximum depth centered at Station 2 (1800 meters). From Station 4 to 6 lies Dewakang Sill that makes seabed topography becomes more complex. Later on Station 6 to 9 lies Selayar Slope with an average depth of 350 meters.

Water mass of Southern Makassar Strait (figure 2) is divided into three layers : mixed layer, thermocline layer and deep layer. Mixed layer with a relatively high and homogeneous high temperature (>25 °C), low salinity (34 PSU) and low density (22 kgm^{-1}) is in depth 0 to 100 m. Thermocline layer at depth 100 to 200 m is characterized by isotherm 20. In the thermocline layer, there is isohaline 34.5 as a marker for North Pacific thermocline waters. Still on the thermocline layer, there is some undulations of isopicnal 24 and 25 that indicate internal waves activity. The deep layer located below the thermocline that characterized by homogeneous low temperature (<5 °C), high salinity (<34.5 PSU) and high density (>27 kgm^{-1}). Through the same figure, it appears that the isotherm 25 (figure 2a), isohaline 34.5 (figure 2b) and isopicnal 23 (figure 2c) increase from the thermocline to the mixed layer. These three parameters move to the upper layer possibly due to upwelling that centered near Dewakang Sill around station 5.

The structure of water masses in Southern Makassar Strait described by using T-S diagram (figure 3). The water mass on the upper layer of the thermocline (isopicnal σ_0 = 24) with a maximum salinity (>34.6 PSU) is characteristic of North Pacific Subtropical Water (NPSW). The section on the bottom layer of the thermocline (isopicnal σ_0 = 26.5) with minimum salinity (34.4 PSU) is characteristic of North Pacific Intermediate Water (NPIW). It is known that the North Pacific waters enter Makassar Strait through the western path from its entrance in the northeastern Sulawesi Sea. The maximum salinity of NPSW in this study decrease gradually from the north (station 3) to the south (Station 6). Instead, the minimum salinity of NPIW in the northern part was 34.40, then increased to 34.52 in the south. Changes in salinity may be due to mixing with more fresh waters or mixing with South Pacific waters that came from eastern part in the Banda Sea [26,27].

IOP Conf. Series: Earth and Environmental Science **54** (2017) 012087 doi:10.1088/1755-1315/54/1/012087

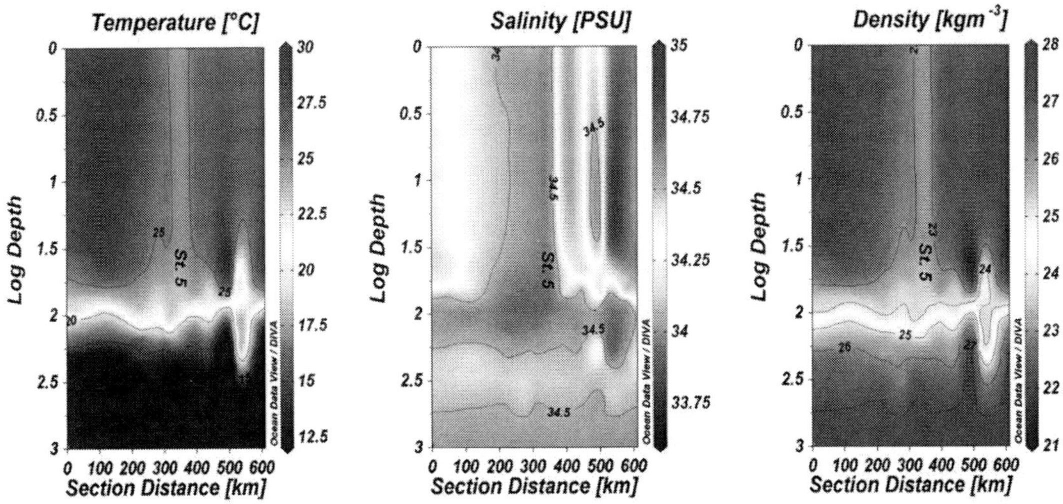

(a) (b) (c)

Figure 2. Distribution of (a) temperature, (b) salinity and (c) density of Southern Makassar Strait during Southeast Monsoon

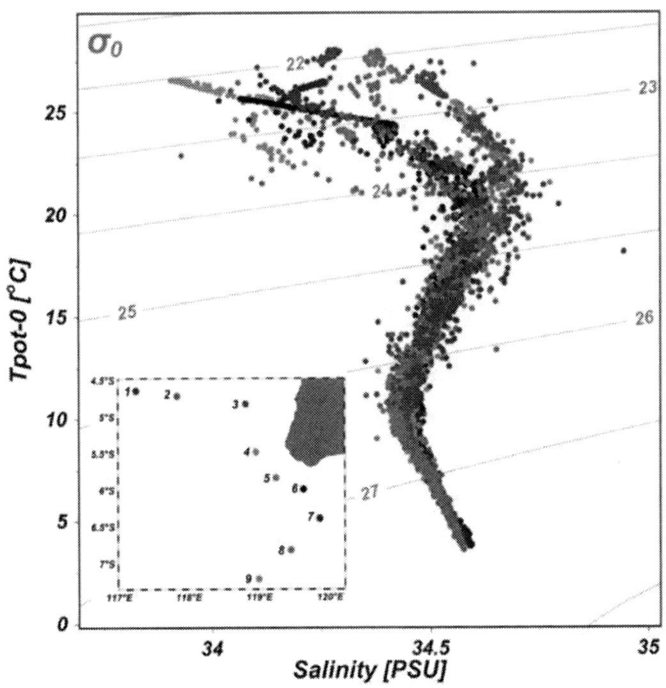

Figure 3. T-S diagram of Southern Makassar Strait during Southeast Monsoon. Map in the diagram shows symbol of T-S for each stations. Tpot-0 on the Y axis is potential temperature at water pressure = 0. Isopycnal is symbolized by σ_0.

3.2 Nutrient Characteristics

The nutrients distributions in Southern Makassar Strait are generally varied within depth (**Figure 4**). Concentration of nutrients in the surface tends to be low, although at some stations increased. Nutrients in the mixed layer to the upper thermocline (depth <200 meters) is also low. The

IOP Conf. Series: Earth and Environmental Science **54** (2017) 012087 doi:10.1088/1755-1315/54/1/012087

concentration of nutrients then increase gradually from the center of thermocline to the deep layer. Maximum nutrient concentration averaged found below the depth of 230 meters. From that distribution, we agree that the Indonesian throughflow influences Indian Ocean biogeochemistry by supplying an effective or net flux of nutrients primarily to thermocline waters [28]. Variation of nutrients concentration is consequence of different biological processes and water mass characteristics [28]. Nutrients from the surface water get utilized by the phytoplankton, thereby depleting their level in the upper layer. Nutrient itself or through dead cell of organisms then sink and accumulate in the thermocline layer. Some particle trapped in the thermocline, while the other part sink to the deeper layer. In deep layer, some bacterias degradate organic matter so that nutrient regenerates, thus make nutrient higher at bottom [29].

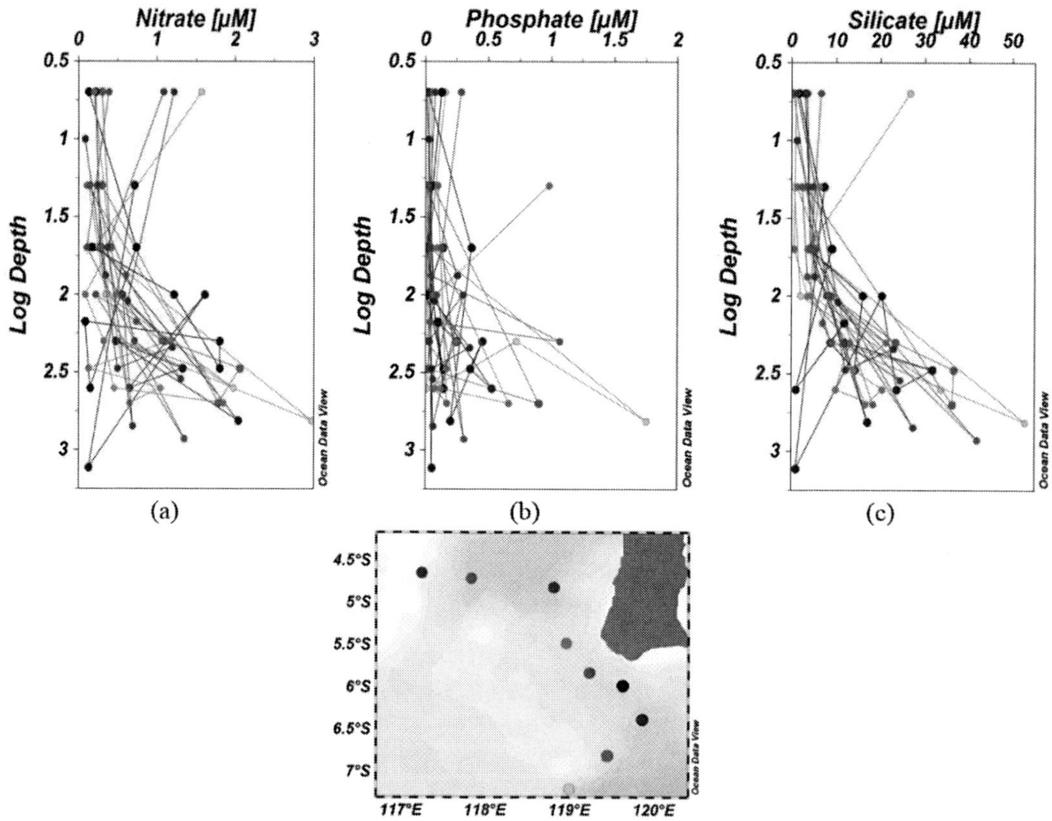

Figure 4. Scatter of (a) nitrate, (b) phosphate and (c) silicate concentration in Southern Makassar Strait during Southeast Monsoon. Map below the scatter diagrams show color symbol for each station.

Upwelling generally trigger nutrient to the surface layer. It was mentioned before that center of upwelling is in station 5 near Dewakang Sill (**Figure 2**). However, from this study we found that center of upwelling is low nutrient (**Figure 5**). Nutrients lower at station 5 allegedly due to its use by the autothrops. Increase of chlorophyll-a in the surface station 5 (figure 6) indicates higher photosynthesis rate than the other stations. Relationships between chlorophyll a and nutrients were opposite, suggesting that in the short term phytoplankton controlled nutrient concentrations [30].

High concentration of nutrients in the surface varied within location. Nitrate was high at Station 1 to 3 (**Figure 5**a), while phosphate (figure 4b) high at station 2 and silicates (figure 4c) high at station 9. Shfting of high nutrient concentration is associated with environmental condition. Station with higher nutrient in the surface apparently had lower salinity (**Figure 2**b). The lower salinity can be indicates an input of fresh water. Then we can assume that another input of nutrient to the Makassar Strait is came from river run-off or human activity near the coast of Sulawesi and Kalimantan [16].

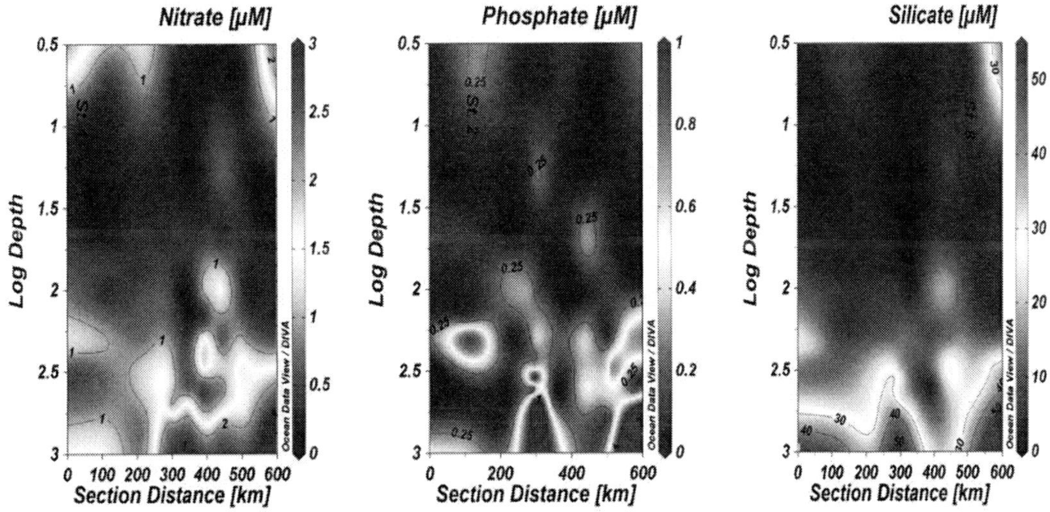

(a) (b) (c)

Figure 5. Concentration of (a) nitrate, (b) phosphate and (c) silicate in Southern Makassar Strait during Southeast Monsoon

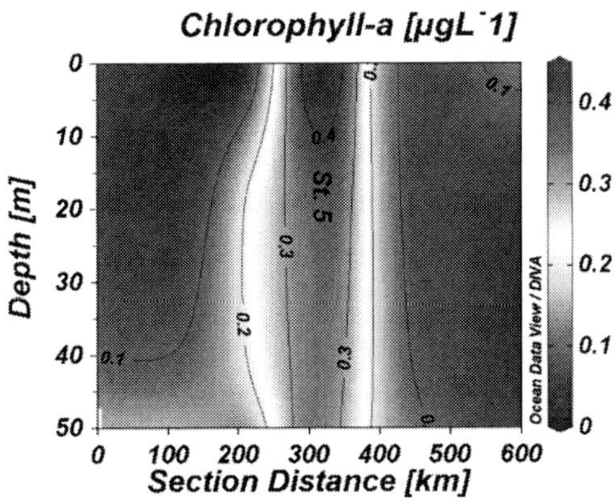

Figure 6. Concentration of chlorophyll-a in Southern Makassar Strait during Southeast Monsoon

Figure 7 shows the distribution of molar ratios between Si: N: P that gathered in quadrant N and P. Quadrant N is indicating the nitrate deficient, whereas quadrant F indicating phosphat deficient. Various biological processes in water such as death, predation, autocatalysist and remineralization of plankton cells can affect the ratios [3]. In **Figure 4**, range of nitrate (0.022-2.978 μM) and phosphate (0.001-1.89 μM) remained lower than silicate (0.157-54.533 μM). This higher silicate concentration possibly reflecting the small rate of dissolution of biogenic silica in the upper ocean that mediated by diatoms [18]. Some reason is that waters upwelling to the surface are generally slightly depleted in N relative to P (below Redfield) [1]. The persistence of high Si:N ratios also carry the signature of biologically spent water in the upper mixed layer [32].

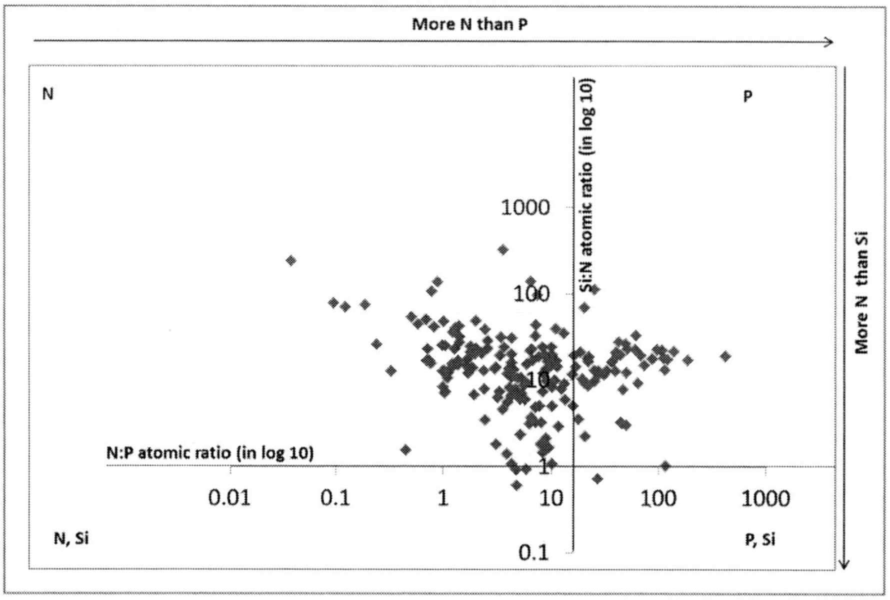

Figure 7. Diagram of the molar ratios between atomic Si: N: P in the Southern Makassar Strait during Southeast Monsoon. The Si:N and N:P atomic ratios of 1:1 and 16:1 are indicated by the horizontal and vertical lines, respectively. Nutrient limitation (assuming sufficient concentration) within each quadrant are indicated by the lettering. [24,25]

4. Conclusion

Physical properties such as temperature, salinity and density during observation show the upwelling signal near Dewakang Sill. Apparently, nutrient profiles in this study areas show a common oceanic nutrient type that lower in the upper and higher in the deeper layer, suggesting their involvement in the biogeochemical processes. At the center of upwelling, nutrient tend to be low but the chlorophyll-a is high. This condition shows that nutrient from deeper layer had been used for photosynthesis. Slighly lower N and P also indicate dominant activity of organism especially for diatoms in the ocean.

Acknowledgments

This research was part of partnership between Bogor Agricultural University, Marine Geological Institute and Marine Fisheries Research Center, Ministry of Sea and Fisheries of Indonesia. We would like to thanks P2O LIPI for analyzing nutrient and chlorophyll-a in this research. We also thank to the Crew of RV Geomarin III who accompany us in this research. We appreciated comments from reviewers to improve the quality of this paper.

5. References

[1] Arrigo K R 2005 Marine microorganisms and global nutrient cycles *Nature* **437** 349-355

[2] D'Croz L and O'Dea A 2007 Variability in upwelling along the Pacific shelf of Panama and implications for the distribution of nutrients and chlorophyll. Estuarine *Coastal and Shelf Science* **73** 325-340

[3] Falkowski P G, Laws E A, Barber R T and Murray J W 2003 Phytoplankton and their role in primary, new, and export production In *Ocean biogeochemistry* (Heidelberg: Springer Berlin) pp 99-121

[4] Fiedler P C, 2002 The annual cycle and biological effects of the Costa Rica Dome *Deep-Sea Research* **49** 321-338

[5] Weber T S 2013 *Biological and physical regulation of the oceanic fixed nitrogen reservoir* [Doctoral dissertation] (Los Angeles: University of California)

[6] Gordon A L and Fine R A 1996 Pathways of water between the Pacific and Indian Oceans in the Indonesian Seas *Nature* **379** 146–149

[7] Ffield and Gordon 1992 Vertical mixing in the indonesian thermocline *Journal of Physical Oceanography* **22** 184-195

[8] Gordon A L, Susanto R D, Ffield A, Huber B A, Pranowo W and Wirasantosa S 2008 Makassar Strait throughflow 2004 to 2006 *Geophysical Research Letters* **35** L036372

[9] Susanto R D, Ffield A, Gordon A L and Adi T R 2012 Variability of Indonesian throughflow within Makassar Strait 2004–2009 *Journal of Geophysical Research- Oceans* 117

[10] Illahude AG 1978 On The Effecting The Productivity of The Southern Makassar Strait *Marine Research in Indonesia* **21** 81-107

[11] Kinkade C, Marra J, Langdon C, Knudson C and Ilahude AG 1997 Monsoonal differences in phytoplankton biomass and production in the Indonesian Seas: Tracing vertical mixing using temperature Deep Sea Research Part I: *Oceanographic Research Papers* **44** 581-592

[12] Habibi A, Setiawan R Y and Zuhdy A Y 2012 Wind-driven coastal upwelling along South of Sulawesi Island *ILMU KELAUTAN: Indonesian Journal of Marine Sciences* **15** 113-118

[13] Atmadipoera A S and Widyastuti P 2015 A Numerical Modeling Study on Upwelling Mechanism in Southern Makassar Strait *Jurnal Ilmu dan Teknologi Kelautan Tropis* **6** 355-371

[14] Setiawan R Y and Kawamura H 2011 Summertime phytoplankton bloom in the South Sulawesi Sea Selected Topics in Applied Earth Observations and Remote Sensing *IEEE Journal* **4** 241-244

[15] Syahdan M, Atmadipoera A S, Susilo S B and Gaol J L 2014 Variability of surface chloropyll-a in Makassar Strait-Java Sea, Indonesia *International Journal of Sciences: Basic and Applied Research (IJSBAR)* **14** 103-116

[16] Afdal dan Riyono SH 2004 Sebaran Klorofil-a dan Kaitannya dengan Kondisi Hidrologi di Selat Makassar *Oseanologi dan Limnologi di Indonesia* **36** 69-82

[17] Nababan B, Rosyadi N, Manurung D, Natih N M and Hakim R 2016 Sea surface temperature and chlorophyll-a concentration in the south of Makassar Strait *Procedia Environmental Sciences* **33** 583-599

[18] Louanchi F and Najjar R G 2000 A global monthly climatology of phosphate, nitrate, and silicate in the upper ocean- Spring-summer export production and shallow remineralization *Global Biogeochemical Cycles* **14** 957-977

[19] Hodgkiss IJ and Lu SH 2004 The effects of nutrients and their ratios on phytoplankton abundance in Junk Bay, Hong Kong *Hydrobiologia* **512** 215-229

[20] Sadhotomo B and Durand JR 1997 The Java Sea Environment in Potier M and S Nurhakim (Ed.) *BIODYNEX: Biology, Dynamics, Exploitation of the Small Pelagic Fishes in the Java Sea* AARD/ORSTOM 14-38

[21] U.S. Environmental Protection Agency (EPA) 1983 Sample Preservation *In Methods for Chemical Analysis of Water and Wastes* EPA-600/4-79-020 (Cincinnati, Ohio, USA: U.S.E.P.A)

[22] Strickland J D and Parsons T R 1972 *A Practical Handbook of Seawater Analysis* (Ottawa: Fisheries Research Board of Canada) p 45-137

[23] Cochlan W and Hendorn J 2012 *Water Quality Methods* (Tiburon, CSA, USA: Cochlan Phytoplankton Ecophysiology Laboratory Romberg Tiburon Center for Environmental Studies San Fransisco State University) 317 pp

[24] Turner R E, Rabalais N N and Dortch Q 2003 Future aquatic nutrient limitations *Marine Pollution Bulletin* **46** 1032-1034

[25] Wu M L, Wang Y S, Wang Y T, Sun F L, Sun C C, Jiang Z Y and Cheng H 2014 Influence of environmental changes on phytoplankton pattern in Daya Bay, South China Sea *Rev. biol. mar. oceanogr* **49** 323-337

[26] Ilahude A G and Gordon A L 1996 Thermocline stratification within the Indonesian Seas *Journal of Geophysical Research: Oceans* **101** 12401-12409
[27] Atmadipoera A, Molcard R, Madec G, Wijffels S, Sprintall J, Koch-Larrouy A, Jaya I and Supangat A 2009 Characteristics and variability of the Indonesian throughflow water at the outflow straits *Deep Sea Research Part I: Oceanographic Research Papers* **56** 1942-1954
[28] Chester R and Jickells T D 2003 The transport of material to the oceans: Relative flux magnitudes *Marine Geochemistry* **1** 92-124
[29] Ayers J M, Strutton P G, Coles V J, Hood R R and Matear R J 2014 Indonesian throughflow nutrient fluxes *Geophysical Research Letters* **41** 5060-5067
[30] Satpathy K K, Mohanty A K, Sahu G, Sarguru S, Sarkar S K and Natesan U 2011 Spatio-temporal variation in physicochemical properties of coastal waters off Kalpakkam, southeast coast of India, during summer, pre-monsoon and post-monsoon period *Environmental monitoring and assessment* **180** 41-62
[31] Perez-Ruzafa A, Fernández A I, Marcos C, Gilabert J, Quispe J I and García-Charton J A 2005 Spatial and temporal variations of hydrological conditions, nutrients and chlorophyll a in a Mediterranean coastal lagoon (Mar Menor, Spain) *Hydrobiologia* **550** 11-27
[32] Tremblay J É, Simpson K, Martin J, Miller L, Gratton Y, Barber D and Price N M 2008 Vertical stability and the annual dynamics of nutrients and chlorophyll *Journal of Geophysical Research* **113** C7

Spatial distribution and ecological risk assessment of heavy metal on surface sediment in west part of Java Sea

Hefni Effendi[1,2*], Yusli Wardiatno[b], Mujizat Kawaroe[3], Mursalin[1], Dea Fauzia Lestari[1]
[1] Center for Environmental Research, Bogor Agricultural University
[2] Department of Aquatic Resources Management, Faculty of Fisheries and Marine Science, Bogor Agricultural University
[3] Department of Marine Science and Technology, Faculty of Fisheries and Marine Science, Bogor Agricultural University

E-mail: hefni_effendi@yahoo.com

Abstract. The surface sediments were identified from west part of Java Sea to evaluate spatial distribution and ecological risk potential of heavy metals (Hg, As, Cd, Cr, Cu, Pb, Zn and Ni). The samples were taken from surface sediment (<0.5 m) in 26 m up to 80 m water depth with Eikman grab. The average material composition on sediment samples were clay (9.86%), sand (8.57%) and mud sand (81.57%). The analysis showed that Pb (11.2%), Cd (49.7%), and Ni (59.5%) exceeded of Probably Effect Level (PEL). Base on ecological risk analysis, Cd (E_r^i:300.64) and Cr (E_r^i:0.02) were categorized to high risk and low risk criteria. The ecological risk potential sequences of this study were Cd>Hg>Pb>Ni>Cu>As>Zn>Cr. Furthermore, the result of multivariate statistical analysis shows that correlation among heavy metals (As/Ni, Cd/Ni, and Cu/Zn) and heavy metals with Risk Index (Cd/Ri and Ni/Ri) had positive correlation in significance level $p<0.05$. Total variance of analysis factor was 80.04% and developed into 3 factors (eigenvalues >1). On the cluster analysis, Cd, Ni, Pb were identified as fairly high contaminations level (cluster 1), Hg as moderate contamination level (cluster 2) and Cu, Zn, Cr with lower contamination level (cluster 3).

1. Introduction

Many human activities are conducted in the west part of Java Sea such as fishing, shipping and industrial activites. Improperly management of environmental risk of those activities could contribute to quite important effect of water contamination and pollution. The contaminants are distributed through suspended material intake carried out from the mainland by rivers flow or run off, especially the rivers are close by the coastal areas. In the marine environment, anthropogenic sources are influenced from several sources such as river discharge, deposition from the atmosphere, industrial waste input and shipping activities [1]. The contaminants movement pattern base on the surface current circulation before sinking and accumulated on the bottom sediment. The contaminant on the water column will be accumulated through the settling process into sediment, therefore sediment may be regarded as one of the pollutant storage.

Sediments act as both carriers and sinks for contaminants, reflecting the history of pollution, and could be bioindicator for aquatic ecosystems [2-4]. Investigation of heavy metal concentration on the sediment is one of method to analyze ecological risk on the sediment. Heavy metals which have low solubility properties, will be absorbed and accumulated on the bottom sediments [5, 6]. Accumulation of heavy metals can be harmful for organism that lives on the sediment. The assessment of ecological risk is used for determining the level of heavy metals contamination on the sediment. The research aims were to identify and evaluate concentrations level, to describe spatial distribution, and to assess the ecological risk potential of heavy metals pollution in west part of Java Sea.

2. Methods

2.1. Sampling locations
Sediment samples were collected from 14 locations in west part of Java Sea. The samples were taken by Ekman grab from < 0.5 m surface layer in water depth of 26 m up to 80 meters. The maps of sampling locations are shown in figure 1.

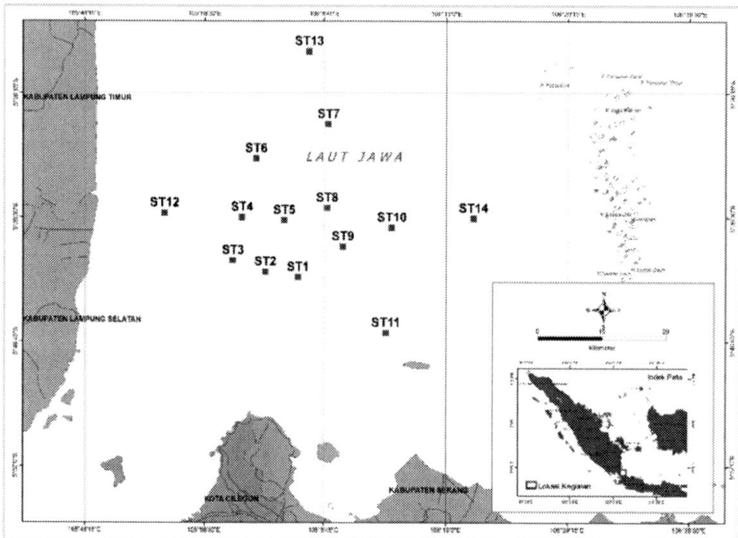

Figure 1. Sampling locations of sediment in the west part of Java Sea

2.2. Sample analysis
Samples of sediment were analyzed at Aquatic Environment Productivity Laboratory, Faculty of Fisheries and Marine Science, Bogor Agricultural University, which has been accredited by National Accreditation Committee based on ISO 17025.

2.3. Data analysis

2.3.1 Materials composition and heavy metals concentration
Sediment materials have been identified and clarified by measuring the composition of sand and mud according to sediment grain size nomenclature of Folk triangle diagram [7]. The effect of heavy metals toxicity on sediment referred to heavy metal quality assessment based on Consensus-Based Sediment Quality Guidelines [8]. Assessment was carried out by comparing Threshold Effect Level (TEL) and Probable Effect Level (PEL) with heavy metals concentration. TEL is a threshold where the heavy metals concentration cause adverse biological effects and occur rarely, while PEL is the threshold where the heavy metals concentration cause adverse biological effects and occur frequently.

2.3.2 Spatial distribution of material composition and heavy metals
Spatial distribution of sediment composition and heavy metals concentration were visualized by software Surfer version 11. Surfer visualizes the map contour and 3 dimension base on grid (horizontal

and vertical line). This software plots tubular data xyz from each composition sediment and heavy metals concentration.

2.3.3 Ecological risk index

Ecological risk potential assessment of heavy metals referred to Hakanson method [9] with the following formula:

$$RI = \sum_{i=1}^{n} E_r^i \,; E_r^i = T_f^i \times C_f^i ; C_f^i = \frac{C_s^i}{C_r^i} \quad (1)$$

C_f^i: contamination level of heavy metal; C_s^i : concentration of heavy metal on sediment; C_r^i : reference value of heavy metal in study location; E_r^i : ecological risk potential of heavy metal; T_f^i : toxicity response factor of heavy metal; RI (Risk Index): ecological risk potential of environment.

The factor scores on each of heavy metals according Hakanson approach [10] were: As (10), Cd (30), Cr (2), Cu (5), Pb (5), Ni (5), and Zn (1). Then, the criteria for ecological risk potential of heavy metals and environment were presented in table 1.

Table 1. Potential ecological risk value and criteria [9]

E_r^i	Ecological risk criteria of heavy metal	ERI	Ecological risk criteria of environment
$E_r^i < 40$	Low risk	RI < 150	Low risk
$40 < E_r^i < 80$	Moderate risk	150 < ERI < 300	Moderate risk
$80 < E_r^i < 160$	Considerable risk	300 < ERI < 600	Considerable risk
$160 < E_r^i < 320$	High risk	RI > 600	Very high risk
> 320	Very high risk		

2.3.4 Multivariate statistical analysis

Factor and cluster analysis were used for statistical multivariate analysis with software SPSS version 16.0. Correlation analysis identified correlation among heavy metals, while factor analysis simplified complex information of several variables based on similarity. Besides that, this analysis could be used for identifying the source pollution of heavy metals on the sediment, from natural source or anthropogenic contribution [10-12]. Cluster analysis is usually coupled with factor analysis to check results and to group individual parameters and variables [10]. The proximity and interrelationship among heavy metals were visualized by dendogram.

3. Results

3.1 Material Composition of Surface Sediment

The compositions of surface sediment were muddy sand (81.57%), clay (9.86%), and sand (8.57%). In general, the muddy sand distribution on the surface layer spread evenly on each station. Wave effect on surface sediment was quite low in 26-80 meters depth, then the sediment particles were accumulated in the same area.

3.2 Spatial Distribution of Heavy Metals Concentration in West Part of Java Sea

Spatial distribution pattern of heavy metals on the sediment were visualized on contour map based on concentration of each location (figure 2). There were eight heavy metals identified, those were Hg, As, Cd, Cr, Cu, Pb, Zn, and Ni. Based on the average value Ni exceeded PEL threshold, Cd and PB exceeded TEL threshold, while Hg, As, Cr, Cu, and Zn were less then TEL threshold. On the range effect, As/Cr/Cu/Zn were distributed 100%, (figure 3 and 4), Hg was distributed 70.62% under range effect of TEL, and the others were distributed 29.4% between range effect of TEL and PEL (figure 5). The other heavy metals (Cd, Pb, Ni) were distributed 44.1%, 88.8% and 40.5% between TEL-PEL range effect, while on the range of >PEL distributed 47%, 11.2% and 59.5%.

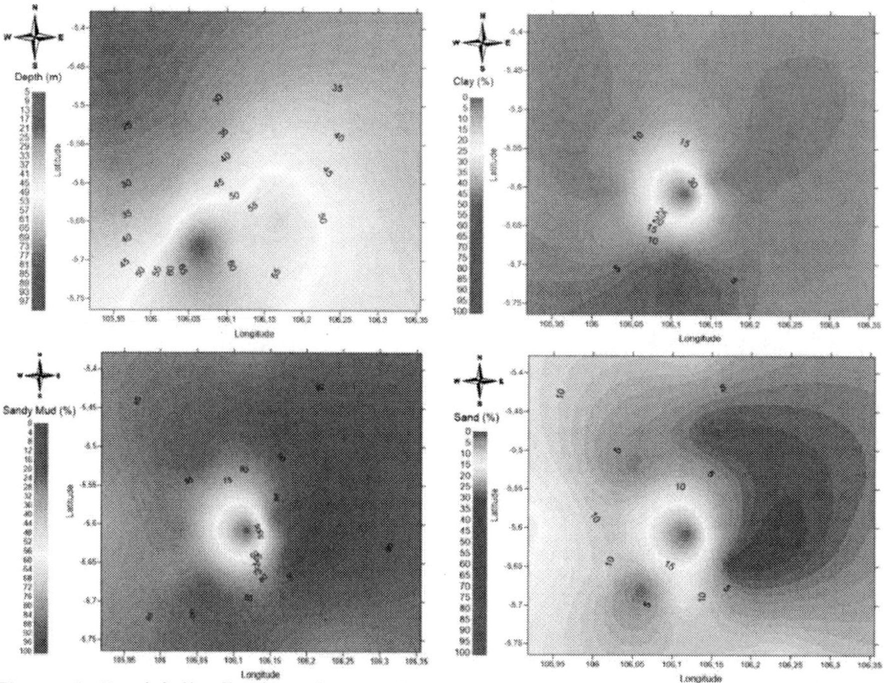

Figure 2. Spatial distribution of water depth (m), clay, sandy mud and sand percentage in west part of Java Sea

Range effect distribution between TEL and PEL for Hg occurred in ST1, ST5 and ST8. Cd was distributed in ST2, ST3, ST4, ST5, ST7, ST11 and ST13. Pb was distributed in ST1, ST2, ST3, ST5, ST7, ST8, ST9, ST10, ST11, ST12, ST13 and ST14. Ni was distributed in ST2, ST3, ST5, ST8, ST9, ST11 and ST12. Range effect >PEL for Cd occurred in ST1, ST6 and ST14. Pb was distributed in ST6 and Ni was distributed in ST1, ST4, ST6, ST7, ST10, ST13 and ST14 (figure 4 and figure 5).

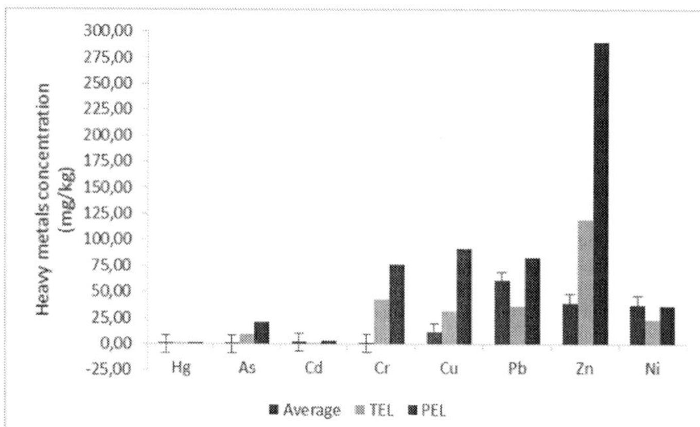

Figure 3. Heavy metals concentration average on the surface sediment in west part of Java Sea

Twenty heavy metals have been identified as essential elements for both humans and organisms at low quantities but toxic at slightly higher quantities. These include Fe, Cr, Ni, Cu, and Zn. Thereafter

Pb, Cd, Hg, As are known as toxic metals and considered the top twenty hazardous substances in the priority [1,14].

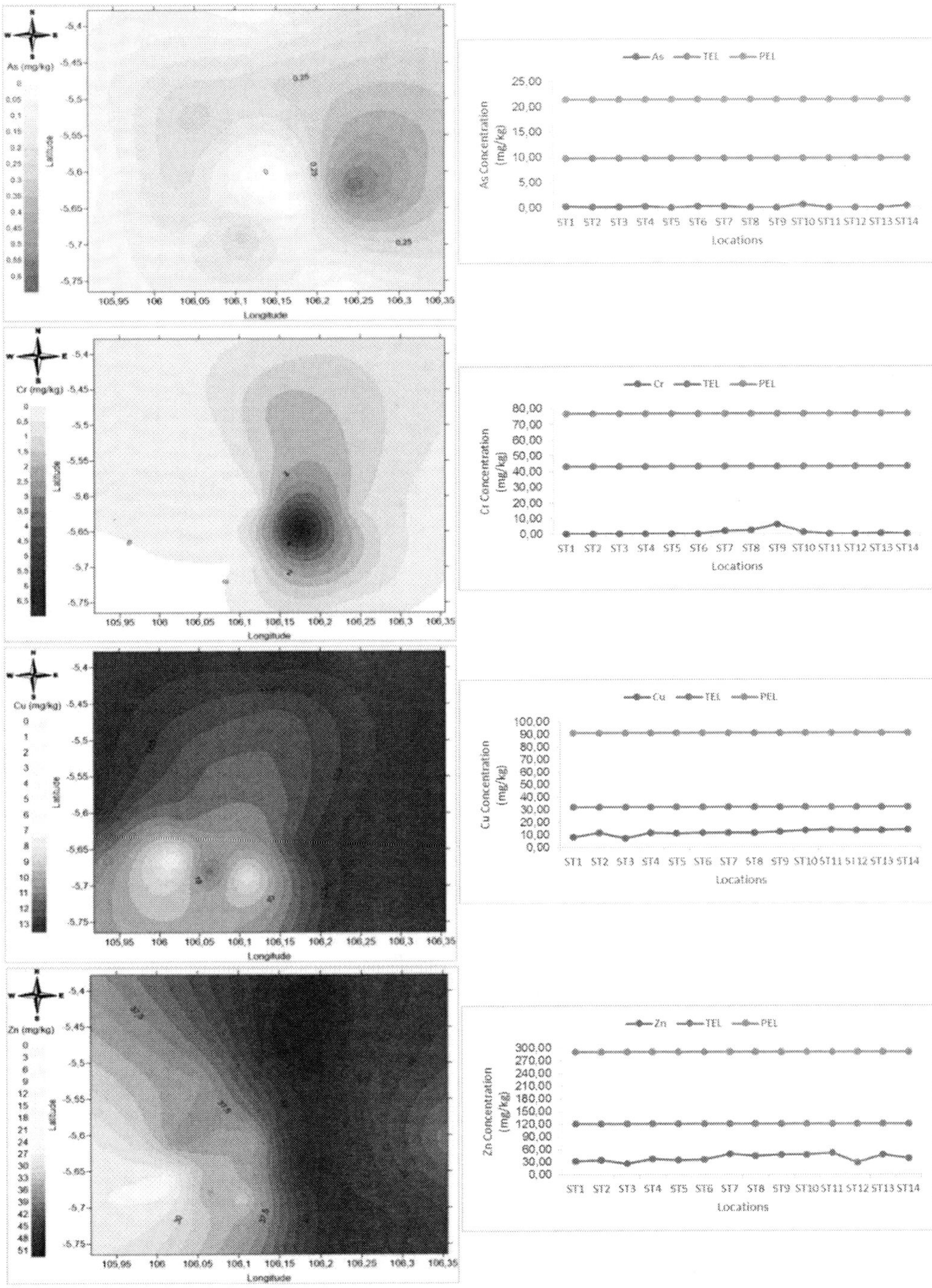

Figure 4. Spatial distribution of heavy metals As, Cr, Cu and Zn on the surface sediment in west part of Java Sea

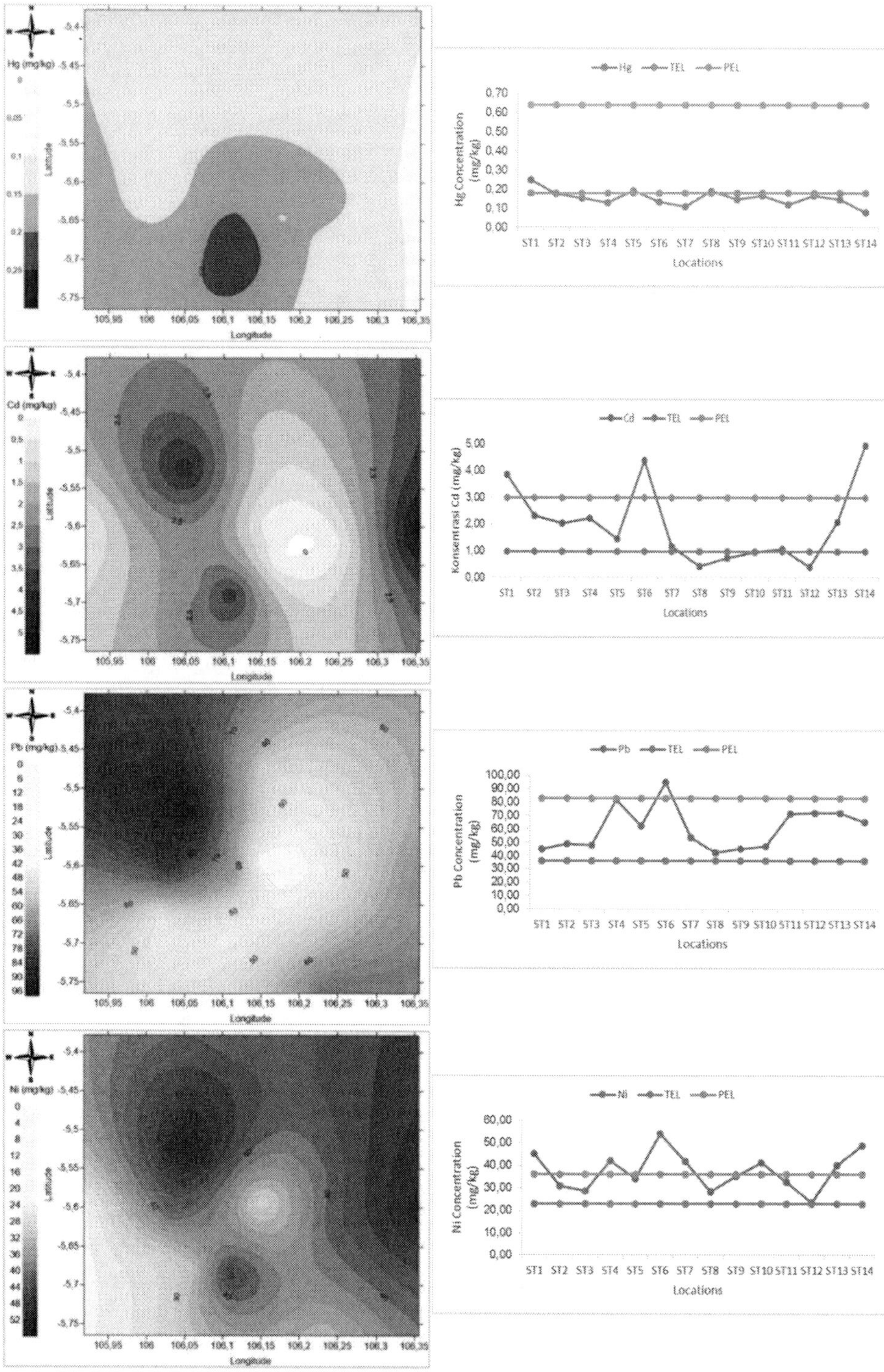

Figure 5. Spatial distribution of heavy metals Hg, Cd, Pb and Ni on the surface sediment in west part of Java Sea

On the waters column Ni appears in colloidal form such as nickel nitrate, chloride and ammonium sulfate which are dissolved nickel. Concentration of Ni was more than PEL concentration (>36mg/kg) and the difference value of concentration was 1.58 kg/mg. However, this value was still less than Ni concentration in the nature (75 mg/kg) [15, 16]. The concentration of Cd was more than TEL concentration (>0.99 mg/kg), the difference value of concentration was 1.01 kg/mg. This value was greater than Cd concentration in the nature (0.20 mg/kg) [15]. Heavy metal Cd is usually mixed with other heavy metals such as Zn and Sn, for example in the mining process, Cd concentration ranged 0.2-0.4%. The natural resources of Cd are derived from volcanic activities, forest fires, and also from the activities on the sea itself. In addition, human activities also increase concentration in the atmosphere. Generally, mining activities, waste combustion, iron, steel, and phosphate fertilizer industries contribute to Cd contamination through water flow and air.

Average concentration of Pb was 60.63 mg/kg, more than TEL (>23 mg/kg). The difference value of Pb and TEL concentration was 24.63 mg/kg, that was more than available concentration on the nature (12.5 mg/kg) [15]. Around 0.1% of Pb is contained on the earth mantel. Anthropogenic activities such as mining dumping, iron and steel fusion, and agriculture waste gain the concentration of Pb through water flow. Besides that, the sources could be produced by oil mining and storage, seabed pipeline, and unused batteries disposal to the sea.

According to sediment quality assessment for aquatic organism conservation [17], high amount of heavy metals which are exceeding PEL concentration could increase possibility of biological effect to aquatic organism. Frequently effect (>50%) occurs at concentration >PEL and rarely effect (25%) occurs at concentration between TEL and PEL or <TEL.

3.3 Ecological Risk Index

The sequences of potential risk index of heavy metals from the highest to the lowest were Cd> Hg> Pb> Ni> Cu> As> Zn> Cr. Heavy metal Cd was categorized to high risk (HR) with the average value $E_r^i = 284.44$, Hg was categorized to moderate risk (MR) with the average value $E_r^i = 77.00$, the other heavy metals were categorized to low risk (LR) with the range of average value $E_r^i = 0.02\text{-}24.25$. The toxicity of Cd has a role in the assessment of ecological risk potential. The toxicity value of Cd equals to 30 means the largest toxicity value than Pb/Ni/Cu/As/Zn/Cr, except Hg [9].

According to sampling locations, 40.58% area were categorized to very high risk (VHR), 36.24% area were categorized to considerable risk (CR) and 23.19% were categorized to moderate risk (MR). The high potential ecological risk occurred in ST14, ST6 and ST1. The sequences of potential ecological risk from the highest to the lowest were T14> ST6> ST1> ST2> ST4> ST13> ST3> ST5> ST7> ST11> ST10> ST9> ST8> ST12 (figure 6). The contribution of Cd, Hg, and the others to ecological contribution to ecological risks of heavy metals on surface sediments were 73.88%, 18.71%, and < 6%. Based on these percentages, Cd contribution was categorized to considerable risks.

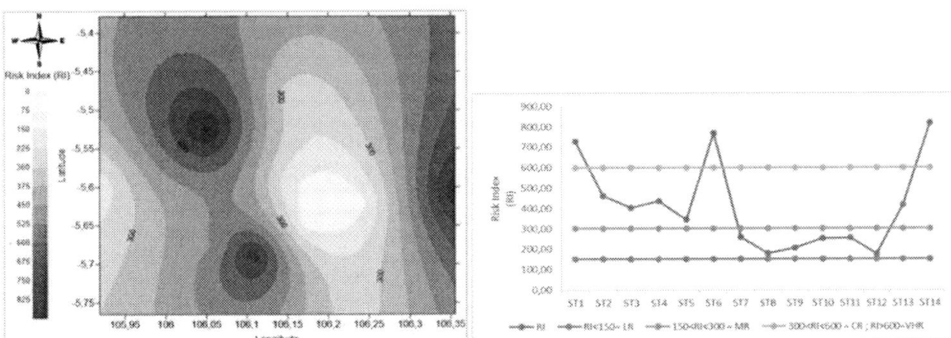

Figure 6. Spatial distribution ecological risk on RI of heavy metals on the surface sediment in west part of Java Sea

3.4 Multivariate Statistical Analysis

3.4.1 Correlation Analysis

The correlation among heavy metals (As/Ni, Cd/Ni, and Cu/Zn) and heavy metals with Risk Index (Cd/Ri and Ni/Ri) had positive correlation in the significance level $p < 0.05$ (table 2). The positive correlation shows the same source between heavy metals, dependency, and identic role during the transport [10, 18].

Table 2. Correlation matrix for heavy metals concentration on the surface sediment in west part of Java Sea

	Hg	As	Cd	Cr	Cu	Pb	Zn	Ni	Ri
Hg	1.00								
As	-0.09	1.00							
Cd	-0.13	0.28	1.00						
Cr	-0.45	-0.19	-0.51	1.00					
Cu	-0.57	-0.06	-0.19	0.12	1.00				
Pb	-0.45	0.06	0.39	-0.45	0.37	1.00			
Zn	-0.38	0.02	-0.37	0.49	**0.61***	-0.09	1.00		
Ni	-0.27	**0.59***	**0.71***	-0.13	0.08	0.39	0.17	1.00	
Ri	-0.05	0.28	**0.99***	-0.52	-0.24	0.38	-0.41	**0.69***	1.00

* Correlation is significant at $P<0.05$

3.4.2 Factor Analysis

Factor analysis had been applied on 9 variables for 14 sampling locations to identify the possibilities of heavy metals source on surface sediment especially for Kaiser Normalization rotation. Total variant and rotation component of heavy metals matrix explained 80.04% from total variant (figure 7). Factor 1 explained 30.20% and was dominated by Cd and Pb. Factor 2 explained 25.30% and was dominated by Cu, Pb, and Zn. Factor 3 explained 24.54% and was dominated by As, Cd, and Ni. Generally, the result of factor analysis were similar to correlation analysis result. In this case, the result of Pb was not similar between factor and correlation analysis result (coefficient correlation $Pb < 0.5$) (table 2).

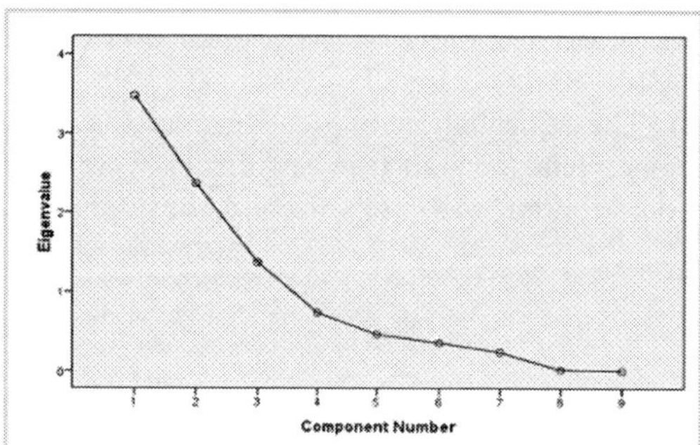

Component	Initial Eigenvalues			Rotation Sum of Squared Loading		
	Total	% of Variance	Cumulative %	Total	% of Variance	Cumulative %
1	3,48	38,61	38,61	2,72	30,20	30,20
2	2,36	26,25	64,86	2,28	25,30	55,50
3	1,37	15,17	80,03	2,21	24,54	80,04

Figure 7. Screen plot of eigen values and variance total on three component factor

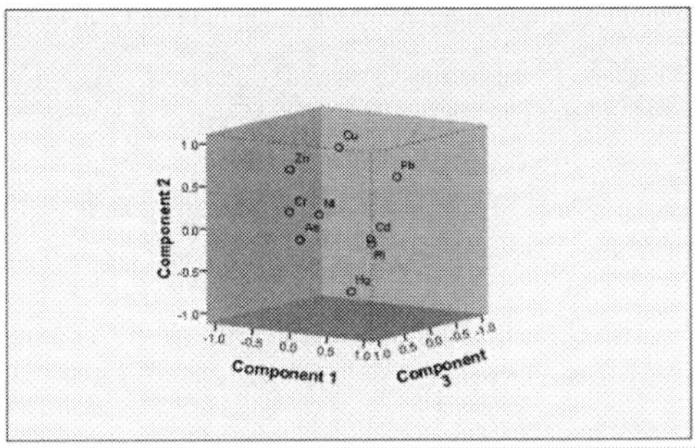

Heavy Metal	Component		
	1	2	3
Hg	-0,05	-0,82	-0,14
As	-0,07	-0,07	**0,81**
Cd	**0,73**	-0,02	**0,59**
Cr	-0,80	-0,06	-0,03
Cu	-0,18	**0,87**	-0,09
Pb	**0,67**	**0,62**	-0,01
Zn	-0,66	**0,60**	0,15
Ni	0,24	0,27	**0,89**
Ri	**0,74**	-0,07	**0,59**

Figure 8. Plot of component factor in rotated component matrix

3.4.3 Cluster Analysis

Average linkage in cluster analysis was used to group heavy metals and sampling locations. Based on the type of heavy metals, Cd/Ni/Pb/As were identified as high contaminant level (cluster 1), Hg was categorized to moderate contaminant level (cluster 2), and Cu/Zn/Cr were categorized to low contaminant level (cluster3) (figure 9a). Based on locations, cluster 1 consisted of ST1, ST2, ST3, ST4, ST5, ST6, ST7, ST11, ST13, and ST14, cluster 2 consisted of ST9 and ST10, while cluster 3 consisted of ST8 and ST12 (figure 9b).

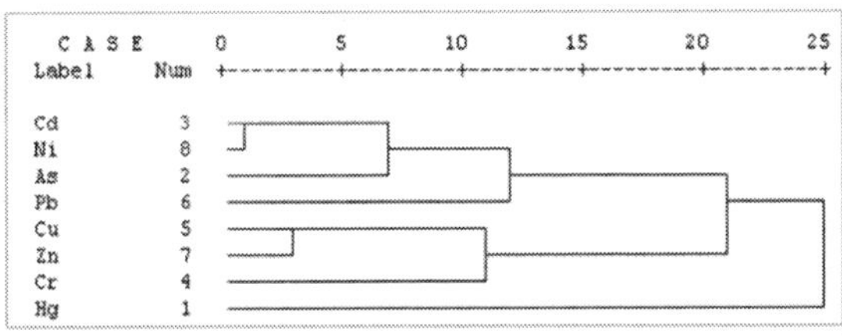

(a)

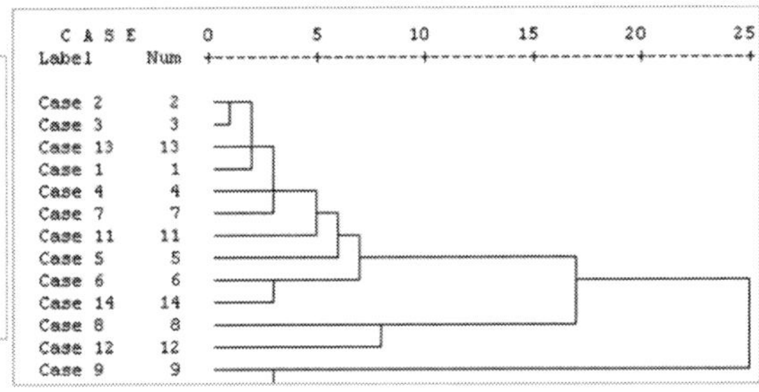

(b)

Figure 9. Dendrogram of heavy metals concentration (a) and sampling locations on the surface sediment in west part of Java Sea (b)

The observation areas were national and international shipping route, seabed pipeline installation, and mining area. Besides that, many industry activities are conducted especially in the coastal area of Banten Bay. Those activities may contribute to heavy metals input to the water and sediment. The availability of surface sediment (<0.5 m) which consist by mud may control heavy metal concentration on the sediment itself. Fine particles have a high ability to absorb dissolved heavy metals and deposited on the surface sediment [1].

4. Conclusions

The concentration of Ni was the highest on the surface sediment (<0.5 m) in west part of Java Sea, exceed the threshold of range effect level (>PEL). The potential risk sequences of heavy metals from the highest to the lowest were Cd>Hg>Pb>Ni>Cu>As>Zn>Cr. Cluster 1 (Cd/Ni/Pb), cluster 2 (Hg), and cluster 3 (Cu/Zn/Cr) were categorized to high, moderate, and low contaminant level. Based on locations, cluster 1 consisted of ST1, ST2, ST3, ST4, ST5, ST6, ST7, ST11, ST13, and ST14, cluster 2 consisted of ST9 and ST10, while cluster 3 consisted of ST8 and ST12.

5. References

[1] Sany SBT, Salleh A, Sulaiman AH, Sasekumar A, Tehrani G and Rezayi M 2012 Distribution characteristics and ecological risk of heavy metals in surface sediments of West Port Malaysia *Environmental Protection Engineering* **38** 139-154

[2] Singh KP, Mohan D, Singh VK and Malik A 2005 Studies on distribution and fractionation of heavy metals in Gomti River sediments A tributary of the Ganges, India *Journal of Hydrology* **312** 14–27

[3] Cevik F, Goksu MZL, Derici OB and Findik O 2009 An assessment of metal pollution in surface sediments of seyhan dam by using enrichment factor, geoaccumulation index and statistical analyses *Environ. Monitoring Assessment.* **152** 309–317

[4] Devesa-Rey R, Diaz-Fierros F and Barral MT 2010 Trace metals in river bed sediments: An assessment of their partitioning and bioavailability by using multivariate exploratory analysis. *Journal of Environmental Management* **91** 2471–2477

[5] Bai JH, Cui BS, Chen B, Zhang KJ, Deng W, Gao HF and Xiao R 2011 Spatial distribution and ecological risk assessment of heavy metals in surface sediments from a typical plateau lake wetland China. *Ecological Modelling.* **222** 301-30

[6] Suresh G, Sutharsan P, Ramasamy V and Venkatachalapathy R 2012 Assessment of spatial distribution and potential ecological risk of the heavy metals in relation to granulometric contents of Veeranam lake sediments, India. *Ecotoxicology and Environmental Safety.* **84** 117–24

[7] Folk RL 1980 *Petrology of Sedimentary Rocks*. (Austin: Hemphill Publishing) 184p

[8] CBSQG 2003 *Consensus-based sediment quality guidelines Recommendations for use & application interim guidance*. (Contaminated Sediment Standing Team) WT-732 40 p

[9] Hakanson L 1980 An ecological risk index for aquatic pollution control, a sedimentological approach. *Water Research* **14** 975-1001

[10] Lu XW, Wang LJ, Li LY, Lei K, Huang L and Kang D 2010 Multivariate statistical analysis of heavy metals in street dust of Baoji, NW China *Hazardous Materials* **173** 744-749

[11] Chen XD, Lu XW and Yang G 2012 Sources identification of heavy metals in urban topsoil from inside the Xi'an Second Ringroad, NW China using multivariate statistical methods *Catena* **98** 73–78

[12] Yang ZP, Lu WX, Long YQ, Bao XH and Yang QC 2011 Assessment of heavy metals contamination in urban topsoil from Changchun City, China *Journal of Geochemical Exploration* **108** 27-38

[13] Li F, Huang J, Zeng G, Yuan X, Li X, Liang J, Wang X, Tang X and Bai B 2013 Spatial risk assessment and source identification of heavy metal in surface sediment from the Dongting Lake, Middle China *Journal of Geochemical Exploration* **132** 75-83

[14] US EPA (Environmental Protection Agency) 2004 *Contamination sediment science priorities*. Prepared for U.S. Environmental Protection Agency by Members of the Contaminated Sediments Science Priorities Workgroup 172p

[15] Tylor SR 1964 Abundance of chemical elements in the continental crust: a new table *Geochimica et Cosmochimica Acta.* **28** 1273-1285

[16] Moore JW 1991 *Inorganic contaminant of surface water: research and monitoring priorities* (New York: Springer-Verlag) 334p

[17] Canadian Council of Minister of the Environment (CCME) 2001 *Canadian sediment quality guidelines for the protection of aquatic life: Introduction*. (Canada: Winnipeg)

[18] Saeedi M, Li LY and Salmanzadeh M 2012 Heavy metals and polycyclic aromatic hydrocarbons: pollution and ecological risk assessment in street dust of Tehran *Journal of Hazardous Materials* **227–228** 9–17

The Existence of Phytoplankton and ZooplanktonDuring Solar Eclipse in A Single Spot of Pramuka Island Waters, Seribu Islands

NTM Pratiwi*, RZulmi, DMulyawati, and GSASulaiman
Department of Aquatic Resources Management,
Faculty of Fisheries and Marine Sciences, Bogor Agricultural University
Jl. AgatisKampus, Dramaga, Bogor 16680

E-mail: niken_tmpratiwi@yahoo.com

Abstract. The observation of phytoplankton and zooplankton in Karang Lebar Lagoon during solar eclipse periods was held in order to understand the existence of the community in this special and unusual phenomenon of nature. There were 13 temporally sampling to take plankton and water quality samples, one day before (twice), during (eight times), and one day after the eclipse (three times). The results show that the number of taxa and abundance of plankton during solar eclipse for both phytoplankton and zooplankton were lower than the daylight normal condition. Furthermore, the highest value of Shannon diversity and Evenness index, and also the succession rate of phytoplankton species were reached just after the solar eclipse over. Most of this pattern was followed by zooplankton species at the next sampling periods. As a whole, there were different excistence pattern of plankton performes between solar eclipse period and normal condition.

1. Introduction

The solar eclipse being a rare natural phenomenon gives an opportunity to investigate how the ionising radiations react to the material surface of the earth due to the fast solar radiation changes. A huge quantity of shorter wavelength radiations are expected to reach the earth's surface during solar eclipse since the disturbance of the heat balance along the supersonic travel of the trajectory of the Moon's shadow could generate eclipse induced gravity waves [1].

Solar eclipse, especially total solar eclipse, is one of long periodical phenomenon, therefore there are lack information about relationship between the phenomenon and its effect, relatively, especially related to the aquatic organisms. Solar eclipse in the area study was a total eclipse with 350 years period. However, another type of solar eclipse in this research location had occurred in 1901,1926,1929,1934,1962,1983, 1984, 1988, and 1995.The total eclipse made a rapid change of light intensity, that presumably leads change of another environment condition [2].

The environmental effects of a solar eclipse have been mainly focused on meteorological parameters [1], photochemistry [3], boundary layer physics [4], total columnar ozone [5], gravity waves [6], ionospheric parameters [7] and also plants [8] and animals [9]. Temperature, relative humidity, wind and cloudiness are among the most common meteorological parameters measured in experimental campaigns during solar eclipses [3].

Radiations at shorter wavelengths are generally influenced more by the eclipse. Various solar eclipse effects on plants mainly related to the abrupt solar light "switch off" such as transient aberrations in the chromosomal structure of root meristems and a delaying seed germination, effects on photosynthesis and evapotranspiration of crop plants, etc. [3].

Light effects on the photochemical phase of photosynthesis may be responsible for the observed depression in photosynthetic rates. Field studies addressing the migratory responses of marine zooplankton (microzooplankton (ciliates), and meso-zooplankton) due to the rapid changes in underwater light intensity were also performed. The light intensity attenuation was simulated with the use of accurate underwater radiative transfer modelling techniques. Ciliates, responded to the rapid decrease in light intensity during the eclipse adopting night-time behaviour. From the meso-zooplankton assemblage, various vertical migratorybehaviours were adopted by different species [4]. The aim of the observation was to study the existance of phytoplankton and zooplankton in Secchi depth layer of Karang Lebar Lagoon during solar eclipse.

2. Materials and Methods

The observation was carried out at wide coral reefs lagoon surround Karang Lebar Lagoon, in the area of Seribu (Thousand) Islands, Jakarta (Figure 1). The sampling site was restricted and protected from anthropogenic activities in order to get clear information about the respond of plankton towards solar eclipse phenomenon.

Vertical sampling of plankton was done compositely from Secchi depth to water surface. There were 13 temporally sampling to take plankton and water quality samples; one day before (twice:12.00 and 16.00 o'clock), during (eight times: 06.00, 08.00,.10.00,.12.00,14.00, 16.00, 18.00, and 20.00 o'clock), and one day after the eclipse (three times: 06.30, 12.00, and 16.30 o'clock). The observation of water quality comprised of temperature, pH, salinity, and nutrients that were analyzed in the Water Quality Laboratory of Aquatic Productivity and Environment Division, Department of Aquatic Resources Management, Faculty of Fisheries and Marine Sciences, Bogor Agricultural University.

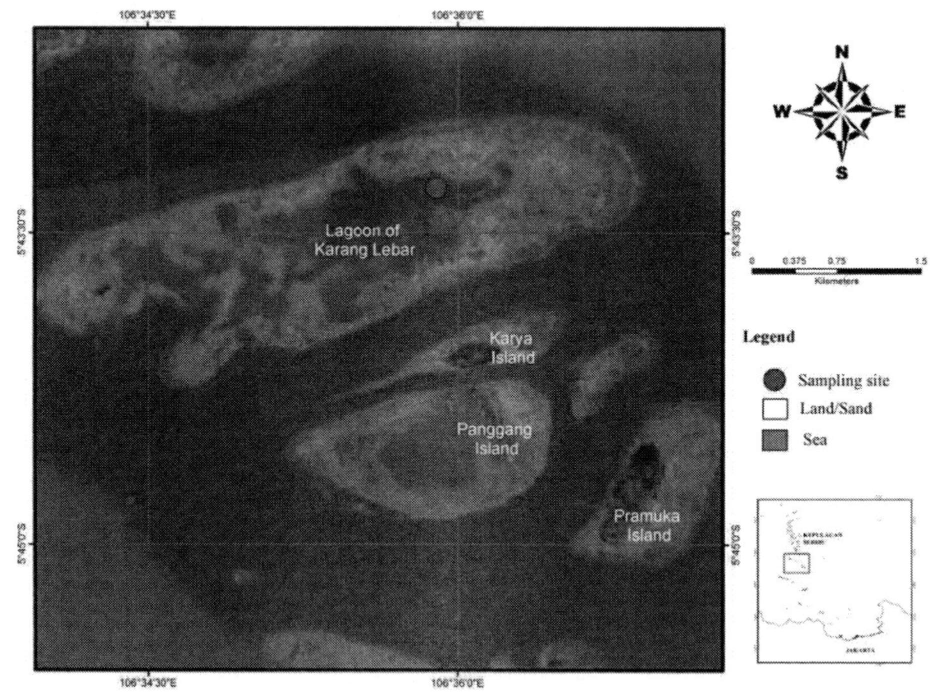

Figure 1. Research location and sampling site at Karang Lebar Lagoon

2.1. Sample collection

The plankton samples were collected with plankton net that drawn vertically (hauling) from Secchi depth to water surface. The samples were preserved with 1% Lugol's solution and identified using plankton identification guide [5], [6], and [7]. The counting of plankton was based on SRC (Sedgwick Rafter Counting cell) observation [8]. The specimens were documented in photograph that was supported by trinocular microscope Zeiss Primo Star with camera and software AxioVision Rel 4.8.

In situ water quality measurements were carried out for temperature, dissolved oxygen/DO, salinity, pH, water depth, water transparency, and light intensity. Nutrients that consist of nitrate-N, nitrite-N, ammonia-N, ammonium-N, and orthophosphate were analyzed in laboratory [8].

2.2. Data analysis

The diversity of plankton along the solar eclipse were determined as Shannon-Wiener diversity index (H'), Evenness index (E), and Simpson diversity or Dominance index (C) [9]. Those indices show the stability of plankton community in the circumstantial condition.

The data analysis was used to determine succession of plankton during unusual and rapid change of light intensity. Summed Difference Index or SDI indicated the rate of succession [10]. The index was formulated as follows.

Summed difference index (SDI; σ_S) is a formulation to calculate the movement, change, or succession rate of plankton. SDI is formulated as follows.

$$\sigma_s = \sum_i \left| d[b_i(t)/B(t)] \right| / dt \quad (1)$$

The formulation for specific time interval is:

$$\sigma_S = \frac{\sum_i \left| [b_i(t_1)/B(t_1)] - [b_i(t_2)/B(t_2)] \right|}{t_2 - t_1} \quad (2)$$

with:

$b_i(t)b_i(t)$: Species density of each observed time

$B(t)B(t)$: Total density of each observed time.

The difference excixtence pattern of plankton performances along the solar eclipse towards the other periods were analyzed by Student t-test [11]

$$t = \frac{\bar{x} - \mu 0}{s/\sqrt{n}} \quad (3)$$

Information:

t: t test

$\bar{x}$: Average sample

$\mu 0$: Average Average specific or particular (which becomes the comparison)

s: standard deviation of the sample

n: number of samples.

3. Results and Discussion

The results showed that the abundance and number of species of phytoplankton were relatively low during the time of solar eclipse. The increasing was shown at 12.00 o'clock of the solar eclipse day, and the extremely spike in abundance value was shown at 12.00 o'clock of the next day. The fluctuation pattern was quite similar between abundance and number of species (Figure 2a and 2b). Bacillariophyceae was the dominance group, mostly, both in abundance and the number of species.

The different condition was shown at noon and afternoon of the next day that the Cyanophyceae had higher density than Bacilloriaphyceae.

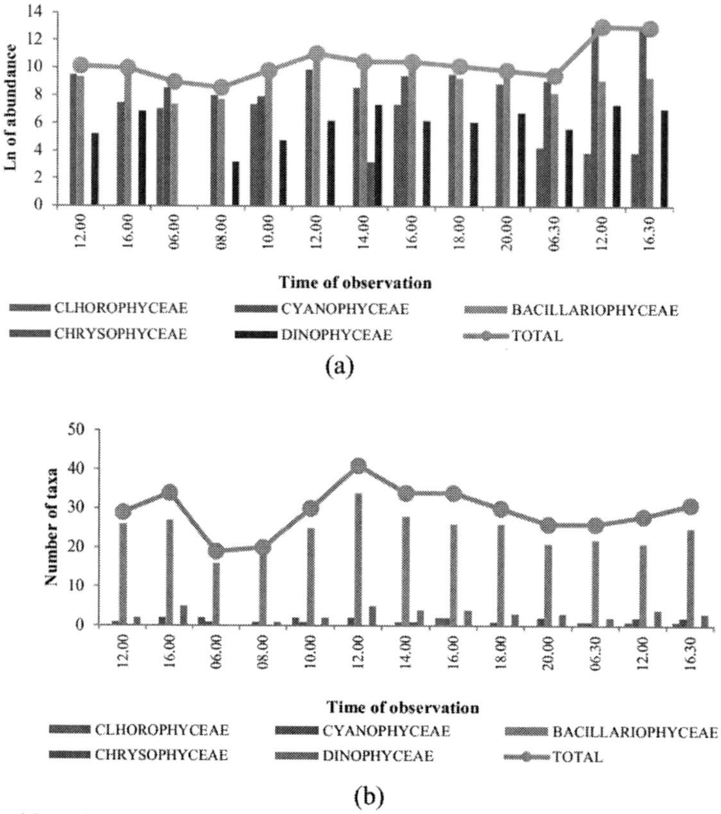

Figure 2. Composition of phytoplankton as abundance (a) and number of species in period of solar eclipse (b)

The amount of light (or solar radiation) strongly affects both the amount and rate of photosynthesis. Thus the photosynthesis occurring in a water sample is proportional to the light intensity; photosynthesis increases with the increasing light intensity up to some maximal value (Pmax). At still higher light intensities, there may be a significant decrease in photosynthesis (called photoinhibition) that is caused by a number of physiological reactions such as shrinkage of chloroplasts in bright light [12].

The drop in solar irradiance and the increase in mesophyll CO_2-concentration during the eclipse did not induce stomatal closure thus not blocking CO_2 uptake by plants. Light effects on the photochemical phase of photosynthesis may be responsible for the observed depression in photosynthetic rates [13].

The zooplankton community also showed a unique pattern. There were low density and species number during solar eclipse, but with different group of dominance. The Crustacea was always in high level of abundance during the observation. But the Protozoan that usually had high number of species was replaced by Crustacea at the time of solar eclipse (Figure 3a and 3b).

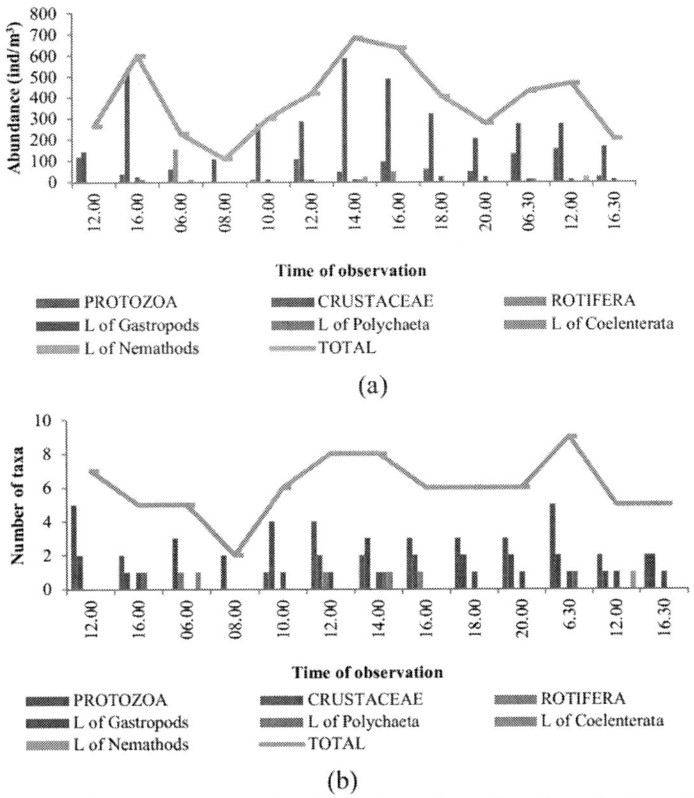

Figure 3. Composition of zoooplankton as abundance (a) and number of species in period of solar eclipse (b)

Animals alter their behaviour [14] and some planktonic Crustacea are vertically redistributed [15] [16]. The evidence suggests that certain species of epipelagic oceanic copepods and euphausiaceans in the Gulf of Mexico responded to the noontime total solar eclipse of March 1970 by migrating to the surface [17].

Light intensity as well as other environmental factors such as temperature, hydrostatic pressure, and pH can affect the direction and intensity of swimming of zooplankters. The vertical migratory performance can be most easily interpreted by the hypothesis zooplankters were responding exclusively toconcurrent light intensity [18].

Field studies addressing the migratory responses of marine zooplankton (microzooplankton (ciliates), and meso-zooplankton) due to the rapid changes in underwater light intensity were also performed. The light intensity attenuation was simulated with the use of accurate underwater radiative transfer modelling techniques. Ciliates, responded to the rapid decrease in light intensity during the eclipse adopting night-time behaviour. From the meso-zooplankton assemblage, various vertical migratory behaviours were adopted by different species [1].

The community structure of phytoplankton indicated that the diversity was low at the beginning of eclipse phenomenon, and increased to moderate at the end of the day. The unique things were appeared at the following day that there were shown declining values of diversity and Evennes index to minimum level. Meanwhile, the opposite thing was occured in dominance index that showed maximum level. The zooplankton showed a different condition. All diversity indices were relatively constant (Figure 4a and 4b).

It was clearly shown that the pattern of diversity was in line with the Eveness value, along the observation. Furthermore, the dominance value was in contrast with the value of those two indices. It means that the relationship between ecological index of the plankton community structure was well illustrated.

The SDI showed the rate of succession or the change of species in certain time interval. The highest rate of plankton occurred in the time of solar eclipse. From Figure 5a and 5b, it is shown that zooplankton community was changed earlier that phytoplankton community.

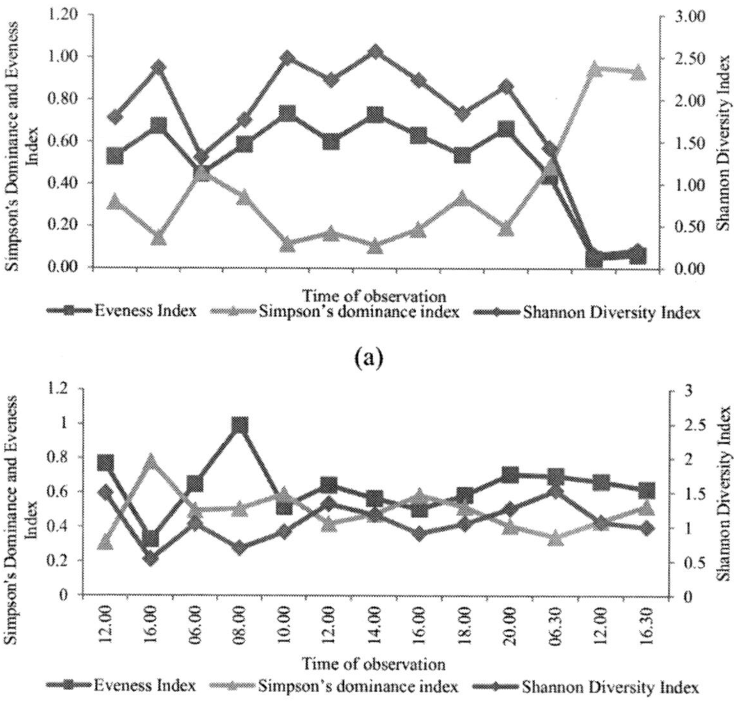

(b)

Figure 4. Diversity indices of phytoplankton (a) and zooplankton (b) in time of solar eclipse phenomenon

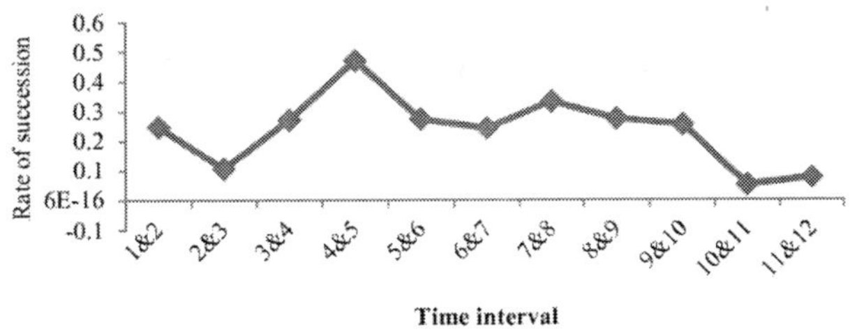

(a)

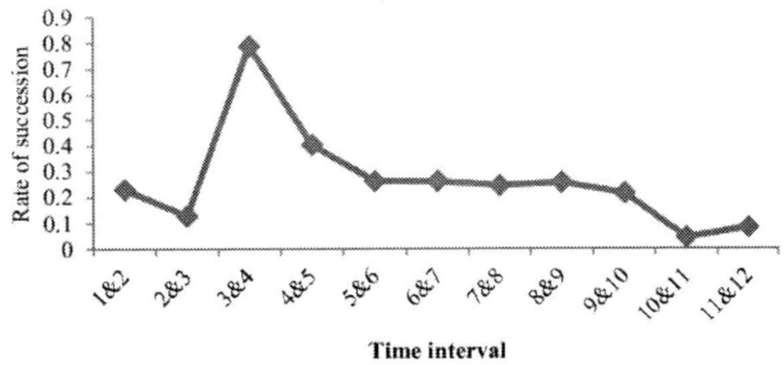

(b)

Figure 5. The rate of succession of phytoplankton (a) and zooplankton (b) in time of solar eclipse phenomenon

There were some descriptions of the excistence of plankton community, such as density, number of species or richness, Shannon diversity index, eveness index, and Simpson's domonance index, in time of solar eclipse phenomenon. The pattern of phytoplankton density was relatively different from the zooplankton.

The density and number of species of phytoplankton along the solar eclipse were significantly different ($p<0.05$) from the density at the same time of one day before and after solar eclipse, but quite similar ($p>0.05$) to the density of dark time, of one day after eclipse. Meanwhile, the density of zooplankton was only significantly different ($p<0.05$) from the density at the same time of one day after solar eclipse.

There were not any significantly differences ($p>0.05$) among all of indices for the phytoplankton community. On the contrary, and the interestingly was shown by the Sannon diversity and the Simpson's dominance indices of zooplankton that significantly different ($p<0.05$) from the condition along the eclipse with the same time of one day after solar eclipse and to the condition of dark time, of one day after eclipse.

As a whole, it was shown that the excistence of phytoplankton along the solar eclipse very similar to those of the dark periods. This was different with the zooplankton that showed relatively similar performance as daylight excistence, along the solar eclipse, and showed a different performance at dark situation. It means that there were different performance pattern of phytoplankton and zooplankton community in responding the solar eclipse phenomenon.

In essence, a high change of intensity produced an alteration of photonegative and photopositive phases and the net result had relatively little effect on the depth at which the animal was located. When changes in light intensity were slow. However, the animals simply followed the movement of the original optimum zone [12].

The environmental condition was represented by the measurement result of *in situ* observation and nutrient analysis as laid in Table 1 and 2. It was indicated that the most different condition at the same time of observation was the light intensity.

There was only 18.25% of light intensity from the normal condition at 06.00 before it became totally dark at 07.00. It was relatively similar with another observation that the light intensity of eclipse day was 1002 lux with the minimum value of the eclipse day as 10 lux [3].

The other water quality parameter that influenced by solar eclipse was pH. As seen in Table 1, the pH condition during eclipse period to the midday of 9 March was decreased. The gamma rays are reaching the earth's surface during eclipse since the pH value of seawater is reduced when exposed to solar radiation because gamma irradiation reduces the pH value of water [1].

Table 1 Water quality in time of solar eclipse at single spot of Pramuka Island

Parameter	Units	8 March		9 March								10 March		
		12.00	16.00	06.00	08.00	10.00	12.00	14.00	16.00	18.00	20.00	06.00	12.00	16.00
Depth	m	11	11	11	11	11	11	11	11	11	11	11	11	11
Transparency	m	6	6	6	6	6	6	6	6	6	6	6	6	6
DO	mg/L	6,9	7,5	3,2	5,2	6,1	7,4	9,1	8,8	8	7,6	3,6	6,3	7,6
Temperature	°C	29,7	30,7	28,6	29,4	29,8	29,9	31,6	32,6	32,4	31,3	29,2	31,4	33,1
pH	-	8,06	8,03	7,44	7,86	7,92	7,96	8,1	8,17	8,22	8,11	7,63	8,11	8,3
Salinity	ppt	29,8	29,8	29,8	30	30	29,9	30,1	30,1	30,2	30,2	30,2	30,2	30,3
Conductivity	µmHos/cm	46200	46200	46100	46400	46400	46300	46600	46700	46800	46800	46600	46700	46900
TDS	g/L	23,56	23,57	23,5	23,67	23,66	23,6	23,77	23,83	23,87	23,87	23,78	23,83	23,92
Light intensity	lux	54700	48473	1248	26220	39600	52980	102060	65900	951	-	6838	85920	40800

There was not any significance difference condition of nutrients, before, during, and after solar eclipse occurred. As a whole, all of nutrients were relatively adequate for the living and growth of phytoplankton.

Table 2 Water nutrients in time of solar eclipse at single spot of Pramuka Island

Parameters	Units	8 March	9 March	10 March
Ammonia (NH_3-N) $^+$	mg/L	0,082	0,050	0,102
Nitrate (NO_3-N)$^-$	mg/L	0,056	0,046	0,062
Nitrite (NO_2-N)$^-$	mg/L	0,002	0,002	0,002
Orthophosphate (PO_4-P) $^+$	mg/L	0,010	0,011	0,011

The results showed that the number of taxa and abundance of plankton during solar eclipse for both phytoplankton and zooplankton were lower than the daylight normal condition. Furthermore, the highest value of Shannon diversity and Evenness index, and also the succession rate of phytoplankton species were reached just after the solar eclipse over. Most of this pattern was followed by zooplankton species at the next sampling periods

4. Conclusion

As a whole, there were different pattern of richness, diversity, and abundance of plankton between solar eclipse period and normal condition.

Reference

[1] Kumar SS, R Rengaiyan. 2011. Influence of solar eclipse on seawater. *Natural Science.* **3**(1): 69-74.

[2] Littmann, Espenak, Willcox. 2008. Totality: Eclipses of The Sun.Oxford University Press

[3] Sambandan K, DeviKS, KumarSS, NancharaiahM, DhatchanamoorthyN. 2012. Effects of solar eclipse on photosynthesis of Portulacaoleracea and *Phyla nodiflora* in coastal wild conditions. *Journal of Phytology*. **4**(2): 34-40.
[4] Economou G, ChristouED, GiannakourouA, GerasopoulosE, GeorgopoulosD, KotoulasV, LyraD, TsakalisN, TzortziouM, VahamidisP, PapathanassiouE, KaramanosA. 2008. Eclipse effects on field crops and marine zooplankton: the 29 March 2006 total solar eclipse. *Atmospheric Chemistryand Physics*.**8**: 4665-4676.
[5] Davis CC. 1955. *The Marine and Freshwater Plankton*. Michigan State University Press. USA.
[6] Yamaji IE. 1979. *Illustration of The Marine Plankton of Japan*. Hoikusha Publishing Co. Osaka.
[7] Tomas CR. 1977. *Identifying Marine Phytoplankton*. Academic Press. California. USA.
[8] APHA, AWWA, WEF. 2012. *Standard Methods for examination of water andwastewater*. 22nd ed. Washington. American Public Health Association.
[9] Krebs CJ. 1972. Ecology: *The Scientific Study of the Interactions that Determine the Distribution and Abundance of Organisms*. Harper & Row Publisher. New York.
[10] LewisWMJr. 1978. Analysis Of Succession In A Tropical Phytoplankton Community and A New Measure of Succession Rate. *The American Naturalist*. **112**(984): 401-414.
[11] Walpole, R.E. dan R.H. Myers. 1995. Ilmu Peluang dan Statistika untuk Insinyur dan Ilmuwan. Edisi keempat. Penerbit ITB. Bandung.
[12] Lalli M, XR Parsons. 2006. *Biological Oceanography: an Introduction. - 2nd ed. Chapter 3: Radiation and Photosynthesis*. Elsevier Butterworth-Heinemann Linacre House, Jordan Hill, Oxford.
[13] SkudBE. 1967.Responses of Marine Organisms during The Solar Eclipse of July 1963. *US Fish and Wildlife Service Fishery Bulletin*. **66:** 259-271.
[14] Jennings S, Bustamante R, Collins HK,Mallinson J. 1998. Reef fish behaviour during a total solar eclipse at Pinta Island, Galapagos, *Journal of Fish Biology*. **53**: 683-686.
[15] VecchioneM, MaplesRS, Donahoe R.1987.Changes in chlorophyll a concentrations in a shallow water column during a solar eclipse. *Smithsonian Contributionto the Marine Sciences*. **29**: 37-44.
[16] Giroud C, BalvayG.1999.The solar eclipse and the migration of some planktonic crustacea in Lake Geneva. *Archaeological Science*.**52**: 199-208.
[17] Bright T, F Ferrari, D Martin. 1972. Effects of a total solar eclipse on the vertical distribution of certain oceanic zooplankters. *Limnology and Oceanography*. **17**(2): 296-301.
[18] Enright JT. 1967. Vertical Diurnal Migration and Endogenous Rhythmicity. *The Journal of Science*. **157**: 937-943.

Diversity and Spatial Distribution of Plankton in Connected Waters of Bali Strait, Between Eastern Part of Java and Western Part of Bali Island

NTM Pratiwi, DY Wulandari, IP Ayu, and A Iswantari
Department of Aquatic Resources Management,
Faculty of Fisheries and Marine Sciences, Bogor Agricultural University
Jl. Agatis, Kampus IPB Dramaga, Bogor 16680

E-mail: niken_tmpratiwi@yahoo.com

Abstract. Bali Strait waters located between Java and Bali Island contain high productivity, which indicated by the high diversity of phytoplankton and zooplankton. The purpose of the research was to understand the local (α), intra-space (β), and regional or total (ϒ) diversity values (that expressed as richness) of plankton taken from surface and thermocline layers. The surface observation was carried out at three locations, two locations, with 5 stations of each, nearshore (Banyuwangi/Jawa Island and Southern part of Bali Island) and one location, with 9 stations, offshore (Bali Strait). The thermocline was observed offshore at all 9 stations as the surface. The diversity of each stations (α) were 11-20 for phytoplankton and 7-12 for zooplankton at the surface, then 9-11 and 7-13, respectively at the thermocline. The total diversity (ϒ) of phytoplankton was 27 and 13 of zooplankton. Based on the β values, the plankton of nearshore were also found at offshore locations. Furthermore, some species of plankton at the surface were not found in thermocline layer. As a whole, phytoplankton and zooplankton were well distributed at the surface, but showed as specific distribution at thermocline layer, especially for phytoplankton.

1. Introduction

Bali Strait located between Java Island and Bali Island connects two waters, i.e. Bali Sea in the north and Indian Ocean in the south. It contain high productivity of fish [1] which is also indicated by a high the diversity of plankton, both phytoplankton and zooplankton [2].

Plankton is microscopic organism, both animals and plants, which drift with the flow [3]. Some types of plankton can swim only passively, not at all move, and others swam quite active [4]. Plankton consists of phytoplankton and zooplankton. Phytoplankton is an autotrophic microorganism which able to produce organic matter from inorganic material through the process of photosynthesis with the help of light. Approximately 95% of primary production in the ocean comes from phytoplankton [5]. Furthermore, zooplankton plays an important role in aquatic food webs, both as a resource for consumers on higher trophic levels and as a conduit for packaging the organic material in the biological pump [6].

Phytoplankton and zooplankton communities are distributed either horizontally or vertically in all waters. Distribution of phytoplankton is influenced by physical factors such as the mass movement of

IOP Conf. Series: Earth and Environmental Science **54** (2017) 012090 doi:10.1088/1755-1315/54/1/012090

water and chemicals, such as nutrients. In addition, zooplankton can rely on water currents to move any great distance.

The presence of phytoplankton is highly correlated to photic and warmer water layer, while zooplankton might be present more at lower water layer. Meanwhile, the water quality and mass water movement could influence the presence of those plankton communities in the water layer.

The thermocline is a unique layer that shows a colder and thicker water, that gives specific condition, such as organic matterials accumulation and nutrients availability that could serve as living area for both phytoplankton and zooplankton.

Study of diversity and spatial distribution of plankton are important to develop. Therefore, the purpose of this research was to understand the local, intra-space diversity, and regional (total) diversity or α, β, and γ diversity.

2. Material and Methods

2.1. Study area

This research was conducted in February-March 2011 using the Research Vessel Baruna Jaya VIII of Oceanography Research Center, Indonesian Institute of Sciences (P2O-LIPI). Sampling of plankton and water quality parameters were taken once at a predetermined location (Figure 1). At the surface water, the observation was carried out at three locations, two nearshore (Banyuwangi/Eastern part of Java Island and Western part of Bali Island) each of 10 stations and offshore (Bali Strait) location with 9 stations. The offshore consist of two locations, the surface and thermocline water column.

2.2. Samples collection

Data were collected at 9 stations in offshore area by taking samples of plankton using plankton net and water samples using a CTD (Conductivity Temperature Depth) that has 12 tubes. At 10 nearshore stations, plankton and water samples were taken directly on the surface.

Plankton at off shore stations was taken by plankton net hauling from 10 meters depth up to the surface, whereas water samples were taken by using a tube on the CTD at 10 meters depth. Plankton at in shore stations was taken by taking water directly from the surface and then it was filtered with plankton net. Plankton sampling at thermocline water column was taken by tube on CTD. When the CTD was lowered, CTD measured the water column temperature of each depth. The depth of thermocline was determined from the temperature-depth profile.

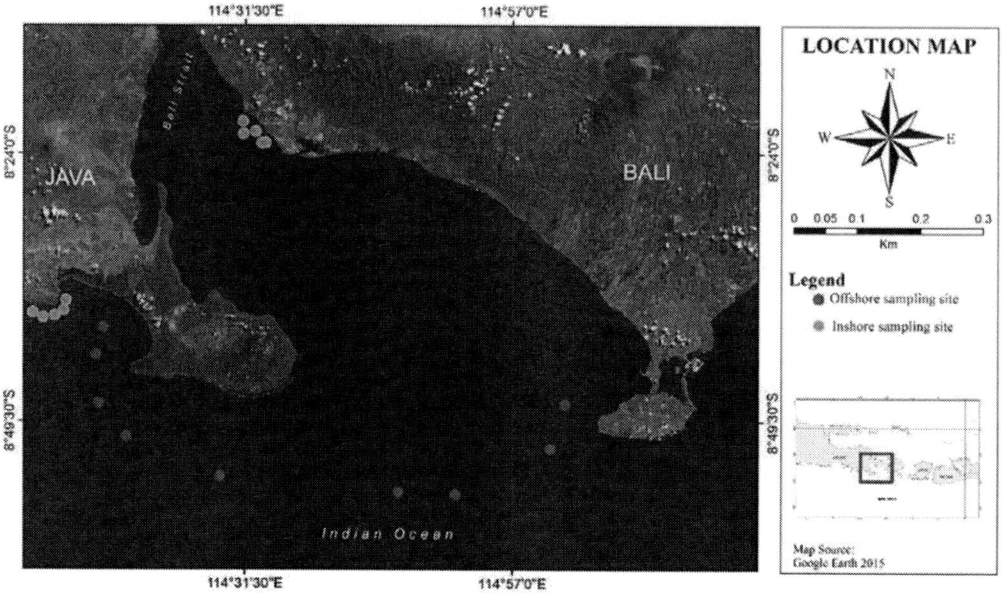

Figure1. Research sites in connected waters of Bali Strait

2.3. Data analysis

Plankton abundance was observed using a Sedgewick Rafter Counting Chamber (SRC). The morphological identification of phytoplankton was defined by using reference book [7] [8] [9] [10]. Plankton abundance was expressed in individuals per m^3 with by the following formula [11]:

$$N = n \times \frac{V_t}{V_{src}} \times \frac{A_{src}}{A_a} \times \frac{1}{V_d}$$

Description:

N	= Abundance of phytoplankton (cell/m^3)
n	= Organisms observed (cell)
V_d	= Volume of filtered water (m^3)
V_t	= Volume of water filtered (ml)
V_{src}	= Volume of the SRC (1 ml)
A_{src}	= Cross-sectional area of SRC
Aa	= Area of observation

The observation of the phytoplankton abundance was carried out on the water samples. Diversity of plankton was analysed employing the species richness of the plankton. Diversity α is the number of species found in a locality (stations), diversity β is the difference in diversity between localities were observed (intra-space/location), and the diversity of γ is the total diversity of species contained in the scope of a wider (regional) [12].

$$\beta_{cc} = [(b+c)/(a+b+c)], \text{ or}$$
$$\beta_{cc} = \beta_3 + \beta_{rich}$$
$$\beta_{-3} = 2\times[(\min(b,c)/(a+b+c)]$$
$$\beta_{rich} = [(b-c)/(a+b+c)]$$

Description:

β_{cc} = Difference in diversity between localities
β_3 = Turnover between localities
β_{rich} = Difference in the proportion of intra-space diversity (*nestedness*)
a = Number of species that can be found in both localities
b = Number of species that are found only in one locality
c = Number of species found only in the two locality

Determination diversity α, β, and γ in this study was to describe the diversity of each location or locality (α), intra-space diversity (β), and the total diversity of plankton species found in this research (γ). Accordingly, local diversity (α_i) is a value richness of plankton at each station, while the total diversity (γ) are the richness of the overall value of stations observed. Meanwhile, the difference in intra-space diversity (β) is obtained based on the results of a particular calculation. Determination of intra-space diversity can be made by Jaccard calculation formula which has two components, namely β_3 and β_{rich}. Components of β_3 are the difference in the proportion of intra-space diversity (nestedness), while the turnover kind intra-space diversity β_{rich} is being compared (turnover) [13]. Here is the formula for Jaccard and its components.

3. Result and Discussion

3.1 Abundance and number of species

The result showed that composition of phytoplankton based on number of species in the waters of the Bali Strait consists of three classes, i.e. Bacillariophyceae, Dinophyceae, and Cyanophyceae. Composition of phytoplankton based on group is presented in (Figure 2). Group of Bacillariophyceae had the highest composition in offshore location with 73-85% and 71-78% for nearshore location. Phytoplankton was dominated by Bacillariophyceae. Bacillariophyceae class (diatoms) was mostly found in this research. Generally, phytoplankton in the sea is composed by diatom species (Bacillariophyceae), followed by dinoflagellates (Dinophyceae) and blue green algae (Cyanophyceae) [14].

Total abundance of phytoplankton in each location was showed in Figure 3. Total abundance ranged from 151476-4285209 cell/m^3. Location that had the highest abundance of phytoplankton was surface, while the lowest was thermocline. Based on abundance, the highest phytoplankton composition was also obtained for Bacillariophyceae. The common phytoplankton in the ocean that is usually found in large abundance consists of two groups that dominate i.e diatoms (class Bacillariophyceae) and dinoflagellates [15] [16].

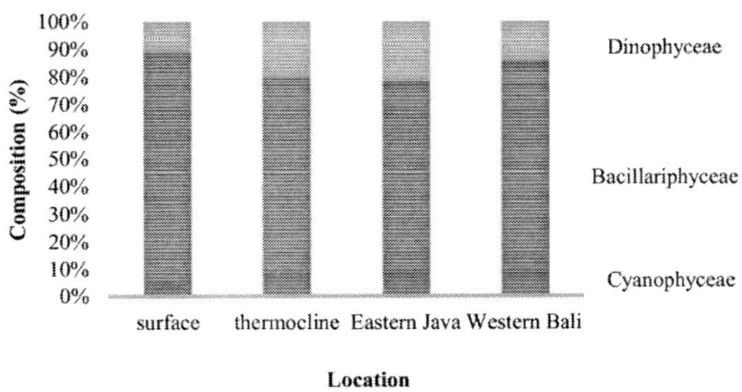

Figure 2. Composition of phytoplankton

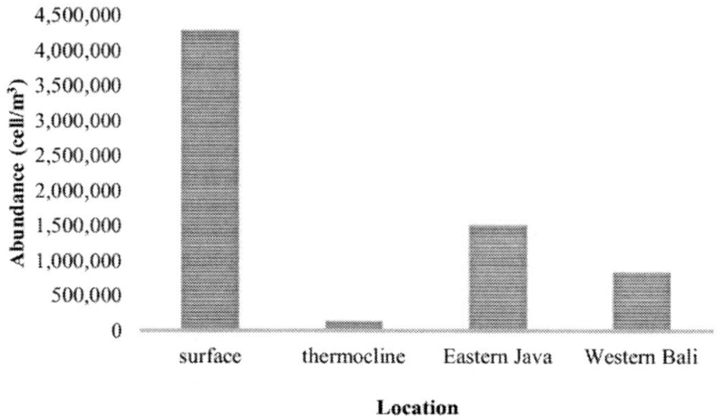

Figure 3. Total abundance of phytoplankton

Zooplankton community in waters of Bali Strait consisted of various zooplankton populations. A total of 5 groups of zooplankton had identified during research period. Based on composition number of zooplankton species (Figure 4), it was known that zooplankton of Crustacea group had the highest number of species in all locations (38 to 45%). Crustacea is the dominant zooplankton found in the sea [14].

Abundance of zooplankton during the study is presented in Figure 5. The abundance ranged from 33098-1413069 ind/m3. Location that had the highest abundance of zooplankton was surface, while the lowest was thermocline. This pattern is similar with the pattern of phytoplankton abundance. Zooplankton abundance is influenced by the abundance of phytoplankton [17].

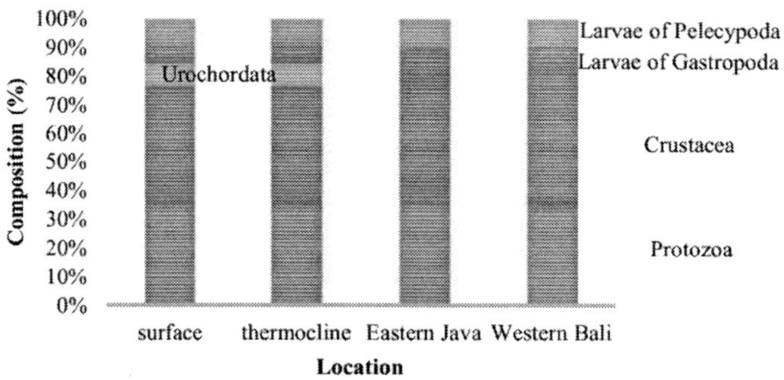

Figure 4. Composition of zooplankton

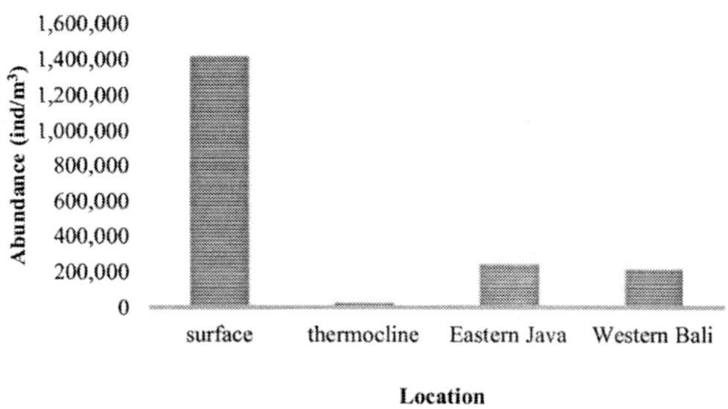

Figure 5. Total abundance of zooplankton

3.2 Local diversity (α), intra-space diversity (β), and regional (total) diversity (γ) phytoplankton

Local diversity (α) and total diversity (γ) of phytoplankton in offshore and nearshore locations (Figure 6a and 6b) showed that the highest diversity was in the α2 for surface offshore (20 genera) and α2 for Eastern Java nearshore (14 genera). Total phytoplankton diversity (γ) in the offshore was 27 and 15 genera. Meanwhile, there were 14 genera in both nearshore locations, Western Bali and Eastern Java. Total diversity (γ) in surface and thermocline location, Eastern Java, and Western Bali are not much different from the phytoplankton species that found in every station (locality). It indicated that phytoplankton species found in every station was almost similar.

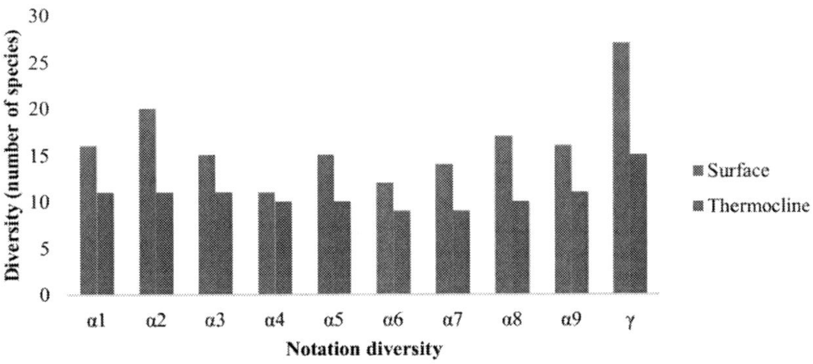

(a)

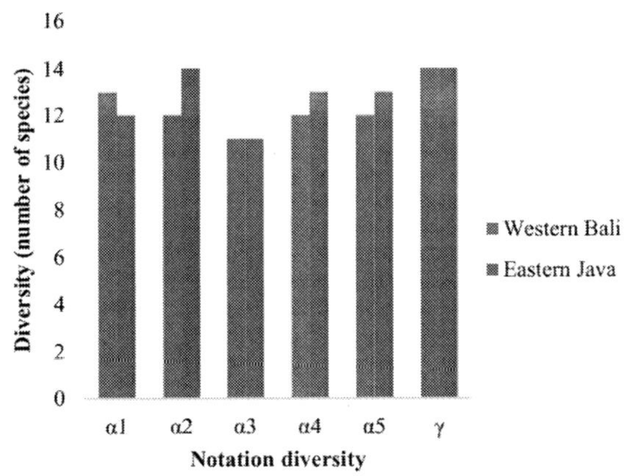

(b)

Figure 6. α and γ diversity of phytoplankton

Table 1 showed the intra-space (β) diversity of phytoplankton in the offshore waters between surface and thermocline location. It showed that the difference in total diversity (β_{cc}) was high during research period (value was close to 1). Diversity of phytoplankton (number of species) on surface location was higher than thermocline location. Generally, the total diversity difference (β_{cc}) between surface location was more affected by the high value of β_3 (turnover) than the value of β_{rich} (nestedness). High value of β_3 in waters indicates that any phytoplankton species in the surface location different from phytoplankton species in the thermocline. When a high value of β_{cc} is affected by β_3, turnover species between localities might occur. Instead, when a high value of β_{cc} is affected by

β_{rich}, phytoplankton in locality that has lower diversity is a part of phytoplankton in locality that has higher diversity, for example on α_{S2} & α_{T2} and α_{S8} & α_{T8}.

The intra-space (β) diversity of zooplankton in the nearshore waters between Eastern Java and Western Bali location was showed in Table 2. Generally, the difference in total diversity (β_{cc}) in both locations tends to have a low value, which means phytoplankton in both locations was almost similar.

Table 1. Intra-space (β) diversity of phytoplankton in the offshore waters

Diversity β	Station								
	α_{S1} & α_{T1}	α_{S2} & α_{T2}	α_{S3} & α_{T3}	α_{S4} & α_{T4}	α_{S5} & α_{T5}	α_{S6} & α_{T6}	α_{S7} & α_{T7}	α_{S8} & α_{T8}	α_{S9} & α_{T9}
β_{cc}	0,714	0,591	0,579	0,765	0,684	0,833	0,647	0,684	0,650
β_3	0,476	0,182	0,421	0,706	0,421	0,667	0,353	0,211	0,400
βrich	0,238	0,409	0,158	0,059	0,263	0,167	0,294	0,474	0,250

Table 2. Intra-space (β) diversity of phytoplankton in the inshore waters

Diversity β	Station				
	α_{WB1} & α_{EJ1}	α_{WB2} & α_{EJ2}	α_{WB3} & α_{EJ3}	α_{WB4} & α_{EJ4}	α_{WB5} & α_{EJ5}
β_{cc}	0,214	0,267	0,429	0,214	0,214
β_3	0,143	0,133	0,429	0,143	0,143
βrich	0,071	0,133	0,000	0,071	0,071

3.3. Local diversity (α), intra-space diversity (β), and regional (total) diversity (γ) zooplankton

Alpha (α) and gamma (γ) diversity of zooplankton in offshore and nearshore locations (Figure 7a and 7b) showed that the highest diversity was in the α9 for surface offshore (13 genera) and α4 for Eastern Java nearshore (14 genera). Total phytoplankton diversity (γ) in offshore location was 27 and 15 genera. Meanwhile, there were 14 genera in both nearshore locations, Western Bali and Eastern Java. Total diversity (γ) in surface and thermocline location, Eastern Java, and Western Bali were not much different from zooplankton that was found in every station (locality). It indicated that species zooplankton found in every station was almost the same.

Table 3 showed the intra-space (β) diversity of zooplankton in the offshore waters between surface and thermocline location. It showed that the difference in total diversity (β_{cc}) was low during research period (value was close to 0). Difference in total diversity (β_{cc}) of zooplankton in offshore waters was more affected by the value of β_{rich}. It means zooplankton in locality that has lower diversity is part of zooplankton in locality that has higher diversity, like in α_{S3} & α_{T3}, α_{S4} & α_{T4}, and α_{S6} & α_{T6}. Similar to zooplankton in offshore waters, the difference in total diversity (β_{cc}) in nearshore waters was generally low. The value of β_{cc} is affected by β_3, which means turnover species between localities might occurr.

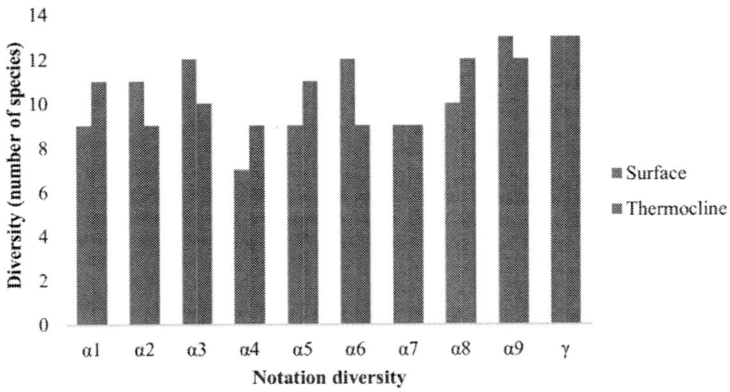

(a)

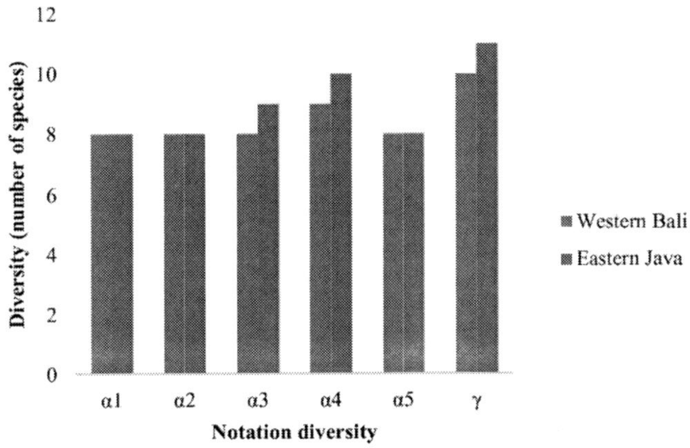

(b)

Figure 7. α and γ diversity of zooplankton

Tabel 3. Intra-space (β) diversity of zooplankton in the offshore waters

Diversity β	Station								
	α_{S1} & α_{T1}	α_{S2} & α_{T2}	α_{S3} & α_{T3}	α_{S4} & α_{T4}	α_{S5} & α_{T5}	α_{S6} & α_{T6}	α_{S7} & α_{T7}	α_{S8} & α_{T8}	α_{S9}& α_{T9}
β_{cc}	0,333	0,333	0,167	0,222	0,462	0,385	0,364	0,308	0,077
β_3	0,167	0,167	0,000	0,000	0,308	0,154	0,364	0,154	0,000
β_{rich}	0,167	0,167	0,167	0,222	0,154	0,231	0,000	0,154	0,077

Table 4. Intra-space (β) diversity of zooplankton in the inshore waters

Diversity β	Station				
	α_{WB1} & α_{EJ1}	α_{WB2} & α_{EJ2}	α_{WB3} & α_{EJ3}	α_{WB4} & α_{EJ4}	α_{WB5} & α_{EJ5}
β_{cc}	0,400	0,222	0,455	0,100	0,222
β_3	0,400	0,222	0,364	0,000	0,222
βrich	0,000	0,000	0,091	0,100	0,000

As a whole, there were found 27 species of phytoplankton and 13 kinds of zooplankton, as each of γ diversity. At the surface water, all species of phytoplankton were found in offshore stations, and 14 of them were found in both nearshore locations. All 13 kinds of zooplankton were also found in offshore stations, and 10 to 11 of them were found in nearshore locations.

4. Conclusion

Phytoplankton and zooplankton were well distributed at the surface, but showed as spesific distribution at thermocline layer, especially for phytoplankton.

5. Acknowledgments

Authors wishing to acknowledge assistance or encouragement from colleagues, special work by technical staff, and financial support from organizations should do so in an unnumbered.

References

[1] Hartoyo D, Purwanto H, Wahyono IB. 1998. Sebaran densitas ikan pelagik di selat bali pada musim timur september 1998, Prosiding seminar riptek kelautan nasional, Bali
[2] Khasanah RI, Sartimbul A, Herawati EY. 2013. Kelimpahan Plankton di perairan selat bali. Jurnal Ilmu Kelautan vol 18(4):193-202
[3] Odum EP. 1971. *Fundamentals of Ecology. Third Ed.* W.B. Saunders Company. Philadelphia.
[4] Chandy JP, Al-Tisan I, Munshi HA, El Reheim HA. 1991. Marine phytoplankton: A study on seasonal abundance and distribution in Al-Jubail. *Issued as Technical Report No. SWCC (RDC)* 17 in December, 1991.
[5] Nielsen SE. 1975. Marine Photosynthesis with Special Emphasis on the Ecological Aspect. *Elsevier Sci. Publ. Co*. Amsterdam.
[6] Montemezzani V, Duggan IC, Hogg ID, Craggs RJ. 2015. A review of potential methods for zooplankton control in wastewater treatment High Rate Algal Ponds and algal production raceways. *Algal research*. 11: 211-261.
[7] Davis CC. 1955. *The Marine and Freshwater Plankton*. Michigan: Michigan State University Press.
[8] Newell GE, Newell RC. 1977. *Marine Plankton: A Practical Guide*. Hutchinson & Co. London.
[9] Yamaji I. 1979. *Illustration of The Marine Plankton of Japan*. Hoikusha Publishing Co. Ltd. Osaka. Japan.
[10] Tomas CR. 1997. *Identifying Marine Phytoplankton*. Academic Press. California, USA.
[11] Rice EW, Baird RB, Eaton AD, Clesceri LS. 2012. *APHA (American Public Health Association): Standard Method for The Examination of Water and Wastewater 22th ed*. Washington DC: AWWA (*American Water Works Association*) and WEF (*Water Environment Federation*).
[12] Anderson MJ, Crist TO, Chase JM, Vellend M, Inouye BD, Freestone AL, Sanders NJ, Cornell HV, Comita LS, Davies KF, Harrison SP, Kraft NJB, Stegen JC, Swenson NG. 2011. Navigating the multiple meanings of b diversity: a roadmap for the practicing ecologist. *Ecology letters*. 14: 19-28.
[13] Carvalho JC, Cardoso P, Gomes P. 2012. Determining the relative roles of species replacement and species richness differences in generating beta-diversity patterns. *Global ecology and biogeography*. 21(7): 760-771.
[14] Nontji A. 2008. *Plankton Laut*. LIPI Press. Jakarta.
[15] Kennish MJ. 1990. *Ecology of estuary. Volume II*. Biological Aspect. CRC Press, Inc. United State.
[16] Skaloud P and M Rezacova. 2004. *Spatial Distribution Phytoplankton in the Eastern Part of the North Sea*. Department of Phycology. Institute of Biology. University of Copenhagen.
[17] Heneash AMM, HRZ Tadrose, MA Maged. Hussein, SK Hamdona, N Abdel-Aziz, SM Gharib. 2015. Potential effects of abiotic factors on the abundance and distribution of the plankton in the Western Harbour, south-eastern Mediterranean Sea, Egypt. *Oceanologia*. 57: 61-70.

Spatial distribution of plankton in Riau islands province, Indonesia

I P Ayu [1*], N T M Pratiwi[1], A Iswantari[1], S Hariyadi[1], D Mulyawati[1], B Subhan[2], D Arafat[2], P Santoso[2] and M Sastria[2]

[1] Department of Aquatic Resources Management, Faculty of Fisheries and Marine Sciences, Bogor Agricultural University, Indonesia
[2] Department of Marine Science and Technology, Faculty of Fisheries and Marine Sciences, Bogor Agricultural University, Indonesia

E-mail: innapuspa@gmail.com

Abstract. Riau Islands which is located at 4°LU – 1°LS and 104°BT – 107°BT, consist of around 3200 islands. It has high marine biodiversity, especially micro-plankton. Biodiversity of marine phytoplankton is usually dominated by diatom and zooplankton by micro-crustacean and early stage of marine biota. Nowadays, biodiversity of micro-plankton is an important study to identify their origin and potential as alien and invasive species. The aim of this research was to determine the biodiversity of marine micro-plankton in Riau Islands. This research was conducted in 14 small islands (Karanggerih, Pemping, Panjang, Melur, Palantuah, Dendun, Mantang, Bunut, Kelong, Mercusuar, Tokong Hiu Kecil, Tokong Hiu Besar, Karimun, Penyengat) in Riau Islands Province. Samples of micro-plankton were collected from surface water using plankton net. Samples were observed under light microscope and identified morphologically. Biodiversity index was calculated. There were found 20-34 taxa of phytoplankton and 10-17 taxa of zooplankton in all sites. Phytoplankton was dominated by Bacillariophyceae group and zooplankton by Crustacean and Protozoa groups. This result is expected for biodiversity bank information and further research.

1. Introduction

Riau Islands Province is a fast growing province since becoming part of a Demilitarized Zone with Singapore in 2006. It has 252.601 km^2 large and consists of 2.408 islands. Seawater ecosystem is becoming a popular destination in tourism sector in Riau islands. Possibilities of marine biodiversity could find in Riau Islands. Indonesia probably has the most diverse assemblage of marine habitats in the world [1].

Nowadays natural resources information, especially biodiversity in marine waters ecosystems has an important role. Alien and invasive species belong to trending topics biodiversity, especially micro-plankton diversity related to various aspect of aquatic ecosystem function [2]. Plankton in micro cell size has varied characteristic among genera and define diversity of organism in ecosystem. Riau islands may have diverse aquatic biota; include micro-plankton which varied in spatial distribution. A quantitative understanding of the abundance, distribution, and diversity ofplankton is essential for estimating their biogeochemical impact, particularly under a changing environment [3]. The aim of this research was to determine biodiversity of micro-plankton in Riau Islands Province, Indonesia.

2. Materials and Methods

Plankton sampling were carried out in 2015 at 14 Islands of Riau Islands (Figure 1). There were 19 stations consist of Karanggerih (B1, B2), Pemping (B3), Panjang (B4), Melur (B5), Palantuah (B6), Dendun (B7), Mantang (B8), Dendun (B9), Bunut (B10), Kelong (B11, B12), Mercusuar (B13), Tokong Hiu Kecil (B14), Tokong Hiu Besar (O3), and Karimun (O4, O5, O6), and Penyengat (O7).

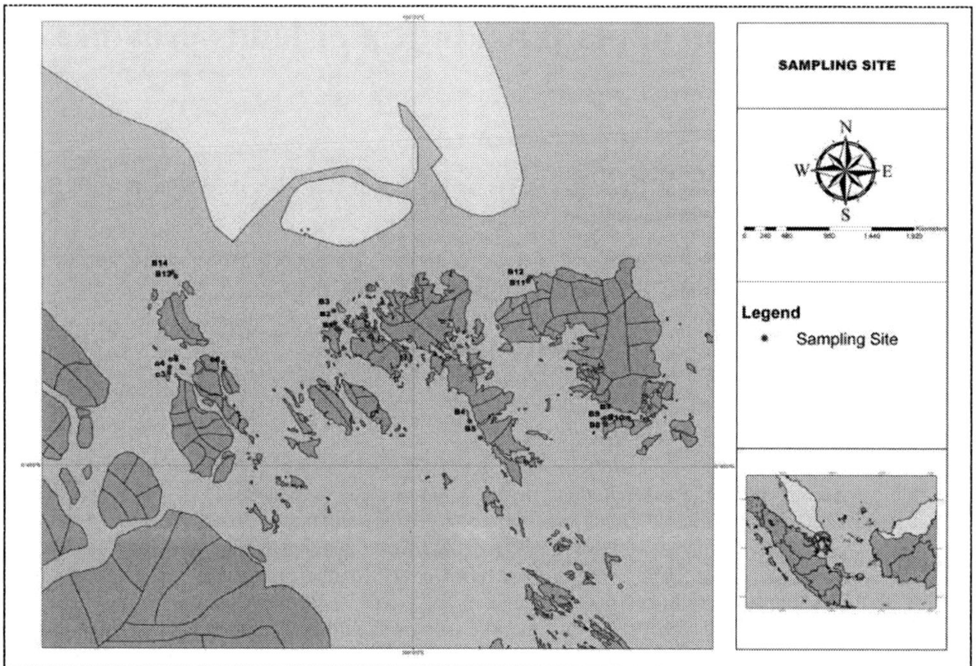

Figure 1. Sampling location of plankton in Riau Islands, Indonesia

Water sample were collected using plankton net with 23 μm mesh size to filter 100 L at surface water. Plankton samples were preserved in situ with Lugol solution [4] methods. Plankton samples were analyzed with microscope following APHA (2005) methods and identified followed [5,6,7, 8, 9, and 10]. Ecological indices, the Shannon-Wiener diversity index (H') [11], Evenness index (E) [11], and domination index (C) [12] were calculated to understand the community structure of plankton. Clustering analysis was calculated by Bray Curtis similarity index. Clustered of dendrogram will show cluster of spatial distribution of plankton. The equations of ecological indices are described as follows.

$$H' = -\sum_{i=1}^{S} p_i \ln p_i \quad (1)$$

with:

H' : Diversity index
p_i : n_i/N
n : total number of organism of particular species
N :total number of organism of all species

$$E = \frac{H'}{H_{maks}} = \frac{H'}{\ln(S)} \quad (2)$$

with:

E : Evenness index
H' : Diversity index

H_{max} : ln S
S : number of genus

$$C= \sum_{n=1}^{n} \left(\frac{n_i}{N}\right)^2 \quad (3)$$

with:
C : Simpson Domination index
n_i : total number of organism of particular species
N : total number of organism

3. Result and Discussion

Structure of plankton community in Riau province island was described by ecological indices, both of phytoplankton and zooplankton show different spatial distribution, which described by clustering analysis. All of phytoplankton taxa were belong to group of Cyanophyceae (2 genera), Bacillariophyceae (17 genera), Dinophyceae (5 genera), Chlorophyceae (2 genera), and Chrysophyceae (1 genera). The number of taxa of each site ranged from 14 to 31 species and the composition of phytoplankton was dominated by Diatom or Bacillariophyceae. The least number of taxa was found in Karimun 4, and the most was found in P. Tokong Hiu Kecil. Karimun 4 also showed the lowest abundance of phytoplankton. The highest abundance was found in P. Dendun. The total of phytoplankton abundance ranged from 61,800 to 11,554, 665 ind/l. The composition of phytoplankton abundance was shown in Figure 2.

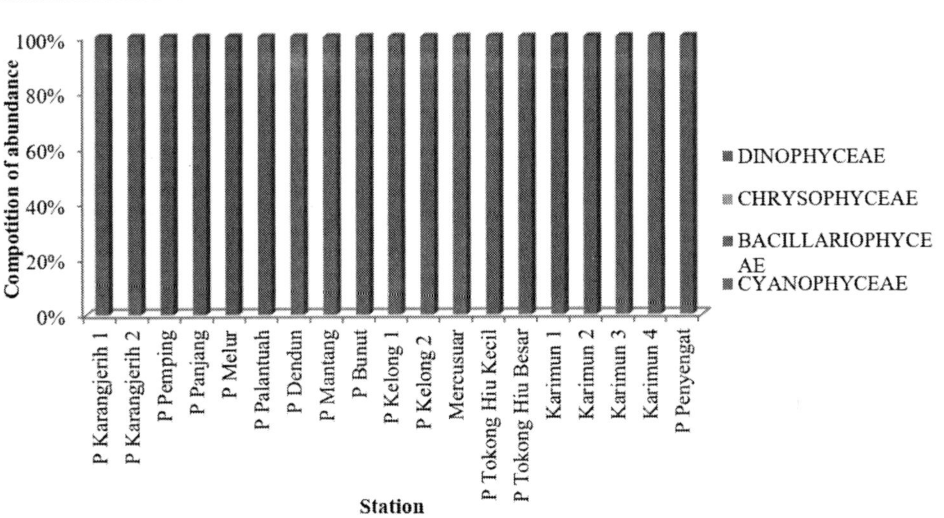

Figure 2. Composition of phytoplankton abundance

Biodiversity within a group of organisms in each site was described by diversity index. Biodiversity may be determined numerically in relation to a variety of parameters (referred to as importance values) including biomass, productivity, and organism count. Most estimates of species diversity are determined from species counts, which can be used to generate bio-indices of three main types species richness, species evenness/dominance, and a combined index of biodiversity [2]. Biodiversity index of phytoplankton and zooplankton was shown in Table 1 and 2.

Table 1. Biodiversity index of phytoplankton in sampling sites

Indices	Stations																		
	B1	B2	B3	B4	B5	B6	B7	B8	B9	B10	B11	B12	B13	B14	O3	O4	O5	O6	O7
H'	2,13	1,32	0,94	1,58	1,40	1,56	1,67	1,75	1,53	1,96	1,80	1,95	2,69	2,47	2,77	2,54	2,70	2,38	1,01
E	0,64	0,43	0,31	0,51	0,44	0,48	0,56	0,57	0,47	0,60	0,57	0,61	0,78	0,73	0,85	0,84	0,81	0,90	0,31
C	0,21	0,50	0,67	0,36	0,39	0,34	0,32	0,31	0,35	0,21	0,23	0,21	0,10	0,16	0,08	0,10	0,09	0,11	0,64

Table 2. Biodiversity index of zooplankton in sampling sites

Indices	Stations																		
	B1	B2	B3	B4	B5	B6	B7	B8	B9	B10	B11	B12	B13	B14	O3	O4	O5	O6	O7
H'	1.46	1.55	1.88	1.66	1.90	1.82	1.82	1.43	1.87	1.80	1.43	1.69	1.86	1.47	1.80	1.91	1.26	1.02	1.57
E	0.63	0.80	0.78	0.67	0.79	0.64	0.67	0.58	0.73	0.72	0.74	0.71	0.72	0.64	0.75	0.73	0.55	0.52	0.63
C	0.33	0.26	0.20	0.29	0.22	0.29	0.25	0.39	0.23	0.25	0.35	0.26	0.22	0.35	0.22	0.21	0.39	0.54	0.33

P. Tokong Hiu besar, which has high number of species, showed high evenness and low dominancy of phytoplankton species. The lowest dominancy was shown in Karimun 1, which has the greatest evenness of all sites. Different with zooplankton, the greatest diversity was shown in P. Melur with low dominancy and the lowest dominancy of zooplankton species was shown in P Palangtuah. The values of Evennes index showed a high similarity of proportion among species which compose the community. The low values of domination index shows that there are not any predominant species in the plankton community.

On the other hand, all of the zooplankton taxa were belong to the group of Protozoa (12 genera), Crustaceae (6 genera), Urochordata (1 genera), Pelecypoda (1 genera), Gastropoda (1 genera), and Echinodermata (1 genera). The abundance of zooplankton was relatively high. The number of taxa of each site ranged from 7 to 17 and the composition of zooplankton was dominated by Crustacea and Protozoa. As on phytoplankton, the least number of taxa was found in Karimun 4. The highest taxa number of zooplankton was found in Palangtuah. Palangtuah also had the highest abundance. The lowest abundance was found in Karangjerih 2. The total zooplankton densities ranged from 2.205 to 40.907 ind/L. The composition of zooplankton abundance was shown in Figure 3.

Spatial distribution of plankton, which shown by Bray Curtis similarity index was use to cluster the stations. There were developed four groups, based on number of species of phytoplankton. The first group consist of P. Karangjerih 1 and P. Dendun; group two consist of P. Karangjerih 2, Kelong 1, Kelong 2, Penyengat, and P. Tokong Hiu Besar; group three consist of P. Pemping, Karimun 2, Mercusuar, P. Panjang, Karimun 3, P. Melur, Palantuah, P. Bunut, and Karimun 4, group four consist of P. Mantang, Karimun 1, and P. Tokong Hiu Kecil (Figure 4). Different with zooplankton, there were developed three groups, based on number of zooplankton species. The first group consist of Karangjerih 1, Kelong 1, Karimun 2, and Karimun 4; P. Kelong 2 belong to group three; and rest of the stations belong to group two (Figure 5). The similarity between sites also shows similarity of spatial distribution of plankton. Plankton diversity between sites was varied, which could be influenced by physics or chemical water parameters. Plankton diversity, especially in spatial distribution has benefit for biodiversity bank information and further research.

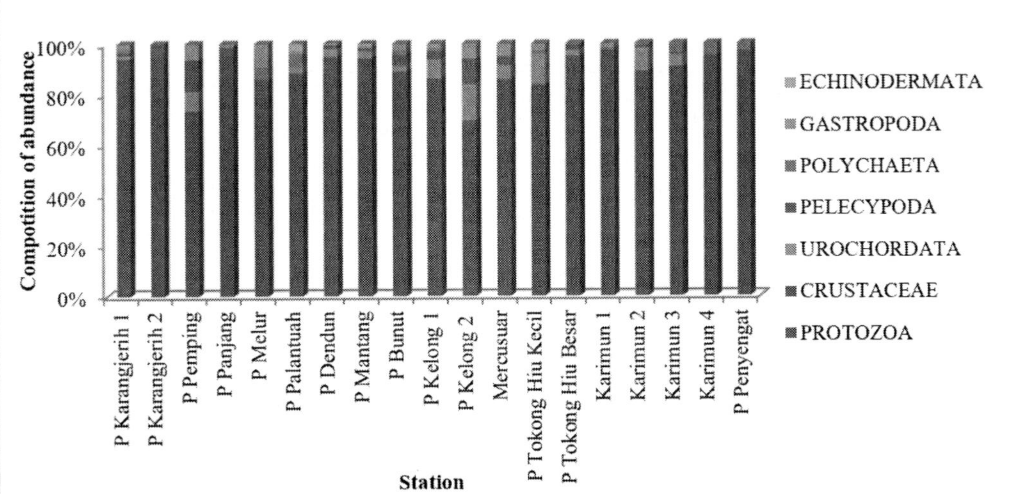

Figure 3. Composition of zooplankton abundance

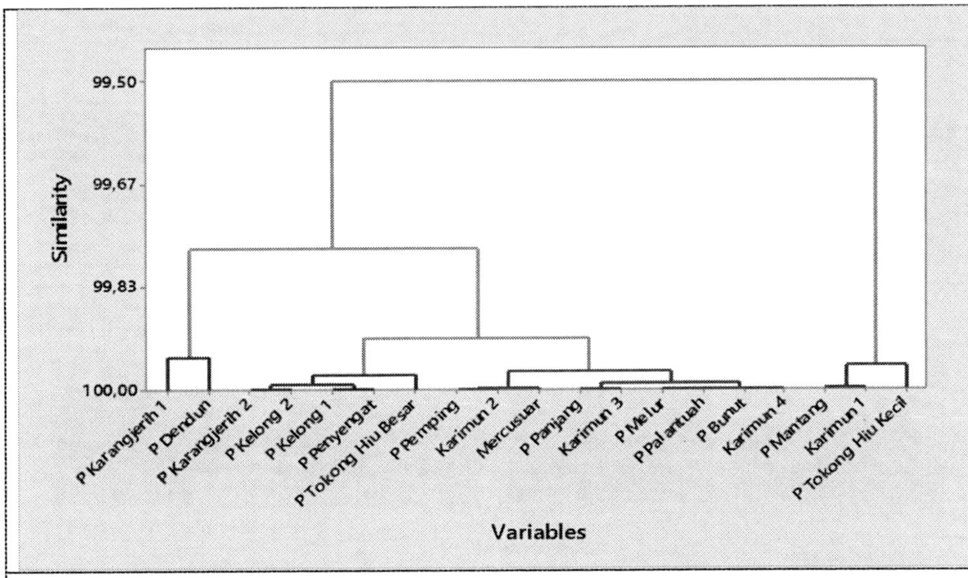

Figure 4. Cluster of stations based on number of phytoplankton species.

IOP Conf. Series: Earth and Environmental Science **54** (2017) 012091 doi:10.1088/1755-1315/54/1/012091

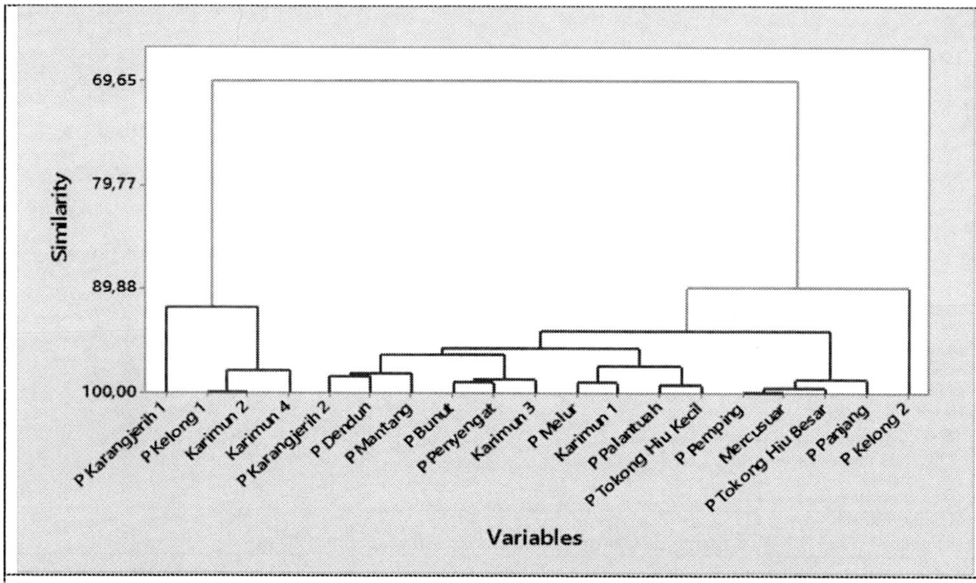

Figure 5. Cluster of stations based on number of zooplankton species.

Some genera of plankton in Riau Islands Province were illustrated in Figure 6

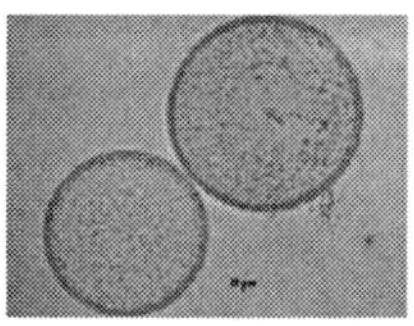	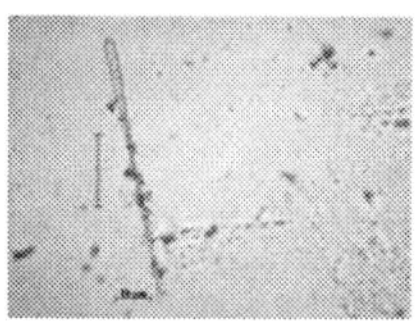	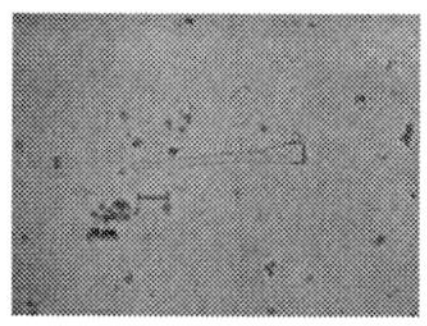	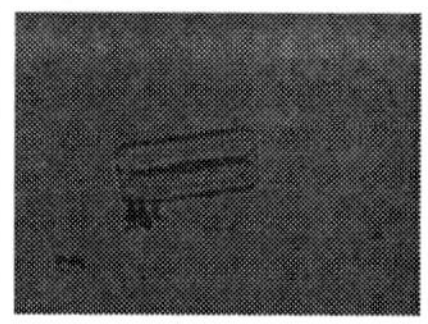
Cyclotella sp.	*Climacosphenia* sp.	*Asterionella* sp.	*Diploneis* sp.
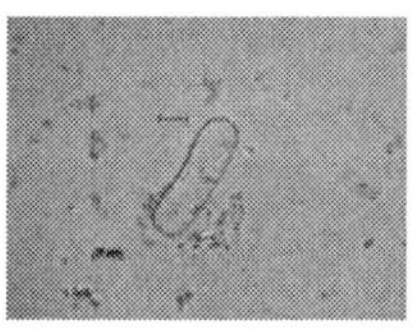	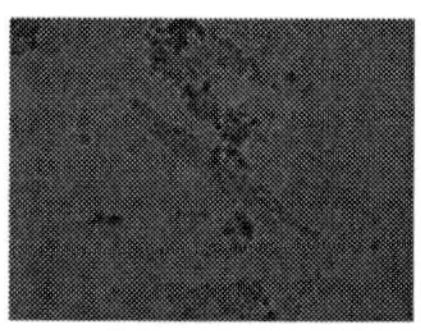	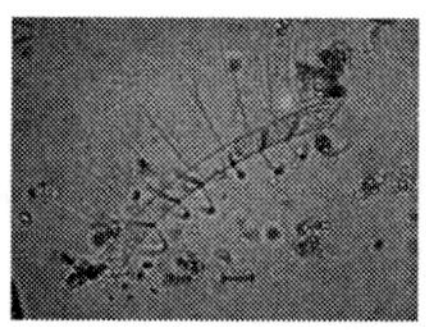	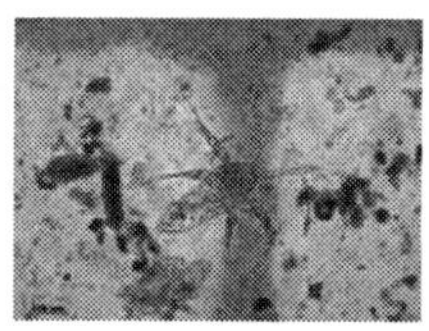
Amphora sp.	*Bacillaria* sp.	*Chaetoceros* sp.	Nauplius
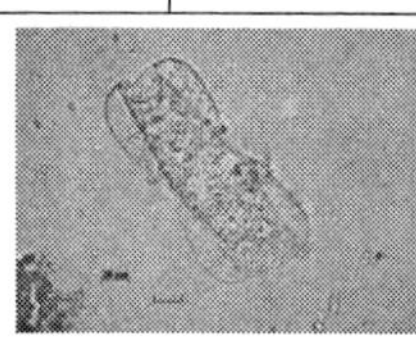		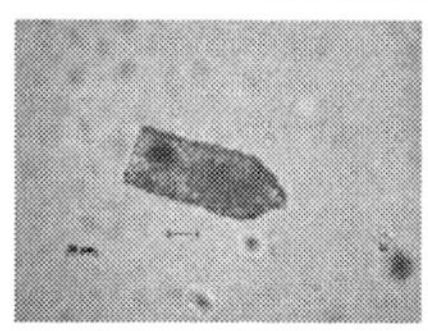	
Amphiprora sp.		*Tintinnopsis* sp.	

Figure 6. Dominant plankton found in sampling site of Riau Islands waters

4. Conclusion

There were found 14-31 taxa of phytoplankton belong to group Cyanophyceae, Bacillariophyceae, Dinophyceae, Chlorophyceae, Chrysophyceae and 7-17 taxa of zooplankton belong to group Protozoa and Crustaceae in Riau Islands Province, Indonesia.

Acknowledgement

We thank to Aldiano Rahmadya, Reza Zulmi, Dudi M Wildan, and Dwi Yuni Wulandari who prepared map layout and plankton photograph.

References

[1] Hutomo M and Moosa M K 2005 Indonesian marine and coastal biodiversity: present status *Indian Journal of Marine Sciences* **34** 88-97
[2] Sigee D C 2005 *Freshwater Microbiology* (England : John Wiley & Sons Ltd)
[3] Rynearson T A and Menden-Deuer S 2016 Drivers that structure biodiversity in the plankton *Aquatic microbial ecology and biogeochemistry: A dual perspective* (Switzerland: Springer International Publishing)
[4] APHA 2012 *Standard method for the examination of water and wastewater 22th ed* Editor Rice E W, Baird R B, Eaton A D, Clesceri L S APHA (American Public Health Association), AWWA (American Water Works Association) and WEF (Water Environment Federation) (Washington DC)
[5] Desikachary TV 1989 *Atlas of Diatom* (Madras: Madras Science Foundation)
[6] Desikachary TV 1988 *Atlas of Diatoms* (Madras: Madras Science Foundation)
[7] Desikachary TV 1986 *Atlas of Diatoms* (Madras: Madras Science Foundation)
[8] Tomas C R 1997 *Identifying marine phytoplankton* Academic Press California
[9] Newell G E and Newell R C 1963 *Marine plankton: a practical guide* Hutchinson educational London
[10] Davis C D 1955 *The Marine and freshwater plankton* (Ohio: Michigan State University Press)
[11] Krebs C J 1972 Ecology: The Scientific study of the interactions that determine the distribution and abundance of organisms (New York: Harper & Row Publisher)
[12] Odum E P 1971 *Fundamental of ecology* 3rd ed (Philadelphia: Saunders Companys)

The optical band gap of $LiTaO_3$ and Nb_2O_5-doped $LiTaO_3$ thin films based on Tauc Plot method to be applied on satellite

R Estrada[1*], N Djohan[1], D Pasole[2], M Dahrul[3], A Kurniawan[3], J Iskandar[3], H Hardhienata[4] and Irzaman[4]

[1] Department of Electrical Engineering, Faculty of Engineering and Computer Science, Krida Wacana Christian University, Jl. Tj. Duren Raya No. 4, Jakarta 11470, Indonesia.
[2] Student of Electrical Engineering, Faculty of Engineering and Computer Science, Krida Wacana Christian University, Jl. Tj. Duren Raya No. 4, Jakarta 11470, Indonesia.
[3] Post Graduate of Biophysics, Faculty of Mathematics and Natural Sciences, Bogor Agricultural University, Jl. Raya Dramaga, Bogor 16680, Indonesia.
[4] Department of Physics, Faculty of Mathematics and Natural Sciences, Bogor Agricultural University, Jl. Raya Dramaga, Bogor 16680, Indonesia.

E-mail: richie.estrada@ukrida.ac.id

Abstract. This research observed the optical band gap of thin films made from $LiTaO_3$ undoped (0%) and doped (5% and 10%) with Nb_2O_5 in the 1 M-solubility deposited on n-type Si (111) substrates. The thin films are manufactured with coating process of substrates by Chemical Solution Deposition (CSD) method using a spin coater device at a rotation speed of 3000 rpm for 30 seconds and annealed in furnace (Nabertherm B180) at a temperature of 850°C for 15 hours. The optical absorption data of thin films are obtained by using an Ocean Optics USB2000 device in the wavelength of visible light. The band gap curve is determined from optical absorption data processing using Tauc Plot method. The Tauc Plot with indirect transition shows that $LiTaO_3$ doped with Nb_2O_5 provides increased optical band gap value in a range less than 3.5 eV. Based on the results of this research, it can be concluded that $LiTaO_3$ and Nb_2O_5-doped $LiTaO_3$ thin films on n-type Si (111) substrate are semiconductor materials and has the potential to be applied on satellite.

1. Introduction

Thin film is a layer of material with the thickness of 10^{-9} m – 10^{-6} m that is grown by using the surface of substrate [1, 2]. The layer of material can be made of organic, inorganic, metal and organometallic which contain conductor, semiconductor, superconductor and insulator properties [3-6]. Basically, atomic structure in the material is divided into two (2) groups of energy band. The band with lower energy is called valence band, whilst the other with higher energy is called conduction band. The area between valence band and conduction band is separated by forbidden band. The amount of energy required to discharge electrons from valence band to conduction band is called the 'band gap' (unit: eV) [7, 8].

The crystal structure in the form of chemical formula ABO_3 is called perovskite [9, 10]. Lithium tantalate ($LiTaO_3$) is an inorganic material [11] that has 4.7 eV of band gap [12] and one of alkali tantalate type perovskite family rhombohedral crystal structure with space group of R3c [13-15] (see Figure 1). Previous studies argued that the doping concentration could change crystalline structure of perovskite [16, 17] that affected the shifts in band gap [1, 18, 19]. Thus, the focus of this research is to investigate the optical band gap of $LiTaO_3$ and Nb_2O_5-doped $LiTaO_3$ thin films on n-type Si (111) substrates using Tauc Plot method from UV-Vis Ocean Optics USB2000 as the measurement device.

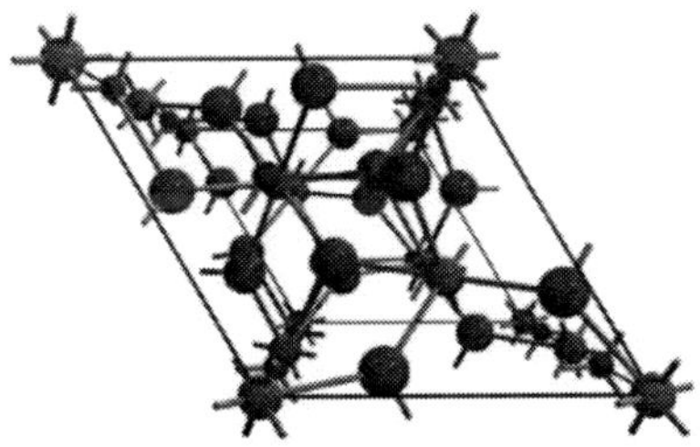

Figure 1. Crystal structure of $LiTaO_3$ [10]

2. Methodology

Lithium tantalate ($LiTaO_3$) was obtained from the result of mixing chemicals powder between lithium acetate ($Li(CH_3COO)$) and tantalum pentoxide (Ta_2O_5) according to balanced chemical equation [18]:

$$2Li(CH_3COO) + Ta_2O_5 + 4O_2 \rightarrow 2LiTaO_3 + 4CO_2 + 3H_2O \quad (1)$$

The solubility of this research was regulated at 1 M to calculate mass composition from lithium acetate powder ($Li(CH_3COO)$, 99.5% purity) and tantalum pentoxide powder (Ta_2O_5, 99.9% purity) corresponding with balanced chemical equation. The result of mass composition calculation indicates the exact amount of lithium acetate powder ($Li(CH_3COO)$, 99.5% purity) is 0.1650 gram and tantalum pentoxide powder (Ta_2O_5, 99.9% purity) is 0.5524 gram. In this research, the niobium oxide powder (Nb_2O_5, 99.9% purity) was used as a doping for lithium tantalate ($LiTaO_3$) with a mass composition of 0.0295 gram (5% from the mass of $LiTaO_3$) and 0.0590 gram (10% from the mass of $LiTaO_3$). The entire calculation of mass composition from chemicals powder (lithium acetate, tantalum pentoxide and niobium oxide) was weighed using analytical scale (AND GR-200).

A thin layer in form of liquid was made by mixing the chemicals powder into the solvent 2-methoxyethanol ($CH_3OCH_2CH_2OH$) with 2.5 ml volume to produce solution formulations of $LiTaO_3$, $LiTaO_3$ + 5% Nb_2O_5 and $LiTaO_3$ + 10% Nb_2O_5. All solution formulation was sonicated by using ultrasonicator device (BRANSON 2510) for 90 minutes to obtain the homogeneous solutions. The coating process of each homogeneous solutions on a n-type Si (111) substrates (half surface from dimensions of 1×1 cm^2) were done with Chemical Solution Deposition (CSD) method using a spin coater device at speed of 3000 rpm for 30 seconds [20]. Each coating process (dripping and rotating) was repeated three (3) times with interruption time for one (1) minute, then annealed in furnace (Nabertherm B180) at a temperature of 850°C for 15 hours.

Research on this thin films referred to the measurement of absorption spectra (optical characterization) from a thin layer that was formed on n-type Si (111) substrates. UV-Vis spectroscopy (Ocean Optics USB2000) was used as a device to measure the absorption spectra (optical characterization) on thin films in the form of absorbance value at a wavelength of visible light. Absorbance value indicates the magnitude of photon energy absorbed by a thin layer due to the influence of excitation of electrons from valence band to conduction band. The optical band gap value for each thin film was determined through extrapolation by withdrawing straight line on curve that was formed from correlation between $(\alpha h\nu)^n$ and $h\nu$ (called Tauc Plot method).

3. Result and Discussion

In this research, the measurement of optical absorbance from thin layer that deposited on n-type Si (111) substrate was presented in absorption spectra (see Figure 2). Related to optical absorbance, the coordinate axes for optical band gap of thin films were determined from the calculation using the following formula (equation) [1, 21, 22]:

$$d = \frac{m}{\rho A} \tag{2}$$

$$\alpha = 2.303 \frac{\text{optical absorbance}}{d} \tag{3}$$

$$\text{y-axis} = (\alpha h\nu)^{0.5} \tag{4}$$

$$\text{x-axis} = h\nu \tag{5}$$

with: m = mass of thin layer on n-type Si (111) substrate (gram)
ρ = density of $LiTaO_3$ (7.46 gram/cm^3)
A = surface area of thin layer on n-type Si (111) substrate (cm^2)
d = thickness of thin layer on n-type Si (111) substrate (cm)
optical absorbance = data values (from measurement using an Ocean Optics USB2000)
α = absorption coefficient (cm^{-1})
$h\nu$ = photon energy (eV)

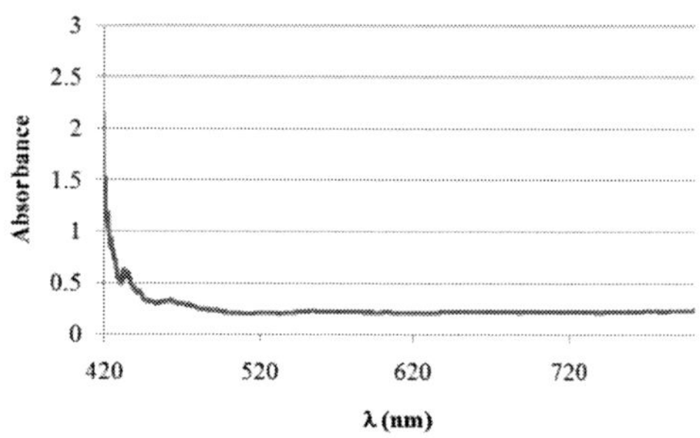

(a)

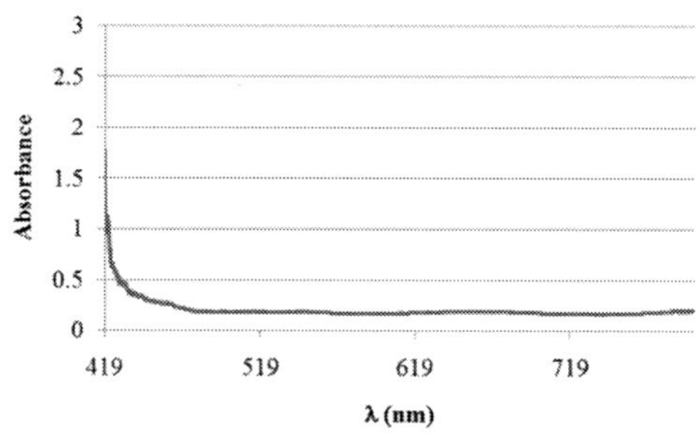

(b)

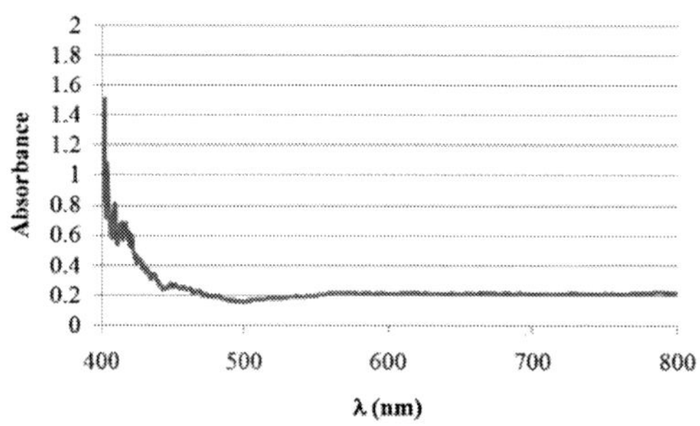

(c)

Figure 2. Absorption spectra: (a) Undoped (b) Doped with 5% Nb_2O_5 (c) Doped with 10% Nb_2O_5

The Tauc Plot for undoped and Nb_2O_5-doped $LiTaO_3$ thin films on n-type Si (111) substrate were presented in Figure 3. The band gap values were determined through extrapolation by withdrawing straight line (as shown in Figure 3) on coordinate axes (($(\alpha h\nu)^{0.5}$ – $h\nu$) and listed in Table 1. The wavelengths of thin films were obtained from the calculation (listed in Table 2) using the following formula (equation) [1]:

$$\lambda = \frac{\text{h c}}{\left(1.602 \times 10^{-19}\ \text{J}\right)\left(\text{optical band gap}\right)} \quad (6)$$

with: h = Planck's constant (6.626×10^{-34} J·s)
c = speed of light (2.998×10^{8} m·s^{-1})
optical band gap = from listed in Table 1

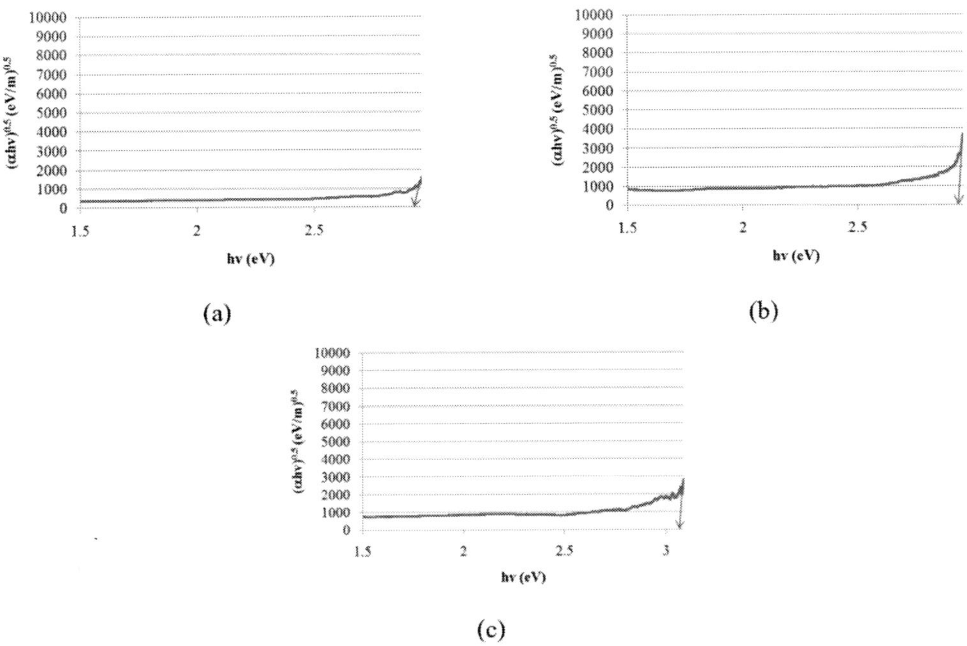

Figure 3. Optical band gap: (a) Undoped (b) Doped with 5% Nb_2O_5 (c) Doped with 10% Nb_2O_5

Table 1. The optical band gap of thin films

	Optical band gap (eV)
$LiTaO_3$ + 0% Nb_2O_5	2.93
$LiTaO_3$ + 5% Nb_2O_5	2.95
$LiTaO_3$ + 10% Nb_2O_5	3.07

Table 2. The wavelength and sensitivity to specific spectra color of light on each thin film

	Wavelength of light (nm)	**Spectra color of light (Based on literature [23])**
$LiTaO_3$ + 0% Nb_2O_5	424	Violet
$LiTaO_3$ + 5% Nb_2O_5	421	Violet
$LiTaO_3$ + 10% Nb_2O_5	404	Violet

4. Conclusion

In this research, the thin film shows increasing optical band gap values through the addition of Nb_2O_5 (5% and 10%) into $LiTaO_3$. Based on the optical band gap values of this research, it can be concluded that the grown of $LiTaO_3$ and Nb_2O_5-doped $LiTaO_3$ thin films on n-type Si (111) substrate are classified into semiconductor materials and has the potential to be applied on satellite.

Acknowledgment
This research is funded by the Ministry of Research, Technology and Higher Education of the Republic of Indonesia through PEKERTI Research Grant No. 771/K3/KM/SPK.LT/2016.

References

[1] Ohring M 1992 *The Materials Science of Thin Films* (San Diego: Academic Press)
[2] Hong X, Gan Y and Wang Y 2011 *Surf. Interface Anal.* **43** 1299
[3] Mercereau J E and Notarys H A 1973 *J. Vac. Sci. Technol.* **10** 646
[4] Fraxedas J 2008 *Molecular Organic Materials: From Molecules to Crystalline Solids* (Cambridge: Cambridge University Press)
[5] Schlueter J A 2009 *Top. Organomet. Chem.* **27** 1
[6] Wang X L, Dou S X and Zhang C 2010 *NPG Asia Mater.* **2** 31
[7] Shur M 2005 *The Electrical Engineering Handbook* ed W K Chen (San Diego: Academic Press) p 153
[8] Wang W, Tadé M O and Shao Z 2015 *Chem. Soc. Rev.* **44** 5371
[9] Inaguma Y, Matsui Y, Shan Y J, Itoh M and Nakamura T 1995 *Solid State Ionics* **79** 91
[10] Su Y, Lang J, Du C and Wang X 2016 *Perovskite Materials - Synthesis, Characterisation, Properties, and Applications* ed L Pan and G Zhu (Rijeka: InTech) **Chapter 16** pp 489–510
[11] Brehmer L, Kaminorz Y, Dietel R, Grasnick G and Herkner G 1997 *Frontiers in Biosensorics I: Fundamental Aspects* ed F W Scheller, F Schubert and J Fedrowitz (Basel: Birkhäuser Verlag) pp 155–166
[12] Nuraje N, Kudaibergenov S and Asmatulu R 2013 *Producing Fuels and Fine Chemicals from Biomass Using Nanomaterials* ed R Luque and A M Balu (Boca Raton: CRC Press) **Chapter 4** pp 95–118
[13] Tarafder A, Annapurna K, Chaliha R S, Tiwari V S, Gupta P K and Karmakar B 2010 *J. Alloy. Compd.* **489** 281
[14] Khalil A, Masaif N, Jennane A and Maaider K 2011 *J. Mater. Environ. Sci.* **2** 196
[15] Tahiri M, Masaif N, Jennane A and Lotfi E M 2012 *Ukr. J. Phys.* **57** 834
[16] Shen K, Sun H L, Ji G, Yang Y, Jiang Z and Song F 2016 *Nanoelectronics and Materials Development* ed A Kar (Rijeka: InTech) **Chapter 6** pp 95–116
[17] Zhang T, Dong Y T, Geng T, Dai Q and Xu Y H 2009 *J. Mater. Chem. Phys.* **114** 257
[18] Rundupadang G C, Estrada R, Kurniawan A, Rohaeti E dan Irzaman 2015 *Prosiding Seminar Nasional Fisika dan Aplikasinya (Jatinangor)* **1** (Sumedang: Universitas Padjadjaran) p FM-07
[19] Irzaman, Pebriyanto Y, Apipah E R, Noor I and Alkadri A 2015 *Integrated Ferroelectrics* **167** 137
[20] Estrada R, Djohan N, Rundupadang G C, Kurniawan A, Iskandar J, Dahrul M, Hardhienata H and Irzaman 2016 *presented in 2nd Int. Conf. on Sci. Tech. and Interdiscip. Res. 2016 (Lampung)* (Bandar Lampung: University of Lampung) [Unpublished]
[21] Ismangil A, Jenie R P, Irmansyah and Irzaman 2015 *J. Procedia Environ. Sci.* **24** 329
[22] Triloki, Rai R and Singh B K 2013 *Proc. Int. Symp. on Nuclear Phys. (Mumbai)* **58** (India: Bhabha Atomic Research Centre) p 838
[23] Bharadwaj V 2014 *Proc. Natl. Conf. Comp. of Colours (Indore)* (Madhya Pradesh: Granthaalayah) pp 1–6

The effect of annealing temperature variation on the optical properties test of $LiTaO_3$ thin films based on Tauc Plot method for satellite technology

N Djohan[1*], R Estrada[1], D Sari[2], M Dahrul[3], A Kurniawan[3], J Iskandar[3], H Hardhienata[4] and Irzaman[4]

[1] Department of Electrical Engineering, Faculty of Engineering and Computer Science, Krida Wacana Christian University, Jl. Tj. Duren Raya No. 4, Jakarta 11470, Indonesia.
[2] Student of Electrical Engineering, Faculty of Engineering and Computer Science, Krida Wacana Christian University, Jl. Tj. Duren Raya No. 4, Jakarta 11470, Indonesia.
[3] Post Graduate of Biophysics, Faculty of Mathematics and Natural Sciences, Bogor Agricultural University, Jl. Raya Dramaga, Bogor 16680, Indonesia.
[4] Department of Physics, Faculty of Mathematics and Natural Sciences, Bogor Agricultural University, Jl. Raya Dramaga, Bogor 16680, Indonesia.

E-mail: nani.djohan@ukrida.ac.id

Abstract. The purpose of the present research is to observe the energy gap of thin films made from $LiTaO_3$ in 1 M-solubility deposited on n-type Si (111) substrates with annealing temperature variation. The manufacture of thin films has been formed by Chemical Solution Deposition (CSD) method using spin coater on 3000 rpm speed for 30 seconds and performed annealing process using furnace (Nabertherm type B180) at a temperature of 750°C, 800°C and 850°C for 15 hours. The absorbance of thin films is measured by using an Ocean Optics USB2000 device and processed into the energy gap curve using Tauc Plot method. The result shows that the energy gap of thin films associated with indirect transitions are increased from 2.78 eV to 2.93 eV with the rise of annealing temperature. The research shows that the thin films on n-type Si (111) substrates made of $LiTaO_3$ produces sensitivity to violet light spectrum and have the potential to be developed as a sensor on satellite technology.

1. Introduction

Ferroelectric thin films are used for a variety of devices such as ferroelectric memories, infrared pyroelectric sensors and other integrated technologies [1-3]. Lithium tantalate ($LiTaO_3$) is currently one of the most interesting ferroelectric materials due to the high dielectric constant and high load storage capacity [4, 5]. $LiTaO_3$ is the object studied intensively over the last few years because it has unique properties [6].

Due to the excellent ferroelectric and pyroelectric characteristics, $LiTaO_3$ ferroelectric materials are used in electronical devices and optical devices, such as high temperature transducers, lasers, electronic filters and non-linear optical crystals [7, 8]. Ferroelectric materials also have a special characteristic as they are spontaneously polarized dielectric and able to change the direction of the internal electricity [9, 10].

$LiTaO_3$ thin film can be created with a number of techniques, e.g. Chemical Solution Deposition (CSD), Pulsed Laser Deposition (PLD), sputtering and Metal Organic Chemical Vapour Deposition (MOCVD), combustion method [11], the Czochralski method [10] and liquid phase epitaxy [12]. These techniques have demonstrated that $LiTaO_3$ thin films are prepared by the CSD method on n-type Si (111) substrate can function as a sensor [4, 6, 9]. The purpose of the present research determines the energy gap of thin films made from lithium tantalate ($LiTaO_3$) on n-type Si (111) substrate with the annealing temperature of 750°C, 800°C and 850°C.

2. Experimental Method

The thin film preparation was started by cutting the n-type of Si (111) substrate using a glass cutter with the size of 1×1 cm^2, then it was cleaned using aqua bidest and then dried [13]. In this experiment, an analytical scale type AND GR-200, Branson 2510 ultrasonicator device, a spin coater, furnace Nabertherm type B180 and an Ocean Optics USB2000 were used as the research devices. $LiTaO_3$ solution was obtained by mixing chemicals according to reaction equalization [14, 15]:

$$2Li(CH_3COO) + Ta_2O_5 + 4O_2 \rightarrow 2LiTaO_3 + 4CO_2 + 3H_2O \quad (1)$$

Based on the reaction equalization, the chemical solution of $LiTaO_3$ at 1 M-solubility can be produced from a mass composition (by weighing with an analytical scale type AND GR-200) of lithium acetate ($Li(CH_3COO)$, 99.5%) with 0.1650 gram, tantalum pentoxide ((Ta_2O_5), 99.9%) with 0.5524 gram and also using 2.5 ml of 2-methoxyethanol ($CH_3OCH_2CH_2OH$) as solvent. The procedure was continued with sonication the chemical solution of $LiTaO_3$ for 60 minutes by using BRANSON 2510 as the ultrasonicator device to obtain the homogeneous solution.

The Chemical Solution Deposition (CSD) method is the coating process of $LiTaO_3$ on the surface of n-type Si (111) substrates which is placed on the rotatable disc [14, 15]. The coating process on the surface of each substrate and rotating process (using a spin coater device) at speed of 3000 rpm during 30 seconds were repeated three (3) times with one (1) minute interruption time [14, 15]. The annealing process (by furnace Nabertherm type B180) with a temperature of 750°C, 800°C, 850°C for 15 hours was used to shape crystal of $LiTaO_3$ solution on n-type Si (111) substrates [16].

In this research, Ocean Optics USB2000 was used to measure the absorption spectrum of visible light of thin films. The results of measurement obtained in form of the recorded of absorbance data at a wavelength of visible light which indicates electron excitation in the crystal structure of $LiTaO_3$. Related to it, the Tauc Plot method was used to estimate the magnitude of energy band by withdrawing the line with a sloping curve from the relationship between hv (x-axis) and $(\alpha hv)^n$ (y-axis).

3. Result and Discussion

The absorbance data at a wavelength of visible light of thin film was obtained through measurement using spectroscopy (Ocean Optics USB2000) and presented in the curve form (see Figure 1). The result shows the values shown as seen in Figure 2 (x-axis and y-axis) which were determined from the following formula [4, 17, 18]:

$$\alpha = 2.303 \frac{\text{absorbance data}}{d} \quad (2)$$

$$\text{y-axis} = (\alpha hv)^{0.5} \quad (3)$$

$$\text{x-axis} = hv \quad (4)$$

with: d = thickness of $LiTaO_3$ on substrate (cm)
absorbance data = the values from measurement by using spectroscopy
α = absorption coefficient (cm^{-1})
hv = photon energy (eV)

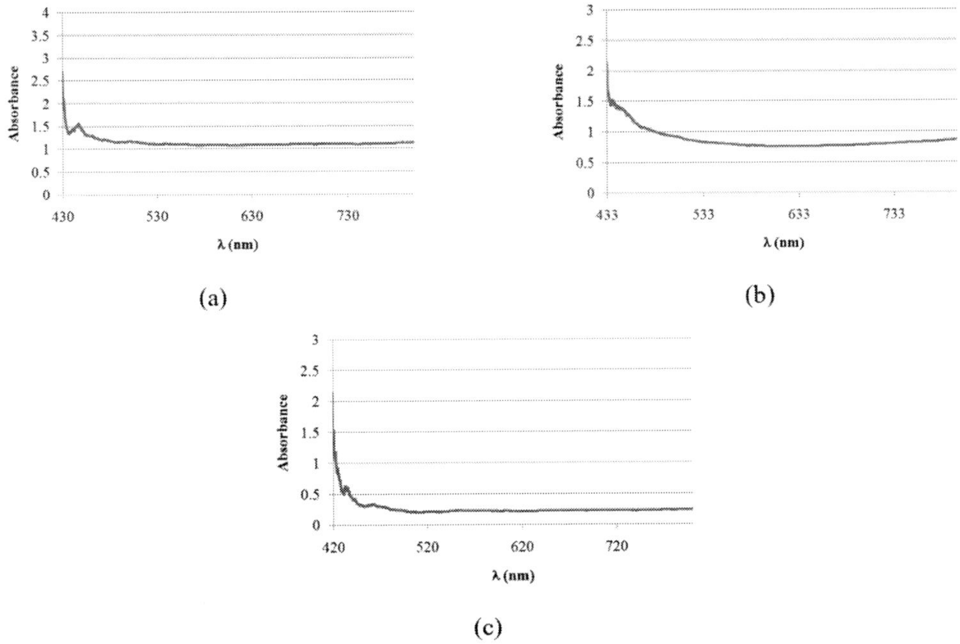

Figure 1. Absorption spectrum: (a) 750°C (b) 800°C (c) 850°C

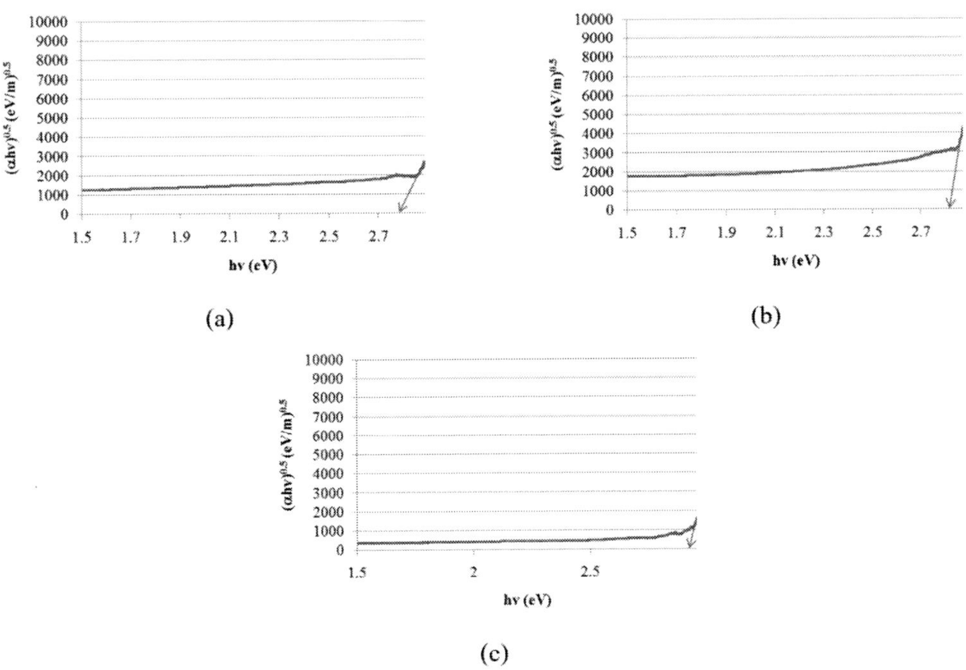

Figure 2. Energy gap: (a) 750°C (b) 800°C (c) 850°C

The energy gap values (Tabel 1) were estimated from Tauc Plot method by withdrawing the line with a sloping curve which was in correspondence between x-axis and y-axis (see Figure 2). The wavelength refers to energy gap value of thin films were determined from the following formula [17, 19]:

$$\lambda = \frac{h\,c}{\left(1.602\times 10^{-19}\ \mathrm{J}\right)\left(\text{energy gap}\right)} \tag{5}$$

with: h = Planck's constant (6.626×10^{-34} J·s)
c = speed of light (2.998×10^{8} m·s^{-1})
energy gap = the values from Table 1

Table 1. The energy gap of thin films

Annealing temperature (°C)	Energy gap (eV)
750	2.78
800	2.82
850	2.93

Table 2. The wavelength and sensitivity to specific light spectrum on each thin film

Annealing temperature (°C)	Wavelength of light (nm)	Spectrum color of light (Based on literature [20])
750	446	Violet
800	440	Violet
850	424	Violet

4. Conclusion

In this research, $LiTaO_3$ thin films was successfully made with spin coating. The energy gap of thin films associated with indirect transition shows increasing from 2.78 eV to 2.93 eV with the rise of annealing temperature. The absorbance of thin films was obtained by using an Ocean Optics USB2000 device in the wavelength of visible light in the range between 424 nm and 446 nm. The research concludes that the thin films on n-type Si (111) substrates made from $LiTaO_3$ produce sensitivity to violet light spectrum and have the potential to be developed as a sensor on satellite technology.

Acknowledgment

This research is funded by the Ministry of Research, Technology and Higher Education of the Republic of Indonesia through PEKERTI Research Grant No. 771/K3/KM/SPK.LT/2016.

References

[1] Harsono B, Liman J, Iskandar J, Rohaeti E and Irzaman 2015 *Intg. Sci. Tech: The Interdiscip. Res. Apch.* ed A Ulvan and I Sukmana (Bandar Lampung: University of Lampung) **Chapter 5** pp 37–43
[2] Gonzales A H M, Simoes A Z, Zaghete M A and Varela J A 2003 *J. Mater. Charact.* **50** 233
[3] Kostritskii S M, Sevostyanov O G, Bourson P, Aillerie M, Fontana M D and Kip D 2007 *J. Ferroelectr.* **352** 61
[4] Ismangil A, Jenie R P, Irmansyah and Irzaman 2015 *J. Procedia Environ. Sci.* **24** 329
[5] Uchino K 2000 *Ferroelectric Devices* (New York: Marcel Dekker, Inc.)
[6] Ismangil A, Irmansyah and Irzaman 2016 *J. Procedia Environ. Sci.* **33** 668
[7] Wendong Z, Qiulin T, Jun L, Chenyang X, Jijun X and Xiujian C 2010 *J. Opt. Laser Technol.* **42** 1223
[8] Stenger V, Shnider M and Sriram S 2012 *Optoelectronic Materials and Devices* **8** 8261-27

[9] Misbakhusshudur M, Ismangil A, Aminullah, Irmansyah and Irzaman 2016 *J. Procedia Environ. Sci.* **33** 615
[10] Irzaman, Maddu A, Syafutra H dan Ismangil A 2010 *Prosiding Seminar Nasional Fisika* 175
[11] Chung S L and Wang C M 2012 *J. Mater. Sci. Technol.* **28** 713
[12] Li N, Katase T, Zhu Y, Matsumoto T, Umemura T, Ikuhara Y and Ohta H 2016 *Appl. Phys. Express* **9** 125501
[13] Irzaman, Pebriyanto Y, Apipah E R, Noor I and Alkadri A 2015 *Integrated Ferroelectrics* **167** 137
[14] Sari F I W, Djohan N, Kurniawan A, Rohaeti E dan Irzaman 2015 *Prosiding Seminar Nasional Fisika dan Aplikasinya (Jatinangor)* **1** (Sumedang: Universitas Padjadjaran) p FM-12
[15] Djohan N, Estrada R, Sari F I W, Kurniawan A, Iskandar J, Dahrul M, Hardhienata H and Irzaman 2016 *presented in 2nd Int. Conf. on Sci. Tech. and Interdiscip. Res. 2016 (Lampung)* (Bandar Lampung: University of Lampung) [Unpublished]
[16] Vilarinho P M, Barroca N, Zlotnik S, Félix P, Fernandes M H 2014 *J. Mater. Sci. Eng. C* **39** 395
[17] Ohring M 1992 *The Materials Science of Thin Films* (San Diego: Academic Press)
[18] Triloki, Rai R and Singh B K 2013 *Proc. Int. Symp. on Nuclear Physics (Mumbai)* **58** (India: Bhabha Atomic Research Centre) p 838
[19] Tauc J, Grigorovici R and Vancu A 1966 *J. Phys. Stat. Sol.* **15** 627
[20] Bharadwaj V 2014 *Proc. Natl. Conf. Comp. of Colours (Indore)* (Madhya Pradesh: Granthaalayah) pp 1–6

Barium Strontium Titanate Thin Film Growth with rotational speed variation as a satellite temperature sensor prototype

Mulyadi[1*], W Rika[1], Sulidah[2], Irzaman[3], Hendradi Hardhienata[3]
[1]Department of Electrical Engineering, Borneo University, Indonesia
[2]Department of Nursery, Borneo University, Indonesia
[3]Department of Physics, Bogor Agriculture University, Indonesia

E-mail: mulyadi@borneo.ac.id

Abstract. Barium Strontium Titanate(BST) is a promising material for sensor devices such as temperature and infrared sensor. $Ba_xSr_{1-x}TiO_3$ thin films with affordable Si substrate were prepared by chemical solution deposition method and spin coating technique for 30 seconds with variation in rotation speed (3000 rpm, 5500 rpm and 8000 rpm). A high baking temperature at 850^0C has been used for 15 hours during the annealing process. The thickness of BST film was calculated via gravimetric calculation. USB 2000 VIS-NIR was used to characterize the optical properties of BST thin film. The obtained reflectance curve showed that the most reflected wavelengths were in the range of 408-452 nm respectively. The result of the optical film characterization is very important for further development as a sensor in satellite technology.

1. Introduction

Barium Strontium Titanate (BST) material is one material which in recent years have intensively studied and developed [1-3]. $Ba_xSr_{1-x}TiO_3$(BST) is a material that has a dielectric constant and a high charge storage capacity.Barium Strontium Titanate having a dielectric constant in excess of 1000 in bulk form, a key challenge for the semiconductor industry is to maintain the high dielectric constant properties of BST while integrating the material in thin film form into IC's [4]. Ferroelectric material has the ability to change the direction of its internal electricity, can be spontaneously polarized and show hysteresis effects associated with the shift dielectric in response to an internal electric field. Piezoelectric properties can be applied as microactuator and sensors, such as infrared sensors and electro-optical properties to be applied to the light sensor [5], polarizability properties can be applied as a non-volatile random accessmemory [6]. In the process of preparing the film, there are several methods used [7], such as sol-gel method [8], pulsed laser ablation deposition method [9], sputtering method [10], MOCVD method [11] and chemical solution deposition methods [12]. Barium strontium titanate preparation methods may be using equipment that is quite simple, cost-effective and carried out in a relatively short time. In this work, we chose a chemical solution deposition method because of its superiority to control the films stoichiometry with good quality and the procedure is relatively easy [13].

2. Materials and Methods

The substrate used is a p-type silicon. The substrate cut into a square with a size of 1 cm x 1 cm 24 pieces. After the cutting process, followed by washing using distilled water. In the experimental work, barium acetate (99 %, Aldrich-243671), strontium acetate (99.995 %, Aldrich - 41283) and titanium isopropoxide oxide (99 %, Aldrich-21642200) were used as precursor materials for the BST synthesis and 2.5 ml of 2-methoxy ethanol as a solvent to obtain 1 M. Then all the solution were stirred with a Branson 2510 ultrasonic stirrer for 90 minutes [14].

The chemical formula of BST thin film as follow:

$$_{0,5}Ba(Ch_3COO)_2 + _{0,5}Sr(CH_3COO)_2 + Ti(C_{12}H_{28}O_4) + _{22}O_2 \longrightarrow Ba_{0,5}Sr_{0,5}TiO_3 + _{17}H_2O + _{16}CO_2$$

2.1. *Thin film growth and annealing process*

The hatching process from a homogeneous solution was done using chemical solution deposition methods (CSD) which has been developed for the growth of perovskite thin films [15]. Si substrate (100) p-type that has been poured by $Ba_{0,5}Sr_{0,5}TiO_3$ solution was rotated using a spin coater at 3000 rpm, for 30 seconds. The hatching process and Si (100) p-type substrate rotation performed three times droplets with a pause for one minute in each hatching [16]. Thin films that have gone through the process chemical solution deposition, further in-annealing using Nabertherm furnace at a temperature of 850^{0}C for 15 hours [17]. This procedure was repeated for rotational speed of 5500 rpm and 8000 rpm.

2.2. *Optical characterization of thin film BST*

Evaluation of optical properties such as reflectance values conducted to observe the light reflection spectrum of the thin film Barium Strontium Titanate. The light source used was visible light, while the tool used in this test was a UV-Vis Spectroscopy Ocean Optics USB 2000. Analysis of the reflectance values in the form of calculation of the energy gap and the refractive index. Energy gap is obtained by creating plotting the relationship between:

$$\left[\ln\left(\frac{R_{max}-R_{min}}{R-R_{min}}\right)\right]^2 \quad to \quad hv \tag{1}$$

Extrapolation performed on the curve that has the highest gradient and cutting hv axis. Values were cut hv axis represents the value of energy gap. Meanwhile, the refractive index is obtained through equation:

$$n = \frac{1+\sqrt{R}}{1-\sqrt{R}} \tag{2}$$

3. Result and Discussion

From the plots in Fig.1 it can be observed that the thickness of BST was obtained using gravimetric methods. The film thickness (d) can be calculated from actual (m) mass deposited onto substrate, (ρ) density of BST (6.02 g/cm3), and the area of the film (A) [18, 19].

$$d = \frac{m}{A(\rho)} \tag{3}$$

The variation of spin coater rotation speed in thin films preparation process led to changes in the thickness of BST film. Although the changes that caused by rotation speed was not significant, but small changes affected the absorption coefficient (α) of the thin film. The value of the energy gap was affected by the value of coefficient absorption [20].

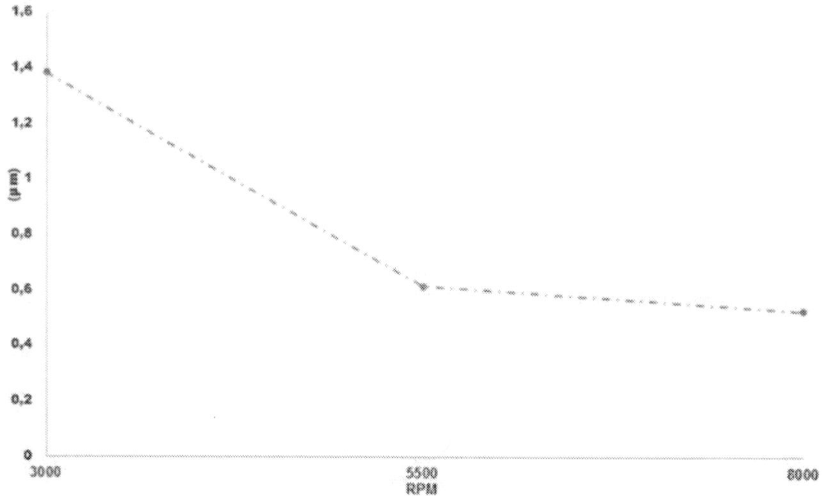

Figure 1. Film thickness

According to Irzaman, the duration of annealing hold time could affect the crystal structure, crystal morphology regularity compilers, and thickness of BST film [21].

The optical properties are measured in the light wavelength region of 400 to 1000 nm. The principal optical properties measured are the reflectance spectrum. It is used to calculate the band gap range of the BST. The energy band gap value calculated using Tauc method from the reflectance data.

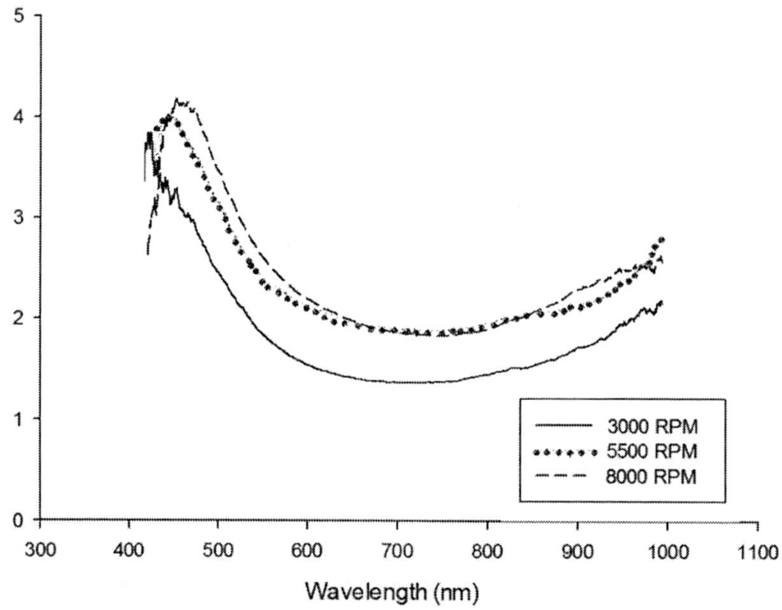

Figure 2. Reflectance spectrum of BST

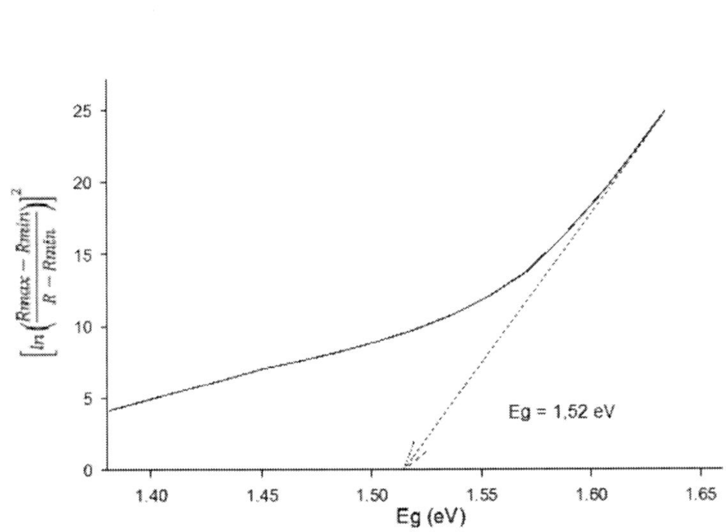

Figure 3. BST energy gap for 3000 rpm

Figure 3, 4 and 5 shows the relationship of optical properties to band gap energy of each sample to prove the film was semiconductor material. Energy gap is obtained by creating plotting the relationship between. The method is just plotting a curve of $(\alpha h\nu)^{1/2}$ to hv. Extrapolation performed on the curve to obtain the corresponding value of energy gap.

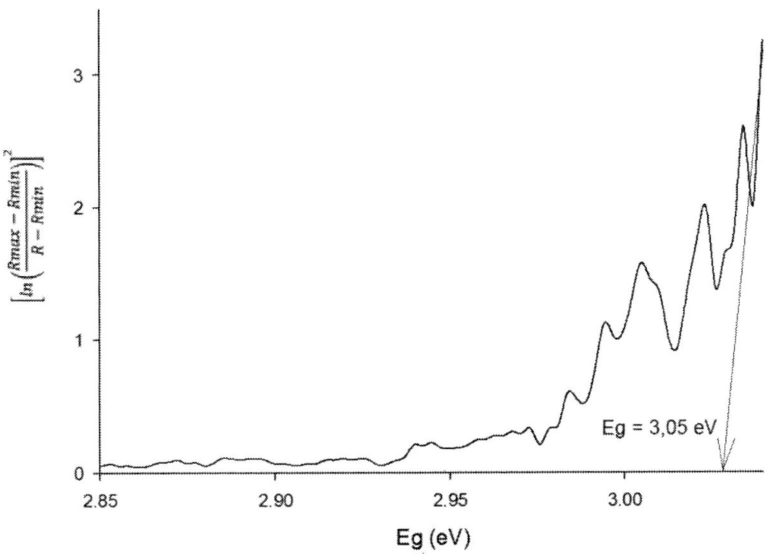

Figure 4. BST energy gap for 5500 rpm

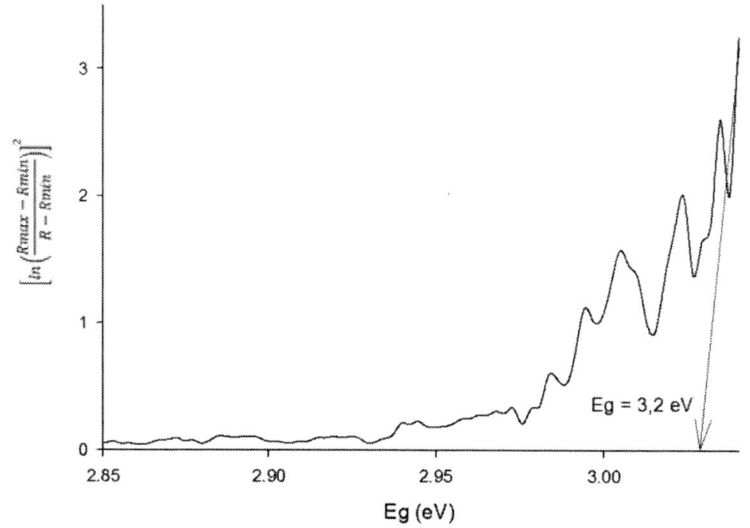

Figure 5. BST energy gap for 8000 rpm

The relationship between semiconductor photon absorption coefficient and photon energy incident on the semiconductor film. Assuming that n=1/2 it could be used to calculate the energy band gap of $Ba_{0,5}Sr_{0,5}TiO_3$ that is grown on Si substrates. Several studies that have been carried out previously indicate that the absorption coefficient is equal to ln $[(R_{max}-R_{min})/(R-R_{min})]$ as shown in equation (1). From the graph in figure 3, 4 and 5 it could be seen that the energy band gap values for BST uses varying the rotational speed of 3000 rpm, 5500 rpm and 8000 rpm are 1.52 eV, 3.05 eV and 3.2 eV respectively. The discrepancy of energy band gap widths are probably due to the varying the rotational speed of chemical solution deposition methods process that make the crystal expand. These values are in accordance to high photon energy in the short wavelength (about 2 to 3 eV). This result is not too different from earlier study of $Ba_{0,5}Sr_{0,5}TiO_3$ energy band gap [22]. Figure 3, 4 and 5 shows the results of the band gap energy is still within in the range of semiconductor energy band gap value. From these results, it seems that if the rotational speed is increased there will be an increase in the band gap energy with the value in the range of semiconductors value range.

4. Conclusions

$Ba_{0,5}Sr_{0,5}TiO_3$ thin film (BST) which is grown on the substrate Si (100) p-type by CSD method uses varying the rotational speed of 3000 rpm, 5500 rpm, 8000 rpm, has been successfully created. Variation of spin coating rotational speed serves to establish the orientation of BST film crystal structure that coincide with the thickness of the crystal substrate. The crystal structure is denser and the energy band gap become narrower which makes it suitable to be applied as a temperature sensor. The energy gap and the refractive index have been successfully analyzed using the reflectance method. The results of the analysis using reflectance method produces a wavelength range of 450 to 750 nm with energy gap are obtained in the range of 1.52 eV to 3.2 eV. Band gap energy values of the films showed that $Ba_{0,5}Sr_{0,5}TiO_3$ was in the range of semiconductors value. The designed BST thin film proved to be sufficient a promising potential as a sensor in terms of optical properties characterization for the satellite technology.

Acknowledgments

The research underlying this paper is partially supported by Penelitian Kerjasama antar Perguruan Tinggi Ministry of Research, Technology and Higher Education of the Republic of Indonesia.

5. References

[1] Friederich A, Kohler C, Nikfalazar M, Wiens A, Sazegar M, Jakoby R, Bauer W and Binder J.R 2014 *J. Eur. Ceram. Soc.* **34** 2925–32
[2] Thongchanthep C and Thountom S 2015 *Ceram. Int.* **41** S95–S99
[3] Tagantsev A K, Sherman V O, Astafiev K F, Venkatesh J and Setter N, 2003 *J. Electroceram.* **11** 5
[4] Remmel T, Schulberg M, Fujimura S, Honma H, Kobayashi H, Kohno H, Owens S, Deslattes R and Pedulla J 2000 *Denver X-ray Conferenceon Applications of X-ray Analysis*; Denver USA
[5] Iskandar J, Syafutra H, Juansah J and Irzaman 2015 *Procedia Environmental Sciences.* **24** 324
[6] Zhang W, Xue S, Liu S, Wang J, Shen B and Zhai J 2014 *J. Alloys Compd.* **617** 740–45
[7] Zuo X H, Deng X Y, Chen Y, Ruan M, Li W, Liu B, Qu Y and Xu B 2010 *Mater. Lett.* **64** 1150
[8] Wang J, Zhang T J, Xiang J H and Zhang B S 2008 *Mater. Chem. Phys.* **108** 445
[9] Craciun V and Singh R K 2000 *Appl. Phys. Lett.* **76** 1932-40
[10] Zhu X H, Guigues B, Defay E, Dubarry C and Aid M 2009 *J. Appl. Phys.* **105** 044108
[11] Sbrockey N M, Cole M W, Kalkur T S, Luong M, Spanier J E and Tompa G S 2011 *Integr.Ferroelectr.* **126** 21
[12] Irzaman, Marwan, Arief A, Hamdani R and Komaro MKM 2009 *Indones. J. Phys.* **19** 119–121
[13] Kao C F and Yan W D 1999 *Applied Organometallic Chemistry.* **13** 383-397
[14] Ubagaramary D and Neeraja P 2012 *Journal of Applied Chemistry.* **2** 5
[15] Chen P C, Tsai M C, Huang Y J, Chiu H T and Lee C Y 2012 *Cryst Eng Comm.* **14** 1990–1993
[16] Seo Y J and Park S W 2004 *J. of the Korean Phy. Society.* **45** 769-772
[17] Morîntale E, Scari N, Dinescu M and Rotaru P 2010 *J. Physics.* **20** 83-87
[18] Irzaman, Syafutra H, Darmasetiawan H, Hardhienata H, Erviansyah R and Huriawati F 2011 *Atom Indonesia.* **37**133-138
[19] Jundale D, Pawar S, Chougule M, Godse P and Patil S 2011 *Scientific Research.* **1** 34-36
[20] Setiawan A, Aminullah, Juansah J and Irzaman 2016 *Procedia Environmental Sciences.* **33** 620
[21] Irzaman, Darmasetiawan H, Hardhienata H, Hikam M, Arifin P, Jusoh N S, Taking S, Jamal Z, Idris M A. 2009 *Atom Indonesia.* **35** 57-62
[22] Irzaman, Syafutra H, Arif A, Alatas H, Hilaluddin M N, Kurniawan A, Iskandar A, Dahrul M, Ismangil A, Yosman D, Aminullah, Prasetyo L B, Yusuf A and Kadri T M 2014 *AIP Conference Proceedings* **1586** 24

Preliminary of Optical Lens Design for Micro-Satellite

Elvira Rachim, Andi Mukhtar Tahir, Agus Herawan

Indonesian National Institute of Aeronautics and Space (LAPAN)

E-mail: elvira.rachim@lapan.go.id

Abstract. The development of micro satellites for the last two decades is emerging rapidly as the need of satellite communication usage is increasing. Earth observation is one of the example of how satellites are on demand. Most observation satellites consist of sensors and imaging system on-board. One of the key element to have a good imaging system is a special optical lens system design. Such lens is designed specifically by calculating every parameter such as refractive, reflective indexes, type of surface, distance and many more. Manufactured lenses sometimes do not match the requirement of an imager system hence the special lens design is needed. This paper will first briefly describe the history of optic, theory related to lens system, then the design and the analysis of lens system for micro-satellites generally and LAPAN A4 particularly.

1. Introduction

The origins of optical technology date back to remote antiquity. Early mirrors were made of polished copper, bronze and later on speculum, a copper alloy rich in tin. At the seventeenth century Willebrord Snell (1591-1626) Professor at Leyden, empirically discovered the long-hidden Law of Refraction in 1621, this was one of the great moments in Optics. By learning precisely how rays of light are redirected on traversing a boundary between two media, Snell in one swoop swung open the door to modern applied Optics. [1]

In the twentieth century, applied Optics had become a phenomenal, more complex design and technology as the advent of the high speed computer was used to design such Optics. The demand of a high speed communication had also taken the Optics technology into another level, and there was also the infrared technology which is highly used in surveillance systems these days. Before a lens can be constructed it must be designed, that is to say, the radii of curvature of the surfaces, the thickness, the air spaces, the diameters of the various components, and the types of glass to be used must all be determined and specified. [2]

Two of the basic elements of Optics which influential are Numerical Aperture (NA) and Reflection. Numerical Aperture is the amount of light from the Field of View which a lens can accept.

$$NA = n \sin \theta \qquad (1)$$

where n is the index of refraction of the medium in which the lens is working and θ is the half-angle of the maximum cone of light that can enter or exit the lens. [3]

There are three kinds of Reflections, Prism, Cataoptric, and Catadioptric. The type which LAPAN satellite used is Catadioptric. Catadioptric is a combination of Refraction and Reflection using lenses and curved mirrors.

This paper will describe mostly on the Optics design for the micro-satellite for LAPAN A4, which requirement will be similar to the LAPAN A3 imager system.

The next generation of LAPAN satellite would be applying an infrared (IR) micrometer Bolometer, whereas the design would much more complex than the regular lens used for microsatellite, everything had been initiated by spectral radiance analysis adjusted by catastrophes sources in Indonesia, mainly wild fire (forest fire) and active volcano [12]. When it comes to the terms of IR micrometer Bolometer. There would be thermal analysis needed.

2. Methodology and Goals

This research was done by studying several literatures and simulation using a lens design software. The goals of this research is so that it will become a reference for further study on Optical Lens Design for satellites, and to find a compatible design for LAPAN A4 imager system.

3. Designing an optical lens

Optical design or lens design is the process used to find the best set of lens construction parameters (e.g., radii of curvature, thicknesses, airspaces, and materials) that optimizes the overall performance (including the manufacturability) of an optical system [4]

There are also some key elements worth to notice in designing a lens, the stray light analysis, and distortion.

3.1. Stray light analysis

Stray light analysis is any unwanted light that reduces performances by striking the image plane of an optical system [4] In designing an optical lens, there is always the possibility of stray lights, hence there is ray tracing to trace the unwanted lights and baffle is commonly used to fix this such error. Baffle functions as a shield of a light not from the field of view.

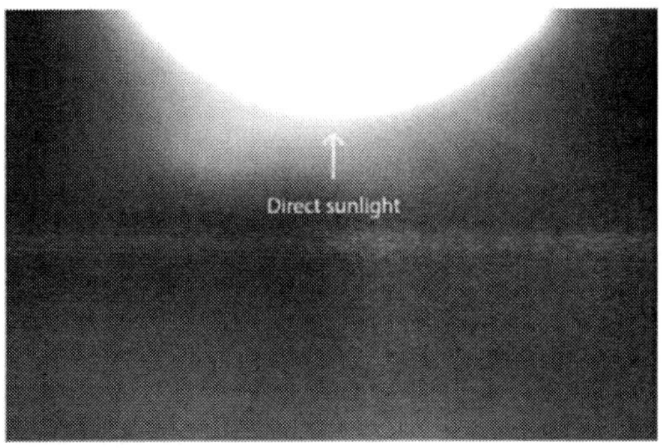

Figure 1. Stray light from the sun just outside the lower edge of FOV, shining through the hole in the primary mirror of a Maksutov–Cassegrain telescope whose baffles are undersized [5]

3.1.1. Distortion

Distortion is an aberration in which the magnification varies over the image height. Object points are imaged to perfect points with no blur, but straight lines in the object become curved lines in the image [4]

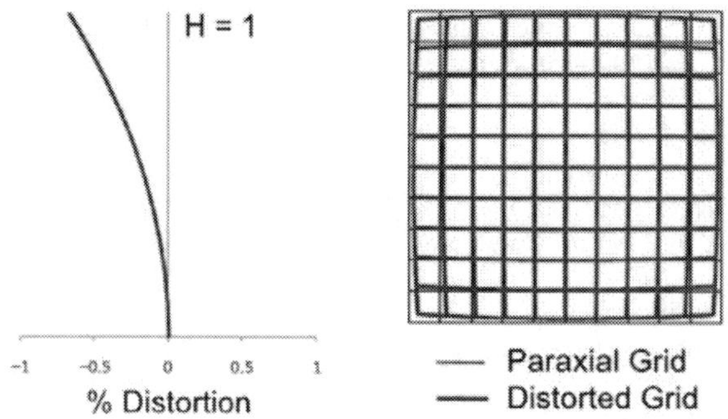

Figure 2. Distortion

In is simplest expression, the distortion is a transverse deformation of an image. If a grid is used as an object, its image in a system with distortion will be a grid of curved lines. This is due to the change in the transverse magnification in a system as a function of the distance from the optical axis. When the magnification decreases as the distance from the axis increases, a barrel shaped image occurs. If the magnification increases with the distance, a pincushion image is obtained [6]

3.1.2. Lens Design Software

There are three most commonly lens design software used these days, Optics Software for Layout and Optimization (OSLO) and TracePro by Lambdares, OpticStudio by Zemax, Code V Optical Design Software by Synopsys. These softwares will mostly help in ray tracing the stray lights and distortion, and optimization.

3.1.3. Basic Optical Design Process

The basic optical design process is as follows:

- A specification document for the optical system is established, and the appropriate type of lens form (e.g., microscope, telescope, camera lens) is selected.
- A starting-point design is chosen from previous projects, historical databases, or patent literature (primarily based on the aperture, field, wavelength, and packaging requirements).
- If an appropriate starting-point design does not exist, designers can choose to start from a first-order solution that identifies focal lengths and object/image/pupil locations and sizes using thin lenses and paraxial rays. The first-order solution can be converted to a third-order thin lens solution where thin lens parameters such as radius of curvature and index of refraction are chosen to minimize third-order aberrations. Thickness is then introduced into the thin lenses to arrive at a thick lens starting point.
- The starting-point design is assessed by tracing real rays at multiple wavelengths through the system to analyze the chromatic and higher-order aberrations. Compliance to performance requirements (e.g., image quality, packaging restrictions, fabrication and assembly tolerances, cost) is evaluated.
- If the performance is not acceptable, optimization is used to improve the design. A set of variables and constraints is defined, and computer software is used to find design solutions as defined by a merit function.

- If the solution is still unacceptable, designers must return to a previous step and iterate until the design meets specifications. This process may require the selection of a different starting-point design and/or requesting a significant change in one or more of the specifications.

The manufacturability of the design is just as important as the final image quality. Therefore, a complete design process and performance evaluation includes tolerancing, stray light analysis, and thermal and other environmental analyses.

Meanwhile the design process for IR micrometer Bolometer is as follows:

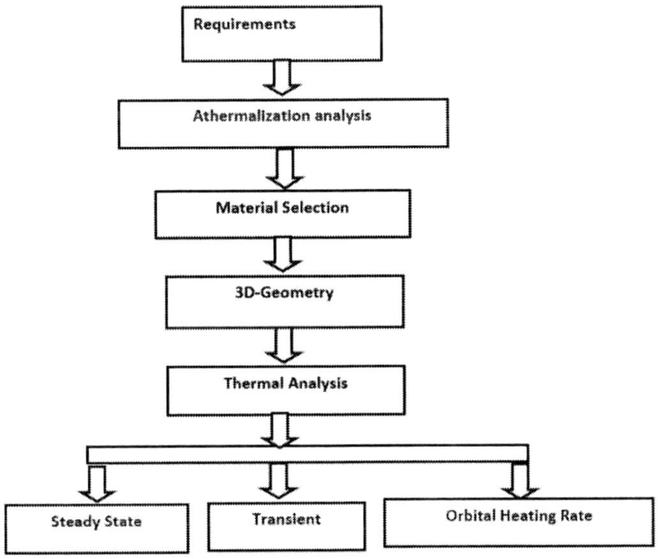

Figure 3. Design process of thermal analysis

The work is initiated by determining requirements of the design. Then, it is continued by doing an analysis of athermalization, either for lens or housing of the sensor. From that process, we've got the material selection for lens and also for the housing which fulfils the requirements. Based on the prior result, we create three dimensional (3D) geometry, yet we separate between the housing and the lens in order to facilitate analysis process. Moreover, this is in accordance with the capacity that we have. After constructing process finished, the analysis goes through the thermal analysis for each geometry using commercial software applications, i.e. Finite element and Thermal Desktop 5.0 / Sinda Fluint 5.0. There are three simulation process that we do throughout the analysis, i.e. steady state, transient analysis, and orbital heating rate using Monte Carlo approach. All three result will be used to analysis what the appropriate thermal control is for IR Microbolometer optical design, so that the sensor can operate in severe condition such as space environment. [13]

For applications prone to temperature fluctuations, it is important to develop an athermal optical system: an optical system that is insensitive to an environment's thermal change and the resulting system defocus. Developing an athermal design, which is dependent on the Coefficient of Thermal Expansion (CTE) of the materials and the change in index with temperature (dn/dT), is especially critical in the infrared. The dn/dT of most IR materials is orders of magnitude higher than those of visible glasses, creating large changes in the refractive index. Additionally, while optical systems are often designed in air, the housing material is also sensitive to thermal change, and should be addressed when considering an athermalized design. [14]

4. The Optical Lens Design for Micro-satellite

The demand of Earth observation is emerging rapidly these days, everyone wants the technology. For the developing countries with low research and development budget have to considered a way to obtain and actually build the technology, hence the answer is micro-satellites.

One of the possible approaches is taking full advantage of the ongoing technology developments leading to further miniaturization of engineering components, development of micro-technologies for sensors and instruments which allow designing dedicated, well-focused Earth observation missions. At the extreme end of the miniaturization, the integration of micro-electromechanical systems (MEMS) with microelectronics for data processing, signal conditioning, power conditioning, and communications leads to the concept of application specific integrated micro-instruments (ASIM). [7]

The lens design depends on the micro-satellite mission. Earth observation or surveillance mission for example, need a proper resolution, there are three kinds of resolutions: spatial resolution, spectral resolution and temporal resolution. Each resolution shall has Ground Sample Distance (GSD) and the Field of View (FOV) measured precisely. The imaging system payload is basically consists of an Optical Head and Electronic Unit. [6]

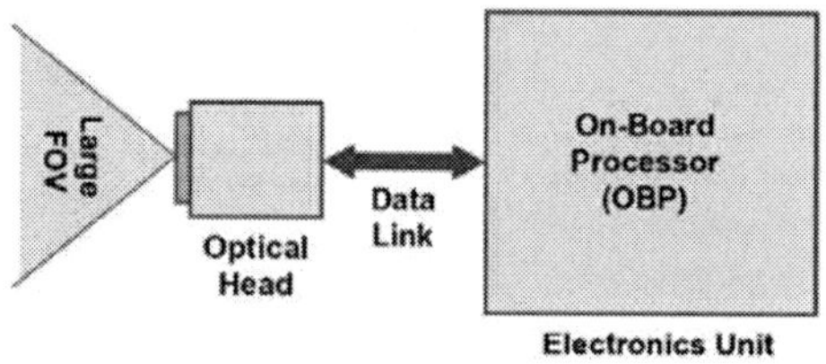

Figure 4.. The Optical Head and Electronic Unit block diagram

The primary purpose of the optical module is to record wide and broadband images of a scene around the satellite. A key requirement of the optical module is therefore to be able to use the field coverage properties to distinguish true element that can be used for star tracking, earth horizon sensing and related tracking functionalities. The optical module must provide all usable telemetric information for the satellite [6]. To obtain that, usually a custom made lens is highly recommended as it will provide the closest specification to imaging system requirement. LAPAN A4 satellite will use catadioptric lens, as we mentioned above, catadioptric is a combination of both refractive and reflective (lenses and mirrors).

Optical systems can be refractive (lenses), reflective (mirrors), or catadioptric (both lenses and mirrors). Because many refractive design forms have analogous reflective forms (e.g., the telephoto and the Cassegrain, the retrofocus and the Schwarzchild, or the Cooke triplet and the reflective triplet), it is useful to understand the advantages and disadvantages of one type over the other. [4]

Some of the advantageous of the reflective systems are no chromatic aberrations hence the mirror can be focused in the visible and used in any wavelength region [2] and better performance over a wider thermal range. [4] Although there are also some of the disadvantageous such as limited FOV and stray lights are harder to manage. One of the advantageous of using a catadioptric system is it can decreased aberrations on lenses.

The method used in designing optics here was the same as to process the optical design in general because it is the common way for all applications. The first process starts with the development of the draft specification that includes first-order properties (Effective Focal Length, Field of View, F / number, spectral band), size and weight restrictions, limitations of material and shape of the surface. Next is to assign one or more starting point. In setting the starting point of the optical engineers rarely developed the early design of the draft itself. Based on the design specifications, engineers specifically used the literature that has been patented or some other database to get the configuration succedeed in meeting the design specifications for their specific applications. One technique that has been used successfully in the developing a starting point for complex systems is to break into the system functional subsystems of a form that can be recognized. The aim at this stage is not a process to produce a complete draft final, it is more an early form with unique characteristics. As it is still based on the satellite bus until Lapan Lapan-A1-A3, the optical system selected as the starting point was close to the existing optical system specifications on the satellite as lens length of about 30 cm and 10 cm diameter lens. Here the selected model was Maksutov Cassegrain as a starting point as in Figure 5. The Maksutov-Cassegrain is a catadioptric design, which means the use of lenses and mirrors. The system consists of a parabolic primary mirror and a second mirror hyperboloid and lenses are used as lenses to eliminate aberration corrector (aberration) generated from the design that only uses only mirrors

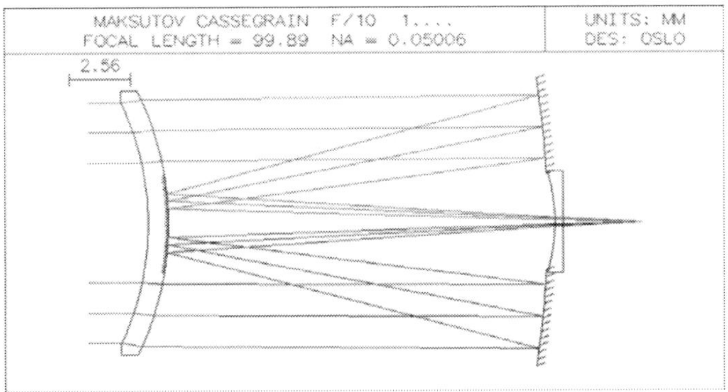

Figure 5. Maksutov Cassegrain, an example of a catadioptric system.

From this design We modificate and optimize it to obtain an optical system to get the desired specifications such as the effective focal length (EFL) 1000 mm NA (numerical aperture) of 0.05, 1.27 degrees FOV, as shown in Figure 6.

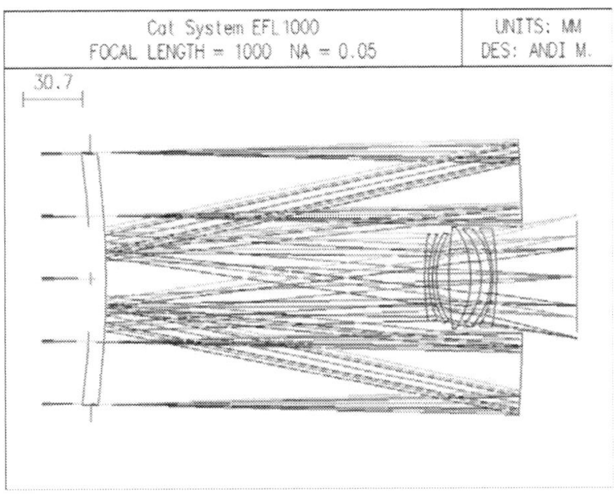

Figure 6. Modificated Lens

When the designing work is done, comes another complex task to do, manufacturing. We shall decide which material will be used, whether it is going to use coating or not. High-reflection

(HR) coatings are needed on reflective optics to improve the base metal reflectivity (e.g., an enhanced dielectric coating can boost the reflectance of an aluminum mirror from 90% to over 97%). Dichroics are short- or long-wave-pass filters used for color balancing or color splitting. Cold and hot mirrors are special dichroics that transmit or reflect, respectively, thermal (long-wave) radiation to protect thermally sensitive optics. [8]

Thermal evaporation is the most commonly used coating method. Source material is heated by resistance heating or by electron beam bombardment to either the sublimation or melting point. Another point of our athermalization design that has become our concern is the housing that holds the lenses in place since its material is affected by thermal change [13]

The Aluminum material does not obtain a high merit comparison of mechanical figure versus thermal stability. Since it has the lowest assembly cost, we finally decide to use it, instead of other materials. Besides, many housing lens use it as the housing material. For this case, Aluminum 2024 T3 has been chosen due to its high strength and excellent fatigue resistance compare to others. [15]

5. Conclusion

Designing an optical lens system for micro-satellite is a complex work to do. We must measure and calculate every aspects. LAPAN might choose to use a catadioptric system because it has advantegous for less expensive and more manageable or flexible control over lenses and mirrors aberrations and errors to fulfill the satellite mission requirement.

References

[1] Eugene H and Ganesan AR 2008 Optics, Dorling Kindersley (India)
[2] Rudolph Kingslake, R. Barry Johnson: Lens Design Fundamentals, Academic Press is an imprint of Elsevier, 2010
[3] UCLA -Brain Research Institute Microscopy Core Facilities (website) : http://www.gonda.ucla.edu/bri_core/na.htm, 2007, last accessed September 2015
[4] Julie Bentley, Craig Olson : Field Guide to Lens Design, SPIE Press, 2012
[5] Eric C. Fest : Stray Light Analysis and Control, SPIE Press, 2013
[6] Simon Thibault, Eric Harvey: Space Telemetric Panomorph Imaging System for Micro/Nano Satellite, SPIE vol.6958,2008
[7] Rainer Sandau: Status and trends of small satellite missions for Earth observation, Elsevier, 2009
[8] Shiuh Chao: Coating Technology, Optical Coatings And Thermal Noise In Precision Measurement, Cambridge University Press, 2012
[9] David Hasenauer: Application-Specific Optical Design, White Paper, Optical SolutionsGroup, Synopsys, Inc. 2012
[10] Andi Mukhtar Tahir: Rancangan Optik untuk Aplikasi Sistem Pencitraan pada Satelit Mikro Menggunakan Perangkat Lunak Optik, Siptekgan XII-2008, ISBN: 978-979-1458-19-1
[11] Wang Wen-fang, Yang Xiao-xu, Zhang Hen-jin, L Gang: Design of Catadioptric Infrared Optical System With Wide Field, Procedia Engineering 15(2011) 4568-4573, Elsevier 2011, doi:10.1016/j.proeng.2011.08.858
[12] Bustanul Arifin, Irwan Priyanto, Andi Mukhtar Tahir: Illumination analysis of LAPAN's IR micro bolometer
[13] Bustanul Arifin, Ahmad Fauzi: The Thermal Analysis Of Lapan's Ir Micro Bolometer Optical Design, International Astronautical Congress (IAC) 2016
[14] Edmund Optics, An Introduction to Passive Athermalization, http://www.edmundoptics.com/resources/application-notes/optics/an-introduction-topassive-athermalization/ (accessed 05.12.16)
[15] Aircraft Spruce and Specialty Co., General Aluminum Information, www.aircraftspruce.com/catalog/mepages/aluminfo.php, (accessed 05.12.16).

Analysis of LAPAN-IPB image lossless compression using differential pulse code modulation and huffman coding

P R Hakim* **and R Permala**

Satellite Technology Centre, Indonesian Aeronautics and Space Agency (LAPAN), Rancabungur, Bogor 16310, Indonesia

E-mail: patriarachmanhakim@yahoo.com

Abstract. LAPAN-A3/IPB satellite is the latest Indonesian experimental microsatellite with remote sensing and earth surveillance missions. The satellite has three optical payloads, which are multispectral push-broom imager, digital matrix camera and video camera. To increase data transmission efficiency, the multispectral imager data can be compressed using either lossy or lossless compression method. This paper aims to analyze Differential Pulse Code Modulation (DPCM) method and Huffman coding that are used in LAPAN-IPB satellite image lossless compression. Based on several simulation and analysis that have been done, current LAPAN-IPB lossless compression algorithm has moderate performance. There are several aspects that can be improved from current configuration, which are the type of DPCM code used, the type of Huffman entropy-coding scheme, and the use of sub-image compression method. The key result of this research shows that at least two neighboring pixels should be used for DPCM calculation to increase compression performance. Meanwhile, varying Huffman tables with sub-image approach could also increase the performance if on-board computer can support for more complicated algorithm. These results can be used as references in designing Payload Data Handling System (PDHS) for an upcoming LAPAN-A4 satellite.

1. Introduction

LAPAN-A3/IPB satellite brings one multispectral push-broom imager with four color channel (RGB-NIR), one digital matrix camera with Bayer color filter array (CFA) filter, and also one analog video camera [1]. With current satellite altitude of 505 km, the multispectral imager will have approximately 15 meter ground sampling distance (GSD) with swath-width of 100 km, while the matrix camera will have around 3 meter resolution with 6 km square coverage area. Special attention must be addressed on the multispectral push-broom imager data transmission, since each channel has 8002 detector, where each detector is represented by 16-bit digital number. Therefore, LAPAN-IPB has 105Mbps data-rate transmission to accommodate real-time transmission of the multispectral imager data. Even though of having quite high transmission rate, LAPAN-IPB multispectral image data still need to be compressed to further optimize bandwidth saving without sacrificing decompressed image quality. In LAPAN-IPB satellite on-board data processing, the multispectral image can be compressed in lossy or lossless method. In general, lossless compression will reduce the image size without affecting its quality, while lossy compression will reduce the size more efficient, but it reduces the quality of the decompressed image [2].

There are several methods that can be used for image compression. Consultative Committee of Satellite Data System (CCSDS) recommends the use of Discrete Wavelet Transform (DWT) for both satellite image lossy and lossless compression [3]. However, DWT algorithm implementation in satellite on-board computer is quite challenging, in term of algorithm itself and in hardware implementation. Therefore, LAPAN-IPB uses Fast-Fourier Transform (FFT) for lossy compression and Differential Pulse Code Modulation (DPCM) for lossless compression, where both methods use Huffman entropy-coding. The analysis of LAPAN-IPB image lossy compression has been done [4], and this paper deals with the analysis of lossless compression one. In general, LAPAN-IPB lossless compression scheme uses DPCM to de-correlate image pixel values while Huffman table is used to encode the resulted DPCM values, using basic concept of representing a more frequent value with fewer number of bit [5].

There are several potential issues that can be discussed in the lossless compression algorithm that has been implemented in LAPAN-IPB. First thing that will be analyzed in this paper is the use of one single neighboring pixel in DPCM algorithm of LAPAN-IPB. In DPCM algorithm, there are several options that can be used to calculate the value of de-correlated pixels before entropy-coded by Huffman table [6]. It consists of the simplest case of taking differences between current and one previous pixel in the same line, which is the method used in LAPAN-IPB, and the more complex form that taking average of two or more neighboring pixels. In general, the trade-off between compression performance and hardware complexity should be treated carefully to decide which type of DPCM should be used in compression algorithm.

The other important thing in LAPAN-IPB lossless compression method that will be discussed is the use of static Huffman table to encode every image captured, although the table can be updated for any acquisition period. One particular static table might be good to compress some particular images, but might produce quite bad performance when applied to other different type of images [7]. For example, the camera could capture town, river, mountain, desert, beach or ocean in one single acquisition period, where each of them has different image characteristic that should be encoded using different Huffman table to produce the best performance. However, the implementation of static Huffman table is much easier compare to variable Huffman table that changes continuously between images. Again, the trade-off between performance and computational-cost should be considered when designing the Huffman coding scheme that gives the best result.

In general, the Huffman entropy-coding will work optimally on homogenous data, and in case of image data, the histogram of pixel value should have small standard deviation. In LAPAN-IPB image lossless compression scheme, due to on-board computer limitation, the Huffman algorithm is applied to one single image line. This is in fact, has made DPCM values compressed by Huffman coding be less homogenous. The third issue that will be investigated in this paper is the possibility of using sub-image data, instead of one single line, when applying Huffman coding algorithm. Given the same data buffer to process DPCM values, it is interested to see whether bigger-image data will give significant improvement or not.

This paper aims to analyze LAPAN-IPB lossless compression algorithm, particularly with current LAPAN-IPB on-board computer limitation. Based on the analysis done, this paper will also propose an improved compression scheme, considering several above-mentioned compression aspects, which are the type of DPCM code used to de-correlate pixel values, the type of Huffman entropy-coding used to compress the result of de-correlation process, and also the use of sub-image approach for entropy coding. Apart of being used as references to optimize LAPAN-IPB multispectral imager real-time data acquisition, these analysis results will also used as reference for designing Payload Data Handling System (PDHS) that is currently being developed for the next LAPAN-A4 satellite.

2. Methodology

First step on this research is study literature about image compression in general with lossless method in particular. In general, image compression consists of image de-correlation process and followed by entropy-coding process [8].De-correlation process is done to exploit the similarity value between neighboring pixels, so that the de-correlated values will have more uniform pattern compare to original image data, therefore can be encoded more efficient by the entropy-coding. There are many image de-correlation algorithms that can be used for image compression purpose, such as Discrete-Cosine Transform (DCT), Fast-Fourier Transform (FFT) and Discrete Wavelet Transform (DWT) [9]. Meanwhile, entropy-coding process is done to compress the resulted de-correlated values, where more frequent values will be encoded with fewer number of bit compare to less frequent values. Some common examples of entropy-coding algorithm are Huffman and arithmetic coding [10]. Figure 1 shows general block diagram of image compression process. De-correlation process is the heart of image compression algorithm since it distinguishes image data compression from any other ordinary data compression.

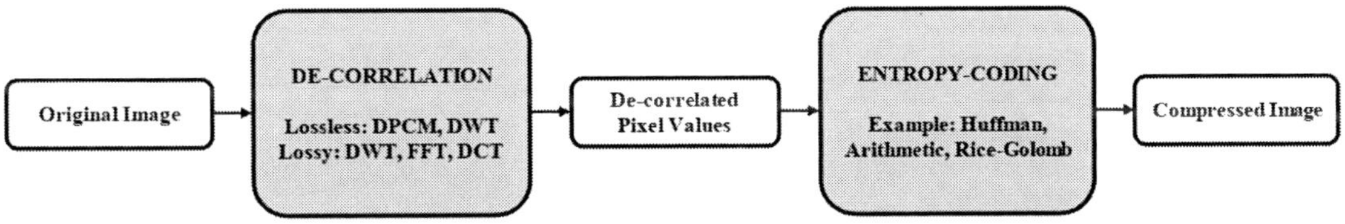

Figure 1. General block diagram of image compression process

Next, LAPAN-IPB lossless compression algorithm is modeled and implemented using MATLAB software to determine the quality of compression algorithm. In LAPAN-IPB lossless compression, DPCM is used to de-correlate pixel values and static Huffman table is used to encode (compress) the de-correlated pixel values. To provide robustness to accommodate various types of image, special-character is added to Huffman table to limit number of bits generated by the entropy-coding process. Figure 2 shows flowchart of image lossless compression that is used in LAPAN-IPB satellite.

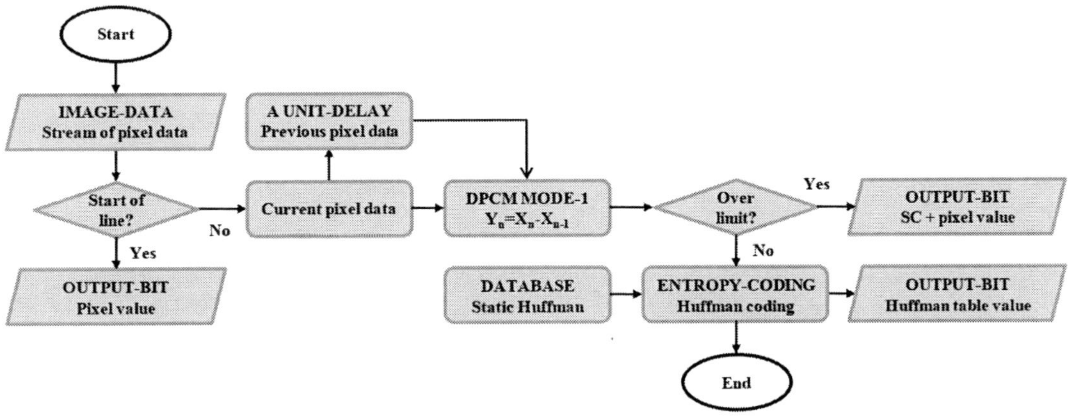

Figure 2. Flowchart of LAPAN-IPB image lossless compression

Finally, several tests and simulations are done to determine the quality of LAPAN-IPB lossless compression, such as the averaged nominal compression ratio produced and how well single static Huffman table adapt to various type of image. Since lossless compression does not reduce the quality of decompressed image, the only important parameter that will be considered is image compression ratio, which is ratio of number of bits of original image compare to the compressed image [11]. Based on these simulation results, several additional simulations are done to see if it is possible to improve the compression quality by changing the type of DPCM and Huffman table used, as well as the use of sub-image as the input-image for entropy-coding process. Figure 3 shows flowchart of experiment and simulation done in this research.

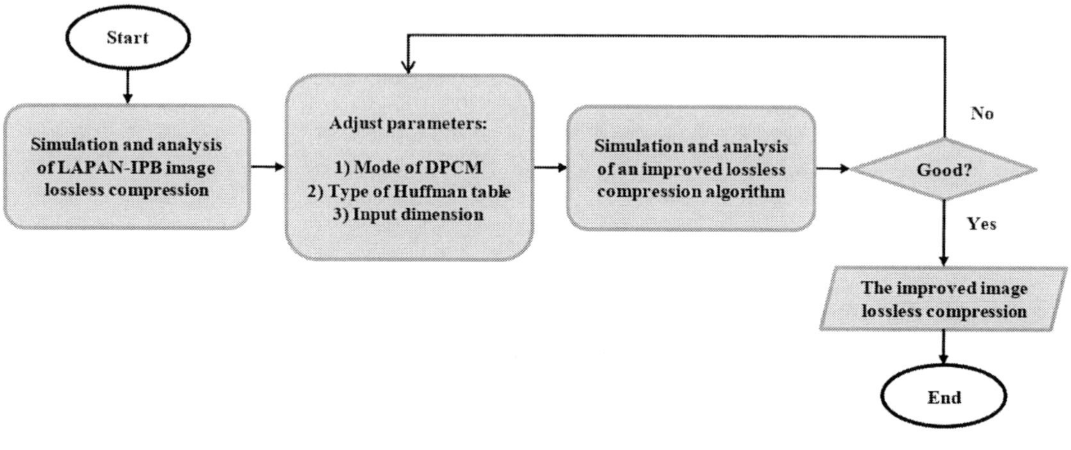

Figure 3. Flowchart of experiment and simulation in this research

3. Simulation Results

Based on figure 3, the experiment and simulation done in this research can be divided into two parts. In the first part, LAPAN-IPB image lossless compression performance will be simulated and analyzed using MATLAB and JAVA software, both for encoding and decoding process. Meanwhile in the second part, several algorithm improvements will be analyzed using MATLAB software, including the different type of DPCM and Huffman-table used in compression algorithm, and multi-dimensional image-input for entropy coding.

3.1. LAPAN-IPB lossless compression

As mentioned earlier in figure 2, LAPAN-IPB image lossless compression uses the simplest method of DPCM algorithm, where it takes a simple difference between current pixel and previous pixel value to be feed for Huffman-coding. Static Huffman-table is then used to encode resulted differenced value. The compression algorithm is simulated by using six images, which consists of three artificial images and three images taken during LAPAN-IPB payload flight test campaign. The flight test images used in this simulation are shown in figure 4. Table 1 shows simulation results, where the first three are for artificial images and the last three are for flight test images. Ideal compression ratio can be achieved when each image is encoded using ideal Huffman-table corresponds to the image, which gives a better compression ratio compare to when static Huffman-table is used to encode all images. In worst-case scenarios, where the static-table has significant characteristic difference to the image, the compression ratio can be reduced significantly.

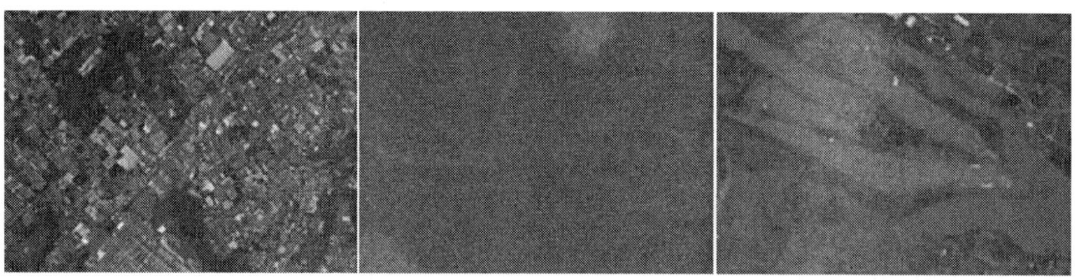

Figure 4. LAPAN-IPB payload flight test images used in simulation: urban area (image-4) in the left, farmland (image-5) in center, and river area (image-6) in the right

Table 1. Compression ratio of LAPAN-IPB image lossless compression

Image Compressed	Ideal Ratio	Static Huffman (mean)		Worst-Case Scenario	
		Ratio	Reduction (%)	Ratio	Reduction (%)
Image-1	1.432	1.259	13.74	1.109	29.13
Image-2	1.904	1.747	8.99	1.658	14.84
Image-3	1.979	1.861	6.34	1.763	12.25
Image-4	1.762	1.703	3.46	1.640	7.44
Image-5	1.774	1.703	4.17	1.641	8.10
Image-6	2.346	2.163	8.46	1.886	24.39
Performance	**1.826**	**1.695**	**7.73**	**1.570**	**16.31**

LAPAN-IPB image lossless compression algorithm uses special-character to minimize the above problem. If the resulted differenced DPCM value is beyond a pre-determined range value, then it produces special-character bits appended to its pixel value. The Huffman table is only applied if the resulted differenced DPCM value is inside the range. The special-character bits are chosen carefully so that it can be differentiated from all other Huffman-table output. The use of this special-character can improve compression ratio slightly, as can be seen in Table 2.

Table 2. Influence of special-character to improve compression ratio

Image Compressed	Ideal Ratio	Static Huffman (mean)		Worst-Case Scenario	
		Ratio	Reduction (%)	Ratio	Reduction (%)
Image-1	1.432	1.290	11.01	1.156	23.88
Image-2	1.904	1.820	4.62	1.723	10.50
Image-3	1.979	1.898	4.27	1.765	12.12
Image-4	1.762	1.704	3.40	1.641	7.37
Image-5	1.774	1.703	4.17	1.641	8.10
Image-6	2.346	2.163	8.46	1.886	24.39
Performance	**1.826**	**1.720**	**6.16**	**1.595**	**14.48**

3.2. Differential Pulse Code Modulation (DPCM)

In DPCM algorithm, there are seven modes that can be used to calculate de-correlated value that will be feed to Huffman-coding [6]. Usually, the neighboring pixels that are used in DPCM calculation can be labeled as W (west), N (north), and NW (north-west), corresponding to its location based on current pixel under consideration. In standard image processing, pixel data is processed by line from north to south and by column from west to east. For example, LAPAN-IPB lossless compression uses DPCM mode-1 that differentiates current pixel with previous pixel in the west, therefore can be labeled as Y=P-W, where Y is de-correlated value, P is current pixel value and W is west-side pixel value. With the same naming convention, DPCM mode-2 uses Y=P-N and mode-3 uses Y=P-NW. Meanwhile, the rest DPCM modes uses more complex form, where DPCM mode-4 uses Y=P-(W+N-NW), mode-5 uses Y=P-(W+(N-NW)/2), mode-6 uses Y=P-(N+(W-NW)/2) and mode-7 uses Y=P-(W+N)/2. Table 3 shows compression ratio produced by each of these DPCM modes, using the same six images that are used in previous LAPAN-IPB lossless compression analysis. It can be seen that the last four modes which are use more than one neighboring pixels, give a much better compression ratio compare to the first three modes which are only use one single neighboring pixel. This shows that LAPAN-IPB lossless compression performance can be improved by approximately 10 percent by using one of these multi-neighboring pixel modes, which are mode-4, mode-5, mode-6 and mode-7. DPCM mode-7 in particular has slight advantage to the other multi-neighboring pixel modes since this mode has the simplest equation and therefore lowest computation complexity.

Table 3. Performance of different type of DPCM calculation

DPCM	Image Compression Ratio						
	Artificial Dummy-Image			Digital Camera Image			Average
	Image-1	Image-2	Image-3	Image-4	Image-5	Image-6	
Mode-1	1.432	1.905	1.979	1.762	1.774	2.346	1.826
Mode-2	1.428	1.833	1.928	1.789	1.803	2.370	1.818
Mode-3	1.337	1.717	1.796	1.642	1.638	2.128	1.678
Mode-4	1.513	1.903	2.096	1.948	1.973	2.567	1.952
Mode-5	1.536	1.991	2.146	1.955	1.985	2.652	1.992
Mode-6	1.533	1.961	2.116	1.970	2.002	2.666	1.989
Mode-7	1.527	1.999	2.130	1.947	1.961	2.662	1.984

3.3. Different Huffman table implementation

Based on Table 1, ideal compression ratio can be produced if the image is encoded using its own Huffman-table that represents its histogram characteristic. This can be done by calculating histogram distribution of de-correlated values for each image before entropy-coding. This histogram distribution is then used to form Huffman-table corresponds to the de-correlated image, therefore compression ratio produced can be considered ideal (best). However, this method needs massive memory to store entire pixel image data, as well as complex on-board circuit to calculate Huffman table value. Note that this method is different from adaptive-Huffman coding, where the value of Huffman-table is progressively changing respect to new pixel data calculated by the algorithm. Adaptive-Huffman coding is beyond the scope of this research, in this research different Huffman-coding implementation in entropy-coding are investigated.

The idea is to use the best possible Huffman-table to encode each image. Since the ideal Huffman-table is expensive to calculate, then this research proposes the use of parallel encoding using a set of different static Huffman-table for each image. Each image is encoded several times, separately but simultaneously, using several pre-determined table, and the best compressed image is then produced as the output. The index of which the table is used can be inserted in the header of the image, to provide information for decoder to decompress the compressed image. Table 4 shows the compression ratio produced by using this parallel encoding compare to original LAPAN-IPB lossless compression algorithm. It can be seen that the compression ratio can be improved by approximately 5 percent. Note that the result can be different if using another set of static Huffman table, but this simulation shows the idea of how parallel Huffman-coding could improve the compression performance.

Table 4. Performance of parallel Huffman-coding

Huffman-coding	Image Compression Ratio						
	Artificial Dummy-Image			Digital Camera Image			Average
	Image-1	Image-2	Image-3	Image-4	Image-5	Image-6	
Ideal	1.432	1.904	1.979	1.762	1.774	2.346	1.826
Static Huffman	1.259	1.747	1.861	1.703	1.703	2.163	1.695
Parallel encoding	1.359	1.840	1.933	1.741	1.745	2.292	1.775
Improvement	7.91	5.31	3.89	2.22	2.48	5.98	4.74

Another aspect that was investigated was about dimension of image data that will be encoded by Huffman-coding. In LAPAN-IPB lossless compression, and also in the proposed parallel Huffman-coding, single Huffman-table is applied to entire image. Actually this is not the best condition to exploit the similarity between pixels. In complete images, anything can be captured by the imager, ranging from homogenous object like ocean or desert, to heterogeneous object such as city with traffic and buildings. This diverge object characteristic usually cannot be encoded optimally to produce the best compression ratio. The idea is to divide the complete image into several sub-images, hoping that each sub-image might have homogenous characteristic, thus can be encoded optimally using parallel Huffman-coding proposed above. Therefore, each sub-image might be encoded by using several pre-determined Huffman-table, each one is best table corresponds to its sub-image, and table-indexing is still same as before, inserted in the beginning of each sub-image data. However, based on several simulations that have been done, this approach gives insignificant performance improvement, only under 5 percent compression ratio increase.

4. Analysis

All of simulations that have been done in this research show that there are some aspects of LAPAN-IPB lossless compression algorithm that can be improved to increase compression ratio. The use of DPCM mode 4-7 could increase the compression ratio about 10 percent, while parallel Huffman table and sub-image approach could further increase the ratio by 5 percent each. If all of these proposed improved methods can be realized, then the lossless compression algorithm could produce about 20 percent better compression ratio compare to LAPAN-IPB algorithm. However, some of the proposed method needs additional on-board processing resources and power, and depending on Payload data Handling System (PDHS) that the satellite has, these methods might not be able to be implemented. This section discusses the trade-off analysis between the compression performance improvement achieved by the proposed method and additional on-board processing complexity needed, in order to see which proposed improvement technically viable to be implemented.

Mode-1 of DPCM algorithm that is used in LAPAN-IPB lossless compression is very simple. It consists of single operation by subtracting current pixel value with previous pixel value, thus only needs one byte-register to store previous pixel value and one addition-subtraction circuit. In the other hand, Mode 4-7 of DPCM algorithm which produce significant compression ratio increase, need more complicated on-board processing. Since it needs more than one neighboring pixels in calculation, at least one of them coming from previous line, it means that more byte-registers are needed to store all pixel values of previous line. In case of LAPAN-IPB multispectral imager where there are 4-bands of 8002 pixels of 2 bytes data each line, then additional 64-kbyte registers are needed. This additional complexity is quite significant, especially if the on-board processor has limited resources. Meanwhile since in mode 4-7 there are one division by two and two or three addition-subtraction operation, then one bit-shift register and some additional addition-subtraction circuit are needed, but these additional complexity are not significant compare to previous additional memory needed to store all pixel values of previous line.

In parallel Huffman-coding scheme, it practically encodes source data simultaneously using several Huffman-tables. However, these parallel Huffman-coding use the same de-correlated values produced by DPCM algorithm, so the additional hardware complexity only comes from the needs of several additional memory-buffer to temporarily store produced bits by each Huffman-table. How significant are these additional complexity depends on the size of image that being compressed. In LAPAN-IPB lossless compression, all bits produced by Huffman-coding are temporarily stored in small data buffer, and then processed to produce byte-data before sending the compressed byte-data to RAM memory. In parallel Huffman-coding, this is not the case. Since the algorithm needs to compare compression ratio of all Huffman-table result, then all bits produced by each Huffman-table must be stored in different data buffer. In standard LAPAN-IPB multispectral image, each image has roughly 50 thousand of lines, that was 50k-lines x 4-bands x 8002-pixels x 2-bytes, or approximately 3-Giga bytes data. Thus, using five Huffman-tables in parallel-coding needs additional 15-GB memory as data buffer, which is quite astonishing.

The needs of extra-ordinary additional data buffer by parallel Huffman-coding scheme can be reduced by using sub-image compression scheme proposed above. By using sub-image approach, the algorithm does not need to compare several compressed full-image, instead the algorithm can compare the result of each Huffman-table compression for pre-determined number of lines. For example, the LAPAN-IPB multispectral image captured can be divided into several sub-images which has 100 lines each, which corresponds to around 250 ms acquisition period. This way, parallel-coding only needs

64-MB data buffer for each table. This requirement is still enormous, but it is significantly easier to implement in low-cost on-board processor.

Table 5 shows several methods that have been proposed to improve the quality of image lossless compression of LAPAN-IPB multispectral imager, with their performance in terms of compression ratio and their feasibility in terms of on-board processor implementation. The method proposed initially is designed for multispectral image but can be easily adopted for matrix image.

Table 5. Trade-off between performance improvement and additional on-board complexity

Method Proposed	Compression-Ratio Improvement	Additional-Hardware Complexity	Remark
DPCM Mode 4-7	10%	Small memory registers and small arithmetic circuit	Very feasible to implement
Parallel Huffman-coding	5%	Very huge RAM memory and small comparative circuit	Not feasible for full-image, must employ sub-image approach
Sub-Image Approach	Under 5%	Small RAM memory and small comparative circuit	Minor compression improvement

5. Conclusions

LAPAN-A3/IPB image lossless compression method produces good compression ratio of around 1.5 to 2 when using ideal Huffman table, however the use of static Huffman table in LAPAN-IPB could reduce the compression ratio for about 10%. To mitigate this performance degradation, the algorithm uses special-character to limit number of bits produced by Huffman table, which could improve the compression ratio up to 5%. Based on simulations done, algorithm performance can be improved by modifying DPCM mode and Huffman-table type that are used. The use of two-or-more neighboring pixels in DPCM calculation could increase compression ratio for about 10%, while the use of parallel Huffman-table with sub-image approach could also increase the ratio for another 10%. All of these improvement shows that it is possible for LAPAN-A4 satellite to produce lossless image compression ratio up to 2.5, or 20% better compare to current LAPAN-IPB, provided that Payload Data Handling System (PDHS) of LAPAN-A4 has enough resources to handle the algorithm proposed.

Acknowledgments

The authors would like to thank Mr. Abdul Rahman as director of LAPAN Satellite Technology Center and Mr. Wahyudi Hasbi as chief engineer of LAPAN-IPB satellite, for their support and assistance so that this work can be well completed.

References

[1] Hasbi W and Suhermanto 2013 Development of LAPAN-A3/IPB satellite an experimental remote sensing microsatellite *ACRS* Bali

[2] Buchman A, Buchman A and Sabou S 2010 Image processing in transform domain: a survey *Carpathian Journal of Electronic and Computer Engineering* **3** pp. 1-4

[3] The Consultative Committee for Space Data Systems 2005 *Image data compression* Blue Book CCSDS **122**.0-B-1

[4] Hakim P R and Permala R 2015 Compression algorithm performance analysis of LAPAN-A3 satellite image using Fast-Fourier Transform *ISAST* Bali

[5] Sharma M 2010 Compression using Huffman coding *International Journal of Computer Science and Network Security* **10** pp. 133-141

[6] Pennebaker W B and Mitchell J L 1993 *JPEG Still Image Data Compression Standard* (New York: Van Nostrand Reinhold)

[7] Salomon D 2008 *A concise introduction to data compression* (London Springer: Verlag London Limited)

[8] Kaur N 2013 A review of image compression using pixel correlation and image decomposition

with results *International Journal of Application of Innovation in Engineering and Management* Vol. **2**

[9] Sharma N, Garg I, Sharma P K and Sharma D 2012 Analysis of transform techniques for 2D image compression *International Journal of Engineering and Technology* Vol. **1** pp. 78-82

[10] Patel D, Bhogan V and Janson A 2013 Simulation and comparison of various lossless data compression techniques based on compression ratio and processing delay *International Journal of Computer Application* Vol. **81**

[11] Anitha S 2015 Lossless image compression and decompression using Huffman coding, *International Research Journal of Engineering and Technology* Vol. **2**

Utilization of LAPAN Satellite (TUBSAT, A2, and A3) in supporting Indonesia's potential as maritime center of the world

A Julzarika[1*]
[1]Research Group of Energy and Mineral Resources, Spatial Plan, and Water Resources Indonesian National Institute of Aeronautics and Space

E-Mail: verbhakov@yahoo.com

Abstract. Indonesia has archipelago area of 2.8 million km^2, territorial sea area of 0.4 km^2. Indonesia have number of 13.466 islands. Coastline length of Indonesia reached 99.093 km^2. Large areas can be monitored using remote sensing technology. Currently, Indonesia have research remote sensing satellites, namely LAPAN TUBSAT, LAPAN A2, LISAT (A3). All of these satellites could be used to monitor Indonesia. These satellites can be used to make the DSM using videogrammetry and depth cue perceptive methods. They also can be used for identification of geobiophysic parameter. Indonesian maritime territory which has sea highway planning can also be monitored using this satellites combination. AIS sensor on LAPAN A2 can be used to identify ships that pass in the territorial waters of Indonesia. It diagonally across the Indonesian region of west to east as much as 14 times a day. At this point it will have detection radius of over 100 km and has the ability to receive signals from maximum of 2000 vessels in the coverage area. Utilization of this satellites is expected to be helpful in supporting the ships cruise monitoring and their support sea highway also in making Indonesia as maritime center of the world.

I. Introduction

Indonesia is a maritime country has an area of 2.8 million km^2 archipelago, territorial sea area of 0.4 km^2, as well as claims on the Continental Shelf beyond the 200 mile area of 3500 km^2 in west Aceh (according to the documents LKRI submissions to the Committee for the limits of the Continental Shelf (CLCS, United Nation). Indonesia has 13,466 islands pieces named and coordinated. Coastline length of Indonesia reached 99,093 km^2, placing it in second place as the country with the longest coastline in the world after Canada. Maritime conditions can be monitored using remote sensing technology and hydrographic.

Remote sensing is the science, engineering and the arts to obtain information about object, area or phenomenon by analyzing data obtained from a device without directly related to the object, area or phenomenon being studied by a device remotely (eg from airplane, spacecraft, satellites, ships or other equipment). Remote sensing support in precision, data acquisition, efficiency cost, and effective time.

According to the philosophy of knowledge, Hydrography also call Marine Geodesy. Etymologically, Hydrography is picture of earth surrounded by water. Definition of old Hydrography in 1960, is limited to the definition bathymetry surveying and mapping, positioning bathymetry [1]. Hydrography activities include coastal, offshore, oceanic. Hydrography simply intended to describe the seabed relief, includes all the elements of the natural and man-made, in principle, similar to the map of the land, in this case the topography [2]. Hydrography is not only associated with marine mapping and positioning, but also with the law of the sea and the physical aspects of integrated coastal area management[3,4]. This shift is due to the advancement of computing technology and measurement instruments. These technologies include photogrammetry, remote sensing, global navigation systems, geoinformatics. In addition, this shift caused by the demands of society and industrial users of hydrography products as a result of increased activity in the waters.

Hydrography definition agreed by the International Hydrography Organization (IHO) in Special Publication Number 32 (SP-32) in 1970 and the Group of Experts on Hydrography Surveying and Nautical Charting in the Second United Nations Regional Cartographic Conference for the Americas in Mexico City in 1979. IHO argued that Hydrography is the branch of applied science which deals with the measurement and description of physical features of the navigable position of the earth's surface and joining coastal areas, with special reference to their use for the purpose of navigation [5,6].

Users of hydrography products can come from variety of sectors, such as transportation and marine navigation, coastal management, exploration and exploitation of marine resources[7], management of the marine environment, engineering offshore, as well as the determination of the law of the sea and the Exclusive Economic Zone (EEZ), Two major information generated by hydrography is depth data for making seabed map and coastline. Both of these data can be obtained by sounding, tides observations, ocean current measurements, and the determination of obstruction under the water as constituent component of the seabed [8].

Hydrography survey aims to obtain depth data and seabed topography configuration (bathymetry), including the location and size of objects as well as the information described in geospatial [9]. Hydrography surveys conducted along the corridor include surveys with varying width [10]. The main lanes should be run at intervals of 100 meters and cross line at intervals of 1000 m. The main strip is run at intervals of 50 m and cross line at intervals of 500 m [11]. Bathymetry survey work should not be conducted in the state of wave with height of more than 1.5 m if no heave compensator or up to 2.5 m when using heave compensator [12]. Remote sensing is used to cope with such conditions. Remote sensing is also useful in efficient and effective of survey mapping.

Shipping lanes and signs that the current monitoring and maintenance needs to be done regularly to maintain the safety and smoothness of the cruise ships do. Maintenance of navigation channel can be done by implementing Hydrography surveys periodically by using Global Navigation Satellite System (GNSS) method of differential real time kinematic (dRTK). This condition can help accurately surveying, precision, and faster when compared to conventional equipment such as arc sextan, theodolites, and other tools [13].Navigation channel has function to give way to the vessel to enter the port area and the pool safely and easily [14]. Another function of the navigation channel is to eliminate the difficulties that occur due to movement of the ship towards the top (minimum ships maneuver activity) and natural disturbances. The world economy will lead to increased use of sea transport is increasingly congested, especially straits and canals, port and intersection crossings or traffic navigation. This research aims to study the use of LAPAN satellite in supportting Indonesia as maritime center of the world.

1.2. 3D modeling Satellite Imagery

Satellite imagery is a product of imaging were performed using a satellite vehicle with passive, active, radio, microwave, and/or sonar sensors. Satellite images with passive sensors have the disadvantage of not free cloud effect but have the look naturally. Satellite imagery with active sensors have advantages of free cloud effect but require special ability to analyze the data. Satellite image sensors commonly referred to as passive optical image sensor is active while the satellite image with an image called Synthetic Aperture Radar (SAR). Image sensor with radio, microwave, and multifunction sonar has advantages but requires special algorithms to analyze the data.

Satellite imagery is available in 2D where the condition is similar to the condition of a single aerial photograph. Satellite imagery can be modeled stereo if at least two satellite images with different viewpoints angle, ie on nadir, backward, and forward. The stereo model must have at least 70% overlay on the front and rear satellites fly and 30% on the fly sideways direction of the satellite and the condition of base height to ratio near to one value. There was also a model of reverse stereo using a single satellite image with few specific methods, one of which depth cue perceptive.

Height model or also known as the 3D model is view of a model with the 3D coordinate system (polar, geodetic, raster and Cartesian) to the reference plane defined on the projection and datum. Height model can be created from optical and radar data. Besides height model can also be defined as a digital model that gives information form of the earth's surface (topography) in the form of raster, vector or other forms of data.

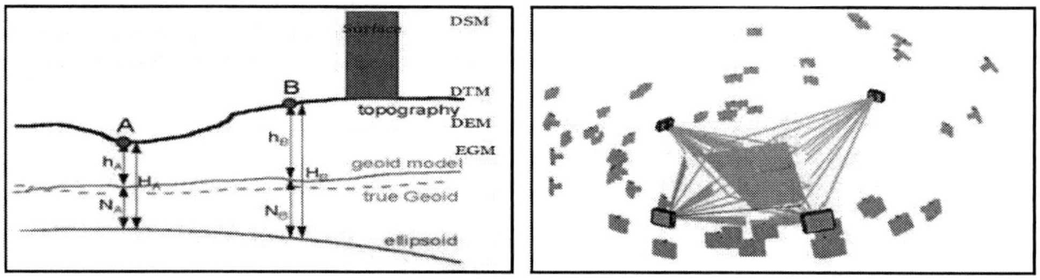

Figure 1. Differences DSM, DEM, DTM, and Geoid (left) and the videogrammetry concept (right)

Height model consists of two pieces of information, namely: height data and height position coordinates on the earth's surface. In some references, the term height model often associated with some other term, namely: Digital Elevation Model (DEM), Digital Terrain Model (DTM) and Digital Surface Model (DSM), see Figure 1. DEM is height surface information displayed without any object earth surface, in a state of bare earth. DTM is the height surface information of the earth without the earth object surface, but is equipped with natural features such as rivers. DSM is a land cover information from the earth surface and above the earth's surface objects (for example, 3D urban areas). Height data storage using GeoTIFF format with the file type 32 bit floating point samples.

Videogrammetry is the development of photogrammetry. Photogrammetry is the art and science of measurement procedures in the object based on the location and shape. This measurement is carried out without direct contact with the object but directly on the image or the image of an object. Photogrammetry has proximity understanding with remote sensing. Videogrammetry is a coordinate measuring technology in three-dimensional form of points on an object is determined by measuring the source of two or more video images in which they were taken from a different angle. Images can be obtained from two angles simultaneously display the objects or derived from sequential images were captured the same video with the look of an object. Videogrammetry expand close-range photogrammetry method and apply it into a sequence of images to present or produce series of three-dimensional models are manufactured using the concept of photogrammetry. Depth Cue Perceptive method is one of the formation of 3D models from photographs/single image with the concept see the depth of an object with a particular viewpoint. The depth of an object this will produce two photographs/images come from photographs/single image with the condition of the base height to a certain ratio. This method is the opposite of the concept of stereo model.

There are several methods of Depth Cue Perceptive that used to make reverse stereo model, namely Averaging, Geometric, Red Search, Green Search, Blue Search, Search Cyan, Yellow Search, Search Yellow, Magenta Search, Neighbor Search, Search Luminance, and Saturation Search. Figure 2 displays the defining of depth technical object from the image or the original video, the concept model of 2D and 3D models, as well as the concept of depth scale use of pictures/video. Both models are only differentiated by the use of the parallax barrier. 3D models enable parallax barrier system that can generate images/video to 3D shape, while the 2D models do not enable parallax barrier system that can not produce images/video to 3D conditions. The greater the number value, the level of depth of stereo 3D model will be more obvious.

Figure 2. The concept of depth images and video

Reverse stereo model is a view in 3D with formation of two photos/images from single image/photo, and is one of the methods for the formation of 3D models. On the reverse stereo models need to be checked at some point/object. It is useful to ascertain the degree of correlation and the condition of base height to ratio of the right images and left images resulting from the Depth Cue Perceptive method.

2. Indonesia as the maritime center of the world

Currently, the need Means Navigation Aid Sailing (SBNP) met only about 3,541 units (66.96%), necessitating the addition of approximately 1750 units SBNP (Tower Suar, signs Beacon, Beacon buoys, signs Signs Lunch, Kids Floats), but the development of additional SBNP number of year 2010-2013 tend to be small, ie, 93 units SBNP for 3 years (1.44% per year). The number of navigation vessels 64 units (already meet the requirements up to the year 2020) but most his age is quite old and inadequate operational reliability.

Great value can be achieved by a program that is World Maritime Axis/Center, which could be achieved if there are policies and programs supporting the Right, Effective and Competitive. Toll sea is the sea of effective connectivity in the form of their ships and sail regularly scheduled from west to east Indonesia [15]. Sea Toll function in Indonesia as a pivot support maritime world in the year 2045. The main line toll ocean uses six major ports in Indonesia, namely the port of Belawan, Tanjung Priok, Tanjung Perak, Makassar, Bitung, and Bintuni Bay. Lane expressway proposed use of this sea route using four sea plan. The path is proposed in the form of Belawan-Tanjung Priok-Makassar-Bitung, Belawan and Tanjung Priok-Makassar-Bintuni Bay, Belawan and Tanjung Perak-Makassar-Bitung, and Belawan-Tanjung Perak-Makassar-Bintuni Bay. The main line should also be supported by a companion track. The line was supported by many seaports around Indonesia [15].

According to the 2013-2014 world economic forum, the condition of the national sea transport is still not functioning significantly. Indonesia connectivity index ranking in the marine transport sector in 2014 increased to 77 compared to the year 2013, which was ranked 104. The rating is lower than Thailand and Malaysia [16]. According [15] there are several indices in the determination of sea transport, namely the provinces connectivity index and the index of sea transport connectivity. Connectivity index is measured by a factor province registered vessel, carrying container capacity, the maximum size of the ship, the number of ship visits, and shipping companies registered. Based on the index of sea transport connectivity, Jakarta has a strong connectivity in Indonesia. Value indices so far compared to the eastern region of Indonesia. Equitable development is needed to address this.

3. Result and Discussion

Based on the statistics of national territorial assets, an area of Indonesian waters reached 5.9 million km^2 including the Exclusive Economic Zone (EEZ). This large area can be monitored using remote sensing technology. Remote sensing can be used for preliminary hydrography surveys. Currently, Indonesia has remote sensing satellite, namely LAPAN TUBSAT (A1), LAPAN A2, LAPAN A3 [17]. The third satellite could be used to monitor Indonesia.

IOP Conf. Series: Earth and Environmental Science **54** (2017) 012097 doi:10.1088/1755-1315/54/1/012097

Utilization of remote sensing technology in supporting the development of hydrography sea toll is on the use of satellite imagery. The satellite imagery can be optical, Synthethic Aperture Radar (SAR), microwave, sonar and lidar. One recommendation sensing data used in the development of toll marine hydrography is satellite imagery LAPAN A2, A3, Sentinel 1, Sentinel 2, Aster and Landsat 8. Satellite of A1, A2, A3 can be used to make height model such as the Digital Surface Model (DSM) using videogrammetry and depth cue perceptive. It also can be used for identification of land cover, geology and mining potential, land changes, coastal, and ocean dynamics, identification of the cruise ship, the Earth's magnetic field, and others. Indonesian maritime territory which has sea toll planning can also be monitored with the use of combination of satellite stretcher. Sea toll is planned to make Indonesia as a maritime center of the world in 2045.

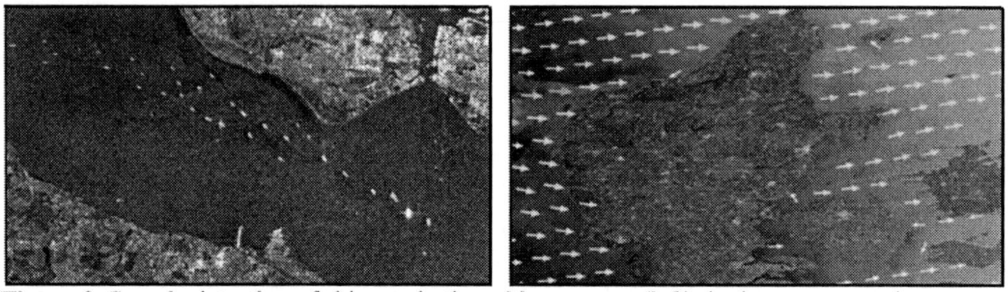

Figure 3. Sample detection of ships and other objects at sea (left), bathymetry mapping, ocean currents (right), oceanographic parameters, the dynamics of the coastal and marine remote sensing data (Sentinel 1). LAPAN imagery could potentially be used for the detection of ships and other objects in the sea

Satellite imagery can be used to make bathymetric map, see Figure 3. This map can be used to a preliminary survey in bathymetric mapping in hydrography. It also can be as the integration and fusion of various bathymetric data. LAPAN A2 image using optical and microwave imagery. LAPAN A2 image can be used to ship identification that will cross the region range [18]. AIS sensor in orbit has a detection radius of over 100 km and has the ability to receive signals from a maximum of 2000 vessels in the area of coverage. As for high performance surveillance mission system, LAPAN A2 digital camera adopts earth observation with 4 band multispectral camera scanning [19]. The camera resolution of 18 m with a coverage of 120 km and 6 m resolution camera with 12 x 12 km coverage. The main camera is equipped with 1000 mm lens is a charge coupled device (CCD) 2000 x 2000 pixels. The main camera has a ground resolution of up to 3.5 meters. Then a second camera with a lens color CCD 1000 mm with a resolution of 752 x 582 pixels. The second camera has a ground resolution of up to 5 m [20].

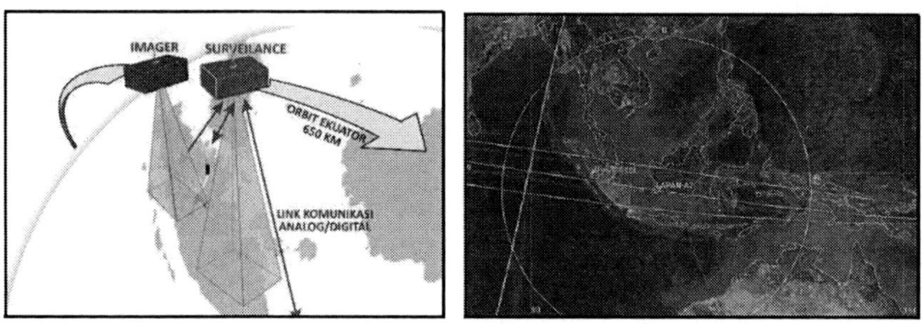

Figure 4. Acquisition system (left) and image coverage LAPAN A2 (right)

IOP Conf. Series: Earth and Environmental Science **54** (2017) 012097 doi:10.1088/1755-1315/54/1/012097

LAPAN A2 equatorial orbits at an altitude of 650 km. Judging from the height of the orbit, the satellite entered to Low Earth Orbit (LEO). Because it orbits at the equator, the satellite will continue to move around the earth's rotation. LAPAN A2 diagonally across the territory of Indonesia as many as 14 times a day, with an orbital period of 100 minutes. LAPAN A2 equatorial satellites circling the Earth through the equator, reaching across the face of the Earth from 6 degrees north latitude and 6 degrees South latitude, AIS device shall be installed in all large and medium ships, see Figure 4. AIS signals emitted from ships and satellites captured determined the ship's position every time, so the route is known. By observing the movement of the ship, can be identified whether the ship merchant ships, fishing or illegal fishing boats.

Sensor Automatic Identification System (AIS) on LAPAN A2 satellite can be used to identify ships that pass in the territorial waters of Indonesia. LAPAN A2 diagonally across the Indonesian region of west to east as much as 14 times a day (noon and night seven times) with a long pass about 20 minutes. At this point the AIS sensor of LAPAN A2 detection will have a radius of over 100 km and has the ability to receive signals from a maximum of 2000 vessels in the coverage area. LAPAN satellite utilization is expected to be useful in supporting the monitoring of vessels at sea and shipping lane expressway support and in making Indonesia as maritime center of the world.

In this paper used depth cue Perceptive value of 10. Using the value 10, the resulting model is already significant in describing the 3D condition. The model results show the level of depth vision is in the inside condition. Image (left) and image (right) has an overlay of more than 70%. This will cause the formation of stereo model so that it can be made 3D models. Examples of 3D images at this paper is in the region of Mount Bromo, see Figure 5.

Image (left)	Image (right)	3D image
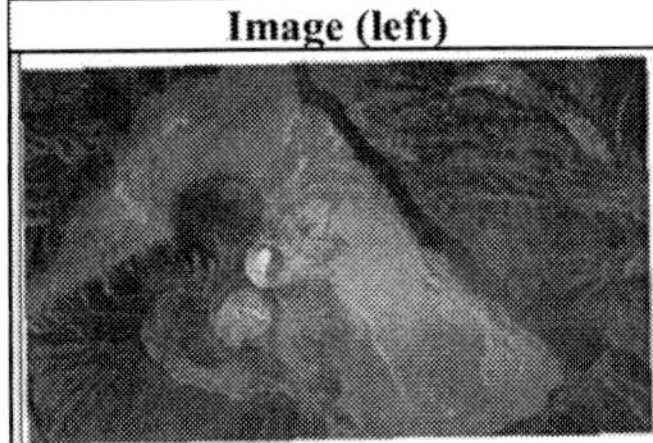	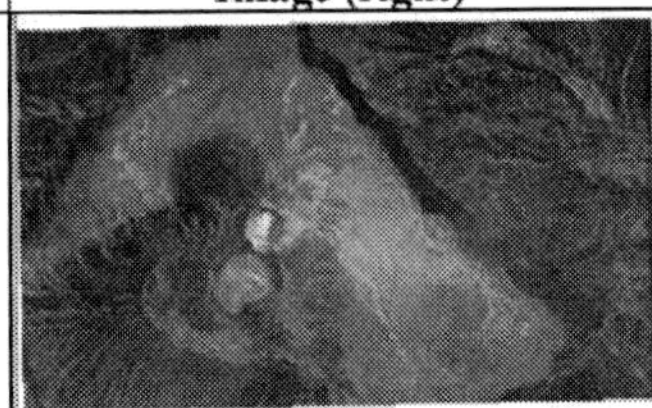	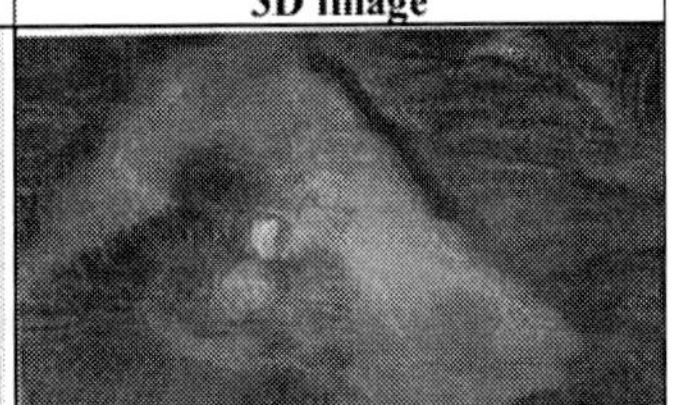
Image (left) is in left side from the image center and have overlay > 70 %	Image (right) is in right side from the image center and have overlay > 70 %	3D image is made from overlay > 70% of image (left) and image (right)

Figure 5. 3D LAPAN imagery in Bromo volcano

Visually, the formation of 3D images at the crater of Mount Bromo is already under optimal conditions with minimal error rate on reverse stereo models. After checking visually, it can be checked by geostatistics. Checking this form of correlation checking the same object in both photo/image with two different viewpoints. The correlation is above 0.9 between the photo/image with a confidence level of 95%. This is one alternative in the formation of a 3D image from a photo/single image. In addition to effective use of the number of photos/images, also produces 3D images with better accuracy as well.

Potential of height model from LAPAN A2 satellite imagery can refer to the height model on LAPAN TUBSAT satellite imagery. Examples of this paper uses height model from LAPAN TUBSAT image of Mount Merapi, see Figure 6. This process includes the creation of stereo image, creating DEM, and the transverse profile [21]. Figure 6 are height model of LAPAN TUBSAT satellite imagery. The left image and right image has an overlay of more than 70%. These conditions will allow the formation of stereo images that can be made 3D models.

This height model can be used for various applications such as geology, mining, spatial, deformation, bathymetry, coastal mapping, currents and waves, as well as a variety of engineering and non-engineering applications. DEM LAPAN image created has done terrain correction, height error correction (bull eyes) and geoid undulation correction. Terrain correction is to transform DSM into DEM. Height error correction is eliminating anomalies occur in pixels height value or height point relative to its surrounding neighbors. Geoid undulation correction for DEM LAPAN image using EGM 2008. In addition to the transverse profile test, the DEM LAPAN also geostatistical tests have been conducted with the least squares method which includes global test, data snooping (outlier), and significance test. It also conducted a height different test, the results obtained are a total height difference of the minimum (zero). This means that the model has height altitude point relative to the plane of the same datum. It is also useful in removing the systematic error that still exists on the height model.

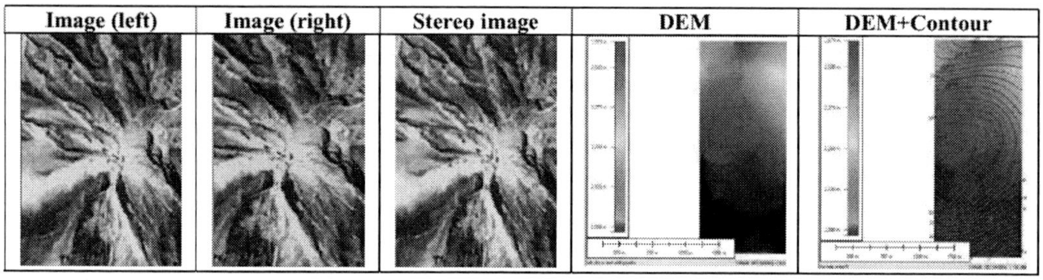

Figure 6. Height Model from LAPAN satellite Imagery

Acknowledgement
Thank you very much to LAPAN for the data of LAPAN imagery and has been included in the LISAT seminar 2016.

Conclusion
Remote sensing and hydrography can play an important role in the development of marine tolls to support the realization of Indonesia as the maritime center of the world. Remote sensing technology can be used to support the development of hydrography in the sea toll. Indonesia has many potential and great resources in creating a world maritime in 2045.

References
[1] IHO, 1960 *Hydrography*. International Hydrographic Bureau
[2] Ingham AE, 1992 *Hydrography for the surveyor and engineer* 3rd edition. (Oxford: BSP)
[3] De Jong et al., 2002 *Hydrography* (Belanda: Delft University Press)
[4] Dyer, 1979 *Estuarine Hydrography and Sedimentation* (Cambridge: Cambridge University Press) ISBN: 0 521 22435 7
[5] IHO, 1979 *Hydrography* (USA: United Nation)
[6] IHO, 2006 *Technical Aspects of the Law of the Sea (TALOS)* (USA: United Nation)
[7] EMR, 1982 *Surveying Offshore Canada Lands for Mineral Resource Development* 3rd edition EMR Surveying and Mapping Branch Report No. M52-43/1983E.
[8] Whitman EC, 1996 Laser airborne bathymetry-lifting the littoral *Sea Technology* p 95-98.
[9] Loweh, 1997 *Manual of Offshore Surveying for Geoscientists and Engineers* (Springer Science+Bussiness)
[10] Thomson DB, Wells DE, Falkenberg WH, 1986 *Hydrographic surveying Surveying Engineering* (University of Calgary) Report No. 10002
[11] Hare R, 1995 Depth and position error budgets for multibeam echosounding *International Hydrographic Review* LXXII **2** p 37-69.
[12] Renouf JK, 1987 Heave compensation for hydrographic survey *UNB Survey Eng* Grad. Seminar
[13] Geng X and Zielinski A, 1999 Precise multibeam acoustic bathymetry *Marine* Geodesy **22** p

157-167.

[14] Halim RA, Seck M., Diouf, dan Tavasszy LA, 2012 Modeling the global freight transportation system: a multi level modeling perpective

[15] Prihartono B, 2015 Pengembangan Tol Laut dalam RPJMN 2015-2019 dan Implementasi (Jakarta: Bappenas)

[16] Schwab, 2015 The Global Competitiveness Report 2013-2014 *world economic forum* (Genewa, Swiss)

[17] Raimadoya M A, Trisasongko BH, Zain A, 2011. Analisis misi dan rancangan LAPAN-IPB Satellite (LISAT) untuk pemantauan kemandirian pangan. *Jurnal Ilmu Pertanian Indonesia* 16 3 p 173-178 ISSN 0853 – 4217

[18] Haryono, 2015 Multitasking programming of OBDH satellite based on PC-104. *International Journal of advanced studies in Computer Science and Engineering (IJASCSE)* **4** 8

[19] Triharjanto RH, Poetro RE., Muhammad H, dan Hardhienata S, 2016 Evaluating deployable solar panel option in earth observation micro-satellite design optimization. *Int'l Journal of Advances in Mechanical & Automobile Engg. (IJAMAE)* **3** 1 ISSN 2349-1485 EISSN 2349-1493

[20] Hasbi W M, Mukhayadi, PA, Budiantoro AH, Syafrudin AM, Tahir dan Kadri TM, 2010 *Penentuan Spesifikasi Optik pada Kamera Pengamatan Bumi di Satelit LAPAN-A2 dan Satelit LAPAN-ORARI* (LAPAN: Pustekelegan)

[21] Karim A, 2010 Design Simulator for the Development of Satellite OBDH case study of LAPAN-TUBSAT (Bandung: Bandung Technology Institute)

Performance of LAPAN-A2 satellite data to classify land cover/land use in Semarang, Central Java

J T Nugroho[1*], Zylshal[1], G A Chulafak[1] and D Kushardono[1]

[1]Remote Sensing Application Center, National Institute of Aeronautics and Space (LAPAN), Indonesia

E-mail: imeljalu@yahoo.com

Abstract. LAPAN-A2 is a 2nd generation of microsatellite developed by the Indonesian National Institute of Aeronautics and Space (LAPAN), Indonesia. This satellite has been launched on September 28, 2015, with several primary missions are: Earth observation using an RGB camera, maritime traffic monitoring using Automatic Identification System (AIS), and supporting disaster mitigation through amateur radio automatic packet reporting system (APRS) and a voice repeater. The installed camera on the satellite can provide imagery with 3.5 m spatial resolution and able to view a swath 7 km wide. This study is attempt to classify and evaluate the accuracy of land cover and land use (LCLU) map classified from the LAPAN-A2 satellite data. As a reference we used the Pleiades-1A Orthorectified imagery data with 0.5 m spatial resolution. Through visual interpretation method, we classified the LCLU into five classess as follows: water body, building, vegetation, road, and bare land. The confusion matrix was applied to evaluate the classification results for three selected area of interest (AOI). We obtained the overall accuracy, producer's accuracy, and user's accuracy are about 61.77%, 61.02%, and 83.78% respectively.

1. Introduction

There have been many studies about the use of high resolution remotely sensed data within numerous fields, such as environment-green area monitoring [1,2], disaster mitigation monitoring for foresty [3,4], landsubsidence [5], volcano eruption [6-8], floods, agriculture monitoring, urban modelling and urban planning. Related to the application of microsatellite, it is shown that the water body can be effectively extracted from Beijing-1 microsatellite images with higher accuracy [9]. This satellite is also one microsatellite of the international Disaster Monitoring Constellation which has rich spectrum information, clear texture structure and good geometry performance for land use classification [10]. Another microsatellite namely MISTIGRI is designed to observe the Earth in the thermal infrared (TIR). The scientific goals and expected applications of the mission are agriculture, hydrology, urban and coastal area [11]. Gaining experience in operating microsatellite, LAPAN Indonesia will continue this project in the future by develop the 3rd satellite, named LAPAN-A3/IPB satellite. This satellite will become the first Indonesian remote sensing experimental satellite with some importance payloads such as an experimental line scan imager payload, high resolution digital space camera with 4 megapixel images, automatic identification systems (AIS), earth magnetic field sensor and automatic packet relay system (APRS) for digital amateur communication [12].

As the 2nd generation of microsatellite developed by the Indonesian National Institute of Aeronautics and Space (LAPAN), LAPAN-A2 has been launched on September 28, 2015. This satellite has three primary missions as follows: Earth observation using an RGB camera, maritime traffic monitoring using Automatic Identification System (AIS), and supporting disaster mitigation using amateur radio automatic packet reporting system (APRS) and a voice repeater [13]. On the satellite, the matrix camera installed and it provides imagery with 3.5 m spatial resolution and able to view a swath 7 km wide. The LAPAN-A2 orbits in a 650 km with an inclination of 6°. It takes about 110 minutes to orbit the Earth and should pass over Indonesia and other near equatorial locations 14 times a day.

Semarang is the fifth-largest city in Indonesia has experience in rapid urbanization, industrialization, and transportation that cause changes in land use. Some agricultural lands was converted into build-up area over 16 years (1994-2010). The conversion of rice fields and upland about 384.40 ha and 414.41 ha respectively, has changed the social economic condition in the study area [14].

This study was inspired by the fact that most of the high resolution imagery are intended for commercial purposes at a costly price and LAPAN-A2 expected to act as complement to the need of high resolution data. On the other side, we consider that it necessary to evaluate the LAPAN-A2 data quality by carried out the accuracy assessment of land cover land use (LCLU) mapping.

2. Data and Method

2.1. Data

The study area situated in Semarang area Central Java (373.7 km^2) with total population of 1.584.906 inhabitants (Semarang City Central Bureau of Statistics, 2014). In this study we used the LAPAN-A2 satellite imagery acquired on 24 February 2016. We divided the study area into three area of interest (AOI), namely AOI-1 (6°57'23.73''S-6°57'56.34''S latitude and 110°23'29.84''E-110°24'2.38''E longitude), AOI-2 (6°59'4.84''S – 6°59'37.43''S latitude and 110°30'40.78''E – 110°31'13.34''E longitude), and AOI-3 (6°57'22.26''S – 6°57'54.86''S latitude and 110°25'4.61''E – 110°25'37.16''E longitude). Wide area for each AOI is 1 km. Figure 1 presents the location for each AOI.

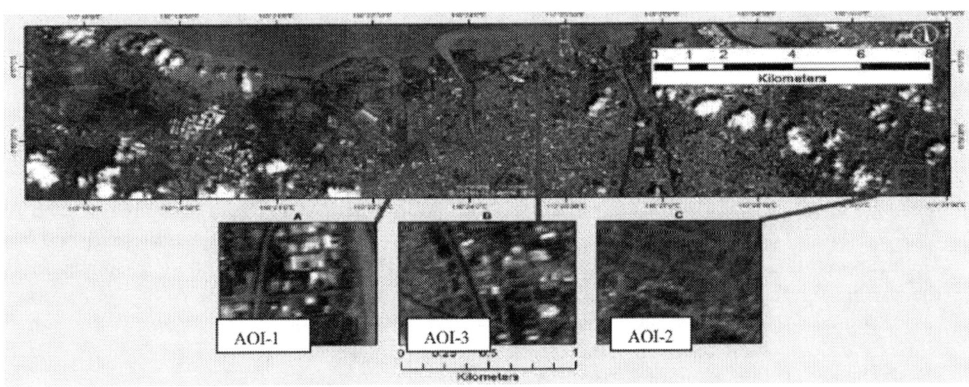

Figure 1. Study area for each AOI

For a comparison, we used the Pleiades-1A Orthorectified imagery as reference data with 0.5 m spatial resolution (acquired on 2013). The LAPAN-A2 data was obtained from the Center for Satellite Technology (Pusteksat), while Pleiades-1A data was obtained from Remote Sensing Technology and Data Center (Pustekdata), LAPAN.

2.2. Method

The accuracy of a classification assessed by a comparison of the classification with some reference data that is believed to accurately reflect the true land cover. Sources of reference data include among other things ground truth, higher resolution satellite images, and maps derived from aerial photo interpretation. Presently LCLU classification accuracy assessments are based solely on the used of the confusion matrix, which is a simple cross-tabulation of the mapped class against that observed in the reference data at a set of validation pixels providing a summary of commision (type I) errors and omission (type II) errors [15]. Confusion matrix set up according to category can calculate the accuracy of each class and reflect the degree of confusion between classes directly. Besides, the confusion matrix is a concise and straightforward method [16]. Confusion matrix use the classified and reference data to compare and set up a matrix. The indexes, such as the overall accuracy, producer's accuracy and user's accuracy, can be generated from the matrix.

Overall matrix is a very effective way to calculate the map accuracy. Overall accuracy is the sum of the major diagonal (the number of correctly classified sample units) divided by the total number of sample units. Producer's accuracy and user's accuracy are computed for individual category accuracy by dividing the total number of sample units in one category by the total number of sample units in this column or row. A commission error represents an area in a category which it does not belong to that category. An omission error reveals the excluded area from the category to which it belongs to. Producer's accuracy plus omission error equals 1 and user's accuracy plus commission error equals 1 [16,17]. Kappa reflects the difference between actual agreement and the agreement ecpected by chance [18].

The results of an accuracy assessment are usually summarized in a confusion matrix. A measure for the overall classification accuracy can be derived from this table by counting how many pixels were classified the same in the satellite image and on the ground and dividing this by the total number of pixels. The total accuracy is measured by calculating the proportional class pixel relative to total tested number of pixel (Total= total corrected/Total tested) [18].

The steps of the methodology applied in the current study in order to investigate the accuracy of land cover/land is implemented in the following manner. First step is geometric correction of the LAPAN-A2 images. In the next step we carried out the visual interpretation or delineation to the LAPAN-A2 imagery for each AOI. The accuracy calculations is conducted by overlain both of data the LAPAN-A2 and Pleiades using union process, then we classified the intersecting area (corrected), error of commision (inclusion), and error of omission (exclusion). Last, the assessment accuracy of land cover land use is examined using confusion matrix.

3. Result and Discussion

To get an accurate reading for the specific locations for samples positions first we selected ground control points (GCP). Geometric correction of the LAPAN-A2 images was done using 25 control points extracted with total RMSE about 5.62. The next step we selected area of interest (AOI) of study area. We have selected three AOI, namely AOI-1, AOI-2, and AOI-3 with total area of each AOI is 1 km. We classify the LCLU into five classess i.e. water body, building, vegetation, road, and bare land. Delineation of LAPAN-A2 data for AOI-1 by visually interpreting method is shown in figure 2 meanwhile figure 3 shows the delineation of Pleiades-1A data for the same AOI. Delineation process is performed for two other AOI, that is AOI-2 and AOI-3 respectively. Figure 4 through figure 5 shows the delineation result of LAPAN-A2 and Pleiades-1A for AOI-2. Last figure 6 and figure 7 shows the delineation result of LAPAN-A2 and Pleiades-1A data for AOI-3.

Both of images are classified as the high resolution products with 3.5 m and 0.5 m for LAPAN-A2 and Pleiades-1A, respectively. A main different between both of satellite is that the Pleiades-1A produces multispectral imagery, meanwhile the LAPAN-A2 data is product of the digital matrix camera formed by red, green, and blue filter [19]. As shown in figure 2 up to figure 7, we found that LAPAN-A2 image was sensitive relatively to certain objects i.e building in particular with bright roof. We also noted that the difficulties appear when extracting objects from the image and identifying the object shapes and object boundary in general.

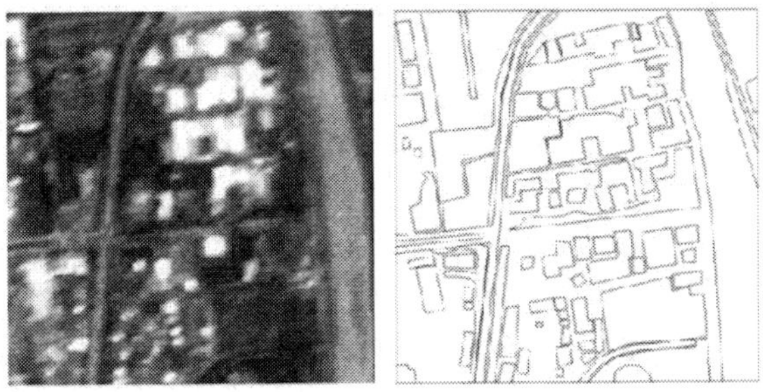

Figure 2. Delineation of LAPAN-A2 data for AOI-1

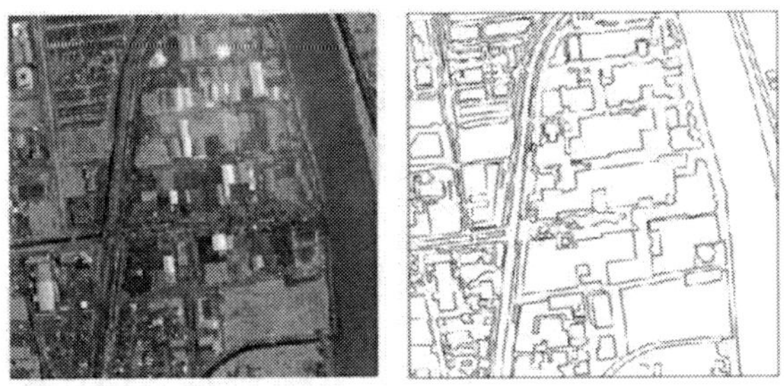

Figure 3. Delineation of Pleiades-1A data for AOI-1

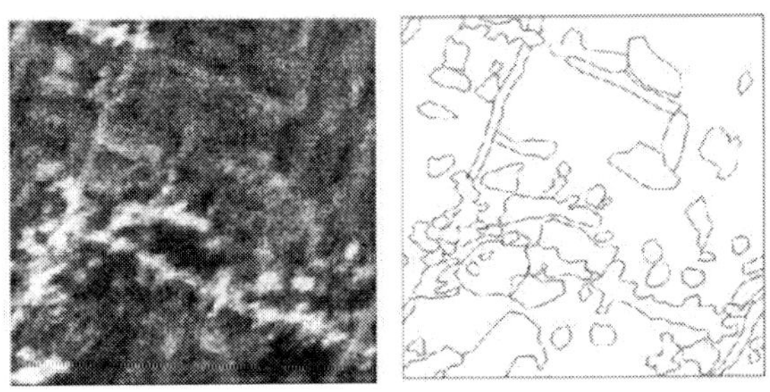

Figure 4. Delineation of LAPAN-A2 data for AOI-2

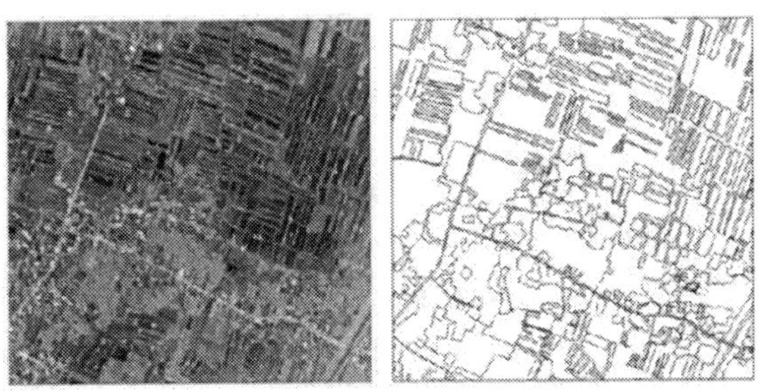

Figure 5. Delineation of Pleiades-1A data for AOI-2

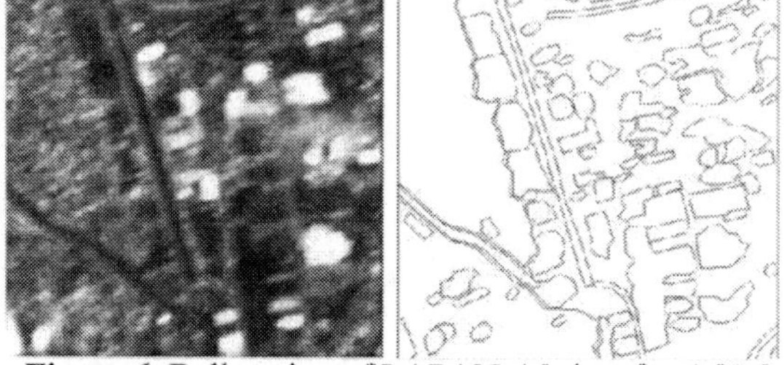

Figure 6. Delineation of LAPAN-A2 data for AOI-3

Figure 7. Delineation of Pleiades-1A data for AOI-3

Data processing for three AOI are carried out to test the overall accuracy and kappa coefficient (Landis and Koch 1977) using confusion matrix [20]. A total of 300 points are generated by stratified random with 60 samples for each LCLU class. Tabel 1 shows the result a measure for overall accuracy of LCLU in Semarang, Central Java using LAPAN-A2 satellite data for each AOI.

Tabel 1. Accuracy assessment of LCLU in Semarang, Central Java using LAPAN-A2 satellite data for three AOI

AOI	Summary Statistics			Overall Accuracy
	FN (m^2)	TP (m^2)	FP (m^2)	
AOI-1	1247.09	621074.36	383183.75	61.77%
AOI-2	23.37	555731.98	448404.27	55.34%
AOI-3	1722.00	538640.84	465169.82	53.57%

where FN is False Negative (omission), TP is True Positive (corrected), and FP is False Positive (comission).

As given in table 1, the overall accuracy of error matrix for AOI-1, AOI-2, and AOI-3 are: 61.77%, 55.34%, and 53.57% respectively. As described in previous section, the inclination orbit of LAPAN-A2 is about 6° and the located (geographical position) of each AOI is upper than 6°. It explains the data acquired at off-nadir condition which has possibility to affecting the data quality and in the end affecting the accuracy. This proved from two AOI, i.e AOI-1 (6°57'23.73''S-6°57'56.34''S latitude and 110°23'29.84''E-110°24'2.38''E longitude) and AOI-2 2 (6°59'4.84''S – 6°59'37.43''S latitude and 110°30'40.78''E – 110°31'13.34''E longitude) where the latitute coordinate of AOI-1 is lower than AOI-2. The overall accuracy difference between two AOI is about 6.43%. In addition it is known that the orbital velocity needed to maintain a stable low Earth orbit is about 7.8 km/s. Satellite imageries are produced by sensor/camera that mounted on moving object. With such speed probably affect the quality of data in particular sensor/camera as the case of LAPAN-A2. Weather conditions are another factor that can't be neglected if we are talking about the quality of satellite data. Overall accuracy usually increases with an increase in the number of fields surveyed [21].

To get more information about how well individual classes were classified presently we carried out the producer's accuracy and user's accuracy test. Tabel 2 portrays in detail the evaluation of overall accuracy and kappa coefficient using confusion matrix for five object classess, i.e water body, building, vegetation, road, and bare land for all AOI. The producer's accuracy of five classess varies from 48.33% (bare land) to 61.02% (water body) whereas for the user's accuracy the value varies from 34.52% (bare land) to 83.78% (road). There is indicate that some classes can be estimated accurately and confidently with fewer numbers of samples, whereas others require larger reference data sets to achieve satisfactory results [21]. Kappa accuracy of 0.44 indicates that there is 44% better agreement.

Tabel 2. Accuracy evaluation result of LCLU in Semarang, Central Java using LAPAN-A2 satellite data for five object classess for three AOI in detail

	Reference						Producer's
Classified	Water body	Building	Vegetation	Road	Bare land	Total	Accuracy
Water body	36	2	15	1	5	**59**	61.02%
Building	0	33	10	2	15	**60**	55.00%
Vegetation	3	4	36	0	17	**60**	60.00%
Road	2	2	7	31	18	**60**	51.67%
Bare land	5	10	13	3	29	**60**	48.33%
Total	**46**	**51**	**81**	**37**	**84**	**299**	
User's Accuracy	78.26%	64.71%	44.44%	83.78%	34.52%		

Kappa	0.44
Overall Accuracy	55.18%

4. Conclusion

This study assesses a performance of land cover/land use map classified from the LAPAN-A2 satellite data. The LCLU was classified into five classes as follows: water body, building, vegetation, road, and bare land. We obtained the overall accuracy, producer's accuracy, and user's accuracy were about 61.77%, 61.02%, and 83.78% respectively. Further research work is needed to determine the appropriate method in processing LAPAN-A2 imagery and hence optimizing the benefit of data utilizations. Moreover, we also conclude that some factors probably affected LAPAN-A2 data quality, i.e acquisition at off-nadir pointing, type and quality of camera, camera setting, and weather condition. The latter this research may usefull as reference in development program of LAPAN satellite in the future.

References

[1] Beiranvand A, Bonyad A E, and Sousani J 2013 Evaluation of changes in per capita green space through remote sensing data *International Journal of Advanced Biological and Biomedical Research* 1 Issue **4** 321-330

[2] Hofmann P, Strobl J, and Nazarkulova A 2011 Mapping green spaces in Bishkek—How reliable can spatial analysis be? *Remote Sens.* **3** 1088-1103

[3] de Waroux Y P and Lambin E F 2012 Monitoring degradation in arid and semi-arid forests and woodlands: The case of the argan woodlands (Morocco) *Applied Geography* **32** 777-786

[4] Souza C Jr, Firestonea L, Silva L M, and Roberts D 2003 Mapping forest degradation in the eastern Amazon from SPOT 4 through spectral mixture models *Remote Sensing of Environment* **87** 494–506

[5] Pasaribu J M, Nugroho J T, and Wiweka 2014 Pemanfaatan penginderaan jauh untuk pemantauan penurunan muka tanah (land subsidence) *Bunga Rampai Pemanfaatan Penginderaan Jauh untuk Pemantauan, Deteksi dan Kajian Lingkungan* (Bogor: Crestpent

Press)
[6] Ebmeier S K, Biggs J, Mather T A, Elliot J R, Wadge G, and Amelung F 2012 Measuring large topographic change with InSAR: lava thicknesses, extrusion rate and subsidence rate at Santiaguito volcano, Guatemala *Earth and Planetary Science Letters* p 335-336 p 216-225
[7] Pinel V, Hooper A, De la Cruz-Reyna S, Reyes-Davila G, Doin M P, and Bascou P 2011 The challenging retrieval of the displacement field from InSAR data for andesitic stratovolcanoes; Case study of Popocatepetl and Colima Volcano, Mexico *Journal of Volcanology and Geothermal Research* p 49-61
[8] Stevens N F, Manville V, and Heron D W 2002 The sensitivity of a volcanic low model to digital elevation model accuracy: experiments with digitised map contours and interferometric SAR at Ruapehu and Taranaki volcanoes, New Zealand *Journal of Volcanology and Geothermal Research* **119** 89-105
[9] Yang C, He R, and Wang S 2008 Extracting waterbody from Beijing-1 micro-satellite images based on knowledge discovery *Geoscience and Remote Sensing Symposium IGARSS IEEE International* DOI: 10.1109/IGARSS.2008.4779856
[10] Qian W 2008 Research on "Beijing 1" Micro-satellite image quality and land use classification precision *The International Archives of the Photogrammetry, Remote Sensing and Spatial Information Sciences* XXXVII B1 941-944
[11] Laguarde J P, Bach M, Boulet G, Briottet X, Cherchali S, Dadou I, Dedieu G, Hagolle O, Jacob F, Nerry F, Olioso A, Ottle C, Pascal V, Roujen J L, Sobrino J A, and Garcia-Moreno T 2010 Combining high spatial resolution and revisit capabilities in the thermal infrared: the MISTIGRI Mission Project *Remote Sensing for Science, Education, and Natural and Cultural Heritage* p 165-172
[12] Hasbi W and Suhermanto 2013 Development of LAPAN-A3/IPB satellite an experimental remote sensing microsatellite *34th Asian Conference on Remote Sensing* p 1508-1515
[13] Hardhienata S, Triharjanto R H and Mukhayadi M 2011 LAPAN-A2 : Indonesian near-equatorial surveilance satellite *18th Asia-Pacific Regional Space Agency Forum (APRSAF)*
[14] Dewi N K and Rudiarto I 2013 Identifikasi alih fungsi lahan pertanian dan kondisi sosial ekonomi masyarakat daerah pinggiran di Kecamatan Gunungpati Kota Semarang *Jurnal Wilayah dan Lingkungan* 1 **2** 175-188
[15] Carneiro J T and Pereira M J 2014 Beyond the Confusion Matrix: Geostatistical error assessment for Landsat landcover maps of the Portuguese landscape *IEEE International on Geoscience and Remote Sensing Symposium (IGARSS)* Quebec City
[16] Ji X and Niu X 2014 The Attribute accuracy assessment of land cover data in the national geographic conditions survey *ISPRS Annals of the Photogrammetry, Remote Sensing and Spatial Information Sciences*, II-435-40 DOI 10.5194/isprsannals-II-4-35-2014
[17] Congalton R G 1991 A Review of assessing the accuracy of classifications of remotely sensed data *Remote Sensing of Environment* 37 **1** 35-46
[18] Adam A H M, Elhag A M H and Abdelrahim M 2013 Accuracy assessment of land use & land cover classification (LU/LC) "Case study of Shomadi area-Renk County-Upper Nile State, South Sudan" *International Journal of Scientific and Research Publications* 3 **5**
[19] Hubel P M, Liu J and Guttosch R J 2004 Spatial frequency response of color image sensors: Bayer color filters and foveon X3 *Proc. of SPIE*, **5301** 402-407
[20] Congalton R G and Green K 2008 *Assessing the Accuracy of Remotely Sensed Data: Principles and Practices* 2nd Edition (CRC Press)
[21] Champagne C, McNairn H, Daneshfar B, and Shang J 2014 A bootstrap method for assessing classification accuracy and confidence for agricultural land use mapping in Canada *Int. J. Appl. Earth Obs. Geoinf.* **29** 44-52

IOP Conf. Series: Earth and Environmental Science **54** (2017) 012099 doi:10.1088/1755-1315/54/1/012099

Identification of typology related to the coastal line changes in Bantul

Guridno Bintar Saputro[1], Mone Iye Cornelia Marschiavelli[1], Farid Ibrahim[2], Edwin Maulana[2]

[1]Geospatial Information Agency, Jl. Raya Jakarta-Bogor KM.46, Bogor, West Java, Indonesia
[2]Parangtritis Geomaritime Science Park, Jl. Parangtritis, Kec. Kretek, Bantul, Daerah Istimewa Yogyakarta, Indonesia

Email: bintar_s@yahoo.com, edwinmaulana35@yahoo.com

Abstract. Typology of coastal associated with landforms and shoreline changes. This study was conducted to determine the relationship typology of coastal especially on the parent material of a coastal with changing shoreline due to changes in the flow of ocean waves. Bantul coastal typology identified using remote sensing data to find the coastal character and patterns. Field surveys conducted for calibration areas undoubtedly result interpretation. The results of this study indicate coastal in Bantul Regency entirely included sandy beaches. The main factor of shoreline change caused by coastal erosion and sedimentation, frequent anomalies shoreline due to erosion and sedimentation. Abrasion occurred in Bantul district south coast will greatly affect reducing the shoreline because of its major parent material is sand. Sedimentation also play a dominant role in influencing increasing in the shoreline in Bantul because many large rivers that flow into the southern coast of Bantul.

1. Introduction

Typology is a beach characteristic in terms of the formation process, material, designation, and the shape of the beach slope. The study of the typology of the beach is very important to do because it is based on the identification of the typology, we can know the potential and the problems that occur on a beach. One of the problems that occur in the coastal region is shoreline change [1-5]. In recent years, changes in the shoreline become a serious issue related to the coastal management [6]. Shoreline change profile changing shoreline is in the process of stability (either forward or backward) caused by various phenomena such as abrasion, accretion, sedimentation, reclaimed or natural disasters. Controlling factor shoreline changes include wind, waves, currents, water levels, sediment supply and coastal structures [7].

Shoreline change pose environmental problems [8-11]. Some issues related to changes in the shoreline of which is the destruction of the environment, the terrain is narrowed, damage to infrastructure, disruption of human activities and administrative boundaries problems. The problem of shoreline change requires serious handling because it can cause a variety of problems including vertical and horizontal conflicts. Real example of the conflict due to changes in the shoreline in

Indonesia is shoreline change caused problems of land management in several regions in Indonesia such as West Java, Yogyakarta, East Java, Ambon and some other areas in Indonesia [12, 13, 14].

One area that is vulnerable to changes in the shoreline in Indonesia is the Special Region of Yogyakarta (DIY). Yogyakarta region vulnerable to shoreline change due to three main reasons: 1) typology sandy beach; 2) direction toward the ocean; and 3) supply of material from upstream. One district in the province that is vulnerable to shoreline changes is Bantul. Shoreline changes in Bantul caused by erosion, sedimentation, wind, big waves and land use changes. Several previous studies indicate that the impact of shoreline change on public economic activities [15-18]. One example abrasion disrupt economic activities in Bantul Regency is abrasion in Kwaru Beach which caused dozens of stalls collapsed. Conditions in Kwaru Beach stalls after abrasion can be seen in Figure 1.

Figure 1. Building collapsed by abrasion in Kuwaru
Source: Field Survey, 2016

The risk reduction due to shoreline changes can be done by several methods. Mapping of the shoreline changes and typology of the beach is one method that can be used to minimize the shoreline change risk (Formerly, et al., 2013; Kumar, et al., 2010; Adamo, et al., 2014; Misra and Balaji, 2015; Shetty, et al., 2015) [19-21]. Based on the results of risk mapping shoreline changes and morphological analysis, can be specified mitigation efforts that can be done to reduce the impact of the shoreline changes. The purpose of this study was to identify the typology beach related to coastal changes of Bantul District so it can be on reducing the shoreline change risk.

2. Data and Methodology

The data used in this study is the Topographic Map of Indonesia, Landsystem Map, BingMaps Imagery and SRTM 30m. The field survey was conducted to collect typology data, aerial photographs and analysis of changes in the shoreline due to erosion and sedimentation. Furthermore, remote sensing data in the form of photos taken during the survey also inclined the field to reinforce the findings of the post-field [23]. Photos taken using the inclined DJI Phantom 4. In-depth interviews were conducted in areas whose interpretation is still doubtful.

Analysis of the data used in the form of spatial analysis. Based on the spatial analysis data can be determined typology in the southern coastal of Bantul District. Spatial data that were analyzed with descriptive methods described explorative. Analysis of exploratory description used for the results of a deeper study by linking between one variable with another variable.

3. Study Area

In general, most of Bantul Regency is located on alluvial plains, karst hills, and coastal landforms. Alluvial plains scattered in the central part of Bantul. This area is influenced by the presence of volcanic sediment material derived from Merapi Volcano. Located in the hilly area that is a path that slopes steep hills with dominant steep slope of more than 70% with a height of 400 meters above sea level. This area is formed by the formation Nglanggran and Wonosari. The beach area is located in the southern part of Bantul. In this area, there is a beach and sand dunes (fluviomarine) with a gentle slope. This area is formed by loose material with a size sand gravels. While in the western part of the district of Bantul are low hills with a gentle slope-steep and located at an altitude of 150 meters above sea level. Study area can be seen in Figure 2

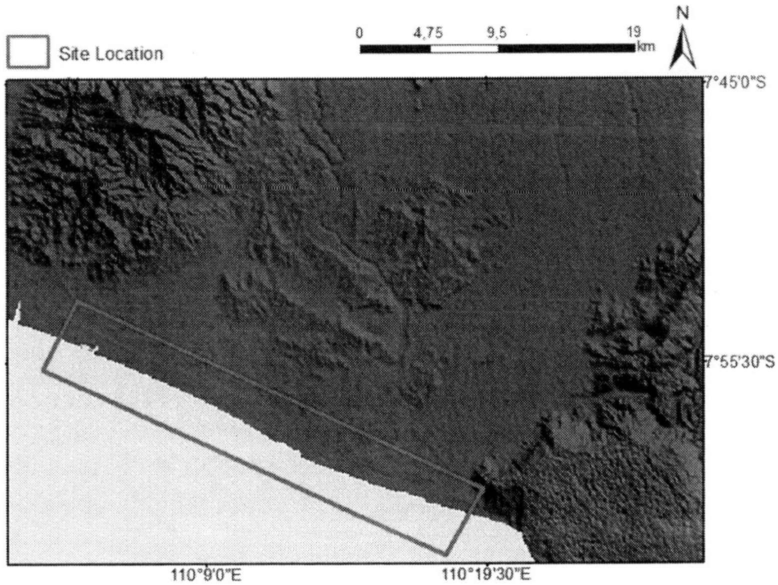

Figure. 2 Study Area
Source: Analysis 2016

Bantul Regency is one of regencies in DIY. Based on 2010-2020's data, population projections, the population of Bantul Regency in 2015 was 971.511 inhabitants spread over 75 villages and 17 subdistricts [24]. Bantul Regency has so many potency. From the broad range of potential, such as tourism potential [25], the potential of agriculture [26], until the potential of culture and tradition that is still awake. Bantul Regency has potential in agriculture in the form of rice, onion, and chili [27]. Potential traditions and cultures that exist among them is the art Ketoprak, Jathilan art and the art of puppetry which is still maintained by the community. In addition to the potential for tourism, agriculture, and cultural traditions that had been mentioned, the Bantul district south section immediately adjacent to the waters of the Indian Ocean region, needs special attention. Because it also has the potential of coastal disasters such as abrasion [28] and wave [29] Spatial planning should pay attention to aspects of existing natural vulnerability.

4. Results and Discussion

4.1. South Coastal of Bantul Typology

Based on the analysis can be seen that the height of the class with the most extensive deployment lies at an altitude of between 25 to 100 meters with an area of 27,709 ha, or about 54.67% of the total Bantul. Grades lowest height that is less than seven meters has an area narrower than most other classes, namely 3,228 ha or approximately 6.37% of the total area throughout Bantul. In general, low-altitude region is an area directly adjacent to the Indian Ocean. In addition, the region located at an altitude of more than 100 meters above sea level are the most Dlingo, Imogiri, Piyungan, and Pajangan

(Bantul Regency, 2013). Bantul has three major rivers and three other rivers, namely (1) Oya River in Sub Dlingo and Imogiri with a length of 35.75 km, (2) the Progo River which flows in the District Sedayu, Pajangan, Pandak, and Srandakan with length of 24 km, (3) the Opak River flowing in the District Piyungan, Banguntapan, Pleret, Jetis, Imogiri, Pundong, and Kretek with a length of 18.75 km, (4) Winongo flowing in the District of Sewon, Bantul, Jetis, Pundong, Kretek with a length of 18.75 km, (5) Bedog River flowing in Kasihan, Pajangan, Bantul, and Pandak with a length of 9.50 km, (6) the Code River which flows in the District Banguntapan, Pleret, Sewon, and Jetis with a length of 7 km.

Typology beach in Bantul consists of three kinds, namely sandy beach, rocky beach and the muddy sand. The sandy beach is characterized by low relief, smooth curved up rough, there is rubble (for the type of white sandy beaches), and comes from a volcanic eruption (for the type of black sand). Several sandy beaches are derived from volcanic material flows through the river. White sandy beach material comes from rocks that experienced the rapture.

The process of formation that occurred beach consists of ocean processes derived from wave energy in areas close to the sea and aeolian processes in the region more toward land. The form of the coastal region is derived from these processes is the physical form of the coast as a result of the sea and the sand dunes as a result of aeolian processes. Coastal formed by deposition of land has the characteristics of a flat relief to choppy and material composed of mud. The process of fluvial processes that occur are due to the flow of the river [30].

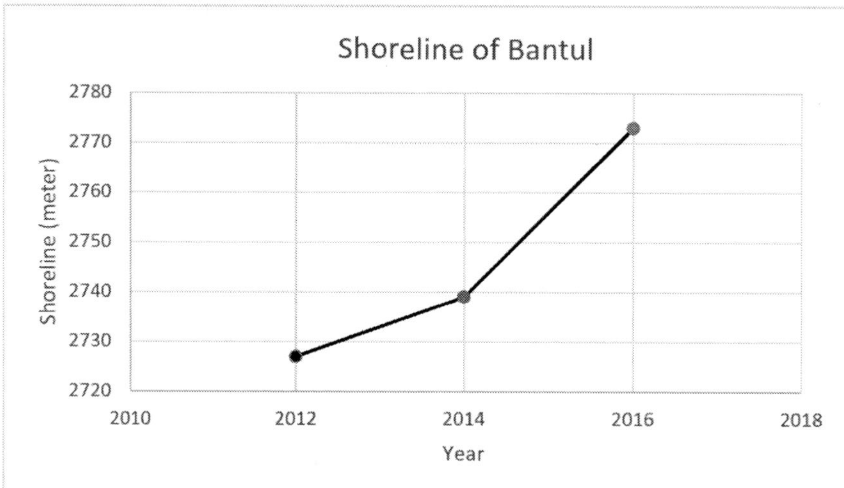

Figure. 3 Shoreline of Bantul
Source: Analysis 2016

Muddy lots formed then in the region ramps and associated with mangrove and seagrass ecosystems. Sometimes to distinguish between muddy and sloping sandy beach is quite difficult, because the sloping sandy beaches tend to be composed of fine sand that can be mixed and accumulated mud so it will be more difficult to overcome [31]. Bantul district has little muddy, the beach average lies at the mouth of the river or are in the delta estuaries, such as the Opak River Estuary and Progo River Estuary because, meetings sandy beach with a river delta or estuary of the river carrying sediment will lead to the muddy shore. Samas beach is just one example of the beach is in flow Opak River estuary, around the mouth of the river Opak is a substrate of the beach turned into muddy. Samas beach is a sandy beach with black sand (Sandy Beach), in addition to the slopes on ramps as well Samas beach and coastal Samas beach area with a long shoreline.

Tirtohargo muddy mangrove area has a uniqueness in terms of constituent materials. Mangrove areas generally have a constituent material in the form of mud. The base substrate owned mangrove areas Tirtohargo substrate is muddy and mixed with fine sandy material. The base substrate sandy mud in mangrove areas is caused due to the mixing of sediment material in the form of sand and silt carried by the Opak River. The blow waves are not far away in the mangrove areas indirectly contributes to refine the material carried by the Opak River. The sand in the area of mangrove came from the eruption of Merapi Volcano is then carried away by the river and anchored to the ocean. Opak estuarine waters is one estuary in Bantul, Yogyakarta. These waters have a dynamic nature with a variety of physical and chemical factors are high. Opak in estuarine waters are mangroves with muddy base substrate, and the open waters immediately adjacent to the sea has a substrate category muddy waters, sandy and a little rocky [32].

Coastal region of Bantul, generally has a sandy beach typology. Rocky beach located on a small coverage area of Bantul, but do not look at a map scale of 1: 25,000. Rocky beach is in the majority most of the coast in the district of Gunung Kidul. Some species on the rocky coast rocky like mussels and oysters are a source of food for coastal communities. Lots of rocky beach in the tropics consisting of coral or other type of limestone rocks that have holes and gaps [33].

Typology rocky beach located on the coastal area of Bantul little coverage. The location is in the rocky coast of the eastern border of the coastal Bantul. Typology rocky beach located on Parangendog Coast region. Parangendog has main constituent material in the form of limestone reefs. Gap of bluff contained in Coastal waterfall Parangendog allow annual found in the region. Reason waterfall named as an annual because of the waterfall appear only once a year during the rainy season. Origin of the word Parangendog name taken from the local language. Parang means stone and endog means eggs. If interpreted as a whole, Parangendog can be interpreted as a stone that resembles an egg. Characters Parangendog Coast has rugged relief with a little sand. The composition of sand is less than the rock.

4.2. The relationship between typology and Shoreline Changes in Bantul

Each beach has a unique typology and peculiarities. The ability to withstand the blow of a wave is affected by the material making up the region. According to local regulations Yogyakarta Special Region No. 16 of 2011 explains that the direction of the management of coastal border to Bantul at least 200 meters [34]. The regulation is based on the condition of the southern waters has waves and a strong current. In addition to the currents and the waves, the material making up the region is dominated by sandy material.

Sandy material has a structure unravel so easily eroded. The existence of the southern coast as an open coastal waves and currents cause the condition has great energy. Serves as a medium flow and sediment transport as erosion agent. Flow is affected by the pounding waves. Waves coming towards the shore can cause beach currents (nearshore current) that can affect the process of sedimentation or erosion on the beach [35].

The sequence of events always hit coastal abrasion Bantul, despite that the coastal state has its own peculiarities. Uniqueness is happening on the South Shore Bantul visible from the time of accretion. Does not need a long time to experience the accretion. Accretion is a change in the shoreline towards sedimentation. If traced between the hours of erosion in September 2013, accretion that occurred at the same location in November 2013. The time required for the recovery of the region for two months after the event [15]. Accretion that occurred in Bantul can be seen in Figure 2.

Figure 2. Kuwaru Beach, Bantul while experiencing accretion (Source: Field Survey, 2016)

Typology muddy beach is in Opak River estuary area and Progo River estuary area. Shoreline changes also take place periodically in the estuary area. Characteristics of the river mouth facilitates the pounding waves landward with the runoff from rivers and the sea. Substrate mud that comes from river runoff does not dominate the region. Sandy material is also available on the Bantul Coastal mangrove areas. Beaches in mangrove areas have greater power to retain its shoreline compared with the typology of sandy beaches. Plant material is accompanied by a muddy mangrove stronger against the pounding waves that hit coastal areas.

Typology rocky beach located on the coastal Parangendog shorelines are relatively stable. Pounding waves directly hit the rocks along the coast. The structure of the solid rock cause cannot rewind the pounding waves of the beach in a short span of time. Characterized rocky beach has a rugged relief with composing materials such as limestone reefs. Coastal disasters that often occur on the rocky shoreline with rugged relief is landslides. Typology rocky beach has a smaller risk to the tsunami and abrasion at risk compared with the sandy and muddy coastal typology.

5. Conclusions

There are three typologies of costal in Bantul, i.e. the sandy beach typology, typology muddy and rocky beach typologies. Sandy typology has two kinds, they are the coastal comes from karst areas and coastal that come from volcanoes. Muddy typology is located in the estuary Opak-Progo River Estuary. The base substrate muddy beach is sandy mud. Typology of rocky beach in Bantul are on the small coverage Bantul. Parangendog rocky beach located on the side of the easternmost district of Bantul. Rocky beach is characterized by the main material in the form of limestone reefs.

Every typology beach in Bantul have a vulnerability to changes in the shoreline. Changes caused by shoreline erosion and sedimentation beach. Beaches that have higher susceptibility to changing shoreline is sandy beach typology then muddy, and rocky. Typology of sandy beaches have the highest vulnerability to changes in the shoreline because it has a material unravel. Open beach conditions resulting in larger waves and currents. Relief sloping sandy beach on the typology of the more trigger the vulnerability to changes in the shoreline.

6. Acknowledgement

I would like to dedicate this paper to Dr. Junun Sartohadi, M.Sc and Syamsul Bahri, Ph.D for his guidance over the years. We also thank to the colleagues in Parangtritis Geomaritime Science Park and Geospatial Information Agency that always helps the writer.

References

[1] Natesan, U., Parthasarathy, A., Vishnunath, R., Kumar, G.E.J., Ferrer, V.A 2015 Monitoring long term shoreline changes along Tamil Nadu, India using geospatial techniques. *Aquatic Procedia* **4** 325-332.

[2] Testut, L., Duvat, V., Ballu, V., Fernandes, R.M.S., Pouget, F.E., Salmon, C., Dyment, J 2016 Shoreline changes in a rising sea level context: The example of Grande Glorieuse, Scattered Islands, Western Indian Ocean *Acta Oecologica* **72** 110-119.

[3] Kermani, S., Boutiba, M., Guendouz, M., Guettouche, M.S., Khelfani, D. 2016. Detection and analysis of shoreline changes using geospatial tools and automatic computation: Case of jijelian sandy coast (East Algeria) *Ocean & Coastal Management* **132** 46-58.

[4] Hapke, C.J., Plant, N.G., Henderson, R.E., Schwab, W.C., Nelson, T.R 2016 Decoupling processes and scales of shoreline morphodynamics. *Marine Geology* **381** 42-53.

[5] Acciarri, A., Bisci, C., Cantalamessa, G., Pancrazio, G.D 2016 Anthropogenic influence on recent evolution of shorelines between the Conero Mt. and the Tronto R. mouth (southern Marche, Central Italy) *Catena* **147** 545-555.

[6] Al-Hatrushi, S.M 2012 Monitoring of the shoreline change using remote sensing and GIS: a case study of Al Hawasnah tidal inlet, Al Batinah coast, Sultanate of Oman *Arab J Geosci* **6** 1479-1484

[7] Lee, C.E 1998 Prediction of Shoreline Change Behind Detached Breakwater *KSCE Journal of Civil Engineering* **2**

[8] Umar, H., Rahman, S., Baeda, A.Y., Klara, S 2015 Identification of Coastal Problem and Prediction of Coastal Erosion Sedimentation in South Sulawesi *Procedia Engineering* **116** 125-133.

[9] Jonah, F.E 2015 Managing coastal erosion hotspots along the Elmina, Cape Coast and Moree area of Ghana *Ocean & Coastal Management* **109** 9-16.

[10] Cellone, F., Carol, E., Tosi, L 2016 Coastal erosion and loss of wetlands in the middle Río de la Plata estuary (Argentina) *Applied Geography* **76** 37-48.

[11] Chao, R.B., Alberti, A.P., Trenhaile, A.S., Casais, M.C., Díaz, M.V 2007 Shore platform abrasion in a para-periglacial environment, Galicia, northwestern Spain *Geomorphology* **83** 136-151.

[12] Munibah, K., Iswati, A., Tjahjono, B 2010 Perubahan Garis Pantai dan Regulasi Pengelolaan Lahan Baru di Delta Cipunagara, Subang, Jawa Barat. *Globë* **12** 151-159.

[13] Christina, Joseph 2005 Studi Perubahan Garis Pantai Selatan Teluk Ambon Luar Dengan Metode Komar Dan Bikjer *Jurnal Teknologi* **2** 23.

[14] Fatimah, I., Haryanto, B 2015 Studi Adaptasi Perilaku Masyarakat Akibat Reklamasi Pantai Di Desa Sugihwaras Kecamatan Jenu Kabupaten Tuba *Swara Bhumi* **2**

[15] Wahyuningsih, D.S., Putra, M.D., Wulan, T.R., Putra, A.S. Maulana, E., Ibrahim, F 2016 Mitigasi Bencana Erosi Kepesisiran di Pantai Kuwaru dan Samas, Kabupaten Bantul, Daerah Istimewa Yogyakarta. *Prosiding Seminar Nasional Geografi UMS 2016*

[16] Putra, A.S., Maulana, E., Wulan, T.R., Nurhidayah, P., Sanjaya, M.D.A., Swastiko, F.A 2016 Uji Akuisisi Data dengan UAV untuk Monitoring Kondisi Mangrove dalam Mencegah Abrasi Air Laut (Studi Kasus: Pesisir Baros, Tirtoargo, Kecamatan Kretek, Kabupaten Bantul) *Prosiding Seminar Nasional Geografi UMS 2016*

[17] Raharjo, I.R., Rosyidie, A 2015 Kesesuaian Kegiatan Wisata di Pantai Baru Pandansimo Sebagai Tujuan Ekowisata *Jurnal Perencanaan Wilayah dan Kota B SAPPK V4N1*.

[18] Pinto, Z 2015 Kajian Perilaku Masyarakat Pesisir yang Mengakibatkan Kerusakan Lingkungan (Studi Kasus di Pantai Kuwaru, Desa Poncosari, Kecamatan Srandakan, Kabupaten Bantul, Provinsi DIY). *Jurnal Wilayah Dan Lingkungan* **3** 163-174

[19] Kumar, A., Narayana, A.C., Jayappa, K.S 2010 Shoreline changes and morphology of spits along southern Karnataka, west coast of India: A remote sensing and statistics-based approach *Geomorphology* **120** 133-152

[20] Adamo, F., Capua, C.D., Filianoti, P., Lanzolla, A.M.L., Morello, R 2014 A coastal erosion model to predict shoreline changes. *Measurement* **47** 734-740

[21] Misra, A., Balaji, R 2015 A study on the shoreline changes and Land-use/land-cover along the South Gujarat shoreline. 8th International Conference on Asian and Pacific Coasts (APAC 2015) *Procedia Engineering* **116** 381-389

[22] Shetty, Jayappaa, K.S., Mitrab, D 2015 Shoreline Change Analysis of Mangalore Coast and Morphometric Analysis of Netravathi-Gurupur and Mulky-Pavanje Spits. International Conference on Water Resources, Coastal and Ocean Engineering (ICWRCOE 2015) *Aquatic Procedia* **4** 182-189

[23] Maulana, E., Wulan, T.R. 2015 Identifikasi Agihan Barkhan pada Zona Inti Gumuk Pasir Parangtritis dengan Menggunakan Data UAV *Simposium Nasional Sains Geoinformasi IV 2015: Penguatan Peran Sains Informasi Geografi dalam Mendukung Penanganan Isyu-Isyu Strategis Nasional*

[24] BPS 2016 Bantul In Figures 2016 Publised by Statistics of Bantul Regency. Publication Number: 34020.1606. ISSN: 0215-5184.

[25] Nawawi, A 2013 Partisipasi Masyarakat dalam Pengelolaan Wisata Pantai Depok di Desa Kretek Parangtritis *Jurnal Nasional Pariwisata*. **5** (2) 103-109

[26] Giyarsih, S. R., Andri K 2001 Regionalisasi Wilayah Kabupaten Bantul (Suatu Kajian untuk Kepentingan Perencanaan Pengembangan Wilayah *Jurnal PWK* **12** (4)

[27] Wulan, T. R., Edwin M., Nita M., Wiwin., Tri R., Farid I., Mega D. P., Dwi S. W., Zheni W 2016 Strategi Penghidupan Masyarakat Pada Periode Krisis Bencana Banjir Pada Lahan Pertanian di Pesisir Kabupaten Bantul (Studi Kasus Masyarakat Dusun Depok, Desa Parangtritis, Kecamatan Kretek, Kabupaten Bantul, Daerah Istimewa Yogyakarta) *Prosiding Seminar Nasional Kelautan 2016*. Universitas Trunojoyo Madura

[28] Umar 2011 Kajian Pengaruh Gelombang Terhadap Kerusakan Pantai Matang Danau Kabupaten Sambas. *Jurnal Teknik Sipil UNTAN* **11** (1)

[29] Benumof, B. F., Curt D. S., Richard J., Seymour., Gary B. G 2000 The Relationship Between Incicent

Wave Energy and Seacliff Erosion Rates: San Diego Country, California *Journal of Coastal Research* **16** (4) 1162-1178.

[30] Khakhim, N 2008 *Analisis Preferensi Visual Lanskap Pesisir Daerah Istimewa Yogyakarta Untuk Pengembangan Pariwisata Pesisir Menuju Pada Pengelolaan Wilayah Pesisir Berkelanjutan.*

[31] Menteri Lingkungan Hidup 2014 *Pedoman Panangulangan Dampak Lingkungan Akibat Tumpahan Minyak Di Laut. Jakarta.*

[32] Pambuko 2015 *Aspek Reproduksi Udang Krosok (Palaemon concinnus DANA 1852) di Estuari Opak Kabupaten Bantul* Universitas Hasanuddin Makasar.

[33] Kementrian Lingkungan Hidup 2014 *Draft Pedoman Penanggulangan Dampak Lingkungan Akibat Tumpahan Minyak di Laut.* Jakarta: Kementrian Lingkungan Hidup

[34] Pemerintah Daerah Istimewa Yogyakarta 2011 *Peratutan Daerah Provinsi Daerah Istimewa Yogyakarta Nomor 16 Tahun 2011.* Yogyakarta: Pemerintah Daerah Istimewa Yogyakarta

[35] Opa, Esry Tommy 2011 Perubahan Garis Pantai Desa Bentenan Kecamatan Pusomaen, Minahasa Tenggara. *Jurnal Perikanan dan Kelautan Tropis* **VII** (3) 109-114.

A study of the partial acquisition technique to reduce the amount of SAR data

Rahmat Arief[1*], Dodi Sudiana[2], Kalamullah Ramli[2]
[1] Remote Sensing Division, Indonesian National Institute of Aeronautics and Space (LAPAN), Jakarta 13710, Indonesia
[2] Department of Electrical Engineering, Universitas Indonesia, Depok 16424, Indonesia

E-mail : rahmat.arief@lapan.go.id

Abstract. Synthetic Aperture Radar (SAR) technology is capable to provide high resolution image data of earth surfaces from a moving vehicle. This causes large volumes of raw data. Many researchs were proposed about compressed radar imaging, which can reduce the sampling rate of the analog digital converter (ADC) on the receiver and eliminate the need of match filter on the radar receiver. Besides the advantages, there is a major problem that produces a large measurement matrix, which causes a very intensive matrix calculation. In this paper is studied a new approach to partial acquisition technique to reduce the amount of raw data using compressed sampling in both the azimuth and range and to reduce the computational load. The results showed that the reconstruction of SAR image using partial acquisition model has good resolution comparable to the conventional method (Range Doppler Algorithm). On a target of a ship, that represents a low level sparsity, a good reconstruction image could be achieved from a fewer number measurement. The method can speed up the computation time by a factor of 2.64 to 4.49 times faster than with a full acquisition matrix.

1. Intoduction

One of main challenges to obtain high resolution synthetic aperture radar (SAR) images is the acquisition mechanism of a backscatter signal using a high rate of analog digital converter (ADC) according Shannon / Nyquist theorem [1]. This causes the volume of SAR raw data larger. The conventional approach is not only complicated and expensive, but also led to work onboard components becomes heavy. But on the other hand, the capacity of onboard memory and downlink transmission is limited. To solve this problem, many techniques have been proposed to compress SAR data. Scalar compression technique was first used is block adaptive quantization (BAQ). BAQ technique aims to estimate the input signal statistics and match quantizer adaptively according to the statistics of input signal and adopt on onboard satellite such as SIR-C [2], FFT-BAQ [3], ALOS PALSAR 2 [4], Flexible Dinamic BAQ (FDBAQ) on Sentinel-1 [5].

Unlike conventional compression methods above, the theory of compressive sensing (CS) [6–8] proposed a new approach, where CS can recover certain signals from the measurement/sampling much less than the Nyquist sampling rate theory. Scheme of CS for radar imaging system was introduced from reseachers [9,10] and which states that the radar system with CS can eliminate the need of match filter on the radar receiver and reduce the sampling rate of the ADC on the receiver. Liu [11] proposed the use random sampling for SAR signal transmission without any changes to the system hardware. All the above research requires the radar signal is sparse and compressible.

Sparse representation model of SAR signals stated that the raw data can be represented as a sparse signal in a certain basis. Herman [12] proposed a sparse representation model in the form of a linear equation with Alltop sequence. Wei [13] described the SAR signal by separating the sparse target and the acquisition matrix of SAR signal. Another approach is the establishment of the linear model of the SAR raw data based on the Born Approximation [14–16]. This paper proposes a model of partial SAR data acquisition to reduce the dimension of the acquisition matrix of SAR signal based on the linear model of the SAR raw data, so that the processing load can be reduced.

2. Method

2.1 Linear model of recieved SAR signal

Pulse radar systems using stop-go approach [16] where the radar antenna transmits chirp signal at time t and the position of the antenna x repeatedly on repetition interval. The chirp signal has a pulsed LFM radar transmitted waveform and can be written as follows:

$$\mathcal{E}^{in}(t,x) = \mathrm{G_a}.\,\mathrm{a(t)}.\,\mathrm{e}^{i(\omega_o(t)+\pi\alpha(t)^2)} \tag{1}$$

where $\mathrm{G_a}$ is the amplitude of the transmitter signal and $\mathrm{a(t)} = \mathrm{rect}\big((t - T_p/2)/T_p\big)$ is a rectangular gate function with T_p as the pulse duration time. The $\omega_o = 2\pi f_c$ is the carrief frequency and LFM pulse chirp rate. When the signal hits an object, it will induce currents hence the object emits the scattered field which is the same signal, but weaker and time delayed. The scattered field $\mathcal{E}^{sc}(t,x)$ is formed from the interaction between the target and the incident field $\mathcal{E}^{in}(t,x)$. Thus its value is the response target depends on the geometry and material properties of the target and of the shape. The equation of scattered signal according [14] can be written as follows:

$$\mathcal{E}^{sc}(t,x) = -\int \frac{\omega_0^{\,2}}{16\pi^2 R^2}\mathrm{V(z)}.\,\mathrm{G_a}\,.\,\mathrm{a(t)}.\,\mathrm{e}^{i(\omega_o(t-\tau)+\pi\alpha(t-\tau)^2)}dz \tag{2}$$

where $R(z) = |x - z|$ is the distance between the antenna and the target and $\tau = 2R(z)/c$ is the time delay, which is the travel time of chirp signal from the antenna to the target and back to the antenna.

For radar imaging, the scattered field can be measured at the antenna and the reflectivity $V(z)$ is a function that must be resolved. We assume the value of the coefficient $V_k \in \mathbb{C}^{Nx1}(N = N_a \times N_r)$ is the coefficient value of the backscattered signal from sparse targets, where k is an index of sparse targets and N_a and N_r are the sampling number of slow time and fast time signal. The linear equation of the SAR signal (2) is formed by separating the components reflectivity V_k and acquisition matrix Ψ SAR signal in the form of discrete [16] written as follows:

$$s_{RT}(t_n,\eta_i) = \sum_{k=1}^{N} \psi_k(t_n,\eta_i)\,.\,V_k \quad \text{or} \quad S = \Psi\,.\,V_k \tag{3}$$

where $\psi_k(t_n,\eta_i) = A_k\,.\,e^{-j\varphi(t_n,\eta_i)}$

$$\psi_k(t_n,\eta_i) = [A_k e^{-j\varphi_1(1,1)},\ldots,A_k e^{-j\varphi_1(1,N_r)},A_k\, e^{-j\varphi_1(2,1)},\ldots,A_k e^{-j\varphi_1(N_a,N_r)}]^T$$

$$A_k = \frac{\omega_0^{\,2}}{16\pi^2 R_{\eta_i k}^{\,2}}\left|a\left(t_n - \frac{2R_{\eta_i k}}{c}\right)\right|^2.\,G_a \text{ and } \varphi(t_n,\eta_i) = 4\pi f_c \frac{R_{\eta_i k}}{c} - \pi K_r\left(t_n - \frac{2R_{\eta_i k}}{c}\right)^2$$

The time scale of fast time and slow time signal are indexed by $t_n = 1,\ldots,N_r$ and $\eta_i = 1,\ldots,N_a$. where N_r and N_a are the amount of sampling number of fast time and slow time signal. Suppose the scatter coefficient vector $V_k = [v_1,v_2,\cdots,v_N]^T$, the new mathematical model of general SAR signal acquisition Ψ is interpreted as basis vector at the fast-time t_n and slow-time η_i and can be written as:

$$\Psi = [\psi_1(t_n,\eta_i);\psi_2(t_n,\eta_i);\;\ldots;\psi_N(t_n,\eta_i)] \tag{4}$$

A measured SAR echo S is obtained by using high rate analog digital converter (ADC) as required by the Nyquist theorem. The goal of SAR image reconstruction is to determine the target reflectivity V_k from the measured SAR echo S.

2.2 Partial acquisition model of SAR signal

In this section proposed a new method to reduce the dimension of the matrix Ψ in formula (7) by dividing the matrix per block in order to reduce computational load. The matrix Ψ as shown in Figure 1(a) has a large size of $(N_a N_r \times N_{target})$, where N_a and N_r are the maximum number of sampling of slow time and fast time signal. This causes the inverse solution of target reflectivity $V_k = \text{inverse}(f(S, \Psi))$ becomes complex. To find the target reflectivity V_k, the number of equations is not required the number of rows $(N_a \times N_r)$ of the matrix Ψ. For this reason, the dimension of the matrix Ψ can be reduced. So the computational load to resolve inverse problems can be reduced as well. The new matrix $\boldsymbol{\Psi_B}$ is formed by dividing the original Ψ into several blocks as shown in Figure 1(b), so that the dimension $\boldsymbol{\Psi_B}$ becomes smaller.

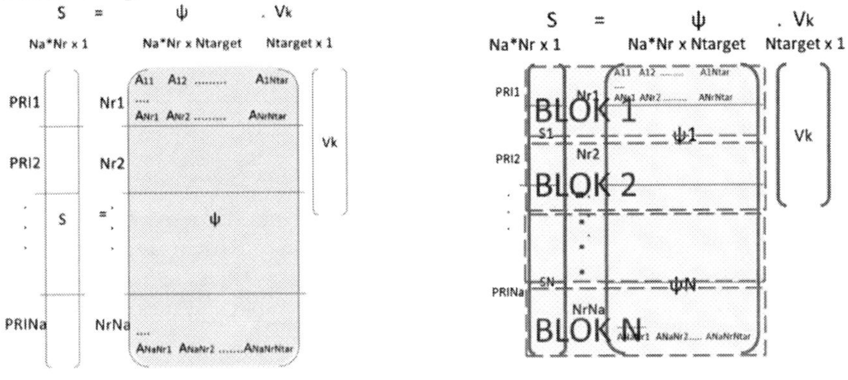

(a) (b)

Figure 1 (a) Fully acquisition matrix Ψ (b) partially $\boldsymbol{\Psi_B}$ by deviding in several blocks

2.3 Low sampling method

One important step in the algorithm CS is randomly low sampling on recieved radar signal s_{RT} (3) is required. A low sampling model in form of fewer random measurement is needed to reduce the SAR raw data. It represent incomplete matrix. The new incomplete radar signal is formulated as follows:

$$y = \Phi S_{RT} = \Phi \Psi V_k + n \quad (5)$$

where Φ is a randomly low sampling measurement matrix with size of $M \times N$, and n is noise matrix. The noise can be stochastic or deterministic. The number of measurements M must be at least greater than the number of K non-zero value, but significantly smaller than the total entries $(K < M \ll N)$. The undersampling ratio become $r = M/N$. The low sampling of slow time (azimuth) signal is obtained by random arrangements of transmitted radar pulses [11,17] and the low sampling of fast time (range) signal is obtained by using lower rate ADC than received signal [15,18].

2.4 Reconstruction algorithm

The scheme of the proposed partially SAR acquisition method states that a new partially matrix $\boldsymbol{\Psi_B}$ is obtained by deviding the acquitition matrix (7) in several N blocks in the same size. Each block produce a new matrix $\boldsymbol{\Psi_{Bi}}$, which has diffirent value compared to other blocks. Thus, the linear equation of each block can be formulated as follows:

$$Blok\ i \rightarrow S_i = \Psi_{Bi} \cdot V_{ki} \qquad \text{for } i = 1,2, \dots, N \quad (6)$$

IOP Conf. Series: Earth and Environmental Science **54** (2017) 012100 doi:10.1088/1755-1315/54/1/012100

where i is an index of each block. The reflectivity target $V_{k1}, V_{k2}, \dots, V_{kN}$ should ideally have the same value. But because the value S_i and Ψ_{Bi} of every blocks are different, the results of V_{ki} are solved using L1 algorithm [19] and obtained different magnitudes. The best reconstructed value from V_{ki} is obtained by comparing the PSNR value of each blocks and the highest PSNR is choosen as the final reconstructed reflectifity target. Figure 2 showed the proposed algorithm.

```
Algorithm
Input : SAR raw data s, fully acquisition matrix Ψ_Full, number of blocks N
to be find  : reconstructed target vk'
Procedure  :
  for i=1 to N
    create partially acquisition matrix Ψ_Bi
    create low sampling matrix Φ
    calculate y_i = ΦΨ_B V_Ki
    calculate V_Ki using L1 algorithm
    calculate PSNR value of V'_K1, V'_K2, ..., V'_KN
  end
  compare V'_K1, V'_K2, ..., V'_KN
  choose the best PSNR value of  V_Ki
end
```

Figure 2 Reconstruction algorithm for partially SAR acquisition

3. Result and discussion

This section describes the performance evaluation of the new model of partially SAR acquisition which are the basis for reconstructing the sparse target on CS-based SAR imaging. Experiments were conducted by evaluating the performance of the partially SAR acquisition and compared to fully SAR acquisition. Experiments were performed on the two input data SAR: an ideal target in the form of point target and ship target from Radarsat-1 as shown in Figure 3.

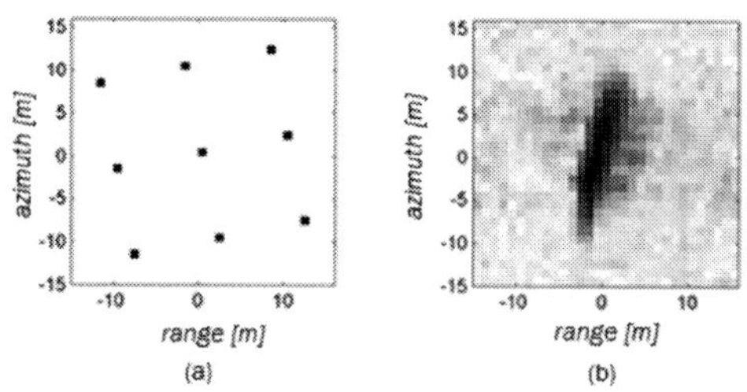

Figure 3 Input raw data (a) point target and (b) a ship target of Radarsat-1 as a complex scene

Both these targets have different pixel intensities and illustrated in one area with a size of 31x31 pixels. SAR parameters used to generate image are as follows: stripmap mode, the frequency of 5.3 GHz, azimuth and range resolution 1.00 m respectively. The total numbers of samples are $N_S = Na \times Nr$, where $N_a = 96$ are $N_r = 126$ ($N_S = 12096$). Low sampling were carried out for all blocks by randomly sampling with M = 50-90 samples at the point target and M = 720-1020 samples at ship.

3.1 Experiment on point targets

Scheme as described in Figure 1(a) divides the acquisition matrix Ψ_{Full} into several blocks each be 1/2, 1/3 1/4, 1/5 1/6 section. The low random sampling is performed on each block a number of measurements M = 90, with 10 samples in azimuth and 9 samples in the range direction. The result of the reconstruction of the experiment is to distinguish between the fully and partially SAR acquisition.

In the first simulation, the full acquisition matrix is divided into 2 blocks of the same size. Reconstructed results were calculated using algorithm as shown in figure 2 for each block and the best reconstruction result is obtained by comparing the value of its PSNR. The PSNR values of block 1 and block 2 are 49.030 dB and 58.595 dB, RMSE values are 0.036 and 0.011, and the calculation times are

3.719 seconds and 3.519 seconds. From the both calculation is obtained best reconstruction result of block 2 with PSNR 58.595 dB, RMSE 0,011 and a calculation time of 3.519 seconds. The next simulation were performed also at 1/3, 1/4, 1/5 and 1/6 of the full block. The PSNR and RMSE values of the reconstructed target point using the proposed method can be seen in Figure 4 and Table 1.

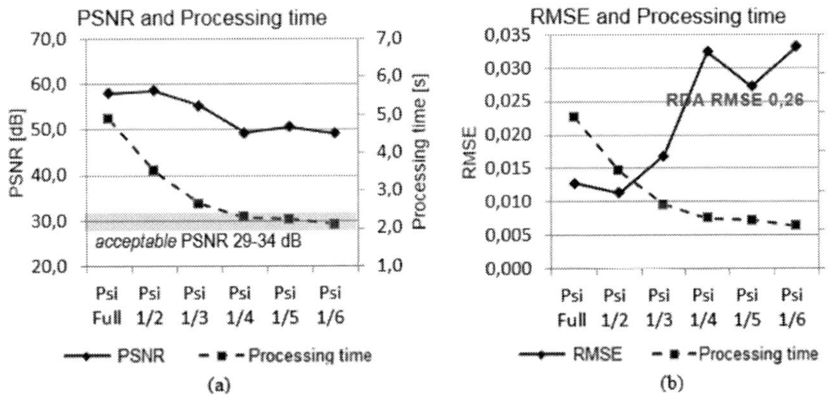

Figure 4 The PSNR, RMSE values and processing time of target point using CS with Ψ_{Full} and Ψ_{B1}=1/2,1/3,1/4,1/5,1/6 of full block

Table 1 Reconstruction result of target point using CS with with Ψ_{Full} and Ψ_{B1}=1/2,1/3,1/4,1/5,1/6 of full block

Parameter	Size of block	CS full	Block 1	Block 2	Block 3	Block 4	Block 5	Block 6	Best	Mean
PSNR [dB]		58,066	49,030	58,595					58,595	53,812
RMSE	1/2	0,013	0,036	0,011					0,011	0,024
Rec. time [s]		4,877	3,719	3,519					3,519	3,619
PSNR [dB]		58,066	49,062	52,948	55,322				55,322	52,444
RMSE	1/3	0,013	0,035	0,021	0,017				0,017	0,024
Rec. time [s]		4,877	2,844	2,657	2,639				2,639	2,714
PSNR [dB]		58,066	47,863	49,274	42,616	45,684			49,274	46,359
RMSE	1/4	0,013	0,040	0,032	0,066	0,047			0,032	0,046
Rec. time [s]		4,877	2,540	2,283	2,316	2,414			2,283	2,388
PSNR [dB]		58,066	48,494	50,006	37,229	37,323	50,576		50,576	44,726
RMSE	1/5	0,013	0,036	0,030	0,132	0,129	0,027		0,027	0,071
Rec. time [s]		4,877	2,412	2,284	2,138	2,074	2,231		2,231	2,228
PSNR [dB]		58,066	49,226	48,686	33,364	36,773	39,535	49,230	49,230	42,802
RMSE	1/6	0,013	0,033	0,034	0,208	0,140	0,104	0,033	0,033	0,092
Rec. time [s]		4,877	2,263	2,082	1,900	1,951	1,943	2,096	2,096	2,039

In this experiment, the reconstruction results on partial SAR acquisition model with blocks 1/2 and 1/3 showed equivalent quality with full matrix acquisition. The reconstruction of the target point with 1/6 partial matrix acquisition of the full block (smallest building blocks of the simulation) have good image quality of best PSNR 49.230 dB, RMSE 0.033 or average (mean) PSNR 42.802 dB, RMSE 0.092.

The image quality of the experimental results of the proposed method are still better than those from the conventional method RDA and has a PSNR value of 23.206 dB and RMSE 0.26. PSNR value is still above the threshold of acceptable PSNR is between 29-34 dB according to [20].

There are some facts that PSNR decreased or RMSE value increases with the division of the smaller blocks. The proposed method resulted in the target point reconstruction with a 1/6 block is still very good, well above the required PSNR and can reduce the computational load. The calculation time of point target using the partial acquisition matrix is faster, because the size of the partial acquisition matrix $\boldsymbol{\Psi}_{\boldsymbol{B}}$ is getting smaller. The complexity of matrix multiplication becomes smaller. Compared with the calculation of the full block, a calculation of 1/6 block is accelerated 2.32 times (see table 1).

3.2 Experiment on ship target

The next experiment is performed on the ship target. This target represents the real target that has a lower level of sparsity and more complex scene compared with the target point. As the previous section, the block dividing scheme as described in Figure 1(b) states that the full acquisition matrix is divided into several blocks equally of 1/2, 1/3 1/4 of the full block. Randomly low sampling of the radar signal is performed on each block with more number of measurements than point target as M = 1000 samples, with details of 20 samples in azimuth and 50 samples in the range direction.

Results reconstruction targets calculated using an algorithm in Figure 2, which is looking for the best reconstruction results by comparing the value of PSNR. Figure 5 shows the results of reconstruction with partial acquisition matrix.

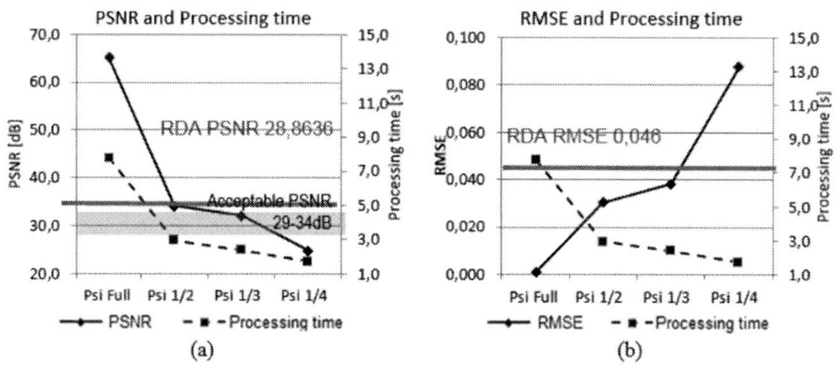

Figure 5 The PSNR, RMSE values and processing time of target ship using CS with Ψ_{Full} and Ψ_{B1}=1/2,1/3,1/4 of full block

The PSNR value of full block decreased from 65.047 dB to 34.104 dB and 32.075 dB at partial acquisition matrix with size 1/2 and 1/3 of full block. The PSNR value decreases or RMSE value increases with the division of the smaller blocks. The reconstruction result of 1/4 of full block indicates PSNR value of 24.847 dB. It shows a lower quality than conventional methods RDA. SAR image reconstruction error rate increases, if the dimension of the partial acquisition matrix gets smaller. The advantage of the partial acquisition model is that the image of the target vessel can be reconsructed using a partial matrix with 1/2, 1/3, 1/4 of full blok faster than the full matrix by a factor of 2.64 to 4.49 times.

4. Conclusion

The new method of partial acquisition techniques have been proposed, performed experiments and analyzed. The experimental results show the performance of the method is better than the conventional method of RDA. It can suppress the side lobe and improve the quality of SAR images from fewer numbers of measurement of SAR signals and can speed up the computation time by a factor of 2.64 to 4.49 times faster than with a full acquisition matrix. The fewer measurement numbers of received radar signals is conducted by low sampling of slow time signals (azimuth) by setting the transmitted radar pulses at random and by low sampling of fast time signals (range) below the Nyquist frequency. Partial acquisition model generates good quality SAR image almost equal with to the full acquisition model at few numbers of measurement to a target point. While on the target ship, this method provides good image quality above acceptable PSNR value.

Acknowledgements
The authors would like to thank Mahdi Kartasasmita for many valuable comments and suggestions which significantly improved this paper.

5. References

[1] Cumming I G and Wong F H 2005 *Digital Processing of Synthetic Aperture Radar Data: Algorithms and Implementation* (Norwood, Massachusetts, USA: Artech House)
[2] Kwok R and Johnson W 1989 Block adaptive quantization of Magellan SAR data *IEEE Trans. Geosci. Remote Sens.* **27**
[3] Fischer J, Benz U and Moreira A 1999 A Efficient SAR raw data compression in frequency domain *Proceedings of International Geoscience and Remote Sensing Symposium, 1999. (IGARSS'99)* vol 4(Hamburg, Germany: IEEE)pp 2261–3
[4] Okada Y, Nakamura S, Iribe K, Yokota Y, Tsuji M, Hariu K, Kankaku Y, Suzuki S, Osawa Y and Shimada M 2013 System characteristics for wide swath L-band SAR onboard ALOS-2/PALSAR-2 *Proc. Asia-Pacific Conference on Synthetic Aperture Radar (APSAR), 2013* (Tsukuba, Japan) pp 141–3
[5] Attema E, Cafforio C, Gottwald M, Guccione P, Guarnieri A M, Rocca F and Snoeij P 2010 Flexible Dynamic Block Adaptive Quantization for Sentinel-1 SAR Missions *IEEE Geosci. Remote Sens. Lett.* **7** 766–70
[6] Candes E J and Tao T 2006 Near-optimal signal recovery from random projections: Universal encoding strategies? *IEEE Trans. Inf. Theory* **52** 5406–25
[7] Candes E J and Wakin M B 2008 An Introduction To Compressive Sampling *IEEE Signal Process. Mag.* **25** 21–30
[8] Donoho D L 2006 Compressed sensing *IEEE Trans. Inf. Theory* **52** 1289–306
[9] Baraniuk R and Steeghs P 2007 Compressive Radar Imaging *Proc. IEEE Radar Conf* pp 128–3
[10] Patel V M, Easley G R, Healy Jr D M and Chellappa R 2010 Compressed Synthetic Aperture Radar *IEEE J. Sel. Top. Signal Process.* **4** 244–54
[11] Liu D and Boufounos P T 2011 High resolution SAR imaging using random pulse timing *Proc. IEEE Int. Geoscience and Remote Sensing Symp. (IGARSS) 2011* (Vancouver, Canada: IEEE) pp 4091–4
[12] Herman M A and Strohmer T 2009 High-Resolution Radar via Compressed Sensing *IEEE Trans. Signal Process.* **57**, **n** 2275–2284
[13] Wei S-J, Zhang X-L, Shi J and Xiang G 2010 Sparse reconstruction for SAR imaging based on compressed sensing *Prog. Electromagn. Res.* **109** 63–81
[14] Cheney M and Borden B 2009 *Fundamentals of Radar Imaging* (Philadelphia, USA: Society for Industrial and Applied Mathematics)
[15] Sun B, Cao Y, Chen J, Li C and Qiao Z 2014 Compressive sensing imaging for general synthetic aperture radar echo model based on Maxwell's equations *EURASIP J. Adv. Signal Process.* **2014** 153
[16] Arief R, Sudiana D and Ramli K 2016 Compressed Synthetic Aperture Radar Imaging Based on Maxwell Equation *J. Teknol.* **78** 15–22
[17] Yang D, Liao G, Zhu S, Yang X and Zhang X 2014 SAR imaging with undersampled data via matrix completion *IEEE Geosci. Remote Sens. Lett.* **11** 1539–43
[18] Arief R, Sudiana D and Ramli K 2013 Noise effects on compressed SAR raw data *Proceedings of 34th Asian Conference on Remote Sensing (ACRS), 20 – 24 October 2013* (Bali, Indonesia) p SC02.1053-1060

[19] Candes E and Romberg J 2005 *L1-magic: Recovery of Sparse Signals via Convex Programming* (California, US: Caltech)

[20] Zain J M, Mohd W M W and El-Qawasmeh E 2011 Software Engineering and Computer Systems, Part I *Proceedings of 2nd International Conference in Computer and Information Science, Kuantan, Malaysia, June 27-29, 2011* (Springer Berlin Heidelberg) p 764

Identification of atmospheric boundary layer thickness using doppler radar datas and WRF - ARW model in Merauke

R J A Putri[1*] and T Setyawan[1]
High School of Meteorology, Climatology, and Geophysics, Indonesia

E-mail: diaputri.2422@gmail.com

Abstract. In the synoptic scale, one of the important meteorological parameter is the atmospheric boundary layer. Aside from being a supporter of the parameters in weather and climate models, knowing the thickness of the layer of the atmosphere can help identify aerosols and the strength of the vertical mixing of pollutants in it. The vertical wind profile data from C-band Doppler radar Mopah-Merauke which is operated by BMKG through Mopah-Merauke Meteorological Station can be used to identify the peak of Atmospheric Boundaryu Layer (ABL). ABL peak marked by increasing wind shear over the layer blending. Samples in January 2015 as a representative in the wet and in July 2015 as the representation of a dry month, shows that ABL heights using WRF models show that in July (sunny weather) ABL height values higher than in January (cloudy)

1. Introduction

Indonesian territory has a diversity of complex topography in form of mountains and hills on the mainland, while the waters that dominate in form of straits, headlands, bays, and so forth. This topographic diversity affect weather formation in all parts of Indonesia. Weather phenomena often occur in Indonesia as heavy rain and thunderstorms are related to mesoscale weather systems. Mesoscale weather dynamics include the study of motion system that has a horizontal scale in the range of 10 to 1000 km [1]. One of the phenomena associated with meso-scale weather systems is the atmospheric boundary layer (ABL) or commonly abbreviated with ABL. ABL is the bottom part of the atmosphere where the flow field influenced directly by the interaction with the Earth's surface and sensitive to variety of conditions in the Earth's surface for a short time scale [2]. Physical and dynamical processes in ABL related to exchange process of mass, momentum, and heat between the earth's surface with the troposphere. Not only that, the process of cloud formation and hydrological cycle are influenced by ABL.

The structure of the ABL can be complex and variable [2-5], and the height (or depth) of the ABL is commonly used to characterize the vertical extents of mixing within the boundary layer and the level at which exchange with the free troposphere occurs [6-9]. ABL heights is one of important and fundamental parameter to knowing the characteristics of the ABL structure such as turbulence, atmospheric stability and diurnal pattern that related to vertical transport process or convection [10] Although the influence of ABL is not felt immediately by humans, but ABL is very influential in weather formation in a region. ABL heights can vary significantly over time as result of several factors including the dynamics of large-scale, cloud, convective mixing and diurnal variation of solar radiation [11]. In addition, the topography also affects the variation of the height of ABL.

Generally, ABL measured in local scale and short periods of time, because ABL has diurnal variation (daily). ABL heights can not be calculated directly but through specific approaches and simulation. ABL heights can be measured using some weather instruments, such as weather radar and radiosonde. ABL measurement by radar performed by [12] at the Bavarian Alps region showed that ABL structure is strongly influenced by the terrain and topography, in this case the mountains. ABL measurement using radar is also done by [10] which shows that the average height of ABL in January was smaller than in July although the difference is not big enough. Nocturnal Boundary Layer (NBL) peak marked by significant wind shear due to the maximum speed (low-level jet) at ground level, while the peak of Convective Boundary Layer (CBL) marked by increasing wind shear at the top of the mixing layer. Not only that, the simulation model of the WRF-ARW model also be performed to determine ABL profile, as did [13] who analyzed the diurnal variation height of ABL based observations and using simulation models of single-column WRF, then compared with calculations using radiometer and wind profiler. The results of simulation-based ABL heights gives a higher yield of about 500-1000 m of ABL heights based observations. A similar study conducted by [14], which resulted in that measurements using WRF is always lower (underestimated) a few hundred meters. In the country have also been ABL measurements using simulation WRF by [15]. Her research using weather radar and comparing turbulence index and richardson numbers calculated from the results of WRF simulations. The results indicate that wind shear was good to identify ABL heights when the weather is sunny, while the spectral width was good to identify ABL heights when the weather is cloudy. While the Richardson number and turbulence index is unable to identify the ABL heights when the weather is sunny or cloudy.Additionally [16] also studied the Climatology LBA by using ECMWF data. The result shows that LBA heights is low around the beach and Characterized by stratus clouds. LBA can reach 3 km, especially in desert areas, and sometimes in the tropics.

In this study the authors will try to identify the height of ABL in Merauke using weather radar to knowing vertical wind profile and WRF-ARW model simulation. Identification of ABL heights performed on two different weather conditions when sunny and rainy day, to determine the height difference of ABL in different weather conditions.

2. Introduction

2.1 Research Location and Time

The research location is in Merauke with location point is Meteorological Station Mopah Merauke with coordinate 08^{0}31'03 "LS and 140^{0}25'01" BT and height 3 m above sea level. Merauke region dominated by lowlands and the hilly area near the coast. Research time was on January 13, 2015 during cloudy and rainy weather conditions, and on July 8, 2015 during cloudy weather conditions.

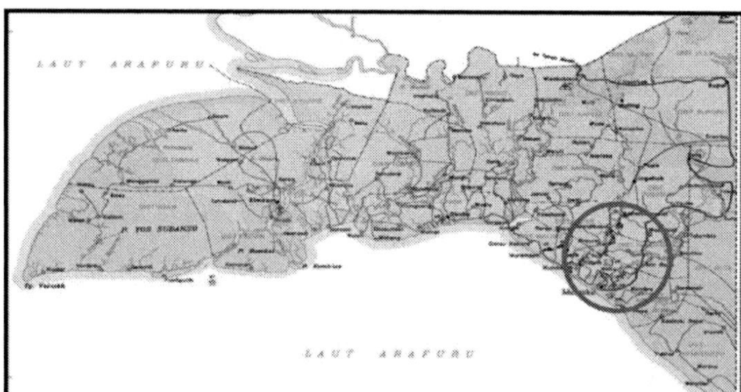

Figure 1. South Papua Island (Merauke is shown with blue circle)

1.2. Data

The data used in this study include:

1. C-Band Doppler radar Data with the format *.RAW from Mopah merauke meteorological station, which then processed using Rainbow TM application for radar gematronic
2. FNL data as input for model WRF-ARW with 10x1o spatial resolution and temporal resolution of 6 hours.

1.3. Step Work

1. Data collection
2. Processing of radar data using Rainbow app. Products shown are Volume Velocity Processing (VVP) that depicting vertically meteorological parameters, such as divergence, direction and wind speed of 3-dimensional (u, v, w). From these products can be seen as a marker peak wind shear atmospheric boundary layer.
3. Running WRF-ARW models with three domain downscaling with the center point of the meteorological station Mopah Merauke (domain 3). Schemes used in the process of running WRF-ARW is the default scheme with the following details:

Table 1. WRF default parameterization scheme

Parameterization Scheme	Number	Selection Scheme
Cumulus	1	Kain-Fritsch
Mycrophysics	3	WSM-3
Shortwave Radiation	1	Dudhia
Longwave radiation	1	RRTM
Planetary Boundary Layer	1	YSU

4. Then running model is done to get .ctl and .dat file. Then we do visualization outcome parameter models which are a boundary layer (pblh).
5. After that, the analysis and discussion based on the results of data processing, and the conclusions drawn.

3. Results

3.1. Identification ABL Heights Using Doppler Radar Data

Identification of ABL heights by using Doppler radar is done by looking the vertical wind profile and the identification result shows that the peak height of ABL elevation change in a day. During the day, ABL heights characterized by maximum wind speeds due to reduced turbulence intensity above the LBA. At night, ABL heights characterized by maximum wind speeds on the bottom layer or so-called nocturnal jet which is a kind of low-level jet. Here is the result of the identification of ABL heights by using weather radar data (marked with a red line).

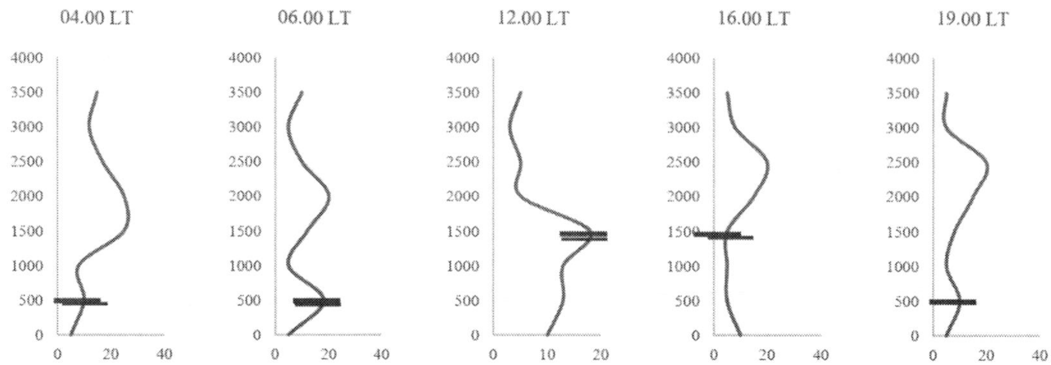

Figure 2. The vertical wind profile using weather radar data on January 13, 2015. ABL heights indicated by red line

Figure 2 shows vertical wind profiler by radar weather on January 13, 2015 during cloudy and rainy weather conditions. ABL heights varies with time. In the early morning at 04.00 LT (Local time = GMT + 9 h) and 06.00 LT, ABL heights are low in the range of 500 m. During the day at 12.00 LT, ABL maximum heights within the value range of 1200 m to 1500 m and it last until afternoon at 16.00 LT. At night, ABL heights decreased again at 500 m.The height of the ABL during the daytime is higher than the height of ABL at night, and the difference ABL heights in the mornings and afternoons tend to be insignificant. ABL layer formation is as a result of the interaction of the atmosphere with the surface of the earth to produce ABL thickness values varying depending on heating / cooling factors of the earth's surface, and also the surface roughness characteristics of the surface topography. ABL greater heights at noon because of high rates of warming at the daytime.

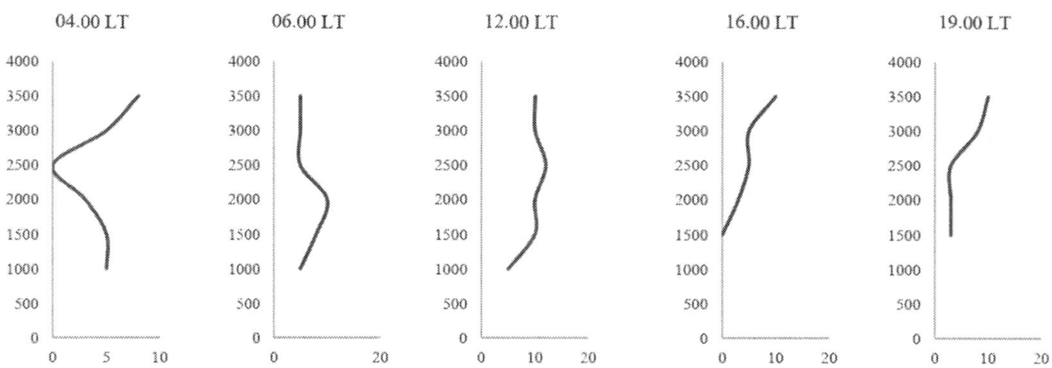

Figure 3. The vertical wind profile using weather radar data on July 8, 2015. ABL heights indicated by red line

Figure 3 shows the vertical wind profiler by radar weather on July 8, 2015 when the weather was sunny and cloudy conditions. But ABL heights on July 8, can not be identified because of the weather radar wind profiler irregular.

3.2. Identification ABL Heights Using WRF-ARW Model Simulation

In addition to using radar, ABL heights can also be identified by the WRF-ARW model simulation. The parameters used for identification ABL heights is pblh. Furthermore, this parameter will displayed in graphical form. Here's a graph of ABL heights every hour using a model WRF-ARW.

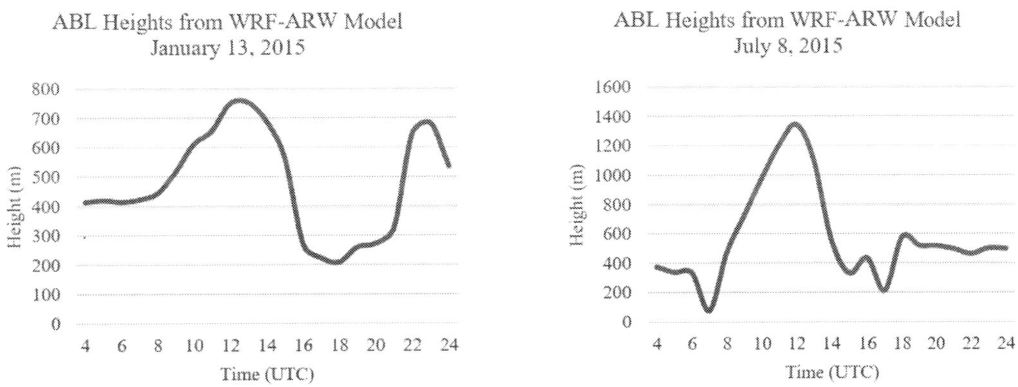

Figure 4. The vertical wind profile using weather radar data on July 8, 2015. ABL heights indicated by red line

Based on WRF-ARW model results on January 13, 2015 when the weather is cloudy and rainy, ABL heights ranging between 208 m to 752 m. In the morning from 04.00 LT to 08.00 LT, ABL heights ranges between 410 to 440 m. ABL heights increase during the day and reach maximum value at 13:00 LT with height about 752 m. In the afternoon until evening ABL heights decreasing and minimum value of ABL heights at 18.00 LT with heights about 208 m.

Based on WRF-ARW model results on July 8, 2015 when the weather was sunny and cloudy, BL heights ranging between 77 m to 1344 m. in the morning from 04.00 LT to 06.00 LT, ABL heights ranges between 332 m to 374 m. differ on January 13, 2015, the minimum height of ABL on July 8, 2015 occurred in the morning at 07.00 LT with height of 77 m. ABL heights begin increase during the day and reach maximum value at 12:00 LT with height about 1344 m. in the afternoon until the evening ABL heights decreases.

ABL heights using WRF models show that in July (sunny weather) ABL height values higher than in January (cloudy). From the data of the average humidity in January and July in Merauke show that the average air humidity in January (wet season) is higher than in July (dry season). In addition, the average temperature in January was lower than in July. This has led to an altitude of ABL in July (sunny weather) is higher than the height of ABL in January (cloudy). Humidity higher in January makes the surface warming during the day is not as intense as in July.

3.3. ABL heights in different weather condition

ABL heights varies with time. Variation of ABL heights caused by daily cycle of solar radiation. ABL heights tends to be low during morning and evening where air condition is stable and radiative cooling of the Earth's surface. The stability ABL withstand turbulence so ABL heights become lower. Unlike the situation during sunshines, when maximum solar radiation occur that generally is going convection activity. Convection causes the radiative warming occurs in the earth's surface so that ABL heights increases.

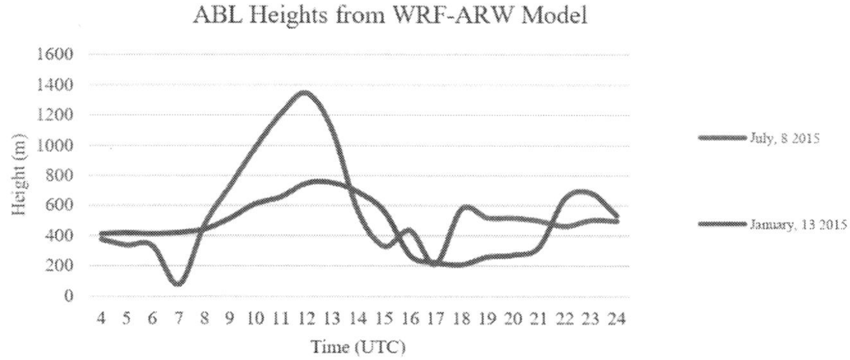

Figure 5. ABL heights in different weather condition (July 8, 2015 is sunny weather and January 13, 2015 is rainy weather)

In addition, weather conditions factors may also affect the ABL height variation. Figure 5 shows the difference in ABL heights in different weather conditions. When cloudy and rainy weather conditions, ABL is higher than when the cloudy and rainy weather conditions. This is due to cloud cover and rain blocking incoming solar radiation so that played a role in decreasing surface temperature. As a consequence of energy that supports the growth of ABL reduced and become lower heights. Unlike when the weather is sunny, the earth's surface receives solar radiation is intense enough so that the earth's surface heat increases as the energy of formation of ABL. This causes ABL heights when sunny conditions are higher.

4. Conclusion

The conclusion of this study are:

1. Identification of ABL heights by using Doppler radar is done by looking the vertical wind profile and the identification result shows that the peak height of ABL elevation change in a day.
2. ABL heights using WRF models show that in July (sunny weather) ABL height values higher than in January (cloudy). This situation is caused by differences in humidity and temperature in the second month, in January of air humidity is higher and temperatures lower than in July makes the surface warming during the day is not as intense as in July.
3. The determination of the height of the ABL method of wind profiles can be done, but there are some conditions in which the height of ABL difficult identified include irregular weather radar wind profiler.

Acknowledgments

This study will not be completed without weather radar data dopler C-Band belongs to the Meteorological Station Mopah, Merauke. Therefore, the authors would like to thank friends BMKG Mopah Meteorological Station for the support data and support to the author.

References

[1] Holton J R 2004 *An Introduction to Dynamic Meteorology* (USA: Elsevier Academic Press)
[2] Stull R B 1988 *An Introduction to Boundary Layer Meteorology* pp 666 (Dordrecht: Kluwer)
[3] Oke T R 1988 *Boundary Layer Climates* (New York: Halsted Press) pp.435
[4] Sorbjan Z 1989 *Structure of the Atmospheric Boundary Layer* pp.317 (Prentice Hall: Englewood Cliffs)
[5] Garratt J R 1992 *The Atmospheric Boundary Layer* pp.335 (Cambridge Atmospheric and Space Science Series: Cambridge Univ. Press)

[6] Bhumralkar C M 1976 Parameterization of the planetary boundary layer in atmospheric general circulation models *Rev Geophys* **14** 215–226
[7] Seibert P F, Beyrich S E, Gryning S, Joffre A, Rasmussen and Tercier P 2000 Review and intercomparison of operational methods for the determination of the mixing height. *Atmos. Environ* **34** 1001–1027
[8] Medeiros B A, Hall and B Stevens 2005 What controls the climatological depth of the PBL? *Journal of Climate* **18** 2877–2892, doi:10.1175/JCLI3417
[9] Seidel D J, Ao C O and Li K *2010* Estimating climatological planetary boundary layer heights from radiosonde observations: Comparison of methods and uncertainty *Analysis Journal of Geophysical Research* **115** 1-15, doi:10.1029/2009JD013680
[10] Dewi S 2012 *Identification of Atmospheric Boundary Layer Thickness Using Doppler Radar Wind Profile in Tangerang and Surrounding Areas Course of Meteorology* (Bandung: Bandung Technology University)
[11] Rani A P 2016 *Study of Atmospheric Boundary Layer Height On Clear and Rainy Weather Conditions*. Course of Meteorology STMKG (Jakarta: School of Meteorology Climatology and Geophysics)
[12] Staudt K 2006 *Determination of the Atmospheric Boundary Layer Height in Complex Terrain During SALSA* 2005 Thesis Department of Micrometeorology (Bayreuth, University of Bayreuth)
[13] Breuer H, Acs F, Horvath A., Nemeth P, dan Rajkai K 2014 Diurnal Course Analysis of the WRF-Simulated and Observation-Based Planetary Boundary Layer Height *Advances in Science and Research* **11** 83-88
[14] Bloch L A 2008 Evaluation *of Daytime Boundary Layer Heights From a Mesoscale Model Using Profilers/Rass Measurements* Thesis Naval Postgraduate School (Monterey, California)
[15] Dewita A 2015 Identification of atmospheric boundary layer thickness using velocity and spectral width datas and turbulence parameter from WRF-ARW in Jakarta and surrounding areas *Journal of Meteorology Climatology dan Geophysics* **2** 161-170
[16] vonEngeln A and Teixeira J 2013 A planetary boundary layer height climatology derived from ECMWF reanalysis data. *Journal of Climate 26 American Meteorological Society* **26** 6575-6590

Satellite image processing for precision agriculture and agroindustry using convolutional neural network and genetic algorithm

Firdaus[1*], Y Arkeman[1,2], A Buono[1] and I Hermadi[1]

[1] Departement of Computer Science, Bogor Agricultural University, Indonesia
[2] Departement of Agroindustrial Technology, Bogor Agricultural University, Indonesia

E-mail: firdaus.prawiradisastra@gmail.com

Abstract. Translating satellite imagery to a useful data for decision making during this time are usually done manually by human. In this research, we are going to translate satellite imagery by using artificial intelligence method specifically using convolutional neural network and genetic algorithm to become a useful data for decision making, especially for precision agriculture and agroindustry. In this research, we are focused on how to made a sustainable land use planning with 3 objectives. The first is maximizing economic factor. Second is minimizing CO_2 emission and the last is minimizing land degradation. Results show that by using artificial intelligence method, can produced a good pareto optimum solutions in a short time.

1. Introduction

Bogor Agricultural University (IPB) and Indonesia National Institute of Aeronautics and Space (LAPAN) today had been cooperation launched a satellite called as LAPAN-IPB Satellite. The challenge now is how to translate image from satellite to become a useful data for decision making, especially for precision agriculture and agroindustry.

During this time, translating satellite imagery to a useful data for decision making usually done manually by human. In this research, we are going to translate satellite imagery by using artificial intelligence (AI) method specifically using convolutional neural network (CNN) and genetic algorithm (GA). One example application in satellite image processing is to develop green and sustainable biodiesel industry. The design process need to be start from top level such as land preparation, middle level such as industry and also in bottom level such as consumer.

In this research, we are focused on how to made a sustainable land use planning with 3 objectives. The first is maximizing economic factor. Second is minimizing CO_2 emission and the last is minimizing land degradation. The input of land use planning is satellite imagery, and this input will be give CNN and GA and those AI method will process intelligently. Through the process we hope, we can obtain a good solution that meets the three criteria above.

Many works on using CNN for satellite image processing can achieve high accuracy between 75% until 91% [1]. CNN basically inspired by biological neural network. It is reproduce some of its functions by using simple but massively interconnected processing units neurons.

Genetic Algorithms is a searching and optimization technique that mimics the process of evolution and genetic structural change in living organisms [2][3][4]. It is, therefore, often called as evolutionary algorithms. In this study, multiobjective (multicriteria) optimization was done. Multiobjective optimization is an optimization of more than one objective functions where conflicts among those functions is likely to occur [5][6].

Continuous use of fossil fuel in the world has caused its scarcity. This has led to a situation where a search for alternative renewable fuel sources from plants (bioenergy) needs to be done [7]. Bioenergy is a source of renewable energy potential to reduce carbon emission compared to fossil fuel. However, increasing bioenergy production might cause land use change (LUC) and an increase in CO2 emission to the atmosphere [8]. Indonesia has a great potential of bioenergy production as this country has a total land size of 188.20 million hectares with 100.80 hectares of it is suitable for farming land [9]. Therefore, good and optimal planning of land utilization is really needed in order to improve bioenergy productivity in Indonesia.

Oil palm is one of the potential plants for bioenergy production. Oil palm plantations have been extensively developed not only by state-owned companies but also by private companies and smallholder estate owners [10]. According to [10] and [11], Indonesia is the biggest palm oil producer in the world. Therefore, planning in land utilization for biodiesel industry is important in order to reach the objectives which do not only focus on the economic aspect but also pay attention to environmental conservation.

Good and right land utilization could help economic development. Land utilization planning is needed for optimal land utilization. The problem of land utilization is so complex that a smart computation technique is needed for optimal land utilization planning [10][12][13].

Genetic algorithm has been used to solve problems in land utilization. [14] used multi-objective evolutionary algorithm (MOEA) for land use allocation. [15] also used non-dominated sorting genetic algorithms (NSGA-II) to optimize spatial multiobjective in land use allocation [16] also used NSGA-II for solving multiobjective landuse problems in Tongzhou, China. [13] also had been done a landuse optimization in region in China, but with different objective function from [16].

Results with numerical example shows CNN and GA can produced a good pareto optimum solutions in a short time. In our previous research [17] we had not considered a spatial aspect.But, in this research we added a spatial aspect such as height of land. So the present research become more complex.

2. Methodology

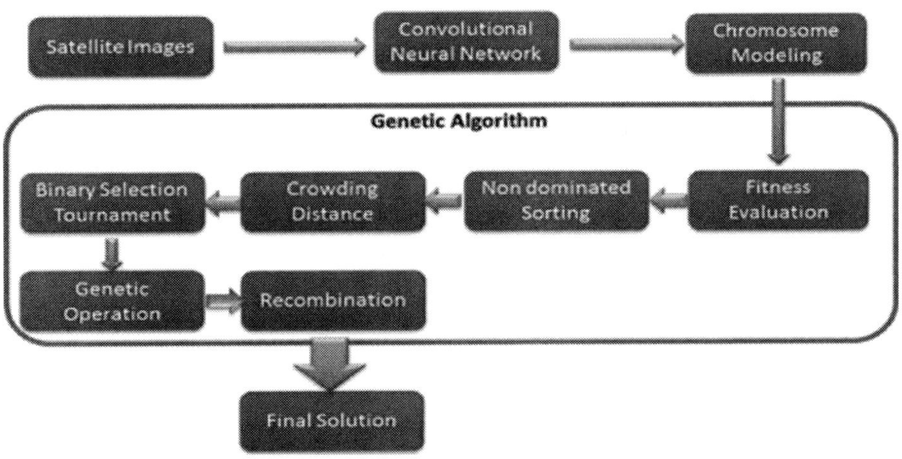

Figure 1. Methodology

First, Image from satellite was processed by convolutional neural network for clasifying plant types in land area. After that, data was modeled as a grid form and land types and land high data also added to grid form. Next, the grid form model was evaluated to assess the existing value of 3 objectives. After an assessment, genetic algorithm will take an optimization process to produce a final best solutions.

2.1. Chromosome modeling

The Chromosome are modeled in 2 dimensonal matrix form according to grid form which had been created before. There are 3 chromosome which had been modeled, land type chromosome, land high chromosome and plant type chromosome. For land type and land high chromosome, there are several types of value which positions are not change. For plant types chromosome there are 3 value which is random between 1 to 3 as in figure 3 below.

1-3	1-3	1-3	1-3	1-3	1-3	1-3	1-3	1-3	1-3
1-3	1-3	1-3	1-3	1-3	1-3	1-3	1-3	1-3	1-3
1-3	1-3	1-3	1-3	1-3	1-3	1-3	1-3	1-3	1-3
1-3	1-3	1-3	1-3	1-3	1-3	1-3	1-3	1-3	1-3
1-3	1-3	1-3	1-3	1-3	1-3	1-3	1-3	1-3	1-3
1-3	1-3	1-3	1-3	1-3	1-3	1-3	1-3	1-3	1-3
1-3	1-3	1-3	1-3	1-3	1-3	1-3	1-3	1-3	1-3
1-3	1-3	1-3	1-3	1-3	1-3	1-3	1-3	1-3	1-3
1-3	1-3	1-3	1-3	1-3	1-3	1-3	1-3	1-3	1-3
1-3	1-3	1-3	1-3	1-3	1-3	1-3	1-3	1-3	1-3

Figure 2. Plant types chromosome

2.2. Objective functions

In this paper, there were three objective functions on which optimizations were done. These included profit factor obtained and modified from the study of Shaygan [15], CO2 emission absorption function modified from [14], and soil fertility function obtained from the modification of [15] and [14]. Chromosome representation was adopted from the ones of Shaygan [15]. Plant types, land types and land high types were limited into 3, 5 and 4, respectively. In order to show system's capability, data in the form of numerical examples were used.

2.2.1 Profit factor function

In this function formulation, Z1 was defined as maximizing profit factor where $Ec_{lu,lh,i,j}$ was the profit factor from the use of *lu* land type and *lh* land high applied in *(i, j)* unit. *Ru* was the number of rows in land use unit and *Cu* was the number of columns in land use unit.

$$Z1 = Max\sum_{lu=1}^{LU}\sum_{lh=1}^{LH}\sum_{i=1}^{Ru}\sum_{j=1}^{Cu} Ec_{lu,lh,i,j} \tag{1}$$

2.2.2 Carbon sequestration factor function

Second function is minimizing carbon emission, so it mean we have to maximizing carbon sequestration, In this function formulation, Z2 was defined as maximizing CO_2 sequestration factor where $Car_{lu,lh,i,j}$ was the sequestration factor from the use of *lu* land type and *lh* land high applied in *(i, j)* unit. *Ru* was the number of rows in land use unit and *Cu* was the number of columns in land use unit.

$$Z2 = Max\sum_{lu=1}^{LU}\sum_{lh=1}^{LH}\sum_{i=1}^{Ru}\sum_{j=1}^{Cu} Car_{lu,lh,i,j} \tag{2}$$

2.2.3 Soil fertility factor function

Third function is minimizing land degradation, so it mean we have to maximizing land fertility, In this function formulation, Z3 was defined as maximizing land fertility factor where $Fer_{lu,lh,i,j}$ was the land fertility factor from the use of *lu* land type and *lh* land high applied in *(i, j)* unit. *Ru* was the number of rows in land use unit and *Cu* was the number of columns in land use unit.

$$Z3 = Max\sum_{lu=1}^{LU}\sum_{lh=1}^{LH}\sum_{i=1}^{Ru}\sum_{j=1}^{Cu} Fer_{lu,lh,i,j} \tag{3}$$

2.3. Non dominatd sorting

Non dominated sorting is one of the multi-objective genetic algorithm process which give a rank to each individual based on their objective functions [14].

2.4. Crowding distance

After rank has been given to each individual, crowding distance value also added to each individual to know the diversity of populatioin [14] [18].

2.5. Binary selection tournament

After the process which has been done before, the selection process will be conducted according to following criteria :

- If individual are located in same front then choose the bigger value of crowding distance.
- If Individual are located in different rank then choose the lower rank (better rank).

2.6. Crossover

Crossover type which has been used in this research is uniform crossover [19] But, it had been modified for 2 dimension chromosome case, which has been adapted based on Bui and Moon crossover principle [20].

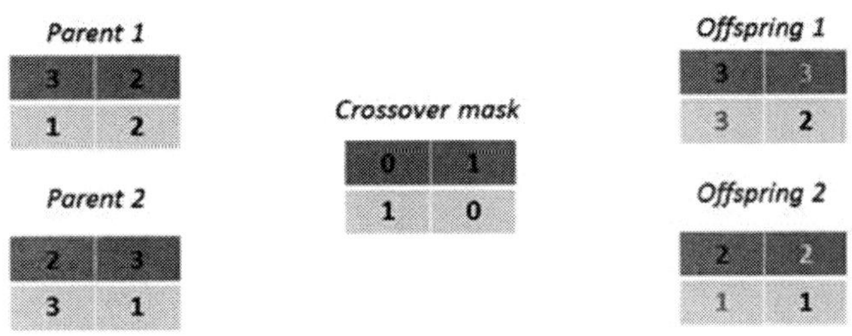

Figure 3. Crossover operator

2.7. Mutation

Mutation which has been used in this research is swap mutation [21]. But, it has been modified by change the swap to generate random number.

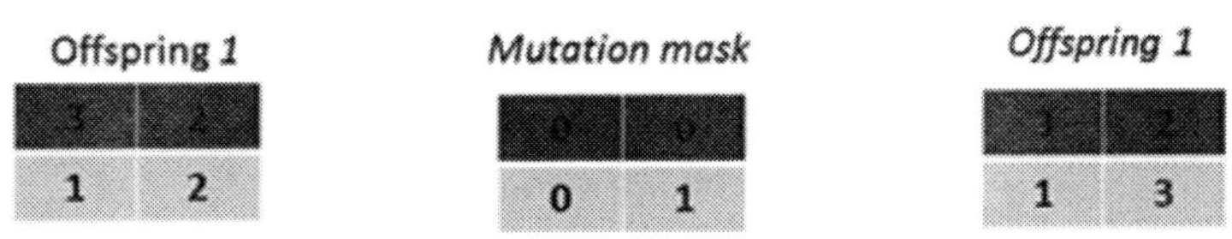

Figure 4. Mutation operator

2.8. Recombination

Recombination is a process to store all best individual in multi-objective genetic algorithm generation process. Parent and offspring are combined, so the size of population become 2 times larger. After that, each individual are given a rank and crowding distance value, then population will be cut half. Cutting at the last individual, considering a crowding distance value. Individual that contain higher value will be take to the next population [18].

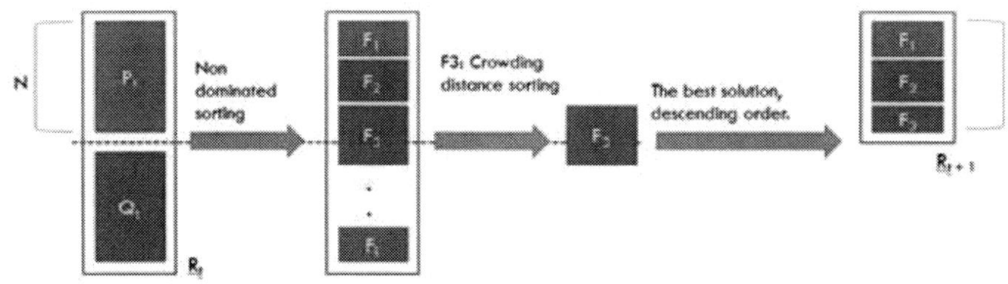

Figure 5. Recombination

3. Results and Disucussions

The following figures shows traveling of individual in multiobjective genetic algorithm generation. Black filled circles are the non dominated individual in generation, black rings are indicates a dominated individual in generation.

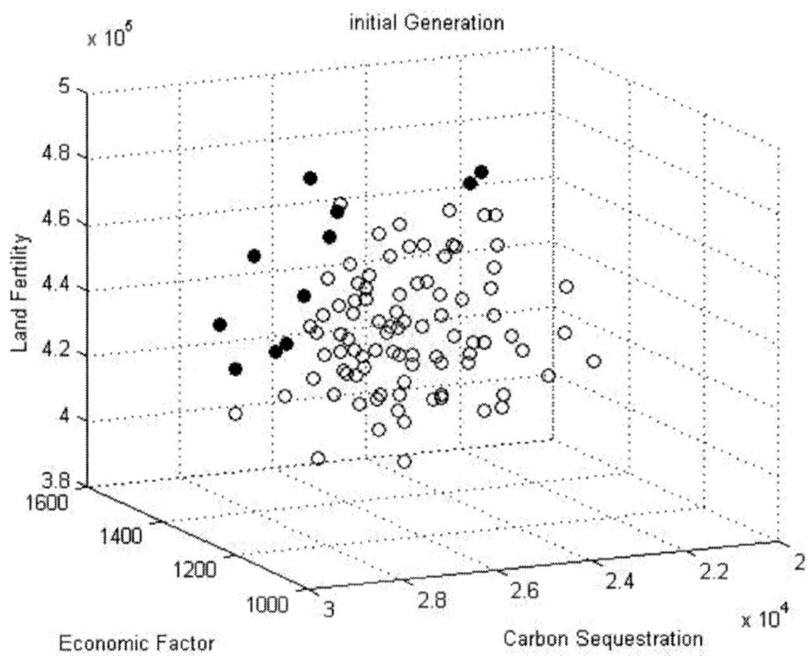

Figure 6. Plot initial population

The following figures are comparing initial population with final population. So we can see the population movement. Black filled circles are indicates final generation, black rings are indicates initial population.

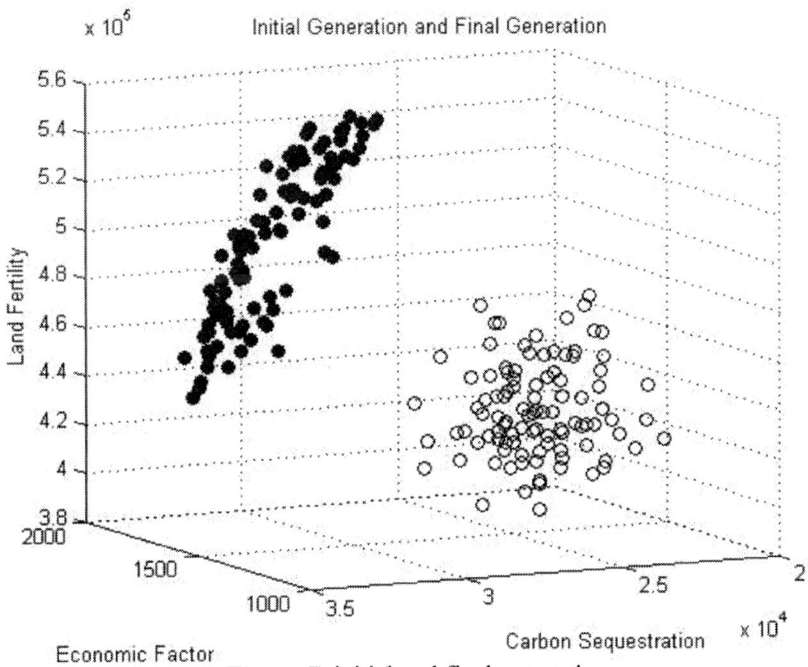

Figure 7. initial and final generation

4. Conclusions

Based on results we got, at the first generated initial population, we can see most of individual still spread randomly and number of non dominated individual still smaller than dominated individual. At the end of generation, population move intelligently to optimum position and we can see all individual are non dominated in final generation, it means genetic algorithm succeed optimizing a problems.

So it can be concluded that satellite imagery for precision agriculture and agroindustry problem can be well solved using convolutional neural network and genetic algorithm

References

[1] Langkvist M, Kiselev A, Alirezaie M and Loutfi A 2016 Classification and segmentation of satellite orthoimagery using convolutional neural network *Remote Sensing* **8**

[2] Arkeman Y, Seminar K B and Gunawan H 2012 *Algoritma genetika Teori dan Aplikasinya untuk Bisnis dan Industri* (Bogor: IPB Press)

[3] Jong K A D 2006 *Evolutionary Computation A Unified Approach* (London: MIT Press)

[4] Bagchi T P 1999 *Multiobjective Scheduling by Genetic Algorithm* (India: Kluwer Academic Publisher)

[5] Arkeman Y, Herdiyeni Y, Hermadi I and Laxmi G F 2013 *Algoritma genetika tujuan jamak (multi-objective genetic algorithms): teori dan aplikasinya untuk bisnis dan agroindustri* (Bogor: IPB Press)

[6] Ratono J, Seminar K B, Arkeman Y and Suroso A I 2015 ERP selection using fuzzy-MOGA approach: A food enterprise case study *TELKOMNIKA* **12**

[7] Zhiyuan H, Piqiang T, Gengqiang P 2005 Multi-objective optimization of cassava-based fuel ethanol used as an alternative automotive fuel in Guangxi, China *ELSEVIER* **83**

[8] Ren Z and Lu X 2011 Using GA for land use planning *IEEE*

[9] Mulyani A and Las I 2008 Potensi sumber daya lahan dan optimalisasi pengembangan komoditas penghasil bioenergi di Indonesia *Jurnal Litbang Pertanian*
[10] Amin M, Hanum C and Charloq 2015 Kandungan hara tanah dan tanaman kelapa sawit menghasilkan terhadap pemberian tandan kosong kelapa sawit dan kedalaman biopori *Jurnal Online Agroekoteknologi*
[11] Astuti E 2008 Pengaruh konsentrasi katalisator dan rasio bahan terhadap kualitas biodiesel dari minyak kelapa *Jurnal Rekayasa Proses*
[12] Zhang H H, Zeng Y N and Bian L 2010 Simulating multi-objective spatial optimization allocation of land use based on the integration of multi-agent system and genetic algorithm *International Journal Environ* **4**
[13] Stewart T J, Janssen R and Herwijenen M V 2004 A genetic algorithm approach to multiobjective land use planning *ELSEVIER* **31**
[14] Datta D, Deb K, Fonsecca C M, Lobo F and Condado P 2007 Multiobjective evolutionary algorithm for land-use management problem *International Journal Computational Intelligence Research* **3**
[15] Shaygan M, Alimohammadi A, Mansourian A, Govara S and Kalami S M 2014 Spatial multi-objective optimization approach for land use allocation using nsga-ii *IEEE*
[16] Cao K, Batty M, Huang B, Liu Y, Yu L and Chen J 2011 Spatial multi-objective land use optimization:extensions to the non-dominated sorting genetic algorithm-II *International Journal of Geographical Information Science* **25**
[17] Firdaus, Arkeman Y, Buono A 2016 A multiobjective genetic algorithm for land use planning for sustainable biodiesel agroindustry *WCCA.AFITA*
[18] Deb K , Pratap A, Agarwal S and Meyarivan T 2002 A fast and elitist multiobjective genetic algorithm: NSGA-II *IEEE* **6**
[19] Hu X B and Paolo E D 2007 An Efficient genetic algorithm with uniform crossover for air traffict control *ELSEVIER*
[20] Hwan I C and Jung H K 2003 Hybrid genetic algorithm for electromagnetic topology optimization *IEEE*
[21] Hyun J, Kim Y H and Ryou H B 2008 Optimal sensor deployment for wireless surveilance sensor networks by a hybrid steady-state genetic algorithm *IEICE*
[22] Deb K 2011 *Multi-objective evolutionari optimisation for product design and manufacturing* (London: Springer)
[23] Engelbrecht A P 2007 *Computational intelligence an introduction* (South Africa: WILEY)
[24] Haupt R L and Haupt S E 2004 *Practical genetic algorithms* (Ner Jersey: WILEY)
[25] Wise M, Hodson E L, Mignone B K, Clarke L, Waldhoff S and Luckow P 2015 An approach to computing marginal land use change carbon intensities for bioenergy in policy applications *ELSEVIER*

IOP Conf. Series: Earth and Environmental Science **54** (2017) 012103 doi:10.1088/1755-1315/54/1/012103

Modelling land use/cover changes with markov-cellular automata in Komering Watershed, South Sumatera

E Kusratmoko*, S D Y Albertus and Supriatna
Magister Program Department of Geography FMIPA Universitas Indonesia Gedung H Kampus UI Depok, Indonesia

E-mail: eko.kusratmoko@ui.ac.id

Abstract. This research has a purpose to study and develop a model that can representing and simulating spatial distribution pattern of land use change in Komering watershed. The Komering watershed is one of nine sub Musi river basin and is located in the southern part of Sumatra island that has an area of 8060,62 km^2. Land use change simulations, achieved through Markov-cellular automata (CA) methodologies. Slope, elevation, distance from road, distance from river, distance from capital sub-district, distance from settlement area area were driving factors that used in this research. Land use prediction result in 2030 also shows decrease of forest acreage up to -3.37%, agricultural land decreased up to -2.13%, and open land decreased up to -0.13%. On the other hand settlement area increased up to 0.07% , and plantation land increased up to 5.56%. Based on the predictive result, land use unconformity percentage to RTRW in Komering watershed is 18.62 % and land use conformity is 58.27%. Based on the results of the scenario, where forest in protected areas and agriculture land are maintained, shows increase the land use conformity amounted to 60.41 % and reduce unconformity that occur in Komering watershed to 17.23 %.

1. Introduction

Komering watershed is one sub-basin of nine sub basin in the Musi basin catchment, and in line with the developments in Komering watershed, the various orders of life is changing rapidly following the various needs of the community. One impact of these changes is that the pattern of use of natural resources which are located around the community. In addition to the growth in population and settlement area will urge the land use/cover pattern in Komering watershed change that is usually converted from the use of agricultural land to non-agricultural so that erosion and sedimentation is high enough. The condition of the river and the land area is seen that the upstream Komering watershed has been impaired or decline of ecosystems and Environment [1]. The setback is particularly indicated environmental quality include the existence of illegal logging for the area of aquaculture, plantations and settlement areas do not pay attention to the principles of environmental sustainability and the turbidity of the water in a river estuary Komering watershed. To prevent such degradation, monitoring of land use/cover change in Komering watershed is very important. Therefore this study was intended to employ Geographycal Information System and remote sensing techniques to detect land use/cover change in Komering watershed between 1990 and 2016, and Markov-cellular automata (CA) methodologies to simulating land use/cover change in year 2030 using Land Change Modeler tools on Idrisi Selva software.

2. Method

2.1. Study Area

The research location is situated in Komering watershed, which is administratively includes three regency, which Ogan Ulu Timur, Ogan Ulu and Ogan Ilir regencies (figure 1).

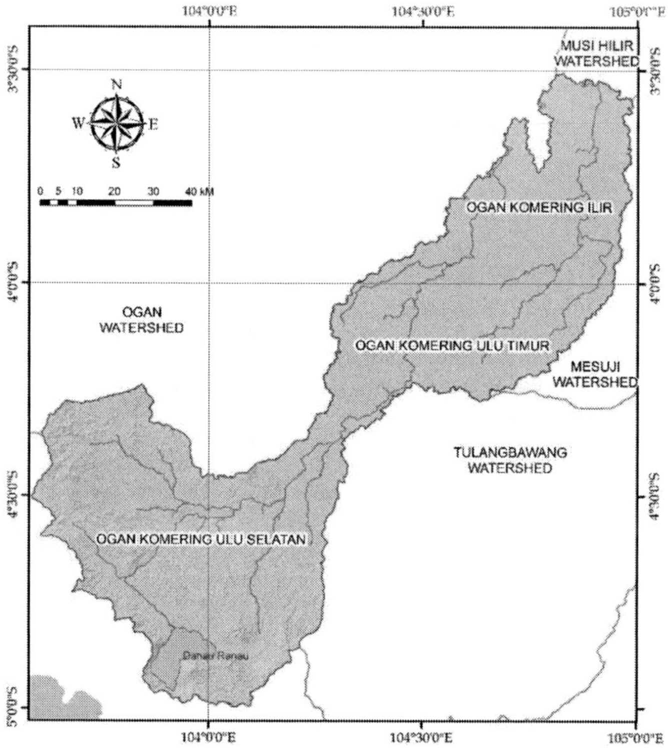

Figure 1. Map of the Komering watershed

2.2. Materials

- Satellite data : Landsat images of path/row 124/63, Landsat 5 TM for 1990 and 2000, Landsat 7 ETM for 2010 and Landsat 8 OLI for 2016. The satellite acquisition date for the images was May to August. DEM data with 30 meter resolution to create watershed boundary.
- Software : Idrisi selva 17, ENVI 5.1, and Arc GIS 10.1.

2.3. Data analysis

- Image classification : Classification of land use/cover was conducted using supervised classification in ENVI 5.1 software. The classification process produced land use/cover maps for the years 1990, 2000, 2010 and 2016. The six land use classes were generated which included; forest, water body, settlement area, plantation, bare-land and agricultural land.
- Land use change detection : land use/cover change detection was conducted by post-classification comparison change detection using ARC GIS 10.1. This was achieved by comparative analysis of the years 1990-2000, 2000-2010, and 2010-2016 (figure 2 and 3).

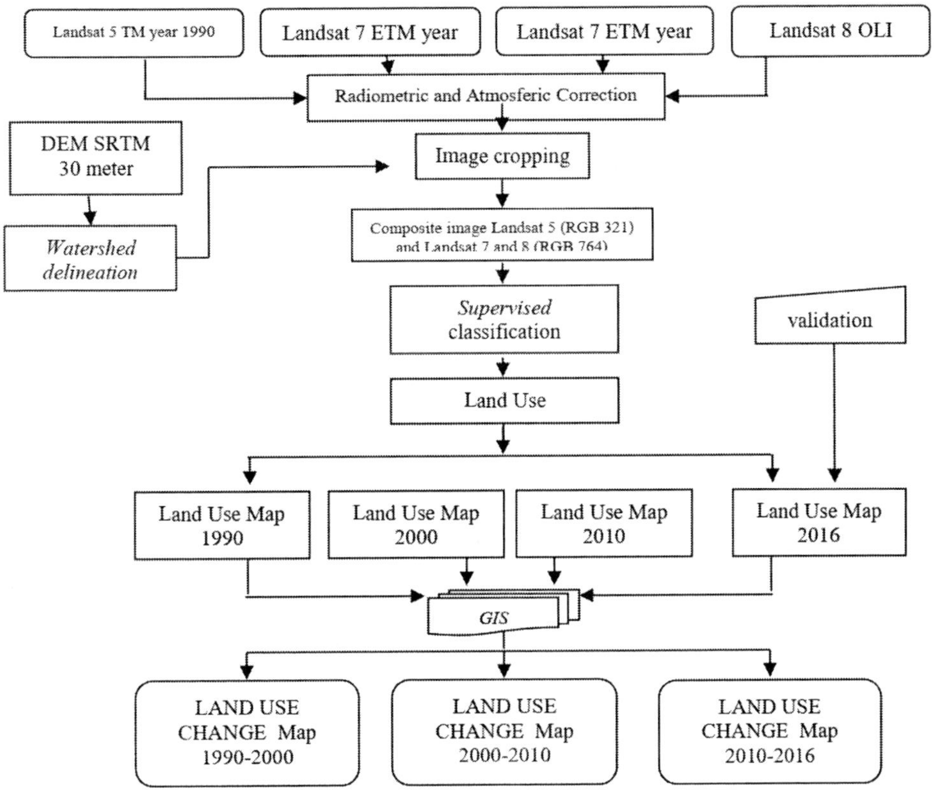

Figure 2. Image analysis and detection process

- CA –Markov Model: Markov chain and CA both are the discrete dynamic model in time and state [2]. One inherent problem with Markov is that it provides no sense of geography. The transition probabilities may be accurate on per category basis, but there is no knowledge of the spatial distribution of occurrences within each landuse category [3]. CA was used to add spatial character into the model. The Idrisi Selva 17 integrates CA with Markov very well [4]. In this research, we take the Idrisi GIS software to simulate the land use/cover in Komering watershed. The driving factors that used in this research are slope, elevation, distance from road, distance from river, distance from capital sub-district, distance from settlement area.

3. Results and Discussions

3.1. Change in Komering Watershed from 1990 to 2016

Six land use/cover classes were produced. These included; forest, water body, settlement area, agriculture land, bare-land and plantation. The results of testing the accuracy of interpretation of Landsat imagery in 2016 visually gives kappa coefficient value by 87% and overall accuracy of 92%. The value is compliant validation of interpretation so that proper land use/cover maps used in the study. A map of the land use/cover interpretation of the results can be good if it has an accuracy of more than 70% [5]. Land use/cover change from 1990 to 2016 is shown in figure 4 and figure 5. The result showed that in Komering watershed revealed a significant change in the proportions of the various land use/cover change types of the study area from the year 1990 to 2016, forest decreased up to -1.66%, agricultural land decreased up to -1.52%, and open land decreased up to -0.12%. On the other hand settlement area increased up to 0.09% , and plantation land increased up to 3.21% (table 1 and 2).

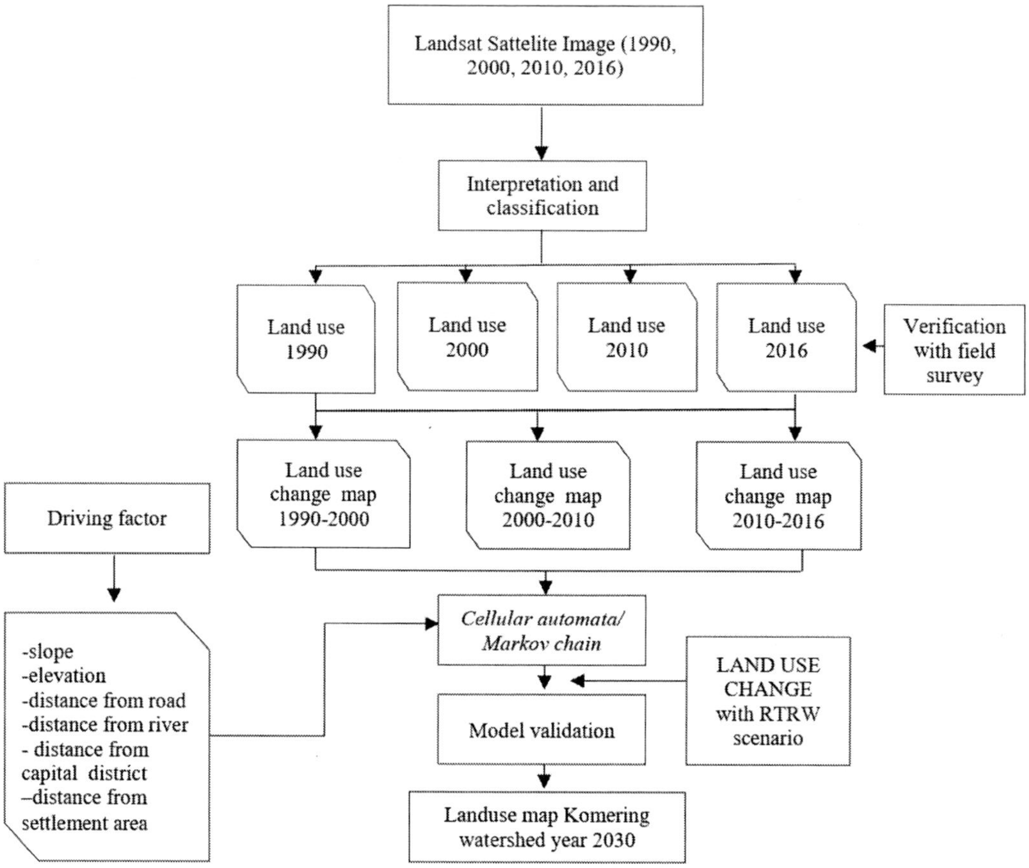

Figure 3. Landuse modelling Process

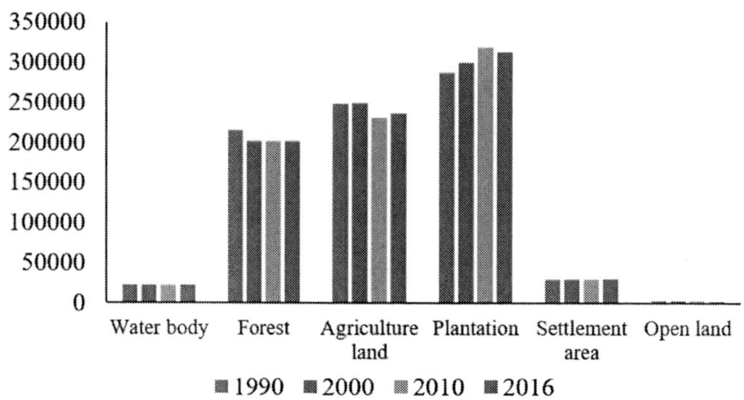

Figure 4. Komering watershed land use/cover in years 1990, 2000, 2010 and 2016

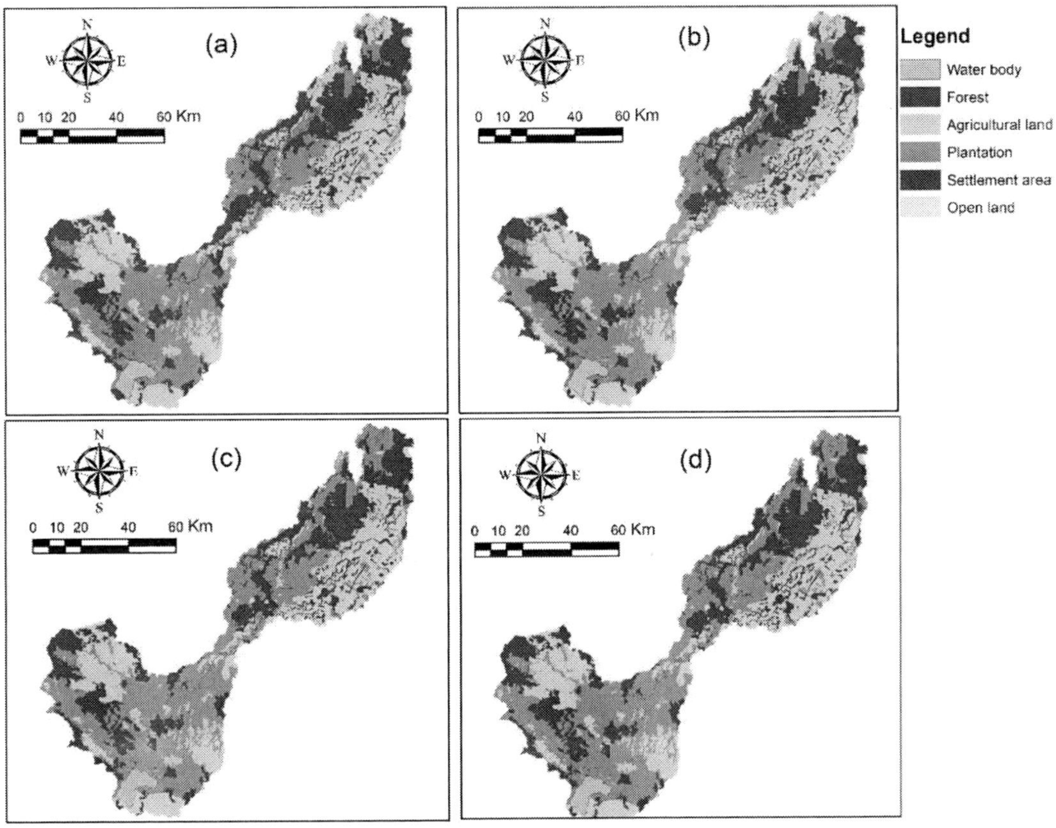

Figure 5. Land use/cover Komering watershed map year 1990 (5a), 2000(5b), 2010(5c), 2016(5d)

Table 1. Areas from different land use/cover classes in 1990, 2000, 2010, 2016.

Landuse	Area							
	1990		2000		2010		2016	
	Hectare	%	Hectare	%	Hectare	%	Hectare	%
Water body	21.817	2,71	21.817	2,71	21.817	2,71	21.814	2,71
Forest	215.188	26,70	201.663	25,02	201.662	25,02	201.812	25,04
Agriculture land	248.858	30,87	249.851	31,00	230.860	28,64	236.574	29,35
Plantation	287.936	35,72	300.256	37,25	319.247	39,61	313.781	38,93
Settlement area	29.232	3,63	29.422	3,65	29.422	3,65	29.993	3,72
Open land	3.031	0,38	3.054	0,38	3.054	0,38	2.089	0,26

Table 2. Areas changes from different land use/cover classes in 1990, 2000, 2010, 2016.

Landuse	Change							
	1990-2000		2000-2010		2010-2016		1990-2016	
	Hectare	%	Hectare	%	Hectare	%	Hectare	%
Water body	0	0,00	0	0,00	-3	0,00	-3	0,00
Forest	-13.525	-1,68	-1	0,00	151	0,02	-13.375	-1,66
Agriculture land	992	0,12	-18.990	-2,36	5.713	0,71	-12.284	-1,52
Plantation	12.320	1,53	18.991	2,36	-5466	-0,68	25.844	3,21
Settlement area	189	0,02	-	0,00	571	0,07	760	0,09
Open land	23	0,00	-	0,00	-966	-0,12	-942	-0,12

3.2. Land use/cover simulation

Model validation is done by comparing the prediction of land use/cover in 2016 with the actual land use/cover maps in 2016 (table 3). Validation test results obtained by the 81% which indicates that the model simulations were done very well. The land use/cover predictive result (figure 6 and 7) in 2030 also shows decrease of forest area up to -3.37%, agricultural land decreased up to -2.13%, and open land decreased up to -0.13% (table 4). On the other hand settlement area increased up to 0.07% , and plantation land increased up to 5.56%. Based on the predictive result, land use/cover unconformity percentage to RTRW in Komering watershed is 18.62 % and land use/cover conformity is 58.27%.

Table 3. Major land use/cover conversions from 1990 to 2016

1990 \ 2016	Water body (hectares)	Forest (hectares)	Agriculture Land (hectares)	Plantation (hectares)	Settlement area (hectares)	Open land (hectares)
Water body	22,088	2	1	2	3	0
Forest	64	207,128	1,360	12,725	301	0
Agriculture land	0	0	246,838	14,148	422	23
Plantation	0	0.	0.	295,763	256	0
Settlement area	0	0	0	0	30,187	0
Open land		150	208	527	79	2,199

Table 4. Land use/cover classes and land use/cover change between 2016 and 2030.

Landuse	2016		2030		% change	
	hectare	%	hectare	%	hectare	%
Water body	21,814	2.71	21,816	2.71	1	0.00
Forest	201,812	25.04	174,670	21.67	-27,141	-3.37
Agriculture land	236,573	29.35	219,404	27.22	-17,169	-2.13
Plantation	313,780	38.93	358,612	44.49	44,831	5.56
Settlement area	29,992	3.72	30,540	3.79	547	0.07
Open land	2,088	0.26	1,017	0.13	-1,070	-0.13

Based on The RTRW scenario, where forest in protected areas and agriculture land are maintained, shows increase the land use/cover conformity amounted to 60.41 % and reduce unconformity that occur in Komering watershed to 17.23 % (table 5). Similarly, the use of land agricultural land has not achieved the expected objective because it does not change the broad decline of agricultural land changes, however, this scenario is very possible to be applied.

Table 5. Land use/cover classes and land use/cover change between 2016 and 2030 with RTRW scenario

	2016		2030		% change	
Landuse	Hectare	%	hectare	%	hectare	%
Water body	21,814	2.71	21,816	2.71	1.	0.00
Forest	201,812	25.04	189,580	23.52	-12,231	-1.52
Agriculture land	236,573	29.35	219,353	27.21	-17,219	-2.14
Plantation	313,780	38.93	343,703	42.64	29,922	3.71
Settlement area	29,992	3.72	30,541	3.79	548	0.07
Open land	2,088	0.26	1,067	0.13	-10,20	-0.13

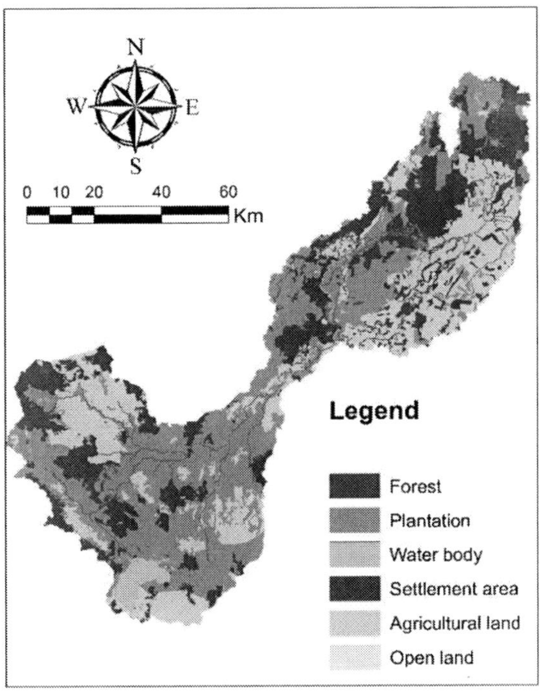

Figure 6. Land use/cover change Komering watershed map year 2030 without skenario

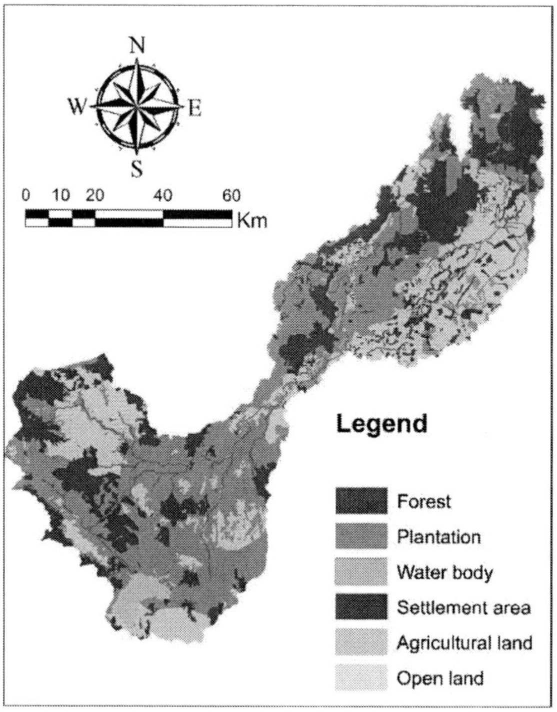

Figure 7. Land use/cover change Komering watershed map year 2030 with scenario

4. Conclusion

This research presents the results for analysing and predicting land use/cover change by using remote sensing data and CA-Markov model. From this study of land use/cover changes in Komering watershed from 1990 to 2016, some conclusions were obtain the decrease in forest areas from 26.7 % to 25.04% is closed related with increase in plantation areas from 35.72% to 38.93%. The simulation of land use/cover in the year 2030 indicated that the forest and agricultural land decreased rapidly and change to plantation area.

References

[1] BPDAS 2009 *Laporan Hasil Identifikasi dan Karakteristik Sub Das Komering. Sumatera Selatan*
[2] Tiur Vera Damayanti Peruge, Samsu Arief H, Sakka 2012 *Model Of Land Use Change Of Use* Cellular *Automata - Markov Chain In The Mamminasata* (Makassar: Universitas Hassanudin)
[3] Eastman JR 2003 *IDRISI Kilimanjaro Guide to GIS and Image Processing Worcester* (MA: Clark University)
[4] Clark Labs 2006 *IDRISI Andes software Worcester* (MA: Clark University)
[5] Purwadhi, Hardiyanti S F 2001 *Interpretasi Citra Digital* (Jakarta : Grasindo)

IOP Conf. Series: Earth and Environmental Science **54** (2017) 012104 doi:10.1088/1755-1315/54/1/012104

Indonesia coverage simulation of SAR satellite at near-equatorial orbit

Harry Septanto, Satriya Utama, Robertus Heru Triharjanto and Suhermanto

Satellite Technology Center, National Institute of Aeronautics and Space (LAPAN), Indonesia

E-mail: harry.septanto@lapan.go.id

Abstract. Properties of SAR (Synthetic Aperture Radar) that able to penetrate the cloud and does not depend on the sunlight are a number of advantages when utilized for monitoring tropical region like the IMC (Indonesian Maritime Continent). Moreover, since having areas along equatorial belt, the IMC is at a shortcoming from perspective of highly inclined LEO (Low Earth Orbit) satellite. It would result shorter and infrequent pass times when compared with a near-equatorial LEO satellite whose low inclination. This paper reports on the investigation of a near-equatorial LEO SAR satellite coverage property through simulations. The simulations is run in nine scenarios of orbit parameter that consist of combinations of attitude {500 km, 600 km, 700 km} and inclination {80, 90, 100}. The target area is defined as 50 km x 50 km around Jakarta. Meanwhile, the SAR sensor simulation is run with swath width of 40 km, incidence angle around 25^0-29^0 and Stripmap mode. Minimum, Maximum and Mean Access Revisit of the target for each scenarios are resulted.

1. Introduction

SAR (synthetic aperture radar) is a type of radar system installed on a moving platform, such as UAV (unmanned aerial vehicle), aircraft or satellite. The objective of the radar is to acquire images of phenomena over the surface of the Earth. Earth observation capabilities provided by a SAR that is carried by a satellite as the payload have been enthused by researchers for years. Unlike optical camera payload that depend on Sun illumination, SAR payload can be operated in all-weather and day-night time. Therefore, it is very valueable for the monitoring of areas with high cloud coverage such as Indonesia. Images acquired from SAR satellites are employed over a wide range of applications including to classify tropical natural forest and plantation and to track plantation dynamics [1], to detect ships [2] and to monitor small land deformation that indicates environmental degradation at urban area [3][4]. In addition, another applications are to generate actual oil spill distribution maps for the Northern Caspian and the Eastern Black Sea [5], to study accurateness burned area mapping in relating to the soil and snag moisture [6], to quantify the amount of snow mass [7], for tomographic analysis of forest volumes [8], to classify coastal zone terrains [9], to generate to identify rice fields in Vietnam [10], to retrieve agricultural crop [11], to monitor glacier surface motion [12], to get better understanding and quantification of surface roughness of sea ice [13], to provide flood mapping in urban areas [14], to name a few.

SAR satellites have been operating since 1978 with the launch of US SAR satellite SEASAT. At the moment, many nations operate SAR satellite system such as Radarsat (Canada), ALOS (Japan),

TerraSAR-X (Germany) and many others. X, C, and L-band are the most popular choice for the radar frequency and the resolution are varied from 3 m, for vehicle class observation, to 100 m for large area observation.

Most of satellites including the satellites stated above are at highly inclined LEO (Low Earth Orbit). Since having areas along equatorial belt, the IMC (Indonesian Maritime Continent) is at a shortcoming from perspective of highly inclined LEO (Low Earth Orbit) satellite. It would result shorter and infrequent pass times when compared with a near-equatorial LEO satellite whose low inclination. Meanwhile, near-equatorial orbit is a new trend in Earth observation satellites recent years, since more countries in equatorial belts in owning Earth observation satellites. In July 2009, Malaysia launched a near-equatorial orbit satellite called Razaksat-1 which has 400 kg mass. The satellite has multispectral imager with 5 m resolution and panchromatic camera with 2.5 m resolution. It is launched to 685 km orbit with 9° inclination. From such orbit the camera was operated from Malaysia six passes per day, until its radio failed in one year after launched [15]. In September 2015, a 78 kg near-equatorial orbit satellite is launched by Indonesia that is called LAPAN-A2/ LAPAN-ORARI satellite [16]. The satellite is launched to 638 km orbit with 6° inclination. It passes Indonesia nine times per day with brings multi-mission including 24 hours AIS (automated identification system) receiver run to track vessels and ships. In December 2015, Singapore launched a near- equatorial orbit satellite which has 500 kg mass to 550 km orbit with 15^0 inclination. The satellite carries panchromatic camera with 1 m resolution and has mean revisit time of 12-16 hours [17].

Currently, there is none of SAR satellite operates at near-equatorial LEO even though the satellite might provide many advantages to nations that have areas along equatorial belt. Hence, studies on the design of the SAR satellite that operates at near-equatorial LEO is challenging, besides would be valuable, especially in the context of the IMC. This paper reports on the investigation of a coverage properties aspect of a near-equatorial LEO SAR satellite.

2. Simulation Setup

A commercial simulation software called STK (System Tool Kit) Pro version 11 from AGI (Analytical Graphics, Inc.) [18] is utilized to run the simulation. Orbit parameters of the satellite used in the simulation is presented in Table 1. In addition, the simulation implements J4 orbit perturbation propagator model which considers the Earth oblateness effects. It is well-known that the Earth oblateness effects is the major orbit perturbation effect for most SAR satellites [20].

Table 1. Satellite's orbit parameters

Orbit's Shape	Circular
Argument of Perigee	90^0
RAAN (Right Ascension of Ascending Node)	0^0

There are nine scenarios that accommodate combination values between altitude and inclination as shown in Table 2. Meanwhile, for all of scenarios, the SAR system is run such that it has about 25^0-29^0 incidence angle and about 40 km swath width. The SAR sensor model used in the simulation is Complex Conic type. This sensor type selection follows the consideration in [19].

Inner (outer) incidence angle of the SAR system, as depicted in Figure 1, is represented by the Inner (Outer) Half Angle parameter of the sensor model. Equation (1) represents the mathematical relation between the incidence angle, θ, and the Half Angle, α, [19].

$$\alpha = \sin^{-1}\left(\frac{R_E}{R_E + h}\sin(\theta)\right) \quad (1)$$

where R_E is radius of the Earth and h is the satellite's altitude.

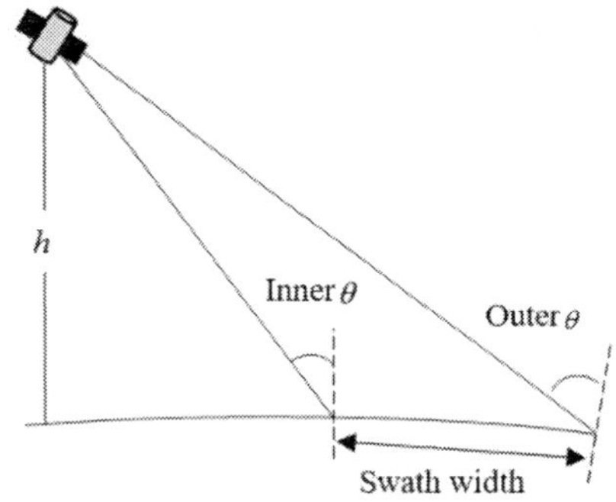

Figure 1. SAR system parameters

Table 2. Combinations of altitude and inclination

Scenario No.	Altitude (km)	Inclination (^{0})
1	500	8
2	600	8
3	700	8
4	500	9
5	600	9
6	700	9
7	500	10
8	600	10
9	700	10

The radar projection on ground (sensor footprint) is formed by defining an inner incidence angle and tuning the outer incidence angle parameter as well as the Minimum and Maximum Clock Angle parameters of the Complex Conic sensor model. The inner incidence angle is defined by 25^0 for all scenarios. It yields a set of outer incidence angle that varies from 28^0 to 29^0. For each scenario, these values are converted to Inner and Outer Half Angle through equation (1) as appeared in Table 3. Moreover, for each scenario, the Minimum Clock Angle is 89.9^0 and the Maximum Clock Angle is 90.1^0 to represent a side-looking radar with a narrow beam width. For each scenario, the SAR system is assumed run Stripmap mode with the defined fixed parameters.

Table 3. Incidence Angles and sensor model's Half Angle equivalent value

Scenario No.	Incidence Angle (^{0})		Half Angle (^{0})	
	Inner	Outer	Inner	Outer
1	25	29	23.0706	26.7135
2	25	28.5	22.7209	25.8546
3	25	28	22.3821	25.0239
4	25	29	23.0706	26.7135

Scenario No.	Incidence Angle (^{0})		Half Angle (^{0})	
	Inner	Outer	Inner	Outer
5	25	28.5	22.7209	25.8546
6	25	28	22.3821	25.0239
7	25	29	23.0706	26.7135
8	25	28.5	22.7209	25.8546
9	25	28	22.3821	25.0239

The target area is defined as 50 km to 50 km north side of Jakarta, Indonesia. This area is depicted in Figure 2.

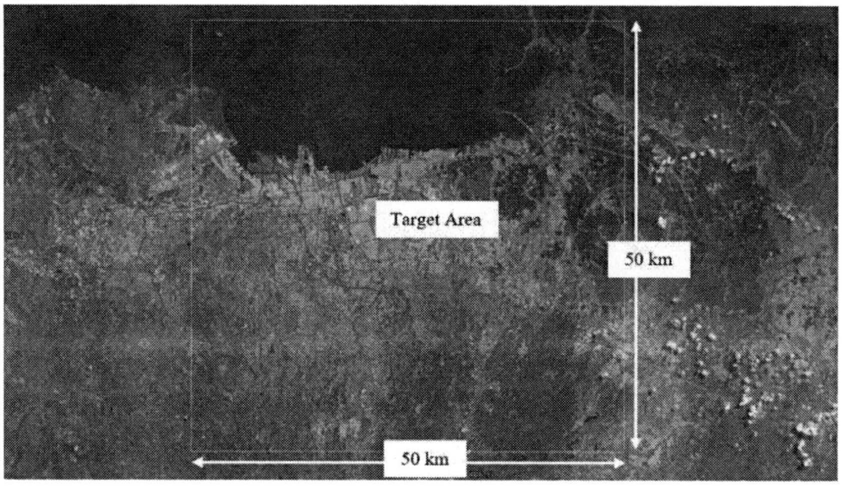

Figure 2. Target area

3. Results and Discussion

The following terminologies that represents coverage properties are necessary to be defined to further discuss the simulation result:

- *Access to Target* occurs when there is an intersection between the sensor footprint and the target.
- *Start Time of Access to Target* is a time when the Access to Target began.
- *Access Revisit* is the time elapsed between two consecutive start times of Access to Target.
- *Mean Access Revisit* is average of access revisit in one year.

Table 4. Coverage properties in 1 year

	Minimum		Maximum	
Total Access to Target	145 times	Scenario 9	201 times	Scenario 1
Access Revisit	8.4 hours	Scenario 4/ 7	15.1 days	Scenario 9
Mean Access Revisit	1.8 days	Scenario 9	2.5 days	Scenario 1

Data presented in Table 4 describes a brief result obtained from the simulation run. These coverage properties are obtained for 1 year simulation time. The Access Revisit and Mean Access Revisit data in Table 4 are roundness values. The orbit's parameters of Scenario 1 yields the maximum times of Access to Target and Mean Access Revisit, i.e. 201 times and 2.5 days, respectively. Meanwhile, simulation results using Scenario 9 presents the longest Access Revisit, i.e. 15.1 days. In addition, simulation results of each scenario are presented in Figure 3, Figure 4 and Figure 5.

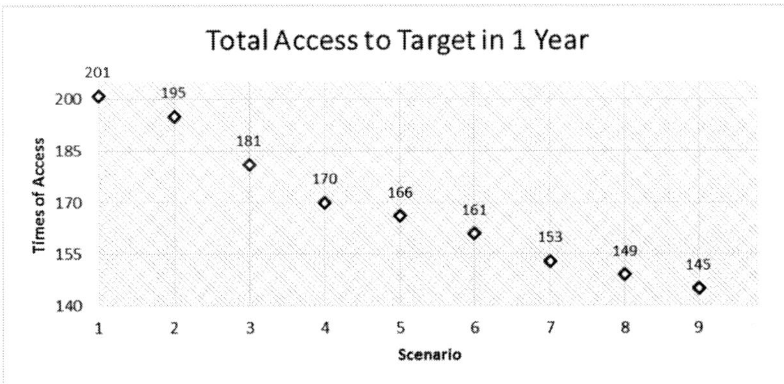

Figure 3. Total Access to Target in 1 year of each scenario

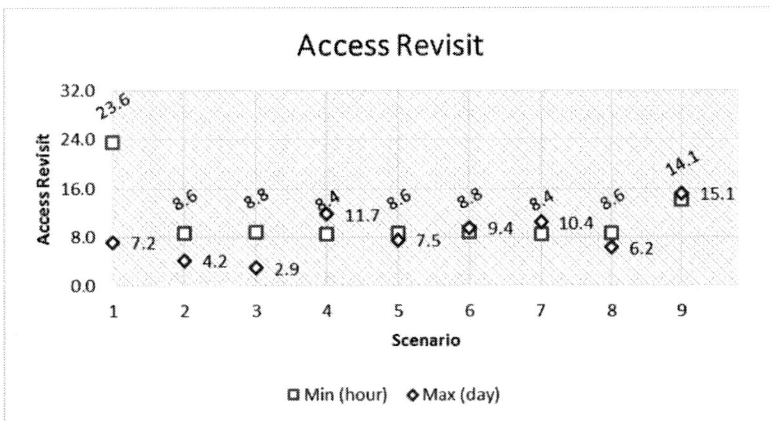

Figure 4. Minimum and maximum Access Revisit of each scenario

Figure 3 describes that, for a same inclination setting, total Access to Target decreases when altitude increases. This trend also appear in Figure 5 that for a same inclination setting, Mean Access Revisit decreases when altitude increases. Meanwhile, differs from these figures, Figure 4 shows fluctuate values. An interest finding is that the orbit parameters of Scenario 2-8 yield similar values of the minimum Access Revisit.

In polar or near-polar orbit, Revisit Time is a terminology that is often used to represent a coverage property, i.e. the time elapsed between observations of the exact same location on Earth. Its value represents temporal resolution of the SAR-based remote sensing system which is an important consideration in many applications [22]. However, a near-equatorial orbit case produces different behavior of path than polar or near-polar orbit case that implies in difficulty of Revisit Time calculation. Instead, especially in this paper, the temporal resolution is represented by (Mean) Access Revisit. Simulation results show that the near-equatorial SAR satellite has higher temporal resolution (Mean Access Revisit), i.e. about two days, than a polar or near-polar orbit SAR satellite that has more than a week of Revisit Time, e.g. TerraSAR-X has 11 days [21].

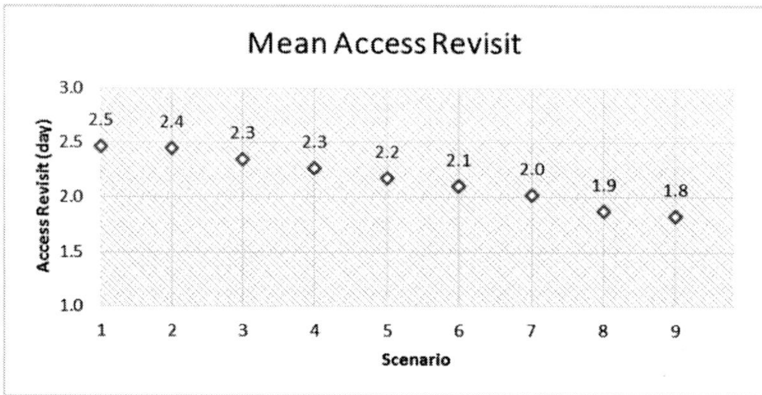

Figure 5. Mean Access Revisit of each scenario

4. Concluding Remarks

Study on SAR-based remote sensing satellite at a near-equatorial orbit is challenging. Through simulation, coverage properties with respect to a 50 km x 50 km target area at north side of Jakarta, Indonesia, are resulted. Each scenario (Mean) Access Revisit that is considered as the temporal resolution of the near-equatorial orbit satellite is presented.

These research results can be further developed with respect to entire area of Indonesia Maritime Contingent. Moreover, a more realistic SAR system model and another SAR modes, i.e. spotlight and scanSAR, are challenging to be implemented in the simulation to obtain more comprehensive coverage properties analysis. Eventually, these research results may become an important part in designing a SAR-based remote sensing satellite at a near-equatorial orbit.

Acknowledgement

This work is supported by Program Insentif Riset Sistem Inovasi Nasional Gelombang Kedua Tahun Anggaran 2016, Kemenristekdikti. The authors would like to thank director of Satellite Technology Center (Pusteksat) LAPAN for providing research support facilities and to director of Space Technology Application Center for providing administration support.

References

[1] Dong, X., Quegan, S., Yumiko, U., Hu, C., and Zeng, T., Feasibility Study of C- and L-band SAR Time Series Data in Tracking Indonesian Plantation and Natural Forest Cover Changes. *IEEE Journal of Selected Topics in Applied Earth Observations and Remote Sensing,* Vol. 8, No. 7, July 2015.

[2] Gambardella, A., Nunziata, F., and Migliaccio, M., A Physical Full-Resolution SAR Ship Detection Filter. *IEEE Geoscience and Remote Sensing Letters*, Vol. 5, No. 4, October 2008.

[3] Sumantyo, J. T. S., Shimada, M., Mathieu, Pierre-P., Abidin, H. Z., Long-Term Consecutive DInSAR for Volume Change Estimation of Land Deformation. *IEEE Transactions on Geoscience and Remote Sensing,* Vol. 50, No. 1, January 2012.

[4] Joyce, K. E., Belliss, S. E., Samsonov, S. V., McNeill, S. J., and Glassey, P. J., A review of the status of satellite remote sensing and image processing techniques for mapping natural

hazards and disasters. *Progress in Physical Geography*.

[5] Ivanov, A.Y. and Kucheiko, A.A., Distribution of oil spills in inland seas based on SAR image analysis: a comparison between the Black Sea and the Caspian Sea. *International Journal of Remote Sensing*, 37(9), pp.2101-2114.

[6] Kalogirou, V., Ferrazzoli, P., Della Vecchia, A. and Foumelis, M., On the SAR backscatter of burned forests: A model-based study in C-Band, over burned pine canopies. *IEEE Transactions on Geoscience and Remote Sensing*, 52(10), pp.6205-6215.

[7] Montomoli, F., Macelloni, G., Brogioni, M., Lemmetyinen, J., Cohen, J. and Rott, H., Observations and Simulation of Multifrequency SAR Data Over a Snow-Covered Boreal Forest. *IEEE Journal of Selected Topics in Applied Earth Observations and Remote Sensing*, 9(3), pp.1216-1228.

[8] Schmitt, M. and Zhu, X.X., Demonstration of Single-Pass Millimeterwave SAR Tomography for Forest Volumes. *IEEE Geoscience and Remote Sensing Letters*, 13(2), pp.202-206.

[9] Gou, S., Li, X. and Yang, X., Coastal Zone Classification With Fully Polarimetric SAR Imagery. *IEEE Geoscience and Remote Sensing Letters*, 13(11), pp.1616-1620.

[10] Hoang, H.K., Bernier, M., Duchesne, S. and Tran, Y.M., Rice Mapping Using RADARSAT-2 Dual-and Quad-Pol Data in a Complex Land-Use Watershed: Cau River Basin (Vietnam). *IEEE Journal of Selected Topics in Applied Earth Observations and Remote Sensing*, 9(7), pp.3082-3096.

[11] Erten, E., Lopez-Sanchez, J.M., Yuzugullu, O. and Hajnsek, I., Retrieval of agricultural crop height from space: A comparison of SAR techniques. *Remote Sensing of Environment*, 187, pp.130-144.

[12] Fang, L., Xu, Y., Yao, W. and Stilla, U., Estimation of glacier surface motion by robust phase correlation and point like features of SAR intensity images. *ISPRS Journal of Photogrammetry and Remote Sensing*, 121, pp.92-112.

[13] Fors, A.S., Brekke, C., Gerland, S., Doulgeris, A.P. and Beckers, J.F., Late Summer Arctic Sea Ice Surface Roughness Signatures in C-Band SAR Data. *IEEE Journal of Selected Topics in Applied Earth Observations and Remote Sensing*, *9*(3), pp.1199-1215.

[14] Giustarini, L., Hostache, R., Matgen, P., Schumann, G.J.P., Bates, P.D. and Mason, D.C., A change detection approach to flood mapping in urban areas using TerraSAR-X. *IEEE Transactions on Geoscience and Remote Sensing*, 51(4), pp.2417-2430.

[15] Ahmad, A., Classification simulation of RazakSAT satellite. *Procedia Engineering for Malaysian Technical Universities Conference on Engineering and Technology (MUCET) 2012*, vol. 53, pp. 472-482, 2013.

[16] Hasbi, W., and Karim, A., Lapan-A2 System Design For Equatorial Survaillance Missions. In *9th International Symposium of The International Academy of Astronautics (IAA) Berlin, April 2013*.

[17] Seah, P. H., Maverick Tan, T. C. and Meurer, R. H., TeLEOS-1 in multi-source maritime security and safety applications," *ACRS (Asian Conference on Remote Sensing),* Nay Pyi Taw, Myanmar, Oct. 27-31, 2014.

[18] http://www.agi.com/products/stk/modules/default.aspx/id/stk-pro

[19] Mattar K E. Modelling of the TerraSAR-X sensor modes in CSIAPS. *Technical Memorandum of DDRC Ottawa TM 2008-284.* Defence R & D Canada-Ottawa, 2009.

[20] Jie, C., Yin-qing, Z., and Chun-sheng, Li., Spaceborne synthetic aperture radar raw data simulation of three dimensional natural terrain. *Proceedings of 2001 CIE International Conference on Radar*. IEEE, 2001.

[21] Toth, C. and Jozkow, G., Remote sensing platform and sensors: A survey. *ISPRS Journal of Photogrammetry and Remote Sensing,* Vol. 115, pp. 22-36, 2016.

[22] Campbell, J. B., *Introduction to Remote Sensing, 4th Ed.* The Guilford Press, 2007.

Noise removal using thresholding and segmentation for random noise Sentinel-1 data

Haris Suka Dyatmika, Katmoko Ari Sambodo, Marendra Eko Budiono and Hendayani

Remote sensing Technology and Data Center, Indonesian National Institute of Aeronautics and Space (LAPAN)

Email: haris.suka@lapan.go.id

Abstract. Sentinel-1 constellation will cover the entire world's land area continuously. Although Sentinel-1 data show consistency and stability, several noise have been observed in the data i.e. random noise on the right and left of the scene. The noise exists on the right and left of the scene of the SAR data that should be no data value. The noise is quite disturbing and interferes the data especially on mosaic product of some scene data sets. The mosaic product have a seam line that separate a scene and the neighbor that should be disappear after the mosaic process. This paper shows study on how to remove the random noise without losing the information contained. Some Sentinel-1 Level-1 GRD (Ground Range Detected) data in Kalimantan area were used in this study. Principally, the methods used for the noise removal were thresholding and segmentation. If the noise removal process using thresholding only, many noise still exist in the data. Region on the right and left of the scene filtered by a certain value of intensity and segmentation area. Generally, improvement of the data was evaluated both after each scene noise removal process and mosaic product. The noise in each scene Sentinel-1 data disappear and the mosaic product look seamless after applying the noise removal.

1. Introduction

Sentinel-1 satellite constellation operate day and night at C-band synthetic aperture radar make them possible to acquire imagery independent of the weather condition. The Sentinel-1 satellite constellation consist of two satellites in a near-polar and sun-synchronous orbit with a 12 day repeat cycle for a single satellite. The constellation of two satellites offers a 6 day repeat cycle. Sentinel-1 data acquisition data processing should be done consistently in order to receive a good observation or monitoring. All global landmasses are regularly covered in VV-VH polarization [1].

A processing step Sentinel-1 products leading to random noise on the right and left of the scene of the Sentinel-1 data [2]. These noise affect the target at the edges much more than those close to the center [3]. Combining the data of the individual sub-swath into consistent mosaic require a good performance [4].

Many methods have been proposed to remove various noise in various field application. Donoho [5] shown work on de-noising by soft-thresholding. The theorem has two components, one concerning about statistical decision theory and the other concerning about wavelet bases and their properties.

The work reported here was a study to applying a general de-noising theorem to remove the random noise in a Sentinel-1 SAR data above. The random noise affect the data both single scene data and more over the mosaic products.

2. Data and method

Almost all of the data Sentinel-1 in Level-1 Ground Range Detected (GRD) products affected by the noise. This paper used some data Sentinel-1 Level-1 GRD products in Kalimantan area to test the proposed methodology. The data were adjacent to another so the mosaic of the data could also evaluate.

The data used for this study were dual polarization i.e. VV and VH polarization Sentinel-1 data, acquired on December 13th and December 19th 2015 with Interferometric Wide (IW) swath mode and descending pass both of them. Each data have amplitude band and intensity virtual band (Figure 1).

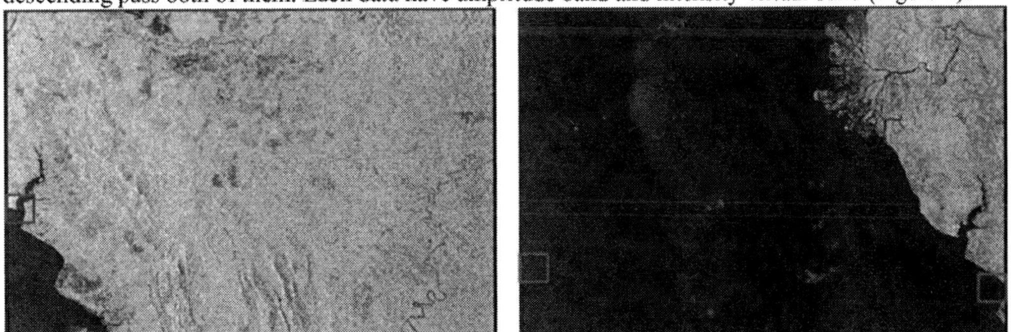

Figure 1. The Sentinel-1 Level-1 GRD data, with color composite combinations (Red=VV, Green=VV, and Blue=VH).

The noise could observed visually after saturate or enhance the data in the noise value. In the other hand, after make mosaic some adjacent data, the noise generate seam line that separate a scene and the neighbor that visible clearly.

The noise removal have two steps i.e. masking the noise then remove the masked noise. The noise masking algorithm perform some kind of classification using thresholding and segmentation principally. To masking the noise, some series of processing step performed i.e. masking the area that close to the left and right border, filter the masked area with the amplitude threshold to get the noise temporary, segmented the noise and filter the noise with the segmentation size.

The IW swath mode acquires data with a 250 km swath at 5 m by 20 m spatial resolution (single look). In Level-1 GRD products, the resolution changed dependent upon of multi-looking performed [6]. One of the data used in this research was 10×10 m pixel spacing and 25326×16831 pixel dimension and the other was similar. In the other hand the noise mainly only fraction on the left and right border, so the first step would be restrict the working area. Processing whole area would be adding many computation load and processing time. The noise left and right fraction covered by only 200 pixels each of them from the edge.

After restrict the area, the noise masking continue with classified the noise by set the threshold to differ the noise and the data. It would be hard to set a perfect threshold to differ the noise and the data. If the threshold is too high, many pixels data contained information would be lost especially on the dark object like water and the opposite. The threshold would be different between each polarization i.e. 37 for VV and 25 for VH. The differences come from range of the backscatter each polarization. The thresholding would be only a process step and would be continue in the next step.

The latter step mainly about segmentation of the noise. The idea is fraction of noise usually only small segment of the data. So the process is set a threshold of area sized i.e. 20000 pixels square. A segment bigger than the threshold would be classified as a data. Finally the noise have been masked with the processing series above. And the last process remain is remove the masked noise.

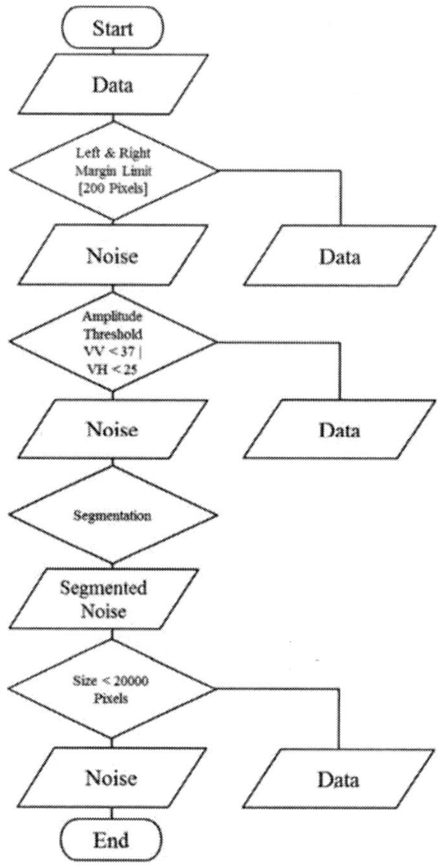

Figure 2. Noise masking algorithm flowchart.

Evaluation on single data could be directly observed but the mosaic products need to process a geometric correction before, because the Level-1 data were not yet geometric corrected. The single data evaluation only need to saturate or enhance the display into the noise value range.

3. Results and discussion

Figure 3 shows data subset with noisy image before enhance, noisy image after enhance and removed noise image after enhance. The subset were area in Figure 1 with red box marked. In the noisy image before enhance, the noise can't be seen because the value of the noise was very low and seen as a dark like 0 value. After enhance to the noise value, the noise is visible clearly. Finally the removed noise image show that the noisy fragment in the previous data were removed.

Also on the mosaic data (Figure 4) the noise on single data affect to the mosaic product. The mosaic product have two seam line that separate a scene and the neighbor that should be disappear after the mosaic process. The first line on the left affected by the noise in image shown on subset 3 and the second line on the right affected by the noise in image shown on subset 1 and 2 (Figure 3). The mosaic product look seamless after applying the noise removal (Figure 4). The mosaic product shown that this method could remove the noise significantly.

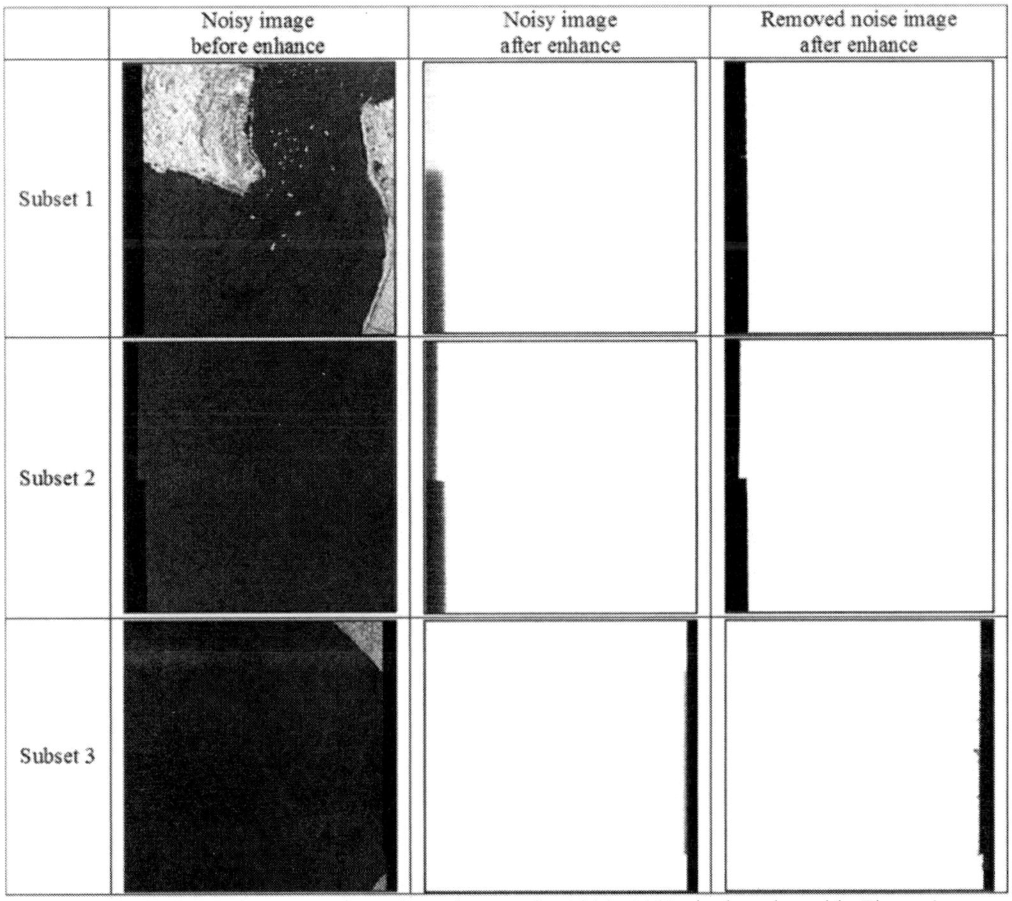

Figure 3. Data subset zoom in to the noise area by 1000×1000 pixels, selected in Figure 1.

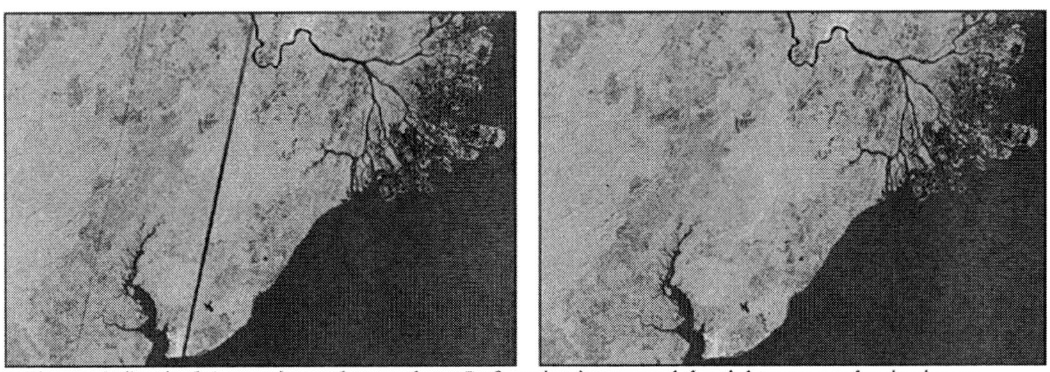

Figure 4. Sentinel-1 mosaic products subset. Left, noisy image and the right, removed noise image.

4. Conclusion

The single scene data Sentinel-1 noise could observed after enhance range value on the noise range value while the mosaic product could be directly observed. The noise removal using thresholding and segmentation could remove the noise on the right and left of the scene Sentinel-1 data significantly.

5. Acknowledgement

The authors would like to thank Remote sensing Technology and Data Center, LAPAN (Indonesian National Institute of Aeronautics and Space) for supporting the research.

References

[1] European Space Agency. 2015. Sentinel High Level Operations Plan (HLOP). 1–66
[2] Guillaume H. 2015. Masking " No-value " Pixels on GRD Products generated by the Sentinel-1 ESA IPF. 1–9
[3] De Zan F and Guarnieri A M. 2006. TOPSAR: Terrain observation by progressive scans. *IEEE Trans. Geosci. Remote Sens.* **44** 2352–60
[4] Wegmüller U, Werner C, Strozzi T, Wiesmann A, Frey O and Santoro M. 2014. SENTINEL-1 SUPPORT IN THE GAMMA SOFTWARE. **42** 5194
[5] Donoho D L. 1995. De-Noising by Soft-Thresholding. *IEEE Trans. Inf. Theory.* **41** 613–27
[6] European Space Agency. 2013. Sentinel-1 User Handbook. **1** 1–80

Preliminary Detection Model of Rapid Mapping Technique for Landslide Susceptibility Zone Using Multi Sensor Imagery (Case Study in Banjarnegara Regency)

I Yanuarsyah[1], R N Khairiah[2]

[1]Geoinformatic of Informatics Engineering Department, Faculty of Engineering, Ibn Khaldun University, Jalan KH Soleh Iskandar KM 2, Tanah Sereal, Bogor, West Jawa, Indonesia
[2]Post Graduate Student, Tropical Biodiversity Conservation, Faculty of Forestry, Bogor Agricultural University (IPB), Bogor

Email: iksal.80@gmail.com

Abstract. This study as a preliminary stage as part of disaster mitigation landslide in Banjarnegara Regency, by utilizing a combination of multi-sensor image to overview the pattern forest cover changes with supported by other parameters such as rainfall, slope, aspect, curvature patterns hill (curvature). The objective is how to develop detection model in rapid mapping technique for detection landslide susceptibility zone. This information is used as basis an early detection for estimating landslide potentially happen in the future. there are four main processes which are optical image processing, SAR image processing, DEM processing and Scoring Geoprocessing. The final zone might be verified by particular landslide event location whether it exist on the result map. It obtain "big five" district with higher prone landslide susceptibility zone such as Batur district, Pejawaran district, Wanayasa district, Kalibening district and Rakit district. Total susceptibility zone in Banjarnegara regency approximately 604.79 Ha with 15,250 prone point location. Thus, it classified as 14.16 Ha of low zone, 286.41 Ha of moderate zone and 304.22 Ha of high zone. This study demonstrates as rapid mapping the enormous potential landslide occurrences investigated by susceptibility zone. In term of landslide prone point, the combination optical image and SAR image quite enough to perform post forest cover changes and it also can overlay with another causative parameter.

1. Introduction

More than hundred victims were buried by landslide at hilly Karangkobar Village, Banjarnegara Regency, in December two years ago. Material losses reached nearly two billion rupiah accompanied by the destruction of irrigation channels, curve of roads and damage to the bridges, which connecting between villages. According to the landslide map occurrences issued by National Disaster Relief Agency shows the location affected by landslides that occurred in three villages in Banjarnegara Regency, which are Tunggoro Village, Sidengkok Village and Sampang Village. The biggest landslide occurred in Jemblung, Sampang Village, Karangkobar district. The disaster resulted in the deaths of 108 people unaccounted (missing), 54 families affected, two people died, four people were seriously injured, 11 people were injured lightly and the number of refugees still in accounted. Damage and loss of material in the form of 30 houses and some vehicles buried by landslide material. Compared with other natural hazards such as volcanic eruptions and floods, landslides cause considerable damage to human beings and massive economic losses.

Data from the Indonesian Agency for Meteorology Climatology and Geophysics shows that 11 days before the disaster occur the area was hit by huge rain, around 50% from its annual rainfall. The Indonesian Institute of Sciences later explained that the ground on the area were soaked by the rain, hence the ground to moved.

Geologically, Indonesia is located at the confluence of three tectonic plates, the Indo-Australian plate, the Eurasian plate and the Pacific plate. Therefore, Indonesia is known as the state is prone to geological disasters, such as earthquakes, volcanic eruptions and tsunamis. Landslide often occur due to higher geological and climatological potential risk with large numbers impact of swallow losses and casualties. Landslide occurred mainly in the hilly areas and during the rainy season. Geological condition of Indonesia is prone to seismic vibrations or volcanic vibration. The vibrations can lead to landslide. Based on events Research Group of Gajah Mada University (UGM), around 60% of the land in Indonesia is an area prone to landslides.

A landslide is a movement of a mass of rock, debris, or earth down a slope, under the influence of gravity [1] in [2]. Landslides involve flowing, sliding, toppling, falling, or spreading, and many landslides exhibit a combination of different types of movements, at the same time or during the lifetime of the land slide. Landslides are present in all continents, and play an important role in the evolution of landscapes. In many areas they also pose a serious threat to the population [3]. For instance, landslides in Java, the most densely populated island in Indonesia, caused 2,095 casualties and 522 injuries in 1981-2007 [4].

Basic disaster management concepts and basic knowledge of landslide phenomena were introduced in order to improve understanding, awareness, motivation and empowerment and to develop effective disaster management measures and mitigation. Therefore, mitigation system become crucial step towards the lead human resources to guarantee the sustainability of life and the environment an areas susceptible to landslides. Disaster risk reduction is part of an effort to maintain the sustainability of human well-being. Resistance and resilience are necessary to adapt to the danger of landslide, as well as to survive the devastating of landslide impacts. Integration between disaster risk reduction and sustainable development becomes urgent in order to reduce and eliminate future damage and loss [4].

Several factors that cause landslide such as settlement of geology, slope and vegetation distribution in identification. Remote sensing techniques well used to determine the zoning of areas susceptible to ground movement, especially to observe areas that are difficult to reach [5].

On satellite images and SRTM 30m DEM, land movement symptom expression is shown by its distinctive shape like a horseshoe shape, a steep escarpment, fracture pattern parallel to the cliff landslides, soil moisture on the slope below the cliff, steps topography along the river bank, and so on. Although these types of avalanches are not always recognizable in the images, preliminary estimates can still be expected from the avalanche product form that shows the appearance of ground motion formation on it [6].

Landslide Susceptibility Zone Map is a map that contain visual information to region tendency of ground movement occurrences. Landslide susceptibility zone is a zone that has the same relative susceptibility for ground movement occurrences. Determination of the susceptibility is based on the parameters, the magnitude of slope, soil and rock types, rainfall, and the amount of movement of soil, land use, seismicity, the numerical values of slope stability, etc. Classification of landslide susceptibility zone is divided into four, namely high, medium, low and very low [7].

This study as a preliminary stage as part of disaster mitigation landslide in Banjarnegara Regency, by utilizing a combination of multi-sensor image to overview the pattern forest cover changes with supported by other parameters such as rainfall, slope, aspect, curvature patterns hill (curvature). The objective is how to develop detection model in rapid mapping technique for detection landslide susceptibility zone. This information is used as basis an early detection for estimating landslide potentially happen in the future.

2. Method

The indirect analysis emphasis in desk study by the overlap between the map of the distribution of landslide occurrences and several parameters (geological, slope and land use), then do the estimation using the data unit geology, grade slope, and units of land use effect on the incidence of ground movement. This method is based on a model calculation of density (density) of ground movement and the weight value (weight value) per unit of geology, grade slope, and units of land use (on each map parameter). The weight values obtained are summed and divided into a maximum of four classes using the upper limit value (upper bound) for each class, which susceptibility zones is very low, the zone of susceptibility is low, the zone of susceptibility medium, and the zone of zone of susceptibility High [7]. This research used indirect analysis to combine optical image and radar image with applying some overlay analysis and scoring for particular parameters. The threshold of vegetation indices and water indices might be used in order to sharp slicing forest cover changes. Elevation model might also need due to significant parameter in obtaining the susceptibility.

2.1. Location

This study located at 109°29'00" - 109°45'50" E and 7°12'00" - 7°31'00" S in Mensiku miniwatershed of Banjarnegara Regency covers 1,70 Km^2 approximately 3.1 % of Central Java Province. It consists of 19 districts (Banjarmangu, Banjarnegara, Batur, Bawang, Kalibening, Karangkobar, Madukara, Mandiraja, Pagentan, Pandanarum, Pejawaran, Punggelan, Purwonegoro, Purworejo Klampok, Rakit, Sigaluh, Susukan, Wanadadi, and Wanayasa).

2.2. Spatial Dataset

This research required spatial data such as i) Landsat 8 OLI images path 120 Row 065 (1 scene before landslide captured in August 30, 2014 and 1 scene after landslide captured in February 22, 2015) gathered from USGS Global Visualization Viewer (http://glovis.usgs.gov/); ii) Sentinel-1A C-band Synthetic Aperture Radar (SAR) image (1 scene before landslide captured in November 24, 2014 and 1 scene after landslide captured in December 18, 2014) gathered from National Institute of Aeronautics and Space; iii) Digital Elevation Model (DEM) derived from Shuttle Radar Topography Mission (SRTM) Arc-second 30 m gathered from The CGIAR Consortium for Spatial Information (CGIAR-CSI); iv) Rainfall annual data (acquition in 2014) produced by Ministry of Agriculture scale 1:250.000; v) Landslide coordinate survey on October 2014. Beyond that also need Banjarnegara administrative boundary in district scale. Data processing required geoprocessing software such as Quantum GIS 2.12.3 Lyon.

2.3. Process Flow

The process flow of this research described in Figure 1, where all the data will be confined by study location (AOI) thus data layout focused on district level. Commonly there are four main processes which are optical image processing, SAR image processing, DEM processing and Scoring Geoprocessing.

Performing first step is Landsat images starting from preprocessing for reducing radiometric and geometric effects, thus to generate vegetation indices (NDVI) and water indices (NDWI) for both images (pre and post landslide) in order to overview forest cover changes during landslide occurrences.

The normalized difference vegetation index (NDVI) is the most widely used index for remote sensing of vegetation in the past two decades. It is equal to $(\rho_{NIR} - \rho_{RED})/(\rho_{NIR} + \rho_{RED})$, where ρ_{RED} is the radiance (in reflectance units) of a red channel near 0.66 μm, and ρ_{NIR} the radiance (in reflectance units) of a near-IR channel around 0.86 μm. Remote sensing of vegetation liquid water has important applications in agriculture and forestry. A normalized difference water index (NDWI) that uses two near-IR channels centered approximately at 0.86 μm and 1.24 μm for remote sensing of vegetation liquid water from space is proposed. The 1.24 μm channel has not been used previously in the formation of vegetation indices [8]. NDWI extraction from Landsat 8 by channel 3 (ρ_{GREEN}) and channel 5 (ρ_{NIR}) through formula $(\rho_{GREEN} - \rho_{NIR})/(\rho_{GREEN} + \rho_{NIR})$ [9]. A new vegetation index, the normalized difference water index (NDWI), is proposed for remote sensing of vegetation liquid water from space. This index uses two narrow channels centered near 0.86 μm and 1.24 μm. Both channels sense similar depths through vegetation canopies, unlike the two channels used in NDVI. NDWI is a measure of liquid water molecules in vegetation canopies that interacted with the incoming solar radiation. It is less sensitive to atmospheric scattering effects than NDVI [8]. Threshold deviation result of the vegetation indices and water indices deviation will be determined such for NDVI (forest changed area > 0.1) and NDWI (forest water content changed < -0.06).

Performing second step is SAR images starting from preprocessing for reducing radiometric and geometric effects, thus to emphasis cover changes by Red-Green Different (RG Different) where post Sentinel-1A as red layer and pre Sentinel-1A as green layer. Yellow color image will be produced without any changes of both layers, contrary some red color (post-changed area) and green color (pre-changed area) appearance if there has land cover changes. Segmentation with object based analysis is used to perform clear object in supervised classification with required some training area. Furthermore, post-changed area from SAR image will be interacted with forest cover changed area from Landsat image (post forest changed area).

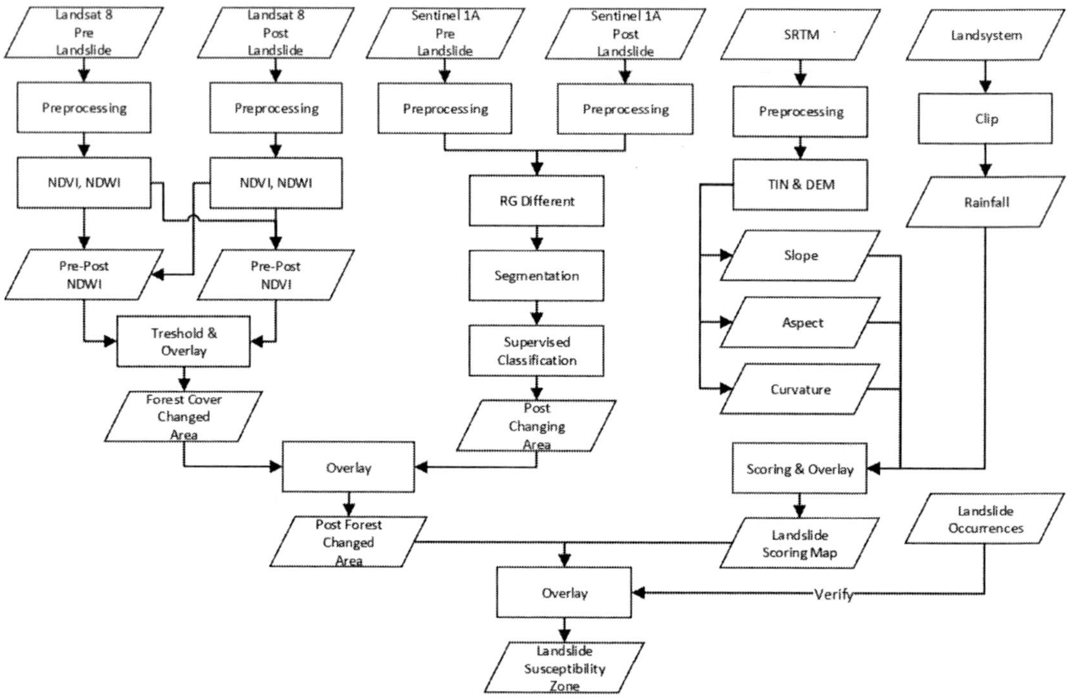

Figure 1. Data processing flow

Performing third step by using DEM extraction method is needed to generate slope data, aspect and curvature data. Rainfall annual data which derived from landsystem map as also causative factors of landslide. Slope map could be classified as flat (< 8 %), gently slope (8 - 15 %), moderate (15 - 25 %), steeply slope (25 - 40 %, and very steeply slope (> 40 %). Aspect map could be classified as flat, north (0 - 23°), northeast (23 - 68°), east (68 - 113°), southeast (113 - 158°), south (158 - 203°), southwest (203 - 248°), west (248 - 293°), northwest (293 - 338°), and north (338 - 360°). Curvature map could be classified as plus (+), minus (-) and zero (0). Curvature map combined with planform curvature and profile curvature. Scoring technique is needed to deploy the detection model of landslide susceptibility zone as shown on Figure 2.

Plan curvature is described as the curvature of a contour line formed by intersecting a horizontal plane with the surface, and therefore controls the convergence or divergence of landslide debris and water in the direction of landslide motion. The positive, zero and negative value indicate convexity, concavity and flat surface, respectively. Profile curvature relates to the convergence and divergence of flow across a surface, and affects the acceleration or deceleration of flow across the surface (concave and convex). In locations with convex (negative) profile curvature the erosion will prevail and in locations with concave (positive) curvature the deposition [10].

Landslide susceptibility map generated by the sum of weight value from the map results alignment between ground motion distribution maps with geological aspect, slope and land use maps by using spatial analysis using GIS [5]. Performing the fourth step by using some overlay analysis of each raster layers and spatial features which are clip, dissolve, raster conversion, intersect to achieve susceptibility zone. The final zone might be verified by particular landslide event location whether it exist on the result map.

No	Changed Area	Score
1	Post Changed Area	1
2	Pre Change Area	0

No	Slope	Score
1	< 8 %	1
2	8 %- 15 %	2
3	15 %- 25 %	2
4	25 %- 40 %	3
5	> 40%	3

No	Rainfall	Score
1	< 2,000 mm/year	1
2	2,000– 3,000 mm/year	2
3	> 3,000 mm/year	3

No	Planform Curvature	Score
1	+ Plus	3
2	- Minus	2
3	o Zero	1

No	Aspect	Score
1	North	3
2	Northeast	3
3	East	1
4	Southeast	1
5	South	3
6	Southwest	3
7	West	2
8	Northwest	2

No	Profile Curvature	Score
1	+ Plus	2
2	- Minus	3
3	o Zero	1

No	Landslide Susceptibility Zone	Scores
1	LOW	<5
2	MODERATE	6- 10
3	HIGH	>11

Figure 2. Scoring deployment of detection model

3. Result and Discussion

Landslides, ground settlement and avalanche interfere greatly and persistently with mass activities. It occurs when hill side or valley side slopes falls using to specific geological, climatic and biotic factors. They are bringing about major disruptions of towns and cities, communication systems and large structure including dams and bridges. Mitigation of disasters due to landslides can be successful only with detailed knowledge about the expected frequency, character and magnitude of mass movements in an area. To forecast possibilities of the future landslides in an area, comprehensive knowledge of causative factors of land sliding is necessary. The wide applicability of geospatial technologies are using in solving various environmental tasks. This technology can be used as an effective aid in natural hazard investigation, as well as for the purpose of environmental planning. Drainage map, contour map, digital elevation model, slope angle map, land use / land cover map, relative relief map, thrust (buffer) map, photolineament [11].

According to the analysis (Figure 3), it obtain "big five" district with higher prone landslide susceptibility zone such as Batur district, Pejawaran district, Wanayasa district, Kalibening district and Rakit district. Total susceptibility zone in Banjarnegara regency approximately 604.79 Ha with 15,250 prone point location. Thus, it classified as 14.16 Ha of low zone, 286.41 Ha of moderate zone and 304.22 Ha of high zone. More detailed explanation as shown in Table 1, susceptible zone in Batur district about 117.86 Ha and become the highest susceptible risk contained equal of high zone and moderate zone (49 %) with almost four thousand prone points spreading in all part of the area (Figure 4). Pejawaran district has 101.66 Ha dominant with 62 % of high susceptible zone with two point five thousand prone point spreading in north part, west and east part of the area. Wanayasa district has 93.85 Ha dominant with 66 % of moderate susceptible zone with two point three thousand prone point spreading in center part and south part of the area. In addition, whole districts of Banjarnegara regency might aware with landslides risk potentially occur further.

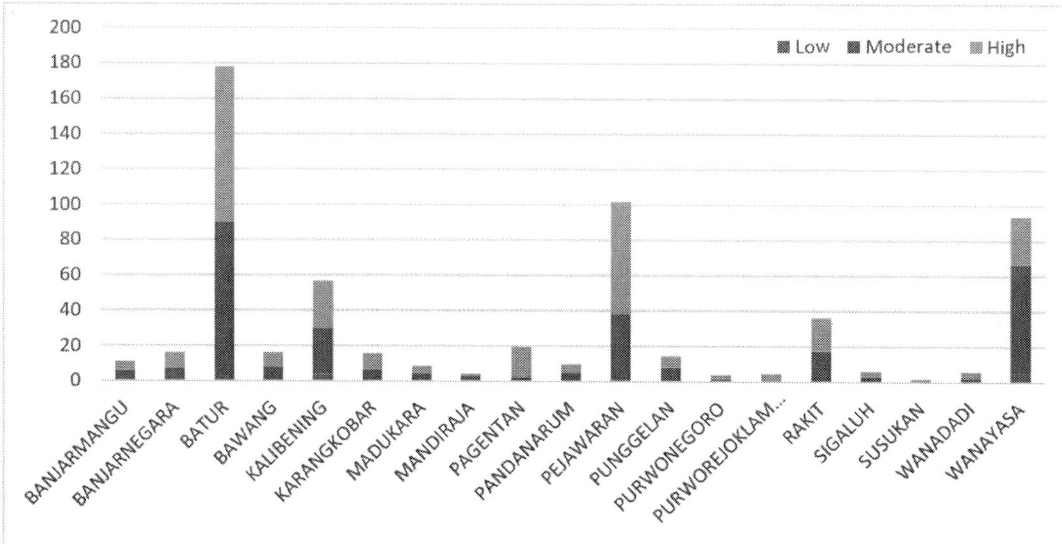

Figure 3. Landslide susceptibility classes in Banjarnegara regency (Ha)

Table 1. Percentage of landslide susceptibility classes in Banjarnegara regency (Ha)

District	Susceptible Area (Ha)	Percentage			Prone Location
		Low	**Moderate**	**High**	
Banjarmangu	11.20	1%	54%	45%	359
Banjarnegara	16.35	0%	44%	56%	517
Batur	177.86	1%	49%	49%	3,887
Bawang	16.21	1%	49%	50%	472
Kalibening	56.98	7%	45%	48%	1,314
Karangkobar	15.66	0%	42%	58%	624
Madukara	8.54	0%	50%	50%	254
Mandiraja	4.10	13%	51%	37%	133
Pagentan	19.53	0%	11%	89%	486
Pandanarum	9.74	0%	49%	51%	293
Pejawaran	101.66	1%	36%	62%	2,449
Punggelan	14.32	3%	55%	42%	578
Purwonegoro	3.89	0%	50%	50%	123
Purworejo Klampok	4.76	0%	17%	83%	132
Rakit	36.20	1%	48%	51%	942
Sigaluh	6.12	0%	45%	55%	171
Susukan	1.86	0%	21%	79%	67
Wanadadi	5.94	0%	34%	65%	176
Wanayasa	93.85	6%	66%	29%	2,273
Total	604.79	2%	47%	50%	**15,250**

When evaluating the probability of landsliding within a specific period of time and within a certain area, it is of major importance to recognize the conditions that can cause the landslide and the process that could trigger the movement. The correlation between landslide areas and associated factors that cause landslides can be allocated from the connections between areas without past landslides and the landslide-related parameters. In order to prepare the landslide susceptibility map quantitatively, the frequency ratio method was implemented using GIS techniques [12].

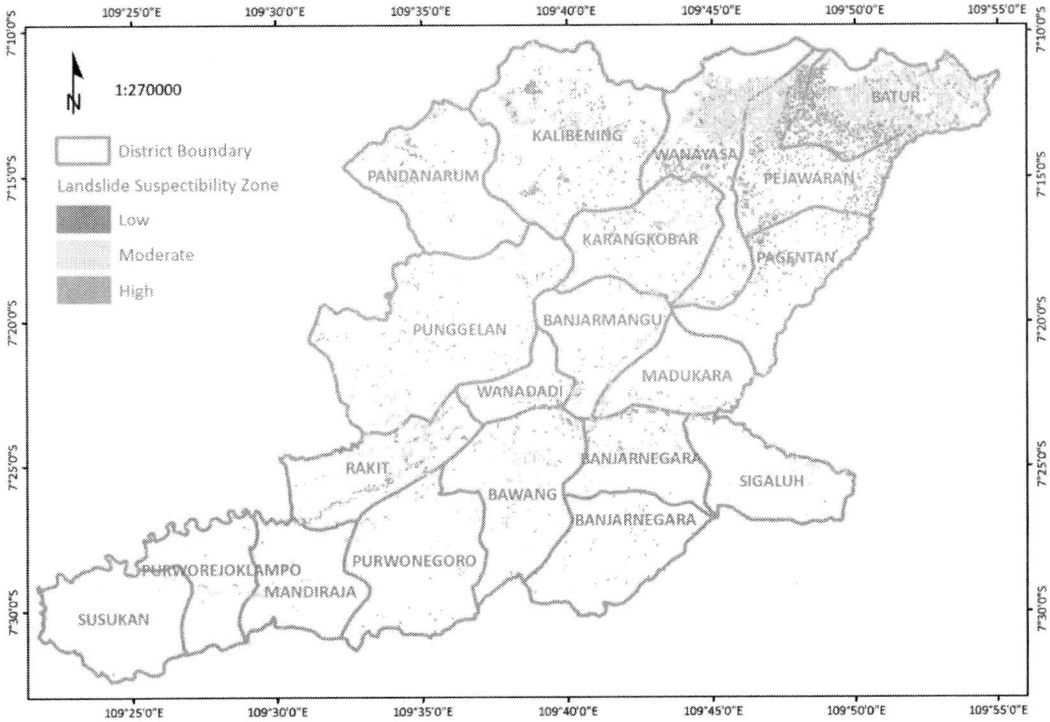

Figure 4. Landslide susceptibility zone Banjarnegara regency

Old landslide motion may still be active again mainly due to the high rainfall in a long time and strong erosion. The ground movement occurs on the slope of a rather steep to steep. Generally, slope formed by rained paddy field and forest cover. On Landsat imagery and SRTM 30 meter DEM, avalanche sightings can be characterized by the shape of a steep escarpment, fracture pattern parallel to the cliff landslides and soil moisture on the slopes below the cliffs. The avalanche shape visually delineated both escarpment avalanches, landslides direction, fracture pattern, and the sliding plane, and the landslide area, and eventually formed a distribution map avalanches [5].

The analysis of this study regarding on the creation of causative parameters and scoring model. Some interpretations might be influenced by the experiences and literature used. As advanced stages, relevant information on additional parameters such as geological factor, weather dynamics, social factors are needed to float a robust model as rapid mapping technique of landslide detection.

4. Conclusion

This study demonstrates as rapid mapping the enormous potential landslide occurrences investigated by susceptibility zone. In term of landslide prone point, the combination optical image and SAR image quite enough to perform post forest cover changes and it also can overlay with another causative parameter. Furthermore, it need to elaborate with quantitative and qualitative parameter in order to obtain the optimal result of landslide susceptibility zone. Eventually, this preliminary study for developing landslide detection model is used as recommendation and part decision support system to local government of Banjarnegara Regency.

References

[1] Cruden, D. M. and Varnes, D. J. 1996. Landslide types and processes. In Turner, A. K. and Schuster, R. L. (Editors), Landslides Investigation and Mitigation: Transportation Research Board, US National Research Council, Special Report 247, Washington, DC, pp 36-75

[2] Hungr, O., Evans, S.G., Bovis, M., and Hutchinson, J.N. 2001. Review of the classification of

landslides of the flow type. *Environmental and Engineering Geoscience*.7: 221-238

[3] Petley, D. N. 2012. Global patterns of loss of life from landslides. *Geology*, 40: 927-930

[4] Nugraha, H., Wacano D., Dipayana G.A., Cahyadi A., Mutaqin B.A., Larasati A. 2015. Geomorphometric characteristics of landslides in the Tinalah Watershed, Menoreh Mountains, Yogyakarta, Indonesia. *Procedia Environmental Sciences*, 28: 578-586

[5] Yunarto. 2012. Remote Sensing Technique And Geographic Information System For Mapping Movement Susceptibility Zones With Indirect Method At Kuningan Regency. *Bulletin of Environmental Geology. Vol. 22, 2: 75-86* pp 4, 7, 10

[6] Noor, D. 2005. Geologi Lingkungan. Graha Ilmu, Yogyakarta

[7] Indonesian National Standard. 2005. Standard of Landslide Susceptibility Zone Map. *SNI 13-7124-2005*. pp 3, 5

[8] Gao, B. C. 1996. NDWI A Normalized Difference Water Index for Remote Sensing of Vegetation Liquid Water From Space. *Remote Sens. Environ*. 58:257-266

[9] LAPAN. 2015. Pedoman Pemanfaatan Data Landsat-8 Untuk Deteksi Daerah Tergenang Banjir (*Inundated Area*). National Institute of Aeronautics and Space. p 18

[10] Huang, J. 2014. Investigation on landslide susceptibility using remote sensing and GIS methods. *Open Access Theses and Dissertations*. Paper 33. pp 40-41

[11] Rai, P. K., Mohan, K., Kumra, V. K.. 2014. Landslide Hazard And Its Mapping Using Remote Sensing And GIS. *Journal of Scientific Research Banaras Hindu University, Varanasi*. 58 : 1-13 ISSN : 0447-9483 p 12

[12] Sivakami C. and Sundaram A 2014 Landslide Susceptibility Zone using Frequency Ratio Model, Remote Sensing & GIS –A Case Study of Western Ghats, India (Part of Kodaikanal Taluk). *Journal of Environment and Earth Science*. ISSN 2225-0948 4:.22. p 6